Unit Conversions (Equivalents)

Length

1 in. = 2.54 cm
1 cm = 0.394 in.
1 ft = 30.5 cm
1 m = 39.37 in. = 3.28 ft
1 mi = 5280 ft = 1.61 km
1 km = 0.621 mi
1 nautical mile (U.S.) = 1.15 mi = 6076 ft = 1.852 km
1 fermi = 1 femtometer (fm) = 10^{-15} m
1 angstrom (Å) = 10^{-10} m
1 light-year (ly) = 9.46×10^{15} m
1 parsec = 3.26 ly = 3.09×10^{16} m

Volume

1 liter (L) = 1000 mL = 1000 cm^3 = $1.0 \times 10^{-3} m^3$ =
    1.057 quart (U.S.) = 54.6 $in.^3$
1 gallon (U.S.) = 4 qt (U.S.) = 231 $in.^3$ = 3.78 L =
    0.83 gal (Imperial)
1 m^3 = 35.31 ft^3

Speed

1 mi/h = 1.47 ft/s = 1.609 km/h = 0.447 m/s
1 km/h = 0.278 m/s = 0.621 mi/h
1 ft/s = 0.305 m/s = 0.682 mi/h
1 m/s = 3.28 ft/s = 3.60 km/h
1 knot = 1.151 mi/h = 0.5144 m/s

Angle

1 radian (rad) = 57.30° = 57°18′
1° = 0.01745 rad
1 rev/min (rpm) = 0.1047 rad/s

Time

1 day = 8.64×10^4 s
1 year = 3.156×10^7 s

Mass

1 atomic mass unit (u) = 1.6605×10^{-27} kg
1 kg = 0.0685 slug
[1 kg has a weight of 2.20 lb where $g = 9.81$ m/s².]

Force

1 lb = 4.45 N
1 N = 10^5 dyne = 0.225 lb

Energy and Work

1 J = 10^7 ergs = 0.738 ft·lb
1 ft·lb = 1.36 J = 1.29×10^{-3} Btu = 3.24×10^{-4} kcal
1 kcal = 4.18×10^3 J = 3.97 Btu
1 eV = 1.602×10^{-19} J
1 kWh = 3.60×10^6 J = 860 kcal

Power

1 W = 1 J/s = 0.738 ft·lb/s = 3.42 Btu/h
1 hp = 550 ft·lb/s = 746 W

Pressure

1 atm = 1.013 bar = 1.013×10^5 N/m²
      = 14.7 lb/in.² = 760 torr
1 lb/in.² = 6.90×10^3 N/m²
1 Pa = 1 N/m² = 1.45×10^{-4} lb/in.²

SI Derived Units and Their Abbreviations

Quantity	Unit	Abbreviation	In Terms of Base Units[†]
Force	newton	N	kg·m/s²
Energy and work	joule	J	kg·m²/s²
Power	watt	W	kg·m²/s³
Pressure	pascal	Pa	kg/(m·s²)
Frequency	hertz	Hz	s^{-1}
Electric charge	coulomb	C	A·s
Electric potential	volt	V	kg·m²/(A·s³)
Electric resistance	ohm	Ω	kg·m²/(A²·s³)
Capacitance	farad	F	A²·s⁴/(kg·m²)
Magnetic field	tesla	T	kg/(A·s²)
Magnetic flux	weber	Wb	kg·m²/(A·s²)
Inductance	henry	H	kg·m²/(s²·A²)

[†]kg = kilogram (mass), m = meter (length), s = second (time), A = ampere (electric current).

Metric (SI) Multipliers

Prefix	Abbreviation	Value
exa	E	10^{18}
peta	P	10^{15}
tera	T	10^{12}
giga	G	10^9
mega	M	10^6
kilo	k	10^3
hecto	h	10^2
deka	da	10^1
deci	d	10^{-1}
centi	c	10^{-2}
milli	m	10^{-3}
micro	μ	10^{-6}
nano	n	10^{-9}
pico	p	10^{-12}
femto	f	10^{-15}
atto	a	10^{-18}

PHYSICS
for
SCIENTISTS & ENGINEERS

Volume II
PHYSICS
for
SCIENTISTS & ENGINEERS
with Modern Physics
Third Edition

DOUGLAS C. GIANCOLI

PRENTICE HALL
Upper Saddle River, New Jersey 07458

Editor-in-Chief: Paul F. Corey
Production Editor: Susan Fisher
Executive Editor: Alison Reeves
Development Editor: David Chelton
Director of Marketing: John Tweedale
Senior Marketing Manager: Erik Fahlgren
Assistant Vice President of Production and Manufacturing: David W. Riccardi
Executive Managing Editor: Kathleen Schiaparelli
Manufacturing Manager: Trudy Pisciotti
Art Manager: Gus Vibal
Director of Creative Services: Paul Belfanti
Advertising and Promotions Manager: Elise Schneider
Editor in Chief of Development: Ray Mullaney
Project Manager: Elizabeth Kell
Photo Research: Mary Teresa Giancoli
Photo Research Administrator: Melinda Reo
Copy Editor: Jocelyn Phillips
Editorial Assistant: Marilyn Coco
Cover photo: Onne van der Wal/Young America
Composition: Emilcomp srl / Preparé Inc.

© 2000, 1989, 1984 by Douglas C. Giancoli
Published by Prentice Hall
Upper Saddle River, NJ, 07458

Printed in the United States of America

10 9 8 7 6 5 4 3

ISBN 0-13-021519-8

Prentice-Hall International (UK) Limited, *London*
Prentice-Hall of Australia Pty. Limited, *Sydney*
Prentice-Hall Canada Inc., *Toronto*
Prentice-Hall Hispanoamericana, S.A., *Mexico City*
Prentice-Hall of India Private Limited, *New Delhi*
Prentice-Hall of Japan, Inc., *Tokyo*
Prentice-Hall (*Singapore*) Pte. Ltd.
Editora Prentice-Hall do Brasil, Ltda., *Rio de Janeiro*

CONTENTS OF VOLUME I

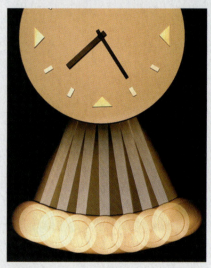

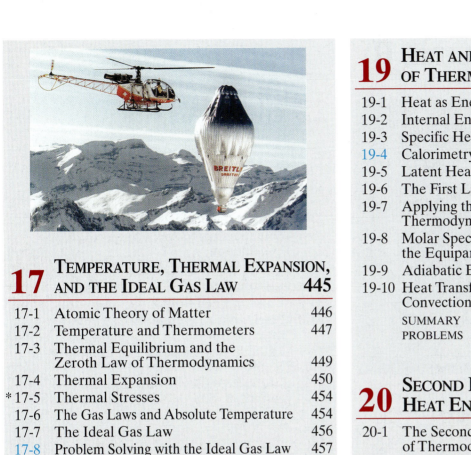

CONTENTS—VOLUME II

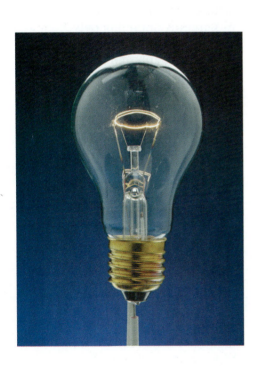

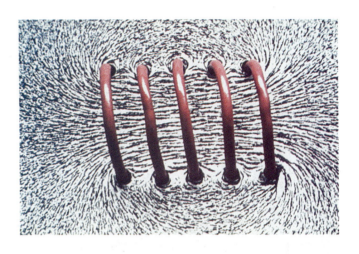

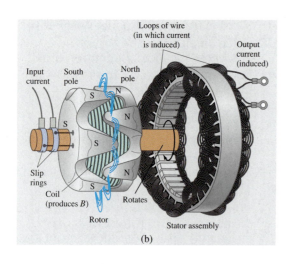

Loops of wire
(in which current
is induced)

Output
current
(induced)

Input
current

South
pole

North
pole

Slip
rings

Coil
(produces *B*)

Rotor

Rotates

Stator assembly

(b)

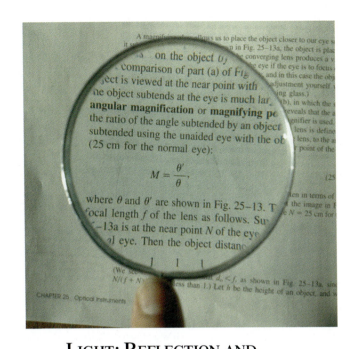

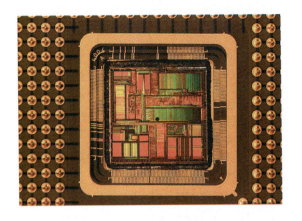

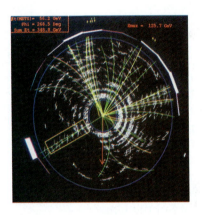

45 ASTROPHYSICS AND COSMOLOGY — 1140

PREFACE

A Brand New Third Edition

It has been more than ten years since the second edition of this calculus-based introductory physics textbook was published. A lot has changed since then, not only in physics itself, but also in how physics is presented. Research in how students learn has provided textbook authors new opportunities to help students learn physics and learn it well.

This third edition comes in two versions. The standard version covers all of classical physics plus a chapter on special relativity and one on the early quantum theory. The extended version, with modern physics, contains a total of nine detailed chapters on modern physics, ending with astrophysics and cosmology. This book retains the original approach: in-depth physics, concrete and nondogmatic, readable.

This new third edition has many improvements in the physics and its applications. Before discussing those changes in detail, here is a list of some of the overall changes that will catch the eye immediately.

Full color throughout is not just cosmetic, although fine color photographs do help to attract the student readers. More important, full color diagrams allow the physics to be displayed with much greater clarity. We have not stopped at a 4-color process; this book has actually been printed in 5 pure colors (5 passes through the presses) to provide better variety and definition for illustrating vectors and other physics concepts such as rays and fields. I want to emphasize that color is used pedagogically to bring out the physics. For example, different types of vectors are given different colors—see the chart on page xxxi.

Many more diagrams, almost double the number in the previous edition, have all been done or redone carefully using full color; there are many more graphs and many more photographs throughout. See for example in optics where new photographs show lenses and the images they make.

Marginal notes have been added as an aid to students to (i) point out what is truly important, (ii) serve as a sort of outline, and (iii) help students find details about something referred to later that they may not remember so well. Besides such "normal" marginal notes, there are also marginal notes that point out brief *problem solving* hints, and others that point out interesting *applications*.

The great laws of physics are emphasized by giving them a marginal note all in capital letters and enclosed in a rectangle. The most important equations, especially those expressing the great laws, are further emphasized by a tan-colored screen behind them.

Chapter opening photographs have been chosen to illustrate aspects of each chapter. Each was chosen with an eye to writing a caption which could serve as a kind of summary of what is in that chapter, and sometimes offer a challenge. Some chapter-opening photos have vectors or other analysis superimposed on them.

Page layout: complete derivations. Serious attention has been paid to how each page was formatted, especially for page turns. Great effort has been made to keep important derivations and arguments on facing pages. Students then don't have to turn back to check. More important, readers repeatedly see before them, on two facing pages, an important slice of physics.

Two kinds of Examples: Conceptual Examples and **Estimates.**

New Physics

The whole idea of a new edition is to improve, to bring in new material, and to delete material that is verbose and only makes the book longer or is perhaps too advanced and not so useful. Here is a brief summary of a few of the changes involving the physics iself. These lists are selections, not complete lists.

New discoveries:
- planets revolving around distant stars
- Hubble Space Telescope
- updates in particle physics and cosmology, such as inflation and the age of the universe

New physics topics added:
- new treatment of how to make estimates (Chapter 1), including new Estimating Examples throughout (in Chapter 1, estimating the volume of a lake, and the radius of the Earth)
- symmetry used much more, including for solving problems
- new Tables illustrating the great range of lengths, time intervals, masses, voltages
- gravitation as curvature of space, and black holes (Chapter 6)
- engine efficiency (Chapter 8 as well as Chapter 20)
- rolling with and without slipping, and other useful details of rotational motion (Chapter 10)
- forces in structures including trusses, bridges, arches, and domes (Chapter 12)
- square wave (Chapter 15)
- using the Maxwell distribution (Chapter 18)
- Otto cycle (Chapter 20)
- statistical calculation of entropy change in free expansion (Chapter 20)
- effects of dielectrics on capacitor connected and not (Chapter 24)
- grounding to avoid electric hazards (Chapter 25)
- three phase ac (Chapter 31)
- equal energy in $\mathbf{E}$ and $\mathbf{B}$ of EM wave (Chapter 32)
- radiation pressure, EM wave (Chapter 32)
- photos of lenses and mirrors with their images (Chapter 33)
- detailed outlines for ray tracing with mirrors and lenses (Chapters 33, 34)
- lens combinations (Chapter 34)
- new radiation standards (Chapter 43)
- Higgs boson, supersymmetry (Chapter 44)

Modern physics. A number of modern physics topics are discussed in the framework of classical physics. Here are some highlights:
- gravitation as curvature of space, and black holes (Chapter 6)
- planets revolving around distant stars (Chapter 6)
- kinetic energy at relativistic speeds (Chapter 7)
- nuclear collisions (Chapter 9)
- star collapse (Chapter 10)
- galaxy red shift, Doppler (Chapter 16)
- atoms, theory of (Chapters 17, 18, 21)
- atomic theory of thermal expansion (Chapter 17)
- mass of hydrogen atom (Chapter 17)
- atoms and molecules in gases (Chapters 17, 18)
- molecular speeds (Chapter 18)
- equipartition of energy; molar specific heats (Chapter 19)
- star size (Chapter 19)
- molecular dipoles (Chapters 21, 23)
- cathode ray tube (Chapters 23, 27)
- electrons in a wire (Chapter 25)
- superconductivity (Chapter 25)
- discovery and properties of the electron, e/m, oil drop experiment (Chapter 27)
- Hall effect (Chapter 27)

- magnetic moment of electrons (Chapter 27)
- mass spectrometer (Chapter 27)
- velocity selector (Chapter 27)
- electron spin in magnetic materials (Chapter 28)
- light and EM wave emission (Chapter 32)
- spectroscopy (Chapter 36)

Many other examples of modern physics are found as Problems, even in early chapters. Chapters 37 and 38 contain the modern physics topics of Special Relativity, and an introduction to Quantum Theory and Models of the Atom. The longer version of this text, "with Modern Physics," contains an additional seven chapters (for a total of nine) which present a detailed and extremely up-to-date treatment of modern physics: Quantum Mechanics of Atoms (Chapters 38 to 40); Molecules and Condensed Matter (Chapter 41); Nuclear Physics (Chapter 42 and 43); Elementary Particles (Chapter 44); and finally Astrophysics, General Relativity, and Cosmology (Chapter 45).

Revised physics and reorganizations. First of all, a major effort has been made to not throw everything at the students in the first few chapters. The basics have to be learned first; many aspects can come later, when the students are more prepared. Secondly, a great part of this book has been rewritten to make it clearer and more understandable to students. Clearer does not always mean simpler or easier. Sometimes making it "easier" actually makes it harder to understand. Often a little more detail, without being verbose, can make an explanation clearer. Here are a few of the changes, big and small:

- new graphs and diagrams to clarify velocity and acceleration; deceleration carefully treated.
- unit conversion now a new Section in Chapter 1, instead of interrupting kinematics.
- circular motion: Chapter 3 now gives only the basics, with more complicated treatment coming later: non-uniform circular motion in Chapter 5, angular variables in Chapter 10.
- Newton's second law now written throughout as $m\mathbf{a} = \Sigma\mathbf{F}$, to emphasize inclusion of all forces acting on a body.
- Newton's third law follows the second directly, with inertial reference frames placed earlier. New careful discussions to head off confusion when using Newton's third law.
- careful rewriting of chapters on Work and Energy, especially potential energy, conservative and nonconservative forces, and the conservation of energy.
- renewed emphasis that $\Sigma\tau = I\alpha$ is not always valid: only for an axis fixed in an inertial frame or if axis is through the CM (Chapters 10 and 11).
- rolling motion introduced early in Chapter 10, with more details later, including rolling with and without slipping.
- rotating frames of reference, and Coriolis, moved later, to Chapter 11, shortened, optional, but still including why an object does not fall straight down on Earth.
- fluids reduced to a single chapter (13); some topics and details dropped or greatly shortened.
- clearer details on how an object floats (Chapter 13).
- distinction between wave interference in space, and in time (beats) (Chapter 16).
- thermodynamics reduced to four chapters; the old chapters on Heat and on the First Law of Thermodynamics have been combined into one (19), with some topics shortened and a more rational sequence of topics achieved.
- heat transfer now follows the first law of thermodynamics (Chapter 19).
- electric potential carefully rewritten for accuracy (Chapter 23).
- CRT, computer monitors, TV, treated earlier (Chapter 23).
- use of Q_{encl} and I_{encl} for Gauss's and Ampère's laws, with subscripts meaning "enclosed".
- Ohm's law and definition of resistance carefully redone (Chapter 25).
- sources of magnetic field, Chapter 28, reorganized for ease of understanding, with some new material, and deletion of the advanced topic on magnetization vector.
- circuits with L, C, and/or R now introduced via Kirchhoff's loop rule, and clarified in other ways too (Chapters 30, 31).
- streamlined Maxwell's equations, with displacement current downplayed (Chapter 32).
- optics reduced to four chapters; polarization is now placed in the same chapter as diffraction.

New Pedagogy

All of the above mentioned revisions, rewritings, and reorganizations are intended to help students learn physics better. They were done in response to contemporary research in how students learn, as well as to kind and generous input from professors who have read, reviewed, or used the previous editions. This new edition also contains some new elements, especially an increased emphasis on conceptual development:

Conceptual Examples, typically 1 or 2 per chapter, sometimes more, are each a sort of brief Socratic question and answer. It is intended that students will be stimulated by the question to think, or reflect, and come up with a response—before reading the Response given. Here are a few:

- using symmetry (Chapters 1, 44, and elsewhere)
- ball moving upward: misconceptions (Chapter 2)
- reference frames and projectile motion: where does the apple land? (Chapter 3)
- what exerts the force that makes a car move? (Chapter 4)
- Newton's third law clarification: pulling a sled (Chapter 4)
- free-body diagram for a hockey puck (Chapter 4)
- advantage of a pulley (Chapter 4), and of a lever (Chapter 12)
- to push or to pull a sled (Chapter 5)
- which object rolls down a hill faster? (Chapter 10)
- moving the axis of a spinning wheel (Chapter 11)
- tragic collapse (Chapter 12)
- finger at top of a full straw (Chapter 13)
- suction cups on a spacecraft (Chapter 13)
- doubling amplitude of SHM (Chapter 14)
- do holes expand thermally? (Chapter 17)
- simple adiabatic process: stretching a rubber band (Chapter 19)
- charge inside a conductor's cavity (Chapter 22)
- how stretching a wire changes its resistance (Chapter 25)
- series or parallel (Chapter 26)
- bulb brightness (Chapter 26)
- spiral path in magnetic field (Ch. 27)
- practice with Lenz's law (Chapter 29)
- motor overload (Chapter 29)
- emf direction in inductor (Chapter 30)
- photo with reflection—is it upside down? (Chapter 33)
- reversible light rays (Chapter 33)
- how tall must a full-length mirror be? (Chapter 33)
- diffraction spreading (Chapter 36)

Estimating Examples, roughly 10% of all Examples, also a new feature of this edition, are intended to develop the skills for making order-of-magnitude estimates, even when the data are scarce, and even when you might never have guessed that any result was possible at all. See, for example, Section 1–6, Examples 1–5 to 1–8.

Problem Solving, with New and Improved Approaches

Learning how to approach and solve problems is a basic part of any physics course. It is a highly useful skill in itself, but is also important because the process helps bring understanding of the physics. Problem solving in this new edition has a significantly increased emphasis, including some new features.

Problem-solving boxes, about 20 of them, are new to this edition. They are more concentrated in the early chapters, but are found throughout the book. They each outline a step-by-step approach to solving problems in general, and/or specifically for the material being covered. The best students may find these separate "boxes" unnecessary (they can skip them), but many students will find it helpful to be reminded of the general approach and of steps they can take to get started; and, I think, they help to build confidence. The general problem solving box in Section 4–8 is placed there, after students have had some experience wrestling with problems, and so may be strongly motivated to read it with close attention. Section 4–8 can, of course, be covered earlier if desired.

Problem-solving Sections occur in many chapters, and are intended to provide extra drill in areas where solving problems is especially important or detailed.

Examples. This new edition has many more worked-out Examples, and they all now have titles for interest and for easy reference. There are even two new categories of Example: Conceptual, and Estimates, as described above. Regular Examples serve as "practice problems". Many new ones have been added, some of the old ones have been dropped, and many have been reworked to provide greater clarity and detail: more steps are spelled out, more of "why we do it this way", and more discussion of the reasoning and approach. In sum, the idea is "to think aloud with the students", leading them to develop insight. The total number of worked-out Examples is about 30% greater than in the previous edition, for an average of 12 to 15 per chapter. There is a significantly higher concentration of Examples in the early chapters, where drill is especially important for developing skills and a variety of approaches. The level of the worked-out Examples for most topics increases gradually, with the more complicated ones being on a par with the most difficult Problems at the end of each chapter, so that students can see how to approach complex problems. Many of the new Examples, and improvements to old ones, provide relevant applications to engineering, other related fields, and to everyday life.

Problems at the end of each chapter have been greatly increased in quality and quantity. There are over 30% more Problems than in the second edition. Many of the old ones have been replaced, or rewritten to make them clearer, and/or have had their numerical values changed. Each chapter contains a large group of Problems arranged by Section and graded according to difficulty: level I Problems are simple, designed to give students confidence; level II are "normal" Problems, providing more of a challenge and often the combination of two different concepts; level III are the most complex, typically combining different issues, and will challenge even superior students. The arrangement by Section number means only that those Problems depend on material up to and including that Section: earlier material may also be relied upon. The ranking of Problems by difficulty (I, II, III) is intended only as a guide.

General Problems. About 70% of Problems are ranked by level of difficulty (I, II, III) and arranged by Section. New to this edition are General Problems that are unranked and grouped together at the end of each chapter, and account for about 30% of all problems. The average total number of Problems per chapter is about 90. Answers to odd-numbered Problems are given at the back of the book.

Complete Physics Coverage, with Options

This book is intended to give students the opportunity to obtain a thorough background in all areas of basic physics. There is great flexibility in choice of topics so that instructors can choose which topics they cover and which they omit. Sections marked with an asterisk can be considered optional, as discussed more fully on p. xxv. Here I want to emphasize that topics not covered in class can still be read by serious students for their own enrichment, either immediately or later. Here is a partial list of physics topics, not the standard ones, but topics that might not usually be covered, and that represent how thorough this book is in its coverage of basic physics. Section numbers are given in parentheses.

- use of calculus; variable acceleration (2–8)
- nonuniform circular motion (5–4)
- velocity-dependent forces (5–5)
- gravitational versus inertial mass; principle of equivalence (6–8)
- gravitation as curvature of space; black holes (6–9)
- kinetic energy at very high speed (7–5)
- potential energy diagrams (8–9)
- systems of variable mass (9–10)
- rotational plus translational motion (10–11)
- using $\Sigma \tau_{CM} = I_{CM} \alpha_{CM}$ (10–11)
- derivation of $K = K_{CM} + K_{rot}$ (10–11)
- why does a rolling sphere slow down? (10–12)
- angular momentum and torque for a system (11–4)
- derivation of $d\mathbf{L}_{CM}/dt = \Sigma \boldsymbol{\tau}_{CM}$ (11–4)
- rotational imbalance (11–6)
- the spinning top (11–8)
- rotating reference frames; inertial forces (11–9)
- coriolis effect (11–10)
- trusses (12–7)
- flow in tubes: Poiseuille's equation (13–11)
- surface tension and capillarity (13–12)
- physical pendulum; torsion pendulum (14–6)
- damped harmonic motion: finding the solution (14–7)
- forced vibrations; equation of motion and its solution; Q-value (14–8)
- the wave equation (15–5)
- mathematical representation of waves; pressure wave derivation (16–2)
- intensity of sound related to amplitude (16–3)
- interference in space and in time (16–6)
- atomic theory of expansion (17–4)
- thermal stresses (17–5)
- ideal gas temperature scale (17–10)
- calculations using the Maxwell distribution of molecular speeds (18–2)
- real gases (18–3)
- vapor pressure and humidity (18–4)
- van der Waals equation of state (18–5)
- mean free path (18–6)
- diffusion (18–7)
- equipartition of energy (19–8)
- energy availability; heat death (20–8)
- statistical interpretation of entropy and the second law (20–9)
- thermodynamic temperature scale; absolute zero and the third law (20–10)
- electric dipoles (21–11, 23–6)
- experimental basis of Gauss's and Coulomb's laws (22–4)
- general relation between electric potential and electric field (23–2, 23–8)

- electric fields in dielectrics (24–5)
- molecular description of dielectrics (24–6)
- current density and drift velocity (25–8)
- superconductivity (25–9)
- RC circuits (26–4)
- use of voltmeters and ammeters; effects of meter resistance (26–5)
- transducers (26–6)
- magnetic dipole moment (27–5)
- Hall effect (27–8)
- operational definition of the ampere and coulomb (28–3)
- magnetic materials—ferromagnetism (28–7)
- electromagnets and solenoids (28–8)
- hysteresis (28–9)
- paramagnetism and diamagnetism (28–10)
- counter emf and torque; eddy currents (29–5)
- Faraday's law—general form (29–7)
- force due to changing $\mathbf{B}$ is nonconservative (29–7)
- LC circuits and EM oscillations (30–5)
- AC resonance; oscillators (31–6)
- impedance matching (31–7)
- three phase AC (31–8)
- changing electric fields produce magnetic fields (32–1)
- speed of light from Maxwell's equations (32–5)
- radiation pressure (32–8)
- fiber optics (33–7)
- lens combinations (34–3)
- aberrations of lenses and mirrors (34–10)
- coherence (35–4)
- intensity in double-slit pattern (35–5)
- luminous intensity (35–8)
- intensity for single-slit (36–2)
- diffraction for double–slit (36–3)
- limits of resolution, the λ limit (36–4, 36–5)
- resolution of the human eye and useful magnification (36–6)
- spectroscopy (36–8)
- peak widths and resolving power for a diffraction grating (36–9)
- x-rays and x-ray diffraction (36–10)
- scattering of light by the atmosphere (36–12)
- time–dependent Schrödinger equation (39–6)
- wave packets (39–7)
- tunneling through a barrier (39–9)
- free-electron theory of metals (41–6)
- semiconductor electronics (41–9)
- standard model, symmetry, QCD, GUT (44–9, 44–10)
- astrophysics, cosmology (Ch. 45)

New Applications

Relevant applications to everyday life, to engineering, and to other fields such as geology and medicine, provide students with motivation and offer the instructor the opportunity to show the relevance of physics. Applications are a good response to students who ask "Why study physics?" Many new applications have been added in this edition. Here are some highlights:

- airbags (Chapter 2)
- elevator and counterweight (Chapter 4)
- antilock brakes and skidding (Chapter 5)
- geosynchronous satellites (Chapter 6)
- hard drive and bit speed (Chapter 10)
- star collapse (Chapter 10)
- forces within trusses, bridges, arches, domes (Chapter 12)
- the Titanic (Chapter 12)
- Bernoulli's principle: wings, sailboats, TIA, plumbing traps and bypasses (Chapter 13)
- pumps (Chapter 13)
- car springs, shock absorbers, building dampers for earthquakes (Chapter 14)
- loudspeakers (Chapters 14, 16, 27)
- autofocusing cameras (Chapter 16)
- sonar (Chapter 16)
- ultrasound imaging (Chapter 16)
- thermal stresses (Chapter 17)
- R-values, thermal insulation (Ch. 19)
- engines (Chapter 20)
- heat pumps, refrigerators, AC; coefficient of performance (Chapter 20)
- thermal pollution (Chapter 20)
- electric shielding (Chapters 21, 28)
- photocopier (Chapter 21)
- superconducting cables (Chapter 25)
- jump starting a car (Chapter 26)
- aurora borealis (Chapter 27)
- solenoids and electromagnetics (Ch. 28)
- computer memory and digital information (Chapter 29)
- seismograph (Chapter 29)

- tape recording (Chapter 29)
- loudspeaker cross-over network (Ch. 31)
- antennas, for **E** or **B** (Chapter 32)
- TV and radio; AM and FM (Chapter 32)
- eye and corrective lenses (Chapter 34)
- mirages (Chapter 35)
- liquid crystal displays (Chapter 36)
- CAT scans, PET, MRI (Chapter 43)

Some old favorites retained (and improved):

- pressure gauges (Chapter 13)
- musical instruments (Chapter 16)
- humidity (Chapter 18)
- CRT, TV, computer monitors (Ch. 23, 27)
- electric hazards (Chapter 25)
- power in household circuits (Chapter 25)
- ammeters and voltmeters (Chapter 26)
- microphones (Chapters 26, 29)
- transducers (Chapter 26, and elsewhere)
- electric motors (Chapter 27)
- car alternator (Chapter 29)
- electric power transmission (Chapter 29)
- capacitors as filters (Chapter 31)
- impedance matching (Chapter 31)
- fiber optics (Chapter 33)
- cameras, telescopes, microscopes, other optical instruments (Chapter 34)
- lens coatings (Chapter 35)
- spectroscopy (Chapter 36)
- electron microscopes (Chapter 38)
- lasers, holography, CD players (Ch. 40)
- semiconductor electronics (Chapter 41)
- radioactivity (Chapters 42 and 43)

Deletions

Something had to go, or the book would have been too long. Lots of subjects were shortened—the detail simply isn't necessary at this level. Some topics were dropped entirely: polar coordinates; center-of-momentum reference frame; Reynolds number (now a Problem); object moving in a fluid and sedimentation; derivation of Poiseuille's equation; Stoke's equation; waveguide and transmission line analysis; electric polarization and electric displacement vectors; potentiometer (now a Problem); negative pressure; combinations of two harmonic motions; adiabatic character of sound waves; central forces.

Many topics have been shortened, often a lot, such as: velocity-dependent forces; variable acceleration; instantaneous axis; surface tension and capillarity; optics topics such as some aspects of light polarizarion. Many of the brief historical and philosophical issues have been shortened as well.

General Approach

This book offers an in-depth presentation of physics, and retains the basic approach of the earlier editions. Rather than using the common, dry, dogmatic approach of treating topics formally and abstractly first, and only later relating the material to the students' own experience, my approach is to recognize that physics is a description of reality and thus to start each topic with concrete observations and experiences that students can directly relate to. Then we move on to the generalizations and more formal treatment of the topic. Not only does this make the material more interesting and easier to understand, but it is closer to the way physics is actually practiced.

This new edition, even more than previous editions, aims to explain the physics in a readable and interesting manner that is accessible and clear. It aims to teach students by anticipating their needs and difficulties, but without oversimplifying. Physics is all about us. Indeed, it is the goal of this book to help students "see the world through eyes that know physics."

As mentioned above, this book includes of a wide range of Examples and applications from technology, engineering, architecture, earth sciences, the environment, biology, medicine, and daily life. Some applications serve only as examples of physical principles. Others are treated in depth. But applications do not dominate the text—this is, after all, a physics book. They have been carefully chosen and integrated into the text so as not to interfere with the development of the physics but rather to illuminate it. You won't find essay sidebars here. The applications are integrated right into the physics. To make it easy to spot the applications, a new *Physics Applied* marginal note is placed in the margin (except where diagrams in the margin prevent it).

It is assumed that students have started calculus or are taking it concurrently. Calculus is treated gently at first, usually in an optional Section so as not to burden students taking calculus concurrently. For example, using the integral in kinematics, Chapter 2, is an optional Section. But in Chapter 7, on work, the integral is discussed fully for all readers.

Throughout the text, *Système International* (SI) units are used. Other metric and British units are defined for informational purposes. Careful attention is paid to significant figures. When a certain value is given as, say, 3, with its units, it is meant to be 3, not assumed to be 3.0 or 3.00. When we mean 3.00 we write 3.00. It is important for students to be aware of the uncertainty in any measured value, and not to overestimate the precision of a numerical result.

Rather than start this physics book with a chapter on mathematics, I have instead incorporated many mathematical tools, such as vector addition and multiplication, directly in the text where first needed. In addition, the Appendices contain a review of many mathematical topics such as trigonometric identities, integrals, and the binomial (and other) expansions. One advanced topic is also given an Appendix: integrating to get the gravitational force due to a spherical mass distribution.

It is necessary, I feel, to pay careful attention to detail, especially when deriving an important result. I have aimed at including all steps in a derivation, and have tried to make clear which equations are general, and which are not, by explicitly stating the limitations of important equations in brackets next to the equation, such as

$$x = x_0 + v_0 t + \tfrac{1}{2} a t^2. \qquad \text{[constant acceleration]}$$

The more detailed introduction to Newton's laws and their use is of crucial pedagogic importance. The many new worked-out Examples include initially fairly simple ones that provide careful step-by-step analysis of how to proceed in solving dynamics problems. Each succeeding Example adds a new element or a new twist that introduces greater complexity. It is hoped that this strategy will enable even less-well-prepared students to acquire the tools for using Newton's laws correctly. If students don't surmount this crucial hurdle, the rest of physics may remain forever beyond their grasp.

Rotational motion is difficult for most students. As an example of attention to detail (although this is not really a "detail"), I have carefully distinguished the position vector (**r**) of a point and the perpendicular distance of that point from an axis, which is called R in this book (see Fig. 10–2). This distinction, which enters particularly in connection with torque, moment of inertia, and angular momentum, is often not made clear—it is a disservice to students to use $\mathbf{r}$ or r for both without distinguishing. Also, I have made clear that it is not always true that $\Sigma\tau = I\alpha$. It depends on the axis chosen (valid if axis is fixed in an inertial reference frame, or through the CM). To not tell this to students can get them into serious trouble. (See pp. 250, 283, 284.) I have treated rotational motion by starting with the simple instance of rotation about an axis (Chapter 10), including the concepts of angular momentum and rotational kinetic energy. Only in Chapter 11 is the more general case of rotation about a point dealt with, and this slightly more advanced material can be omitted if desired (except for Sections 11–1 and 11–2 on the vector product and the torque vector). The end of Chapter 10 has an optional subsection containing three slightly more advanced Examples, using $\Sigma\tau_{CM} = I_{CM}\alpha_{CM}$: car braking distribution, a falling yo-yo, and a sphere rolling with and without slipping.

Among other special treatments is Chapter 28, Sources of Magnetic Field: here, in one chapter, are discussed the magnetic field due to currents (including Ampère's law and the law of Biot-Savart) as well as magnetic materials, ferromagnetism, paramagnetism, and diamagnetism. This presentation is clearer, briefer, and more of a whole, and all the content is there.

Organization

The general outline of this new edition retains a traditional order of topics: mechanics (Chapters 1 to 12); fluids, vibrations, waves, and sound (Chapter 13 to 16); kinetic theory and thermodynamics (Chapters 17 to 20). In the two-volume version of this text, volume I ends here, after Chapter 20. The text continues with electricity and magnetism (Chapters 21 to 32), light (Chapters 33 to 36), and modern physics (Chapters 37 and 38 in the short version, Chapters 37 to 45 in the extended version "with Modern Physics"). Nearly all topics customarily taught in introductory physics courses are included. A number of topics from modern physics are included with the classical physics chapters as discussed earlier.

The tradition of beginning with mechanics is sensible, I believe, because it was developed first, historically, and because so much else in physics depends on it. Within mechanics, there are various ways to order topics, and this book allows for considerable flexibility. I prefer, for example, to cover statics after dynamics, partly because many students have trouble working with forces without motion. Besides, statics is a special case of dynamics—we study statics so that we can prevent structures from becoming dynamic (falling down)—and that sense of being at the limit of dynamics is intuitively helpful. Nonetheless statics (Chapter 12) can be covered earlier, if desired, before dynamics, after a brief introduction to vector addition. Another option is light, which I have placed after electricity and magnetism and EM waves. But light could be treated immediately after the chapters on waves (Chapters 15 and 16). Special relativity is Chapter 37, but could instead be treated along with mechanics—say, after Chapter 9.

Not every chapter need be given equal weight. Whereas Chapter 4 might require $1\frac{1}{2}$ to 2 weeks of coverage, Chapter 16 or 22 may need only $\frac{1}{2}$ week.

Some instructors may find that this book contains more material than can be covered completely in their courses. But the text offers great flexibility in choice of topics. Sections marked with a star (asterisk) are considered optional. These Sections contain slightly more advanced physics material, or material not usually covered in typical courses, and/or interesting applications. They contain no material needed in later chapters (except perhaps in later optional Sections). This does not imply that all nonstarred Sections must be covered: there still remains considerable flexibility in the choice of material. For a brief course, all optional material could be dropped as well as major parts of Chapters 11, 13, 16, 26, 30, 31, and 36 as well as selected parts of Chapters 9, 12, 19, 20, 32, 34, and the modern physics chapters. Topics not covered in class can be a valuable resource for later study; indeed, this text can serve as a useful reference for students for years because of its wide range of coverage.

Thanks

Some 60 physics professors provided input or direct feedback on every aspect of this textbook. The reviewers and contributors to this third edition are listed below. I owe each a debt of gratitude.

Ralph Alexander, University of Missouri at Rolla

Zaven Altounian, McGill University

Charles R. Bacon, Ferris State University

Bruce Birkett, University of California, Berkeley

Art Braundmeier, Southern Illinois University at Edwardsville

Wayne Carr, Stevens Institute of Technology

Edward Chang, University of Massachusetts, Amherst

Charles Chiu, University of Texas at Austin

Lucien Crimaldi, University of Mississippi

Robert Creel, University of Akron

Alexandra Cowley, Community College of Philadelphia

Timir Datta, University of South Carolina

Gary DeLeo, Lehigh University

John Dinardo, Drexel University

Paul Draper, University of Texas, Arlington

Alex Dzierba, Indiana University

William Fickinger, Case Western University

Jerome Finkelstein, San Jose State University

Donald Foster, Wichita State University

Gregory E. Frances, Montana State University

Lothar Frommhold, University of Texas at Austin

Thomas Furtak, Colorado School of Mines

Edward Gibson, California State University, Sacramento

Christopher Gould, University of Southern California

John Gruber, San Jose State University

Martin den Boer, Hunter College

Greg Hassold, General Motors Institute

Joseph Hemsky, Wright State University

Laurent Hodges, Iowa State University

Mark Holtz, Texas Tech University

James P. Jacobs, University of Montana

James Kettler, Ohio University Eastern Campus

Jean Krisch, University of Michigan

Mark Lindsay, University of Louisville

Eugene Livingston, University of Notre Dame

Bryan Long, Columbia State Community College

Daniel Mavlow, Princeton University

Pete Markowitz, Florida International University

John McCullen, University of Arizona, Tucson

Peter Nemeth, New York University

Hon-Kie Ng, Florida State University

Eugene Patroni, Georgia Institute of Technology

Robert Pelcovits, Brown University

William Pollard, Valdosta State University

Joseph Priest, Miami University

Carl Rotter, West Virginia University

Lawrence Rees, Brigham Young University

Peter Riley, University of Texas at Austin

Roy Rubins, University of Texas at Arlington

Mark Semon, Bates College

Robert Simpson, University of New Hampshire

Mano Singham, Case Western University

Harold Slusher, University of Texas at El Paso

Don Sparks, Los Angeles Pierce Community College

Michael Strauss, University of Oklahoma

Joseph Strecker, Wichita State University

William Sturrus, Youngstown State University

Arthur Swift, University of Massachusetts, Amherst

Leo Takahasi, The Pennsylvania State University

Edward Thomas, Georgia Institute of Technology

Som Tyagi, Drexel University

John Wahr, University of Colorado

Robert Webb, Texas A & M University

James Whitmore, The Pennsylvania State University

W. Steve Quon, Ventura College

I owe special thanks to Irv Miller, not only for many helpful physics discussions, but for having worked out all the Problems and managed the team that also worked out the Problems, each checking the other, and finally for producing the Solutions Manual and all the answers to the odd-numbered Problems at the end of this book. He was ably assisted by Zaven Altounian and Anand Batra.

I am particularly grateful to Robert Pelcovits and Peter Riley, as well as to Paul Draper and James Jacobs, who inspired many of the new Examples, Conceptual Examples, and Problems.

Crucial for rooting out errors, as well as providing excellent suggestions, were the perspicacious Edward Gibson and Michael Strauss, both of whom carefully checked all aspects of the physics in page proof.

Special thanks to Bruce Birkett for input of every kind, from illuminating discussions on pedagogy to a careful checking of details in many sections of this book. I wish also to thank Professors Howard Shugart, Joe Cerny, Roger Falcone and Buford Price for helpful discussions, and for hospitality at the University of California, Berkeley. Many thanks also to Prof. Tito Arecchi at the Istituto Nazionale di Ottica, Florence, Italy, and to the staff of the Institute and Museum for the History of Science, Florence, for their hospitality.

Finally, I wish to thank the superb editorial and production work provided by all those with whom I worked directly at Prentice Hall: Susan Fisher, Marilyn Coco, David Chelton, Kathleen Schiaparelli, Trudy Pisciotti, Gus Vibal, Mary Teresa Giancoli, and Jocelyn Phillips.

The biggest thanks of all goes to Paul Corey, whose constant encouragement and astute ability to get things done, provided the single strongest catalyst.

The final responsibility for all errors lies with me, of course. I welcome comments and corrections.

D.C.G.

AVAILABLE SUPPLEMENTS

For the Student

Student Study Guide and Solutions Manual
Douglas Brandt, Eastern Illinois University. (0-13-021475-2)
Contains chapter objectives, summaries with additional examples, self-study quizzes, key mathematical equations, and complete worked-out solutions to alternate odd problems in the text.

Doing Physics with Spreadsheets: A Workbook
Gordon Aubrecht, T. Kenneth Bolland, and Michael Ziegler, all of The Ohio State University.
(0-13-021474-4)
Designed to introduce students to the use of spreadsheets for solving simple and complex physics problems. Students are either provided with spreadsheets or must construct their own, then use the model to most closely approximate natural behavior. The amount of spreadsheet construction and the complexity of the spreadsheet increases as the student gains experience.

Science on the Internet: A Student's Guide, 1999
Andrew Stull and Carl Adler (0-13-021308-X)
The perfect tool to help students take advantage of the *Physics for Scientists and Engineers, Third Edition* Web page. This useful resource gives clear steps to access Prentice Hall's regularly updated physics resources, along with an overview of general World Wide Web navigation strategies. Available FREE for students when packaged with the text.

Prentice Hall/*New York Times* Themes of the Times — Physics
This unique newspaper supplement brings together a collection of the latest physics-related articles from the pages of *The New York Times*. Updated twice per year and available FREE to students when packaged with the text.

For the Instructor

Instructor's Solutions Manual
Irvin A. Miller, Drexel University.
Print version (0-13-021381-0); Electronic (CD-ROM) version (0-13-021481-7)
Contains detailed worked solutions to every problem in the text. Electronic versions are available in CD-ROM (dual platform for both Windows and Macintosh systems) for instructors with Microsoft Word or Word-compatible software.

Test Item File
Robert Pelcovits, Brown University; David Curott, University of North Alabama; and Edward Oberhofer, University of North Carolina at Charlotte (0-13-021482-5)
Contains over 2200 multiple choice questions, about 25% conceptual in nature. All are referenced to the corresponding Section in the text and ranked by difficulty.

Prentice Hall Custom Test Windows (0-13-021477-9); Macintosh (0-13-021476-0)
Based on the powerful testing technology developed by Engineering Software Associates, Inc. (ESA), Prentice Hall Custom Test includes all questions from the Test Item File and allows instructors to create and tailor exams to their own needs. With the Online Testing Program, exams can also be administered on line and data can then be automatically transferred for evaluation. A comprehensive desk reference guide is included along with online assistance.

Transparency Pack (0-13-021470-1)
Includes approximately 400 full color transparencies of images from the text.

Media Supplements

Physics for Scientists and Engineers Web Site www.prenhall.com/giancoli
A FREE innovative online resource that provides students with a wealth of activities and exercises for each text chapter. Features on the site include:

- Practice Questions, Destinations (links to related sites), NetSearch keywords and algorithmically generated numeric Practice Problems by Carl Adler of East Carolina University.
- Physlet Problems (Java-applet simulations) by Wolfgang Christian of Davidson College.
- Warmups and Puzzles essay questions and Applications from Gregor Novak and Andrew Gavrin at Indiana University-Purdue University, Indianapolis.
- Ranking Task Exercises edited by Tom O'Kuma of Lee College, Curtis Hieggelke of Joliet Junior College and David Maloney of Indiana University-Purdue University, Fort Wayne.

Using Prentice Hall CW '99 technology, the website grades and scores all objective questions, and results can be automatically e-mailed directly to the instructors if so desired. Instructors can also create customized syllabi online and link directly to activities on the Giancoli website.

Presentation Manager CD-ROM
Dual Platform (Windows/Macintosh; 0-13-214479-5)
This CD-ROM enables instructors to build custom sequences of Giancoli text images and Prentice Hall digital media for playback in lecture presentations. The CD-ROM contains all text illustrations, digitized segments from the Prentice Hall *Physics You Can See* videotape as well as additional lab and demonstration videos and animations from the Prentice Hall *Interactive Journey Through Physics* CD-ROM. Easy to navigate with Prentice Hall Presentation Manager software, instructors can preview, sequence, and play back images, as well as perform keyword searches, add lecture notes, and incorporate their own digital resources.

Physics You Can See *Video*
(0-205-12393-7)
Contains eleven two- to five-minute demonstrations of classical physics experiments. It includes segments such as "Coin and Feather" (acceleration due to gravity), "Monkey and Gun" (projectile motion), "Swivel Hips" (force pairs), and "Collapse a Can" (atmospheric pressure).

CAPA: A Computer-Assisted Personalized Approach to Assignments, Quizzes, and Exams
CAPA is an on-line homework system developed at Michigan State University that instructors can use to deliver problem sets with randomized variables for each student. The system gives students immediate feedback on their answers to problems, and records their participation and performance. Prentice Hall has arranged to have half of the even-numbered problems of Giancoli, *Physics for Scientists and Engineers, Third Edition*, coded for use with the CAPA system. For additional information about the CAPA system, please visit the web site at http://www.pa.msu.edu/educ/CAPA/.

WebAssign
WebAssign is a web-based homework delivery, collection, grading, and recording service developed and hosted by North Carolina State University. Prentice Hall will arrange for end-of-chapter problems from Giancoli, *Physics for Scientists and Engineers, Third Edition* to be coded for use with the *WebAssign* system for instructors who wish to take advantage of this service. For more information on the *WebAssign* system and its features, please visit http://webassign.net/info or e-mail webassign@ncsu.edu.

NOTES TO STUDENTS AND INSTRUCTORS ON THE FORMAT

1. Sections marked with a star (*) are considered optional. They can be omitted without interrupting the main flow of topics. No later material depends on them except possibly later starred sections. They may be fun to read though.

2. The customary conventions are used: symbols for quantities (such as m for mass) are italicized, whereas units (such as m for meter) are not italicized. Boldface (**F**) is used for vectors.

3. Few equations are valid in all situations. Where practical, the limitations of important equations are stated in square brackets next to the equation. The equations that represent the great laws of physics are displayed with a tan background, as are a few other equations that are so useful that they are indispensable.

4. The number of significant figures (see Section 1–3) should not be assumed to be greater than given: if a number is stated as (say) 6, with its units, it is meant to be 6 and not 6.0 or 6.00.

5. At the end of each chapter is a set of Questions that students should attempt to answer (to themselves at least). These are followed by Problems which are ranked as level I, II, or III, according to estimated difficulty, with level I Problems being easiest. These Problems are arranged by Section, but Problems for a given Section may depend on earlier material as well. There follows a group of General Problems, which are not arranged by Section nor ranked as to difficulty. Questions and Problems that relate to optional Sections are starred.

6. Being able to solve problems is a crucial part of learning physics, and provides a powerful means for understanding the concepts and principles. This book contains many aids to problem solving: (a) worked-out Examples and their solutions in the text, which are set off with a vertical blue line in the margin, and should be studied as an integral part of the text; (b) special "Problem-solving boxes" placed throughout the text to suggest ways to approach problem solving for a particular topic—but don't get the idea that every topic has its own "techniques," because the basics remain the same; (c) special problem-solving Sections (marked in blue in the Table of Contents); (d) "Problem solving" marginal notes (see point 8 below) which refer to hints for solving problems within the text; (e) some of the worked-out Examples are Estimation Examples, which show how rough or approximate results can be obtained even if the given data are sparse (see Section 1–6); and finally (f) the Problems themselves at the end of each chapter (point 5 above).

7. Conceptual Examples look like ordinary Examples but are conceptual rather than numerical. Each proposes a question or two, which hopefully starts you to think and come up with a response. Give yourself a little time to come up with your own response before reading the Response given.

8. Marginal notes: brief notes in the margin of almost every page are printed in blue and are of four types: (a) ordinary notes (the majority) that serve as a sort of outline of the text and can help you later locate important concepts and equations; (b) notes that refer to the great laws and principles of physics, and these are in capital letters and in a box for emphasis; (c) notes that refer to a problem-solving hint or technique treated in the text, and these say "Problem Solving"; (d) notes that refer to an application of physics, in the text or an Example, and these say "Physics Applied."

9. This book is printed in full color. But not simply to make it more attractive. The color is used above all in the Figures, to give them greater clarity for our analysis, and to provide easier learning of the physical principles involved. The Table on the next page is a summary of how color is used in the Figures, and shows which colors are used for the different kinds of vectors, for field lines, and for other symbols and objects. These colors are used consistently throughout the book.

10. Appendices include useful mathematical formulas (such as derivatives and integrals, trigonometric identities, areas and volumes, expansions), and a table of isotopes with atomic masses and other data. Tables of useful data are located inside the front and back covers.

Vectors

A general vector

    resultant vector (sum) is slightly thicker

    components of any vector are dashed

Displacement (**D, r**)

Velocity (**v**)

Acceleration (**a**)

Force (**F**)

    Force on second or

    third object in same figure

Momentum (**p** or m**v**)

Angular momentum (**L**)

Angular velocity (ω)

Torque (τ)

Electric field (**E**)

Magnetic field (**B**)

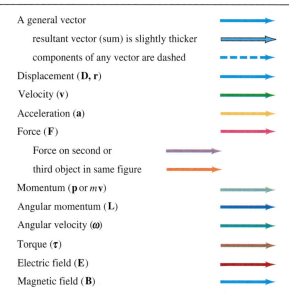

Electricity and magnetism

Electric field lines

Equipotential lines

Magnetic field lines

Electric charge (+)   +   or   • +

Electric charge (–)   –   or   • –

Electric circuit symbols

Wire

Resistor

Capacitor

Inductor

Battery

Optics

Light rays

Object

Real image (dashed)

Virtual image (dashed and paler)

Other

Energy level (atom, etc.)

Measurement lines   ⊢—1.0 m—⊣

Path of a moving object

Direction of motion or current

This comb has been passed through hair, or rubbed by a cloth or paper towel, which gave it a static electric charge. The electrical charge on the comb induces a polarization (separation of charge) in all those scraps of paper, and thus attracts them.

Our introduction to electricity in this chapter covers conductors and insulators, and Coulomb's law which relates the force between two point charges as a function of their distance apart. We introduce the concept of electric field, which is easy to understand and powerful to use.

Electric Charge and Electric Field

The word "electricity" may evoke an image of complex modern technology: computers, lights, motors, electric power. But the electric force would seem to play an even deeper role in our lives. According to atomic theory, electric forces between atoms and molecules hold them together to form liquids and solids, and electric forces are also involved in the metabolic processes that occur within our bodies. Many of the forces we have dealt with so far, such as elastic forces, the normal force, and other contact forces (pushes and pulls) are now considered to result from electric forces acting at the atomic level. Gravity, on the other hand, is a separate force.[†]

The earliest studies on electricity date back to the ancients, but it has been only in the past two centuries that electricity was studied in detail. We will discuss the development of ideas about electricity, including practical devices, as well as the relation to magnetism, in the next twelve chapters.

[†] As we discussed in Section 6–7, physicists in this century came to recognize four different fundamental forces in nature: (1) gravitational force, (2) electromagnetic force (we will see later that electric and magnetic forces are intimately related), (3) strong nuclear force, and (4) weak nuclear force. The last two forces operate at the level of the nucleus of an atom. A recent theory has combined the electromagnetic and weak nuclear forces so they are now considered to have a common origin known as the electroweak force.

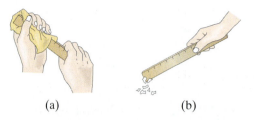

FIGURE 21–1 Rub a plastic ruler (a) and bring it close (b) to some tiny pieces of paper.

(a)   (b)

FIGURE 21–2 Unlike charges attract, whereas like charges repel one another.

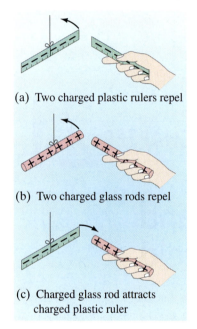

(a) Two charged plastic rulers repel

(b) Two charged glass rods repel

(c) Charged glass rod attracts charged plastic ruler

Likes repel; unlikes attract

21–1 Static Electricity; Electric Charge and Its Conservation

The word *electricity* comes from the Greek word *elektron*, which means "amber." Amber is petrified tree resin, and the ancients knew that if you rub an amber rod with a piece of cloth, the amber attracts small pieces of leaves or dust. A piece of hard rubber, a glass rod, or a plastic ruler rubbed with a cloth will also display this "amber effect," or **static electricity** as we call it today. You can readily pick up small pieces of paper with a plastic comb or ruler that you've just vigorously rubbed with even a paper towel. See the photo on the previous page and Fig. 21–1. You have probably experienced static electricity when combing your hair or when taking a synthetic blouse or shirt from a clothes dryer. And you may have felt a shock when you touched a metal doorknob after sliding across a car seat or walking across a nylon carpet. In each case, an object becomes "charged" due to a rubbing process and is said to possess a net **electric charge**.

Is all electric charge the same, or is it possible that there is more than one type? In fact, there are *two* types of electric charge, as the following simple experiments show. A plastic ruler is suspended by a thread and rubbed vigorously with a cloth to charge it. When a second ruler, which has also been charged in the same way, is brought close to the first, it is found that the one ruler *repels* the other. This is shown in Fig. 21–2a. Similarly, if a rubbed glass rod is brought close to a second charged glass rod, again a repulsive force is seen to act, Fig. 21–2b. However, if the charged glass rod is brought close to the charged plastic ruler, it is found that they *attract* each other, Fig. 21–2c. The charge on the glass must therefore be different from that on the plastic. Indeed, it is found experimentally that all charged objects fall into one of two categories. Either they are attracted to the plastic and repelled by the glass, just as glass is; or they are repelled by the plastic and attracted to the glass, just as the plastic ruler is. Thus there seem to be two, and only two, types of electric charge. Each type of charge repels the same type but attracts the opposite type. That is: **unlike charges attract; like charges repel.**

The two types of electric charge were referred to as *positive* and *negative* by the American statesman, philosopher, and scientist Benjamin Franklin (1706–1790). The choice of which name went with which type of charge was of course arbitrary. Franklin's choice set the charge on the rubbed glass rod to be positive charge, so the charge on a rubbed plastic ruler (or amber) is called negative charge. We still follow this convention today.

Franklin argued that whenever a certain amount of charge is produced on one body in a process, an equal amount of the opposite type of charge is produced on another body. The positive and negative are to be treated *algebraically*, so that during any process, the net change in the amount of charge produced is zero. For example, when a plastic ruler is rubbed with a paper towel, the plastic acquires a negative charge and the towel an equal amount of positive charge. The charges are separated, but the sum of the two is zero. This is an example of a law that is now well established: the **law of conservation of electric charge**, which states that

the net amount of electric charge produced in any process is zero.

If one object or one region of space acquires a positive charge, then an equal amount of negative charge will be found in neighboring areas or objects. No viola-

LAW OF CONSERVATION OF ELECTRIC CHARGE

tions have ever been found, and this conservation law is as firmly established as those for energy and momentum.

21–2 | Electric Charge in the Atom

Only within the past century has it become clear that an understanding of electricity begins inside the atom itself. In later chapters we will discuss atomic structure and the ideas that led to our present view of the atom in more detail. But it will help our understanding of electricity if we discuss it briefly now.

A simplified model of an atom shows it as having a tiny but heavy, positively charged nucleus surrounded by one or more negatively charged electrons (Fig. 21–3). The nucleus contains protons, which are positively charged, and neutrons, which have no net electric charge. All protons and all electrons have exactly the same magnitude of electric charge, but their signs are opposite. Hence, neutral atoms, having no net charge, contain equal numbers of protons and electrons. Sometimes, as we shall see, an atom may lose one or more of its electrons, or may gain extra electrons. In this case the atom will have a net positive or negative charge, and is called an **ion**.

In solid materials the nuclei tend to remain close to fixed positions, whereas some of the electrons move quite freely. The charging of a solid object by rubbing can be explained by the transfer of electrons from one material to the other. When a plastic ruler becomes negatively charged by rubbing with a paper towel, the transfer of electrons from the towel to the plastic leaves the towel with a positive charge equal in magnitude to the negative charge acquired by the plastic. In liquids and gases, nuclei or ions can move as well as electrons.

Normally when objects are charged by rubbing, they hold their charge only for a limited time and eventually return to the neutral state. Where does the charge go? In some cases it is neutralized by charged ions in the air (formed, for example, by collisions with charged particles known as cosmic rays that reach the Earth from space). Often more importantly, the charge can "leak off" onto water molecules in the air. This is because water molecules are **polar**—that is, even though they are neutral, their charge is not distributed uniformly, Fig. 21–4. Thus the extra electrons on, say, a charged plastic ruler can "leak off" into the air because they are attracted to the positive end of water molecules. A positively charged object, on the other hand, can be neutralized by transfer of loosely held electrons from water molecules in the air. On dry days, static electricity is much more noticeable since the air contains fewer water molecules to allow leakage. On humid or rainy days, it is difficult to make any object hold a net charge for long.

21–3 | Insulators and Conductors

Suppose we have two metal spheres, one highly charged and the other electrically neutral (Fig. 21–5a). If we now place a metal object, such as a nail, so that it touches both the spheres (Fig. 21–5b), it is found that the previously uncharged sphere quickly becomes charged. If, instead, we connect the two spheres together by a wooden rod or a piece of rubber (Fig. 21–5c), the uncharged ball does not become noticeably charged. Materials like the iron nail are said to be **conductors** of electricity, whereas wood and rubber are **nonconductors** or **insulators**.

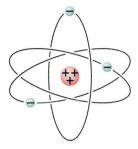

FIGURE 21–3 Simple model of the atom.

FIGURE 21–4 Diagram of a water molecule. Because it has opposite charges on different ends, it is called a "polar" molecule.

Charged   Neutral

(a)

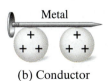

Metal

(b) Conductor

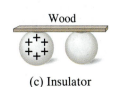

Wood

(c) Insulator

FIGURE 21–5 (a) A charged metal sphere and a neutral metal sphere. (b) The two spheres connected by a metal nail, which conducts charge from one sphere to the other. (c) The two spheres connected by an insulator (wood); almost no charge is conducted.

Metals are generally good conductors whereas most other materials are insulators (although even insulators conduct electricity very slightly). It is interesting that nearly all natural materials fall into one or the other of these two quite distinct categories. There are some materials, however (notably silicon, germanium, and carbon), that fall into an intermediate (but distinct) category known as **semiconductors**.

From the atomic point of view, the electrons in an insulating material are bound very tightly to the nuclei. In a good conductor, on the other hand, some of the electrons are bound very loosely and can move about freely within the material (although they cannot *leave* the object easily) and are often referred to as *free electrons* or *conduction electrons*. When a positively charged object is brought close to or touches a conductor, the free electrons in the conductor are attracted by this positive charge and move quickly toward it. On the other hand, the free electrons move swiftly away from a negative charge that is brought close. In a semiconductor, there are very few free electrons, and in an insulator, almost none.

(a) Neutral metal rod

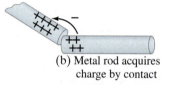

(b) Metal rod acquires charge by contact

FIGURE 21–6 (a) Neutral metal rod acquires a charge (b) when placed in contact with a charged metal object.

FIGURE 21–7 Charging by induction.

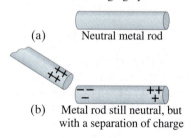

(a)   Neutral metal rod

(b)   Metal rod still neutral, but with a separation of charge

FIGURE 21–8 Inducing a charge on an object connected to ground.

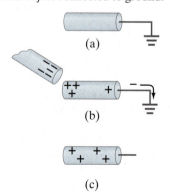

(a)

(b)

(c)

21–4   Induced Charge; the Electroscope

Suppose a positively charged metal object is brought close to an uncharged metal object. If the two touch, the free electrons in the neutral one are attracted to the positively charged object and some will pass over to it, Fig. 21–6. Since the second object is now missing some of its negative electrons, it will have a net positive charge. This process is called "charging by conduction," or "by contact," and the two objects end up with the same sign of charge.

Now suppose a positively charged object is brought close to a neutral metal rod, but does not touch it. Although the electrons of the metal rod do not leave the rod, they still move within the metal toward the charged object, which leaves a positive charge at the opposite end, Fig. 21–7. A charge is said to have been *induced* at the two ends of the metal rod. Of course no net charge has been created in the rod; charges have merely been *separated*. The net charge on the metal rod is still zero. However, if the metal were broken into two pieces, we could have two charged objects, one charged positively and one charged negatively.

Another way to induce a net charge on a metal object is to connect it with a conducting wire to the ground (or a conducting pipe leading into the ground) as shown in Fig. 21–8a ($\perp$ means "ground"). The object is then said to be "grounded" or "earthed." Now the Earth, since it is so large and can conduct, can easily accept or give up electrons; hence it acts like a reservoir for charge. If a charged object—say negative this time—is brought up close to the metal, free electrons in the metal are repelled and many of them move down the wire into the Earth, Fig. 21–8b. This leaves the metal positively charged. If the wire is now cut, the metal will have a positive induced charge on it (Fig. 21–8c). If the wire were cut after the negative object is moved away, the electrons would all have moved back into the metal and it would be neutral.

An **electroscope** is a device that can be used for detecting charge. As shown in Fig. 21–9, inside of a case are two movable metal leaves, often made of gold. (Sometimes only one leaf is movable.) The leaves are connected by a conductor to a metal knob on the outside of the case, but are insulated from the case itself. If a positively charged object is brought close to the knob, a separation of charge is induced, as electrons are attracted up into the knob, leaving the leaves positively charged, Fig. 21–10a. The two leaves repel each other as shown. If, instead, the

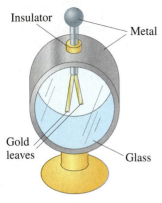

FIGURE 21–9 Electroscope.

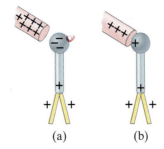

(a)   (b)

FIGURE 21–10 Electroscope charged
(a) by induction, (b) by conduction.

knob is charged by conduction, the whole apparatus acquires a net charge as shown in Fig. 21–10b. In either case, the greater the amount of charge, the greater the separation of the leaves.

Note, however that you cannot tell the sign of the charge in this way, since a negative charge will cause the leaves to separate just as much as an equal magnitude positive charge—in either case the two leaves repel each other. An electroscope can, however, be used to determine the sign of the charge if it is first charged by conduction, say negatively, as in Fig. 21–11a. Now if a negative object is brought close, as in Fig. 21–11b, more electrons are induced to move down into the leaves and they separate further. On the other hand, if a positive charge is brought close, the electrons are induced to flow upward, leaving the leaves less negative and their separation is reduced, Fig. 21–11c.

The electroscope was much used in the early studies of electricity. The same principle, aided by some electronics, is used in much more sensitive modern **electrometers**.

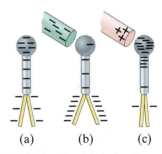

(a)   (b)   (c)

FIGURE 21–11 A previously charged electroscope can be used to determine the sign of a given charge.

21–5   Coulomb's Law

We have seen that an electric charge exerts a force on other electric charges. What factors affect the magnitude of this force? To answer this, the French physicist Charles Coulomb (1736–1806) investigated electric forces in the 1780s using a torsion balance (Fig. 21–12) much like that used by Cavendish for his studies of the gravitational force (Chapter 6).

Although precise instruments for the measurement of electric charge were not available in Coulomb's time, he was able to prepare small spheres with different magnitudes of charge in which the *ratio* of the charges was known. He reasoned that if a charged conducting sphere is placed in contact with an identical uncharged sphere, the charge on the first would be shared equally by the two of them because of symmetry. He thus had a way to produce charges equal to $\frac{1}{2}, \frac{1}{4}$, and so on, of the original charge. Although he had some difficulty with induced charges, Coulomb was able to argue that the force one tiny charged object exerted on a second tiny charged object is directly proportional to the charge on each of them. That is, if the charge on either one of the objects was doubled, the force was doubled; and if the charge on both of the objects was doubled, the force increased to four times the original value. This was the case when the distance between the two charges remained the same. If the distance between them was allowed to increase, he found that the force decreased with the *square of the distance* between

FIGURE 21–12 Principle of Coulomb's apparatus. It is similar to Cavendish's, which was used for the gravitational force. When a charged sphere is placed close to the one on the suspended bar, the bar rotates slightly. The suspending fiber resists the twisting motion and the angle of twist is proportional to the force applied. By the use of this apparatus, Coulomb investigated how the electric force varies as a function of the magnitude of the charges and of the distance between them.

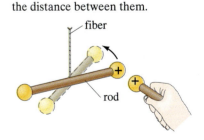

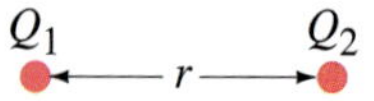

FIGURE 21–13 Coulomb's law, Eq. 21–1, gives the force between two point charges, Q_1 and Q_2, a distance r apart.

COULOMB'S LAW

them. That is, if the distance was doubled, the force fell to one-fourth of its original value. Thus, Coulomb concluded, the force one tiny charged object exerts on a second one is proportional to the product of the magnitude of the charge on one, Q_1, times the magnitude of the charge on the other, Q_2, and inversely proportional to the square of the distance r between them (Fig. 21–13). As an equation, we can write **Coulomb's law** as

$$F = k\frac{Q_1 Q_2}{r^2}, \qquad (21\text{–}1)$$

where k is a proportionality constant.

Equation 21–1 gives the *magnitude* of the electric force that either object exerts on the other. The *direction* of the electric force *is always along the line joining the two objects*. If the two charges have the same sign, the force on either object is directed away from the other. If the two charges have opposite signs, the force on one is directed toward the other, Fig. 21–14. Notice that the force one charge exerts on the second is equal but opposite to that exerted by the second on the first, in accord with Newton's third law.

The validity of Coulomb's law today rests on precision measurements that are much more sophisticated than Coulomb's original experiment. The exponent, 2, in Coulomb's law has been shown to be accurate to 1 part in 10^{16} [that is, $2 \pm (1 \times 10^{-16})$].

Since we are dealing here with a new quantity (electric charge), we could choose its unit so that the proportionality constant k in Eq. 21–1 would be one. Indeed, such a system of units was once common.[†] However, the most widely used unit now is the **coulomb** (C), which is the SI unit. The precise definition of the coulomb today is in terms of electric current and magnetic field, and will be discussed later (Section 28–3). In SI units, k has the value

$$k = 8.988 \times 10^9 \, \text{N} \cdot \text{m}^2/\text{C}^2 \approx 9.0 \times 10^9 \, \text{N} \cdot \text{m}^2/\text{C}^2.$$

*Unit for charge:
the coulomb*

FIGURE 21–14 Direction of the force depends on whether the charges have the same sign, (a) and (b), or opposite signs (c).

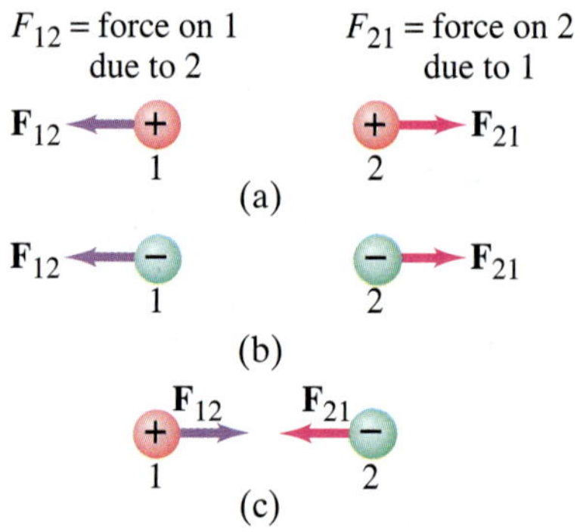

F_{12} = force on 1 due to 2 F_{21} = force on 2 due to 1

(a)

(b)

(c)

*Charge on electron
(the elementary charge)*

Thus, 1 C is that amount of charge which, if placed on each of two point objects that are 1.0 m apart, will result in each object exerting a force of $(9.0 \times 10^9 \, \text{N} \cdot \text{m}^2/\text{C}^2)(1.0 \, \text{C})(1.0 \, \text{C})/(1.0 \, \text{m})^2 = 9.0 \times 10^9 \, \text{N}$ on the other. This would be an enormous force, equal to the weight of almost a million tons. We don't normally encounter charges as large as a coulomb.

Charges produced by rubbing ordinary objects (such as a comb or plastic ruler) are typically around a microcoulomb $(1 \, \mu\text{C} = 10^{-6} \, \text{C})$ or less. Objects that carry a positive charge have a deficit of electrons, whereas negatively charged objects have an excess of electrons. The magnitude of the charge on one electron has been determined to be about $1.602 \times 10^{-19} \, \text{C}$, and its sign is negative. This is the smallest charge found in nature,[‡] and because of its fundamental nature, it is given the symbol e and is often referred to as the *elementary charge*:

$$e = 1.602 \times 10^{-19} \, \text{C}.$$

Note that e is defined as a positive number, so the charge on the electron is $-e$. (The charge on a proton, on the other hand, is $+e$). Since an object cannot gain or lose a fraction of an electron, the net charge on any object must be an integral multiple of this charge. Electric charge is thus said to be **quantized** (existing only in discrete amounts: $1e$, $2e$, $3e$, etc.). Because e is so small, however, we normally don't notice this discreteness in macroscopic charges $(1 \, \mu\text{C}$ requires about 10^{13} electrons), which thus seem continuous.

Electric charge is quantized

[†] This is a cgs system of units, and the unit of electric charge is called the *electrostatic unit* (esu) or the statcoulomb. One esu is defined as that charge, on each of two point objects 1 cm apart, that gives rise to a force of 1 dyne.

[‡] According to the standard model of elementary particle physics, subnuclear particles called quarks have a smaller charge than that on the electron, equal to $\frac{1}{3}e$ or $\frac{2}{3}e$. Quarks have not been detected directly as isolated objects, and theory indicates that free quarks may not be detectable.

Note the similarity of Coulomb's law to the law of universal gravitation, Eq. 6–1. Both are inverse square laws $\left(F \propto 1/r^2\right)$. Both also have a proportionality to a product of a property of each body—mass for gravity, electric charge for electricity. A major difference between the two laws is that gravity is always an attractive force, whereas the electric force can be either attractive or repulsive.

The constant k in Eq. 21–1 is often written in terms of another constant, ϵ_0, called the **permittivity of free space**. It is related to k by $k = 1/4\pi\epsilon_0$. Coulomb's law can then be written

$$F = \frac{1}{4\pi\epsilon_0}\frac{Q_1 Q_2}{r^2},$$ (21–2)

where

$$\epsilon_0 = \frac{1}{4\pi k} = 8.85 \times 10^{-12}\,\text{C}^2/\text{N}\cdot\text{m}^2.$$

COULOMB'S LAW (in terms of ϵ_0)

Equation 21–2 looks more complicated than Eq. 21–1, but other fundamental equations we haven't seen yet are simpler in terms of ϵ_0 rather than k. It doesn't matter which form we use, of course, since Eqs. 21–1 and 21–2 are equivalent.

It should be recognized that Eqs. 21–1 and 21–2 apply to objects whose size is much smaller than the distance between them. Ideally, it is precise for **point charges** (spatial size negligible compared to other distances). For finite-sized objects, it is not always clear what value to use for r, particularly since the charge may not be distributed uniformly on the objects. If the two objects are spheres and the charge is known to be distributed uniformly on each, then r is the distance between their centers.

Coulomb's law describes the force between two charges when they are at rest. Additional forces come into play when charges are in motion, and these will be discussed in later chapters. In this chapter we discuss only charges at rest, the study of which is called **electrostatics**.

When calculating with Coulomb's law, we can usually ignore the signs of the charges and determine direction based on whether the force is attractive or repulsive.

EXAMPLE 21–1 **Electric force on electron by proton.** Determine the magnitude of the electric force on the electron of a hydrogen atom exerted by the single proton $\left(Q_2 = +e\right)$ that is its nucleus. Assume the electron "orbits" the proton at its average distance of $r = 0.53 \times 10^{-10}$ m, Fig. 21–15.

SOLUTION We use Coulomb's law, $F = k Q_1 Q_2/r^2$ (Eq. 21–1), with $r = 0.53 \times 10^{-10}$ m, and $Q_1 = Q_2 = 1.6 \times 10^{-19}$ C (ignoring the signs of the charges):

$$F = \frac{\left(9.0 \times 10^9\,\text{N}\cdot\text{m}^2/\text{C}^2\right)\left(1.6 \times 10^{-19}\,\text{C}\right)\left(1.6 \times 10^{-19}\,\text{C}\right)}{\left(0.53 \times 10^{-10}\,\text{m}\right)^2}$$

$$= 8.2 \times 10^{-8}\,\text{N}.$$

The direction of the force on the electron is toward the proton, since the charges have opposite signs and the force is attractive.

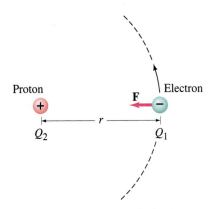

FIGURE 21–15 Example 21–1.

CONCEPTUAL EXAMPLE 21–2 **Which charge exerts the greater force?** Two positive point charges, $Q_1 = 50\,\mu\text{C}$ and $Q_2 = 1\,\mu\text{C}$, are separated by a distance l, Fig. 21–16. Which is larger in magnitude, the force that Q_1 exerts on Q_2, or the force that Q_2 exerts on Q_1?

RESPONSE From Coulomb's law, the force on Q_1 exerted by Q_2 is:

$$F_{12} = k\frac{Q_1 Q_2}{l^2}.$$

The force on Q_2 exerted by Q_1 is the same except that Q_1 and Q_2 are reversed. The equation is symmetric with respect to the two charges, so $F_{21} = F_{12}$. Newton's third law also tells us that these two forces must have equal magnitude.

FIGURE 21–16 Example 21–2.

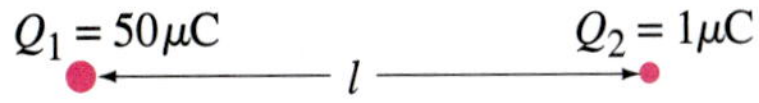

$Q_1 = 50\,\mu\text{C}$ $Q_2 = 1\,\mu\text{C}$

l

It is very important to keep in mind that Eq. 21–1 (or 21–2) gives the force on a charge due to only *one* other charge. If several (or many) charges are present, the *net force on any one of them will be the vector sum of the forces due to each of the others*. This **principle of superposition** is based on experiment, and tells us that electric force vectors add like any other vector. For continuous distributions of charge, the sum becomes an integral.

Superposition principle: electric forces add as vectors

When dealing with several charges, it is helpful to use double subscripts on each of the forces involved. The first subscript refers to the particle *on* which the force acts; the second refers to the particle that exerts the force. For example, if we have three charges, $\mathbf{F}_{31}$ means the force exerted *on* particle 3 *by* particle 1.

As in all problem solving, it is very important to draw a diagram, in particular a free-body diagram (Chapter 4) for each body, showing all the forces acting on that body. In applying Coulomb's law, we can deal with charge magnitudes only (leaving out minus signs) to get the magnitude of each force. Then determine the direction of the force physically (along the line joining the two particles: like charges repel, unlike attract), and show the force on the diagram. Finally, add all the forces on one object together as vectors.

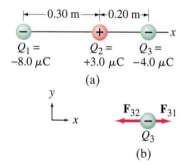

FIGURE 21–17 Diagram for Example 21–3.

EXAMPLE 21–3 Three charges in a line. Three charged particles are arranged in a line, as shown in Fig. 21–17a. Calculate the net electrostatic force on particle 3 (the $-4.0\ \mu\text{C}$ on the right) due to the other two charges.

SOLUTION The net force on particle 3 will be the vector sum of the force $\mathbf{F}_{31}$ exerted by particle 1 and the force $\mathbf{F}_{32}$ exerted by particle 2: $\mathbf{F} = \mathbf{F}_{31} + \mathbf{F}_{32}$. The magnitudes of these two forces are

$$F_{31} = \frac{(9.0 \times 10^9\ \text{N} \cdot \text{m}^2/\text{C}^2)(4.0 \times 10^{-6}\ \text{C})(8.0 \times 10^{-6}\ \text{C})}{(0.50\ \text{m})^2} = 1.2\ \text{N}$$

$$F_{32} = \frac{(9.0 \times 10^9\ \text{N} \cdot \text{m}^2/\text{C}^2)(4.0 \times 10^{-6}\ \text{C})(3.0 \times 10^{-6}\ \text{C})}{(0.20\ \text{m})^2} = 2.7\ \text{N}.$$

Since we were calculating the magnitudes of the forces, we omitted the signs of the charges; but we must be aware of them to get the direction of each force. Let the line joining the particles be the x axis, and we take it positive to the right. Then, because $\mathbf{F}_{31}$ is repulsive and $\mathbf{F}_{32}$ is attractive, the directions of the forces are as shown in Fig. 21–17b: F_{31} points in the positive x direction and F_{32} points in the negative x direction. The net force on particle 3 is then

$$F = -F_{32} + F_{31} = -2.7\ \text{N} + 1.2\ \text{N} = -1.5\ \text{N}.$$

The magnitude of the net force is 1.5 N, and it points to the left.

Notice in this Example that the charge in the middle (Q_2) in no way blocks the effect of the other charge (Q_1).

EXAMPLE 21–4 Electric force using vector components. Calculate the net electrostatic force on charge Q_3 shown in Fig. 21–18a due to the charges Q_1 and Q_2.

SOLUTION The forces $\mathbf{F}_{31}$ and $\mathbf{F}_{32}$ have the directions shown in the diagram since Q_1 exerts an attractive force and Q_2 a repulsive force. The magnitudes of $\mathbf{F}_{31}$ and $\mathbf{F}_{32}$ are (ignoring signs since we know the directions)

$$F_{31} = \frac{(9.0 \times 10^9\ \text{N} \cdot \text{m}^2/\text{C}^2)(6.5 \times 10^{-5}\ \text{C})(8.6 \times 10^{-5}\ \text{C})}{(0.60\ \text{m})^2} = 140\ \text{N},$$

$$F_{32} = \frac{(9.0 \times 10^9\ \text{N} \cdot \text{m}^2/\text{C}^2)(6.5 \times 10^{-5}\ \text{C})(5.0 \times 10^{-5}\ \text{C})}{(0.30\ \text{m})^2} = 330\ \text{N}.$$

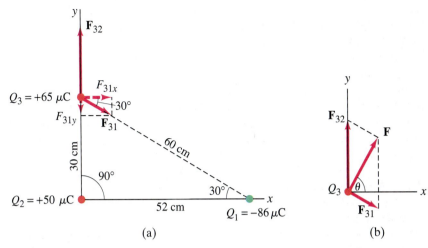

FIGURE 21–18 Determining the forces for Example 21–4.

We resolve $\mathbf{F}_{31}$ into its components along the x and y axes, as shown in Fig. 21–18a:

$$F_{31x} = F_{31} \cos 30° = 120 \,\text{N},$$

$$F_{31y} = -F_{31} \sin 30° = -70 \,\text{N}.$$

The force $\mathbf{F}_{32}$ has only a y component. So the net force $\mathbf{F}$ on Q_3 has components

$$F_x = F_{31x} = 120 \,\text{N}$$

$$F_y = F_{32} + F_{31y} = 330 \,\text{N} - 70 \,\text{N} = 260 \,\text{N}.$$

Thus the magnitude of the net force is

$$F = \sqrt{F_x^2 + F_y^2} = \sqrt{(120 \,\text{N})^2 + (260 \,\text{N})^2} = 290 \,\text{N};$$

and it acts at an angle θ (see Fig. 21–18b) given by $\tan \theta = F_y/F_x =$ 260 N/120 N = 2.2, so $\theta = 65°$.

* Vector Form of Coulomb's Law

Coulomb's law can be written in vector form (as we did for Newton's law of universal gravitation in Chapter 6, Section 6–2) as

$$\mathbf{F}_{12} = k \frac{Q_1 Q_2}{r_{21}^2} \hat{\mathbf{r}}_{21},$$

where $\mathbf{F}_{12}$ is the vector force on charge Q_1 due to Q_2 and $\hat{\mathbf{r}}_{21}$ is the unit vector pointing from Q_2 toward Q_1. That is $\mathbf{r}_{21}$ points from the "source" charge (Q_2) toward the charge on which we want to know the force (Q_1). See Fig. 21–19. The charges Q_1 and Q_2 can be either positive or negative, and this will affect the direction of the electric force. If Q_1 and Q_2 have the same sign, the product $Q_1 Q_2 > 0$ and the force on Q_1 points away from Q_2—that is, it is repulsive. If Q_1 and Q_2 have opposite signs, $Q_1 Q_2 < 0$ and $\mathbf{F}_{12}$ points toward Q_2—that is, it is attractive.

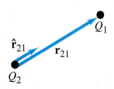

FIGURE 21–19 Determining the force on Q_1 due to Q_2, showing the direction of the unit vector $\hat{\mathbf{r}}_{21}$.

21–6 | The Electric Field

FIGURE 21–20 An electric field surrounds every charge. P is an arbitrary point.

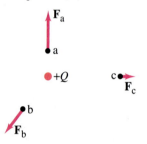

FIGURE 21–21 Force exerted by charge $+Q$ on a small test charge, q, placed at points a, b, and c.

Both the gravitational force and the electrical force act over a distance: there is a force even when the two objects are not touching. The idea of a force *acting at a distance* was a difficult one for early thinkers. Newton himself felt uneasy with this idea when he published his law of universal gravitation. A helpful way to look at the situation uses the idea of the **field**, developed by the British scientist Michael Faraday (1791–1867). In the electrical case, according to Faraday, an *electric field* extends outward from every charge and permeates all of space (Fig. 21–20). When a second charge is placed near the first charge, it feels a force because of the electric field that is there (say, at point P in Fig. 21–20). The electric field at the location of the second charge is considered to interact directly with this charge to produce the force.

We can in principle investigate the electric field surrounding a charge or group of charges by measuring the force on a small positive **test charge**. By a test charge we mean a charge so small that the force it exerts does not significantly alter the distribution of the charges that create the field being measured. The force on a tiny positive test charge q placed at various locations in the vicinity of a single positive charge Q would be as shown in Fig. 21–21. The force at b is less than at a because the distance is greater (Coulomb's law); and the force at c is smaller still. In each case, the force is directed radially outward from Q. The electric field is defined in terms of the force on such a positive test charge. In particular, the **electric field**, **E**, at any point in space is defined as the force **F** exerted on a tiny positive test charge at that point divided by the magnitude of the test charge q:

Definition of electric field

$$\mathbf{E} = \frac{\mathbf{F}}{q}. \tag{21–3}$$

Ideally, **E** is defined as the limit of $\mathbf{F}/q$ as q is taken smaller and smaller, approaching zero. From this definition (Eq. 21–3), we see that the electric field at any point in space is a vector whose direction is the direction of the force on a positive test charge at that point, and whose magnitude is the *force per unit charge*. Thus **E** is measured in units of newtons per coulomb (N/C).

The reason for defining **E** as $\mathbf{F}/q$ (with $q \to 0$) is so that **E** does not depend on the magnitude of the test charge q. This means that **E** describes only the effect of the charges creating the electric field at that point.

The electric field at any point in space can be measured, based on the definition, Eq. 21–3. For simple situations involving one or several point charges, we can calculate what **E** will be. For example, the electric field at a distance r from a single point charge Q would have magnitude

$$E = \frac{F}{q} = \frac{kqQ/r^2}{q}$$

$$= k\frac{Q}{r^2}; \qquad \text{[single point charge]} \quad \textbf{(21–4a)}$$

or, in terms of ϵ_0 as in Eq. 21–2 $\left(k = 1/4\pi\epsilon_0\right)$:

Electric field due to one point charge

$$E = \frac{1}{4\pi\epsilon_0}\frac{Q}{r^2}. \qquad \text{[single point charge]} \quad \textbf{(21–4b)}$$

Notice that E is independent of q—that is, it depends only on the charge Q which produces the field, and not on the value of the test charge q. Equations 21–4 could be referred to as the electric field form of Coulomb's law.

EXAMPLE 21–5 **Electrostatic copier.** An electrostatic copier works by selectively arranging positive charges (in a pattern to be copied) on the surface of a nonconducting drum, then gently sprinkling negatively charged dry toner (ink) particles onto the drum. The toner particles temporarily stick to the pattern on the drum and are later transferred to paper and "melted" to produce the copy. Suppose each toner particle has a mass of 9.0×10^{-16} kg and carries an average of 20 extra electrons to provide an electric charge. Assuming that the electric force on a toner particle must exceed twice its weight in order to ensure sufficient attraction, compute the required electric field strength near the surface of the drum. See Fig. 21–22.

SOLUTION The minimum value of electric field satisfies the relation

$$qE = 2\,mg$$

where $q = 20e$. Hence

$$E = \frac{2\,mg}{q} = \frac{2(9.0 \times 10^{-16}\,\text{kg})(9.8\,\text{m/s}^2)}{20(1.6 \times 10^{-19}\,\text{C})}$$

$$= 5.5 \times 10^{3}\,\text{N/C}.$$

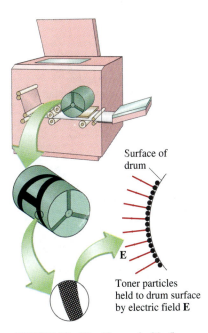

Surface of drum

E

Toner particles held to drum surface by electric field **E**

FIGURE 21–22 Example 21–5.

EXAMPLE 21–6 **Electric field of a single point charge.** Calculate the magnitude and direction of the electric field at a point P which is 30 cm to the right of a point charge $Q = -3.0 \times 10^{-6}$ C.

SOLUTION The magnitude of the electric field due to a single point charge is given by Eq. 21–4:

$$E = k\frac{Q}{r^2} = \frac{(9.0 \times 10^9\,\text{N}\cdot\text{m}^2/\text{C}^2)(3.0 \times 10^{-6}\,\text{C})}{(0.30\,\text{m})^2} = 3.0 \times 10^5\,\text{N/C}.$$

The direction of the electric field is *toward* the charge Q as shown in Fig. 21–23a since we defined the direction as that of the force on a positive test charge. If Q had been positive, the electric field would have pointed away, as in Fig. 21–23b.

FIGURE 21–23 Example 21–6. Electric field at point P (a) due to a negative charge Q, and (b) due to a positive charge Q.

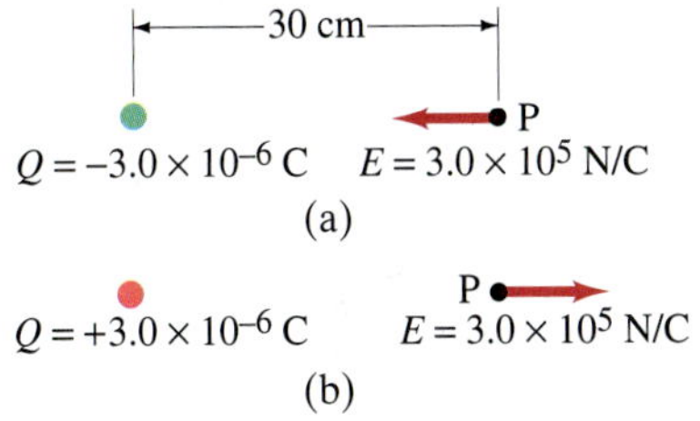

|←——30 cm——→|

$Q = -3.0 \times 10^{-6}$ C $E = 3.0 \times 10^5$ N/C   P
(a)

$Q = +3.0 \times 10^{-6}$ C   P $E = 3.0 \times 10^5$ N/C
(b)

This Example illustrates a general result: The electric field due to a positive charge points away from the charge, whereas **E** due to a negative charge points toward that charge.

If the field is due to more than one charge, the individual fields (call them $\mathbf{E}_1$, $\mathbf{E}_2$, etc.) due to each charge are added vectorially to get the total field at any point:

$$\mathbf{E} = \mathbf{E}_1 + \mathbf{E}_2 + \cdots. \qquad (21\text{–}5)$$

The validity of this **superposition principle** for electric fields is fully confirmed by experiment.

If we are given the electric field **E** at a given point in space, then we can calculate the force **F** on any charge q (even if not small) placed at that point by writing (see Eq. 21–3):

$$\mathbf{F} = q\mathbf{E}.$$

If q is positive, **F** and **E** will point in the same direction. If q is negative, **F** and **E** point in opposite directions. See Fig. 21–24.

FIGURE 21–24 (a) Electric field at a given point in space. (b) Force on a positive charge. (c) Force on a negative charge.

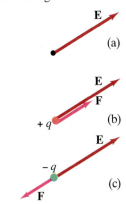

E (a)

E F $+q$ (b)

E $-q$ F (c)

$Q_1 = -25\,\mu C$ P $Q_2 = +50\,\mu C$

$\leftarrow r_1 = 2.0\,\text{cm} \rightarrow\!\leftarrow$ ——— $r_2 = 8.0\,\text{cm}$ ———$\rightarrow$

(a)

Q_1 $\mathbf{E}_1$ Q_2

$\mathbf{E}_2$ (b)

FIGURE 21–25 Example 21–7.
In (b), we don't know the relative
lengths of $\mathbf{E}_1$ and $\mathbf{E}_2$ until we do the
calculation.

EXAMPLE 21–7 **E in between two point charges.** Two point charges are separated by a distance of 10.0 cm. One has a charge of $-25\,\mu C$ and the other $+50\,\mu C$. (*a*) What is the direction and magnitude of the electric field at a point P in between them, that is 2.0 cm from the negative charge (Fig. 21–25a)? (*b*) If an electron is placed at rest at P, what will its acceleration (direction and magnitude) be initially?

SOLUTION (*a*) The field will be a combination of two fields both pointing to the left: the field due to the negative charge Q_1 points toward Q_1, and the field due to the positive charge Q_2 points away from Q_2, again to the left, Fig. 21–25b. Thus, we can add the magnitudes of the two fields together algebraically:

$$E \;=\; k\frac{Q_1}{r_1^2} + k\frac{Q_2}{r_2^2} \;=\; k\left(\frac{Q_1}{r_1^2} + \frac{Q_2}{r_2^2}\right) \;=\; k\frac{Q_1}{r_1^2}\left[1 + \frac{(Q_2/Q_1)}{(r_2^2/r_1^2)}\right]$$

where in the last step we factored out (Q_1/r_1^2). We substitute in $r_1 = 2.0\,\text{cm} = 2.0 \times 10^{-2}\,\text{m}$ and $r_2 = 8.0 \times 10^{-2}\,\text{m}$:

$$E \;=\; (9.0 \times 10^9\,\text{N}\cdot\text{m}^2/\text{C}^2)\frac{(25 \times 10^{-6}\,\text{C})}{(2.0 \times 10^{-2}\,\text{m})^2}\left[1 + \frac{(50/25)}{(8.0/2.0)^2}\right]$$

$$=\; 5.6 \times 10^8\left[1 + \tfrac{1}{8}\right]\text{N/C} \;=\; 6.3 \times 10^8\,\text{N/C}.$$

Notice how factoring out Q_1/r_1^2 on the first line allowed us to see the relative strengths of the two contributing fields—namely that Q_2's field is only $\tfrac{1}{8}$ of Q_1's (or $\tfrac{1}{9}$ of the total).

(*b*) The electron will feel a force to the *right* since it is negatively charged and the acceleration will therefore be to the right. From the definition of electric field, Eq. 21–3, the force on any charge q (even if not small) placed in an electric field E is given by $F = qE$. Hence the magnitude of the acceleration is

$$a \;=\; \frac{F}{m} \;=\; \frac{qE}{m} \;=\; \frac{(1.60 \times 10^{-19}\,\text{C})(6.3 \times 10^8\,\text{N/C})}{9.1 \times 10^{-31}\,\text{kg}} \;=\; 1.1 \times 10^{20}\,\text{m/s}^2.$$

EXAMPLE 21–8 **E above two point charges.** Calculate the total electric field (*a*) at point A and (*b*) at point B in Fig. 21–26 due to both charges, Q_1 and Q_2.

SOLUTION (*a*) The calculation is much like that of Example 21–4, but now we are dealing with electric fields. The electric field at A is the vector sum of the fields $\mathbf{E}_{A1}$ due to Q_1, and $\mathbf{E}_{A2}$ due to Q_2; for each point charge $E = kQ/r^2$, so

$$E_{A1} \;=\; \frac{(9.0 \times 10^9\,\text{N}\cdot\text{m}^2/\text{C}^2)(50 \times 10^{-6}\,\text{C})}{(0.60\,\text{m})^2} \;=\; 1.25 \times 10^6\,\text{N/C},$$

$$E_{A2} \;=\; \frac{(9.0 \times 10^9\,\text{N}\cdot\text{m}^2/\text{C}^2)(50 \times 10^{-6}\,\text{C})}{(0.30\,\text{m})^2} \;=\; 5.0 \times 10^6\,\text{N/C}.$$

The directions are as shown, so the total electric field at A, $\mathbf{E}_A$, has components

$$E_{Ax} \;=\; E_{A1}\cos 30° \;=\; 1.1 \times 10^6\,\text{N/C},$$
$$E_{Ay} \;=\; E_{A2} - E_{A1}\sin 30° \;=\; 4.4 \times 10^6\,\text{N/C}.$$

➡ **PROBLEM SOLVING**

Ignore signs of charges and determine direction physically, showing directions on diagram

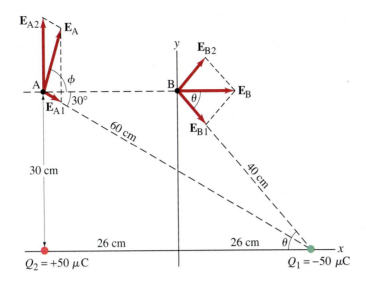

FIGURE 21–26 Calculation of the electric field at points A and B for Example 21–8.

Thus the magnitude of $\mathbf{E}_A$ is

$$E_A = \sqrt{(1.1)^2 + (4.4)^2} \times 10^6\,\text{N/C} = 4.5 \times 10^6\,\text{N/C},$$

and its direction is ϕ given by $\tan\phi = E_{Ay}/E_{Ax} = 4.4/1.1 = 4.0$, so $\phi = 76°$.
(b) Because B is equidistant (40 cm by the Pythagorean theorem) from the two equal charges, the magnitudes of E_{B1} and E_{B2} are the same; that is,

$$E_{B1} = E_{B2} = \frac{kQ}{r^2} = \frac{(9.0 \times 10^9\,\text{N·m}^2/\text{C}^2)(50 \times 10^{-6}\,\text{C})}{(0.40\,\text{m})^2}$$

$$= 2.8 \times 10^6\,\text{N/C}.$$

Also, because of the symmetry, the y components are equal and opposite. Hence the total field E_B is horizontal and equals $E_{B1}\cos\theta + E_{B2}\cos\theta = 2E_{B1}\cos\theta$; from the diagram, $\cos\theta = 26\,\text{cm}/40\,\text{cm} = 0.65$. Then

$$E_B = 2E_{B1}\cos\theta = 2(2.8 \times 10^6\,\text{N/C})(0.65) = 3.6 \times 10^6\,\text{N/C},$$

and the direction of $\mathbf{E}_B$ is along the $+x$ direction.

➡ **PROBLEM SOLVING**

Use symmetry to save work, when possible

PROBLEM SOLVING Electrostatics

Solving electrostatics problems follows, to a large extent, the general problem-solving procedure discussed in Section 4–8. In particular,

1. Draw a careful diagram—namely, a free-body diagram for each object, showing all the forces acting on that object, or the electric field at a point due to all sources.

2. Apply Coulomb's law to get the magnitude of the force that each contributing charge exerts on a charged object, or the electric field at a point. Deal only with magnitudes of charges (leaving out minus signs), and obtain the magnitude of each force or electric field. Then determine the direction of each force or electric field physically (like charges repel each other, unlike charges attract). Show and label each vector force or field on your diagram. Then add vectorially all the forces on an object, or the contributing fields at a point, to get the resultant.

3. Use symmetry (say, in the geometry) whenever possible.

21–7 Electric Field Calculations for Continuous Charge Distributions

In many cases we can treat[†] charge as being distributed continuously. We can divide up a charge distribution into infinitesimal charges dQ, each of which will act as a tiny point charge. The contribution to the electric field at a distance r from each dQ is

$$dE = \frac{1}{4\pi\epsilon_0}\frac{dQ}{r^2}. \qquad (21\text{–}6a)$$

Then the electric field, **E**, at any point is obtained by summing over all the infinitesimal contributions, which is the integral

$$\mathbf{E} = \int d\mathbf{E}. \qquad (21\text{–}6b)$$

Note that $d\mathbf{E}$ is a vector (Eq. 21–6a gives its magnitude). [In situations where Eq. 21–6b is difficult to evaluate, other techniques (discussed in the next two chapters) can often be used instead to determine **E**. Numerical integration can also be used in many cases.]

EXAMPLE 21–9 **A ring of charge.** A thin, ring-shaped object of radius a holds a total charge Q distributed uniformly around it. Determine the electric field at a point P on its axis, a distance x from the center. See Fig. 21–27. Let λ be the charge per unit length (C/m).

SOLUTION The electric field, $d\mathbf{E}$, due to a particular segment of the ring of length dl has magnitude

$$dE = \frac{1}{4\pi\epsilon_0}\frac{dQ}{r^2}.$$

The whole ring has length (circumference) of $2\pi a$, so the charge on a length dl is

$$dQ = Q\left(\frac{dl}{2\pi a}\right) = \lambda\,dl$$

where $\lambda = Q/2\pi a$ is the charge per unit length. Now we write dE as

$$dE = \frac{1}{4\pi\epsilon_0}\frac{\lambda\,dl}{r^2}.$$

The vector $d\mathbf{E}$ has components dE_x along the x axis and $dE_\perp$ perpendicular to the x axis (Fig. 21–27). We are going to sum (integrate) around the entire ring. We note that an equal-length segment diametrically opposite the dl shown will produce a $d\mathbf{E}$ whose component perpendicular to the x axis will just cancel the $dE_\perp$ shown. This is true for all segments of the ring, so by symmetry **E** will be

→ **PROBLEM SOLVING**

Use symmetry when possible

[†] Because we believe there is a minimum charge (e), the treatment here is only for convenience; it is nonetheless useful and accurate since e is usually very much smaller than macroscopic charges.

FIGURE 21–27 Example 21–9.

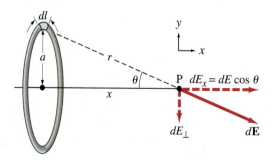

directed along the x axis and we need only sum the x components, dE_x. The total field is then

$$E = E_x = \int dE_x = \int dE \cos\theta = \frac{1}{4\pi\epsilon_0} \lambda \int \frac{dl}{r^2} \cos\theta.$$

Since $\cos\theta = x/r$, where $r = (x^2 + a^2)^{\frac{1}{2}}$, we have

$$E = \frac{\lambda}{(4\pi\epsilon_0)} \frac{x}{(x^2 + a^2)^{\frac{3}{2}}} \int_0^{2\pi a} dl = \frac{1}{4\pi\epsilon_0} \frac{\lambda x (2\pi a)}{(x^2 + a^2)^{\frac{3}{2}}}$$

$$= \frac{1}{4\pi\epsilon_0} \frac{Qx}{(x^2 + a^2)^{\frac{3}{2}}}.$$

At great distances, $x \gg a$, this reduces to $E = Q/4\pi\epsilon_0 x^2$. We would expect this result since at great distances the ring would appear to be a point charge $(1/r^2$ dependence$)$.

➡ **PROBLEM SOLVING**

Check result by noting that at a great distance the ring looks like a point charge

Note in this Example three important problem solving techniques or "tricks" that can be used elsewhere: (1) The use of symmetry to reduce the complexity of the problem; (2) expressing the charge dQ in terms of a charge density (here linear, $\lambda = Q/2\pi a$); and (3) the checking of the answer at the limit of large r, which serves as an indication (but not proof) of the correctness of the answer—if the result did not check at large r, the result no doubt would be wrong entirely.

➡ **PROBLEM SOLVING**

Hints

EXAMPLE 21–10 **Long line of charge.** Determine the magnitude of the electric field at any point P a distance x from a very long line (a wire, say) of uniformly distributed charge, Fig. 21–28. Assume x is much smaller than the length of the wire, and let λ be the charge per unit length (C/m).

SOLUTION We set up a coordinate system so the wire is on the y axis with origin 0 as shown. A segment of wire dy has charge $dQ = \lambda\,dy$. The field $d\mathbf{E}$ at P due to such a length of wire at y has magnitude

$$dE = \frac{1}{4\pi\epsilon_0} \frac{dQ}{r^2} = \frac{1}{4\pi\epsilon_0} \frac{\lambda\,dy}{(x^2 + y^2)},$$

where $r = (x^2 + y^2)^{\frac{1}{2}}$ as shown in Fig. 21–28. The vector $d\mathbf{E}$ has components dE_x and dE_y as shown where $dE_x = dE\cos\theta$ and $dE_y = dE\sin\theta$. If the wire is extremely long in both directions (so distant contributions have little effect compared to nearby ones), or if 0 is at the midpoint of the wire (even if the wire is short), the y component of $\mathbf{E}$ will be zero since there will be equal contributions to $E_y = \int dE_y$ from above and below point 0, so

$$E_y = \int dE \sin\theta = 0.$$

Then we have

$$E = E_x = \int dE \cos\theta = \frac{\lambda}{4\pi\epsilon_0} \int \frac{\cos\theta\,dy}{x^2 + y^2}.$$

The integration here is over y, along the wire, with x treated as constant. We must now write θ as a function of y, or y as a function of θ. We do the latter: since $y = x\tan\theta$, $dy = x\,d\theta/\cos^2\theta$ and $(x^2 + y^2) = x^2/\cos^2\theta$. Then

$$E = \frac{\lambda}{4\pi\epsilon_0} \frac{1}{x} \int_{-\pi/2}^{\pi/2} \cos\theta\,d\theta = \frac{\lambda}{4\pi\epsilon_0 x} (\sin\theta)\Big|_{-\pi/2}^{\pi/2} = \frac{1}{2\pi\epsilon_0} \frac{\lambda}{x},$$

where we have assumed the wire is extremely long in both directions ($y \to \pm\infty$) which corresponds to the limits $\theta = \pm\pi/2$. Thus the field of a long straight line of charge decreases inversely as the first power of the distance from the wire. This result, obtained for an infinite wire, is a good approximation for a wire of finite length as long as x is small compared to the distance of P from the ends of the wire.

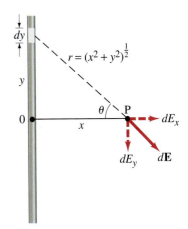

FIGURE 21–28 Example 21–10.

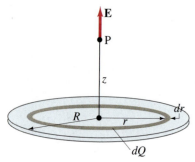

FIGURE 21–29 Example 21–11: a uniformly charged flat disk of radius R.

EXAMPLE 21–11 **Uniformly charged disk.** Charge is distributed uniformly over a thin circular disk of radius R. The charge per unit area (C/m^2) is σ. Calculate the electric field at a point P on the axis of the disk, a distance z above its center, Fig. 21–29.

SOLUTION We can think of the disk as a set of concentric rings. We can then apply the result of Example 21–9 to each of these rings, and then sum over all the rings. For the ring of radius r shown in Fig. 21–29, the electric field has magnitude

$$dE = \frac{1}{4\pi\epsilon_0} \frac{z\,dQ}{(z^2 + r^2)^{\frac{3}{2}}}$$

where we have used the result of Example 21–9 and have written dE (instead of E) for this thin ring of total charge dQ. The ring has area $(dr)(2\pi r)$ and we insert the charge per unit area $\sigma = dQ/(2\pi r\,dr)$:

$$dE = \frac{1}{4\pi\epsilon_0} \frac{z\sigma 2\pi r\,dr}{(z^2 + r^2)^{\frac{3}{2}}} = \frac{z\sigma r\,dr}{2\epsilon_0(z^2 + r^2)^{\frac{3}{2}}}.$$

Now we sum over all the rings, starting at $r = 0$ out to the largest with $r = R$:

$$E = \frac{z\sigma}{2\epsilon_0} \int_0^R \frac{r\,dr}{(z^2 + r^2)^{\frac{3}{2}}} = \frac{z\sigma}{2\epsilon_0} \left[-\frac{1}{(z^2 + r^2)^{\frac{1}{2}}} \right]_0^R$$

$$= \frac{\sigma}{2\epsilon_0} \left[1 - \frac{z}{(z^2 + R^2)^{\frac{1}{2}}} \right].$$

This gives the magnitude of **E** at any point z along the axis of the disk. The direction of each $d\mathbf{E}$ due to each ring is along the z axis (as in Example 21–9), and therefore the direction of **E** is along z. If Q (and σ) are positive, **E** points away from the disk; if Q (and σ) are negative, **E** points toward the disk.

If the radius of the disk in Example 21–11 is much greater than the distance of our point P from the disk (i.e., $z \ll R$) then we can obtain a very useful result: the second term in the solution above becomes very small, so

Electric field near a thin large surface

$$E = \frac{\sigma}{2\epsilon_0}. \qquad\qquad \text{[infinite plane]} \quad \textbf{(21–7)}$$

This result is valid for any point above (or below) an infinite plane of any shape holding a uniform charge density σ. It is also valid for points close to a finite plane, as long as the point is close to the plane compared to the distance to the plane's edge. Thus the field near a large uniformly charged plane is uniform, and directed outward if the plane is positively charged.

It is interesting to compare here the distance dependence of the electric field due to a point charge $(E \sim 1/r^2)$, due to a very long uniform line of charge $(E \sim 1/r)$, and due to a very large uniform plane of charge (E does not depend on r).

EXAMPLE 21–12 **Two parallel plates.** Determine the electric field between two large parallel plates or sheets, which are very thin and are separated by a distance d which is small compared to their height and width. One plate carries a uniform surface charge density σ and the other carries a uniform surface charge density $-\sigma$, as shown in Fig. 21–30.

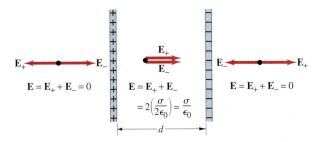

FIGURE 21–30 Example 21–12.

SOLUTION From Eq. 21–7, each plate sets up an electric field of magnitude $E = \pm\sigma/2\epsilon_0$. The field due to the positive plate points away from that plate whereas the field due to the negative plate points toward that plate. Hence, in the region between the plates, the fields add together as shown:

$$E = E_+ + E_- = \frac{\sigma}{2\epsilon_0} + \frac{\sigma}{2\epsilon_0} = \frac{\sigma}{\epsilon_0}.$$

Electric field between two large oppositely charged parallel plates

The field is uniform, since the plates are very large compared to their separation, so this result is valid for any point, whether near one or the other of the plates, or midway between them as long as the point is far from the ends. Outside the plates, the fields cancel,

$$E = E_+ + E_- = \frac{\sigma}{2\epsilon_0} - \frac{\sigma}{2\epsilon_0} = 0,$$

as shown in the diagram. These results are valid ideally for infinitely large plates; they are a good approximation for finite plates if the separation is much less than the dimensions of the plate and for points not too close to the edge. These useful and extraordinary results illustrate the principle of superposition and its great power.

21–8 Field Lines

Since the electric field is a vector, it is sometimes referred to as a *vector field*. We could indicate the electric field with arrows at various points in a given situation, such as at a, b, and c in Fig. 21–31. The directions of $\mathbf{E}_a$, $\mathbf{E}_b$, and $\mathbf{E}_c$ are the same as that of the forces shown earlier in Fig. 21–21, but the lengths (magnitudes) are different since we divide by q. However, the relative lengths of $\mathbf{E}_a$, $\mathbf{E}_b$, and $\mathbf{E}_c$ are the same as for the forces since we divide by the same q each time. To indicate the electric field in such a way at *many* points, however, would result in many arrows, which might appear complicated or confusing. To avoid this, we use another technique, that of field lines.

In order to visualize the electric field, we draw a series of lines to indicate the direction of the electric field at various points in space. These **electric field lines** (sometimes called *lines of force*) are drawn so that they indicate the direction of the force due to the given field on a positive test charge. The lines of force due to a single isolated positive charge are shown in Fig. 21–32a and for a single isolated negative charge in Fig. 21–32b. In part (a) the lines point radially outward from the charge, and in part (b) they point radially inward toward the charge because that is the direction the force would be on a positive test charge in each case (as in Fig. 21–23). Only a few representative lines are shown. One could just as well draw lines in between those shown since the electric field exists there as well. We can always draw the lines so that the *number of lines starting on a positive charge, or ending on a negative charge, is proportional to the magnitude of the charge*. Notice that near the charge, where the electric field is greatest, the lines are closer together. This is a general property of electric field lines: *the closer the lines are together, the stronger the electric field in that region*. In fact the lines can always be drawn so that the number of lines crossing unit area perpendicular to **E** is proportional to the magnitude of the electric field.

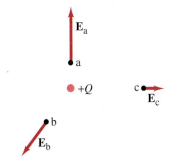

FIGURE 21–31 Electric field vector shown at three points, due to a single point charge Q. (Compare to Fig. 21–21.)

FIGURE 21–32 Electric field lines (a) near a single positive point charge, (b) near a single negative point charge.

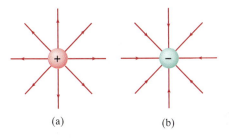

(a)   (b)

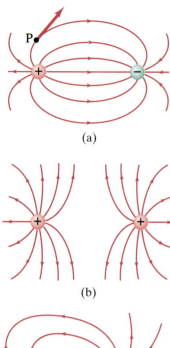

(a)

(b)

-Q ++ +2Q

(c)

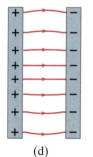

(d)

FIGURE 21–33 Electric field lines for four arrangements of charges.

FIGURE 21–34 The Earth's gravitational field.

Figure 21–33a shows the electric field lines due to two charges of opposite sign. The electric field lines are curved in this case and they are directed from the positive charge to the negative charge. The direction of the field at any point is directed tangentially as shown by the arrow at point P. To satisfy yourself that this is the correct pattern for the electric field lines, you can make a few calculations such as those done in Example 21–8 for just this case (see Fig. 21–26). Figures 21–33b and c show the electric field lines (b) for two equal positive charges, and (c) for unequal charges, $+2Q$ and $-Q$; note that twice as many lines leave $+2Q$ as there are lines entering $-Q$ (number of lines is proportional to magnitude of Q). Finally, in (d), we see the field between two oppositely charged parallel plates. Notice that the electric field lines between the two plates start out perpendicular to the surface of the metal plates (we'll see why this is always true in the next Section) and go directly from one plate to the other, as we expect because a positive test charge placed between the plates would feel a strong repulsion from the positive plate and a strong attraction to the negative plate. The field lines between the plates are parallel and equally spaced in the central region far from the edges (as in Example 21–12), but fringe outward near the edges. Thus, in the central region, the electric field has the same magnitude at all points and we can write (see Example 21–12)

$$ E = \text{constant} = \frac{\sigma}{\epsilon_0}. \quad \text{[between two closely spaced parallel plates]} \quad \textbf{(21–8)} $$

The fringing of the field near the edges can often be ignored, particularly if the separation of the plates is small compared to their size.

We summarize the properties of field lines as follows:

1. The field lines indicate the direction of the electric field; the field points in the direction tangent to the field line at any point.
2. The lines are drawn so that the magnitude of the electric field, E, is proportional to the number of lines crossing unit area perpendicular to the lines. The closer the lines, the stronger the field.
3. Electric field lines start on positive charges and end on negative charges; and the number starting or ending is proportional to the magnitude of the charge.

Also note that field lines never cross. Why not? Because it would not make sense for the electric field to have two values at the same point.

The field concept can also be applied to the gravitational force as mentioned in Chapter 6. Thus we can say that a **gravitational field** exists for every object that has mass. One object attracts another by means of the gravitational field. The Earth, for example, can be said to possess a gravitational field (Fig. 21–34) which is responsible for the gravitational force on objects. The *gravitational field* is defined as the *force per unit mass*. The magnitude of the Earth's gravitational field at any point is then (GM_E/r^2), where M_E is the mass of the Earth, r is the distance of the point from the Earth's center, and G is the gravitational constant (Chapter 6). At the Earth's surface, r is simply the radius of the Earth and the gravitational field is simply equal to g, the acceleration due to gravity (since $F/m = mg/m = g$). Beyond the Earth, the gravitational field can be calculated at any point as a sum of terms due to Earth, Sun, Moon, and other bodies that contribute significantly.

21–9 Electric Fields and Conductors

We now discuss some properties of conductors. First, *the electric field inside a conductor is zero in the static situation*—that is, when the charges are at rest. If there were an electric field within a conductor, there would be a force on its free electrons. The electrons would move until they reached positions where the electric field, and therefore the electric force on them, did become zero.

This reasoning has some interesting consequences. For one, *any net charge on a conductor distributes itself on the surface.* For a negatively charged conductor, you can imagine that the negative charges repel one another and race to the surface to get as far from one another as possible. Another consequence is the following. Suppose that a positive charge Q is surrounded by an isolated uncharged metal conductor whose shape is a spherical shell, Fig. 21–35. Because there can be no field within the metal, the lines leaving the positive charge must end on negative charges on the inner surface of the metal. Thus an equal amount of negative charge, $-Q$, is induced on the inner surface of the spherical shell. Then, since the shell is neutral, a positive charge of the same magnitude, $+Q$, must exist on the outer surface of the shell. Thus, although no field exists in the metal itself, an electric field exists outside of it, as shown in Fig. 21–35, as if the metal were not even there.

A related property of static electric fields and conductors is that *the electric field is always perpendicular to the surface outside of a conductor.* If there were a component of **E** parallel to the surface (Fig. 21–36), electrons at the surface would move along the surface in response to this force until they reached positions where no net force was exerted on them parallel to the surface—that is, until the electric field was perpendicular to the surface.

These properties pertain only to conductors. Inside a nonconductor, which does not have free electrons, an electric field can exist as we will see later, in Chapter 24. And the electric field outside a nonconductor does not necessarily make an angle of 90° to the surface.

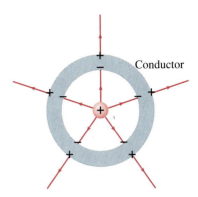

FIGURE 21–35 A charge placed inside a spherical shell. Charges are induced on the conductor surfaces. The electric field exists even beyond the shell but not within the conductor itself.

FIGURE 21–36 If the electric field **E** at the surface of a conductor had a component parallel to the surface, $\mathbf{E}_\parallel$, the latter would accelerate electrons into motion. In the static case (no charges are in motion), $\mathbf{E}_\parallel$ must be zero, and so the electric field must be perpendicular to the conductor's surface: $\mathbf{E} = \mathbf{E}_\perp$.

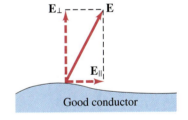

| **CONCEPTUAL EXAMPLE 21–13** | **Shielding, and safety in a storm.** A hollow metal box is placed between two parallel charged plates as shown in Fig. 21–37a. What is the field like inside the box? |

RESPONSE If our metal box were solid, and not hollow, the electrons in the box, even if it were neutral overall, would redistribute themselves along the surface so that the field lines would not penetrate the conducting metal of the box. For a hollow box, the external field is not changed since the electrons in the metal can move just as freely as before to the surface. Hence we conclude that the field inside the hollow metal box is zero. So the field lines are something like those shown in Fig. 21–37b. A conducting box used in this way is an effective device for shielding delicate instruments and electronic circuits from unwanted external electric fields. We also can see that a relatively safe place to be during a lightning storm is inside a car, surrounded by metal. See also Fig. 21–38, where we see that a person inside a porous "cage" is protected from a strong electric discharge.

FIGURE 21–37 Example 21–13.

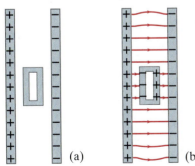

(a)   (b)

FIGURE 21–38 A strong electric field exists in the vicinity of this "Faraday cage," so strong that electrons are pulled off atoms of the air, and charge flows to (or from) the metal cage. Yet the person inside the cage is not affected.

21–10 Motion of a Charged Particle in an Electric Field

If an object having an electric charge q is at a point in space where the electric field is **E**, the force on the object is given by

$$\mathbf{F} = q\mathbf{E}$$

(see Eq. 21–3). In the past few Sections we have seen how to determine **E** for some particular situations. Now let us suppose we know **E** and we want to find the force on a charged object and its subsequent motion.

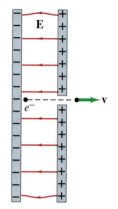

FIGURE 21–39 Example 21–14.

EXAMPLE 21–14 **Electron accelerated by electric field.** An electron (mass $m = 9.1 \times 10^{-31}$ kg) is accelerated in the uniform field **E** $(E = 2.0 \times 10^4$ N/C) between two parallel charged plates. The separation of the plates is 1.5 cm. The electron is accelerated from rest near the negative plate and passes through a tiny hole in the positive plate, Fig. 21–39. (*a*) With what speed does it leave the hole? (*b*) Show that the gravitational force can be ignored. Assume the hole is so small that it does not affect the uniform field between the plates.

SOLUTION (*a*) The magnitude of the force on the electron is

$$F = qE$$

and is directed to the right. The magnitude of the electron's acceleration is

$$a = \frac{F}{m} = \frac{qE}{m}.$$

Between the plates **E** is uniform so the electron undergoes uniformly accelerated motion with acceleration

$$a = \frac{(1.6 \times 10^{-19}\,\text{C})(2.0 \times 10^4\,\text{N/C})}{(9.1 \times 10^{-31}\,\text{kg})} = 3.5 \times 10^{15}\,\text{m/s}^2.$$

It travels a distance $x = 1.5 \times 10^{-2}$ m before reaching the hole, and since its initial speed was zero, we can use the kinematic equation, $v^2 = v_0^2 + 2ax$ (Eq. 2–12c), with $v_0 = 0$:

$$v = \sqrt{2ax} = \sqrt{2(3.5 \times 10^{15}\,\text{m/s}^2)(1.5 \times 10^{-2}\,\text{m})} = 1.0 \times 10^7\,\text{m/s}.$$

There is no electric field outside the plates, so after passing through the hole, the electron moves with this speed, which is now constant.
(*b*) The magnitude of the electric force on the electron is

$$qE = (1.6 \times 10^{-19}\,\text{C})(2.0 \times 10^4\,\text{N/C}) = 3.2 \times 10^{-15}\,\text{N}.$$

The gravitational force is

$$mg = (9.1 \times 10^{-31}\,\text{kg})(9.8\,\text{m/s}^2) = 8.9 \times 10^{-30}\,\text{N},$$

which is 10^{14} times smaller! Note that the electric field due to the electron does not enter the problem (since a particle cannot exert a force on itself).

EXAMPLE 21–15 **Electron moving perpendicular to E.** Suppose an electron (say from Example 21–14) traveling with speed $v_0 = 1.0 \times 10^7$ m/s enters a uniform electric field $\mathbf{E}$ at right angles to $\mathbf{v}_0$ as shown in Fig. 21–40. Describe its motion by giving the equation of its path while in the electric field. Ignore gravity.

SOLUTION When the electron enters the electric field (at $x = y = 0$) it has velocity $\mathbf{v}_0 = v_0\mathbf{i}$ in the x direction. The electric field $\mathbf{E}$, pointing vertically upward, imparts a uniform vertical acceleration to the electron of

$$a_y = \frac{F}{m} = \frac{qE}{m} = -\frac{eE}{m},$$

where we set $q = -e$ for the electron. Its vertical position is given by

$$y = \frac{1}{2}a_y t^2 = -\frac{eE}{2m}t^2$$

since the motion is at constant acceleration. The horizontal position is given by

$$x = v_0 t$$

since $a_x = 0$. We eliminate t between these two equations and obtain

$$y = -\frac{eE}{2mv_0^2}x^2,$$

which is the equation of a parabola (just as in projectile motion, Section 3–7).

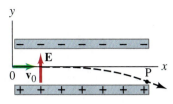

FIGURE 21–40 Example 21–15.

21–11 Electric Dipoles

The combination of two equal charges of opposite sign, $+Q$ and $-Q$, separated by a distance l, is referred to as an **electric dipole**. The quantity Ql is called the **dipole moment** and is represented[†] by the symbol p. The dipole moment can be considered to be a vector $\mathbf{p}$, of magnitude Ql, that points from the negative to the positive charge as shown in Fig. 21–41. Many molecules, such as the diatomic molecule CO, have a dipole moment (C has a small positive charge and O a small negative charge of equal magnitude), and are referred to as **polar molecules**. Even though the molecule as a whole is neutral, there is a separation of charge that results from an uneven sharing of electrons by the two atoms.[‡] (Symmetric diatomic molecules, like O_2, have no dipole moment.) The water molecule, with its uneven sharing of electrons (O is negative, the two H are positive), also has a dipole moment—see Figs. 21–4 and 21–42.

Dipole moment

FIGURE 21–41 A dipole consists of equal but opposite charges, $+Q$ and $-Q$, separated by a distance l. The dipole moment $\mathbf{p} = Ql$ and points from the negative to the positive charge.

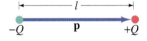

[†] Be careful not to confuse this p for dipole moment with p for momentum.

[‡] The value of the separated charges may be a fraction of e (say $\pm 0.2e$ or $\pm 0.4e$) but note that such charges don't violate what we said about e being the smallest charge. These charges less than e cannot be isolated and merely represent how much time electrons spend around one atom or the other.

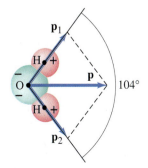

FIGURE 21–42 In the water molecule (H_2O), the electrons spend more time around the oxygen atom than around the two hydrogen atoms. The net dipole moment $\mathbf{p}$ can be considered as the vector sum of two dipole moments $\mathbf{p}_1$ and $\mathbf{p}_2$ that point from the O to each H as shown: $\mathbf{p} = \mathbf{p}_1 + \mathbf{p}_2$.

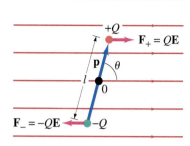

FIGURE 21–43 An electric dipole in a uniform electric field.

Dipole in an External Field

First let us consider a dipole, of dipole moment $p = Ql$, that is placed in a uniform electric field **E**, as shown in Fig. 21–43. If the field is uniform, the force QE on the positive charge and the force $-QE$ on the negative charge result in no net force on the dipole. There will, however, be a *torque* on the dipole which has magnitude (calculated about the center, 0, of the dipole)

$$\tau = QE\frac{l}{2}\sin\theta + QE\frac{l}{2}\sin\theta = pE\sin\theta. \qquad \textbf{(21–9a)}$$

This can be written in vector notation as

$$\boldsymbol{\tau} = \mathbf{p} \times \mathbf{E}. \qquad \textbf{(21–9b)}$$

The effect of the torque is to try to turn the dipole so **p** is parallel to **E**. The work done on the dipole by the electric field to change the angle θ from θ_1 to θ_2, is (see Eq. 10–22)

$$W = \int_{\theta_1}^{\theta_2} \tau \, d\theta.$$

We need to write the torque as $\tau = -pE\sin\theta$ because its direction is opposite to the direction of increasing θ (right hand rules). Then

$$W = \int_{\theta_1}^{\theta_2} \tau \, d\theta = -pE\int_{\theta_1}^{\theta_2}\sin\theta \, d\theta = pE\cos\theta \Big|_{\theta_1}^{\theta_2} = pE(\cos\theta_2 - \cos\theta_1).$$

Positive work done by the field decreases the potential energy, U, of the dipole in this field. If we choose $U = 0$ when **p** is perpendicular to **E** (that is, choosing $\theta_1 = 90°$ so $\cos\theta_1 = 0$), and setting $\theta_2 = \theta$, then

Potential energy of dipole in electric field

$$U = -W = -pE\cos\theta = -\mathbf{p}\cdot\mathbf{E}. \qquad \textbf{(21–10)}$$

If the electric field is *not* uniform, the force on the $+Q$ of the dipole may not have the same magnitude as on the $-Q$, so there may be a net force as well as a torque.

Electric Field Produced by a Dipole

We have just seen how an external electric field affects an electric dipole. Now let us suppose that there is no external field, and we want to determine the electric field produced *by* the dipole. For brevity, we restrict ourselves to points that are on the perpendicular bisector of the dipole, such as point P in Fig. 21–44 which is a distance r above the midpoint of the dipole. Note that r in Fig. 21–44 is not the distance from either charge to P; the latter distance is $(r^2 + l^2/4)^{\frac{1}{2}}$ and this is what must be used in Eq. 21–4. The total field at P is

$$\mathbf{E} = \mathbf{E}_+ + \mathbf{E}_-,$$

FIGURE 21–44 Electric field due to an electric dipole.

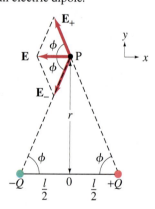

where $\mathbf{E}_+$ and $\mathbf{E}_-$ are the fields due to the $+$ and $-$ charges respectively. The magnitudes E_+ and E_- are equal:

$$E_+ = E_- = \frac{1}{4\pi\epsilon_0}\frac{Q}{r^2 + l^2/4}.$$

Their y components cancel at point P, so the magnitude of the total field **E** is

$$E = 2E_+\cos\phi = \frac{1}{2\pi\epsilon_0}\left(\frac{Q}{r^2 + l^2/4}\right)\frac{l}{2(r^2 + l^2/4)^{\frac{1}{2}}}$$

or

$$E = \frac{1}{4\pi\epsilon_0}\frac{p}{(r^2 + l^2/4)^{\frac{3}{2}}}. \qquad \begin{bmatrix}\text{on perpendicular bisector}\\\text{of dipole}\end{bmatrix} \quad \textbf{(21–11)}$$

Far from the dipole, $r \gg l$, this reduces to

$$E = \frac{1}{4\pi\epsilon_0}\frac{p}{r^3}. \qquad \begin{bmatrix}\text{on perpendicular bisector}\\\text{of dipole; } r \gg l\end{bmatrix} \quad \textbf{(21–12)}$$

So the field decreases more rapidly for a dipole than for a single point charge $(1/r^3$ versus $1/r^2)$, which we expect since at large distances the two opposite charges appear so close together as to neutralize each other. This $1/r^3$ dependence also applies for points not on the perpendicular bisector (see the Problems).

EXAMPLE 21–16 **Dipole in a field.** The dipole moment of a water molecule is 6.1×10^{-30} C·m. A water molecule is placed in a uniform electric field with magnitude 2.0×10^5 N/C. (a) What is the magnitude of the maximum torque that the field can exert on the molecule? (b) What is the potential energy when the torque is at its maximum? (c) In what position will the potential energy take on its greatest value? Why is this different than the position where the torque is maximized?

SOLUTION (a) From Eq. 21–9 we see that τ is maximized when θ is 90°. Then $\tau = pE = (6.1 \times 10^{-30} \text{ C·m})(2.0 \times 10^5 \text{ N/C}) = 1.2 \times 10^{-24}$ N·m.
(b) The potential energy is given by Eq. 21–10. For $\theta = 90°$ the potential energy is zero. Note that the potential energy is negative for smaller values of θ, so U is not a minimum for $\theta = 90°$.
(c) The potential energy will be a maximum when $\cos\theta = -1$, so $\theta = 180°$, meaning **E** and **p** are antiparallel. The potential energy is maximized when the dipole is oriented so that it has to rotate through the largest angle, 180°, to reach the equilibrium position at $\theta = 0°$. The torque on the other hand is maximized when the electric forces are perpendicular to **p**.

Summary

There are two kinds of **electric charge**, positive and negative. These designations are to be taken algebraically— that is, any charge is plus or minus so many coulombs (C), in SI units.

Electric charge is **conserved**: if a certain amount of one type of charge is produced in a process, an equal amount of the opposite type is also produced; thus the *net* charge produced is zero.

We can better understand electricity by appealing to atomic theory, which theorizes an atom as consisting of a positively charged nucleus surrounded by negatively charged electrons. Each electron has a charge $-e = -1.6 \times 10^{-19}$ C.

Conductors are those materials in which many electrons are relatively free to move, whereas electric **insulators** are those in which very few electrons are free to move.

An object is negatively charged when it has an excess of electrons, and positively charged when it has less than its normal amount of electrons. The charge on any object is thus a whole number times $+e$ or $-e$. That is, charge is **quantized**.

An object can become charged by rubbing (in which electrons are transferred from one material to another), by conduction (which is transfer of charge from one charged object to another by touching), or by induction (the separation of charge within an object because of the close approach of another charged object but without touching).

Electric charges exert a force on each other. If two charges are of opposite types, one positive and one negative, they each exert an attractive force on the other. If the two charges are the same type, each repels the other.

The magnitude of the force one point charge exerts on another is proportional to the product of their charges, and inversely proportional to the square of the distance between them:

$$F = k\frac{Q_1 Q_2}{r^2};$$

this is **Coulomb's law.** In SI units, k is often written as $1/4\pi\epsilon_0$.

We think of an **electric field** as existing in space around any charge or group of charges. The force on another charged object is then said to be due to the electric field present at its location.

The *electric field*, **E**, at any point in space due to one or more charges, is defined as the force per unit charge that would act on a test charge q placed at that point:

$$\mathbf{E} = \frac{\mathbf{F}}{q}.$$

Electric fields are represented by **electric field lines** that start on positive charges and end on negative charges. Their direction indicates the direction the force would be on a tiny positive test charge placed at a point. The lines can be drawn so that the number per unit area is proportional to the magnitude of E.

The static electric field (that is, no charges moving) inside a conductor is zero, and the electric field lines just outside a charged conductor are perpendicular to its surface.

An **electric dipole** is a combination of two equal but opposite charges, $+Q$ and $-Q$, separated by a distance l. The **dipole moment** is $p = Ql$. A dipole placed in a uniform electric field feels no net force but does feel a net torque (unless **p** is parallel to **E**). The electric field produced by a dipole decreases as the third power of the distance r from the dipole $(E \propto 1/r^3)$ for r large compared to l.

Questions

1. If you charge a pocket comb by rubbing it with a silk scarf, how can you determine if the comb is positively or negatively charged?

2. Why does a shirt or blouse taken from a clothes dryer sometimes cling to your body?

3. Explain why fog or rain droplets tend to form around ions or electrons in the air.

4. A positively charged rod is brought close to a neutral piece of paper, which it attracts. Draw a diagram showing the separation of charge and explain why attraction occurs.

5. Why does a plastic ruler that has been rubbed with a cloth have the ability to pick up small pieces of paper? Why is this difficult to do on a humid day?

6. Contrast the *net charge* on a conductor to the "free charges" in the conductor.

7. Figures 21–7 and 21–8 show how a charged rod placed near an uncharged metal object can attract (or repel) electrons. There are a great many electrons in the metal, yet only some of them move as shown. Why not all of them?

8. When an electroscope is charged, the two leaves repel each other and remain at an angle. What balances the electric force of repulsion so that the leaves don't separate further?

9. The form of Coulomb's law is very similar to that for Newton's law of universal gravitation. What are the differences between these two laws? Compare also gravitational mass and electric charge.

10. We are not normally aware of the gravitational or electric force between two ordinary objects. What is the reason in each case? Give an example where we are aware of each one and why.

11. Is the electric force a conservative force? Why or why not? (See Chapter 8.)

12. What experimental observations mentioned in the text rule out the possibility that the numerator in Coulomb's law contains the sum $(Q_1 + Q_2)$ rather than the product $Q_1 Q_2$?

13. When a charged ruler attracts small pieces of paper, sometimes a piece jumps quickly away after touching the ruler. Explain.

14. Explain why we use *small* test charges when measuring electric fields.

15. When determining an electric field, must we use a *positive* test charge, or would a negative one do as well? Explain.

16. Draw the electric field lines surrounding two negative electric charges a distance l apart.

17. Assume that the two opposite charges in Fig. 21–33a are 12.0 cm apart. Consider the magnitude of the electric field 2.5 cm from the positive charge. On which side of this charge—top, bottom, left, or right—is the electric field the strongest? The weakest?

18. Consider the electric field at the three points indicated by the letters A, B, and C in Fig. 21–45. First draw an arrow at each point indicating the direction of the net force that a positive test charge would experience if placed at that point, then list the letters in order of *decreasing* field strength (strongest first).

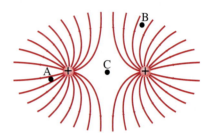

FIGURE 21–45   Question 18.

19. Why can electric field lines never cross?

20. Show, using the three rules for field lines given in Section 21–8, that the electric field lines starting or ending on a single point charge must be symmetrically spaced around the charge.

21. Given two point charges, Q and $2Q$, a distance l apart, is there a point along the straight line that passes through them where $E = 0$ when their signs are (a) opposite, (b) the same? If yes, state roughly where this point will be.

22. Suppose the ring of Fig. 21–27 has a uniformly distributed negative charge Q. What is the magnitude and direction of **E** at point P?

23. Consider a small positive test charge located on an electric field line at some point, such as point P in Fig. 21–33a. Is the direction of the velocity and/or acceleration of the test charge along this line? Discuss.

24. We wish to determine the electric field at a point near a positively charged metal sphere (a good conductor). We do so by bringing a small test charge, q_0, to this point and measure the force F_0 on it. Will F_0/q_0 be greater than, less than, or equal to, the electric field **E** as it was at that point before the test charge was present?

25. In what ways does the electron motion in Example 21–15 resemble projectile motion (Section 3–7)? In which ways not?

26. Describe the motion of the dipole shown in Fig. 21–43 if it is released from rest at the position shown.

27. Explain why there can be a net force on an electric dipole placed in a nonuniform electric field.

Problems

Section 21–5

1. (I) Calculate the magnitude of the force between two 2.50-C point charges 3.0 m apart.

2. (I) How many electrons make up a charge of $-30.0\,\mu C$?

3. (I) What is the magnitude of the electric force of attraction between an iron nucleus ($q = +26e$) and its innermost electron if the distance between them is 1.5×10^{-12} m?

4. (I) What is the repulsive electrical force between two protons in a nucleus that are 5.0×10^{-15} m apart from each other?

5. (I) What is the magnitude of the force a $+25$-μC charge exerts on a $+3.0$-mC charge 35 cm away? $\left(1\,\mu C = 10^{-6}\,C, 1\,mC = 10^{-3}\,C.\right)$

6. (II) Two charged smoke particles exert a force of 4.2×10^{-2} N on each other. What will be the force if they are moved so they are only one eighth as far apart?

7. (II) Two charged balls are 15.0 cm apart. They are moved, and the force on each of them is found to have been tripled. How far apart are they now?

8. (II) A person scuffing her feet on a wool rug on a dry day accumulates a net charge of $-40\,\mu C$. How many excess electrons does this person get, and by how much does her mass increase?

9. (II) What is the total charge of all the electrons in 1.0 kg of H_2O?

10. (II) Particles of charge $+70$, $+48$, and $-80\,\mu C$ are placed in a line (Fig. 21–46). The center one is 0.35 m from each of the others. Calculate the net force on each charge due to the other two.

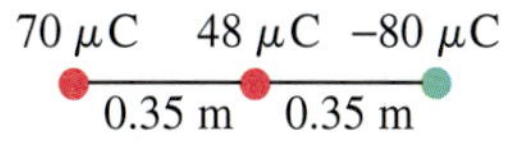

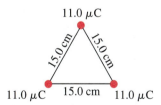

FIGURE 21–46   Problem 10.

11. (II) Three positive particles of charges $11.0\,\mu C$ are located at the corners of an equilateral triangle of side 15.0 cm (Fig. 21–47). Calculate the magnitude and direction of the net force on each particle.

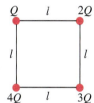

FIGURE 21–47   Problem 11.

12. (II) A charge of 6.00 mC is placed at each corner of a square 0.100 m on a side. Determine the magnitude and direction of the force on each charge.

13. (II) Repeat Problem 12 for the case when two of the positive charges, on opposite corners, are replaced by negative charges of the same magnitude (Fig. 21–48).

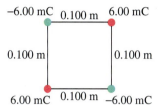

FIGURE 21–48   Problem 13.

14. (II) At each corner of a square of side l there are point charges of magnitude Q, $2Q$, $3Q$, and $4Q$ (Fig. 21–49). Determine the force on each charge due to the other three charges.

FIGURE 21–49   Problem 14.

15. (II) Repeat Problem 14 assuming the $3Q$ charge is replaced by a $-3Q$ charge (i.e., its sign is changed).

16. (II) Compare the electric force holding the electron in orbit $\left(r = 0.53 \times 10^{-10}\,m\right)$ around the proton nucleus of the hydrogen atom, with the gravitational force between the same electron and proton. What is the ratio of these two forces?

17. (II) Two positive point charges are a fixed distance apart. The sum of their charges is Q_T. What charge must each have in order to (a) maximize the electric force between them, and (b) minimize it?

18. (II) Two point charges have a total charge of $560\,\mu C$. When placed 1.10 m apart, the force each exerts on the other is 22.8 N and is repulsive. What is the charge on each?

19. (II) Two charges, $-Q_0$ and $-3Q_0$, are a distance l apart. These two charges are free to move but do not because there is a third charge nearby. What must be the charge and placement of the third charge for the first two to be in equilibrium?

20. (III) A $+7.7\,\mu C$ and a $-3.5\,\mu C$ charge are placed 18.5 cm apart. Where can a third charge be placed so that it experiences no net force?

21. (III) Two small nonconducting spheres have a total charge of $90.0\,\mu C$. When placed 1.16 m apart, the force each exerts on the other is 12.0 N and is repulsive. What is the charge on each? What if the force were attractive?

22. (III) Two small charged spheres hang from cords of equal length l as shown in Fig. 21–50 and make small angles θ_1 and θ_2 with the vertical. (a) If $Q_1 = Q$, $Q_2 = 2Q$, and $m_1 = m_2 = m$, determine the ratio θ_1/θ_2. (b) If $Q_1 = Q$, $Q_2 = 2Q$, $m_1 = m$, and $m_2 = 2m$, determine the ratio θ_1/θ_2. (c) Estimate the distance between the spheres for each case.

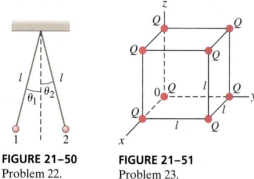

FIGURE 21–50
Problem 22.

FIGURE 21–51
Problem 23.

23. (III) On each corner of a cube of side l there is a point charge Q. What is the force on the charge at the origin 0 due to the others? Give answer in vector notation for the charge at the origin in Fig. 21–51.

Sections 21–6 to 21–8

24. (I) What is the magnitude of the acceleration experienced by an electron in an electric field of 600 N/C? How does the direction of the acceleration depend on the direction of the field at that point?

25. (I) What is the magnitude and direction of the electric force on an electron in a uniform electric field of strength 1360 N/C that points due east?

26. (I) A proton is released in a uniform electric field, and it experiences an electric force of 2.75×10^{-14} N toward the south. What are the magnitude and direction of the electric field?

27. (I) What is the magnitude and direction of the electric field 20.0 cm directly above an isolated 33.0×10^{-6}-C charge?

28. (I) What is the magnitude and direction of the electric field at a point midway between a -8.0-μC and a $+7.0$-μC charge 8.0 cm apart? Assume no other charges are nearby.

29. (I) The electric force on a $+4.20$-μC charge is $\mathbf{F} = 5.85 \times 10^{-4}\,\mathbf{Nj}$. What is the electric field at the position of the charge?

30. (I) What is the electric field at a point when the force on a 1.25-μC charge placed at that point is $\mathbf{F} = (3.0\mathbf{i} - 5.0\mathbf{j}) \times 10^{-3}$ N?

31. (II) Electric field lines can always be drawn so that their number per unit area perpendicular to the lines is proportional to the electric field magnitude E. However, if Coulomb's law were not valid—that is, if the field due to a single point did not fall off precisely as $1/r^2$ (so the exponent on r was not precisely 2)—this property of field lines would not be true. Show why. [*Hint*: Take the example of a single point charge.]

32. (II) Draw, approximately, the electric field lines about two point charges, $+Q$ and $-3Q$, which are a distance l apart.

33. (II) Draw, approximately, the electric field lines emanating from a uniformly charged straight wire whose length l is not great. The spacing between lines near the wire should be somewhat less than l. [*Hint*: Also consider points very far from the wire.]

34. (II) What is the electric field strength at a point in space where a proton $(m = 1.67 \times 10^{-27}$ kg$)$ experiences an acceleration of 1 million "g's"?

35. (II) An electron is released from rest in a uniform electric field and accelerates to the north at a rate of 145 m/s². What is the magnitude and direction of the electric field?

36. (II) The electric field midway between two equal but opposite point charges is 845 N/C, and the distance between the charges is 16.0 cm. What is the magnitude of the charge on each?

37. (II) Use Coulomb's law to determine the magnitude and direction of the electric field at points A and B in Fig. 21–52 due to the two positive charges $(Q = 7.0\,\mu$C$)$ shown. Is your result consistent with Fig. 21–33b?

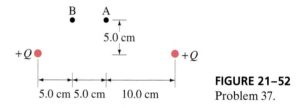

FIGURE 21–52
Problem 37.

38. (II) Calculate the electric field at the center of a square 52.5 cm on a side if one corner is occupied by a $+45.0$-μC charge and the other three are occupied by -27.0-μC charges.

39. (II) Calculate the electric field at one corner of a square 1.00 m on a side if the other three corners are occupied by 3.25×10^{-6}-C charges.

40. (II) (a) Determine the electric field $\mathbf{E}$ at the origin 0 in Fig. 21–53 due to the two charges at A and B. (b) Repeat, but let the charge at B be reversed in sign.

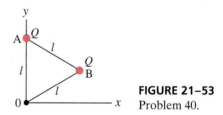

FIGURE 21–53
Problem 40.

41. (II) You are given two unknown point charges, Q_1 and Q_2. At a point on the line joining them, one-third of the way from Q_1 to Q_2, the electric field is zero (Fig. 21–54). What is the ratio Q_1/Q_2?

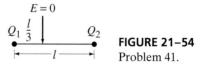

FIGURE 21–54
Problem 41.

42. (II) Two parallel circular rings of radius R have their centers on the x axis separated by a distance l as shown in Fig. 21–55. If each ring carries a uniformly distributed charge Q, find the electric field, $\mathbf{E}(x)$, at points along the x axis.

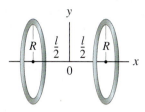

FIGURE 21–55
Problem 42.

43. (II) (a) Two equal charges Q are positioned at points $(x = l, y = 0)$ and $(x = -l, y = 0)$. Determine the electric field as a function of y for points along the y axis. (b) Show that the field is a maximum at $y = \pm l/\sqrt{2}$.

44. (II) At what position, $x = x_M$, is the magnitude of the electric field along the axis of the ring of Example 21–9 a maximum?

45. (II) Suppose the charge Q on the ring of Fig. 21–27 was all distributed uniformly on only the upper half of the ring, and no charge was on the lower half. Determine the electric field $\mathbf{E}$ at P. (Take y vertically upward.)

46. (II) Estimate the electric field at a point 2.8 cm perpendicular to the midpoint of a 2.0-m-long thin wire carrying a total charge of 4.75 μC.

47. (II) The uniformly charged straight wire in Fig. 21–28 has the length L, where point 0 is at the midpoint. Show that the field at point P, a perpendicular distance x from 0, is given by

$$E = \frac{\lambda}{2\pi\epsilon_0} \frac{L}{x(L^2 + 4x^2)^{\frac{1}{2}}},$$

where λ is the charge per unit length.

48. (III) A thin rod of length l carries a total charge Q distributed uniformly along its length. See Fig. 21–56. Determine the electric field along the axis of the rod starting at one end—that is, find $E(x)$ for $x \geq 0$ in Fig. 21–56.

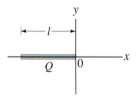

FIGURE 21–56
Problem 48.

49. (III) A thin rod bent into the shape of an arc of a circle of radius R carries a uniform charge per unit length λ. The arc subtends a total angle $2\theta_0$, symmetric about the x axis, as shown in Fig. 21–57. Determine the electric field $\mathbf{E}$ at the origin 0.

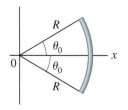

FIGURE 21–57
Problems 49 and 50.

50. (III) (a) Suppose the circular arc shown in Fig. 21–57, carries a charge per unit length λ that varies with θ as $\lambda = \lambda_0 \cos \theta$, where θ is measured from the x axis. Determine the electric field $\mathbf{E}$ at the origin 0. (b) Repeat, assuming $\lambda = \lambda_0 \sin \theta$.

51. (III) Suppose the uniformly charged wire of Fig. 21–28 starts at point 0 and rises vertically to a length L. (a) Determine the components of the electric field E_x and E_y at point P, a distance x from 0. (That is, calculate $\mathbf{E}$ near one end of a long wire, in the plane perpendicular to the wire.) (b) If the wire extends from $y = 0$ to $y = \infty$, so that $L = \infty$, show that $\mathbf{E}$ makes a 45° angle to the horizontal for any x.

52. (III) Suppose in Example 21–10 that $x = 0.250$ m, $Q = 3.15\ \mu$C, and that the uniformly charged wire is only 6.0 m long and extends along the y axis from $y = -4.0$ m to $y = +2.0$ m. (a) Calculate E_x and E_y at point P. (b) Determine what the error would be if you simply used the result of Example 21–10, $E = \lambda/2\pi\epsilon_0 x$. Express this error as $(E_x - E)/E$ and E_y/E.

53. (III) *Uniform plane of charge.* Charge is distributed uniformly over a large square plane of side L, as shown in Fig. 21–58. The charge per unit area (C/m^2) is σ. Calculate the electric field at a point P a distance z above the center of the plane, in the limit $L \to \infty$. [*Hint:* Divide the plane into long narrow strips of width dy, and use the result of Example 21–10 to sum the fields due to each strip to get the total field.]

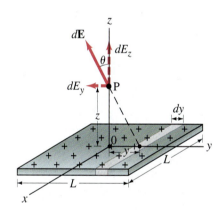

FIGURE 21–58 Problem 53.

Section 21–10

54. (II) An electron with speed $v_0 = 21.5 \times 10^6$ m/s is traveling parallel to an electric field $(\mathbf{v_0} \| \mathbf{E})$ of magnitude $E = 11.4 \times 10^3$ N/C. (a) How far will it travel before it stops? (b) How much time will elapse before it returns to its starting point?

55. (II) An electron has an initial velocity $v_0 = 8.0 \times 10^4$ m/s $\mathbf{i}$. It enters a region where $\mathbf{E} = (2.0\mathbf{i} + 8.0\mathbf{j}) \times 10^4$ N/C. (a) Determine the vector acceleration of the electron as a function of time. (b) At what angle θ is it moving (relative to its initial direction) at $t = 1.0$ ns?

56. (II) A water droplet of radius 0.020 mm remains stationary, in the air. If the electric field of the Earth is 150 N/C downward, how many excess electron charges must the water droplet have?

57. (II) At what angle will the electrons in Example 21–15 leave the uniform electric field at the end of the parallel plates (point P in Fig. 21–40)? Assume the plates are 6.0 cm long and $E = 5.0 \times 10^3$ N/C. Ignore fringing of the field.

58. (II) Suppose electrons enter the uniform electric field midway between two plates, moving at an upward 45° angle as shown in Fig. 21–59. What maximum speed can the electrons have if they are to avoid striking the upper plate? Ignore fringing of the field.

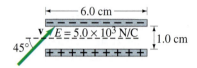

FIGURE 21–59 Problem 58.

59. (III) A positive charge q is placed at the center of a circular ring of radius R. The ring carries a uniformly distributed negative charge of total magnitude $-Q$. (a) If the charge q is displaced from the center a small distance x as shown in Fig. 21–60, show that it will undergo simple harmonic motion when released. (b) If its mass is m, what is its period?

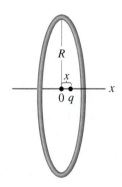

FIGURE 21–60 Problem 59.

General Problems

66. How close must two electrons be if the electric force between them is equal to the weight of either at the Earth's surface?

67. Imagine that space invaders could deposit extra electrons in equal amounts on the Earth and on your car, which has a mass of 1050 kg. Note that the rubber tires would provide some insulation. How much charge Q would need to be placed on your car (same amount on the Earth) in order to levitate it (overcome gravity)? [*Hint:* Assume that the Earth's charge is spread uniformly so it acts as if it were located at the Earth's center, and then the separation distance is the radius of the Earth.]

Section 21–11

60. (II) A dipole consists of charges $+e$ and $-e$ separated by 0.68 nm. It is in an electric field $E = 2.7 \times 10^4$ N/C. (a) What is the value of the dipole moment? (b) What is the torque on the dipole when it is perpendicular to the field? (c) What is the torque on the dipole when it is at an angle of 45° to the field? (d) What is the work required to rotate the dipole from being oriented parallel to the field to being antiparallel to the field?

61. (II) The HCl molecule has a dipole moment of about 3.4×10^{-30} C·m. The two atoms are separated by about 1.0×10^{-10} m. (a) What is the net charge on each atom? (b) Is this equal to an integral multiple of e? If not, explain. (c) What maximum torque would this dipole experience in a 2.5×10^4 N/C electric field? (d) How much energy would be needed to rotate one molecule 45° from its equilibrium position of lowest potential energy?

62. (II) Suppose both charges in Fig. 21–44 were positive. (a) Show that the field on the perpendicular bisector, for $r \gg l$, is given by $(1/4\pi\epsilon_0)(2Q/r^2)$. (b) Explain why the field decreases as $1/r^2$ here whereas for a dipole it decreases as $1/r^3$.

63. (II) An electric dipole, of dipole moment p and moment of inertia I, is placed in a uniform electric field **E**. (a) If displaced by an angle θ as shown in Fig. 21–43 and released, under what conditions will it oscillate in simple harmonic motion? (b) What will be its frequency?

64. (III) Suppose a dipole **p** is placed in a nonuniform electric field $\mathbf{E} = E\mathbf{i}$ that points along the x axis. If E depends only on x, show that the net force on the dipole is

$$\mathbf{F} = \left(\mathbf{p} \cdot \frac{d\mathbf{E}}{dx} \right)\mathbf{i},$$

where $d\mathbf{E}/dx$ is the gradient of the field in the x direction.

65. (III) (a) Show that at points along the axis of a dipole (along the same line that contains $+Q$ and $-Q$), the electric field has magnitude

$$E = \frac{1}{4\pi\epsilon_0} \frac{2p}{r^3}$$

for $r \gg l$ (Fig. 21–44), where r is the distance from the point to the center of the dipole. (b) In what direction does **E** point?

68. A 3.0-g copper penny has a positive charge of 5.5 μC. What fraction of its electrons has it lost?

69. Suppose that electrical attraction, rather than gravity, were responsible for holding the Moon in orbit around the Earth. If equal and opposite charges Q were placed on the Earth and the Moon, what should be the value of Q to maintain the present orbit? Use these data: mass of Earth $= 5.97 \times 10^{24}$ kg, mass of Moon $= 7.35 \times 10^{22}$ kg, radius of orbit $= 3.84 \times 10^8$ m. Treat the Earth and Moon as point particles.

70. One type of *electric quadrupole* consists of two dipoles placed end to end with their negative charges (say) overlapping; that is, in the center is $-2Q$ flanked (on a line) by a $+Q$ to either side (Fig. 21–61). Determine the electric field **E** at points along the perpendicular bisector and show that E decreases as $1/r^4$.

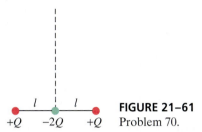

FIGURE 21–61
Problem 70.

71. Three charged particles are placed at the corners of an equilateral triangle of side 1.20 m (Fig. 21–62). The charges are $+4.0\,\mu C$, $-8.0\,\mu C$, and $-6.0\,\mu C$. Calculate the magnitude and direction of the net force on each due to the other two.

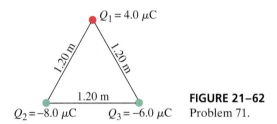

FIGURE 21–62
Problem 71.

72. A proton $(m = 1.67 \times 10^{-27}\,kg)$ is suspended at rest in a uniform electric field **E**. Take into account gravity and determine **E**.

73. Calculate the magnitude of the electric field at the center of a square with sides 35 cm long if the corners, taken in rotation, have charges of $1.0\,\mu C$, $2.0\,\mu C$, $3.0\,\mu C$, and $4.0\,\mu C$ (all positive).

74. In a simple model of the hydrogen atom, the electron revolves in a circular orbit around the proton with a speed of $1.1 \times 10^6\,m/s$. What is the radius of the electron's orbit?

75. Two charges, $-Q_0$ and $-4Q_0$, are a distance l apart. These two charges are free to move but do not because there is a third charge nearby. What must be the magnitude of the third charge and its placement in order for the first two to be in equilibrium?

76. A point charge $(m = 1.0\,g)$ at the end of an insulating string of length 55 cm is observed to be in equilibrium in a uniform horizontal electric field of 10,000 N/C, when the pendulum's position is as shown in Fig. 21–63, with the charge 12 cm above the lowest (vertical) position. If the field points to the right in Fig. 21–63, determine the magnitude and sign of the point charge.

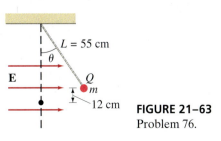

FIGURE 21–63
Problem 76.

77. A positive point charge $Q_1 = 1.85 \times 10^{-5}\,C$ is fixed at the origin of coordinates, and a negative charge $Q_2 = -7.65 \times 10^{-6}\,C$ is fixed to the x axis at $x = +2.00\,m$. Find the location of the place(s) along the x axis where the electric field due to these two charges is zero.

78. Estimate the net force between the CO group and the HN group shown in Fig. 21–64. The C and O have effective charges $\pm 0.40e$ and the H and N have effective charges $\pm 0.20e$ where $e = 1.6 \times 10^{-19}\,C$. [*Hint*: Do not include the "internal" forces between C and O, or between H and N.]

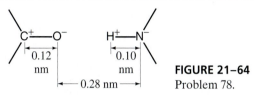

FIGURE 21–64
Problem 78.

79. An electron moving with a speed of $2.0 \times 10^6\,m/s$ to the right enters a uniform electric field region where the field is known to be parallel to its direction of motion. If the electron is to be brought to rest in the space of 5.4 cm, (*a*) in what direction is the electric field, and (*b*) what is the magnitude of the field?

80. The two strands of the helix-shaped DNA molecule (the genetic material in living cells) are held together by electrostatic forces as shown in Fig. 21–65. Assume that the net average charge indicated on H and N atoms is effectively $0.2e$, and on the indicated C and O atoms is $0.4e$, that atoms on each molecule are separated by $1.0 \times 10^{-10}\,m$, and that all relevant angles are 120°. Estimate the net force between: (*a*) a thymine and an adenine; and (*b*) a cytosine and a guanine. (*c*) Estimate the total force for a DNA molecule containing 10^5 pairs of such molecules.

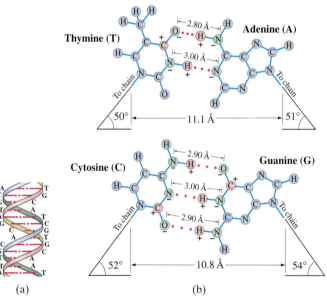

FIGURE 21–65 Problem 80. (a) Section of a DNA double helix. (b) "Close-up" view of the helix, showing how A and T attract each other and how G and C attract each other through electrostatic forces, to hold the double helix together. The red dots are used to indicate the electrostatic attraction (often called a "weak bond" or "hydrogen bond"). Note that there are two weak bonds between A and T, and three between C and G. The distance unit is the angstrom $(1\,Å = 10^{-10}\,m)$.

81. Suppose electrons enter the electric field midway between two plates at an angle θ_0 to the horizontal, as shown in Fig. 21–66. The path is symmetrical, so they leave at the same angle θ_0 and just barely miss the top plate. What is θ_0? Ignore fringing of the field.

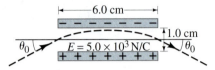

FIGURE 21–66 Problem 81.

82. Two point charges, $Q_1 = -6.7\,\mu\text{C}$ and $Q_2 = 1.3\,\mu\text{C}$, are located between two oppositely charged parallel plates, as shown in Fig. 21–67. The two point charges are separated by a distance of $x = 0.34\,\text{m}$. Assume that the electric field produced by the charged plates is uniform and equal to $E = 73{,}000\,\text{N/C}$. Calculate the net electrostatic force on Q_1 and give its direction.

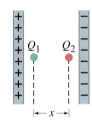

FIGURE 21–67 Problem 82.

83. A small lead ball is encased in insulating plastic and suspended vertically from an ideal spring $(k = 126\,\text{N/m})$ above a lab table, Fig. 21–68. The total mass of the coated ball is 0.800 kg, and its center lies 15.0 cm above the tabletop when in equilibrium. The ball is pulled down 5.00 cm below equilibrium, an electric charge $Q = -3.00 \times 10^{-6}\,\text{C}$ is deposited on the ball, and the system is released. Using what you know about harmonic oscillation, write an expression for the electric field strength as a function of time that would be measured at the point on the tabletop (P) directly below the ball.

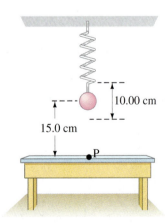

FIGURE 21–68 Problem 83.

84. A large electroscope is made with "leaves" that are 75-cm-long wires with 22-g balls at the ends. When charged, nearly all the charge resides on the balls. If the wires each make a 30° angle with the vertical (Fig. 21–69), what total charge Q must have been applied to the electroscope?

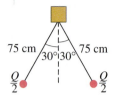

FIGURE 21–69 Problem 84.

85. Three very large square planes of charge are arranged as shown (on edge) in Fig. 21–70. From left to right, the planes have charge densities per unit area of $-0.50\,\mu\text{C/m}^2$, $+0.10\,\mu\text{C/m}^2$, and $-0.35\,\mu\text{C/m}^2$. Find the total electric field (direction and magnitude) at the points A, B, C, and D. Assume the plates are much larger than the distance AD.

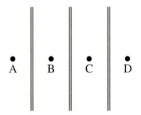

FIGURE 21–70 Problem 85.

86. What is the total charge of all the electrons in a 15-kg bar of aluminum? What is the net charge of the bar? (Aluminum has 13 electrons per atom and an atomic mass of 27 u.)

87. Given the two charges shown in Fig. 21–71, at what position(s) x is the electric field zero? Is the field zero at any other points, not on the x axis?

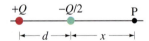

FIGURE 21–71 Problem 87.

88. An electron moves in a circle of radius r around a very long uniformly charged wire in a vacuum chamber, as shown in Fig. 21–72. The charge density on the wire is $\lambda = 0.14\,\mu\text{C/m}$. (a) What is the electric field at the electron (magnitude and direction)? (b) What is the speed of the electron?

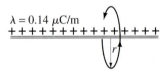

FIGURE 21–72 Problem 88.

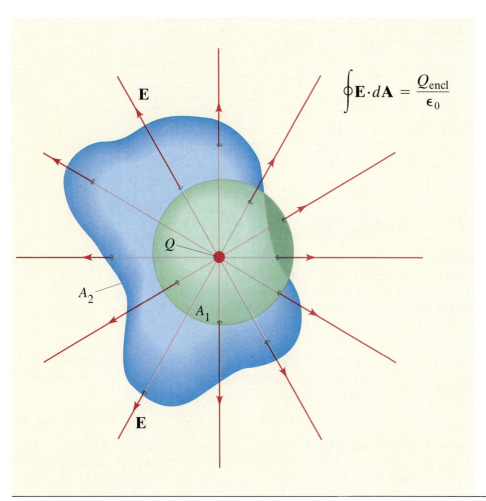

$$\oint \mathbf{E} \cdot d\mathbf{A} = \frac{Q_{\text{encl}}}{\epsilon_0}$$

Gauss's law is an elegant relation between electric charge and electric field. It is more general than Coulomb's law. Gauss's law involves an integral of the electric field $\mathbf{E}$ at each point on a closed surface. The surface is only imaginary, and we choose the shape and placement of the surface so that we can figure out the integral. In this drawing, two different surfaces are shown, both enclosing a point charge Q. Gauss's law states that the product $\mathbf{E} \cdot d\mathbf{A}$, where $d\mathbf{A}$ is an infinitesimal area of the surface, integrated over the entire surface, equals the charge enclosed by the surface Q_{encl} divided by ϵ_0. Both surfaces here enclose the same charge Q. Hence $\oint \mathbf{E} \cdot d\mathbf{A}$ will give the same result for both.

CHAPTER 22

Gauss's Law

G auss's law, which we develop and discuss in this chapter, is a statement of the relation between electric charge and electric field. It is a more general and elegant form of Coulomb's law.

We can, in principle, determine the electric field due to any given distribution of electric charge using Coulomb's law. The total electric field at any point will be the vector sum (or integral) of contributions from all charges present (see Eqs. 21–5 and 21–6). Except for some simple cases, the sum or integral can be quite complicated to evaluate. For situations in which an analytic solution (such as we carried out in the Examples of Sections 21–6 and 21–7) is not possible, a computer can be used.

In some cases, however, the electric field due to a given charge distribution can be calculated more easily or more elegantly using Gauss's law, as we shall see in this chapter. But the major importance of Gauss's law is that it gives us additional insight into the nature of electrostatic fields, and a more general relationship between charge and field.

Before discussing Gauss's law itself, we first discuss the concept of *flux*.

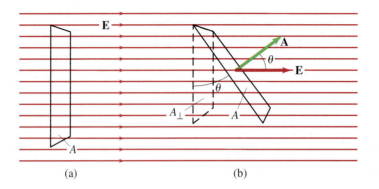

FIGURE 22–1 A uniform field **E** (indicated by the parallel field lines) passing through a surface of area A: (a) perpendicular to **E**; (b) not perpendicular to **E**. The dashed surface of area $A_\perp$ in (b) is the projection of **A** perpendicular to the field **E**.

(a)   (b)

22–1 Electric Flux

Imagine a surface of area A through which a uniform electric field **E** passes, Fig. 22–1. The surface could be a rectangle (as shown), a circle, or any other shape. If the electric field direction is perpendicular to the surface, as in Fig. 22–1a, the **electric flux**, Φ_E, through this surface is defined as the product

$$\Phi_E = EA.$$

If the area A is not perpendicular to **E**, but rather makes an angle θ as shown in Fig. 22–1b, fewer field lines will pass through the area. In this case we define the electric flux through the surface as

$$\Phi_E = EA_\perp = EA\cos\theta, \qquad\qquad \text{[E uniform]} \quad \textbf{(22–1a)}$$

where $A_\perp$ is the projection of the area A on a surface perpendicular to **E** as shown. The area A of a surface can be represented by a vector **A** whose magnitude is A and whose direction is perpendicular to the surface, as shown in Fig. 22–1b. The angle θ is the angle between **E** and **A**, so the electric flux can also be written

$$\Phi_E = \mathbf{E} \cdot \mathbf{A}. \qquad\qquad \text{[E uniform]} \quad \textbf{(22–1b)}$$

Because of how we defined it, the electric flux has a simple intuitive interpretation in terms of field lines. We saw in Section 21–8 that field lines can always be drawn so that the number (N) passing through unit area perpendicular to the field $(A_\perp)$ is proportional to the magnitude of the field (E): that is, $E \propto N/A_\perp$. Hence,

Electric flux is proportional to the number of field lines passing through the area

$$N \propto EA_\perp = \Phi_E,$$

so the flux through an area is proportional to the number of lines passing through that area.

EXAMPLE 22–1 **Electric flux.** (*a*) Calculate the electric flux through the rectangle shown in Fig. 22–1a. The rectangle is 10 cm by 20 cm and the electric field is uniform at 200 N/C. (*b*) What is the flux in Fig. 22–1b if the angle θ is 30°?

SOLUTION (*a*) The electric flux is

$$\begin{aligned}\Phi_E &= EA\cos\theta \\ &= (200\,\text{N/C})(0.10\,\text{m} \times 0.20\,\text{m})\cos 0° = 4.0\,\text{N}\cdot\text{m}^2/\text{C}.\end{aligned}$$

(*b*) In this case the flux is

$$\Phi_E = (200\,\text{N/C})(0.10\,\text{m} \times 0.20\,\text{m})\cos 30° = 3.5\,\text{N}\cdot\text{m}^2/\text{C}.$$

Now let us consider the more general case, when the electric field **E** is not uniform and the surface is not flat, Fig. 22–2. We divide up the chosen surface into n small elements of surface whose areas are $\Delta A_1, \Delta A_2, \ldots \Delta A_n$. We choose the division so that each ΔA_i is small enough that (1) it can be considered flat, and (2) the electric field varies so little over this small area that it can be considered uniform over this tiny area. Then the electric flux through the entire surface is approximately

$$\Phi_E \approx \sum_{i=1}^{n} \mathbf{E}_i \cdot \Delta \mathbf{A}_i,$$

where $\mathbf{E}_i$ is the field passing through $\Delta \mathbf{A}_i$. In the limit as we let $\Delta \mathbf{A}_i \to 0$, the sum becomes an integral over the entire surface and the relation becomes mathematically exact:

$$\Phi_E = \int \mathbf{E} \cdot d\mathbf{A}. \qquad (22\text{–}2)$$

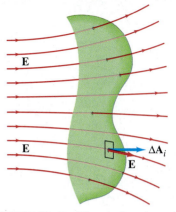

FIGURE 22–2 Electric flux through a curved surface. One small area of the surface, $\Delta \mathbf{A}_i$, is indicated.

Electric flux defined

In many cases (in particular, for Gauss's law) we deal with the flux through a *closed* surface—that is, a surface that completely encloses a volume (like a sphere or the surface of a football), Fig. 22–3. In this case, the net flux through the surface is given by

$$\Phi_E = \oint \mathbf{E} \cdot d\mathbf{A}, \qquad (22\text{–}3)$$

Electric flux through a closed surface

where the integral sign is written $\oint$ to indicate that the integral is over the value of **E** on an enclosing surface.

Up to now we have not been concerned with the fact that there is an ambiguity in the direction of the vector **A** that represents a surface. For example, in Fig. 22–1, the vector **A** could point upward and to the right (as shown) or downward to the left and still be perpendicular to the surface. For a closed surface, we define (arbitrarily) the direction of **A**, or of $d\mathbf{A}$, to point *outward* from the enclosed volume, Fig. 22–4. For a line leaving the enclosed volume (on the right in Fig. 22–4), the angle θ between **E** and $d\mathbf{A}$ must be less than $\pi/2$ ($= 90°$), so $\cos\theta > 0$. For a line entering the volume (on the left in Fig. 22–4) $\theta > \pi/2$, so $\cos\theta < 0$. Hence, *flux entering the enclosed volume is negative* ($\int E \cos\theta \, dA < 0$), whereas *flux leaving the volume is positive*. Consequently, Eq. 22–3 gives the net flux *out* of the volume. If Φ_E is negative, there is a net flux *into* the volume.

Flux entering is negative
Flux leaving is positive

In Figs. 22–3 and 22–4, each field line that enters the volume also leaves the volume. Hence $\Phi_E = \oint \mathbf{E} \cdot d\mathbf{A} = 0$. There is no net flux into or out of this surface. The flux, $\oint \mathbf{E} \cdot d\mathbf{A}$, will be nonzero only if one or more lines start or end within the surface.

FIGURE 22–3 Electric flux through a closed surface.

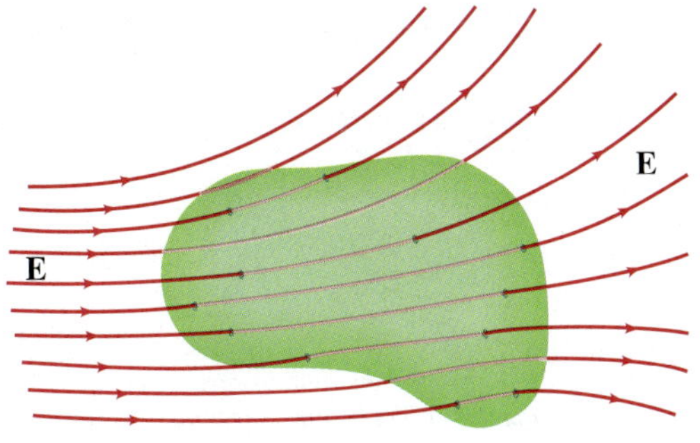

FIGURE 22–4 The direction of an element of area $d\mathbf{A}$ is taken to point outward from an enclosed surface.

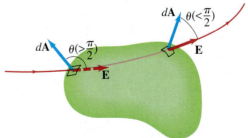

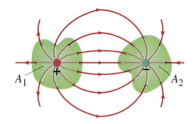

FIGURE 22–5 An electric dipole. Flux through surface A_1 is positive. Flux through A_2 is negative.

FIGURE 22–6 Net flux through surface A is negative.

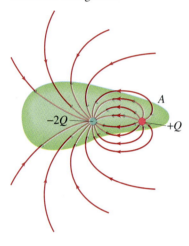

FIGURE 22–7 A single point charge Q at the center of an imaginary sphere of radius r (our "gaussian surface"—that is, the closed surface we choose to use for applying Gauss's law in this case).

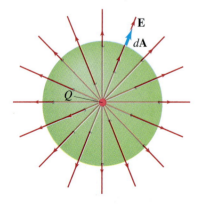

Since electric field lines start and stop only on electric charges, the flux will be nonzero only if the surface encloses a net charge. For example, the surface labeled A_1 in Fig. 22–5 encloses a positive charge and there is a net outward flux through this surface ($\Phi_E > 0$). The surface A_2 encloses an equal magnitude negative charge and there is a net inward flux ($\Phi_E < 0$). For the configuration shown in Fig. 22–6, the flux through the surface shown is negative (count the lines). The value of Φ_E depends on the charge enclosed by the surface, and this is what Gauss's law is all about.

[The concept of flux applies equally well to fluid flow, and thus makes an interesting analogy. (Indeed, the word "flux" comes from the Latin word for "flow.") The electric field **E** at each point corresponds to the fluid flow velocity **v**, so the electric field lines correspond to streamlines of a fluid flow. The flux, Φ, through a surface for a fluid is the volume rate of flow and is given by $\Phi = \int \mathbf{v} \cdot d\mathbf{A}$. In Figs. 22–1, 22–2, and 22–3, the lines can correspond to the steady streamline flow of a fluid with no sources (such as a faucet) or sinks (such as a leak or drain). In this case, the net flux through a closed surface, as in Fig. 22–3, is zero: what flows in also flows out. In Figs. 22–5 and 22–6, there are sources (corresponding to a positive charge) where flow lines start, and also sinks (corresponding to negative charge) where flow lines end. Although this comparison between electric flux and fluid flux is interesting, and perhaps offers some insight, do not get them confused—an electric flux is not a flow of any substance. Flux can be defined for any vector field, and we will later use it also for the magnetic field.]

22–2 Gauss's Law

The precise relation between the electric flux through a closed surface and the net charge Q_{encl} enclosed within that surface is given by **Gauss's law**:

$$\oint \mathbf{E} \cdot d\mathbf{A} \; = \; \frac{Q_{\text{encl}}}{\epsilon_0}, \tag{22–4}$$

where ϵ_0 is the same constant (permittivity of free space) that appears in Coulomb's law. The integral on the left is over the value of **E** on a closed surface, which we choose for our convenience in any given situation. The charge Q_{encl} is the net charge *enclosed* by that surface. It doesn't matter where or how the charge is distributed within the surface. Any charge outside this surface must not be included. A charge outside the chosen surface may affect the position of the electric field lines, but will not affect the net number of lines entering or leaving the surface. For example, Q_{encl} for the gaussian surface A_1 in Fig. 22–5 would be the positive charge enclosed by A_1; the negative charge does contribute to the electric field at A_1 but it is *not* enclosed by surface A_1 and so is not included in Q_{encl}.

Before we discuss the validity of Gauss's law, we note that the integral is often rather difficult to carry out in practice. We rarely need to do it except for some fairly simple situations that will be discussed shortly (Section 22–3).

Now let us see how Gauss's law is related to Coulomb's law.[†] First, we show that Coulomb's law follows from Gauss's law. In Fig. 22–7 we have a single charge Q.

[†] Note that Gauss's law would look more complicated in terms of the constant $k = 1/4\pi\epsilon_0$ that we originally used in Coulomb's law (Eq. 21–1 or 21–4a):

Coulomb's law	*Gauss's law*
$E = k\dfrac{Q}{r^2}$	$\oint \mathbf{E} \cdot d\mathbf{A} = 4\pi k Q$
$E = \dfrac{1}{4\pi\epsilon_0}\dfrac{Q}{r^2}$	$\oint \mathbf{E} \cdot d\mathbf{A} = \dfrac{Q}{\epsilon_0}.$

Gauss's law has a simpler form using ϵ_0; Coulomb's law is simpler using k. The normal convention is to use ϵ_0 rather than k because Gauss's law is considered more general and therefore it is preferable to have it in simpler form.

For our "gaussian surface," we choose an imaginary sphere of radius r centered on the charge. Because Gauss's law is supposed to be valid for any surface, we have chosen one that will make our calculation easy. Because of the symmetry of this (imaginary) sphere about the charge at its center, we know that $\mathbf{E}$ must have the same magnitude at any point on the surface, and that $\mathbf{E}$ points radially outward (or inward) parallel to $d\mathbf{A}$, an element of the surface area. Hence we write the integral in Gauss's law as

$$\oint \mathbf{E} \cdot d\mathbf{A} = \oint E \, dA = E \oint dA = E(4\pi r^2)$$

since the surface area of a sphere of radius r is $4\pi r^2$, and the magnitude of $\mathbf{E}$ is the same at all points on this spherical surface. Then Gauss's law becomes, with $Q_{encl} = Q$,

$$\frac{Q}{\epsilon_0} = \oint \mathbf{E} \cdot d\mathbf{A} = E(4\pi r^2).$$

Solving for E we obtain

$$E = \frac{Q}{4\pi\epsilon_0 r^2},$$

which is the electric field form of Coulomb's law, Eq. 21–4b.

Now let us do the reverse, and derive Gauss's law from Coulomb's law for static electric charges. First we consider a single point charge Q surrounded by an imaginary spherical surface as in Fig. 22–7. Coulomb's law tells us that the electric field at the spherical surface is $E = (1/4\pi\epsilon_0)(Q/r^2)$. Reversing the argument we just used, we have

$$\oint \mathbf{E} \cdot d\mathbf{A} = \oint \frac{1}{4\pi\epsilon_0} \frac{Q}{r^2} \, dA = \frac{Q}{4\pi\epsilon_0 r^2}(4\pi r^2) = \frac{Q}{\epsilon_0}.$$

This is Gauss's law, with $Q_{encl} = Q$, and we derived it for the special case of a spherical surface enclosing a point charge at its center. But what about some other surface, such as the irregular surface labeled A_2 in Fig. 22–8? The same number of field lines (due to our charge Q) pass through surface A_2, as pass through the spherical surface, A_1. Therefore, because the flux through a surface is proportional to the number of lines through it as we saw in Section 22–1, the flux through A_2 is the same as through A_1:

$$\oint_{A_2} \mathbf{E} \cdot d\mathbf{A} = \oint_{A_1} \mathbf{E} \cdot d\mathbf{A} = \frac{Q}{\epsilon_0}.$$

Hence we can expect that

$$\oint \mathbf{E} \cdot d\mathbf{A} = \frac{Q}{\epsilon_0}$$

would be valid for *any* surface surrounding a single point charge Q.

Finally, let us look at the case of more than one charge. For each charge, Q_i, enclosed by the chosen surface,

$$\oint \mathbf{E}_i \cdot d\mathbf{A} = \frac{Q_i}{\epsilon_0},$$

where $\mathbf{E}_i$ refers to the electric field produced by Q_i alone. By the superposition principle for electric fields (Eq. 21–5), the total field $\mathbf{E}$ is equal to the sum of the fields due to each separate charge, $\mathbf{E} = \Sigma\mathbf{E}_i$. Hence

$$\oint \mathbf{E} \cdot d\mathbf{A} = \oint (\Sigma\mathbf{E}_i) \cdot d\mathbf{A} = \Sigma \frac{Q_i}{\epsilon_0} = \frac{Q_{encl}}{\epsilon_0},$$

where $Q_{encl} = \Sigma Q_i$ is the total net charge enclosed within the surface. Thus we see, based on this simple argument, that Gauss's law follows from Coulomb's law for any distribution of electric charge enclosed within a closed surface of any shape.

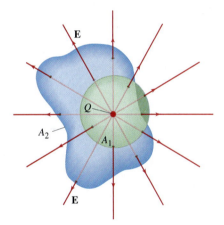

FIGURE 22–8 A single point charge surrounded by a spherical surface A_1, and an irregular surface, A_2.

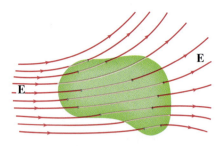

FIGURE 22–9 Electric flux through a closed surface. (Same as Fig. 22–3.) No electric charge is enclosed by this surface ($Q_{encl} = 0$).

The derivation of Gauss's law from Coulomb's law is valid for electric fields produced by static electric charges. We will see later that electric fields can also be produced by changing magnetic fields. Coulomb's law cannot be used to describe such electric fields. But Gauss's law *is* found to hold also for electric fields generated in this way. Hence *Gauss's law is a more general law than Coulomb's law.* It holds for any electric field whatsoever.

Even for the case of static electric fields we are treating in this chapter, it is important to recognize that **E** on the left side of Gauss's law is not necessarily due only to the charge Q_{encl} that appears on the right. For example, in Fig. 22–9 there is an electric field **E** at all points on the imaginary gaussian surface, but it is not due to the charge enclosed by the surface (which is $Q_{encl} = 0$ in this case). The electric field **E** which appears on the left side of Gauss's law is the *total* electric field at each point, on the gaussian surface chosen, not just that due to the charge Q_{encl}, which appears on the right side. Gauss's law has been found to be valid for the total field at any surface. It tells us that any *difference* between the input and output flux of the electric field over any surface is due to charge within that surface.

CONCEPTUAL EXAMPLE 22–2 | **Flux from Gauss's law.** Consider the two gaussian surfaces, A_1 and A_2, shown in Fig. 22–10. The only charge present is the charge Q at the center of surface A_1. What is the net flux through each surface, A_1 and A_2?

RESPONSE The surface A_1 encloses the charge $+Q$. By Gauss's law, the net flux through A_1 is then Q/ϵ_0. For surface A_2, the charge $+Q$ is outside the surface. Surface A_2 encloses zero net charge, so the net electric flux through A_2 is zero, by Gauss's law. Note that all field lines that enter the volume enclosed by surface A_2 also leave it.

FIGURE 22–10 Example 22–2. Two gaussian surfaces.

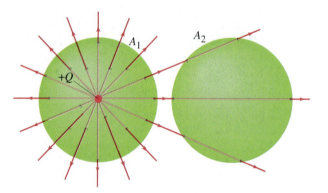

22–3 | Applications of Gauss's Law

Gauss's law is a very compact and elegant way to write the relation between electric charge and electric field. It also offers a simple way to determine the electric field when the charge distribution is simple and/or possesses a high degree of symmetry. In order to apply Gauss's law however, we must choose the "gaussian" surface very carefully (for the integral on the left side of Gauss's law) so we can determine **E**. We normally try to think of a surface that has just the symmetry needed so that E will be constant on all or on parts of its surface. Sometimes we choose a surface so the flux through part of the surface is zero.

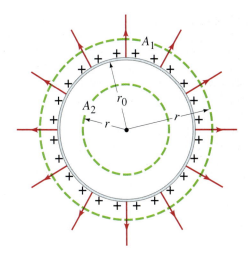

FIGURE 22–11 Cross sectional drawing of a thin spherical shell of radius r_0, carrying a net charge Q uniformly distributed. A_1 and A_2 represent two Gaussian surfaces we use to determine **E**.

EXAMPLE 22–3 **Spherical conductor.** A thin spherical shell of radius r_0 possesses a total net charge Q that is uniformly distributed on it, Fig. 22–11. Determine the electric field at points (a) outside the shell, and (b) inside the shell. (c) What if the conductor were a solid sphere?

SOLUTION (a) Because the charge is distributed symmetrically, the electric field must also be symmetric. Thus the field must be directed radially outward (inward if $Q < 0$) and must depend only on r, not on angle (spherical coordinates). First we want to find **E** outside the spherical shell, so we choose our imaginary gaussian surface to be a sphere of radius r $(r > r_0)$ concentric with the shell, and shown in Fig. 22–11 as a dashed circle A_1 outside the shell. The electric field **E** then has the same magnitude at all points on the surface, and because **E** is perpendicular to this surface, the cosine of the angle between **E** and $d\mathbf{A}$ is always 1. Gauss's law then gives $\left(\text{with } Q_{\text{encl}} = Q\right)$

$$\oint \mathbf{E} \cdot d\mathbf{A} = E(4\pi r^2) = \frac{Q}{\epsilon_0}$$

or

$$E = \frac{1}{4\pi\epsilon_0} \frac{Q}{r^2}. \qquad\qquad [r > r_0]$$

Field outside spherical shell is same as for point charge at center

Thus the field outside a uniform spherical shell of charge is the same as if all the charge were concentrated at the center as a point charge.

(b) Inside the shell, the field must also be symmetric. So E must again have the same value at all points on a spherical gaussian surface $\left(A_2 \text{ in Fig. 22–11}\right)$ concentric with the shell. Thus E can be factored out of the integral and, with $Q_{\text{encl}} = 0$, we have

$$\oint \mathbf{E} \cdot d\mathbf{A} = E(4\pi r^2) = 0.$$

Hence

$$E = 0 \qquad\qquad [r < r_0]$$

E = 0 inside uniformly charged spherical shell

inside a uniform spherical shell of charge.

(c) These same results also apply to a uniformly charged solid spherical conductor, since all the charge would lie in a thin layer at the surface.

Above results apply also for solid conducting sphere

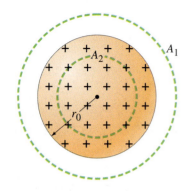

FIGURE 22–12 A solid sphere of uniform charge density.

Uniform sphere produces same field as a point charge at center

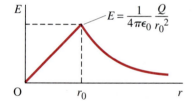

FIGURE 22–13 Magnitude of the electric field as a function of the distance r from the center of a uniformly charged solid sphere.

Electric field inside a uniformly charged nonconducting sphere

EXAMPLE 22–4 **Solid sphere of charge.** An electric charge Q is distributed uniformly throughout a nonconducting sphere of radius r_0, Fig. 22–12. Determine the electric field (*a*) outside the sphere $(r > r_0)$ and (*b*) inside the sphere $(r < r_0)$.

SOLUTION Since the charge is distributed symmetrically in the sphere, the electric field at all points must again be symmetric. **E** depends only on r and is directed radially outward (or inward if $Q < 0$).
(*a*) For our gaussian surface we choose a sphere of radius r $(r > r_0)$, labeled A_1 in Fig. 22–12. Since E depends only on r, Gauss's law gives, with $Q_{encl} = Q$,

$$\oint \mathbf{E} \cdot d\mathbf{A} = E(4\pi r^2) = \frac{Q}{\epsilon_0}$$

or

$$E = \frac{1}{4\pi\epsilon_0} \frac{Q}{r^2}.$$

Again, the field outside a spherically symmetric distribution of charge is the same as that for a point charge of the same magnitude located at the center of the sphere.
(*b*) Inside the sphere, we choose for our gaussian surface a concentric sphere of radius r $(r < r_0)$, labeled A_2 in Fig. 22–12. From symmetry, the magnitude of **E** is the same at all points on A_2, and **E** is perpendicular to the surface, so

$$\oint \mathbf{E} \cdot d\mathbf{A} = E(4\pi r^2).$$

We must equate this to Q_{encl}/ϵ_0 where Q_{encl} is the charge enclosed by A_2. Q_{encl} is not the total charge Q but only a portion of it. We define the **charge density**, ρ_E, as the charge per unit volume $(\rho_E = dQ/dV)$, and here we are given that $\rho_E = $ constant. So the charge enclosed by the gaussian surface A_2, a sphere of radius r, is

$$Q_{encl} = \left(\frac{\frac{4}{3}\pi r^3 \rho_E}{\frac{4}{3}\pi r_0^3 \rho_E} \right) Q = \frac{r^3}{r_0^3} Q.$$

Hence, from Gauss's law,

$$E(4\pi r^2) = \frac{Q_{encl}}{\epsilon_0} = \frac{r^3}{r_0^3} \frac{Q}{\epsilon_0}$$

or

$$E = \frac{1}{4\pi\epsilon_0} \frac{Q}{r_0^3} r. \qquad\qquad [r < r_0]$$

Thus the field increases linearly with r, until $r = r_0$. It then decreases as $1/r^2$, as plotted in Fig. 22–13.

The results above would have been difficult to obtain from Coulomb's law by integrating over the sphere. Using Gauss's law and the symmetry of the situation, this result is obtained rather easily, and shows the great power of Gauss's law. However, its use in this way is limited mainly to cases where the charge distribution has a high degree of symmetry. In such cases, we *choose* a simple surface on which $E = $ constant, so the integration is simple. Gauss's law holds, of course, for any surface. The next two Examples are symmetric cases that we did treat before, using Coulomb's law, but we get the result more easily using Gauss's law.

EXAMPLE 22–5 **Long uniform line of charge.** A very long straight wire possesses a uniform positive charge per unit length, λ. Calculate the electric field at points near (but outside) the wire, far from the ends.

SOLUTION Because of the symmetry, we expect the field to be directed radially outward and to depend only on the perpendicular distance, r, from the wire. Because of the cylindrical symmetry, the field will be the same at all points on a gaussian surface that is a cylinder with the wire along its axis, Fig. 22–14. **E** is perpendicular to this surface at all points. For Gauss's law, we need a closed surface, so we include the flat ends of the cylinder. Since **E** is parallel to the ends, there is no flux through the ends (the cosine of the angle between **E** and $d\mathbf{A}$ on the ends is $\cos 90° = 0$). So Gauss's law tells us

$$\oint \mathbf{E} \cdot d\mathbf{A} = E(2\pi r l) = \frac{Q_{\text{encl}}}{\epsilon_0} = \frac{\lambda l}{\epsilon_0},$$

where l is the length of our chosen gaussian surface ($l \ll$ length of wire), and $2\pi r$ is its circumference. Hence

$$E = \frac{1}{2\pi\epsilon_0}\frac{\lambda}{r}.$$

This is the same result as we got in Example 21–10 using Coulomb's law (we used x there instead of r), but here it took much less effort. Again we see the great power of Gauss's law.[†]

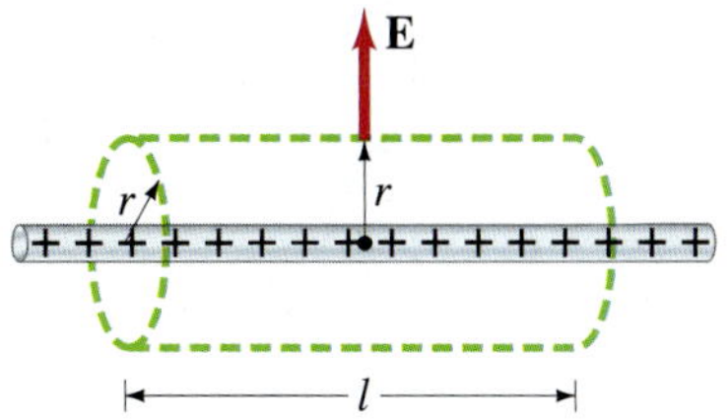

FIGURE 22–14 Calculation of **E** due to a very long line of charge.

EXAMPLE 22–6 **Infinite plane of charge.** Charge is distributed uniformly, with a surface charge density σ ($\sigma =$ charge per unit area $= dQ/dA$), over a very large but very thin nonconducting flat plane surface. Determine the electric field at points near the plane.

SOLUTION We choose as our gaussian surface a small, closed cylinder whose axis is perpendicular to the plane and which extends through the plane as shown in Fig. 22–15. Because of the symmetry, we expect **E** to be directed perpendicular to the plane on both sides as shown, and to be uniform over the end caps of the cylinder, each of whose area is A. Since no flux passes through the curved sides of the cylinder, all the flux is through the two end caps. So Gauss's law gives

$$\oint \mathbf{E} \cdot d\mathbf{A} = 2EA = \frac{Q_{\text{encl}}}{\epsilon_0} = \frac{\sigma A}{\epsilon_0},$$

where $Q_{\text{encl}} = \sigma A$ is the charge enclosed by our gaussian cylinder. The electric field is then

$$E = \frac{\sigma}{2\epsilon_0}.$$

This is the same result we obtained much more laboriously in Chapter 21, Eq. 21–7. The field is uniform for points far from the ends of the plane, and close to its surface.

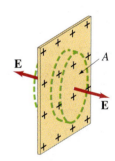

FIGURE 22–15 Calculation of the electric field outside a large uniformly charged nonconducting plane surface.

Electric field near a thin uniformly charged plane

[†] But note that the method of Example 21–10 allows calculation of E also for a short line of charge by using the appropriate limits for the integral, whereas Gauss's law is not readily adapted due to lack of symmetry.

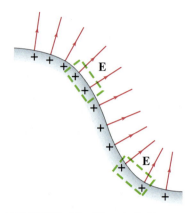

FIGURE 22–16 Electric field near surface of a conductor.

EXAMPLE 22–7 **Electric field near any conducting surface.** Show that the electric field just outside the surface of any good conductor of arbitrary shape is given by

$$E = \frac{\sigma}{\epsilon_0},$$

where σ is the surface charge density on the conductor's surface at that point.

SOLUTION We choose as our gaussian surface a small cylindrical box, as we did in the previous example. We choose the cylinder to be very small in height, so that one of its circular ends is just above the conductor (Fig. 22–16). The other end is just below the conductor's surface, and the sides are perpendicular to it. The electric field is zero inside a conductor and is perpendicular to the surface just outside it (Section 21–9), so electric flux passes only through the outside end of our cylindrical box. We choose the area A (of the flat cylinder end) small enough so that E is essentially uniform over it. Then Gauss's law gives

$$\oint \mathbf{E} \cdot d\mathbf{A} = EA = \frac{Q_{\text{encl}}}{\epsilon_0} = \frac{\sigma A}{\epsilon_0},$$

so that

$$E = \frac{\sigma}{\epsilon_0}. \qquad \text{[at surface of conductor]} \quad \text{(22–5)}$$

Electric field at surface of a conductor

This is a useful result which applies for a conductor of any shape.

When is $E = \sigma/\epsilon_0$ and when is $E = \sigma/2\epsilon_0$

FIGURE 22–17 Thin flat charged conductor with surface charge density σ at each surface, but for the conductor as a whole, the charge density is $\sigma' = 2\sigma$.

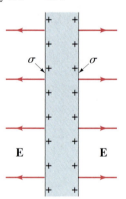

Why is it that the field outside a large plane nonconductor is $E = \sigma/2\epsilon_0$ (Example 22–6) whereas outside a conductor it is $E = \sigma/\epsilon_0$ (Example 22–7)? The reason for the factor of 2 comes not so much from conductor versus nonconductor as from what we mean by the surface charge density σ. For a conductor, the charge lies at the surface and all the electric field lines leave on one side of the surface. For a thin plane nonconductor, the lines leave both sides, Fig. 22–15. If we had a large, thin, flat conducting plane, the charge would accumulate on both surfaces (Fig. 22–17) and the field would emanate from both sides. If we called σ' the surface charge for the plane as a *whole*, each face of the plane would have surface charge $\sigma = \sigma'/2$, and so the result of Example 22–7 would give $E = (\sigma'/2)/\epsilon_0 = \sigma'/2\epsilon_0$, the same as for a nonconducting plane. Normally, however, we use σ to apply to *each* face of a conducting plane and then we would get $E = \sigma/\epsilon_0$. Thus the factor of 2 between Examples 22–6 and 22–7 comes from two different ways of defining σ.

We saw in Section 21–9 that in the static situation, the electric field inside any conductor must be zero even if it has a net charge. (Otherwise, the free charges in the conductor would move—until the net force on each, and hence $\mathbf{E}$, were zero.) We also mentioned there that any net electric charge on a conductor must all reside on its outer surface. This is readily shown using Gauss's law. Consider any charged conductor of any shape, such as that shown in Fig. 22–18, which carries a net charge Q. We choose the gaussian surface, shown dashed in the diagram, so that it lies just below the surface of the conductor. Our gaussian surface can be arbitrarily close to the surface, but still *inside* the conductor. The electric field is zero at all points on this gaussian surface since it is inside the conductor. Hence, from Gauss's law, Eq. 22–4, the net charge within the surface must be zero. Hence, there can be no net charge within the conductor. Any net charge must lie on the surface of the conductor.

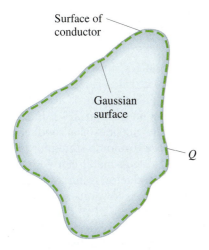

FIGURE 22–18 An insulated charged conductor of arbitrary shape, showing gaussian surface (dashed) just below the surface of the conductor.

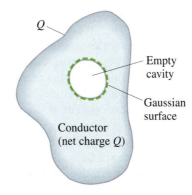

FIGURE 22–19 An empty cavity inside a charged conductor carries zero net charge.

If there is an empty cavity inside a conductor, can charge accumulate on that (inner) surface too? As shown in Fig. 22–19, if we imagine a gaussian surface (shown dashed) just inside the conductor above the cavity, we know that **E** must be zero everywhere on this surface since it is inside the conductor. Hence, by Gauss's law, *there can be no net charge at the surface of the cavity*.

But what if the cavity is not empty and there is a charge inside it?

CONCEPTUAL EXAMPLE 22–8 **Conductor with charge inside a cavity.** Suppose a conductor carries a net charge $+Q$ and contains a cavity, inside of which resides a point charge $+q$. What can you say about the charges on the inner and outer surfaces of the conductor?

RESPONSE As shown in Fig. 22–20, a gaussian surface just inside the conductor surrounding the cavity must contain zero net charge ($E = 0$ in a conductor). Thus a net charge of $-q$ must exist on the cavity surface. The conductor itself carries a net charge $+Q$, so its outer surface must carry a charge equal to $Q + q$.

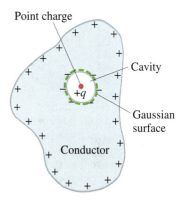

FIGURE 22–20 Example 22–8.

PROBLEM SOLVING **Gauss's Law for Symmetric Charge Distributions**

1. First identify the symmetry of the charge distribution: spherical, cylindrical, planar. This identification should suggest a gaussian surface for which **E** will be constant and/or zero on all or on parts of the surface: a sphere for spherical symmetry, a cylinder for cylindrical symmetry and a small cylinder or "pillbox" for planar symmetry.

2. Draw the appropriate gaussian surface making sure it passes through the point where you want to know the electric field.

3. Use the symmetry of the charge distribution to determine the direction of **E** on the gaussian surface.

4. Evaluate the flux, $\oint \mathbf{E} \cdot d\mathbf{A}$. With the appropriate gaussian surface, the dot product $\mathbf{E} \cdot d\mathbf{A}$ should be zero or equal to $\pm E\, dA$, with the magnitude of E being constant.

5. Calculate the charge *enclosed* by the gaussian surface. Remember it's the enclosed charge that matters. Ignore all the charge outside the gaussian surface.

6. Equate the flux to the enclosed charge and solve for E.

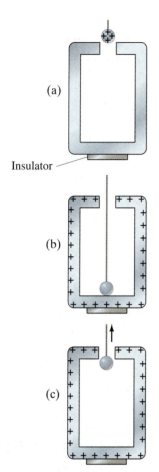

(a)

Insulator

(b)

(c)

FIGURE 22–21 (a) A charged conductor (metal ball) is lowered into an insulated metal can (a good conductor) carrying zero net charge. (b) The charged ball is touched to the can and all of its charge quickly flows to the outer surface of the can. (c) When the ball is then removed, it is found to carry zero net charge.

* 22–4 Experimental Basis of Gauss's and Coulomb's Law

Gauss's law predicts that any net charge on a conductor must lie only on its surface. But is this true in real life? Let us see how it can be verified experimentally. And in confirming this prediction of Gauss's law, Coulomb's law is also confirmed since the latter follows from Gauss's law, as we saw in Section 22–2. Indeed, the earliest observation that charge resides only on the outside of a conductor was recorded by Benjamin Franklin some 30 years before Coulomb stated his law.

A simple experiment is illustrated in Fig. 22–21. A metal can with a small opening at the top rests on an insulator. The can, a conductor, is initially uncharged (Fig. 22–21a). A charged metal ball (also a conductor) is lowered by an insulating thread into the can, and is allowed to touch the can (Fig. 22–21b). The ball and can now form a single conductor. Gauss's law, as discussed above, predicts that all the charge will flow to the outer surface of the can. (The flow of charge in such situations does not occur instantaneously, but the time involved is usually negligible.) These predictions are confirmed in experiments by (1) connecting an electroscope to the can, which will show that the can is charged, and (2) connecting an electroscope to the ball after it has been withdrawn from the can (Fig. 22–21c), which will show that the ball carries zero charge.

The precision with which Coulomb's and Gauss's laws hold can be stated quantitatively by writing Coulomb's law as

$$F = k\frac{Q_1 Q_2}{r^{2+\delta}}.$$

For a perfect inverse-square law, $\delta = 0$. The most recent and precise experiments (1971) give $\delta = (2.7 \pm 3.1) \times 10^{-16}$. Thus Coulomb's and Gauss's laws are found to be valid to an extremely high precision!

Summary

The **electric flux** passing through a flat area A for a uniform electric field $\mathbf{E}$ is

$$\Phi_E = \mathbf{E} \cdot \mathbf{A}.$$

If the field is not uniform, the flux is determined from the integral

$$\Phi_E = \int \mathbf{E} \cdot d\mathbf{A}.$$

The direction of the vector $\mathbf{A}$ or $d\mathbf{A}$ is chosen to be perpendicular to the surface whose area is A or dA, and points outward from an enclosed surface. The flux through a surface is proportional to the number of field lines passing through it.

Gauss's law states that the net flux passing out of any closed surface is equal to the net charge Q_{encl} enclosed by the surface divided by ϵ_0:

$$\oint \mathbf{E} \cdot d\mathbf{A} = \frac{Q_{encl}}{\epsilon_0}.$$

Gauss's law can in principle be used to determine the electric field due to a given charge distribution, but its usefulness is mainly limited to a small number of cases, usually where the charge distribution displays much symmetry. The real importance of Gauss's law is that it is a more general and elegant statement (than Coulomb's law) for the relation between electric charge and electric field. It is one of the basic equations of electromagnetism.

Questions

1. If the electric flux through a closed surface is zero, is the electric field necessarily zero at all points on the surface? What about the converse: If $\mathbf{E} = 0$ at all points on the surface is the flux through the surface zero?

2. Is the electric field $\mathbf{E}$ in Gauss's law, $\oint \mathbf{E} \cdot d\mathbf{A} = Q_{encl}/\epsilon_0$, that due only to the charge Q_{encl}?

3. A point charge is surrounded by a spherical gaussian surface of radius r. If the sphere is replaced by a cube of side r, will Φ_E be larger, smaller, or the same?

4. What can you say about the flux through a closed surface that encloses an electric dipole?

5. The electric field $\mathbf{E}$ is zero at all points on a closed surface; is there necessarily no net charge within the surface? If a surface encloses zero net charge, is the electric field necessarily zero at all points on the surface?

6. Define gravitational flux in analogy to electric flux. Are there "sources" and "sinks" for the gravitational field as there are for the electric field? Discuss.

7. Would Gauss's law be helpful in determining the electric field due to an electric dipole?

8. A spherical basketball (a nonconductor) is given a charge Q distributed uniformly over its surface. What can you say about the electric field inside the ball? A person now steps on the ball, collapsing it, and forcing most of the air out without altering the charge. What can you say about the field inside now?

9. In Example 22–5, it may seem that the electric field calculated is due only to the charge on the wire that is enclosed by the cylinder chosen as our gaussian surface. In fact, the entire charge along the whole length of the wire contributes to the field. Explain how the charge outside the cylindrical gaussian surface of Fig. 22–14 contributes to E at the gaussian surface. [*Hint*: Compare to what the field would be due to a short wire.]

10. Suppose the line of charge in Example 22–5 extended only a short way beyond the ends of the cylinder shown in Fig. 22–14. How would the result of Example 22–5 be altered?

11. A point charge Q is surrounded by a spherical surface of radius r_0, whose center is at Q. Later, the charge is moved to the right a distance $\frac{1}{2}r_0$, but the sphere remains where it was, Fig. 22–22. How is the electric flux Φ_E through the sphere changed? Is the electric field at the surface of the sphere changed? For each "yes" answer, describe the change.

FIGURE 22–22
Question 11.

FIGURE 22–23
Question 12.

12. A conductor carries a net positive charge Q. There is a hollow cavity within the conductor, at whose center is a negative point charge $-q$ (Fig. 22–23). What is the charge on (*a*) the outer surface of the conductor and (*b*) the inner surface of the conductor?

13. A point charge q is placed at the center of the cavity of a thin metal shell which is neutral. Will a charge Q placed outside the shell feel an electric force? Explain.

14. In Fig. 22–24, two objects, O_1 and O_2, have charges $+1.0 \ \mu C$ and $-2.0 \ \mu C$ respectively, and a third object, O_3, is electrically neutral. (*a*) What is the electric flux through the surface A_1 that encloses all the three objects? (*b*) What is the electric flux through the surface A_2 that encloses the third object only?

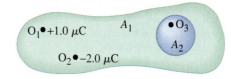

FIGURE 22–24
Question 14.

Problems

Section 22–1

1. (I) A flat circle of radius 15 cm is placed in a uniform electric field of magnitude $5.8 \times 10^2 \ N/C$. What is the electric flux through the circle when its face is (*a*) perpendicular to the field lines, (*b*) at $45°$ to the field lines, and (*c*) parallel to the field lines.

2. (I) The Earth possesses an electric field of (average) magnitude $150 \ N/C$ near its surface. The field points radially inward. Calculate the net electric flux outward through a spherical surface surrounding, and just beyond, the Earth's surface.

3. (II) A cube of side l is placed in a uniform field $E = 6.50 \times 10^3 \ N/C$ with edges parallel to the field lines. What is the net flux through the cube? What is the flux through each of its six faces?

4. (II) A uniform field $\mathbf{E}$ is parallel to the axis of a hollow hemisphere of radius R, Fig. 22–25. (*a*) What is the electric flux through the hemispherical surface? (*b*) What is the result if $\mathbf{E}$ is instead perpendicular to the axis?

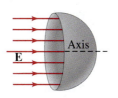

FIGURE 22–25
Problem 4.

Section 22–2

5. (I) The total electric flux from a cubical box 28.0 cm on a side is $1.45 \times 10^3 \, \mathrm{N \cdot m^2/C}$. What charge is enclosed by the box?

6. (I) Figure 22–26 shows five closed surfaces that surround various charges as indicated. Determine the electric flux through each surface, S_1, S_2, S_3, S_4, and S_5. The surfaces are flat "pill-box" surfaces that extend only slightly above and below the plane of the paper.

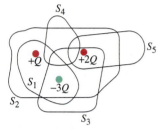

FIGURE 22–26   Problem 6.

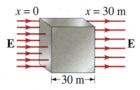

FIGURE 22–27   Problem 7.

7. (II) In a certain region of space, the electric field is constant in direction (say horizontally, in the x direction), but its magnitude decreases from $E = 560 \, \mathrm{N/C}$ at $x = 0$ to $E = 410 \, \mathrm{N/C}$ at $x = 30 \, \mathrm{m}$. Determine the charge within a cubical box of side $l = 30 \, \mathrm{m}$, where the box is oriented so that four of its sides are parallel to the field lines (Fig. 22–27).

8. (II) A point charge Q is placed at the center of a cube of side l. What is the flux through one face of the cube?

Section 22–3

9. (I) The field just outside a 3.50-cm-radius metal ball is $2.75 \times 10^2 \, \mathrm{N/C}$ and points toward the ball. What charge resides on the ball?

10. (I) Starting from the result of Example 22–3, show that the electric field just outside a uniformly charged spherical conductor is $E = \sigma/\epsilon_0$, consistent with Example 22–7.

11. (I) A long thin wire, hundreds of meters long, carries a uniformly distributed charge of $-2.8 \, \mu\mathrm{C}$ per meter of length. What are the magnitude and direction of the electric field at points (a) 5.0 m and (b) 2.0 m from the wire?

12. (II) A solid metal sphere of radius 3.00 m carries a total charge of $-3.50 \, \mu\mathrm{C}$. What is the magnitude of the electric field at a distance from the sphere's center of (a) 0.15 m, (b) 2.90 m, (c) 3.10 m, and (d) 6.00 m? How would the answers differ if the sphere were (e) a thin shell, or (f) a solid nonconductor uniformly charged throughout?

13. (II) A 15.0-cm-diameter nonconducting sphere carries a total charge of $12.0 \, \mu\mathrm{C}$ distributed uniformly throughout its volume. Graph the electric field E as a function of the distance r from the center of the sphere from $r = 0$ to $r = 30 \, \mathrm{cm}$.

14. (II) A flat square sheet of thin aluminum foil, 25 cm on a side, carries a uniformly distributed 35 nC charge. What, approximately, is the electric field (a) 1.0 cm above the sheet and (b) 20 m above the sheet?

15. (II) A spherical cavity of radius 4.50 cm is at the center of a metal sphere of radius 18.0 cm. A point charge $Q = 5.50 \, \mu\mathrm{C}$ rests at the very center of the cavity, whereas the metal conductor carries no net charge. Determine the electric field at a point (a) 3.0 cm from the center of the cavity and (b) 6.0 cm from the center of the cavity.

16. (II) A point charge Q rests at the center of an uncharged thin spherical conducting shell. What is the electric field E as a function of r (a) for r less than the radius of the shell, (b) inside the shell, and (c) beyond the shell? (d) Does the shell affect the field due to Q alone? Does the charge Q affect the shell?

17. (II) A solid metal cube has a spherical cavity at its center as shown in Fig. 22–28. At the center of the cavity there is a point charge $Q = +8.00 \, \mu\mathrm{C}$. The metal cube carries a net charge $q = -7.00 \, \mu\mathrm{C}$ (not including Q). Determine (a) the total charge on the surface of the spherical cavity and (b) the total charge on the outer surface of the cube.

FIGURE 22–28   Problem 17.

FIGURE 22–29
Problems 18, 19, and 20.

18. (II) Two large, flat metal plates are separated by a distance that is very small compared to their height and width. The conductors are given equal but opposite uniform surface charge densities $\pm\sigma$. Ignore edge effects and use Gauss's law to show that for points far from the edges, (a) the electric field between the plates is $E = \sigma/\epsilon_0$ and (b) that outside the plates on either side the field is zero. (c) How would your results be altered if the two plates were nonconductors? (See Fig. 22–29.)

19. (II) Suppose the two conducting plates in Problem 18 have the *same* sign of charge. What then will be the electric field (a) between them and (b) outside them on either side? (c) What if the planes are nonconducting?

20. (II) The electric field between two square metal plates is $100 \, \mathrm{N/C}$. The plates are 1.0 m on a side and are separated by 3.0 cm, as in Fig. 22–29. What is the charge on each plate? Neglect edge effects.

21. (II) Two thin concentric spherical shells of radii r_1 and r_2 $(r_1 < r_2)$ contain uniform surface charge densities σ_1 and σ_2, respectively (see Fig. 22–30). Determine the electric field for (a) $r < r_1$, (b) $r_1 < r < r_2$, and (c) $r > r_2$. (d) Under what conditions will $E = 0$ for $r > r_2$? (e) Under what conditions will $E = 0$ for $r_1 < r < r_2$?

FIGURE 22–30   Two spherical shells (Problem 21).

FIGURE 22–31
Problems 22, 23, and 34.

22. (II) Suppose the nonconducting sphere of Example 22–4 has a spherical cavity of radius r_1 centered at the sphere's center (Fig. 22–31). Assuming the charge Q is distributed uniformly in the "shell" (between $r = r_1$ and $r = r_0$), determine the electric field as a function of r for (a) $r < r_1$, (b) $r_1 < r < r_0$, and (c) $r > r_0$.

23. (II) Suppose in Fig. 22–31, Problem 22, there is also a charge q at the center of the cavity. Determine the electric field for (a) $0 < r < r_1$, (b) $r_1 < r < r_0$, and (c) $r > r_0$.

24. (II) Suppose the thick spherical shell of Problem 22 is a conductor. It carries a total net charge Q and at its center there is a point charge q. What total charge is found on (a) the inner surface of the shell and (b) the outer surface of the shell? Determine the electric field for (c) $0 < r < r_1$, (d) $r_1 < r < r_0$, and (e) $r > r_0$.

25. (II) Suppose that at the center of the cavity inside the shell (charge Q) of Fig. 22–11 (and Example 22–3), there is a point charge $q \; (\neq Q)$. Determine the electric field for (a) $r < r_0$ and (b) $r > r_0$. What are your answers if (c) $q = Q$ and (d) $q = -Q$?

26. (II) A spherical rubber balloon carries a total charge Q uniformly distributed on its surface. At $t = 0$ the nonconducting balloon has radius r_0 and the balloon is then slowly blown up so that r increases linearly to $2r_0$ in a time T. Determine the electric field as a function of time (a) just outside the balloon surface and (b) at $r = 4r_0$.

27. (II) A long cylindrical shell of radius R_0 and length L $(R_0 \ll L)$ possesses a uniform surface charge density (charge per unit area) σ (Fig. 22–32). Determine the electric field at points (a) outside the cylinder $(r > R_0)$ and (b) inside the cylinder $(r < R_0)$; assume the points are far from the ends and not too far from the shell $(r \ll L)$. (c) Compare to the result for a long line of charge, Example 22–5.

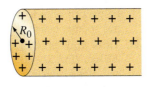

FIGURE 22–32 Problem 27.

FIGURE 22–33 Problem 28.

28. (II) A very long solid nonconducting cylinder of radius R_0 and length L $(R_0 \ll L)$ possesses a uniform volume charge density ρ_E (C/m^3), Fig. 22–33. Determine the electric field at points (a) outside the cylinder $(r > R_0)$ and (b) inside the cylinder $(r < R_0)$. Do only for points far from the ends and for which $r \ll L$.

29. (II) A thin cylindrical shell of radius R_1 is surrounded by a second concentric cylindrical shell of radius R_2 (Fig. 22–34). The inner shell has a total charge $+Q$ and the outer shell $-Q$. Assuming the length L of the shells is much greater than R_1 or R_2, determine the electric field as a function of r (the perpendicular distance from the common axis of the cylinders) for (a) $r < R_1$, (b) $R_1 < r < R_2$, and (c) $r > R_2$. (d) What is the kinetic energy of an electron if it moves between (and concentric with) the shells in a circular orbit of radius $(R_1 + R_2)/2$?

FIGURE 22–34 Problems 29, 30, 31, and 32.

30. (II) In Problem 29 (a) under what conditions will $E = 0$ for $r > R_2$? (b) Under what conditions will $E = 0$ for $R_1 < r < R_2$?

31. (II) A thin cylindrical shell of radius $R_1 = 5.0$ cm is surrounded by a second cylindrical shell of radius $R_2 = 9.0$ cm, as in Fig. 22–34. Both cylinders are 5.0 m long and the inner one carries a total charge $Q_1 = -3.8\,\mu C$ and the outer one $Q_2 = +3.2\,\mu C$. For points far from the ends of the cylinders, determine the electric field at a radial distance r from the central axis of (a) 3.0 cm, (b) 6.0 cm, and (c) 12.0 cm.

32. (II) (a) If an electron $(m = 9.1 \times 10^{-31}\,kg)$ escaped from the surface of the inner cylinder in Problem 31 (Fig. 22–34) with negligible speed, what would be its speed when it reached the outer cylinder? (b) If a proton $(m = 1.67 \times 10^{-27}\,kg)$ revolves in a circular orbit of radius $r = 6.0$ cm about the axis (i.e., between the cylinders), what must be its speed?

33. (II) A very long solid nonconducting cylinder of radius R_1 is uniformly charged with a charge density ρ_E. It is surrounded by a concentric cylindrical tube of inner radius R_2 and outer radius R_3 as shown in Fig. 22–35, and it too carries a uniform charge density ρ_E. Determine the electric field as a function of the distance r from the center of the cylinders for (a) $0 < r < R_1$, (b) $R_1 < r < R_2$, (c) $R_2 < r < R_3$, and (d) $r > R_3$. (e) If $\rho_E = 15\,\mu C/m^3$ and $R_1 = \frac{1}{2}R_2 = \frac{1}{3}R_3 = 5.0$ cm, plot E as a function of r from $r = 0$ to $r = 20.0$ cm. Assume the cylinders are very long compared to R_3.

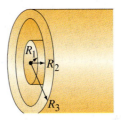

FIGURE 22–35 Problem 33.

34. (III) Suppose the density of charge between r_1 and r_0 of the hollow sphere of Problem 22 (Fig. 22–31) varies as $\rho_E = \rho_0 r_1/r$. Determine the electric field as a function of r for (a) $r < r_1$, (b) $r_1 < r < r_0$, and (c) $r > r_0$. (d) Plot E versus r from $r = 0$ to $r = 2r_0$.

35. (III) A point charge Q is on the axis of a cylinder at its center. The diameter of the cylinder is equal to its length L (Fig. 22–36). What is the total flux through the curved sides of the cylinder? [Hint: First calculate the flux through the ends.]

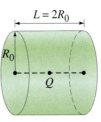

FIGURE 22–36 Problem 35.

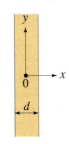

FIGURE 22–37 Problem 36.

36. (III) A flat slab of nonconducting material (Fig. 22–37) carries a uniform charge per unit volume, ρ_E. The slab has thickness d which is small compared to the height and breadth of the slab. Determine the electric field as a function of x (a) inside the slab and (b) outside the slab (at distances much less than the slab's height or breadth). Take the origin at the center of the slab.

General Problems

37. Write Gauss's law for the gravitational field **g** (see Section 6–6).

38. The Earth is surrounded by an electric field, pointing inward at every point, of magnitude $E \approx 150$ N/C near the surface. (a) What is the net charge on the Earth? (b) How many excess electrons per square meter on the Earth's surface does this correspond to?

39. A cube of side l has one corner at the origin of coordinates, and extends along the positive x, y, and z axes. Suppose the electric field in this region is given by $E = (a + by)\mathbf{j}$. Determine the charge inside the cube.

40. A solid nonconducting sphere of radius r_0 has a total charge Q which is distributed according to $\rho_E = br$, where ρ_E is the charge per unit volume, or charge density (C/m^3), and b is a constant. Determine (a) b in terms of Q, (b) the electric field at points inside the sphere, and (c) the electric field at points outside the sphere.

41. A point charge of 3.50 nC is located at the origin and a second charge of -5.00 nC is located on the x axis at $x = 1.50$ m. Calculate the electric flux through a sphere centered at the origin with radius 1.00 m. Repeat the calculation for a sphere of radius 2.00 m.

42. A point charge produces an electric flux of $+500$ N·m²/C through a gaussian sphere of radius 15.0 cm centered on the charge. (a) What is the flux through a gaussian sphere with a radius 35.0 cm? (b) What is the magnitude and sign of the charge?

43. A point charge Q is placed a distance $r_0/2$ above the surface of an imaginary spherical surface of radius r_0 (Fig. 22–38). (a) What is the electric flux through the sphere? (b) What range of values does E have at the surface of the sphere? (c) Is **E** perpendicular to the sphere at all points? (d) Is Gauss's law useful for obtaining E at the surface of the sphere?

FIGURE 22–38 Problem 43.   **FIGURE 22–39** Problem 44.

44. Three large but thin charged sheets are parallel to each other as shown in Fig. 22–39. Sheet I has a total surface charge density of 9.0 nC/m², sheet II a charge of -2.0 nC/m², and sheet III a charge of 5.0 nC/m². What is the force per unit area on each sheet, in N/m²?

45. Neutral hydrogen can be modeled as a positive point charge $+e$ surrounded by a distribution of negative charge with volume density given by $\rho_E(r) = -Ae^{-2r/a_0}$ where $a_0 = 0.53 \times 10^{-10}$ m is called the *Bohr radius*, and A is a constant such that the total amount of negative charge is $-e$. (a) What is the net charge inside a sphere of radius a_0? (b) What is the strength of the electric field at a distance a_0 from the nucleus?

46. A very large thin plane has uniform surface charge density σ. Touching it on the right (Fig. 22–40) is a long and wide slab of thickness d with uniform volume charge density ρ_E. Determine the electric field (a) to the left of the plane, (b) to the right of the slab, and (c) everywhere inside the slab.

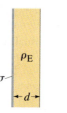

FIGURE 22–40 Problem 46.   **FIGURE 22–41** Problem 47.

47. A sphere of radius r_0 carries a volume charge density ρ_E (Fig. 22–41). A spherical cavity of radius $r_0/2$ is then scooped out and left empty, as shown. (a) What is the magnitude and direction of the electric field at point A? (b) What is the direction and magnitude of the electric field at point B?

48. Dry air will break down and generate a spark if the electric field exceeds about 3×10^6 N/C. How much charge could be packed onto a green pea (diameter 0.75 cm) before the pea spontaneously discharges?

49. Three very large sheets are separated by equal distances of 20.0 cm (Fig. 22–42). The first and third sheets are very thin and nonconducting and have surface charge densities of $+5.00 \ \mu C/m^2$ and $-5.00 \ \mu C/m^2$ respectively. The middle sheet is conducting but has no net charge. (a) What is the electric field inside the middle sheet? What is the electric field (b) between the left and middle sheets, and (c) between the middle and right sheets? (d) What is the charge density on the surface of the middle sheet facing the left sheet, and (e) on the surface facing the right sheet?

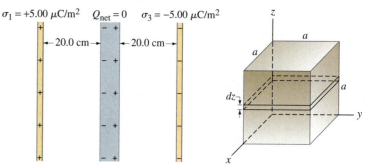

FIGURE 22–42 Problem 49.   **FIGURE 22–43** Problem 50.

50. Careful measurement of the electric field in the ground can help provide useful information about charge. In a particular region, a technician has determined that the electric field in a cubical volume, 1.00 m on a side, is

$$E = E_0\left(1 + \frac{z}{a}\right)\mathbf{\hat{i}} + E_0\left(\frac{z}{a}\right)\mathbf{\hat{j}}$$

where $E_0 = 1.00$ N/C and $a = 1.00$ m. The cube has its sides parallel to the coordinate axes, Fig. 22–43. Determine the net charge within the cube.

Lightning: The potential difference (voltage) between clouds and the Earth can become so high that electrons are pulled off atoms of the air by the large electric field. The air becomes a conductor as the ionized atoms and freed electrons flow rapidly, colliding with more atoms, and causing more ionization. The massive flow of charge reduces the potential difference and the "discharge" quickly ceases. The light represents energy released when the ions and electrons recombine to form atoms.

CHAPTER 23

Electric Potential

We saw in Chapters 7 and 8 that the concept of energy was extremely valuable in dealing with mechanical problems. For one thing, energy is a conserved quantity and is thus an important tool for understanding nature. Furthermore, we saw that many problems could be solved using the energy concept even though a detailed knowledge of the forces involved was not possible, or when a calculation involving Newton's laws would have been too difficult.

The energy point of view can be used in electricity, and it is especially useful. It not only extends the law of conservation of energy, but it gives us another way to view electrical phenomena; and it is a tool in solving problems more easily, in many cases, than by using forces and electric fields.

23–1 | Electric Potential and Potential Difference

To apply conservation of energy, we need to define electric potential energy as for other types of potential energy (Chapter 8). As we saw, potential energy can be defined only for a conservative force. Recall that the work done by a conservative force in moving an object between any two positions is independent of the path

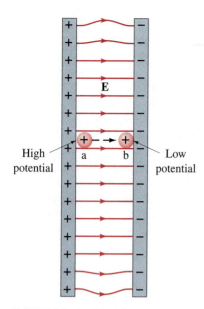

High potential

Low potential

FIGURE 23–1 Work is done by the electric field in moving the positive charge from position a to position b.

Potential is potential energy per unit charge

Potential difference

The volt (1 V = 1 J/C)

Voltage = potential difference

V = 0 chosen arbitrarily

taken. It is easy to see that the electrostatic force between any two charges $(F = kQ_1Q_2/r^2)$ is conservative: the dependence on position is $1/r^2$ just as for the gravitational force, which we saw in Section 8–7 is conservative. Hence the electrostatic force given by Coulomb's law is conservative and we can define potential energy U for it.

We define the change in electric potential energy, $U_b - U_a$, when a charge q moves from some point a to a second point b, as the negative of the work done by the electric force to move the charge from a to b. For example, consider the electric field between two equally but oppositely charged parallel plates whose separation is small compared to their width and height, so the field **E** will be uniform over most of the region, Fig. 23–1. Now consider a tiny positive point charge q placed at point a very near the positive plate as shown. This charge q is so small it doesn't affect **E**. If this charge q at point a is released, the electric force will do work on the charge and accelerate it toward the negative plate. In the process, the charged particle will have its kinetic energy K increased. By conservation of energy, the potential energy will decrease by an equal amount, equal to the negative of the work done by the electric force. In accord with the conservation of energy, electric potential energy is transformed into kinetic energy, and the total energy is conserved. Note that the positive charge q has its greatest potential energy at point a, near the positive plate,[†] so $(U_b - U_a) < 0$. The reverse is true for a negative charge: its potential energy is greatest near the negative plate.

We defined the electric field (Chapter 21) as the force per unit charge. Similarly, it is useful to define the **electric potential** (or simply the **potential** when "electric" is understood) as the *potential energy per unit charge*. Electric potential is given the symbol V. If a positive test charge q has electric potential energy U_a at some point a (relative to some zero potential energy), the electric potential V_a at this point is

$$V_a = \frac{U_a}{q}.$$

As we discussed in Chapter 8, only differences in potential energy are physically meaningful. Hence only the **difference in potential**, or the **potential difference**, between two points a and b (such as between a and b in Fig. 23–1) is measurable. When the electric force does positive work on a charge, the kinetic energy increases and the potential energy decreases. The difference in potential energy, $U_b - U_a$, is equal to the negative of the work, W_{ba}, done by the electric force to move the charge from point a to point b, so the potential difference V_{ba} is

$$V_{ba} = V_b - V_a = \frac{U_b - U_a}{q} = -\frac{W_{ba}}{q}. \tag{23–1}$$

Note that electric potential, like electric field, does not depend on our test charge q. V depends on the other charges that create the field, not on q; q acquires potential energy by being in the potential V due to the other charges.

We can see from our definition that the positive plate in Fig. 23–1 is at a higher potential than the negative plate. Thus a positively charged object moves naturally from a high potential to a low potential. A negative charge does the reverse.

The unit of electric potential, and of potential difference, is joules/coulomb and is given a special name, the **volt**, in honor of Alessandro Volta (1745–1827; he is best known for having invented the electric battery). The volt is abbreviated V, so 1 V = 1 J/C. Potential difference, since it is measured in volts, is often referred to as **voltage**.

If we wish to speak of the potential, V_a, at some point a, we must be aware that V_a depends on where the potential is chosen to be zero. The zero point for electric potential in a given situation, just as for potential energy, can be chosen arbitrarily

[†] At this point it has its greatest ability to do work (on some other object or system).

since only differences in potential energy can be measured. Often the ground, or a conductor connected directly to the ground (the Earth), is taken as zero potential, and other potentials are given with respect to ground. (Thus, a point where the voltage is 50 V is one where the difference of potential between it and ground is 50 V.) In other cases, as we shall see, we may choose the potential to be zero at an infinite distance ($r = \infty$).

CONCEPTUAL EXAMPLE 23–1 | **A negative charge.** Suppose a negative charge, such as an electron, is placed at point b in Fig. 23–1. If the electron is free to move, will its electric potential energy increase or decrease? How will the electric potential change?

RESPONSE An electron placed at point b will move toward the positive plate. (An electron placed at point a would not move.) As the electron moves to the left, its potential energy *decreases* as its kinetic energy gets larger. But note that the electron moves from a point b at low potential to a point a at higher potential: $\Delta V = V_a - V_b > 0$. (The potentials V_a and V_b are due to the charges on the plates, not due to the electron.)

Since the electric potential difference is defined as the potential energy difference per unit charge, then the change in potential energy of a charge q when moved between two points a and b is

$$U_b - U_a = q(V_b - V_a) = qV_{ba}. \tag{23–2}$$

Electric potential and potential energy

That is, if an object with charge q moves through a potential difference V_{ba}, its potential energy changes by an amount qV_{ba}. For example, if the potential difference between the two plates in Fig. 23–1 is 6 V, then a +1 C charge moved (say by an external force) from b to a will gain $(1\,\text{C})(6\,\text{V}) = 6\,\text{J}$ of electric potential energy. (And it will lose 6 J of electric potential energy if it moves from a to b.) Similarly, a 2-C charge will gain 12 J, and so on. Thus, electric potential difference is a measure of how much energy an electric charge can acquire in a given situation. And, since energy is the ability to do work, the electric potential difference is also a measure of how much work a given charge can do. The exact amount depends both on the potential difference and on the charge.

To better understand electric potential, let's make a comparison to the gravitational case when a rock falls from the top of a cliff. The greater the height, h, of a cliff, the more potential energy ($= mgh$) the rock has at the top of the cliff, relative to the bottom, and the more kinetic energy it will have when it reaches the bottom. The actual amount of kinetic energy it will acquire, and the amount of work it can do, depends both on the height of the cliff and the mass m of the rock. A large rock and a small rock can be at the same height h (Fig. 23–2a) and thus have the same "gravitational potential," but the larger rock has the greater potential energy. The electrical case is similar (Fig. 23–2b): the potential energy change, or the work that can be done, depends both on the potential difference (corresponding to the height of the cliff) and on the charge (corresponding to mass), Eq. 23–2. [But note a significant difference: electric charge comes in two types, + and −, whereas gravitational mass is always +.]

Potential likened to height of a cliff

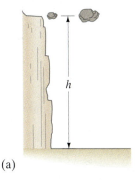

(a)

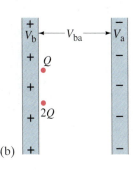

(b)

FIGURE 23–2 (a) Two rocks are at the same height. The larger rock has more potential energy. (b) Two charges have the same electric potential. The 2Q charge has more potential energy.

TABLE 23–1
Some Typical Voltages

Source	Voltage (approx.)
Thundercloud to ground	10^8 V
High-voltage power line	10^6 V
Power supply for TV tube	10^4 V
Automobile ignition	10^4 V
Household outlet	10^2 V
Automobile battery	12 V
Flashlight battery	1.5 V
Resting potential across nerve membrane	10^{-1} V
Potential changes on skin (EKG and EEG)	10^{-4} V

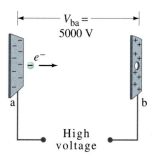

FIGURE 23–3 Electron accelerated in TV picture tube. Example 23–2.

Practical sources of electrical energy such as batteries and electric generators are meant to maintain a potential difference. The actual amount of energy used or transformed depends on how much charge flows. For example, consider an automobile headlight connected to a 12.0-V battery. The amount of energy transformed (into light and thermal energy) is proportional to how much charge flows, which in turn depends on how long the light is on. If over a given period of time 5.0 C of charge flows through the light, the total energy transformed is $(5.0 \text{ C})(12.0 \text{ V}) = 60 \text{ J}$. If the headlight is left on twice as long, 10.0 C of charge will flow and the energy transformed is $(10.0 \text{ C})(12.0 \text{ V}) = 120 \text{ J}$.

Table 23–1 presents some typical voltages.

EXAMPLE 23–2 **Electron in TV tube.** Suppose an electron in the picture tube of a television set is accelerated from rest through a potential difference $V_{ba} = +5000 \text{ V}$ (Fig. 23–3). (*a*) What is the change in potential energy of the electron? (*b*) What is the speed of the electron $(m = 9.1 \times 10^{-31} \text{ kg})$ as a result of this acceleration? (*c*) Repeat for a proton $(m = 1.67 \times 10^{-27} \text{ kg})$ that accelerates through a potential difference of $V_{ba} = -5000 \text{ V}$.

SOLUTION (*a*) The charge on an electron is $e = -1.6 \times 10^{-19} \text{ C}$. Therefore its change in potential energy (Eq. 23–2) is equal to

$$\Delta U = qV_{ba} = (-1.6 \times 10^{-19} \text{ C})(+5000 \text{ V})$$
$$= -8.0 \times 10^{-16} \text{ J}.$$

The minus sign indicates that the potential energy decreases. (The potential difference, V_{ba}, has a positive sign since the final potential is higher than the initial potential; that is, negative electrons are attracted from a negative electrode to a positive one.)

(*b*) The potential energy lost by the electron becomes kinetic energy $(= K)$. From conservation of energy (Eq. 8–9), $\Delta K + \Delta U = 0$, so

$$\Delta K = -\Delta U$$
$$\tfrac{1}{2}mv^2 - 0 = -qV_{ba},$$

where the initial kinetic energy is zero since we assume the electron started from rest. We solve for v and put in the mass of the electron $m = 9.1 \times 10^{-31} \text{ kg}$:

$$v = \sqrt{-\frac{2qV_{ba}}{m}}$$
$$= \sqrt{-\frac{2(-1.6 \times 10^{-19} \text{ C})(5000 \text{ V})}{9.1 \times 10^{-31} \text{ kg}}}$$
$$= 4.2 \times 10^7 \text{ m/s}.$$

[Note: For such a high speed, which is $\tfrac{1}{7}$ the speed of light, we really should use the theory of relativity, Chapter 37, to get a more precise result.]

(*c*) The proton has the same magnitude of charge as the electron, though of opposite sign. Hence for the same magnitude of V_{ba} we expect the same change in U, but a lesser speed since the proton's mass is greater. Thus:

$$\Delta U = qV_{ba} = (+1.6 \times 10^{-19} \text{ C})(-5000 \text{ V}) = -8.0 \times 10^{-16} \text{ J},$$

and

$$v = \sqrt{-\frac{2qV_{ba}}{m}} = \sqrt{-\frac{2(1.6 \times 10^{-19} \text{ C})(-5000 \text{ V})}{(1.67 \times 10^{-27} \text{ kg})}}$$
$$= 9.8 \times 10^5 \text{ m/s}.$$

Note that the energy doesn't depend on the mass, only on the charge and voltage. The speed *does* depend on m.

23–2 Relation between Electric Potential and Electric Field

The effects of any charge distribution can be described either in terms of electric field or in terms of electric potential. Electric potential is often easier to use because it is a scalar, as compared to electric field which is a vector. There is a crucial connection between the electric potential produced by a given arrangement of charges and the electric field due to those charges, which we now examine.

We start by recalling the relation between a conservative force $\mathbf{F}$ and the potential energy U associated with that force. As discussed in Section 8–2, the difference in potential energy between any two points in space, a and b, is given by Eq. 8–4:

$$U_b - U_a = -\int_a^b \mathbf{F} \cdot d\mathbf{l},$$

where $d\mathbf{l}$ is an infinitesimal increment of displacement, and the integral is taken along any path in space from point a to point b. For the electrical case, we are more interested in the potential difference, given by Eq. 23–1, $V_{ba} = V_b - V_a = (U_b - U_a)/q$, rather than in the potential energy itself. Also, the electric field $\mathbf{E}$ at any point in space is defined as the force per unit charge (Eq. 21–3): $\mathbf{E} = \mathbf{F}/q$. Putting these two relations in the above equation gives us

$$V_{ba} = V_b - V_a = -\int_a^b \mathbf{E} \cdot d\mathbf{l}. \qquad \text{(23–3)}$$

V related to $\mathbf{E}$

This is the general relation between electric field and potential difference. See Fig. 23–4. If we are given the electric field due to some arrangement of electric charge, we can use Eq. 23–3 to determine V_{ba}.

A simple special case is when the field is uniform. In Fig. 23–1, for example, a path parallel to the electric field lines from point a at the positive plate to point b at the negative plate gives (since $\mathbf{E}$ and $d\mathbf{l}$ are in the same direction at each point),

$$V_b - V_a = -\int_a^b \mathbf{E} \cdot d\mathbf{l} = -E\int_a^b dl = -Ed$$

or

$$V_{ba} = -Ed \qquad \text{[only if E is uniform]} \qquad \text{(23–4)}$$

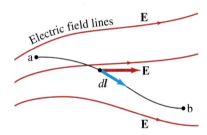

FIGURE 23–4 Integrating $\mathbf{E} \cdot d\mathbf{l}$ from point a to point b in a nonuniform electric field $\mathbf{E}$.

where d is the distance, parallel to the field lines, between points a and b. Be careful not to use Eq. 23–4 unless you are sure the electric field is uniform.

From either Eq. 23–3 or 23–4 we can see that the units for electric field intensity can be written as volts per meter (V/m) as well as newtons per coulomb (N/C). These are equivalent in general, since $1\,\text{N/C} = 1\,\text{N} \cdot \text{m}/\text{C} \cdot \text{m} = 1\,\text{J/C} \cdot \text{m} = 1\,\text{V/m}$.

EXAMPLE 23–3 **Special case: Uniform electric field obtained from voltage.** Two parallel plates are charged to a voltage of 50 V. If the separation between the plates is 5.0 cm, calculate the electric field between them, ignoring any fringing.

SOLUTION We have from Eq. 23–4, dealing only with magnitudes for convenience,

$$E = \frac{V_{ba}}{d} = \frac{50\,\text{V}}{0.050\,\text{m}} = 1000\,\text{V/m}.$$

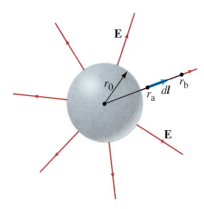

E

r_0

r_a dl r_b

E

FIGURE 23–5 Example 23–4: integrating $\mathbf{E} \cdot d\mathbf{l}$ for the field outside a spherical conductor.

FIGURE 23–6 (a) E versus r, and (b) V versus r, for a uniformly charged solid conducting sphere of radius r_0 (the charge distributes itself on the surface); r is the distance from the center of the sphere.

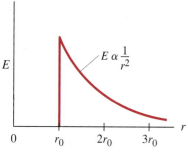

$E \propto \dfrac{1}{r^2}$

E

0 r_0 $2r_0$ $3r_0$ r

(a)

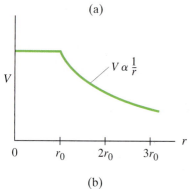

$V \propto \dfrac{1}{r}$

V

0 r_0 $2r_0$ $3r_0$ r

(b)

EXAMPLE 23–4 Charged conducting sphere. Determine the potential at a distance r from the center of a uniformly charged conducting sphere of radius r_0 for (a) $r > r_0$, (b) $r = r_0$, (c) $r < r_0$. The total charge on the sphere is Q.

SOLUTION (a) The charge Q is distributed over the surface of the sphere since it is a conductor. We saw in Example 22–3 that the electric field outside a conducting sphere is

$$E = \frac{1}{4\pi\epsilon_0}\frac{Q}{r^2} \qquad\qquad [r > r_0]$$

and points radially outward (inward if $Q < 0$). Since we know $\mathbf{E}$, we can use Eq. 23–3 and integrate along a radial line with $d\mathbf{l}$ parallel to $\mathbf{E}$ (Fig. 23–5) between two points which are distances r_a and r_b from the sphere's center:

$$V_b - V_a = -\int_{r_a}^{r_b}\mathbf{E}\cdot d\mathbf{l} = -\frac{Q}{4\pi\epsilon_0}\int_{r_a}^{r_b}\frac{dr}{r^2} = \frac{Q}{4\pi\epsilon_0}\left(\frac{1}{r_b} - \frac{1}{r_a}\right).$$

If we let $V = 0$ for $r = \infty$ (say $V_b = 0$ at $r_b = \infty$), then at any other point r (for $r > r_0$) we have

$$V = \frac{1}{4\pi\epsilon_0}\frac{Q}{r}. \qquad\qquad [r > r_0]$$

We will see in the next Section that this same equation applies for the potential a distance r from a single point charge. Thus the electric potential outside a spherical conductor with a uniformly distributed charge is the same as if all the charge were at its center.

(b) As r approaches r_0, we see that

$$V = \frac{1}{4\pi\epsilon_0}\frac{Q}{r_0} \qquad\qquad [r = r_0]$$

at the surface of the conductor.

(c) For points within the conductor, $E = 0$. Thus the integral, $\int \mathbf{E}\cdot d\mathbf{l}$, between $r = r_0$ and any point within the conductor gives zero change in V. Hence V is constant within the conductor:

$$V = \frac{1}{4\pi\epsilon_0}\frac{Q}{r_0}. \qquad\qquad [r \leq r_0]$$

The whole conductor, not just its surface, is at this same potential. Plots of both E and V as a function of r are shown in Fig. 23–6 for a conducting sphere.

EXAMPLE 23–5 Breakdown voltage. In many kinds of equipment, very high voltages are used. A problem with high voltage is that the air can become ionized due to the high electric fields: free electrons in the air (produced by cosmic rays, for example) can be accelerated by such high fields to speeds sufficient to ionize O_2 and N_2 molecules by collision, knocking out one or more of their electrons. The air then becomes conducting and the high voltage cannot be maintained as charge flows. The breakdown of air occurs for electric fields of about 3×10^6 V/m. (a) Show that the breakdown voltage for a spherical conductor in air is proportional to the radius of the sphere, and (b) estimate the breakdown voltage in air for a sphere of diameter 1.0 cm.

SOLUTION (a) The electric potential at the surface of a spherical conductor of radius r_0 (Example 23–4), and the electric field just outside its surface, are

$$V = \frac{1}{4\pi\epsilon_0}\frac{Q}{r_0} \quad\text{and}\quad E = \frac{1}{4\pi\epsilon_0}\frac{Q}{r_0^2}.$$

Combining these we obtain

$$V = r_0 E. \qquad\qquad \text{[at surface of spherical conductor]}$$

(b) For $r_0 = 5 \times 10^{-3}$ m, the breakdown voltage in air is

$$V = (5 \times 10^{-3}\,\text{m})(3 \times 10^6\,\text{V/m}) \approx 15,000\,\text{V}.$$

Example 23–5 makes clear why large rounded terminals without sharp edges are used for high-voltage equipment. It also explains why breakdown, or sparks, occur at rough edges or points (regions with small radius of curvature) on a conductor, and why conductors are usually made very smooth.

23–3 Electric Potential Due to Point Charges

The electric potential at a distance r from a single point charge Q can be derived directly from Eq. 23–3, $V_b - V_a = -\int \mathbf{E} \cdot d\mathbf{l}$. The electric field due to a single point charge has magnitude (Eq. 21–4)

$$E = \frac{1}{4\pi\epsilon_0} \frac{Q}{r^2} \quad \text{or} \quad E = k \frac{Q}{r^2}$$

(where $k = 1/4\pi\epsilon_0 = 8.99 \times 10^9 \, \text{N·m}^2/\text{C}^2$), and is directed radially outward from the charge (inward if $Q < 0$). We take the integral in Eq. 23–3 along a (straight) field line (Fig. 23–7) from point a, a distance r_a from Q, to point b, a distance r_b from Q. Then $d\mathbf{l}$ will be parallel to $\mathbf{E}$ and $dl = dr$. Thus

$$V_b - V_a = -\int_{r_a}^{r_b} \mathbf{E} \cdot d\mathbf{l} = -\frac{Q}{4\pi\epsilon_0} \int_{r_a}^{r_b} \frac{1}{r^2} \, dr = \frac{1}{4\pi\epsilon_0} \left(\frac{Q}{r_b} - \frac{Q}{r_a} \right).$$

As mentioned earlier, only differences in potential have physical meaning. We are free, therefore, to choose the value of the potential at some one point to be whatever we please. It is common to choose the potential to be zero at infinity (let $V_b = 0$ at $r_b = \infty$). Then the electric potential V at a distance r from a single point charge is

$$V = \frac{1}{4\pi\epsilon_0} \frac{Q}{r}. \qquad \left[\begin{array}{l} \text{single point charge;} \\ V = 0 \text{ at } r = \infty \end{array}\right] \quad \textbf{(23–5)}$$

We can think of V here as representing the absolute potential, where $V = 0$ at $r = \infty$, or we can think of V as the potential difference between r and infinity. Notice that the potential V decreases with the first power of the distance, whereas the electric field (Eq. 21–4) decreases as the *square* of the distance. The potential near a positive charge is large, and it decreases toward zero at very large distances (Fig. 23–8). For a negative charge, the potential is negative and increases toward zero at large distances (Fig. 23–9).

In Example 23–4 we found that the potential due to a uniformly charged sphere is given by the same relation, Eq. 23–5, for points outside the sphere. Thus we see that the potential outside a uniformly charged sphere is the same as if all the charge were concentrated at its center.

EXAMPLE 23–6 **Work to force two + charges close together.** What minimum work is required by an external force to bring a charge $q = 3.00 \, \mu\text{C}$ from a great distance away (take $r = \infty$) to a point 0.500 m from a charge $Q = 20.0 \, \mu\text{C}$?

SOLUTION The work done by the electric field is equal to the negative of the change in potential energy:

$$W = -qV_{ba} = -\frac{q}{4\pi\epsilon_0} \left(\frac{Q}{r_b} - \frac{Q}{r_a} \right),$$

where $r_b = 0.500 \, \text{m}$ and $r_a = \infty$. The second term is zero ($1/\infty = 0$), so

$$W = -(3.00 \times 10^{-6} \, \text{C}) \frac{(8.99 \times 10^9 \, \text{N·m}^2/\text{C}^2)(2.00 \times 10^{-5} \, \text{C})}{(0.500 \, \text{m})} = -1.08 \, \text{J}.$$

The electric field does negative work in this case. In order to bring the charge to this point, an *external* force would have to do work $W = +1.08 \, \text{J}$, assuming no acceleration of the charges.

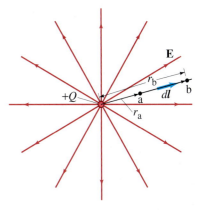

FIGURE 23–7 We integrate Eq. 23–3 along the straight line (shown in black) from point a to point b. The line ab is parallel to a field line.

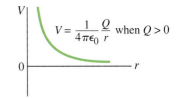

FIGURE 23–8 Potential V as a function of distance r from a single point charge Q when the charge is positive.

FIGURE 23–9 Potential V as a function of distance r from a single point charge Q when the charge is negative.

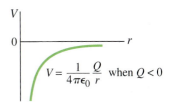

To determine the electric field surrounding a collection of two or more point charges requires adding up the electric fields due to each charge. Since the electric field is a vector, this can often be a chore. To find the electric potential due to a collection of point charges is far easier, since the electric potential is a scalar, and hence you only need to add numbers together without concern for direction. This is a major advantage in using electric potential. We do have to include the signs of charges, however.

Potentials add as scalars
(Fields add as vectors)

EXAMPLE 23–7 **Potential above two charges.** Calculate the electric potential at points A and B in Fig. 23–10 due to the two charges shown. (This is the same situation as Example 21–8, Fig. 21–26, where we calculated the electric field at these points.) Assume $V = 0$ at $r = \infty$.

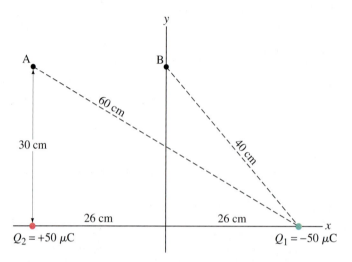

FIGURE 23–10 Example 23–7. (See also Example 21–8, Fig. 21–26.)

SOLUTION The potential at point A is the sum of the potentials due to the + and − charges, and we use Eq. 23–5 for each:

$$V_A = V_{A2} + V_{A1}$$
$$= \frac{(9.0 \times 10^9 \,\text{N} \cdot \text{m}^2/\text{C}^2)(5.0 \times 10^{-5}\,\text{C})}{0.30\,\text{m}}$$
$$+ \frac{(9.0 \times 10^9 \,\text{N} \cdot \text{m}^2/\text{C}^2)(-5.0 \times 10^{-5}\,\text{C})}{0.60\,\text{m}}$$
$$= 1.50 \times 10^6 \,\text{V} - 0.75 \times 10^6 \,\text{V}$$
$$= 7.5 \times 10^5 \,\text{V}.$$

At point B:

$$V_B = V_{B2} + V_{B1}$$
$$= \frac{(9.0 \times 10^9 \,\text{N} \cdot \text{m}^2/\text{C}^2)(5.0 \times 10^{-5}\,\text{C})}{0.40\,\text{m}}$$
$$+ \frac{(9.0 \times 10^9 \,\text{N} \cdot \text{m}^2/\text{C}^2)(-5.0 \times 10^{-5}\,\text{C})}{0.40\,\text{m}}$$
$$= 0\,\text{V}.$$

It should be clear that the potential will be zero everywhere on the plane equidistant between the two charges. Thus this plane is an equipotential surface with $V = 0$.

A summation like these can be easily performed for any number of point charges.

23–4 Potential Due to Any Charge Distribution

If we know the electric field in a region of space due to any distribution of electric charge, we can determine the difference in potential between two points in the region using Eq. 23–3, $V_{ba} = -\int_a^b \mathbf{E} \cdot d\mathbf{l}$. In many cases we don't know $\mathbf{E}$ as a function of position, and it may be difficult to calculate. We can calculate the potential V due to a given charge distribution in another way, often easier, using the potential due to a single point charge, Eq. 23–5:

$$V = \frac{1}{4\pi\epsilon_0}\frac{Q}{r},$$

where $V = 0$ at $r = \infty$. Then we can sum over all the charges. If we have n individual point charges, the potential at some point a (relative to $V = 0$ at $r = \infty$) is

$$V_a = \sum_{i=1}^{n} V_i = \frac{1}{4\pi\epsilon_0}\sum_{i=1}^{n}\frac{Q_i}{r_{ia}}, \qquad \textbf{(23–6a)}$$

Potential due to many point charges

where r_{ia} is the distance from the ith charge (Q_i) to the point a. (We already used this approach in Example 23–7.) If the charge distribution can be considered continuous, then

$$V = \frac{1}{4\pi\epsilon_0}\int\frac{dq}{r}, \qquad \textbf{(23–6b)}$$

Potential due to a continuous charge distribution

where r is the distance from a tiny element of charge, dq, to the point where V is being determined.

EXAMPLE 23–8 **Potential due to a ring of charge.** A thin circular ring of radius R carries a uniformly distributed charge Q. Determine the electric potential at a point P on the axis of the ring a distance x from its center, Fig. 23–11.

SOLUTION Each point on the ring is equidistant from point P, and this distance is $(x^2 + R^2)^{\frac{1}{2}}$. So the potential at P is:

$$V = \frac{1}{4\pi\epsilon_0}\int\frac{dq}{r} = \frac{1}{4\pi\epsilon_0}\frac{1}{(x^2 + R^2)^{\frac{1}{2}}}\int dq = \frac{1}{4\pi\epsilon_0}\frac{Q}{(x^2 + R^2)^{\frac{1}{2}}}.$$

Note that for points very far away from the ring, $x \gg R$, this result reduces to $(1/4\pi\epsilon_0)(Q/x)$, the potential of a point charge, as we should expect.

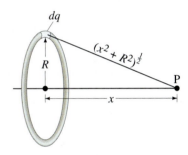

FIGURE 23–11 Calculating the potential at point P, a distance x from the center of a uniform ring of charge (Example 23–8).

EXAMPLE 23–9 **Potential due to a charged disk.** A thin flat disk, of radius R, carries a uniformly distributed charge Q, Fig. 23–12. Determine the potential at a point P on the axis of the disk, a distance x from its center.

SOLUTION Divide the disk into thin rings of radius r and thickness dr. The charge Q is distributed uniformly, so the charge contained in each ring is proportional to its area. The disk has area πR^2 and each thin ring has area $dA = (2\pi r)(dr)$. Hence

$$\frac{dq}{Q} = \frac{2\pi r\, dr}{\pi R^2}$$

so

$$dq = Q\frac{(2\pi r)(dr)}{\pi R^2} = \frac{2Qr\, dr}{R^2}.$$

Then the potential at P, using Eq. 23–6b in which r is replaced by $(x^2 + r^2)^{\frac{1}{2}}$, is

$$V = \frac{1}{4\pi\epsilon_0}\int\frac{dq}{(x^2 + r^2)^{\frac{1}{2}}} = \frac{2Q}{4\pi\epsilon_0 R^2}\int_0^R\frac{r\, dr}{(x^2 + r^2)^{\frac{1}{2}}} = \frac{Q}{2\pi\epsilon_0 R^2}(x^2 + r^2)^{\frac{1}{2}}\Big|_{r=0}^{r=R}$$

$$= \frac{Q}{2\pi\epsilon_0 R^2}\left[(x^2 + R^2)^{\frac{1}{2}} - x\right].$$

FIGURE 23–12 Calculating the electric potential at point P on the axis of a uniformly charged thin disk. Example 23–9.

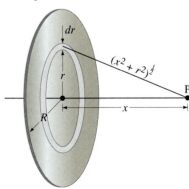

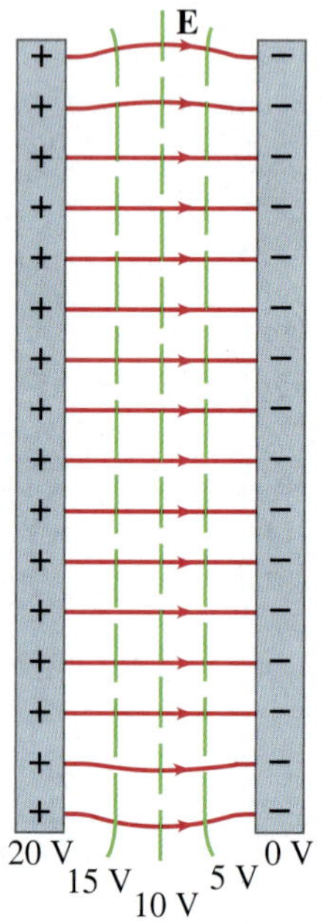

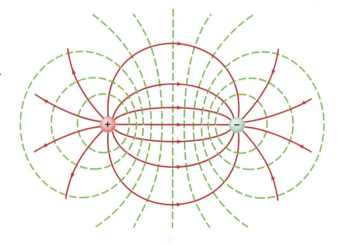

FIGURE 23–13 Equipotential lines (the green dashed lines) between two oppositely charged parallel plates. Note that they are perpendicular to the electric field lines (solid red lines).

The electric potential can be represented graphically by drawing **equipotential lines** or, in three dimensions, **equipotential surfaces**. An equipotential surface is one on which all points are at the same potential. That is, the potential difference between any two points on the surface is zero, and no work is required to move a charge from one point to the other. An *equipotential surface must be perpendicular to the electric field* at any point. If this were not so—that is, if there were a component of **E** parallel to the surface—it would require work to move the charge along the surface against this component of **E**; and this would contradict the idea that it is an equipotential surface. This can also be seen from Eq. 23–3, $\Delta V = -\int \mathbf{E} \cdot d\mathbf{l}$. On a surface where V is constant, $\Delta V = 0$, so we must have either $\mathbf{E} = 0$, $d\mathbf{l} = 0$, or $\cos\theta = 0$ where θ is the angle between **E** and $d\mathbf{l}$. Thus in a region where **E** is not zero, the path $d\mathbf{l}$ along an equipotential must have $\cos\theta = 0$, meaning $\theta = 90°$ and **E** is perpendicular to the equipotential.

The fact that the electric field lines and equipotential surfaces are mutually perpendicular helps us locate the equipotentials when the electric field lines are known. In a normal two-dimensional drawing, we show equipotential *lines*, which are the intersections of equipotential surfaces with the plane of the drawing. In Fig. 23–13, a few of the equipotential lines are drawn (dashed green lines) for the electric field (red lines) between two parallel plates at a potential difference of 20 V. The negative plate is arbitrarily chosen to be zero volts and the potential of each equipotential line is indicated. Note that **E** points toward lower values of V. The equipotential lines for the case of two equal but oppositely charged particles are shown in Fig. 23–14 as green dashed lines. Equipotential lines and surfaces, unlike field lines, are always continuous and never end, and so continue beyond the borders of Figs. 23–13 and 23–14.

We saw in Section 21–9 that there can be no electric field within a conductor in the static case, for otherwise the free electrons would feel a force and would move. Indeed, the entire volume of *a conductor must be entirely at the same potential in the static case*, and the surface of a conductor is then an equipotential surface. (If it weren't, the free electrons at the surface would move, since whenever there is a potential difference between two points, free charges will move.) This is fully consistent with our result, discussed earlier, that the electric field at the surface of a conductor must be perpendicular to the surface.

Conductors are equipotential surfaces

FIGURE 23–14 Equipotential lines (green, dashed) are always perpendicular to the electric field lines (solid red) shown here for two equal but oppositely charged particles.

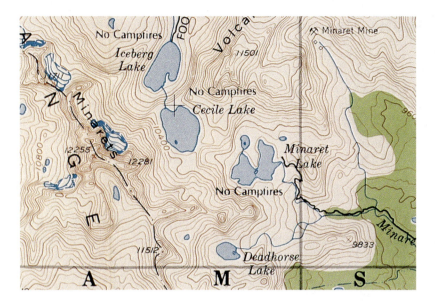

FIGURE 23–15 A topographic map (here, a portion of the Sierra Nevada in California) shows continuous contour lines, each of which is at a fixed height above sea level. Here they are at 80 ft (25 m) intervals. If you walk along one contour line, you neither climb nor descend. If you cross lines, and especially (maximally), if you climb perpendicular to the lines, you will be changing your gravitational potential (rapidly, if the lines are close together).

A useful analogy for equipotential lines is a topographic map: the contour lines are essentially gravitational equipotential lines (Fig. 23–15).

23–6 Electric Dipoles

Two equal point charges Q, of opposite sign, separated by a distance l, are called an **electric dipole**, as we already saw in Section 21–11. Also, the two charges we saw in Figs. 23–10 and 23–14 constitute an electric dipole, and the latter shows the electric field lines and equipotential surfaces for a dipole. Because electric dipoles occur often in physics, as well as in other fields, it is useful to examine them more closely.

Let us calculate the electric potential at an arbitrary point P due to a dipole, as shown in Fig. 23–16. As usual, we take $V = 0$ at $r = \infty$. Since V is the sum of the potentials due to each of the two charges, we have

$$ V = \frac{1}{4\pi\epsilon_0} \frac{Q}{r} + \frac{1}{4\pi\epsilon_0} \frac{(-Q)}{r + \Delta r} = \frac{1}{4\pi\epsilon_0} Q \left(\frac{1}{r} - \frac{1}{r + \Delta r} \right) = \frac{Q}{4\pi\epsilon_0} \frac{\Delta r}{r(r + \Delta r)}, $$

where r is the distance from P to the positive charge and $r + \Delta r$ is the distance to the negative charge. This equation becomes simpler if we consider points P whose distance from the dipole is much larger than the separation of the two charges—that is, for $r \gg l$. From the diagram we can see that in this case, $\Delta r \approx l \cos\theta$. Since $r \gg \Delta r$, we can neglect Δr in the denominator as compared to r. Therefore, we obtain

$$ V = \frac{1}{4\pi\epsilon_0} \frac{Ql \cos\theta}{r^2} = \frac{1}{4\pi\epsilon_0} \frac{p \cos\theta}{r^2} \qquad \text{[dipole; } r \gg l \text{]} \quad \textbf{(23–7)} $$

where $p = Ql$ is called the **dipole moment**. When θ is between $0°$ and $90°$, V is positive. If θ is between $90°$ and $180°$, V is negative (since $\cos\theta$ is then negative). This makes sense since in the first case P is closer to the positive charge and in the second case it is closer to the negative charge. At $\theta = 90°$, the potential is zero ($\cos 90° = 0$), in agreement with the result of Example 23–7 (point B). From Eq. 23–7, we see that the potential decreases as the *square* of the distance from the dipole, whereas for a single point charge the potential decreases with the first power of the distance (Eq. 23–5). It is not surprising that the potential should fall off faster for a dipole; for when you are far from a dipole, the two equal but opposite charges appear so close together as to tend to neutralize each other.

Table 23–2 gives the dipole moments for several molecules. The + and − signs indicate on which atoms these charges lie. The last two entries are a part of many organic molecules and play an important role in molecular biology.

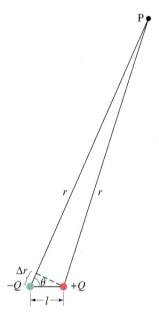

FIGURE 23–16 Electric dipole. Calculation of potential V at point P.

TABLE 23–2 Dipole Moments of Selected Molecules

Molecule	Dipole Moment (C·m)
$H_2^{(+)}O^{(-)}$	6.1×10^{-30}
$H^{(+)}Cl^{(-)}$	3.4×10^{-30}
$N^{(-)}H_3^{(+)}$	5.0×10^{-30}
$>N^{(-)}{-}H^{(+)\ddagger}$	$\approx 3.0 \times 10^{-30}$
$>C^{(+)}{=}O^{(-)\ddagger}$	$\approx 8.0 \times 10^{-30}$

$\ddagger$These groups often appear on larger molecules; hence the value for the dipole moment will vary somewhat, depending on the rest of the molecule.

EXAMPLE 23–10 **The C=O group dipole.** The distance between the carbon (+) and oxygen (−) atoms in the group C=O which occurs in many organic molecules is about 1.2×10^{-10} m and the dipole moment of this group is about 8.0×10^{-30} C·m. Calculate (a) the effective charge Q on the C (carbon) and O (oxygen) atoms, and (b) the potential 9.0×10^{-10} m from the dipole along its axis, with the oxygen being the nearer atom (that is, to the left in Fig. 23–16, so $\theta = 180°$). (c) What would the potential be at this point if only the oxygen (O) were charged?

SOLUTION (a) The dipole moment $p = Ql$. Therefore

$$Q = \frac{p}{l} = \frac{8.0 \times 10^{-30} \text{ C·m}}{1.2 \times 10^{-10} \text{ m}} = 6.7 \times 10^{-20} \text{ C}.$$

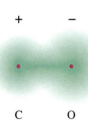

+ −

C   O

FIGURE 23–17 Electron "cloud" around C and O in the C=O group. The C=O group has a dipole moment because two electrons originally on the carbon atom spend some of their time in the vicinity of the oxygen atom.

Although this charge is less than e, the smallest known charge, it is not a charge that can be isolated, but is the effective charge that results from unequal sharing of the electrons, Fig. 23–17.

(b) Since $\theta = 180°$, we have, using Eq. 23–7:

$$V = \frac{1}{4\pi\epsilon_0} \frac{p \cos\theta}{r^2}$$

$$= \frac{(9.0 \times 10^9 \text{ N·m}^2/\text{C}^2)(8.0 \times 10^{-30} \text{ C·m})(-1.00)}{(9.0 \times 10^{-10} \text{ m})^2} = -0.089 \text{ V}.$$

(c) If we assume that the oxygen has charge $Q = -6.7 \times 10^{-20}$ C, as in part (a) above, and that the carbon is not charged, we use the formula for a single charge:

$$V = \frac{1}{4\pi\epsilon_0} \frac{Q}{r} = \frac{(9.0 \times 10^9 \text{ N·m}^2/\text{C}^2)(-6.7 \times 10^{-20} \text{ C})}{9.0 \times 10^{-10} \text{ m}} = -0.67 \text{ V}.$$

Of course, we expect the potential of a single charge to have greater magnitude than that of a dipole of equal charge at the same distance.

23–7 | E Determined from V

We can use Eq. 23–3, $V_b - V_a = -\int_a^b \mathbf{E} \cdot d\mathbf{l}$, to determine the difference in potential between two points if the electric field is known in the region between those two points. By inverting Eq. 23–3, we can write the electric field in terms of the potential. Then the electric field can be determined from a knowledge of V. Let us see how to do this.

We write Equation 23–3 in differential form as

$$dV = -\mathbf{E} \cdot d\mathbf{l} = -E_l \, dl,$$

where dV is the infinitesimal difference in potential between two points a distance dl apart, and E_l is the component of the electric field in the direction of the infinitesimal displacement $d\mathbf{l}$. We can then write

$$E_l = -\frac{dV}{dl}. \tag{23–8}$$

Thus *the component of the electric field in any direction is equal to the negative of the rate of change of the electric potential with distance in that direction.* The quantity dV/dl is called the gradient of V in a particular direction. If the direction is not specified, the term *gradient* refers to that direction in which V changes most rapidly; this would be the direction of $\mathbf{E}$ at that point, so we can write

$$E = -\frac{dV}{dl}. \qquad [\text{if } d\mathbf{l} \parallel \mathbf{E}]$$

If $\mathbf{E}$ is written as a function of x, y, and z, and we let l refer to the x, y, and z axes, then Eq. 23–8 becomes

$$E_x = -\frac{\partial V}{\partial x}, \qquad E_y = -\frac{\partial V}{\partial y}, \qquad E_z = -\frac{\partial V}{\partial z}. \qquad \text{(23–9)} \qquad \textit{E related to V}$$

Here, $\partial V / \partial x$ is the "partial derivative" of V with respect to x, with y and z held constant.[†]

EXAMPLE 23–11 E for ring and disk. Determine the electric field at point P on the axis of (*a*) a circular ring of charge (Fig. 23–11) and (*b*) a uniformly charged disk (Fig. 23–12).

SOLUTION (*a*) From Example 23–8,

$$V = \frac{1}{4\pi\epsilon_0} \frac{Q}{\left(x^2 + R^2\right)^{\frac{1}{2}}}.$$

Then

$$E_x = -\frac{\partial V}{\partial x} = \frac{1}{4\pi\epsilon_0} \frac{Qx}{\left(x^2 + R^2\right)^{\frac{3}{2}}}$$

$$E_y = E_z = 0.$$

This is the same result we obtained in Example 21–9, but here we didn't have to break the vector electric field into components and then integrate.

(*b*) From Example 23–9,

$$V = \frac{Q}{2\pi\epsilon_0 R^2} \left[\left(x^2 + R^2\right)^{\frac{1}{2}} - x\right],$$

so

$$E_x = -\frac{\partial V}{\partial x} = \frac{Q}{2\pi\epsilon_0 R^2} \left[1 - \frac{x}{\left(x^2 + R^2\right)^{\frac{1}{2}}}\right]$$

$$E_y = E_z = 0.$$

For points very close to the disk, $x \ll R$, this can be approximated by

$$E_x \approx \frac{Q}{2\pi\epsilon_0 R^2} = \frac{\sigma}{2\epsilon_0}$$

where $\sigma = Q/\pi R^2$ is the surface charge density. We also obtained these results in Chapter 21, Example 21–11 and Eq. 21–7.

If we compare this last Example with Examples 21–9 and 21–11, we see that here, as for many charge distributions, it is easier to calculate V first, and then $\mathbf{E}$ from Eq. 23–9, than to calculate $\mathbf{E}$ due to each charge from Coulomb's law. This is because V due to many charges is a scalar sum, whereas $\mathbf{E}$ is a vector sum.

23–8 Electrostatic Potential Energy; the Electron Volt

Suppose a point charge q is moved between two points in space, a and b, where the electric potential due to other charges is V_a and V_b, respectively. The change in electrostatic potential energy of q in the field of these other charges is, according to Eq. 23–2,

$$\Delta U = U_b - U_a = q(V_b - V_a) = qV_{ba}.$$

[†] Equation 23–9 can be written as a vector equation,

$$\mathbf{E} = -\text{grad } V = -\nabla V = -\left(\mathbf{i}\frac{\partial}{\partial x} + \mathbf{j}\frac{\partial}{\partial y} + \mathbf{k}\frac{\partial}{\partial z}\right)V$$

where the symbol ∇ is called the *del* or *gradient operator*: $\nabla = \mathbf{i}\frac{\partial}{\partial x} + \mathbf{j}\frac{\partial}{\partial y} + \mathbf{k}\frac{\partial}{\partial z}.$

Now suppose we have a system of several point charges. What is the electrostatic potential energy of the system? It is most convenient to choose the electric potential energy to be zero when the charges are very far (ideally infinitely far) apart. A single point charge, Q_1, in isolation, has no potential energy, because if there are no other charges around, no electric force can be exerted on it. If a second point charge Q_2 is brought close to Q_1, the potential due to Q_1 at the position of this second charge is

$$V = \frac{1}{4\pi\epsilon_0}\frac{Q_1}{r_{12}},$$

where r_{12} is the distance between the two. The potential energy of the two charges, relative to $V = 0$ at $r = \infty$, is

Potential energy, 2 point charges

$$U = Q_2 V = \frac{1}{4\pi\epsilon_0}\frac{Q_1 Q_2}{r_{12}}. \qquad \textbf{(23–10)}$$

This represents the work that needs to be done by an external force to bring Q_2 from infinity ($V = 0$) to a distance r_{12} from Q_1. It is also the negative of the work needed to separate them to infinity.

If the system consists of three charges, the total potential energy will be the work needed to bring all three together. Equation 23–10 represents the work needed to bring Q_2 close to Q_1; to bring a third charge Q_3 so that it is a distance r_{13} from Q_1 and r_{23} from Q_2 requires work equal to

$$\frac{1}{4\pi\epsilon_0}\frac{Q_1 Q_3}{r_{13}} + \frac{1}{4\pi\epsilon_0}\frac{Q_2 Q_3}{r_{23}}.$$

So the potential energy of a system of three point charges is

Potential energy, 3 point charges

$$U = \frac{1}{4\pi\epsilon_0}\left(\frac{Q_1 Q_2}{r_{12}} + \frac{Q_1 Q_3}{r_{13}} + \frac{Q_2 Q_3}{r_{23}}\right). \qquad [V = 0 \text{ at } r = \infty]$$

For a system of four charges, the potential energy would contain six such terms, and so on. (Caution must be used when making such sums to avoid double counting of the different pairs.)

The Electron Volt Unit

The joule is a very large unit for dealing with energies of electrons, atoms, or molecules (see Example 23–2), and for this purpose, the unit **electron volt** (eV) is used. One electron volt is defined as the energy acquired by a particle carrying a charge equal to that on an electron ($q = e$) when it moves through a potential difference of 1 V. Since $e = 1.6 \times 10^{-19}$ C, and since the change in potential energy equals qV, 1 eV is equal to $(1.6 \times 10^{-19}\text{ C})(1.0\text{ V}) = 1.6 \times 10^{-19}$ J:

Electron volt (energy unit)

$$1\,\text{eV} = 1.6 \times 10^{-19}\,\text{J}.$$

An electron that accelerates through a potential difference of 1000 V will lose 1000 eV of potential energy and will thus gain 1000 eV or 1 keV (kilo-electron volt) of kinetic energy. On the other hand, if a particle has a charge equal to twice the charge on the electron ($= 2e = 3.2 \times 10^{-19}$ C), when it moves through a potential difference of 1000 V its energy will change by 2000 eV.

Although the electron volt is handy for *stating* the energies of molecules and elementary particles, it is not a proper SI unit. For calculations it should be converted to joules using the conversion factor given above. In Example 23–2, for example, the electron acquired a kinetic energy of 8.0×10^{-16} J. We normally would quote this energy as 5000 eV $(= 8.0 \times 10^{-16}\text{ J}/1.6 \times 10^{-19}\text{ J/eV})$. But in determining its speed in SI units we had to use the kinetic energy in J.

EXAMPLE 23–12 **Disassembling a hydrogen atom.** Calculate the work needed to "disassemble" a hydrogen atom. Assume that the proton and electron are initially separated by a distance equal to the "average" radius of the hydrogen atom in its ground state, 0.529×10^{-10} m, and that they end up an infinite distance apart from each other.

SOLUTION From Eq. 23–10 we have initially

$$U = \frac{1}{4\pi\epsilon_0}\frac{Q_1 Q_2}{r} = -\frac{1}{4\pi\epsilon_0}\frac{e^2}{r} = \frac{-(8.99 \times 10^9\,\text{N}\cdot\text{m}^2/\text{C}^2)(1.60 \times 10^{-19}\,\text{C})^2}{(0.529 \times 10^{-10}\,\text{m})}$$
$$= -27.2(1.60 \times 10^{-19})\,\text{J} = -27.2\,\text{eV}.$$

This represents the potential energy. The total energy must include also the kinetic energy of the electron moving in an orbit of radius $r = 0.529 \times 10^{-10}$ m. From $F = ma$ for centripetal acceleration, we have $(1/4\pi\epsilon_0)(e^2/r^2) = mv^2/r$. Then

$$K = \tfrac{1}{2}mv^2 = \tfrac{1}{2}\left(\frac{1}{4\pi\epsilon_0}\right)\frac{e^2}{r}$$

which equals $-\tfrac{1}{2}U$ (as calculated above), so $K = +13.6\,\text{eV}$. The total energy initially is $E = K + U = 13.6\,\text{eV} - 27.2\,\text{eV} = -13.6\,\text{eV}$. To separate a stable hydrogen atom into a proton and an electron at rest very far apart ($U = 0$ at $r = \infty$, $K = 0$ because $v = 0$) requires $+13.6\,\text{eV}$. This is, in fact, the measured ionization energy for hydrogen.

*23–9 Cathode Ray Tube: TV and Computer Monitors, Oscilloscope

An important device that makes use of voltage, and that allows us to "visualize" voltages in the sense of displaying graphically how a voltage changes in time, is the **cathode ray tube (CRT)**. A CRT used in this way is an *oscilloscope*—but an even more common use of a CRT is as the picture tube of television sets and computer monitors.

The operation of a CRT depends first of all on the phenomenon of **thermionic emission**, discovered by Thomas Edison (1847–1931) in the course of experiments on developing the electric light bulb. To understand how thermionic emission occurs, consider two small plates (electrodes) inside an evacuated "bulb" or "tube" as shown in Fig. 23–18, to which is applied a potential difference (by a battery, say). The negative electrode is called the **cathode**, the positive one the **anode**. If the negative cathode is heated (usually by an electric current, as in a lightbulb) so that it becomes hot and glowing, it is found that negative charge leaves the cathode and flows to the positive anode. These negative charges are now called electrons, but originally they were called **cathode rays** since they seemed to come from the cathode.

We can understand how electrons might be "boiled off" a hot metal plate if we treat electrons like molecules in a gas. This makes sense if electrons are relatively free to move about inside a metal, which is consistent with metals being good conductors. However, electrons don't readily escape from the metal. If an electron were to escape outside the metal surface, a net positive charge would remain behind, and this would attract the electron back. To escape, an electron needs a certain minimum kinetic energy, just as molecules in a liquid must have a minimum kinetic energy to "evaporate" into the gaseous state. We saw in Chapter 18 that the average kinetic energy ($\bar{K}$) of molecules in a gas is proportional to the absolute temperature T. We can apply this idea, but only very roughly, to free electrons in a metal as if they made up an "electron gas." Of course, some electrons have more kinetic energy than average and others less. At room temperature, very few electrons would have sufficient energy to escape. At high temperature, $\bar{K}$ is larger and many electrons escape—just as molecules evaporate from liquids, which occurs more readily at high temperatures. Thus, significant thermionic emission occurs only at elevated temperatures.

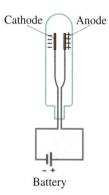

FIGURE 23–18 If the cathode inside the evacuated glass tube is heated to glowing, negatively charged "cathode rays" (electrons) are "boiled off" and flow across to the anode (+) to which they are attracted.

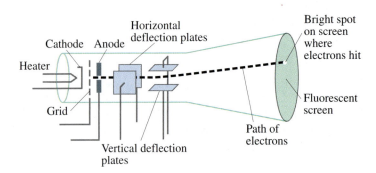

PHYSICS APPLIED

CRT

PHYSICS APPLIED

TV and computer monitors

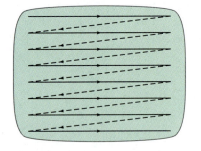

FIGURE 23–20 Electron beam sweeps across a television screen in a succession of horizontal lines.

FIGURE 23–21 An electrocardiogram (ECG) trace displayed on a CRT.

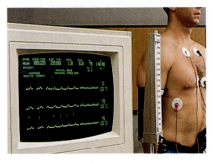

The **cathode-ray tube** (CRT) derives its name from the fact that inside an evacuated glass tube, a beam of cathode rays (electrons) is directed to various parts of a screen to produce a "picture." A simple CRT is diagrammed in Fig. 23–19. Electrons emitted by the heated cathode are accelerated by a high voltage (5,000–50,000 V) applied to the anode. The electrons pass out of this "electron gun" through a small hole in the anode. The inside of the tube face is coated with a fluorescent material that glows when struck by electrons. A tiny bright spot is thus visible where the electron beam strikes the screen. Two horizontal and two vertical plates deflect the beam of electrons when a voltage is applied to them. The electrons are deflected toward whichever plate is positive. By varying the voltage on the deflection plates, the bright spot can be placed at any point on the screen. Today many CRTs use magnetic deflection coils (Chapter 27) instead of electric plates.

In the picture tube or monitor for a computer or television set, the electron beam is made to sweep over the screen in the manner shown in Fig. 23–20. The beam is swept horizontally by the horizontal deflection plates or coils. When the horizontal deflecting field is maximum in one direction, the beam is at one edge of the screen. As the field decreases to zero, the beam moves to the center; and as the field increases to a maximum in the opposite direction, the beam approaches the opposite edge. When the beam reaches this edge, the voltage or current abruptly changes to return the beam to the opposite side of the screen. Simultaneously, the beam is deflected downward slightly by the vertical deflection plates (or coils), and then another horizontal sweep is made. For standard television in the United States, 525 lines constitutes a complete sweep over the entire screen. (High-definition TV provides more than double this number of lines, giving greater picture sharpness.) The complete picture of 525 lines is swept out in $\frac{1}{30}$ s. Actually, a single vertical sweep takes $\frac{1}{60}$ s and involves every other line. The lines in between are then swept out over the next $\frac{1}{60}$ s (called interlacing). We see a picture because the image is retained by the fluorescent screen and by our eyes for about $\frac{1}{20}$ s. The picture we see consists of the varied brightness of the spots on the screen. The brightness at any point is controlled by the grid (a "porous" electrode, such as a wire grid, that allows passage of electrons) which can limit the flow of electrons by means of the voltage applied to it: the more negative this voltage, the more electrons are repelled and the fewer pass through. The voltage on the grid is determined by the video signal (a voltage) sent out by the TV station and received by the TV set. Accompanying this signal are signals that synchronize the grid voltage to the horizontal and vertical sweeps.

An **oscilloscope** is a device for amplifying, measuring, and visually observing an electrical signal (a "signal" is usually a time-varying voltage), especially rapidly changing signals. The signal is displayed on the screen of a CRT. In normal operation, the electron beam is swept horizontally at a uniform rate in time by using a time varying potential difference applied to the horizontal deflection plates. The signal to be displayed is applied, after amplification, to the vertical deflection plates. The visible "trace" on the screen, which could be an ECG (Fig. 23–21), a voltage in an electronic device being repaired, or a signal from an experiment, is thus a plot of the signal voltage (vertically) versus time (horizontally).

Summary

Electric potential is defined as electric potential energy per unit charge. That is, the **electric potential difference** between any two points in space is defined as the difference in potential energy of a test charge q placed at those two points, divided by the charge q:

$$V_{ba} = \frac{U_b - U_a}{q}.$$

Potential difference is measured in volts ($1\text{ V} = 1\text{ J/C}$) and is sometimes referred to as **voltage**.

The change in potential energy of a charge q when it moves through a potential difference V_{ba} is

$$\Delta U = qV_{ba}.$$

The potential difference V_{ba} between two points, a and b, is given by the relation

$$V_{ba} = -\int_a^b \mathbf{E} \cdot d\mathbf{l}.$$

Thus V_{ba} can be found in any region where $\mathbf{E}$ is known. If the electric field is uniform, the integral is easy: $V_{ba} = -Ed$, where d is the distance (parallel to the field lines) between the two points.

When V is known, the components of $\mathbf{E}$ can be found from the inverse of the above relation, namely

$$E_x = -\frac{\partial V}{\partial x}, \quad E_y = -\frac{\partial V}{\partial y}, \quad E_z = -\frac{\partial V}{\partial z}.$$

An **equipotential line** or **surface** is all at the same potential, and is perpendicular to the electric field at all points.

The electric potential due to a single point charge Q, relative to zero potential at infinity, is given by

$$V = \frac{1}{4\pi\epsilon_0}\frac{Q}{r}.$$

The potential due to any charge distribution can be obtained by summing (or integrating) over the potentials for all the charges.

Questions

1. If two points are at the same potential, does this mean that no work is done in moving a test charge from one point to the other? Does this imply that no force must be exerted?

2. If a negative charge is initially at rest in an electric field, will it move toward a region of higher potential or lower potential? What about a positive charge? How does the potential energy of the charge change in each of these two instances?

3. State clearly the difference (*a*) between electric potential and electric field, (*b*) between electric potential and electric potential energy.

4. An electron is accelerated by a potential difference of, say, 0.10 V. How much greater would its final speed be if it were accelerated with four times as much voltage?

5. Can a particle ever move from a region of low electric potential to one of high potential and yet have its electric potential energy decrease? Explain.

6. If $V = 0$ at a point in space, must $\mathbf{E} = 0$? If $\mathbf{E} = 0$ at some point, must $V = 0$ at that point? Explain. Give examples for each.

7. When dealing with practical devices, we often take the ground (the Earth) to be 0 V. (*a*) If instead we said the ground was -10 V, how would this affect V and E at other points? (*b*) Does the fact that the Earth carries a net charge affect the choice of V at its surface?

8. Can two equipotential lines cross? Explain.

9. Draw in a few equipotential lines in Fig. 21–33b and c.

10. What can you say about the electric field in a region of space that has the same potential throughout?

11. A satellite orbits the Earth along a gravitational equipotential line. What shape must the orbit be?

12. Suppose the charged ring of Example 23–8 was not uniformly charged, so that the density of charge was twice as great near the top as near the bottom. Assuming the total charge Q is unchanged, would this affect the potential at point P on the axis (Fig. 23–11)? Would it affect the value of $\mathbf{E}$ at that point? Is there a discrepancy here? Explain.

13. Consider a metal conductor in the shape of a football. If it carries a total charge Q, where would you expect the charge density σ to be greatest, at the ends or along the flatter sides? Explain. (*Hint:* Near the surface of a conductor, $E = \sigma/\epsilon_0$.)

14. A conducting sphere carries a charge Q and a second identical conducting sphere is neutral. The two are initially insulated, but then they are placed in contact. (*a*) What can you say about the potential of each when they are in contact? (*b*) Will charge flow from one to the other? If so, how much? (*c*) If the spheres do not have the same radius, how are your answers to parts (*a*) and (*b*) altered?

15. At a particular point, the electric field points due north. In what direction(s) will the rate of change of potential be (*a*) greatest, (*b*) least, and (*c*) zero?

16. If you know V at a point in space, can you calculate $\mathbf{E}$ at that point? If you know $\mathbf{E}$ at a point can you calculate V at that point? If not, what else must be known in each case?

17. Equipotential lines are spaced 1.00 V apart. Does the distance between the lines in different regions of space tell you anything about the relative strengths of $\mathbf{E}$ in those regions? If so, what?

18. If the electric field $\mathbf{E}$ is uniform in a region, what can you infer about the electric potential V? If V is uniform in a region of space, what can you infer about $\mathbf{E}$?

19. Is the electric potential energy of two unlike charges positive or negative? What about two like charges? What is the significance of the sign of the potential energy in each case?

Problems

Section 23–1

1. (I) How much work is needed to move a -7.0-μC charge from ground to a point whose potential is $+6.00$ V higher?

2. (I) How much work is needed to move a proton from a point with a potential of $+100$ V to a point where it is -50 V?

3. (I) How much kinetic energy will an electron gain (in joules) if it falls through a potential difference of 21,000 V in a TV picture tube?

4. (I) An electron acquires 16.4×10^{-16} J of kinetic energy when it is accelerated by an electric field from plate A to plate B. What is the potential difference between the plates, and which plate is at the higher potential?

5. (II) The work done by an external force to move a -8.10-μC charge from point a to point b is 8.00×10^{-4} J. If the charge was started from rest and had 2.10×10^{-4} J of kinetic energy when it reached point b, what must be the potential difference between a and b?

Section 23–2

6. (I) The electric field between two parallel plates connected to a 45-V battery is 1500 V/m. How far apart are the plates?

7. (I) An electric field of 640 V/m is desired between two parallel plates 11.0 mm apart. How large a voltage should be applied?

8. (I) How strong is the electric field between two parallel plates 5.0 mm apart if the potential difference between them is 110 V?

9. (I) What is the maximum amount of charge that a spherical conductor of radius 5.0 cm can hold in air?

10. (I) What minimum radius must a large conducting sphere of an electrostatic generating machine have if it is to carry 30,000 V without discharge into the air? How much charge will it carry?

11. (II) A uniform electric field $\mathbf{E} = -300$ N/Ci points in the negative x direction as shown in Fig. 23–22. The x and y coordinates of points A, B, and C are given on the diagram (in meters). Determine the differences in potential (a) V_{BA}, (b) V_{CB}, and (c) V_{CA}.

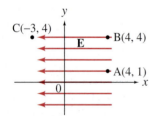

FIGURE 23–22
Problem 11.

12. (II) The electric potential of a very large flat metal plate is V_0. It carries a uniform distribution of charge of surface density $\sigma (C/m^2)$. Determine V at a distance x from the plate. Consider the point x to be far from the edges and assume x is much smaller than the plate dimensions.

13. (II) The Earth produces an inwardly directed electric field of magnitude 150 V/m near its surface. (a) What is the potential of the Earth's surface relative to $V = 0$ at $r = \infty$? (b) If the potential of the Earth is chosen to be zero, what is the potential at infinity? (Ignore the fact that positive charge in the ionosphere approximately cancels the Earth's net charge; how would this affect your answer?)

14. (II) A 32-cm-diameter conducting sphere is charged to 500 V relative to $V = 0$ at $r = \infty$. (a) What is the surface charge density σ? (b) At what distance will the potential due to the sphere be only 10 V?

15. (II) An insulated spherical conductor of radius r_1 carries a charge Q. A second conducting sphere of radius r_2 and initially uncharged is then connected to the first by a long conducting wire. (a) After the connection, what can you say about the electric potential of each sphere? (b) How much charge is transferred to the second sphere? Assume the connected spheres are far apart compared to their radii. (Why make this assumption?)

16. (II) Determine the difference in potential between two points that are distances R_a and R_b from a very long $(\gg R_a$ or $R_b)$ straight wire carrying a uniform charge per unit length λ.

17. (II) A nonconducting sphere of radius r_0 carries a total charge Q distributed uniformly throughout its volume. Determine the electric potential as a function of the distance r from the center of the sphere for (a) $r > r_0$ and (b) $r < r_0$. Take $V = 0$ at $r = \infty$. (c) Plot V versus r and E versus r.

18. (III) Repeat Problem 17 assuming the charge density ρ_E increases as the square of the distance from the center of the sphere, and $\rho_E = 0$ at the center.

19. (III) A very long conducting cylinder (length L) of radius R_0 $(R_0 \ll L)$ carries a uniform surface charge density $\sigma (C/m^2)$. The cylinder is at an electric potential V_0. What is the potential, at points far from the end, at a distance r from the center of the cylinder? Determine for (a) $r > R_0$ and (b) $r < R_0$. (c) Is $V = 0$ at $r = \infty$ (assume $L = \infty$)? Explain.

20. (III) A hollow spherical conductor, carrying a net charge $+Q$, has inner radius r_1 and outer radius $r_2 = 2r_1$ (Fig. 23–23). At the center of the sphere is a point charge $+Q/2$. (a) Write the electric field strength E in all three regions as a function of r. Then determine the potential as a function of r, the distance from the center, for (b) $r > r_2$, (c) $r_1 < r < r_2$, and (d) $0 < r < r_1$. (e) Plot both V and E as a function of r from $r = 0$ to $r = 2r_2$.

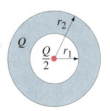

FIGURE 23–23
Problem 20.

Section 23–3

21. (I) (a) What is the electric potential 0.50×10^{-10} m from a proton (charge $+e$)? Let $V = 0$ at $r = \infty$. (b) What is the potential energy of an electron at this point?

22. (I) A charge Q creates an electric potential of $+125$ V at a distance of 15 cm. What is Q?

23. (II) A $+25$-μC charge is placed 6.0 cm from an identical $+25$-μC charge. How much work would be required by an external force to move a $+0.10$-μC test charge from a point midway between them to a point 1.0 cm closer to either of the charges?

24. (II) A 3.0-μC and a -2.0-μC charge are placed 4.0 cm apart. At what points along the line joining them is (a) the electric field zero and (b) the potential zero? Let $V = 0$ at $r = \infty$.

25. (II) How much voltage must be used to accelerate a proton (radius 1.2×10^{-15} m) so that it has sufficient energy to just penetrate a silicon nucleus? A silicon nucleus has a charge of $+14e$ and its radius is about 3.6×10^{-15} m. Assume the potential is that for point charges.

26. (II) Consider point a which is 70 cm north of a -3.8-μC point charge, and point b which is 80 cm west of the charge (Fig. 23–24). Determine (a) $V_{ba} = V_b - V_a$, and (b) $\mathbf{E}_b - \mathbf{E}_a$ (magnitude and direction).

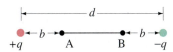

FIGURE 23–24
Problem 26.

27. (II) An electron starts from rest 72.5 cm from a fixed point charge with $Q = -0.125\ \mu$C. How fast will the electron be moving when it is very far away?

28. (II) Two identical $+7.5$-μC point charges are initially spaced 5.5 cm from each other. If they are released at the same instant from rest, how fast will they be moving when they are very far away from each other? Assume they have identical masses of 1.0 mg.

29. (II) Two equal but opposite charges are separated by a distance d, as shown in Fig. 23–25. Determine a formula for $V_{BA} = V_B - V_A$ for points B and A on the line between the charges situated as shown.

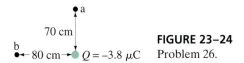

FIGURE 23–25   Problem 29.

Section 23–4

30. (II) Three point charges are arranged at the corners of a square of side L as shown in Fig. 23–26. What is the potential at the fourth corner (point A), taking $V = 0$ at a great distance.

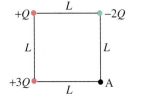

FIGURE 23–26
Problem 30.

31. (II) A flat ring of inner radius R_1 and outer radius R_2, Fig. 23–27, carries a uniform surface charge density σ. Determine the electric potential at points along the axis (the x axis).

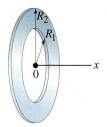

FIGURE 23–27
Problem 31.

32. (II) A thin rod of length $2L$ is centered on the x axis as shown in Fig. 23–28. The rod carries a uniformly distributed charge Q. Determine the potential V as a function of y for points along the y axis. Let $V = 0$ at infinity.

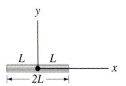

FIGURE 23–28   Problems 32, 33, 34, and 48.

33. (III) Determine the potential $V(x)$ for points along the x axis outside the rod of Fig. 23–28 (Problem 32).

34. (III) The charge on the rod of Fig. 23–28 has a nonuniform linear charge distribution, $\lambda = ax$. Determine the potential V for (a) points along the y axis and (b) points along the x axis outside the rod.

35. (III) Suppose the flat circular disk of Fig. 23–12 (Example 23–9) has a nonuniform surface charge density $\sigma = ar^2$, where r is measured from the center of the disk. Find the potential $V(x)$ at points along the x axis, relative to $V = 0$ at $x = \infty$.

Section 23–5

36. (I) Draw a conductor in the shape of a football. This conductor carries a net negative charge, $-Q$. Draw in a dozen or so electric field lines and equipotential lines.

37. (II) Equipotential surfaces are to be drawn 100 V apart near a very large uniformly charged plate carrying a charge density $\sigma = 0.55\ \mu$C/m^2. How far apart (in space) are the equipotential surfaces?

38. (II) A metal sphere of radius $r_0 = 0.30$ m carries a charge $Q = 0.50\ \mu$C. Equipotential surfaces are to be drawn for 100-V intervals outside the sphere. Determine the radius r of (a) the first, (b) the tenth, and (c) the 100th equipotential from the surface.

Section 23–6

39. (I) An electron and a proton are 0.53×10^{-10} m apart. (a) What is their dipole moment if they are at rest? (b) What is the average dipole moment if the electron revolves about the proton in a circular orbit?

40. (II) Calculate the electric potential due to a dipole whose dipole moment is 4.8×10^{-30} C·m at a point 1.1×10^{-9} m away if this point is far from the dipole, and: (*a*) along the axis of the dipole nearer the positive charge; (*b*) 45° above the axis but nearer the positive charge; (*c*) 45° above the axis but nearer the negative charge.

41. (II) (*a*) In Example 23–10 part *b*, calculate the electric potential without using the dipole approximation, Eq. 23–7; that is, don't assume $r \gg l$. (*b*) What is the percent error in this case when the dipole approximation is used?

42. (III) Show that if an electric dipole is placed in a uniform electric field, then a torque is exerted on it equal to $pE \sin \phi$, where ϕ is the angle between the dipole moment vector and the direction of the electric field as shown in Fig. 23–29. What is the net force on the dipole? How are your answers affected if the field is nonuniform? Note that the dipole moment vector **p** is defined so that its magnitude is Ql and its direction is pointing from the negative end to the positive end as shown.

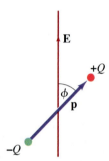

FIGURE 23–29
Problem 42.

43. (III) The dipole moment, considered as a vector, points from the negative to the positive charge. The water molecule, Fig. 23–30, has a dipole moment **p** which can be considered as the vector sum of the two dipole moments $\mathbf{p}_1$ and $\mathbf{p}_2$ as shown. The distance between each H and the O is about 0.96×10^{-10} m; the lines joining the center of the O atom with each H atom make an angle of 104° as shown, and the net dipole moment has been measured to be $p = 6.1 \times 10^{-30}$ C·m. (*a*) Determine the effective charge q on each H atom. (*b*) Determine the electric potential, far from the molecule, due to each dipole, $\mathbf{p}_1$ and $\mathbf{p}_2$, and show that

$$V = \frac{1}{4\pi\epsilon_0} \frac{p \cos\theta}{r^2},$$

where p is the magnitude of the net dipole moment, $\mathbf{p} = \mathbf{p}_1 + \mathbf{p}_2$, and V is the total potential due to both $\mathbf{p}_1$ and $\mathbf{p}_2$. Take $V = 0$ at $r = \infty$.

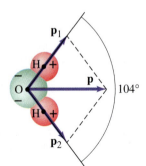

FIGURE 23–30
Problem 43.

Section 23–7

44. (I) What is the potential gradient just outside the surface of a uranium nucleus $(Q = +92e)$ whose diameter is about 15×10^{-15} m?

45. (I) Show that the electric field of a single point charge (Eq. 21–4) follows from Eq. 23–5, $V = (1/4\pi\epsilon_0)(Q/r)$.

46. (II) The electric potential in a region of space varies as $V = ay/(b^2 + y^2)$. Determine **E**.

47. (II) In a certain region of space, the electric potential is given by $V = y^2 + 2xy - 4xyz$. Determine the electric field vector, **E**, in this region.

48. (III) Use the results of Problems 32 and 33 to determine the electric field due to the uniformly charged rod of Fig. 23–28 for points (*a*) along the *y* axis and (*b*) along the *x* axis.

Section 23–8

49. (I) Determine the mutual electrostatic potential energy (in electron volts) of two protons in a uranium (^{235}U) nucleus (*a*) if they are at the surface, on opposite sides of the nucleus, and (*b*) if one is at the center and the other is at the surface. The diameter of a ^{235}U nucleus is about 15×10^{-15} m. Ignore the other protons for this calculation.

50. (I) How much work must be done to bring three electrons from a great distance apart to within 1.0×10^{-10} m from one another?

51. (I) What potential difference is needed to give a helium nucleus $(Q = 3.2 \times 10^{-19}$ C) 48 keV of kinetic energy?

52. (I) What is the speed of (*a*) a 3.5 keV (kinetic energy) electron and (*b*) a 3.5 keV proton?

53. (II) Write the total electrostatic potential energy, U, for (*a*) four point charges and (*b*) five point charges. Draw a diagram defining all quantities.

54. (II) An alpha particle (which is a helium nucleus, $Q = +2e$, $m = 6.64 \times 10^{-27}$ kg) is emitted in a radioactive decay with kinetic energy 5.53 MeV. What is its speed?

55. (II) An electron starting from rest acquires 2.0 keV of kinetic energy in moving from point A to point B. (*a*) How much kinetic energy would a proton acquire, starting from rest at B and moving to point A? (*b*) Determine the ratio of their speeds at the end of their respective trajectories.

56. (II) Four equal point charges, Q, are fixed at the corners of a square of side b. (*a*) What is their total electrostatic potential energy? (*b*) How much potential energy will a fifth charge, Q, have at the center of the square (relative to $V = 0$ at $r = \infty$)? (*c*) If constrained to remain in that plane, is the fifth charge in stable or unstable equilibrium? If unstable, what maximum kinetic energy could it acquire? (*d*) Repeat part (*c*) for a negative $(-Q)$ charge.

57. (II) Repeat Problem 56, parts *a* and *b*, assuming that two of the charges, on diagonally opposite corners, are replaced by $-Q$ charges.

58. (II) Determine the total electrostatic potential energy of a conducting sphere of radius r_0 that carries a total charge Q distributed uniformly on its surface.

59. (III) Determine the total electrostatic potential energy of a nonconducting sphere of radius r_0 carrying a total charge Q distributed uniformly throughout its volume.

* **60.** (I) Use the ideal gas as a model to estimate the rms speed of a free electron in a metal at 300 K, and at 2500 K (the typical temperature of the cathode in a CRT).

* **61.** (III) In a given CRT, electrons are accelerated horizontally by 15 kV. They then pass through a uniform electric field E for a distance of 2.8 cm which deflects them upward so they reach the top of the screen 22 cm away, 11 cm above the center. Estimate the value of E.

General Problems

62. Sketch the electric field and equipotential lines for two charges of the same sign and magnitude separated by a distance d.

63. By rubbing a nonconducting material, a charge of 10^{-8} C can readily be produced. If this is done to a sphere of radius 10 cm, estimate the potential produced at the surface. Let $V = 0$ at $r = \infty$.

64. If the electrons in a single raindrop, 3 mm in diameter, could be removed from the Earth (without removing the atomic nuclei), by how much would the potential of the Earth increase?

65. A lightning flash transfers 4.0 C of charge and 4.2 MJ of energy to the Earth. (*a*) Between what potential difference did it travel? (*b*) How much water could this boil, starting from room temperature?

66. At each corner of a cube of side l there is a point charge Q, Fig. 23–31. (*a*) What is the potential at the center of the cube ($V = 0$ at $r = \infty$)? (*b*) What is the potential at each corner due to the other seven charges? (*c*) What is the total potential energy of this system?

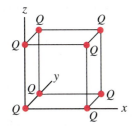

FIGURE 23–31 Problem 66.

67. How much voltage must be used to accelerate a proton so that it has just sufficient energy to touch the surface of an iron nucleus? An iron nucleus has a charge 26 times that of the proton ($= e$) and its radius is about 4.0×10^{-15} m, whereas the proton has a radius of about 1.2×10^{-15} m. Assume the nucleus is spherical and uniformly charged.

* **68.** Electrons are accelerated by 14 kV in a CRT. The screen is 30 cm wide and is 34 cm from the 2.6-cm-long deflection plates. Over what range must the horizontally deflecting electric field vary to sweep the beam fully across the screen?

69. Suppose a uniform layer of electrons was held near the surface of the Earth by the Earth's gravity. What is the maximum number of electrons that could be held in this manner? Ignore all other electric charges and fields except those of the electrons themselves.

70. Four point charges are located at the corners of a square that is 8.0 cm on a side. The charges, going in rotation around the square, are Q, $2Q$, $-3Q$, and $2Q$, where $Q = 4.8\ \mu C$ (Fig. 23–32). What is the total electric potential energy stored in the system, relative to $U = 0$ at infinite separation?

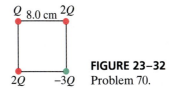

FIGURE 23–32
Problem 70.

71. In a television picture tube, electrons are accelerated by thousands of volts through a vacuum. If a television set were laid on its back, would electrons be able to move upward against the force of gravity? What potential difference, acting over a distance of 3.0 cm, would be needed to balance the downward force of gravity so that an electron would remain stationary? Assume that the electric field is uniform.

72. A proton starting from rest acquires 5.2 keV of kinetic energy in moving from point P to point Q. (*a*) How much kinetic energy would an electron acquire, starting from rest at Q and moving to point P? (*b*) Determine the ratio of their speeds at the end of their respective trajectories.

73. In a photocell, ultraviolet (UV) light provides enough energy to some electrons in barium metal to eject them from the surface at high speed. See Fig. 23–33. To measure the maximum energy of the electrons, another plate above the barium surface is kept at a negative enough potential that the emitted electrons are slowed down and stopped, and return to the barium surface. If the plate voltage is -3.02 V (compared to the barium) when the fastest electrons are stopped, what was the speed of these electrons when they were emitted?

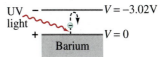

FIGURE 23–33 Problem 73.

74. Near the surface of the Earth there is an electric field of about 150 V/m which points downward. Two identical balls with mass $m = 0.540$ kg are dropped from a height of 2.00 m, but one of the balls is positively charged with $q_1 = 550\ \mu C$, and the second is negatively charged with $q_2 = -550\ \mu C$. Use conservation of energy to determine the difference in the speed of the two balls when they hit the ground. (Neglect air resistance.)

75. Three charges are at the corners of an equilateral triangle (side L) as shown in Fig. 23–34. Determine the potential at the midpoint of each of the sides.

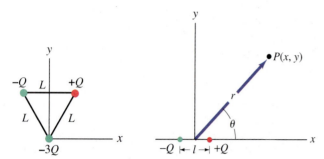

FIGURE 23–34 Problem 75. **FIGURE 23–35** Problem 76.

76. Determine the components of the electric field, E_x and E_y, at any point P in the xy plane due to a dipole, Fig. 23–35, starting with Eq. 23–7. Assume $r = (x^2 + y^2)^{\frac{1}{2}} \gg l$.

77. (a) What is the electric potential a distance of 2.5×10^{-15} m away from a proton? Let $V = 0$ at $r = \infty$. (b) What is the electric potential energy of a system that consists of two protons 2.5×10^{-15} m apart—as might occur inside a typical nucleus?

78. A thin flat nonconducting disk, with radius R and charge Q, has a hole with a radius $R/2$ in its center. Find the electric potential $V(x)$ at points along the symmetry (x) axis of the disk (a line perpendicular to the disk, passing through its center). Let $V = 0$ at $x = \infty$.

79. A Geiger counter is used to detect charged particles emitted by radioactive nuclei. It consists of a thin, positively charged central wire of radius R_a surrounded by a concentric conducting cylinder of radius R_b with an equal negative charge (Fig. 23–36). The charge per unit length on the inner wire is λ (units C/m). The cylindrical assembly is filled with low-pressure inert gas. Charged particles ionize some of these gas atoms; the resulting free electrons are attracted toward the central wire. If the radial electric field is strong enough, the freed electrons gain enough energy to ionize other atoms, causing an "avalanche" of electrons to strike the central wire, generating an electric "signal." Find the expression for the electric field between the wire and the cylinder, and show that the potential difference between R_a and R_b is

$$V_a - V_b = \Delta V = \frac{\lambda}{2\pi\epsilon_0 \ln(R_b/R_a)}.$$

FIGURE 23–36 Problem 79.

80. A Van de Graaff generator (Fig. 23–37) can develop a very large potential difference, even millions of volts. Electrons are pulled off the belt by the high voltage pointed electrode at A, leaving the belt positively charged. The belt carries the positive charge up inside the spherical shell where it passes to the pointed conductor at B, and races to the outer surface of the conducting sphere. As more charge is brought up, the sphere reaches extremely high voltage. Consider a Van de Graaff generator with a sphere of radius 0.15 m. (a) What is the electric potential on the surface of the sphere when electrical breakdown occurs? (Assume that $V = 0$ at $r = \infty$.) (b) What is the charge on the sphere for the potential found in part (a)?

GENERATOR

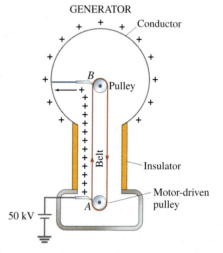

FIGURE 23–37
Problem 80.

81. Show that if two dipoles with dipole moments p_1 and p_2 are in line with one another (Fig. 23–38), the potential energy of one in the presence of the other (their "interaction energy") is given by

$$U = -\frac{1}{2\pi\epsilon_0} \frac{p_1 p_2}{r^3}$$

where r is the distance between the two dipoles. Assume that r is much greater than the length of either dipole.

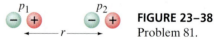

FIGURE 23–38
Problem 81.

82. Show that the electrostatic potential energy of two dipoles in a plane as shown in Fig. 23–39 is given by

$$U = \frac{1}{4\pi\epsilon_0} \frac{p_1 p_2}{r^3} \left[\cos(\theta_1 - \theta_2) - 3\cos\theta_1 \cos\theta_2\right].$$

Assume r is much larger than the length of each dipole. The vector dipole moments, $\mathbf{p}_1$ and $\mathbf{p}_2$, point from the negative charge toward the positive charge of the dipole.

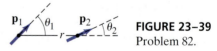

FIGURE 23–39
Problem 82.

83. A nonconducting sphere of radius r_2 contains a concentric spherical cavity of radius r_1. The material between r_1 and r_2 carries a uniform charge density ρ_E (C/m^3). Determine the electric potential V, relative to $V = 0$ at $r = \infty$, as a function of the distance r from the center for (a) $r > r_2$, (b) $r_1 < r < r_2$, and (c) $r < r_1$. Is V continuous at r_1 and r_2?

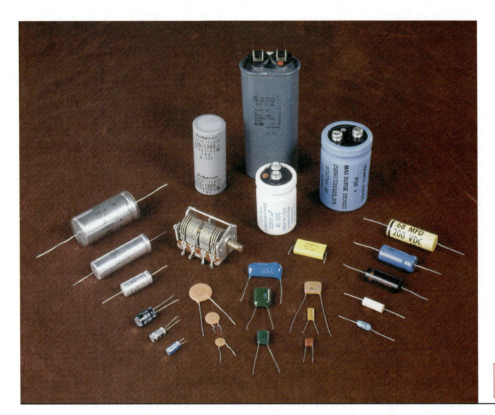

Capacitors come in a wide range of sizes and shapes, only a few of which are shown here. A capacitor is basically two conductors that do not touch, and which therefore can store charge of opposite sign on its two conductors. Capacitors are used in a wide variety of circuits, as we shall see in this and later chapters.

Capacitance, Dielectrics, Electric Energy Storage

This chapter will complete our study of electrostatics. It deals first of all with an important device, the capacitor, which is used in nearly all electronic circuits. We will also discuss electric energy storage and the effects of an insulator, or dielectric, on electric fields and potential differences.

24–1 Capacitors

A **capacitor**, sometimes called a *condenser*, is a device that can store electric charge, and usually consists of two conducting objects (usually plates or sheets) placed near each other but not touching. Capacitors are widely used in electronic circuits: they store charge for later use, as in a camera flash, and as energy backup in computers if the power fails; capacitors block surges of charge and energy to protect circuits; they form part of the tuner of a radio; very tiny capacitors serve as memory for the "ones" and "zeroes" of the binary code in the random access memory (RAM) of computers; and capacitors serve many other applications, some of which we will discuss.

➡ PHYSICS APPLIED

Uses of capacitors

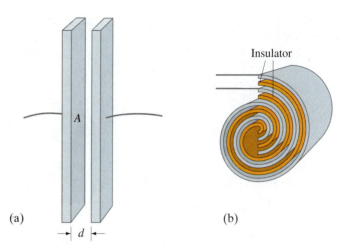

(a)   (b)

FIGURE 24–1 Capacitors: Diagrams of (a) parallel plate, (b) cylindrically shaped (rolled up parallel plate).

A simple capacitor consists of a pair of parallel plates of area A separated by a small distance d (Fig. 24–1a). Often the two plates are rolled into the form of a cylinder with paper or other insulator separating the plates, Fig. 24–1b. On the previous page is a photo of some actual capacitors used for various applications. In a diagram, a capacitor is represented by the symbol

   [capacitor symbol]

Another symbol for a capacitor you may encounter is ⊣⊢. A battery, which is a source of voltage, is indicated by the symbol:

   [battery symbol]

with unequal arms.

If a voltage is applied to a capacitor, say by connecting the capacitor to a battery as in Fig. 24–2, it quickly becomes charged. One plate acquires a negative charge, the other an equal amount of positive charge. Each battery terminal, connecting wire, and plate of the capacitor are conductors and at the same potential; hence the full battery voltage appears across the capacitor. For a given capacitor, it is found that the amount of charge Q acquired by each plate is proportional to the magnitude of the potential difference V_{ba} between them:

$$Q = CV_{ba}. \qquad (24–1)$$

Capacitance

The constant of proportionality, C, in the above relation is called the **capacitance** of the capacitor. The unit of capacitance is coulombs per volt and this unit is called a **farad** (F). Most capacitors have capacitance in the range of 1 pF $\left(\text{picofarad} = 10^{-12}\,\text{F}\right)$ to 1 μF $\left(\text{microfarad} = 10^{-6}\,\text{F}\right)$. The relation, Eq. 24–1, was first suggested by Volta in the late eighteenth century. The capacitance C does not in general depend on Q or V. Its value depends only on the size, shape, and relative position of the two conductors, and also on the material that separates them.

Unit is farad $(1\,\text{F} = 1\,\text{C/V})$

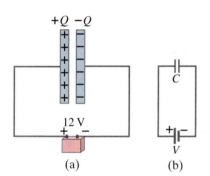

(a)   (b)

FIGURE 24–2 (a) Parallel-plate capacitor connected to a battery. (b) Same circuit shown using symbols.

24–2 | Determination of Capacitance

The capacitance of a given capacitor can be determined experimentally directly from Eq. 24–1, by measuring the charge Q on either conductor for a given potential difference V_{ba}.

For capacitors whose geometry is simple, we can determine C analytically, and in this Section we assume the conductors are separated by a vacuum or air. To illustrate this, we now determine C for a parallel-plate capacitor, Fig. 24–3. Each plate has area A and the two plates are separated by a distance d. We assume d is small compared to the dimensions of each plate so that the electric field $\mathbf{E}$ is uniform between them and we can ignore fringing (lines of $\mathbf{E}$ not straight) at the edges. We saw earlier (Example 21 – 12) that the electric field between two closely spaced parallel plates has magnitude $E = \sigma/\epsilon_0$ and its direction is perpendicular to the plates. Since σ is the charge per unit area, $\sigma = Q/A$, then the field between the plates is

$$E = \frac{Q}{\epsilon_0 A}.$$

The relation between electric field and electric potential, as given by Eq. 23 – 4, is

$$V_{ba} = -\int_a^b \mathbf{E} \cdot d\mathbf{l}.$$

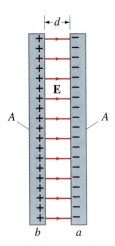

FIGURE 24–3 Parallel-plate capacitor, each of whose plates has area A. Fringing of the field is ignored.

We can take the line integral along a path antiparallel to the field lines, from one plate to the other; then $\theta = 180°$ and $\cos 180° = -1$, so

$$V_{ba} = V_b - V_a = -\int_a^b E\, dl \cos 180° = +\int_a^b E\, dl = \frac{Q}{\epsilon_0 A}\int_a^b dl = \frac{Qd}{\epsilon_0 A}.$$

This relates Q to V_{ba}, and from it we can get the capacitance C in terms of the geometry of the plates:

$$C = \frac{Q}{V_{ba}} = \epsilon_0 \frac{A}{d}. \qquad \text{[parallel-plate capacitor]} \quad \textbf{(24–2)}$$

Parallel plate capacitance formula

Note also from Eq. 24 – 2 that the value of C does not depend on Q or V_{ba}, so Q is predicted to be proportional to V_{ba} as is found experimentally.

EXAMPLE 24–1 **Capacitor calculations.** (a) Calculate the capacitance of a capacitor whose plates are 20 cm × 3.0 cm and are separated by a 1.0-mm air gap. (b) What is the charge on each plate if the capacitor is connected to a 12-V battery? (c) What is the electric field between the plates? (d) Estimate the area of the plates needed to achieve a capacitance of 1 F, given the same air gap d.

SOLUTION (a) The area $A = (20 \times 10^{-2}\,\text{m})(3.0 \times 10^{-2}\,\text{m}) = 6.0 \times 10^{-3}\,\text{m}^2$. The capacitance C is then

$$C = \epsilon_0 \frac{A}{d} = (8.85 \times 10^{-12}\,\text{C}^2/\text{N}\cdot\text{m}^2)\frac{6.0 \times 10^{-3}\,\text{m}^2}{1.0 \times 10^{-3}\,\text{m}} = 53\,\text{pF}.$$

(b) The charge on each plate (where we write simply V in place of V_{ba}) is:

$$Q = CV = (53 \times 10^{-12}\,\text{F})(12\,\text{V}) = 6.4 \times 10^{-10}\,\text{C}.$$

(c) From Eq. 23–4 for a uniform electric field, the magnitude of E is

$$E = \frac{V}{d} = \frac{12\,\text{V}}{1.0 \times 10^{-3}\,\text{m}} = 1.2 \times 10^4\,\text{V/m}.$$

(d) We solve for A in Eq. 24–2 and find that to make a 1-F capacitor with a 1.0-mm air gap requires plates with an area

$$A = \frac{Cd}{\epsilon_0} \approx \frac{(1\,\text{F})(1.0 \times 10^{-3}\,\text{m})}{(9 \times 10^{-12}\,\text{C}^2/\text{N}\cdot\text{m}^2)} \approx 10^8\,\text{m}^2.$$

This is the area of a square $10^4\,\text{m} = 10\,\text{km}$ on a side: the size of a city like San Francisco or Boston!

Ten or fifteen years ago, a capacitance greater than $1 \mu F$ was unusual. Today there are capacitors available that are 1 or 2 F, yet they are physically small, a few cm on a side. Such capacitors are used as power backups for low voltage applications, such as computer memory and VCR's where the time and date can be maintained through tiny charge flow. [Capacitors are superior to rechargable batteries for this purpose because they can be recharged more than 10^5 times with no degradation.] How are these high capacitance capacitors made? One type uses activated carbon which has very high porosity, so that the surface area is very large; one tenth of a gram of activated carbon can have a surface area of $100 \, m^2$. Furthermore, the equal and opposite charges exist in an electric "double layer" which is a layer of charge at the interface between the carbon particles and the sulfuric acid surrounding them. The positive charge resides at the edge of the carbon and the negative charge at the edge of the acid with a spacing between them of about 10^{-9} m. Thus, the capacitance of 0.1 g of activated carbon, whose internal area can be $10^2 \, m^2$, is $C = \epsilon_0 A/d = (8.85 \times 10^{-12} \, C^2/N \cdot m^2)(10^2 \, m^2)/(10^{-9} \, m) \approx 1 \, F$.

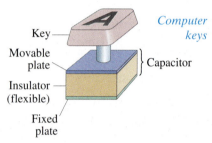

Key
Movable plate
Insulator (flexible)
Fixed plate

$\left.\begin{array}{c}\end{array}\right\}$ Capacitor

Computer keys

One type of computer keyboard operates by capacitance. As shown in Fig. 24–4, each key is connected to the upper plate of a capacitor. The upper plate moves down when the key is pressed, reducing the spacing between the capacitor plates, and increasing the capacitance (Eq. 24–2: smaller d, larger C). The *change* in capacitance becomes an electric signal that is detected by an electronic circuit.

The proportionality, $C \propto A/d$ in Eq. 24–2, is valid also for a parallel-plate capacitor that is rolled up into a spiral cylinder, as in Fig. 24–1b. However, the constant factor, ϵ_0, must be replaced if an insulator such as paper separates the plates, as is usual, and this is discussed in Section 24–5. For a true cylindrical capacitor—consisting of two long coaxial cylinders—the result is somewhat different as the next Example shows.

FIGURE 24–4 Key on a computer keyboard. Pressing the key reduces the capacitor spacing thus increasing the capacitance which can be detected electronically.

FIGURE 24–5 (a) Cylindrical capacitor consists of two coaxial cylindrical conductors. (b) The electric field lines are shown in cross-sectional view.

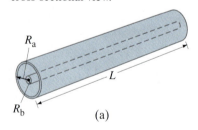

R_a

L

R_b

(a)

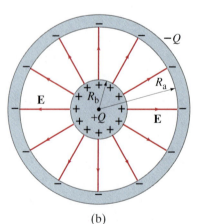

$-Q$

R_a

R_b

$+Q$

E

E

(b)

EXAMPLE 24–2 **Cylindrical capacitor.** A cylindrical capacitor consists of a cylinder (or wire) of radius R_b surrounded by a coaxial cylindrical shell of inner radius R_a, Fig. 24–5a. Both cylinders have length L which we assume is much greater than the separation of the cylinders, $R_a - R_b$, so we can neglect end effects. The capacitor is charged (say, by connecting it to a battery) so that one cylinder has a charge $+Q$ (say, the inner one) and the other one a charge $-Q$. Determine a formula for the capacitance.

SOLUTION To obtain $C = Q/V_{ba}$, we need to determine the potential difference between the cylinders, V_{ba}, in terms of Q. We can use our earlier result (Example 21–10 or 22–5) that the electric field outside a long wire is directed radially outward and has magnitude $E = (1/2\pi\epsilon_0)(\lambda/r)$, where r is the distance from the axis and λ is the charge per unit length, Q/L; then $E = (1/2\pi\epsilon_0)(Q/Lr)$ for points between the cylinders.

To obtain V_{ba} in terms of Q, we use this result in Eq. 23–3, $V_{ba} = -\int_a^b \mathbf{E} \cdot d\mathbf{l}$, and write the line integral from the outer cylinder to the inner one (so $V_{ba} > 0$) along a radial line[†]:

$$V_{ba} = -\int_a^b \mathbf{E} \cdot d\mathbf{l} = -\frac{Q}{2\pi\epsilon_0 L}\int_{R_a}^{R_b}\frac{dr}{r} = -\frac{Q}{2\pi\epsilon_0 L}\ln\frac{R_b}{R_a} = \frac{Q}{2\pi\epsilon_0 L}\ln\frac{R_a}{R_b}.$$

Q and V_{ba} are proportional, and the capacitance C is

$$C = \frac{Q}{V_{ba}} = \frac{2\pi\epsilon_0 L}{\ln(R_a/R_b)}. \qquad \text{[cylindrical capacitor]}$$

Does the dependence on L, R_a, and R_b make sense intuitively? (See discussion immediately after Eq. 24–2.)

[†] Note that **E** points outward in Fig. 24–5b, but $d\mathbf{l}$ points inward for our chosen direction of integration; the angle between **E** and $d\mathbf{l}$ is 180° and cos 180° = −1. Also, $dl = -dr$ because dr increases outward. These two minus signs cancel.

EXAMPLE 24–3 **Spherical capacitor.** A spherical capacitor consists of two thin concentric spherical conducting shells, of radius r_a and r_b as shown in Fig. 24–6. The inner shell carries a uniformly distributed charge Q on its surface, and the outer shell an equal but opposite charge $-Q$. Determine the capacitance of the two shells.

SOLUTION In Example 22–3 we used Gauss's law to show that the electric field outside a uniformly charged conducting sphere is $E = Q/4\pi\epsilon_0 r^2$ as if all the charge were concentrated at the center. Now we use Eq. 23–3, $V_{ba} = -\int_a^b \mathbf{E} \cdot d\mathbf{l}$, and integrate along a radial line to obtain the potential difference between the two conducting shells:

$$V_{ba} = -\int_a^b \mathbf{E} \cdot d\mathbf{l} = -\frac{Q}{4\pi\epsilon_0} \int_{r_a}^{r_b} \frac{1}{r^2}\, dr$$

$$= \frac{Q}{4\pi\epsilon_0}\left(\frac{1}{r_b} - \frac{1}{r_a}\right) = \frac{Q}{4\pi\epsilon_0}\left(\frac{r_a - r_b}{r_a r_b}\right).$$

Finally,

$$C = \frac{Q}{V_{ba}} = 4\pi\epsilon_0\left(\frac{r_a r_b}{r_a - r_b}\right).$$

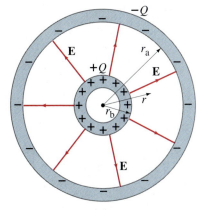

FIGURE 24–6 Cross section of a spherical capacitor. The thin inner shell has radius r_b and the thin outer shell has radius r_a.

A single isolated conductor can also be said to have a capacitance, C. In this case, C can still be defined as the ratio of the charge to absolute potential V on the conductor (relative to $V = 0$ at $r = \infty$), so that the relation

$$Q = CV$$

remains valid. For example, the potential of a single conducting sphere of radius r_b can be obtained from our results in Example 24–3 by letting r_a become infinitely large. As $r_a \to \infty$, then

$$V = \frac{Q}{4\pi\epsilon_0}\left(\frac{1}{r_b} - \frac{1}{r_a}\right) = \frac{1}{4\pi\epsilon_0}\frac{Q}{r_b};$$

so its capacitance is

$$C = \frac{Q}{V} = 4\pi\epsilon_0 r_b.$$

But note that a single conductor alone is not considered a capacitor. In practical cases, a single conductor may be near other conductors or the Earth (which can be thought of as the other "plate" of a capacitor), and these will affect the value of the capacitance.

24–3 | Capacitors in Series and Parallel

Capacitors are found in many electric circuits. By electric circuit we mean a closed path of conductors, usually wires connecting capacitors and/or other devices, in which charge can flow and which includes a source of voltage such as a battery. The battery voltage is usually given the symbol V, which means that V represents a potential *difference*. Capacitors can be connected together in various ways. Two common ways are in *series*, or in *parallel*, and we now discuss both.

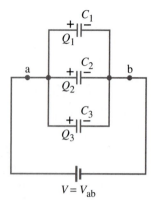

FIGURE 24–7 Capacitors in parallel: $C = C_1 + C_2 + C_3$.

A circuit containing three capacitors connected in **parallel** is shown in Fig. 24–7. They are in "parallel" because when a battery of voltage V is connected to points a and b, this voltage $V = V_{ab}$ exists across each of the capacitors. That is, since the left-hand plates of all the capacitors are connected by conductors, they all reach the same potential V_a when connected to the battery; and the right-hand plates each reach potential V_b. Each capacitor plate acquires a charge given by $Q_1 = C_1 V$, $Q_2 = C_2 V$, and $Q_3 = C_3 V$. The total charge Q that must leave the battery is then

$$Q = Q_1 + Q_2 + Q_3 = C_1 V + C_2 V + C_3 V.$$

Let us try to find a single equivalent capacitor that will hold the same charge Q at the same voltage $V = V_{ab}$. It will have a capacitance C_{eq} given by

$$Q = C_{eq} V.$$

Combining the two previous equations, we have

$$C_{eq} V = C_1 V + C_2 V + C_3 V = (C_1 + C_2 + C_3)V$$

or

Capacitors in parallel

$$C_{eq} = C_1 + C_2 + C_3. \qquad \text{[parallel]} \quad \textbf{(24–3)}$$

The net effect of connecting capacitors in parallel is thus to *increase* the capacitance. This makes sense because we are essentially increasing the area of the plates where charge can accumulate (see, for example, Eq. 24–2).

Capacitors can also be connected in **series**. That is, end to end, as shown in Fig. 24–8. A charge $+Q$ flows from the battery to one plate of C_1, and $-Q$ flows to one plate of C_3. The regions A and B between the capacitors were originally neutral; so the net charge there must still be zero. The $+Q$ on the left plate of C_1 attracts a charge of $-Q$ on the opposite plate. Because region A must have a zero net charge, there is thus $+Q$ on the left plate of C_2. The same considerations apply to the other capacitors, so we see the charge on each capacitor is the same value Q. A single capacitor that could replace these three in series without affecting the circuit (that is, Q and V the same) would have a capacitance C_{eq} where

FIGURE 24–8 Capacitors in series: $\dfrac{1}{C} = \dfrac{1}{C_1} + \dfrac{1}{C_2} + \dfrac{1}{C_3}$.

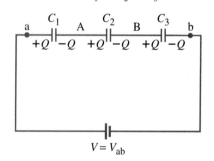

$$Q = C_{eq} V.$$

Now the total voltage V across the three capacitors in series must equal the sum of the voltages across each capacitor:

$$V = V_1 + V_2 + V_3.$$

We also have $Q = C_1 V_1$, $Q = C_2 V_2$, and $Q = C_3 V_3$, so we substitute for V, V_1, V_2, and V_3 into the last equation and get

$$\frac{Q}{C_{eq}} = \frac{Q}{C_1} + \frac{Q}{C_2} + \frac{Q}{C_3} = Q\left(\frac{1}{C_1} + \frac{1}{C_2} + \frac{1}{C_3}\right)$$

or

Capacitors in series

$$\frac{1}{C_{eq}} = \frac{1}{C_1} + \frac{1}{C_2} + \frac{1}{C_3}. \qquad \text{[series]} \quad \textbf{(24–4)}$$

Note that the equivalent capacitance C_{eq} is smaller than the smallest contributing capacitance.

Other connections of capacitors can be analyzed similarly using charge conservation, and often simply in terms of series and parallel connections.

EXAMPLE 24–4 **Equivalent capacitance.** Determine the capacitance of a single capacitor that will have the same effect as the combination shown in Fig. 24–9. Take $C_1 = C_2 = C_3 = C$.

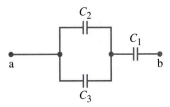

FIGURE 24–9 Example 24–4.

SOLUTION C_2 and C_3 are connected in parallel, so they are equivalent to a single capacitor having capacitance

$$C_{23} = C_2 + C_3 = 2C.$$

C_{23} is in series with C_1, so the equivalent capacitance, C_{eq}, is given by

$$\frac{1}{C_{eq}} = \frac{1}{C_1} + \frac{1}{C_{23}} = \frac{1}{C} + \frac{1}{2C} = \frac{3}{2C}.$$

Hence $C_{eq} = \frac{2}{3}C$, which is the equivalent capacitance of the entire combination.

➡ **PROBLEM SOLVING**

Remember to take the reciprocal

EXAMPLE 24–5 **Capacitor combination.** (a) Determine the equivalent capacitance of the combination shown in Fig. 24–10a (that is, the capacitance between points a and b). Take $C_1 = 6.0\,\mu F$, $C_2 = 4.0\,\mu F$, and $C_3 = 8.0\,\mu F$. (b) If the capacitors are charged by a 12-V battery placed between a and b, determine the charge on each capacitor and the potential difference across each.

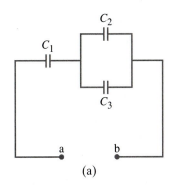

(a)

SOLUTION (a) C_2 and C_3 are connected in parallel, so they are equivalent to a single capacitor of capacitance (Eq. 24–3)

$$C_{23} = C_2 + C_3 = 4.0\,\mu F + 8.0\,\mu F = 12.0\,\mu F.$$

C_{23} is in series with C_1 as shown in Fig. 24–10b. So the equivalent capacitance C of the entire combination is given by (Eq. 24–4)

$$\frac{1}{C_{eq}} = \frac{1}{C_1} + \frac{1}{C_{23}} = \frac{1}{6.0\,\mu F} + \frac{1}{12.0\,\mu F} = \frac{3}{12.0\,\mu F}.$$

Hence $C = 12.0\,\mu F/3 = 4.0\,\mu F$.

(b) The total charge that flows from the battery is

$$Q = CV = (4.0 \times 10^{-6}\,F)(12\,V) = 4.8 \times 10^{-5}\,C.$$

Both C_1 and C_{23} carry this charge Q. The voltage across C_1 is then

$$V_1 = \frac{Q}{C_1} = \frac{4.8 \times 10^{-5}\,C}{6.0 \times 10^{-6}\,F} = 8.0\,V.$$

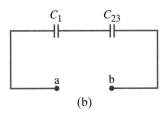

(b)

FIGURE 24–10 Example 24–5.

The voltage across the combination C_{23} is

$$V_{23} = \frac{Q}{C_{23}} = \frac{4.8 \times 10^{-5}\,C}{12.0 \times 10^{-6}\,F} = 4.0\,V.$$

Since the actual capacitors C_2 and C_3 are in parallel, this represents the voltage across each of them:

$$V_2 = V_3 = 4.0\,V.$$

The charges on C_2 and C_3 are

$$Q_2 = C_2 V_2 = (4.0 \times 10^{-6}\,F)(4.0\,V) = 1.6 \times 10^{-5}\,C$$
$$Q_3 = C_3 V_3 = (8.0 \times 10^{-6}\,F)(4.0\,V) = 3.2 \times 10^{-5}\,C.$$

To summarize:

$$V_1 = 8.0\,V \qquad Q_1 = 48\,\mu C$$
$$V_2 = 4.0\,V \qquad Q_2 = 16\,\mu C$$
$$V_3 = 4.0\,V \qquad Q_3 = 32\,\mu C.$$

Note that $Q_2 + Q_3 = Q_1 = Q$, as it should.

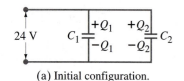

24 V

(a) Initial configuration.

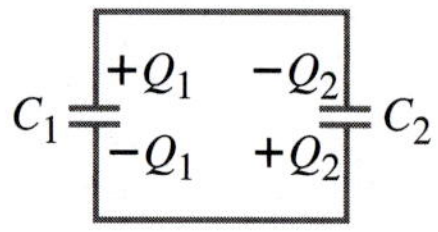

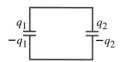

(b) At the instant of reconnection only.

(c) A short time later.

FIGURE 24–11 Example 24–6.

EXAMPLE 24–6 Capacitors reconnected. Two capacitors, $C_1 = 2.2\,\mu\text{F}$ and $C_2 = 1.2\,\mu\text{F}$ are connected in parallel to a 24-volt source as shown in Fig. 24–11a. After they are charged they are disconnected from the source and from each other and reconnected directly to each other, with plates of opposite sign connected together (see Fig. 24–11b). Find the charge on each capacitor and the potential across each after equilibrium is established.

SOLUTION First we need to calculate how much charge has been placed on each capacitor after the power source has charged them fully. Using Eq. 24–1 for each capacitor we find:

$$Q_1 = C_1 V = (2.2\,\mu\text{F})(24\,\text{V}) = 52.8\,\mu\text{C},$$
$$Q_2 = C_2 V = (1.2\,\mu\text{F})(24\,\text{V}) = 28.8\,\mu\text{C}.$$

Now we examine Fig. 24–11b. The capacitors are connected in parallel, and the potential difference across each must quickly reach the same value. Thus, the charge cannot remain as shown in Fig. 24–11b, but the charge must rearrange itself so that the upper plates at least have the same sign of charge, with the lower plates having the opposite charge as shown in Fig. 24–11c. Equation 24–1 applies for each:

$$q_1 = C_1 V' \qquad \text{and} \qquad q_2 = C_2 V'$$

where V' is the voltage across each capacitor after the charges have rearranged themselves. We don't know q_1, q_2, or V', so we need a third equation. This is provided by charge conservation. The charges have rearranged themselves between Figs. 24–11b and c. The total charge on the upper plates in those two figures must be the same, so we have

$$q_1 + q_2 = Q_1 - Q_2 = 24.0\,\mu\text{C}.$$

Combining the last three equations we find:

$$V' = (q_1 + q_2)/(C_1 + C_2) = 24.0\,\mu\text{C}/3.4\,\mu\text{F} = 7.06\,\text{V} \approx 7.1\,\text{V}$$
$$q_1 = C_1 V' = (2.2\,\mu\text{F})(7.06\,\text{V}) = 15.5\,\mu\text{F} \approx 16\,\mu\text{F}$$
$$q_2 = C_2 V' = (1.2\,\mu\text{F})(7.06\,\text{V}) = 8.5\,\mu\text{F}$$

where we have kept only two significant figures in our final answers.

24–4 Electric Energy Storage

A charged capacitor stores electrical energy. The energy stored in a capacitor will be equal to the work done to charge it. The net effect of charging a capacitor is to remove charge from one plate and add it to the other plate. This is what a battery does when it is connected to a capacitor. A capacitor does not become charged instantly. It takes time (Section 26–4). Initially, when the capacitor is uncharged, it requires no work to move the first bit of charge over. When some charge is on each plate, it requires work to add more charge of the same sign because of the electric repulsion. The more charge already on a plate, the more work is required to add additional charge. The work needed to add a small amount of charge dq, when a potential difference V is across the plates, is $dW = V\,dq$. Since $V = q/C$ at any moment (Eq. 24–1), where C is the capacitance, the work needed to store a total charge Q is

$$W = \int_0^Q V\,dq = \frac{1}{C}\int_0^Q q\,dq = \frac{1}{2}\frac{Q^2}{C}.$$

Thus we can say that the energy "stored" in a capacitor is

$$U = \frac{1}{2}\frac{Q^2}{C}$$

when the capacitor C carries charges $+Q$ and $-Q$ on its two conductors. Since $Q = CV$, where V is the potential difference across the capacitor, we can also write

Energy stored in a capacitor

$$U = \frac{1}{2}\frac{Q^2}{C} = \frac{1}{2}CV^2 = \frac{1}{2}QV. \qquad (24\text{–}5)$$

EXAMPLE 24–7 **Energy stored in a capacitor.** A camera flash unit stores energy in a $150\,\mu\text{F}$ capacitor at 200 V. How much electric energy can be stored?

➡ **PHYSICS APPLIED**

Camera flash

SOLUTION From Eq. 24–5, we have

$$U = \text{energy} = \tfrac{1}{2}CV^2 = \tfrac{1}{2}(150 \times 10^{-6}\,\text{F})(200\,\text{V})^2 = 3.0\,\text{J}.$$

Notice how the units work out: $\text{FV}^2 = \left(\dfrac{\text{C}}{\text{V}}\right)(\text{V}^2) = \text{CV} = \text{C}\left(\dfrac{\text{J}}{\text{C}}\right) = \text{J}.$

If this energy could be released in $\frac{1}{1000}$ of a second $(10^{-3}\,\text{s})$ the power output would be equivalent to 3000 W.

It is useful to think of the energy stored in a capacitor as being stored in the electric field between the plates. As an example let us calculate the energy stored in a parallel-plate capacitor in terms of the electric field.

We have seen (Eq. 23–4) that the electric field **E** between two close parallel plates is (approximately) uniform and its magnitude is related to the potential difference by $V = Ed$ where d is the separation. Also, Eq. 24–2 tells us $C = \epsilon_0 A/d$ for a parallel-plate capacitor. Thus

$$U = \tfrac{1}{2}CV^2 = \tfrac{1}{2}\left(\frac{\epsilon_0 A}{d}\right)(E^2 d^2)$$

$$= \tfrac{1}{2}\epsilon_0 E^2 Ad.$$

The quantity Ad is the volume between the plates in which the electric field E exists. If we divide both sides by the volume, we obtain an expression for the energy per unit volume or **energy density**, u:

$$u = \text{energy density} = \tfrac{1}{2}\epsilon_0 E^2. \qquad (24\text{–}6)$$

Energy stored per unit volume in electric field

The *electric energy stored per unit volume in any region of space is proportional to the square of the electric field* in that region. We derived Eq. 24–6 for the special case of a parallel-plate capacitor. But it can be shown to be true for any region of space where there is an electric field.

24–5 | Dielectrics

In most capacitors there is an insulating sheet of material (such as paper or plastic) called a **dielectric** between the plates. This serves several purposes. First of all, dielectrics break down (allowing electric charge to flow) less readily than air, so higher voltages can be applied without charge passing across the gap. Furthermore, a dielectric allows the plates to be placed closer together without touching, thus allowing an increased capacitance because d is less in Eq. 24–2. Finally, it is found experimentally that if the dielectric fills the space between the two conductors, it increases the capacitance by a factor K which is known as the **dielectric constant**. Thus

$$C = KC_0, \qquad (24\text{–}7)$$

Dielectric constant (defined)

where C_0 is the capacitance when the space between the two conductors of the capacitor is a vacuum and C is the capacitance when the space is filled with a material whose dielectric constant is K.

TABLE 24–1
Dielectric constants (at 20° C)

Material	Dielectric constant K	Dielectric strength (V/m)
Vacuum	1.0000	
Air (1 atm)	1.0006	3×10^6
Paraffin	2.2	10×10^6
Polystyrene	2.6	24×10^6
Rubber, neoprene	6.7	12×10^6
Vinyl (plastic)	2–4	50×10^6
Paper	3.7	15×10^6
Quartz	4.3	8×10^6
Oil	4	12×10^6
Glass, Pyrex	5	14×10^6
Porcelain	6–8	5×10^6
Mica	7	150×10^6
Water (liquid)	80	
Strontium titanate	300	8×10^6

Energy density in electric field inside a dielectric

The values of the dielectric constant for various materials are given in Table 24–1. Note that for air (at 1 atm pressure) $K = 1.0006$, which differs so little from 1.0000 that the capacitance for an air-filled capacitor hardly differs from that for a vacuum. Also shown in Table 24–1 is the **dielectric strength**, the maximum electric field before breakdown (charge flow) occurs.

For a parallel-plate capacitor (see Eq. 24–2),

$$C = K\epsilon_0 \frac{A}{d} \qquad \text{[parallel-plate capacitor]} \quad \textbf{(24–8)}$$

when the space between the plates is completely filled with a dielectric whose dielectric constant is K. (The situation when the dielectric only partially fills the space will be discussed shortly in Example 24–9.) The quantity $K\epsilon_0$ appears so often in formulas that we define a new quantity

$$\epsilon = K\epsilon_0 \qquad \textbf{(24–9)}$$

called the **permittivity** of a material. Then the capacitance of a parallel-plate capacitor becomes

$$C = \epsilon \frac{A}{d}.$$

Note that ϵ_0 represents the permittivity of free space (a vacuum).

The energy density stored in an electric field E (Section 24–4) in a dielectric is given by (see Eq. 24–6)

$$u = \tfrac{1}{2}K\epsilon_0 E^2 = \tfrac{1}{2}\epsilon E^2.$$

Two simple experiments illustrate the effect of a dielectric. In the first, Fig. 24–12a, a battery of voltage V_0 is kept connected to a capacitor as a dielectric is inserted between the plates. If the charge on the plates without dielectric is Q_0, then when the dielectric is inserted, it is found experimentally (first by Faraday) that the charge Q on the plates is increased by a factor K,

$$Q = KQ_0. \qquad \text{[voltage constant]}$$

The capacitance has increased to $C = Q/V_0 = KQ_0/V_0 = KC_0$, which is Eq. 24–7. In a second experiment, Fig. 24–12b, a battery V_0 is connected to a capacitor C_0 which then holds a charge $Q_0 = C_0V_0$. The battery is then disconnected, leaving the capacitor isolated with charge Q_0 and still at voltage V_0. Next a dielectric is inserted between the plates of the capacitor. The charge remains Q_0 (there is nowhere for the charge to go) but the voltage is found experimentally to drop by a factor K:

$$V = \frac{V_0}{K}. \qquad \text{[charge constant]}$$

Note that the capacitance changes to $C = Q_0/V = Q_0/(V_0/K) = KQ_0/V_0 = KC_0$, so this experiment too confirms Eq. 24–7.

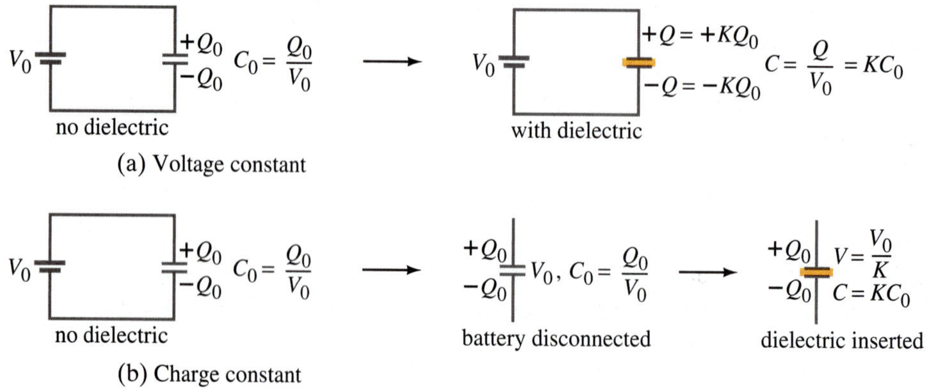

FIGURE 24–12 Two experiments with a capacitor. Dielectric inserted with (a) voltage held constant, (b) charge held constant.

(a) Voltage constant

(b) Charge constant

The electric field within a dielectric is also altered. When no dielectric is present, the electric field between the plates of a parallel-plate capacitor is given by Eq. 23–4:

$$E_0 = \frac{V_0}{d},$$

where V_0 is the potential difference between the plates and d is their separation. If the capacitor is isolated so that the charge remains fixed on the plates when a dielectric is inserted, filling the space between the plates, the potential difference drops to $V = V_0/K$. So the electric field in the dielectric is now

$$E = E_D = \frac{V}{d} = \frac{V_0}{Kd}$$

or

$$E_D = \frac{E_0}{K}. \qquad \text{[in a dielectric]} \quad \textbf{(24–10)}$$

The electric field within a dielectric is thus also reduced by a factor equal to the dielectric constant. Although the field is reduced in a dielectric (or insulator), it is not reduced all the way to zero as in a conductor.

EXAMPLE 24–8 **Dielectric removal.** A parallel-plate capacitor, filled with a dielectric with $K = 3.4$, is connected to a 100-V battery (Fig. 24–13a). After the capacitor is fully charged, the battery is disconnected. The plates have area $A = 4.0\,\text{m}^2$, and are separated by $d = 4.0\,\text{mm}$. (a) Find the capacitance, the charge on the capacitor, the electric field strength, and the energy stored in the capacitor. (b) The dielectric is carefully removed, without changing the plate separation nor does any charge leave the capacitor (Fig. 24–13b). Find the new values of capacitance, electric field strength, voltage between the plates, and the energy stored on the capacitor.

SOLUTION (a) First we find the capacitance, with dielectric:

$$C = \frac{K\epsilon_0 A}{d} = \frac{3.4(8.85 \times 10^{-12}\,\text{C}^2/\text{N}\cdot\text{m}^2)(4.0\,\text{m}^2)}{4.0 \times 10^{-3}\,\text{m}} = 3.0 \times 10^{-8}\,\text{F}.$$

The charge Q on the plates is

$$Q = CV = (3.0 \times 10^{-8}\,\text{F})(100\,\text{V}) = 3.0 \times 10^{-6}\,\text{C}.$$

The electric field between the plates is

$$E = \frac{V}{d} = \frac{100\,\text{V}}{4.0 \times 10^{-3}\,\text{m}} = 25\,\text{kV/m}.$$

Finally, the total energy stored in the capacitor is

$$U = \tfrac{1}{2}CV^2 = \tfrac{1}{2}(3.0 \times 10^{-8}\,\text{F})(100\,\text{V})^2 = 1.5 \times 10^{-4}\,\text{J}.$$

(b) The capacitance without dielectric is

$$C_0 = \frac{C}{K} = \frac{(3.0 \times 10^{-8}\,\text{F})}{(3.4)} = 8.8 \times 10^{-9}\,\text{F}.$$

The charge Q doesn't change, so $V = Q/C$ increases by a factor $K = 3.4$ to 340 V. The electric field is

$$E = \frac{V}{d} = \frac{340\,\text{V}}{4.0 \times 10^{-3}\,\text{m}} = 85\,\text{kV/m}.$$

The energy stored is

$$U = \tfrac{1}{2}CV^2 = \tfrac{1}{2}(8.8 \times 10^{-9}\,\text{F})(340\,\text{V})^2 = 5.1 \times 10^{-4}\,\text{J}.$$

Where did all this extra energy come from? The energy increased because work had to be done to remove the dielectric. The work required was $W = 5.1 \times 10^{-4}\,\text{J} - 1.5 \times 10^{-4}\,\text{J} = 3.6 \times 10^{-4}\,\text{J}$. (We will see in the next Section that work is required because of the force of attraction between induced charge on the dielectric and the charges on the plates—see Fig. 24–14c.)

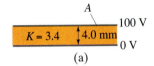

FIGURE 24–13 Example 24–8.

Molecular Description of Dielectrics

Molecular description of dielectrics

Let us now examine, from the molecular point of view, why the capacitance of a capacitor should increase when a dielectric is inserted between the plates. Consider a capacitor whose plates are separated by an air gap. This capacitor has a charge $+Q$ on one plate and $-Q$ on the other (Fig. 24–14a). The capacitor is isolated (not connected to a battery) so charge cannot flow to or from the plates. The potential difference between the plates, V_0, is given by Eq. 24–1: $Q = C_0 V_0$; the subscripts $(_0)$ refer to the situation when only air is between the plates. Now we insert a dielectric between the plates (Fig. 24–14b). The molecules of the dielectric may be *polar*. That is, although the molecules are neutral, they may have a permanent dipole moment (as water does). Because of the electric field between the plates, the molecules will tend to become oriented as shown in Fig. 24–14b; they won't be perfectly aligned because of thermal motion (Chapter 18), but they will usually be at least partially aligned (the stronger the electric field the more alignment).

Even if the molecules are not polar, the electric field between the plates will induce some separation of charge in the molecules (induced dipole moment). Although the electrons do not leave the molecules, they will move slightly within the molecules toward the positive plate. So the situation is still as illustrated in Fig. 24–14b.

The net effect in either case is as if there were a net negative charge on the outer edge of the dielectric facing the positive plate, and a net positive charge on the opposite side, as shown in Fig. 24–14c.

We can visualize that some of the electric field lines do not pass through the dielectric but instead end on charges induced on the surface of the dielectric as shown. Hence the electric field within the dielectric is less than in air. Now imagine a positive test charge within the dielectric. Because the electric field is less, the force a test charge feels is reduced by some factor K (equal, as we shall see, to the dielectric constant). Because the force on our test charge is reduced by a factor K, the work needed to move it from one plate to the other is reduced by a factor K. (We assume

FIGURE 24–14 Molecular view of the effects of a dielectric.

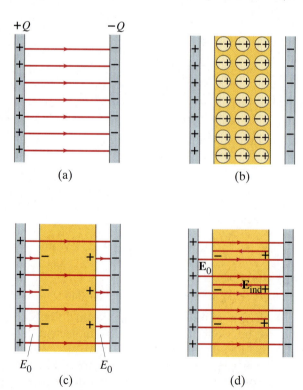

that the dielectric fills all the space between the plates, even though Fig. 24–14 leaves a space so we can show the field there.) The voltage, which is the work done per unit charge, must therefore also have decreased by the factor K. That is, the voltage between the plates is now

$$V = \frac{V_0}{K}.$$

Now the charge Q on the plates has not changed, because they are isolated. So we have

$$Q = CV,$$

where C is the capacitance when the dielectric is present. When we combine this with the relation, $V = V_0/K$, we obtain

$$C = \frac{Q}{V} = \frac{Q}{V_0/K} = \frac{QK}{V_0} = KC_0,$$

since $C_0 = Q/V_0$. Thus we see why the capacitance is increased by a factor K.

As shown in Fig. 24–14d, the electric field within the dielectric E_D can be considered as the vector sum of the electric field $\mathbf{E}_0$ due to the "free" charges on the conducting plates, and the field $\mathbf{E}_{ind}$ due to the induced charge on the surfaces of the dielectric. Since these two fields are in opposite directions, the net field within the dielectric, $E_0 - E_{ind}$, is less than E_0. The precise relationship is given by Eq. 24–10:

$$E_D = E_0 - E_{ind} = \frac{E_0}{K},$$

so

$$E_{ind} = E_0\left(1 - \frac{1}{K}\right).$$

The electric field between two parallel plates is related to the surface charge density, σ (see Section 22–3 and Example 22–7), by $E = \sigma/\epsilon_0$. Thus

$$E_0 = \frac{\sigma}{\epsilon_0}$$

where $\sigma = Q/A$ is the surface charge density on the conductor; Q is the net charge on the conductor and is often called the **free charge** (since charge is free to move in a conductor). Similarly, we define an equivalent induced surface charge density σ_{ind} on the dielectric; then

$$E_{ind} = \frac{\sigma_{ind}}{\epsilon_0}$$

where E_{ind} is the electric field due to the induced charge $Q_{ind} = \sigma_{ind} A$ on the surface of the dielectric, Fig. 24–14d. Q_{ind} is often called the **bound charge**, since it is on an insulator and is not free to move. Since $E_{ind} = E_0(1 - 1/K)$ as shown above, we now have

$$\sigma_{ind} = \sigma\left(1 - \frac{1}{K}\right) \tag{24–11a}$$

and

$$Q_{ind} = Q\left(1 - \frac{1}{K}\right). \tag{24–11b}$$

Since K is always greater than 1, we see that the charge induced on the dielectric is always less than the free charge on each of the capacitor plates.

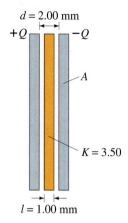

$d = 2.00$ mm

$+Q$ $-Q$

A

$K = 3.50$

$l = 1.00$ mm

FIGURE 24–15 Example 24–9.

EXAMPLE 24–9 **Dielectric partially fills capacitor.** A parallel-plate capacitor has plates of area $A = 250\,\text{cm}^2$ and separation $d = 2.00\,\text{mm}$. The capacitor is charged to a potential difference $V_0 = 150\,\text{V}$. Then the battery is disconnected (the charge Q on the plates then won't change), and a dielectric sheet ($K = 3.50$) of the same area A but thickness $l = 1.00\,\text{mm}$ is placed between the plates as shown in Fig. 24–15. Determine (a) the initial capacitance of the air-filled capacitor, (b) the charge on each plate before the dielectric is inserted, (c) the charge induced on each face of the dielectric after it is inserted, (d) the electric field in the space between each plate and the dielectric, (e) the electric field in the dielectric, (f) the potential difference between the plates after the dielectric is added, and (g) the capacitance after the dielectric is in place.

SOLUTION (a) Before the dielectric is in place, the capacitance is

$$C_0 = \epsilon_0 \frac{A}{d} = (8.85 \times 10^{-12}\,\text{C}^2/\text{N}\cdot\text{m}^2)\left(\frac{2.50 \times 10^{-2}\,\text{m}^2}{2.00 \times 10^{-3}\,\text{m}}\right) = 111\,\text{pF}.$$

(b) The charge on each plate is

$$Q = C_0 V_0 = (1.11 \times 10^{-10}\,\text{F})(150\,\text{V}) = 1.66 \times 10^{-8}\,\text{C}.$$

(c) From Eq. 24–11b,

$$Q_{\text{ind}} = Q\left(1 - \frac{1}{K}\right) = (1.66 \times 10^{-8}\,\text{C})\left(1 - \frac{1}{3.50}\right) = 1.19 \times 10^{-8}\,\text{C}.$$

(d) The electric field in the gaps between the plates and the dielectric (see Fig. 24–14c) is the same as in the absence of the dielectric since the charge on the plates has not been altered. Gauss's law, as applied in Example 22–7, could be used here, which gives $E_0 = \sigma/\epsilon_0$. Or we can note that, in the absence of the dielectric, $E_0 = V_0/d = Q/C_0 d$ (since $V_0 = Q/C_0$) $= Q/\epsilon_0 A$ (since $C_0 = \epsilon_0 A/d$) which is the same result. Thus

$$E_0 = \frac{Q}{\epsilon_0 A} = \frac{1.66 \times 10^{-8}\,\text{C}}{(8.85 \times 10^{-12}\,\text{C}^2/\text{N}\cdot\text{m}^2)(2.50 \times 10^{-2}\,\text{m}^2)} = 7.50 \times 10^4\,\text{V/m}.$$

(e) In the dielectric the electric field is (Eq. 24–10)

$$E_D = \frac{E_0}{K} = \frac{7.50 \times 10^4\,\text{V/m}}{3.50} = 2.14 \times 10^4\,\text{V/m}.$$

(f) To obtain the potential difference in the presence of the dielectric we use Eq. 23–3, and integrate along a straight line parallel to the field lines:

$$V = -\int \mathbf{E} \cdot d\mathbf{l} = E_0(d - l) + E_D l,$$

which can be simplified to

$$V = E_0\left(d - l + \frac{l}{K}\right)$$

$$= (7.50 \times 10^4\,\text{V/m})\left(1.00 \times 10^{-3}\,\text{m} + \frac{1.00 \times 10^{-3}\,\text{m}}{3.50}\right)$$

$$= 96.4\,\text{V}.$$

(g) In the presence of the dielectric, the capacitance is

$$C = \frac{Q}{V} = \frac{1.66 \times 10^{-8}\,\text{C}}{96.4\,\text{V}} = 172\,\text{pF}.$$

Note that if the dielectric filled the space between the plates, the answers to (f) and (g) would be 42.9 V and 387 pF, respectively.

Summary

A **capacitor** is a device used to store charge and consists of two separated conductors. The two conductors generally carry equal and opposite charges, Q, and the ratio of this charge to the potential difference V between the conductors is called the **capacitance**, C; so

$$Q = CV.$$

The capacitance of a parallel-plate capacitor is proportional to the area A of each plate and inversely proportional to their separation d:

$$C = \epsilon_0 \frac{A}{d}.$$

The space between the conductors contains a nonconducting material such as air, paper, or plastic. The latter materials are referred to as **dielectrics**, and the capacitance is proportional to a property of dielectrics called the **dielectric constant**, K (nearly equal to 1 for air). For a parallel-plate capacitor

$$C = K\epsilon_0 \frac{A}{d} = \epsilon \frac{A}{d}$$

where $\epsilon = K\epsilon_0$ is called the **permittivity** of the dielectric material.

When capacitors are connected in **parallel**, the equivalent capacitance is the sum of the individual capacitances:

$$C_{eq} = C_1 + C_2 + \cdots.$$

When capacitors are connected in **series**, the reciprocal of the equivalent capacitance equals the sum of the reciprocals of the individual capacitances:

$$\frac{1}{C_{eq}} = \frac{1}{C_1} + \frac{1}{C_2} + \cdots.$$

A charged capacitor stores an amount of electric energy given by

$$U = \tfrac{1}{2}QV = \tfrac{1}{2}CV^2 = \tfrac{1}{2}\frac{Q^2}{C}.$$

This energy can be thought of as stored in the electric field between the plates. In any electric field $\mathbf{E}$ in free space the **energy density** u (energy per unit volume) is

$$u = \tfrac{1}{2}\epsilon_0 E^2.$$

If a dielectric is present, the energy density is

$$u = \tfrac{1}{2}K\epsilon_0 E^2 = \tfrac{1}{2}\epsilon E^2.$$

Questions

1. Suppose two nearby conductors carry the same negative charge. Can there be a potential difference between them? If so, can the definition of capacitance, $C = Q/V$, be used here?

2. Suppose the separation of plates d in a parallel-plate capacitor is not very small compared to the dimensions of the plates. Would you expect Eq. 24–2 to give an overestimate or underestimate of the true capacitance? Explain.

3. Suppose one of the plates of a parallel-plate capacitor was moved so that the area of overlap was reduced by half, but they are still parallel. How would this affect the capacitance?

4. Explain how the relation for the capacitance of a cylindrical capacitor, Example 24–2, makes sense intuitively. Use arguments such as those just after Eq. 24–2.

5. Describe a simple method of measuring ϵ_0 using a capacitor.

6. When a battery is connected to a capacitor, why do the two plates acquire charges of the same magnitude? Will this be true if the two conductors are different sizes or shapes?

7. A large copper sheet of thickness l is placed between the parallel plates of a capacitor, but does not touch the plates. How will this affect the capacitance?

8. Suppose three identical capacitors are connected to a battery. Will they store more energy if connected in series or in parallel?

9. The parallel plates of an isolated capacitor carry opposite charges, Q. If the separation of the plates is increased, is a force required? Is the potential difference changed? What happens to the work done in the pulling process?

10. How does the energy in a capacitor change if (a) the potential difference is doubled, (b) the charge on each plate is doubled, and (c) the separation of the plates is doubled, as the capacitor remains connected to a battery?

11. For dielectrics consisting of polar molecules, how would you expect the dielectric constant to change with temperature?

12. An isolated charged capacitor has horizontal plates. If a thin dielectric is inserted a short way between the plates, Fig. 24–16, how will it move when it is then released?

Dielectric

FIGURE 24–16
Question 12.

13. Suppose a battery remains connected to the capacitor in Question 12. What then will happen when the dielectric is released?

14. A dielectric is pulled from between the plates of a capacitor which remains connected to a battery. What changes occur to the capacitance, charge on the plates, potential difference, energy stored, and electric field?

15. How does the energy stored in a capacitor change when a dielectric is inserted if (a) the capacitor is isolated so Q doesn't change; (b) the capacitor remains connected to a battery so V doesn't change?

16. We have seen that the capacitance C depends on the size, shape, and position of the two conductors, as well as on the dielectric constant K. What then did we mean when we said that C is a constant in Eq. 24–1?

17. What value might we assign to the dielectric constant for a good conductor? Explain.

18. *Dissolving Power of Water.* The very high dielectric constant of water, $K = 80$ (Table 24–1), has a profound effect on materials in that it allows many of them to be dissolved in water. For example, ordinary table salt, NaCl (sodium chloride), whose crystal structure (Fig. 24–17a) is held together by the attractive forces between the ions Na^+ and Cl^-, is easily dissolved when placed in water. Explain why we would expect that the electric field produced by each ion would be reduced by a factor equal to the dielectric constant; that is, discuss the extension of Eq. 24–10 to the field of a point charge in a dielectric, and thus explain (using this simple model) how salt is dissolved (see Fig. 24–17b).

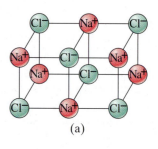

(a)

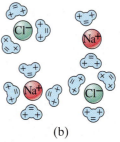

(b)

FIGURE 24–17 (a) Sodium chloride crystal; (b) sodium chloride dissolving in water. Question 18.

Problems

Section 24–1

1. (I) The two plates of a capacitor hold $+2500\,\mu C$ and $-2500\,\mu C$ of charge, respectively, when the potential difference is 950 V. What is the capacitance?

2. (I) A 12,000-pF capacitor holds $28.0 \times 10^{-8}\,C$ of charge. What is the voltage across the capacitor?

3. (I) The potential difference between two parallel wires in air is 12.0 V. They carry equal and opposite charge of magnitude 75 pC. What is the capacitance of the two wires?

4. (I) How much charge flows from a 12-V battery when it is connected to a 15.6-μF capacitor?

5. (I) The charge on a capacitor increases by $16\,\mu C$ when the voltage across it increases from 28 V to 48 V. What is the capacitance of the capacitor?

6. (II) A capacitor C_1 carries a charge Q_0. It is then connected directly to a second, initially uncharged, capacitor C_2. What charge will each carry now? What will be the potential difference across each?

7. (II) It takes 25 J of energy to move a 0.20-mC charge from one plate of a 16-μF capacitor to the other. How much charge is on each plate?

8. (II) A 2.40-μF capacitor is charged to 880 V and a 4.00-μF capacitor is charged to 560 V. (*a*) These capacitors are then disconnected from their batteries, and the positive plates are now connected to each other and the negative plates are connected to each other. What will be the potential difference across each capacitor and the charge on each? (*b*) What is the voltage and charge for each capacitor if plates of opposite sign are connected?

Section 24–2

9. (I) A 0.40-μF capacitor is desired. What area must the plates have if they are to be separated by a 4.0-mm air gap?

10. (I) What is the capacitance per unit length (F/m) of a coaxial cable whose inner conductor has a 1.0-mm diameter and the outer cylindrical sheath has a 5.0-mm diameter? Assume the space between is filled with air.

11. (I) Determine the capacitance of the Earth, assuming it to be a spherical conductor.

12. (II) Use Gauss's law to show that $\mathbf{E} = 0$ inside the inner conductor of a cylindrical capacitor (see Fig. 24–5 and Example 24–2) as well as outside the outer cylinder.

13. (II) Dry air will break down if the electric field exceeds about 3.0×10^6 V/m. What amount of charge can be placed on a capacitor if the area of each plate is 8.5 cm^2?

14. (II) An electric field of 2.80×10^5 V/m is desired between two parallel plates each of area 21.0 cm^2 and separated by 0.250 cm of air. What charge must be on each plate?

15. (II) In the limit of a very small separation of the two cylinders of a cylindrical capacitor $(R_a - R_b \ll R_a$ in Fig. 24–5) show that the relation derived in Example 24–2 reduces to that of a parallel-plate capacitor (Eq. 24–2).

16. (II) Suppose a capacitor carries a charge of $\pm 4.2\,\mu C$, and an electric field of 2.0 kV/mm is desired between the plates which are separated by 4.0 mm of air. What must each plate's area be?

17. (II) How strong is the electric field between the plates of a 0.80-μF air-gap capacitor if they are 2.0 mm apart and each has a charge of $72\,\mu C$?

18. (II) Show that for a spherical capacitor (Example 24–3), if the spacing between shells is very small $(r_a - r_b \ll r_a)$, the formula reduces to that for a parallel-plate capacitor.

19. (II) A large metal sheet of thickness l is placed between, and parallel to, the plates of the parallel-plate capacitor of Fig. 24–3. It does not touch the plates, and extends beyond their edges. (a) What is now the net capacitance in terms of A, d, and l? (b) If $l = \frac{2}{3}d$, by what factor does the capacitance change when the sheet is inserted?

Section 24–3

20. (I) Six 1.8-μF capacitors are connected in parallel. What is the equivalent capacitance? What is their equivalent capacitance if connected in series?

21. (I) The capacitance of a portion of a circuit is to be reduced from 3600 pF to 1600 pF. What capacitance can be added to the circuit to produce this effect without removing existing circuit elements? Must any existing connections be broken to accomplish this?

22. (II) Suppose three parallel-plate capacitors, whose plates have areas A_1, A_2, and A_3 and separations d_1, d_2, and d_3, are connected in parallel. Show, using only Eq. 24–2, that Eq. 24–3 is valid.

23. (II) (a) Determine the equivalent capacitance of the circuit shown in Fig. 24–18. (b) If $C_1 = C_2 = 2C_3 = 14.0\ \mu$F, how much charge is stored on each capacitor when $V = 25.0$ V?

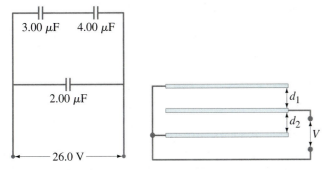

FIGURE 24–18 Problems 23, 24, and 43.

24. (II) In Fig. 24–18, suppose $C_1 = C_2 = C_3 = 16.0\ \mu$F. If the charge on C_2 is $Q_2 = 24.0\ \mu$C, determine the charge on each of the other capacitors, the voltage across each capacitor, and the voltage V_{ab} across the entire combination.

25. (II) A 3.00-μF and a 4.00-μF capacitor are connected in series and this combination is connected in parallel with a 2.00-μF capacitor (see Fig. 24–19). (a) What is the net capacitance? (b) If 26.0 V is applied across the whole network, calculate the voltage across each capacitor.

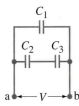

FIGURE 24–19 Problem 25.   **FIGURE 24–20** Problem 26.

26. (II) Three conducting plates, each of area A, are connected as shown in Fig. 24–20. (a) Are the two capacitors thus formed connected in series or in parallel? (b) Determine C as a function of d_1, d_2, and A. Assume $d_1 + d_2$ is much less than the dimensions of the plates. (c) The middle plate can be moved (changing the values of d_1 and d_2), so as to vary the capacitance. What are the minimum and maximum values of the net capacitance?

27. (II) Consider three capacitors, of capacitance 3000 pF, 5000 pF, and 0.010 μF. What are the maximum and minimum capacitances that you can form from these? How do you make the connection in each case?

28. (II) A 0.20-μF and a 0.30-μF capacitor are connected in series to a 9.0-V battery. Calculate (a) the potential difference across each capacitor and (b) the charge on each. (c) Repeat parts (a) and (b) assuming the two capacitors are in parallel.

29. (II) In Fig. 24–21, suppose $C_1 = C_2 = C_3 = C_4 = C$. (a) Determine the equivalent capacitance between points a and b. (b) Determine the charge on each capacitor and the potential difference across each if $V_{ba} = V$.

30. (II) Suppose in Fig. 24–21 that $C_1 = C_2 = C_3 = 16.0\ \mu$F and $C_4 = 36.0\ \mu$F. If the charge on C_2 is $Q_2 = 12.4\ \mu$C, determine the charge on each of the other capacitors, the voltage across each capacitor, and the voltage V_{ab} across the entire combination.

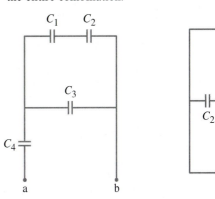

FIGURE 24–21 Problems 29, 30, and 44.   **FIGURE 24–22** Problem 31.

31. (II) The switch S in Fig. 24–22 is connected downward so that capacitor C_2 becomes fully charged by the battery of voltage V_0. If the switch is then connected upward, determine the charge on each capacitor after the switching.

32. (II) (a) Determine the equivalent capacitance between points a and b for the combination of capacitors shown in Fig. 24–23. (b) Determine the charge on each capacitor and the voltage across each if $V_{ab} = V$.

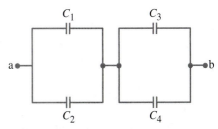

FIGURE 24–23 Problems 32 and 33.

33. (II) Suppose in Problem 32, Fig. 24–23, that $C_1 = C_3 = 8.0\ \mu$F, $C_2 = C_4 = 16\ \mu$F, and $Q_3 = 30\ \mu$C. Determine (a) the charge on each of the other capacitors, (b) the voltage across each capacitor, and (c) the voltage V_{ba} across the combination.

34. (II) Two capacitors connected in parallel produce an equivalent capacitance of 35.0 μF but when connected in series the equivalent capacitance is only 4.0 μF. What is the individual capacitance of each capacitor?

Problems   **629**

35. (II) In the *capacitance bridge* shown in Fig. 24–24, a voltage V_0 is applied and the variable capacitor C_1 is adjusted until there is zero voltage between points a and b as measured on the voltmeter (•–Ⓥ–•). Determine the unknown capacitance C_x if $C_1 = 8.9\,\mu F$ and the fixed capacitors have $C_2 = 18.0\,\mu F$ and $C_3 = 6.0\,\mu F$.

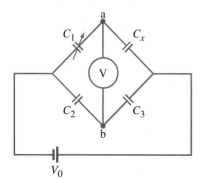

FIGURE 24–24
Problem 35.

36. (II) Two capacitors, $C_1 = 3200\,pF$ and $C_2 = 2200\,pF$, are connected in series to a 12.0-V battery. The capacitors are later disconnected from the battery and connected directly to each other, positive plate to positive plate, and negative plate to negative plate. What then will be the charge on each capacitor?

37. (III) Suppose one plate of a parallel-plate capacitor were tilted so it made a small angle θ with the other plate, as shown in Fig. 24–25. Determine a formula for C in terms of A, d, and θ, where A is the area of each plate and θ is small. Assume the plates are square. [*Hint*: Imagine the capacitor as many infinitesimal capacitors in parallel.]

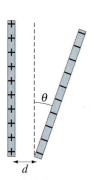

FIGURE 24–25
Problem 37.

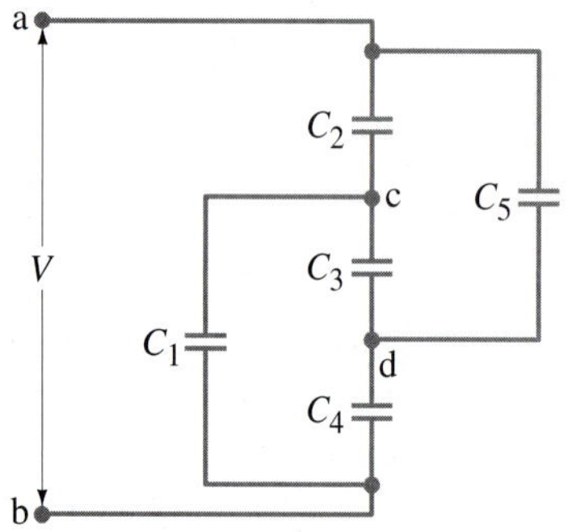

FIGURE 24–26 Problem 38.

38. (III) A voltage V is applied to the capacitor network shown in Fig. 24–26. (*a*) What is the equivalent capacitance? [*Hint*: Assume a potential difference V_{ab} exists across the network as shown; write potential differences for various pathways through the network from a to b in terms of the charges on the capacitors and the capacitances.] (*b*) Determine the equivalent capacitance if $C_2 = C_4 = 8.0\,\mu F$ and $C_1 = C_3 = C_5 = 6.0\,\mu F$.

Section 24–4

39. (I) 1200 V is applied to a 2800-pF capacitor. How much electric energy is stored?

40. (I) There is an electric field near the Earth's surface whose intensity is about 150 V/m. How much energy is stored per cubic meter in this field?

41. (I) How much energy is stored by the electric field between two square plates, 8.0 cm on a side, separated by a 1.5 mm air gap? The charges on the plates are equal and opposite and of magnitude 420 μC.

42. (II) A parallel-plate capacitor has fixed charges $+Q$ and $-Q$. The separation of the plates is then doubled. (*a*) By what factor does the energy stored in the electric field change? (*b*) How much work must be done if the separation of the plates is doubled from d to $2d$? The area of each plate is A.

43. (II) In Fig. 24–18, let $V = 10.0\,V$ and $C_1 = C_2 = C_3 = 2200\,pF$. How much energy is stored in the capacitor network?

44. (II) What is the total energy stored in the network of capacitors in Fig. 24–21, Problem 29?

45. (II) How much energy must a 12-V battery expend to fully charge a 0.15-μF and a 0.20-μF capacitor when they are placed (*a*) in parallel, (*b*) in series? (*c*) How much charge flowed from the battery in each case?

46. (II) (*a*) Suppose the outer radius R_a of a cylindrical capacitor was doubled, but the charge was kept constant. By what factor would the stored energy change? Where would the energy come from? (*b*) Repeat, assuming the voltage remains constant.

47. (II) A 3.0-μF capacitor is charged by a 12-V battery. It is disconnected from the battery and then connected to an uncharged 5.0-μF capacitor. Determine the total stored energy (*a*) before the two capacitors are connected and (*b*) after they are connected. (*c*) What is the change in energy?

48. (II) How much work would be required to remove the metal sheet from between the plates of the capacitor in Problem 19, assuming: (*a*) the battery remains connected so the voltage remains constant; (*b*) the battery is disconnected so the charge remains constant?

49. (II) (*a*) Show that each plate of a parallel-plate capacitor exerts a force

$$ F = \frac{1}{2} \frac{Q^2}{\epsilon_0 A} $$

on the other, by calculating dW/dx where dW is the work needed to increase the separation by dx. (*b*) Why does using $F = QE$, with E being the electric field between the plates, give the wrong answer?

50. (II) Show that the electrostatic energy stored in the electric field outside an isolated spherical conductor of radius R carrying a net charge Q is

$$U = \frac{1}{8\pi\epsilon_0} \frac{Q^2}{R}.$$

Do this in three ways: (a) Use Eq. 24–6 for the energy density in an electric field [Hint: Consider spherical shells of thickness dr]; (b) use Eq. 24–5 together with the capacitance of an isolated sphere (Section 24–2); (c) by calculating the work needed to bring all the charge Q up from infinity in infinitesimal bits dq.

Section 24–5

51. (I) What is the capacitance of a pair of circular plates with a radius of 5.0 cm separated by 3.2 mm of mica?

52. (I) What is the capacitance of two square parallel plates 5.5 cm on a side that are separated by 1.8 mm of paraffin?

53. (II) A 3500-pF air-gap capacitor is connected to a 22-V battery. If a piece of mica is placed between the plates, how much charge will flow from the battery?

54. (II) Suppose the capacitor in Example 24–8 remains connected to the battery as the dielectric is removed. What will be the work required to remove the dielectric in this case?

55. (II) How much energy would be stored in the capacitor of Problem 41 if a mica dielectric is placed between the plates? Assume the mica is 1.5 mm thick (and therefore fills the space between the plates).

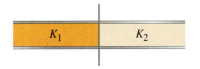

FIGURE 24–27
Problem 56.

56. (II) Two different dielectrics each fill half the space between the plates of a parallel-plate capacitor as shown in Fig. 24–27. Determine a formula for the capacitance in terms of K_1, K_2, the area A of the plates, and the separation d. [Hint: Can you consider this capacitor as two capacitors in series or in parallel?]

57. (II) Two different dielectrics fill the space between the plates of a parallel-plate capacitor as shown in Fig. 24–28. Determine a formula for the capacitance in terms of K_1, K_2, the area A, of the plates, and the separation $d_1 = d_2 = d/2$. [Hint: Can you consider this capacitor as two capacitors in series or in parallel?]

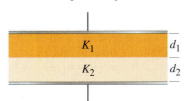

FIGURE 24–28
Problems 57 and 58.

58. (II) Repeat Problem 57 (Fig. 24–28) but assume the separation $d_1 \neq d_2$.

59. (II) Two identical capacitors are connected in parallel and each acquires a charge Q_0 when connected to a source of voltage V_0. The voltage source is disconnected and then a dielectric ($K = 4.0$) is inserted to fill the space between the plates of one of the capacitors. Determine (a) the charge now on each capacitor, and (b) the voltage now across each capacitor.

60. (III) A slab of width d and dielectric constant K is inserted a distance x between the square parallel plates (of side l) of a capacitor as shown in Fig. 24–29. Determine, as a function of x, (a) the capacitance, (b) the energy stored if the potential difference is V_0, and (c) the magnitude and direction of the force exerted on the slab (assume V_0 is constant).

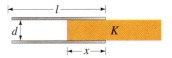

FIGURE 24–29 Problem 60.

* Section 24–6

* 61. (II) Repeat Example 24–9 assuming the battery remains connected when the dielectric is inserted. Also, what is the free charge on the plates after the dielectric is added (let this be part (h) of the problem)?

* 62. (II) Show that the capacitor in Example 24–9 with dielectric inserted can be considered as equivalent to three capacitors in series, and using this assumption show that the same value for the capacitance is obtained as was obtained in part (g) of the Example.

* 63. (II) In Example 24–9 what percent of the stored energy is stored in the electric field in the dielectric?

* 64. (II) Using Example 24–9 as a model, derive a formula for the capacitance of a parallel-plate capacitor whose plates have area A, separation d, with a dielectric of thickness $l (l < d)$ with dielectric constant K placed between the plates.

* 65. (III) The capacitor shown in Fig. 24–30 is connected to a 90.0-V battery. Calculate (and sketch) the electric field everywhere between the capacitor plates. Find both the free charge on the capacitor plate and the induced charge on the faces of the glass dielectric plate.

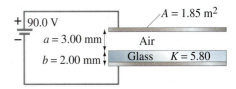

FIGURE 24–30 Problem 65.

▮ General Problems

66. An electric circuit was accidentally constructed using a 5.0-μF capacitor instead of the required 16-μF value. What can a technician add to correct this circuit?

67. A *cardiac defibrillator* is used to shock a heart that is beating erratically. A capacitor in this device is charged to 6000 V and stores 200 J of energy. What is its capacitance?

68. A homemade capacitor is assembled by placing two 9-inch pie pans 10 cm apart and connecting them to the opposite terminals of a 9-V battery. Estimate (*a*) the capacitance, (*b*) the charge on each plate, (*c*) the electric field halfway between the plates, (*d*) the work done by the battery to charge the plates. (*e*) Which of the above values change if a dielectric is inserted?

69. How does the energy stored in a capacitor change if (*a*) the potential difference is doubled, (*b*) the charge on each plate is doubled, and (*c*) the separation of the plates is doubled, as the capacitor remains connected to a battery?

70. A huge 7.0-F capacitor has enough stored energy to heat 2.5 kg of water from 20°C to 95°C. What is the potential difference across the plates?

71. An uncharged capacitor is connected to a 24.0-V battery until it is fully charged, after which it is disconnected from the battery. A slab of paraffin is then inserted between the plates. What will now be the voltage between the plates?

72. It takes 18.5 J of energy to move a 13.0-mC charge from one plate of a 12.0-μF capacitor to the other. How much charge is on each plate?

73. A coaxial cable, Fig. 24–31, consists of an inner cylindrical conducting wire of radius R_b surrounded by a dielectric insulator. Surrounding the dielectric insulator is an outer conducting sheath of radius R_a, which is usually "grounded." (*a*) Determine an expression for the capacitance per unit length of a cable whose insulator has dielectric constant K. (*b*) For a given cable, $R_b = 3.5$ mm and $R_a = 9.0$ mm. The dielectric constant of the dielectric insulator is $K = 2.6$. Suppose that there is a potential of 1.0 kV between the inner conducting wire and the outer conducting sheath. Find the capacitance per meter of the cable.

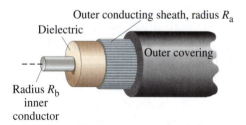

Outer conducting sheath, radius R_a
Dielectric
Outer covering
Radius R_b inner conductor

FIGURE 24–31 Problem 73.

74. The electric field between the plates of a paper-separated ($K = 3.75$) capacitor is 9.21×10^4 V/m. The plates are 1.95 mm apart and the charge on each plate is 0.475 μC. Determine the capacitance of this capacitor and the area of each plate.

75. A parallel-plate capacitor is isolated with a charge $\pm Q$ on each plate. If the separation of the plates is halved and a dielectric (constant K) is inserted in place of air, by what factor does the energy storage change? To what do you attribute the change in stored potential energy? How does the new value of the electric field between the plates compare with the original value?

76. A 3.5-μF capacitor is charged by a 12.4-V battery and then is disconnected from the battery. When this capacitor (C_1) is then connected to a second (initially uncharged) capacitor, C_2, the voltage on the first drops to 5.2 V. What is the value of C_2?

77. The power supply for a pulsed nitrogen laser has a 0.060-μF capacitor with a maximum voltage rating of 25 kilovolts. (*a*) Estimate how much energy could be stored in this capacitor. (*b*) If 10 percent of this stored electrical energy is converted to light energy in a pulse that is 10 microseconds long, what is the power of the laser pulse?

78. The variable capacitance of an old radio tuner consists of four plates connected together placed alternately between four other plates, also connected together (Fig. 24–32). Each plate is separated from its neighbor by 1.0 mm of air. One set of plates can move so that the area of overlap of each plate varies from 2.0 cm² to 9.0 cm². (*a*) Are these seven capacitors connected in series or in parallel? (*b*) Determine the range of capacitance values.

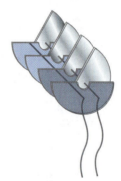

FIGURE 24–32
Problems 78 and 79.

79. A high-voltage supply can be constructed from a variable capacitor with interleaving plates which can be rotated as in Fig. 24–32. A version of this type of capacitor with more plates has a capacitance which can be varied from 10 pF to 1 pF. (*a*) Initially, this capacitor is charged by a 10,000-volt power supply when the capacitance is 10 pF. It is then disconnected from the power supply and the capacitance reduced to 1.0 pF by rotating the plates. What is the voltage across the capacitor now? (*b*) What is a major disadvantage of this as a high-voltage power supply?

80. A 150-pF capacitor is connected in series with an unknown capacitor, and as a series combination they are connected to a 25.0-V battery. If the 150-pF capacitor stores 125 pC of charge on its plates, what is the unknown capacitance?

81. A circuit contains a single 330-pF capacitor hooked across a battery. It is desired to store three times as much energy by adding a single capacitor to this one. How would you hook it up and what would its value be?

82. What are the values of effective capacitance which can be obtained by connecting four identical capacitors, each having a capacitance C?

83. In the circuit shown in Fig. 24–33, $C_1 = 1.0\,\mu\text{F}$, $C_2 = 2.0\,\mu\text{F}$, $C_3 = 3.0\,\mu\text{F}$, and a voltage $V_{ab} = 24\,\text{V}$ is applied across points a and b. After C_1 is fully charged the switch is thrown to the right. What is the final charge and potential difference on each capacitor?

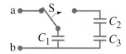

FIGURE 24–33 Problem 83.

84. The long cylindrical capacitor shown in Fig. 24–34 consists of 4 concentric cylinders, with respective radii R_a, R_b, R_c, and R_d. The cylinders b and c are joined by metal strips, as indicated. Determine the capacitance per unit length of this arrangement. (Assume equal and opposite charges are placed on the innermost and outermost cylinders.)

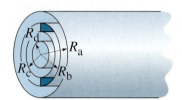

FIGURE 24–34 Problem 84.

85. A parallel plate capacitor has plate area A, plate separation x, and has a charge Q stored on its plates (Fig. 24–35). Find the amount of work required to double the plate separation to $2x$, assuming the charge remains constant at Q. Show that your answer is consistent with the change in energy stored by the capacitor.

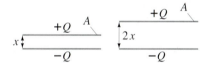

FIGURE 24–35 Problem 85.

86. Consider the use of capacitors as memory cells. A charged capacitor would represent a one and an uncharged capacitor a zero. Suppose these capacitors were fabricated on a silicon chip and had a capacitance of 30 femto-farads each $\left(1\,\text{fF} = 10^{-15}\,\text{F}.\right)$ The dielectric filling the space between the parallel plates has dielectric constant $K = 1.00 \times 10^4$ and a dielectric strength of $5.0 \times 10^7\,\text{V/m}$. (a) If the operating voltage is 3.0 volts, how many electrons would be stored on one of these capacitors when charged? (b) If no safety factor is allowed, how thin a dielectric layer could we use for operation at 3.0 volts? (c) Using the layer thickness from your answer to part (b), what would be the area of the capacitor plates?

87. A parallel plate capacitor with plate area $A = 2.5\,\text{m}^2$ and plate separation $d = 3.0\,\text{mm}$ is connected to a 45 V battery (Fig. 24–36). (a) Determine the charge on the capacitor, the electric field, the capacitance, and the energy stored in the capacitor. (b) With the capacitor still connected to the battery, a slab of plastic with dielectric strength $K = 3.6$ is placed between the plates of the capacitor, so that the gap is completely filled with the dielectric. What are the new values of charge, electric field, capacitance, and the energy U stored in the capacitor?

FIGURE 24–36 Problem 87.

88. A smooth conducting sphere of radius r_0 carries a charge Q. Half of the energy stored in its electric field is contained in a volume of what radius?

89. Paper has a dielectric constant of $K = 3.7$ and a dielectric strength of $15 \times 10^6\,\text{V/m}$. Suppose that a typical sheet of paper has a thickness of 0.030 mm. You make a "home made" capacitor by placing a sheet of 8.5×11 inch paper between two aluminum foil sheets (Fig. 24–37). The thickness of the aluminum foil is 0.040 mm. (a) What is the capacitance C_0 of your device? (b) About how much charge could you store on your capacitor before it would break down? (c) Show in a sketch how you could overlay sheets of paper and aluminum for a parallel combination. If you made 100 such capacitors, and connected the ends of the sheets in parallel, so that you have a single large capacitor of capacitance $100\,C_0$, how thick would your new large capacitor be? (d) What is the maximum voltage you can apply to this $100\,C_0$ capacitor without breakdown?

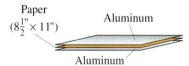

FIGURE 24–37 Problem 89.

90. In lightning storms, the potential difference between the Earth and the bottom of the thunderclouds can be as high as 35,000,000 V. The bottoms of the thunderclouds are typically 1500 m above the Earth, and can have an area of $110\,\text{km}^2$. For the purpose of this problem, model the Earth-cloud system as a huge capacitor and calculate (a) the capacitance of the Earth-cloud system, (b) the charge stored in the "capacitor," and (c) the energy stored in the "capacitor."

The glow of the thin wire filament of a light bulb is caused by the electric current passing through it. Electric energy is transformed to thermal energy (via collisions between moving electrons and atoms of the wire), which causes the wire's temperature to become so high that it glows. Electric current and electric power in electric circuits are of basic importance in everyday life. We examine both dc and ac in this chapter, and include the microscopic analysis of electric current, as well as a look at electric hazards.

CHAPTER 25

Electric Currents and Resistance

FIGURE 25–1 Alessandro Volta. In this portrait, Volta exhibits his battery to Napoleon in 1801.

In the previous four chapters we have been studying static electricity: electric charges at rest. In this chapter we begin our study of charges in motion, and we call a flow of charge an electric current.

In everyday life we are familiar with electric currents flowing in wires and other conductors. Indeed, most practical electrical devices depend on electric current: current flowing through a light bulb, current in the heating element of a stove or electric heater, and of course currents in electronic devices. Electric currents can exist in conductors such as wires and also in other devices, such as the CRT of a television or computer monitor whose charged electrons flow through space (Section 23–9).

In electrostatic situations, we saw (Sections 21–9 and 22–3) that the electric field must be zero inside a conductor (if it weren't, the charges would move). But when charges are moving in a conductor, there can be an electric field in the conductor. Indeed, an electric field is needed to get charges into motion, and to keep them in motion in any normal conductor.

We first look at electric current from a macroscopic point of view: that is, current as measured in a laboratory. Later in the chapter we look at currents from a microscopic (theoretical) point of view as a flow of electrons in a wire.

We can control the flow of charge using electric fields and electric potential (voltage), concepts we have just been discussing. In order to have a current in a wire, a potential difference is needed, which can be provided by a battery.

Until the year 1800, the technical development of electricity consisted mainly of producing a static charge by friction. It all changed in 1800 when Alessandro Volta (1745–1827; Fig. 25–1) invented the electric battery, and with it produced the first steady flow of electric charge—that is, a steady electric current.

25–1 The Electric Battery

The events that led to the discovery of the battery are interesting. For not only was this an important discovery, but it also gave rise to a famous scientific debate.

In the 1780s, Luigi Galvani (1737–1798), professor at the University of Bologna, carried out a series of experiments on the contraction of a frog's leg muscle through electricity produced by static electricity. Galvani found that contraction of the muscle could also be produced when dissimilar metals were inserted into the frog. Galvani believed that the source of the electric charge was in the frog muscle or nerve itself, and that the metal merely transmitted the charge to the proper points. When he published his work in 1791, he termed this charge "animal electricity." Many wondered, including Galvani himself, if he had discovered the long-sought "life-force."

Volta, at the University of Pavia 200 km away, was skeptical of Galvani's results, and came to believe that the source of the electricity was not in the animal itself, but rather in the *contact between the dissimilar metals*. Volta realized that a moist conductor, such as a frog muscle or moisture at the contact point of two dissimilar metals, was necessary in the circuit if it was to be effective. He also saw that the contracting frog muscle was a sensitive instrument for detecting electric "tension" or "electromotive force" (his words for what we now call potential), in fact more sensitive than the best available electroscopes that he and others had developed.[†]

Volta's research found that certain combinations of metals produced a greater effect than others, and, using his measurements, he listed them in order of effectiveness. (This "electrochemical series" is still used by chemists today.) He also found that carbon could be used in place of one of the metals.

Volta then conceived his greatest contribution to science. Between a disc of zinc and one of silver, he placed a piece of cloth or paper soaked in salt solution or dilute acid and piled a "battery" of such couplings, one on top of another, as shown in Fig. 25–2. This "pile" or "battery" produced a much increased potential difference. Indeed, when strips of metal connected to the two ends of the pile were brought close, a spark was produced. Volta had designed and built the first electric battery.

A battery produces electricity by transforming chemical energy into electrical energy. Today a great variety of electric cells and batteries are available, from flashlight batteries to the storage battery of a car. The simplest batteries contain two plates or rods made of dissimilar metals (one can be carbon) called **electrodes**. The electrodes are immersed in a solution, such as a dilute acid, called the **electrolyte**. Such a device is properly called an **electric cell**, and several cells connected together is a **battery**, although today even a single cell is called a battery. The chemical reactions involved in most electric cells are quite complicated. Here we describe how one very simple cell works, emphasizing the physical aspects.

The cell shown in Fig. 25–3 uses dilute sulfuric acid as the electrolyte. One of the electrodes is made of carbon, the other of zinc. That part of each electrode outside the solution is called the **terminal**, and connections to wires and circuits are made here. The acid reacts with the zinc electrode and tends to dissolve it. Each zinc atom leaves two electrons behind and enters the solution as a positive ion. The zinc electrode thus acquires a negative charge. As the electrolyte becomes positively charged, electrons are pulled off the carbon electrode. Thus the carbon electrode becomes positively charged. Because there is an opposite charge on the two electrodes, there is a potential difference between the two terminals. In a cell

FIGURE 25–2 A voltaic battery: Taken from Volta's original publication.

Cells and batteries

FIGURE 25–3 Simple electric cell.

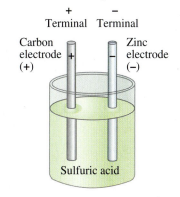

[†] Volta's most sensitive electroscope measured about 40 V per degree (angle of leaf separation). Nonetheless, he was able to estimate the potential differences produced by dissimilar metals in contact: for a silver-zinc contact he got about 0.7 V, remarkably close to today's value of 0.78 V.

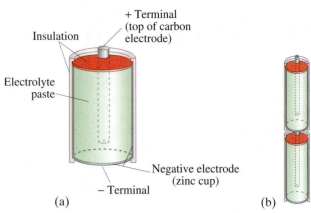

FIGURE 25–4 (a) Diagram of an ordinary dry cell (like a D-cell or AA). The cylindrical zinc cup is covered on the sides; its flat bottom is the negative terminal. (b) Two dry cells (AA type) connected in series. Note that the positive terminal of one cell pushes against the negative terminal of the other.

+ Terminal (top of carbon electrode)

Insulation

Electrolyte paste

Negative electrode (zinc cup)

− Terminal

(a)

(b)

whose terminals are not connected, only a small amount of the zinc is dissolved, for as the zinc electrode becomes increasingly negative, any new positive zinc ions produced are attracted back to the electrode. Thus, a particular potential difference or voltage is maintained between the two terminals. If charge is allowed to flow between the terminals, say, through a wire (or a lightbulb), then more zinc can be dissolved. After a time, one or the other electrode is used up and the cell becomes "dead."

The voltage that exists between the terminals of a battery depends on what the electrodes are made of and their relative ability to be dissolved or give up electrons.

When two or more cells are connected so that the positive terminal of one is connected to the negative terminal of the next, they are said to be connected in *series* and their voltages add up. Thus, the voltage between the ends of two 1.5-V flashlight batteries connected in series is 3.0 V, whereas the six 2-V cells of an automobile storage battery give 12 V. Figure 25–4 shows (a) a diagram of a common "dry cell" or "flashlight battery" used in portable radios, Walkmans, flashlights, etc., and (b) shows two of them in series.

25–2 | Electric Current

Electric circuit

Battery

When a continuous conducting path is connected between the terminals of a battery, we have an electric **circuit**, Fig. 25–5a. On any diagram of a circuit, as in Fig. 25–5b, we represent a battery by the symbol

$$\dashv\!\!\underset{+}{\mid}\!\underset{-}{\mid}\!\vdash.$$ [battery symbol]

The device powered by the battery could be a lightbulb (which is just a fine wire inside an evacuated glass bulb), a heater, a radio, or whatever. When such a circuit is formed, charge can flow through the wires of the circuit, from one terminal of the battery to the other. Any flow of charge such as this is called an **electric current**. Electric current can flow whenever there is a potential difference between the ends of a conductor—or, more simply, if there are opposite charges at the two ends of the conductor, or even in empty space.

FIGURE 25–5 (a) A simple electric circuit. (b) Schematic drawing of the same circuit.

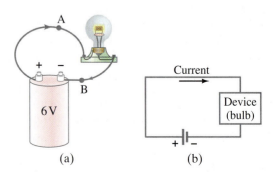

More precisely, the electric current in a wire is defined as the net amount of charge that passes through the wire's full cross section at any point per unit time. Thus, the average current $\bar{I}$ is defined as

$$\bar{I} = \frac{\Delta Q}{\Delta t},$$ (25–1a) *Electric current*

where ΔQ is the amount of charge that passes through the conductor at any location during the time interval Δt. The instantaneous current is defined by the differential limit

$$I = \frac{dQ}{dt}.$$ (25–1b)

Electric current is measured in coulombs per second; this is given a special name, the **ampere** (abbreviated amp or A), after the French physicist André Ampère (1775–1836). Thus, $1\,A = 1\,C/s$. Smaller units of current are often used, such as the milliampere $(1\,mA = 10^{-3}\,A)$ and microampere $(1\,\mu A = 10^{-6}\,A)$. *The ampere* $(1\,A = 1\,C/s)$

In any single circuit, with only a single path for current to follow such as in Fig. 25–5, a steady current at any instant is the same at one point (say point A) as at any other point (such as B). This follows from the conservation of electric charge (charge doesn't disappear).

EXAMPLE 25–1 **Current is flow of charge.** A steady current of 2.5 A flows in a wire for 4.0 min. (*a*) How much charge passed by any point in the circuit? (*b*) How many electrons would this be?

SOLUTION (*a*) Since the current was 2.5 A, or 2.5 C/s, then in 4.0 minutes (= 240 seconds) the total charge that flowed was, from Eq. 25–1,

$$\Delta Q = I\,\Delta t$$
$$= (2.5\,C/s)(240\,s) = 600\,C.$$

(*b*) The charge on one electron is $1.60 \times 10^{-19}\,C$, so 600 C would consist of

$$\frac{600\,C}{1.6 \times 10^{-19}\,C/electron} = 3.8 \times 10^{21}\text{ electrons.}$$

We saw in Chapter 21 that conductors contain many free electrons. Thus, if a continuous conducting wire is connected to the terminals of a battery, negatively charged electrons flow in the wire. When the wire is first connected, the potential difference between the terminals of the battery sets up an electric field inside the wire and parallel to it. Thus free electrons at one end of the wire are attracted into the positive terminal, and at the same time, electrons leave the negative terminal of the battery and enter the wire at the other end. There is a continuous flow of electrons through the wire that begins as soon as the wire is connected to *both* terminals. However, when the conventions of positive and negative charge were devised two centuries ago, it was assumed that positive charge flowed in a wire. For nearly all purposes, positive charge flowing in one direction is exactly equivalent to negative charge flowing in the opposite direction,[†] as shown in Fig. 25–6. Today, we still use the historical convention of positive current when discussing the direction of a current. So when we speak of the current in a circuit, we mean the direction positive charge would flow. This is sometimes referred to as **conventional current**. When we want to speak of the direction of electron flow, we will specifically state it is the electron current. In liquids and gases, both positive and negative charges (ions) can move.

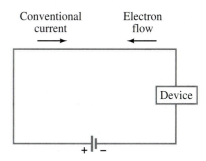

FIGURE 25–6 Conventional current from + to − is equivalent to a negative (electron) current flowing from − to +.

Conventional current

[†] An exception is discussed in Section 27–8.

SECTION 25–2 Electric Current **637**

25–3 Ohm's Law: Resistance and Resistors

To produce an electric current in a circuit, a difference in potential is required. One way of producing a potential difference is by a battery. It was Georg Simon Ohm (1787–1854) who established experimentally that the current in a metal wire is proportional to the potential difference V applied to its two ends:

$$I \propto V.$$

If, for example, we connect a wire to a 6-V battery, the current flow will be twice what it would be if the wire were connected to a 3-V battery. For simplicity, we are using V, rather than V_{ba}, to represent potential difference or voltage. It is also found that reversing the sign of the voltage does not affect the magnitude of the current.

Water analogy

It is helpful to compare an electric current to the flow of water in a river or a pipe acted on by gravity. If the pipe (or river) is nearly level, the flow rate is small. But if one end is somewhat higher than the other, the flow rate—or current—is greater. The greater the difference in height, the swifter the current. We saw in Chapter 23 that electric potential is analogous, in the gravitational case, to the height of a cliff. This applies in the present case to the height through which the fluid flows. Just as an increase in height can cause a greater flow of water, so a greater electric potential difference, or voltage, causes a greater electric current flow.

Exactly how much current flows in a wire depends not only on the voltage, but also on the resistance the wire offers to the flow of electrons. The walls of a pipe, or the banks of a river and rocks in the middle, offer resistance to the flow of current. Similarly, electrons are slowed down because of interactions with the atoms of the wire. The higher this resistance, the less the current for a given voltage V. We then define *resistance* so that the current is inversely proportional to the resistance: that is,

$$R = \frac{V}{I} \tag{25–2a}$$

where R is the **resistance** of a wire or other device, V is the potential difference across the device, and I is the current that flows through it. Eq. 25–2a is often written as

$$V = IR. \tag{25–2b}$$

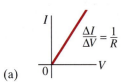

(a)

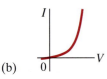

(b)

FIGURE 25–7 Graphs of current vs. voltage for (a) a metal conductor which obeys Ohm's law, and (b) for a nonohmic device, in this case a semiconductor diode.

As mentioned above, Ohm found experimentally that in metal conductors R is a constant independent of V, a result known as **Ohm's law**. Also, Eq. 25–2b, $V = IR$, is itself sometimes called Ohm's law, but only when referring to materials or devices for which R is a constant independent of V. But R is not a constant for many substances, nor for devices such as diodes, vacuum tubes, transistors, and so on. Thus "Ohm's law" is not a fundamental law, but rather a description of a certain class of materials (metal conductors). Materials or devices that do not follow Ohm's law (R = constant) are said to be *nonohmic*. See Fig. 25–7.

The ohm $(1\,\Omega = 1\,V/A)$

The unit for resistance is called the **ohm** and is abbreviated Ω (Greek capital omega). Because $R = V/I$, we see that $1.0\,\Omega$ is equivalent to $1.0\,V/A$.

FIGURE 25–8 Example 25–2.

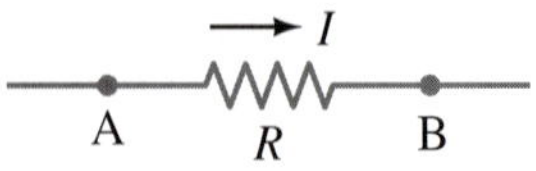

CONCEPTUAL EXAMPLE 25–2 Current and potential. Current I enters a resistor R as shown in Fig. 25–8. (a) Is the potential higher at point A or at point B? (b) Is the current greater at point A or at point B?

RESPONSE (a) Positive charge always flows from + to −, from high potential to low potential. Think again of the gravitational analogy: a mass will fall down from high gravitational potential to low. So for positive current I, point A is at a higher potential than point B. (b) Conservation of charge requires that whatever current flows into the resistor at point A, emerges at point B. Current does not get "used up" by a resistor, just as an object that falls through a gravitational potential difference does not gain or lose mass. So the current is the same at A and B.

EXAMPLE 25–3 Flashlight bulb resistance. A small flashlight bulb (Fig. 25–9) draws 300 mA from its 1.5-V battery. (*a*) What is the resistance of the bulb? (*b*) If the voltage dropped to 1.2 V, how would the current change?

SOLUTION (*a*) From Eq. 25–2,

$$R = \frac{V}{I} = \frac{1.5\,V}{0.30\,A} = 5.0\,\Omega.$$

(*b*) If the resistance didn't change, the current would be

$$I = \frac{V}{R} = \frac{1.2\,V}{5.0\,\Omega} = 0.24\,A,$$

or a drop of 60 mA. Actually, resistance does depend on temperature (Section 25–4), so this is only a rough approximation.

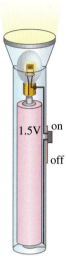

FIGURE 25–9 Flashlight (Example 25–3). Note how circuit is completed along the side strip.

All electric devices, from heaters to lightbulbs to stereo amplifiers, offer resistance to the flow of current. The filaments of lightbulbs and electric heaters are special types of wires whose resistance results in their becoming very hot. Generally, the connecting wires have very low resistance in comparison to the resistance of the wire filaments or coils. In many circuits, particularly in electronic devices, **resistors** are used to control the amount of current. Resistors have resistances from less than an ohm to millions of ohms (see Figs. 25–10 and 25–11). The main types are "wire-wound" resistors which consist of a coil of fine wire; "composition" resistors which are usually made of the semiconductor carbon; and thin metal films.

When we draw a diagram of a circuit, we indicate a resistance with the symbol

–∿∿– .   [resistor symbol]

Wires whose resistance is negligible, however, are shown simply as straight lines.

FIGURE 25–10 Photo of resistors (mostly).

Resistor Color Code			
Color	**Number**	**Multiplier**	**Tolerance (%)**
Black	0	1	
Brown	1	10^1	
Red	2	10^2	
Orange	3	10^3	
Yellow	4	10^4	
Green	5	10^5	
Blue	6	10^6	
Violet	7	10^7	
Gray	8	10^8	
White	9	10^9	
Gold		10^{-1}	5%
Silver		10^{-2}	10%
No color			20%

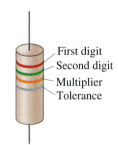

First digit
Second digit
Multiplier
Tolerance

FIGURE 25–11 The resistance value of a given resistor is written on the exterior, or may be given as a color code, as shown above and in the table: the first two colors represent the first two digits in the value of the resistance, the third color represents the power of ten that it must be multiplied by, and the fourth is the manufactured tolerance. For example, a resistor whose four colors are red, green, orange, and silver has a resistance of 25,000 Ω (25 kΩ), give or take 10 percent.

TABLE 25–1 Resistivity and Temperature Coefficients (at 20°C)

Material	Resistivity, ρ ($\Omega \cdot$ m)	Temperature Coefficient, α (C°)$^{-1}$
Conductors		
Silver	1.59×10^{-8}	0.0061
Copper	1.68×10^{-8}	0.0068
Gold	2.44×10^{-8}	0.0034
Aluminum	2.65×10^{-8}	0.00429
Tungsten	$5.6 \ \times 10^{-8}$	0.0045
Iron	9.71×10^{-8}	0.00651
Platinum	$10.6 \ \times 10^{-8}$	0.003927
Mercury	$98 \ \ \times 10^{-8}$	0.0009
Nichrome (alloy of Ni, Fe, Cr)	$100 \ \ \times 10^{-8}$	0.0004
Semiconductors[†]		
Carbon (graphite)	$(3-60) \times 10^{-5}$	−0.0005
Germanium	$(1-500) \times 10^{-3}$	−0.05
Silicon	0.1−60	−0.07
Insulators		
Glass	$10^9 - 10^{12}$	
Hard rubber	$10^{13} - 10^{15}$	

[†] Values depend strongly on presence of even slight amounts of impurities.

25–4 Resistivity

It is found experimentally that the resistance R of a metal wire is directly proportional to its length l and inversely proportional to its cross-sectional area A. That is,

$$R = \rho \frac{l}{A}, \qquad (25\text{–}3)$$

Resistivity

where ρ, the constant of proportionality, is called the **resistivity** and depends on the material used. Typical values of ρ, whose units are $\Omega \cdot$ m (see Eq. 25–3), are given for various materials in the middle column of Table 25–1. The values depend somewhat on purity, heat treatment, temperature, and other factors. Notice that silver has the lowest resistivity and is thus the best conductor (although it is expensive). Copper is close and much less expensive, so it is clear why most wires are made of copper. Aluminum, although it has a higher resistivity, is much less dense than copper; it is thus preferable to copper in some situations, such as transmission lines, because its resistance for the same weight is less than that for copper.

The reciprocal of the resistivity, called the **conductivity** σ, is

Conductivity

$$\sigma = \frac{1}{\rho}, \qquad (25\text{–}4)$$

and has units of $(\Omega \cdot$ m$)^{-1}$.

EXAMPLE 25–4 Speaker wires. Suppose you want to connect your stereo to remote speakers (Fig. 25–12). (*a*) If each wire must be 20 m long, what diameter copper wire should you use to keep the resistance less than 0.10 Ω per wire? (*b*) If the current to each speaker is 4.0 A, what is the voltage drop across each wire?

SOLUTION (a) We solve Eq. 25–3 for the area A and use Table 25–1:

$$A = \rho \frac{l}{R} = \frac{(1.68 \times 10^{-8}\,\Omega\cdot m)(20\,m)}{(0.10\,\Omega)} = 3.4 \times 10^{-6}\,m^2.$$

The cross-sectional area A of a circular wire is related to its diameter d by $A = \pi d^2/4$. The diameter must then be at least

$$d = \sqrt{\frac{4A}{\pi}} = 2.1 \times 10^{-3}\,m = 2.1\,mm.$$

(b) From $V = IR$ we have

$$V = IR = (4.0\,A)(0.10\,\Omega) = 0.40\,V.$$

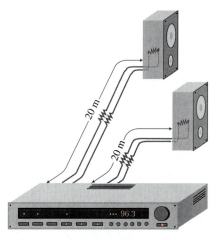

FIGURE 25–12 Example 25–4.

| CONCEPTUAL EXAMPLE 25–5 | **Stretching changes resistance.** A wire of resistance R is stretched uniformly until it is twice its original length. What happens to its resistance?

RESPONSE If the length l doubles then the cross-sectional area A halves, so that the volume $(V = Al)$ of the wire remains the same. From Eq. 25–3 we see that the resistance would increase by a factor of four $\left(2/\frac{1}{2} = 4\right)$.

Temperature Dependence of Resistivity

The resistivity of a material depends somewhat on temperature. In general, the resistance of metals increases with temperature. This is not surprising, for at higher temperatures, the atoms are moving more rapidly and are arranged in a less orderly fashion. So they might be expected to interfere more with the flow of electrons. If the temperature change is not too great, the resistivity of metals usually increases nearly linearly with temperature. That is,

$$\rho_T = \rho_0\left[1 + \alpha(T - T_0)\right] \qquad \textbf{(25–5)}$$

Effect of temperature

where ρ_0 is the resistivity at some reference temperature T_0 (such as 0°C or 20°C), ρ_T is the resistivity at a temperature T and α is the *temperature coefficient of resistivity*. Values for α are given in Table 25–1. Note that the temperature coefficient for semiconductors can be negative. Why? It seems that at higher temperatures, some of the electrons that are not normally free in a semiconductor become free and can contribute to the current. Thus, the resistance of a semiconductor can decrease with an increase in temperature, although this is not always the case.

| EXAMPLE 25–6 | **Resistance thermometer.** The variation in electrical resistance with temperature can be used to make precise temperature measurements. Platinum is commonly used since it is relatively free from corrosive effects and has a high melting point. Suppose at 20°C the resistance of a platinum resistance thermometer is 164.2 Ω. When placed in a particular solution, the resistance is 187.4 Ω. What is the temperature of this solution?

SOLUTION Since the resistance R is directly proportional to the resistivity ρ, we can combine Eq. 25–3 with Eq. 25–5:

$$R = R_0\left[1 + \alpha(T - T_0)\right].$$

Here $R_0 = \rho_0 L/A$ is the resistance of the wire at $T_0 = 20$°C. We solve this equation for T and find (see Table 25–1 for α)

$$T = T_0 + \frac{R - R_0}{\alpha R_0} = 20°C + \frac{187.4\,\Omega - 164.2\,\Omega}{(3.927 \times 10^{-3}(\text{C}°)^{-1})(164.2\,\Omega)} = 56.0°C.$$

FIGURE 25–13 A thermistor shown next to a millimeter ruler for scale.

More convenient for some applications is a *thermistor* (Fig. 25–13), which consists of a metal oxide or semiconductor whose resistance also varies in a repeatable way with temperature. They can be made quite small and respond very quickly to temperature changes. Resistance thermometers have another advantage in that they can be used at very high or low temperatures where gas or liquid thermometers would be useless.

The value of α in Eq. 25–5 itself can depend on temperature, so it is important to check the temperature range of validity of any value (say, in a handbook of physical data). If the temperature range is wide, Eq. 25–5 is not adequate and terms proportional to the square and cube of the temperature are needed, but they are generally very small except when $T - T_0$ is large.

25–5 | Electric Power

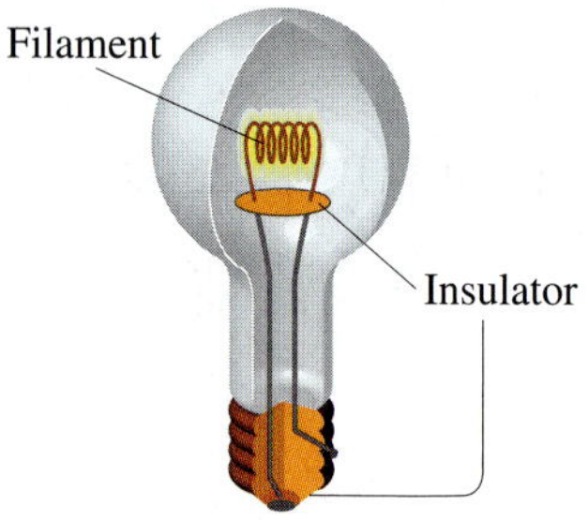

Electric energy is useful to us because it can be easily transformed into other forms of energy. Motors, whose operation we will examine in Chapter 27, transform electric energy into mechanical work.

In other devices such as electric heaters, stoves, toasters, and hair dryers, electric energy is transformed into thermal energy in a wire resistance known as a "heating element." And in an ordinary lightbulb, the tiny wire filament (Fig. 25–14) becomes so hot it glows; only a few percent of the energy is transformed into visible light, and the rest, over 90 percent, into thermal energy. Lightbulb filaments and heating elements in household appliances have resistances typically of a few ohms to a few hundred ohms.

Electric energy is transformed into thermal energy or light in such devices, and there are many collisions between the moving electrons and the atoms of the wire. In each collision, part of the electron's kinetic energy is transferred to the atom with which it collides. As a result, the kinetic energy of the wire's atoms increases and hence the temperature of the wire element increases. The increased thermal energy can be transferred as heat by conduction and convection to the air in a heater or to food in a pan, by radiation to bread in a toaster, or radiated as light.

To find the power transformed by an electric device, recall that the energy transformed when an infinitesimal charge dq moves through a potential difference V is $dU = dq\, V$ (Eq. 23–2). If dt is the time required for an amount of charge dq to move through the potential difference V, the power P, which is the rate energy is transformed, is

$$P = \frac{dU}{dt} = \frac{dq}{dt} V.$$

The charge that flows per second, dq/dt, is the electric current I. Thus we have

Electric power (general)

$$P = IV. \tag{25–6}$$

This general relation gives us the power transformed by any device, where I is the current passing through it and V is the potential difference across it. It also gives the power delivered by a source such as a battery. The SI unit of electric power is the same as for any kind of power, the **watt** (1 W = 1 J/s).

The rate of energy transformation in a resistance R can be written, using $V = IR$, in two other ways:

Electric power (in resistance R)

$$P = IV \tag{25–7a}$$
$$= I(IR) = I^2 R \tag{25–7b}$$
$$= \left(\frac{V}{R}\right)V = \frac{V^2}{R}. \tag{25–7c}$$

Equations 25–7b and c apply only to resistors, whereas Eq. 25–7a, $P = IV$, applies to any device.

FIGURE 25–14   Incandescent lightbulb.

Filament

Insulator

EXAMPLE 25–7 **Headlights.** Calculate the resistance of a 40-W automobile headlight designed for 12 V (Fig 25–15).

12 V

SOLUTION Since we are given $P = 40$ W and $V = 12$ V, we can use Eq. 25–7c and solve for R:

$$R = \frac{V^2}{P} = \frac{(12\ \text{V})^2}{(40\ \text{W})} = 3.6\ \Omega.$$

40-W Headlight

FIGURE 25–15 Example 25–7.

This is the resistance when the bulb is burning brightly at 40 W. When the bulb is cold, the resistance is much lower, as we saw in Eq. 25–5. Since the current is high when the resistance is low, lightbulbs burn out most often when first turned on.

It is energy, not power, that you pay for on your electric bill. Since power is the *rate* energy is transformed, the total energy used by any device is simply its power consumption multiplied by the time it is on. If the power is in watts and the time is in seconds, the energy will be in joules since $1\ \text{W} = 1\ \text{J/s}$. Electric companies usually specify the energy with a much larger unit, the **kilowatt-hour** (kWh). One kWh $= (1000\ \text{W})(3600\ \text{s}) = 3.60 \times 10^6$ J.

You pay for energy

Kilowatt-hour (unit of energy)

EXAMPLE 25–8 **Electric heater.** An electric heater draws a steady 15.0 A on a 120-V line. How much power does it require and how much does it cost per month (30 days) if it operates 3.0 h per day and the electric company charges 10.5 cents per kWh?

SOLUTION The power is

$$P = IV = (15.0\ \text{A})(120\ \text{V}) = 1800\ \text{W}$$

or 1.80 kW. To operate it for $(3.0\ \text{h/d})(30\ \text{d}) = 90$ h would cost $(1.80\ \text{kW})(90\ \text{h})(\$0.105) = \$17$.

EXAMPLE 25–9 **ESTIMATE** **Lightning bolt.** Lightning is a spectacular example of electric current in a natural phenomenon (Fig. 25–16). There is much variability to lightning bolts, but a typical event can transfer 10^9 J of energy across a potential difference of perhaps 5×10^7 V during a time interval of about 0.2 s. Use this information to estimate the total amount of charge transferred, the current, and the average power over the 0.2 s.

➡ **PHYSICS APPLIED**

Lightning

SOLUTION From Eq. 23–2, energy $= QV$, so

$$Q \approx \frac{10^9\ \text{J}}{5 \times 10^7\ \text{V}} = 20\ \text{C}.$$

The current over the 0.2 s is about

$$I = \frac{Q}{t} \approx \frac{20\ \text{C}}{0.2\ \text{sec}} = 100\ \text{A}.$$

Since most lightning bolts consist of several stages, it is possible that individual parts could carry currents much higher than this. The average power delivered is

$$\overline{P} = \frac{\text{energy}}{\text{time}} = \frac{10^9\ \text{J}}{0.2\ \text{sec}} = 5 \times 10^9\ \text{W} = 5\ \text{GW}.$$

We can also use Eq. 25–7a:

$$P = IV = (100\ \text{A})(5 \times 10^7\ \text{V}) = 5\ \text{GW}.$$

FIGURE 25–16 Example 25–9: a lightning bolt. See caption for photo at start of Chapter 23.

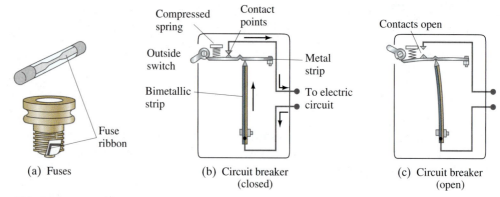

Compressed spring

Contact points

Contacts open

Outside switch

Metal strip

Bimetallic strip

To electric circuit

Fuse ribbon

(a) Fuses

(b) Circuit breaker (closed)

(c) Circuit breaker (open)

FIGURE 25–17 (a) Fuses. When the current exceeds a certain value, the metallic ribbon melts and the circuit opens. Then the fuse must be replaced. (b) A circuit breaker. Electric current passes to a circuit via a bimetallic strip. When the current is great enough (i.e., too great to be safe), the bimetallic strip is heated and bends so far to the left that the notch in the spring-loaded metal strip drops down over the end of the bimetallic strip; (c) the circuit then opens at the contact points (one is attached to the metal strip) and the outside switch is also flipped. As soon as the bimetallic strip cools down, it can be reset using the outside switch.

25–6 Power in Household Circuits

PHYSICS APPLIED

Safety—wires getting hot

Fuses, circuit breakers, and shorts

The electric wires that carry electricity to lights and other electric appliances have some resistance, although usually it is quite small. Nonetheless, if the current is large enough, the wires will heat up and produce thermal energy at a rate equal to I^2R, where R is the wire's resistance. One possible hazard is that the current-carrying wires in the wall of a building may become so hot as to start a fire. Thicker wires, of course, have less resistance (see Eq. 25–3) and thus can carry more current without becoming too hot. When a wire carries more current than is safe, it is said to be "overloaded." To prevent overloading, fuses or circuit breakers are installed in circuits. They are basically switches (Fig. 25–17) that open the circuit when the current exceeds some particular value. A 20-A fuse or circuit breaker, for example, opens when the current passing through it exceeds 20 A. If a circuit repeatedly burns out a fuse or opens a circuit breaker, there are two possibilities: there may be too many devices drawing current in that circuit; or there is a fault somewhere, such as a "short." A short, or "short circuit," means that two wires have crossed (perhaps because the insulation has worn down) so the path of the current is shortened. The resistance of the circuit is then very small, so the current will be very large. Short circuits, of course, should be remedied immediately.

Household circuits are designed with the various devices connected so that each receives the standard voltage (usually 120 V in the United States) from the electric company (Fig. 25–18). Circuits with the devices arranged as in Fig. 25–18 are called *parallel circuits*, as we will discuss more fully in the next chapter. When a fuse blows or circuit breaker opens, the total current being drawn on that circuit should be checked.

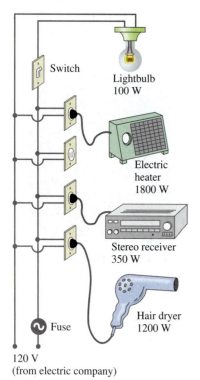

Switch

Lightbulb 100 W

Electric heater 1800 W

Stereo receiver 350 W

Hair dryer 1200 W

Fuse

120 V (from electric company)

FIGURE 25–18 Connection of household appliances.

EXAMPLE 25–10 Will a fuse blow? Determine the total current drawn by all the devices in the circuit of Fig. 25–18.

SOLUTION The circuit in Fig. 25–18 draws the following currents: the lightbulb draws $I = P/V = 100 \text{ W}/120 \text{ V} = 0.8 \text{ A}$; the heater draws $1800 \text{ W}/120 \text{ V} = 15.0 \text{ A}$; the stereo draws a maximum of $350 \text{ W}/120 \text{ V} = 2.9 \text{ A}$; and the hair dryer draws $1200 \text{ W}/120 \text{ V} = 10.0 \text{ A}$. The total current drawn, if all devices are used at the same time, is

$$0.8 \text{ A} + 15.0 \text{ A} + 2.9 \text{ A} + 10.0 \text{ A} = 28.7 \text{ A}.$$

If the circuit in Fig. 25–18 is designed for a 20-A fuse, the fuse should blow, and we hope it will, to prevent overloaded wires from getting hot enough to start a fire. Something will have to be turned off to get this circuit below 20 A. (Houses and apartments usually have several circuits, each with its own fuse or circuit breaker; try moving one of the devices to another circuit.) If the circuit is designed for a 30-A fuse, it shouldn't blow, so if it does, a short may be the problem. (The most likely place is in the cord of one of the devices.) Proper fuse size is selected according to the wire used to supply the current; a properly rated fuse should *never* be replaced by a higher-rated one. A fuse blowing or a circuit breaker opening is acting like a switch, making an "open circuit." By an open circuit, we mean that there is no longer a complete conducting path, so no current can flow; it is as if $R = \infty$.

25–7 | Alternating Current

When a battery is connected to a circuit, the current flows steadily in one direction. This is called a **direct current**, or **dc**. Electric generators at electric power plants, however, produce **alternating current**, or **ac**. (Sometimes capital letters are used, DC and AC.) An alternating current reverses direction many times per second and is commonly sinusoidal, as shown in Fig. 25–19. The electrons in a wire first move in one direction and then in the other. The current supplied to homes and businesses by electric companies is ac throughout virtually the entire world. We will discuss and analyze ac circuits in detail in Chapter 31. But because ac circuits are so common in real life, we will discuss some of their simpler aspects here.

DC and AC

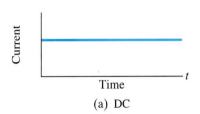

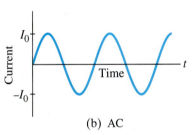

FIGURE 25–19 (a) Direct current. (b) Alternating current.

The voltage produced by an ac electric generator is sinusoidal, as we shall see later. The current it produces is thus sinusoidal (Fig. 25–19b). We can write the voltage as a function of time as

$$V = V_0 \sin 2\pi f t = V_0 \sin \omega t.$$

The potential V oscillates between $+V_0$ and $-V_0$. V_0 is referred to as the **peak voltage**. The frequency f is the number of complete oscillations made per second, and $\omega = 2\pi f$. In most areas of the United States and Canada, f is 60 Hz (the unit "hertz," as we saw in Chapter 14, means cycles per second). In many countries, 50 Hz is used.

From Eq. 25–2, $V = IR$, if a voltage V exists across a resistance R, then the current I is

$$I = \frac{V}{R} = \frac{V_0}{R} \sin \omega t = I_0 \sin \omega t. \tag{25–8}$$

The quantity $I_0 = V_0/R$ is the **peak current**. The current is considered positive when the electrons flow in one direction and negative when they flow in the opposite direction. It is clear from Fig. 25–19b that an alternating current is as often positive as it is negative. Thus, the average current is zero. This does not mean, however, that no power is needed or that no heat is produced in a resistor. Electrons do move back and forth, and do produce heat. Indeed, the power delivered to a resistance R at any instant is

$$P = I^2 R = I_0^2 R \sin^2 \omega t.$$

Because the current is squared, we see that the power is always positive, Fig. 25–20. The quantity $\sin^2 \omega t$ varies between 0 and 1; and it is not too difficult to show[†] that its average value is $\frac{1}{2}$ as can be seen graphically in the figure. Thus, the *average power* developed, $\overline{P}$, is

$$\overline{P} = \frac{1}{2} I_0^2 R.$$

FIGURE 25–20 Power delivered to a resistor in an ac circuit.

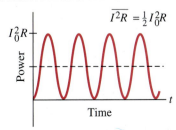

[†] A graph of $\cos^2 \omega t$ versus t is identical to that for $\sin^2 \omega t$ in Fig. 25–20 except that the points are shifted $\left(\text{by } \frac{1}{4} \text{ cycle} \right)$ on the time axis. Hence the average value of $\sin^2$ and $\cos^2$, averaged over one or more full cycles, will be the same: $\overline{\sin^2 \omega t} = \overline{\cos^2 \omega t}$. From the trigonometric identity $\sin^2 \theta + \cos^2 \theta = 1$, we can write

$$\overline{(\sin^2 \omega t)} + \overline{(\cos^2 \omega t)} = 2\overline{(\sin^2 \omega t)} = 1.$$

Hence the average value of $\sin^2 \omega t$ is $\frac{1}{2}$.

Since power can also be written $P = V^2/R = (V_0^2/R)\sin^2 \omega t$, we also have that the average power is

$$\overline{P} = \frac{1}{2}\frac{V_0^2}{R}.$$

The average or mean value of the *square* of the current or voltage is thus what is important for calculating average power: $\overline{I^2} = \frac{1}{2}I_0^2$ and $\overline{V^2} = \frac{1}{2}V_0^2$. The square root of each of these is the **rms** (root-mean-square) value of the current or voltage:

rms current
$$I_{rms} = \sqrt{\overline{I^2}} = \frac{I_0}{\sqrt{2}} = 0.707 I_0, \qquad \textbf{(25–9a)}$$

rms voltage
$$V_{rms} = \sqrt{\overline{V^2}} = \frac{V_0}{\sqrt{2}} = 0.707 V_0. \qquad \textbf{(25–9b)}$$

The rms values of V and I are sometimes called the "effective values." They are useful because they can be substituted directly into the power formulas, Eqs. 25–7, to get the average power:

$$\overline{P} = \frac{1}{2} I_0^2 R = I_{rms}^2 R \qquad \textbf{(25–10a)}$$

$$\overline{P} = \frac{1}{2}\frac{V_0^2}{R} = \frac{V_{rms}^2}{R}. \qquad \textbf{(25–10b)}$$

Thus, a direct current whose values of I and V equal the rms values of I and V for an alternating current will produce the same power. Hence it is usually the rms value of current that is specified or measured. For example, in the United States and Canada, standard line voltage[†] is 120 V ac. The 120 V is V_{rms}; the peak voltage V_0 is

$$V_0 = \sqrt{2}\,V_{rms} = 170\text{ V}.$$

In most of Europe the rms voltage is 240 V, so the peak voltage is 340 V.

EXAMPLE 25–11 **Hair dryer.** (*a*) Calculate the resistance and the peak current in a 1000-W hair dryer (Fig. 25–21) connected to a 120-V line. (*b*) What happens if it is connected to a 240-V line in Britain?

SOLUTION (*a*) We can apply Eq. 25–7a using rms values. Then the rms current is

$$I_{rms} = \frac{\overline{P}}{V_{rms}} = \frac{1000\text{ W}}{120\text{ V}} = 8.33\text{ A}.$$

Thus $I_0 = \sqrt{2}\,I_{rms} = 11.8$ A. Then the resistance is

$$R = \frac{V_{rms}}{I_{rms}} = \frac{120\text{ V}}{8.33\text{ A}} = 14.4\ \Omega.$$

The resistance could equally well be calculated using peak values: $R = V_0/I_0 = 170\text{ V}/11.8\text{ A} = 14.4\ \Omega$.
(*b*) When connected to a 240-V line, more current would flow and the resistance would change with the increased temperature (Section 25–4). But let us make an estimate based on the same 14.4 Ω resistance. The average power delivered would be

$$\overline{P} = \frac{V_{rms}^2}{R} = \frac{(240\text{ V})^2}{(14.4\ \Omega)} = 4000\text{ W}.$$

This is four times the dryer's power rating and would undoubtedly melt the heating element or the wire coils of the motor. Be sure your hair dryer (or electric shaver) has a 120/240 V switch before traveling too far from home, or carry a transformer (Section 29–6) for travelers.

Motor
Fan
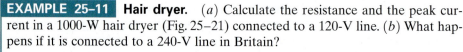
Heating coils
Switch
Cord

FIGURE 25–21 A hair dryer. Most of the current goes through the heating coils, a pure resistance; a small part goes to the motor to turn the fan. Example 25–11.

[†]The line voltage can vary, depending on the total load; the frequency of 60 Hz, however, remains extremely steady.

25–8 Microscopic View of Electric Current: Current Density and Drift Velocity

Up to now in this chapter we have dealt mainly with a macroscopic, everyday-world, view of electric current. We did mention, however, that according to atomic theory, the electric current in metal wires is carried by negatively charged electrons, and that in liquid solutions current can also be carried by positive and/or negative ions. Let us now look at this microscopic picture in more detail.

When a potential difference is applied to the two ends of a wire of uniform cross-section, the direction of the electric field **E** is parallel to the walls of the wire (Fig. 25–22). The existence of **E** within the conducting wire does not contradict our earlier result that **E** = 0 inside a conductor in the electrostatic case. We are no longer dealing with the static case. Charges are free to move in a conductor, and hence can move under the action of the electric field. If all the charges are at rest, then **E** must be zero (electrostatics).

We now define a new microscopic quantity, the **current density**, **j**. It is defined as the *electric current per unit cross-sectional area* at any point in space. If the current density **j** in a wire of cross-sectional area A is uniform over the cross section, then j is related to the electric current by

$$ j = \frac{I}{A} \quad \text{or} \quad I = jA. \tag{25–11} $$

If the current density is not uniform, then the general relation is

$$ I = \int \mathbf{j} \cdot d\mathbf{A}, \tag{25–12} $$

where $d\mathbf{A}$ is an element of surface and I is the current through the surface over which the integration is taken. The direction of the current density at any point is the direction that a positive charge would move when placed at that point—that is, the direction of **j** at any point is generally the same as the direction of **E**, Fig. 25–22. (Inertial effects can usually be ignored.) The current density exists for any *point* in space. The current I, on the other hand, refers to a conductor as a whole, and hence is a macroscopic quantity.

The direction of **j** is chosen to represent the direction of flow of positive charge. In a conductor, it is negatively charged electrons that move, so they move in the direction of −**j**, or −**E** (to the left in Fig. 25–22). We can imagine the free electrons as moving about randomly at high speeds, bouncing off the atoms of the wire (somewhat like the molecules of a gas—Chapter 18). When an electric field exists in the wire, Fig. 25–23, the electrons feel a force and initially begin to accelerate. But they soon reach a more or less steady average speed (due to collisions with atoms in the wire), known as their **drift velocity**, v_d. The drift velocity is normally very much smaller than the electrons' average random speed.

We can relate v_d to the macroscopic current I in the wire. In a time Δt, the electrons will travel a distance $l = v_d \Delta t$ on average. Suppose the wire has cross-sectional area A. Then in time Δt, electrons in a volume $V = Al = Av_d \Delta t$ will pass through the cross section A of wire, as shown in Fig. 25–24. If there are n free electrons (each of charge $-e$) per unit volume ($n = N/V$), then the total charge ΔQ that passes through the area A in a time Δt is

$$ \Delta Q = (\text{no. of charges, } N) \times (\text{charge per particle}) $$
$$ = (nV)(-e) = -(nAv_d \Delta t)(e). $$

The current I in the wire is thus

$$ I = \frac{\Delta Q}{\Delta t} = -neAv_d. \tag{25–13} $$

FIGURE 25–22 Electric field **E** in a uniform wire of cross-sectional area A carrying a current I. The current density $j = I/A$.

Current

density

defined

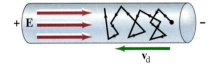

FIGURE 25–23 Electric field **E** in a wire gives electrons in random motion a drift velocity v_d.

Drift velocity

FIGURE 25–24 Electrons in the volume Al will all pass through the cross-section indicated in a time Δt, where $l = v_d \Delta t$.

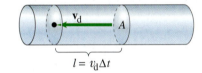

Current (microscopic variables)

The current density, $j = I/A$, is

Current density

in terms of

drift velocity

$$j = -nev_d. \tag{25-14}$$

In vector form, this is written

$$\mathbf{j} = -ne\mathbf{v}_d, \tag{25-15}$$

where the minus sign indicates that the direction of (positive) current flow is opposite to the drift velocity of electrons.

We can generalize Eq. 25–15 to any type of charge flow, such as flow of ions in an electrolyte. If there are several types of ions (which can include free electrons), each of density n_i (number per unit volume), charge q_i ($q_i = -e$ for electrons) and drift velocity $\mathbf{v}_{di}$, then the net current density at any point is

$$\mathbf{j} = \sum_i n_i q_i \mathbf{v}_{di}. \tag{25-16}$$

The total current I passing through an area A perpendicular to a uniform $\mathbf{j}$ is then

$$I = \sum_i n_i q_i v_{di} A.$$

EXAMPLE 25–12 **Electron speeds in a wire.** A copper wire 3.2 mm in diameter, carries a 5.0-A current. Determine (a) the current density in the wire, and (b) the drift velocity of the free electrons. (c) Estimate the rms speed of electrons assuming they behave like an ideal gas at 20°C. Assume that one electron per Cu atom is free to move (the others remain bound to the atom).

SOLUTION (a) The cross-sectional area of the wire is

$$A = \pi r^2 = (3.14)(1.60 \times 10^{-3}\,\text{m})^2 = 8.0 \times 10^{-6}\,\text{m}^2.$$

The current density is then

$$j = \frac{I}{A} = \frac{5.0\,\text{A}}{8.0 \times 10^{-6}\,\text{m}^2} = 6.2 \times 10^5\,\text{A/m}^2.$$

(b) Since we assume there is one free electron per atom, the density of free electrons, n, is the same as the density of Cu atoms. The atomic mass of Cu is 63.5 u (see Periodic Table inside the back cover), so 63.5 g of Cu contains one mole or 6.02×10^{23} free electrons. The mass density of copper (Table 13–1) is $\rho_D = 8.9 \times 10^3\,\text{kg/m}^3$, where $\rho_D = m/V$. (We use ρ_D to distinguish it here from ρ for resistivity.) So the number of free electrons per unit volume is

$$n = \frac{N}{V} = \frac{N}{m/\rho_D} = \frac{N(1\,\text{mole})}{m(1\,\text{mole})}\rho_D$$

$$n = \left(\frac{6.02 \times 10^{23}\,\text{electrons}}{63.5 \times 10^{-3}\,\text{kg}}\right)(8.9 \times 10^3\,\text{kg/m}^3) = 8.4 \times 10^{28}\,\text{m}^{-3}.$$

Then, by Eq. 25–14, the drift velocity is

$$v_d = \frac{j}{ne} = \frac{6.2 \times 10^5\,\text{A/m}^2}{(8.4 \times 10^{28}\,\text{m}^{-3})(1.6 \times 10^{-19}\,\text{C})} = 4.6 \times 10^{-5}\,\text{m/s},$$

which is only about 0.05 mm/s.

(c) If we model the free electrons as an ideal gas (a rather rough approximation), we use Eq. 18–5 to estimate the random rms speed of an electron as it darts around:

$$v_{rms} = \sqrt{\frac{3kT}{m}} = \sqrt{\frac{3(1.38 \times 10^{-23}\,\text{J/K})(293\,\text{K})}{9.11 \times 10^{-31}\,\text{kg}}} = 1.2 \times 10^5\,\text{m/s}.$$

Thus we see that the drift velocity (average speed in the direction of the current) is very much less than the rms thermal speed of the electrons, by a factor of about 10^9. [Note: The result in (c) is an underestimate. Quantum theory calculations, and experiments, give the rms speed in copper to be about $1.6 \times 10^6\,\text{m/s}$.]

The drift velocity of electrons in a wire is clearly very slow, only about 0.05 mm/s for the example above, which means it takes an electron 20×10^3 s, or $5\frac{1}{2}$ h, to travel only 1 m. This is not, of course, how fast "electricity travels": when you flip a light switch, the light—even if many meters away—goes on nearly instantaneously, for electric fields travel essentially at the speed of light $(3 \times 10^8 \text{ m/s})$. We can think of electrons in a wire as being like a pipe full of water: when a little water enters one end of the pipe, almost immediately some water issues forth at the other end.

Electricity's "speed"

Equation 25–2b, $V = IR$, can be written in terms of microscopic quantities as follows. We write the resistance R in terms of the resistivity ρ:

$$R = \rho \frac{l}{A};$$

and we write V and I as

$$I = jA$$

and

$$V = El.$$

The last relation follows from Eq. 23–3, where we assume the electric field is uniform within the wire and l is the length of the wire (or a portion of the wire) between whose ends the potential difference is V. Thus, from $V = IR$, we have

$$V = IR$$

$$El = (jA)\left(\rho \frac{l}{A}\right) = j\rho l$$

so

$$j = \frac{1}{\rho} E = \sigma E, \qquad (25\text{–}17)$$

where $\sigma = 1/\rho$ is the *conductivity* (Eq. 25–4). For a metal conductor, ρ and σ do not depend on V (and hence not on E). Therefore the current density $\mathbf{j}$ is proportional to the electrical field $\mathbf{E}$ in the conductor. This is the "microscopic" statement of Ohm's law. Equation 25–17, which can be written in vector form as

$$\mathbf{j} = \sigma \mathbf{E} = \frac{1}{\rho} \mathbf{E},$$

is sometimes taken as the definition of conductivity σ and resistivity ρ.

EXAMPLE 25–13 **Electric field inside a wire.** What is the electric field inside the wire of Example 25–12?

SOLUTION Table 25–1 gives $\rho = 1.68 \times 10^{-8} \, \Omega \cdot \text{m}$ for copper. Since $j = 6.2 \times 10^5 \text{ A/m}^2$,

$$E = \rho j = (1.68 \times 10^{-8} \, \Omega \cdot \text{m})(6.2 \times 10^5 \text{ A/m}^2)$$

$$= 1.0 \times 10^{-2} \text{ V/m}.$$

For comparison, the electric field between the plates of a capacitor is often much larger; in Example 24–1, for example, E is on the order of 10^4 V/m. Thus we see that only a modest electric field is needed for current flow in practical cases.

SECTION 25–8 Microscopic View of Electric Current: Current Density and Drift Velocity **649**

* 25–9 Superconductivity

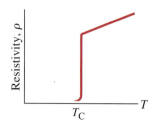

FIGURE 25–25 A superconducting material has zero resistivity when its temperature is below T_C, its "transition temperature." At T_C, the resistivity jumps to a "normal" non-zero value and increases with temperature as most materials do (Eq. 25–5).

High-temperature superconductors

FIGURE 25–26 An experimental train in Japan, supported by the magnetic field produced by current in coils beneath the tracks (in the red containers).

At very low temperatures, near absolute zero, the resistivity (Section 25–4) of certain metals and certain compounds or alloys becomes zero as measured by the highest-precision techniques. Materials in such a state are said to be **superconducting**. This phenomenon was first observed by H. K. Onnes (1853–1926) in 1911 when he cooled mercury below 4.2 K (−269°C). He found that at this temperature, the resistance of mercury suddenly dropped to zero. In general, superconductors become superconducting only below a certain *transition temperature* T_C, which is usually within a few degrees of absolute zero. Current in a ring-shaped superconducting material has been observed to flow for years in the absence of a potential difference, with no measurable decrease. Measurements show that the resistivity ρ of superconductors is less than $4 \times 10^{-25}\ \Omega \cdot m$, which is over 10^{16} times smaller than that for copper, and is considered to be zero in practice. See Fig. 25–25.

Much research has been done on superconductivity in recent years to try to understand why it occurs, and to find materials that superconduct at more reasonable temperatures to reduce the cost and inconvenience of refrigeration at the required very low temperature. Before 1986 the highest temperature at which a material was found to superconduct was 23 K, and this required liquid helium to keep the material cold. In 1987, a compound of yttrium, barium, copper, and oxygen was developed that can be superconducting at 90 K. Since this is above the temperature of liquid nitrogen, 77 K, boiling liquid nitrogen is sufficiently cold to keep the material superconducting. This was an important breakthrough since liquid nitrogen is much more easily and cheaply obtained than is the liquid helium needed for previous superconductors. Since then, superconductivity at temperatures in the vicinity of 160 K have been reported, though in fragile compounds.

Considerable research is being done to develop high-T_C superconductors as wires that can carry currents strong enough to be practical. Most applications today use a bismuth-strontium-calcium-copper oxide, known (for short) as BSCCO. A major use today of superconductors is for carrying the current in electromagnets (we shall see in Chapter 27 that electric currents produce magnetic fields). In large non-superconducting magnets, a great amount of energy is needed just to maintain the current, and this energy is wasted as heat.

A major problem is how to make a useable, bendable wire out of the BSCCO, which is very brittle. One solution is to embed tiny filaments of the high-T_C superconductor in a metal alloy matrix. The first major commercial use of high-T_C superconductors embeds the filaments in silver; the wires are formed into a cable to carry very high currents for the city of Detroit's electric distribution grid. The superconducting wire is wrapped around a tube carrying liquid nitrogen to keep the BSCCO below T_C. The wire is not resistanceless, because of the silver connections, but the resistance is much less than that of a conventional copper cable. Indeed, the 100 kg of this 130-m-long superconducting cable can carry as much current as 8000 kg of the copper cable it replaces.

Electric motors, generators, and transformers using superconductors are also being worked on, and will be much smaller and lighter than conventional ones. Prototype motors under development are half the size and weight of non-superconducting motors.

Superconductors could make electric cars more practical, make computers much faster, and have great potential in devices to store energy for use at peak demand. Superconductors are being studied for use in high-speed ground transportation: the magnetic fields produced by superconducting magnets would be used to "levitate" vehicles over tracks so there is essentially no friction (Fig. 25–26). The levitation arises from the repulsive force between the magnet (say, on the train) and the eddy currents produced in the track below (or, vice versa).

25–10 | Electric Hazards; Leakage Currents

An electric shock can damage the body or may even be fatal. The severity of a shock depends on the magnitude of the current, how long it acts, and through what part of the body it passes. A current passing through vital organs such as the heart or brain is especially serious. Electric current heats tissue and can cause burns. A current also stimulates nerves and muscles, and we feel a "shock."

Most people can "feel" a current of about 1 mA. Currents of a few mA cause pain but rarely much damage in a healthy person. However, currents above 10 mA cause severe contraction of the muscles, and a person may not be able to release the source of the current (say, a faulty appliance or wire). Death from paralysis of the respiratory system can occur. Artificial respiration, however, can sometimes revive a victim. If a current above about 70 mA passes across the torso so that a portion passes through the heart for a second or more, the heart muscles will begin to contract irregularly and blood will not be properly pumped. This condition is called "ventricular fibrillation." If it lasts for long, death results. Strangely enough, if the current is much larger, on the order of 1 A, the damage may be less and death by heart failure may be less likely[†] under some conditions.

The seriousness of a shock depends on the effective resistance of the body. Living tissue has quite low resistance since the fluid of cells contains ions that can conduct quite well. However, the outer layers of skin, when dry, offer much resistance. The effective resistance between two points on opposite sides of the body when the skin is dry is in the range of 10^4 to 10^6 Ω. However, when the skin is wet, the resistance may be 10^3 Ω or less. A person in good contact with the ground who touches a 120-V dc line with wet hands can suffer a current

$$I = \frac{120 \text{ V}}{1000 \text{ }\Omega} = 120 \text{ mA}.$$

As we saw above, this could be lethal.

Figure 25–27 shows how the circuit is completed when a person touches an electric wire. One side of a 120-V source is connected to ground by a wire connected to a buried conductor, such as a water pipe. Thus the current passes from the high-voltage wire, through the person, to the ground; it passes through the ground back to the other terminal of the source to complete the circuit. If the person in Fig. 25–27 stands on a good insulator—thick-soled shoes or a dry wood floor—there will be much more resistance in the circuit and consequently much less current will flow. However, if the person stands with bare feet on the ground, or is sitting in a bathtub, there is considerable danger because the resistance is much less. In a bathtub, not only are you wet, but the water is in contact with the drain pipe that leads to the ground. That is why it is strongly recommended not to touch anything electrical in such a situation.

A principal danger comes from touching a bare wire whose insulation has worn off, or from a bare wire inside an appliance when you're tinkering with it. (Always unplug an electrical device before investigating its insides!) Sometimes a wire inside a device breaks or loses its insulation and comes in contact with the case. If the case is metal, it will conduct electricity. A person could then suffer a

→ PHYSICS APPLIED

Electric shock

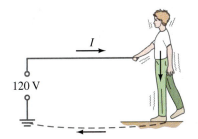

FIGURE 25–27 A person receives an electric shock when the circuit is completed.

→ PHYSICS APPLIED

Grounding and shocks

[†] Apparently, larger currents bring the entire heart to a standstill. Upon release of the current, the heart returns to its normal rhythm. This may not happen when fibrillation occurs since it is often hard to stop once it starts. Fibrillation may also occur as a result of a heart attack or during heart surgery. A device known as a *defibrillator* can apply a brief high current to the heart; this causes complete heart stoppage and is often followed by resumption of normal beating.

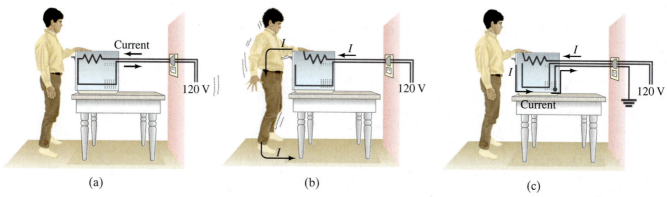

(a)   (b)   (c)

FIGURE 25–28 (a) An electric appliance operating normally with a two-prong plug. (b) Short to the case with ungrounded case: shock. (c) Short to the case with the case grounded with third prong.

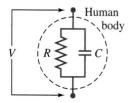

FIGURE 25–29 Human body modeled electrically as resistance and capacitor in parallel when a voltage is applied.

Leakage current

severe shock merely by touching the case, as shown in Fig. 25–28b. To prevent an accident, metal cases are supposed to be connected directly to ground, so they cannot become "hot." Then if a "hot" wire touches the grounded case, a short circuit to ground immediately occurs internally, as shown in Fig. 25–28c; most of the current passes through the low resistance ground wire rather than through the person. Furthermore, the high current immediately opens the fuse or circuit breaker in the circuit. Grounding a metal case is best done by a separate ground wire connected to the third (round) prong of a 3-prong plug. It can also be done by connecting the case to the larger prong of a so-called "polarized" 2-prong plug. Of course not only the device, but also the outlets must be wired correctly to ground.

The human body acts as if it had capacitance in parallel with its resistance (Fig. 25–29). A dc current can pass through the resistance, but not the capacitance. An ac current, like the changing currents discussed in Section 25–7, can exist also in the capacitive branch. Because of the additional path allowing current flow, the ac current for a given V_{rms} will be greater than for the same dc voltage. Thus an ac voltage is more dangerous than an equal dc voltage.

Another danger is *leakage current*, by which we mean a current along an unintended path. Leakage currents are often capacitively coupled. For example, a wire in a lamp forms a capacitor with the metal case; charges moving in one conductor attract or repel charge in the other, so there is a current. Typical electrical codes limit leakage currents to 1 mA for any device. A 1-mA leakage current is usually harmless. It can be very dangerous, however, to a hospital patient with implanted electrodes connected to ground through the apparatus. This is because the current can pass directly through the heart as compared to the usual situation where the current enters at the hands and spreads out through the body. Although 70 mA may be needed to cause heart fibrillation when entering through the hands (very little of it actually passes through the heart), as little as 0.02 mA has been known to cause fibrillation when passing directly to the heart. Thus, a "wired" patient is in considerable danger from leakage current even from as simple an act as touching a lamp.

Summary

An electric battery serves as a source of potential difference by transforming chemical energy into electric energy. A simple battery consists of two electrodes made of different metals immersed in a solution or paste known as an electrolyte.

Electric current, I, refers to the rate of flow of electric charge and is measured in **amperes** (A): 1 A equals a flow of 1 C/s past a given point.

The direction of **conventional current** flow is that of positive charge. In a wire, it is actually negatively charged electrons that move, so they flow in a direction opposite to the direction of the conventional current. Positive conventional current always flows from a high potential to a low potential.

The **resistance** R of a device is defined by the relation

$$V = IR,$$

where I is the current in the device when a potential difference V is applied across it. For materials such as metals, R is a constant independent of V (thus $I \propto V$), a result known as Ohm's law.

The unit of resistance is the **ohm** (Ω), where $1\,\Omega = 1\,\text{V/A}$. See Table 25–2.

TABLE 25–2	Summary of Units
Current	$1\,\text{A} = 1\,\text{C/s}$
Potential difference	$1\,\text{V} = 1\,\text{J/C}$
Power	$1\,\text{W} = 1\,\text{J/s}$
Resistance	$1\,\Omega = 1\,\text{V/A}$

The resistance R of a wire is inversely proportional to its cross-sectional area A, and directly proportional to its length l and to a property of the material called its resistivity: $R = \rho l/A$. The **resistivity**, ρ, increases with temperature for metals, but for semiconductors it may decrease.

The rate at which energy is transformed in a resistance R from electric to other forms of energy (such as heat and light) is equal to the product of current and voltage. That is, the **power** transformed, measured in watts, is given by

$$P = IV$$

and for resistors can be written as

$$P = I^2R = \frac{V^2}{R}.$$

The total electric energy transformed in any device equals the product of power and the time during which the device is operated. In SI units, energy is given in joules ($1\,\text{J} = 1\,\text{W} \cdot \text{s}$), but electric companies use a larger unit, the **kilowatt-hour** $(1\,\text{kWh} = 3.6 \times 10^6\,\text{J})$.

Electric current can be **direct current (dc)**, in which the current is steady in one direction; or it can be **alternating current (ac)**, in which the current reverses direction at a particular frequency f, typically 60 Hz. Alternating currents are often sinusoidal in time, $I = I_0 \sin \omega t$, where $\omega = 2\pi f$, and are produced by an alternating voltage.

The rms values of sinusoidally alternating currents and voltages are given by

$$I_{\text{rms}} = \frac{I_0}{\sqrt{2}} \quad \text{and} \quad V_{\text{rms}} = \frac{V_0}{\sqrt{2}},$$

respectively, where I_0 and V_0 are the peak values. The power relationship, $P = IV = I^2R = V^2/R$, is valid for the average power in alternating currents when the rms values of V and I are used.

Current density $\mathbf{j}$ is the current per cross-sectional area. From a microscopic point of view, the current density is related to the number of charge carriers per unit volume, n, their charge, q, and their drift velocity, $\mathbf{v}_d$, by $\mathbf{j} = nq\mathbf{v}_d$. The electric field within a wire is related to $\mathbf{j}$ by $\mathbf{j} = \sigma\mathbf{E}$ where $\sigma = 1/\rho$ is the **conductivity**.

At very low temperatures certain materials become **superconducting**, which means their electrical resistance becomes zero.

Electric shocks are caused by current passing through the body. To avoid shocks, the body must not become part of a circuit by allowing different parts of the body to touch objects at different potentials.

Questions

1. Car batteries can be rated in ampere-hours ($A \cdot h$). What aspect of the battery is being rated?

2. When an electric cell is connected to a circuit, electrons flow away from the negative terminal in the circuit. But within the cell, electrons flow *to* the negative terminal. Explain.

3. Develop an analogy between blood circulation and an electrical circuit. Discuss what plays the role of the heart for the electric case, and so on.

4. In a car, one terminal of the battery is said to be connected to "ground." Since it is not really connected to the ground, what is meant by this expression?

5. When you turn on a water faucet, the water usually flows immediately. You don't have to wait for water to flow from the faucet valve to the spout. Why not? Is the same thing true when you connect a wire to the terminals of a battery?

6. Can a copper wire and an aluminum wire of the same length have the same resistance? Explain.

7. If a rectangular solid made of carbon has sides of length a, $2a$, $3a$, how would you connect the wires from a battery so as to obtain (a) the least resistance, (b) the greatest resistance?

8. The equation $P = V^2/R$ indicates that the power dissipated in a resistor decreases if the resistance is increased, whereas the equation $P = I^2R$ implies the opposite. Is there a contradiction here? Explain.

9. What happens when a lightbulb burns out?

10. Explain why lightbulbs almost always burn out just as they are turned on and not after they have been on for some time.

11. Which draws more current, a 100-W lightbulb or a 75-W bulb? Which has the higher resistance?

12. Electric power is transferred over large distances at very high voltages. Explain how the high voltage reduces power losses in the transmission lines.

13. Why is it dangerous to replace a 15-A fuse that blows repeatedly with a 25-A fuse?

14. When electric lights are operated on low-frequency ac (say, 10 Hz) they flicker noticeably. Why?

15. Driven by ac power, the same electrons pass back and forth through your reading lamp over and over again. Explain why the light stays lit instead of going out after the first pass of electrons.

16. The heating element in a toaster is made of Nichrome wire. Immediately after the toaster is turned on, is the current (I_{rms}) in the wire increasing, decreasing, or staying constant? Explain.

17. Is current used up in a resistor?

18. A voltage V is connected across a wire of length l and radius r. How is the electron drift velocity affected if (a) l is doubled, (b) r is doubled, (c) V is doubled.

19. Compare the drift velocities and electric currents in two wires that are geometrically identical and the density of atoms is similar, but the number of free electrons per atom in the material of one wire is twice that in the other.

20. Why is it more dangerous to turn on an electric appliance when you are standing outside in bare feet than when you are inside wearing shoes with thick soles?

Problems

Sections 25–2 and 25–3

1. (I) A current of 1.50 A flows in a wire. How many electrons are flowing past any point in the wire per second? The charge on one electron is 1.60×10^{-19} C.

2. (I) A service station charges a battery using a current of 5.7 A for 7.0 h. How much charge passes through the battery?

3. (I) What is the current in amperes if 1000 Na^+ ions were to flow across a cell membrane in 7.5 μs? The charge on the sodium is the same as on an electron, but positive.

4. (I) What is the resistance of a toaster if 110 V produces a current of 4.2 A?

5. (I) What voltage will produce 0.25 A of current through a 3000-Ω resistor?

6. (II) An electric device draws 5.50 A at 110 V. (a) If the voltage drops by 10 percent, what will be the current, assuming nothing else changes? (b) If the resistance of the device were reduced by 10 percent, what current would be drawn at 110 V?

7. (II) A 9.0-V battery is connected to a bulb whose resistance is 1.6 Ω. How many electrons leave the battery per minute?

8. (II) If a 12-V battery pushes a current of 0.50 A through a resistor, what is its resistance, and how many joules of energy does the battery lose in a minute?

9. (II) A hair dryer draws 7.5 A when plugged into a 120-V line. (a) What is its resistance? (b) How much charge passes through it in 15 min? (Assume direct current.)

10. (II) A bird stands on a dc electric transmission line carrying 2500 A (Fig. 25–30). The line has $2.5 \times 10^{-5}\ \Omega$ resistance per meter and the bird's feet are 4.0 cm apart. What potential difference does the bird feel?

FIGURE 25–30 Problem 10.

Section 25–4

11. (I) What is the diameter of a 1.00-m length of tungsten wire whose resistance is 0.22 Ω?

12. (I) What is the resistance of a 3.5-m length of copper wire 1.5 mm in diameter?

13. (II) Compare the resistance of 10.0 m of aluminum wire 2.0 mm in diameter with 20.0 m of copper filament wire 2.5 mm in diameter.

14. (II) Can a 2.5-mm-diameter copper wire have the same resistance as a tungsten wire of the same length? Give numerical details.

15. (II) A certain copper wire has a resistance of 10.0 Ω. At what point along the length must the wire be cut so that the resistance of one piece is 5.0 times the resistance of the other? What is the resistance of each piece.

16. (II) How much would you have to raise the temperature of a copper wire (originally at 20°C) to increase its resistance by 20 percent?

17. (II) A length of aluminum wire is connected to a precision 10.00-V power supply, and a current of 0.4212 A is precisely measured at 20.0°C. The wire is placed in a new environment of unknown temperature where the measured current is 0.3618 A. What is the unknown temperature?

18. (II) Estimate at what temperature copper will have the same resistivity as tungsten does at 20°C.

19. (II) A 100-W lightbulb has a resistance of about 12 Ω when off (20°C) and 140 Ω when on (hot). Estimate the temperature of the filament when "on" assuming an average temperature coefficient of resistivity $\alpha = 0.0060 \ (C°)^{-1}$.

20. (II) A rectangular solid made of carbon has sides lying along the x, y, and z axes, whose lengths are 1.0 cm, 2.0 cm, and 4.0 cm, respectively (Fig. 25–31). Determine the resistance for current that flows through the solid in (a) the x direction, (b) the y direction, and (c) the z direction. Assume the resistivity is $\rho = 3.0 \times 10^{-5} \ \Omega \cdot m$.

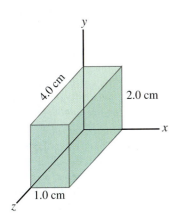

FIGURE 25–31
Problem 20.

21. (II) A length of wire is cut in half and the two lengths are wrapped together side by side to make a thicker wire. How does the resistance of this new combination compare to the resistance of the original wire?

22. (II) For some applications, it is important that the value of a resistance not change with temperature. For example, suppose you made a 4.70-kΩ resistor from a carbon resistor and a Nichrome wire-wound resistor connected together so the total resistance is the sum of their separate resistances. What value should each of these resistors have (at 0°C) so that the combination is temperature independent?

23. (II) (a) Show that if a straight wire of cross-sectional area A lies along the x axis, the rate at which charge flows is given by

$$\frac{dq}{dt} = -\sigma A \frac{dV}{dx},$$

where dV/dx is the potential gradient, and σ is the conductivity. (b) Make an analogy to heat conduction (Chapter 19). Would you expect σ and k (thermal conductivity) to be related?

24. (III) A hollow cylindrical resistor with inner radius r_1 and outer radius r_2, and length l, is made of a material whose resistivity is ρ (Fig. 25–32). (a) Show that the resistance is given by

$$R = \frac{\rho}{2\pi l} \ln \frac{r_2}{r_1}$$

for current that flows radially outward. [Hint: Divide the resistor into concentric cylindrical shells and integrate.] (b) Evaluate the resistance R for such a resistor made of carbon whose inner and outer radii are 1.0 mm and 1.8 mm and whose length is 3.0 cm. (c) What is the resistance in part (b) for current flowing parallel to the axis?

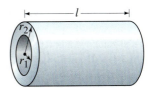

FIGURE 25–32   Problem 24.

25. (III) Determine a formula for the total resistance of a spherical shell made of material whose conductivity is σ and whose inner and outer radii are r_1 and r_2. Assume the current flows radially outward.

26. (III) The filament of a light bulb has a resistance of 12 Ω at 20°C and 140 Ω when hot (as in Problem 19). (a) Calculate the temperature of the filament when it is hot, and take into account the change in length and area of the filament due to thermal expansion (assume tungsten for which the thermal expansion coefficient is ≈5 × $10^{-6} \ C°^{-1}$). (b) In this temperature range, what is the percentage change in resistance due to thermal expansion, and what is the percentage change in resistance due solely to the change in ρ? Use Eq. 25–5.

Sections 25–5 and 25–6

27. (I) What is the maximum power consumption of a 9.0-V portable cassette player that draws a maximum of 350 mA of current?

28. (I) The element of an electric oven is designed to produce 3.1 kW of heat when connected to a 240-V source. What must be the resistance of the element?

29. (I) What is the maximum voltage that can be applied to a 5.4-kΩ resistor rated at $\frac{1}{4}$ watt?

30. (I) A hair dryer has two settings: 600 W and 1200 W. (a) At which setting do you expect the resistance to be higher? After making a guess, determine the resistance at (b) the lower setting, (c) the higher setting.

31. (I) (a) What is the resistance and current through a 60-W lightbulb if it is connected to its proper source voltage of 120 V? (b) Repeat for a 150-W bulb.

32. (I) You buy a 60-W lightbulb in Europe, where electricity is delivered to homes at 240 V. If you use the lightbulb in the United States at 120 V (assume its resistance doesn't change), how bright will it be relative to 60-W 120-V bulbs? Estimate how much power it will consume.

33. (II) How many kWh of energy does a 550-W toaster use in the morning if it is in operation for a total of 10 min? At a cost of 12 cents/kWh, how much would this add to your monthly electric energy bill if you made toast five mornings per week?

34. (II) At $0.110 per kWh, what does it cost to leave a 60-W porch light on day and night for a year?

35. (II) What is the total amount of energy stored in a 12-V, 90-A·h car battery when it is fully charged?

36. (II) A transistor to be used in a circuit is rated for a maximum current of 28 mA if operated at 9.0 V. (a) What is the maximum acceptable power to the transistor, and (b) what would be the limiting current if the voltage applied were actually only 7.0 V?

37. (II) How many 100-W lightbulbs, connected to 120 V as in Fig. 25–18, can be used without blowing a 2.5-A fuse?

38. (II) What is the efficiency of a 0.50-hp electric motor that draws 4.6 A from a 120-V line?

39. (II) A power station delivers 520 kW of power to a factory through wires of total resistance of 3.0 Ω. How much less power is wasted if the electricity is delivered at 50,000 V rather than 12,000 V?

40. (II) A 2800-W oven is hooked to a 240-V source. (a) What is the resistance of the oven? (b) How long will it take to boil 100 mL of water assuming 80 percent efficiency? (c) How much will this cost at 10 cents/kWh?

41. (III) The current in an electromagnet connected to a 240-V line is 14.5 A. At what rate must cooling water pass over the coils if the water temperature is to rise by no more than 6.50 C°?

42. (III) A small immersion heater can be used in a car to heat a cup of water for coffee. If the heater can heat 150 mL of water from 5°C to 95°C in 6.0 min, approximately how much current does it draw from the 12-V battery, and what is its resistance? Assume the manufacturer's claim of 60 percent efficiency.

Section 25–7

43. (I) Calculate the peak current in a 1.8-kΩ resistor connected to a 120-V rms ac source.

44. (I) An ac voltage, whose peak value is 180 V, is across a 330-Ω resistor. What is the value of the peak and rms currents in the resistor?

45. (I) What is the resistance of the circuits in your house as seen by the power company, when (a) everything electrical is turned off, and (b) there is a lone 75-W lightbulb burning on the porch?

46. (II) The peak value of an alternating current passing through a 1500-W device is 6.0 A. What is the rms voltage across it?

47. (II) Calculate the peak voltage across, and peak current through, an 1800-W arc welder connected to a 450-V ac line.

48. (II) What is the maximum instantaneous power dissipated by, and maximum current passing through, a 3.0-hp pump connected to a 240-V ac power source?

49. (II) A heater coil connected to a 240-V ac line has a resistance of 38 Ω. (a) What is the average power used? (b) What are the maximum and minimum values of the instantaneous power?

50. (II) Suppose a current is given by the equation $I = 1.80 \sin 210t$, where I is in amperes, and t in seconds. (a) What is the frequency? (b) What is the rms value of the current? (c) If this is the current through a 42.0-Ω resistor, what is the equation that describes the voltage as a function of time?

Section 25–8

51. (II) A 0.55-mm-diameter copper wire carries a tiny current of 2.5 μA. Estimate (a) the electron drift velocity in the wire, (b) the current density, and (c) the electric field.

52. (II) A 5.00-m length of 2.0-mm-diameter wire carries a 750-mA current when 22.0 mV is applied to its ends. If the drift velocity has been measured (by the Hall effect— Section 27–8) to be 1.7×10^{-5} m/s, determine (a) the resistance R of the wire, (b) the resistivity ρ, (c) the current density j, (d) the electric field inside the wire, and (e) the number n of free electrons per unit volume.

53. (III) At a point high in the Earth's atmosphere, He^{2+} ions in a concentration of $2.8 \times 10^{12}/m^3$ are moving due north at a speed of 2.0×10^6 m/s. Also, an $8.0 \times 10^{11}/m^3$ concentration of O_2^- ions is moving due south at a speed of 7.2×10^6 m/s. Determine the magnitude and direction of the current density **j** at this point.

General Problems

54. How many coulombs are there in 1.00 ampere-hour?

55. A person accidentally leaves a car with the lights on. If each of the two front lights uses 40 W and each of the two rear lights 6 W, for a total of 92 W, how long will a fresh 12-V battery last if it is rated at 90 A·h? Assume the full 12 V appears across each bulb.

56. What is the average current drawn by a 1.5-hp 120-V motor?

57. The *conductance G* of an object is defined as the reciprocal of the resistance R: that is, $G = 1/R$. The unit of conductance is the mho $(= ohm^{-1})$, which is also called the siemens (S). What is the conductance (in siemens) of an object that draws 800 mA of current at 12.0 V?

58. 10.0 m of wire consists of 5.0 m of copper followed by 5.0 m of aluminum of equal diameter (both 1.0 mm). A potential difference of 25 V is placed across the composite wire. (a) What is the total resistance of the wire? (b) What is the current flow through the wire? (c) What are the voltages across the aluminum part and across the copper part?

59. An ordinary flashlight uses 2 D-cell 1.5 V batteries connected in series as in Fig. 25–4b. The bulb draws 350 mA when turned on. (a) Calculate the resistance of the bulb and the power dissipated. (b) By what factor would the power increase if 4 D-cells in series were used with the same bulb? (Neglect heating effects of the filament.) Why shouldn't you try this?

60. The heating element of a 110-V, 900-W heater is 5.4 m long. If it is made of iron, what must its diameter be?

61. (a) A particular household uses a 1.8-kW heater 3.0 h/day ("on" time), four 100-W lightbulbs 6.0 h/day, a 3.0-kW electric stove element for a total of 1.4 h/day, and miscellaneous power amounting to 2.0 kWh/day. If electricity costs $0.105 per kWh, what will be their monthly bill (30 days)? (b) How much coal (which produces 7000 kcal/kg) must be burned by a 35-percent-efficient power plant to provide the yearly needs of this household?

62. A small city requires about 10 MW of power. Suppose that instead of using high-voltage lines to supply the power, the power is delivered at 120 V. Assuming a two-wire line of 0.50-cm-diameter copper wire, estimate the cost of the energy lost to heat per hour per meter. Assume the cost of electricity is about 10 cents per kWh.

63. A 1200-W hair dryer is designed for 117 V. (a) What will be the percentage change in power output if the voltage drops to 105 V? Assume no change in resistance. (b) How would the actual change in resistivity with temperature affect your answer?

64. The wiring in a house must be thick enough so it doesn't become so hot as to start a fire. What diameter must a copper wire be if it is to carry a maximum current of 30 A and produce no more than 1.6 W of heat per meter of length?

65. An air conditioner draws 12 A at 120 V ac. The connecting cord is copper wire which has a diameter of 1.628 mm. (a) How much power does the air conditioner draw? (b) If the total length of wire is 15 m, how much power is dissipated in the wiring? (c) If no. 12 wire, with a diameter of 2.053 mm were used instead, how much power would be dissipated? (d) Assuming that the air conditioner is run 12 hours per day, how much money per day would be saved by using no. 12 wire? Assume that the cost of electricity is 10 cents per kWh.

66. In a "brownout" situation, the voltage supplied by the electric company falls. Assuming the percent drop is small, show that the power output of a given appliance falls by approximately twice that percent, assuming the resistance does not change. How much of a voltage drop does it take for a 60-W lightbulb to begin acting like a 50-W bulb?

67. A microwave oven running at 60 percent efficiency delivers 900 W of energy per second to the interior. Determine (a) the power drawn from the source, and (b) the rms current drawn. Assume a source voltage of 120 Vrms.

68. A 1.00-Ω wire is drawn out to 3.00 times its original length. What is its resistance now?

69. 220 V is applied to two different conductors made of the same material. One conductor is twice as long and twice as thick as the second. What is the ratio of the power transformed in the first relative to the second?

70. An electric heater is used to heat a room of volume 68 m³. Air is brought into the room at 5°C and is changed completely twice per hour. Heat loss through the walls amounts to approximately 850 kcal/h. If the air is to be maintained at 20°C, what minimum wattage must the heater have? (The specific heat of air is about 0.17 kcal/kg·C°.)

71. A 6.50-Ω resistor is made from a coil of copper wire whose total mass is 18.0 g. What is the diameter of the wire and how long is it?

72. The new EV-1 electric car makes use of storage batteries as its source of energy. Its mass is 1300 kg and it is powered by 26 batteries, each 12 V, 52 A·h. Assume that the car is driven on the level at an average speed of 40 km/h, and the average retarding force is 240 N. Assume 100 percent efficiency and neglect energy used for acceleration. Note that no energy is consumed when the vehicle is stopped since the engine doesn't need to idle. (a) Determine the horsepower required. (b) After approximately how many kilometers must the batteries be recharged?

73. A 100-W, 120-V lightbulb has a resistance of 12 Ω when cold (20°C) and 140 Ω when on (hot). Calculate its power consumption at (a) the instant it is turned on, and (b) after a few moments when it is hot.

74. A capacitor is often used in electronics to keep energy flowing even if there is a momentary loss of power from the electric company. What capacitance would be required for a television, operating at an internal dc voltage of 120 V at 150 W, to provide sufficient energy during a 0.10 s lapse in power?

75. The Tevatron accelerator at Fermilab (Illinois) is designed to carry an 11 mA beam of protons traveling at very nearly the speed of light $(3.0 \times 10^8 \text{ m/s})$ around a ring 6300 m in circumference. How many protons are stored in the beam?

76. How far does an average electron move along the wires of a 500-W toaster during an alternating current cycle? The power cord has copper wires of diameter 1.8 mm and is plugged into a standard 60 Hz 120 V ac outlet. [Hint: The maximum current in the cycle is related to the maximum drift velocity. The maximum velocity in an oscillation is related to the maximum displacement; see Chapter 14.]

77. For the wire in Fig. 25–33, whose diameter varies uniformly from a to b as shown, suppose a current $I = 2.0$ A enters at a. If $a = 3.0$ mm and $b = 4.0$ mm, what is the current density (assume uniform) at each end?

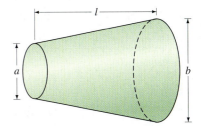

FIGURE 25–33 Problems 77 and 78.

78. The cross section of a portion of wire increases uniformly as shown in Fig. 25–34 so it has the shape of a truncated cone. The diameter at one end is a and at the other it is b, and the total length along the axis is l. If the material has resistivity ρ, determine the resistance R between the two ends in terms of a, b, l, and ρ. Assume that the current flows uniformly through each section, and that the taper is small, i.e., $(b - a) \ll l$.

This "Walkman" contains circuits within it that are dc, at least in part. (The audio signal is ac.) The circuit diagram below shows a possible amplifier circuit (actually, two identical circuits are used, one for each stereo channel). Although the large triangle is an amplifier circuit board containing transistors (not discussed in this chapter), the other circuit elements are ones we have met, resistors and capacitors, and we discuss them in circuits in this chapter. We also discuss voltmeters and ammeters, and how they are built and used.

CHAPTER 26

DC Circuits

TABLE 26–1 Symbols for Circuit Elements	
Symbol	**Device**
⊣⊢	Battery
⊣⊢	Capacitor
-⋁⋁⋁-	Resistor
——	Wire with negligible resistance
⏚ or ⏚	Ground

Electric circuits are basic parts of all electronic gear from radio and TV sets to computers and automobiles. Scientific measurements, from physics to biology and medicine, make use of electric circuits. In Chapter 25, we discussed the basic principles of electric current. Now we will apply these principles to analyze dc circuits and to understand the operation of a number of useful instruments.[†]

When we draw a diagram for a circuit, we represent batteries, capacitors, and resistors by the symbols shown in Table 26–1. Wires whose resistance is negligible compared to other resistance in the circuit are drawn simply as straight lines. Some circuit diagrams show a ground symbol (⏚ or ⏚) which may mean a real connection to the ground, perhaps via a metal pipe, or it may simply mean a common connection, such as the frame of a car.

For the most part in this chapter, except in Section 26–4, we will be interested in circuits operating in their steady state—that is, we won't be looking at a circuit at the moment a change is made in it, such as when a battery or resistor is connected or disconnected, but rather a short time later when the currents have reached their steady values.

[†] Ac circuits that contain only a voltage source and resistors can be analyzed like the dc circuits in this chapter. However, ac circuits that contain capacitors and other circuit elements are more complicated, and we discuss them in Chapter 31.

26–1 EMF and Terminal Voltage

To have current in an electric circuit, we need a device such as a battery or an electric generator that transforms one type of energy (chemical, mechanical, light, and so on) into electric energy. Such a device is called a *source* of *electromotive force* or of *emf*. (The term "electromotive force" is a misnomer since it does not refer to a "force" that is measured in newtons. Hence, to avoid confusion, we prefer to use the abbreviation, emf.) The potential difference between the terminals of such a source, when no current flows to an external circuit, is called the **emf** of the source. The symbol $\mathscr{E}$ is usually used for emf (don't confuse it with E for electric field).

emf

Why battery voltage isn't constant

You may have noticed in your own experience that when a current is drawn from a battery, the voltage across its terminals drops below its rated emf. For example, if you start a car with the headlights on, you may notice the headlights dim. This happens because the starter draws a large current, and the battery voltage drops as a result. The voltage drop occurs because the chemical reactions in a battery cannot supply charge fast enough to maintain the full emf. For one thing, charge must flow (within the electrolyte) between the electrodes of the battery, and there is always some hindrance to completely free flow. Thus, a battery itself has some resistance, which is called its **internal resistance**; it is usually designated r. A real battery is then modeled as if it were a perfect emf $\mathscr{E}$ in series with a resistor r, as shown in Fig. 26–1. Since this resistance r is inside the battery, we can never separate it from the battery. The two points a and b in the diagram represent the two terminals of the battery. What we measure is the **terminal voltage** $V_{ab} = V_a - V_b$. When no current is drawn from the battery, the terminal voltage equals the emf, which is determined by the chemical reactions in the battery: $V_{ab} = \mathscr{E}$. However, when a current I flows naturally from the battery there is an internal drop in voltage equal to Ir. Thus the terminal voltage (the actual voltage delivered) is[†]

$$V_{ab} = \mathscr{E} - Ir.$$

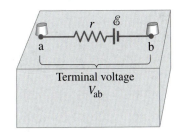

FIGURE 26–1 Diagram for an electric cell or battery.

Terminal voltage

For example, if a 12-V battery has an internal resistance of 0.1 Ω, then when 10 A flows from the battery, the terminal voltage is $12\,V - (10\,A)(0.1\,\Omega) = 11\,V$. The internal resistance of a battery is usually small. For example, an ordinary flashlight battery when fresh may have an internal resistance of perhaps 0.05 Ω. (However, as it ages and the electrolyte dries out, the internal resistance increases to many ohms.) Car batteries have even lower internal resistance.

EXAMPLE 26–1 Battery with internal resistance. A 65.0-Ω resistor is connected to the terminals of a battery whose emf is 12.0 V and whose internal resistance is 0.5 Ω, Fig. 26–2. Calculate (*a*) the current in the circuit, (*b*) the terminal voltage of the battery, V_{ab}, and (*c*) the power dissipated in the resistor R and in the battery's internal resistance r.

SOLUTION (*a*) From the equation above relating emf $\mathscr{E}$ to terminal voltage, we have

$$V_{ab} = \mathscr{E} - Ir,$$

where $V_{ab} = IR$ (Eq. 25–2). Hence $IR = \mathscr{E} - Ir$ or $\mathscr{E} = I(R + r)$, and so

$$I = \frac{\mathscr{E}}{R + r} = \frac{12.0\,V}{65.5\,\Omega} = 0.183\,A.$$

(*b*) The terminal voltage is

$$V_{ab} = \mathscr{E} - Ir = 12.0\,V - (0.183\,A)(0.5\,\Omega) = 11.9\,V.$$

(*c*) The power dissipated is

$$P_R = I^2R = (0.183\,A)^2(65.0\,\Omega) = 2.18\,W$$
$$P_r = I^2r = (0.183\,A)^2(0.5\,\Omega) = 0.02\,W.$$

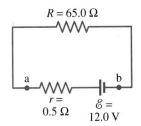

FIGURE 26–2 Example 26–1.

[†] When a battery is being charged, a current is forced to pass through it and we have to write $V_{ab} = \mathscr{E} + Ir$. See Example 26–10.

In much of what follows, unless stated otherwise, we assume that the battery's internal resistance is negligible, and that the battery voltage given is its terminal voltage, which we will usually write simply as V rather than V_{ab}.

26–2 Resistors in Series and in Parallel

When two or more resistors are connected end to end as shown in Fig. 26–3, they are said to be connected in **series**. The resistors could be simple resistors as were pictured in Fig. 25–10, or they could be lightbulbs, heating elements, or other resistive devices. Any charge that passes through R_1 in Fig. 26–3a will also pass through R_2 and then R_3. Hence the same current I passes through each resistor. (If it did not, this would imply that charge was accumulating at some point in the circuit, which does not happen in the steady state.) We let V represent the voltage across all three resistors. We assume all other resistance in the circuit can be ignored, and so V equals the terminal voltage of the battery. We let V_1, V_2, and V_3 be the potential differences across each of the resistors, R_1, R_2, and R_3, respectively, as shown in Fig. 26–3a. From $V = IR$, we can write $V_1 = IR_1$, $V_2 = IR_2$, and $V_3 = IR_3$. Because the resistors are connected end to end, energy conservation tells us that the total voltage V is equal to the sum of the voltages across each resistor:

Series circuit: voltages add; current the same in each R

$$V = V_1 + V_2 + V_3 = IR_1 + IR_2 + IR_3. \qquad \text{[series]} \quad \textbf{(26–1)}$$

To see in more detail why this is true, note that an electric charge q passing through R loses potential energy by qV_1. In passing through R_2 and R_3, the potential energy U decreases by qV_2 and qV_3, for a total $\Delta U = qV_1 + qV_2 + qV_3$; this sum must equal the energy given to q by the battery, qV, so that energy is conserved. Hence $qV = q(V_1 + V_2 + V_3)$, and so $V = V_1 + V_2 + V_3$, which is Eq. 26–1.

Now let us determine the equivalent single resistance R_{eq} that would draw the same current as our combination; see Fig. 26–3c. Such a single resistance R_{eq} would be related to V by

$$V = IR_{eq}.$$

We equate this expression with Eq. 26–1, $V = I(R_1 + R_2 + R_3)$, and find

Resistances in series

$$R_{eq} = R_1 + R_2 + R_3. \qquad \text{[series]} \quad \textbf{(26–2)}$$

This is, in fact, what we expect. When we put several resistances in series, the total resistance is the sum of the separate resistances. This applies to any number of resistances in series. Note that when you add more resistance to the circuit, the current will decrease. For example, if a 12-V battery is connected to a 4-Ω resistor, the current will be 3 A. But if the 12-V battery is connected to three 4-Ω resistors in series, the total resistance is 12 Ω and the current will be only 1 A.

FIGURE 26–3 (a) Resistances connected in series: $R_{eq} = R_1 + R_2 + R_3$. (b) Resistances could be lightbulbs, or any other type of resistance. (c) Equivalent single resistance R_{eq} that draws the same current.

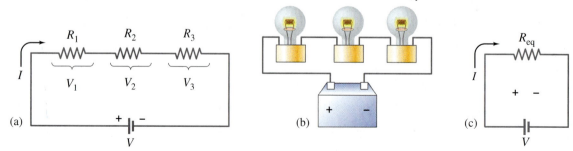

Another simple way to connect resistors is in **parallel**, so that the current from the source splits into separate branches, as shown in Fig. 26–4. The wiring in houses and buildings is arranged so all electric devices are in parallel, as we already saw in Chapter 25, Fig. 25–18. With parallel wiring, if you disconnect one device $\left(\text{say } R_1 \text{ in Fig. 26–4}\right)$, the current to the others is not interrupted. But in a series circuit, if one device $\left(\text{say } R_1 \text{ in Fig. 26–3}\right)$ is disconnected, the current *is* stopped to all the others.

In a parallel circuit, Fig. 26–4a, the total current I that leaves the battery breaks into three branches. We let I_1, I_2, and I_3 be the currents through each of the resistors, R_1, R_2, and R_3, respectively. Because electric charge is conserved, the current flowing into a junction (where different wires or conductors meet) must equal the current flowing out of the junction. Thus, in Fig. 26–4a,

$$I = I_1 + I_2 + I_3. \qquad \text{[parallel]}$$

Parallel circuit:
currents add;
voltage the same across each R

When resistors are connected in parallel, each experiences the same voltage. (Indeed, any two points in a circuit connected by a wire of negligible resistance are at the same potential.) Hence the full voltage of the battery is applied to each resistor in Fig. 26–4a, so

$$I_1 = \frac{V}{R_1}, \qquad I_2 = \frac{V}{R_2}, \qquad \text{and} \qquad I_3 = \frac{V}{R_3}.$$

Let us now determine what single resistor R_{eq} (Fig. 26–4c) will draw the same current I as these three resistances in parallel. This equivalent resistance R_{eq} must satisfy

$$I = \frac{V}{R_{\text{eq}}}.$$

We now combine the equations above:

$$I = I_1 + I_2 + I_3,$$

$$\frac{V}{R_{\text{eq}}} = \frac{V}{R_1} + \frac{V}{R_2} + \frac{V}{R_3}.$$

When we divide out the V from each term, we have

$$\frac{1}{R_{\text{eq}}} = \frac{1}{R_1} + \frac{1}{R_2} + \frac{1}{R_3}. \qquad \text{[parallel]} \quad \textbf{(26–3)}$$

Resistances in parallel

For example, suppose you connect two 4-Ω loudspeakers to a single set of output terminals of your stereo amplifier or receiver. (Ignore the other channel for a moment—our two speakers are both connected to the left channel, say.) The equivalent resistance will then be found from

$$\frac{1}{R_{\text{eq}}} = \frac{1}{4\,\Omega} + \frac{1}{4\,\Omega} = \frac{2}{4\,\Omega} = \frac{1}{2\,\Omega}$$

and so $R_{\text{eq}} = 2\,\Omega$. Thus the net resistance is *less* than that of each single resistance. This may at first seem surprising. But remember that when you put resistors in parallel, you are giving the current additional paths to follow. Hence the net resistance will be less.

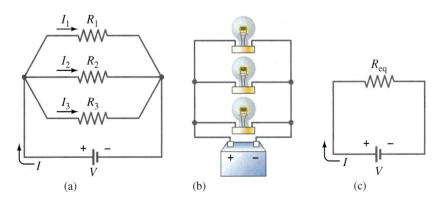

(a) (b) (c)

FIGURE 26–4 (a) Resistances connected in parallel: $1/R_{\text{eq}} = 1/R_1 + 1/R_2 + 1/R_3$, which could be (b) lightbulbs; (c) shows the equivalent circuit with R_{eq} obtained from Eq. 26–3.

FIGURE 26–5 Water pipes in parallel—analogy to electric currents in parallel.

FIGURE 26–6 Example 26–2.

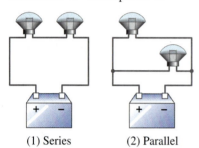

(1) Series   (2) Parallel

FIGURE 26–7 Example 26–3.

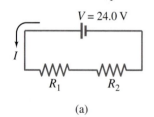

(a)

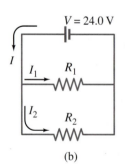

(b)

An analogy may help here. Consider two pipes taking in water near the top of a dam and releasing it below as shown in Fig. 26–5. The gravitational potential difference, proportional to the height h, is the same for both pipes, just as in the electrical case of parallel resistors. If both pipes are open, rather than only one, twice as much water current will flow through. That is, with two equal pipes open, the net resistance to the flow of water will be reduced, by half. Note that if both pipes are closed, the dam offers infinite resistance to the flow of water. This corresponds in the electrical case to an open circuit—when no current flows—so the electrical resistance is infinite.

Notice that the forms of the equations for resistors, Eqs. 26–2 and 26–3, are just the reverse of their counterparts for capacitors, Chapter 24, Eqs. 24–3 and 24–4. That is, the formula for resistors in series has the same form as the formula for capacitors in parallel, and vice versa.

CONCEPTUAL EXAMPLE 26–2 | **Series or Parallel?** (a) The lightbulbs in Fig. 26–6 are identical and have identical resistance R. Which configuration produces more light? (b) Which way do you think the headlights of a car are wired?

RESPONSE (a) The parallel combination has lower resistance ($= R/2$) than the series combination ($= 2R$). There will be more total current in configuration (2). The total power transformed, which is proportional to the light produced, is $P = IV$, so the greater current in (2) means more light.
(b) In parallel (2), because if one bulb goes out, the other bulb can stay lit. If they were in series (1), when one bulb burned out (the filament broke), the circuit would be open and no current would flow, so even the good bulb would not light.

EXAMPLE 26–3 **Series and parallel resistors.** Two 100-Ω resistors are connected (a) in series, and (b) in parallel, to a 24.0-V battery. See Fig. 26–7. What is the current through each resistor and what is the equivalent resistance of each circuit?

SOLUTION (a) All the current that flows out of the battery passes first through R_1 and then R_2. So the current I is the same in both resistors; and the potential difference across the battery, V, equals the total change in potential across the two resistors:

$$V = V_1 + V_2 = IR_1 + IR_2.$$

Hence

$$I = \frac{V}{R_1 + R_2} = \frac{24.0\text{ V}}{100\ \Omega + 100\ \Omega} = 0.120\text{ A.}$$

The equivalent resistance, using Eq. 26–2, is $R_{eq} = R_1 + R_2 = 200\ \Omega$. We could also get R_{eq} by thinking from the point of view of the battery: the total resistance R_{eq} must equal the battery voltage divided by the current it puts out: $R_{eq} = V/I = 24.0\text{ V}/0.120\text{ A} = 200\ \Omega$.
[Note that the voltage across R_1 is $V_1 = IR_1 = (0.120\text{ A})(100\ \Omega) = 12.0\text{ V}$, and that across R_2 is $V_2 = IR_2 = 12.0\text{ V}$, each being half of the battery voltage. A simple circuit like Fig. 26–7a is thus often called a simple "voltage divider."]
(b) Any given charge (or electron) can flow through only one or the other of the two resistors. Just as a river may break into two streams when going around an island, here too the total current I from the battery (Fig. 26–7b) equals the sum of the separate currents through the two resistors:

$$I = I_1 + I_2.$$

The potential difference across each resistor is the battery voltage $V = 24.0\text{ V}$. Hence

$$I = I_1 + I_2 = \frac{V}{R_1} + \frac{V}{R_2} = \frac{24.0\text{ V}}{100\ \Omega} + \frac{24.0\text{ V}}{100\ \Omega} = 0.24\text{ A} + 0.24\text{ A} = 0.48\text{ A.}$$

The equivalent resistance is

$$R_{eq} = \frac{V}{I} = \frac{24.0\,\text{V}}{0.48\,\text{A}} = 50\,\Omega.$$

We could also have obtained this result from Eq. 26–3:

$$\frac{1}{R_{eq}} = \frac{1}{100\,\Omega} + \frac{1}{100\,\Omega} = \frac{2}{100\,\Omega} = \frac{1}{50\,\Omega},$$

and so $R_{eq} = 50\,\Omega$.

EXAMPLE 26–4 **Circuit with series and parallel.** How much current flows from the battery shown in Fig. 26–8a?

SOLUTION The current I that flows out of the battery all passes through the 400-Ω resistor, but then it splits into I_1 and I_2 passing through the 500-Ω and 700-Ω resistors. The latter two are in parallel. We look for simplicity, something that we already know how to treat. So let's start by finding the equivalent resistance, R_P, of the parallel resistors, 500-Ω and 700-Ω:

$$\frac{1}{R_P} = \frac{1}{500\,\Omega} + \frac{1}{700\,\Omega} = 0.0020\,\Omega^{-1} + 0.0014\,\Omega^{-1} = 0.0034\,\Omega^{-1}.$$

This is $1/R_P$, so we take the reciprocal to find R_P. (It is a common mistake to forget to take this reciprocal. Notice that the units of reciprocal ohms, Ω^{-1}, are a reminder.) Thus

$$R_P = \frac{1}{0.0034\,\Omega^{-1}} = 290\,\Omega.$$

This 290 Ω is the equivalent resistance of the two parallel resistors, and is in series with the 400-Ω resistor as shown in the equivalent circuit of Fig. 26–8b. To find the total equivalent resistance R_{eq}, we add the 400-Ω and 290-Ω resistances together, since they are in series, and find

$$R_{eq} = 400\,\Omega + 290\,\Omega = 690\,\Omega.$$

The total current flowing from the battery is then

$$I = \frac{V}{R_{eq}} = \frac{12.0\,\text{V}}{690\,\Omega} = 0.017\,\text{A} = 17\,\text{mA}.$$

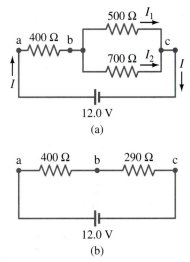

FIGURE 26–8 (a) Circuit for Examples 26–4 and 26–5. (b) Equivalent circuit, showing the equivalent resistance of 290 Ω for the two parallel resistors in (a).

EXAMPLE 26–5 **Current in one branch.** What is the current flowing through the 500-Ω resistor in Fig. 26–8a?

SOLUTION We need to find the voltage V_{bc} across the 500-Ω resistor, and then apply $V = IR$ to get the current. First we find the voltage across the 400-Ω resistor, V_{ab}, since we know that 17 mA passes through it; V_{ab} can be found using $V = IR$:

$$V_{ab} = (0.017\,\text{A})(400\,\Omega) = 6.8\,\text{V}.$$

Since the total voltage across the network of resistors is $V_{ac} = 12.0\,\text{V}$, then V_{bc} must be $12.0\,\text{V} - 6.8\,\text{V} = 5.2\,\text{V}$. Then Ohm's law tells us that the current I_1 through the 500-Ω resistor is

$$I_1 = \frac{5.2\,\text{V}}{500\,\Omega} = 1.0 \times 10^{-2}\,\text{A} = 10\,\text{mA}.$$

This is the answer we wanted. We can also calculate the current I_2 through the 700-Ω resistor since the voltage across it is also 5.2 V:

$$I_2 = \frac{5.2\,\text{V}}{700\,\Omega} = 7\,\text{mA}.$$

Notice that when I_1 combines with I_2 to form the total current I (at point c in Fig. 26–8a), their sum is $10\,\text{mA} + 7\,\text{mA} = 17\,\text{mA}$. This is, of course, the total current as calculated in Example 26–4.

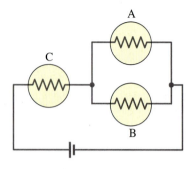

FIGURE 26–9 Example 26–6, three identical lightbulbs.

FIGURE 26–10 Example 26–7.

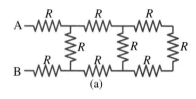

(a)

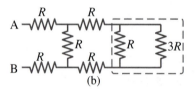

(b)

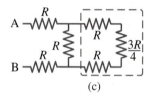

(c)

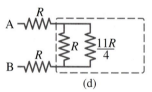

(d)

FIGURE 26–11 Currents can be calculated using Kirchhoff's rules.

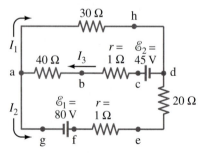

CONCEPTUAL EXAMPLE 26–6 **Bulb brightness in a circuit.** The circuit shown in Fig. 26–9 has three identical lightbulbs, each of resistance R. How will the brightness of bulbs A and B compare with that of bulb C?

RESPONSE The current that passes through bulb C splits into two equal parts when it reaches the junction leading to bulbs A and B, because the resistance of bulbs A and B are equal. Thus, bulbs A and B receive half the current, and will be less bright than bulb C.

EXAMPLE 26–7 **Resistor "ladder."** (a) Estimate the equivalent resistance of the "ladder" of equal 100-Ω resistors shown in Fig. 26–10a. In other words, what resistance would an ohmmeter read if connected between points A and B? (b) What is the current through each of the three resistors on the left if a 50.0 V battery is connected between points A and B?

SOLUTION It may seem that none of these resistors is in series or parallel. But there is a place to start: the three resistors on the far right are definitely in series with one another. Each has resistance $R(= 100\,\Omega)$, so those last three have a net resistance of $3R(= 300\,\Omega)$. Next we can see that this combination is in parallel with the next resistor to the left, as shown in the dashed box in Fig. 26–10b. The equivalent resistance of the resistors in the dashed box (b), call it R_{eq1}, is given by

$$\frac{1}{R_{\text{eq1}}} = \frac{1}{R} + \frac{1}{3R} = \frac{4}{3R}, \qquad \text{so} \qquad R_{\text{eq1}} = \frac{3R}{4}$$

which numerically is $300\,\Omega/4 = 75\,\Omega$. Next, this equivalent resistance of $3R/4$ is in series with the next two (Fig. 26–10c). The resistors in the dashed box (c) are in series, and are equivalent to $2R + 3R/4 = 11R/4$ (which equals $1100\,\Omega/4 = 275\,\Omega$). Now this $11R/4$ is in parallel with the next step on the ladder (Fig. 26–10d). The resistance in the dashed box (d) is equivalent to

$$\frac{1}{R_{\text{eq2}}} = \frac{1}{R} + \frac{4}{11R} = \frac{15}{11R}, \qquad \text{so} \qquad R_{\text{eq2}} = \frac{11R}{15}.$$

This in turn is in series with two more, yielding the final equivalent resistance of

$$R_{\text{eq}} = \frac{11R}{15} + R + R = \frac{41}{15}R.$$

We are given that $R = 100\,\Omega$, so $R_{\text{eq}} = 273\,\Omega$. Notice again the value of using algebra all along: we can have an answer that enables us to calculate the overall resistance regardless of the value of the individual resistor.

(b) If a 50.0 V battery is placed between points A and B, then the current issuing from the battery, and passing through the two resistors on the left, is $I = 50.0\,\text{V}/273\,\Omega = 0.183\,\text{A}$. The current through the first cross resistance is less than this since this R is in parallel with a net resistance of $2\frac{3}{4}R = \frac{11}{4}R$ (Fig. 26–10d). Let I_1 be the current through this R and I_2 the current in the $\frac{11}{4}R$, with $I_1 + I_2 = I$. The potential difference across R is the same as that across the $\frac{11}{4}R$, so $I_1 R = I_2(\frac{11}{4}R)$, and so $I_2 = \frac{4}{11}I_1$. Then

$$I = I_1 + I_2 = \left(1 + \tfrac{4}{11}\right)I_1 = \tfrac{15}{11}I_1.$$

Hence $I_1 = \frac{11}{15}I = \frac{11}{15}(0.183\,\text{A}) = 0.134\,\text{A}$.

26–3 Kirchhoff's Rules

In the last few Examples we have been able to find the currents flowing in circuits by combining resistances in series and parallel. This technique can be used for many circuits. However, some circuits are too complicated for that analysis. For example, we cannot find the currents flowing in each part of the circuit shown in Fig. 26–11 simply by combining resistances as we did before.

To deal with such complicated circuits, we use Kirchhoff's rules, devised by G. R. Kirchhoff (1824–1887) in the mid-nineteenth century. There are two of them, and they are simply convenient applications of the laws of conservation of charge and energy. **Kirchhoff's first** or **junction rule** is based on the conservation of charge, and we already used it in deriving the rule for parallel resistors. It states that

> at any junction point, the sum of all currents entering the junction must equal the sum of all currents leaving the junction.

Junction rule
(conservation of charge)

That is, what goes in must come out. For example, at the junction point a in Fig. 26–11, I_3 is entering whereas I_1 and I_2 are leaving. Thus Kirchhoff's junction rule states that $I_3 = I_1 + I_2$. We already saw an instance of this at the end of Example 26–5.

Kirchhoff's **second** or **loop rule** is based on the conservation of energy. It states

> the sum of the changes in potential around any closed path of a circuit must be zero.

Loop rule
(conservation of energy)

To see why this should hold, consider a rough analogy with the potential energy of a roller coaster on its track. When it starts from the station, it has a particular potential energy. As it climbs the first hill, its potential energy increases and reaches a peak at the top. As it descends the other side, its potential energy decreases and reaches a local minimum at the bottom of the hill. As the roller coaster continues on its path, its potential energy goes through more changes. But when it arrives back at the starting point, it has exactly as much potential energy as it had when it started at this point. Another way of saying this is that there was as much uphill as there was downhill.

Similar reasoning can be applied to an electric circuit. We will do the circuit of Fig. 26–11 shortly but first we consider the simpler circuit in Fig. 26–12. We have chosen it to be the same as the equivalent circuit of Fig. 26–8b already discussed. The current in this circuit is $I = (12.0\,\text{V})/(690\,\Omega) = 0.017\,\text{A}$, as we calculated in Example 26–4. The positive side of the battery, point e in Fig. 26–12a, is at a high potential compared to point d at the negative side of the battery. That is, point e is like the top of a hill for a roller coaster. We can now follow the current around the circuit starting at any point we choose. Let us start at point e and follow a positive test charge completely around this circuit. As we go, we will note all changes in potential. When the test charge returns to point e, the potential there will be the same as when we started, so the total change in potential will be zero. It is useful to plot the changes in voltage around the circuit, and we do this in Fig. 26–12b; point d is arbitrarily taken as zero. As our positive test charge goes from point e to point a, there is no change in potential since there is no source of potential nor any resistance. However, as the charge passes through the 400-Ω resistor to get to point b, there is a decrease in potential of $V = IR = (0.017\,\text{A})(400\,\Omega) = 6.8\,\text{V}$. In effect, the positive test charge is flowing "downhill" since it is heading toward the negative terminal of the battery. This is indicated in the graph of Fig. 26–12b. The decrease in potential between the two ends of a resistor ($= IR$) is called a **voltage drop**. Because this is a *decrease* in potential, we use a *negative* sign when applying Kirchhoff's loop rule; that is,

$$V_{\text{ba}} = V_{\text{b}} - V_{\text{a}} = -6.8\,\text{V}.$$

As the charge proceeds from b to c there is another voltage drop of $(0.017\,\text{A}) \times (290\,\Omega) = 5.2\,\text{V}$, and since this is a decrease in potential, we write

$$V_{\text{cb}} = -5.2\,\text{V}.$$

There is no change in potential as our test charge moves from c to d. But when it moves from d, which is the negative or low potential side of the battery, to point e which is the positive terminal, the potential *increases* by 12.0 V. That is,

$$V_{\text{ed}} = +12.0\,\text{V}.$$

The sum of all the changes in potential in going around the circuit of Fig. 26–12 is

$$-6.8\,\text{V} - 5.2\,\text{V} + 12.0\,\text{V} = 0.$$

And this is exactly what Kirchhoff's loop rule said it would be.

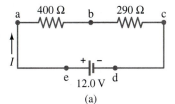

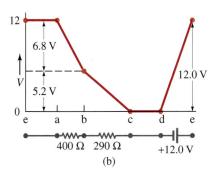

FIGURE 26–12 Changes in potential around the circuit in (a) are plotted in (b).

➡ **PROBLEM SOLVING**

Be consistent with signs

When using Kirchhoff's rules, we will designate the current in each separate branch of the given circuit by a different subscript, such as I_1, I_2, and I_3 in Fig. 26–13 (this is the same circuit as in Fig. 26–11). You do not have to know in advance in which direction these currents actually are moving. You make a guess and calculate the potentials around the circuit for that direction. If the current actually flows in the opposite direction, your answer will have a negative sign.

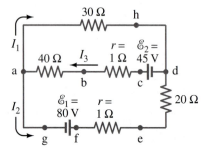

FIGURE 26–13 Currents can be calculated using Kirchhoff's rules. See Example 26–8.

EXAMPLE 26–8   Using Kirchhoff's rules.   Calculate the currents I_1, I_2, and I_3 in each of the branches of the circuit in Fig. 26–13.

SOLUTION   Since (positive) current tends to move away from the positive terminal of a battery, we assume I_2 and I_3 to have the directions shown in Fig. 26–13. The direction of I_1 is not obvious in advance, so we arbitrarily chose the direction indicated. We have three unknowns and therefore we need three equations. We first apply Kirchhoff's junction rule to the currents at point a, where I_3 enters and I_2 and I_1 leave:

$$I_3 = I_1 + I_2. \tag{a}$$

This same equation holds at point d, so we get no new information there. We now apply Kirchhoff's loop rule to two different closed loops. First we apply it to the loop ahdcba. We start (and end) at point a. From a to h we have a voltage drop $V_{ha} = -(I_1)(30\,\Omega)$. From h to d there is no change, but from d to c the potential increases by 45 V: that is, $V_{cd} = +45$ V. From c to a the voltage drops through the two resistances by an amount $V_{ac} = -(I_3)(40\,\Omega + 1\,\Omega)$. Thus we have $V_{ha} + V_{cd} + V_{ac} = 0$, or

$$-30I_1 + 45 - (40 + 1)I_3 = 0 \tag{b}$$

where we have omitted the units. For our second loop, we take the complete circuit ahdefga. (We could have just as well taken abcdefg instead.) Again we start at point a and have $V_{ha} = -(I_1)(30\,\Omega)$, and $V_{dh} = 0$. But when we take our positive test charge from d to e, it actually is going uphill, against the flow of current—or at least against the *assumed* direction of the current, which is what counts in this calculation. Thus $V_{ed} = I_2(20\,\Omega)$ has a *positive* sign. Similarly, $V_{fe} = I_2(1\,\Omega)$. From f to g there is a decrease in potential of 80 V since we go from the high potential terminal of the battery to the low. Thus $V_{gf} = -80$ V. Finally, $V_{ag} = 0$, and the sum of the potential charges around this loop is then

$$-30I_1 + (20 + 1)I_2 - 80 = 0. \tag{c}$$

The physics is now done. The rest is algebra. We have three equations—labeled (a), (b), and (c)—in three unknowns. From Eq. (c) we have

$$I_2 = \frac{80 + 30I_1}{21} = 3.8 + 1.4I_1. \tag{d}$$

From Eq. (b) we have

$$I_3 = \frac{45 - 30I_1}{41} = 1.1 - 0.73I_1. \tag{e}$$

We substitute these into Eq. (a) and solve for I_1:

$$I_1 = I_3 - I_2 = 1.1 - 0.73I_1 - 3.8 - 1.4I_1$$
$$3.1I_1 = -2.7$$
$$I_1 = -0.87 \text{ A}.$$

The negative sign indicates that the direction of I_1 is actually opposite to that initially assumed and shown in Fig. 26–13. Note that the answer automatically comes out in amperes because all values were in volts and ohms. From Eq. (d) we have

$$I_2 = 3.8 + 1.4I_1 = 2.6 \text{ A},$$

and from Eq. (e)

$$I_3 = 1.1 - 0.73I_1 = 1.7 \text{ A}.$$

This completes the solution.

Kirchhoff's Rules

1. Label + and − for each battery. The long side of a battery symbol is +.

2. Label the current in each branch of the circuit with a symbol and an arrow (as in Fig. 26–13): The direction of the arrow can be chosen arbitrarily. If the current is actually in the opposite direction, it will come out with a minus sign in the solution.

3. Apply Kirchhoff's junction rule at one or more junctions, and the loop rule for one or more loops. You will need as many independent equations as there are unknowns. You may write down more equations than this, but you will find that some of the equations will be redundant (that is, not be independent in the sense of providing new information). You may use $V = IR$ for each resistor, which sometimes will reduce the number of unknowns.

4. In applying the loop rule, follow each loop in one direction only. Pay careful attention to subscripts, and to signs:

   (a) For a resistor, the sign of the potential difference is negative if your chosen loop direction is the same as the chosen current direction through that resistor; the sign is positive if you are moving opposite to the chosen current direction.

   (b) For a battery, the sign of the potential difference is positive if your loop direction moves from the negative terminal toward the positive; the sign is negative if you are moving from the positive terminal toward the negative terminal.

5. Solve the equations algebraically for the unknowns. Be careful in manipulating equations not to err with signs. At the end, check your answers by plugging them into the original equations, or even by using any additional equations not used previously (either loop or junction rule equations).

EXAMPLE 26–9 Wheatstone bridge. A wheatstone bridge is a type of "bridge circuit" used to make measurements of resistance. The unknown resistance to be measured, R_x, is placed in the circuit with accurately known resistances R_1, R_2, and R_3. One of these, R_3, is a variable resistor which is adjusted so that when the switch is closed momentarily, the ammeter Ⓐ shows zero current flow. (We will see later in this Chapter how an ammeter works.) (a) Determine R_x in terms of R_1, R_2, and R_3. (b) If a Wheatstone bridge is "balanced" when $R_1 = 630\,\Omega$, $R_2 = 972\,\Omega$, and $R_3 = 42.6\,\Omega$, what is the value of the unknown resistance?

SOLUTION (a) We are told that R_3 has been adjusted until no current flows through the ammeter. Hence points B and D in Fig. 26–14 are at the same potential, so $V_{AB} = V_{AD}$ or

$$I_3 R_3 = I_1 R_1.$$

I_1 is the current that passes through R_1 and also through R_2 when the bridge is balanced; I_3 is the current through R_3 and R_x. When the bridge is balanced the voltage across R_x equals that across R_2, so

$$I_3 R_x = I_1 R_2.$$

We divide these two equations and find

$$R_x = \frac{R_2}{R_1} R_3.$$

In practice, the ammeter is very sensitive, and so when R_3 is being adjusted the switch is closed only momentarily to check if the current is zero or not.

(b)
$$R_x = \frac{R_2}{R_1} R_3 = \left(\frac{972\,\Omega}{630\,\Omega}\right)(42.6\,\Omega) = 65.7\,\Omega.$$

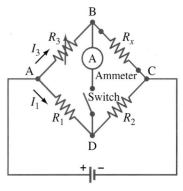

FIGURE 26–14 Example 26–9, Wheatstone bridge.

EMFs in Series and in Parallel; Charging a Battery

When two or more sources of emf, such as batteries, are arranged in series, as in Fig. 26–15a, the total voltage is the algebraic sum of their respective voltages. On the other hand, when a 20-V and a 12-V battery are connected oppositely, as shown in Fig. 26–15b, the net voltage V_{ca} is 8 V. That is, a positive test charge moved from a to b gains in potential by 20 V, but when it passes from b to c it drops by 12 V. So the net change is $20\,\text{V} - 12\,\text{V} = 8\,\text{V}$. You might think that connecting batteries in reverse like this would be wasteful. And for most purposes that would be true. But such a reverse arrangement is precisely how a battery charger works. In Fig. 26–15b, the 20-V source is charging up the 12-V battery. Because of its greater voltage, the 20-V source is forcing charge back into the 12-V battery: electrons are being forced into its negative terminal and removed from its positive terminal. An automobile alternator keeps the car battery charged in the same way. A voltmeter placed across the terminals of a (12-V) car battery with the engine running fairly fast can tell you whether or not the alternator is charging the battery. If it is, the voltmeter reads 13 or 14 V. If the battery is not being charged, the voltage will be 12 V, or less if the battery is discharging. Car batteries can be recharged, but other batteries may not be rechargeable, since the chemical reactions in many cannot be reversed. In such cases, the arrangement of Fig. 26–15b would simply waste energy.

FIGURE 26–15 Batteries in series, (a) and (b), and in parallel, (c).

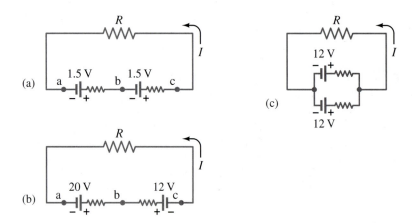

Sources of emf can also be arranged in parallel, Fig. 26–15c, which is useful normally only if the emfs are the same. A parallel arrangement is not used to increase voltage, but rather to provide more energy when large currents are needed. Each of the cells in parallel has to produce only a fraction of the total current, so the loss due to internal resistance is less than for a single cell; and the batteries will go dead less quickly.

FIGURE 26–16 Example 26–10, a jump start.

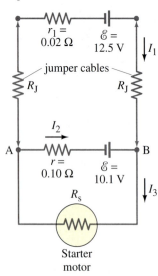

EXAMPLE 26–10 **Jump starting a car.** A good car battery is being used to jump start a car with a weak battery. The good battery has an emf of 12.5 V and internal resistance $0.020\,\Omega$. Suppose the weak battery has an emf of 10.1 V and internal resistance $0.10\,\Omega$. Each copper jumper cable is 3.0 m long and 0.50 cm in diameter, and can be attached as shown in Fig. 26–16. Assume the starter motor can be represented as a resistor $R_s = 0.15\,\Omega$. Determine the current through the starter motor (a) if only the weak battery is connected to it, and (b) if the good battery is also connected, as shown in Fig. 26–16.

SOLUTION (a) The circuit is simple: an emf of 10.1 V connected to two resistances in series, $0.10\,\Omega + 0.15\,\Omega = 0.25\,\Omega$. Hence the current is $I = V/R = (10.1\,\text{V})/(0.25\,\Omega) = 40\,\text{A}$.

(b) We need to find the resistance of the jumper cables that connect the good battery. From Eq. 25–3, each has resistance $R_J = \rho L/A =$ $(1.68 \times 10^{-8}\,\Omega\cdot\text{m})(3.0\,\text{m})/(\pi)(0.25 \times 10^{-2}\,\text{m})^2 = 0.0026\,\Omega$. Kirchhoff's loop rule for the full outside loop gives

$$12.5\,\text{V} - I_1(2R_J + r) - I_3 R_S = 0$$
$$12.5\,\text{V} - I_1(0.025\,\Omega) - I_3(0.15\,\Omega) = 0. \qquad (a)$$

The loop rule for the lower loop, including the weak battery and the starter, gives

$$10.1\,\text{V} - I_3(0.15\,\Omega) - I_2(0.10\,\Omega) = 0. \qquad (b)$$

The junction rule at point B gives

$$I_1 + I_2 = I_3. \qquad (c)$$

We have three equations in three unknowns. We combine equations (a) and (c), eliminating I_1, to obtain

$$12.5\,\text{V} - (I_3 - I_2)(0.025\,\Omega) - I_3(0.15\,\Omega) = 0$$
$$12.5\,\text{V} - I_3(0.175\,\Omega) + I_2(0.025\,\Omega) = 0.$$

Combining this last with (b) gives $I_3 = 71\,\text{A}$. Quite a bit better than in (a). The other currents are $I_2 = -6.2\,\text{A}$ and $I_1 = 77\,\text{A}$.

The circuit shown in Fig. 26–16, without the starter motor, is how a battery can be charged. The stronger battery pushes charge back into the weaker battery. Note in this Example that $I_2 = -6.2\,\text{V}$ and so is in the opposite direction from that assumed in Fig. 26–16. The terminal voltage of the weak 10.1 V battery is thus $V_{BA} = 10.1\,\text{V} + (6.2\,\text{A})(0.10\,\Omega) = 10.7\,\text{V}$.

EXAMPLE 26–11 **Jumper cables reversed.** What would happen if the jumper cables of Example 26–10 were mistakenly connected in reverse, the positive terminal of each battery connected to the negative terminal of the other battery? Why could this be dangerous?

SOLUTION The circuit is shown in Fig. 26–17. Even before the starter motor is engaged (the switch S in Fig. 26–17 is open) there is trouble if the batteries are connected in this way. By Kirchhoff's loop rule in the single (upper) loop carrying current I, we have

$$12.5\,\text{V} - I(2R_J + 0.10\,\Omega + 0.02\,\Omega) + 10.1\,\text{V} = 0$$

where each $R_J = 0.0026\,\Omega$. We solve for I:

$$I = \frac{22.6\,\text{V}}{0.125\,\Omega} = 180\,\text{A}.$$

The extremely high current through the batteries could cause them to become very hot and explode. For example, the power dissipated in the weak battery would be $P = I^2 r = (180\,\text{A})^2(0.10\,\Omega) = 3200\,\text{W}!$

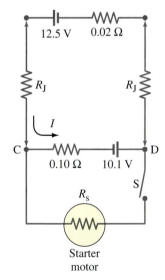

FIGURE 26–17 Example 26–11.

26–4 Circuits Containing Resistor and Capacitor (RC Circuits)

Our study of circuits in this chapter has, until now, dealt with steady currents that don't change in time. Now we examine circuits that contain both resistance and capacitance. Such a circuit is called an **RC circuit**. RC circuits are common in everyday life: they are used to control the speed of a car's windshield wiper, and the timing of the change of a traffic light from red to green. They are used in camera flashes and in heart pacemakers.

RC circuit

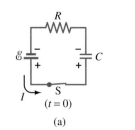

Charging the capacitor

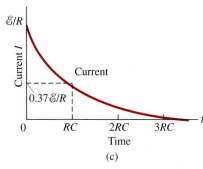

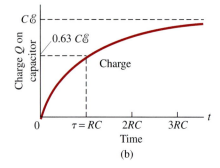

FIGURE 26–18 For the *RC* circuit shown in (a), the charge on the capacitor increases with time as shown in (b), and the current through the resistor decreases with time as shown in (c).

Charge on charging capacitor

Voltage across charging capacitor

Time constant = RC

Let us now examine the simple *RC* circuit shown in Fig. 26–18a. When the switch S is closed, current immediately begins to flow through the circuit. Electrons will flow out from the negative terminal of the battery, through the resistor *R*, and accumulate on the upper plate of the capacitor. And electrons will flow into the positive terminal of the battery, leaving a positive charge on the other plate of the capacitor. As charge accumulates on the capacitor, the potential difference across it increases; and the current is reduced until eventually the voltage across the capacitor equals the emf of the battery, $\mathcal{E}$. There is then no potential difference across the resistor, and no further current flows. The charge *Q* on the capacitor thus increases gradually as shown in Fig. 26–18b and reaches a maximum value equal to $C\mathcal{E}$ (Eq. 24–1, $Q_{max} = CV_{ba} = C\mathcal{E}$). The mathematical form of this curve—that is, *Q* as a function of time—can be derived using conservation of energy (or Kirchhoff's loop rule). The emf $\mathcal{E}$ of the battery will equal the sum of the voltage drops across the resistor (*IR*) and the capacitor (*Q/C*):

$$\mathcal{E} = IR + \frac{Q}{C}. \quad (26\text{–}4)$$

The resistance *R* includes all resistance in the circuit, including the internal resistance of the battery; *I* is the current in the circuit at any instant, and *Q* is the charge on the capacitor at that same instant. Although $\mathcal{E}$, *R*, and *C* are constants, both *Q* and *I* are functions of time. The rate at which charge flows through the resistor ($I = dQ/dt$) is equal to the rate at which charge accumulates on the capacitor. Thus we can write

$$\mathcal{E} = R\frac{dQ}{dt} + \frac{1}{C}Q.$$

This equation can be solved by rearranging it:

$$\frac{dQ}{C\mathcal{E} - Q} = \frac{dt}{RC}.$$

We now integrate from $t = 0$, when $Q = 0$, to time *t* when a charge *Q* is on the capacitor:

$$\int_0^Q \frac{dQ}{C\mathcal{E} - Q} = \frac{1}{RC}\int_0^t dt$$

$$-\ln(C\mathcal{E} - Q) - (-\ln C\mathcal{E}) = \frac{t}{RC}$$

or

$$\ln(C\mathcal{E} - Q) - \ln(C\mathcal{E}) = -\frac{t}{RC}$$

so

$$\ln\left(1 - \frac{Q}{C\mathcal{E}}\right) = -\frac{t}{RC}.$$

We take the exponential of both sides

$$1 - \frac{Q}{C\mathcal{E}} = e^{-t/RC}$$

or

$$Q = C\mathcal{E}(1 - e^{-t/RC}). \quad (26\text{–}5a)$$

The potential difference across the capacitor is $V_C = Q/C$, so

$$V_C = \mathcal{E}(1 - e^{-t/RC}). \quad (26\text{–}5b)$$

From Eqs. 26–5 we see that the charge *Q* on the capacitor, and the voltage V_C across it, increase from zero at $t = 0$ to maximum values $Q_{max} = C\mathcal{E}$ and $V_C = \mathcal{E}$ after a very long time. The quantity *RC* that appears in the exponent is called the **time constant** τ of the circuit:

$$\tau = RC.$$

(The units of *RC* are $\Omega \cdot F = (V/A)(C/V) = C/(C/s) = s$.) It represents the time required for the capacitor to reach $(1 - e^{-1}) = 0.63$ or 63 percent of its full charge. Thus the product RC is a measure of how quickly the capacitor gets charged. In a circuit, for example, where $R = 200\,\text{k}\Omega$ and $C = 3.0\,\mu\text{F}$, the time constant is $(2.0 \times 10^5\,\Omega)(3.0 \times 10^{-6}\,\text{F}) = 0.60\,\text{s}$. If the resistance is much lower, the time constant is much smaller. This makes sense, since a lower resistance will retard the flow of charge less. All circuits contain some resistance (if only in the connecting wires), so a capacitor never can be charged instantaneously when connected to a battery.

From Eqs. 26–5, it appears that Q and V_C never quite reach their maximum values within a finite time. However, they reach 86 percent of maximum in $2RC$, 95 percent in $3RC$, 98 percent in $4RC$, and so on. Q and V_C approach their maximum values asymptotically. For example, if $R = 20\,\text{k}\Omega$ and $C = 0.30\,\mu\text{F}$, the time constant is $(2.0 \times 10^4\,\Omega)(3.0 \times 10^{-7}\,\text{F}) = 6.0 \times 10^{-3}\,\text{s}$. So the capacitor is more than 98 percent charged in less than $\frac{1}{40}$ of a second.

The current I through the circuit of Fig. 26–18a at any time t can be obtained by differentiating Eq. 26–5a:

$$I = \frac{dQ}{dt} = \frac{\mathscr{E}}{R}e^{-t/RC}. \qquad\qquad (26\text{--}6) \qquad \textit{Current in resistor}$$

Thus, at $t = 0$, the current is $I = \mathscr{E}/R$, as expected for a circuit containing only a resistor (there is not yet a potential difference across the capacitor). The current then drops exponentially in time with a time constant equal to RC. This is shown in Fig. 26–18c. The time constant RC represents the time required for the current to drop to $1/e \approx 0.37$ of its initial value.

EXAMPLE 26–12 *RC* **circuit, with emf.** The capacitance in the circuit of Fig. 26–18a is $C = 0.30\,\mu\text{F}$, the total resistance is $20\,\text{k}\Omega$, and the battery emf is 12 V. Determine (*a*) the time constant, (*b*) the maximum charge the capacitor could acquire, (*c*) the time it takes for the charge to reach 99 percent of this value, (*d*) the current I when the charge Q is half its maximum value, (*e*) the maximum current, and (*f*) the charge Q when, the current I is 0.20 its maximum value.

SOLUTION (*a*) The time constant is $RC = (2.0 \times 10^4\,\Omega)(3.0 \times 10^{-7}\,\text{F}) = 6.0 \times 10^{-3}\,\text{s}$.
(*b*) The maximum charge would be $Q = C\mathscr{E} = (3.0 \times 10^{-7}\,\text{F})(12\,\text{V}) = 3.6\,\mu\text{C}$.
(*c*) In Eq. 26–5a, we set $Q = 0.99C\mathscr{E}$:

$$0.99C\mathscr{E} = C\mathscr{E}(1 - e^{-t/RC}),$$

or

$$e^{-t/RC} = 1 - 0.99 = 0.01.$$

Then

$$\frac{t}{RC} = -\ln(0.01) = 4.6$$

so

$$t = 4.6RC = 28 \times 10^{-3}\,\text{s}$$

or 28 ms (less than $\frac{1}{30}$ s).
(*d*) From part (*b*) the maximum charge is $3.6\,\mu\text{C}$. When the charge is half this value, $1.8\,\mu\text{C}$, the current I in the circuit can be found using the original differential equation, or Eq. 26–4:

$$I = \frac{1}{R}\left(\mathscr{E} - \frac{Q}{C}\right) = \frac{1}{2.0 \times 10^4\,\Omega}\left(12\,\text{V} - \frac{1.8 \times 10^{-6}\,\text{C}}{0.30 \times 10^{-6}\,\text{F}}\right) = 300\,\mu\text{A}.$$

(*e*) The current is a maximum when there is no charge on the capacitor ($Q = 0$):

$$I_{\text{max}} = \frac{\mathscr{E}}{R} = \frac{12\,\text{V}}{2.0 \times 10^4\,\Omega} = 600\,\mu\text{A}.$$

(*f*) Again using Eq. 26–4, with $I = 0.20I_{\text{max}} = 120\,\mu\text{A}$, we have

$$Q = C(\mathscr{E} - IR) = (3.0 \times 10^{-7}\,\text{F})\left[12\,\text{V} - (1.2 \times 10^{-4}\,\text{A})(2.0 \times 10^4\,\Omega)\right] = 2.9\,\mu\text{C}.$$

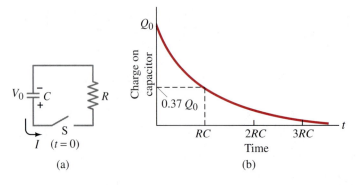

FIGURE 26–19 For the *RC* circuit shown in (a), the charge *Q* on the capacitor decreases with time, as shown in (b), after the switch S is closed at $t = 0$. The voltage across the capacitor follows the same curve since $V \propto Q$.

(a)

(b)

The circuit just discussed involved the *charging* of a capacitor by a battery through a resistance. Now let us look at another situation: when a capacitor is already charged (say to a voltage V_0), and it is allowed to *discharge* through a resistance *R* as shown in Fig. 26–19a. (In this case there is no battery.) When the switch S is closed, charge begins to flow through resistor *R* from one side of the capacitor toward the other side, until the capacitor is fully discharged. The voltage across the resistor at any instant equals that across the capacitor:

Capacitor discharges

$$IR = \frac{Q}{C}.$$

The rate at which charge leaves the capacitor equals the negative of the current in the resistor, $I = -dQ/dt$, because the capacitor is discharging (*Q* is decreasing). So we write the above equation as

$$-\frac{dQ}{dt} R = \frac{Q}{C}.$$

We rearrange this to

$$\frac{dQ}{Q} = -\frac{dt}{RC}$$

and integrate it from $t = 0$ when the charge on the capacitor is Q_0, to some time *t* when the charge is *Q*:

$$\ln \frac{Q}{Q_0} = -\frac{t}{RC}$$

or

Charge on discharging capacitor

$$Q = Q_0 e^{-t/RC}. \qquad (26\text{–}7)$$

Thus the charge on the capacitor decreases exponentially in time with a time constant *RC*. This is shown in Fig. 26–19b. The current is

Current in resistor

$$I = -\frac{dQ}{dt} = \frac{Q_0}{RC} e^{-t/RC} = I_0 e^{-t/RC}, \qquad (26\text{–}8)$$

and it too is seen to decrease exponentially in time with the same time constant *RC*. Both the charge on the capacitor and the voltage across it ($V_C = Q/C$), as well as the current in the resistor, decrease to 37 percent of their original value in one time constant $t = \tau = RC$.

EXAMPLE 26–13 **Discharging *RC* circuit.** In the *RC* circuit shown in Fig. 26–20, the battery has fully charged the capacitor, so $Q_0 = C\mathscr{E}$. Then at $t = 0$ the switch is thrown from position a to b. The battery emf is 20.0 V, and the capacitance $C = 1.02\ \mu\text{F}$. The current *I* is observed to decrease to 0.50 of its initial value in 40 μs. (*a*) What is the value of *R*? (*b*) What is the value of *Q*, the charge on the capacitor, at $t = 0$? (*c*) What is *Q* at $t = 60\ \mu$s?

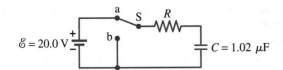

FIGURE 26–20   Example 26–13.

SOLUTION (*a*) At $t = 0$, the battery is removed from the circuit and the capacitor begins discharging through the resistor, as in Fig. 26–19. At any time t later (Eq. 26–7) we have

$$Q = Q_0 e^{-t/RC} = C\mathscr{E}e^{-t/RC},$$

and

$$I = -\frac{dQ}{dt} = \frac{\mathscr{E}}{R}e^{-t/RC} = I_0 e^{-t/RC}.$$

At $t = 40\,\mu s$, $I = 0.50I_0$. Hence

$$0.50I_0 = I_0 e^{-t/RC}$$

or, taking natural logs on both sides ($\ln 0.50 = -0.693$):

$$0.693 = \frac{t}{RC}$$

and

$$R = \frac{t}{(0.693)C} = \frac{(40 \times 10^{-6}\,\text{s})}{(0.693)(1.02 \times 10^{-6}\,\text{F})} = 57\,\Omega.$$

(*b*) At $t = 0$,

$$Q = Q_0 = C\mathscr{E} = (1.02 \times 10^{-6}\,\text{F})(20.0\,\text{V}) = 20.4\,\mu C.$$

(*c*) At $t = 60\,\mu s$,

$$Q = Q_0 e^{-\frac{60 \times 10^{-6}\,\text{s}}{(57\,\Omega)(1.02 \times 10^{-6}\,\text{F})}} = 7.3\,\mu C.$$

Applications of *RC* Circuits

The charging and discharging in an *RC* circuit can be used to produce voltage pulses at a regular frequency. The charge on the capacitor increases to a particular voltage, and then discharges. A simple way of initiating the discharge is by the use of a gas-filled tube that breaks down when the voltage across it reaches a certain value V_0. After the discharge is finished, the tube no longer conducts current and the recharging process repeats itself, starting at V_0'. Figure 26–21 shows a possible circuit, and the "sawtooth" voltage it produces.

An automobile turn signal indicator can be an application of a sawtooth oscillator circuit. Here the emf is supplied by the car battery ($\mathscr{E} = 12\,\text{V}$); the neon bulb, which flashes on at a rate of perhaps 2 cycles per second, is the turn signal indicator. The main component of the "flasher unit" is a moderately large capacitor.

The intermittent windshield wipers of a car can also use an *RC* circuit. The *RC* time constant, which can be changed using a multi-positioned switch for different values of R with fixed C, determines the rate at which the wipers come on.

Another interesting use of an *RC* circuit is the electronic heart pacemaker, which can make a stopped heart start beating again by applying an electric stimulus through electrodes attached to the chest. The stimulus can be repeated at the normal heartbeat rate if necessary. The heart itself contains *pacemaker* cells, which send out tiny electric pulses at a rate of 60 to 80 per minute. These signals induce the start of each heartbeat. In some forms of heart disease, the natural pacemaker fails to function properly, and the heart loses its beat. People suffering from this ailment now commonly make use of *electronic pacemakers* which produce a regular voltage pulse that starts and controls the frequency of the heartbeat. The electrodes are implanted in or near the heart and the circuit usually contains a capacitor and a resistor. The charge on the capacitor increases to a certain point and then discharges. Then it starts charging again. The pulsing rate depends on the values of R and C.

⇒ **PHYSICS APPLIED**

Sawtooth voltage; turn signals and wipers; pacemakers

FIGURE 26–21 (a) An *RC* circuit, coupled with a gas-filled tube as a switch, can produce a repeating "sawtooth" voltage, as shown in (b).

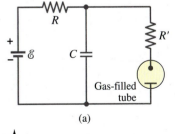

(a)

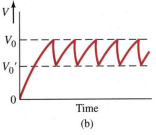

(b)

FIGURE 26–22 A multimeter used as a voltmeter.

Ammeter uses shunt resistor

An **ammeter** is used to measure current, and a **voltmeter** measures potential difference or voltage. The crucial part of an analog ammeter or voltmeter, in which the reading is by a pointer on a scale (Fig. 26–22), is a *galvanometer*. The galvanometer works on the principle of the force between a magnetic field and a current-carrying coil of wire, and will be discussed in Chapter 27. For now, we merely need to know that the deflection of the needle of a galvanometer is proportional to the current flowing through it. The *full-scale current sensitivity*, I_m, of a galvanometer is the current needed to make the needle deflect full scale.

A galvanometer can be used directly to measure small dc currents. For example, a galvanometer whose sensitivity I_m is 50 μA can measure currents from about 1 μA (currents smaller than this would be hard to read on the scale) up to 50 μA. To measure larger currents, a resistor is placed in parallel with the galvanometer. Thus an ammeter, represented by the symbol •—Ⓐ—•, consists of a galvanometer (•—Ⓖ—•) in parallel with a resistor called the **shunt resistor**, ("shunt" is a synonym for "in parallel"), as shown in Fig. 26–23. The shunt resistance is R_{sh}, and the resistance of the galvanometer coil, through which current passes, is r. The value of R_{sh} is chosen according to what full-scale deflection is desired and is normally very small, giving an ammeter a very small internal resistance.

FIGURE 26–23 An ammeter is a galvanometer in parallel with a (shunt) resistor with low resistance, R_{sh}.

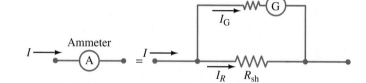

EXAMPLE 26–14 **Ammeter design.** Design an ammeter to read 1.0 A at full scale using a galvanometer with a full-scale sensitivity of 50 μA and a resistance $r = 30\ \Omega$. Check if the scale is linear.

SOLUTION When the total current I entering the ammeter is 1.0 A, we want the current I_G through the galvanometer to be precisely 50 μA (to give full-scale deflection). See Fig. 26–23. Thus, when 1.0 A flows into the meter, we want 0.999950 A $\left(= I_R\right)$ to pass through the shunt resistor R_{sh}. Since the potential difference across the shunt is the same as across the galvanometer,

$$I_R R_{sh} = I_G r,$$

then

$$R_{sh} = \frac{I_G r}{I_R} = \frac{\left(5.0 \times 10^{-5}\,\text{A}\right)(30\ \Omega)}{(0.999950\,\text{A})} = 1.5 \times 10^{-3}\ \Omega,$$

or 0.0015 Ω. The shunt resistor must thus have a very low resistance so that most of the current passes through it.

If the current I into the meter is 0.50 A, this will produce a current to the galvanometer equal to $I_G = I_R R_{sh}/r = (0.50\,\text{A})\left(1.5 \times 10^{-3}\ \Omega\right)/30\ \Omega = 25\ \mu$A, which gives a deflection half of full scale, as required.

Voltmeter uses series resistor

A voltmeter (•—Ⓥ—•) also consists of a galvanometer and a resistor. But the resistor R_{ser} is connected in series, Fig. 26–24, and it is usually large, giving a voltmeter a high internal resistance.

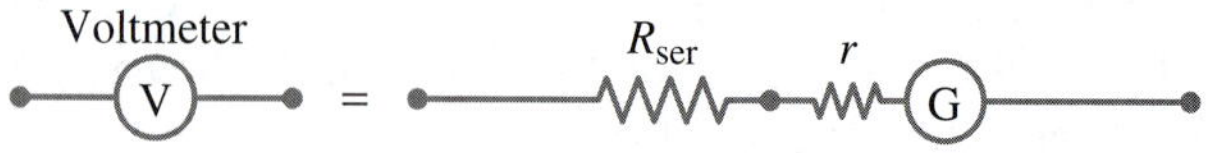

FIGURE 26–24 A voltmeter is a galvanometer in series with a resistor with high resistance, R_{ser}.

EXAMPLE 26–15 **Voltmeter design.** Using the same galvanometer with internal resistance $r = 30\,\Omega$ and full-scale current sensitivity of $50\,\mu\text{A}$, design a voltmeter that reads from 0 to 15 V. Is the scale linear?

SOLUTION When a potential difference of 15 V exists across the terminals of our voltmeter, we want $50\,\mu\text{A}$ to be passing through it so as to give a full-scale deflection. From Ohm's law, $V = IR$, we have (see Fig. 26–24)

$$15\,\text{V} = (50\,\mu\text{A})(r + R_{\text{ser}}),$$

so

$$R_{\text{ser}} = \frac{15\,\text{V}}{5.0 \times 10^{-5}\,\text{A}} - r = 300\,\text{k}\Omega - 30\,\Omega = 300\,\text{k}\Omega.$$

Notice that $r = 30\,\Omega$ is so small compared to the value of R_{ser} that it doesn't influence the calculation significantly.

The scale will again be linear: if the voltage to be measured is 6.0 V, the current passing through the voltmeter will be $(6.0\,\text{V})/(3.0 \times 10^5\,\Omega) = 2.0 \times 10^{-5}\,\text{A}$, or $20\,\mu\text{A}$. This will produce two-fifths of full-scale deflection, as required $(6.0\,\text{V}/15.0\,\text{V} = 2/5)$.

The meters discussed above are for direct current. A dc meter can be modified to measure ac with the addition of diodes which allow current to flow in one direction only. An ac meter can be calibrated to read rms or peak values.

* Use of Voltmeters and Ammeters

Suppose you wish to determine the current I in the circuit shown in Fig. 26–25a, and the voltage V across the resistor R_1. How exactly are ammeters and voltmeters connected to the circuit being measured?

Because an ammeter is used to measure the current flowing in the circuit, it must be inserted directly into the circuit, in series with the other elements, as shown in Fig. 26–25b. The smaller its internal resistance, the less it will affect the circuit.

A voltmeter, on the other hand, is connected in parallel with the circuit element across which the voltage is to be measured. It is used to measure the potential difference between two points and its two wire leads (connecting wires) are connected to the two points, as shown in Fig. 26–25c where the voltage across R_1 is being measured. The larger its internal resistance $(R_{\text{ser}} + r$ in Fig. 26–24) the less it affects the circuit being measured.

Voltmeters and ammeters can have several series or shunt resistors to offer a choice of range. **Multimeters** can measure voltage, current, and resistance. Sometimes they are called VOMs (Volt-Ohm-Meter). Meters with digital readout are called digital voltmeters (DVM) or digital multimeters (DMM), Fig. 26–26.

To measure resistance, the meter must contain a battery of known voltage connected in series to a resistor (R_{ser}) and to an ammeter. This makes an **ohmmeter**, Fig. 26–27. The resistor whose resistance is to be measured completes the circuit. The deflection is inversely proportional to the resistance. The calibration of the scale depends on the value of the series resistor. Since an ohmmeter sends a current through the device whose resistance is to be measured, it should not be used on very delicate devices that could be damaged by the current.

The **sensitivity** of a meter is generally specified on the face. It may be given as so many ohms per volt, which indicates how many ohms of resistance there are in the meter per volt of full-scale reading. For example, if the sensitivity is $30{,}000\,\Omega/\text{V}$, this means that on the 10-V scale the meter has a resistance of $300{,}000\,\Omega$. The full-scale current sensitivity, I_{m}, discussed earlier, is just the reciprocal of the sensitivity in Ω/V.

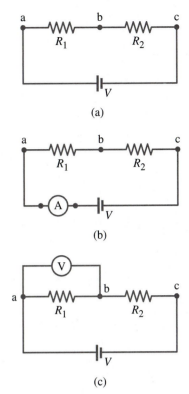

(a)

(b)

(c)

FIGURE 26–25 Measuring current and voltage.

FIGURE 26–26 A digital multimeter being used to measure resistance.

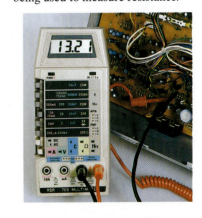

FIGURE 26–27 An ohmmeter.

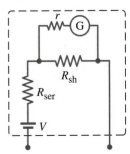

* Effects of Meter Resistance

It is important to know the sensitivity of a meter, for in many cases the resistance of the meter can seriously affect your results. Take the following Example.

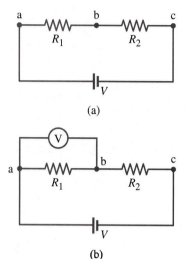

(a)

(b)

FIGURE 26–28 Example 26–16.

EXAMPLE 26–16 **Voltage reading versus true voltage.** Suppose you are testing an electronic circuit which has two resistors, R_1 and R_2, each 15 kΩ, connected in series as shown in Fig. 26–28a. The battery maintains 8.0 V across them and has negligible internal resistance. A voltmeter whose sensitivity is 10,000 Ω/V is put on the 5.0-V scale. What voltage does the meter read when connected across R_1, and what error is caused by the finite resistance of the meter?

SOLUTION On the 5.0-V scale, the voltmeter has an internal resistance of $(5.0\,\text{V})(10{,}000\,\Omega/\text{V}) = 50{,}000\,\Omega$. When connected across R_1, as in Fig. 26–28b, we have this 50 kΩ in parallel with $R_1 = 15$ kΩ. The net resistance R_{eq} of these two is given by

$$\frac{1}{R_{eq}} = \frac{1}{50\,\text{k}\Omega} + \frac{1}{15\,\text{k}\Omega} = \frac{13}{150\,\text{k}\Omega};$$

so $R_{eq} = 11.5$ kΩ. This $R_{eq} = 11.5$ kΩ is in series with $R_2 = 15$ kΩ, so the total resistance of the circuit is now 26.5 kΩ. Hence the current from the battery is

$$I = \frac{8.0\,\text{V}}{26.5\,\text{k}\Omega} = 0.30\,\text{mA}.$$

Then the voltage drop across R_1, which is the same as that across the voltmeter, is $(3.0 \times 10^{-4}\,\text{A})(11.5 \times 10^3\,\Omega) = 3.5$ V. [The voltage drop across R_2 is $(3.0 \times 10^{-4}\,\text{A})(15 \times 10^3\,\Omega) = 4.5$ V, for a total of 8.0 V.] If we assume the meter is precise, it will read 3.5 V. In the normal circuit, without the meter, $R_1 = R_2$ so the voltage across R_1 is half that of the battery, or 4.0 V. Thus the voltmeter, because of its internal resistance, gives a low reading. In this case it is off by 0.5 V, or more than 10 percent.

Example 26–16 illustrates how seriously a meter can affect a circuit and give a misleading reading. If the resistance of a voltmeter is much higher than the resistance of the circuit, however, it will have little effect and its readings can be trusted, at least to the manufactured precision of the meter, which for ordinary analog meters is typically 3 to 4 percent of full-scale deflection. An ammeter also can interfere with a circuit, but the effect is minimal if its resistance is much less than that of the circuit as a whole. For both voltmeters and ammeters, the more sensitive the galvanometer the less effect it will have. A 50,000-Ω/V meter is far better than a 1,000-Ω/V meter.

Electronic voltmeters using transistors, including digital meters, have very high input resistance (usually specified in ohms) in the range 10^6 to $10^8\,\Omega$, and even higher. Hence they have very little effect on most circuits and their readings are reliable for most circuits. The precision of digital meters is typically one part in 10^4 (= 0.01 percent) or better. State-of-the-art instruments reach a precision of one part in 10^6.

* 26–6 Transducers and the Thermocouple

A *transducer* is a device that converts one type of energy into another. A high fidelity loudspeaker is one kind of transducer—it transforms electric energy into sound energy (see Chapter 27). So is a microphone, which changes sound into an electrical signal.

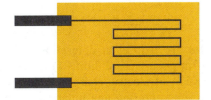

FIGURE 26–29 Wire strain gauge.

Transducers are often used for measuring particular quantities. In this case they can be thought of as changing one kind of signal into another kind, usually electrical. The *resistance thermometer* and the *thermistor* mentioned in the previous chapter are examples; basically they translate a change in temperature into a change in an electrical property, resistance.

A *strain gauge*, Fig. 26–29, uses the elasticity of a wire that will stretch an amount proportional to the stress (or force) applied to it. When stretched, its resistance must increase since it is longer and its cross-sectional area is reduced (see Eq. 25–3). The fine wire of a strain gauge is usually bonded to a flexible backing material. When fixed tightly to a structure, the resistance of the strain gauge changes in direct proportion to any change in stress on the structure. Since the wire must not be strained beyond its elastic limit, the change in length is generally quite small. Hence the change in resistance is quite small (less than one part per thousand) and a very sensitive Wheatstone bridge is used to measure it. Strain gauges must be carefully calibrated and are used in many applications, such as by architects and engineers on models of proposed structures to determine the stress at critical points. A wire strain gauge can be attached to a membrane or diaphragm. A change in pressure against the membrane causes strain in the wire. Thus a strain gauge can be used to measure pressure, and is then called a *pressure transducer*.

Another kind of pressure transducer makes use of the *piezoelectric effect*. This effect occurs in certain crystals, such as quartz, that become polarized (Section 24–5) when a mechanical force is applied and produce an emf proportional to the force.

The *thermocouple* is a device that produces an electrical signal when subjected to differing temperatures. Unlike the resistance thermometer, in which it is the resistance that changes with temperature, a thermocouple produces an emf and is based on the "thermoelectric effect." When two dissimilar metals, say iron and copper, are joined at the ends, as shown in Fig. 26–30a, it is found that an emf is produced if the two junctions are at different temperatures.[†] The magnitude of this emf depends on the temperature difference. In operation, one junction of a thermocouple is kept at a known temperature. This "reference temperature" is often 0°C. The other junction, called the "test junction," is placed where the desired temperature is to be measured. The emf is measured by some precise means, Fig. 26–30b, whose terminals need to be made of the same metal, and kept at the same temperature, so that no additional emfs are produced. Often the thermocouple junctions are connected to lead-in wires, and these additional connections must also be kept at the same temperature.

A *microphone* is a transducer that changes a sound wave into an electrical signal. One type of microphone transducer is the capacitor (or condenser) microphone, Fig. 26–31. The changing air pressure in a sound wave causes one plate of the capacitor C to move back and forth. We saw in Chapter 24 that the capacitance is inversely proportional to the separation of the plates. Thus a sound wave causes the capacitance to change. This in turn causes the charge Q on the plates to change ($Q = CV$) so that an electric current is generated at the same frequencies as the incoming sound wave.

➡ **PHYSICS APPLIED**
_____

Strain gauge, piezoelectric, thermocouple, microphones

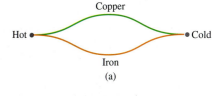

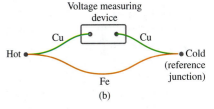

FIGURE 26–30 Thermocouple.

FIGURE 26–31 Diagram of a capacitor microphone.

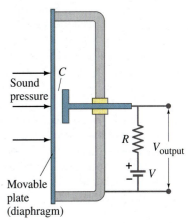

[†] Part of the theoretical explanation is that electrons in one metal occupy lower energy states than in the other, so some of them flow across the junction. This leaves one metal slightly more positive than the other, and so a *contact potential* exists between them. If the two junctions are at the same temperature, the same contact potential exists at each; they balance each other and no current flows. But if one of the junctions is at a higher temperature, the energy states are altered and the contact potentials will be different. In this case there will be a net emf and a current will flow.

Summary

A device that transforms one type of energy into electrical energy is called a **source** of **emf**. A battery behaves like a source of emf in series with an **internal resistance**. The emf is the potential difference determined by the chemical reactions in the battery and equals the terminal voltage when no current is drawn. When a current is drawn, the voltage at the battery's terminals is less than its emf by an amount equal to the Ir drop across the internal resistance.

When resistances are connected in **series** (end to end), the equivalent resistance is the sum of the individual resistances:

$$R_{eq} = R_1 + R_2 + \cdots.$$

When resistors are connected in **parallel**, the reciprocal of the total resistance equals the sum of the reciprocals of the individual resistances:

$$\frac{1}{R_{eq}} = \frac{1}{R_1} + \frac{1}{R_2} + \cdots.$$

In a parallel connection, the net resistance is less than any of the individual resistances.

Kirchhoff's rules are helpful in determining the currents and voltages in circuits. Kirchhoff's **junction rule** is based on conservation of electric charge and states that the sum of all currents entering any junction equals the sum of all currents leaving that junction. The second, or **loop rule**, is based on conservation of energy and states that the algebraic sum of the voltage changes around any closed path of the circuit must be zero.

When an **RC circuit** containing a resistor R in series with a capacitance C is connected to a dc source of emf, the voltage across the capacitor rises gradually in time characterized by an exponential of the form $\left(1 - e^{-t/RC}\right)$ where the **time constant**, $\tau = RC$, is the time it takes for the voltage to reach 63 percent of its maximum value. The current through the resistor decreases as $e^{-t/RC}$.

A capacitor discharging through a resistor is characterized by the same time constant: in a time $\tau = RC$, the voltage across the capacitor drops to 37 percent of its initial value. The charge on the capacitor, and voltage across it, decreases as $e^{-t/RC}$, as does the current.

Questions

1. Explain why birds can sit on power lines safely, while leaning a metal ladder up against one to fetch a stuck kite is extremely dangerous.

2. Discuss the advantages and disadvantages of Christmas tree lights connected in parallel versus those connected in series.

3. If all you have is a 120-V line, would it be possible to light several 6-V lamps without burning them out? How?

4. Two lightbulbs of resistance R_1 and R_2 $(>R_1)$ are connected in series. Which is brighter? What if they are connected in parallel?

5. Describe carefully the difference between emf and potential difference.

6. Household outlets are often double outlets. Are these connected in series or parallel? How do you know?

7. With two identical lightbulbs and two identical batteries, how would you arrange the bulbs and batteries in a circuit in order to get the maximum possible total power out. (Assume that the batteries have negligible internal resistance.)

8. Explain why Kirchhoff's junction rule is based on conservation of electric charge.

9. Explain why Kirchhoff's loop rule is a result of the conservation of energy.

10. How does the overall resistance of your room's electric circuit change when instead of having a single 60-W lightbulb on, you turn on an additional 100-W bulb?

11. Given the circuit shown in Fig. 26–32, use the words "increases," "decreases," or "stays the same" to complete the following statements:

    (a) If R_7 increases, the potential difference between A and E (assume no resistance in Ⓐ and ℰ) _____.

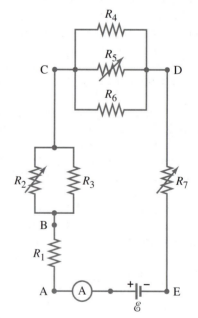

FIGURE 26–32
Question 11.

    (b) If R_7 increases, the potential difference between A and E (assume Ⓐ and ℰ have resistance) _____.

    (c) If R_7 increases, the voltage drop across R_4 _____.

    (d) If R_2 decreases, the current through R_1 _____.

    (e) If R_2 decreases, the current through R_6 _____.

    (f) If R_2 decreases, the current through R_3 _____.

    (g) If R_5 increases, the voltage drop across R_2 _____.

    (h) If R_5 increases, the voltage drop across R_4 _____.

    (i) If R_2, R_5, and R_7 increase, ℰ _____.

12. Why are batteries connected in series? Why in parallel? Does it matter if the batteries are nearly identical or not in either case?

13. Can the terminal voltage of a battery ever exceed its emf? Explain.

14. The 18-V source in Fig. 26–33 is "charging" the 12-V battery. Explain how it does this.

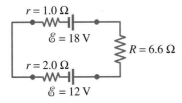

FIGURE 26–33 Questions 14 and 18; and Problem 24.

15. Explain in detail how you could measure the internal resistance of a battery.

16. Compare and discuss the formulas for the equivalent values for resistors and for capacitors when connected in series and in parallel.

17. Suppose that three identical capacitors are connected to a battery. Will they store more energy if connected in series or in parallel?

18. When applying Kirchhoff's loop rule (such as in Fig. 26–33), does the sign (or direction) of a battery's emf depend on the direction of current through the battery?

19. In an RC circuit, current flows from the battery until the capacitor is completely charged. Is the total energy supplied by the battery equal to the total energy stored by the capacitor? If not, where does the extra energy go?

20. Design a circuit in which two different switches of the type shown in Fig. 26–34 can be used to operate the same lightbulb from opposite sides of a room.

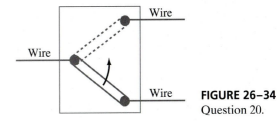

FIGURE 26–34
Question 20.

* **21.** What is the main difference between a voltmeter and an ammeter?

* **22.** What would happen if you mistakenly used an ammeter where you needed to use a voltmeter?

* **23.** Explain why an ideal ammeter would have zero resistance and an ideal voltmeter infinite resistance.

Problems

Section 26–1

1. (I) Calculate the terminal voltage for a battery with an internal resistance of $0.900\ \Omega$ and an emf of 8.50 V when the battery is connected in series with (*a*) a $68.0\text{-}\Omega$ resistor, and (*b*) a $680\text{-}\Omega$ resistor.

2. (I) Four 2.0-V cells are connected in series to a $12\text{-}\Omega$ lightbulb. If the resulting current flow is 0.62 A, what is the internal resistance of each cell, assuming they are identical and neglecting the wires?

3. (II) A 1.5-V dry cell can be tested by connecting it to a low-resistance ammeter. It should be able to supply at least 25 A. What is the internal resistance of the cell in this case?

4. (II) What is the internal resistance of a 12.0-V car battery whose terminal voltage drops to 9.8 V when the starter draws 60 A? What is the resistance of the starter?

Section 26–2

In the following Problems neglect the internal resistance of a battery unless the Problem refers to it.

5. (I) Four $90\text{-}\Omega$ lightbulbs are connected in series. What is the total resistance of the circuit? What is their resistance if they are connected in parallel?

6. (I) Three $40\text{-}\Omega$ lightbulbs and three $80\text{-}\Omega$ lightbulbs are connected in series. (*a*) What is the total resistance of the circuit? (*b*) What is their resistance if all six are wired in parallel?

7. (I) Given only one $25\text{-}\Omega$ and one $70\text{-}\Omega$ resistor, list all possible values of resistance that can be obtained.

8. (I) Suppose that you have a $500\text{-}\Omega$, a $900\text{-}\Omega$, and a $1.40\text{-k}\Omega$ resistor. What is (*a*) the maximum, and (*b*) the mininium resistance you can obtain by combining these?

9. (II) Suppose that you have a 6.0-V battery and you wish to apply a voltage of only 4.0 V. Given an unlimited supply of $1.0\text{-}\Omega$ resistors, how could you connect them so as to make a "voltage divider" that produced a 4.0-V output for a 6.0-V input?

10. (II) Three $1.20\text{-k}\Omega$ resistors can be connected together in four different ways, making combinations of series and/or parallel circuits. What are these four ways and what is the net resistance in each case?

11. (II) What is the net resistance of the circuit connected to the battery in Fig. 26–35? Each resistance has $R = 2.8\ \text{k}\Omega$.

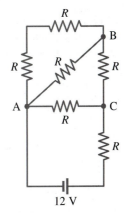

FIGURE 26–35
Problems 11 and 18.

12. (II) Eight lights are connected in series across a 110-V line. (*a*) What is the voltage across each bulb? (*b*) If the current is 0.60 A, what is the resistance of each bulb and the power dissipated in each?

13. (II) Eight lights are connected in parallel to a 110-V source by two long leads of total resistance 1.5 Ω. If 340 mA flows through each bulb, what is the resistance of each, and what fraction of the total power is wasted in the leads?

14. (II) Eight 7.0-W Christmas tree lights are connected in series to each other and to a 110-V source. What is the resistance of each bulb?

15. (II) A close inspection of an electric circuit reveals that a 480-Ω resistor was inadvertently soldered in the place where a 320-Ω resistor is needed. How can this be fixed without removing anything from the existing circuit?

16. (II) Two resistors when connected in series to a 110-V line use one fourth the power that is used when they are connected in parallel. If one resistor is 1.6 kΩ, what is the resistance of the other?

17. (II) A 75-W, 110-V bulb is connected in parallel with a 40-W, 110-V bulb. What is the net resistance?

18. (II) Calculate the current through each resistor in Fig. 26–35 if each resistance $R = 2.20$ kΩ. What is the potential difference between points A and B?

19. (II) Consider the network of resistors shown in Fig. 26–36. Answer qualitatively: (*a*) What happens to the voltage across each resistor when the switch S is closed? (*b*) What happens to the current through each when the switch is closed? (*c*) What happens to the power output of the battery when the switch is closed? (*d*) Let $R_1 = R_2 = R_3 = R_4 = 100$ Ω and $V = 45.0$ V. Determine the current through each resistor before and after closing the switch. Are your qualitative predictions confirmed?

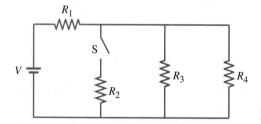

FIGURE 26–36
Problem 19.

20. (II) Three equal resistors (*R*) are connected to a battery, as shown in Fig. 26–37. Qualitatively, what happens to (*a*) the voltage drop across each of these resistors, (*b*) the current flow through each, and (*c*) the terminal voltage of the battery, when the switch S is opened, after having been closed for a long time? (*d*) If the emf of the battery is 12.0 V, what is its terminal voltage when the switch is closed if the internal resistance is 0.50 Ω and $R = 5.50$ Ω? (*e*) What is the terminal voltage when the switch S is open?

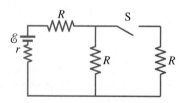

FIGURE 26–37
Problem 20.

21. (II) A battery with an emf of 12.0 V shows a terminal voltage of 11.8 V when operating in a circuit with two lightbulbs, each rated at 3.0 W (at 12.0 V) which are connected in parallel with it. What is the battery's internal resistance?

22. (III) A 3.8-kΩ and a 2.1-kΩ resistor are connected in parallel; this combination is connected in series with a 1.8-kΩ resistor. If each resistor is rated at $\frac{1}{2}$ W, what is the maximum voltage that can be applied across the whole network?

Section 26–3

23. (I) Calculate the current in the circuit of Fig. 26–38 and show that the sum of all the voltage changes around the circuit is zero.

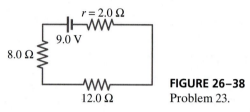

FIGURE 26–38
Problem 23.

24. (II) Determine the terminal voltage of each battery in Fig. 26–33.

25. (II) Determine the magnitudes and directions of the currents through R_1 and R_2 in Fig. 26–39.

26. (II) Repeat Problem 25, now assuming that each battery has an internal resistance $r = 1.2$ Ω.

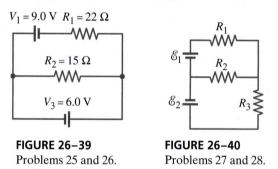

FIGURE 26–39
Problems 25 and 26.

FIGURE 26–40
Problems 27 and 28.

27. (II) Determine the magnitudes and directions of the currents through each resistor shown in Fig. 26–40. The batteries have emfs of $\mathscr{E}_1 = 9.0$ V and $\mathscr{E}_2 = 12.0$ V, and the resistors have values of $R_1 = 15$ Ω, $R_2 = 20$ Ω, and $R_3 = 40$ Ω.

28. (II) Repeat Problem 27 assuming each battery has internal resistance $r = 1.0$ Ω.

29. (II) Determine the current through each of the resistors in Fig. 26–41.

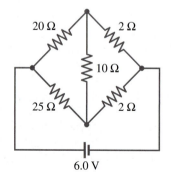

FIGURE 26–41
Problems 29 and 30.

30. (II) If the 20-Ω resistor in Fig. 26–41 is shorted out (resistance = 0), what then would be the current through the 10-Ω resistor?

31. (II) The current through the 4.0-kΩ resistor in Fig. 26–42 is 3.50 mA. What is the terminal voltage V_{ba} of the "unknown" battery? (There are two answers. Why?)

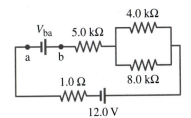

FIGURE 26–42
Problem 31.

32. (II) Suppose the 10-Ω resistor in Fig. 26–43 were replaced by an unknown resistance R. If the current through this unknown resistance is measured to be $I_2 = 0.90$ A to the right, what is the value of R? Assume $r = 1.0 \, \Omega$.

33. (II) Suppose the 6.0-V battery in Fig. 26–43 is replaced by an unknown emf $\mathscr{E}$. If the current through the 10-Ω resistor is $I_2 = 0.30$ A to the left, what is $\mathscr{E}$? Assume $r = 1.0 \, \Omega$.

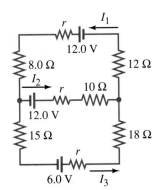

FIGURE 26–43
Problems 32, 33, 34, and 35.

34. (III) Determine the currents I_1, I_2, and I_3 in Fig. 26–43. Assume the internal resistance of each battery is $r = 1.0 \, \Omega$. What is the terminal voltage of the 6.0-V battery?

35. (III) What would the current I_1 be in Fig. 26–43 ($r = 1.0 \, \Omega$) if the 18-Ω resistor were shorted out?

36. (III) Determine the net resistance of the network shown in Fig. 26–44 (*a*) between points a and c, and (*b*) between points a and b. Assume $R' = R$. [*Hint:* Use symmetry.]

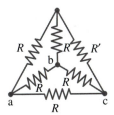

FIGURE 26–44
Problems 36 and 39.

37. (III) A voltage V is applied to n resistors connected in parallel. If the resistors are instead all connected in series with the applied voltage, show that the power transformed is decreased by a factor n^2.

38. (III) For the circuit shown in Fig. 26–45, determine (*a*) the current through the 14-V battery and (*b*) the potential difference between points a and b.

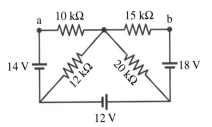

FIGURE 26–45   Problem 38.

39. (III) Determine the net resistance in Fig. 26–44 (*a*) between points a and c, and (*b*) between points a and b. Assume $R' \neq R$. [*Hint:* Apply an emf and determine currents; use symmetry at junctions.]

40. (III) Twelve resistors, each of resistance R, are connected as the edges of a cube as shown in Fig. 26–46. Determine the equivalent resistance (*a*) between points a and b, the ends of a side; (*b*) between points a and c, the ends of a face diagonal; (*c*) between points a and d, the ends of the volume diagonal. [*Hint:* Apply an emf and determine currents; use symmetry at junctions.]

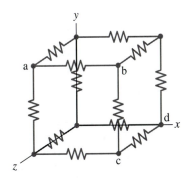

FIGURE 26–46
Problem 40.

Section 26–4

41. (II) In Fig. 26–18a, the total resistance is 15 kΩ, and the battery's emf is 24.0 V. If the time constant is measured to be 55 μs, calculate (*a*) the total capacitance of the circuit and (*b*) the time it takes for the voltage across the resistor to reach 16.0 V

42. (II) The RC circuit of Fig. 26–19a has $R = 6.7$ kΩ and $C = 6.0 \, \mu$F. The capacitor is at voltage V_0 at $t = 0$, when the switch is closed. How long does it take the capacitor to discharge to 1.0 percent of its initial voltage?

43. (II) How long does it take for the energy stored in a capacitor in a series RC circuit (Fig. 26–18a) to reach half its maximum value? Express answer in terms of the time constant $\tau = RC$.

44. (II) Two 6.0-μF capacitors, two 2.2-kΩ resistors, and a 12.0-V source are connected in series. Starting from the uncharged state, how long does it take for the current to drop from its initial value to 1.50 mA?

45. (III) Determine the time constant for charging the capacitor in the circuit of Fig. 26–47. [*Hint*: Use Kirchhoff's rules.] (*b*) What is the maximum charge on the capacitor?

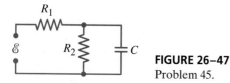

FIGURE 26–47
Problem 45.

46. (III) Two resistors and two uncharged capacitors are arranged as shown in Fig. 26–48. Then a potential difference of 24 V is applied across the combination as shown. (*a*) What is the potential at point a with *S* open? (Let *V* = 0 at the negative terminal of the source.) (*b*) What is the potential at point b with the switch open? (*c*) When the switch is closed, what is the final potential of point b? (*d*) How much charge flows through the switch S after it is closed?

47. (III) Suppose the switch S in Fig. 26–48 is closed. What is the time constant (or time constants) for charging the capacitors after the 24 V is applied.

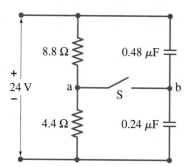

FIGURE 26–48
Problems 46 and 47.

* **Section 26–5**

* **48.** (I) What is the resistance of a voltmeter on the 250-V scale if the meter sensitivity is 50,000 Ω/V?

* **49.** (I) An ammeter has a sensitivity of 20,000 Ω/V. What current passing through the galvanometer produces full-scale deflection?

* **50.** (II) A galvanometer has an internal resistance of 30 Ω and deflects full scale for a 50-μA current. Describe how to use this galvanometer to make (*a*) an ammeter to read currents up to 30 A, and (*b*) a voltmeter to give a full-scale deflection of 1000 V.

* **51.** (II) A galvanometer has a sensitivity of 35 kΩ/V and internal resistance 20.0 Ω. How could you make this into (*a*) an ammeter that reads 2.0 A full scale, or (*b*) a voltmeter reading 1.00 V full scale?

* **52.** (II) A milliammeter reads 20 mA full scale. It consists of a 0.20-Ω resistor in parallel with a 30-Ω galvanometer. How can you change this ammeter to a voltmeter giving a full-scale reading of 10 V without taking the ammeter apart? What will be the sensitivity (Ω/V) of your voltmeter?

* **53.** (II) A 45-V battery of negligible internal resistance is connected to a 37-kΩ and a 28-kΩ resistor in series. What reading will a voltmeter, of internal resistance 100 kΩ, give when used to measure the voltage across each resistor? What is the percent inaccuracy due to meter resistance for each case?

* **54.** (II) An ammeter whose internal resistance is 60 Ω reads 4.25 mA when connected in a circuit containing a battery and two resistors in series whose values are 700 Ω and 400 Ω. What is the actual current when the ammeter is absent?

* **55.** (II) A battery with $\mathscr{E}$ = 12.0 V and internal resistance r = 1.0 Ω is connected to two 9.0-kΩ resistors in series. An ammeter of internal resistance 0.50 Ω measures the current and at the same time a voltmeter with internal resistance 11.5 kΩ measures the voltage across one of the 9.0-kΩ resistors in the circuit. What do the ammeter and voltmeter read?

* **56.** (II) A 12.0-V battery (assume the internal resistance = 0) is connected to two resistors in series. A voltmeter whose internal resistance is 15.0 kΩ measures 5.5 V and 4.0 V, respectively, when connected across each of the resistors. What is the resistance of each resistor?

* **57.** (II) Two 8.4-kΩ resistors are placed in series and connected to a battery. A voltmeter of sensitivity 1000 Ω/V is on the 3.0-V scale and reads 2.0 V when placed across either of the resistors. What is the emf of the battery? (Ignore its internal resistance.)

* **58.** (III) The voltage across a 120-kΩ resistor in a circuit containing additional resistance (R_2) in series with a battery (V) is measured, by a 20,000-Ω/V meter on the 100-V scale, to be 25 V. On the 30-V scale, the reading is 23 V. What is the actual voltage in the absence of the voltmeter? What is R_2?

* **59.** (III) (*a*) A voltmeter and an ammeter can be connected as shown in Fig. 26–49a to measure a resistance *R*. The value of *R* will not quite be V/I where *V* is the voltmeter reading and *I* is the ammeter reading since some of the current actually goes through the voltmeter. Show that the actual value of *R* is given by

$$\frac{1}{R} = \frac{I}{V} - \frac{1}{R_v}$$

where R_v is the voltmeter resistance. Note that $R \approx V/I$ if $R_v \gg R$. (*b*) A voltmeter and an ammeter can also be connected as shown in Fig. 26–49b to measure a resistance *R*. Show in this case that

$$R = \frac{V}{I} - R_A$$

where *V* and *I* are the voltmeter and ammeter readings and R_A is the resistance of the ammeter. Note that $R \approx V/I$ if $R_A \ll R$.

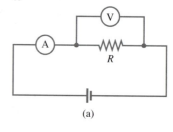

(a)

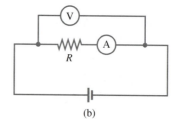

(b)

FIGURE 26–49
Problem 59.

* 60. (I) A copper-constantan thermocouple produces emfs of about $40\,\mu\text{V}/\text{C}°$. If the reference temperature is 25°C, what must be the temperature of the test junction (assume a temperature increase) if the emf produced is 1.72 mV?

* 61. (II) For iron-copper junctions near room temperature, the emf produced by a thermocouple is about $14\,\mu\text{V}/\text{C}°$. If emfs can be detected as low as $0.50\,\mu\text{V}$, to what accuracy can the temperature be read?

* 62. (II) The *strain factor* K of a strain gauge is defined as the fractional change in resistance $(\Delta R/R)$ divided by the fractional change in length:

$$K = \frac{\Delta R/R}{\Delta L/L}$$

and is relatively constant at a given temperature. In other words, the change in resistance ΔR is proportional to the change in length of the wire ΔL. (a) Show that this linear relationship is reasonable for small $\Delta L/L$. (b) A strain gauge with a strain factor of 1.8 is attached to a small muscle about 4.5 mm wide. The gauge is connected as the unknown arm in a Wheatstone bridge (Example 26-9). When the muscle is relaxed, the Wheatstone bridge is balanced when $R_2/R_1 = 1.4800$ and $R_3 = 40.700\,\Omega$. When the muscle contracts, balance occurs for $R_3 = 40.736\,\Omega$. How much has the muscle widened?

General Problems

63. Suppose that you wish to apply a 0.25-V potential difference between two points on the body. The resistance is about $2000\,\Omega$, and you only have a 6.0-V battery. How can you connect up one or more resistors so that you can produce the desired voltage?

64. A three-way lightbulb can produce 50 W, 100 W, or 150 W, at 120 V. Such a bulb contains two filaments that can be connected to the 120 V individually or in parallel. Describe how the connections to the two filaments are made to give each of the three wattages. What must be the resistance of each filament?

65. Suppose you want to run some apparatus that is 115 m from an electric outlet. Each of the wires connecting your apparatus to the 120-V source has a resistance per unit length of $0.0065\,\Omega/\text{m}$. If your apparatus draws 3.0 A, what will be the voltage drop across the connecting wires and what voltage will be applied to your apparatus?

66. Electricity can be a hazard in hospitals, particularly to patients who are connected to electrodes, such as an ECG. For example, suppose that the motor of a motorized bed shorts out to the bed frame, and the bed frame's connection to a ground has broken (or was not there in the first place). If a nurse touches the bed and the patient at the same time, she becomes a conductor and a complete circuit can be made through the patient to ground through the ECG apparatus. This is shown schematically in Fig. 26-50. Calculate the current through the patient.

67. A heart pacemaker is designed to operate at 72 beats/min using a 7.5-μF capacitor in a simple RC circuit. What value of resistance should be used if the pacemaker is to fire (capacitor discharge) when the voltage reaches 45 percent of maximum?

68. The internal resistance of a 1.35-V mercury cell is $0.030\,\Omega$, whereas that of a 1.5-V dry cell is $0.35\,\Omega$. Explain why three mercury cells can more effectively power a 2-W hearing aid that requires 4.0 V than can three dry cells.

69. Suppose that a person's body resistance is $1100\,\Omega$. (a) What current passes through the body when the person accidentally is connected to 110 V? (b) If there is an alternative path to ground whose resistance is $40\,\Omega$, what current passes through the person? (c) If the voltage source can produce at most 1.5 A, how much current passes through the person in case (b)?

70. An unknown length of platinum wire 0.920 mm in diameter is placed as the unknown resistance in a Wheatstone bridge (Example 26-9 and Fig. 26-51). Arms 1 and 2 have resistance of $38.0\,\Omega$ and $46.0\,\Omega$, respectively. Balance is achieved when R_3 is $3.48\,\Omega$. How long is the platinum wire?

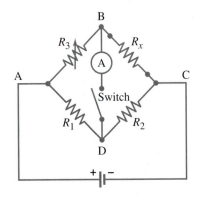

FIGURE 26-51 Wheatstone bridge. Problem 70.

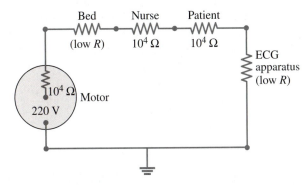

FIGURE 26-50 Problem 66.

71. Suppose two batteries, with unequal emfs of 2.0 V and 3.0 V, are connected as shown in Fig. 26–52. If each internal resistance is $r = 0.10\,\Omega$, and $R = 4.0\,\Omega$, what is the voltage across the resistor R?

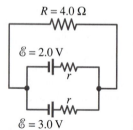

$R = 4.0\,\Omega$

$\mathscr{E} = 2.0\,V$

r

r

$\mathscr{E} = 3.0\,V$

FIGURE 26–52
Problem 71.

72. Electrocardiographs are often connected as shown in Fig. 26–53. The leads are said to be capacitively coupled. A time constant of 3.0 s is typical and allows rapid changes in potential to be accurately recorded. If $C = 3.0\,\mu F$, what value must R have?

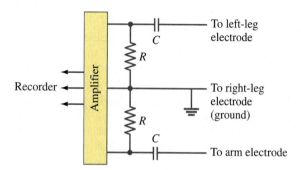

To left-leg electrode

C

R

Recorder

Amplifier

R

C

To right-leg electrode (ground)

To arm electrode

FIGURE 26–53   Problem 72.

73. A battery produces 40.8 V when 7.40 A are drawn from it and 44.5 V when 2.20 A are drawn. What is the emf and internal resistance of the battery?

74. How many $\frac{1}{2}$-W resistors, each of the same resistance, must be used to produce an equivalent 1.2-kΩ, 5-W resistor? What is the resistance of each, and how must they be connected?

75. Some light dimmer switches use a variable resistor as shown in Fig. 26–54. The slide moves from position $x = 0$ to $x = 1$, and the resistance up to slide position x is proportional to x (the total resistance is $R_{pot} = 100\,\Omega$). What is the power expended in the lightbulb if (a) $x = 1.00$, (b) $x = 0.50$, (c) $x = 0.25$?

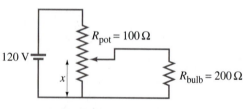

$R_{pot} = 100\,\Omega$

120 V

x

$R_{bulb} = 200\,\Omega$

FIGURE 26–54   Problem 75.

76. A **potentiometer** is a device to precisely measure potential differences or emf, using a "null" technique. In the simple potentiometer circuit shown in Fig. 26–55, R' represents the total resistance of the resistor from A to B (which could be a long uniform "slide" wire), whereas R represents the resistance of only the part from A to the movable contact at C. When the unknown emf to be measured, $\mathscr{E}_x$, is placed into the circuit as shown, the movable contact C is moved until the galvanometer G gives a null reading (i.e., zero) when the switch S is closed. The resistance between A and C for this situation we call R_x. Next, a standard emf, $\mathscr{E}_s$, which is known precisely, is inserted into the circuit in place of $\mathscr{E}_x$ and again the contact C is moved until zero current flows through the galvanometer when the switch S is closed. The resistance between A and C now is called R_s. (a) Show that the unknown emf is given by

$$\mathscr{E}_x = \left(\frac{R_x}{R_s}\right)\mathscr{E}_s$$

where R_x, R_s, and $\mathscr{E}_s$ are all precisely known. The working battery is assumed to be fresh and give a constant voltage. (b) A slide-wire potentiometer is balanced against a 1.0182-V standard cell when the slide wire is set at 25.4 cm out of a total length of 100.0 cm. For an unknown source, the setting is 45.8 cm. What is the emf of the unknown? (c) The galvanometer of a potentiometer has an internal resistance of 30 Ω and can detect a current as small as 0.015 mA. What is the minimum uncertainty possible in measuring an unknown voltage? (d) Explain the advantage of using this "null" method of measuring emf.

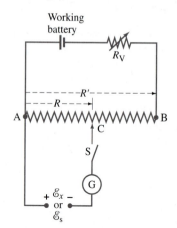

Working battery

R_V

R'

R

A   B

C

S

G

$+\ \mathscr{E}_x\ -$
or
$\mathscr{E}_s$

FIGURE 26–55
Potentiometer circuit.
Problem 76.

77. Electronic devices often use an RC circuit to protect against power outages as shown in Fig. 26–56. (a) If the device is supposed to keep the supply voltage at least 70 percent of nominal for as long as 0.20 s, how big a resistance is needed? The capacitor is 14 μF. (b) Between which two terminals should the device be connected, a and b, b and c, or a and c?

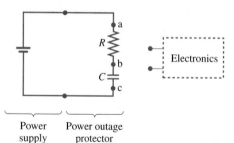

a

R

b

C

c

Electronics

Power supply

Power outage protector

FIGURE 26–56
Problem 77.

78. For the circuit shown in Fig. 26–19a, show that the decrease in energy stored in the capacitor from $t = 0$ until one time constant has elapsed equals the energy dissipated as heat in the resistor.

79. A solar cell, 3.0 cm square, has an output of 350 mA at 0.80 V when exposed to full sunlight. A solar panel that delivers 1.0 A of current at an emf of 100 V to an external load is needed. How many cells will you need? How big a panel will you need, and how should you connect the cells to one another? How can you optimize the output of your solar panel?

80. Determine the current in each resistor of the circuit shown in Fig. 26–57.

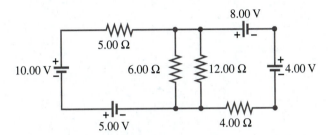

FIGURE 26–57 Problem 80.

81. In the circuit shown in Fig. 26–58, switch S is closed at time $t = 0$. (a) What is the current I_0 leaving the battery at $t = 0$, immediately after the switch is closed? (b) What is the current I a "long time" later? (c) What charge has accumulated on the capacitor after this long time? (d) If, finally, switch S is opened again, how long will it take after the switch is opened for the capacitor to lose 80 percent of its charge?

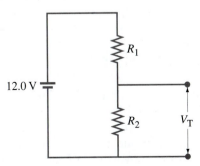

FIGURE 26–58 Problem 81.

82. A power supply has a fixed output voltage of 12.0 V, but you need $V_T = 3.0$ V for an experiment. (a) Using the voltage divider shown in Fig. 26–59, what should R_2 be if R_1 is 10.0 Ω? (b) What will the terminal voltage V_T be if you connect a load to the 3.0-volt terminal, which has a resistance of 7.0 Ω?

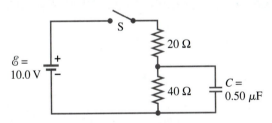

FIGURE 26–59 Problem 82.

83. In the circuit shown in Fig. 26–60, switch S is closed at time $t = 0$. (a) After the capacitor is fully charged, what is the voltage across it? How much charge is on it? (b) Switch S is now opened. How long does it now take for the capacitor to discharge until it has only 5.0 percent of its initial charge?

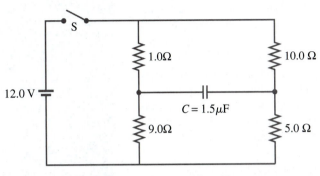

FIGURE 26–60 Problem 83.

84. Figure 26–61 shows the circuit for a simple *sawtooth oscillator*. At time $t = 0$, its switch S is closed. The neon bulb has initially infinite resistance until the voltage across it reaches 90.0 V, and then it begins to conduct with very little resistance (essentially zero). It stops conducting (its resistance becomes essentially infinite) when the voltage drops down to 70.0 V. (a) At what time t_1 does the neon bulb reach 90.0 V and start conducting? (b) At what time t_2 does the bulb reach 90.0 V for a second time and again become conducting? (c) Sketch the sawtooth waveform between $t = 0$ and $t = 0.50$ seconds.

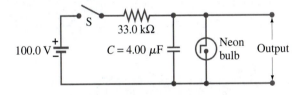

FIGURE 26–61 Problem 84.

Magnets produce magnetic fields, but so do electric currents. An electric current flowing in this straight wire produces a magnetic field which causes the tiny pieces of iron (iron "filings") to align in the field. We will see in this chapter how magnetic field is defined, and that the magnetic field direction is along the iron filings. The magnetic field lines due to the electric field in this long wire form circles around the wire. We will also discuss how magnetic fields exert forces on electric currents and on charged particles, as well as useful applications of the interaction between magnetic fields and electric currents and moving electric charges.

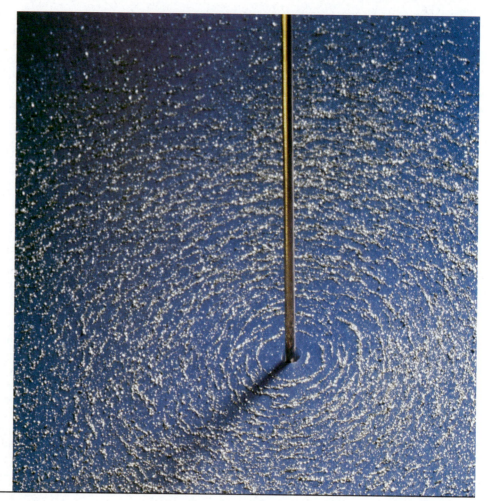

CHAPTER 27

Magnetism

The history of magnetism begins thousands of years ago, with the ancient civilizations in Asia Minor. It was in a region of Asia Minor known as Magnesia that rocks were found that could attract each other. These rocks were called "magnets" after their place of discovery.

Not until the nineteenth century was it seen that magnetism and electricity are closely related. A crucial discovery was that electric currents produce magnetic effects (we will say "magnetic fields") like magnets do. All kinds of practical devices depend on magnetism, as we shall see in this and later chapters: from compasses to motors, loudspeakers, computer memory, and electric generators.

27–1 | Magnets and Magnetic Fields

Poles of a magnet

We have all observed a magnet attract paper clips, nails, and other objects made of iron. Any magnet, whether it is in the shape of a bar or a horseshoe, has two ends or faces, called **poles**, which is where the magnetic effect is strongest. If a magnet is suspended from a fine thread, it is found that one pole of the magnet will always point toward the north. It is not known for sure when this fact was discovered, but it is known that the Chinese were making use of it as an aid to navigation by the eleventh century and perhaps earlier. This is, of course, the principle of a compass. A compass needle is simply a magnet which is supported at its center of gravity so that it can rotate freely. The pole of a freely suspended magnet that points toward geographic north is called the **north pole** of the magnet. The other pole points toward the south and is called the **south pole**.

It is a familiar fact that when two magnets are brought near one another, each exerts a force on the other. The force can be either attractive or repulsive and can be felt even when the magnets don't touch. If the north pole of one magnet is brought near the north pole of a second magnet, the force is repulsive. Similarly, if two south poles are brought close, the force is repulsive. But when a north pole is brought near a south pole, the force is attractive. These results are shown in Fig. 27–1, and are reminiscent of the force between electric charges; like poles repel, and unlike poles attract. *But do not confuse magnetic poles with electric charge*. They are not the same thing. One important difference is that a positive or negative electric charge can easily be isolated. But the isolation of a single magnetic pole seems impossible. If a bar magnet is cut in half, you do not obtain isolated north and south poles. Instead, two new magnets are produced, Fig. 27–2. If the cutting operation is repeated, more magnets are produced, each with a north and a south pole. Physicists have searched for isolated single magnetic poles (monopoles), but no magnetic monopole has ever been observed.

Only iron and a few other materials such as cobalt, nickel, gadolinium and certain alloys show strong magnetic effects. They are said to be **ferromagnetic** (from the Latin word *ferrum* for iron). All other materials show some slight magnetic effect, but it is extremely weak and can be detected only with delicate instruments. We will look in more detail at ferromagnetism in Sections 28–7 and 28–9.

We found it useful to speak of an electric field surrounding an electric charge. In the same way, we can imagine a **magnetic field** surrounding a magnet. The force one magnet exerts on another can then be described as the interaction between one magnet and the magnetic field of the other. Just as we drew electric field lines, we can also draw **magnetic field lines**. They can be drawn, as for electric field lines, so that (1) the direction of the magnetic field is tangent to a line at any point, and (2) the number of lines per unit area is proportional to the strength of the magnetic field.

The *direction* of the magnetic field at a given point can be defined as the direction that the north pole of a compass needle would point when placed at that point. (A more precise definition will be given in Section 27–3.) Figure 27–3a shows how one magnetic field line around a bar magnet is found using compass needles. The magnetic field determined in this way for the field outside a bar magnet is shown in Fig. 27–3b. Notice that because of our definition, the lines always point out from the north pole toward the south pole of a magnet (the north pole of a magnetic compass needle is attracted to the south pole of another magnet). Figure 27–4 shows how thin iron filings reveal the magnetic field lines by lining up like compass needles.

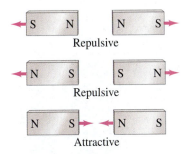

FIGURE 27–1 Like poles of a magnet repel; unlike poles attract.

FIGURE 27–2 If you break a magnet in half, you do not obtain isolated north and south poles; instead, two new magnets are produced, each with a north and a south pole.

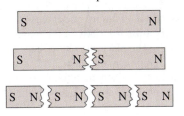

Magnetic field lines

FIGURE 27–4 Thin iron filings indicate the magnetic field lines around a bar magnet.

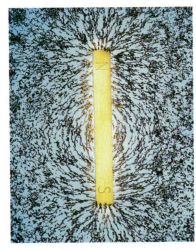

FIGURE 27–3 (a) Plotting a magnetic field line of a bar magnet. (b) Magnetic field lines for a bar magnet.

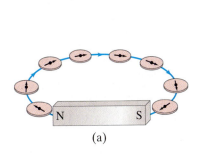

(a)

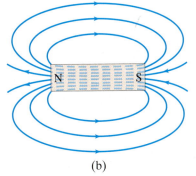

(b)

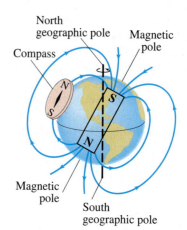

FIGURE 27-5 The Earth acts like a huge magnet; but its magnetic poles are not at the geographic poles, which are defined to be on the Earth's rotation axis.

FIGURE 27-6 Using a map and compass in the wilderness. First you align the compass case so the needle points away from true north (N) exactly the number of degrees of declination as stated on the map: 15° for the place shown on this topographic map of a part of California. Then align the map with true north, as shown, *not* with the compass needle.

➡ **PHYSICS APPLIED**

Use of a compass.

Magnetic field lines continue inside a magnet, as indicated in Fig. 27–3b. Indeed, given the lack of single magnetic poles, magnetic field lines form closed loops, unlike electric field lines that begin on positive charges and end on negative charges.

The Earth's magnetic field is shown in Fig. 27–5. The pattern of field lines is as if there were an (imaginary) bar magnet inside the Earth. Since the north pole of a compass needle points north, the magnetic pole which is in the geographic north is actually a south pole magnetically, as indicated in Fig. 27–5 by the S on the schematic bar magnet inside the Earth. (Remember that the north pole of one magnet is attracted to the south pole of a second.) Nonetheless, this pole is still often called the "north magnetic pole," or "geomagnetic north," simply because it is in the north. Similarly, the Earth's southern magnetic pole, near the geographic south pole, is magnetically a north pole. The Earth's magnetic poles do not coincide with the *geographic* poles, which are on the Earth's axis of rotation. The north magnetic pole, for example, is in northern Canada, about 1300 km from the geographic north pole, or "true north." This must be taken into account when using a compass (Fig. 27–6). The angular difference between magnetic north, as indicated by a compass, and true (geographical) north, is called the **magnetic declination**. In the U.S. it varies from 0° to perhaps 25°, depending on location.

Notice in Fig. 27–5 that the Earth's magnetic field is not tangent to the Earth's surface at all points. The angle that the Earth's magnetic field makes with the horizontal at any point is referred to as the **angle of dip**.

The simplest magnetic field is one that is uniform—it doesn't change from one point to another. A perfectly uniform field over a large area is not easy to produce. But the field between two flat parallel pole pieces of a magnet is nearly uniform if the area of the pole faces is large compared to their separation, as shown in Fig. 27–7. At the edges, the field "fringes" out somewhat and is no longer uniform. The parallel evenly spaced field lines in the drawing indicate that the field is uniform at points not too near the edge, much like the electric field between two parallel plates (Fig. 23–1).

FIGURE 27-7 Magnetic field between two large poles of a magnet is nearly uniform except at the edges.

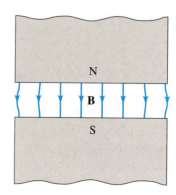

FIGURE 27–8 Deflection of compass needles near a current-carrying wire, showing the presence and direction of the magnetic field.

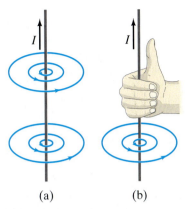

(a)   (b)

FIGURE 27–9 (a) Magnetic field lines around an electric current in a straight wire. (b) Right-hand rule for remembering the direction of the magnetic field: when the thumb points in the direction of the conventional current, the fingers wrapped around the wire point in the direction of the magnetic field.

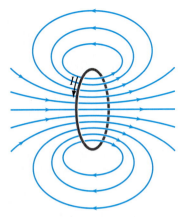

FIGURE 27–10 Magnetic field lines due to a circular loop of wire.

27–2 Electric Currents Produce Magnetism

During the eighteenth century, many scientists sought to find a connection between electricity and magnetism. A stationary electric charge and a magnet were shown not to have any influence on each other. But in 1820, Hans Christian Oersted (1777–1851) found that when a compass needle is placed near an electric wire, the needle deflects as soon as the wire is connected to a battery and a current flows. As we have seen, a compass needle can be deflected by a magnetic field. What Oersted found was that **an electric current produces a magnetic field**. He had found a connection between electricity and magnetism.

A compass needle placed near a straight section of current-carrying wire aligns itself so it is tangent to a circle drawn around the wire, Fig. 27–8. Thus, the magnetic field lines produced by a current in a straight wire are in the form of circles with the wire at their center, Fig. 27–9a. The direction of these lines is indicated by the north pole of the compass in Fig. 27–8. There is a simple way to remember the direction of the magnetic field lines in this case. It is called a **right-hand rule**: you grasp the wire with your right hand so that your thumb points in the direction of the conventional (positive) current; then your fingers will encircle the wire in the direction of the magnetic field, Fig. 27–9b. The magnetic field lines due to a circular loop of current-carrying wire can be determined in a similar way using a compass. The result is shown in Fig. 27–10. Again the right-hand rule can be used, as shown in Fig. 27–11.

Electric currents produce magnetic fields

Right-hand rule for magnetic field direction

FIGURE 27–11 Right-hand rule for determining the direction of the magnetic field relative to the current.

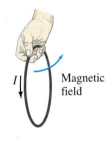

Magnetic field

27–3 Force on an Electric Current in a Magnetic Field; Definition of B

In Section 27–2 we saw that an electric current exerts a force on a magnet, such as a compass needle. By Newton's third law, we might expect the reverse to be true as well: we should expect that *a magnet exerts a force on a current-carrying wire.* Experiments indeed confirm this effect, and it too was first observed by Oersted.

Magnet exerts a force on an electric current

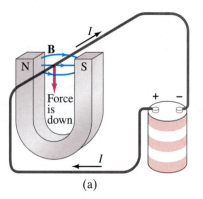

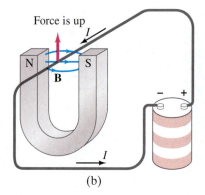

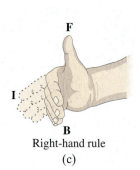

(a)   (b)   (c)

FIGURE 27–12 (a) Force on a current-carrying wire placed in a magnetic field **B**; (b) same, but current reversed; (c) right-hand rule for setup in (b).

Let us look at the force exerted on a current-carrying wire in detail. Suppose a straight wire is placed between the poles of a horseshoe magnet as shown in Fig. 27–12. When a current flows in the wire, a force is exerted on the wire. But this force is *not* toward one or the other pole of the magnet. Instead, the force is directed at right angles to the magnetic field direction, downward in Fig. 27–12a. If the current is reversed in direction, the force is in the opposite direction, Fig. 27–12b. It is found that *the direction of the force is always perpendicular to the direction of the current and also perpendicular to the direction of the magnetic field,* **B**. This statement does not completely describe the direction, however: the force could be either up or down in Fig. 27–12b and still be perpendicular to both the current and to **B**. Experimentally, the direction of the force is given by another **right-hand rule**, as illustrated in Fig. 27–12c. Orient your right hand so that outstretched fingers can point in the direction of the current, and when you bend your fingers they point in the direction of the magnetic field lines. Then your thumb points in the direction of the force on the wire.

Right-hand rule
for force on current due to **B**

This describes the direction of the force. What about its magnitude? It is found experimentally that the magnitude of the force is directly proportional to the current I in the wire, and to the length l of wire in the magnetic field (assumed uniform). Furthermore, if the magnetic field is made stronger, the force is proportionally greater. The force also depends on the angle θ between the current direction and the magnetic field (Fig. 27–13). When the current is perpendicular to the field lines, the force is strongest. When the wire is parallel to the magnetic field lines, there is no force at all. At other angles, the force is proportional to $\sin\theta$ (Fig. 27–13). Thus for a current I in a wire, with length l in a uniform magnetic field B, we have

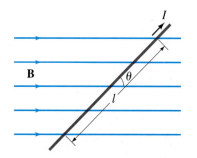

FIGURE 27–13 Current-carrying wire in a magnetic field. Force on the wire is directed into the page.

$$F \propto IlB \sin\theta.$$

Up to now we have not defined the magnetic field strength precisely. In fact, the magnetic field B can be conveniently defined in terms of the above proportion so that the proportionality constant is precisely 1. Thus we have

Force on electric current
in a uniform magnetic field

$$F = IlB \sin\theta. \qquad (27\text{–}1)$$

If the direction of the current is perpendicular to the field ($\theta = 90°$), then the force is

$$F_{\text{max}} = IlB. \qquad [\mathbf{I} \perp \mathbf{B}] \qquad (27\text{–}2)$$

Definition of
magnetic field

If the current is parallel to the field ($\theta = 0°$), the force is zero. The magnitude of **B** can then be defined as $B = F_{\text{max}}/Il$ where F_{max} is the magnitude of the force on a straight length l of wire carrying a current I when the wire is perpendicular to **B**.

The relation between the force **F** on a wire carrying current I, and the magnetic field **B** that causes the force, can be written as a vector equation. To do so, we recall that the direction of **F** is given by the right-hand rule (Fig. 27–13c), and the magnitude by Eq. 27–1. This is consistent with the definition of the vector cross product (see Section 11–1), so we can write

$$\mathbf{F} = I\boldsymbol{l} \times \mathbf{B};\qquad(27\text{–}3)$$

Force on electric current in a uniform magnetic field

here, $\boldsymbol{l}$ is a vector whose magnitude is the length of the wire and its direction is along the wire (assumed straight) in the direction of the conventional (positive) current.

The above discussion applies if the magnetic field is uniform and the wire is straight. If **B** is not uniform, or if the wire does not everywhere make the same angle θ with **B**, then Eq. 27–3 can be written

$$d\mathbf{F} = I\,d\boldsymbol{l} \times \mathbf{B},\qquad(27\text{–}4)$$

Force on electric current segment in any magnetic field

where $d\mathbf{F}$ is the infinitesimal force acting on a differential length $d\boldsymbol{l}$ of the wire. The total force on the wire is then found by integrating.

Equation 27–4 can serve (just as well as Eq. 27–2 or 27–3) as a practical definition of **B**. An equivalent way to define **B**, in terms of the force on a moving electric charge, is discussed in the next Section.

The SI unit for magnetic field B is the **tesla** (T). From Eqs. 27–1, 2, 3, or 4, it is clear that $1\,\mathrm{T} = 1\,\mathrm{N/A\cdot m}$. An older name for the tesla is the "weber per meter squared" $(1\,\mathrm{Wb/m^2} = 1\,\mathrm{T})$. Another unit commonly used to specify magnetic field is a cgs unit, the **gauss** (G): $1\,\mathrm{G} = 10^{-4}\,\mathrm{T}$. A field given in gauss should always be changed to teslas before using with other SI units. To get a "feel" for these units, we note that the magnetic field of the Earth at its surface is about $\frac{1}{2}\,\mathrm{G}$ or $0.5 \times 10^{-4}\,\mathrm{T}$. On the other hand, strong electromagnets can produce fields on the order of several teslas and superconducting magnets over 10 T.

The tesla and the gauss (units)

On a diagram, when we want to represent a magnetic field that is pointing out of the page (toward us) or into the page, we use $\odot$ or $\times$. The $\odot$ is meant to resemble the tip of an arrow pointing directly toward the reader, whereas the $\times$ or $\otimes$ resembles the tail of an arrow going away. (See Figs. 27–14 and 27–15.)

EXAMPLE 27–1 **Measuring a magnetic field.** A rectangular loop of wire hangs vertically as shown in Fig. 27–14. A magnetic field **B** is directed horizontally, perpendicular to the wire, and points out of the page at all points as represented by the symbol $\odot$. The magnetic field **B** is very nearly uniform along the horizontal portion of wire ab (length $l = 10.0\,\mathrm{cm}$) which is near the center of a large magnet producing the field. The top portion of the wire loop is free of the field. The loop hangs from a balance which measures a downward force (in addition to the gravitational force) of $F = 3.48 \times 10^{-2}\,\mathrm{N}$ when the wire carries a current $I = 0.245\,\mathrm{A}$. What is the magnitude of the magnetic field B at the center of the magnet?

SOLUTION The magnetic forces on the two vertical sections of the wire loop point to the left and right, respectively. They are equal and in opposite directions and so add up to zero. Hence, the net magnetic force on the loop is that on the horizontal section ab whose length is $l = 0.100\,\mathrm{m}$ (and $\theta = 90°$ so $\sin\theta = 1$); thus

$$B = \frac{F}{Il} = \frac{3.48 \times 10^{-2}\,\mathrm{N}}{(0.245\,\mathrm{A})(0.100\,\mathrm{m})} = 1.42\,\mathrm{T}.$$

This technique is a highly precise means of determining magnetic fields.

FIGURE 27–14 Measuring a magnetic field **B**. Example 27–1.

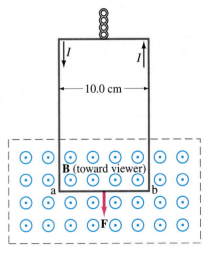

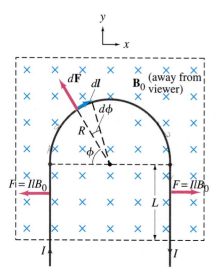

y ↳ x

× × × × × × × ×
× dF × dl × × B_0 (away from viewer) ×
× × × × × × × ×
× × $d\phi$ × × × ×
× × R × × × ×
× × ϕ × × × ×
$F = IlB_0$ × × × × × $F = IlB_0$
× × × × × L × ×
× × × × × × × ×
I I

FIGURE 27–15 Example 27–2.

EXAMPLE 27–2 **Magnetic force on a semicircular wire.** A rigid wire, carrying a current I, consists of a semicircle of radius R and two straight portions as shown in Fig. 27–15. The wire lies in a plane perpendicular to a uniform magnetic field $\mathbf{B}_0$. The straight portions each have length l within the field. Determine the net force on the wire due to the magnetic field $\mathbf{B}_0$.

SOLUTION The forces on the two straight sections are equal $(= IlB_0)$ and in opposite directions, so they cancel. Hence the net force is that on the semicircular portion. We divide the semicircle into short lengths $dl = R\,d\phi$ as indicated and make use of Eq. 27–4, $d\mathbf{F} = I\,d\mathbf{l} \times \mathbf{B}$, to find

$$dF = IB_0 R\,d\phi,$$

where dF is the force on the length $dl = R\,d\phi$, and the angle between $d\mathbf{l}$ and $\mathbf{B}_0$ is 90° (so $\sin\theta = 1$ in the cross product). The x component of the force $d\mathbf{F}$ on the segment dl shown, and the x component of $d\mathbf{F}$ for a symmetrically located dl on the other side of the semicircle, will cancel each other. Thus for the entire semicircle there will be no x component of force. Hence we need be concerned only with the y components, each equal to $dF \sin\phi$, and the total force will have magnitude

$$F = \int_0^\pi dF \sin\phi = IB_0 R \int_0^\pi \sin\phi\,d\phi = -IB_0 R \cos\phi \Big|_0^\pi = 2IB_0 R,$$

with direction vertically upward along the y axis in Fig. 27–15.

27–4 Force on an Electric Charge Moving in a Magnetic Field

We have seen that a current-carrying wire experiences a force when placed in a magnetic field. Since a current in a wire consists of moving electric charges, we might expect that freely moving charged particles (not in a wire) would also experience a force when passing through a magnetic field. Indeed, this is the case.

From what we already know we can predict the force on a single moving electric charge. If N such particles of charge q pass by a given point in time t, they constitute a current $I = Nq/t$. We let t be the time for a charge q to travel a distance L in a magnetic field $\mathbf{B}$; then $\mathbf{l} = \mathbf{v}t$ where $\mathbf{v}$ is the velocity of the particle. Thus, the force on these N particles is, by Eq. 27–3, $\mathbf{F} = I\mathbf{l} \times \mathbf{B} = (Nq/t)(\mathbf{v}t) \times \mathbf{B} = Nq\mathbf{v} \times \mathbf{B}$. The force on *one* of the N particles is then

$$\boxed{\mathbf{F} = q\mathbf{v} \times \mathbf{B}.} \tag{27–5a}$$

This basic and important result can be considered as an alternative way of defining the magnetic field $\mathbf{B}$, in place of Eq. 27–4 or 27–3. The magnitude of the force in Eq. 27–5a is

$$F = qvB \sin\theta. \tag{27–5b}$$

This gives the magnitude of the force on a particle of charge q moving with velocity $\mathbf{v}$ at a point where the magnetic field has magnitude B. The angle between $\mathbf{v}$ and $\mathbf{B}$ is θ. The force is greatest when the particle moves perpendicular to $\mathbf{B}$ ($\theta = 90°$):

$$F_{max} = qvB. \qquad [\mathbf{v} \perp \mathbf{B}]$$

The force is *zero* if the particle moves *parallel* to the field lines ($\theta = 0°$). The *direction* of the force is perpendicular to the magnetic field $\mathbf{B}$ and to the velocity $\mathbf{v}$ of the particle. It is given again by a right-hand rule, as for any cross product (for $q > 0$): orient your right hand so that your outstretched fingers point along the direction of motion of the particle ($\mathbf{v}$) and when you bend your fingers they point along the direction of $\mathbf{B}$. Then your thumb will point in the direction of the force. This is true only for *positively* charged particles, and will be "down" for the situation shown in Fig. 27–16. For negatively charged particles, the force is in exactly the opposite direction, "up" in Fig. 27–16.

Force on moving charge in magnetic field

FIGURE 27–16 Force on charged particles due to a magnetic field is perpendicular to the magnetic field direction.

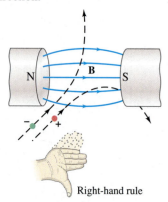

N $\mathbf{B}$ S

Right-hand rule

EXAMPLE 27–3 **Magnetic force on a proton.** A proton having a speed of 5.0×10^6 m/s in a magnetic field feels a force of 8.0×10^{-14} N toward the west when it moves vertically upward. When moving horizontally in a northerly direction, it feels zero force. What is the magnitude and direction of the magnetic field in this region? (The charge on a proton is $q = +e = 1.6 \times 10^{-19}$ C.)

SOLUTION Since the proton feels no force when moving north, the field must be in a north–south direction. In order to produce a force to the west when the proton moves upward, the right-hand rule tells us that **B** must point toward the north. (Your thumb points west and the outstretched fingers of your right hand point upward only when your bent fingers point north.) The magnitude of **B**, from Eq. 27–5 with $\theta = 90°$, is

$$B = \frac{F}{qv} = \frac{8.0 \times 10^{-14}\,\text{N}}{(1.6 \times 10^{-19}\,\text{C})(5.0 \times 10^6\,\text{m/s})} = 0.10\,\text{T}.$$

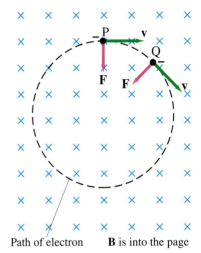

The path of a charged particle moving in a plane perpendicular to a uniform magnetic field is a circle. See Fig. 27–17, where the magnetic field is directed *into* the paper, as represented by ×'s. An electron at point P is moving to the right, and the force on it at this point is downward as shown (use the right-hand rule and reverse the direction for negative charge). The electron is thus deflected downward. A moment later, say when it reaches point Q, the force is still perpendicular to the velocity and is in the direction shown. Since the force is always perpendicular to **v**, the magnitude of **v** does not change—it moves at constant speed. But the particle changes direction, and moves in a circular path with constant centripetal acceleration (see Example 27–4) due to the magnetic force directed toward the center of this circle at all points. The electron moves clockwise in Fig. 27–17. A positive particle would feel a force in the opposite direction and would thus move counterclockwise.

Path of electron   **B** is into the page

FIGURE 27–17 Force exerted by a uniform magnetic field on a moving charged particle (in this case, an electron) produces a circular path.

EXAMPLE 27–4 **Electron's path in a uniform magnetic field.** An electron travels at 2.0×10^7 m/s in a plane perpendicular to a 0.010-T magnetic field. Describe its path.

SOLUTION The electron moves at speed v in a curved path whose radius of curvature is found using Newton's second law, $F = ma$. We have a centripetal acceleration $a = v^2/r$ (Eq. 3–14). The force is given by Eq. 27–5 with $\sin\theta = 1$, $F = qvB$, so we have

$$F = ma$$
$$qvB = \frac{mv^2}{r}.$$

We solve for r and find

$$r = \frac{mv}{qB}.$$

Since **F** is perpendicular to **v**, the magnitude of **v** doesn't change. From this equation we see that if **B** = constant, then r = constant, and the curve must be a circle as we claimed above. To get r we put in the numbers:

$$r = \frac{(9.1 \times 10^{-31}\,\text{kg})(2.0 \times 10^7\,\text{m/s})}{(1.6 \times 10^{-19}\,\text{C})(0.010\,\text{T})} = 1.1 \times 10^{-2}\,\text{m} = 1.1\,\text{cm}.$$

The time T required for a particle of charge q moving with constant speed v to make one circular revolution in a uniform magnetic field **B** ($\perp$ **v**) is $T = 2\pi r/v$, where $2\pi r$ is the circumference of its circular path. From Example 27–4, $r = mv/qB$, so

$$T = \frac{2\pi m}{qB}.$$

Since T is the period of rotation, the frequency of rotation is

$$f = \frac{1}{T} = \frac{qB}{2\pi m}.$$ (27–6)

This is often called the **cyclotron frequency** of a particle in a field because this is the rotation frequency of particles in a cyclotron (see Problem 59). Note that f does not depend on the speed v; if v is large, r is large ($r = mv/qB$) for a given B, but the frequency is independent of v and r, as long as v is not near the speed of light (see Section 7–5 or Chapter 37).

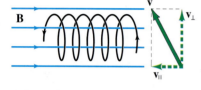

FIGURE 27–18
Conceptual Example 27–5.

PHYSICS APPLIED

The aurora borealis

| **CONCEPTUAL EXAMPLE 27–5** | **A spiral path.** |

A spiral path. What is the path of a charged particle in a uniform magnetic field if its velocity is *not* perpendicular to the magnetic field?

RESPONSE The velocity vector can be broken down into components parallel and perpendicular to the field. The velocity component parallel to the field lines experiences no force, and so this component remains constant. The velocity component perpendicular to the field results in circular motion about the field lines. Putting these two motions together produces a helical (spiral) motion around the field lines as shown in Fig. 27–18.

CONCEPTUAL EXAMPLE 27–6 **Aurora borealis.** Charged ions approach the Earth from the Sun (the "solar wind") and enter the atmosphere mainly near the poles, sometimes causing a phenomenon called the *aurora borealis* or "northern lights" in northern latitudes. Why toward the poles?

RESPONSE A glance at Fig. 27–19 (see also Fig. 27–18) provides the answer. Imagine a stream of charged particles approaching the Earth as shown. The velocity component perpendicular to the field for each particle becomes a circular orbit around the field lines, whereas the velocity component parallel to the field carries the particle along the field lines toward the poles. The high concentration of charged particles ionizes the air, and the recombining of electrons with atoms emits light (Chapter 38), which is the aurora, especially during periods of high sunspot activity when the solar wind is greater.

FIGURE 27–19 (a) Diagram showing a charged particle approaching the Earth which is "captured" by the magnetic field of the Earth. Such particles follow the field lines toward the poles as shown. (b) Photo of aurora borealis.

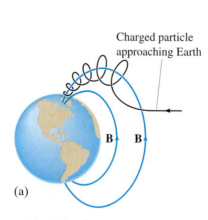

Charged particle approaching Earth

(a)

(b)

If a particle of charge q moves with velocity $\mathbf{v}$ in the presence of both a magnetic field $\mathbf{B}$ and an electric field $\mathbf{E}$, it will feel a force

Lorentz equation

$$\mathbf{F} = q(\mathbf{E} + \mathbf{v} \times \mathbf{B})$$ (27–7)

where we have made use of Eqs. 21–3 and 27–5. Equation 27–7 is often called the **Lorentz equation** and is considered one of the basic equations in physics.

EXAMPLE 27–7 **Velocity selector, or filter: Crossed E and B fields.** Some electronic devices and experiments need a beam of charged particles all moving at nearly the same velocity. This can be achieved using both a uniform electric field and a uniform magnetic field, arranged so they are at right angles to each other. As shown in Fig. 27–20a, particles of charge q pass through slit S_1 and enter the region where **B** points into the page and **E** points from the positive plate toward the negative plate. If the particles enter with different velocities, show how this device "selects" a particular velocity, and determine what this velocity is.

SOLUTION After passing through slit S_1, each particle is subject to two forces as shown in Fig. 27–20b. If q is positive, the magnetic force is upwards and the electric force downwards. (Vice versa if q is negative.) The exit slit, S_2, is assumed to be directly in line with S_1 and the particles' velocity **v**. Depending on the magnitude of **v**, some particles will be bent upwards and some downwards. The only ones to make it through the slit S_2 will be those for which the net force is zero: $\sum F = qvB - qE = 0$. Hence this device selects particles whose velocity is

$$v = \frac{E}{B}. \qquad (27\text{–}8)$$

This result doesn't depend on the sign of the charge q.

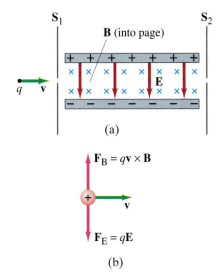

FIGURE 27–20 A velocity selector: if $v = E/B$, the particles make it through S_1 and S_2.

27–5 Torque on a Current Loop; Magnetic Dipole Moment

When an electric current flows in a closed loop of wire placed in a magnetic field, as shown in Fig. 27–21, the magnetic force on the current can produce a torque. This is the basic principle behind a number of important practical devices, including voltmeters, ammeters, and motors. (We discuss these applications in the next Section.) The interaction between a current and a magnetic field is important in other areas as well, including atomic physics.

When current flows through the loop in Fig. 27–21a, whose face we assume is parallel to **B** and is rectangular, the magnetic field exerts a force on both vertical sections of wire as shown, $\mathbf{F}_1$ and $\mathbf{F}_2$ (see also top view, Fig. 27–21b). Notice that, by the right-hand rule, the direction of the force $\mathbf{F}_1$ on the upward current on the left is in the opposite direction from the equal magnitude force $\mathbf{F}_2$ on the descending current on the right. These forces give rise to a net torque that tends to rotate the coil about its vertical axis.

Let us calculate the magnitude of this torque. From Eq. 27–2, the force has magnitude $F = IaB$, where a is the length of the vertical arm of the coil. The lever arm for each force is $b/2$, where b is the width of the coil and the "axis" is at the midpoint. The total torque is the sum of the torques due to each of the forces, so

$$\tau = IaB\frac{b}{2} + IaB\frac{b}{2} = IabB = IAB,$$

where $A = ab$ is the area of the coil. If the coil consists of N loops of wire, the torque on N wires becomes

$$\tau = NIAB.$$

If the coil makes an angle θ with the magnetic field, as shown in Fig. 27–21c, the forces are unchanged, but each lever arm is reduced from $\frac{1}{2}b$ to $\frac{1}{2}b\sin\theta$. Note that the angle θ is chosen to be the angle between **B** and the perpendicular to the face of the coil, Fig. 27–21c. So the torque becomes

$$\tau = NIAB\sin\theta. \qquad (27\text{–}9)$$

This formula, derived here for a rectangular coil, is valid for any shape of flat coil.

FIGURE 27–21 Calculating the torque on a current loop in a magnetic field **B**. (a) Loop face parallel to **B** field lines; (b) top view; (c) loop makes an angle to **B**, reducing the torque since the lever arm is reduced.

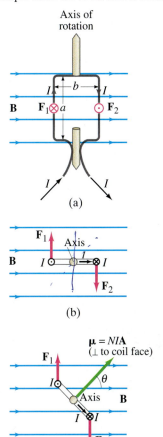

The quantity *NIA* is called the **magnetic dipole moment** of the coil and is considered a vector:

Magnetic dipole moment

$$\boldsymbol{\mu} = NI\mathbf{A}, \tag{27-10}$$

where the direction of **A** (and therefore of **μ**) is *perpendicular* to the plane of the coil (the green arrow in Fig. 27–21c) consistent with the right-hand rule (cup your right hand so your fingers wrap around the loop in the direction of current flow, then your thumb points in the direction of **μ** and **A**). With this definition of **μ**, we can rewrite Eq. 27–9 in vector form:

$$\boldsymbol{\tau} = NI\mathbf{A} \times \mathbf{B}$$

or

$$\boldsymbol{\tau} = \boldsymbol{\mu} \times \mathbf{B}, \tag{27-11}$$

which gives the correct magnitude and direction for **τ**.

Equation 27–11 has the same form as Eq. 21–9b for an electric dipole (with electric dipole moment **p**) in an electric field **E**, which is $\boldsymbol{\tau} = \mathbf{p} \times \mathbf{E}$. And just as an electric dipole has potential energy given by $U = -\mathbf{p} \cdot \mathbf{E}$ when in an electric field, we expect a similar form for a magnetic dipole in a magnetic field. In order to rotate a current loop (Fig. 27–21) so as to increase θ, we must do work against the force due to the magnetic field. Hence the potential energy depends on angle (see Eq. 10–22, the work-energy principle for rotational motion) as

$$U = \int \tau \, d\theta = \int NIAB \sin\theta \, d\theta = -\mu B \cos\theta + C.$$

If we choose $U = 0$ at $\theta = \pi/2$, then the arbitrary constant C is zero and the potential energy is

$$U = -\mu B \cos\theta = -\boldsymbol{\mu} \cdot \mathbf{B}, \tag{27-12}$$

as expected. Bar magnets and compass needles, as well as current loops, can be considered as magnetic dipoles. Note the striking similarities of the fields produced by a bar magnet and a current loop, Figs. 27–3b and 27–10.

EXAMPLE 27–8 **Torque on a coil.** A circular coil of wire has a diameter of 20.0 cm and contains 10 loops. The current in each loop is 3.00 A, and the coil is placed in a 2.00-T magnetic field. Determine the maximum and minimum torque exerted on the coil by the field.

SOLUTION Equation 27–9 is valid for any shape of coil, including circular, where the area is

$$A = \pi r^2 = \pi (0.100 \, \text{m})^2 = 3.14 \times 10^{-2} \, \text{m}^2.$$

The maximum torque occurs when the coil's face is parallel to the magnetic field, so $\theta = 90°$ in Fig. 27–21c, and $\sin\theta = 1$ in Eq. 27–9:

$$\tau = NIAB \sin\theta = (10)(3.00 \, \text{A})(3.14 \times 10^{-2} \, \text{m}^2)(2.00 \, \text{T})(1) = 1.88 \, \text{m} \cdot \text{N}.$$

The minimum torque occurs if $\sin\theta = 0$, for which $\theta = 0°$, and then $\tau = 0$ from Eq. 27–9.

EXAMPLE 27–9 **Magnetic moment of a hydrogen atom.** Determine the magnetic dipole moment of the electron orbiting the proton of a hydrogen atom, assuming (in the Bohr model) it is in its ground state with a circular orbit of radius 0.529×10^{-10} m. [Note: This is a very rough picture of atomic structure, but nonetheless gives an accurate result.]

SOLUTION From Newton's second law, $F = ma$, we have, since the electron is held in its orbit by the coulomb force,

$$\frac{e^2}{4\pi\epsilon_0 r^2} = \frac{mv^2}{r};$$

so

$$v = \sqrt{\frac{e^2}{4\pi\epsilon_0 mr}} = \sqrt{\frac{(8.99 \times 10^9 \, \text{N} \cdot \text{m}^2/\text{C}^2)(1.60 \times 10^{-19} \, \text{C})^2}{(9.11 \times 10^{-31} \, \text{kg})(0.529 \times 10^{-10} \, \text{m})}} = 2.19 \times 10^6 \, \text{m/s}.$$

Since current is the electric charge that passes a given point per unit time, the revolving electron is equivalent to a current

$$I = \frac{e}{T} = \frac{ev}{2\pi r},$$

where $T = 2\pi r/v$ is the time required for one orbit. Since the area of the orbit is $A = \pi r^2$, the magnetic dipole moment is

$$\mu = IA = \frac{ev}{2\pi r}(\pi r^2) = \tfrac{1}{2}evr$$

$$= \tfrac{1}{2}(1.60 \times 10^{-19} \, \text{C})(2.19 \times 10^6 \, \text{m/s})(0.529 \times 10^{-10} \, \text{m}) = 9.27 \times 10^{-24} \, \text{A} \cdot \text{m}^2,$$

or 9.27×10^{-24} J/T.

*27–6 Applications: Galvanometers, Motors, Loudspeakers

The basic component of analog meters (those with pointer and dial), including analog ammeters, voltmeters, and ohmmeters, is a galvanometer. We have already seen how these meters are designed (Section 26–5), and now we can examine how the crucial element, a galvanometer, works. As shown in Fig. 27–22, a **galvanometer** consists of a coil of wire (with attached pointer) suspended in the magnetic field of a permanent magnet. When current flows through the loop of wire, the magnetic field exerts a torque on the loop, as given by Eq. 27–9, $\tau = NIAB \sin\theta$. This torque is opposed by a spring which exerts a torque τ_s approximately proportional to the angle ϕ through which it is turned (Hooke's law). That is, $\tau = k\phi$, where k is the stiffness constant of the spring. Thus the coil and the attached pointer will rotate only to the point where the spring torque balances the torque due to the magnetic field. From Eq. 27–9 we then have $k\phi = NIAB \sin\theta$ or

$$\phi = \frac{NIAB \sin\theta}{k}.$$

Thus the deflection of the pointer, ϕ, is directly proportional to the current I flowing in the coil. But it also depends on the angle θ the coil makes with **B**. For a useful meter we need ϕ to depend only on the current I, independent of θ. To solve this problem, curved pole pieces are used and the galvanometer coil is wrapped around a cylindrical iron core as shown in Fig. 27–23. The iron tends to concentrate the magnetic field lines so that **B** always points parallel to the face of the coil at the wire outside the core. The force is then always perpendicular to the face of the coil and the torque will not vary with angle. Thus ϕ will be proportional to I, as required.

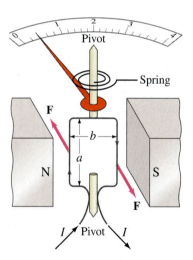

FIGURE 27–22 Galvanometer.

FIGURE 27–23 Galvanometer coil wrapped on an iron core.

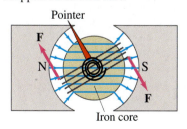

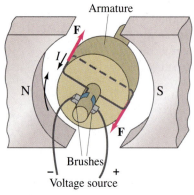

FIGURE 27–24 Diagram of a simple dc motor.

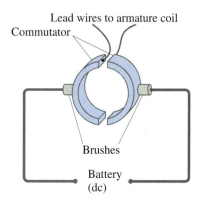

FIGURE 27–25 The commutator-brush arrangement in a dc motor assures alternation of the current in the armature to keep rotation continuous. The commutators are attached to the motor shaft and turn with it, whereas the brushes remain stationary.

➡ P H Y S I C S   A P P L I E D

Motor

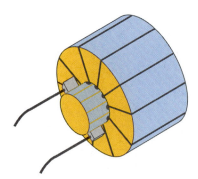

FIGURE 27–26 Motor with many windings.

FIGURE 27–27 Loudspeaker.

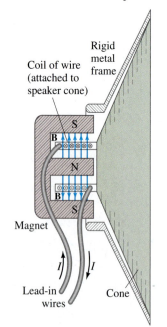

An **electric motor** changes electric energy into (rotational) mechanical energy. A motor works on the same principle as a galvanometer, except that there is no spring so the coil can rotate continuously in one direction. The coil is larger and is mounted on a large cylinder called the **rotor** or **armature**, Fig. 27–24. Actually, there are several coils, although only one is indicated in the Figure. The armature is mounted on a shaft or axle. At the moment shown in Fig. 27–24, the magnetic field exerts forces on the current in the loop as shown. However, when the coil, which is rotating clockwise in Fig. 27–24, passes beyond the vertical position the forces would then act to return the coil back to vertical if the current remained the same. But if the current could somehow be reversed at that critical moment, the forces would reverse, and the coil would continue rotating in the same direction. Thus, alternation of the current is necessary if a motor is to turn continuously in one direction. This can be achieved in a **dc motor** with the use of **commutators** and **brushes**: as shown in Fig. 27–25, the brushes are stationary contacts that rub against the conducting commutators mounted on the motor shaft. At every half revolution, each commutator changes its connection to the other brush. Thus the current in the coil reverses every half revolution as required for continuous rotation. Most motors contain several coils, called "windings," each located in a different place on the armature, Fig. 27–26. Current flows through each coil only during a small part of a revolution, at the time when its orientation results in the maximum torque. In this way, a motor produces a much steadier torque than can be obtained from a single coil. An **ac motor**, with ac current as input, can work without commutators since the current itself alternates. Many motors use wire coils to produce the magnetic field (electromagnets) instead of a permanent magnet. Indeed the design of most motors is more complex than described here, but the general principles remain the same.

A **loudspeaker** also works on the principle that a magnet exerts a force on a current-carrying wire. The electrical output of a stereo or TV set is connected to the wire leads of the speaker. The speaker leads are connected internally to a coil of wire, which is itself attached to the speaker cone, Fig. 27–27. The speaker cone is usually made of stiffened cardboard and is mounted so that it can move back and forth freely. A permanent magnet is mounted directly in line with the coil of wire. When the alternating current of an audio signal flows through the wire coil, which is free to move within the magnet, the coil experiences a force due to the magnetic field of the magnet. As the current alternates at the frequency of the audio signal, the coil and attached speaker cone move back and forth at the same frequency, causing alternate compressions and rarefactions of the adjacent air, and sound waves are produced. A speaker thus changes electrical energy into sound energy, and the frequencies and intensities of the emitted sound waves can be an accurate reproduction of the electrical input.

27-7 | Discovery and Properties of the Electron

The electron plays a basic role in our understanding of electricity and magnetism today. But its existence was not suggested until the 1890s. We discuss it here because magnetic fields were crucial for measuring its properties.

Toward the end of the nineteenth century, studies were being done on the discharge of electricity through rarefied gases. One apparatus, diagrammed in Fig. 27–28, was a glass tube fitted with electrodes and evacuated so only a small amount of gas remained inside. When a very high voltage was applied to the electrodes, a dark space seemed to extend outward from the cathode (negative electrode) toward the opposite end of the tube; and that far end of the tube would glow. If one or more screens containing a small hole were inserted as shown, the glow was restricted to a tiny spot on the end of the tube. It seemed as though something being emitted by the cathode traveled to the opposite end of the tube. These "somethings" were named **cathode rays**.

There was much discussion at the time about what these rays might be. Some scientists thought they might resemble light. But the observation that the bright spot at the end of the tube could be deflected to one side by an electric or magnetic field suggested that cathode rays could be charged particles; and the direction of the deflection was consistent with a negative charge. Furthermore, if the tube contained certain types of rarefied gas, the path of the cathode rays was made visible by a slight glow.

Estimates of the charge e of the (assumed) cathode-ray particles, as well as of their charge-to-mass ratio e/m, had been made by 1897. But in that year, J. J. Thomson (1856–1940) was able to measure e/m directly, using the apparatus shown in Fig. 27–29. Cathode rays are accelerated by a high voltage and then pass between a pair of parallel plates built into the tube. The voltage applied to the plates produces an electric field, and a pair of coils produces a magnetic field. When only the electric field is present, say with the upper plate positive, the cathode rays are deflected upward as in path a in Fig. 27–29. If only a magnetic field exists, say inward, the rays are deflected downward along path c. These observations are just what is expected for a negatively charged particle. The force on the rays due to the magnetic field is $F = evB$, where e is the charge and v is the velocity of the cathode rays. In the absence of an electric field, the rays are bent into a curved path, so we have, from $F = ma$,

$$evB = m\frac{v^2}{r},$$

and thus

$$\frac{e}{m} = \frac{v}{Br}.$$

The radius of curvature r can be measured and so can B. The velocity v can be found by applying an electric field in addition to the magnetic field. The electric field E is adjusted so that the cathode rays are undeflected and follow path b in Fig. 27–29. This

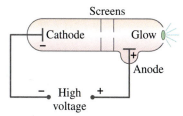

FIGURE 27–28 Discharge tube. In some models, one of the screens is the anode (positive plate).

FIGURE 27–29 Cathode rays deflected by electric and magnetic fields.

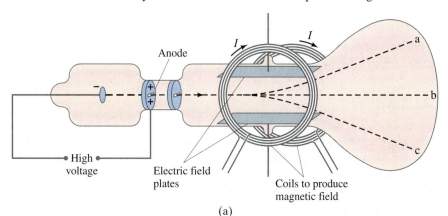

(a)

is just like the velocity selector of Example 27–7 where the force due to the electric field, $F = eE$, is balanced by the force due to the magnetic field, $F = evB$. Thus $eE = evB$ and $v = E/B$. Combining this with the above equation we have

e/m measured

$$\frac{e}{m} = \frac{E}{B^2 r} . \qquad\qquad \textbf{(27–13)}$$

The quantities on the right side can all be measured so that although e and m could not be determined separately, the ratio e/m could be determined. The accepted value today is $e/m = 1.76 \times 10^{11}$ C/kg. Cathode rays soon came to be called **electrons**.

"Discovery" of the electron

It is worth noting that the "discovery" of the electron, like many others in science, is not quite so obvious as discovering gold or oil. Should the discovery of the electron be credited to the person who first saw a glow in the tube? Or to the person who first called them cathode rays? Perhaps neither one, for they had no conception of the electron as we know it today. In fact, the credit for the discovery is generally given to Thomson, but not because he was the first to see the glow in the tube. Rather it is because he believed that this phenomenon was due to tiny negatively charged particles and made careful measurements on them. Furthermore he argued that these particles were constituents of atoms, and not ions or atoms themselves as many thought, and he developed an electron theory of matter. His view is close to what we accept today, and this is why Thomson is credited with the "discovery." Note, however, that neither he nor anyone else ever actually saw an electron itself. We discuss this briefly, for it illustrates the fact that discovery in science is not always a clear-cut matter. In fact some philosophers of science think the word "discovery" is often not appropriate, such as in this case.

Thomson believed that an electron was not an atom, but rather a constituent, or part, of an atom. Convincing evidence for this came soon with the determination of the charge and the mass of the cathode rays. Thomson's student J. S. Townsend made the first direct (but rough) measurements of e in 1897. But it was the more refined **oil-drop experiment** of Robert A. Millikan (1868–1953) that yielded a precise value for the charge on the electron and showed that charge comes in discrete amounts. In this experiment, tiny droplets of mineral oil carrying an electric charge were allowed to fall under gravity between two parallel plates, Fig. 27–30. The electric field E between the plates was adjusted until the drop was suspended in midair. The downward pull of gravity, mg, was then just balanced by the upward force due to the electric field. Thus $qE = mg$, so the charge $q = mg/E$. The mass of the droplet was determined by measuring its terminal velocity in the absence of the electric field. Sometimes the drop was charged negatively, and sometimes positively, suggesting that the drop had acquired or lost electrons (by friction, leaving the atomizer). Millikan's painstaking observations and analysis presented convincing evidence that any charge was an integral multiple of a smallest charge, e, that was ascribed to the electron, and that the value of e was 1.6×10^{-19} C. (Today's value of e, as mentioned in Chapter 21, is $e = 1.602 \times 10^{-19}$ C.) This value of e, combined with the measurement of e/m, gives the mass of the electron to be $(1.6 \times 10^{-19}\,\text{C})/(1.76 \times 10^{11}\,\text{C/kg}) = 9.1 \times 10^{-31}$ kg. This mass is less than a thousandth the mass of the smallest atom, and thus confirmed the idea that the electron is only a part of an atom. The accepted value today for the mass of the electron is $m_e = 9.11 \times 10^{-31}$ kg.

Millikan oil-drop experiment

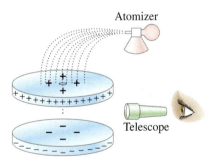

Atomizer

Telescope

FIGURE 27–30 Millikan's oil-drop experiment.

CRT, Revisited

The cathode ray tube (CRT), which is the picture tube of TV sets, oscilloscopes, and computer monitors, was discussed in Chapter 23. There, in Fig. 23–19, we saw a design using electric deflection plates to maneuver the electron beam. Many CRTs, however, make use of the magnetic field produced by coils to maneuver the electron beam. They operate much like the coils shown in Fig. 27–29. Both types—electrostatic deflection and magnetic deflection—are in use today.

* 27-8 The Hall Effect

When a current-carrying conductor is held firmly in a magnetic field, the field exerts a sideways force on the charges moving in the conductor. For example, if electrons move to the right in the rectangular conductor shown in Fig. 27–31a, the inward magnetic field will exert a downward force on the electrons $\mathbf{F}_B = -e\mathbf{v}_d \times \mathbf{B}$, where $\mathbf{v}_d$ is the drift velocity of the electrons (Section 25–8). So the electrons will tend to move nearer face D than face C. There will thus be a potential difference between faces C and D of the conductor. This potential difference builds up until the electric field $\mathbf{E}_H$ it produces exerts a force, $e\mathbf{E}_H$, on the moving charges that is equal and opposite to the magnetic force. This effect is called the **Hall effect** after E. H. Hall, who discovered it in 1879. The difference of potential produced is called the **Hall emf**.

The electric field due to the separation of charge is called the *Hall field*, $\mathbf{E}_H$, and points downward in Fig. 27–31a, as shown. In equilibrium, the force due to this electric field is balanced by the magnetic force $ev_d B$, so

$$eE_H = ev_d B.$$

Hence $E_H = v_d B$. The Hall emf is then (assuming the conductor is long and thin so E_H is uniform)

$$\mathcal{E}_H = E_H l = v_d B l, \qquad (27\text{–}14)$$

where l is the width of the conductor.

A current of negative charges moving to the right is equivalent to positive charges moving to the left, at least for most purposes. But the Hall effect can distinguish these two. As can be seen in Fig. 27–31b, positive particles moving to the left are deflected downward, so that the bottom surface is positive relative to the top surface. This is the reverse of part (a). Indeed, the direction of the emf in the Hall effect first revealed that it is negative particles that move in metal conductors.

The magnitude of the Hall emf is proportional to the strength of the magnetic field. The Hall effect can thus be used to measure magnetic field strengths. First the conductor, called a *Hall probe*, is calibrated with known magnetic fields. Then, for the same current, its emf output will be a measure of B. Hall probes can be made very small and are convenient and accurate to use.

The Hall effect can also be used to measure the drift velocity of charge carriers when the external magnetic field B is known. Such a measurement also allows us to determine the density of charge carriers in the material.

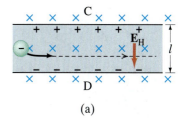

(a)

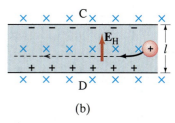

(b)

FIGURE 27–31 The Hall effect. (a) Negative charges moving to the right as the current. (b) Positive charges moving to the left as the current.

EXAMPLE 27–10 Drift velocity using the Hall effect. A long copper strip 1.8 cm wide and 1.0 mm thick is placed in a 1.2-T magnetic field as in Fig. 27–31a. When a steady current of 15 A passes through it, the Hall emf is measured to be 1.02 μV. Determine the drift velocity of the electrons and the density of free (conducting) electrons (number per unit volume) in the copper.

SOLUTION The drift velocity (Eq. 27–14) is

$$v_d = \frac{\mathcal{E}_H}{Bl} = \frac{1.02 \times 10^{-6}\,\text{V}}{(1.2\,\text{T})(1.8 \times 10^{-2}\,\text{m})} = 4.7 \times 10^{-5}\,\text{m/s}.$$

The density of charge carriers n is obtained from Eq. 25–13, $I = nev_d A$, where A is the cross-sectional area through which the current I flows. Then

$$n = \frac{I}{ev_d A} = \frac{15\,\text{A}}{(1.6 \times 10^{-19}\,\text{C})(4.7 \times 10^{-5}\,\text{m/s})(1.8 \times 10^{-2}\,\text{m})(1.0 \times 10^{-3}\,\text{m})}$$

$$= 11 \times 10^{28}\,\text{m}^{-3}.$$

This value for the density of free electrons in copper, $n = 11 \times 10^{28}$ per m³, is the experimentally measured value. It represents *more* than one free electron per atom, which as we saw in Example 25–12 is $8.4 \times 10^{28}\,\text{m}^{-3}$.

Mass Spectrometer

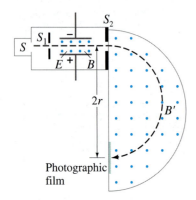

FIGURE 27–32 Bainbridge mass spectrometer. The magnetic fields B and B' point out of the paper (indicated by the dots).

Various methods were developed in the early part of this century to measure the masses of atoms. One of the most accurate was the **mass spectrometer** of Fig. 27–32. Ions are produced by heating, or by an electric current, in the source S. Often the particles are accelerated and then pass through slit S_1 into a velocity selector of crossed electric and magnetic fields (as in Example 27–7). Only those ions whose speed is $v = E/B$ will pass through undeflected and emerge through slit S_2. In the second region, after S_2, there is only a magnetic field B' so the ions follow a circular path. The radius of their path can be measured because the ions darken the photographic plate where they strike. Since $qvB' = mv^2/r$ and $v = E/B$, then we have

$$m = \frac{qB'r}{v} = \frac{qBB'r}{E}.$$

By measuring the quantities on the right, m can be determined. Note that for ions of the same charge, the mass of each is proportional to the radius of its path.

The masses of many atoms were measured in this way. When a pure substance was used, it was sometimes found that two or more closely spaced marks would appear on the film. For example, neon produced two marks whose radii corresponded to atoms of mass 20 and 22 atomic mass units (u). Impurities were ruled out and it was concluded that there must be two types of neon with different masses. These different forms were called **isotopes**. It was soon found that most elements are mixtures of isotopes, and that the difference in mass is due to different numbers of neutrons. Mass spectrometers can be used to separate not only different elements and isotopes, but different molecules as well.

EXAMPLE 27–11 Mass spectrometry. Carbon atoms of atomic mass 12.0 u are found to be mixed with another, unknown, element. In a mass spectrometer, the carbon traverses a path of radius 22.4 cm and the unknown's path has a 26.2 cm radius. What is the unknown element? Assume they have the same charge.

SOLUTION Since mass is proportional to the radius, we have

$$\frac{m_x}{m_C} = \frac{26.2 \text{ cm}}{22.4 \text{ cm}} = 1.17.$$

Thus $m_x = 1.17 \times 12.0 \text{ u} = 14.0 \text{ u}$. The other element is probably nitrogen (see the Periodic Table, inside the back cover). However, it could also be an isotope of carbon or oxygen. Further physical or chemical analysis would be needed.

Summary

A magnet has two **poles**, north and south. The north pole is that end which points toward geographic north when the magnet is freely suspended. Unlike poles of two magnets attract each other, whereas like poles repel.

We can imagine that a **magnetic field** surrounds every magnet. The SI unit for magnetic field is the **tesla** (T).

Electric currents produce magnetic fields. For example, the lines of magnetic field due to a current in a straight wire form circles around the wire and the field exerts a force on magnets placed near it.

A magnetic field exerts a force on an electric current. The force on an infinitesimal length of wire $d\mathbf{l}$ carrying a current I in a magnetic field $\mathbf{B}$ is

$$d\mathbf{F} = I \, d\mathbf{l} \times \mathbf{B}.$$

If the field $\mathbf{B}$ is uniform over a straight length $\mathbf{l}$ of wire, then the force is

$$\mathbf{F} = I\mathbf{l} \times \mathbf{B}$$

which has magnitude

$$F = IlB \sin \theta$$

where θ is the angle between magnetic field $\mathbf{B}$ and the wire. The direction of the force is perpendicular to the wire and to the magnetic field, and is given by the right-hand rule. This relation serves as the definition of magnetic field $\mathbf{B}$.

Similarly, a magnetic field **B** exerts a force on a charge q moving with velocity **v** given by

$$\mathbf{F} = q\mathbf{v} \times \mathbf{B}.$$

The magnitude of the force is

$$F = qvB \sin \theta,$$

where θ is the angle between **v** and **B**. The path of a charged particle moving perpendicular to a uniform magnetic field is a circle.

If both electric and magnetic fields are present, the force is

$$\mathbf{F} = q\mathbf{E} + q\mathbf{v} \times \mathbf{B}.$$

The torque on a current loop in a magnetic field **B** is

$$\boldsymbol{\tau} = \boldsymbol{\mu} \times \mathbf{B},$$

where $\boldsymbol{\mu}$ is the **magnetic dipole moment** of the loop:

$$\boldsymbol{\mu} = NI\mathbf{A}.$$

Here N is the number of coils carrying current I in the loop and **A** is a vector perpendicular to the plane of the loop (use right-hand rule) and has magnitude equal to the area of the loop.

The measurement of the charge-to-mass ratio (e/m) of the electron was done using magnetic and electric fields. The charge e on the electron was first measured in the Millikan oil-drop experiment and then its mass was obtained from the measured value of the e/m ratio.

Questions

1. A compass needle is not always balanced parallel to the Earth's surface, but one end may dip downward. Explain.

2. Draw the magnetic field lines around a straight section of wire carrying a current horizontally to the left.

3. In what direction are the magnetic field lines surrounding a straight wire carrying a current that is moving directly toward you?

4. A horseshoe magnet is held vertically with the north pole on the left and south pole on the right. A wire passing between the poles, equidistant from them, carries a current directly away from you. In what direction is the force on the wire?

5. In the relation, $\mathbf{F} = I\mathbf{l} \times \mathbf{B}$, which pairs of the vectors (**F**, **l**, **B**) are always at 90°? Which can be at other angles?

6. The magnetic field due to current in wires in your home can affect a compass. Discuss the effect in terms of currents, including if they are ac or dc.

7. If a negatively charged particle enters a region of uniform magnetic field which is perpendicular to the particle's velocity, will the kinetic energy of the particle increase, decrease, or stay the same. Explain your answer. (Neglect gravity.)

8. In Fig. 27–33, charged particles move in the vicinity of a current-carrying wire. For each charged particle the arrow indicates the direction of motion of the particle and the + or − indicates the sign of the charge. For each of the particles, indicate the direction of the magnetic force due to the magnetic field produced by the wire.

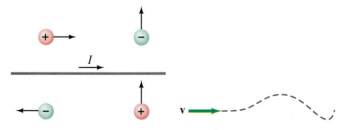

FIGURE 27–33 Question 8.   **FIGURE 27–34** Question 9.

9. A positively charged particle in a nonuniform magnetic field follows the trajectory shown in Fig. 27–34. Indicate the direction of the magnetic field everywhere in space, assuming the path is always in the plane of the page, and indicate the relative magnitudes of the field in each region.

10. Note that the pattern of magnetic field lines surrounding a bar magnet is similar to that of the electric field around an electric dipole. From this fact, predict how the magnetic field will change with distance (*a*) when near one pole of a very long bar magnet, and (*b*) when far from a magnet as a whole.

11. Explain why a strong magnet held near a television screen causes the picture to become distorted. Also, explain why the picture sometimes goes completely black where the field is the strongest. [But don't risk damage to your TV by trying this.]

12. Describe the trajectory of a negatively charged particle in the velocity filter of Fig. 27–20 if its speed exceeds E/B. What is its trajectory if $v < E/B$? Would it make any difference if the particle were positively charged?

13. Can you set a resting electron into motion with a magnetic field? With an electric field?

14. A charged particle is moving in a circle under the influence of a uniform magnetic field. If an electric field that points in the same direction as the magnetic field is turned on, describe the path the charged particle will take.

15. The force on a particle in a magnetic field is the idea behind *electromagnetic pumping*. It is used to pump metallic fluids (such as sodium) and more recently to pump blood in artificial heart machines. The basic design is shown in Fig. 27–35. An electric field is applied perpendicular to a blood vessel and to a magnetic field. Explain how ions are caused to move. Do positive and negative ions feel a force in the same direction?

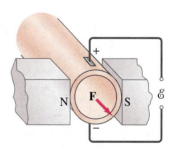

FIGURE 27–35 Electromagnetic pumping in a blood vessel. Question 15.

16. A beam of electrons is directed toward a horizontal wire carrying a current from left to right (Fig. 27–36). In what direction is the beam deflected?

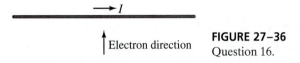

FIGURE 27–36
Question 16.

17. What kind of field or fields surround a moving electric charge?

18. Could we have defined the direction of the magnetic field **B** to be in the direction of the force on a moving charged particle? Explain.

19. A charged particle moves in a straight line through a particular region of space. Could there be a nonzero magnetic field in this region? If so, give two possible situations.

20. If a moving charged particle is deflected sideways in some region of space, can we conclude, for certain, that $\mathbf{B} \neq 0$ in that region? Explain.

21. In a particular region of space there is a uniform magnetic field **B**. Outside this region, $B = 0$. Can you inject an electron from outside into the field perpendicularly so it will move in a closed circular path in the field? What if the electron is injected near the center?

22. How could you tell whether moving electrons in a certain region of space are being deflected by an electric field or by a magnetic field (or by both)?

23. How can you make a compass without using iron or other ferromagnetic material?

24. Describe how you could determine the dipole moment of a bar magnet or compass needle.

25. In what positions (if any) will a current loop placed in a uniform magnetic field be in (*a*) stable equilibrium, and (*b*) unstable equilibrium?

* **26.** A rectangular piece of semiconductor is inserted in a magnetic field and a battery is connected to its ends as shown in Fig. 27–37. When a sensitive voltmeter is connected between points a and b, it is found that point a is at a higher potential than b. What is the sign of the charge carriers in this semiconductor material?

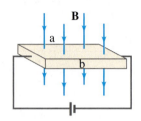

FIGURE 27–37   Question 26.

* **27.** Two ions have the same mass, but one is singly ionized and the other is doubly ionized. How will their positions on the film of the mass spectrometer of Fig. 27–32 differ?

Problems

Section 27–3

1. (I) (*a*) What is the force per meter of length on a straight wire carrying a 7.40-A current when perpendicular to a 0.90-T uniform magnetic field? (*b*) What if the angle between the wire and field is 45.0°?

2. (I) Calculate the magnetic force on a 240-m length of wire stretched between two towers carrying a 150-A current. The Earth's magnetic field of 5.0×10^{-5} T makes an angle of 60° with the wire.

3. (I) How much current is flowing in a wire 4.20 m long if the maximum force on it is 0.900 N when placed in a uniform 0.0800-T field?

4. (I) A 1.5-m length of wire carrying 4.5 A of current is oriented horizontally. At that point on the Earth's surface, the dip angle of the Earth's magnetic field makes an angle of 40° to the wire. Estimate the magnetic force on the wire due to the Earth's magnetic field of 5.5×10^{-5} T at this point.

5. (I) The force on a wire carrying 8.75 A is a maximum of 1.18 N when placed between the pole faces of a magnet. If the pole faces are 55.5 cm in diameter, what is the approximate strength of the magnetic field?

6. (II) The magnetic force per meter on a wire is measured to be only 45 percent of its maximum possible value. Sketch the relationship of the wire and the field if the force were a maximum, and sketch the relationship as it actually is, calculating the angle between the wire and the magnetic field.

7. (II) The force on a wire is a maximum of 5.30 N when placed between the pole faces of a magnet. The current flows horizontally to the right and the magnetic field is vertical. The wire is observed to "jump" toward the observer when the current is turned on. (*a*) What type of magnetic pole is the top pole face? (*b*) If the pole faces have a diameter of 10.0 cm, estimate the current in the wire if the field is 0.15 T. (*c*) If the wire is tipped so that it now makes an angle of 10° with the horizontal, what force will it now feel?

8. (II) Suppose the straight wires connected to the conductor bent into a semicircle in Fig. 27–15 were bent outwardly, still in the plane of the page, so they were horizontal at the base of the semicircle. If a length L of each remained in the field **B**, what would be the total force on the conductor as a whole?

9. (II) A straight 2.0-mm-diameter copper wire can just "float" horizontally in air because of the force of the Earth's magnetic field **B**, which is horizontal, perpendicular to the wire, and of magnitude 5.0×10^{-5} T. What current does the wire carry?

10. (II) A long wire stretches along the *x* axis and carries a 3.0-A current to the right (+*x*). The wire passes through a uniform magnetic field $\mathbf{B} = (0.20\mathbf{i} - 0.30\mathbf{j} + 0.25\mathbf{k})$T. Determine the components of the force on the wire per cm of length.

11. (III) A curved wire, connecting two points a and b, lies in a plane perpendicular to a uniform magnetic field **B** and carries a current I. Show that the resultant magnetic force on the wire, no matter what its shape, is the same as that on a straight wire connecting the two points carrying the same current I. See Fig. 27–38.

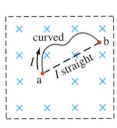

FIGURE 27–38 Problem 11.

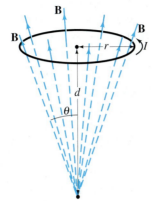

FIGURE 27–39 Problem 12.

12. (III) A circular loop of wire, of radius r, carries current I. It is placed in a magnetic field whose straight lines seem to diverge from a point a distance d below the ring on its axis. (That is, the field makes an angle θ with the loop at all points, Fig. 27–39, where $\tan\theta = r/d$.) Determine the force on the loop.

Section 27–4

13. (I) Determine the magnitude and direction of the force on an electron traveling 7.75×10^5 m/s horizontally to the east in a vertically upward magnetic field of strength 0.85 T.

14. (I) Find the direction of the force on a negative charge for each diagram shown in Fig. 27–40, where **v** is the velocity of the charge and **B** is the direction of the magnetic field. ($\otimes$ means the vector points inward. $\odot$ means it points outward, toward the viewer.)

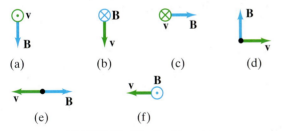

FIGURE 27–40 Problem 14.

15. (I) Determine the direction of **B** for each case in Fig. 27–41, where **F** represents the force on a positively charged particle moving with velocity **v**.

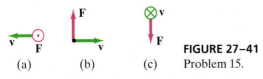

FIGURE 27–41
Problem 15.

16. (I) An electron is projected vertically upward with a speed of 1.80×10^6 m/s into a uniform magnetic field of 0.250 T that is directed horizontally away from the observer. Describe the electron's path in this field.

17. (I) A particle of charge q moves in a circular path of radius r in a uniform magnetic field **B**. Show that its momentum is $p = qBr$.

18. (II) What is the velocity of a beam of electrons that go undeflected when passing through crossed electric and magnetic fields of magnitude 8.8×10^3 V/m and 3.5×10^{-3} T, respectively? What is the radius of the electron orbit if the electric field is turned off?

19. (II) For a particle of mass m and charge q moving in a circular path in a magnetic field B, (a) show that its kinetic energy is proportional to r^2, the square of the radius of curvature of its path, and (b) show that its angular momentum is $L = qBr^2$, about the center of the circle.

20. (II) An electron moves with velocity $\mathbf{v} = (4.0\mathbf{i} - 6.0\mathbf{j}) \times 10^4$ m/s in a magnetic field $\mathbf{B} = (-0.80\mathbf{i} + 0.60\mathbf{j})$T. Determine the magnitude and direction of the force on the electron.

21. (II) A 5.0-MeV (kinetic energy) proton enters a 0.20-T field, in a plane perpendicular to the field. What is the radius of its path?

22. (II) An electron experiences the greatest force as it travels 2.9×10^6 m/s in a magnetic field when it is moving northward. The force is upward and of magnitude 7.2×10^{-13} N. What is the magnitude and direction of the magnetic field?

23. (II) A doubly charged helium atom whose mass is 6.6×10^{-27} kg is accelerated by a voltage of 2100 V. (a) What will be its radius of curvature if it moves in a plane perpendicular to a uniform 0.340-T field? (b) What is its period of revolution?

24. (II) A 3.40-g bullet moves with a speed of 160 m/s perpendicular to the Earth's magnetic field of 5.00×10^{-5} T. If the bullet possesses a net charge of 13.5×10^{-9} C, by what distance will it be deflected from its path due to the Earth's magnetic field after it has traveled 1.00 km?

25. (II) Suppose the Earth's magnetic field at the equator has magnitude 0.40×10^{-4} T and a northerly direction at all points. How fast must a singly ionized uranium ion ($m = 238$ u, $q = e$) move so as to circle the Earth 5.0 km above the equator? Can you ignore gravity?

26. (II) A proton (mass m_p), a deuteron ($m = 2m_p$, $Q = e$), and an alpha particle ($m = 4m_p$, $Q = 2e$) are accelerated by the same potential difference V and then enter a uniform magnetic field **B** where they move in circular paths perpendicular to **B**. Determine the radius of the paths for the deuteron and alpha particle in terms of that for the proton.

27. (II) A proton moves through a region of space where there is a magnetic field, $\mathbf{B} = (0.45\mathbf{i} + 0.20\mathbf{j})$T and an electric field $\mathbf{E} = (3.0\mathbf{i} - 4.2\mathbf{j}) \times 10^3$ V/m. At a given instant, the proton's velocity is $\mathbf{v} = (6.0\mathbf{i} + 3.0\mathbf{j} - 5.0\mathbf{k}) \times 10^3$ m/s. Determine the components of the total force on the proton.

28. (II) An electron experiences a force $\mathbf{F} = (3.8\mathbf{i} - 2.7\mathbf{j}) \times 10^{-13}$ N when passing through a magnetic field $\mathbf{B} = (0.35\text{ T})\mathbf{k}$. Determine the electron's velocity.

29. (II) An electron enters a uniform magnetic field $B = 0.23$ T at a 45° angle to **B**. Determine the radius r and pitch p (distance between loops) of the electron's helical path assuming its speed is 3.0×10^6 m/s. See Fig. 27–42.

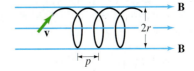

FIGURE 27–42
Problem 29.

30. (II) The path of protons emerging from an accelerator must be bent by 90° by a "bending magnet" so as not to strike a barrier in their path a distance l from their exit hole in the accelerator. Show that the field **B** in the bending magnet, which we assume is uniform and can extend over an area $l \times l$, must have magnitude of at least $B \geq (2mK/e^2l^2)^{\frac{1}{2}}$, where m is the mass of a proton and K is its kinetic energy.

31. (II) A proton moving with speed $v = 2.0 \times 10^5$ m/s in a field-free region abruptly enters an essentially uniform magnetic field $B = 0.850$ T as shown in Fig. 27–43 (**B** $\perp$ **v**). If the proton enters the magnetic field region at a 45° angle as shown, (a) at what angle does it leave and (b) at what distance x does it exit from the field?

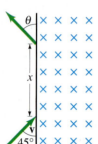

FIGURE 27–43
Problem 31.

Section 27–5

32. (I) A 13.0-cm-diameter circular loop of wire is placed with its face parallel to the uniform magnetic field between the pole pieces of a large magnet. When 7.10 A flows in the coil, the torque on it is 0.185 m·N. What is the magnetic field strength?

33. (I) How much work is required to rotate the current loop (Fig. 27–21) in a uniform magnetic field **B** from (a) $\theta = 0°$ ($\boldsymbol{\mu} \| \textbf{B}$) to $\theta = 180°$, (b) $\theta = 90°$ to $\theta = -90°$?

34. (II) Show that the magnetic dipole moment $\boldsymbol{\mu}$ of an electron orbiting the proton nucleus of a hydrogen atom is related to the orbital momentum L of the electron by

$$\mu = \frac{e}{2m} L.$$

35. (II) A circular coil 17.0 cm in diameter and containing twelve loops lies flat on the ground. The Earth's magnetic field at this location has magnitude 5.50×10^{-5} T and points into the Earth at an angle of 66.0° below a line pointing due north. If a 7.10-A clockwise current passes through the coil, (a) determine the torque on the coil, and (b) which edge of the coil rises up, north, east, south, or west.

36. (II) A 20-loop circular coil 20 cm in diameter lies in the xy plane. The current in each loop of the coil is 7.6 A clockwise, and an external magnetic field $\textbf{B} = (0.80\textbf{i} + 0.60\textbf{j} - 0.65\textbf{k})$ T passes through the coil. Determine: (a) the magnetic moment of the coil, $\boldsymbol{\mu}$; (b) the torque on the coil due to the external magnetic field; (c) the potential energy U of the coil in the field (take the same zero for U as we did in our discussion of Fig. 27–21).

37. (III) Suppose a nonconducting rod of length l carries a uniformly distributed charge Q. It is rotated with angular velocity ω about an axis perpendicular to the rod at one end. Show that the magnetic dipole moment of this rod is $\frac{1}{6}Q\omega l^2$. [*Hint*: Consider the motion of each infinitesimal length of the rod.]

* 38. (I) A galvanometer needle deflects full scale for a 63.0-μA current. What current will give full-scale deflection if the magnetic field weakens to 0.860 of its original value?

* 39. (I) If the restoring spring of a galvanometer weakens by 20 percent over the years, what current will give full-scale deflection if it originally required 36 μA?

* 40. (I) If the current to a motor drops by 18 percent, by what factor does the output torque change?

Section 27–7

41. (I) What is the value of q/m for a particle that moves in a circle of radius 8.0 mm in a 0.46-T magnetic field if a crossed 200-V/m electric field will make the path straight?

42. (II) An oil drop whose mass is determined to be 3.3×10^{-15} kg is held at rest between two large plates separated by 1.0 cm as in Fig. 27–30. If the potential difference between the plates is 340 V, how many excess electrons does this drop have?

* Section 27–8

* 43. (II) A rectangular sample of a metal is 3.0 cm wide and 500 μm thick. When it carries a 42-A current and is placed in a 0.80-T magnetic field it produces a 6.5-μV Hall emf. Determine: (a) the Hall field in the conductor; (b) the drift speed of the conduction electrons; (c) the density of free electrons in the metal.

* 44. (II) In a probe that uses the Hall effect to measure magnetic fields, a 12.0-A current passes through a 1.50-cm-wide 1.30-mm-thick strip of sodium metal. If the Hall emf is 2.42 μV, what is the magnitude of the magnetic field (take it perpendicular to the flat face of the strip)? Assume one free electron per atom of Na, and take its specific gravity to be 0.971.

* 45. (II) The Hall effect can be used to measure blood flow rate because the blood contains ions that constitute an electric current. (a) Does the sign of the ions influence the emf? (b) Determine the flow velocity in an artery 3.3 mm in diameter if the measured emf is 0.10 mV and B is 0.070 T. (In actual practice, an alternating magnetic field is used.)

* Section 27–9

* 46. (I) In a mass spectrometer, germanium atoms have radii of curvature equal to 21.0, 21.6, 21.9, 22.2, and 22.8 cm. The largest radius corresponds to an atomic mass of 76 u. What are the atomic masses of the other isotopes?

* 47. (II) Suppose the electric field between the electric plates in the mass spectrometer of Fig. 27–32 is 2.48×10^4 V/m and the magnetic fields $B = B' = 0.58$ T. The source contains carbon isotopes of mass numbers 12, 13, and 14 from a long-dead piece of a tree. (To estimate atomic masses, multiply by 1.66×10^{-27} kg.) How far apart are the lines formed by the singly charged ions of each type on the photographic film? What if the ions were doubly charged?

*48. (II) A mass spectrometer is being used to monitor air pollu-
tants. It is difficult, however, to separate molecules with
nearly equal mass such as CO (28.0106 u) and N_2 (28.0134 u).
How large a radius of curvature must a spectrometer have if
these two molecules are to be separated on the film by
0.50 mm?

*49. (II) One form of mass spectrometer accelerates ions by a
voltage V before they enter a magnetic field B. The ions are
assumed to start from rest. Show that the mass of an ion is
$m = qB^2R^2/2V$, where R is the radius of the ions' path in
the magnetic field and q is their charge.

General Problems

50. Protons move in a circle of radius 5.10 cm in a 0.725-T magnetic field. What value of electric field could make their paths
straight? In what direction must the electric field point?

51. Protons with momentum 4.8×10^{-16} kg·m/s are magnetically steered clockwise in a circular path 2.0 km in diameter
at Fermi National Accelerator Laboratory in Illinois. What
is the magnitude and direction of the field in the magnets
surrounding the beam pipe?

52. A proton and an electron have the same kinetic energy
upon entering a region of constant magnetic field. What is
the ratio of the radii of their circular paths?

53. Near the equator, the Earth's magnetic field points almost
horizontally to the north and has magnitude
$B = 0.50 \times 10^{-4}$ T. What should be the magnitude and
direction for the velocity of an electron if its weight is to be
exactly balanced by the magnetic force?

54. Calculate the force on an airplane which has acquired a net
charge of 1550 μC and moves with a speed of 120 m/s perpendicular to the Earth's magnetic field of 5.0×10^{-5} T.

55. The power cable for an electric trolley carries a horizontal
dc current of 330 A toward the east. The Earth's magnetic
field has a strength 5.0×10^{-5} T and makes an angle of dip
of 22° at this location. Calculate the magnitude and direction of the magnetic force on a 10-m length of this cable.

56. Two stiff parallel wires a distance l apart in a horizontal plane
act as rails to support a light metal rod of mass m (perpendicular to each rail), Fig. 27–44. A magnetic field $\mathbf{B}$, directed vertically upward (outward in diagram), acts throughout. At
$t = 0$, wires connected to the rails are connected to a constant current source and a current I begins to flow through
the system. Determine the speed of the rod, which starts from
rest at $t = 0$, as a function of time (a) assuming no friction
between the rod and the rails, and (b) if the coefficient of
friction is μ_k. (c) In which direction does the rod move, east
or west, if the current through it heads north?

57. Suppose the rod in Fig. 27–44 (Problem 56) has mass
$m = 0.40$ kg and length 22 cm and the current through it is
$I = 40$ A. If the coefficient of static friction is $\mu_s = 0.50$,
determine the minimum magnetic field $\mathbf{B}$ (not necessarily
vertical) that will just cause the rod to slide. Give the magnitude of $\mathbf{B}$ and its direction relative to the vertical.

58. Estimate the approximate maximum deflection of the electron beam near the center of a TV screen due to the Earth's
5.0×10^{-5} T field. Assume the screen is 20 cm from the electron gun where the electrons are accelerated (a) by 2.0 kV,
or (b) by 30 kV. Note that in color TV sets, the beam must
be directed accurately to within less than 1 mm in order to
strike the correct phosphor. Because the Earth's field is significant here, mu-metal shields are used to reduce the
Earth's field in the CRT. (See Section 23–9.)

59. The **cyclotron** (Fig. 27–45) is a device used to accelerate elementary particles such as protons to high speeds. Particles
starting at point A with some initial velocity travel in circular orbits in the magnetic field B. The particles are accelerated to higher speeds each time they pass in the gap
between the metal "dees," where there is an electric field E.
(There is no electric field within the cavity of the metal
dees.) The electric field changes direction each half-cycle,
owing to an ac voltage $V = V_0 \sin 2\pi ft$, so that the particles are increased in speed at each passage through the gap.
(a) Show that the frequency f of the voltage must be
$f = Bq/2\pi m$, where q is the charge on the particles and m
their mass. (b) Show that the kinetic energy of the particles
increases by $2qV_0$ each revolution, assuming that the gap is
small. (c) If the radius of the cyclotron is 2.0 m and the magnetic field strength is 0.50 T, what will be the maximum
kinetic energy of accelerated protons in MeV?

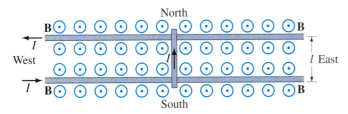

FIGURE 27–44 Looking down on a rod sliding on
rails. Problems 56 and 57.

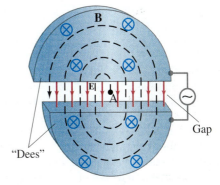

FIGURE 27–45 A cyclotron. Problem 59.

60. The rectangular loop of wire shown in Fig. 27–21 has mass m and carries current I. Show that if the loop is oriented at an angle $\theta \ll 1$ (in radians), then when it is released it will execute simple harmonic motion about $\theta = 0$. Calculate the period of the motion.

61. Magnetic fields are very useful in particle accelerators for "beam steering"; that is, the magnetic fields can be used to change the beam's direction without altering its speed (Fig. 27–46). Discuss how this could work with a beam of protons. What happens to protons that are not moving with the speed that the magnetic field is designed for? If the field extends over a region 5.0 cm wide and has a magnitude of 0.33 T, by approximately what angle will a beam of protons traveling at 0.75×10^7 m/s be bent?

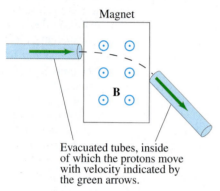

Evacuated tubes, inside of which the protons move with velocity indicated by the green arrows.

FIGURE 27–46 Problem 61.

62. A square loop of aluminum wire is 20.0 cm on a side. It is to carry 25.0 A and rotate in a uniform 1.65-T magnetic field as shown in Fig. 27–47. (*a*) Determine the minimum diameter of the wire so that it will not fracture from tension or shear. Assume a safety factor of 10. (See Table 12–2.) (*b*) What is the resistance of a single loop of this wire?

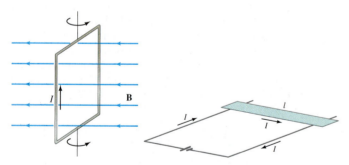

FIGURE 27–47
Problem 62.

FIGURE 27–48   Problem 63.

63. A sort of "projectile launcher" is shown in Fig. 27–48. A large current moves in a closed loop composed of fixed rails, a power supply, and a very light, almost frictionless bar touching the rails. A magnetic field is perpendicular to the plane of the circuit. If the bar has a length of 20 cm, a mass of 1.5 g, and is placed in a field of 1.7 T, what constant current flow is needed in order for it to accelerate to 30 m/s in a distance of 1.0 m? In what direction must the field point?

64. (*a*) What value of magnetic field would make a beam of electrons, traveling to the right at a speed of 4.8×10^6 m/s, go undeflected through a region where there is a uniform electric field of 10,000 V/m pointing vertically up? (*b*) What is the direction of the magnetic field if it is known to be perpendicular to the electric field? (*c*) What is the frequency of the circular orbit of the electrons if the electric field is turned off?

65. In a certain cathode ray tube, electrons are accelerated horizontally by 25 kV. They then pass through a uniform magnetic field B for a distance of 3.5 cm, which deflects them upward so they reach the top of the screen 22 cm away, 11 cm above the center. Estimate the value of B.

66. **Zeeman effect.** In the Bohr model of the hydrogen atom, the electron is held in its circular orbit of radius r about its proton nucleus by electrostatic attraction. If the atoms are placed in a weak magnetic field **B**, the rotation frequency of electrons rotating in a plane perpendicular to **B** is changed by an amount

$$\Delta f = \pm \frac{eB}{4\pi m}$$

where e and m are the charge and mass of an electron. (*a*) Derive this result, assuming the force due to **B** is much less than that due to electrostatic attraction of the nucleus. (*b*) What does the ± sign indicate?

67. A proton follows a spiral path through a gas in a magnetic field of 0.010 T, perpendicular to the plane of the spiral, as shown in Fig. 27–49. In two successive loops, P and Q, the radii are 10.0 mm and 8.5 mm, respectively. Calculate the change in the kinetic energy of the proton as it travels from P to Q.

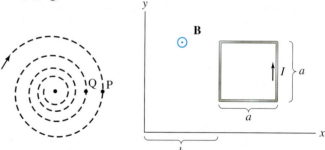

FIGURE 27–49
Problem 67.

FIGURE 27–50   Problem 68.

68. The net force on a current loop in a uniform magnetic field is zero, since contributions to the net force from opposite sides of the loop cancel. However, if the field varies in magnitude from one side of the loop to the other, then a net force can be generated on the loop. Consider a square loop with sides whose length is a, located with one side at $x = b$ in the xy plane (Fig. 27–50). A magnetic field is directed along z, with a magnitude that varies with x according to

$$B = B_0 \left(1 - \frac{x}{d} \right).$$

If the current in the loop circulates counter-clockwise (that is, the magnetic dipole moment of the loop is along the z axis), find an expression for the net force on the loop.

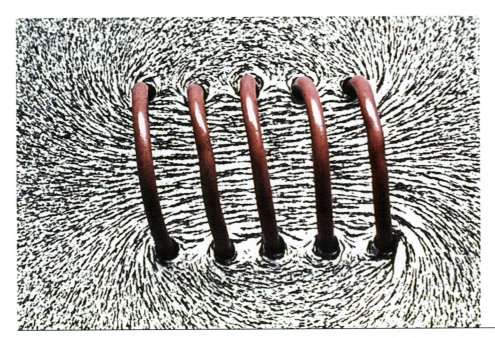

A long coil of wire with many closely spaced loops is called a solenoid. When a solenoid carries an electric current, a nearly uniform magnetic field is produced within the loops as suggested by the alignment of the iron filings in this photo. The magnitude of the field inside a solenoid is readily found using Ampère's law, one of the great general laws of electromagnetism, relating magnetic fields and electric currents. We examine these connections in detail in this chapter, as well as other means for producing magnetic fields.

CHAPTER 28

Sources of Magnetic Field

In the previous chapter, we discussed primarily the effects (forces and torques) that a magnetic field has on electric currents and on moving electric charges. We did see, however, that magnetic fields are produced not only by magnets but also by electric currents (Oersted's great discovery). It is this aspect of magnetism, the production of magnetic fields, that we discuss in this chapter. We will now see how magnetic field strengths are determined for some simple situations, and discuss some general relations between magnetic fields and their sources. We begin with the simplest case, the magnetic field created by a long straight wire carrying a steady electric current. We then look at how such a field, created by one wire, exerts a force on a second current-carrying wire. Interestingly enough, this interaction is used for the precise definitions of both the units of electric current and electric charge, the ampere and the coulomb.

Then we develop an elegant general approach to finding the connection between current and magnetic field known as Ampère's law, one of the fundamental equations of physics. We also examine a second technique for determining the magnetic field due to a current, known as the Biot-Savart law; though harder to visualize intuitively, the Biot-Savart law does allow us to solve problems more readily in many cases than Ampère's law.

Finally, we end this chapter with a discussion of what we understand about iron and other magnetic materials and how they produce magnetic fields.

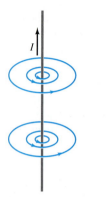

FIGURE 28–1 Same as Fig. 27–9a, magnetic field lines around a long straight wire carrying an electric current I.

Magnetic field due to current in a long straight wire

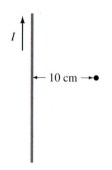

FIGURE 28–2 Example 28–1.

➡ **PHYSICS APPLIED**

A compass, near a current, may not point North

FIGURE 28–3 (a) Two parallel conductors carrying currents I_1 and I_2. (b) Magnetic field produced by I_1. (Field produced by I_2 is not shown.)

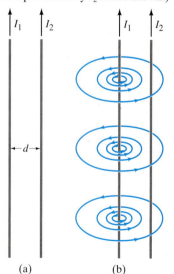

28–1 Magnetic Field Due to a Straight Wire

We saw in Section 27–2, Fig. 27–9, that the magnetic field due to the electric current in a long straight wire is such that the field lines are circles with the wire at the center (Fig. 28–1). You might expect that the field strength at a given point would be greater if the current flowing in the wire were greater; and that the field would be less at points farther from the wire. This is indeed the case. Careful experiments show that the magnetic field B at a point near a long straight wire is directly proportional to the current I in the wire and inversely proportional to the distance r from the wire:

$$B \propto \frac{I}{r}.$$

This relation is valid as long as r, the perpendicular distance to the wire, is much less than the distance to the ends of the wire (i.e., the wire is long).

The proportionality constant is written[†] as $\mu_0/2\pi$; thus,

$$B = \frac{\mu_0}{2\pi}\frac{I}{r}. \qquad \text{[outside a long straight wire]} \quad \textbf{(28–1)}$$

The value of the constant μ_0, which is called the **permeability of free space**, is $\mu_0 = 4\pi \times 10^{-7}\,\text{T·m/A}$ (see Section 28–3).

EXAMPLE 28–1 **Calculation of B near a wire.** A vertical electric wire in the wall of a building carries a dc current of 25 A upward. What is the magnetic field at a point 10 cm due north of this wire (Fig. 28–2)?

SOLUTION According to Eq. 28–1:

$$B = \frac{\mu_0 I}{2\pi r} = \frac{(4\pi \times 10^{-7}\,\text{T·m/A})(25\,\text{A})}{(2\pi)(0.10\,\text{m})} = 5.0 \times 10^{-5}\,\text{T},$$

or 0.50 G. By the right-hand rule (Fig. 27–9b), the field points to the west (into the page in Fig. 28–2) at this point. Since this field has about the same magnitude as Earth's, a compass placed at this location would not point north, but in a northwesterly direction.

28–2 Force between Two Parallel Wires

We have seen that a wire carrying a current produces a magnetic field (magnitude given by Eq. 28–1 for a long straight wire), and furthermore that such a wire feels a force when placed in a magnetic field (Section 27–3, Eq. 27–1). Thus, we expect that two current-carrying wires would exert a force on each other.

Consider two long parallel conductors separated by a distance d, as in Fig. 28–3a. They carry currents I_1 and I_2, respectively. Each current produces a magnetic field that is "felt" by the other so that each must exert a force on the other, as Ampère first pointed out. For example, the magnetic field B_1 produced by I_1 is given by Eq. 28–1. At the location of the second conductor, the magnitude of the magnetic field produced by I_1 is

$$B_1 = \frac{\mu_0}{2\pi}\frac{I_1}{d}.$$

See Fig. 28–3b where the field due *only* to I_1 is shown. According to Eq. 27–2, the

[†]The constant is chosen in this complicated way so that Ampère's law (Section 28–4), which is considered more fundamental, will have a simple and elegant form.

force F per unit length l on the conductor carrying current I_2, when the field and current are perpendicular, is

$$\frac{F}{l} = I_2 B_1.$$

Note that the force on I_2 is due only to the field produced by I_1. Of course I_2 also produces a field, but it does not exert a force on itself. We substitute in the above formula for B_1 and find that the force per unit length on I_2 is

$$\frac{F}{l} = \frac{\mu_0}{2\pi} \frac{I_1 I_2}{d}. \qquad \textbf{(28-2)}$$

If we use the right-hand rule of Fig. 27–9b, we see that the lines of B_1 are as shown in Fig. 28–3b. Then using the right-hand rule of Fig. 27–12c, we see that the force exerted on I_2 will be to the left in Fig. 28–3b. That is, I_1 exerts an attractive force on I_2 (Fig. 28–4a). This is true as long as the currents are in the same direction. If I_2 is in the opposite direction, the right-hand rule indicates that the force is in the opposite direction. That is, I_1 exerts a repulsive force on I_2 (Fig. 28–4b).

Reasoning similar to that above shows that the magnetic field produced by I_2 exerts an equal but opposite force on I_1. We expect this to be true also, of course, from Newton's third law. Thus, as shown in Fig. 28–4, parallel currents in the same directions attract each other, whereas parallel currents in opposite directions repel.

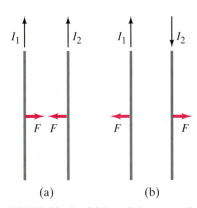

FIGURE 28–4 (a) Parallel currents in the same direction exert an attractive force on each other. (b) Antiparallel currents (in opposite directions) exert a repulsive force on each other.

EXAMPLE 28–2 Force between two current carrying wires. The two wires of a 2.0-m-long appliance cord are 3.0 mm apart and carry a current of 8.0 A dc. Calculate the force between these wires.

SOLUTION Equation 28–2 gives us

$$F = \frac{(2.0 \times 10^{-7}\,\text{T·m/A})(8.0\,\text{A})^2(2.0\,\text{m})}{(3.0 \times 10^{-3}\,\text{m})} = 8.5 \times 10^{-3}\,\text{N},$$

where we have written $\mu_0/2\pi = 2.0 \times 10^{-7}\,\text{T·m/A}$. Since the currents are in opposite directions, the force would tend to spread them apart.

EXAMPLE 28–3 Suspending a current with a current. A horizontal wire carries a current $I_1 = 80$ A dc. A second parallel wire 20 cm below it (Fig. 28–5) must carry how much current I_2 so that it doesn't fall due to gravity? The lower wire has a mass of 0.12 g per meter of length.

FIGURE 28–5 Example 28–3.

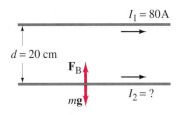

SOLUTION The force of gravity on wire 2 is downward, and per each meter of length has magnitude

$$\frac{F}{l} = \frac{mg}{l} = \frac{(0.12 \times 10^{-3}\,\text{kg})(9.8\,\text{m/s}^2)}{1.0\,\text{m}} = 1.18 \times 10^{-3}\,\text{N/m}.$$

The magnetic force on wire 2 must be upward (hence I_2 must have the same direction as I_1) and, with $d = 0.20$ m and $I_1 = 80$ A, has magnitude

$$\frac{F}{l} = \frac{\mu_0}{2\pi} \frac{I_1 I_2}{d}.$$

We solve for I_2 and find

$$I_2 = \frac{2\pi d}{\mu_0 I_1}\left(\frac{F}{l}\right) = \frac{2\pi(0.20\,\text{m})}{(4\pi \times 10^{-7}\,\text{T·m/A})(80\,\text{A})}\left(1.18 \times 10^{-3}\,\text{N/m}\right) = 15\,\text{A}.$$

28–3 Operational Definitions of the Ampere and the Coulomb

You may have wondered how the constant μ_0 in Eq. 28–1 could be exactly $4\pi \times 10^{-7}\,\text{T}\cdot\text{m/A}$. Here is how it happened. With an older definition of the ampere, μ_0 was measured experimentally to be very close to this value. Today, however, μ_0 is *defined* to be exactly $4\pi \times 10^{-7}\,\text{T}\cdot\text{m/A}$. This, of course, could not be done if the ampere were defined independently. The ampere, the unit of current, is now defined in terms of the magnetic field B it produces using the defined value of μ_0.

In particular, we use the force between two parallel current-carrying wires, Eq. 28–2, to define the ampere precisely. If $I_1 = I_2 = 1\,\text{A}$ exactly, and the two wires are exactly 1 m apart, then

$$\frac{F}{l} = \frac{\mu_0}{2\pi}\frac{I_1 I_2}{d} = \frac{(4\pi \times 10^{-7}\,\text{T}\cdot\text{m/A})\,(1\,\text{A})(1\,\text{A})}{(2\pi)\,(1\,\text{m})} = 2 \times 10^{-7}\,\text{N/m}.$$

Definitions

of ampere

and coloumb

Thus, *one **ampere** is defined as that current flowing in each of two long parallel conductors 1 m apart, which results in a force of exactly $2 \times 10^{-7}\,\text{N/m}$ of length of each conductor.*

This is the precise definition of the ampere. The **coulomb** is then defined as being *exactly* one ampere-second: $1\,\text{C} = 1\,\text{A}\cdot\text{s}$. The value of k or ϵ_0 in Coulomb's law (Section 21–5) is obtained from experiment.

This may seem a rather roundabout way of defining quantities. The reason behind it is the desire for **operational definitions** of quantities—that is, definitions of quantities that can actually be measured given a definite set of operations to carry out. For example, the unit of charge, the coulomb, could be defined in terms of the force between two equal charges after defining a value for ϵ_0 or k in Eqs. 21–1 or 21–2. However, to carry out an actual experiment to measure the force between two charges is very difficult. For one thing, any desired amount of charge is not easily obtained precisely; and charge tends to leak from objects into the air. The amount of current in a wire, on the other hand, can be varied accurately and continuously (by putting a variable resistor in a circuit). Thus the force between two current-carrying conductors is far easier to measure precisely. And this is why the ampere is defined first and then the coulomb in terms of the ampere. At the National Institute of Standards and Technology in Maryland, precise measurement of current is made using circular coils of wire rather than straight lengths because it is more convenient and accurate.

Electric and magnetic field strengths are also defined operationally: the electric field in terms of the measurable force on a charge, via Eq. 21–3; and the magnetic field in terms of the force per unit length on a current-carrying wire, via Eq. 27–2.

28–4 Ampère's Law

In Section 28–1 we saw that Eq. 28–1 gives the relation between the current in a long straight wire and the magnetic field it produces. This equation is valid only for a long straight wire. The following question arises: is there a general relation between a current in a wire of any shape and the magnetic field around it? The answer is yes: the French scientist André Marie Ampère (1775–1836), proposed such a relation shortly after Oersted's discovery. Consider an arbitrary closed path around a current as shown in Fig. 28–6, and imagine this path as being made up of short segments each of length Δl. First, we take the product of the length of each segment times the component of **B** parallel to that segment (call this component $B_\parallel$). If we now sum all these terms, according to Ampère, the result will be equal to μ_0 times the net current I_{encl} that passes through the surface enclosed by the path:

$$\sum B_\parallel \Delta l = \mu_0 I_{\text{encl}}.$$

The lengths Δl are chosen so that $B_\parallel$ is essentially constant along each length. The

FIGURE 28–6 Arbitrary path enclosing a current, for Ampère's law. The path is broken down into segments of equal length Δl.

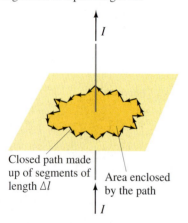

I

Closed path made up of segments of length Δl

Area enclosed by the path

I

sum must be made over a *closed path*; and I_{encl} is the net current passing through the surface bounded by this closed path. In the limit $\Delta l \rightarrow 0$, this relation becomes

$$\oint \mathbf{B} \cdot d\mathbf{l} = \mu_0 I_{encl},\qquad(28\text{-}3)$$

AMPÈRE'S LAW

where $d\mathbf{l}$ is an infinitesimal length vector and the vector dot product assures that the parallel component of $\mathbf{B}$ is taken. Equation 28–3 is known as **Ampère's law**. The integrand in Eq. 28–3 is taken around a closed path, and I_{encl} is the current passing through the area enclosed by the chosen path.

To understand Ampère's law better, let us apply it to the simple case of a long straight wire carrying a current I which we've already examined, and which served as an inspiration for Ampère himself. Suppose we want to find the magnitude of $\mathbf{B}$ at some point A which is a distance r from the wire (Fig. 28–7). We know the magnetic field lines are circles with the wire at their center. So to apply Eq. 28–3 we choose as our path of integration a circle of radius r. The choice of path is ours, so we choose one that will be convenient: at any point on this circular path, $\mathbf{B}$ will be tangent to the circle. Furthermore, since all points on the path are the same distance from the wire, by symmetry we expect B to have the same magnitude at each point. Thus for any short segment of the circle (Fig. 28–7), $\mathbf{B}$ will be parallel to that segment, and hence

$$\mu_0 I = \oint \mathbf{B} \cdot d\mathbf{l} = \oint B\,dl = B \oint dl = B(2\pi r),$$

where $\oint dl = 2\pi r$, the circumference of the circle, and $I_{encl} = I$. We solve for B and obtain

$$B = \frac{\mu_0 I}{2\pi r}.$$

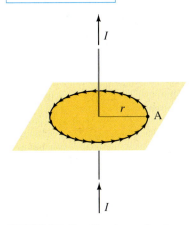

FIGURE 28–7 Circular path of radius r.

B for straight wire using Ampère's law

This is just Eq. 28–1 for the field near a long straight wire as discussed earlier.

Ampère's law thus works for this simple case. A great many experiments indicate that Ampère's law is valid in general. However, as with Gauss's law for the electric field, its practical value as a means to calculate the magnetic field is limited mainly to simple or symmetric situations. Its importance is that it relates the magnetic field to the current in a direct and mathematically elegant way. Ampère's law is thus considered one of the basic laws of electricity and magnetism. It is valid for any situation where the currents and fields are steady and not changing in time, and no magnetic materials are present.

We now can see why the constant in Eq. 28–1 is written $\mu_0/2\pi$. This is done so that only μ_0 appears in Eq. 28–3, rather than, say, $2\pi k$ if we had used k in Eq. 28–1. In this way, the more fundamental equation, Ampère's law, has the simpler form.

It should be noted, that the $\mathbf{B}$ in Ampère's law is not necessarily due only to the current I_{encl}. Ampère's law, like Gauss's law for the electric field, is valid in general. $\mathbf{B}$ is the field at each point in space along the chosen path due to all sources—including the current I enclosed by the path, but also due to any other sources. For example, the field surrounding two parallel current-carrying wires is the vector sum of the fields produced by each, and the field lines are shown in Fig. 28–8. If the path chosen for the integral (Eq. 28–3) is a circle centered on one of the wires with radius less than the distance between the wires (the dashed line in Fig. 28–8), only the current (I_1) in the encircled wire is included on the right side of Eq. 28–3. $\mathbf{B}$ on the left side of the equation must be the total $\mathbf{B}$ at each point due to both wires. Note also that $\oint \mathbf{B} \cdot d\mathbf{l}$ for the path shown in Fig. 28–8 is the same whether the second wire is present or not (in both cases, it equals $\mu_0 I_1$). How can this be? It can be so because although the fields due to the two wires tend to cancel one another at points between them, such as point N in the diagram $(\mathbf{B} = 0$ at a point midway between the wires if $I_1 = I_2)$, at a point such as M in the Figure, the fields add together to produce a larger field. In the *sum*, $\oint \mathbf{B} \cdot d\mathbf{l}$, these effects just balance so that $\oint \mathbf{B} \cdot d\mathbf{l} = \mu_0 I_1$, whether the second wire is there or not. The integral $\oint \mathbf{B} \cdot d\mathbf{l}$ will be the same in each case, even though $\mathbf{B}$ will not be the same at every point for each of the two cases.

FIGURE 28–8 Magnetic field lines around two long parallel wires whose equal currents, I_1 and I_2, are coming out of the paper toward the viewer.

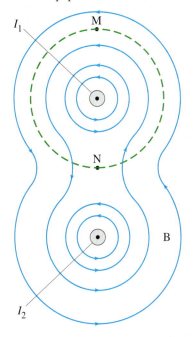

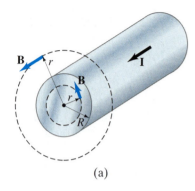

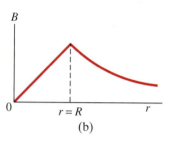

(a)

(b)

FIGURE 28–9 Magnetic field inside and outside a cylindrical conductor (Example 28–4).

EXAMPLE 28–4 **Field inside and outside a wire.** A long straight cylindrical wire conductor of radius R carries a current I of uniform current density in the conductor. Determine the magnetic field at (a) points outside the conductor ($r > R$), and (b) points inside the conductor ($r < R$). See Fig. 28–9. Assume that r, the radial distance from the axis, is much less than the length of the wire. (c) If $R = 2.0$ mm and $I = 60$ A, what is B at $r = 1.0$ mm, $r = 2.0$ mm, and $r = 3.0$ mm?

SOLUTION (a) Because the wire is long, straight, and cylindrical, we expect from the symmetry of the situation that the magnetic field must be the same at all points that are the same distance from the center of the conductor—there is no reason why any such point should have preference over the others at the same distance from the wire (they are physically equivalent), and so B must have the same value at all points the same distance from the center. We also expect **B** to be tangent to circles around the wire (Fig. 28–1), so let us choose a circular path of integration outside the wire ($r > R$), but concentric with it, as we did in Fig. 28–7. Then $I_{encl} = I$, so

$$\oint \mathbf{B} \cdot d\mathbf{l} = B(2\pi r) = \mu_0 I_{encl}$$

or

$$B = \frac{\mu_0 I}{2\pi r}. \qquad [r > R]$$

which is the same result as for a thin wire.

(b) Inside the wire ($r < R$), we again choose a circular path concentric with the cylinder; we expect **B** to be tangential to this path, and again, because of the symmetry, it will have the same magnitude at all points on the circle. The current enclosed in this case is less than I by a factor of the ratio of the areas:

$$I_{encl} = I \frac{\pi r^2}{\pi R^2}.$$

So Ampère's law gives

$$\oint \mathbf{B} \cdot d\mathbf{l} = \mu_0 I_{encl}$$

$$B(2\pi r) = \mu_0 I \left(\frac{\pi r^2}{\pi R^2} \right)$$

so

$$B = \frac{\mu_0 I r}{2\pi R^2}. \qquad [r < R]$$

The field is zero at the center of the conductor and increases linearly with r until $r = R$; beyond $r = R$, B decreases as $1/r$. This is shown in Fig. 28–9b. Note that these results are valid only for points close to the center of the conductor as compared to its length. For a current to flow, there must be connecting wires (to a battery, say), and the field due to these conducting wires, if not very far away, will destroy the assumed symmetry.

(c) At $r = 2.0$ mm, the surface of the wire, $r = R$, so

$$B = \frac{\mu_0 I}{2\pi R} = \frac{(4\pi \times 10^{-7}\,\text{T·m/A})(60\,\text{A})}{(2\pi)(2.0 \times 10^{-3}\,\text{m})} = 6.0 \times 10^{-3}\,\text{T}.$$

We saw in (b) that inside the wire B is linear in r. So at $r = 1.0$ mm, B will be half what it is at $r = 2.0$ mm or 3.0×10^{-3} T. Outside the wire, B falls off as $1/r$, so at $r = 3.0$ mm it will be two-thirds as great as at $r = 2.0$ mm, or $B = 4.0 \times 10^{-3}$ T. To check, we use our result in (a), $B = \mu_0 I/2\pi r$, which gives the same result.

CONCEPTUAL EXAMPLE 28–5 **Coaxial cable.** A *coaxial cable* is a single wire surrounded by a cylindrical metallic braid, as shown in Fig. 28–10. The two conductors are separated by an insulator. The central wire carries current to the other end of the cable, and the outer braid carries the return current and is usually considered ground. Describe the magnetic field (a) in the space between the conductors, and (b) outside the cable.

RESPONSE (a) In the space between the conductors, we can apply Ampère's law for a circular path around the center wire, just as we did for the case shown in Fig. 28–7, and the magnitude is as given by Eq. 28–1. The current in the outer conductor has no bearing on this result. (Ampère's law uses only the current enclosed *inside* the path; as long as the currents outside the path don't affect the symmetry of the field, they do not contribute to the field along the path at all).
(b) Outside the cable, we can draw a similar circular path, for we expect the field to have the same circular symmetry. Now, however, there are two currents enclosed by the path, and they add up to zero. The field outside the cable is zero.

The nice feature of coaxial cables is that they are self-shielding: no stray magnetic fields escape outside the cable. The outer cylindrical conductor also shields external electric fields from coming in (see also Example 21–13). This makes them ideal for carrying signals near sensitive equipment. Audiophiles use coaxial cables between stereo equipment components and even to the loudspeakers.

➡ **PHYSICS APPLIED**

Coaxial cable (shielding)

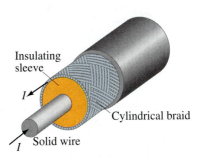

FIGURE 28–10 Coaxial cable. Conceptual Example 28–5.

EXAMPLE 28–6 **A nice use for Ampère's law.** Use Ampère's law to show that in any region of space where there are no currents the magnetic field cannot be both unidirectional and nonuniform as shown in Fig. 28–11a.

SOLUTION The wider spacing of lines near the top of Fig. 28–11a indicates the field has a smaller magnitude at the top than it does lower down. We now apply Ampère's law to the rectangular path abcd shown dashed in the diagram. Since no current is enclosed by this path,

$$\oint \mathbf{B} \cdot d\mathbf{l} = 0.$$

The integral along sections ab and cd is zero, since $\mathbf{B} \perp d\mathbf{l}$. Thus

$$\oint \mathbf{B} \cdot d\mathbf{l} = B_{bc}l - B_{da}l = (B_{bc} - B_{da})l,$$

which is not zero since the field B_{bc} along the path bc is less than the field B_{da} along path da. Hence we have a contradiction: $\oint \mathbf{B} \cdot d\mathbf{l}$ cannot be both zero (since $I = 0$) and nonzero. Thus we have shown that a nonuniform unidirectional field is not consistent with Ampère's law. A nonuniform field whose direction also changes, as in Fig. 28–11b, is consistent with Ampère's law (convince yourself this is so), and possible. The fringing of a permanent magnet's field (Fig. 27–7) has this shape.

FIGURE 28–11 Example 28–6.

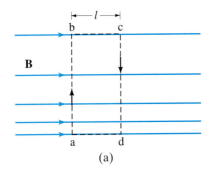

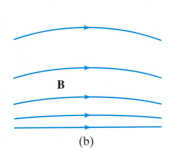

(a)

(b)

1. Ampère's Law, like Gauss' law, is always a valid statement. But as a calculation tool it is limited primarily to systems with a high degree of symmetry. The first step in applying Ampère's Law is to identify any useful symmetry.

2. Choose an integration path that reflects the symmetry (see the Examples to get a feel for this). In particular, search for paths where B has constant magnitude along the entire path or along segments of the path. Make sure your integration path passes through the point where you wish to evaluate the magnetic field.

3. Use symmetry to determine the direction of **B** along the integration path. With a smart choice of path, **B** will be either parallel or perpendicular to the path.

4. Evaluate the right-hand side of Ampère's Law by determining the enclosed current. Be careful with signs. Let the fingers of your right hand curl along the direction of **B** so that your thumb shows the direction of positive current. If the problem involves a solid conductor and your integration path does not enclose the full current, calculate the enclosed current using the current density (current per unit area) multiplied by the enclosed area (as in Example 28–4).

28–5 | Magnetic Field of a Solenoid and a Toroid

A long coil of wire consisting of many loops is called a **solenoid**. Each loop produces a magnetic field as was shown in Fig. 27–10. In Fig. 28–12a, we see the field due to a solenoid when the coils are far apart. Near each wire, the field lines are very nearly circles as for a straight wire (that is, at distances that are small compared to the curvature of the wire). Between any two wires, the fields due to each loop tend to cancel. Toward the center of the solenoid, the fields add up to give a field that can be fairly large and fairly uniform. For a long solenoid, with closely packed coils, the field is nearly uniform and parallel to the solenoid axes within the entire cross section, as shown in Fig. 28–12b. The field outside the solenoid is very small compared to the field inside, except near the ends. Note that the same number of field lines that are concentrated inside the solenoid, spread out into the vast open space outside.

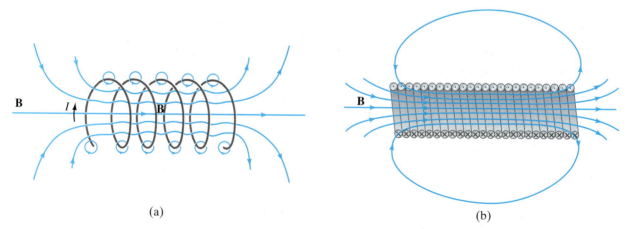

(a)

(b)

FIGURE 28–12 Magnetic field due to a solenoid: (a) loosely spaced turns, (b) closely spaced turns.

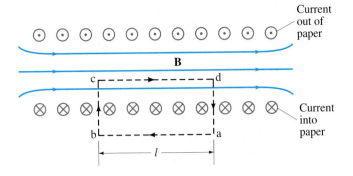

FIGURE 28–13 Magnetic field inside a long solenoid is uniform. Dashed lines indicate the path chosen for use in Ampère's law.

We now use Ampère's law to determine the magnetic field inside a very long (ideally, infinitely long) closely packed solenoid. We choose the path abcd shown in Fig. 28–13, far from either end, for applying Ampère's law. We will consider this path as made up of four segments, the sides of the rectangle: ab, bc, cd, da. Then the left side of Eq. 28–3, Ampère's law, becomes

$$\oint \mathbf{B} \cdot d\mathbf{l} = \int_a^b \mathbf{B} \cdot d\mathbf{l} + \int_b^c \mathbf{B} \cdot d\mathbf{l} + \int_c^d \mathbf{B} \cdot d\mathbf{l} + \int_d^a \mathbf{B} \cdot d\mathbf{l}.$$

The field outside the solenoid is so small as to be negligible compared to the field inside. Thus the first term in this sum will be zero. Furthermore, $\mathbf{B}$ is perpendicular to the segments bc and da inside the solenoid, and is nearly zero between and outside the coils, so these terms too are zero. Therefore we have reduced the integral to the segment cd where $\mathbf{B}$ is the nearly uniform field inside the solenoid, and is parallel to $d\mathbf{l}$, so

$$\oint \mathbf{B} \cdot d\mathbf{l} = \int_c^d \mathbf{B} \cdot d\mathbf{l} = Bl,$$

where l is the length cd. Now we determine the current enclosed by this loop for the right side of Ampère's law, Eq. 28–3. If a current I flows in the wire of the solenoid, the total current enclosed by our path abcd is NI where N is the number of loops our path encircles (five in Fig. 28–13). Thus Ampère's law gives us

$$Bl = \mu_0 NI.$$

If we let $n = N/l$ be the *number of loops per unit length*, then

$$B = \mu_0 nI. \qquad \text{[solenoid]} \qquad \textbf{(28–4)}$$

Magnetic field inside a solenoid

This is the magnitude of the magnetic field within a solenoid. Note that B depends only on the number of loops per unit length, n, and the current I. The field does not depend on position within the solenoid, so B is uniform. This is strictly true only for an infinite solenoid, but it is a good approximation for real ones for points not close to the ends.

EXAMPLE 28–7 **Field inside a solenoid.** A thin 10-cm-long solenoid used for fast electromechanical switching has a total of 400 turns of wire and carries a current of 2.0 A. Calculate the field inside near the center.

SOLUTION The number of turns per unit length is $n = 400/0.10\,\text{m} = 4.0 \times 10^3\,\text{m}^{-1}$. Thus

$$B = \mu_0 nI = (12.57 \times 10^{-7}\,\text{T·m/A})(4.0 \times 10^3\,\text{m}^{-1})(2.0\,\text{A})$$
$$= 1.0 \times 10^{-2}\,\text{T}.$$

A close look at Fig. 28–12 shows that the field outside of a solenoid is much like that of a bar magnet (Fig. 27–3). Indeed, a solenoid acts like a magnet, with one end acting as a north pole and the other as south pole, depending on the direction of the current in the loops. Since magnetic field lines leave the north pole of a magnet, the north poles of the solenoids in Fig. 28–12 are on the right.

Solenoids have many practical applications, and we discuss some of them later in the chapter, in Section 28–8.

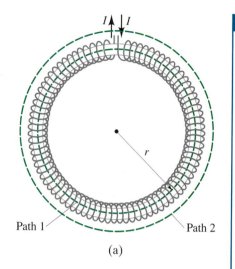

Path 1   Path 2

(a)

(b)

FIGURE 28–14 (a) A toroid. (b) A section of the toroid showing direction of the current for three loops: ⊙ means current toward viewer, ⊗ means current away from viewer.

EXAMPLE 28–8 **Toroid.** Use Ampère's law to determine the magnetic field (a) inside and (b) outside a toroid, which is like a solenoid bent into the shape of a circle as shown in Fig. 28–14a.

SOLUTION The magnetic field lines inside the toroid will be circles concentric with the toroid. (If you think of the toroid as a solenoid bent into a circle, the field lines bend along with the solenoid.) We choose as our path of integration one of these field lines of radius r inside the toroid as shown by the dashed line labeled "path 1" in Fig. 28–14a. We make this choice to use the symmetry of the situation, so B must be the same at all points along the path (although it is not necessarily the same across the whole cross-section of the toroid); so Ampère's law

$$\oint \mathbf{B} \cdot d\mathbf{l} = \mu_0 I_{encl}$$

becomes

$$B(2\pi r) = \mu_0 NI,$$

where N is the total number of coils and I is the current in each of the coils. Thus

$$B = \frac{\mu_0 NI}{2\pi r}.$$

The magnetic field B is not uniform within the toroid: it is largest along the inner edge (where r is smallest) and smallest at the outer edge. However, if the toroid is large, but thin (so that the difference between the inner and outer radii is small compared to the average radius) the field will be essentially uniform within the toroid. In this case, the formula for B reduces to that for a straight solenoid $B = \mu_0 nI$ where $n = N/(2\pi r)$ is the number of coils per unit length. (b) Outside the toroid, we choose as our path of integration a circle concentric with the toroid, "path 2" in Fig. 28–14a. This path encloses N loops carrying current I in one direction and N loops carrying the same current in the opposite direction. (Figure 28–14b shows the directions of the current for the parts of the loop on the inside and outside of the toroid.) Thus the net current enclosed by path 2 is zero. For a very tightly packed toroid, all points on path 2 are equidistant from the toroid and equivalent, so we expect B to be the same at all points along the path. Hence, Ampère's law gives

$$\oint \mathbf{B} \cdot d\mathbf{l} = \mu_0 I_{encl}$$

$$B(2\pi r) = 0$$

or

$$B = 0.$$

The same is true for a path taken at a radius smaller than that of the toroid. So there is no field exterior to a very tightly wound toroid. It is all inside the loops.

28–6 Biot-Savart Law

The usefulness of Ampère's law for determining the magnetic field **B** due to particular electric currents is restricted to situations where the symmetry of the given currents allows us to evaluate $\oint \mathbf{B} \cdot d\mathbf{l}$ readily. This does not, of course, invalidate Ampère's law nor does it reduce its fundamental importance. Recall the electric case, where Gauss's law is considered fundamental but is limited in its use for actually calculating **E**. We must often determine the electric field **E** by another method summing over contributions due to infinitesimal charge elements dq via Coulomb's law: $dE = (1/4\pi\epsilon_0)(dq/r^2)$. A magnetic equivalent to this infinitesimal form of Coulomb's law would be helpful for currents that do not have great symmetry. Such a law was developed by Jean Baptiste Biot (1774–1862) and Felix Savart (1791–1841) shortly after Oersted's discovery in 1820 that a current produces a magnetic field.

According to Biot and Savart, a current I flowing in any path can be considered as many tiny (infinitesimal) current elements, such as in the wire of Fig. 28–15. If $d\mathbf{l}$ represents any infinitesimal length along which the current is flowing, then the magnetic field, $d\mathbf{B}$, at any point P in space, due to this element of current, is given by

$$d\mathbf{B} = \frac{\mu_0 I}{4\pi} \frac{d\mathbf{l} \times \hat{\mathbf{r}}}{r^2}, \tag{28–5}$$

Biot-Savart law

where **r** is the displacement vector from the element $d\mathbf{l}$ to the point P, and $\hat{\mathbf{r}} = \mathbf{r}/r$ is the unit vector in the direction of **r** (see Fig. 28–15). Equation 28–5 is known as the **Biot-Savart law**. The magnitude of $d\mathbf{B}$ is

$$dB = \frac{\mu_0 I\, dl \sin\theta}{4\pi r^2}, \tag{28–6}$$

where θ is the angle between $d\mathbf{l}$ and **r** (Fig. 28–15). The total magnetic field at point P is then found by summing (integrating) over all current elements:

$$\mathbf{B} = \int d\mathbf{B}.$$

Note that this is a *vector* sum. The Biot-Savart law is the magnetic equivalent of Coulomb's law in its infinitesimal form. It is even an inverse square law, like Coulomb's law.

An important difference between the Biot-Savart law and Ampère's law (Eq. 28–3) is that in Ampère's law $\left[\oint \mathbf{B} \cdot d\mathbf{l} = \mu_0 I_{\text{encl}}\right]$, **B** is not necessarily due only to the current enclosed by the path of integration. But in the Biot-Savart law the field $d\mathbf{B}$ in Eq. 28–5 is due only, and entirely, to the current element $I\, d\mathbf{l}$. To find the total **B** at any point in space, it is necessary to include *all* currents.

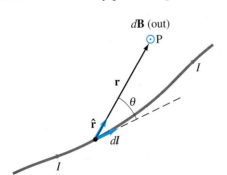

FIGURE 28–15 Biot-Savart law: the field at P due to current element $I\, d\mathbf{l}$ is $d\mathbf{B} = (\mu_0 I/4\pi)(d\mathbf{l} \times \hat{\mathbf{r}}/r^2)$.

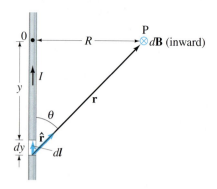

FIGURE 28–16 Determining **B** due to a long straight wire using the Biot-Savart law.

EXAMPLE 28–9 *B* **due to current *I* in straight wire.** For the field near a long straight wire carrying a current I, show that the Biot-Savart law gives the same result as Eq. 28–1, $B = \mu_0 I / 2\pi r$.

SOLUTION We calculate the magnetic field in Fig. 28–16 at point P, which is a perpendicular distance R from an infinitely long wire. The current is moving upwards, and both $d\boldsymbol{l}$ and $\hat{\mathbf{r}}$, which appear in the cross product of Eq. 28–5, are in the plane of the page. Hence the direction of the field $d\mathbf{B}$ due to each element of current must be directed into the plane of the page as shown (right-hand rule for the cross product $d\boldsymbol{l} \times \hat{\mathbf{r}}$). Thus all the $d\mathbf{B}$ have the same direction at point P, and add up to give **B** the same direction consistent with our previous results (Figs. 28–1, 28–7, and 28–9). The magnitude of **B** will be

$$B = \frac{\mu_0 I}{4\pi} \int_{y=-\infty}^{+\infty} \frac{dy \sin\theta}{r^2},$$

where $dy = dl$ and $r^2 = R^2 + y^2$. Note that we are integrating over y (the length of the wire) so R is considered constant. Both y and θ are variables, but they are not independent. In fact, $y = -R/\tan\theta$. Note that we measure y as positive upward from point 0, so for the current element we are considering $y < 0$. Then

$$dy = +R \csc^2\theta \, d\theta = \frac{R \, d\theta}{\sin^2\theta} = \frac{R \, d\theta}{(R/r)^2} = \frac{r^2 \, d\theta}{R}.$$

So our integral becomes

$$B = \frac{\mu_0 I}{4\pi} \frac{1}{R} \int_{\theta=0}^{\pi} \sin\theta \, d\theta = -\frac{\mu_0 I}{4\pi R} \cos\theta \Big|_0^\pi = \frac{\mu_0 I}{2\pi R}.$$

This is just Eq. 28–1 for the field near a long wire, where R has been used instead of r.

FIGURE 28–17 Determining **B** due to a current loop.

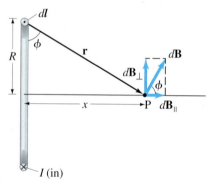

EXAMPLE 28–10 **Current loop.** Determine **B** for points on the axis of a circular loop of wire of radius R carrying a current I, Fig. 28–17.

SOLUTION For an element of current at the top of the loop, the magnetic field $d\mathbf{B}$ at point P on the axis has the direction shown, and magnitude (Eq. 28–5)

$$dB = \frac{\mu_0 I \, dl}{4\pi r^2}$$

since $d\boldsymbol{l}$ is perpendicular to **r** so $|d\boldsymbol{l} \times \hat{\mathbf{r}}| = dl$. We can break $d\mathbf{B}$ down into components $dB_\parallel$ and $dB_\perp$, which are parallel and perpendicular to the axis as shown. When we sum over all the elements of the loop symmetry tells us that the perpendicular components will cancel on opposite sides, so $B_\perp = 0$. Hence, the total **B** will point along the axis, and will have magnitude

$$B = B_\parallel = \int dB \cos\phi = \int dB \frac{R}{r} = \int dB \frac{R}{(R^2 + x^2)^{\frac{1}{2}}},$$

where x is the distance of P from the center of the ring, and $r^2 = R^2 + x^2$. Now we put in dB from the equation above and integrate around the current loop, noting that all segments $d\boldsymbol{l}$ of current are the same distance, $(R^2 + x^2)^{\frac{1}{2}}$, from point P:

$$B = \frac{\mu_0 I}{4\pi} \frac{R}{(R^2 + x^2)^{\frac{3}{2}}} \int dl = \frac{\mu_0 I R^2}{2(R^2 + x^2)^{\frac{3}{2}}}$$

since $\int dl = 2\pi R$, the circumference of the loop. At the very center of the loop (where $x = 0$) the field has its maximum value

$$B = \frac{\mu_0 I}{2R}. \qquad \text{[at center of loop]}$$

Recall from Section 27–5 that a current loop, such as that just discussed (Fig. 28–17) is considered a *magnetic dipole*. We saw there that a magnetic dipole has a dipole moment

$$\mu = NIA,$$

where A is the area of the loop and N is the number of coils in the loop, each carrying current I. We also saw in Chapter 27 that a magnetic dipole placed in an external magnetic field experiences a torque and possesses potential energy, just like an electric dipole. In Example 28–10, we have looked at another aspect of a magnetic dipole: the magnetic field *produced by* a magnetic dipole has magnitude, along the dipole axis, of

$$B = \frac{\mu_0 IR^2}{2(R^2 + x^2)^{\frac{3}{2}}}.$$

We can write this in terms of the magnetic dipole moment $\mu = IA = I\pi R^2$ (for a single loop $N = 1$):

$$B = \frac{\mu_0}{2\pi} \frac{\mu}{(R^2 + x^2)^{\frac{3}{2}}}. \qquad \text{[magnetic dipole]} \quad \textbf{(28–7a)}$$

(Be careful to distinguish μ for dipole moment from μ_0, the magnetic permeability constant.) For distances far from the loop, $x \gg R$, this becomes

$$B \approx \frac{\mu_0}{2\pi} \frac{\mu}{x^3}. \qquad \left[\begin{matrix}\text{on axis,} \\ \text{magnetic dipole, } x \gg R\end{matrix}\right] \quad \textbf{(28–7b)}$$

The magnetic field on the axis of a magnetic dipole decreases with the cube of the distance, just as for an electric dipole. B decreases as the cube of the distance also for points not on the axis, although the multiplying factor is not the same. The magnetic field due to a current loop can be determined at various points using the Biot-Savart law and the results are in accord with experiment. The field lines around a current loop are shown in Fig. 28–18.

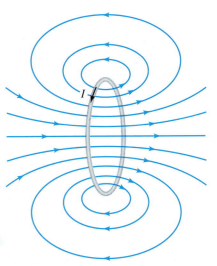

FIGURE 28–18 Magnetic field due to a circular loop of wire. (Same as Fig. 27–10.)

EXAMPLE 28–11 B due to a wire segment. One quarter of a circular loop of wire carries a current I as shown in Fig. 28–19. The current I enters and leaves on straight segments of wire, as shown; the straight wires are along the radial direction from the center C of the circular portion. Find the magnetic field at point C.

FIGURE 28–19 Example 28–11.

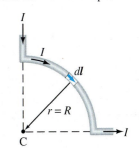

SOLUTION The current in the straight sections produces no magnetic field at point C because $d\boldsymbol{l}$ and $\hat{\mathbf{r}}$ in the Biot-Savart law (Eq. 28–5) are parallel and therefore $d\boldsymbol{l} \times \hat{\mathbf{r}} = 0$. (Said another way, the magnitude of $d\boldsymbol{l} \times \mathbf{r}$ is $(dl)(r)\sin\theta$ where $\theta = 0$.) Each piece $d\boldsymbol{l}$ of the curved section of wire produces a field $d\mathbf{B}$ that points into the paper at C (right-hand rule). The magnitude of each $d\mathbf{B}$ is (Eq. 28–6)

$$dB = \frac{\mu_0 I \, dl}{4\pi R^2}$$

where $r = R$ is the radius of the curved section, and $\sin\theta$ in Eq. 28–6 is $\sin 90° = 1$. Because $r = R$ for all pieces $d\boldsymbol{l}$, then

$$B = \int dB = \frac{\mu_0 I}{4\pi R^2} \int dl = \frac{\mu_0 I}{4\pi R^2} \left(\frac{1}{4} 2\pi R\right) = \frac{\mu_0 I}{8R}.$$

*28–7 Magnetic Materials—Ferromagnetism

Magnetic fields can be produced (a) by magnetic materials (magnets) and (b) by electric currents. We have studied the latter extensively in this chapter. Now we take a brief look at magnetic materials, which are so present in everyday life: ordinary magnets, iron cores in motors and electromagnets, magnetic recording tape and computer data storage disks, even the magnetic stripe on credit cards. We saw in Section 27–1 that iron (and a few other materials) can be made into strong magnets. These materials are said to be **ferromagnetic**. We now look more deeply into the sources of ferromagnetism.

A bar magnet, with its two opposite poles near either end, resembles an electric dipole (equal-magnitude positive and negative charges separated by a distance). Indeed, a bar magnet is sometimes referred to as a "magnetic dipole." There are opposite "poles" separated by a distance. And the magnetic field lines of a bar magnet form a pattern much like that for the electric field of an electric dipole: compare Fig. 21–33a with Fig. 27–3b.

Microscopic examination reveals that a magnet is actually made up of tiny regions known as **domains**, which are at most about 1 mm in length or width. Each domain behaves like a tiny magnet with a north and a south pole. In an unmagnetized piece of iron, these domains are arranged randomly, as shown in Fig, 28–20a. The magnetic effects of the domains cancel each other out, so this piece of iron is not a magnet. In a magnet, the domains are preferentially aligned in one direction as shown in Fig. 28–20b (downward in this case). A magnet can be made from an unmagnetized piece of iron by placing it in a strong magnetic field. (You can make a needle magnetic, for example, by stroking it with one pole of a strong magnet.) Careful observations show in this case that the magnetization of domains may actually rotate slightly so as to be more nearly parallel to the external field. Or, more commonly, the borders of domains move so that those domains whose magnetic orientation is parallel to the external field grow in size at the expense of other domains, as can be seen by comparing Figs. 28–20a and b. This explains how a magnet can pick up unmagnetized pieces of iron like paper clips or bobby pins. The magnet's field causes a slight alignment of the domains in the unmagnetized object so that the object becomes a temporary magnet with its north pole facing the south pole of the permanent magnet, and vice versa; thus, attraction results. In the same way, elongated iron filings will arrange themselves in a magnetic field just as a compass needle does, and will reveal the shape of the magnetic field, Fig. 28–21.

An iron magnet can remain magnetized for a long time, and thus is referred to as a "permanent magnet." However, if you drop a magnet on the floor or strike it with a hammer, you may jar the domains into randomness. The magnet can thus lose some or all of its magnetism. Heating a magnet too can cause a loss of magnetism, for raising the temperature increases the random thermal motion of the atoms which tends to randomize the domains. Above a certain temperature known as the **Curie temperature** (1043 K for iron), a magnet cannot be made at all.[†]

The striking similarity between the fields produced by a bar magnet (Fig. 27–3b) and by a loop of electric current (Fig. 28–18) suggests that the magnetic field produced by a current may have something to do with ferromagnetism, an idea proposed by Ampère in the nineteenth century. According to modern atomic theory, the atoms that make up any material can be roughly visualized as containing electrons that orbit around a central nucleus. Since the electrons are charged, they constitute an electric current and therefore produce a magnetic

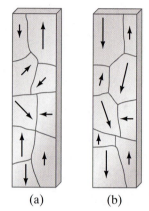

FIGURE 28–20 (a) An unmagnetized piece of iron is made up of domains that are randomly arranged. Each domain is like a tiny magnet; the arrows represent the magnetization direction, with the arrowhead being the N pole. (b) In a magnet, the domains are preferentially aligned in one direction, and may be altered in size by the magnetization process.

FIGURE 28–21 Iron filings line up along magnetic field lines.

[†] Iron, nickel, cobalt, gadolinium, and certain alloys are ferromagnetic at room temperature; several other elements and alloys have low Curie temperature and thus are ferromagnetic only at low temperatures.

field. But if there is no external field, the electron orbits in different atoms are arranged randomly, so the magnetic effects due to the many orbits in all the atoms in a material cancel out. However, electrons by themselves produce an additional intrinsic magnetic field—they have an intrinsic magnetic moment referred to as their "spin" magnetic moment.[†] It is the magnetic field due to electron spin that is now believed to produce ferromagnetism. In most materials, the magnetic fields due to electron spin cancel out because they are oriented at random. But in iron and other ferromagnetic materials, a complicated cooperative mechanism, known as "exchange coupling," operates. The result is that the spin of the electrons contributing to the ferromagnetism in a domain point in the same direction. Thus the tiny magnetic fields due to each of these electrons add up to give the magnetic field of a domain. And when the domains are aligned, as we have seen, a strong magnet results.

*28–8 Electromagnets and Solenoids

A long coil of wire consisting of many loops of wire, as discussed in Section 28–5, is called a solenoid. The magnetic field within a solenoid can be fairly large since it will be the sum of the fields due to the current in each loop (see Fig. 28–22). The solenoid acts like a magnet; one end can be considered the north pole and the other the south pole, depending on the direction of the current in the loops (use the right-hand rule). Since the magnetic field lines leave the north pole of a magnet, the north pole of the solenoid in Fig. 28–22 is on the right.

If a piece of iron is placed inside a solenoid, the magnetic field is increased greatly because the domains of the iron are aligned by the magnetic field produced by the current. The resulting magnetic field is the sum of that due to the current and that due to the iron, and can be hundreds or thousands of times that due to the current alone (see Section 28–9). This arrangement is called an **electromagnet**. The iron used in electromagnets acquires and loses its magnetism quite readily when the current is turned on or off, and so is referred to as "soft iron." (It is "soft" only in a magnetic sense.) Iron that holds its magnetism even when there is no externally applied field is called "hard iron." Hard iron is used in permanent magnets. Soft iron is usually used in electromagnets so that the field can be turned on and off readily. Whether iron is hard or soft depends on heat treatment and other factors.

Electromagnets find use in many practical applications, from use in motors and generators to producing large magnetic fields for research. For some applications an iron core is not present—the magnetic field comes only from the current in the wire coils. When the current flows continuously in a normal electromagnet, a great deal of waste heat (I^2R power) can be produced. Cooling coils, which are tubes carrying water, must be used to absorb the heat in bigger installations. For some applications, superconducting magnets are used. The current-carrying wires are made of superconducting material (Section 25–9) kept below the transition temperature. Very high fields can be produced in the absence of an iron core, higher than what could be reached with an iron core (due to saturation of the iron—see the next Section). No electric power is needed to maintain large current in the superconducting coils, which means large savings of electricity, nor must huge amounts of heat be dissipated. Of course, energy is needed to keep the superconducting coils at the necessary low temperature.

Another useful device consists of a solenoid into which a rod of iron is partially inserted. This combination is also referred to as a solenoid. One simple use is

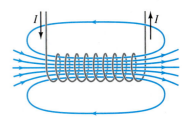

FIGURE 28–22 Magnetic field of a solenoid. The north pole of this solenoid, thought of as a magnet, is on the right, and the south pole is on the left.

→ PHYSICS APPLIED

Electromagnets and solenoids

[†]The name "spin" comes from an early suggestion that this intrinsic magnetic moment arises from the electron "spinning" on its axis (as well as "orbiting" the nucleus) to produce the extra field. However this view of a spinning electron is oversimplified and not valid.

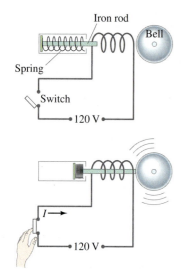

FIGURE 28–23 Solenoid used as a doorbell.

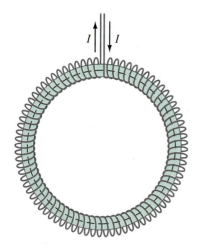

FIGURE 28–24 Iron-core toroid.

FIGURE 28–25 Total magnetic field B in an iron-core toroid as a function of the external field B_0 (B_0 is caused by the current I in the coil).

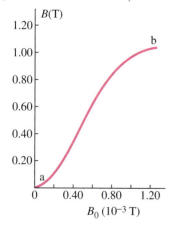

as a doorbell (Fig. 28–23). When the circuit is closed by pushing the button, the coil effectively becomes a magnet and exerts a force on the iron rod. The rod is pulled into the coil and strikes the bell. A larger solenoid is used in the starters of cars; when you engage the starter, you are closing a circuit that not only turns the starter motor, but activates a solenoid that first moves the starter into direct contact with the gears on the engine's flywheel. Solenoids are used as switches in many other devices, such as tape recorders. They have the advantage of moving mechanical parts quickly and accurately.

* 28–9 Magnetic Fields in Magnetic Materials; Hysteresis

The field of a long solenoid is directly proportional to the current. Indeed, Eq. 28–4 tells us that the field B_0 inside a solenoid is given by

$$B_0 = \mu_0 n I.$$

This is valid if there is only air inside the coil. If we put a piece of iron or other ferromagnetic material inside the solenoid, the field will be greatly increased, often by hundreds or thousands of times. This occurs because the domains in the iron become preferentially aligned by the external field. The resulting magnetic field is the sum of that due to the current and that due to the iron. It is sometimes convenient to write the total field in this case as a sum of two terms:

$$\mathbf{B} = \mathbf{B}_0 + \mathbf{B}_M. \tag{28–8}$$

Here, $\mathbf{B}_0$ refers to the field due only to the current in the wire (the "external field"). It is the field that would be present in the absence of a ferromagnetic material. Then $\mathbf{B}_M$ represents the additional field due to the ferromagnetic material itself; often $\mathbf{B}_M \gg \mathbf{B}_0$.

The total field inside a solenoid in such a case can also be written by replacing the constant μ_0 in Eq. 28–4 by another constant, μ, characteristic of the material inside the coil:

$$B = \mu n I; \tag{28–9}$$

μ is called the **magnetic permeability** of the material (do not confuse it with $\boldsymbol{\mu}$ for magnetic moment). For ferromagnetic materials, μ is much greater than μ_0. For all other materials, its value is very close to μ_0 (Section 28–10). The value of μ, however, is not constant for ferromagnetic materials; it depends on the value of the external field B_0, as the following experiment shows.

Measurements on magnetic materials are generally done using a toroid, which is essentially a long solenoid bent into the shape of a circle (Fig. 28–24), so that practically all the lines of $\mathbf{B}$ remain within the toroid. Suppose the toroid has an iron core that is initially unmagnetized and there is no current in the windings of the toroid. Then the current I is slowly increased, and B_0 increases linearly with I. The total field B also increases, but follows the curved line shown in the graph of Fig. 28–25. (Note the different scales: $B \gg B_0$.) Initially, point a, the domains (Section 28–7) are randomly oriented. As B_0 increases, the domains become more and more aligned until at point b, nearly all are aligned. The iron is said to be approaching **saturation**. Point b is typically 70 percent of full saturation. (If B_0 is increased further, the curve continues to rise very slowly, and reaches 98 percent saturation only when B_0 reaches a value about a thousandfold above that at point b; the last few domains are very difficult to align.) Next, suppose the external field B_0 is reduced by decreasing the current in the coils. As the current is reduced to zero, shown as point c in Fig. 28–26, the domains do not become completely random. Some permanent magnetism remains. If the current is then reversed in direction, enough domains can be turned around so $B = 0$ (point d). As the reverse current is increased further, the iron approaches saturation in the opposite direction (point e). Finally, if the current is again reduced to zero and then increased in the original direction, the total field follows the path efgb, again approaching saturation at point b.

Notice that the field did not pass through the origin (point a) in this cycle. The fact that the curves do not retrace themselves on the same path is called **hysteresis**. The curve bcdefgb is called a **hysteresis loop**. In such a cycle, much energy is transformed to thermal energy (friction) due to realigning of the domains. It can be shown that the energy dissipated in this way is proportional to the area of the hysteresis loop.

At points c and f, the iron core is magnetized even though there is no current in the coils. These points correspond to a permanent magnet. For a permanent magnet, it is desired that ac and af be as large as possible. Materials for which this is true are said to have high **retentivity**.

Materials with a broad hysteresis curve as in Fig. 28–26 are said to be magnetically "hard." On the other hand, a hysteresis curve such as that in Fig. 28–27 occurs for "soft" iron. This is preferred for *electromagnets* and transformers (Section 29–6) since the field can be more readily switched off, and the field can be reversed with less loss of energy.

A ferromagnetic material can be demagnetized—that is, made unmagnetized. This can be done by reversing the magnetizing current repeatedly while decreasing its magnitude. This results in the curve of Fig. 28–28. The heads of a tape recorder are demagnetized in this way. The alternating magnetic field acting at the heads due to a demagnetizer is strong when the demagnetizer is placed near the heads and decreases as it is moved slowly away. Video and audio tapes themselves can be erased and ruined by a magnetic field, as can computer disks.

*28–10 Paramagnetism and Diamagnetism

All materials are magnetic to some extent. Nonferromagnetic materials fall into two principal classes: *paramagnetic*, in which the magnetic permeability μ is slightly greater than μ_0; and *diamagnetic*, in which, μ is slightly less than μ_0. The ratio of μ to μ_0 for any material is called the **relative permeability** K_m:

$$K_m = \frac{\mu}{\mu_0}.$$

Another useful parameter is the **magnetic susceptibility** X_m defined as

$$X_m = K_m - 1.$$

Paramagnetic substances have $K_m > 1$ and $X_m > 0$, whereas diamagnetic substances have $K_m < 1$ and $X_m < 0$. See Table 28–1.

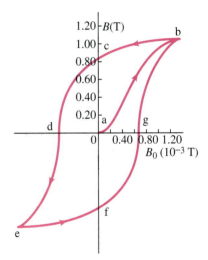

FIGURE 28–26 Hysteresis curve.

FIGURE 28–27 Hysteresis curve for soft iron.

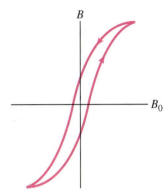

FIGURE 28–28 Successive hysteresis loops during demagnetization.

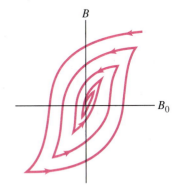

TABLE 28–1 Paramagnetism and Diamagnetism: Magnetic Susceptibilities			
Paramagnetic substance	X_m	**Diamagnetic substance**	X_m
Aluminum	2.3×10^{-5}	Copper	-9.8×10^{-6}
Calcium	1.9×10^{-5}	Diamond	-2.2×10^{-5}
Magnesium	1.2×10^{-5}	Gold	-3.6×10^{-5}
Oxygen (STP)	2.1×10^{-6}	Lead	-1.7×10^{-5}
Platinum	2.9×10^{-4}	Nitrogen (STP)	-5.0×10^{-9}
Tungsten	6.8×10^{-5}	Silicon	-4.2×10^{-6}

The difference between paramagnetic and diamagnetic materials can be understood theoretically at the molecular level on the basis of whether or not the molecules have a permanent magnetic dipole moment. **Paramagnetism** occurs in materials whose molecules (or ions) have a permanent magnetic dipole moment.[†] In the absence of an external field, the molecules are randomly oriented and no magnetic effects are observed. However, when an external magnetic field is applied, say, by putting the material in a solenoid, the applied field exerts a torque on the magnetic dipoles (Section 27–5), tending to align them parallel to the field. The total magnetic field (external plus that due to aligned magnetic dipoles) will be slightly greater than B_0. The thermal motion of the molecules reduces the alignment, however. A useful quantity is the **magnetization vector**, **M**, defined as the magnetic dipole moment per unit volume,

$$\mathbf{M} = \frac{\boldsymbol{\mu}}{V},$$

where $\boldsymbol{\mu}$ is the magnetic dipole moment of the sample and V its volume. It is found experimentally that M is directly proportional to the external magnetic field (tending to align the dipoles) and inversely proportional to the Kelvin temperature T (tending to randomize dipole directions). This is called *Curie's law*, after Pierre Curie (1859–1906), who first noted it:

$$M = C \frac{B}{T},$$

where C is a constant. If the ratio B/T is very large (B very large or T very small) Curie's law is no longer accurate; as B is increased (or T decreased), the magnetization approaches some maximum value, M_{max}. This makes sense, of course, since M_{max} corresponds to complete alignment of all the permanent magnetic dipoles. However, even for very large magnetic fields, $\approx 2.0\,\text{T}$, deviations from Curie's law are normally noted only at very low temperatures, on the order of a few kelvins.

Ferromagnetic materials, as mentioned in Section 28–7, are no longer ferromagnetic above a characteristic temperature called the Curie temperature (1043 K for iron). Above this Curie temperature, they generally are paramagnetic.

Paramagnetism

Diamagnetism

Diamagnetic materials (for which μ_m is slightly less than μ_0) are made up of molecules that have no permanent magnetic dipole moment. When an external magnetic field is applied, magnetic dipoles are induced, but the induced magnetic dipole moment is in the direction opposite to that of the field. Hence the total field will be slightly less than the external field. The effect of the external field—in the crude model of electrons orbiting nuclei—is to increase the "orbital" speed of electrons revolving in one direction, and to decrease the speed of electrons revolving in the other direction; the net result is a net dipole moment opposing the external field. Diamagnetism is present in all materials, but is weaker even than paramagnetism and so is overwhelmed by paramagnetic and ferromagnetic effects in materials that display these other forms of magnetism.

[†] Other types of paramagnetism also occur whose origin is different from that described here, such as in metals where free electrons can contribute.

Summary

Ampère's law states that the line integral of the magnetic field **B** around any closed loop is equal to μ_0 times the total net current I_{encl} enclosed by the loop:

$$\oint \mathbf{B} \cdot d\mathbf{l} = \mu_0 I_{\text{encl}}.$$

The magnetic field B at a distance r from a long straight wire is directly proportional to the current I in the wire and inversely proportional to r. The magnetic field lines are circles centered at the wire.

The magnetic field inside a long tightly wound solenoid is $B = \mu_0 n I$ where n is the number of coils per unit length and I is the current in each coil.

The force that one long current-carrying wire exerts on a second parallel current-carrying wire a distance l away serves as the definition of the ampere unit, and ultimately of the coulomb as well.

The **Biot-Savart law** is useful for determining the magnetic field due to a known arrangement of currents. It states that

$$d\mathbf{B} = \frac{\mu_0 I}{4\pi} \frac{d\mathbf{l} \times \hat{\mathbf{r}}}{r^2},$$

where $d\mathbf{B}$ is the contribution to the total field at some point P due to a current I along an infinitesimal length $d\mathbf{l}$ of its path, and $\hat{\mathbf{r}}$ is the unit vector along the direction of the displacement vector **r** from $d\mathbf{l}$ to P. The total field **B** will be the integral over all $d\mathbf{B}$.

Iron and a few other materials can be made into strong permanent magnets. They are said to be **ferromagnetic**. Ferromagnetic materials are made up of tiny **domains**—each a tiny magnet—which are preferentially aligned in a permanent magnet, but randomly aligned in a nonmagnetized sample.

When a ferromagnetic material is placed in a magnetic field B_0 due to a current, say inside a solenoid or toroid, the material becomes magnetized. When the current is turned off, however, the material remains magnetized, and when the current is increased in the opposite direction (and then again reversed), a graph of the total field B versus B_0 is a **hysteresis loop**, and the fact that the curves do not retrace themselves is called **hysteresis**.

Questions

1. The magnetic field due to current in wires in your home can affect a compass. Discuss the problem in terms of currents, depending on whether they are ac or dc, and their distance away.

2. Compare and contrast the magnetic field due to a long straight current and the electric field due to a long straight line of electric charge at rest (Section 21–7).

3. Two long wires carrying equal currents I are at right angles to each other, but don't quite touch. Describe the magnetic force one exerts on the other.

4. A horizontal wire carries a large current. A second wire carrying a current in the same direction is suspended below. Can the current in the upper wire hold the lower wire in suspension against gravity? Under what conditions will it be in equilibrium?

5. A horizontal current-carrying wire, free to move in Earth's gravitational field, is suspended directly above a second, parallel, current-carrying wire. (a) In what direction is the current in the lower wire? (b) Can the upper wire be held in stable equilibrium due to the magnetic force of the lower wire? Explain.

6. (a) Write Ampère's law for a path that surrounds both conductors in Fig. 28–8. (b) Repeat, assuming the lower current I_2, is in the opposite direction $(I_2 = -I_1)$.

7. Suppose the cylindrical conductor of Fig. 28–9a has a concentric cylindrical hollow cavity inside it (so it looks like a pipe). What can you say about **B** in the cavity?

8. Explain why a field such as that shown in Fig. 28–11b is consistent with Ampère's law. Could the lines curve upward instead of downward?

9. What would be the effect on B inside a long solenoid if (a) the diameter of all the loops was doubled, or (b) the spacing between loops was doubled, or (c) the solenoid's length was doubled along with a doubling in the total number of loops.

10. Use the Biot-Savart law to convince yourself that the field of the current loop in Fig. 28–18 is correct as shown for points off the axis.

11. Do you think **B** will be the same for all points in the plane of the current loop of Fig. 28–18?

12. Why does twisting the lead-in wires to electrical devices reduce the magnetic effects of the leads?

13. Compare the Biot-Savart law with Coulomb's law: What are the similarities and differences?

14. A type of magnetic switch similar to a solenoid is a **relay**. A relay is an electromagnet (the iron rod inside the coil doesn't move) which, when activated, attracts a piece of soft iron on a pivot. Design a relay (a) to make a doorbell and (b) to close an electrical switch. A relay is used in the latter case when you need to switch on a circuit carrying a very large current but you do not want that large current flowing through the main switch. For example the starter switch of a car is connected to a relay so that the large current needed for the starter doesn't pass to the dashboard switch.

15. How might you measure the magnetic dipole moment of the Earth?

16. How might you define or determine the magnetic pole strength (the magnetic equivalent of a single electric charge) for (*a*) a bar magnet, (*b*) a current loop?

17. A heavy magnet attracts, from rest, a heavy block of iron. Before striking the magnet the block has acquired considerable kinetic energy. (*a*) What is the source of this kinetic energy? (*b*) When the block strikes the magnet, some of the latter's domains are jarred into randomness; describe the energy transformations.

18. Will a magnet attract any metallic object, or only those made of iron? (Try it and see.) Why is this so?

19. An unmagnetized nail will not attract an unmagnetized paper clip. However, if one end of the nail is in contact with a magnet, the other end *will* attract a paper clip. Explain.

20. How do you suppose the first magnets found in Magnesia were formed?

21. Why will either pole of a magnet attract an unmagnetized piece of iron?

22. Suppose you have three iron rods, two of which are magnetized but the third is not. How would you determine which two are the magnets without using any additional objects?

23. Two iron bars attract each other no matter which ends are placed close together. Are both magnets? Explain.

*24. Describe the magnetization curve for (*a*) paramagnetic substance and (*b*) a diamagnetic substance, and compare to that for a ferromagnetic substance (Fig. 28–26).

*25. Can all materials be considered (*a*) diamagnetic, (*b*) paramagnetic, (*c*) ferromagnetic?

Problems

Sections 28–1 and 28–2

1. (I) Jumper cables used to start a stalled vehicle often carry a 65-A current. How strong is the magnetic field 7.5 cm away? Compare to the Earth's magnetic field.

2. (I) If an electric wire is allowed to produce a magnetic field no larger than that of the Earth $(0.55 \times 10^{-4}\,\text{T})$ at a distance of 25 cm, what is the maximum current the wire can carry?

3. (I) What is the magnitude and direction of the force between two parallel wires 45 m long and 6.0 cm apart, each carrying 35 A in the same direction?

4. (I) A vertical straight wire carrying an upward 22-A current exerts an attractive force per unit length of $8.8 \times 10^{-4}\,\text{N/m}$ on a second parallel wire 7.0 cm away. What current (magnitude and direction) flows in the second wire?

5. (I) In Fig. 28–29, a long straight wire carries current *I* out of the page toward the viewer. Indicate, with appropriate arrows, the direction of **B** at each of the points C, D, and E in the plane of the page.

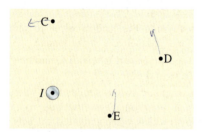

FIGURE 28–29 Problem 5.

6. (II) An experiment on the Earth's magnetic field is being carried out 1.00 m from an electric cable. What is the maximum allowable current in the cable if the experiment is to be accurate to ±1.0 percent?

7. (II) Two long thin parallel wires 15.0 cm apart carry 25-A currents in the same direction. Determine the magnetic field strength at a point 12.0 cm from one wire and 5.0 cm from the other (Fig. 28–30).

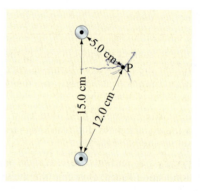

FIGURE 28–30
Problem 7.

8. (II) A horizontal compass is placed 20 cm due south from a straight vertical wire carrying a 40-A current downward. In what direction does the compass needle point at this location? Assume the horizontal component of the Earth's field at this point is $0.45 \times 10^{-4}\,\text{T}$ and the magnetic declination is 0°.

9. (II) A long horizontal wire carries 22.0 A of current due north. What is the net magnetic field 20.0 cm due west of the wire if the Earth's field there points downward, 40° below the horizontal, and has magnitude $5.0 \times 10^{-5}\,\text{T}$?

10. (II) A straight stream of protons passes a given point in space at a rate of 1.5×10^{9} protons/s. What magnetic field do they produce 2.0 m from the beam?

11. (II) Determine the magnetic field midway between two long straight wires 2.0 cm apart in terms of the current *I* in one when the other carries 15 A. Assume these currents are (*a*) in the same direction, and (*b*) in opposite directions.

12. (II) A long pair of wires serves to conduct 25.0 A of dc current to and from an instrument. If the wires are of negligible diameter but are 2.8 mm apart, what is the magnetic field 10.0 cm from their midpoint, in their plane (Fig. 28–31)? Compare to the magnetic field of the Earth.

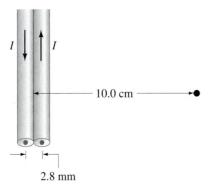

FIGURE 28–31 Problem 12.

13. (II) A compass needle points 20° E of N outdoors. However, when it is placed 12.0 cm to the east of a vertical wire inside a building, it points 55° E of N. What is the magnitude and direction of the current in the wire? The Earth's field there is 0.50×10^{-4} T and is horizontal.

14. (II) A rectangular loop of wire is placed next to a straight wire, as shown in Fig. 28–32. There is a current of 2.5 A in both wires. What is the magnitude and direction of the net force on the loop?

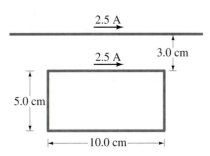

FIGURE 28–32 Problem 14.

15. (II) Let two long parallel wires, a distance d apart, carry equal currents I in the same direction. One wire is at $x = 0$, the other at $x = d$, Fig. 28–33. Determine **B** between the wires as a function of x.

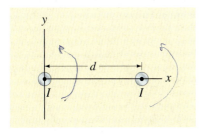

FIGURE 28–33 Problems 15 and 16.

16. (II) Repeat Problem 15 if the wire at $x = 0$ carries twice the current ($2I$) as the other wire, and in the opposite direction.

17. (II) Two long wires are oriented so that they are perpendicular to each other, and at their closest, they are 20.0 cm apart (Fig. 28–34). What is the magnitude of the magnetic field at a point midway between them if the top one carries a current of 20.0 A and the bottom one carries 5.0 A?

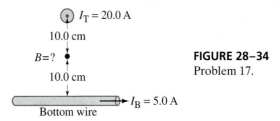

FIGURE 28–34
Problem 17.

18. (II) Two long parallel wires 7.00 cm apart carry 16.5-A currents in the same direction. Determine the magnetic field strength at a point P 12.0 cm from one wire and 13.0 cm from the other. See Fig. 28–35.

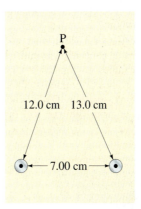

FIGURE 28–35
Problem 18.

19. (III) A very long flat conducting strip of width L and negligible thickness lies in a horizontal plane and carries a uniform current I across its cross section. (a) Show that at points a distance y directly above its center, the field is given by

$$B = \frac{u_0 I}{\pi L} \tan^{-1} \frac{L}{2y},$$

assuming the strip is infinitely long. [*Hint*: Divide the strip into many thin "wires," and sum (integrate) over these.] (b) What value does B approach for $y \gg L$? Does this make sense? Explain.

20. (III) An electron is moving in a plane that also contains a long straight current-carrying wire. The electron is heading at a 45° angle toward the wire, with a speed of 3.4×10^6 m/s, when it is 50 cm away. The electron reaches only as close as 1.0 cm, before being repelled away, always moving in the same plane. What is the current in the wire?

Sections 28–4 and 28–5

21. (I) A 40.0-cm long solenoid 1.35 cm in diameter is to produce a field of 0.385 mT at its center. How much current should the solenoid carry if it has 1000 turns of wire?

22. (I) A 32-cm-long solenoid, 1.8 cm in diameter, is to produce a 0.30-T magnetic field at its center. If the maximum current is 5.7 A, how many turns must the solenoid have?

23. (I) A 2.5-mm-diameter copper wire carries a 40-A current. Determine the magnetic field: (a) at the surface of the wire; (b) inside the wire, 0.50 mm below the surface; (c) outside the wire 2.5 mm from the surface.

24. (II) A toroid (Fig. 28–14) has a 50.0-cm inner diameter and a 54.0-cm outer diameter. It carries a 25.0 A current in its 500 coils. Determine the range of values for B inside the toroid.

25. (II) A 20.0 m long copper wire, 2.00 mm in diameter including insulation, is tightly wrapped in a single layer with adjacent coils touching, to form a solenoid of diameter 2.50 cm. What is (a) the length of the solenoid and (b) the field at the center when the current in the wire is 20.0 A?

26. (II) (a) Use Eq. 28–1, and the vector nature of B, to show that the magnetic field lines around two long parallel wires carrying equal currents $I_1 = I_2$ are as shown in Fig. 28–8. (b) Draw the equipotential lines around two stationary positive electric charges. (c) Are these two diagrams similar? Identical? Why or why not?

27. (II) A coaxial cable consists of a solid inner conductor of radius R_1, surrounded by a concentric cylindrical tube or inner radius R_2 and outer radius R_3 (Fig. 28–36). The conductors carry equal and opposite currents I_0 distributed uniformly across their cross-sections. Determine the magnetic field at a distance R from the axis for: (a) $R < R_1$; (b) $R_1 < R < R_2$; (c) $R_2 < R < R_3$; (d) $R > R_3$.

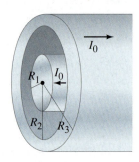

FIGURE 28–36   Problems 27 and 28.

28. (III) Suppose the current in the coaxial cable of Problem 27, Fig. 28–36, is not uniformly distributed, but instead the current density j varies linearly with distance from the center: $j_1 = C_1 R$ for the inner conductor and $j_2 = C_2 R$ for the outer conductor. Each conductor still carries the same total current I_0, in opposite directions. Determine the magnetic field in terms of I_0 in the same four regions of space as in Problem 27.

Section 28–6

29. (I) The Earth's magnetic field is essentially that of a magnetic dipole. If the field near the North Pole is about 1.0×10^{-4} T, what will it be (approximately) 13,000 km above the surface at the North Pole?

30. (II) A wire, in a plane, has the shape shown in Fig. 28–37, two arcs of a circle connected by radial lengths of wire. Determine B at point C in terms of R_1, R_2, θ, and the current I.

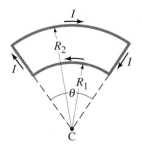

FIGURE 28–37   Problem 30.

31. (II) A circular conducting ring of radius R is connected to two exterior straight wires ending at two ends of a diameter (Fig. 28–38). The current I splits into unequal portions while passing through the ring as shown. What is B at the center of the ring?

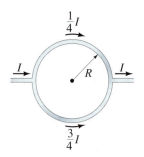

FIGURE 28–38   Problem 31.

32. (II) A small loop of wire of radius 1.8 cm is placed at the center of a 25.0-cm wire loop. The planes of the loops are perpendicular to each other, and a 7.0-A current flows in each. Estimate the torque the large loop exerts on the smaller one. What simplifying assumption did you make?

33. (II) A wire is formed into the shape of two half circles connected by equal-length straight sections as shown in Fig. 28–39. A current I flows in the circuit clockwise as shown. Determine (a) the magnitude and direction of the magnetic field at the center, C, and (b) the magnetic dipole moment of the circuit.

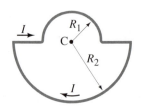

FIGURE 28–39   Problem 33.

34. (II) Use the Biot-Savart law to show that the magnetic field **B**, due to a single point charge q moving with velocity **v**, at a point P whose position vector relative to the charge is **r** (Fig. 28–40) is given by

$$\mathbf{B} = \frac{\mu_0}{4\pi} \frac{q\mathbf{v} \times \mathbf{r}}{r^3}.$$

(Assume v is much less than the speed of light.)

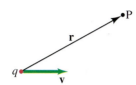

FIGURE 28–40 Problem 34.

35. (II) A nonconducting circular disk, of radius R, carries a uniformly distributed electric charge Q. The plate is set spinning with angular velocity ω about an axis perpendicular to the plate through its center (Fig. 28–41). Determine (*a*) its magnetic dipole moment and (*b*) the magnetic field at points on its axis a distance x from its center; (*c*) does Eq. 28–7b apply in this case for $x \gg R$?

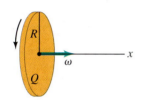

FIGURE 28–41 Problem 35.

36. (II) Consider a straight section of wire of length l, as in Fig. 28–42, which carries a current I. (*a*) Show that the magnetic field at a point P a distance R from the wire along its perpendicular bisector is

$$B = \frac{\mu_0 I}{2\pi R} \frac{l}{(l^2 + 4R^2)^{\frac{1}{2}}}.$$

(*b*) Show that this is consistent with Example 28–9 for an infinite wire.

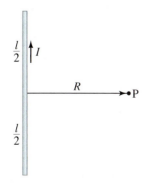

FIGURE 28–42 Problem 36.

37. (II) A segment of wire of length l carries a current I as shown in Fig. 28–43. (*a*) Show that for points along the positive x axis (the axis of the wire), such as point Q, the magnetic field **B** is zero. (*b*) Determine a formula for the field at points along the y axis, such as point P.

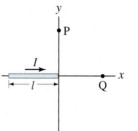

FIGURE 28–43 Problem 37.

38. (II) Use the result of Problem 37 to find the magnetic field at point P in Fig. 28–44 due to the current in the square loop.

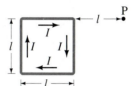

FIGURE 28–44 Problem 38.

39. (II) A wire is bent into the shape of a regular polygon with n sides whose vertices are a distance R from the center. (See Fig. 28–45, which shows the special case of $n = 6$.) If the wire carries a current I_0, (*a*) determine the magnetic field at the center; (*b*) if n is allowed to become very large ($n \to \infty$), show that the formula in part (*a*) reduces to that for a circular loop (Example 28–10).

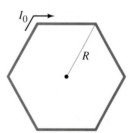

FIGURE 28–45 Problem 39.

40. (III) Start with the result of Example 28–10 for the magnetic field along the axis of a single loop to obtain the field inside a very long solenoid (Eq. 28–4).

41. (III) A single rectangular loop of wire, with sides a and b, carries a current I. An xy coordinate system has its origin at the lower left corner of the rectangle with the x axis parallel to side b (Fig. 28–46). Determine the magnetic field B at all points (x, y) within the loop.

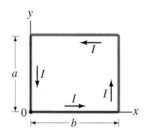

FIGURE 28–46 Problem 41.

42. (III) A square loop of wire, of side l, carries a current I. (*a*) Determine the magnetic field B at points on a line perpendicular to the plane of the square which passes through the center of the square (Fig. 28–47). Express B as a function of x, the distance along the line from the center of the square. (*b*) For $x \gg l$, does the square appear to be a magnetic dipole? If so, what is its dipole moment?

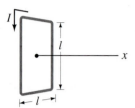

FIGURE 28–47
Problem 42.

* Section 28–7

* **43. (II)** An iron atom has a magnetic dipole moment of about $1.8 \times 10^{-23}\,\text{A}\cdot\text{m}^2$. (*a*) Determine the dipole moment of an iron bar 12 cm long, 1.2 cm wide, and 1.2 cm thick, if it is 100 percent saturated. (*b*) What torque would be exerted on this bar when placed in a 1.2-T field acting at right angles to the bar?

* Section 28–9

* **44. (I)** The following are some values of B and B_0 for a piece of annealed iron as it is being magnetized:

$B_0(10^{-4}\,\text{T})$	0	0.13	0.25	0.50	0.63	0.78	1.0	1.3
$B(\text{T})$	0	0.0042	0.010	0.028	0.043	0.095	0.45	0.67
$B_0(10^{-4}\,\text{T})$		1.9	2.5	6.3	13.0	130	1,300	10,000
$B(\text{T})$		1.01	1.18	1.44	1.58	1.72	2.26	3.15

Determine the magnetic permeability μ for each value and plot a graph of μ versus B_0.

* **45. (I)** A large thin toroid has 400 loops of wire per meter, and a 20-A current flows through the wire. If the relative permeability of the iron is 3000, what is the total field B inside the toroid?

* **46. (II)** An iron-core solenoid is 38 cm long, 1.8 cm in diameter, and has 600 turns of wire. A magnetic field of 2.2 T is produced when 48 A flows in the wire. What is the permeability μ at this high field strength?

General Problems

47. Three long parallel wires are 38.0 cm from one another. (Looking along them, they are at three corners of an equilateral triangle.) The current in each wire is 8.00 A, but that in wire M is opposite to that in wires N and P (Fig. 28–48). Determine the magnetic force per unit length on each wire due to the other two.

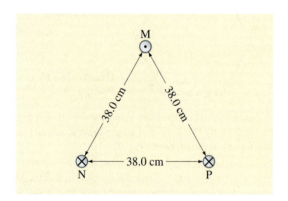

FIGURE 28–48 Problems 47 and 48.

48. In Fig. 28–48 the top wire is 2.00-mm-diameter copper wire and is suspended in air due to the two magnetic forces from the bottom two wires. The current is 20.0 A in each of the two bottom wires. Calculate the required current flow in the suspended wire.

49. An electron enters a large solenoid at a 7.0° angle to the axis. If the field is a uniform $3.3 \times 10^{-2}\,\text{T}$, determine the radius and pitch (distance between loops) of the electron's helical path if its speed is $1.3 \times 10^{7}\,\text{m/s}$.

50. A rectangular loop of wire carries a 2.0-A current and lies in a plane which also contains a very long straight wire carrying a 10.0-A current as shown in Fig. 28–49. Determine (*a*) the net force and (*b*) the net torque on the loop due to the straight wire.

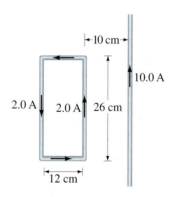

FIGURE 28–49 Problem 50.

51. A very large flat conducting sheet of thickness t carries a uniform current density **j** throughout (Fig. 28–50). Determine the magnetic field (magnitude and direction) at a distance y above the plane. (Assume the plane is infinitely long and wide.)

FIGURE 28–50
Problem 51.

52. A long horizontal wire carries a current of 48 A. A second wire, made of 2.5-mm-diameter copper wire and parallel to the first but 15 cm below it, is held in suspension magnetically (Fig. 28–51). (*a*) What is the magnitude and direction of the current in the lower wire? (*b*) Is the lower wire in stable equilibrium? (*c*) Repeat parts (*a*) and (*b*) if the second wire is suspended 15 cm *above* the first due to the first's magnetic field.

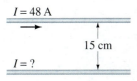

$I = 48$ A

15 cm

$I = ?$

FIGURE 28–51
Problem 52.

53. You have 1.0 kg of copper and want to make a practical solenoid that produces the greatest possible magnetic field. Consider variables such as solenoid diameter, length, and so on, to determine if you should make your copper wire long and thin, short and fat, or something else.

54. For two long parallel wires separated by a distance L, carrying currents I_1 and I_2 as in Fig. 28–8, show that for the circular path of radius r ($r < L$) centered on I_1, that

$$\oint \mathbf{B} \cdot d\mathbf{l} = \mu_0 I_1$$

in accord with Ampère's law. (But do not use Ampère's law.)

55. Near the Earth's poles the magnetic field is about 1 G $(1 \times 10^{-4}\,\text{T})$. Imagine a simple model in which the Earth's field is produced by a single current loop around the equator. What current would this loop carry?

56. A square loop of wire, of side L, carries a current I. Show that the magnetic field at the center of the square is

$$B = \frac{2\sqrt{2}\,\mu_0 I}{\pi L}.$$

[*Hint*: Determine **B** for each segment of length L.]

57. In Problem 56, if you reshaped the square wire into a circle, would B increase or decrease at the center? Explain.

58. Two large coils of wire, each with N turns carrying a current I and separated by a distance equal to the radius R of the coils, are called *Helmholtz coils* (see Fig. 28–52). (*a*) Determine B at points x along the line joining their centers. Let $x = 0$ at the center of one coil, $x = R$ at the center of the other. (*b*) Plot B versus x from $x = 0$ to $x = R$. (*c*) Determine B on the axis midway between the coils $(x = R/2)$ if $R = 20.0$ cm, $I = 35$ A, and each coil contains $N = 350$ turns. (*d*) Show that the field midway between the coils is particularly uniform by showing that when the separation of the coils is R, then $dB/dx = 0$ and $d^2B/dx^2 = 0$ at the midpoint.

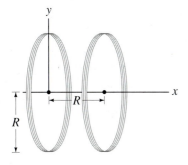

FIGURE 28–52
Problem 58.

59. A 175-g model airplane charged to 18.0 mC and traveling at 2.8 m/s passes within 8.6 cm of a wire, nearly parallel to its path, carrying a 30-A current. What acceleration (in g's) does this interaction give the airplane?

60. Suppose that an electromagnet uses a coil 2.0 m in diameter made from square copper wire 2.0 mm on a side; the power supply produces 50 V at a maximum power output of 1.0 kW. (*a*) How many turns are needed to run the power supply at maximum power? (*b*) What is the magnetic field strength at the center of the coil? (*c*) If you use a greater number of turns and this same power supply, will a greater magnetic field strength result? Explain.

61. Four long straight parallel wires located at the corners of a square of side l carry equal currents I_0 perpendicular to the page as shown in Fig. 28–53. Determine the magnitude and direction of **B** at the center of the square.

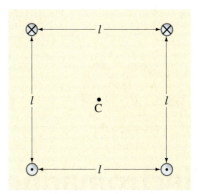

FIGURE 28–53 Problem 61.

62. Determine the magnetic field at the point P due to the long wire with a square bend shown in Fig. 28–54. The point P is halfway between the two corners. [*Hint*: You can use the results of Problems 36 and 37.]

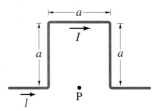

FIGURE 28–54 Problem 62.

63. You want to get an idea of the magnitude of magnetic fields produced by overhead power lines. You estimate that the two wires run about 30 m above the ground about 3 m apart. A call to the local power company provides the information that the lines operate at 10 kilovolts and provide a maximum of 40 MW to the local area. Estimate the maximum magnetic field you might experience walking under these power lines, and compare to the Earth's field. [Note that if the current is ac, the magnetic field will be changing too.]

One of the great laws of physics is Faraday's law of induction, which says that a changing magnetic field produces an induced emf. This photo shows a bar magnet moving inside a coil, and the galvanometer registers an induced current. This phenomenon of electromagnetic induction is the basis for many practical devices, from generators to alternators to transformers, tape recording and reading computer memory.

CHAPTER 29

Electromagnetic Induction and Faraday's Law

In Chapter 27, we discussed two ways in which electricity and magnetism are related: (1) an electric current produces a magnetic field; and (2) a magnetic field exerts a force on an electric current or moving electric charge. These discoveries were made in 1820–1821. Scientists then began to wonder: if electric currents produce a magnetic field, is it possible that a magnetic field can produce an electric current? Ten years later the American Joseph Henry (1797–1878) and the Englishman Michael Faraday (1791–1867) independently found that it was possible. Henry actually made the discovery first. But Faraday published his results earlier and investigated the subject in more detail. We now discuss this phenomenon and some of its world-changing applications such as the electric generator.

29–1 Induced EMF

In his attempt to produce an electric current from a magnetic field, Faraday used an apparatus like that shown in Fig. 29–1. A coil of wire, X, was connected to a battery. The current that flowed through X produced a magnetic field that was intensified by the iron core. Faraday hoped that by using a strong enough battery, a steady current in X would produce a great enough magnetic field to produce a current in a second coil Y. This second circuit, Y, contained a galvanometer to detect any current but contained no battery. He met no success with steady currents.

*Constant **B** induces no emf*

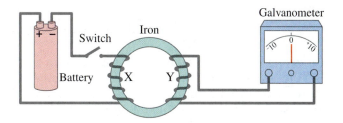

FIGURE 29–1 Faraday's experiment to induce an emf.

But the long-sought effect was finally observed when Faraday saw the galvanometer in circuit Y deflect strongly at the moment he closed the switch in circuit X. And the galvanometer deflected strongly in the opposite direction when he opened the switch. A *steady* current in X had produced *no* current in Y. Only when the current in X was starting or stopping was a current produced in Y.

Faraday concluded that although a steady magnetic field produces no current, a *changing* magnetic field can produce an electric current! Such a current is called an **induced current**. When the magnetic field through coil Y changes, a current flows as if there were a source of emf in the circuit. We therefore say that

an induced emf is produced by a changing magnetic field.

Changing **B** *induces an emf*

Faraday did further experiments on **electromagnetic induction**, as this phenomenon is called. For example, Fig. 29–2 shows that if a magnet is moved quickly into a coil of wire, a current is induced in the wire. If the magnet is quickly removed, a current is induced in the opposite direction. Furthermore, if the magnet is held steady and the coil of wire is moved toward or away from the magnet, or the coil is rotated, again an emf is induced and a current flows. Motion or change is required to induce an emf. It doesn't matter whether the magnet or the coil moves. It is the relative motion that counts.

FIGURE 29–2 (a) A current is induced when a magnet is moved toward a coil. (b) The induced current is opposite when the magnet is moved away from the coil. Note that the galvanometer zero is at the center of the scale and the needle deflects left or right, depending on the direction of the current. In (c) no current is induced if the magnet does not move relative to the coil. It is the relative motion that counts: the magnet can be held steady and the coil moved, which also induces an emf.

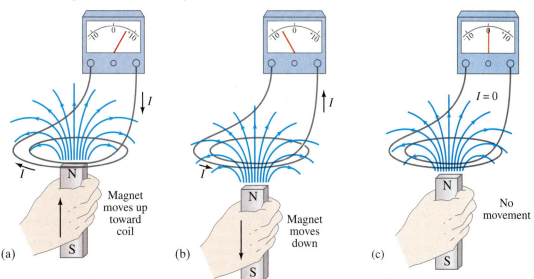

(a) Magnet moves up toward coil

(b) Magnet moves down

(c) No movement $I = 0$

29-2 Faraday's Law of Induction; Lenz's Law

Faraday investigated quantitatively what factors influence the magnitude of the emf induced. He found first of all that the more rapidly the magnetic field changes, the greater the induced emf. But the emf is not simply proportional to the rate of change of the magnetic field, **B**. Rather the emf is proportional to the rate of change of the **magnetic flux**, Φ_B, passing through the circuit or loop of area A. Magnetic flux for a uniform magnetic field is defined as

$$\Phi_B = B_\perp A = BA \cos \theta = \mathbf{B} \cdot \mathbf{A}. \qquad [\mathbf{B} \text{ uniform}] \quad (29\text{-}1\text{a})$$

Here $B_\perp$ is the component of the magnetic field **B** perpendicular to the face of the loop, and θ is the angle between **B** and the vector **A** (representing the area) whose direction is perpendicular to the face of the loop. These quantities are shown in Fig. 29–3 for a square loop of side l whose area is $A = l^2$. If the area is of some other shape, or **B** is not uniform, the magnetic flux can be written[†]

$$\Phi_B = \int \mathbf{B} \cdot d\mathbf{A}. \qquad (29\text{-}1\text{b})$$

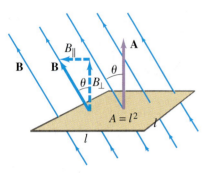

FIGURE 29–3 Determining the flux through a flat loop of wire. This loop is square, of side l and area $A = l^2$.

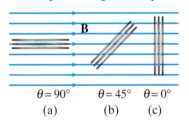

FIGURE 29–4 Magnetic flux Φ_B is proportional to the number of lines of **B** that pass through the loop.

As we saw earlier, the lines of **B** (like lines of **E**) can be drawn such that the number of lines per unit area is proportional to the field strength. Then the flux Φ_B can be thought of as being proportional to the *total number of lines passing through the loop*. This is illustrated in Fig. 29–4, where the loop is viewed from the side (on edge). For $\theta = 90°$, no lines pass through the loop and $\Phi_B = 0$, whereas Φ_B is a maximum when $\theta = 0°$. The unit of magnetic flux is the tesla-meter²; this is called a **weber**: $1 \text{ Wb} = 1 \text{ T} \cdot \text{m}^2$.

With this definition of the flux, we can now write down the results of Faraday's investigations—namely, that the emf induced in a circuit is equal to the rate of change of magnetic flux through the circuit:

FARADAY'S LAW OF INDUCTION

$$\mathscr{E} = -\frac{d\Phi_B}{dt}. \qquad (29\text{-}2\text{a})$$

This fundamental result is known as **Faraday's law of induction**, and is one of the basic laws of electromagnetism.

If the circuit contains N closely wrapped loops, the emfs induced in each add together, so

$$\mathscr{E} = -N\frac{d\Phi_B}{dt}. \qquad (29\text{-}2\text{b})$$

The minus sign in Eq. 29–2 is placed there to remind us in which direction the induced emf acts. Experiments show that:

An induced emf gives rise to a current whose magnetic field opposes the original change in flux.

Lenz's law

This is known as **Lenz's law**. Said another way, valid even if no current can flow (as when a circuit is not complete), is:

An induced emf is always in a direction that opposes the original change in flux that caused it.

Let us apply Lenz's law to the case of relative motion between a magnet and a

---

[†] The integral is taken over an open surface—that is, one bounded by a closed curve such as a circle or square. In the present discussion, the area is that enclosed by the loop under discussion. The area is not an enclosed surface as we used in Gauss's law, Chapter 22.

coil, Fig. 29–2. The changing flux through the coil induces an emf, which produces a current in the coil. And this induced current produces its own magnetic field. In Fig. 29–2a the distance between the coil and the magnet decreases. The magnetic field (and number of field lines), and therefore the flux, through the coil increases. The magnetic field of the magnet points upward. To oppose this upward increase, the field inside the coil produced by the induced current points *downward*. Thus, Lenz's law tells us that the current moves as shown (use the right-hand rule). In Fig. 29–2b, the flux *decreases* (because the magnet is moved away), so the induced current produces an *upward* magnetic field inside the coil that is "trying" to maintain the status quo. Thus the current is as shown.

Let us consider what would happen if Lenz's law were not true, but were just the reverse. The induced current in this imaginary situation would produce a flux in the same direction as the original change. This greater change in flux would produce an even larger current followed by a still greater change in flux, and so on. The current would continue to grow indefinitely, producing power $(= I^2R)$ even after the original stimulus ended. This would violate the conservation of energy. Such "perpetual motion" devices do not exist. Thus, Lenz's law as stated above (and not its opposite) is consistent with the law of conservation of energy.

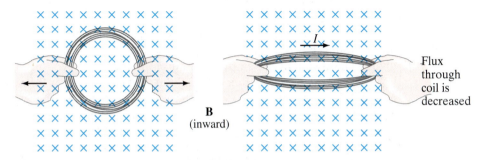

FIGURE 29–5 A current can be induced by changing the area of the coil. In both this case and that of Fig. 29–6, the flux through the coil is reduced. Here the brief induced current acts in the direction shown so as to try to maintain the original flux ($\Phi = BA$) by producing its own magnetic field into the page. That is, as the area A decreases, the current acts to increase B in the original (inward) direction.

It is important to note that an emf is induced whenever there is a change in flux. Since magnetic flux $\Phi_B = \int \mathbf{B} \cdot d\mathbf{A} = \int B \cos\theta \, dA$, we see that an emf can be induced in three ways: (1) by a changing magnetic field B; (2) by changing the area A of the loop in the field; or (3) by changing the loop's orientation θ with respect to the field. Figures 29–1 and 29–2 illustrated case 1. Examples of cases 2 and 3 are illustrated in Figs. 29–5 and 29–6, respectively.

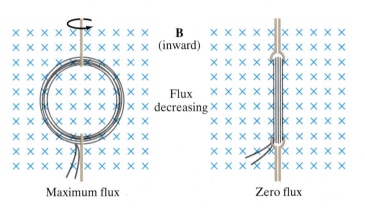

FIGURE 29–6 A current can be induced by rotating a coil in a magnetic field.

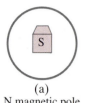

(a)
N magnetic pole
moving toward loop
into the page

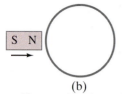

(b)
N magnetic pole moving
toward the loop in the plane
of the page

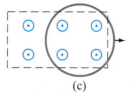

(c)
Pulling the loop to the right
out of a magnetic field
that points out of the page

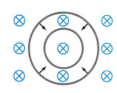

(d)
Shrinking a loop in a
magnetic field pointing
into the page

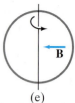

(e)
Rotating the loop about the
vertical diameter by pulling
the left side toward the reader
and pushing the right side
away from the reader in a
magnetic field that points
from right to left in the plane
of the page

FIGURE 29–7 Example 29–1.

CONCEPTUAL EXAMPLE 29–1 | **Practice with Lenz's law.** In which direction is the current induced in the loop for each situation in Fig. 29–7?

RESPONSE (*a*) Magnetic field lines point out from the N pole of a magnet, so as the magnet moves toward the loop, the field points into the page and is getting stronger. The current will be induced in the counterclockwise direction to produce a field **B** *out* of the page so that its own flux counteracts the externally imposed change.
(*b*) The field is in the plane of the page, so the flux through the loop is zero throughout the process; hence there is no change in magnetic flux with time, and there will be no induced emf or current in the loop.
(*c*) Initially, the magnetic flux pointing out of the page passes through the loop. If you pull the loop out of the field, the induced current will be in a direction to make up the deficiency: the current flow will be counterclockwise to produce an outward (toward the reader) magnetic field.
(*d*) The flux is into the page and the coil area shrinks so the flux will decrease; hence the induced current will be clockwise to try to produce its own flux into the page to make up for the flux decrease.
(*e*) Initially there is no flux through the loop (why?). When you start to rotate the loop, the flux begins passing through the loop increasing to the left. To counteract this, the loop will have current induced in a counterclockwise direction so as to produce its own flux to the right.

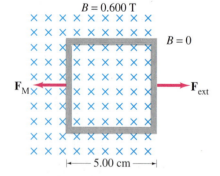

$B = 0.600\,\mathrm{T}$

$B = 0$

$\mathbf{F_M}$ $\mathbf{F}_{\text{ext}}$

5.00 cm

FIGURE 29–8 Example 29–2. The square coil in a magnetic field $B = 0.600\,\mathrm{T}$ is pulled abruptly to the right to a region where $B = 0$.

EXAMPLE 29–2 **Pulling a coil from a magnetic field.** A square coil of wire with side 5.00 cm contains 100 loops and is positioned perpendicular to a uniform 0.600-T magnetic field, as shown in Fig. 29–8. It is quickly and uniformly pulled from the field (moving perpendicular to **B**) to a region where B drops abruptly to zero. At $t = 0$, the right edge of the coil is at the edge of the field. It takes 0.100 s for the whole coil to reach the field-free region. Find (*a*) the rate of change in flux through the coil, (*b*) the emf and current induced, and (*c*) how much energy is dissipated in the coil if its resistance is 100 Ω. (*d*) What was the average force required?

SOLUTION (*a*) First we find how the magnetic flux changes during the time interval $\Delta t = 0.100\,\mathrm{s}$. The area of the coil is $A = (5.00 \times 10^{-2}\,\mathrm{m})^2 = 2.50 \times 10^{-3}\,\mathrm{m}^2$. The flux is initially $\Phi_B = BA = (0.600\,\mathrm{T})(2.50 \times 10^{-3}\,\mathrm{m}^2) = 1.50 \times 10^{-3}\,\mathrm{Wb}$. After 0.100 s, the flux is zero. The rate of change in flux is constant (because the coil is square), equal to

$$\frac{\Delta\Phi_B}{\Delta t} = \frac{0 - 1.50 \times 10^{-3}\,\mathrm{Wb}}{0.100\,\mathrm{s}} = -1.50 \times 10^{-2}\,\mathrm{Wb/s}.$$

(*b*) The emf induced (Eq. 29–2) during this period is

$$\mathcal{E} = -N\frac{d\Phi_B}{dt} = -(100)(-1.50 \times 10^{-2}\,\mathrm{Wb/s}) = 1.50\,\mathrm{V}.$$

The current is

$$I = \frac{\mathscr{E}}{R} = \frac{1.50\text{ V}}{100\,\Omega} = 15.0\text{ mA},$$

and, by Lenz's law, must be clockwise to oppose the decreasing flux into the page. (c) The total energy dissipated is

$$E = Pt = I^2Rt = (1.50 \times 10^{-2}\text{ A})^2(100\,\Omega)(0.100\text{ s}) = 2.25 \times 10^{-3}\text{ J}.$$

(d) We can calculate the force directly using $\mathbf{F} = I\mathbf{l} \times \mathbf{B}$, Eq. 27–3, for constant $\mathbf{B}$. The force the magnetic field exerts on the top and bottom sections of the square loop of Fig. 29–8 are in opposite directions and cancel each other. The magnetic force $\mathbf{F_M}$ exerted on the left vertical section of the square loop acts to the left as shown because the current is up (clockwise). The right side of the loop is in the region where $\mathbf{B} = 0$. Hence the needed external force, to the right, has magnitude

$$F_{\text{ext}} = NIlB = (100)(0.0150\text{ A})(0.0500\text{ m})(0.600\text{ T}) = 0.0450\text{ N},$$

since there are $N = 100$ current-carrying loops.
We can also calculate the average force using the result of part (c): From conservation of energy, the energy dissipated E is equal to the work W needed to pull the coil out of the field. Because $W = \bar{F}d$ where $d = 5.00$ cm, then

$$\bar{F} = \frac{W}{d} = \frac{2.25 \times 10^{-3}\text{ J}}{5.00 \times 10^{-2}\text{ m}} = 0.0450\text{ N},$$

which is the same answer, as we expect.

29–3 EMF Induced in a Moving Conductor

Another way to induce an emf is shown in Fig. 29–9, and this situation helps illuminate the nature of the induced emf. Assume that a uniform magnetic field $\mathbf{B}$ is perpendicular to the area bounded by the U-shaped conductor and the movable rod resting on it. If the rod is made to move at a speed v, it travels a distance $dx = v\,dt$ in a time dt. Therefore, the area of the loop increases by an amount $dA = l\,dx = lv\,dt$ in a time dt. By Faraday's law, there is an induced emf $\mathscr{E}$ whose magnitude is given by

$$\mathscr{E} = \frac{d\Phi_B}{dt} = \frac{B\,dA}{dt} = \frac{Blv\,dt}{dt} = Blv. \quad \text{(29–3)}$$

This equation is valid as long as B, l, and v are mutually perpendicular. (If they are not, we use only the components of each that are mutually perpendicular.) An emf induced on a conductor moving in a magnetic field is sometimes called a **motional emf**.

We can also obtain Eq. 29–3 without using Faraday's law. We saw in Chapter 27 that a charged particle moving perpendicular to a magnetic field B with speed v experiences a force $\mathbf{F} = q\mathbf{v} \times \mathbf{B}$. When the rod of Fig. 29–9a moves to the right with speed v, the electrons in the rod move with this same speed. Therefore, since $\mathbf{v} \perp \mathbf{B}$, each electron feels a force $F = qvB$, which acts upward as shown in Fig. 29–9b. If the rod were not in contact with the U-shaped conductor, free electrons would collect at the upper end of the rod, leaving the lower end positive. There must thus be an induced emf. If the rod does slide on the U-shaped conductor, the electrons will flow into it. There will then be a clockwise (conventional) current flowing in the loop. To calculate the emf, we determine the work W needed to move a charge q from one end of the rod to the other against this potential difference: $W = \text{force} \times \text{distance} = (qvB)(l)$. The emf equals the work done per unit charge, so $\mathscr{E} = W/q = qvBl/q = Blv$, just as above.[†]

FIGURE 29–9 (a) A conducting rod is moved to the right on a U-shaped conductor in a uniform magnetic field $\mathbf{B}$ that points out of the page. (b) Rod only, showing force on one electron.

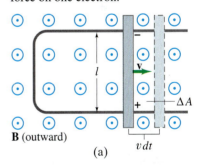

(a)

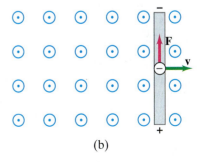

(b)

[†] This argument, which is basically the same as for the Hall effect (Section 27–8), explains this one way of inducing an emf. It does not explain the general case of electromagnetic induction.

This emf produces an electric field E in the rod which moves the electrons along (as in Section 25–8). Assuming a uniform E in the rod, then $E = \mathscr{E}/l = Bv$.

FIGURE 29–10 Example 29 – 3.

EXAMPLE 29–3 **Does a moving airplane develop a large emf?** An airplane travels 1000 km/h in a region where the Earth's magnetic field is 5.0×10^{-5} T and is nearly vertical (Fig. 29–10). What is the potential difference induced between the wing tips that are 70 m apart?

SOLUTION Since $v = 1000$ km/h $= 280$ m/s, and $\mathbf{v} \perp \mathbf{B}$, we have
$$\mathscr{E} = Blv = (5.0 \times 10^{-5}\,\text{T})(70\,\text{m})(280\,\text{m/s}) = 1.0\,\text{V}.$$
Not much to worry about.

EXAMPLE 29–4 **Force on the rod.** To make the rod of Fig. 29–9a move to the right at speed v, you need to apply a force. (a) Explain and determine the magnitude of the required force. (b) What external power is needed to move the rod?

SOLUTION (a) When the rod moves to the right, a current flows downward in the rod, as already discussed. We can see this also from Lenz's law: the upward magnetic flux through the loop is increasing, so the induced current must oppose the increase. Thus the current is clockwise so as to produce a magnetic field into the page (right-hand rule). The force on the moving rod is $\mathbf{F} = I\mathbf{l} \times \mathbf{B}$ for a constant $\mathbf{B}$ (Eq. 27–3). The right-hand rule tells us this force is to the left, and is thus a "drag force" opposing our effort to move the rod to the right.
The magnitude of the external force, to the right, needs to be $F = IlB$, where the current $I = \mathscr{E}/R = Blv/R$. The resistance R is that of the whole circuit, the rod and the U-shaped conductor. The force F required to move the rod is thus
$$F = IlB = \frac{B^2 l^2}{R} v.$$
If B, l, and R are constant, then a constant speed v is produced by a constant force. Constant R implies that all the resistance is in the rod and none in the U-shaped conductor.
(b) The external power needed to move the rod for constant R is
$$P = Fv = \frac{B^2 l^2 v^2}{R}.$$
The power dissipated in the resistance is $P = I^2 R$. With $I = \mathscr{E}/R = Blv/R$,
$$P = I^2 R = \frac{B^2 l^2 v^2}{R}$$
so the power input equals that dissipated in the resistance at any moment.

29–4 Electric Generators

Probably the most important practical result of Faraday's great discovery was the development of the **electric generator** or **dynamo**. A generator transforms mechanical energy into electric energy. This is just the opposite of what a motor does. Indeed, a generator is basically the inverse of a motor. A simplified diagram of an **ac generator** is shown in Fig. 29–11. A generator consists of many coils of wire (only one is shown) wound on an armature that can rotate in a magnetic field. The axle is turned by some mechanical means (falling water, car motor belt), and an emf is induced in the rotating coil. An electric current is thus the *output* of a generator. In Fig. 29–11, the equation $\mathbf{F} = q\mathbf{v} \times \mathbf{B}$ tells us that, with the armature rotating counterclockwise, the (conventional) current in the wire labeled A on the armature is outward; therefore it is outward at brush A, as shown. (Each brush presses against a continuous slip ring.) After one-half revolution, wire A will be where wire C is now in the drawing, and the current then at brush A will be inward. Thus the current produced is alternating. Let us look at this in more detail.

FIGURE 29–11 An ac generator.

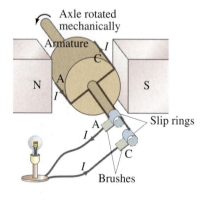

Axle rotated mechanically

Armature

N A S

Slip rings

Brushes

Let us assume the loop is being made to rotate in a uniform magnetic field **B** with constant angular velocity ω. From Faraday's law (Eq. 29–2a), the induced emf is

$$\mathscr{E} = -\frac{d\Phi_B}{dt} = -\frac{d}{dt}\int \mathbf{B} \cdot d\mathbf{A} = -\frac{d}{dt}[BA\cos\theta]$$

where A is the area of the loop and θ is the angle between **B** and **A**. Since $\omega = d\theta/dt$, then $\theta = \theta_0 + \omega t$. We arbitrarily take $\theta_0 = 0$, so

$$\mathscr{E} = -BA\frac{d}{dt}(\cos\omega t) = BA\omega\sin\omega t.$$

If the rotating coil contains N loops,

$$\mathscr{E} = NBA\omega\sin\omega t$$
$$= \mathscr{E}_0\sin\omega t.$$

(29–4)

Thus the output emf is sinusoidal (Fig. 29–12) with amplitude $\mathscr{E}_0 = NBA\omega$. Such a rotating coil in a magnetic field is the basic operating principle of an ac generator.

Over 99 percent of the electricity used in the United States is produced from generators (Fig. 29–13). The frequency $f = \omega/2\pi$ is 60 Hz for general use in the United States and Canada, although 50 Hz is used in many countries. In electric power generating plants, the armature is mounted on a heavy axle connected to a turbine, which is the modern equivalent of a waterwheel. Water pressure at a dam can turn the turbine at a hydroelectric plant. Most of the power generated at present in the United States, however, is done at steam plants, where the burning of fossil fuels (coal, oil, natural gas) boils water to produce high-pressure steam that turns the turbines. Likewise, at nuclear power plants, the nuclear energy released is used to produce steam to turn turbines. Thus, a heat engine (Chapter 20) connected to a generator is the principal means of generating electric power.

The frequency of 60 Hz is maintained very precisely by power companies, and in doing Problems, we will assume it is at least as precise as other numbers given.

EXAMPLE 29–5 **An ac generator.** The armature of a 60-Hz ac generator rotates in a 0.15-T magnetic field. If the area of the coil is $2.0 \times 10^{-2}\,\text{m}^2$, how many loops must the coil contain if the peak output is to be $\mathscr{E}_0 = 170\,\text{V}$?

SOLUTION From Eq. 29–4 we see that the maximum emf is $\mathscr{E}_0 = NBA\omega$. Since $\omega = 2\pi f = (6.28)(60\,\text{s}^{-1}) = 377\,\text{s}^{-1}$, we have

$$N = \frac{\mathscr{E}_0}{BA\omega} = \frac{170\,\text{V}}{(0.15\,\text{T})(2.0 \times 10^{-2}\,\text{m}^2)(377\,\text{s}^{-1})} = 150\,\text{turns}.$$

A **dc generator** is much like an ac generator, except the slip rings are replaced by split-ring commutators, Fig. 29–14a, just as in a dc motor. The output of such a generator is as shown and can be smoothed out by placing a capacitor in parallel with the output (Section 26–4). More common is the use of many armature windings, as in Fig. 29–14b, which produces a smoother output.

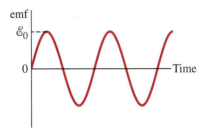

FIGURE 29–12 An ac generator produces an alternating current. The output emf $\mathscr{E} = \mathscr{E}_0\sin\omega t$, where $\mathscr{E}_0 = NAB\omega$ (Eq. 29–4).

➡ **PHYSICS APPLIED**

Power plants

FIGURE 29–13 Water-driven generators at the base of Boulder Dam, Nevada.

➡ **PHYSICS APPLIED**

DC generator

FIGURE 29–14 (a) A dc generator with one set of commutators, and (b) a dc generator with many sets of commutators and windings.

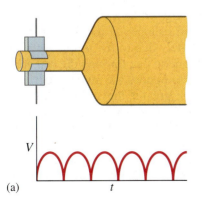

(a)

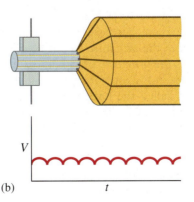

(b)

In the past, automobiles used dc generators. More common now, however, are ac generators or **alternators**, which avoid the problems of wear and electrical arcing (sparks) across the split-ring commutators of dc generators. Alternators differ from the generators discussed above in the following way. In an alternator, current from the battery produces a magnetic field in an electromagnet, called the *rotor*, which is made to rotate by a belt from the engine. Surrounding the rotating rotor are a set of stationary coils called the *stator*, Fig. 29–15. The magnetic field of the rotor passes through the stator coils and, since the rotor is rotating, the field through the fixed stator coils is changing. Hence an alternating current is induced in the stator coils, which is the output. This ac output is changed to dc for charging the battery by the use of semiconductor diodes, which allow current flow in one direction only.

Output emf of alternator

FIGURE 29–15 (a) Simplified schematic diagram of an alternator. The input electromagnet current to the rotor is connected through continuous slip rings. Sometimes the rotor electromagnet is replaced by a permanent magnet. (b) Actual shape of an alternator. The rotor is made to turn by a belt from the engine. The current in the wire coil of the rotor produces a magnetic field inside it on its axis that points horizontally from left to right, thus making north and south poles of the plates attached at either end. These end plates are made with triangular fingers that are bent over the coil—hence there are alternating N and S poles quite close to one another, with magnetic field lines between them as shown by the blue lines. As the rotor turns, these field lines pass through the fixed stator coils (shown on the right for clarity, but in operation the rotor rotates within the stator), inducing a current in them, which is the output.

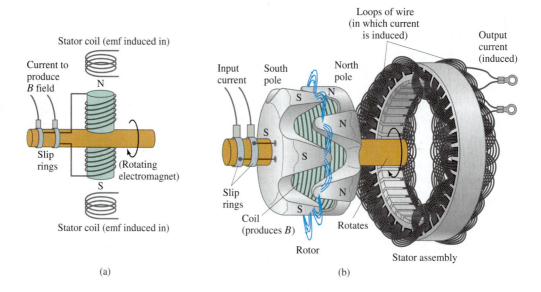

(a) (b)

---

* **29–5** **Counter EMF and Torque; Eddy Currents**

* **Counter EMF**

A motor turns and produces mechanical energy when a current is made to flow in it. From our description in Section 27–6 of a simple dc motor, you might expect that the armature would accelerate indefinitely due to the torque on it. However, as the armature of the motor turns, the magnetic flux through the coil changes and an emf is generated. This induced emf acts to oppose the motion (Lenz's law) and is called the **back emf** or **counter emf**. The greater the speed of the motor, the greater the counter emf. A motor normally turns and does work on something, but if there were no load, the motor's speed would increase until the counter emf equaled the input voltage. When there is a mechanical load, the speed of the motor may be limited also by the load. The counter emf will then be less than the external applied voltage. The greater the mechanical load, the slower the motor rotates and the lower is the counter emf ($\mathscr{E} \propto \omega$, Eq. 29–4).

Back emf

EXAMPLE 29–6 Counter emf in a motor. The armature windings of a dc motor have a resistance of $5.0\,\Omega$. The motor is connected to a 120-V line, and when the motor reaches full speed against its normal load, the counter emf is 108 V. Calculate (*a*) the current into the motor when it is just starting up, and (*b*) the current when it reaches full speed.

SOLUTION (*a*) Initially, the motor is not turning (or turning very slowly), so there is no induced counter emf. Hence, from Ohm's law, the current is

$$I = \frac{V}{R} = \frac{120\,\text{V}}{5.0\,\Omega} = 24\,\text{A}.$$

(*b*) At full speed, the counter emf is a source of emf that opposes the exterior source. We represent this counter emf as a battery in the equivalent circuit shown in Fig. 29–16. In this case, Ohm's law (or Kirchhoff's rule) gives

$$120\,\text{V} - 108\,\text{V} = I\,(5.0\,\Omega).$$

Therefore

$$I = \frac{12\,\text{V}}{5.0\,\Omega} = 2.4\,\text{A}.$$

This result shows that the current can be very high when a motor first starts up. This is why the lights in your house may dim when the motor of the refrigerator (or other large motor) starts up. The large initial current causes the voltage at the outlets to drop, since the house wiring has resistance and there is some voltage drop across it when large currents are drawn.

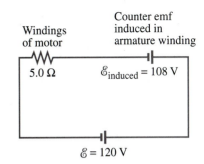

FIGURE 29–16 Circuit of a motor showing induced counter emf.

Effect of back emf on current

CONCEPTUAL EXAMPLE 29–7 Motor overload. When using an appliance such as a blender, electric drill, or sewing machine, if the appliance is overloaded or jammed so that the motor slows appreciably or stops while the power is still connected, the device can burn out and be ruined. Explain why this happens.

RESPONSE The motors are designed to run at a certain speed for a given applied voltage and the designer must take the expected counter emf into account. If the rotation speed is reduced, the counter emf will not be as high as expected ($\mathscr{E} \propto \omega$, Eq. 29–4), and the current will increase, and may become large enough that the windings of the motor heat up to the point of ruining the motor.

➡ **PHYSICS APPLIED**

Burning out a motor

* Counter Torque

In a generator, the situation is the reverse of that for a motor. As we saw, the mechanical turning of the armature induces an emf in the loops, which is the output. If the generator is not connected to an external circuit, the emf exists at the terminals but no current flows. In this case, it takes little effort to turn the armature. But if the generator *is* connected to a device that draws current, then a current flows in the coils of the armature. Because this current-carrying coil is in a magnetic field, there will be a torque exerted on it (as in a motor), and this torque opposes the motion (use the right-hand rule for the force on a wire, in Fig. 29–11). This is called a **counter torque**. The greater the electrical load—that is, the more current that is drawn—the greater will be the counter torque. Hence the external applied torque will have to be greater to keep the generator turning. This of course makes sense from the conservation-of-energy principle. More mechanical-energy input is needed to produce more electrical-energy output.

Counter torque

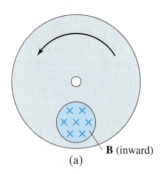

(a)

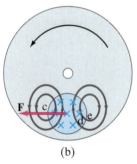

(b)

FIGURE 29–17 Production of eddy currents in a rotating wheel.

FIGURE 29–18 Repairing a step-down transformer on a utility pole.

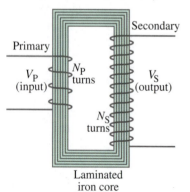

FIGURE 29–19 Step-up transformer ($N_P = 4$, $N_S = 12$).

* Eddy Currents

Induced currents are not always confined to well-defined paths such as in wires. Consider, for example, the rotating metal wheel in Fig. 29–17a. A magnetic field is applied to a limited area as shown and points into the paper. The section of wheel in the magnetic field has an emf induced in it because the conductor is moving, carrying electrons with it. The flow of (conventional) current is upward in the region of the magnetic field (Fig. 29–17b), and the current follows a downward return path outside that region. Why? According to Lenz's law, the induced currents oppose the change that causes them. Consider the part of the wheel labeled c in Fig. 29–17b, where the magnetic field is zero but is just about to enter a region where **B** points into the page. To oppose this change, the induced current is counterclockwise to produce a field pointing out of the page (right-hand rule). Similarly region d is about to move to e, where **B** is zero; hence the current is clockwise to produce an inward field opposed to this change. These currents are referred to as **eddy currents** and can be present in any conductor that is moving across a magnetic field or through which the magnetic flux is changing.

In Fig. 29–17, the magnetic field exerts a force **F** on the induced currents it has created, and that force opposes the rotational motion. Eddy currents can be used in this way as a smooth braking device on, say, a rapid-transit car. In order to stop the car, an electromagnet can be turned on that applies its field either to the wheels or to the moving steel rail below. Eddy currents can also be used to dampen (reduce) the oscillation of a vibrating system. Eddy currents, however, can be a problem. For example, eddy currents induced in the armature of a motor or generator produce heat ($P = I\mathcal{E}$) and waste energy. To reduce the eddy currents, the armatures are *laminated*; that is, they are made of very thin sheets of iron that are well insulated from one another. (See Fig. 29–19 in the next Section.) Thus the total path length of the eddy currents is confined to each slab, which increases the total resistance; hence the current is less and there is less wasted energy.

29–6 | Transformers and Transmission of Power

A transformer is a device for increasing or decreasing an ac voltage. Transformers are found everywhere: in TV sets to give the high voltage needed for the picture tube, in converters for plugging in a portable "Walkman," on utility poles (Fig. 29–18) to reduce the high voltage from the electric company to that usable in houses (110 V or 220 V), and in many other applications. A **transformer** consists of two coils of wire known as the **primary** and **secondary** coils. The two coils can be interwoven (with insulated wire); or they can be linked by a soft iron core which is laminated to prevent eddy-current losses (Section 29–5), as shown in Fig. 29–19. Transformers are designed so that (nearly) all the magnetic flux produced by the current in the primary also passes through the secondary coil, and we assume this is true in what follows. We also assume that energy losses in the resistance of the coils and hysteresis in the iron can be ignored—a good approximation for real transformers, which are often better than 99 percent efficient.

When an ac voltage is applied to the primary, the changing magnetic field it produces will induce an ac voltage of the same frequency in the secondary. However, the voltage will be different according to the number of loops in each coil. From Faraday's law, the voltage or emf induced in the secondary is

$$V_S = N_S \frac{d\Phi_B}{dt},$$

where N_S is the number of turns in the secondary coil, and $d\Phi_B/dt$ is the rate at which the magnetic flux changes.

The input primary voltage, V_P, is also related to the rate at which the flux changes

$$V_P = N_P \frac{d\Phi_B}{dt},$$

where N_P is the number of turns in the primary coil. This follows because the changing flux produces a counter emf, $N_P d\Phi_B/dt$, in the primary that exactly balances the applied voltage V_P if the resistance of the primary can be ignored (Kirchhoff's rules). We divide these two equations, assuming little or no flux is lost, to find

$$\frac{V_S}{V_P} = \frac{N_S}{N_P}.$$

(29–5)   *Transformer equation*

This *transformer equation* tells how the secondary (output) voltage is related to the primary (input) voltage; V_S and V_P in Eq. 29–5 can be the rms values for both, or peak values for both. (DC voltages don't work in a transformer because there would be no changing magnetic flux.)

If N_S is greater than N_P, we have a **step-up transformer**. The secondary voltage is greater than the primary voltage. For example, if the secondary has twice as many turns as the primary, then the secondary voltage will be twice that of the primary. If N_S is less than N_P, we have a **step-down transformer**.

Although ac voltage can be increased (or decreased) with a transformer, we don't get something for nothing. Energy conservation tells us that the power output can be no greater than the power input. A well-designed transformer can be greater than 99 percent efficient, so little energy is lost to heat. The power output thus essentially equals the power input. Since power $P = VI$ (Eq. 25–6), we have

$$V_P I_P = V_S I_S,$$

or

$$\frac{I_S}{I_P} = \frac{N_P}{N_S}.$$

(29–6)   *Transformer equation II*

> **EXAMPLE 29–8** **Portable radio transformer.** A transformer for home use of a portable radio reduces 120-V ac to 9.0-V ac. (Such a device also contains diodes to change the 9.0-V ac to dc.) The secondary contains 30 turns and the radio draws 400 mA. Calculate: (*a*) the number of turns in the primary; (*b*) the current in the primary; and (*c*) the power transformed.
>
> **SOLUTION** (*a*) This is a step-down transformer, and from Eq. 29–5 we have
>
> $$N_P = N_S \frac{V_P}{V_S} = \frac{(30)(120 \text{ V})}{(9.0 \text{ V})} = 400 \text{ turns}.$$
>
> (*b*) From Eq. 29–6,
>
> $$I_P = I_S \frac{N_S}{N_P} = (0.40 \text{ A})\left(\frac{30}{400}\right) = 0.030 \text{ A}.$$
>
> (*c*) The power transformed is
>
> $$P = I_S V_S = (9.0 \text{ V})(0.40 \text{ A}) = 3.6 \text{ W},$$
>
> which is, assuming 100 percent efficiency, the same as the power in the primary, $P = (120 \text{ V})(0.030 \text{ A}) = 3.6 \text{ W}$.

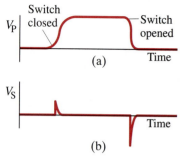

Switch closed V_P Switch opened

(a) Time

V_S

Time

(b)

FIGURE 29–20 A dc voltage turned on and off as shown in (a) produces voltage pulses in the secondary (b). Voltage scales in (a) and (b) are not necessarily the same.

A transformer operates only on ac. A dc current in the primary does not produce a changing flux and therefore induces no emf in the secondary. However, if a dc voltage is applied to the primary through a switch, at the instant the switch is opened or closed there will be an induced current in the secondary. For example, if the dc is turned on and off as shown in Fig. 29–20a, the voltage induced in the secondary is as shown in Fig. 29–20b. Notice that the secondary voltage drops to zero when the dc voltage is steady. This is basically how, in the ignition system of an automobile, the high voltage is created to produce the spark across the gap of a spark plug that ignites the gas-air mixture. The transformer is referred to simply as the "ignition coil," and transforms the 12 V of the battery (when switched off) into a spike of as much as 25 kV.

FIGURE 29–21

The transmission of electric power from power plants to homes makes use of transformers at various stages.

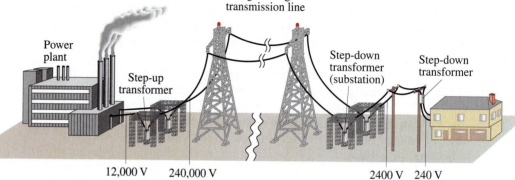

High voltage transmission line

Power plant

Step-up transformer

Step-down transformer (substation)

Step-down transformer

12,000 V   240,000 V   2400 V   240 V

► PHYSICS APPLIED

Transformers help power transmission

Transformers play an important role in the transmission of electricity. Power plants are often situated some distance from metropolitan areas, so electricity must often be transmitted over long distances (Fig. 29–21). There is always some power loss in the transmission lines, and this loss can be minimized if the power is transmitted at high voltage, using transformers, as the following Example shows.

EXAMPLE 29–9 Transmission lines. An average of 120 kW of electric power is sent to a small town from a power plant 10 km away. The transmission lines have a total resistance of 0.40 Ω. Calculate the power loss if the power is transmitted at (*a*) 240 V and (*b*) 24,000 V.

SOLUTION We cannot use $P = V^2/R$ because if R is the resistance of the transmission lines, we don't know the voltage drop along them; the given voltages are applied across the lines plus the load (the town). But for each case we can determine the current I in the lines, and then find the power loss from $P = I^2R$. (*a*) If 120 kW is sent at 240 V, the total current will be

$$I = \frac{P}{V} = \frac{1.2 \times 10^5 \text{ W}}{2.4 \times 10^2 \text{ V}} = 500 \text{ A}.$$

The power loss in the lines, P_L, is then

$$P_L = I^2R = (500 \text{ A})^2 (0.40 \text{ }\Omega) = 100 \text{ kW}.$$

Thus, over 80 percent of all the power would be wasted as heat in the power lines!

(b) If 120 kW is sent at 24,000 V, the total current will be

$$I = \frac{P}{V} = \frac{1.2 \times 10^5 \text{ W}}{2.4 \times 10^4 \text{ V}} = 5.0 \text{ A}.$$

The power loss in the lines is then

$$P_{\text{L}} = I^2 R = (5.0 \text{ A})^2 (0.40 \, \Omega) = 10 \text{ W},$$

which is less than $\frac{1}{100}$ of 1 percent. We see that the greater the voltage, the less the current and thus the less power is wasted in the transmission lines. It is for this reason that power is usually transmitted at very high voltages, as high as 700 kV.

The great advantage of ac, and a major reason it is in nearly universal use, is that the voltage can easily be stepped up or down by a transformer. The output voltage of an electric generating plant is stepped up prior to transmission. Upon arrival in a city, it is stepped down in stages at electric substations prior to distribution. The voltage in lines along city streets is typically 2400 V and is stepped down to 240 V or 120 V for home use by transformers (Figs. 29–18 and 29–21).

29–7 A Changing Magnetic Flux Produces an Electric Field

We have seen in earlier chapters (especially Chapter 25, Section 25–8) that when an electric current flows in a wire, there is an electric field in the wire that does the work of moving the electrons in the wire. In this chapter we have seen that a changing magnetic flux induces a current in the wire, which implies that there is an electric field in the wire induced by the changing magnetic flux. Thus we come to the important conclusion that

a changing magnetic flux produces an electric field.

This result applies not only to wires and other conductors, but is actually a general result that applies to any region in space. Indeed, an electric field will be produced at any point in space where there is a changing magnetic field.

Faraday's Law—General Form

We can put these ideas into mathematical form by generalizing our relation between an electric field and the potential difference between two points a and b: $V_{\text{ab}} = \int_a^b \mathbf{E} \cdot d\mathbf{l}$ (Eq. 23–3) where $d\mathbf{l}$ is an element of displacement along the path of integration. The emf $\mathscr{E}$ induced in a circuit is equal to the work done per unit charge by the electric field, which equals the integral of $\mathbf{E} \cdot d\mathbf{l}$ around the closed path:

$$\mathscr{E} = \oint \mathbf{E} \cdot d\mathbf{l}. \tag{29–7}$$

We combine this with Eq. 29–2a, to obtain a more elegant and general form of Faraday's law

$$\oint \mathbf{E} \cdot d\mathbf{l} = -\frac{d\Phi_B}{dt} \tag{29–8}$$

FARADAY'S LAW
(general form)

which relates the changing magnetic flux to the electric field it produces. The integral on the left is taken around a path enclosing the area through which the magnetic flux Φ_B is changing. This more elegant statement of Faraday's law (Eq. 29–8) is valid not only in conductors, but in any region of space. To illustrate this, let us take an Example.

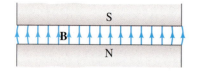

(a)

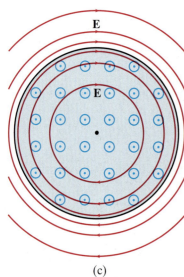

(b)

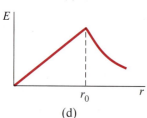

(c)

(d)

FIGURE 29–22 Example 29–10.
(a) Side view of nearly constant **B**.
(b) Top view, for determining the
electric field **E** at point P. (c) Lines of
E produced by increasing **B** (pointing
outward). (d) Graph of E vs. r.

EXAMPLE 29–10 E produced by changing B. A magnetic field **B** between the pole faces of an electromagnet is nearly uniform at any instant over a circular area of radius r_0 as shown in Figs. 29–22a and b. The current in the windings of the electromagnet is increasing in time so that **B** changes in time at a constant rate $d\mathbf{B}/dt$ at each point. Beyond the circular region $(r > r_0)$, we assume **B** = 0 at all times. Determine the electric field **E** at any point P a distance r from the center of the circular area.

SOLUTION The changing magnetic flux through a circle of radius r, shown dashed in Fig. 29–22b, will produce an emf around this circle. Because all points on the dashed circle are equivalent physically, the electric field too will show this symmetry and will be in the plane perpendicular to **B**. Thus we can expect **E** to be perpendicular to **B** and to be tangent to the circle of radius r; the direction of **E** will be as shown in Fig. 29–22b and c, since by Lenz's law the induced **E** needs to be capable of producing a current that generates a magnetic field opposing the original change in **B**. By symmetry, we also expect **E** to have the same magnitude at all points on the circle of radius r. We therefore take this circle as our path of integration in Eq. 29–8 (ignoring the minus sign so we can concentrate on magnitude since we got the direction of **E** from Lenz's law) and obtain

$$E(2\pi r) = (\pi r^2)\frac{dB}{dt}, \qquad\qquad [r < r_0]$$

since $\Phi_B = BA = B(\pi r^2)$ at any instant. We solve for E and obtain

$$E = \frac{r}{2}\frac{dB}{dt}. \qquad\qquad [r < r_0]$$

This expression is valid up to the edge of the circle $(r \leq r_0)$, beyond which **B** = 0. If we now consider a point where $r > r_0$, the flux through a circle of radius r is $\Phi_B = \pi r_0^2 B$. Then Eq. 29–8 gives

$$E(2\pi r) = \pi r_0^2 \frac{dB}{dt} \qquad\qquad [r > r_0]$$

or

$$E = \frac{r_0^2}{2r}\frac{dB}{dt}. \qquad\qquad [r > r_0]$$

Thus the magnitude of the electric field increases linearly from zero at the center of the magnet to $E = (dB/dt)(r_0/2)$ at the edge, and then decreases inversely with distance in the region beyond the edge of the magnetic field. The electric field lines are circles as shown in Fig. 29–22c. A graph of E vs. r is shown in Fig. 29–22d.

* Forces Due to Changing B are Nonconservative

Example 29–10 illustrates an important difference between electric fields produced by changing magnetic fields and electric fields produced by electric charges at rest (electrostatic fields). Electric field lines produced in the electrostatic case (Chapters 21 to 24) start and stop on electric charges. But the electric field lines produced by a changing magnetic field are continuous; they form closed loops. This distinction goes even further and is an important one. In the electrostatic case, the potential difference between two points is given by

$$V_{ab} = V_a - V_b = \int_a^b \mathbf{E} \cdot d\mathbf{l}.$$

If the integral is around a closed loop, so points a and b are the same, then $V_{ab} = 0$. Hence the integral of $\mathbf{E} \cdot d\mathbf{l}$ around a closed path is zero:

$$\oint \mathbf{E} \cdot d\mathbf{l} = 0. \qquad\qquad \text{[electrostatic field]}$$

This followed from the fact that the electrostatic force (Coulomb's law) is a conservative force, and so a potential energy function could be defined. Indeed, the relation above, $\oint \mathbf{E} \cdot d\mathbf{l} = 0$, tells us that the work done per unit charge around any closed path is zero (or the work done between any two points is independent of path—see Chapter 8), which is a property only of a conservative force. But in the nonelectrostatic case, when the electric field is produced by a changing magnetic field, the integral around a closed path is *not* zero, but is given by Eq. 29–8:

$$\oint \mathbf{E} \cdot d\mathbf{l} = -\frac{d\Phi_B}{dt}.$$

We thus come to the conclusion that the forces due to changing magnetic fields are *nonconservative*. We are not able therefore to define a potential energy, or potential function, at a given point in space for the nonelectrostatic case. Although static electric fields are *conservative fields*, the electric field produced by a changing magnetic field is a **nonconservative field**.

* 29–8 Applications of Induction: Sound Systems, Computer Memory, the Seismograph

There are various types of *microphones*, and many operate on the principle of induction. In one form, a microphone is just the inverse of a loudspeaker (Section 27–6). A small coil connected to a membrane is suspended close to a small permanent magnet, as shown in Fig. 29–23. The coil moves in the magnetic field when sound waves strike the membrane. The frequency of the induced emf will be just that of the impinging sound waves, and this emf is the "signal" that can be amplified and sent to loudspeakers, or sent to a tape recorder to be recorded on tape. In a "ribbon" microphone, a thin metal ribbon is suspended between the poles of a permanent magnet. The ribbon vibrates in response to sound waves, and the emf induced in the ribbon is proportional to its velocity.

Tape recording and tape playback is done by *tape heads* inside a tape recorder (or cassette deck). Recording tape for use in audio and video tape recorders contains a thin layer of magnetic oxide on a thin plastic tape. During recording, the audio or video signal voltage is sent to the recording head, which acts as a tiny electromagnet (Fig. 29–24) that magnetizes the tiny section of tape passing over the narrow gap in the head at each instant. In playback, the changing magnetism of the moving tape at the gap causes corresponding changes in the magnetic field within the soft-iron head, which in turn induces an emf in the coil (Faraday's law). This induced emf is the output signal that can be amplified and sent to a loudspeaker (or, in the case of a video signal, to the picture tube). In audio and video recorders, the signals may be *analog*—they vary continuously in amplitude over time. The variation in degree of magnetization of the tape at any point reflects the variation in amplitude of the audio or video signal.

Digital information, such as used on computer disks (floppy disks or hard disks) or on magnetic computer tape and some types of digital tape recorders, is read and written using heads that are basically the same as just described (Fig. 29–24). The essential difference is in the signals, which are not analog, but are digital, and in particular binary, meaning that only two values are possible for each of the extremely high number of predetermined spaces on the tape or disk. The two possible values are usually referred to as 1 and 0. The signal voltage does not vary continuously but rather takes on only two values, say +5 volts and 0 volts, corresponding to the 1 or 0. Thus, information is carried as a series of "bits," each of which can have only one of two values, 1 or 0.

In another field, geophysics, an important device based on electromagnetic induction is one type of *seismograph* or *geophone*. A seismograph is placed in direct contact with the Earth and converts the motion of the Earth—whether due to an earthquake or to an explosion (such as for mineral prospecting or for detecting a bomb test)—into an electrical signal. A seismograph contains a magnet and a

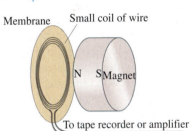

FIGURE 29–23 Diagram of a microphone that works by induction.

FIGURE 29–24 Recording and/or playback head for tape or disk. In recording (or "writing"), the electric input signal to the head, which acts as an electromagnet, magnetizes the passing tape or disk. In playback (or "reading"), the changing magnetic field of the passing tape or disk induces a changing magnetic field in the head, which in turn induces in the coil an emf that is the output signal.

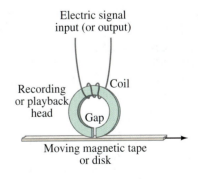

coil of wire, one of which is fixed rigidly to the case which moves as the Earth does where it is planted. The other element is inertial and is suspended from the case by a spring. In the type shown in Fig. 29–25, the coil moves with the Earth, and the relative motion of the magnet and coil produces an induced emf in the coil, which is the output of the device. In many geophones, the coil is inertial and the magnet moves with the Earth.

FIGURE 29–25 (a) A seismograph or geophone. (b) A seismograph reading (Northridge, California earthquake, January 17, 1994).

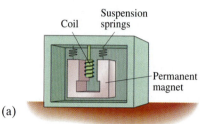

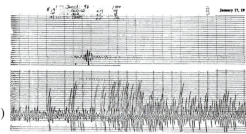

(a)

(b)

Summary

The **magnetic flux** passing through a loop is equal to the product of the area of the loop times the perpendicular component of the (uniform) magnetic field: $\Phi_B = B_\perp A = BA \cos\theta$. If **B** is not uniform, then

$$\Phi_B = \int \mathbf{B} \cdot d\mathbf{A}.$$

If the magnetic flux through a coil of wire changes in time, an emf is induced in the coil. The magnitude of the induced emf equals the time rate of change of the magnetic flux through the loop times the number N of loops in the coil:

$$\mathcal{E} = -N \frac{d\Phi_B}{dt}.$$

This is **Faraday's law of induction**.

The induced emf, and the current it produces, are in a direction that opposes the change in flux which caused them (**Lenz's law**).

We can also see from Faraday's law that a straight wire of length l moving with speed v perpendicular to a magnetic field of strength B has an emf induced between its ends equal to:

$$\mathcal{E} = Blv.$$

Faraday's law also tells us that *a changing magnetic field produces an electric field*. The mathematical relation is

$$\oint \mathbf{E} \cdot d\mathbf{l} = -\frac{d\Phi_B}{dt}$$

and is the general form of Faraday's law. The integral on the left is taken around the loop through which the magnetic flux Φ_B is changing.

An electric **generator** changes mechanical energy into electrical energy. Its operation is based on Faraday's law: a coil of wire is made to rotate uniformly by mechanical means in a magnetic field, and the changing flux through the coil induces a sinusoidal current, which is the output of the generator.

A motor, which operates in the reverse of a generator, acts like a generator in that a **counter emf** is induced in its rotating coil; since this counter emf opposes the input voltage, it can act to limit the current in a motor coil.

Similarly, a generator acts somewhat like a motor in that a **counter torque** acts on its rotating coil.

A **transformer**, which is a device to change the magnitude of an ac voltage, consists of a primary and a secondary coil. The changing flux due to an ac voltage in the primary induces an ac voltage in the secondary. In a 100 percent efficient transformer, the ratio of output to input voltages (V_S/V_P) equals the ratio of the number of turns N_S in the secondary to the number N_P in the primary:

$$\frac{V_S}{V_P} = \frac{N_S}{N_P}.$$

The ratio of secondary to primary current is in the inverse ratio of turns:

$$\frac{I_S}{I_P} = \frac{N_P}{N_S}.$$

Questions

1. What would be the advantage, in Faraday's experiments (Fig. 29–1), of using coils with many turns?

2. Suppose you are holding a circular loop of wire and suddenly thrust a magnet, south pole first, toward the center of the circle. Is a current induced in the wire? Is a current induced when the magnet is held steady within the loop? Is a current induced when you withdraw the magnet? In each case, if your answer is yes, specify the direction.

3. Suppose you are looking along a line through the centers of two circular (but separate) wire loops, one behind the other. A battery is suddenly connected to the front loop, establishing a clockwise current. (a) Will a current be induced in the second loop? (b) If so, when does this current start? (c) When does it stop? (d) In what direction is this current? (e) Is there a force between the two loops? (f) If so, in what direction?

4. The battery discussed in Question 3 is disconnected. Will a current be induced in the second loop? If so, when does it start and stop? In what direction is this current?

5. Two loops of wire are moving in the vicinity of a very long straight wire carrying a steady current as shown in Fig. 29–26. Find the direction of the induced current in each loop.

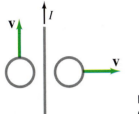

FIGURE 29–26
Question 5.

6. Is there a force between the two loops discussed in Question 5? If so, in what direction?

7. In what direction will the current flow in Fig. 29–9 if the rod moves to the left, which decreases the area of the loop to the left?

8. Some modern stove burners are based on induction. That is, an ac current passes around a coil that is the "burner," a burner that never gets hot. Explain why it will heat a metal pan but not a glass container.

9. A region where no magnetic field is desired is surrounded by a sheet of low-resistivity metal. (*a*) Will this sheet shield the interior from a rapidly changing magnetic field outside? Explain. (*b*) Will it act as a shield to a static magnetic field? (*c*) What if the sheet is superconducting (resistivity = 0)?

10. Show, using Lenz's law, that the emf induced in the moving rod in Fig. 29–9 is positive at the bottom and negative at the top so that the current flows clockwise in the circuit loop on the left.

11. What is the advantage of placing the two insulated electric wires carrying ac close together or even twisted about each other?

12. Explain why, exactly, the lights may dim briefly when a refrigerator motor starts up. When an electric heater is turned on, the lights may stay dimmed as long as it is on. Explain the difference.

* **13.** Use Fig. 29–11 and the right-hand rules to show why the counter torque in a generator *opposes* the motion.

* **14.** Will an eddy-current brake (Fig. 29–17) work on a copper or aluminum wheel, or must it be ferromagnetic?

* **15.** It has been proposed that eddy currents be used to help sort solid waste for recycling. The waste is first ground into tiny pieces and iron removed with a dc magnet. The waste then is allowed to slide down an incline over permanent magnets. How will this aid in the separation of nonferrous metals (Al, Cu, Pb, brass) from nonmetallic materials?

* **16.** The pivoted metal bar with slots in Fig. 29–27 falls much more quickly through a magnetic field than does a solid bar. Explain in detail.

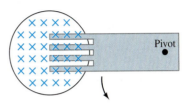

FIGURE 29–27 Question 16.

* **17.** If an aluminum sheet is held between the poles of a large bar magnet, it requires some force to pull it out of the magnetic field even though the sheet is not ferromagnetic and does not touch the pole faces. Explain.

* **18.** A metal bar, pivoted at its upper end, oscillates freely in the absence of a magnetic field; but in a magnetic field, its oscillations are quickly damped out. Explain. (This *magnetic damping* is used in a number of practical devices.)

19. An enclosed transformer has four wire leads coming from it. How could you determine the ratio of turns on the two coils without taking the transformer apart? How would you know which wires paired with which?

20. A transformer designed for a 120-V ac input will often "burn out" if connected to a 120-V dc source. Explain. [*Hint*: The resistance of the primary coil is usually very low.]

* **21.** Since a magnetic microphone is basically like a loudspeaker, could a loudspeaker (Section 27–6) actually serve as a microphone? That is, could you speak into a loudspeaker and obtain an output signal that could be amplified? Explain. Discuss, in light of your response, how a microphone and loudspeaker differ in construction.

Problems

Sections 29–1 and 29–2

1. (I) The magnetic flux through a coil of wire containing two loops changes uniformly from −80 Wb to +58 Wb in 0.72 s. What is the emf induced in the coil?

2. (I) A 26-cm-diameter circular loop of wire lies in a plane perpendicular to a 0.90-T magnetic field. It is removed from the field in 0.15 s. What is the average induced emf?

3. (I) The rectangular loop shown in Fig. 29–28 is pushed into the magnetic field which points inward as shown. In what direction is the induced current? Explain.

FIGURE 29–28 Problem 3.

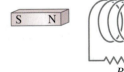

FIGURE 29–29 Problem 4.

4. (I) The north pole of the magnet in Fig. 29–29 is being inserted into the coil. In which direction is the induced current flowing through the resistor *R*? Explain.

5. (I) A 7.2-cm-diameter loop of wire is initially oriented perpendicular to a 1.3-T magnetic field. It is rotated so that its plane is parallel to the field direction in 0.20 s. What is the average induced emf in the loop?

6. (I) A 9.2-cm-diameter wire coil is initially oriented so that its plane is perpendicular to a magnetic field of 0.63 T pointing up. During the course of 0.15 s, the field is changed to one of 0.25 T pointing down. What is the average induced emf in the coil?

7. (II) What is the direction of the induced current in the circular loop due to the current shown in each part of Fig. 29–30? Explain.

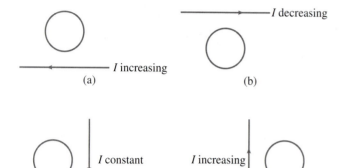

(a)   (b)

(c)   (d)

FIGURE 29–30   Problem 7.

8. (II) (a) If the resistance of the resistor in Fig. 29–31 is slowly increased, what is the direction of the current induced in the small circular loop inside the larger loop? (b) What would it be if the small loop were placed outside and to the left of the larger one? Explain.

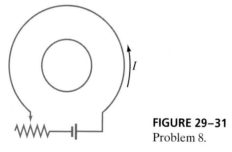

FIGURE 29–31
Problem 8.

9. (II) If the solenoid in Fig. 29–32 is being pulled away from the loop shown, in what direction is the induced current in the loop? Explain.

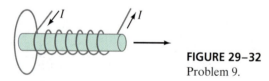

FIGURE 29–32
Problem 9.

10. (II) The magnetic field perpendicular to a circular loop of wire 20 cm in diameter is changed from +0.52 T to −0.45 T in 180 ms, where + means the field points away from an observer and − toward the observer. (a) Calculate the induced emf. (b) In what direction does the induced current flow?

11. (II) An elastic circular loop in the plane of the paper lies in a 0.75 T magnetic field pointing into the paper. If the loop's diameter changes from 20.0 cm to 6.0 cm in 0.50 s, (a) what is the direction of the induced current, (b) what is the magnitude of the average induced emf, and (c) if the loop's resistance is 2.5 Ω, what is the average induced current during the 0.50 s?

12. (II) A square loop of wire 15 cm on a side is rotated uniformly about an axis through its center and parallel to two sides. It rotates 360° in a magnetic field **B** perpendicular to the axis in 45 ms. If the rms induced emf is 70 mV, what is the average value of B?

13. (II) A 20-cm-diameter circular loop of wire has a resistance of 150 Ω. It is initially in a 0.40-T magnetic field, with its plane perpendicular to **B**, but is removed from the field in 100 ms. Calculate the electric energy dissipated in this process.

14. (II) A single rectangular loop of wire with the dimensions shown in Fig. 29–33 is situated so that part is inside a region of uniform magnetic field of 0.450 T and part is outside the field. The total resistance of the loop is 0.230 Ω. Calculate the force required to pull the loop from the field (to the right) at a constant velocity of 3.40 m/s. Neglect gravity.

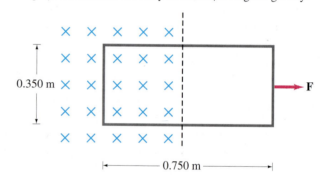

FIGURE 29–33   Problem 14.

15. (II) The magnetic field perpendicular to a single 15.6-cm-diameter circular loop of copper wire decreases uniformly from 0.550 T to zero. If the wire is 2.05 mm in diameter, how much charge moves through the coil during this operation?

16. (II) The magnetic flux through each loop of a 60-loop coil is given by $(8.8t - 0.51t^3) \times 10^{-2}$ T·m², where the time t is in seconds. (a) Determine the emf $\mathscr{E}$ as a function of time. (b) What is $\mathscr{E}$ at $t = 1.0$ s and $t = 5.0$ s?

17. (II) A 35.0-cm-diameter coil consists of 20 turns of circular copper wire 2.0 mm in diameter. A uniform magnetic field, perpendicular to the plane of the coil, changes at a rate of 3.20×10^{-3} T/s. Determine (a) the current in the loop, and (b) the rate at which thermal energy is produced.

18. (II) A single circular loop of wire is placed inside a long solenoid with its plane perpendicular to the axis of the solenoid. The area of the loop is A_1 and that of the solenoid, which has n turns per unit length, is A_2. A current $I = I_0 \cos \omega t$ flows in the solenoid turns. What is the induced emf in the small loop?

19. (II) The area of an elastic circular loop decreases at a constant rate, $dA/dt = -3.50 \times 10^{-2}$ m²/s. The loop is in a magnetic field $B = 0.48$ T whose direction is perpendicular to the plane of the loop. At $t = 0$, the loop has area $A = 0.285$ m². Determine the induced emf at $t = 0$, and at $t = 2.00$ s.

20. (II) Suppose the radius of the elastic loop in Problem 19 increases at a constant rate, $dr/dt = 7.00$ cm/s. Determine the emf induced in the loop at $t = 0$ and at $t = 1.00$ s.

21. (III) Determine the magnetic flux through a square loop of side a (Fig. 29–34) if one side is parallel to, and a distance a from, a straight wire that carries a current I.

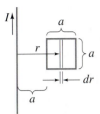

FIGURE 29–34
Problem 21.

Section 29–3

22. (I) The moving rod in Fig. 29–9 is 19.0 cm long and moves with a speed of 25.0 cm/s. If the magnetic field is 0.750 T, calculate the emf developed.

23. (II) In Fig. 29–9 the rod moves with a speed of 1.8 m/s, is 24.0 cm long, and has 2.2 Ω resistance. The magnetic field is 0.35 T and the resistance of the U-shaped conductor is 26.0 Ω at a given instant. Calculate: (a) the emf induced; (b) the current flowing in the U-shaped conductor, and (c) the external force needed to keep the rod's velocity constant at that instant.

24. (II) If the U-shaped conductor in Fig. 29–9 has resistivity ρ, whereas that of the moving rod is negligible, derive a formula for the current I as a function of time. Assume the rod has length l, starts at the bottom of the U at $t = 0$, and moves with uniform speed v in the magnetic field B. The cross-sectional area of the rod and all parts of the U is A.

25. (II) A conducting rod rests on two long frictionless parallel rails in a magnetic field $\mathbf{B}$ ($\perp$ to the rails and rod) as in Fig. 29–35. (a) If the rails are horizontal and the rod is given an initial push, will the rod travel at constant speed even though a magnetic field is present? (b) Suppose at $t = 0$, when the rod has speed $v = v_0$, the two rails are connected electrically by a wire from point a to point b. Assuming the rod has resistance R and the rails have negligible resistance, determine the speed of the rod as a function of time. Discuss your answer.

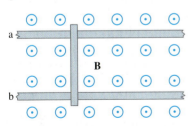

FIGURE 29–35
Problems 25 and 26.

26. (III) Suppose a conducting rod (mass m, resistance R) rests on two frictionless and resistanceless parallel rails a distance l apart in a uniform magnetic field $\mathbf{B}$ ($\perp$ to the rails and the rod) as in Fig. 29–35. At $t = 0$, the rod is at rest and a source of emf is connected to the points a and b. Determine the speed of the rod as a function of time (a) if the source puts out a constant current I, (b) the source puts out a constant emf $\mathscr{E}_0$. (c) Does the rod reach a terminal speed in either case? If so, what is it?

27. (III) A short section of wire, of length a, is moving with velocity $\mathbf{v}$, parallel to a very long wire carrying a current I as shown in Fig. 29–36. The near end of the wire section is a distance b from the long wire. Assuming the vertical wire is very long compared to $a + b$, determine the emf between the ends of the short section. Assume $\mathbf{v}$ is (a) in the same direction as I, (b) in the opposite direction to I.

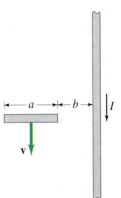

FIGURE 29–36
Problem 27.

Section 29–4

28. (I) The generator of a car idling at 950-rpm produces 12.4 V. What will the output be at a rotation speed of 2500 rpm assuming nothing else changes?

29. (I) Show that the rms output of an ac generator is $V_{\text{rms}} = NAB\omega/\sqrt{2}$.

30. (II) A simple generator has a 420-loop square coil 21.0 cm on a side. How fast must it turn in a 0.350-T field to produce a 120-V peak output?

31. (II) A 350-loop circular armature coil with a diameter of 10.0 cm rotates at 60 rev/s in a uniform magnetic field of strength 0.45 T. What is the rms voltage output of the generator? What would you do to the rotation frequency in order to double the rms voltage output?

* Section 29–5

* 32. (I) A motor has an armature resistance of 3.75 Ω. If it draws 9.20 A when running at full speed and connected to a 120-V line, how large is the counter emf?

* 33. (I) The counter emf in a motor is 72 V when operating at 1800 rpm. What would be the counter emf at 2500 rpm if the magnetic field is unchanged?

* 34. (II) The counter emf in a motor is 100 V when the motor is operating at 1000 rpm. How would you change the motor's magnetic field if you wanted to reduce the counter emf to 75 V when the motor was running at 2500 rpm?

* 35. (II) What will be the current in the motor of Example 29–6 if the load causes it to run at half speed?

* 36. (II) A dc generator is rated at 10 kW, 200 V, and 50 A when it rotates at 1000 rpm. The resistance of the armature windings is 0.40 Ω. (a) Calculate the "no-load" voltage at 1000 rpm (when there is no circuit hooked up to the generator). (b) Calculate the full-load voltage (i.e. at 50 A) when the generator is run at 800 rpm. Assume that the magnitude of the magnetic field remains constant.

37. (I) A transformer is designed to change 120 V into 8500 V, and there are 500 turns in the primary. How many turns are in the secondary, assuming 100 percent efficiency?

38. (I) A transformer has 720 turns in the primary and 120 in the secondary. What kind of transformer is this and, assuming 100 percent efficiency, by what factor does it change the voltage? By what factor does it change the current?

39. (I) A step-up transformer increases 22 V to 120 V. What is the current in the secondary as compared to the primary? Assume 100 percent efficiency.

40. (I) Neon signs require 12 kV for their operation. To operate from a 220-V line, what must be the ratio of secondary to primary turns of the transformer? What would the voltage output be if the transformer were connected backward?

41. (II) A model-train transformer plugs into 120 V ac, and draws 0.65 A while supplying 15 A to the train. (a) What voltage is present across the tracks? (b) Is the transformer step-up or step-down?

42. (II) The output voltage of a 100-W transformer is 12 V and the input current is 26 A. (a) Is this a step-up or a step-down transformer? (b) By what factor is the voltage multiplied?

43. (II) A transformer has 330 primary turns and 1510 secondary turns. The input voltage is 120 V and the output current is 15.0 A. What is the output voltage and input current assuming 100 percent efficiency?

44. (II) If 30 MW of power at 45 kV (rms) arrives at a town from a generator via 3.0-Ω transmission lines, calculate (a) the emf at the generator end of the lines, and (b) the fraction of the power generated that is lost in the lines.

45. (III) If 65 kW is to be transmitted over two 0.100-Ω lines, estimate how much power is saved if the voltage is stepped up from 120 V to 1200 V and then down again, rather than simply transmitting at 120 V. Assume the transformers are each 99 percent efficient.

46. (III) Design a dc transmission line that can transmit 300 MW of electricity 200 km with only a 2 percent loss. The wires are to be made of aluminum and the voltage is 600 kV.

47. (I) Determine the electric field in the moving rod of Problem 22.

48. (II) In Fig. 29–22c, draw two identical small circles, one placed near the center of the magnetic field region and the other near its edge (but not beyond the edge). Show that the emf around each circle is the same, even though E is greater in the region of the outer circle.

49. (II) The *betatron*, a device used to accelerate electrons to high energy, consists of a circular vacuum tube placed in a magnetic field (Fig. 29–37), into which electrons are injected. The electromagnet produces a field that (1) keeps the electrons in their circular orbit inside the tube, and (2) increases the speed of the electrons when B changes. (a) Explain how the electrons are accelerated. (See Fig. 29–37.) (b) In what direction are the electrons moving in Fig. 29–37 (give directions as if looking down from above)? (c) Should B increase or decrease to accelerate the electrons? (d) The magnetic field is actually 60 Hz ac; show that the electrons can be accelerated only during $\frac{1}{4}$ of a cycle ($\frac{1}{240}$ s). (During this time they make hundreds of thousands of revolutions and acquire very high energy.)

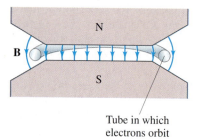

Tube in which electrons orbit

FIGURE 29–37 Problems 49 and 50.

50. (III) Show that the electrons in a betatron, Problem 49 and Fig. 29–37, are accelerated at constant radius if the magnetic field B_0 at the position of the electron orbit in the tube is equal to half the average value of the magnetic field (B_{av}) over the area of the circular orbit at each moment: $B_0 = \frac{1}{2} B_{av}$. (This is the reason the pole faces have a rather odd shape, as indicated in Fig. 29–37.)

51. (III) Find a formula for the net electric field in the moving rod of Problem 26 as a function of time for each case, (a) and (b).

General Problems

52. Suppose you are looking at two current loops in the plane of the page as shown in Fig. 29–38. When the switch is thrown in the left-hand coil, (a) what is the direction of the induced current in the other loop? (b) What is the situation after a "long" time? (c) What is the direction of the induced current in the second loop if the second loop is quickly pulled horizontally to the right?

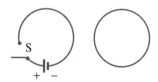

FIGURE 29–38 Problem 52.

53. A simple generator is used to generate a peak output voltage of 24.0 V. The square armature consists of windings that are 7.0 cm on a side and rotates in a field of 0.420 T at a rate of 60 rev/s. How many loops of wire should be wound on the square armature?

54. A square loop 24.0 cm on a side has a resistance of 6.50 Ω. It is initially in a 0.755-T magnetic field, with its plane perpendicular to **B**, but is removed from the field in 40.0 ms. Calculate the electric energy dissipated in this process.

55. Two conducting rails of negligible resistance 30 cm apart rest on a 5.0° ramp. They are joined at the bottom by a 0.60 Ω resistor, and at the top a copper bar of mass 0.040 kg is laid across the rails. The whole apparatus is immersed in a vertical 0.55 T field. What is the terminal (steady) velocity of the bar as it slides frictionlessly down the rails?

56. A pair of power transmission lines each have a 0.80-Ω resistance and carry 700 A over 9.0 km. If the rms input voltage is 42 kV, calculate (a) the voltage at the other end, (b) the power input, (c) power loss in the lines, and (d) the power output.

57. Power is generated at 24 kV at a generating plant which is located 100 km from a town that requires 50 MW of power at 12 kV. The transmission lines from the plant to the town have a total resistance of 0.10 Ω/km. What should the output voltage of the transformer at the generating plant be for an overall transmission efficiency of 98.5 percent, assuming a perfect transformer?

58. A coil with 150 turns, a radius of 5.2 cm and a resistance of 11.0 Ω surrounds a solenoid with 200 turns/cm and a radius of 4.5 cm. The current in the solenoid changes at a constant rate from 0 A to 2.0 A in 0.10 s. Calculate the magnitude and direction of the induced current in the coil.

59. A ring with a radius of 3.0 cm and a resistance of 0.025 Ω is rotated about an axis through its diameter by 90° in a magnetic field of 0.15 T perpendicular to that axis. What is the largest number of electrons that would flow past a fixed point in the ring as this process is accomplished?

60. Calculate the peak output voltage of a simple generator whose square armature windings are 6.60 cm on a side if the armature contains 125 loops and rotates in a field of 0.200 T at a rate of 120 rev/s.

61. A small electric car overcomes a 250-N friction force when traveling 30 km/h. The electric motor is powered by ten 12-V batteries connected in series and is coupled directly to the wheels whose diameters are 50 cm. The 300 armature coils are rectangular, 10 cm by 15 cm, and rotate in a 0.60-T magnetic field. (a) How much current does the motor draw to produce the required torque? (b) What is the back emf? (c) How much power is dissipated in the coils? (d) What percent of the input power is used to drive the car?

62. A **search coil** for measuring B (also called a *flip coil*) is a small coil with N turns, each of cross-sectional area A. It is connected to a so-called **ballistic galvanometer**, which is a device to measure the total charge Q that passes through it in a short time. The flip coil is placed in the magnetic field to be measured with its face perpendicular to the field. It is then quickly rotated 180°. Show that the total charge Q that flows in the induced current during this short "flip" time is proportional to the magnetic field B. In particular, show that B is given by

$$B = \frac{QR}{2NA}$$

where R is the total resistance of the circuit, including that of the coil and that of the ballistic galvanometer which measures the charge Q.

63. The primary windings of a transformer, which has an 80 percent efficiency, are connected to 110 V ac. The secondary windings are connected across a 2.4 Ω, 75 W lightbulb. (a) Calculate the current through the primary windings of the transformer. (b) Calculate the ratio of the number of primary windings of the transformer to the number of secondary windings of the transformer.

64. What is the energy dissipated as a function of time in a circular loop of ten turns of wire having a radius of 10.0 cm and a resistance of 2.0 Ω if the plane of the loop is perpendicular to a magnetic field given by

$$B(t) = B_0 e^{-t/\tau}$$

with $B_0 = 0.50$ T and $\tau = 0.10$ s?

65. A thin metal rod of length L rotates with angular velocity ω about an axis through one end. The rotation axis is perpendicular to the rod and is parallel to a uniform magnetic field **B**. Determine the emf developed between the ends of the rod.

66. High-intensity desk lamps are rated at 40 W but require only 12 V. They contain a transformer that converts 120 V household voltage. (a) Is the transformer step-up or step-down? (b) What is the current in the secondary when the lamp is on? (c) What is the current in the primary? (d) What is the resistance of the bulb when on?

67. Show that the power loss in transmission lines, P_L, is given by $P_L = (P_T)^2 R_L/V^2$, where P_T is the power transmitted to the user, V is the delivered voltage, and R_L is the resistance of the power lines.

* 68. The magnetic field of a "shunt-wound" dc motor is produced by field coils placed in parallel with the armature coils. Suppose that the field coils have a resistance of 36.0 Ω and the armature coils 3.00 Ω. The back emf at full speed is 105 V when the motor is connected to 115 V dc. (a) Draw the equivalent circuit for the situations when the motor is just starting and when it is running full speed. (b) What is the total current drawn by the motor at start up? (c) What is the total current drawn when the motor runs at full speed?

69. Apply Faraday's law, in the form of Eq. 29–8, to show that the static electric field between the plates of a parallel-plate capacitor cannot drop abruptly to zero at the edges, but must, in fact, fringe. Use the path shown dashed in Fig. 29–39.

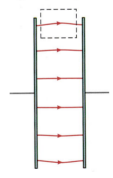

FIGURE 29–39
Problem 69.

70. A circular metal disk of radius R rotates with angular velocity ω about an axis through its center perpendicular to its face. The disk rotates in a uniform magnetic field B whose direction is parallel to the rotation axis. Determine the emf induced between the center and the edges.

71. What is the magnitude and direction of the electric field at each point in the rotating disk of Problem 70?

* 72. Suppose the "magnetic brake" of Fig. 29–17 acts on a circular metal disk of radius R and thickness d, whose electrical resistivity is ρ. The magnetic field **B**, perpendicular to the disk, acts over a small area A whose center is a distance l from the center of the wheel. At a moment when the disk is rotating with angular speed ω about an axis through its center, work out an approximate formula for the torque acting to slow it down.

A spark plug in a car receives a high voltage, which produces a high enough electric field in the air across its gap to pull electrons off the atoms in the air-gasoline mixture and form a spark. The high voltage is produced, from the basic 12 V of the car battery, by an induction coil which is basically a transformer or mutual inductance. Any coil of wire has a self-inductance, and a changing current in it causes an emf to be induced. Such inductors are useful in many circuits.

CHAPTER 30

Inductance; and Electromagnetic Oscillations

We discussed in the last chapter how a changing magnetic flux through a circuit induces an emf in that circuit. Before that we saw that an electric current produces a magnetic field. Combining these two ideas, we expect that a changing current in one circuit ought to induce an emf and a current in a second nearby circuit and even induce an emf in itself. We already saw an example in the previous chapter (transformers), but now we will treat this effect in a more general way in terms of what we will call mutual inductance and self-inductance. The concept of inductance also gives us a springboard to treat energy storage in a magnetic field. This chapter concludes with an analysis of circuits that contain inductance as well as resistance and/or capacitance.

30–1 | Mutual Inductance

If two coils of wire are placed near each other, as in Fig. 30–1, a changing current in one will induce an emf in the other. According to Faraday's law, the emf $\mathscr{E}_2$ induced in coil 2 is proportional to the rate of change of flux passing through it. This flux is due to the current I_1 in coil 1, and it is often convenient to express the emf in coil 2 in terms of the current in coil 1.

We let Φ_{21} be the magnetic flux in each loop of coil 2 created by the current in coil 1. If coil 2 contains N_2 closely packed loops, then $N_2\Phi_{21}$ is the total flux passing through coil 2. If the two coils are fixed in space, $N_2\Phi_{21}$ is proportional to the current I_1 in coil 1; the proportionality constant is called the **mutual inductance**, M_{21}, defined by

$$M_{21} = \frac{N_2\Phi_{21}}{I_1}.$$ **(30–1)**

The emf $\mathscr{E}_2$ induced in coil 2 due to a changing current in coil 1 is, by Faraday's law,

$$\mathscr{E}_2 = -N_2\frac{d\Phi_{21}}{dt}.$$

We combine this with Eq. 30–1 rewritten as $\Phi_{21} = M_{21}I_1/N_2$ (and take its

FIGURE 30–1 A changing current in one coil will induce a current in the second coil.

Mutual inductance

Coil 1   Coil 2

I_1

$\mathscr{E}_2$
(induced)

derivative) and obtain

$$\mathcal{E}_2 = -M_{21} \frac{dI_1}{dt}. \qquad \text{(30–2)}$$

emf due to mutual inductance

This relates the change in current in coil 1 to the emf it induces in coil 2. The mutual inductance of coil 2 with respect to coil 1, M_{21}, is a "constant" in that it does not depend on I_1; M_{21} depends on "geometric" factors such as the size, shape, number of turns, and relative positions of the two coils, and also on whether iron (or other ferromagnetic material) is present. For example, the farther apart the two coils are in Fig. 30–1, the fewer lines of flux can pass through coil 2, so M_{21} will be less. For some arrangements, the mutual inductance can be calculated (see Example 30–1). More often it is determined experimentally.

Suppose, now, we consider the reverse situation: when a changing current in coil 2 induces an emf in coil 1. In this case,

$$\mathcal{E}_1 = -M_{12} \frac{dI_2}{dt}$$

where M_{12} is the mutual inductance of coil 1 with respect to coil 2. It is possible to show, although we will not prove it here, that $M_{12} = M_{21}$. Hence, for a given arrangement we do not need the subscripts and we can let

$$M = M_{12} = M_{21},$$

so that

$$\mathcal{E}_1 = -M \frac{dI_2}{dt} \qquad \text{(30–3a)}$$

and

$$\mathcal{E}_2 = -M \frac{dI_1}{dt}. \qquad \text{(30–3b)}$$

The SI unit for mutual inductance is the henry (H), where $1\,\text{H} = 1\,\text{V}\cdot\text{s/A} = 1\,\Omega\cdot\text{s}$.

EXAMPLE 30–1 **Solenoid and coil.** A long thin solenoid of length l and cross-sectional area A contains N_1 closely packed turns of wire. Wrapped around it is an insulated coil of N_2 turns, Fig. 30–2. Assume all the flux from coil 1 (the solenoid) passes through coil 2, and calculate the mutual inductance.

SOLUTION We first determine the flux produced by the solenoid, all of which passes uniformly through coil N_2: The magnetic field inside the solenoid was derived in Chapter 28, Eq. 28–4:

$$B = \mu_0 \frac{N_1}{l} I_1,$$

where I_1 is the current in the solenoid. The solenoid is close packed, so we assume all the flux in the solenoid links the coil. Then the flux Φ_{21} through coil 2 is

$$\Phi_{21} = BA = \mu_0 \frac{N_1}{l} I_1 A.$$

Hence the mutual inductance is

$$M = \frac{N_2 \Phi_{21}}{I_1} = \frac{\mu_0 N_1 N_2 A}{l}.$$

Note that we calculated M_{21}; if we had tried to calculate M_{12}, it would have been difficult. Given $M_{12} = M_{21} = M$, we can do the simpler calculation to obtain M. Note also that M depends only on geometric factors, and not on the currents.

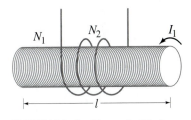

FIGURE 30–2 Example 30–1.

A transformer is an example of mutual inductance in which the coupling is maximized so nearly all flux lines pass through both coils. However, mutual inductance has other uses as well. For example some pacemakers, which are used to regulate the heartbeat so that blood flow is maintained in heart patients, are powered externally. Power in an external coil is transmitted via mutual inductance to a second coil in the pacemaker at the heart. This has the advantage over battery-powered pacemakers in that surgery is not needed to replace a battery when it wears out.

Mutual inductance can sometimes be a problem, however. Any changing current in a circuit can induce an emf in another part of the same circuit or in a different circuit even though the conductors are not in the shape of a coil. The mutual inductance M is usually small unless coils with many turns and/or iron cores are involved. However, in situations where small signals are present, problems due to mutual inductance often arise. Shielded cable, in which an inner conductor is surrounded by a cylindrical grounded conductor, is often used to reduce the problem.

30–2 Self-Inductance

The concept of inductance applies also to a single isolated coil of N turns. When a changing current passes through a coil (or solenoid), a changing magnetic flux is produced inside the coil, and this in turn induces an emf in that same coil. This induced emf opposes the change in flux (Lenz's law). For example, if the current through the coil is increasing, the increasing magnetic flux induces an emf that opposes the original current and tends to retard its increase. If the current is decreasing in the coil, the decreasing flux induces an emf in the same direction as the current, thus tending to maintain the original current.

The magnetic flux Φ_B passing through the N turns of a coil is proportional to the current I in the coil, so we define the **self-inductance** L (in analogy to mutual inductance, Eq. 30–1) as

$$L = \frac{N\Phi_B}{I}. \tag{30–4}$$

Then the emf $\mathscr{E}$ induced in a coil of self-inductance L is, from Faraday's law,

$$\mathscr{E} = -N\frac{d\Phi_B}{dt} = -L\frac{dI}{dt}. \tag{30–5}$$

Like mutual inductance, self-inductance is measured in henrys. The magnitude of L depends on the geometry and on the presence of a ferromagnetic material. Self-inductance can be defined, as above, for any circuit or part of a circuit.

Circuits always contain some inductance, but often it is quite small unless the circuit contains a coil of many turns. A coil that has significant self-inductance L is called an **inductor**. It is shown on circuit diagrams by the symbol

—⦚⦚⦚⦚—.   [inductor symbol]

It can serve a useful purpose in certain circuits. Often, inductance is to be avoided in a circuit. Precision resistors are normally wire wound and thus would have inductance as well as resistance. The inductance can be minimized by winding the insulated wire back on itself in the opposite sense so that the current going in opposite directions produces little net magnetic flux; this is called a "noninductive winding."

If an inductor has negligible resistance, it is the inductance (or induced emf) that controls a changing current. If a source of changing or alternating voltage is applied to the coil, this applied voltage will just be balanced by the induced emf of the coil given by Eq. 30–5. Thus we can see from Eq. 30–5 that, for a given $\mathscr{E}$, if the inductance L is large, the change in the current—and therefore the current itself if it is ac—will be small. The greater the inductance, the less the ac current. An inductance thus acts something like a resistance to impede the flow of alternating current. We use the term *reactance* or *impedance* for this quality of an inductor. We will discuss reactance and impedance more fully in Chapter 31, and we shall see that it depends not only on L, but also on the frequency. Here we mention one example of its importance. The resistance of the primary in a transformer is usually quite small, perhaps less than $1\,\Omega$. If resistance alone limited the current in a transformer, tremendous currents would flow when a high voltage was applied. Indeed, a dc voltage applied to a transformer can burn it out. It is the induced emf (or reactance) of the coil that limits the current to a reasonable value.

Common inductors have inductances in the range from about $1\,\mu H$ to about $1\,H$ (where $1\,H = 1\,\text{henry} = 1\,\Omega \cdot s$).

CONCEPTUAL EXAMPLE 30–2 **Direction of emf in inductor.** (*a*) Suppose current passes through the coil in Fig. 30–3 from left to right as shown, and is increasing with time. In which direction is the induced emf? (*b*) If the current passes through the coil in the same direction, but is decreasing in time, what then is the direction of the induced emf?

RESPONSE (*a*) From Lenz's Law we know that the induced emf must oppose the change in magnetic flux. If the current is increasing, so is the magnetic flux. The induced emf acts to oppose the increasing flux, which means it acts like a source of emf that opposes the outside source of emf driving the current. So the induced emf in the coil acts to oppose I in Fig. 30–3a. In other words the inductor might be thought of as a battery with a positive terminal at point A on the left and negative at point B on the right.
(*b*) If the current is decreasing, then by Lenz's Law the induced emf acts to bolster the flux—like a source of emf reinforcing the external emf. The induced emf acts to increase I in Fig. 30–3b, so in this situation you can think of the induced emf as a battery with its positive terminal on the right, at point B.

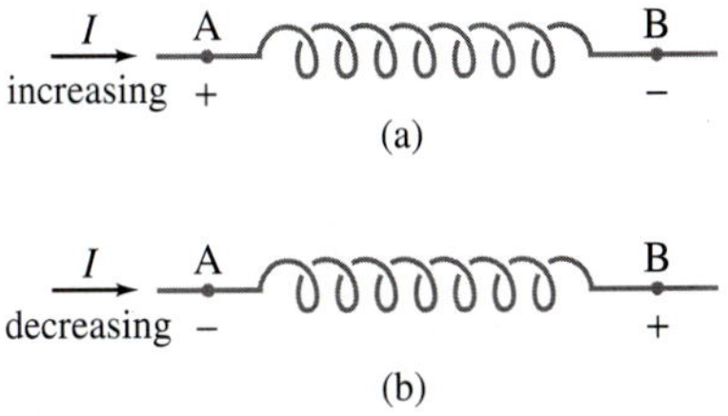

FIGURE 30–3 Example 30–2. The $+$ and $-$ signs refer to the induced emf, due to the changing current, as if points A and B were the terminals of a battery.

EXAMPLE 30–3 **Solenoid inductance.** (*a*) Determine a formula for the self-inductance L of a tightly wrapped solenoid (a long coil) containing N turns of wire in its length l and whose cross-sectional area is A. (*b*) Calculate the value of L if $N = 100$, $l = 5.0\,\text{cm}$, $A = 0.30\,\text{cm}^2$ and the solenoid is air filled. (*c*) Calculate L if the solenoid has an iron core with $\mu = 4000\,\mu_0$.

SOLUTION (*a*) To determine the inductance L, it is usually simplest to start with Eq. 30–4, so we need to first determine the flux. According to Eq. 28–4, the magnetic field inside a solenoid (ignoring end effects) is constant: $B = \mu_0 nI$ where $n = N/l$. The flux is $\Phi_B = BA = \mu_0 NIA/l$, so

$$L = \frac{N\Phi_B}{I} = \frac{\mu_0 N^2 A}{l}.$$

(*b*) Since $\mu_0 = 4\pi \times 10^{-7}\,\text{T}\cdot\text{m/A}$

$$L = \frac{(4\pi \times 10^{-7}\,\text{T}\cdot\text{m/A})(100)^2(3.0 \times 10^{-5}\,\text{m}^2)}{(5.0 \times 10^{-2}\,\text{m})} = 7.5\,\mu H.$$

(*c*) Here we replace μ_0 by $\mu = 4000\,\mu_0$ so L will be 4000 times larger: $L = 0.030\,H = 30\,\text{mH}$.

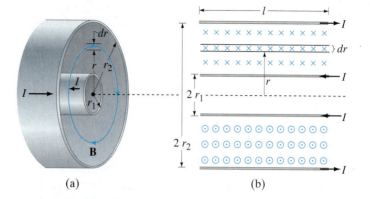

FIGURE 30–4 Example 30–4. Coaxial cable: (a) end view, (b) side view (cross-section).

(a)　　　　　　　(b)

EXAMPLE 30–4 **Coaxial cable inductance.** Determine the inductance per unit length of a coaxial cable whose inner conductor has a radius r_1 and the outer conductor has a radius r_2, Fig. 30–4. Assume the conductors are thin, so that the magnetic field within them can be ignored. The conductors carry equal currents I in opposite directions.

SOLUTION Again we need to find the magnetic flux, $\Phi_B = \int \mathbf{B} \cdot d\mathbf{A}$, this time between the conductors. The lines of $\mathbf{B}$ are circles surrounding the inner conductor (only one is shown in Fig. 30–4a). From Ampère's law, $\oint \mathbf{B} \cdot d\mathbf{l} = \mu_0 I$, the magnitude of the field at a distance r from the center, when the inner conductor carries a current I, is (see also Example 28–4):

$$B = \frac{\mu_0 I}{2\pi r}.$$

The magnetic flux through a rectangle of width dr and length l (along the cable, Fig. 30–4b), a distance r from the center, is

$$d\Phi_B = B(l\,dr) = \frac{\mu_0 I}{2\pi r} l\,dr.$$

The total flux in a length l of cable is

$$\Phi_B = \int d\Phi_B = \frac{\mu_0 Il}{2\pi} \int_{r_1}^{r_2} \frac{dr}{r} = \frac{\mu_0 Il}{2\pi} \ln \frac{r_2}{r_1}.$$

Since the current I all flows in one direction in the inner conductor, and the same current I all flows in the opposite direction in the outer conductor, we have only one turn, so $N = 1$. Hence the self-inductance for a length l is

$$L = \frac{\Phi_B}{I} = \frac{\mu_0 l}{2\pi} \ln \frac{r_2}{r_1}.$$

The inductance per unit length is

$$\frac{L}{l} = \frac{\mu_0}{2\pi} \ln \frac{r_2}{r_1}.$$

Note that L depends only on geometric factors and not on the current I.

30–3 | Energy Stored in a Magnetic Field

When an inductor of inductance L is carrying a current I which is changing at a rate dI/dt, energy is being supplied to the inductor at a rate

$$P = I\mathscr{E} = LI \frac{dI}{dt}$$

where P stands for power and we used[†] Eq. 30–5. Let us calculate the work needed to increase the current in an inductor from zero to some value I. Using this

---

[†]No minus sign here because we are supplying power to oppose the emf of the inductor.

last equation the work dW done in a time dt is

$$dW = P\,dt = LI\,dI.$$

Then the total work done to increase the current from zero to I is

$$W = \int dW = \int_0^I LI\,dI = \tfrac{1}{2}LI^2.$$

This work done is equal to the energy U stored in the inductor when it is carrying a current I (and we take $U = 0$ when $I = 0$):

$$U = \tfrac{1}{2}LI^2. \tag{30-6}$$

Energy stored in inductor

This can be compared to the energy stored in a capacitor, C, when the potential difference across it is V (see Section 24–4):

$$U = \tfrac{1}{2}CV^2.$$

Just as the energy stored in a capacitor can be considered to reside in the electric field between its plates, so the energy in an inductor can be considered to be stored in its magnetic field. To write the energy in terms of the magnetic field, let us use the result of Example 30–3, that the inductance of an ideal solenoid (end effects ignored) is $L = \mu_0 N^2 A / l$. Now the magnetic field B in a solenoid is related to the current I by $B = \mu_0 NI/l$. Thus

$$U = \tfrac{1}{2}LI^2 = \frac{1}{2}\left(\frac{\mu_0 N^2 A}{l}\right)\left(\frac{Bl}{\mu_0 N}\right)^2$$

$$= \frac{1}{2}\frac{B^2}{\mu_0}Al.$$

We can think of this energy as residing in the volume enclosed by the windings, which is Al. Then the energy per unit volume or *energy density* is

$$u = \text{energy density} = \frac{1}{2}\frac{B^2}{\mu_0}. \tag{30-7}$$

Energy density in magnetic field

This formula, which was derived for the special case of a solenoid, can be shown to be valid for any region of space where a magnetic field exists. If a ferromagnetic material is present, μ_0 is replaced by μ. This equation is analogous to that for an electric field, $\tfrac{1}{2}\epsilon_0 E^2$, Eq. 24–6.

EXAMPLE 30–5 **Energy stored in a coaxial cable.** (*a*) How much energy is being stored per unit length in a coaxial cable whose conductors have radii r_1 and r_2 (Example 30–4, Fig. 30–4) and which carry a current I? (*b*) Where is the energy density highest?

SOLUTION (*a*) The inductance per unit length, as we saw in Example 30–4, is

$$\frac{L}{l} = \frac{\mu_0}{2\pi}\ln\frac{r_2}{r_1}.$$

We want to find the energy stored per unit length, which is

$$\frac{U}{l} = \frac{\tfrac{1}{2}LI^2}{l} = \frac{\mu_0 I^2}{4\pi}\ln\frac{r_2}{r_1}.$$

(*b*) Since $B = \mu_0 I/2\pi r$, the field is greatest near the surface of the inner conductor, so the energy density, $u = B^2/2\mu_0$, will be greatest there.

30–4 *LR* Circuits

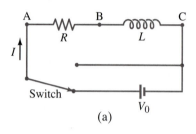

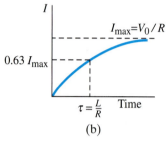

FIGURE 30–5 (a) *LR* circuit; (b) growth of current when connected to battery.

Any inductor will have some resistance. We represent an inductor by drawing its inductance L and its resistance R separately, as in Fig. 30–5a. The resistance R could also include a separate resistor connected in series. Now we ask, what happens when a battery or other source of dc voltage V_0 is connected in series to such an LR circuit? At the instant the switch connecting the battery is closed, the current starts to flow. The current increases starting from zero. It is opposed by the induced emf in the inductor which means point B in Fig. 30–5 is positive relative to point C. However, as soon as current starts to flow, there is also a voltage ($= IR$) across the resistance. Hence the voltage applied across the inductance is reduced and the current increases less rapidly. The current thus rises gradually as shown in Fig. 30–5b, and approaches the steady value $I_{max} = V_0/R$ when all the voltage drop is across the resistance.

We can show this analytically by applying Kirchhoff's loop rule to the circuit of Fig. 30–5a. The emf's in the circuit are the battery voltage V_0, and the emf $\mathcal{E} = -L(dI/dt)$ in the inductor opposing the increasing current. Hence the sum of the potential changes around the loop is

$$V_0 - L\frac{dI}{dt} - IR = 0,$$

where I is the current in the circuit at any instant. We rearrange this to obtain

$$L\frac{dI}{dt} + RI = V_0. \tag{30-8}$$

This is a linear differential equation and can be integrated in the same way we did in Section 26–4 for an RC circuit. We rewrite Eq. 30–8 and then integrate:

$$\int_{I=0}^{I} \frac{dI}{V_0 - IR} = \int_0^t \frac{dt}{L}.$$

Then

$$-\frac{1}{R}\ln\left(\frac{V_0 - IR}{V_0}\right) = \frac{t}{L}$$

or,

$$I = \frac{V_0}{R}\left(1 - e^{-t/\tau}\right) \tag{30-9}$$

where

$$\tau = \frac{L}{R} \tag{30-10}$$

Current in LR circuit when battery connected at $t = 0$

Inductive time constant, $\tau = L/R$

is the **time constant** of the LR circuit. The symbol τ represents the time required for the current I to reach $(1 - 1/e) = 0.63$ or 63 percent of its maximum value (V_0/R). Equation 30–9 is plotted in Fig. 30–5b. (Compare to the RC circuit, Section 26–4.)

Now let us flip the switch in Fig. 30–5a so that the battery is taken out of the circuit, and points A and C are connected together as shown in Fig. 30–6 at the moment when the switching occurs (call it $t = 0$) and the current is I_0. Then the differential equation (Eq. 30–8) becomes (since $V_0 = 0$):

$$L\frac{dI}{dt} + RI = 0.$$

FIGURE 30–6 The switch is flipped quickly so the battery is removed but we still have a circuit. The current at this moment (call it $t = 0$) is I_0.

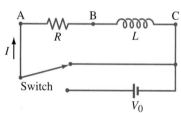

We rearrange this equation and integrate:

$$\int_{I_0}^{I} \frac{dI}{I} = -\int_0^t \frac{R}{L}\,dt$$

where $I = I_0$ at $t = 0$, and $I = I$ at time t.

We integrate this last equation to obtain

$$\ln \frac{I}{I_0} = -\frac{R}{L}t$$

or

$$I = I_0 e^{-t/\tau} \tag{30–11}$$

where again the time constant is $\tau = L/R$. The current thus decays exponentially to zero as shown in Fig. 30–7.

This analysis shows that there is always some "reaction time" when an electromagnet, for example, is turned on or off. We also see that an LR circuit has properties similar to an RC circuit (Section 26–4). Unlike the capacitor case, however, the time constant here is *inversely* proportional to R.

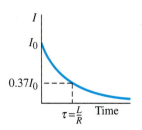

FIGURE 30–7 Decay of the current in Fig. 30–6 in time after the battery is removed from the circuit.

EXAMPLE 30–6 **An *LR* circuit.** At $t = 0$, a 12.0-V battery is connected in series with a 30-Ω resistor and a 220-mH inductor, as shown in Fig. 30–8. (*a*) What is the current at $t = 0$? (*b*) What is the time constant? (*c*) What is the maximum current? (*d*) How long will it take the current to reach half its maximum possible value? (*e*) At this instant, at what rate is energy being delivered by the battery, and (*f*) at what rate is energy being stored in the inductor's magnetic field?

SOLUTION (*a*) The current cannot instantaneously jump from zero to some other value when the switch is closed because the inductor opposes the change $(\mathscr{E}_L = -L(dI/dt))$. Hence just after the switch is closed I is still zero at $t = 0$ and then begins to increase.
(*b*) The time constant is, from Eq. 30–10, $\tau = L/R = (0.22\,\text{H})/(30\,\Omega) = 7.3\,\text{ms}$.
(*c*) The current reaches its maximum steady value after a long time, when $dI/dt = 0$ so $I_{max} = V_0/R = 12.0\,\text{V}/30\,\Omega = 0.40\,\text{A}$.
(*d*) We set $I = \frac{1}{2}I_{max} = V_0/2R$ in Eq. 30–9, which gives us

$$1 - e^{-t/\tau} = \tfrac{1}{2}$$

or

$$e^{-t/\tau} = 1 - \tfrac{1}{2} = \tfrac{1}{2}.$$

We solve for t:

$$t = \tau \ln 2 = (7.3 \times 10^{-3}\,\text{s})(0.69) = 5.0\,\text{ms}.$$

(*e*) At this instant, $I = I_{max}/2 = 200\,\text{mA}$, so the power being delivered by the battery is

$$P = IV = (0.20\,\text{A})(12\,\text{V}) = 2.4\,\text{W}.$$

(*f*) From Eq. 30–6, the energy stored in an inductor L at any instant is

$$U = \tfrac{1}{2}LI^2$$

where I is the current in the inductor at that instant. The *rate* at which the energy changes is

$$\frac{dU}{dt} = LI\frac{dI}{dt}.$$

We can differentiate Eq. 30–9 to obtain dI/dt, or use the differential equation, Eq. 30–8, directly:

$$\frac{dU}{dt} = I\left(L\frac{dI}{dt}\right) = I(V_0 - RI)$$

$$= (0.20\,\text{A})\big[12\,\text{V} - (30\,\Omega)(0.20\,\text{A})\big] = 1.2\,\text{W}.$$

Since only part of the battery's power is feeding the inductor at this instant, where is the rest going?

FIGURE 30–8 Example 30–7.

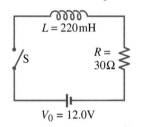

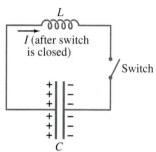

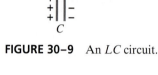

FIGURE 30–9 An *LC* circuit.

In any electric circuit, there can be three basic components: resistance, capacitance, and inductance, in addition to a source of emf. (There can also be more complex components, such as diodes or transistors.) We have previously discussed both *RC* and *LR* circuits. Now we look at an *LC* circuit, one that contains only a capacitance *C* and an inductance, *L*, Fig. 30–9. This is an idealized circuit in which we assume there is no resistance in the inductor; in the next Section we include resistance. Let us suppose the capacitor in Fig. 30–9 is initially charged so that one plate has charge Q_0 and the other plate has charge $-Q_0$. Suppose that at $t = 0$, the switch is closed. The capacitor immediately begins to discharge. As it does so, the current *I* through the inductor increases. We now apply Kirchhoff's loop rule (sum of potential changes around a loop is zero):

$$-L\frac{dI}{dt} + \frac{Q}{C} = 0.$$

Because charge leaves the positive plate on the capacitor to produce the current *I* as shown in Fig. 30–9, the charge *Q* on the (positive) plate of the capacitor is decreasing, so $I = -dQ/dt$. We can then rewrite the above equation as

$$\frac{d^2Q}{dt^2} + \frac{Q}{LC} = 0. \tag{30–12}$$

This is a familiar differential equation. It has the same form as the equation for simple harmonic motion (Chapter 14, Eq. 14–3). The solution of Eq. 30–12 is

$$Q = Q_0\cos(\omega t + \phi) \tag{30–13}$$

where Q_0 and ϕ are constants that depend on the initial conditions. We insert Eq. 30–13 into Eq. 30–12, noting that $d^2Q/dt^2 = -\omega^2 Q_0\cos(\omega t + \phi)$; thus

$$-\omega^2 Q_0\cos(\omega t + \phi) + \frac{1}{LC}Q_0\cos(\omega t + \phi) = 0$$

or

$$\left(-\omega^2 + \frac{1}{LC}\right)\cos(\omega t + \phi) = 0.$$

This relation can be true for all times *t* only if $\left(-\omega^2 + 1/LC\right) = 0$, which tells us that

$$\omega = 2\pi f = \sqrt{\frac{1}{LC}}. \tag{30–14}$$

Equation 30–13 shows that the charge on the capacitor in an *LC* circuit oscillates sinusoidally. The current in the inductor is

$$I = -\frac{dQ}{dt} = \omega Q_0\sin(\omega t + \phi)$$

$$= I_0\sin(\omega t + \phi); \tag{30–15}$$

so the current too is sinusoidal. The maximum value of *I* is $I_0 = \omega Q_0 = Q_0/\sqrt{LC}$. Equations 30–14 and 30–15 for *Q* and *I* when $\phi = 0$, are plotted in Fig. 30–10.

Now let us look at *LC* oscillations from the point of view of energy. The energy stored in the electric field of the capacitor at any time *t* is (see Eq. 24–5):

$$U_E = \frac{1}{2}\frac{Q^2}{C} = \frac{Q_0^2}{2C}\cos^2(\omega t + \phi).$$

LC angular frequency

FIGURE 30–10 Charge *Q* and current *I* in an *LC* circuit. The period $T = \dfrac{1}{f} = \dfrac{2\pi}{\omega} = 2\pi\sqrt{LC}$.

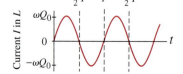

The energy stored in the magnetic field of the inductor at the same instant is (Eq. 30–6)

$$U_B = \frac{1}{2}LI^2 = \frac{L\omega^2 Q_0^2}{2}\sin^2(\omega t + \phi) = \frac{Q_0^2}{2C}\sin^2(\omega t + \phi)$$

where we used Eq. 30–14. If we let $\phi = 0$, then at times $t = 0$, $t = \frac{1}{2}T$, $t = T$, and so on (where T is the period $= 1/f = 2\pi/\omega$), we have $U_E = Q_0^2/2C$ and $U_B = 0$. That is, all the energy is stored in the electric field of the capacitor. But at $t = \frac{1}{4}T, \frac{3}{4}T$, and so on, $U_E = 0$ and $U_B = Q_0^2/2C$, and so all the energy is stored in the magnetic field of the inductor. At any time t, the total energy is

$$U = U_E + U_B = \frac{1}{2}\frac{Q^2}{C} + \frac{1}{2}LI^2$$

$$= \frac{Q_0^2}{2C}\left[\cos^2(\omega t + \phi) + \sin^2(\omega t + \phi)\right] = \frac{Q_0^2}{2C}. \qquad \textbf{(30–16)} \qquad \textit{Energy}$$

Hence the total energy is constant, and energy is conserved.

What we have in this LC circuit is an **LC oscillator** or **electromagnetic oscillation**. The charge Q oscillates back and forth, from one plate of the capacitor to the other, and repeats this continuously. Likewise, the current oscillates back and forth as well. They are also energy oscillations: when Q is a maximum, the energy is all stored in the electric field of the capacitor; but when Q reaches zero, the current I is a maximum and all the energy is stored in the magnetic field of the inductor. Thus the energy oscillates between being stored in the electric field of the capacitor and in the magnetic field of the inductor. See Fig. 30–11.

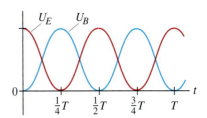

FIGURE 30–11 Energy U_E (red line) and U_B (blue line) stored in the capacitor and the inductor as a function of time. Note how the energy oscillates between electric and magnetic.

EXAMPLE 30–7 **LC circuit.** A 1200-pF capacitor is fully charged by a 500-V dc power supply. It is disconnected from the power supply and is connected, at $t = 0$, to a 75-mH inductor. Determine: (a) the initial charge on the capacitor; (b) the maximum current; (c) the frequency f and period T of oscillation; and (d) the total energy oscillating in the system.

SOLUTION (a) The 500-V power supply, before being disconnected, charged the capacitor to a charge of

$$Q_0 = CV = (1.2 \times 10^{-9}\,\text{F})(500\,\text{V}) = 6.0 \times 10^{-7}\,\text{C}.$$

(b) The maximum current, I_{max}, is (see Eq. 30–15)

$$I_{\text{max}} = \omega Q_0 = \frac{Q_0}{\sqrt{LC}} = \frac{(6.0 \times 10^{-7}\,\text{C})}{\sqrt{(0.075\,\text{H})(1.2 \times 10^{-9}\,\text{F})}} = 63\,\text{mA}.$$

(c) Equation 30–14 gives us the frequency:

$$f = \frac{\omega}{2\pi} = \frac{1}{(2\pi\sqrt{LC})} = 17\,\text{kHz},$$

and the period T is

$$T = \frac{1}{f} = 6.0 \times 10^{-5}\,\text{s}.$$

(d) Finally the total energy (Eq. 30–16) is

$$U = \frac{Q_0^2}{2C} = \frac{(6.0 \times 10^{-7}\,\text{C})^2}{2(1.2 \times 10^{-9}\,\text{F})} = 1.5 \times 10^{-4}\,\text{J}.$$

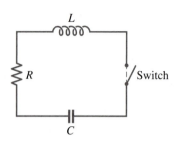

FIGURE 30–12 An *LRC* circuit.

30–6 | *LC* Oscillations with Resistance (*LRC* Circuit)

The *LC* circuit discussed in the previous Section is an idealization. There is always some resistance *R* in any circuit, and so we now discuss such a simple *LRC* circuit, Fig. 30–12.

Suppose again that the capacitor is initially given a charge Q_0 and the battery or other source is then removed from the circuit. The switch is closed at $t = 0$. Since there is now a resistance in the circuit, we expect some of the energy to be converted to thermal energy, and so we don't expect undamped oscillations as in a pure *LC* circuit. Indeed, if we use Kirchhoff's loop rule around this circuit, we obtain

$$-L\frac{dI}{dt} - IR + \frac{Q}{C} = 0,$$

which is the same equation we had in Section 30–5 with the addition of the voltage drop *IR* across the resistor. Since $I = -dQ/dt$, as we saw in Section 30–5, this equation becomes

$$L\frac{d^2Q}{dt^2} + R\frac{dQ}{dt} + \frac{1}{C}Q = 0. \tag{30-17}$$

This second-order differential equation in the variable *Q* has precisely the same form as that for the damped harmonic oscillator, Eq. 14–15:

$$m\frac{d^2x}{dt^2} + b\frac{dx}{dt} + kx = 0.$$

FIGURE 30–13 Charge *Q* on the capacitor in an *LRC* circuit as a function of time: curve A is for underdamped oscillation $(R^2 < 4L/C)$, curve B is for critically damped, and curve C is for overdamped.

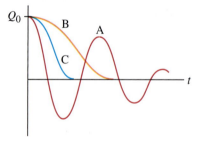

Angular frequency (damped LC)

Hence we can analyze our *LRC* circuit in the same way as for damped harmonic motion, Section 14–7. Our system may undergo damped oscillations, curve A in Fig. 30–13 (underdamped system), or it may be critically damped (curve B), or overdamped (curve C), depending on the relative values of *R*, *L*, and *C*. Using the results of Section 14–7, with *m* replaced by *L*, *b* by *R*, and *k* by C^{-1}, we find that the system will be underdamped when

$$R^2 < \frac{4L}{C},$$

and overdamped for $R^2 > 4L/C$. Critical damping (curve B in Fig. 30–13) occurs when $R^2 = 4L/C$. If *R* is smaller than $\sqrt{4L/C}$, the angular frequency, ω', will be

$$\omega' = \sqrt{\frac{1}{LC} - \frac{R^2}{4L^2}} \tag{30-18}$$

(compare to Eq. 14–20). And the charge *Q* as a function of time will be

$$Q = Q_0 e^{-\frac{R}{2L}t} \cos(\omega't + \phi) \tag{30-19}$$

where ϕ is a phase constant (compare to Eq. 14–19).

Oscillators are an important element in many electronic devices: radios and television sets use them for tuning, tape recorders use them (the "bias frequency") when recording, and so on. Because some resistance is always present, electrical oscillators generally need a periodic input of power to compensate for the energy converted to thermal energy in the resistance.

EXAMPLE 30–8 Damped oscillations. At $t = 0$, a 40-mH inductor is placed in series with a resistance $R = 3.0\,\Omega$ and a charged capacitor $C = 4.8\,\mu\text{F}$. (*a*) Show that this circuit will oscillate. (*b*) Determine the frequency. (*c*) What is the time required for the charge amplitude to drop to half its starting value? (*d*) What is the current amplitude? (*e*) What value of *R* will make the circuit nonoscillating?

SOLUTION (*a*) To be underdamped, we must have $R^2 < 4L/C$. Since $R^2 = 9.0\,\Omega^2$ and $4L/C = 4(0.040\,\text{H})/(4.8 \times 10^{-6}\,\text{F}) = 3.3 \times 10^4\,\Omega^2$, this relation is satisfied, so the circuit will oscillate.

(*b*) We use Eq. 30–18:

$$f' = \frac{\omega'}{2\pi} = \frac{1}{2\pi}\sqrt{\frac{1}{LC} - \frac{R^2}{4L^2}} = 2.3 \times 10^3\,\text{Hz}.$$

(*c*) From Eq. 30–19, the amplitude will be half when

$$e^{-\frac{R}{2L}t} = \tfrac{1}{2}$$

or

$$t = \frac{2L}{R}\ln 2 = 18\,\text{ms}.$$

(*d*) We differentiate Eq. 30–19 to obtain $I = -dQ/dt$ (see comment just before Eq. 30–12), noting that $\phi = 0$ since $Q = Q_0$ at $t = 0$:

$$I = -\frac{dQ}{dt} = Q_0 e^{-\frac{R}{2L}t}\left(\frac{R}{2L}\cos\omega't + \frac{1}{\sqrt{LC}}\sin\omega't\right)$$

where we approximated $\omega' \approx \sqrt{1/LC}$. Because R is much less than $\sqrt{4L/C}$ (see part *a* above), we can ignore the $\cos\omega't$ term, so

$$I \approx \frac{Q_0}{\sqrt{LC}}e^{-\frac{R}{2L}t}\sin\omega't.$$

Hence the initial current amplitude is $Q_0/\sqrt{LC}$. We are not given Q_0, but if it were, say, $1.0\,\mu\text{C}$, then $I_0 = (1.0 \times 10^{-6}\,\text{C})/\sqrt{(40 \times 10^{-3}\,\text{H})(4.8 \times 10^{-6}\,\text{F})} = 2.3\,\text{mA}$.

(*e*) To make the circuit critically damped or overdamped, we must use the criterion $R^2 \geq 4L/C = 3.3 \times 10^4\,\Omega^2$. Hence we must have $R \geq 180\,\Omega$.

Summary

A changing current in a coil of wire will induce an emf in a second coil placed nearby. The **mutual inductance**, M, is defined as the proportionality constant between the induced emf $\mathscr{E}_2$ in the second coil and the time rate of change of current in the first:

$$\mathscr{E}_2 = -M\,dI_1/dt.$$

We can also write M as

$$M = \frac{N\Phi_B}{I}$$

where Φ_B is the magnetic flux through one coil (or circuit) with N loops produced by the current I in a second coil or circuit.

Within a single coil, a changing current induces an opposing emf, $\mathscr{E}$, so a coil has a **self-inductance** L defined by

$$\mathscr{E} = -L\,dI/dt.$$

This induced emf acts as an *impedance* to the flow of an alternating current. We can also write L as

$$L = N\frac{\Phi_B}{I}$$

where Φ_B is the flux through the inductance when a current I flows in its N loops.

When the current in an inductance L is I, the energy stored in the inductance is given by

$$U = \tfrac{1}{2}LI^2.$$

This energy can be thought of as being stored in the mag-

netic field of the inductor. The energy density u in any magnetic field B is given by

$$u = \tfrac{1}{2}\frac{B^2}{\mu},$$

where μ is the magnetic permeability in that region.

When an inductance L and resistor R are connected in series to a source of emf, V_0, the current rises according to an exponential of the form

$$I = \frac{V_0}{R}\left(1 - e^{-t/\tau}\right),$$

where

$$\tau = L/R$$

is the time constant. The current eventually levels out at $I = V_0/R$. If the battery is suddenly switched out of the **LR circuit**, and the circuit remains complete, the current drops exponentially, $I = I_0 e^{-t/\tau}$, with the same time constant τ.

The current in a pure **LC circuit** (or charge on the capacitor) would oscillate sinusoidally. The energy too would oscillate back and forth between electric and magnetic, from the capacitor to the inductor, and back again. The current in a series **LRC** circuit (or charge on the capacitor), in which the capacitor at some instant is charged, can undergo damped oscillations or undergo critically damped or overdamped oscillations.

Questions

1. In situations where a small signal must travel over a distance, a "shielded cable" is used in which the signal wire is surrounded by an insulator and then enclosed by a cylindrical conductor. Why is a "shield" necessary?

2. What is the advantage of placing the two electric wires carrying ac close together?

3. The primary of a transformer on a telephone pole has a resistance of $0.10\,\Omega$ and the input voltage is 2400 V ac. Can you estimate the current that will flow? Will it be 24,000 A? Explain.

4. A transformer designed for a 120-V ac input will often "burn out" if connected to a 120-V dc source. Explain. (*Hint*: The resistance of the primary coil is usually very low.)

5. How would you arrange two flat circular coils so that their mutual inductance was (*a*) greatest, (*b*) least (without separating them by a great distance)?

6. If the two coils in Fig. 30–1 were connected electrically, would there still be mutual inductance?

7. Suppose the second coil of N_2 turns in Fig. 30–2 were moved so it was near the end of the solenoid. How would this affect the mutual inductance?

8. Would two circuits with mutual inductance also have self-inductance? Explain.

9. Is the self-inductance per unit length of a solenoid greater near its center or near its ends?

10. Is the energy density greatest near the ends of a solenoid or near its center?

11. If you are given a fixed length of wire, how would you shape it to obtain the greatest self-inductance? The least?

12. In a battery, when the current is in the same direction as the emf, the energy of the battery decreases, whereas if the current is in the opposite direction, the energy of the battery increases (as in charging a battery). Is this also true for an inductor?

13. Two solenoids have the same length and circular cross-sectional area. Both consist of tightly packed turns of wire, but one uses thicker wire than the other. Which has the greater inductance? Which has the greater inductive time constant?

14. Does the emf of the battery in Fig. 30–5a affect the time needed for the *LR* circuit to reach (*a*) a given fraction of its maximum possible current, (*b*) a given value of current?

15. A circuit with large inductive time constant carries a steady current. If a switch is opened, there can be a very large (and sometimes dangerous) spark or "arcing over." Explain.

16. At the instant the battery is connected into the *LR* circuit of Fig. 30–5a, the emf in the inductor has its maximum value even though the current is zero. Explain.

17. Explain physically why we might expect the time constant of an *LR* circuit to be proportional to *L* and inversely proportional to *R*, Eq. 30–10.

18. Explain how a solenoid whose ends are connected together can by itself oscillate like an *LC* circuit. Where does the capacitance come from?

19. What keeps an *LC* circuit oscillating even after the capacitor has discharged completely?

20. Is the current in the inductor always the same as that in the resistor of the *LRC* circuit of Fig. 30–12?

Problems

Section 30–1

1. (I) Suppose the second coil of Example 30–1 (Fig. 30–2) has twice the diameter of the solenoid, but is still concentric with it. What then will be the mutual inductance? Assume the solenoid is very long.

2. (II) A 2.44-m-long coil containing 300 loops is wound on an iron core $\left(\text{average } \mu = 2000\mu_0\right)$ along with a second coil of 100 loops. The loops of each coil have a radius of 2.00-cm. If the current in the first coil drops uniformly from 12.0 A to zero in 98.0 ms, determine: (*a*) the mutual inductance M; (*b*) the emf induced in the second coil.

3. (II) Determine the mutual inductance per unit length between two long solenoids, one inside the other, whose radii are r_1 and $r_2(r_2 < r_1)$ and whose turns per unit length are n_1 and n_2.

4. (III) A small thin coil with N_2 loops, each of area A_2, is placed inside a long solenoid, near its center. The solenoid has N_1 loops in its length l and has area A_1. Determine the mutual inductance as a function of θ, the angle between the plane of the small coil and the axis of the solenoid.

5. (III) A long straight wire and a small rectangular wire loop lie in the same plane, Fig. 30–14. Determine the mutual inductance in terms of l_1, l_2, and w. Assume the wire is very long compared to l_1, l_2 and w, and that the rest of its circuit is very far away compared to l_1, l_2 and w.

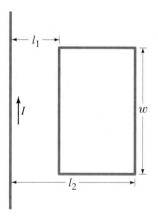

FIGURE 30–14 Problem 5.

Section 30–2

6. (I) If the current in a 180-mH coil changes steadily from 20.0 mA to 38.0 mA in 340 ms, what is the induced emf?

7. (I) Estimate the inductance L of a 0.45-m-long air-filled coil 3.7 cm in diameter containing 20,000 loops.

8. (I) What is the inductance of a coil if it produces an emf of 8.50 V when the current in it changes uniformly from −22.0 mA to +23.0 mA in 21.0 ms?

9. (I) What is the inductance of a coaxial cable 22.0 m long if its inner and outer conductors have diameters of 2.0 mm and 3.5 mm?

10. (I) A 35-V emf is induced in a 150-mH coil by a current that rises uniformly from zero to I_0 in 3.0 ms. What is the value of I_0?

11. (II) If the outer conductor of a coaxial cable has radius 3.0 mm, what should be the radius of the inner conductor so that the inductance per unit length does not exceed 40 nH per meter?

12. (II) (a) Show that if two circuits, such as the coils in Fig. 30–1, carry currents I_1 and I_2, the magnetic flux through each is $\Phi_1 = L_1 I_1 + M I_2$ and $\Phi_2 = L_2 I_2 + M I_1$. (b) Determine a formula for the emf induced in each coil in terms of the rate of change of current in the two coils.

13. (II) The wire of a tightly wound solenoid is unwound and used to make another tightly wound solenoid of 3.0 times the diameter. By what factor does the inductance change?

14. (II) A coil has 2.70-Ω resistance and 0.418-H inductance. If the current is 5.00 A and is increasing at a rate of 4.50 A/s, what is the potential difference across the coil at this moment?

15. (II) Ignoring any mutual inductance, what is the equivalent inductance of two inductors connected (a) in series, (b) in parallel?

16. (II) A toroid has a rectangular cross-section as shown in Fig. 30–15. Show that the self-inductance is

$$ L = \frac{\mu_0 N^2 h}{2\pi} \ln \frac{r_2}{r_1} $$

where N is the total number of turns and r_1, r_2, and h are the dimensions shown in Fig. 30–15. [Hint: Use Ampère's law to get B as a function of r inside the toroid, and integrate.]

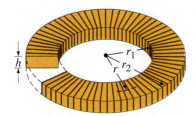

FIGURE 30–15 Problems 16 and 22. A toroid of rectangular cross-section, with N turns carrying a current I.

Section 30–3

17. (I) The magnetic field inside an air-filled solenoid 32.0 cm long and 2.10 cm in diameter is 0.600 T. Approximately how much energy is stored in this field?

18. (I) How much energy is stored in a 400-mH inductor at an instant when the current is 9.0 A?

19. (I) Typical large values for electric and magnetic fields attained in laboratories are about 1.0×10^4 V/m and 2.0 T. (a) Determine the energy density for each field and compare. (b) What magnitude electric field would be needed to produce the same energy density as the 2.0-T magnetic field?

20. (II) What is the energy density at the center of a circular loop of wire carrying a 30-A current if the radius of the loop is 28.0 cm?

21. (II) Calculate the magnetic and electric energy densities at the surface of a 3.0-mm-diameter copper wire carrying a 25-A current.

22. (II) For the toroid of Fig. 30–15, determine the energy density in the magnetic field as a function of $r(r_1 < r < r_2)$ and integrate this over the volume to obtain the total energy stored in the toroid, which carries a current I in each of its N loops.

23. (II) Determine the total energy stored per unit length in the magnetic field between the coaxial cylinders of a coaxial cable by using Eq. 30–7 for the energy density and integrating over the volume. Compare your answer to that obtained in Example 30–5.

Section 30–4

24. (II) After how many time constants does the current in Fig. 30–5 reach within (a) 10 percent, (b) 1.0 percent, and (c) 0.10 percent of its maximum value?

25. (II) How many time constants does it take for the potential difference across the resistor in an LR circuit like that in Fig. 30–6 to drop to 1.0 percent of its original value?

26. (II) In Example 30–6, where is the power delivered by the battery going besides into the inductor? Do a calculation to show that energy is conserved.

27. (II) Determine dI/dt at $t = 0$ (when the battery is connected) for the circuit of Fig. 30–5a and show that if I continued to increase at this rate, it would reach its maximum value in one time constant.

28. (II) It takes 2.56 ms for the current in an LR circuit to increase from zero to half its maximum value. Determine (a) the time constant of the circuit, (b) the resistance of the circuit if $L = 310$ H.

29. (II) (a) Determine the energy stored in the inductor L as a function of time for the LR circuit of Fig. 30–5a. (b) After how many time constants does the stored energy reach 99 percent of its maximum value?

30. (II) In the circuit of Fig. 30–16, determine the current in each resistor (I_1, I_2, I_3) at the moment (a) the switch is closed, (b) a long time after the switch is closed. After the switch has been closed a long time, and reopened, what is each current (c) just after it is opened, (d) after a long time?

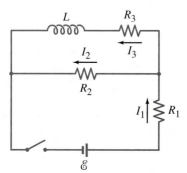

FIGURE 30–16
Problem 30.

Section 30–5

31. (I) The variable capacitor in the tuner of an AM radio has a capacitance of 1800 pF when the radio is tuned to a station at 550 kHz. (a) What must be the capacitance for a station at 1600 kHz? (b) What is the inductance (assumed constant)?

32. (I) (a) If the initial conditions of an LC circuit were $I = I_0$ and $Q = 0$ at $t = 0$, write Q as a function of time. (b) Practically, how could you set up these initial conditions?

33. (I) Use the definitions of the farad and henry to show that $1/\sqrt{LC}$ has units of s^{-1}.

34. (II) A 760-pF capacitor is charged to 135 V and then quickly connected to a 175-mH inductor. Determine (a) the frequency of oscillation, (b) the peak value of the current, and (c) the maximum energy stored in the magnetic field of the inductor.

35. (II) At $t = 0$, $Q = Q_0$ and $I = 0$ in an LC circuit. At the first moment when the energy is shared equally by the inductor and the capacitor, (a) what is the charge on the capacitor? (b) How much time has elapsed (in terms of the period T)?

Section 30–6

36. (II) In an oscillating LRC circuit, how much time does it take for the energy stored in the fields of the capacitor and inductor to fall to half its initial value? (See Fig. 30–12; assume $R \ll \sqrt{4L/C}$.)

37. (II) A damped LC circuit loses 5.5 percent of its electromagnetic energy per cycle to thermal energy. If $L = 65$ mH and $C = 1.00\,\mu F$, what is the value of R?

38. (II) (a) Obtain a differential equation for the current I in the LRC circuit of Fig. 30–12 by differentiating Eq. 30–17. (b) Obtain a general solution for I, assuming the circuit is connnected at $t = 0$ with charge Q_0 on the capacitor. Assume $R \ll \sqrt{4L/C}$. (c) Compare your result to part (d) of Example 30–8, and note any differences. (d) What complications would arise if R were not much less than $\sqrt{4L/C}$, say $R \approx \sqrt{L/C}$?

39. (II) Starting with Eq. 30–19, with $\phi = 0$, show that the current I in a lightly damped LRC circuit is given by

$$I \approx \frac{Q_0}{\sqrt{LC}} e^{-\frac{R}{2L}t} \sin(\omega' t + \delta)$$

where

$$\delta = \tan^{-1}\frac{R}{2L\omega'}.$$

40. (II) Show that the phase constant ϕ in Eq. 30–19 is given by

$$\phi = -\cot^{-1}\left(\frac{4L}{R^2C}\right)^{\frac{1}{2}}$$

if $I = 0$ at $t = 0$ in the circuit of Fig. 30–12. Assume that $R^2 \ll 4L/C$.

41. (III) How much resistance must be added to a pure LC circuit ($L = 300$ mH, $C = 1800$ pF) to change the oscillator's frequency by 0.10 percent? Will it be increased or decreased?

42. (III) Suppose a battery of fixed voltage V_0 is connected in the LRC circuit of Fig. 30–12. At $t = 0$, the switch is closed. Assuming this is a very overdamped circuit ($R^2 \gg 4L/C$), (a) determine the current as a function of time. (b) Plot I vs. t. (c) Compare your result to a pure RC circuit, for which $L = 0$. Real RC circuits always have some inductance, so the result of this Problem is more realistic than a pure RC circuit.

▢ General Problems

43. An air-filled cylindrical inductor has 3000 turns, and it is 2.5 cm in diameter and 28.2 cm long. Ignoring end effects, what is its inductance? How many turns would you need to generate the same inductance if the core were iron-filled instead? Assume the magnetic permeability of iron is about 1000 times that of free space.

44. At $t = 0$, the current through a 60.0 mH inductor is 50.0 mA and is increasing at the rate of 100 mA/s. What is the initial energy stored in the inductor, and how long does it take for the energy to increase by a factor of 10 from the initial value?

45. A 3000-pF capacitor is charged to 120 V and then quickly connected to an inductor. The frequency of oscillation is observed to be 20 kHz. Determine (a) the inductance, (b) the peak value of the current, and (c) the maximum energy stored in the magnetic field of the inductor.

46. Estimate the mutual inductance M between two small loops, of radius r_1 and r_2, which are separated by a distance l that is large compared to r_1 and r_2, Fig. 30–17. Give M as a function of θ, the angle between the planes of the two coils. Assume the line joining their centers is perpendicular to the plane of coil 1.

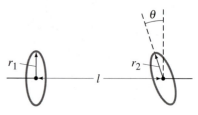

FIGURE 30–17 Problem 46.

47. Approximately how many turns does an air-filled coil have if it is 2.2 cm in diameter and 17.0 cm long and its inductance is 25 mH? How many turns are needed if it has an iron core and $\mu = 10^3 \mu_0$?

48. (a) Show that the self-inductance L of a toroid (Fig. 30–18) of radius r_0 containing N loops each of diameter d is

$$L \approx \frac{\mu_0 N^2 d^2}{8r_0}$$

if $r_0 \gg d$. Assume the field is uniform inside the toroid; is this actually true? Is this result consistent with L for a solenoid? Should it be? (b) Calculate the inductance L of a large toroid if the diameter of the coils is 2.0 cm and the diameter of the whole ring is 50 cm. Assume the field inside the toroid is uniform. There are a total of 550 loops of wire.

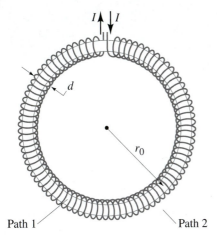

FIGURE 30–18 A toroid, Problem 48.

49. A pair of straight parallel thin wires, such as a lamp cord, each of radius r, are a distance l apart and carry current to a circuit some distance away. Ignoring the field within each wire, show that the inductance per unit length is $(\mu_0/\pi) \ln[(l - r)/r]$.

50. The potential difference across a given coil is 2.55 V at an instant when the current is 360 mA and is changing at a rate of 340 mA/s. At a later instant, the potential difference is 1.82 V while the current is 420 mA and is decreasing at a rate of 180 mA/s. Determine the inductance and resistance of the coil.

51. (a) Show that if two thin coils (inductance L_1 and L_2) are connected in series and placed close to each other, the net self-inductance is

$$L = L_1 + L_2 \pm 2M$$

where M is their mutual inductance. Explain the ± sign. (b) How can M be made zero (or nearly so)? (c) Determine the net inductance if the two coils are connected in parallel, assuming the mutual inductance is ignorable. If M cannot be ignored, how would it affect your answer?

52. Assuming the Earth's magnetic field averages about 0.50×10^{-4} T near the surface of the Earth, estimate the total energy stored in this field in the first 10 km above the Earth's surface.

53. Show that the fraction of electromagnetic energy lost (to thermal energy) per cycle in a lightly damped $(R^2 \ll 4L/C)$ LRC circuit is approximately

$$\frac{\Delta U}{U} = \frac{2\pi R}{L\omega} = \frac{2\pi}{Q}.$$

The quantity Q, defined as $Q = L\omega/R$, is called the Q-value, or quality factor, of the circuit and is a measure of the damping present. A high Q-value means smaller damping and less energy input required to maintain oscillations.

54. Two tightly wound solenoids have the same length and circular cross-sectional area. But solenoid 1 uses wire that is half as thick as solenoid 2. (a) What is the ratio of their inductances? (b) What is the ratio of their inductive time constants (assuming no other resistance in the circuits)?

55. (a) For an underdamped LRC circuit, determine a formula for the energy $U = U_E + U_B$ stored in the electric and magnetic fields as a function of time. Give in terms of the initial charge Q_0 on the capacitor. (b) Show how dU/dt is related to the rate energy is transformed in the resistor, $I^2 R$.

56. An electronic device needs to be protected against sudden surges in current. In particular, after the power is turned on the current should rise to no more than 7.5 mA in the first 100 μs. The device has resistance 150 Ω and is designed to operate at 50 mA. How would you protect this device?

57. The circuit shown in Fig. 30–19a can integrate (in the calculus sense) the input voltage V_{in}, if the time constant L/R is large compared with the time during which V_{in} varies. Explain how this integrator works and sketch its output for the square wave signal input shown in Fig. 30–19b. [Hint: Write Kirchhoff's loop rule for the circuit. Multiply each term in this differential equation (in I) by a factor $e^{Rt/L}$ to make it easier to integrate.]

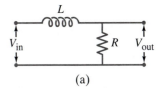

(a)

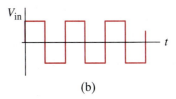

(b)

FIGURE 30–19 Problem 57.

A stereo system is a very complex ac circuit. The power source for home stereo is ac. The audio signal is ac at sound frequencies. The circuits inside a stereo contain resistors, capacitors, and inductors, plus other more complicated elements like diodes and transistors. A loudspeaker also contains an ac circuit, called a crossover (we'll look at this too in this chapter) to divide the signal between woofer and tweeters.

CHAPTER 31

AC Circuits

In earlier chapters we discussed circuits that contain combinations of resistor, capacitor, and inductor (or all three), but only when they are connected to a dc source of emf or to no source as in the discharge of a capacitor in an *RC* circuit or oscillation of an *LC* or *LRC* circuit. See Sections 26–4, 30–4, 30–5, and 30–6. Now we discuss these circuit elements when connected to a source of alternating voltage that produces an alternating current (ac). Such **ac circuits** are important first of all because the output of most generators (Section 29–4) is sinusoidal and most electricity generated and transmitted over wires is ac. Secondly, any voltage that varies in time, no matter how complex it is, can be written as a sum of sine and cosine terms of different frequencies in a Fourier series. Thus, the response of resistors, capacitors, and inductors to a sinusoidal source of emf is of basic importance.

31–1 Introduction: AC Circuits

We briefly discussed alternating currents in Section 25–7, and saw there that for sinusoidally varying currents and voltages, the rms and peak values are related by

$$V_{rms} = \frac{V_0}{\sqrt{2}}, \qquad I_{rms} = \frac{I_0}{\sqrt{2}}.$$

We now examine how a resistor, a capacitor, and an inductor behave when connected to a source of alternating voltage, whose symbol is

●—⊙—●   ,   [symbol for ac source]

which produces a sinusoidal voltage of frequency *f*.

We first look at the *R*, *C*, and *L* separately, and then later all of them together. In each case we assume the voltage gives rise to a current

$$I = I_0 \sin 2\pi f t = I_0 \sin \omega t \tag{31–1}$$

where $\omega = 2\pi f$.

31-2 | AC Circuit Containing only Resistance *R*

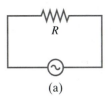

When an ac source is connected to a resistor as in Fig. 31–1a, the current increases and decreases with the alternating voltage, and Kirchhoff's loop rule tells us $V - IR = 0$. Hence

$$V = I_0 R \sin \omega t = V_0 \sin \omega t$$

where $V_0 = I_0 R$ is the peak voltage. Figure 31–1b shows the voltage (red curve) and the current (blue curve). Because the current is zero when the voltage is zero and the current reaches a peak when the voltage does, we say that the current and voltage are **in phase**. Energy is transformed into heat (Section 25–7), at an average rate

$$\overline{P} = \overline{IV} = I_{\text{rms}}^2 R = V_{\text{rms}}^2 / R.$$

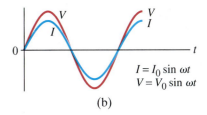

31-3 | AC Circuit Containing only Inductance *L*

In Fig. 31–2a an inductor of inductance *L*, represented by the symbol —, is connected to the ac source. We ignore any resistance it might have (it is usually small). The voltage applied to the inductor will be equal to the emf generated in the inductor by the changing current as given by Eq. 30–5. This is because the sum of the electric potential changes around any closed circuit must add up to zero, as Kirchhoff's rule tells us. Thus

$$V - L\frac{dI}{dt} = 0$$

or

$$V = L\frac{dI}{dt} = \omega L I_0 \cos \omega t. \tag{31–2}$$

Using the identity $\cos \theta = \sin(\theta + 90°)$, we can write

$$V = \omega L I_0 \sin(\omega t + 90°) = V_0 \sin(\omega t + 90°) \tag{31–3a}$$

where

$$V_0 = I_0 \omega L \tag{31–3b}$$

is the peak voltage. The current *I* and voltage *V* across the inductor as a function of time are graphed in Fig. 31–2b. It is clear from this graph, as well as from Eqs. 31–3, that the current and voltage are out of phase by a quarter cycle, which is equivalent to $\pi/2$ radians or 90°. We see from the graph that

in an inductor, the current lags the voltage by 90°.

That is, the current in an inductor reaches its peaks a quarter cycle after the voltage does. Alternatively, we can say that the voltage leads the current by 90°.

Because the current and voltage are out of phase by 90°, no energy is transformed to other forms of energy in an inductor on the average; in particular, no energy is dissipated as thermal energy. This can be seen as follows from Fig. 31–2b. From point c to d, the voltage is increasing from zero to its maximum. The current, however, is in the opposite direction to the voltage and is approaching zero. The average power over this interval, *VI*, is negative. From d to e, however, both *V* and *I* are positive so *VI* is positive; this contribution just balances the negative contribution of the previous quarter cycle. Similar considerations apply to the rest of the cycle. Thus, the average power transformed over one or many cycles is zero. We can see that energy from the source passes into the magnetic field of the inductor, where it is stored temporarily. Then the field decreases and the energy is transferred back to the source. None is dissipated in this process. Compare this to a resistor where the current is always in the same direction as the voltage and energy is transferred out of the source and never back into it. (The product *VI* is never negative.) This energy is not stored in the resistor, but is transformed to thermal energy.

FIGURE 31–1 Resistor connected to ac source. Current is in phase with the voltage across a resistor.

FIGURE 31–2 Inductor connected to an ac source. Current (blue curve) lags voltage (red curve) by a quarter cycle or 90°.

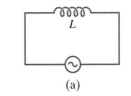

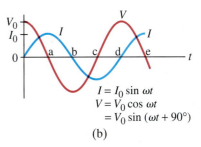

Inductor: current lags voltage

Just as a resistor impedes the flow of charge, so too an inductor impedes the flow of charge in an alternating current due to the back emf produced. For a resistor R, the peak current and peak voltage are related by $V_0 = I_0 R$. We can write a similar relation for an inductor:

$$V_0 = I_0 X_L \qquad \left[\begin{array}{l}\text{maximum values}\\ \text{not an any instant}\end{array}\right] \qquad \textbf{(31–4a)}$$

where, from Eq. 31–3b

$$X_L = \omega L. \qquad \textbf{(31–4b)}$$

Reactance (impedance) of inductor

The term X_L is called the *inductive reactance,* or *impedance,* of the inductor, and it is easy to show it has units of ohms. Normally, we use the term "reactance" to refer solely to the inductive properties. We then reserve the term "impedance" to include the total "impeding" qualities of the coil—its inductance as well as any resistance it may have (more on this in Section 31–5). In the absence of any resistance (or capacitance), the impedance is the same as the reactance.

The quantities V_0 and I_0 in Eq. 31–4a refer to peak values. (It is also valid for rms values, $V_{\text{rms}} = I_{\text{rms}} X_L$.) Note, however, that although this equation relates the peak values, the peak current and voltage are not reached at the same time; so Eq. 31–4a is *not valid at a particular instant,* as is the case for a resistor ($V = IR$).

Eq. 31–4a NOT valid at any instant

Note from Eq. 31–4b that if $\omega = 2\pi f = 0$ (so the current is dc), there is no back emf and no impedance to the flow of charge.

EXAMPLE 31–1 **Reactance of a coil.** A coil has a resistance $R = 1.00\,\Omega$ and an inductance of 0.300 H. Determine the current in the coil if (*a*) 120-V dc is applied to it; (*b*) 120-V ac (rms) at 60.0 Hz is applied.

SOLUTION (*a*) There is no inductive reactance $(X_L = 0$ since $f = 0)$, so we can write for the resistance:

$$I = \frac{V}{R} = \frac{120\,\text{V}}{1.00\,\Omega} = 120\,\text{A}.$$

(*b*) The inductive reactance in this case is:

$$X_L = 2\pi f L = (6.28)(60.0\,\text{s}^{-1})(0.300\,\text{H}) = 113\,\Omega.$$

In comparison to this, the $1.00\,\Omega$ resistance can be ignored. Thus

$$I_{\text{rms}} \approx \frac{V_{\text{rms}}}{X_L} = \frac{120\,\text{V}}{113\,\Omega} = 1.06\,\text{A}.$$

[It might be tempting to say that the total impedance is $113\,\Omega + 1\,\Omega = 114\,\Omega$. This might imply that about 1 percent of the voltage drop is across the resistor, or about 1 V; and that across the inductance is 119 V. Although the 1 Vrms across the resistor is accurate, the other statements are *not* true because of the alteration in phase in an inductor. This will be discussed in Section 31–5.]

31–4 AC Circuit Containing only Capacitance C

When a capacitor is connected to a battery, the capacitor plates quickly acquire equal and opposite charges; but no steady current flows in the circuit. A capacitor prevents the flow of a dc current. However, if a capacitor is connected to an alternating source of voltage, as in Fig. 31–3a, an alternating current will flow continuously. This can happen because when the ac voltage is first turned on, charge begins to flow so that one plate acquires a negative charge and the other a positive charge. But when the voltage reverses itself, the charges flow in the opposite direction. Thus, for an alternating applied voltage, an ac current is present in the circuit continuously.

Let us look at this in more detail. By Kirchhoff's loop rule, the applied source voltage must equal the voltage V across the capacitor at any moment:

$$V = \frac{Q}{C}$$

where C is the capacitance and Q is the charge on the capacitor plates. Now the current I at any instant (given as $I = I_0 \sin \omega t$) is

$$I = \frac{dQ}{dt} = I_0 \sin \omega t.$$

Hence the charge Q on the plates at any instant is given by

$$Q = \int_0^t dQ = \int_0^t I_0 \sin \omega t \, dt = -\frac{I_0}{\omega} \cos \omega t.$$

Then the voltage across the capacitor is

$$V = \frac{Q}{C} = -I_0 \left(\frac{1}{\omega C} \right) \cos \omega t.$$

Using the trigonometric identity $\cos \theta = -\sin(\theta - 90°)$ we can rewrite this as

$$V = I_0 \left(\frac{1}{\omega C} \right) \sin(\omega t - 90°) = V_0 \sin(\omega t - 90°) \qquad \textbf{(31–5a)}$$

where

$$V_0 = I_0 \left(\frac{1}{\omega C} \right) \qquad \textbf{(31–5b)}$$

is the peak voltage. The current $I (= I_0 \sin \omega t)$ and voltage V (Eq. 31–5a) across the capacitor are graphed in Fig. 31–3b. It is clear from this graph, as well as a comparison of Eq. 31–5a with Eq. 31–1, that the current and voltage are out of phase by a quarter cycle or 90° ($\pi/2$ radians):

The current leads the voltage across a capacitor by 90°.

Alternatively we can say that the voltage lags the current by 90°. This is the opposite of what happens for an inductor, where the current lags the voltage by 90°.

Because the current and voltage are out of phase by 90°, the average power dissipated is zero, just as for an inductor. Energy from the source is fed to the capacitor, where it is stored in the electric field between its plates. As the field decreases, the energy returns to the source.

Thus, in an ac circuit, *only a resistance will dissipate energy* to thermal energy.

A relationship between the applied voltage and the current in a capacitor can be written just as for an inductance:

$$V_0 = I_0 X_C \qquad \begin{bmatrix} \text{maximum values} \\ \text{not at any instant} \end{bmatrix} \qquad \textbf{(31–6a)}$$

where X_C is the **capacitive reactance** (or *impedance*) of the capacitor, and has units of ohms. X_C is given by (see Eq. 31–5b)

$$X_C = \frac{1}{\omega C}. \qquad \textbf{(31–6b)}$$

Equation 31–6a relates the peak values of V and I, or the rms values ($V_{\text{rms}} = I_{\text{rms}} X_C$). But it is not valid at a particular instant because I and V are not in phase.

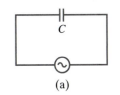

(a)

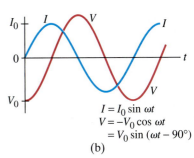

$$I = I_0 \sin \omega t$$
$$V = -V_0 \cos \omega t$$
$$= V_0 \sin(\omega t - 90°)$$

(b)

FIGURE 31–3 Capacitor connected to an ac source. Current leads voltage by a quarter cycle, or 90°.

Note from Eq. 31–6b that for dc conditions, $\omega = 2\pi f = 0$ and X_C becomes infinite. This is as it should be, since a pure capacitor does not pass dc current. Also, note that the reactance of an inductor increases with frequency, but that of a capacitor decreases with frequency.

EXAMPLE 31–2 **Capacitor reactance.** What are the peak and rms currents in the circuit of Fig. 31–3a if $C = 1.0\,\mu\text{F}$ and $V_{\text{rms}} = 120\,\text{V}$? Calculate for (a) $f = 60\,\text{Hz}$, and then for (b) $f = 6.0 \times 10^5\,\text{Hz}$.

SOLUTION (a) $V_0 = \sqrt{2}\,V_{\text{rms}} = 170\,\text{V}$. Then

$$X_C = \frac{1}{2\pi f C} = \frac{1}{(6.28)(60\,\text{s}^{-1})(1.0 \times 10^{-6}\,\text{F})} = 2.7\,\text{k}\Omega.$$

Thus

$$I_0 = \frac{V_0}{X_C} = \frac{170\,\text{V}}{2.7 \times 10^3\,\Omega} = 63\,\text{mA},$$

$$I_{\text{rms}} = \frac{V_{\text{rms}}}{X_C} = \frac{120\,\text{V}}{2.7 \times 10^3\,\Omega} = 44\,\text{mA}.$$

(b) For $f = 6.0 \times 10^5\,\text{Hz}$, X_C will be $0.27\,\Omega$, $I_0 = 630\,\text{A}$, and $I_{\text{rms}} = 440\,\text{A}$. The dependence on f is dramatic.

Capacitors are used for a variety of purposes. some of which have already been described. Two other applications are illustrated in Fig. 31–4. In Fig. 31–4a, circuit A is said to be capacitively coupled to circuit B. The purpose of the capacitor is to prevent a dc voltage from passing from A to B but allowing an ac signal to pass relatively unimpeded. If C is sufficiently large, the ac signal will not be significantly attenuated, whereas dc is filtered out. The capacitor in Fig. 31–4b also passes ac but not dc. In this case, a dc voltage can be maintained between circuits A and B. If the capacitance C is large enough, the capacitor offers little impedance to an ac signal leaving A. Such a signal then passes to ground instead of into B. Thus the capacitor in Fig. 31–4b acts like a *filter* when a constant dc voltage is required; any sharp variation in voltage will pass to ground instead of into circuit B. Capacitors used in these two ways are very common in circuits.

Loudspeakers that have separate speakers for low- and high-frequency sounds can use a simple "cross-over" that uses inductors and capacitors to filter frequencies reaching each speaker. Low-frequency sounds go to a large-diameter speaker, the *woofer*, and high-frequency sounds go to a small-diameter *tweeter*. (Recall standing waves from Chapter 16—small vibrating objects have high resonant frequencies whereas large vibrating objects have lower frequencies.) The simple diagram of Fig. 31–5 shows how the output signal from the amplifier is divided and goes to both tweeter and woofer, in parallel. A capacitor in the tweeter circuit impedes low-frequency signals (much like the capacitor in Fig. 31–4a). An inductor in the woofer circuit impedes high-frequency signals $(X_L = 2\pi f L)$ so mainly low-frequency sounds are emitted by the woofer.

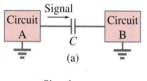
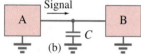

FIGURE 31–4 Two common uses for a capacitor.

FIGURE 31–5
Loudspeaker cross-over.

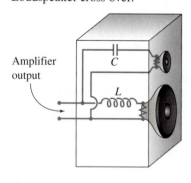

FIGURE 31–6 An *LRC* circuit.

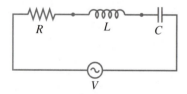

31–5 LRC Series AC Circuit

We now examine a circuit containing all three elements in series, a resistor R, an inductor L, and a capacitor C, Fig 31–6. If a given circuit contains only two of these elements, we can still use the results of this Section by setting $R = 0$, $L = 0$, or $C = \infty$ (infinity) as needed. We let V_R, V_L, and V_C represent the voltage across each element at a *given instant* in time; and V_{R0}, V_{L0}, and V_{C0} represent the *maximum* (peak) values of these voltages. The voltage across each of the elements will follow the phase relations we discussed in the previous Sections. That is, V_R will be in phase with the current; V_L will lead the current by 90°; and V_C will lag behind the current by 90°. And at any instant the voltage V supplied by the source will be, by Kirchhoff's loop rule,

$$V = V_R + V_L + V_C. \qquad \textbf{(31-7)}$$

However, because the various voltages are not in phase, they do not reach their peak values at the same time, and so the peak voltage of the source, V_0, will *not* equal $V_{R0} + V_{L0} + V_{C0}$. Likewise the rms voltages (which are what ac voltmeters usually read) will not simply add up to give the rms voltage of the source.

Let us now examine the circuit in detail. What we would like to find in particular is the impedance of the circuit as a whole, the peak current I_0 that flows, and the phase difference between the source voltage and the current.

First we note that the current at any instant must be the same at all points in the circuit. Thus, *the currents in each element are in phase with each other, although the voltages are not.* We choose our origin in time ($t = 0$) so that the current I at any time t is

$$I = I_0 \sin \omega t$$

just as we did in the previous Sections of this chapter.

We will analyze the LRC circuit using[†] a so-called **phasor diagram**: Arrows (acting somewhat like vectors) are drawn in an xy coordinate system to represent the amplitude of each voltage. The *length of each arrow represents the magnitude of the peak voltage across each element*:

$$V_{R0} = I_0 R, \qquad V_{L0} = I_0 X_L, \qquad V_{C0} = I_0 X_C.$$

The angle of each arrow represents the phase of each voltage, relative to the current, and the arrows rotate at angular frequency ω to take into account the time dependence of the voltages and current. In particular, *the projection of each arrow on the y axis will represent the voltage across each element at a given time.* Let us see how this works.

First we draw the phasor diagram for time $t = 0$. The current at $t = 0$ is $I = I_0 \sin \omega t = 0$, and we draw the arrow representing I_0 along the positive x axis in our phasor diagram, Fig. 31–7a. (Note that the projection of I_0 on the y axis is zero, which corresponds to $I = 0$ at $t = 0$.) The voltage across a resistor is always in phase with the current, so the arrow representing V_{R0} is drawn parallel to I_0 along the x axis as shown. Since the voltage across the inductor, V_L, leads the current by 90°, V_{L0} leads V_{R0} by 90°, and is drawn perpendicular to it as shown. V_C lags the current by 90°, so V_{C0} lags V_{R0} by 90°, and hence V_{C0} is drawn perpendicular to V_{R0}, but downward (Fig. 31–7a). Now if we let this diagram rotate as a whole at angular frequency ω, then we obtain the diagram shown in Fig. 31–7b: after a time t, each arrow has rotated through an angle ωt. Then, as mentioned above, the projections of each arrow on the y axis represent the voltages across each element at the instant t. See Fig. 31–7c. For example, the projection of V_{R0} on the y axis is $V_{R0} \sin \omega t$ ($= I_0 R \sin \omega t = IR$ since $I = I_0 \sin \omega t$). The projections of V_{L0} and V_{C0} on the y axis are $V_L = V_{L0} \cos \omega t = V_L \sin(\omega t + 90°)$, and $V_C = -V_{C0} \cos \omega t = V_{C0} \sin(\omega t - 90°)$. These results are consistent with our earlier results as shown in Figs. 31–1, 31–2, and 31–3. See also Eqs. 31–3a and 31–5a. Maintaining the 90° angle between each vector ensures the correct phase relations. Although these facts show the validity of a phasor diagram, what we are really interested in is how to add the voltages.

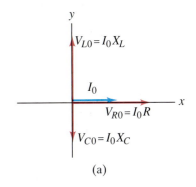

(a)

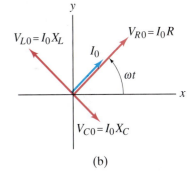

(b)

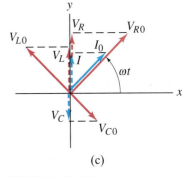

(c)

FIGURE 31–7 Phasor diagram for a series LRC circuit.

[†]We could instead do our analysis by rewriting Eq. 31–7 as a differential equation (setting $V_C = Q/C$, $V_R = IR = (dQ/dt)R$, and $V_L = L\, dI/dt$) and trying to solve the differential equation. The differential equation we would get would look like Eq. 14–21 in Section 14–8 (on forced vibrations), and would be solved in the same way. Phasor diagrams are easier, and at the same time give us some physical insight into the situation.

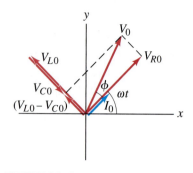

FIGURE 31–8 Phasor diagram for a series *LRC* circuit showing the sum vector, V_0.

The sum of the projections of the three vectors on the y axis is equal to the projection of their sum. But the sum of the projections represents the instantaneous voltage across the whole circuit, which is the source voltage V. We can then let the vector sum of these vectors be the vector that represents the peak source voltage V_0 on a phasor diagram. This is shown in Fig. 31–8, where it is seen that V_0 makes an angle ϕ with V_{R0} and I_0. As time passes, V_0 rotates with the other vectors, so the instantaneous voltage V (projection of V_0 on y axis) is (see Fig. 31–8):

$$V = V_0 \sin(\omega t + \phi).$$

Thus we see that the voltage from the source is out of phase[†] with the current by an angle ϕ.

From this analysis we can now draw some useful conclusions. First we determine the total **impedance** Z of the circuit, which is defined by the relation

$$V_{rms} = I_{rms} Z, \qquad \text{or} \qquad V_0 = I_0 Z. \tag{31–8}$$

From Fig. 31–8, we see, using the Pythagorean theorem (V_0 is the hypoteneuse), that

$$V_0 = \sqrt{V_{R0}^2 + (V_{L0} - V_{C0})^2}$$
$$= \sqrt{I_0^2 R^2 + (I_0 X_L - I_0 X_C)^2}$$
$$= I_0 \sqrt{R^2 + (X_L - X_C)^2}.$$

Thus, from Eq. 31–8, and then Eqs. 31–4b and 31–6b,

Total impedance of LRC ac circuit

$$Z = \sqrt{R^2 + (X_L - X_C)^2} \tag{31–9a}$$
$$= \sqrt{R^2 + \left(\omega L - \frac{1}{\omega C}\right)^2}. \tag{31–9b}$$

This gives the total impedance Z of the circuit. Also from Fig. 31–8 we can find the phase angle ϕ:

Phase angle of LRC ac circuit

$$\tan \phi = \frac{V_{L0} - V_{C0}}{V_{R0}} = \frac{I_0 (X_L - X_C)}{I_0 R} = \frac{X_L - X_C}{R} \tag{31–10a}$$

or

$$\cos \phi = \frac{V_{R0}}{V_0} = \frac{I_0 R}{I_0 Z} = \frac{R}{Z}. \tag{31–10b}$$

Note that Fig. 31–8 was drawn for the case $X_L > X_C$, and the current lags the source voltage by ϕ. If the reverse is true, $X_L < X_C$, then ϕ in Eq. 31–10 is less than zero, and the current leads the source voltage.

Finally, we can determine the power dissipated in the circuit. We saw earlier that power is dissipated only by a resistance; none is dissipated by inductance or capacitance. Therefore, the average power $\overline{P} = I_{rms}^2 R$. But from Eq. 31–10b, $R = Z \cos \phi$. Therefore

Power factor

$$\overline{P} = I_{rms}^2 Z \cos \phi$$
$$= I_{rms} V_{rms} \cos \phi. \tag{31–11}$$

The factor $\cos \phi$ is referred to as the **power factor** of the circuit. For a pure resistor, $\cos \phi = 1$ and $\overline{P} = I_{rms} V_{rms}$. For a capacitor or inductor alone, $\phi = -90°$ or $+90°$, respectively, so $\cos \phi = 0$ and no power is dissipated.

[†] As a check, note that if $R = X_C = 0$, then $\phi = 90°$, and V_0 would lead the current by 90°, as it must for an inductor alone. Similarly, if $R = L = 0$, $\phi = -90°$ and V_0 would lag the current by 90°, as it must for a capacitor alone.

The test of this analysis is, of course, in experiment; and experiment is in full agreement with these results.[†]

EXAMPLE 31–3 *LRC* Circuit. Suppose that $R = 25.0\,\Omega$, $L = 30.0\,\text{mH}$, and $C = 12.0\,\mu\text{F}$ in Fig. 31–6, and that they are connected to a 90.0-V ac (rms) 500-Hz source. Calculate (*a*) the current in the circuit, (*b*) the voltmeter readings (rms) across each element, (*c*) the phase angle ϕ, and (*d*) the power dissipated in the circuit.

SOLUTION (*a*) First, we find the individual impedances at $f = 500\,\text{Hz} = 500\,\text{s}^{-1}$:

$$X_L = 2\pi f L = 94.2\,\Omega,$$

$$X_C = \frac{1}{2\pi f C} = 26.5\,\Omega.$$

Then

$$Z = \sqrt{R^2 + (X_L - X_C)^2}$$
$$= \sqrt{(25.0\,\Omega)^2 + (94.2\,\Omega - 26.5\,\Omega)^2} = 72.2\,\Omega.$$

From Eq. 31–8,

$$I_{\text{rms}} = \frac{V_{\text{rms}}}{Z} = \frac{90.0\,\text{V}}{72.2\,\Omega} = 1.25\,\text{A}.$$

(*b*) The rms voltage across each element is

$$(V_R)_{\text{rms}} = I_{\text{rms}} R = (1.25\,\text{A})(25.0\,\Omega) = 31.2\,\text{V}$$

$$(V_L)_{\text{rms}} = I_{\text{rms}} X_L = (1.25\,\text{A})(94.2\,\Omega) = 118\,\text{V}$$

$$(V_C)_{\text{rms}} = I_{\text{rms}} X_C = (1.25\,\text{A})(26.5\,\Omega) = 33.1\,\text{V}.$$

Notice that these do *not* add up to give the source voltage, 90.0 V (rms). Indeed, the rms voltage across the inductance *exceeds* the source voltage. This can happen because the different voltages are out of phase with each other, and at any instant one voltage can be negative, to compensate for a large positive voltage of another. The rms voltages, however, are always positive by definition. Although the rms voltages do not have to add up to the source voltage, the instantaneous voltages at any time do add up, of course, to the source voltage at that instant.

(*c*) The phase angle ϕ is given by Eq. 31–10b,

$$\cos\phi = \frac{R}{Z} = \frac{25.0\,\Omega}{72.2\,\Omega} = 0.346,$$

so $\phi = 69.7°$. Note that ϕ is positive because $X_L > X_C$ in this case, so $V_{L0} > V_{C0}$ in Fig. 31–8.

(*d*) $\overline{P} = I_{\text{rms}} V_{\text{rms}} \cos\phi = (1.25\,\text{A})(90.0\,\text{V})(25.0\,\Omega/72.2\,\Omega) = 39.0\,\text{W}.$

[†] At the beginning of this Section, we chose the phase of the current so that $I = I_0 \sin\omega t$. This choice is, of course, arbitrary. (What is physically important is the phase difference, ϕ, between current and voltage.) If we had chosen, instead,

$$V = V_0 \sin\omega t,$$

then the current I would be

$$I = I_0 \sin(\omega t - \phi)$$

where ϕ and I_0 have the same values as given by Eqs. 31–8, 31–9, and 31–10.

31–6 Resonance in AC Circuits

The rms current in an LRC series circuit is given by (see Eqs. 31–8 and 31–9b):

$$I_{rms} = \frac{V_{rms}}{Z} = \frac{V_{rms}}{\sqrt{R^2 + \left(\omega L - \frac{1}{\omega C}\right)^2}}. \qquad \textbf{(31–12)}$$

Because the impedance of inductors and capacitors depends on the frequency f ($= \omega/2\pi$) of the source, the current in an LRC circuit will depend on frequency. From Eq. 31–12 we can see that the current will be maximum at a frequency such that

$$\left(\omega L - \frac{1}{\omega C}\right) = 0.$$

We solve this for ω and call the solution ω_0:

Resonance

$$\omega_0 = \sqrt{\frac{1}{LC}}. \qquad \textbf{(31–13)}$$

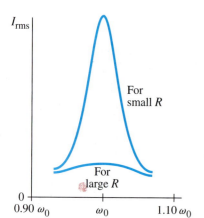

When $\omega = \omega_0$, the circuit is in **resonance**, and $f_0 = \omega_0/2\pi$ is the **resonant frequency** of the circuit. At this frequency, $X_C = X_L$, so the impedance is purely resistive and $\cos\phi = 1$. A graph of I_{rms} versus ω is shown in Fig. 31–9 for particular values of R, L, and C. For smaller R compared to X_L and X_C, the resonance peak will be higher and sharper. When R is very small, the circuit approaches the pure LC circuit we discussed in Section 30–5.

This electrical resonance is analogous to mechanical resonance, which we discussed in Chapter 14. The energy transferred to the system by the source is a maximum at resonance whether it is electrical resonance, the oscillation of a spring, or pushing a child on a swing (Section 14–8). That this is true in the electrical case can be seen from Eq. 31–11. At resonance, $\cos\phi = 1$, and I_{rms} is a maximum. For constant voltage V_{rms}, the power then is a maximum at resonance. A graph of power versus frequency peaks like that for the current, Fig. 31–9.

Electric resonance is used in many circuits. Radio and TV sets, for example, use resonant circuits for tuning in a station. Many frequencies reach the circuit, but a significant current flows only for those at or near the resonant frequency. Either L or C is variable so that different stations can be tuned in.

FIGURE 31–9 Current in LRC circuit as a function of frequency, showing resonance peak at $\omega = \omega_0 = \sqrt{1/LC}$.

EXAMPLE 31–4 **Radio station oscillator.** A radio station is authorized to broadcast on a frequency of 1040 kHz. If you are designing a receiver circuit to pick up this station, and already have a coil with an inductance of 4.0 mH, what capacitance will you need?

SOLUTION The tuned circuit must have a resonant frequency of 1040 kHz. Using an inductor in series with a capacitor will accomplish the job. Since the resonance will be at

$$f_0 = \frac{1}{2\pi}\sqrt{\frac{1}{LC}}$$

you will need to find a capacitor with a value of

$$C = \frac{1}{L(2\pi f_0)^2} = \frac{1}{(4.0 \times 10^{-3}\,\text{H})(2\pi \times 1.04 \times 10^6\,\text{s}^{-1})^2} = 5.85 \times 10^{-12}\,\text{F} = 5.85\,\text{pF}$$

*31-7 | Impedance Matching

It is common to connect one electric circuit to a second circuit. For example, a TV antenna is connected to a TV set; an FM tuner is connected to an amplifier; the output of an amplifier is connected to a speaker; electrodes for an ECG or EEG (electrocardiogram and electroencephalogram—electrical traces of heart and brain signals) are connected to an amplifier or a recorder. In many cases it is important that the maximum power be transferred from one to the other, with a minimum of loss. This can be achieved when the output impedance of the one device matches the input impedance of the second.

To show why this is true, we consider simple circuits that contain only resistance. In Fig. 31–10 the source in circuit 1 could represent a power supply, the output of an amplifier, or the signal from an antenna, a laboratory probe, or a set of electrodes. R_1 represents the resistance of this device and includes the internal resistance of the source. R_1 is called the output impedance (or resistance) of circuit 1. The output of circuit 1 is across the terminals a and b which are connected to the input of circuit 2 which may be very complicated. We let R_2 be the equivalent "input resistance" of circuit 2.

The power delivered to circuit 2 is $P = I^2 R_2$ where $I = V/(R_1 + R_2)$. Thus

$$P = I^2 R_2 = \frac{V^2 R_2}{(R_1 + R_2)^2}.$$

We divide the top and bottom of the right side by R_1^2 and find

$$P = \frac{V^2}{R_1} \frac{\left(\dfrac{R_2}{R_1}\right)}{\left(1 + \dfrac{R_2}{R_1}\right)^2}.$$

The question is, if the resistance of the source is R_1, what value should R_2 have so that the maximum power is transferred to circuit 2? To determine this, we take the derivative of P with respect to R_2 and set it equal to zero:

$$0 = \frac{dP}{dR_2} = \frac{V^2}{R_1^2} \frac{(1 - R_2/R_1)}{(1 + R_2/R_1)^3}.$$

This expression can be zero only if $(1 - R_2/R_1) = 0$, or

$$R_2 = R_1.$$

Impedance matching

Thus, the maximum power is transmitted when the *output impedance* of one device *equals the input impedance* of the second. This is called **impedance matching**.

In an ac circuit that contains capacitors and inductors, the different phases are important and the analysis is more complicated. However, the same result holds: to maximize power transfer it is important to match impedances $(Z_2 = Z_1)$.

In addition, one must be aware that it is possible to seriously distort a signal. For example, when a second circuit is connected, it may put the first circuit into resonance, or take it out of resonance for a certain frequency.

Without proper consideration of the impedances involved, one can make measurements that are completely meaningless. These considerations are normally examined by engineers when designing an integrated set of apparatus. It has happened that researchers have connected several components to one another without regard for impedance matching, and made a "new discovery" that later was found (embarrassingly) to be due to impedance mismatch rather than the natural phenomenon they had thought.

In some cases, a transformer is used to alter an impedance, so it can be matched to that of a second circuit. If Z_s is the secondary impedance and Z_p the

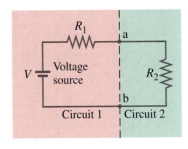

FIGURE 31–10 Output of the circuit on the left is input to the circuit on the right.

primary impedance, then $V_s = I_s Z_s$ and $V_p = I_p Z_p$ (I and V are either peak or rms values of current and voltage). Hence

$$\frac{Z_p}{Z_s} = \frac{V_p I_s}{V_s I_p} = \left(\frac{N_p}{N_s}\right)^2$$

where we have used Eqs. 29–5 and 29–6 for a transformer. Thus the impedance can be changed with a transformer.

Some instruments, such as oscilloscopes, require only a signal voltage but very little power. Maximum power transfer is then not important and such instruments can have a high input impedance, which has the advantage that the instrument draws very little current and disturbs the original circuit as little as possible.

* 31–8 Three-Phase AC

Transmission lines typically consist of four wires, rather than two as you might have guessed. One of these wires is the ground; the remaining three are used to transmit three-phase ac power which is a superposition of three ac voltages 120° out of phase with each other:

$$V_1 = V_0 \sin \omega t$$
$$V_2 = V_0 \sin(\omega t + 2\pi/3)$$
$$V_3 = V_0 \sin(\omega t + 4\pi/3).$$

(See Fig. 31–11.) Why is three-phase power used? We saw in Fig. 25–20 that single-phase ac (i.e., the voltage V_1 by itself) delivers power to the load in pulses. A much smoother flow of power can be delivered if we use three-phase power. Suppose that each of the three voltages making up the three-phase source is hooked up to a resistor R. Then the power delivered is:

$$P = \frac{1}{R}\left(V_1^2 + V_2^2 + V_3^2\right).$$

You can show that this power is a constant equal to $3V_0^2/2R$, which is three times the rms power delivered by a single-phase source. This smooth flow of power makes electrical equipment run smoothly. Although houses use single-phase ac power, most industrial-grade machinery is wired for three-phase power.

FIGURE 31–11 The three voltages, out of phase by 120° $\left(=\tfrac{2}{3}\pi \text{ radians}\right)$, in a three-phase power line.

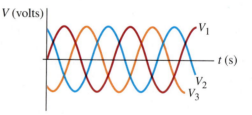

EXAMPLE 31–5 **Three-phase circuit.** In a three-phase circuit there is 266 V rms between line 1 and ground. What is the rms voltage between lines 2 and 3?

SOLUTION We are given $V_{rms} = V_0/\sqrt{2} = 266$ V. Hence $V_0 = 376$ V. Now $V_3 - V_2 = V_0[\sin(\omega t + 4\pi/3) - \sin(\omega t + 2\pi/3)] = 2V_0 \sin\tfrac{1}{2}\left(\tfrac{2\pi}{3}\right)\cos\tfrac{1}{2}(2\omega t)$ where we used the identity: $\sin A - \sin B = 2 \sin\tfrac{1}{2}(A - B)\cos\tfrac{1}{2}(A + B)$. The rms voltage is

$$(V_3 - V_2)_{rms} = \frac{1}{\sqrt{2}}2V_0 \sin\frac{\pi}{3} = \sqrt{2}(376 \text{ V})(0.866) = 460 \text{ Vrms}.$$

Summary

Capacitance and inductance offer *impedance* to the flow of alternating current just as resistance does. This impedance is referred to as **reactance**, X. For capacitance and inductance the reactance is defined, as for resistors, as the proportionality constant between voltage and current (either the rms or peak values). Across a capacitor,

$$V_0 = I_0 X_C,$$

and across an inductor,

$$V_0 = I_0 X_L.$$

The reactance of a capacitor decreases with frequency:

$$X_C = 1/\omega C,$$

where $\omega = 2\pi f$ and f is the frequency.

The reactance of an inductor increases with frequency:

$$X_L = \omega L.$$

Whereas the current through a resistor is always in phase with the voltage across it, this is not true for inductors and capacitors: in an inductor, the current lags the voltage by 90°, and in a capacitor the current leads the voltage by 90°.

In an **LRC series circuit**, the total impedance Z is defined by the equivalent of $V = IR$ for resistance, namely $V_0 = I_0 Z$ or $V_{rms} = I_{rms} Z$; Z is related to R, C, and L by

$$Z = \sqrt{R^2 + (X_L - X_C)^2}.$$

The current in the circuit lags (or leads) the source voltage by an angle ϕ given by $\cos\phi = R/Z$. Only the resistor in an LRC circuit dissipates energy, and at a rate

$$\overline{P} = I_{rms}^2 Z \cos\phi$$

where the factor $\cos\phi$ is referred to as the **power factor**.

An LRC series circuit **resonates** at a frequency given by

$$\omega_0 = \frac{1}{\sqrt{LC}} \quad \text{or} \quad f_0 = \frac{\omega_0}{2\pi} = \frac{1}{2\pi\sqrt{LC}}.$$

The rms current in the circuit is largest when the applied voltage has a frequency equal to f_0. The lower the resistance R, the higher and sharper the resonance peak.

Questions

1. Under what conditions is the impedance in an *LRC* circuit a minimum?

2. Why can we assume that the current in an *LRC* circuit will have the same frequency as the applied emf?

3. In an *LRC* circuit, if $X_L > X_C$, the circuit is said to be predominantly "inductive." And if $X_C > X_L$, the circuit is said to be predominantly "capacitive." Discuss the reasons for these terms. In particular, do they say anything about the relative values of L and C at a given frequency?

4. Do the results of Section 31–5 approach the proper expected results when ω approaches zero? What are the expected results?

5. When an ac generator is connected to an *LRC* circuit, where does the energy come from ultimately? Where does it go? How do the values of L, C, and R affect the energy supplied by the generator?

6. Discuss the validity of both of Kirchhoff's rules (Section 26–3) when applied to ac circuits that contain several loops.

7. Is it possible for the instantaneous power output of an ac generator connected to an *LRC* circuit ever to be negative? Explain.

8. Can you tell whether the current in an *LRC* circuit leads or lags the applied voltage from a knowledge of the power factor, $\cos\phi$?

9. If $\cos\phi$ were less than zero, Eq. 31–11 tells us that $\overline{P} < 0$. Can this happen? Can $\cos\phi$ be made negative? Explain.

10. Does the power factor, $\cos\phi$, depend on frequency? Does the power dissipated in an *LRC* circuit depend on frequency?

11. What is the significance of the sign of $\phi(+$ or $-)$? Is this a convention, or is it a fixed rule?

12. Describe briefly how the frequency of the source emf affects the impedance of (a) a pure resistance, (b) a pure capacitance, (c) a pure inductance, (d) an *LRC* circuit near resonance (R small), (e) an *LRC* circuit far from resonance (R small).

13. Discuss the response of an *LRC* circuit as $R \to 0$ when the frequency is (a) at resonance, (b) near resonance, (c) far from resonance. Is there energy dissipation in each case? Discuss the transformations of energy that occur in each case.

14. Can you tell whether or not a circuit is in resonance if you are given the value of the power factor, $\cos\phi$?

15. An *LRC* resonance circuit is often called an *oscillator* circuit. What is it that oscillates?

16. Compare the oscillations of an *LRC* circuit to the vibration of a mass m on a spring. What do L and C correspond to in the mechanical system?

Problems

1. (I) What is the reactance of a 7.2 μF capacitor at a frequency of (a) 60 Hz, (b) 1.0 MHz?

2. (I) At what frequency will a 22.0-mH inductor have a reactance of 660 Ω?

3. (I) At what frequency will a 2.40-μF capacitor have a reactance of 6.70 kΩ?

4. (I) Plot a graph of the impedance of a 5.8-μF capacitor as a function of frequency from 10 Hz to 1000 Hz.

5. (I) Plot a graph of the impedance of a 5.0-mH inductor as a function of frequency from 100 Hz to 10,000 Hz.

6. (I) Calculate the impedance of, and rms current in, a 36.0-mH radio coil connected to 750-V (rms) 33.3-kHz ac line. Ignore resistance.

7. (II) What is the inductance L of the primary of a transformer whose input is 110 V at 60 Hz and the current drawn is 2.2 A? Assume no current in the secondary.

8. (II) (a) What is the impedance of a well-insulated 0.036-μF capacitor connected to a 22-kV (rms), 600-Hz line? (b) What will be the peak value of the current and its frequency?

9. (II) Instead of starting our analysis of ac circuits with Eq. 31–1, suppose we assumed that the external voltage was given as $V = V_0 \sin \omega t$. Show that when this voltage is (a) connected only to a capacitor C, the current would be $I = \omega C V_0 \cos \omega t = \omega C V_0 \sin(\omega t + 90°)$, and (b) connected only to an inductor L, the current would be $I = -(V_0/\omega L) \cos \omega t = (V_0/\omega L) \sin(\omega t - 90°)$.

10. (II) A current $I = 1.80 \cos 377t$ (I in amps, t in seconds and the "angle" is in radians) flows in a series LR circuit in which $L = 3.85$ mH and $R = 260 \Omega$. What is the average power dissipation?

11. (II) A capacitor is placed in parallel with some device, B, as in Fig. 31–4b, to filter out stray high-frequency signals, but to allow ordinary 60-Hz ac to pass through with little loss. Suppose that circuit B in Fig. 31–4b is a resistance $R = 400 \Omega$ connected to ground, and that $C = 0.35 \mu$F. What percent of the incoming current will pass through C rather than R if (a) it is 60 Hz; (b) 60,000 Hz?

Section 31–5

12. (I) A 1.20-kΩ resistor and a 6.8-μF capacitor are connected in series to an ac source. Calculate the impedance of the circuit if the source frequency is (a) 60 Hz; (b) 60,000 Hz.

13. (I) A 9.0-kΩ resistor is in series with a 26.0-mH inductor and an ac source. Calculate the impedance of the circuit if the source frequency is (a) 50 Hz; (b) 30,000 Hz.

14. (I) For a 120-V, 60-Hz voltage, a current of 70 mA passing through the body for 1.0 s could be lethal. What must be the impedance of the body for this to occur?

15. (II) (a) What is the rms current in a series RC circuit if $R = 6.0$ kΩ, $C = 0.80 \mu$F, and the rms applied voltage is 120 V at 60 Hz? (b) What is the phase angle between voltage and current? (c) What is the power dissipated by the circuit? (d) What are the voltmeter readings across R and C?

16. (II) (a) What is the rms current in a series LR circuit when a 60.0-Hz, 120-V rms ac voltage is applied, where $R = 765 \Omega$ and $L = 250$ mH? (b) What is the phase angle between voltage and current? (c) How much power is dissipated? (d) What are the rms voltage readings across R and L?

17. (II) A 35-mH inductor with 2.0-Ω resistance is connected in series to a 20-μF capacitor and a 60-Hz, 45-V source. Calculate (a) the rms current, (b) the phase angle, and (c) the power dissipated in this circuit.

18. (II) A 40-mH coil whose resistance is 0.80 Ω is connected to a capacitor C and a 360-Hz source voltage. If the current and voltage are to be in phase, what value must C have?

19. (II) What is the resistance of a coil if its impedance is 335 Ω and its reactance is 45.5 Ω?

20. (II) In the LRC circuit of Fig. 31–6, suppose $I = I_0 \sin \omega t$ and $V = V_0 \sin(\omega t + \phi)$. Determine the instantaneous power dissipated in the circuit from $P = IV$ using these equations and show that on the average, $\overline{P} = \frac{1}{2} V_0 I_0 \cos \phi$, which confirms Eq. 31–11.

21. (II) If $V = V_0 \sin \omega t$, what is the average value of V over (a) a whole cycle, (b) each half cycle? How do these compare to V_{rms}?

22. (II) A circuit consists of a 150-Ω resistor in series with a 40.0-mH inductor and a 60.0-V ac generator. The power dissipated by the circuit is 15.5 W. What is the frequency of the generator?

23. (II) What is the total impedance, phase angle, and rms current in an LRC circuit connected to a 10.0-kHz, 800-V (rms) source if $L = 32.0$ mH, $R = 8.70$ kΩ, and $C = 5000$ pF?

Section 31–6

24. (I) A 3200-pF capacitor is connected in series to a 26.0-μH coil of resistance 2.00 Ω. What is the resonant frequency of this circuit?

25. (I) What is the resonant frequency of the LRC circuit of Example 31–3? At what rate is energy taken from the generator, on the average, at this frequency?

26. (II) An LRC circuit has $L = 4.15$ mH and $R = 220 \Omega$. (a) What value must C have to produce resonance at 33.0 kHz? (b) What will be the maximum current at resonance if the peak external voltage is 136 V?

27. (II) What will be the peak current in Problem 26 if the capacitor is chosen so that the resonant frequency is twice the applied frequency of 33.0 kHz?

28. (II) A resonant circuit using a 220-μF capacitor is to resonate at 18.0 kHz. The air-core inductor is to be a solenoid with closely packed coils made from 12.0 m of insulated wire 1.1 mm in diameter. How many loops will the inductor contain?

29. (II) (a) Show that oscillation of charge Q on the capacitor of an LRC circuit has amplitude

$$Q_0 = \frac{V_0}{\sqrt{(\omega R)^2 + \left(\omega^2 L - \dfrac{1}{C}\right)^2}}.$$

(b) At what angular frequency, ω', will Q_0 be a maximum? (c) Compare to a damped harmonic oscillator, and discuss. (See also Question 16 in this chapter.)

30. (II) Show that the width of a sharp resonance peak, defined as the difference in (angular) frequency between the two frequencies where $I = \frac{1}{2}I_0$, is given by $\Delta\omega \approx \sqrt{3}\,R/L$.

31. (II) (a) Determine a formula for the average power $\bar{P}$ dissipated in an LRC circuit in terms of L, R, C, ω, and V_0. (b) At what frequency is the power a maximum? (c) Find an approximate formula for the width of the resonance peak in average power, $\Delta\omega$, which is the difference in the two (angular) frequencies where $\bar{P}$ has half its maximum value. Assume a sharp peak.

* Section 31–7

* 32. (I) The output of an ECG amplifier has an impedance of 35 kΩ. It is to be connected to an 8.0-Ω loudspeaker through a transformer. What should be the turns ratio of the transformer?

* 33. (I) An audio amplifier has output connections for 4 Ω, 8 Ω, and 16 Ω. If two 8-Ω speakers are to be connected in parallel, to which output terminals should they be connected?

General Problems

34. Suppose circuit B in Fig. 31–4a consists of a resistance $R = 800\,\Omega$, and a capacitance $C = 1.2\,\mu\text{F}$. Will this capacitor act to eliminate 60-Hz ac but pass a high-frequency signal of frequency 60,000 Hz? To check this, determine the voltage drop across R for a 130-mV signal of frequency (a) 60 Hz; (b) 60,000 Hz.

35. A 230-mH coil, whose resistance is 18.5 Ω, is connected to a capacitor C and a 3360-Hz source voltage. If the current and voltage are to be in phase, what value must C have?

36. A circuit contains two elements, but it is not known if they are L, R, or C. The current in this circuit when connected to a 120-V 60-Hz source is 5.6 A and lags the voltage by 50°. What are the two elements and what are their values?

37. An inductance coil operates at 240 V and 60 Hz. It draws 22.8 A. What is the coil's inductance?

38. (a) What is the impedance of a well-insulated 0.038-μF capacitor connected to a 4.0-kV (rms) 700-Hz line? (b) What will be the peak value of the current?

39. A 3.5-kΩ resistor in series with a 620-mH inductor is driven by an ac power supply. At what frequency is the impedance double that of the impedance at 60 Hz?

40. (a) What is the rms current in an RC circuit if $R = 8.80\,\text{k}\Omega$, $C = 1.80\,\mu\text{F}$, and the rms applied voltage is 120 V at 60.0 Hz? (b) What is the phase angle between voltage and current? (c) What is the power dissipated by the circuit? (d) What are the voltmeter readings across R and C?

41. An inductance coil draws 2.5 A dc when connected to a 36-V battery. When connected to a 60-Hz 120-V (rms) source, the current drawn is 3.8 A (rms). Determine the inductance and resistance of the coil.

42. The **Q factor** of a resonance circuit can be defined as the ratio of the voltage across the capacitor (or inductor) to the voltage across the resistor, at resonance. The larger the Q factor, the sharper the resonance curve will be and the sharper the tuning. (a) Show that the Q factor is given by the equation $Q = (1/R)\sqrt{L/C}$. (b) At a resonant frequency $f_0 = 1.0\,\text{MHz}$, what must be the value of L and R to produce a Q factor of 1000? Assume that $C = 0.010\,\mu\text{F}$. (c) What is the Q factor of the circuit of Example 31–3? [Note: see also Problem 53 in Chapter 30.]

43. In a series LRC circuit, the inductance is 20 mH, the capacitance is 50 nF, and the resistance is 200 Ω. At what frequencies is the power factor equal to 0.17?

44. In our analysis of a series LRC circuit, Fig. 31–6, suppose we chose $V = V_0 \sin\omega t$. (a) Construct a phasor diagram, like that of Fig. 31–8, for this case. (b) Write a formula for the current I, defining all terms.

45. A voltage $V = 0.95\sin 754t$ is applied to an LRC circuit (I is in amperes, t is in seconds and the "angle" is in radians) which has $L = 22.0\,\text{mH}$, $R = 23.2\,\text{k}\Omega$, and $C = 0.30\,\mu\text{F}$. (a) What is the impedance and phase angle? (b) How much power is dissipated in the circuit? (c) What is the rms current and voltage across each element?

46. *Filter circuit.* Figure 31–12 shows a simple filter circuit designed to pass dc voltages with minimal attenuation and to remove, as much as possible, any ac components (such as 60-Hz line voltage that could cause hum in a stereo receiver, for example). Assume $V_{\text{in}} = V_1 + V_2$ where V_1 is dc and $V_2 = V_{20}\sin\omega t$, and that any resistance is very small. (a) Determine the current through the capacitor: give amplitude and phase (assume $R = 0$ and $X_L > X_C$). (b) Show that the ac component of the output voltage, $V_{2\,\text{out}}$, equals $(Q/C) - V_1$, where Q is the charge on the capacitor at any instant, and determine the amplitude and phase of $V_{2\,\text{out}}$. (c) Show that the attenuation of the ac voltage is greatest when $X_C \ll X_L$, and calculate the ratio of the output to input ac voltage in this case. (d) Compare the dc output to input voltage.

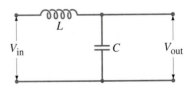

FIGURE 31–12 Problems 46 and 47.

47. Show that if the inductor L in the filter circuit of Fig. 31–12 (Problem 46) is replaced by a large resistor R, there will still be significant attenuation of the ac voltage and little attenuation of the dc voltage if the input dc voltage is high and the current (and power) are low.

48. A resistor R, capacitor C, and inductor L are connected in parallel across an ac generator as shown in Fig. 31–13. The source emf is $V = V_0 \sin \omega t$. Determine the current as a function of time (including amplitude and phase) (a) in the resistor, (b) in the inductor, (c) in the capacitor. (d) What is the total current leaving the source? (Give amplitude I_0 and phase.) (e) Determine the impedance Z defined as $Z = V_0/I_0$. (f) What is the power factor?

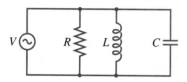

FIGURE 31–13 Problem 48.

49. Suppose a series LRC circuit has two resistors, R_1 and R_2, two capacitors, C_1 and C_2, and two inductors, L_1 and L_2, all in series. Calculate the total impedance of the circuit.

50. Determine the inductance L of the primary of a transformer whose input is 220 V at 60 Hz when the current drawn is 5.8 A. Assume no current in the secondary.

51. You have a small electromagnet that consumes 300 W from a residential circuit operating at 120 V at 60 Hz. Using your ac multimeter, you determine that the unit draws 4.0 A rms. What are the values of the inductance and the internal resistance?

52. An inductor L in series with a resistor R, driven by a sinusoidal voltage source, responds as described by the following differential equation:

$$V_0 \sin \omega t = L\frac{dI}{dt} + RI.$$

Show that a current of the form $I = I_0 \sin(\omega t - \phi)$ flows through the circuit by direct substitution into the differential equation. Determine the amplitude of the current (I_0) and the phase difference ϕ between the current and the voltage source.

53. For the circuit shown in Fig. 31–14, $V = V_0 \sin \omega t$. Calculate the current in each element of the circuit, as well as the total impedance.

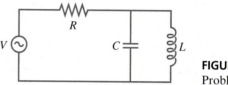

FIGURE 31–14 Problem 53.

54. Show that if the condition $R_1 R_2 = L/C$ is satisfied in the circuit shown in Fig. 31–15, then the potential difference between points a and b is zero for all frequencies.

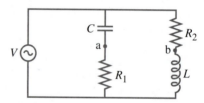

FIGURE 31–15 Problem 54.

These circular disk antennas, each 25 m in diameter, are pointed to receive radio waves from out in space. Radio waves are electromagnetic (EM) waves that have frequencies from a few hundred Hz to about 100 MHz. These antennas are connected together electronically to achieve better detail, and are a part of the Very Large Array in New Mexico searching the heavens for information about the Cosmos.

We will see in this chapter that Maxwell predicted the existence of EM waves from his famous equations. Maxwell's equations themselves are a magnificent summary of electromagnetism. We will also examine how EM waves carry energy and momentum.

Maxwell's Equations and Electromagnetic Waves

The culmination of electromagnetic theory in the nineteenth century was the prediction, and the experimental verification, that waves of electromagnetic fields could travel through space. This achievement opened a whole new world of communication—first the wireless telegraph, then radio and television. And it yielded the spectacular prediction that light is an electromagnetic wave.

The theoretical prediction of electromagnetic waves was the work of the Scottish physicist James Clerk Maxwell (1831–1879; Fig. 32–1), who unified, in one magnificent theory, all the phenomena of electricity and magnetism.

The development of electromagnetic theory in the early part of the nineteenth century by Oersted, Ampère, and others was not actually done in terms of electric and magnetic fields. The idea of the field was introduced somewhat later by Faraday, and was not generally used until Maxwell showed that all electric and magnetic phenomena could be described using only four equations involving electric and magnetic fields. These equations, known as **Maxwell's equations**, are the basic equations for all electromagnetism. They are fundamental in the same sense that Newton's three laws of motion and the law of universal gravitation are for mechanics. In a sense, they are even more fundamental, since they are consistent with the theory of relativity (Chapter 37), whereas Newton's laws are not. Because all of electromagnetism is contained in this set of four equations, Maxwell's equations are considered one of the great triumphs of human intellect.

Before proceeding to a discussion of Maxwell's equations and electromagnetic waves, we first need to discuss a major new prediction of Maxwell's, and, in addition, Gauss's law for magnetism.

FIGURE 32–1 James Clerk Maxwell (1831–1879).

32–1 # Changing Electric Fields Produce Magnetic Fields; Ampère's Law and Displacement Current

Ampère's Law

That a magnetic field is produced by an electric current was discovered by Oersted, and the mathematic relation is given by Ampère's law (Eq. 28–3):

$$\oint \mathbf{B} \cdot d\boldsymbol{l} = \mu_0 I_{\text{encl}}.$$

Is it possible that magnetic fields could be produced in another way as well? For if a changing magnetic field produces an electric field, as discussed in Section 29–7, then perhaps the reverse might be true as well: that *a changing electric field will produce a magnetic field*. If this were true, it would signify a beautiful symmetry in nature.

To back up this idea that a changing electric field might produce a magnetic field, we use an indirect argument that goes something like this. According to Ampère's law, we divide any chosen closed path into short segments $d\boldsymbol{l}$, take the dot product of each $d\boldsymbol{l}$ with the magnetic field $\mathbf{B}$ at that segment, and sum (integrate) all these products over the chosen closed path. That sum will equal μ_0 times the total current I that passes through a surface bounded by the path of the line integral. When we applied Ampère's law to the field around a straight wire (Section 28–4), we imagined the current as passing through the circular area enclosed by our circular loop, and that area is the flat surface 1 shown in Fig. 32–2. However, we could just as well use the sackshaped surface 2 in Fig. 32–2 as the surface for Ampère's law, since the same current I passes through it.

Now consider the closed circular path for the situation of Fig. 32–3, where a capacitor is being discharged. Ampère's law works for surface 1 (current I passes through surface 1), but it does not work for surface 2, since no current passes through surface 2. There is a magnetic field around the wire, so the left side of Ampère's law is not zero; yet no current flows through surface 2, so the right side *is* zero. We seem to have a contradiction of Ampère's law.

There is a magnetic field present in Fig. 32–3, however, only if charge is flowing to or away from the capacitor plates. The changing charge on the plates means that the electric field between the plates is changing in time. Maxwell resolved the problem of no current through surface 2 in Fig. 32–3 by proposing that there needs to be an extra term in Ampère's law involving the changing electric field.

Let us see what this term should be by determining it for the changing electric field between the capacitor plates in Fig. 32–3. The charge Q on a capacitor of capacitance C is $Q = CV$ where V is the potential difference between the plates. Also recall that $V = Ed$ where d is the (small) separation of the plates and E is the (uniform) electric field strength between them, and we ignore any fringing of the field. Also, for a parallel-plate capacitor, $C = \epsilon_0 A/d$, where A is the area of each plate (see Chapter 24). We combine these to obtain:

$$Q = CV = \left(\epsilon_0 \frac{A}{d}\right)(Ed) = \epsilon_0 AE.$$

Now if the charge on the plate changes at a rate dQ/dt, the electric field changes at a proportional rate. That is, by differentiating this expression for Q, we have:

$$\frac{dQ}{dt} = \epsilon_0 A \frac{dE}{dt}.$$

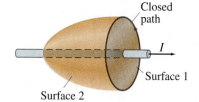

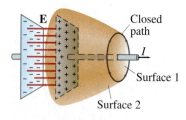

FIGURE 32–2 Ampère's law applied to two different surfaces bounded by the same closed path.

FIGURE 32–3 A capacitor discharging. No conduction current passes through surface 2. An extra term is needed in Ampère's law.

Now dQ/dt is also the current I flowing into or out of the capacitor:

$$I = \frac{dQ}{dt} = \epsilon_0 A \frac{dE}{dt} = \epsilon_0 \frac{d\Phi_E}{dt}$$

where $\Phi_E = EA$ is the **electric flux** through the closed path (surface 2 in Fig. 32–3). In order to make Ampère's law work for surface 2 in Fig. 32–3, as well as for surface 1 (where current I flows), we therefore write:

$$\oint \mathbf{B} \cdot d\mathbf{l} = \mu_0 I_{encl} + \mu_0 \epsilon_0 \frac{d\Phi_E}{dt}. \qquad \text{(32–1)}$$

Ampère's law (generalized)

This equation represents the general form of Ampère's law[†], and embodies Maxwell's idea that a magnetic field can be caused not only by an ordinary electric current, but also by a changing electric field or changing electric flux. Although we arrived at it for a special case, Eq. 32–1 has proved valid in general. The last term on the right in Eq. 32–1 us usually very small, and not easy to measure experimentally.

EXAMPLE 32–1 Charging capacitor. A 30-pF air-gap capacitor has circular plates of area $A = 100\,\text{cm}^2$. It is charged by a 70-V battery through a 2.0-Ω resistor. At the instant the battery is connected, the electric field between the plates is changing most rapidly. At this instant, calculate (a) the current into the plates, and (b) the rate of change of electric field between the plates. (c) Determine the magnetic field induced between the plates. Assume **E** is uniform between the plates at any instant and is zero at all points beyond the edges of the plates.

SOLUTION (a) In Section 26–4 we discussed RC circuits and saw that the charge on a capacitor being charged, as a function of time, is

$$Q = CV_0\left(1 - e^{-t/RC}\right),$$

where V_0 is the voltage of the battery. To find the current at $t = 0$, we differentiate this and substitute the values $V_0 = 70\,\text{V}$, $C = 30\,\text{pF}$, $R = 2.0\,\Omega$:

$$\left.\frac{dQ}{dt}\right|_{t=0} = \left.\frac{CV_0}{RC} e^{-t/RC}\right|_{t=0} = \frac{V_0}{R} = \frac{70\,\text{V}}{2.0\,\Omega} = 35\,\text{A}.$$

This is the rate at which charge accumulates on the capacitor and equals the current flowing in the circuit at this instant.

(b) The electric field between two closely spaced conductors is given by

$$E = \frac{\sigma}{\epsilon_0} = \frac{Q/A}{\epsilon_0}$$

as we saw in Chapter 21 (see Example 21–12). Hence

$$\frac{dE}{dt} = \frac{dQ/dt}{\epsilon_0 A} = \frac{35\,\text{A}}{\left(8.85 \times 10^{-12}\,\text{C}^2/\text{N}\cdot\text{m}^2\right)\left(1.0 \times 10^{-2}\,\text{m}^2\right)} = 4.0 \times 10^{14}\,\text{V/m}\cdot\text{s}.$$

[†] Actually, there is a third term on the right for the case when a magnetic field is produced by magnetized materials. This can be accounted for by changing μ_0 to μ, but we will mainly be interested in cases where no magnetic material is present. In the presence of a dielectric, ϵ_0 is replaced by $\epsilon = K\epsilon_0$ (see Section 24–5).

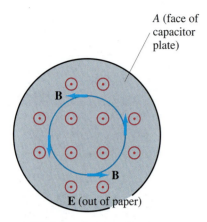

B

B

E (out of paper)

FIGURE 32–4 Frontal view of a circular plate of a parallel-plate capacitor. **E** between plates points out toward viewer; lines of **B** are circles. (Example 32–1.)

A (face of capacitor plate)

(*c*) Although we will not prove it, we might expect the lines of **B** to be perpendicular to **E** and, because of symmetry, to be circles, as shown in Fig. 32–4; this is the same symmetry we saw for the inverse situation of a changing magnetic field producing an electric field (Section 29–7, see Fig. 29–22). To determine B between the plates, we apply Ampère's law, Eq. 32–1, with the current $I_{\text{encl}} = 0$:

$$\oint \mathbf{B} \cdot d\mathbf{l} = \mu_0 \epsilon_0 \frac{d\Phi_E}{dt}.$$

We choose our path to be a circle of radius r, centered at the center of the plate, and thus following a magnetic field line such as the one shown in Fig. 32–4. For $r \leq r_0$ (the radius of plate) the flux through a circle of radius r is $E(\pi r^2)$ since E is assumed uniform within the plates at any moment. So from Ampère's law we have

$$B(2\pi r) = \mu_0 \epsilon_0 \frac{d}{dt}(\pi r^2 E)$$

$$= \mu_0 \epsilon_0 \pi r^2 \frac{dE}{dt}.$$

Hence

$$B = \frac{\mu_0 \epsilon_0}{2} r \frac{dE}{dt}. \qquad [r \leq r_0]$$

We assume $\mathbf{E} = 0$ for $r > r_0$, so for points beyond the edge of the plates all the flux is contained within the plates (area $= \pi r_0^2$) and $\Phi_E = E\pi r_0^2$. Thus Ampère's law gives

$$B(2\pi r) = \mu_0 \epsilon_0 \frac{d}{dt}(\pi r_0^2 E)$$

$$= \mu_0 \epsilon_0 \pi r_0^2 \frac{dE}{dt}$$

or

$$B = \frac{\mu_0 \epsilon_0 r_0^2}{2r} \frac{dE}{dt}. \qquad [r \geq r_0]$$

B has its maximum value at $r = r_0$ which, from either relation above (using $r_0 = \sqrt{A/\pi} = 5.6$ cm), is

$$B = \frac{\mu_0 \epsilon_0 r_0}{2} \frac{dE}{dt}$$

$$= \tfrac{1}{2}(4\pi \times 10^{-7}\,\text{T} \cdot \text{m/A})(8.85 \times 10^{-12}\,\text{C}^2/\text{N} \cdot \text{m}^2)(5.6 \times 10^{-2}\,\text{m})(4.0 \times 10^{14}\,\text{V/m} \cdot \text{s})$$

$$= 1.2 \times 10^{-4}\,\text{T}.$$

This is a very small field and lasts only briefly (the time constant $RC = 6.0 \times 10^{-11}$ s) and so would be very difficult to measure.

Let us write the magnetic field B outside the capacitor plates of Example 32–1 in terms of the current I that leaves the plates. The electric field between the plates can be written $E = \sigma/\epsilon_0 = Q/\epsilon_0 A$, as we just saw (part *b*). Hence B for $r > r_0$ is, with $dQ/dt = I$,

$$B = \frac{\mu_0 \epsilon_0 r_0^2}{2r} \frac{dE}{dt} = \frac{\mu_0 \epsilon_0 r_0^2}{2r} \frac{I}{\epsilon_0 \pi r_0^2} = \frac{\mu_0 I}{2\pi r}.$$

This is the same formula for the field that surrounds a wire (Eq. 28–1). Thus the B field outside the capacitor is the same as that outside the wire. In other words, the magnetic field produced by the changing electric field between the plates is the same as that produced by the current in the wire.

Displacement Current

Maxwell thought it useful to interpret the second term on the right in Eq. 32–1 as being *equivalent* to an electric current. He called it a **displacement current**, I_D. An ordinary current I is then called a **conduction current**. Ampère's law can then be written

$$\oint \mathbf{B} \cdot d\boldsymbol{l} = \mu_0 (I + I_D)_{encl} \qquad (32\text{–}2)$$

where

$$I_D = \epsilon_0 \frac{d\Phi_E}{dt}. \qquad (32\text{–}3)$$

Displacement current

The term "displacement current" was based on an old discarded theory; don't let it confuse you: I_D does not represent a flow of electric charge[†] nor is there a displacement.

32–2 Gauss's Law for Magnetism

We are almost in a position to state Maxwell's equations, but first we need to discuss the magnetic equivalent of Gauss's law. As we saw in Chapter 29, for a magnetic field $\mathbf{B}$, the *magnetic flux* Φ_B through a surface is defined as

$$\Phi_B = \int \mathbf{B} \cdot d\mathbf{A}$$

where the integral is over the area of either an open or a closed surface. The magnetic flux through a closed surface—that is, a surface which completely encloses a volume—is written

$$\Phi_B = \oint \mathbf{B} \cdot d\mathbf{A}.$$

In the electric case, we saw in Section 22–2 that the electric flux Φ_E through a closed surface is equal to the total net charge Q enclosed by the surface, divided by ϵ_0 (Eq. 22–4):

$$\oint \mathbf{E} \cdot d\mathbf{A} = \frac{Q}{\epsilon_0}.$$

This relation is Gauss's law for electricity.

We can write a similar relation for the magnetic flux. We have seen, however, that in spite of intense searches, no isolated magnetic poles (monopoles)—the magnetic equivalent of single electric charges—have ever been observed. Hence, **Gauss's law for magnetism** is

$$\oint \mathbf{B} \cdot d\mathbf{A} = 0. \qquad (32\text{–}4)$$

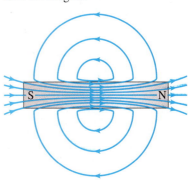

FIGURE 32–5 Magnetic field lines for a bar magnet.

In terms of magnetic field lines, this relation tells us that as many lines enter the enclosed volume as leave it. If, indeed, magnetic monopoles do not exist, then there are no "sources" or "sinks" for magnetic field lines to start or stop on, as electric field lines start on positive charges and end on negative charges. Magnetic field lines must then be continuous. Even for a bar magnet, a magnetic field $\mathbf{B}$ exists inside as well as outside the magnetic material, and the lines of $\mathbf{B}$ are closed loops as shown in Fig. 32–5.

[†]The interpretation of the changing electric field as a displacement current does fit in well with our discussion in Chapter 31 where we saw that an alternating current seems to pass through a capacitor (although charge doesn't). It also means that Kirchhoff's point rule will be valid even at a capacitor plate: for conduction current flows into the plate, but no conduction current flows out of the plate. Instead a "displacement current" flows out of one plate (toward the other).

32–3 Maxwell's Equations

MAXWELL'S

EQUATIONS

With the extension of Ampère's law given by Eq. 32–1, plus Gauss's law for magnetism (Eq. 32–4), we are now ready to state all four of Maxwell's equations. We have seen them all before in the past dozen chapters. In the absence of dielectric or magnetic materials, **Maxwell's equations** are:

$$\oint \mathbf{E} \cdot d\mathbf{A} = \frac{Q}{\epsilon_0} \tag{32–5a}$$

$$\oint \mathbf{B} \cdot d\mathbf{A} = 0 \tag{32–5b}$$

$$\oint \mathbf{E} \cdot d\mathbf{l} = -\frac{d\Phi_B}{dt} \tag{32–5c}$$

$$\oint \mathbf{B} \cdot d\mathbf{l} = \mu_0 I + \mu_0 \epsilon_0 \frac{d\Phi_E}{dt}. \tag{32–5d}$$

The first two of Maxwell's equations are simply Gauss's law for electricity (Chapter 22, Eq. 22–4) and Gauss's law for magnetism (Section 32–2, Eq. 32–4). The third is Faraday's law (Chapter 29, Eq. 29–2) and the fourth is Ampère's law as modified by Maxwell (Eq. 32–1). (We dropped the subscripts for simplicity.)

They can be summarized in words: (1) a generalized form of Coulomb's law relating electric field to its sources, electric charges; (2) the same for the magnetic field, except that if there are no magnetic monopoles, magnetic field lines are continuous—they do not begin or end (as electric field lines do on charges); (3) an electric field is produced by a changing magnetic field; (4) a magnetic field is produced by an electric current or by a changing electric field.

Maxwell's equations are the basic equations for all electromagnetism. All of electromagnetism is contained in this set of four equations. They are as fundamental as Newton's three laws of motion and the law of universal gravitation.

In earlier chapters, we have seen that we can treat electric and magnetic fields separately if they do not vary in time. But we cannot treat them independently if they do change in time. For a changing magnetic field produces an electric field; and a changing electric field produces a magnetic field. An important outcome of these relations is the production of electromagnetic waves.

32–4 Production of Electromagnetic Waves

According to Maxwell, a magnetic field will be produced in empty space if there is a changing electric field. From this, Maxwell derived another startling conclusion. If a changing magnetic field produces an electric field, that electric field will itself be changing. This changing electric field will in turn produce a magnetic field, which itself will be changing and so will produce a changing electric field; and so on. When Maxwell manipulated his equations, he found that the net result of these interacting changing fields was a *wave* of electric and magnetic fields that can actually propagate (travel) through space! We now examine, in a simplified way, how such **electromagnetic waves** can be produced. (In Section 32–5 we will examine them more quantitatively.)

Consider two conducting rods that will serve as an "antenna," Fig. 32–6a. Suppose that these two rods are connected by a switch to the opposite terminals of a battery. As soon as the switch is closed, the upper rod quickly becomes positively charged and the lower one negatively charged. Electric field lines are formed as indicated by the red lines in Fig. 32–6b. While the charges are flowing, a current exists whose direction is indicated by the arrows. A magnetic field is therefore produced near the antenna. The magnetic field lines encircle the wires and therefore, in the plane of the page, **B** points into the paper ($\otimes$) on the right and out of the paper ($\odot$)

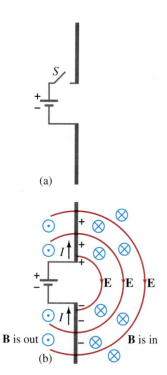

FIGURE 32–6 Fields produced by charge flowing in conductors. It takes time for the **E** and **B** fields to travel outward to distant points.

on the left. Now we ask, how far out do these electric and magnetic fields extend? In the static case, the fields extend outward indefinitely far. However, when the switch in Fig. 32–6 is closed, the fields quickly appear nearby, but it takes time for them to reach distant points. Both electric and magnetic fields store energy, and this energy cannot be transferred to distant points with infinite speed.

Now we look at the situation of Fig. 32–7 where our antenna is connected to an ac generator. In Fig. 32–7a, the connection has just been completed. Charge starts building up and fields form just as in Fig. 32–6. The + and − signs in Fig. 32–7a indicate the net charge on each rod. The black arrows indicate the direction of the current. The electric field is represented by the red lines in the plane of the page; and the magnetic field, according to the right-hand rule, is into ($\otimes$) or out of ($\odot$) the page. In Fig. 32–7b, the voltage of the ac generator has reversed in direction; the current is reversed and the new magnetic field is in the opposite direction. Because the new fields have changed direction, the old lines fold back to connect up to some of the new lines and form closed loops as shown.[†] The old fields, however, don't suddenly disappear; they are on their way to distant points. Indeed, because a changing magnetic field produces an electric field, and a changing electric field produces a magnetic field, this combination of changing electric and magnetic fields moving outward is self-supporting, no longer depending on the antenna charges.

The fields not far from the antenna, referred to as the *near field*, become quite complicated, but we are not so interested in them. We are instead mainly interested in the fields far from the antenna (they are generally what we detect), which we refer to as the **radiation field**. The electric field lines form loops, as shown in Fig. 32–8, and continue moving outward. The magnetic field lines also form closed loops, but are not shown since they are perpendicular to the page. Although the lines are shown only on the right of the source, fields also travel in other directions. The field strengths are greatest in directions perpendicular to the oscillating charges; and they drop to zero along the direction of oscillation—above and below the antenna in Fig. 32–8.

The magnitudes of both **E** and **B** in the radiation field are found to decrease with distance as $1/r$. (Compare this to the static electric field given by Coulomb's law where **E** decreases as $1/r^2$.) The energy carried by the electromagnetic wave is proportional (as for any wave, Chapter 15) to the square of the amplitude, E^2 or B^2, as will be discussed further in Section 32–7, so the intensity of the wave decreases as $1/r^2$.

Several things about the radiation field can be noted from Fig. 32–8. First, *the electric and magnetic fields at any point are perpendicular to each other, and to the direction of motion.* Second, we can see that the fields alternate in direction (**B** is into the page at some points and out of the page at others, similarly for **E**). Thus, the field strengths vary from a maximum in one direction, to zero, to a maximum in the other direction. The electric and magnetic fields are "in phase": that is, they each are zero at the same points and reach their maxima at the same points in space. Finally, very far from the antenna (Fig. 32–8b) the field lines are quite flat over a reasonably large area, and the waves are referred to as **plane waves**.

[†] We are considering waves traveling through empty space, so there are no charges for lines of **E** to start or stop on, so they form closed loops. Magnetic field lines always form closed loops since there seem to be no single (separate) magnetic poles.

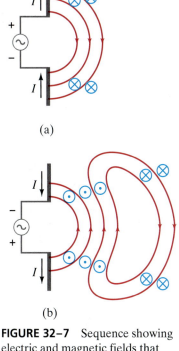

FIGURE 32–7 Sequence showing electric and magnetic fields that spread outward from oscillating charges on two conductors connected to an ac source (see the text).

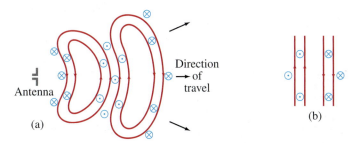

FIGURE 32–8 (a) The radiation fields (far from the antenna) produced by a sinusoidal signal on the antenna. The closed loops represent electric field lines. The magnetic field lines, perpendicular to the page and represented by $\otimes$ and $\odot$, also form closed loops. (b) Very far from the antenna the wave fronts (field lines) are essentially flat over a fairly large area, and are referred to as *plane waves*.

If the source voltage varies sinusoidally, then the electric and magnetic field strengths in the radiation field will also vary sinusoidally. The sinusoidal character of the waves is diagrammed in Fig. 32–9, which shows the field *strengths* plotted as a function of position. Notice that **B** and **E** are perpendicular to each other and to the direction of travel, as pointed out above.

We call these waves electromagnetic (EM) waves. They are *transverse* waves and resemble other types of waves (Chapter 15). However, EM waves are always waves of *fields*, not of matter like waves on water or a rope. Because they are fields, EM waves can propagate in empty space.

We have seen in the above analysis that EM waves are produced by electric charges that are oscillating in an antenna, and hence are undergoing acceleration. In fact, we can say in general that

accelerating electric charges give rise to electromagnetic waves.

Electromagnetic waves can be produced in other ways as well, requiring description at the atomic and nuclear levels, as we will discuss later.

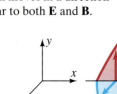

FIGURE 32–9 Electric and magnetic field strengths in an electromagnetic wave. **E** and **B** are at right angles to each other. The entire pattern moves in a direction perpendicular to both **E** and **B**.

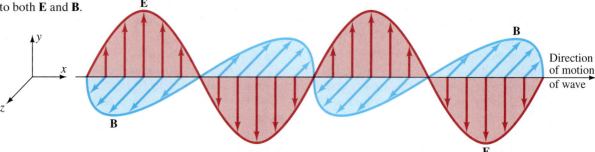

32–5 | Electromagnetic Waves, and Their Speed, from Maxwell's Equations

Let us now examine how the existence of EM waves follows from Maxwell's equations. We will see that Maxwell's prediction of the existence of EM waves was startling. Equally startling was the speed at which they were predicted to travel.

We begin by considering a region of free space, where there are *no charges or conduction currents*—that is, far from the source so that the wave fronts (the field lines in Fig. 32–8) are essentially flat over a reasonable area. They are then called **plane waves**, as we saw, meaning that at any instant, **E** and **B** are uniform over a reasonably large plane perpendicular to the direction of propagation. We also assume, in a particular coordinate system, that the wave is traveling in the x direction with velocity $\mathbf{v} = v\mathbf{i}$, that **E** is parallel to the y axis, and that **B** is parallel to the z axis, as in Fig. 32–9.

Maxwell's equations, with $Q = I = 0$, become

$$\oint \mathbf{E} \cdot d\mathbf{A} = 0 \tag{32–6a}$$

Maxwell's

$$\oint \mathbf{B} \cdot d\mathbf{A} = 0 \tag{32–6b}$$

equations

$$\oint \mathbf{E} \cdot d\mathbf{l} = -\frac{d\Phi_B}{dt} \tag{32–6c}$$

in vacuum

$$\oint \mathbf{B} \cdot d\mathbf{l} = \mu_0 \epsilon_0 \frac{d\Phi_E}{dt}. \tag{32–6d}$$

Notice the beautiful symmetry of these equations. The term on the right in the last equation, conceived by Maxwell, is essential for this symmetry. It is also essential if electromagnetic waves are to be produced, as we will now see.

If the wave is sinusoidal with wavelength λ and frequency f, then, as we saw in Chapter 15, Section 15–4, such a traveling wave can be written as

$$E = E_y = E_0 \sin(kx - \omega t)$$

$$B = B_z = B_0 \sin(kx - \omega t)$$

(32–7)

where

$$k = \frac{2\pi}{\lambda}, \qquad \omega = 2\pi f, \qquad \text{and} \qquad f\lambda = \frac{\omega}{k} = v,$$

(32–8)

with v being the speed of the wave. Although visualizing the wave as sinusoidal is helpful, we will not have to assume this in most of what follows.

Consider now a small rectangle in the plane of the electric field as shown in Fig. 32–10. This rectangle has a finite height Δy, and a very thin width which we take to be the infinitesimal distance dx. In order to show that $\mathbf{E}$, $\mathbf{B}$, and $\mathbf{v}$ are in the orientation shown, we apply Lenz's law to this rectangular loop. The changing magnetic flux through this loop is related to the electric field around the loop by Faraday's law (Maxwell's third equation, Eq. 32–6c). For the case shown, B through the loop is decreasing in time (the wave is moving to the right). So the electric field must be in a direction to oppose this change, meaning E must be greater on the right side of the loop than on the left, as shown (so it could produce a counterclockwise current whose magnetic field would act to oppose the change in Φ_B—but of course there is no current). This brief argument shows that the orientation of $\mathbf{E}$, $\mathbf{B}$, and $\mathbf{v}$ are in the correct relation as shown. That is, $\mathbf{v}$ is in the direction of $\mathbf{E} \times \mathbf{B}$. Now let us apply Faraday's law, which is Maxwell's third equation (Eq. 32–6c),

$$\oint \mathbf{E} \cdot d\mathbf{l} = -\frac{d\Phi_B}{dt}$$

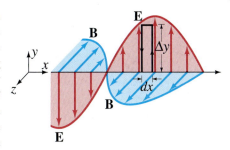

FIGURE 32–10 Applying Faraday's law to the rectangle $(\Delta y)(dx)$.

to the rectangle of height Δy and width dx shown in Fig. 32–10. First we consider $\oint \mathbf{E} \cdot d\mathbf{l}$. Along the short top and bottom sections of length dx, $\mathbf{E}$ is perpendicular to $d\mathbf{l}$, so $\mathbf{E} \cdot d\mathbf{l} = 0$. Along the vertical sides, we let E be the electric field along the left side, and on the right side where it will be slightly larger, it is $E + dE$. Thus, if we take our loop counterclockwise,

$$\oint \mathbf{E} \cdot d\mathbf{l} = (E + dE)\,\Delta y - E\,\Delta y = dE\,\Delta y.$$

For the right side of Faraday's law, the magnetic flux through the loop changes as

$$\frac{d\Phi_B}{dt} = \frac{dB}{dt}\,dx\,\Delta y,$$

since the area of the loop, $(dx)(\Delta y)$, is not changing. Thus, Faraday's law gives us

$$dE\,\Delta y = -\frac{dB}{dt}\,dx\,\Delta y$$

or

$$\frac{dE}{dx} = -\frac{dB}{dt}.$$

Actually, both E and B are functions of position x and time t. We should therefore use partial derivatives:

$$\frac{\partial E}{\partial x} = -\frac{\partial B}{\partial t}$$

(32–9)

where $\partial E/\partial x$ means the derivative of E with respect to x while t is held fixed, and $\partial B/\partial t$ is the derivative of B with respect to t while x is kept fixed.

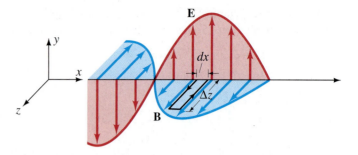

FIGURE 32–11 Applying Maxwell's fourth equation to the rectangle $(\Delta z)(dx)$.

We can obtain another important relation between E and B in addition to Eq. 32–9. To do so, we consider now a small rectangle in the plane of **B**, whose length and width are Δz and dx as shown in Fig. 32–11. To this rectangular loop we apply Maxwell's fourth equation (the extension of Ampère's law):

$$\oint \mathbf{B} \cdot d\boldsymbol{l} = \mu_0 \epsilon_0 \frac{d\Phi_E}{dt}$$

where we have taken $I = 0$ since we assume the absence of conduction currents. Along the short sides (dx), $\mathbf{B} \cdot d\boldsymbol{l}$ is zero since **B** is perpendicular to $d\boldsymbol{l}$. Along the longer sides (Δz), we let B be the magnetic field along the left side of length Δz, and $B + dB$ be the field along the right side. We again integrate counterclockwise, so

$$\oint \mathbf{B} \cdot d\boldsymbol{l} = B\,\Delta z - (B + dB)\,\Delta z = -dB\,\Delta z.$$

The right side of Maxwell's fourth equation is

$$\mu_0 \epsilon_0 \frac{d\Phi_E}{dt} = \mu_0 \epsilon_0 \frac{dE}{dt}\, dx\,\Delta z.$$

Equating the two expressions, we obtain

$$-dB\,\Delta z = \mu_0 \epsilon_0 \frac{dE}{dt}\, dx\,\Delta z$$

or

$$\frac{\partial B}{\partial x} = -\mu_0 \epsilon_0 \frac{\partial E}{\partial t} \qquad\qquad \textbf{(32–10)}$$

where we have replaced dB/dx and dE/dt by the proper partial derivatives as before.

We can use Eqs. 32–9 and 32–10 to obtain a relation between the magnitudes of **E** and **B**, and the speed v. Let E and B be given by Eqs. 32–7 as a function of x and t. When we apply Eq. 32–9, taking the derivatives of E and B as given by Eqs. 32–7, we obtain

$$kE_0 \cos(kx - \omega t) = \omega B_0 \cos(kx - \omega t)$$

or

$$\frac{E_0}{B_0} = \frac{\omega}{k} = v,$$

since $v = \omega/k$ (see Eq. 32–8 or 15–12). Since E and B are in phase, we see that E and B are related by

$$\frac{E}{B} = v \qquad\qquad \textbf{(32–11)}$$

at any point in space, where v is the velocity of the wave.

Now we apply Eq. 32–10 to the sinusoidal fields (Eqs. 32–7) and we obtain

$$kB_0 \cos(kx - \omega t) = \mu_0 \epsilon_0 \omega E_0 \cos(kx - \omega t)$$

or

$$\frac{B_0}{E_0} = \frac{\mu_0 \epsilon_0 \omega}{k} = \mu_0 \epsilon_0 v.$$

But $B_0/E_0 = 1/v$ from Eq. 32–11, so

$$\mu_0 \epsilon_0 v = \frac{1}{v}$$

or

$$v = \frac{1}{\sqrt{\epsilon_0 \mu_0}}. \qquad\qquad\qquad \textbf{(32–12)}$$

Thus the speed of electromagnetic waves in free space is a constant, independent of the wavelength or frequency. If we put in values for ϵ_0 and μ_0 we find

$$v = \frac{1}{\sqrt{\epsilon_0 \mu_0}} = \frac{1}{\sqrt{(8.85 \times 10^{-12}\,\text{C}^2/\text{N}\cdot\text{m}^2)(4\pi \times 10^{-7}\,\text{T}\cdot\text{m}/\text{A})}}$$

$$= 3.00 \times 10^8\,\text{m/s}.$$

Speed of EM waves is speed of light

This is a remarkable result. For this is precisely equal to the measured speed of light!

* Derivation of Speed of Light (General)

We can derive the speed of EM waves without having to assume sinusoidal waves by combining Eqs. 32–9 and 32–10 as follows. We take the derivative with respect to t of Eq. 32–10

$$\frac{\partial^2 B}{\partial t\, \partial x} = -\mu_0 \epsilon_0 \frac{\partial^2 E}{\partial t^2}.$$

We next take the derivative of Eq. 32–9 with respect to x:

$$\frac{\partial^2 E}{\partial x^2} = -\frac{\partial^2 B}{\partial t\, \partial x}.$$

Since $\partial^2 B/\partial t\, \partial x$ appears in both relations, we obtain

$$\frac{\partial^2 E}{\partial t^2} = \frac{1}{\mu_0 \epsilon_0} \frac{\partial^2 E}{\partial x^2}. \qquad\qquad \textbf{(32–13a)}$$

By taking other derivatives of Eqs. 32–9 and 32–10 we obtain the same relation for B:

$$\frac{\partial^2 B}{\partial t^2} = \frac{1}{\mu_0 \epsilon_0} \frac{\partial^2 B}{\partial x^2}. \qquad\qquad \textbf{(32–13b)}$$

Both of Eqs. 32–13 have the form of the *wave equation* for a plane wave traveling in the x direction,

$$\frac{\partial^2 y}{\partial t^2} = v^2 \frac{\partial^2 y}{\partial x^2},$$

as discussed in Section 15–5 (Eq. 15–16). We see that the velocity v is given by

$$v^2 = \frac{1}{\mu_0 \epsilon_0}$$

in agreement with Eq. 32–12. Thus we see that a natural outcome of Maxwell's equations is that E and B obey the wave equation for waves traveling with speed $v = 1/\sqrt{\mu_0 \epsilon_0}$. It was on this basis that Maxwell predicted the existence of electromagnetic waves.

32–6 Light as an Electromagnetic Wave and the Electromagnetic Spectrum

The calculations in Section 32–5 gave the result that Maxwell himself determined: that the speed of EM waves is 3.00×10^8 m/s, the same as the measured speed of light.

Light had been shown some 60 years previously to behave like a wave (we'll discuss this in Chapter 35). But nobody knew what kind of wave it was—that is, what is it that is oscillating in a light wave? Maxwell, on the basis of the calculated speed of EM waves, argued that light must be an electromagnetic wave. This idea soon came to be generally accepted by scientists, but not fully until after EM waves were experimentally detected. EM waves were first generated and detected experimentally by Heinrich Hertz (1857–1894) in 1887, eight years after Maxwell's death. Hertz used a spark-gap apparatus in which charge was made to rush back and forth for a short time, generating waves whose frequency was about 10^9 Hz. He detected them some distance away using a loop of wire in which an emf was produced when a changing magnetic field passed through. These waves were later shown to travel at the speed of light, 3.00×10^8 m/s, and to exhibit all the characteristics of light such as reflection, refraction, and interference. The only difference was that they were not visible. Hertz's experiment was a strong confirmation of Maxwell's theory.

The wavelengths of visible light were measured in the first decade of the nineteenth century, long before anyone imagined that light was an electromagnetic wave. The wavelengths were found to lie between 4.0×10^{-7} m and 7.5×10^{-7} m; or 400 nm to 750 nm $\left(1 \text{ nm} = 10^9 \text{ m}\right)$. The frequencies of visible light can be found using Eq. 15–1, which we rewrite here:

$$f\lambda = c, \qquad (32\text{–}14)$$

c is symbol for speed of light

where f and λ are the frequency and wavelength, respectively, of the wave. Here, c is the speed of light, 3.00×10^8 m/s; it gets the special symbol c because of its universality for all EM waves in free space. Equation 32–14 tells us that the frequencies of visible light are between 4.0×10^{14} Hz and 7.5×10^{14} Hz. $\left(\text{Recall that } 1 \text{ Hz} = 1 \text{ cycle per second} = 1 \text{ s}^{-1}.\right)$

But visible light is only one kind of EM wave. As we have seen, Hertz produced EM waves of much lower frequency, about 10^9 Hz. These are called **radio waves**, since frequencies in this range are used today to transmit radio and TV signals. Electromagnetic waves, or EM radiation as we sometimes call it, have been produced or detected over a wide range of frequencies. They are usually categorized as shown in Fig. 32–12, which is known as the **electromagnetic spectrum**.

EM spectrum

Radio waves and microwaves can be produced in the laboratory using electronic equipment, as we saw in Fig. 32–7. Higher-frequency waves are very difficult to produce electronically. These and other types of EM waves are produced in natural processes, as emission from atoms, molecules, and nuclei (more on this later).

FIGURE 32–12
Electromagnetic spectrum.

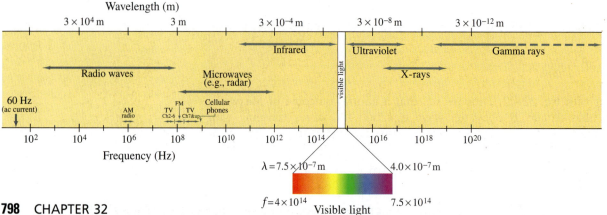

EM waves can be produced by the acceleration of electrons or other charged particles, such as electrons accelerating in the antenna of Fig. 32–7. Another example is X-rays, which are produced (Chapter 36) when fast-moving electrons are rapidly decelerated upon striking a metal target. Even the visible light emitted by an ordinary incandescent light is due to electrons undergoing acceleration within the hot filament. We will meet various types of EM waves later. However, it is worth mentioning here that infrared (IR) radiation (EM waves whose frequency is just less than that of visible light) is mainly responsible for the heating effect of the Sun. The Sun emits not only visible light but substantial amounts of IR and UV (ultraviolet) as well. The molecules of our skin tend to "resonate" at infrared frequencies, so it is these that are preferentially absorbed and thus warm us up. We humans experience EM waves differently depending on their wavelengths: Our eyes detect wavelengths between 4 and 7×10^{-7} m (visible light), whereas our skin detects longer wavelengths (IR). Many EM wavelengths we don't detect directly at all.

EXAMPLE 32–2 **Wavelengths of EM waves.** Calculate the wavelength: (a) of a 60-Hz EM wave, (b) of a 93.3-MHz FM radio wave, and (c) of a beam of visible red light from a laser at frequency 4.74×10^{14} Hz.

SOLUTION (a) Since $c = \lambda f$,

$$\lambda = \frac{c}{f} = \frac{3.0 \times 10^8 \text{ m/s}}{60 \text{ s}^{-1}} = 5.0 \times 10^6 \text{ m},$$

or 5000 km. 60 Hz is the frequency of ac current in the United States, and, as we see here, one wavelength stretches all the way across the country.

(b)

$$\lambda = \frac{3.00 \times 10^8 \text{ m/s}}{93.3 \times 10^6 \text{ s}^{-1}} = 3.22 \text{ m}.$$

The length of an FM antenna is about half this $(\frac{1}{2}\lambda)$.

(c)

$$\lambda = \frac{3.00 \times 10^8 \text{ m/s}}{4.74 \times 10^{14} \text{ s}^{-1}} = 6.33 \times 10^{-7} \text{ m} \, (= 633 \text{ nm}).$$

EXAMPLE 32–3 **Determining E and B in EM waves.** Assume the 60 Hz EM wave in Example 32–2 is a sinusoidal wave propagating in the z direction with **E** pointing in the x direction, and $E_0 = 2.0$ V/m. Write vector expressions for **E** and **B** as functions of position and time.

SOLUTION From Eq. 32–8 we have

$$k = 2\pi/\lambda = \frac{2\pi}{5.0 \times 10^6 \text{ m}} = 1.26 \times 10^{-6} \text{ m}^{-1}$$

$$\omega = 2\pi f = 2\pi(60 \text{ Hz}) = 3.77 \times 10^2 \text{ rad/s}.$$

From Eq. 32–11 with $v = c$, we find that

$$B_0 = \frac{E_0}{c} = \frac{2.0 \text{ V/m}}{3.0 \times 10^8 \text{ m/s}} = 6.7 \times 10^{-9} \text{ T}.$$

The direction of propagation is that of **E** × **B**, as in Fig. 32–9. With **E** pointing in the x direction, and the wave propagating in the z direction, **B** must point in the y direction. Using Eqs. 32–7 we find:

$$\mathbf{E} = \mathbf{i}(2.0 \text{ V/m}) \sin[(1.26 \times 10^{-6} \text{ m}^{-1})z - (3.77 \times 10^2 \text{ rad/s})t]$$

$$\mathbf{B} = \mathbf{j}(6.67 \times 10^{-9} \text{ T}) \sin[(1.26 \times 10^{-6} \text{ m}^{-1})z - (3.77 \times 10^2 \text{ rad/s})t]$$

FIGURE 32–13 Coaxial cable.

Electromagnetic waves can travel along transmission lines as well as in empty space. When a source of emf is connected up to a transmission line—be it two parallel wires or a coaxial cable (Fig. 32–13)—the electric field within the wires is not set up immediately at all points along them, just as we saw in Section 32–4 with reference to Fig. 32–7. Indeed, it can be shown that if the wires are separated by air, the electrical signal travels along the wires at the speed $c = 3.0 \times 10^8$ m/s. For example, when you flip a light switch, the light actually goes on a tiny fraction of a second later. If the wires are in a medium whose electric permittivity is ϵ and magnetic permeability is μ, the speed is not given by Eq. 32–12, but by

$$v = \frac{1}{\sqrt{\epsilon \mu}}.$$

32–7 Energy in EM Waves; the Poynting Vector

Electromagnetic waves carry energy from one region of space to another. This energy is associated with the moving electric and magnetic fields. In Section 24–4, we saw that the energy density (J/m^3) stored in an electric field E is $u_E = \frac{1}{2}\epsilon_0 E^2$, where u_E is the energy per unit volume. The energy stored in a magnetic field B, as we discussed in Section 30–3, is given by $u_B = \frac{1}{2}B^2/\mu_0$. Thus, the total energy stored per unit volume in a region of space where there is an electromagnetic wave is

$$u = \frac{1}{2}\epsilon_0 E^2 + \frac{1}{2}\frac{B^2}{\mu_0}.$$ **(32–15)**

In this equation, E and B represent the electric and magnetic field strengths of the wave at any instant in a small region of space. We can write Eq. 32–15 in terms of the E field only, since from Eq. 32–12 we have $\sqrt{\epsilon_0 \mu_0} = 1/c$, and from Eq. 32–11, $B = E/c$. We insert these into Eq. 32–15 to obtain

$$u = \frac{1}{2}\epsilon_0 E^2 + \frac{1}{2}\frac{\epsilon_0 \mu_0 E^2}{\mu_0}$$

$$= \epsilon_0 E^2.$$ **(32–16a)**

Notice that the energy density associated with the B field is equal to that associated with the E field, so each contributes half to the total energy. We can also write the energy density in terms of the B field only:

$$u = \epsilon_0 E^2 = \epsilon_0 c^2 B^2 = \frac{\epsilon_0 B^2}{\epsilon_0 \mu_0},$$

so

$$u = \frac{B^2}{\mu_0}.$$ **(32–16b)**

We can also write u in one term containing both E and B:

$$u = \epsilon_0 E^2 = \epsilon_0 EcB = \frac{\epsilon_0 EB}{\sqrt{\epsilon_0 \mu_0}},$$

or

$$u = \sqrt{\frac{\epsilon_0}{\mu_0}}\, EB.$$ **(32–16c)**

Equations 32–16 give the energy density in any region of space at any instant.

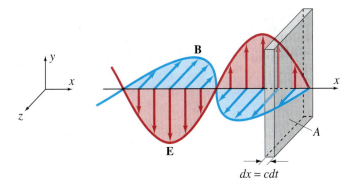

FIGURE 32–14 Electromagnetic wave carrying energy through area A.

$dx = c\,dt$

Now let us determine the energy the wave transports per unit time per unit area. This is given by a vector **S**, which is called the **Poynting vector**.[†] The units of **S** are W/m². The direction of **S** is the direction in which the energy is transported, which is the direction in which the wave is moving. Let us imagine the wave is passing through an area A perpendicular to the x axis as shown in Fig. 32–14. In a short time dt, the wave moves to the right a distance $dx = c\,dt$ where c is the wave speed. The energy that passes through A in the time dt is the energy that occupies the volume $dV = A\,dx = Ac\,dt$. The energy density u is $u = \epsilon_0 E^2$ where E is the electric field in this volume at the given instant. So the total energy dU contained in this volume dV is the energy density u times the volume: $dU = u\,dV = (\epsilon_0 E^2)(Ac\,dt)$. Therefore the energy crossing the area A per time dt is

$$S = \frac{1}{A}\frac{dU}{dt} = \epsilon_0 c E^2. \tag{32–17}$$

Since $E = cB$ and $c = 1/\sqrt{\epsilon_0 \mu_0}$, this can also be written:

$$S = \epsilon_0 c E^2 = \frac{cB^2}{\mu_0} = \frac{EB}{\mu_0}.$$

The direction of **S** is along **v**, perpendicular to **E** and **B**, so the Poynting vector **S** can be written

$$\mathbf{S} = \frac{1}{\mu_0}(\mathbf{E} \times \mathbf{B}). \tag{32–18}$$

Poynting vector

Equation 32–17 or 32–18 gives the energy transported per unit area per unit time at any *instant*. We often want to know the *average* over an extended period of time since the frequencies are usually so high we don't detect the rapid time variation. If E and B are sinusoidal, then $\overline{E^2} = E_0^2/2$, just as for electric currents and voltages (Section 25–7), where E_0 is the *maximum* value of E. Thus we can write for the magnitude of the Poynting vector, on the average,

$$\overline{S} = \frac{1}{2}\epsilon_0 c E_0^2 = \frac{1}{2}\frac{c}{\mu_0}B_0^2 = \frac{E_0 B_0}{2\mu_0}, \tag{32–19a}$$

where B_0 is the maximum value of B. This time averaged value of **S** is the **intensity**, defined as the average power transferred across unit area (Section 15–3). We can also write

Intensity

$$\overline{S} = \frac{E_{\text{rms}} B_{\text{rms}}}{\mu_0} \tag{32–19b}$$

where E_{rms} and B_{rms} are the rms values $\left(E_{\text{rms}} = \sqrt{\overline{E^2}},\ B_{\text{rms}} = \sqrt{\overline{B_{\text{rms}}^2}}\right)$.

[†] After J. H. Poynting (1852–1914).

EXAMPLE 32–4 **E and B from the Sun.** Radiation from the Sun reaches the Earth (above the atmosphere) at a rate of about 1350 W/m². Assume that this is a single EM wave and calculate the maximum values of E and B.

SOLUTION Since $\overline{S} = 1350 \text{ W/m}^2 = \epsilon_0 c E_0^2 / 2$, then

$$E_0 = \sqrt{\frac{2\overline{S}}{\epsilon_0 c}}$$

$$= \sqrt{\frac{2(1350 \text{ W/m}^2)}{(8.85 \times 10^{-12} \text{ C}^2/\text{N} \cdot \text{m}^2)(3.0 \times 10^8 \text{ m/s})}}$$

$$= 1.01 \times 10^3 \text{ V/m}.$$

From Eq. 32–11, $B = E/c$, so

$$B_0 = \frac{E_0}{c} = \frac{1.01 \times 10^3 \text{ V/m}}{3.00 \times 10^8 \text{ m/s}} = 3.37 \times 10^{-6} \text{ T}.$$

This Example illustrates that B has a small numerical value compared to E. This is because of the different units for E and B and the way these units are defined. But, as we saw earlier, B contributes the same energy to the wave as E does.

* 32–8 Radiation Pressure

If electromagnetic waves carry energy, as we have just seen, then we might expect them to also carry linear momentum. When an electromagnetic wave encounters the surface of a material object and is absorbed or reflected, there will be a force exerted on the surface as a result of the momentum transfer ($F = dp/dt$) just as when a moving object strikes a surface. The force per unit area exerted by the waves is called **radiation pressure**, and its existence was predicted by Maxwell. He showed that if a beam of EM radiation (light, for example) is completely absorbed by an object, then the momentum transferred is,

$$\Delta p = \frac{\Delta U}{c}$$

where ΔU is the energy absorbed by the object in a time Δt and c is the speed of light. If instead, the radiation is fully reflected (suppose the object is a mirror) then the momentum transferred is twice as great, just as when a ball bounces elastically off a surface (Chapter 7):

$$\Delta p = \frac{2\Delta U}{c}.$$

If a surface absorbs some of the energy, and reflects some of it, then $\Delta p = k \, \Delta U/c$, where k is a constant between 1 and 2.

Using Newton's second law we can calculate the force and the pressure exerted by radiation on the object. The force F is given by

$$F = \frac{dp}{dt}.$$

The average rate that energy is delivered to the object is related to the Poynting vector by

$$\frac{dU}{dt} = \overline{S}A,$$

where A is the cross-sectional area of the object which intercepts the radiation.

The radiation pressure P (assuming full absorption) is given by

$$P = \frac{F}{A} = \frac{1}{A}\frac{dp}{dt} = \frac{1}{Ac}\frac{dU}{dt} = \frac{\bar{S}}{c}.$$

Radiation pressure
(absorbed)

If the light is fully reflected, the pressure is twice as great:

$$P = \frac{2\bar{S}}{c}.$$

Radiation pressure
(reflected)

EXAMPLE 32–5 ESTIMATE Solar pressure.   Radiation from the sun that reaches the Earth's surface transports energy at a rate of about $1000\ \text{W/m}^2$. Estimate the pressure and force exerted by the Sun on your outstretched hand.

SOLUTION   The radiation is partially reflected and partially absorbed, so let us estimate simply $P = \bar{S}/c$, which gives

$$P \approx \frac{\bar{S}}{c} = \frac{1000\ \text{W/m}^2}{3 \times 10^8\ \text{m/s}} \approx 3 \times 10^{-6}\ \text{N/m}^2.$$

An estimate of the area of your outstretched hand might be about $10\ \text{cm}$ by $20\ \text{cm}$, so $A = 0.02\ \text{m}^2$. Then the force is

$$F = PA \approx (3 \times 10^{-6}\ \text{N/m}^2)(0.02\ \text{m}^2) \approx 6 \times 10^{-8}\ \text{N}.$$

These numbers are tiny. The force of gravity on your hand, for comparison, is maybe a half pound, or with $m = 0.2\ \text{kg}$, $mg \approx (0.2\ \text{kg})(9.8\ \text{m/s}^2) \approx 2\ \text{N}$. We see that the radiation pressure on your hand is imperceptible.

Although you can not directly feel the effects of radiation pressure, the phenomenon is quite dramatic when applied to atoms irradiated by a laser beam. An atom has a mass on the order of $10^{-27}\ \text{kg}$, and a laser can deliver energy at a rate of $1000\ \text{W/m}^2$. This is the same intensity used in Example 32–5, above, but here a radiation pressure of $10^{-6}\ \text{N/m}^2$ would be very significant on a molecule whose mass might be 10^{-23} to $10^{-26}\ \text{kg}$. It is possible, in fact, to move atoms and molecules around by steering them with a laser beam, in a device called "optical tweezers." Optical tweezers have some remarkable applications. They are of great interest to biologists, especially since optical tweezers can manipulate live organisms without damaging them. Optical tweezers have been used to measure the elastic properties of DNA by pulling each end of the molecule with a tweezers.

➡ PHYSICS APPLIED

Optical tweezers

* 32–9 Radio and Television

Electromagnetic waves offer the possibility of transmitting information over long distances. Among the first to realize this and put it into practice was Guglielmo Marconi (1874–1937) who, in the 1890s, invented and developed the wireless telegraph. With it, messages could be sent hundreds of kilometers at the speed of light without the use of wires. The first signals were merely long and short pulses that could be translated into words by a code, such as the "dots" and "dashes" of the Morse code. The next decade saw the development of vacuum tubes. Out of this early work radio and television were born. We now discuss briefly (1) how radio and TV signals are transmitted, and (2) how they are received at home.

➡ PHYSICS APPLIED

"Wireless" transmission

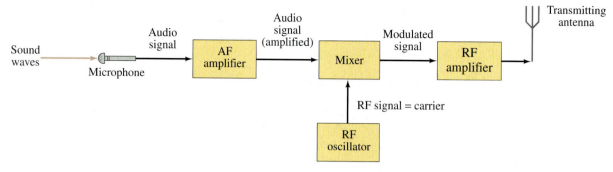

FIGURE 32–15 Block diagram of a radio transmitter.

Transmission of radio waves

Carrier frequency

The process by which a radio station transmits information (words and music) is outlined in Fig. 32–15. The audio (sound) information is changed into an electrical signal of the same frequencies by, say, a microphone or tape recorder head. This electrical signal is called an audiofrequency (AF) signal, since the frequencies are in the audio range (20 to 20,000 Hz). The signal is amplified electronically and is then mixed with a radio-frequency (RF) signal called its **carrier frequency**. AM radio stations have carrier frequencies from about 530 kHz to 1600 kHz. For example, "710 on your dial" means a station whose carrier frequency is 710 kHz. FM radio stations have much higher carrier frequencies, between 88 MHz and 108 MHz. The carrier frequencies for TV stations in the United States lie between 54 MHz and 88 MHz for channels 2 through 6, and between 174 MHz and 216 MHz for channels 7 through 13; UHF (ultra-high-frequency) stations have even higher carrier frequencies, between 470 MHz and 890 MHz.

➡ **PHYSICS APPLIED**

AM and FM

The mixing of the audio and carrier frequencies is done in two ways. In **amplitude modulation** (AM), the amplitude of the higher frequency carrier wave is made to vary in proportion to the amplitude of the audio signal, as shown in Fig. 32–16. It is called "amplitude modulation" because the amplitude of the carrier is altered ("modulate" means to change or alter). In **frequency modulation** (FM), the *frequency* of the carrier wave is made to change in proportion to the audio signal's amplitude, as shown in Fig. 32–17.

The mixed signal is amplified further and is then sent to the antenna, where the complex mixture of frequencies is sent out in the form of EM waves.

A television transmitter works in a similar way, using frequency modulation, except that both audio and video signals are mixed with carrier frequencies.

FIGURE 32–16 In amplitude modulation (AM), the amplitude of the carrier signal is made to vary in proportion to the audio signal's amplitude.

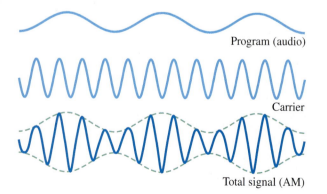

FIGURE 32–17 In frequency modulation (FM), the frequency of the carrier signal is made to change in proportion to the audio signal's amplitude. This method is used by FM radio and television.

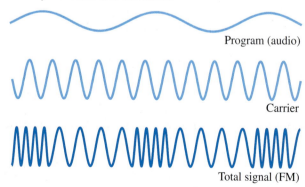

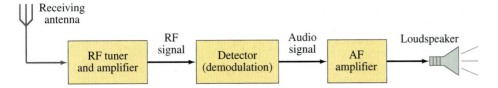

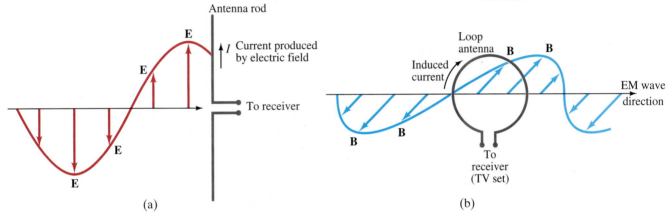

FIGURE 32–18 Block diagram of a simple radio receiver.

Now let us look at the other end of the process, the reception of radio and TV programs at home. A simple radio receiver is diagrammed in Fig. 32–18. The EM waves sent out by all stations are received by the antenna. One kind of antenna consists of one or more conducting rods; the electric field in the EM waves exerts a force on the electrons in the conductor, causing them to move back and forth at the frequencies of the waves (Fig. 32–19a). A second type of antenna consists of a tubular coil of wire, often found in AM radios, or the simple loop of a UHF television antenna. These antennas detect the magnetic field of the wave, for the changing B field induces an emf in the coil (Fig. 32–19b).

➡ **PHYSICS APPLIED**

Radio and TV receivers

Antennas

FIGURE 32–19 Antennas. (a) Electric field of EM wave produces a current in an antenna consisting of straight wire or rods. (b) Changing magnetic field induces an emf and current in a loop antenna.

Antenna rod

Current produced by electric field

To receiver

Loop antenna

Induced current

EM wave direction

To receiver (TV set)

(a) (b)

The signal the antenna detects and sends to the receiver is very small and contains frequencies from many different stations. The receiver selects out a particular RF frequency (actually a narrow range of frequencies) corresponding to a particular station using a resonant LC circuit (Sections 30–5 and 31–6) with a variable capacitor or inductor. A simple example is shown in Fig. 32–20. A particular station is "tuned-in" by adjusting L and C so that the resonant frequency of the circuit equals that of the station's carrier frequency. The signal, containing both audio and carrier frequencies, next goes to the *detector* (Fig. 32–18) where "demodulation" takes place—that is, the RF carrier frequency is separated from the audio signal. The audio signal is amplified and sent to a loudspeaker or headphones.

Modern receivers have more stages than those shown. Various means are used to increase the sensitivity and selectivity (ability to detect weak signals and distinguish them from other stations), and to minimize distortion of the original signal.[†]

A television receiver does similar things to both the audio and the video signals. The audio signal goes finally to the loudspeaker, and the video signal to the picture tube, a *cathode ray tube* (CRT) whose operation was discussed in Sections 23–9 and 27–7.

FIGURE 32–20 Simple tuning stage of a radio.

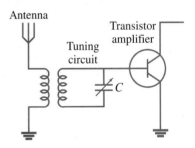

[†] For *FM stereo broadcasting*, two signals are carried by the carrier wave. One of these contains frequencies up to about 15 kHz, which includes most audio frequencies. The other signal includes the same range of frequencies, but 19 kHz is added to it. A stereo receiver subtracts this 19,000-Hz signal and distributes the two signals to the left and right channels. The first signal actually consists of the sum of left and right channels ($L + R$), so mono radios detect all the sound. The second signal is the difference between left and right ($L - R$). Hence the receiver must add and subtract the two signals to get pure left and right signal for each channel.

EXAMPLE 32–6 **Tuning a station.** An FM radio station transmits at 100 MHz. Calculate (a) the transmitting wavelength, and (b) the value of the capacitance in the tuning circuit if $L = 0.40 \, \mu\text{H}$.

SOLUTION (a) The carrier frequency is $f = 100 \, \text{MHz} = 1.0 \times 10^8 \, \text{s}^{-1}$, so

$$\lambda = \frac{c}{f} = \frac{(3.0 \times 10^8 \, \text{m/s})}{(1.0 \times 10^8 \, \text{s}^{-1})} = 3.0 \, \text{m}.$$

The wavelengths of other FM signals (88 MHz to 108 MHz) are close to this. FM antennas are typically 1.5 m long, or about a half wavelength. This length is chosen so that the antenna reacts in a resonant fashion and thus is more sensitive to FM frequencies.

(b) According to Eq. 30–14 or 31–13, the resonant frequency is $f_0 = 1/(2\pi\sqrt{LC})$. Therefore,

$$C = \frac{1}{4\pi^2 f_0^2 L} = \frac{1}{4(3.14)^2(1.0 \times 10^8 \, \text{s}^{-1})^2(4.0 \times 10^{-7} \, \text{H})}$$

$$= 6.3 \, \text{pF}.$$

Of course, the capacitor or inductor is variable, so other stations can be selected too.

The various regions of the radio-wave spectrum are assigned by governmental agencies for various purposes. Besides those mentioned above, there are "bands" assigned for use by ships, airplanes, police, military, amateurs, satellites and space, and radar. Cellular telephones, for example, occupy a band from 824 MHz to 894 MHz.

Summary

James Clerk Maxwell synthesized an elegant theory in which all electric and magnetic phenomena could be described using four equations, now called **Maxwell's equations**. They are based on earlier ideas, but Maxwell added one more—that a changing electric field produces a magnetic field. Maxwell's equations are

$$\oint \mathbf{E} \cdot d\mathbf{A} = \frac{Q}{\epsilon_0} \qquad \oint \mathbf{E} \cdot d\mathbf{l} = -\frac{d\Phi_B}{dt}$$

$$\oint \mathbf{B} \cdot d\mathbf{A} = 0 \qquad \oint \mathbf{B} \cdot d\mathbf{l} = \mu_0 I + \mu_0 \epsilon_0 \frac{d\Phi_E}{dt}.$$

The two on the left are Gauss's laws for electricity and for magnetism; the two on the right are Faraday's law and Ampère's law (as extended by Maxwell).

Maxwell's theory predicted that transverse **electromagnetic (EM) waves** would be produced by accelerating electric charges, and that these waves would propagate through space at the speed of light c, given by

$$c = \frac{1}{\sqrt{\epsilon_0 \mu_0}}.$$

The oscillating electric and magnetic fields in an EM wave are perpendicular to each other and to the direction of propagation.

After EM waves were experimentally detected in the late 1800s, the idea that light is an EM wave (although of much higher frequency than those detected directly) became generally accepted. The **electromagnetic spectrum** includes EM waves of a wide variety of wavelengths, from microwaves and radio waves to visible light to X-rays and gamma rays, all of which travel through space at a speed $c = 3.00 \times 10^8 \, \text{m/s}$.

The energy carried by EM waves can be described by the **Poynting vector**

$$\mathbf{S} = \frac{1}{\mu_0} \mathbf{E} \times \mathbf{B}$$

which gives the rate energy is carried across unit area per unit time when the electric and magnetic fields in an EM wave in free space are $\mathbf{E}$ and $\mathbf{B}$.

Questions

1. Suppose you are looking along the same direction as an electric field **E** that is increasing. Will the induced magnetic field be clockwise or counterclockwise? What if **E** points toward you and is decreasing?

2. What is the direction of the displacement current in Fig. 32–3? (Note: The capacitor is discharging.)

3. Why is it that the magnetic field of a displacement current in a capacitor is so much harder to detect than the magnetic field of a conduction current?

4. Are there any good reasons for calling the term $\mu_0 \epsilon_0 \, d\Phi_E/dt$ in Eq. 32–1 a "current"? Explain.

5. The electric field in an EM wave traveling north oscillates in an east–west plane. Describe the direction of the magnetic field vector in this wave.

6. Is sound an electromagnetic wave? If not, what kind of wave is it?

7. Can EM waves travel through a perfect vacuum? Can sound waves?

8. When you flip a light switch, does the overhead light go on immediately? Explain.

9. Are the wavelengths of radio and television signals longer or shorter than those detectable by the human eye?

10. What does the result of Example 32–2 tell you about the phase of a 60-Hz ac current that starts at a power plant as compared to its phase at a house 200 km away?

11. When you connect two loudspeakers to the output of a stereo amplifier, should you be sure the lead wires are equal in length so that there will not be a time lag between speakers? Explain.

12. In the electromagnetic spectrum, what type of EM wave would have a wavelength of 10^3 km? 1 km? 1 m? 1 cm? 1 mm? 1 μm?

* 13. A lost person may signal by flashing a flashlight on and off using Morse code. This is actually a modulated EM wave. Is it AM or FM? What is the frequency of the carrier, approximately?

* 14. Can two radio or TV stations broadcast on the same carrier frequency? Explain.

* 15. If a radio transmitter has a vertical antenna, should a receiver's antenna (rod type) be vertical or horizontal to obtain best reception?

* 16. The carrier frequencies of FM broadcasts are much higher than for AM broadcasts. On the basis of what you learned about diffraction in Chapter 15, explain why AM signals can be detected more readily than FM signals behind low hills or buildings.

17. Discuss how cordless telephones make use of EM waves. What about cellular telephones?

Problems

Section 32–1

1. (I) Determine the rate at which the electric field changes between the round plates of a capacitor, 6.0 cm in diameter, if the plates are spaced 1.3 mm apart and the voltage across them is changing at a rate of 120 V/s.

2. (I) Calculate the displacement current I_D between the square plates, 3.8 cm on a side, of a capacitor if the electric field is changing at a rate of 2.0×10^6 V/m·s.

3. (II) At a given instant, a 1.8-A current flows in the wires connected to a parallel-plate capacitor. What is the rate at which the electric field is changing between the plates if the square plates are 1.60 cm on a side?

4. (II) A 1200-nF capacitor with circular parallel plates 2.0 cm in diameter is accumulating charge at the rate of 35.0 mC/s at some instant in time. What will be the induced magnetic field strength 10.0 cm radially outward from the center of the plates? What will be the value of the field strength after the capacitor is fully charged?

5. (II) Show that the displacement current through a parallel-plate capacitor can be written $I_D = C \, dV/dt$, where V is the voltage across the capacitor at any instant.

6. (II) Suppose an air-gap capacitor has circular plates of radius $R = 2.5$ cm and separation $d = 2.0$ mm. A 96.0-Hz emf, $\mathscr{E} = \mathscr{E}_0 \cos \omega t$, is applied to the capacitor. The maximum displacement current is 35 μA. Determine (a) the maximum conduction current I, (b) the value of $\mathscr{E}_0$, (c) the maximum value of $d\Phi_E/dt$ between the plates. Neglect fringing.

Section 32–3

7. (II) If magnetic monopoles existed, which of Maxwell's equations would be altered, and what would be their new form? Let Q_m be the strength of a magnetic monopole, analogous to electric charge Q.

Section 32–5

8. (I) If the magnetic field in a traveling EM wave has a peak value of 17.5 nT, what is the peak value of the electric field strength?

9. (I) If the electric field in an EM wave has a peak value of 0.43×10^{-4} V/m, what is the peak value of the magnetic field strength?

10. (I) In an EM wave traveling west, the B field oscillates vertically and has a frequency of 80.0 kHz and an rms strength of 6.75×10^{-9} T. What is the frequency and rms strength of the electric field and what is its direction?

11. (II) The electric field of a plane EM wave is given by $E_x = E_0 \cos(kz + \omega t)$, $E_y = E_z = 0$. Determine (a) the magnitude and direction of **B**, and (b) the direction of propagation.

Section 32-6

12. (I) What is the frequency of a microwave whose wavelength is 1.80 cm?

13. (I) (a) What is the wavelength of a 27.75×10^9 Hz radar signal? (b) What is the frequency of an X-ray with wavelength 0.10 nm?

14. (I) An EM wave has a wavelength of 850 nm. What is its frequency, and how would we classify it?

15. (I) An EM wave has frequency 9.56×10^{14} Hz. What is its wavelength, and how would we classify it?

16. (I) A light-year is a measure of distance (not time). How many meters does light travel in a year?

17. (II) How long would it take a message sent as radio waves from Earth to reach (a) Mars, (b) a spacecraft near Saturn?

18. (II) Pulsed lasers used for science and medicine produce very short bursts of electromagnetic energy. If the laser light wavelength is 1062 nm (this corresponds to a Neodymium-YAG laser), and the pulse lasts for 30 picoseconds, how many wavelengths are found within the laser pulse? How short would the pulse need to be to fit only one wavelength?

Section 32-7

19. (I) The $\mathbf{E}$ field in an EM wave has a peak of 36.5 mV/m. What is the average rate at which this wave carries energy across unit area per unit time?

20. (II) The magnetic field in a traveling EM wave has an rms strength of 32.5 nT. How long does it take to deliver 335 J of energy to 1.00 cm^2 of a wall that it hits perpendicularly?

21. (II) How much energy is transported across a 1.00 cm^2 area per hour by an EM wave whose E field has an rms strength of 28.6 mV/m?

22. (II) A spherically spreading EM wave comes from a 1000-W source. At a distance of 10.0 m, what is the intensity, and what is the rms value of the electric field?

23. (II) What is the energy contained in a 1.00-m^3 volume near the Earth's surface due to radiant energy from the Sun? See Example 32-4.

24. (II) A 12.8-mW laser puts out a narrow beam 2.00 mm in diameter. What are the average (rms) values of E and B in the beam?

25. (II) Estimate the average power output of the Sun, given that about 1350 W/m^2 reaches the upper atmosphere of the Earth.

26. (II) A high-energy pulsed laser emits a 1.0-ns-long pulse of average power 2.5×10^{11} W. The beam is 2.2×10^{-3} m in radius. Determine (a) the energy delivered in each pulse, and (b) the rms value of the electric field.

27. (III) (a) Show that the Poynting vector $\mathbf{S}$ points radially inward toward the center of a circular parallel-plate capacitor when it is being charged as in Example 32-1. (b) Integrate $\mathbf{S}$ over the cylindrical boundary of the capacitor gap to show that the rate at which energy enters the capacitor is equal to the rate at which electrostatic energy is being stored in the electric field of the capacitor (Section 24-4). Ignore fringing of $\mathbf{E}$.

* Section 32-8

* 28. (II) Estimate the radiation pressure due to a 100 W bulb at a distance of 8.0 cm from the center of the bulb. Estimate the force exerted on your fingertip if you place it at this point.

* 29. (III) When the Sun became hot and luminous long ago, it is believed that it ejected dust particles and individual atoms out of the solar system using radiation pressure. Calculate how small the dust particles had to be in order to be ejected by comparing the radiation force with the gravitational force on the particles. Assume the particles are spherical, have a density of $2.0 \times 10^3 \text{ kg/m}^3$, and totally absorb the Sun's radiation. The Sun has an average power output of 3.8×10^{26} W.

* 30. (III) We discussed in the text two different equations for the momentum transfer due to radiation, one valid for total absorption, the other valid for total reflection. The latter equation is identical to the former with the exception of a factor of two. Give a simple mechanical argument for this factor of two by treating light as massless particles (photons) each with momentum p. In the case of absorption, light particles collide completely inelastically with a massive object, whereas in the case of reflection they collide elastically. Analyze these collisions and explain the factor of two.

* Section 32-9

* 31. (I) The variable capacitor in the tuner of an AM radio has a capacitance of 2400 pF when the radio is tuned to a station at 550 kHz. What must the capacitance be for a station at the other end of the dial, 1550 kHz?

* 32. (I) The oscillator of a 96.1-MHz FM station has an inductance of 1.8μH. What value must the capacitance be?

* 33. (II) A certain FM radio tuning circuit has a fixed capacitor $C = 840$ pF. Tuning is done by a variable inductance. What range of values must the inductance have to tune stations from 88 MHz to 108 MHz?

* 34. (II) An amateur radio operator wishes to build a receiver that can tune a range from 14.0 MHz to 15.0 MHz. A variable capacitor has a minimum capacitance of 92 pF. (a) What is the required value of the inductance? (b) What is the maximum capacitance used on the variable capacitor?

General Problems

35. (a) What is the wavelength of an AM station at 680 kHz? (b) An FM station broadcasts at 100.7 MHz. What is the wavelength of this wave?

36. Compare 940 on the AM dial to 94 on the FM dial. Which has the longer wavelength, and by what factor is it larger?

37. Television broadcast frequencies range between 54.0 MHz for Channel 2 to 806 MHz for Channel 69 (and beyond). What are the wavelengths for these two channels?

38. If the Sun were to disappear or somehow radically change its output, how long would it take for us on Earth to learn about it?

39. (a) How long did it take for a message sent from Earth to reach the first astronauts on the Moon? (b) How long will it take for a message from Earth to reach the first astronauts who arrive on Mars; assume Mars is at its closest approach to Earth $(78 \times 10^6 \text{ km})$?

40. A radio voice signal from the Apollo crew on the Moon is beamed to a listening crowd from a radio speaker. If you are standing 50 m from the loudspeaker, what is the total time lag between when you hear the sound and when the sound left the Moon?

41. A point source emits light energy uniformly in all directions at an average rate P_0 with a single frequency f. Show that the peak electric field in the wave is given by

$$E_0 = \sqrt{\frac{\mu_0 c P_0}{2\pi r^2}}.$$

42. What are E_0 and B_0 2.00 m from a 100-W light source? Assume the bulb emits radiation of a single frequency uniformly in all directions.

43. Estimate the rms electric field in the sunlight that hits Mars, knowing that the Earth receives about 1350 W/m² and that Mars is 1.52 times farther away from the Sun (on average) than is the Earth.

44. At a given instant in time, a traveling EM wave is noted to have its maximum magnetic field pointing west and its maximum electric field pointing south. In which direction is the wave traveling? If the rate of energy flow is 500 W/m², what are the maximum values for the two fields?

45. Who will hear the voice of a singer first: a person in the balcony 50 m away from the stage (see Fig. 32–21), or a person 3000 km away at home whose ear is next to the radio? Roughly how much sooner? Assume the microphone is a few centimeters from the singer and the temperature is 20°C.

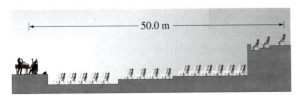

FIGURE 32–21 Problem 45.

46. A 1.80-m-long FM receiving antenna is oriented parallel to the electric field of an EM wave. How large must the **E** field strength be to produce a 1.0-mV rms voltage between the ends of the antenna? What is the rate of energy transport per unit area?

47. Suppose a 50-kW radio station emits EM waves uniformly in all directions. (a) How much energy per second crosses a 1.0-m² area 100 m from the transmitting antenna? (b) What is the rms magnitude of the **E** field at this point, assuming the station is operating at full power? (c) What is the voltage induced in a 1.0-m-long vertical car antenna at this distance?

48. Repeat Problem 47 for a distance of 100 km from the station.

49. Referring to Problem 47, what is the maximum power level of the radio station so as to avoid electrical breakdown of air at a distance of 1.0 m from the antenna? Assume the antenna is a point source. Air breaks down in an electric field of about 3×10^6 V/m. [Hint: See Problem 41.]

50. How large an emf (rms) will be generated in an antenna that consists of a 380-loop circular coil of wire 2.2 cm in diameter if the EM wave has a frequency of 810 kHz and is transporting energy at an average rate of 1.0×10^{-4} W/m² at the antenna? [Hint: You can use Eq. 29–4 for a generator, since it could be applied to an observer moving with the coil so that the magnetic field is oscillating with the frequency $f = \omega/2\pi$.]

* 51. The variable capacitance of a radio tuner consists of six plates connected together placed alternately between six other plates, also connected together (Fig. 32–22). Each plate is separated from its neighbor by 1.1 mm of air. One set of plates can move so that the area of overlap varies from 1.0 cm² to 9.0 cm². (a) Are these capacitors connected in series or in parallel? (b) Determine the range of capacitance values. (c) What value of inductor is needed if the radio is to tune AM stations from 550 kHz to 1600 kHz?

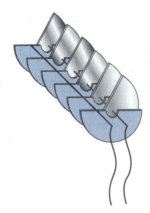

FIGURE 32–22 Problem 51.

52. A cylindrical conductor of radius r and conductivity σ carries a steady current I distributed uniformly across its cross-section. (a) Determine **E** inside the conductor. (b) Determine **B** just outside the conductor. (c) Determine the Poynting vector **S** at the surface of the conductor and show that it is normal to the surface and points inward. (d) Integrate over **S** to show that the rate at which electromagnetic energy enters the sides of the conductor is equal to the rate at which energy is dissipated, I^2R. Thus we can think of the energy as entering the conductor in the form of electromagnetic fields from the sides rather than through the ends.

Reflection from still water, as from a glass mirror, can be analyzed using the ray model of light.
Is this picture right side up? How can you tell? What are the clues? See Example 33–2.
In this first chapter on light and optics, we will use the ray model of light to understand the formation of images by mirrors, both plane and curved (spherical), and we will begin examining the refraction of light in transparent media, which will lead us to a study of lenses in the next chapter.

CHAPTER 33

Light: Reflection and Refraction

The sense of sight is extremely important to us, for it provides us with a large part of our information about the world. How do we see? What is the something called *light* that enters our eyes and causes the sensation of sight? How does light behave so that we can see everything that we do? We saw in Chapter 32 that light can be considered a form of electromagnetic radiation. We now examine the subject of light in detail in the next several chapters.

We see an object in one of two ways: (1) the object may be a source of light, such as a lightbulb, a flame, or a star, in which case we see the light emitted directly from the source; or (2), more commonly, we see an object by light reflected from it. In the latter case, the light may have originated from the sun, artificial lights, or a campfire. An understanding of how bodies *emit* light was not achieved until the 1920s, and this will be discussed in Chapter 38. How light is *reflected* from objects was understood much earlier, and will be discussed in Section 33–3.

33–1 The Ray Model of Light

A great deal of evidence suggests that *light travels in straight lines* under a wide variety of circumstances. For example, a point source of light like the sun casts distinct shadows, and the beam of a flashlight appears to be a straight line. In fact, we infer the positions of objects in our environment by assuming that light moves from the object to our eyes in straight-line paths. Our orientation to the physical world is based on this assumption.

This reasonable assumption has led to the **ray model** of light. This model assumes that light travels in straight-line paths called light **rays**. Actually, a ray is an idealization; it is meant to represent an extremely narrow beam of light. When we see an object, according to the ray model, light reaches our eyes from each point on the object. Although light rays leave each point in many different directions, normally only a small bundle of these rays can enter an observer's eye, as shown in Fig. 33–1. If the person's head moves to one side, a different bundle of rays will enter the eye from each point.

We saw in Chapter 32 that light can be considered as an electromagnetic wave. Although the ray model of light does not deal with this aspect of light (we discuss the wave nature of light in Chapters 35 and 36), the ray model has been very successful in describing many aspects of light such as reflection, refraction, and the formation of images by mirrors and lenses. Because these explanations involve straight-line rays at various angles, this subject is referred to as **geometric optics**. In ignoring the wave properties of light we must be careful that when the light rays pass by objects or through apertures, these must be large compared to the wavelength of the light (so the wave phenomena of interference and diffraction, as discussed in Chapter 15, can be ignored), and we ignore what happens to the light at the edges of objects until we get to Chapters 35 and 36.

Light rays

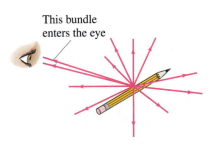

This bundle enters the eye

FIGURE 33–1 Light rays come from each single point on an object. A small bundle of rays leaving one point is shown entering a person's eye.

33–2 The Speed of Light and Index of Refraction

Galileo attempted to measure the speed of light by trying to measure the time required for light to travel a known distance between two hilltops. He stationed an assistant on one hilltop and himself on another and ordered the assistant to lift the cover from a lamp the instant he saw a flash from Galileo's lamp. Galileo measured the time between the flash of his lamp and when he received the light from his assistant's lamp. The time was so short that Galileo concluded it merely represented human reaction time, and that the speed of light must be extremely high.

The first successful determination that the speed of light is finite was made by the Danish astronomer Ole Roemer (1644–1710). Roemer had noted that the carefully measured period of Io, one of Jupiter's moons with an average period of 42.5 h, varied slightly, depending on the relative motion of Earth and Jupiter. When Earth was moving away from Jupiter, the period of the moon was slightly longer, and when Earth was moving toward Jupiter, the period was slightly shorter. He attributed this variation to the extra time needed for light to travel the increasing distance to Earth when Earth is receding, or to the shorter travel time for the decreasing distance when the two planets are approaching one another. Roemer concluded that the speed of light—though great—is finite.

Since then a number of techniques have been used to measure the speed of light. Among the most important were those carried out by the American Albert

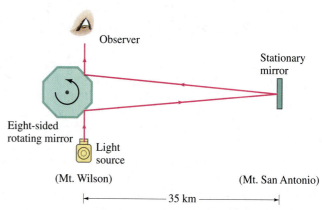

FIGURE 33–2 Michelson's speed-of-light apparatus (not to scale).

Michelson (1852–1931). Michelson used the rotating mirror apparatus diagrammed in Fig. 33–2 for a series of high-precision experiments carried out from 1880 to the 1920s. Light from a source was directed at one face of a rotating eight-sided mirror. The reflected light traveled to a stationary mirror a large distance away and back again as shown. If the rotating mirror was turning at just the right rate the returning beam of light would reflect from one face of the mirror into a small telescope through which the observer looked. If the speed of rotation was only slightly different, the beam would be deflected to one side and would not be seen by the observer. From the required speed of the rotating mirror and the known distance to the stationary mirror, the speed of light could be calculated. In the 1920s, Michelson set up the rotating mirror on the top of Mt. Wilson in southern California and the stationary mirror on Mt. San Antonio, 35 km away. He later measured the speed of light in vacuum using a long evacuated tube.

The accepted value today for the speed of light, c, in vacuum is[‡]

$$c = 2.99792458 \times 10^8 \text{ m/s}.$$

We usually round this off to

$$3.00 \times 10^8 \text{ m/s}$$

when extremely precise results are not required. In air, the speed is only slightly less. In other transparent materials such as glass and water, the speed is always less than that in vacuum. For example, in water it travels at about $\frac{3}{4}c$. The ratio of the speed of light in vacuum to the speed v in a given material is called the **index of refraction** n of that material:

$$n = \frac{c}{v}. \qquad \text{(33–1)}$$

The index of refraction is never less than 1, and its value for various materials is given in Table 33–1. For example, since $n = 2.42$ for diamond, the speed of light in diamond is

$$v = \frac{c}{n} = \frac{(3.00 \times 10^8 \text{ m/s})}{2.42} = 1.24 \times 10^8 \text{ m/s}.$$

As we shall see later, n varies somewhat with the wavelength of the light—except in vacuum—so a particular wavelength is specified; in Table 33–1 it is that of yellow light with wavelength $\lambda = 589$ nm.

TABLE 33–1
Indices of refraction[†]

Material	$n = \dfrac{c}{v}$
Vacuum	1.0000
Air (at STP)	1.0003
Water	1.33
Ethyl alcohol	1.36
Glass	
Fused quartz	1.46
Crown glass	1.52
Light flint	1.58
Lucite or Plexiglas	1.51
Sodium chloride	1.53
Diamond	2.42

[†] $\lambda = 589$ nm

Index of refraction

33–3 | Reflection; Image Formation by a Plane Mirror

When light strikes the surface of an object, some of the light is reflected. The rest is either absorbed by the object (and transformed to thermal energy) or, if the object is transparent like glass or water, part of it is transmitted through. For a very shiny object such as a silvered mirror, over 95 percent of the light may be reflected.

[‡]When the meter was redefined in 1983, the speed of light was given this fixed value, and the meter then defined in terms of it (see Section 1–4).

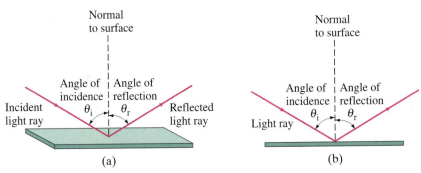

FIGURE 33–3 Law of reflection: (a) Shows an incident ray being reflected at the top of a flat surface; (b) shows a side or "end-on" view, which we will usually use because of its clarity.

When a narrow beam of light strikes a flat surface (Fig. 33–3) we define the **angle of incidence**, θ_i, to be the angle an incident ray makes with the normal to the surface, and the **angle of reflection**, θ_r, to be the angle the reflected ray makes with the normal. It is found that the *incident and reflected rays lie in the same plane with the normal to the surface*, and that

Angle of incidence and Angle of reflection (angle with the normal)

> **the angle of incidence equals the angle of reflection.**

Law of reflection

This is the **law of reflection** and is indicated in Fig. 33–3. It was known to the ancient Greeks, and you can confirm it yourself by shining a narrow flashlight beam at a mirror in a darkened room.

When light is incident upon a rough surface, even microscopically rough such as this page, it is reflected in many directions, Fig. 33–4. This is called **diffuse reflection**. The law of reflection still holds, however, at each small section of the surface. Because of diffuse reflection in all directions, an ordinary object can be seen from many different angles. When you move your head to the side, different reflected rays reach your eye from each point on the object (such as this page), Fig. 33–5a. Let us compare diffuse reflection to reflection from a mirror, which is known as *specular* reflection. ("Speculum" is Latin for mirror.) When a narrow beam of light shines on a mirror, the light will not reach your eye unless your eye is positioned at just the right place where the law of reflection is satisfied, as shown in Fig. 33–5b. This is what gives rise to the special image-forming properties of mirrors.

When you look straight in a mirror, you see what appears to be yourself as well as various objects around and behind you, Fig. 33–6. Your face and the other objects look as if they are in front of you, beyond the mirror; but, of course, they are not. What you see in the mirror is an **image** of the objects.

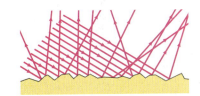

FIGURE 33–4 Diffuse reflection from a rough surface.

Mirrors (specular reflection)

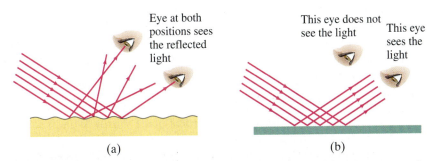

FIGURE 33–5 A beam of light from a flashlight shines on (a) white paper, and (b) a mirror. In part (a), you can see the white light reflected at various points because of diffuse reflection. But in part (b), you see the reflected light only when your eye is placed correctly $(\theta_r = \theta_i)$; this is known as specular reflection. (Galileo, using similar arguments, showed that the Moon must have a rough surface rather than a highly polished surface like a mirror, as some people thought.)

FIGURE 33–6 When you look in a mirror, you see an image of yourself and objects around you. Note that you don't see yourself as others see you, because left and right appear reversed in the image.

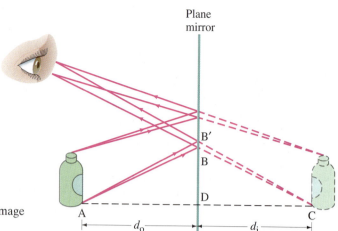

FIGURE 33–7 Formation of a virtual image by a plane mirror.

Figure 33–7 shows how an image is formed by a plane mirror (that is, flat), according to the ray model. We are viewing the mirror, on edge, in the diagram of Fig. 33–7, and the rays are shown reflecting from the front surface. (Good mirrors are generally made by putting a highly reflective metallic coating on one surface of a very flat piece of glass.) Rays from two different points on an object are shown in Fig. 33–7: rays leaving from a point on the top of the bottle, and from a point on the bottom. Rays leave each point on the object going in many directions, but only those that enclose the bundle of rays that reach the eye from the two points are shown. The diverging rays that enter the eye *appear* to come from behind the mirror as shown by the dashed lines. That is, our eyes and brain interpret any rays that enter an eye as having traveled a straight-line path. The point from which each bundle of rays seems to come is one point on the image. For each point on the object, there is a corresponding image point. Let us concentrate on the two rays that leave point A on the object and strike the mirror at points B and B'. The angles ADB and CDB are right angles; and angles ABD and CBD are equal because of the law of reflection. Therefore, the two triangles ABD and CBD are congruent, and the length AD = CD. Thus the image appears as far behind the mirror as the object is in front: the **image distance**, d_i (distance from mirror to image, Fig. 33–7), equals the **object distance**, d_o. From the geometry, we also see that the height of the image is the same as that of the object.

The light rays do not actually pass through the image location itself. It merely *seems* as though the light is coming from the image because our brains interpret any light entering our eyes as having come in a *straight-line path from in front of us*. Because the rays do not actually pass through the image, the image would not appear on paper or film placed at the location of the image. Therefore, it is called a **virtual image**. This is to distinguish it from a **real image** in which the light does pass through the image and which therefore could appear on paper or film placed at the image position. We will see that curved mirrors and lenses can form real images. A movie projector lens, for example, produces a real image that is visible on the screen.

Ray model of how images are formed

Each point of image is where rays intersect

Image distance = object distance

Real and virtual images

CONCEPTUAL EXAMPLE 33–1 | **How tall must a full-length mirror be?** A woman 1.60 m tall stands in front of a vertical plane mirror. What is the minimum height of the mirror, and how high must its lower edge be above the floor, if she is to see her whole body? Assume her eyes are 10 cm below the top of her head.

RESPONSE The situation is diagrammed in Fig. 33–8. First consider the ray from her foot, AB, which upon reflection becomes BE and enters the eye E. The light from point A (her foot) enters the eye after reflecting at B; so the mirror needs to extend no lower than B. Because the angle of reflection equals the angle of incidence, the height BD is half of the height AE. Because AE = 1.60 m − 0.10 m = 1.50 m, then BD = 0.75 m. Similarly, if the woman is to see the top of her head, the top edge of the mirror only needs to reach point F,

How tall a mirror is needed to see your whole reflection?

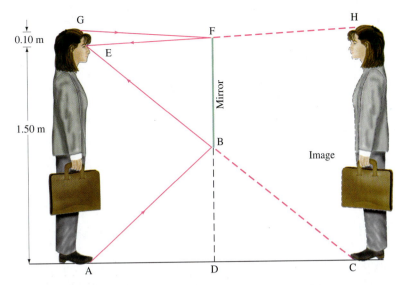

FIGURE 33–8   Seeing oneself in a mirror. Example 33–1.

which is 5 cm below the top of her head (half of GE = 10 cm). Thus, DF = 1.55 m, and the mirror need have a vertical height of only (1.55 m − 0.75 m) = 0.80 m. And the mirror's bottom edge must be 0.75 m above the floor. In general, a mirror need be only half as tall as a person for that person to see all of himself or herself. Does this result depend on the person's distance from the mirror?

CONCEPTUAL EXAMPLE 33–2   Is the photo upside down?   Close examination of the photograph on the first page of this chapter reveals that in the top portion, the image of the Sun is seen clearly, whereas in the lower portion, the image of the Sun is partially blocked by the tree branches. Why isn't the reflection in the water an exact replica of the real scene? Illustrate your answer by drawing a sketch of this situation, showing the Sun, the camera, the branch, and two rays going from the Sun to the camera (one direct and one reflected). Does your sketch tell you if the photograph is right side up?

RESPONSE   We need to draw two diagrams, one assuming the photo on p. 810 is right side up, and another assuming it is upside down. Figure 33–9 is drawn assuming the photo is upside down. In this case, the Sun blocked by the tree would be the direct view, and the full view of the Sun the reflection. Thus, as diagrammed in Fig. 33–9, the ray which reflects off the water and into the camera travels at an angle below the branch, whereas the ray that travels directly to the camera passes through the branches. This works. Try to draw a diagram assuming the photo is right side up (thus assuming that the image of the Sun in the reflection is higher above the horizon than it is as viewed directly). It won't work. The photo on p. 810 is upside down.

   Also, what about the people in the photo? Try to draw a diagram showing why they don't appear in the reflection. [*Hint*: Assume they are not sitting on the edge of poolside, but back from the edge a bit.] Then try to draw a diagram of the reverse (i.e., assume the photo is right side up so the people are visible only in the reflection). In general, reflected images are not perfect replicas when different planes (distances) are involved.

FIGURE 33–9   Example 33–2.

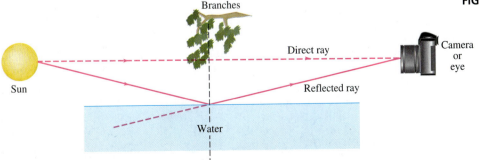

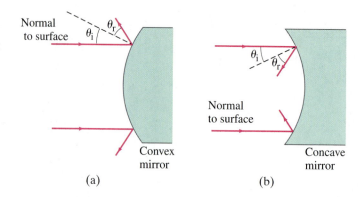

FIGURE 33–10 Mirrors with convex and concave spherical surfaces. Note that $\theta_r = \theta_i$ for each ray.

FIGURE 33–11 (a) A concave cosmetic mirror gives a magnified image. (b) A convex mirror in a store demagnifies and so includes a wide field of view.

(a)

(b)

33–4 Formation of Images by Spherical Mirrors

Reflecting surfaces do not have to be flat. The most common *curved* mirrors are *spherical*, which means they form a section of a sphere. A spherical mirror is called **convex** if the reflection takes place on the outer surface of the spherical shape so that the center of the mirror surface bulges out toward the viewer (Fig. 33–10a). A mirror is called **concave** if the reflecting surface is on the inner surface of the sphere so that the center of the mirror sinks away from the viewer (like a "cave"), Fig. 33–10b. Concave mirrors are used as shaving or cosmetic mirrors (Fig. 33–11a), and convex mirrors are sometimes used on cars and trucks (rearview mirrors) and in shops (to watch for thieves), because they take in a wide field of view (Fig. 33–11b).

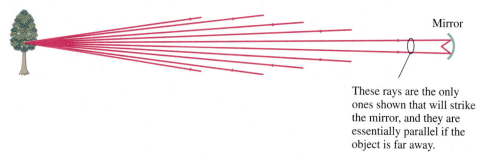

These rays are the only ones shown that will strike the mirror, and they are essentially parallel if the object is far away.

FIGURE 33–12 If the object's distance is large compared to the size of the mirror (or lens), the rays are nearly parallel. They are parallel for an object at infinity (∞).

To see how spherical mirrors form images, we first consider an object that is very far from a concave mirror. For a distant object, as shown in Fig. 33–12, the rays from each point on the object that reach the mirror will be nearly parallel. *For an object infinitely far away* (the Sun and stars approach this), *the rays would be precisely parallel*. Now consider such parallel rays falling on a concave mirror as in Fig. 33–13. The law of reflection holds for each of these rays at the point each strikes the mirror. As can be seen, they are not all brought to a single point. In

FIGURE 33–13 Parallel rays striking a concave spherical mirror do not focus at precisely a single point.

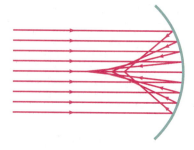

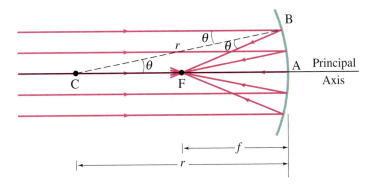

FIGURE 33–14 Rays parallel to the principal axis of a spherical mirror come to a focus at F, called the focal point, as long as the mirror is small in width as compared to its radius of curvature, r, so that the rays are "paraxial"—that is, make only small angles with the axis.

order to form a sharp image, the rays must come to a point. Thus a spherical mirror will not make as sharp an image as a plane mirror will. However, as we show below, if the mirror is small compared to its radius of curvature, so that the reflected rays make only a *small angle* upon reflection, then the rays will cross each other at very nearly a single point, or **focus**, as shown in Fig. 33–14. In the case shown, the rays are parallel to the **principal axis**, which is defined as the straight line perpendicular to the curved surface at its center (line CA in the diagram). The point F, where rays parallel to the principal axis come to a focus, is called the **focal point** of the mirror. The distance between F and the center of the mirror, length FA, is called the **focal length**, f, of the mirror. Another way of defining the focal point is to say that it is the *image point for an object infinitely far away* along the principal axis. The image of the Sun, for example, would be at F.

Now we will show, for a mirror whose reflecting surface is small compared to the radius of curvature, that the rays very nearly meet at a common point, F, and we will also calculate the focal length f. In this approximation, we consider only rays that make a small angle with the principal axis; such rays are called **paraxial rays**, and their angles are exaggerated in Fig. 33–14 only so the labeling will be clear. First we consider a ray that strikes the mirror at B in Fig. 33–14. The point C is the center of curvature of the mirror (the center of the sphere of which the mirror is a part). So the dashed line CB is equal to r, the radius of curvature, and CB is normal to the mirror's surface at B. The incoming ray that hits the mirror at B makes an angle θ with this normal, and hence the reflected ray, BF, also makes an angle θ with the normal. Note also from the geometry that angle BCF is also θ as shown. The triangle CBF is isosceles because two of its angles are equal. Thus length CF = BF. We assume the mirror has a width or diameter that is small compared to its radius of curvature, so the angles are small, and the length FB is nearly equal to length FA. In this approximation, FA = FC. But FA = f, the focal length, and CA = 2 FA = r. Thus the focal length is half the radius of curvature:

$$f = \frac{r}{2}. \tag{33–2}$$

This argument assumed only that the angle θ was small, so the same result applies for all other paraxial rays. Thus all paraxial rays pass through the same point F in the paraxial approximation.

Since it is only approximately true that the rays come to a perfect focus at F, the more curved the mirror, the worse the approximation (Fig. 33–13) and the more blurred the image. This "defect" of spherical mirrors is called **spherical aberration**; we will discuss it more with regard to lenses in Chapter 34. A *parabolic* reflector, on the other hand, will reflect the rays to a perfect focus. However, because parabolic shapes are much harder to make and thus much more expensive, spherical mirrors are used for most purposes. (Many astronomical telescopes use parabolic reflectors.) We consider here only spherical mirrors and we will assume that they are small compared to their radius of curvature so that the image is sharp and Eq. 33–2 holds.

Small angle approximation

Focal point
Focal length

Focal length
of mirror

Parabolic mirror

We saw that for an object at infinity the image is located at the focal point of a concave spherical mirror, where $f = r/2$. But where does the image lie for an object not at infinity? First consider the object shown as an arrow in Fig. 33–15, which is placed at point O between F and C. Let us determine where the image will be for a given point O′ on the object. To do this we can draw several rays and make sure these reflect from the mirror such that the angle of reflection equals the angle of incidence. Determining the image position is simplified if we deal with three particularly simple rays. These are the rays labeled 1, 2, and 3 in Fig. 33–15 and we draw them leaving object point O′ as follows:

➡ R A Y   D I A G R A M

Finding the image position for a curved mirror

Ray 1 is drawn parallel to the axis; therefore after reflection it must pass along a line through F (as we saw in Fig. 33–14, and drawn here in Fig. 33–15a).

Ray 2 leaves O′ and is made to pass through F; therefore it must reflect so it is parallel to the axis (Fig. 33–15b).

Ray 3 is chosen to be perpendicular to the mirror, and so is drawn so that it passes through C, the center of curvature; it is along a radius of the spherical surface, and because it is perpendicular to the mirror it will be reflected back on itself (Fig. 33–15c).

The point at which these rays cross is the image point I′. All other rays from the same object point will pass through this image point. To find the image point for any object point, only these three particular rays need to be used. Actually, only two of these rays are needed, but the third serves as a check.

We have shown the image point in Fig. 33–15 only for a single point on the object. Other points on the object are imaged nearby, so a complete image of the object is formed, as shown by the dashed arrow in Fig. 33–15c. Because the light

Real image

actually passes through the image itself, this is a real image that will appear on a piece of paper or film placed there. This can be compared to the virtual image formed by a plane mirror (the light does not actually pass through that image, Fig. 33–7).

The image in Fig. 33–15 can be seen by the eye when the eye is placed to the left of the image so that some of the rays diverging from each point on the image (as point I′) can enter the eye as shown in Fig. 33–15c. (See also Figs. 33–1 and 33–7.)

FIGURE 33–15 Rays leave point O′ on the object (an arrow). Shown are the three most useful rays for determining where the image I′ is formed. [Note that our mirror's height is not small compared to f, so our diagram will not give the precise position of the image.]

(a) Ray 1 goes out from O′ parallel to the axis and reflects through F.

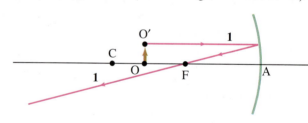

(b) Ray 2 goes through F and then reflects back parallel to the axis.

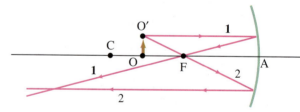

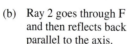

(c) Ray 3 heads out perpendicular to mirror and then reflects back on itself and goes through C (center of curvature)

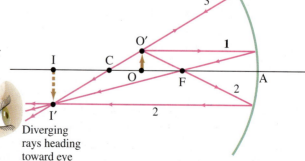

Diverging rays heading toward eye

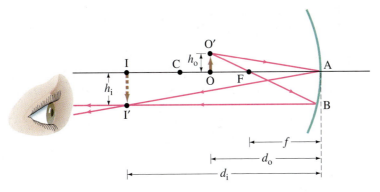

FIGURE 33-16 Diagram for deriving the mirror equation. For the derivation, we assume the mirror size is small compared to its radius of curvature.

Image points can be determined, roughly, by drawing the three rays as described above, but high accuracy is hard to obtain. For one thing it is difficult to draw, maintaining small angles for the "paraxial" rays as we assumed. However, it is possible to derive an equation that gives the image distance if the object distance and radius of curvature of the mirror are known. To do this, we refer to Fig. 33–16. The distance of the object from the center of the mirror, called the **object distance**, is labeled d_o. The **image distance** is labeled d_i. The height of the object OO' is called h_o and the height of the image, $I'I$, is h_i. Two rays are shown, $O'BI'$ (same as ray 2 in the previous figure) and $O'AI'$. The ray $O'AI'$ obeys the law of reflection, of course, so the two right triangles $O'AO$ and $I'AI$ are similar. Therefore, we have

$$\frac{h_o}{h_i} = \frac{d_o}{d_i}.$$

For the other ray, $O'FBI'$, the triangles $O'FO$ and AFB are also similar since the length $AB = h_i$ (in our approximation of a mirror that is small compared to its radius) and $FA = f$, the focal length of the mirror. Therefore,

$$\frac{h_o}{h_i} = \frac{OF}{FA} = \frac{d_o - f}{f}.$$

The left sides of the two preceding expressions are the same, so we can equate the right sides:

$$\frac{d_o}{d_i} = \frac{d_o - f}{f}.$$

We now divide both sides by d_o and rearrange to obtain

$$\frac{1}{d_o} + \frac{1}{d_i} = \frac{1}{f}. \tag{33-3}$$

Mirror equation

This is the equation we were seeking. It is called the **mirror equation** and relates the object and image distances to the focal length f (where $f = r/2$).

The **lateral magnification**, m, of a mirror is defined as the height of the image divided by the height of the object. From our first set of similar triangles above, we can write:

$$m = \frac{h_i}{h_o} = -\frac{d_i}{d_o}. \tag{33-4}$$

Lateral magnification for curved mirror

The minus sign in Eq. 33–4 is inserted as a convention. Indeed, we must be careful about the signs of all quantities in Eqs. 33–3 and 33–4. Sign conventions are chosen so as to give the correct locations and orientations of images, as predicted by ray diagrams. The sign conventions we use are: the image height h_i is positive if the image is upright, and negative if inverted, relative to the object (assuming h_o is taken as positive); d_i and d_o are positive if image and object are on the reflecting side of the mirror (as in Fig. 33–16), but if either image or object is behind the mirror, the corresponding distance is negative (an example can be seen in Fig. 33–17, Example 33–5). Thus the magnification (Eq. 33–4) is positive for an upright image and negative for an inverted image. We will summarize sign conventions more fully after discussing convex mirrors later in this Section.

Sign conventions

EXAMPLE 33–3 **Image in a concave mirror.** A 1.50-cm-high diamond ring is placed 20.0 cm from a concave mirror whose radius of curvature is 30.0 cm. Determine (*a*) the position of the image, and (*b*) its size.

SOLUTION The focal length $f = r/2 = 15.0$ cm. The ray diagram is basically like that shown in Fig. 33–15 and Fig. 33–16, since the object is between F and C. Referring to either Fig. 33–15 or Fig. 33–16, we have CA = r = 30.0 cm, FA = f = 15.0 cm, and OA = d_o = 20.0 cm. (*a*) From Eq. 33–3

$$\frac{1}{d_i} = \frac{1}{f} - \frac{1}{d_o} = \frac{1}{15.0 \text{ cm}} - \frac{1}{20.0 \text{ cm}} = 0.0167 \text{ cm}^{-1}.$$

➡ **P R O B L E M   S O L V I N G**

Remember to take the reciprocal

So $d_i = 1/0.0167 \text{ cm}^{-1} = 60.0$ cm. The image is 60.0 cm from the mirror on the same side as the object.
(*b*) From Eq. 33–4, the lateral magnification is

$$m = -\frac{60.0 \text{ cm}}{20.0 \text{ cm}} = -3.00.$$

Therefore the image height is

$$h_i = mh_o = (-3.00)(1.5 \text{ cm}) = -4.5 \text{ cm}.$$

The minus sign reminds us that the image is inverted, as in Figs. 33–15 and 33–16.

CONCEPTUAL EXAMPLE 33–4 **Reversible rays.** If the object in Example 33–3 is placed instead where the image is (see Fig. 33–16), where will the new image be?

RESPONSE The mirror equation is symmetric in d_o and d_i. Thus the new image will be where the old object was. Indeed, in Fig. 33–16 we need only reverse the direction of the rays to get our new situation.

EXAMPLE 33–5 **Object closer to concave mirror.** A 1.00-cm-high object is placed 10.0 cm from a concave mirror whose radius of curvature is 30.0 cm. (*a*) Draw a ray diagram to locate (approximately) the position of the image. (*b*) Determine the position of the image and the magnification analytically.

SOLUTION (*a*) Since $f = r/2 = 15.0$ cm, the object is between the mirror and the focal point. We draw the three rays as described earlier (Fig. 33–15) and this is shown in Fig. 33–17. Ray 1 leaves the tip of our object heading toward the mirror parallel to the axis, and reflects through F. Ray 2 cannot head toward F because it would not strike the mirror; so ray 2 must head as if it started at F (dashed line) and heads to the mirror and then is reflected parallel to the principal axis. Ray 3 is perpendicular to the mirror, as before. The rays reflected from the mirror diverge and so never meet at a point. They appear, however, to be coming from a point behind the mirror. This point is the image which is thus behind the mirror and *virtual*. (Why?)
(*b*) We use Eq. 33–3 to find d_i when d_o = 10.0 cm:

$$\frac{1}{d_i} = \frac{1}{15.0 \text{ cm}} - \frac{1}{10.0 \text{ cm}} = \frac{2-3}{30.0 \text{ cm}} = -\frac{1}{30.0 \text{ cm}}.$$

Therefore, $d_i = -30.0$ cm. The minus sign means the image is behind the mirror. The lateral magnification is $m = -d_i/d_o = -(-30.0 \text{ cm})/(10.0 \text{ cm}) = +3.00$. So the image is 3.00 times larger than the object; the plus sign indicates that the image is upright (which is consistent with the ray diagram, Fig. 33–17).
[Note that the image distance cannot be obtained accurately by measuring on Fig. 33–17, because our diagram violates the paraxial ray assumption (we had to, to make all rays clearly visible).]

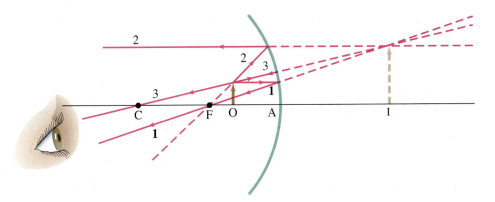

FIGURE 33–17 Object placed within the focal point F. The image is *behind* the mirror and is *virtual*, Example 33–5. [Note that the vertical scale (height of object = 1.0 cm) is different from the horizontal (OA = 10.0 cm) for ease of drawing, and this will affect the precision of the drawing.]

It is useful to compare Figs. 33–15 and 33–17. We can see that if the object is within the focal point, as in Fig. 33–17, the image is virtual, upright, and magnified. This is how a shaving or cosmetic mirror is used—you must place your head within the focal point if you are to see yourself right side up (Fig. 33–11a). If the object is *beyond* the focal point, as in Fig. 33–15, the image is real and inverted (upside down—and hard to use!). Whether the magnification is greater or less than 1.0 in the latter case depends on the position of the object relative to the center of curvature, point C.

The mirror equation also holds for a plane mirror: the focal length is $f = r/2 = \infty$, and Eq. 33–3 gives $d_i = -d_o$.

The analysis used for concave mirrors can be applied to **convex** mirrors. Even the mirror equation (Eq. 33–3) holds for a convex mirror, although the quantities involved must be carefully defined. Figure 33–18a shows parallel rays falling on a convex mirror. Again spherical aberration will be present, but we assume the mirror's size is small compared to its radius of curvature. The reflected rays diverge, but seem to come from point F behind the mirror. This is the **focal point**, and its distance from the center of the mirror is the **focal length**, f. It is easy to show that again $f = r/2$. We see that an object at infinity produces a virtual image in a convex mirror. Indeed, no matter where the object is placed on the reflecting side of a convex mirror, the image will be virtual and erect, as indicated in Fig. 33–18b. To find the image we draw rays 1 and 3 according to the rules used before on the concave mirror, as shown in Fig. 33–18b. Note that although rays 1 and 3 don't actually pass through points F and C, the line along which they are drawn do (shown dashed).

→ **PHYSICS APPLIED**

Magnifying mirror

Convex mirrors—analysis

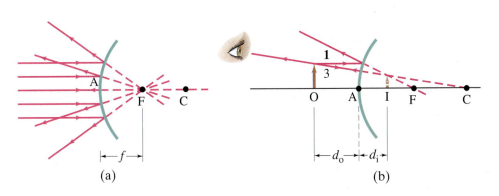

FIGURE 33–18 Convex mirror: (a) the focal point is at F, behind the mirror; (b) the image I of the object at O is virtual, upright, and smaller than the object. [Not to scale for Example 33–6.]

The mirror equation, Eq. 33–3, holds for convex mirrors but the focal length f must be considered negative, as must the radius of curvature. The proof is left as a Problem. It is also left as a Problem to show that Eq. 33–4 for the magnification is also valid.

▶ PHYSICS APPLIED

Convex rearview mirror

EXAMPLE 33–6 **Convex rearview mirror.** A convex rearview car mirror has a radius of curvature of 16 m. Determine the location of the image and its magnification for an object 10.0 m from the mirror.

SOLUTION The ray diagram will be like Fig. 33–18, but the large object distance ($d_o = 10.0$ m) makes a precise drawing difficult. We have a convex mirror, so r is negative by convention. Specifically, $r = -16$ m, so $f = -8.0$ m. The mirror equation gives

$$\frac{1}{d_i} = \frac{1}{f} - \frac{1}{d_o} = \frac{1}{-8.0\text{ m}} - \frac{1}{10.0\text{ m}} = -\frac{18}{80\text{ m}}.$$

So $d_i = -80\text{ m}/18 = -4.4$ m, or 4.4 m behind the mirror. The lateral magnification is $m = -d_i/d_o = -(-4.4\text{ m})/(10.0\text{ m}) = 0.44$.
So the upright image is a little less than half the size of the object. Rearview mirrors sometimes come with warnings that objects look farther away (appear smaller) than they really are; judging distances in a spherical mirror requires skill.

PROBLEM SOLVING **Spherical Mirrors**

1. Always draw a ray diagram even though you are going to make an analytic calculation—the diagram serves as a check, even if not precise. Draw at least two, and preferably three, of the easy-to-draw rays as described in Fig. 33–15. Generally draw the rays starting from a point on the object to the left of the mirror and moving to the right.

2. Use Eqs. 33–3 and 33–4; it is crucially important to follow the sign conventions.

3. **Sign Conventions**
   (a) When the object, image, or focal point is on the reflecting side of the mirror (on the left in all our drawings), the corresponding distance is considered positive. If any of these points is behind the mirror (on the right) the corresponding distance is negative.[†]
   (b) The image height h_i is positive if the image is upright, and negative if inverted, relative to the object (that is, if h_o is taken as positive).

[†] Object distances are positive for material objects, but can be negative in systems with more than one mirror or lens—see Section 34–3.

33–5 Refraction: Snell's Law

When light passes from one medium into another with a different index of refraction (Section 33–2), part of the incident light is reflected at the boundary. The remainder passes into the new medium. If a ray of light is incident at an angle to the surface (other than perpendicular), the ray is bent as it enters the new medium. This bending is called **refraction**. Figure 33–19a shows a ray passing from air into water. The angle θ_1 is the angle the incident ray makes with the perpendicular to the surface and is called the **angle of incidence**. The angle θ_2 is the **angle of refraction**, the angle the refracted ray makes with the perpendicular. Notice that the ray bends toward the normal when entering the water. This is always the case

Angle of refraction

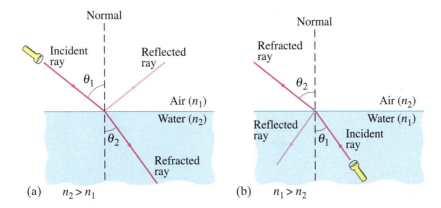

FIGURE 33–19 Refraction.
(a) Light refracted when passing from air (n_1) into water (n_2): $n_2 > n_1$.
(b) Light refracted when passing from water (n_1) into air (n_2): $n_1 > n_2$.

(a) $n_2 > n_1$ (b) $n_1 > n_2$

when the ray enters a medium where the speed of light is *less*. If light travels from one medium into a second where its speed is *greater*, the ray bends away from the normal; this is shown in Fig. 33–19b for a ray traveling from water to air.

Refraction is responsible for a number of common optical illusions. For example, a person standing in waist-deep water appears to have shortened legs. As shown in Fig. 33–20, the rays leaving the person's foot are bent at the surface. The observer's brain assumes the rays to have traveled a straight-line path, and so the feet appear to be higher than they really are. Similarly, when you put a pencil in water, it appears to be bent (Fig. 33–21).

The angle of refraction depends on the speed of light in the two media and on the incident angle. An analytical relation between θ_1 and θ_2 was arrived at experimentally about 1621 by Willebrord Snell (1591–1626). It is known as **Snell's law** and is written:

$$n_1 \sin \theta_1 = n_2 \sin \theta_2. \qquad \textbf{(33–5)}$$

θ_1 is the angle of incidence and θ_2 is the angle of refraction; n_1 and n_2 are the respective indices of refraction in the materials. See Fig. 33–19. The incident and refracted rays lie in the same plane, which also includes the perpendicular to the surface. Snell's law is the basic **law of refraction**. Snell's law can be derived from the wave theory of light (Chapter 35), and in fact we did derive it in Section 15–10 where Eq. 15–19 is just a combination of Eqs. 33–5 and 33–1.

It is clear from Snell's law that if $n_2 > n_1$, then $\theta_2 < \theta_1$. That is, if light enters a medium where n is greater (and its speed less), then the ray is bent toward the normal. And if $n_2 < n_1$, then $\theta_2 > \theta_1$, so the ray bends away from the normal. This is what we saw in Fig. 33–19.

PHYSICS APPLIED

Optical illusions

*Snell's law
(law of refraction)*

FIGURE 33–20 Ray diagram showing why a person's legs look shorter when standing in waist-deep water: the path of light traveling from the bather's foot to the observer's eye bends at the water's surface, and our brain interprets the light as having traveled in a straight line, from higher up (dashed line).

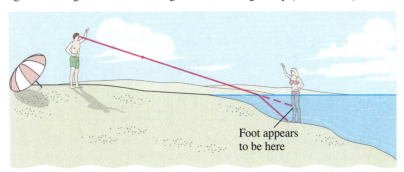

Foot appears to be here

FIGURE 33–21 A pencil in water looks bent even when it isn't.

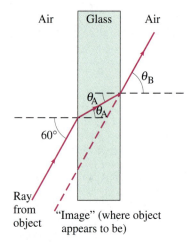

FIGURE 33–22 Light passing through a piece of glass (Example 33–7).

EXAMPLE 33–7 **Refraction through flat glass.** Light strikes a flat piece of uniformly thick glass at an incident angle of 60°, as shown in Fig. 33–22. If the index of refraction of the glass is 1.50, (a) what is the angle of refraction θ_A in the glass; (b) what is the angle θ_B at which the ray emerges from the glass?

SOLUTION (a) We assume the incident ray is in air, so $n_1 = 1.00$ and $n_2 = 1.50$. Then, from Eq. 33–5 we have

$$\sin \theta_A = \frac{1.00}{1.50} \sin 60° = 0.577,$$

so $\theta_A = 35.2°$.
(b) Since the faces of the glass are parallel, the incident angle at the second surface is just θ_A, so $\sin \theta_A = 0.577$. This time $n_1 = 1.50$ and $n_2 = 1.00$. Thus, $\theta_B \,(= \theta_2)$ is

$$\sin \theta_B = \frac{1.50}{1.00} \sin \theta_A = 0.866,$$

and $\theta_B = 60.0°$. The direction of the beam is thus unchanged by passing through a flat piece of glass of uniform thickness. It should be clear that this works for any angle of incidence. The ray is displaced slightly to one side, however. You can observe this by looking through a piece of glass (near its edge) at some object and then moving your head to the side so that you see the object directly.

EXAMPLE 33–8 **Apparent depth of a pool.** A swimmer has dropped her goggles in the shallow end of a pool, marked as 1.0 m deep. But the goggles don't look that deep. Why? How deep do the goggles appear to be when you look straight down into the water?

SOLUTION The ray diagram of Fig. 33–23 shows why the water seems less deep than it actually is. Rays traveling upward from the goggles on the bottom are refracted *away* from the normal as they exit the water. The rays appear to be diverging from a point higher in the water. To calculate the apparent depth d', given a real depth $d = 1.0$ m, we use Snell's law with $n_1 = 1$ for air and $n_2 = 1.33$ for water: $\sin \theta_1 = n_2 \sin \theta_2$. We are considering only small angles, so $\sin \theta \approx \tan \theta \approx \theta$, with θ in radians. So Snell's law becomes

$$\theta_1 \approx n_2 \theta_2.$$

From Fig. 33–23, we see that

$$\theta_1 \approx \tan \theta_1 = \frac{x}{d'} \quad \text{and} \quad \theta_2 \approx \tan \theta_2 = \frac{x}{d}.$$

Putting these into Snell's law, $\theta_1 \approx n_2 \theta_2$, we get

$$\frac{x}{d'} \approx n_2 \frac{x}{d} \quad \text{or} \quad d' \approx \frac{d}{n_2} = \frac{1.0 \text{ m}}{1.33} = 0.75 \text{ m}.$$

The pool seems only three-fourths as deep as it actually is.

FIGURE 33–23 Example 33–8.

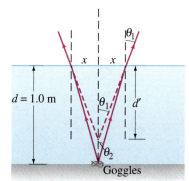

33–6 Visible Spectrum and Dispersion

An obvious property of visible light is its color. Color is related to the wavelengths or frequencies of the light. (How this was discovered will be discussed in Chapter 35.) Visible light—that to which our eyes are sensitive—falls in the wavelength range of about 400 nm to 750 nm.[†] This is known as the **visible spectrum**, and within it lie

[†] Sometimes the angstrom (Å) unit is used when referring to light: $1 \text{ Å} = 1 \times 10^{-10}$ m. Then visible light falls in the wavelength range of 4000 Å to 7500 Å.

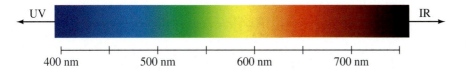

FIGURE 33–24 The spectrum of visible light, showing the range of wavelengths for the various colors.

the different colors from violet to red, as shown in Fig. 33–24. Light with wavelength shorter than 400 nm is called **ultraviolet** (UV), and light with wavelength greater than 750 nm is called **infrared** (IR).[‡] Although human eyes are not sensitive to UV or IR, some types of photographic film do respond to them.

A prism separates white light into a rainbow of colors, as shown in Fig. 33–25. This happens because the index of refraction of a material depends on the wavelength, as shown for several materials in Fig. 33–26. White light is a mixture of all visible wavelengths, and when incident on a prism, as in Fig. 33–27, the different wavelengths are bent to varying degrees. Because the index of refraction is greater for the shorter wavelengths, violet light is bent the most and red the least as indicated. This spreading of white light into the full spectrum is called **dispersion**.

FIGURE 33–25 White light passing through a prism is broken down into its constituent colors.

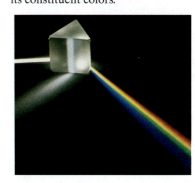

FIGURE 33–26 Index of refraction as a function of wavelength for various transparent solids.

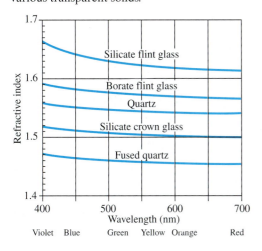

FIGURE 33–27 White light dispersed by a prism into the visible spectrum.

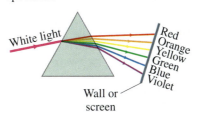

Rainbows are a spectacular example of dispersion—by drops of water. You can see rainbows when you look at falling water droplets with the Sun at your back. Figure 33–28 shows how red and violet rays are bent by spherical water droplets and are reflected off the back surface. Red is bent the least and so reaches the observer's eyes from droplets higher in the sky, as shown in the diagram. Thus the top of the rainbow is red.

➡ **PHYSICS APPLIED**
Rainbows

[‡]The complete electromagnetic spectrum is illustrated in Fig. 32–12.

FIGURE 33–28 (a) Ray diagram explaining how a rainbow (b) is formed.

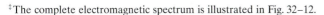

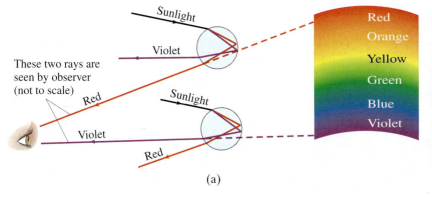

33–7 | Total Internal Reflection; Fiber Optics

When light passes from one material into a second material where the index of refraction is less (say, from water into air), the light bends away from the normal, as for rays I and J in Fig. 33–29. At a particular incident angle, the angle of refraction will be 90°, and the refracted ray would skim the surface (ray K) in this case. The incident angle at which this occurs is called the **critical angle**, θ_C. From Snell's law, θ_C is given by

Critical angle

$$\sin \theta_C = \frac{n_2}{n_1} \sin 90° = \frac{n_2}{n_1}.$$ **(33–6)**

For any incident angle less than θ_C there will be a refracted ray, although part of the light will also be reflected at the boundary. However, for incident angles greater than θ_C, Snell's law would tell us that $\sin \theta_2$ is greater than 1.00. Yet the sine of an angle can never be greater than 1.00. In this case there is no refracted ray at all, and *all of the light is reflected*, as for ray L in Fig. 33–29. This effect is called **total internal reflection**. But note that total internal reflection can occur only when light strikes a boundary where the medium beyond has a lower index of refraction.

Total internal reflection

FIGURE 33–29 Since $n_2 < n_1$, light rays are totally internally reflected if $\theta > \theta_C$ as for ray L. If $\theta < \theta_C$, as for rays I and J, only a part of the light is reflected, and the rest is refracted.

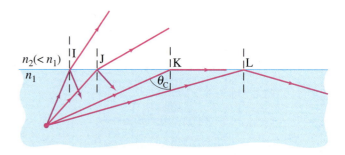

CONCEPTUAL EXAMPLE 33–9 View up from under water. Describe what a person would see who looked up at the world from beneath the perfectly smooth surface of a lake or swimming pool.

RESPONSE For an air–water interface, the critical angle is given by

$$\sin \theta_C = \frac{1.00}{1.33} = 0.750.$$

Therefore, $\theta_C = 49°$. Thus the person would see the outside world compressed into a circle whose edge makes a 49° angle with the vertical. Beyond this angle, the person would see reflections from the sides and bottom of the pool or lake (Fig. 33–30).

FIGURE 33–30 (a) Light rays, and (b) view looking upward from beneath the water (the surface of the water must be very smooth).

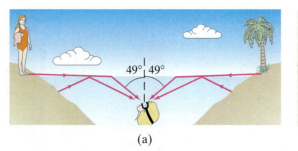

(a)   (b)

Diamonds

Diamonds achieve their brilliance from a combination of dispersion (previous Section) and total internal reflection. Because diamonds have a very high index of refraction of about 2.4, the critical angle for total internal reflection is

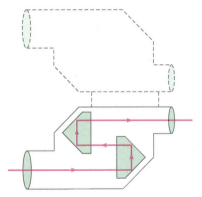

FIGURE 33–31 Total internal reflection of light by prisms in binoculars.

FIGURE 33–32 Light reflected totally at the interior surface of a glass or transparent plastic fiber.

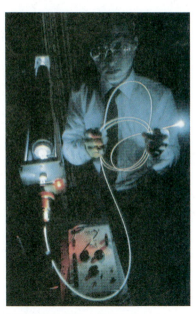

FIGURE 33–33 Total internal reflection within the tiny fibers of this light pipe makes it possible to transmit light in complex paths with minimal loss.

FIGURE 33–34 (a) How a fiber-optic image is made. (b) Example of a fiber-optic device inserted through the nose, and the image seen.

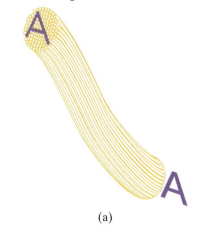

(a)

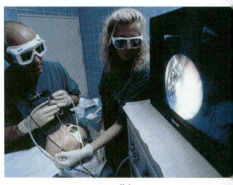

(b)

only 25°. Incident light therefore strikes many of the internal surfaces before it strikes one at less than 25° and emerges. After many such reflections, the light has traveled far enough that the colors have become sufficiently separated to be seen individually and brilliantly by the eye after leaving the crystal.

Many optical instruments, such as binoculars, use total internal reflection within a prism to reflect light. The advantage is that very nearly 100 percent of the light is reflected, whereas even the best mirrors reflect somewhat less than 100 percent. Thus the image is brighter. For glass with $n = 1.50$, $\theta_C = 41.8°$. Therefore, 45° prisms will reflect all the light internally, if oriented as shown in the binoculars of Fig. 33–31.

Total internal reflection is the principle behind **fiber optics**. Glass and plastic fibers as thin as a few micrometers in diameter can now be made. A bundle of such tiny fibers is called a **light pipe** or cable, and light can be transmitted along it with almost no loss because of total internal reflection. Figure 33–32 shows how light traveling down a thin fiber makes only glancing collisions with the walls so that total internal reflection occurs. Even if the light pipe is bent into a complicated shape, the critical angle still won't be exceeded, so light is transmitted practically undiminished to the other end (see Fig. 33–33). Very small losses do occur, mainly by reflection at the ends and absorption within the fiber.

Important applications of fiber-optic cables are in telecommunications and medicine. They are being used to transmit telephone calls, video signals, and computer data. The signal is a modulated light beam (a light beam whose intensity can be varied) and is transmitted at a much higher rate and with less loss and less interference than an electrical signal in a copper wire. Fibers have been developed that can support over one hundred separate wavelengths, each modulated to carry up to 10 gigabits $(10^{10}\ \text{bits})$ of information per second. That amounts to a terabit $(10^{12}\ \text{bits})$ per second for the full one hundred wavelengths. The sophisticated use of fiber optics to transmit a clear picture, is particularly useful in medicine, Fig. 33–34. For example, a patient's lungs can be examined by inserting a light pipe known as a bronchoscope through the mouth and down the bronchial tube. Light is sent down an outer set of fibers to illuminate the lungs. The reflected light returns up a central core set of fibers. Light directly in front of each fiber travels up that fiber. At the opposite end, a viewer sees a series of bright and dark spots, much like a TV screen—that is, a picture of what lies at the opposite end. Lenses are used at each end: at the object end to bring the rays in parallel, and at the viewing end as a telescope. The image may be viewed directly or on a TV monitor or film. The fibers must be optically insulated from one another, usually by a thin coating of material whose refractive index is less than that of the fiber. The fibers must be arranged precisely parallel to one another if the picture is to be clear. The more fibers there are, and the smaller they are, the more detailed the picture. Such instruments, including bronchoscopes, colonoscopes (for viewing the colon) and endoscopes (stomach or other organs) are extremely useful for examining hard-to-reach places.

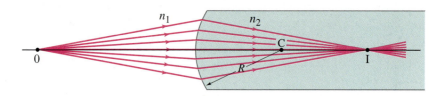

FIGURE 33–35 Rays from a point O on an object will be focused at a single image point I by a spherical boundary between two transparent materials $(n_2 > n_1)$, as long as the rays make small angles with the axis.

* **33–8** **Refraction at a Spherical Surface**

We now examine the refraction of rays at the spherical surface of a transparent material. Such a surface could be one face of a lens or the cornea of the eye. To be general, let us consider an object which is located in a medium whose index of refraction is n_1, and rays from each point on the object can enter a medium whose index of refraction is n_2. The radius of curvature of the spherical boundary is R, and its center of curvature is at point C, Fig. 33–35. We now show that all rays leaving a point O on the object will be focused at a single point I, the image point, if we consider only paraxial rays, rays that make a small angle with the axis. To do so, we consider a single ray that leaves point O as shown in Fig. 33–36. From Snell's law, Eq. 33–5, we have

$$n_1 \sin \theta_1 = n_2 \sin \theta_2.$$

We are assuming that angles $\theta_1, \theta_2, \alpha, \beta,$ and γ are small, so $\sin \theta \approx \theta$ (in radians), and Snell's law becomes, approximately,

$$n_1 \theta_1 = n_2 \theta_2.$$

Also, $\beta + \phi = 180°$ and $\theta_2 + \gamma + \phi = 180°$, so

$$\beta = \gamma + \theta_2.$$

Similarly for triangle OPC,

$$\theta_1 = \alpha + \beta.$$

These three relations can be combined to yield

$$n_1 \alpha + n_2 \gamma = (n_2 - n_1)\beta.$$

Since we are considering only the case of small angles, we can write, approximately,

$$\alpha = \frac{h}{d_o}, \qquad \beta = \frac{h}{R}, \qquad \gamma = \frac{h}{d_i},$$

where d_o and d_i are the object and image distances and h is the height as shown in Fig. 33–36. We substitute these into the previous equation, divide through by h, and obtain

$$\frac{n_1}{d_o} + \frac{n_2}{d_i} = \frac{n_2 - n_1}{R}. \tag{33–7}$$

For a given object distance d_o, this equation tells us d_i, the image distance, does not depend on the angle of a ray. Hence all paraxial rays meet at the same point I. This

FIGURE 33–36 Diagram for proving that all paraxial rays from O focus at the same point I $(n_2 > n_1)$.

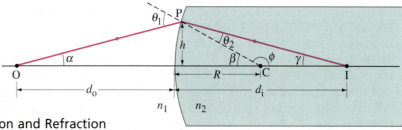

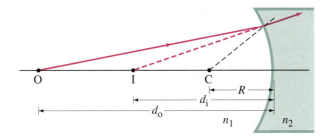

FIGURE 33–37 Rays from O refracted by a concave surface form a virtual image $(n_2 > n_1)$.

is true, of course, only for rays that make small angles with the axis and with each other. This is equivalent to assuming that the width of the refracting spherical surface is small compared to its radius of curvature. If this assumption is not true, the rays will not converge to a point; there will be spherical aberration, just as for a mirror (see Fig. 33–13), and the image will be blurry. (Spherical aberration will be discussed further in Section 34–10.)

We derived Eq. 33–7 using Fig. 33–36 for which the spherical surface is convex (as viewed by the incoming ray). It is also valid for a concave surface—as can be seen using Fig. 33–37—if we make the following conventions:

1. If the surface is convex (so the center of curvature C is on the side of the surface opposite to that from which the light comes), R is positive; if the surface is concave (C on the same side from which the light comes) R is negative.

2. The image distance, d_i, follows the same convention: positive if on the opposite side from where the light comes, negative if on the same side.

3. The object distance is positive if on the same side from which the light comes (this is the normal case, although when several surfaces bend the light it may not be so), otherwise it is negative.

For the case shown in Fig. 33–37 with a concave surface, both R and d_i are negative when used in Eq. 33–7. Note, in this case, that the image is virtual.

EXAMPLE 33–10 Apparent depth II. A person looks vertically down into a 1.0-m-deep pool. How deep does the pool appear to be?

SOLUTION Example 33–8 solved this problem using Snell's law. Here we use Eq. 33–7. A ray diagram is shown in Fig. 33–38. Point O represents a point on the pool's bottom. The rays diverge and appear to come from point I, the image. We have $d_o = 1.0$ m and, for a flat surface $R = \infty$. Then Eq. 33–7 becomes

$$\frac{1.33}{1.0 \text{ m}} + \frac{1.00}{d_i} = \frac{(1.00 - 1.33)}{\infty} = 0$$

Hence $d_i = -(1.0 \text{ m})/(1.33) = -0.75$ m. So the pool appears to be only three-fourths as deep as it actually is, the same result we got in Example 33–8. The minus sign tells us the image point I is on the same side of the surface as O, and the image is virtual. At angles other than vertical, this conclusion must be modified.

FIGURE 33–38 Example 33–10.

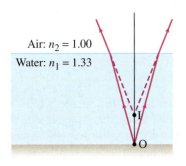

Air: $n_2 = 1.00$

Water: $n_1 = 1.33$

EXAMPLE 33–11 **A spherical "lens."** A point source of light is placed at a distance of 25.0 cm from the center of a glass sphere ($n = 1.5$) of radius 10.0 cm, Fig. 33–39. Find the image of the source.

SOLUTION As shown in Fig. 33–39, there are two refractions, and we treat them successively, one at a time. The light rays from the source first refract from the convex glass surface facing the source. We analyze this refraction ignoring the back side of the sphere, treating it as if it were in the shape shown in Fig. 33–35. Using Eq. 33–7, assuming paraxial rays, with $n_1 = 1.0$, $n_2 = 1.5$, $R = 10.0$ cm and $d_o = 25.0$ cm $- 10.0$ cm $= 15.0$ cm, we solve for the image distance as formed at surface 1, d_{i1}:

$$\frac{1}{d_{i1}} = \frac{1}{1.5}\left(\frac{1.5 - 1.0}{10.0 \text{ cm}} - \frac{1.0}{15.0 \text{ cm}}\right) = -\frac{1}{90.0 \text{ cm}}.$$

Thus, the image of the first refraction is located 90.0 cm *to the left* of the front surface. This image now serves as the object for the refraction occuring at the back surface (surface 2) of the sphere. This surface is concave so $R = -10.0$ cm, and we consider a ray close to the axis. Then the object distance is $d_{o2} = 90.0$ cm $+ 2(10.0 \text{ cm}) = 110.0$ cm, and Eq. 33–7 yields, with $n_1 = 1.5$, $n_2 = 1.0$,

$$\frac{1}{d_{i2}} = \frac{1}{1.0}\left(\frac{1.0 - 1.5}{-10.0 \text{ cm}} - \frac{1.5}{110.0 \text{ cm}}\right) = \frac{4.0}{110.0 \text{ cm}}$$

so $d_{i2} = 28$ cm. Thus, the final image is located a distance 28 cm from the back side of the sphere.

FIGURE 33–39 Example 33–11.

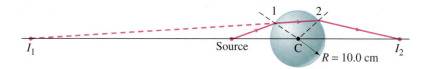

Summary

Light appears to travel along straight-line paths, called **rays**, at a speed v that depends on the **index of refraction**, n, of the material; that is

$$v = \frac{c}{n},$$

where c is the speed of light in vacuum.

When light reflects from a flat surface, *the angle of reflection equals the angle of incidence.* This **law of reflection** explains why mirrors can form **images**.

In a plane mirror, the image is virtual, upright, the same size as the object, and is as far behind the mirror as the object is in front.

A spherical mirror can be concave or convex. A concave spherical mirror focuses parallel rays of light (light from a very distant object) to a point called the **focal point**. The distance of this point from the mirror is the **focal length** f of the mirror and

$$f = \frac{r}{2}$$

where r is the radius of curvature of the mirror.

Parallel rays falling on a convex mirror reflect from the mirror as if they diverged from a common point behind the mirror. The distance of this point from the mirror is the focal length and is considered negative for a convex mirror.

For a given object, the position and size of the image formed by a mirror can be found by ray tracing. Algebraically, the relation between image and object distances, d_i and d_o, and the focal length f, is given by the **mirror equation**:

$$\frac{1}{d_o} + \frac{1}{d_i} = \frac{1}{f}.$$

The ratio of image height to object height, which equals the magnification m, is

$$m = \frac{h_i}{h_o} = -\frac{d_i}{d_o}.$$

If the rays that converge to form an image actually pass through the image, so the image would appear on film or a screen placed there, the image is said to be a **real image**. If the rays do not actually pass through the image, the image is a **virtual image**.

When light passes from one transparent medium into another, the rays bend or refract. The **law of refraction (Snell's law)** states that

$$n_1 \sin \theta_1 = n_2 \sin \theta_2,$$

where n_1 and θ_1 are the index of refraction and angle with the normal to the surface for the incident ray, and n_2 and θ_2 are for the refracted ray.

The wavelength of light determines its color; the **visible spectrum** extends from 400 nm (violet) to about 750 nm (red).

Glass prisms (and other transparent materials) can break white light down into its constituent colors because the index of refraction varies with wavelength, a phenomenon known as **dispersion**.

When light rays reach the boundary of a material where the index of refraction decreases, the rays will be **totally internally reflected** if the incident angle, θ_1, is such that Snell's law would predict $\sin \theta_2 > 1$; this occurs if θ_1 exceeds the critical angle θ_C given by

$$\sin \theta_C = \frac{n_2}{n_1}.$$

Questions

1. What would be the appearance of the Moon if it had (*a*) a rough surface; (*b*) a polished mirrorlike surface?

2. Archimedes is said to have burned the whole Roman fleet in the harbor of Syracuse by focusing the rays of the Sun with a huge spherical mirror. Is this reasonable?

3. If a concave mirror produces a real image, is the image necessarily inverted?

4. When you use a concave mirror, you cannot see an inverted image of yourself unless you place your head beyond the center of curvature C. Yet you can see an inverted image of another object placed between C and F, as in Fig. 33–15. Explain. [*Hint*: You can see a real image only if your eye is behind the image, so that the image can be formed.]

5. What is the focal length of a plane mirror? What is the magnification of a plane mirror?

6. Does the mirror equation, Eq. 33–3, hold for a plane mirror? Explain.

7. How might you determine the speed of light in a solid, rectangular, transparent object?

8. When you look at the Moon's reflection from a ripply sea, it appears elongated (Fig. 33–40). Explain.

9. What is the angle of refraction when a light ray meets the boundary between two materials perpendicularly?

10. When you look down into a swimming pool or a lake, you are likely to underestimate its depth. How does the apparent depth vary with the viewing angle? (Use ray diagrams.)

11. Draw a ray diagram to show why a stick looks bent when part of it is under water (Fig. 33–21).

12. When a wide beam of parallel light enters water at an angle, the beam broadens. Explain.

13. Your eye looks into an aquarium and views a fish inside. One ray of light that emerges from the tank from the fish is shown in Fig. 33–41. Also shown is the apparent position of the fish as seen by the eye ball. In the drawing, indicate the approximate position of the actual fish. Briefly justify your answer.

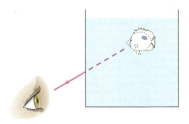

FIGURE 33–41
Question 13.

14. How can you "see" a round drop of water on a table even though the water is transparent and colorless?

15. Can a light ray traveling in air be totally reflected when it strikes a smooth water surface if the incident angle is right?

16. When you look up at an object in air from beneath the surface in a swimming pool, does the object appear to be the same size as when you see it directly in air? Explain.

17. What type of mirror is shown in Fig. 33–42?

FIGURE 33–42
Question 17.

FIGURE 33–40
Question 8.

Problems

Section 33–2

1. (I) What is the speed of light in (a) ethyl alcohol, (b) lucite?

2. (I) The speed of light in ice is 2.29×10^8 m/s. What is the index of refraction of ice?

3. (I) How long does it take light to reach us from the Sun, 1.50×10^8 km away?

4. (I) Our nearest star (other than the Sun) is 4.2 light years away. That is, it takes 4.2 years for the light to reach Earth. How far away is it in meters?

5. (II) Light is emitted from an ordinary light bulb filament in wave-train bursts of about 10^{-8} s in duration. What is the length in space of such wave trains?

6. (II) The speed of light in a certain substance is 92 percent of its value in water. What is the index of refraction of that substance?

7. (II) What is the minimum angular speed at which Michelson's eight-sided mirror would have had to rotate in order that light would be reflected into an observer's eye by succeeding mirror faces (Fig. 33–2)?

Section 33–3

8. (I) Suppose that you want to take a photograph of yourself as you look at your image in a mirror 1.8 m away. For what distance should the camera lens be focused?

9. (II) Stand up two plane mirrors so they form a right angle as in Fig. 33–43. When you look into this double mirror, you see yourself as others see you, instead of reversed as in a single mirror. Make a ray diagram to show how this occurs.

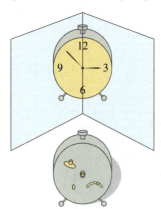

FIGURE 33–43
Problems 9 and 14.

10. (II) A person whose eyes are 1.54 m above the floor stands 2.30 m in front of a vertical plane mirror whose bottom edge is 40 cm above the floor, Fig. 33–44. What is the horizontal distance x to the base of the wall supporting the mirror of the nearest point on the floor that can be seen reflected in the mirror?

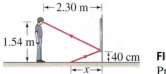

FIGURE 33–44
Problem 10.

11. (II) Two plane mirrors meet at a 135° angle, Fig. 33–45. If light rays strike one mirror at 40° as shown, at what angle ϕ do they leave the second mirror?

FIGURE 33–45
Problem 11.

12. (II) Suppose you are 80 cm from a plane mirror. What area of the mirror is used to reflect the rays entering one eye from a point on the tip of your nose if your pupil diameter is 5.0 mm?

13. (II) Show that if two plane mirrors meet at an angle ϕ, a single ray reflected successively from both mirrors is deflected through an angle of 2ϕ independent of the incident angle. Assume $\phi < 90°$ and that only two reflections, one from each mirror, take place.

14. (III) Suppose a third mirror is placed beneath the two shown in Fig. 33–43, so that all three are perpendicular to each other. (a) Show that for such a "corner reflector," any incident ray will return in its original direction after three reflections. (b) What happens if it makes only two reflections?

Section 33–4

15. (I) A solar cooker, really a concave mirror pointed at the Sun, focuses the Sun's rays 18.2 cm in front of the mirror. What is the radius of the spherical surface from which the mirror was made?

16. (I) How far from a concave mirror (radius 22.0 cm) must an object be placed if its image is to be at infinity?

17. (II) Show with ray diagrams that the magnification of a concave mirror is less than 1 if the object is beyond the center of curvature C, and is greater than 1 if it is within this point.

18. (II) If you look at yourself in a shiny Christmas tree ball with a diameter of 9.0 cm when your face is 25.0 cm away from it, where is your image? Is it real or virtual? Is it upright or inverted?

19. (II) A mirror at an amusement park shows an upright image of any person who stands 1.5 m in front of it. If the image is three times the person's height, what is the radius of curvature of the mirror?

20. (II) A dentist wants a small mirror that, when 2.00 cm from a tooth, will produce a 5.0× upright image. What kind of mirror must be used and what must its radius of curvature be?

21. (II) Some rearview mirrors produce images of cars to your rear that are smaller than they would be if the mirror were flat. Are the mirrors concave or convex? What is a mirror's radius of curvature if cars 20.0 m away appear 0.33 their normal size?

22. (II) A luminous object 3.0 mm high is placed 20 cm from a convex mirror of radius of curvature 20 cm. (a) Show by ray tracing that the image is virtual, and estimate the image distance. (b) Show that to compute this (negative) image distance from Eq. 33–3, it is sufficient to let the focal length be −10 cm. (c) Compute the image size, using Eq. 33–4.

23. (II) (a) Where should an object be placed in front of a concave mirror so that it produces an image at the same location as the object? (b) Is the image real or virtual? (c) Is the image inverted or erect? (d) What is the magnification of the image?

24. (II) The image of a distant tree is virtual and very small when viewed in a curved mirror. The image appears to be 14.0 cm behind the mirror. What kind of mirror is it, and what is its radius of curvature?

25. (II) Use the mirror equation to show that the magnitude of the magnification of a concave mirror is less than 1 if the object is beyond the center of curvature $C(d_o > r)$, and is greater than 1 if the object is within $C(d_o < r)$.

26. (II) Show, using a ray diagram, that the magnification m of a convex mirror is $m = -d_i/d_o$, just as for a concave mirror. [Hint: Consider a ray from the top of the object that reflects at the center of the mirror.]

27. (II) Use ray diagrams to show that the mirror equation, Eq. 33–3, is valid for a convex mirror as long as f is considered negative.

28. (II) The magnification of a convex mirror is $+0.55\times$ for objects 3.2 m from the mirror. What is the focal length of this mirror?

29. (II) A 4.5-cm tall object is placed 28 cm in front of a spherical mirror. It is desired to produce a virtual image that is erect and 3.5 cm tall. (a) What type of mirror should be used? (b) Where is the image located? (c) What is the focal length of the mirror? (d) What is the radius of curvature of the mirror?

30. (II) A shaving/cosmetic mirror is designed to magnify your face by a factor of 1.3 when your face is placed 20.0 cm in front of it. (a) What type of mirror is it? (b) Describe the type of image that it makes of your face. (c) Calculate the required radius of curvature for the mirror.

31. (III) A short thin object (like a short length of wire) of length l is placed on, and parallel to, the principal axis of a spherical mirror. Show that its image has length $l' = m^2 l$ so the longitudinal magnification is equal to $-m^2$ where m is the lateral magnification (Eq. 33–4). Why the minus sign? [Hint: Find the image positions for both ends of the rod, and assume l is very small.]

Section 33–5

32. (I) A flashlight beam strikes the surface of a pane of glass ($n = 1.50$) at a 63° angle to the normal. What is the angle of refraction?

33. (I) A diver shines a flashlight upward from beneath the water at a 32.5° angle to the vertical. At what angle does the light leave the water?

34. (I) A light beam coming from an underwater spotlight exits the water at an angle of 76.0°. At what angle of incidence did it hit the air–water interface from below the surface?

35. (I) Rays of the Sun are seen to make a 43.0° angle to the vertical beneath the water. At what angle above the horizon is the Sun?

36. (II) Light is incident on an equilateral glass prism at a 45.0° angle to one face, Fig. 33–46. Calculate the angle at which light emerges from the opposite face. Assume that $n = 1.50$.

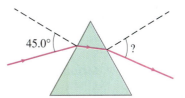

FIGURE 33–46
Problems 36 and 52.

37. (II) In searching the bottom of a pool at night, a watchman shines a narrow beam of light from his flashlight, 1.3 m above the water level, onto the surface of the water at a point 2.7 m from his foot at the edge of the pool (Fig. 33–47). Where does the spot of light hit the bottom of the pool, measured from the wall beneath his foot, if the pool is 2.1 m deep?

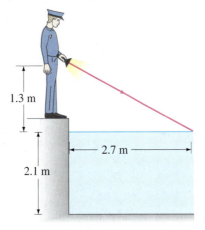

FIGURE 33–47
Problem 37.

38. (II) A beam of light in air strikes a slab of crown glass ($n = 1.52$) and is partially reflected and partially refracted. Find the angle of incidence if the angle of reflection is twice the angle of refraction.

39. (II) If the medium to the left of the glass in Fig. 33–22 is different than that to the right (so there are three different materials of different index of refraction), show that the angle of refraction into the third medium (on the right of the glass) is the same as if the light passed from the first medium to the third directly (as if the glass were of zero thickness).

40. (II) An aquarium filled with water has flat glass sides whose index of refraction is 1.58. A beam of light from outside the aquarium strikes the glass at a 43.5° angle to the perpendicular (Fig. 33–48). What is the angle of this light ray when it enters (a) the glass, and then (b) the water? (c) What would be the refracted angle if the ray entered the water directly?

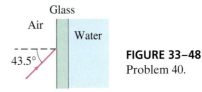

FIGURE 33–48
Problem 40.

41. (II) Prove in general that for a light beam incident on a uniform layer of transparent material, as in Fig. 33–22, the direction of the emerging beam is parallel to the incident beam, independent of the incident angle θ. Assume the medium on the two sides is the same.

42. (III) A light ray is incident on a flat piece of glass as in Fig. 33–22. Show that if the incident angle θ is small, the ray is displaced a distance $d = t\theta(n - 1)/n$, where t is the thickness of the glass and θ is in radians.

Section 33–6

43. (I) By what percent, approximately, does the speed of red light (700 nm) exceed that of violet light (400 nm) in silicate flint glass? (See Fig. 33–26).

44. (I) By what percent is the speed of blue light (450 nm) less than the speed of red light (700 nm), in silicate flint glass (see Fig. 33–26).

45. (II) A light beam strikes a piece of glass at a 60.00° incident angle. The beam contains two wavelengths, 450.0 nm and 700.0 nm, for which the index of refraction of the glass is 1.4820 and 1.4742, respectively. What is the angle between the two refracted beams?

46. (II) A parallel beam of light containing two wavelengths, $\lambda_1 = 450$ nm and $\lambda_2 = 650$ nm, enters the silicate flint glass of an equilateral prism as shown in Fig. 33–49. At what angle does each beam leave the prism (give angle with normal to the face)?

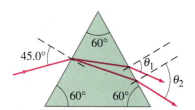

FIGURE 33–49
Problem 46.

Section 33–7

47. (I) What is the critical angle for the interface between water and lucite? To be internally reflected, the light must start in which material?

48. (I) The critical angle for a certain liquid-air surface is 51.3°. What is the index of refraction of the liquid?

49. (II) A beam of light is emitted in a pool of water from a depth of 82.0 cm. Where must it strike the air–water interface, relative to the spot directly above it, in order that the light does *not* exit the water?

50. (II) A ray of light enters a light fiber at an angle of 15.0° with the long axis of the fiber, as in Fig. 33–50. Calculate the distance the light ray travels between successive reflections off the sides of the fiber. Assume that the fiber has an index of refraction of 1.55 and is 1.40×10^{-4} m in diameter.

FIGURE 33–50 Problem 50.

51. (II) A beam of light is emitted 8.0 cm beneath the surface of a liquid and strikes the surface 7.0 cm from the point directly above the source. If total internal reflection occurs, what can you say about the index of refraction of the liquid?

52. (III) Suppose a ray strikes the left face of the prism in Fig. 33–46 at 45° as shown, but is totally internally reflected at the opposite side. If the apex angle (at the top) is $\theta = 75°$, what can you say about the index of refraction of the prism?

53. (III) A beam of light enters the end of an optic fiber as shown in Fig. 33–51. Show that we can guarantee total internal reflection at the side surface of the material (at point a), if the index of refraction is greater than about 1.42. In other words, regardless of the angle α, the light beam reflects back into the material at point a.

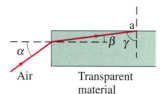

Air   Transparent material   **FIGURE 33–51** Problem 53.

54. (III) (*a*) What is the minimum index of refraction for a glass or plastic prism to be used in binoculars (Fig. 33–31) so that total internal reflection occurs at 45°? (*b*) Will binoculars work if its prisms (assume $n = 1.50$) are immersed in water? (*c*) What minimum n is needed if the prisms are immersed in water?

* Section 33–8

* 55. (II) A 12.0-cm-thick plane piece of glass ($n = 1.50$) lies on the surface of a 12.0-cm-deep pool of water. How far below the top of the glass does the bottom of the pool seem, as viewed from directly above?

* 56. (II) A fish is swimming in water inside a thin spherical glass bowl of uniform thickness. Assuming the radius of curvature of the bowl is 25.0 cm, locate the image of the fish if the fish is located: (*a*) at the center of the bowl; (*b*) 20.0 cm from the side of the bowl between the observer and the center of the bowl. [*Hint*: For (*a*) what is the angle of incidence of the light as it strikes the water–glass interface? Given this, where is the image? Finally, verify this by computation, noticing that R must be negative.]

* 57. (III) Show that Eq. 33–7 is valid for both convex and concave spherical surfaces and for differently located objects and images as long as the conventions discussed in Section 33–8 are adhered to. Show this by using diagrams similar to Fig. 33–36 for all possible cases. Assume both $n_2 > n_1$ and then $n_2 < n_1$.

* 58. (III) A coin lies at the bottom of a 1.00 m deep pool. If a viewer sees it at a 45° angle, where is the image of the coin, relative to the coin. [*Hint*: The image is found by tracing back to the intersection of two rays.]

General Problems

59. Two plane mirrors are facing each other 2.0 m apart as in Fig. 33–52. You stand 1.5 m away from one of these mirrors and look into it. You will see multiple images of yourself. (*a*) How far away from you are the first three images of yourself in the mirror in front of you? (*b*) Which way are these first three images facing, toward you or away from you?

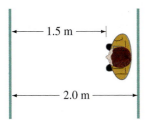

FIGURE 33–52
Problem 59.

60. We wish to determine the depth of a swimming pool filled with water by measuring the width ($x = 5.50$ m) and then noting that the bottom edge of the pool is just visible at an angle of 14.0° above the horizontal as shown in Fig. 33–53. Calculate the depth of the pool.

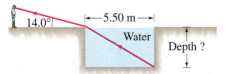

FIGURE 33–53   Problem 60.

61. A 1.70-m-tall person stands 3.80 m from a convex mirror and notices that he looks precisely half as tall as he does in a plane mirror placed at the same distance. What is the radius of curvature of the convex mirror? (Assume that $\sin\theta \approx \theta$.)

62. The critical angle of a certain piece of plastic in air is $\theta_C = 37.3°$. What is the critical angle of the same plastic if it is immersed in water?

63. Each student in a physics lab is assigned to find the location where a bright object may be placed in order that a concave mirror with radius of curvature $r = 40$ cm, will produce an image three times the size of the object. Two students complete the assignment at different times using identical equipment, but when they compare notes later, they discover that their answers for the object distance are not the same. Explain why they do not necessarily need to repeat the lab, and justify your response with a calculation.

64. A kaleidoscope makes symmetric patterns with two plane mirrors having a 60° angle between them as shown in Fig. 33–54. Draw the location of the images (some of them images of images) of the object placed between the mirrors.

FIGURE 33–54
Problem 64.

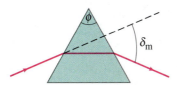

FIGURE 33–55   Problems 65 and 66.

65. If the apex angle of a prism is $\phi = 72°$ (see Fig. 33–55), what is the minimum incident angle for a ray if it is to emerge from the opposite side (i.e., not be totally internally reflected), given $n = 1.56$?

66. When light passes through a prism, the angle that the refracted ray makes relative to the incident ray is called the deviation angle δ, Fig. 33–55. Show that this angle is a minimum when the ray passes through the prism symmetrically, perpendicular to the bisector of the apex angle ϕ, and show that the minimum deviation angle, δ_m, is related to the prism's index of refraction n by

$$n = \frac{\sin\frac{1}{2}(\phi + \delta_m)}{\sin\phi/2}.$$

67. *Fermat's principle* states that "light travels between two points along that path which requires the least time, as compared to other nearby paths." From Fermat's principle derive (*a*) the law of reflection ($\theta_i = \theta_r$) and (*b*) the law of refraction (Snell's law). [*Hint:* Choose two appropriate points so that a ray between them can undergo reflection or refraction. Draw a rough path for a ray between these points, and write down an expression of the time required for light to travel the arbitrary path chosen. Then take the derivative to find the minimum.]

68. The end faces of a cylindrical glass rod ($n = 1.54$) are perpendicular to the sides. Show that a light ray entering an end face at any angle will be totally internally reflected inside the rod when it strikes the sides. Assume the rod is in air. What if it were in water?

69. Suppose Fig. 33–35 shows a cylindrical rod whose end has a radius of curvature $R = 2.0$ cm, and the rod is immersed in water with index of refraction of 1.33. The rod has index of refraction 1.50. Find the location of the image of an object 2.0 mm high located 20 cm away from the rod.

70. An optical fiber is a long transparent cylinder of diameter d and index of refraction n. If this fiber is bent sharply, some light hitting the side of the cylinder may escape rather than reflect back into the fiber (Fig. 33–56). What is the smallest radius of curvature at a short bent section for which total internal reflection will be assured for light initially travelling parallel to the axis of the fiber?

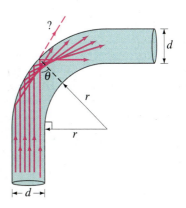

FIGURE 33–56   Problem 70.

Of the many optical devices we discuss in this chapter, the magnifying glass is the simplest. Here it is magnifying a page that describes how it works according to the ray model.

In this Chapter we examine thin lenses in detail, seeing how to determine image position as a function of object position and the focal length of the lens, based on the ray model. We will then examine various optical devices from cameras and eyeglasses to telescopes and microscopes.

Much of the remainder of this chapter will deal with optical devices that are used to produce magnified images of objects. We first discuss the **simple magnifier**, or **magnifying glass**, which is simply a converging lens, Fig. 34–26.

How large an object appears, and how much detail we can see on it, depends on the size of the image it makes on the retina. This, in turn, depends on the angle subtended by the object at the eye. For example, a penny held 30 cm from the eye looks twice as tall as one held 60 cm away because the angle it subtends is twice as great (Fig. 34–27). When we want to examine detail on an object, we bring it up close to our eyes so that it subtends a greater angle. However, our eyes can accommodate only up to a point (the near point), and we will assume a standard distance of 25 cm as the near point in what follows.

A magnifying glass allows us to place the object closer to our eye so that it subtends a greater angle. As shown in Fig. 34–28a, the object is placed at the focal point or just within it. Then the converging lens produces a virtual image, which must be at least 25 cm from the eye if the eye is to focus on it. If the eye is relaxed, the image will be at infinity, and in this case the object is exactly at the focal point. (You make this slight adjustment yourself when you "focus" on the object by mov-

FIGURE 34–26 Photo of a magnifying glass and the image it makes.

FIGURE 34–27 When the same object is viewed at a shorter distance, the image on the retina is greater, so the object appears larger and more detail can be seen. The angle θ that the object subtends in (a) is greater than in (b).

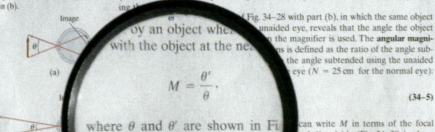

... Fig. 34–28 with part (b), in which the same object ... unaided eye, reveals that the angle the object ... the magnifier is used. The **angular magni-** ... is defined as the ratio of the angle sub- ... the angle subtended using the unaided eye ($N = 25$ cm for the normal eye):

$$M = \frac{\theta'}{\theta},$$

(34–5)

where θ and θ' are shown in Fig ... can write M in terms of the focal length by noting that $\theta = h/N$ ... and $\theta' = h/d_o$ (Fig. 34–28a), where h is the height of the object and ... e angles are small so θ and θ' equal ... fying glass, and (b) with the unaided eye.

CHAPTER 34

Lenses and Optical Instruments

FIGURE 34–1 (a) Converging and (b) diverging lenses, shown in cross section. (c) Photo of a converging lens (on the left) and a diverging lens. (d) Converging lenses (above), and diverging lenses, lying flat, and raised off the paper to form images.

The laws of reflection and refraction, particularly the latter, are the basis for explaining the operation of many optical instruments. In this chapter we discuss and analyze simple lenses using the model of ray optics discussed in the previous chapter. We then analyze a number of optical instruments, from the magnifying glass and the human eye to telescopes and microscopes. Figure 34–1 shows a variety of lenses.

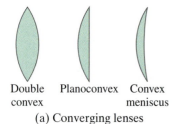

Double convex   Planoconvex   Convex meniscus

(a) Converging lenses

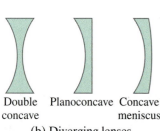

Double concave   Planoconcave   Concave meniscus

(b) Diverging lenses

(c)

(d)

34–1 Thin Lenses; Ray Tracing

The most important simple optical device is no doubt the thin lens. The development of optical devices using lenses dates to the sixteenth and seventeenth centuries, although the earliest record of eyeglasses dates from the late thirteenth century.[†] Today we find lenses in eyeglasses, cameras, magnifying glasses, telescopes, binoculars, microscopes, and many specialized instruments. A thin lens is usually circular in cross section, and its two faces are portions of a sphere. (Although cylindrical surfaces are also possible, we will concentrate on spherical.) The two faces can be concave, convex, or plane; several types are shown in Fig. 34–1, in cross section (previous page). The importance of lenses is that they form images of objects, as shown in Fig. 34–2.

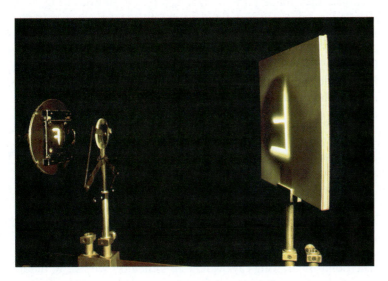

FIGURE 34–2 Converging lens (in holder) forms an image (large "F" on screen at right) of a bright object (glowing "F" at the left).

Consider the rays parallel to the axis of the double convex lens which is shown in cross section in Fig. 34–3a. We assume the lens is made of glass or transparent plastic, so its index of refraction is greater than that of the air outside. The **axis** of a lens is a straight line passing through the very center of the lens and perpendicular to its two surfaces (Fig. 34–3). From Snell's law, we can see that each ray in Fig. 34–3a is bent toward the axis at both lens surfaces (note the dashed lines indicating the normals to each surface for the top ray). If rays parallel to the axis fall on a thin lens, they will be focused to a point called the **focal point**, **F**. This will not be precisely true for a lens with spherical surfaces. But it will be very nearly true—that is, parallel rays will be focused to a tiny region that is nearly a point—if the diameter of the lens is small compared to the radii of curvature of the two lens surfaces. This criterion is satisfied by a **thin lens**, one that is very thin compared to its diameter, and we consider only thin lenses here.

[†] Rounded gemstones used as magnifiers probably date from much earlier.

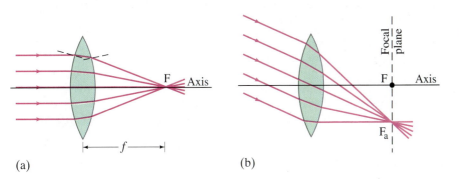

(a)　　　　(b)

FIGURE 34–3 Parallel rays are brought to a focus by a converging thin lens.

The rays from a point on a distant object are essentially parallel—see Fig. 33–12. Therefore we can also say that *the focal point is the image point for an object at infinity on the principal axis*. Thus, the focal point of a lens can be found by locating the point where the Sun's rays (or those of some other distant object) are brought to a sharp image, Fig. 34–4. The distance of the focal point from the center of the lens is called the **focal length**, f. A lens can be turned around so that light can pass through it from the opposite side. The focal length is the same on both sides, as we shall see later, even if the curvatures of the two lens surfaces are different. If parallel rays fall on a lens at an angle, as in Fig. 34–3b, they focus at a point F_a. The plane in which all points such as F and F_a fall is called the **focal plane** of the lens.

Focal length of lens

FIGURE 34–4 Image of the Sun burning a hole, almost, on a piece of paper.

Power of lens

FIGURE 34–5 Diverging lens.

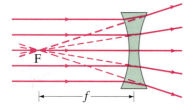

Any lens[†] that is thicker in the center than at the edges will make parallel rays converge to a point, and is called a **converging lens** (see Fig. 34–1a). Lenses that are thinner in the center than at the edges (Fig. 34–1b) are called **diverging lenses** because they make parallel light diverge, as shown in Fig. 34–5. The focal point, F, of a diverging lens is defined as that point from which refracted rays, originating from parallel incident rays, seem to emerge as shown in the Figure. And the distance from F to the lens is called the **focal length**, just as for a converging lens.

Optometrists and ophthalmologists, instead of using the focal length, use the reciprocal of the focal length to specify the strength of eyeglass (or contact) lenses. This is called the **power**, P, of a lens:

$$P = \frac{1}{f}. \tag{34–1}$$

The unit for lens power[‡] is the diopter (D), which is an inverse meter: $1\,\text{D} = 1\,\text{m}^{-1}$. For example, a 20-cm-focal-length lens has a power $P = 1/0.20\,\text{m} = 5.0\,\text{D}$. We will mainly use the focal length here, but we will refer again to the power of a lens when we discuss eyeglass lenses in Section 34–6.

The most important parameter of a lens is its focal length f. For a converging lens, f is easily measured by finding the image point for the Sun or other distant objects. Once f is known, the image position can be found for any object. To find the image point by drawing rays would be difficult if we had to determine all the refractive angles. Instead, we can do it very simply by making use of certain facts we already know, such as that a ray parallel to the axis of the lens passes (after refraction) through the focal point. In fact, to find an image point, we need consider only the three rays indicated in Fig. 34–6, which shows an arrow as the object and a converging lens forming an image to the right. These rays, emanating from a single point on the object, are drawn as if the lens were infinitely thin, and we show only a single sharp bend within the lens instead of the refractions at each surface. These three rays are drawn as follows:

➡ **R A Y   D I A G R A M**

Finding the image position formed by a thin lens

Ray 1 is drawn parallel to the axis; therefore it is refracted by the lens so that it passes along a line through the focal point F, Fig. 34–6a. (See also Fig. 34–3a.)

Ray 2 is drawn on a line passing through the other focal point F′ (front side of lens in Fig. 34–6) and emerges from the lens parallel to the axis, Fig. 34–6b.

Ray 3 is directed toward the very center of the lens, where the two surfaces are essentially parallel to each other; this ray therefore emerges from the lens at the same angle as it entered; as we saw in Example 33–8, the ray would be displaced slightly to one side, but since we assume the lens is thin, we draw ray 3 straight through as shown.

Actually, any two of these rays will suffice to locate the image point, which is the point where they intersect. Drawing the third can serve as a check.

[†] We are assuming the lens has an index of refraction greater than that of the surrounding material, such as a glass or plastic lens in air, which is the usual situation.

[‡] Note that lens power has nothing to do with power as the rate of doing work or transforming energy (Section 8–8).

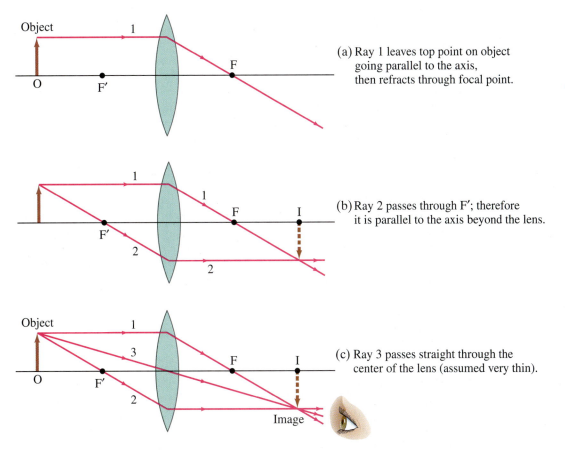

(a) Ray 1 leaves top point on object going parallel to the axis, then refracts through focal point.

(b) Ray 2 passes through F′; therefore it is parallel to the axis beyond the lens.

(c) Ray 3 passes straight through the center of the lens (assumed very thin).

FIGURE 34–6 Finding the image by ray tracing for a converging lens. Rays leave each point on the object. Shown are the three most useful rays, leaving the tip of the object, for determining where the image of that point is formed.

In this way we can find the image point for one point of the object (the top of the arrow in Fig. 34–6). The image points for all other points on the object can be found similarly to determine the complete image of the object. Because the rays actually pass through the image for the case shown in Fig. 34–6, it is a **real image** (see page 814). The image could be detected by film, or actually seen on a white surface placed at the position of the image (Fig. 34–2).

The image can also be seen directly by the eye when the eye is placed behind the image, as shown in Fig. 34–6c, so that some of the rays diverging from each point on the image enter the eye.[†] See Fig. 34–7.

[†] Why, in order to see the image, the rays must be diverging from each point on the image will be discussed in Section 34–6, but is essentially because we see real objects when diverging rays from each point enter the eye as shown in Fig. 33–1.

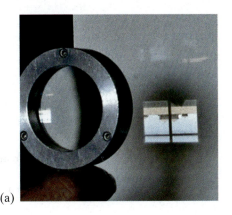

(a)

(b)

FIGURE 34–7 (a) A converging lens can form a real image (here of a distant building) on a screen. (b) That real image is also directly visible to the eye. Figure 34–1d shows images seen by the eye made by both diverging and converging lenses.

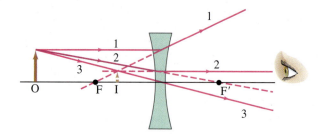

FIGURE 34–8 Finding the image by ray tracing for a diverging lens.

➡ RAY TRACING
For a diverging lens

By drawing the same three rays we can determine the image position for a diverging lens, as shown in Fig. 34–8. Note that ray 1 is drawn parallel to the axis, but does not pass through the focal point F′ behind the lens. Instead it seems to come from the focal point F in front of the lens (dashed line). Ray 2 is directed toward F′ and is refracted parallel by the lens. Ray 3 passes directly through the center of the lens. The three refracted rays seem to emerge from a point on the left of the lens. This is the image, I. Because the rays do not pass through the image, it is a **virtual image**. Note that the eye does not distinguish between real and virtual images—both are visible.

34–2 | The Lens Equation

We now derive an equation that relates the image distance to the object distance and the focal length of the lens. This will make the determination of image position quicker and more accurate than doing ray tracing. Let d_o be the object distance, the distance of the object from the center of the lens, and d_i be the image distance, the distance of the image from the center of the lens; and let h_o and h_i refer to the heights of the object and image. Consider the two rays shown in

FIGURE 34–9 Deriving the lens equation for a converging lens.

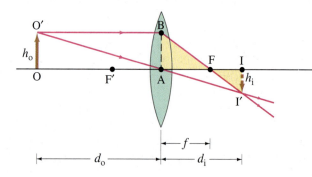

Fig. 34–9 for a converging lens (assumed to be very thin). The triangles FI′I and FBA (highlighted in yellow in Fig. 34–9) are similar because angle AFB equals angle IFI′; so

$$\frac{h_i}{h_o} = \frac{d_i - f}{f},$$

since length $AB = h_o$. Triangles OAO′ and IAI′ are similar. Therefore,

$$\frac{h_i}{h_o} = \frac{d_i}{d_o}.$$

We equate the right sides of these two equations, divide by d_i, and rearrange to obtain

LENS EQUATION

$$\frac{1}{d_o} + \frac{1}{d_i} = \frac{1}{f}.$$

(34–2)

This is called the **lens equation**. It relates the image distance d_i to the object

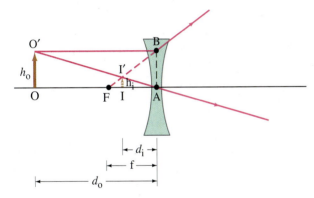

FIGURE 34–10 Deriving the lens equation for a diverging lens.

distance d_o and the focal length f. It is the most useful equation in geometric optics. (Interestingly, it is exactly the same as the mirror equation, Eq. 33–3). Note that if the object is at infinity, then $1/d_o = 0$, so $d_i = f$. Thus the focal length is the image distance for an object at infinity, as mentioned earlier.

We can derive the lens equation for a diverging lens using Fig. 34–10. Triangles IAI′ and OAO′ are similar; and triangles IFI′ and AFB are similar. Thus (noting that length $AB = h_o$)

$$\frac{h_i}{h_o} = \frac{d_i}{d_o} \quad \text{and} \quad \frac{h_i}{h_o} = \frac{f - d_i}{f}.$$

When these are equated and simplified, we obtain

$$\frac{1}{d_o} - \frac{1}{d_i} = -\frac{1}{f}.$$

This equation becomes the same as Eq. 34–2 if we make f and d_i negative. That is, we take f to be *negative for a diverging lens*, and d_i negative when the image is on the same side of the lens as the light comes from. Thus Eq. 34–2 will be valid for both converging and diverging lenses, and for *all* situations, if we use the following **sign conventions**:

1. The focal length is positive for converging lenses and negative for diverging lenses.

2. The object distance is positive if it is on the side of the lens from which the light is coming (this is usually the case, although when lenses are used in combination, it might not be so); otherwise, it is negative.

3. The image distance is positive if it is on the opposite side of the lens from where the light is coming; if it is on the same side, d_i is negative. Equivalently, the image distance is positive for a real image and negative for a virtual image.

4. The height of the image, h_i, is positive if the image is upright, and negative if the image is inverted relative to the object. (h_o is always taken as positive.)

➡ **PROBLEM SOLVING**

Sign conventions for lenses

Magnification

The **lateral magnification**, m, of a lens is defined as the ratio of the image height to object height, $m = h_i/h_o$. From Figs. 34–9 and 34–10 and the conventions just stated, we have

$$m = \frac{h_i}{h_o} = -\frac{d_i}{d_o}. \qquad (34\text{–}3)$$

Lateral magnification of a lens

For an upright image the magnification is positive, and for an inverted image m is negative.

From convention 1 above, it follows that the power (Eq. 34–1) of a converging lens, in diopters, is positive, whereas the power of a diverging lens is negative. A converging lens is sometimes referred to as a **positive lens**, and a diverging lens as a **negative lens**.

1. As always, read and reread the problem.

2. Draw a ray diagram, precise if possible, but even a rough one can serve as confirmation of analytic results. Draw at least two, and preferably three, of the easy-to-draw rays described in Figs. 34–6 and 34–8.

3. For analytic solutions, solve for unknowns in the lens equation (Eq. 34–2) and the magnification (Eq. 34–3). The lens equation (Eq. 34–2) involves reciprocals—avoid the obvious error of forgetting to take the reciprocal.

4. Follow the **Sign Conventions** above.

5. Check that your analytic answers are consistent with your ray diagram.

EXAMPLE 34–1 **Image formed by converging lens.** What is (*a*) the position, and (*b*) the size, of the image of a large 7.6-cm-high flower placed 1.00 m from a +50.0-mm-focal-length camera lens?

SOLUTION Figure 34–11 is a rough ray diagram, showing only rays 1 and 3 for a single point on the flower. We see that the image ought to be a little behind the focal point, F, to the right of the lens. (*a*) We find the image position analytically using the lens equation, Eq. 34–2. The camera lens is converging, with $f = +5.00$ cm, and $d_o = 100$ cm, and so the lens equation gives

$$\frac{1}{d_i} = \frac{1}{f} - \frac{1}{d_o} = \frac{1}{5.00 \text{ cm}} - \frac{1}{100 \text{ cm}} = \frac{20.0 - 1.0}{100 \text{ cm}}.$$

Then

$$d_i = \frac{100 \text{ cm}}{19.0} = 5.26 \text{ cm},$$

or 52.6 mm behind the lens. Notice that the image is 2.6 mm farther from the lens than the image for an object at infinity. Indeed, when focusing a camera lens, the closer the object is to the camera, the farther the lens must be from the film. (*b*) The magnification is

$$m = -\frac{d_i}{d_o} = -\frac{5.26 \text{ cm}}{100 \text{ cm}} = -0.0526;$$

so

$$h_i = mh_o = (-0.0526)(7.6 \text{ cm}) = -0.40 \text{ cm}.$$

The image is 4.0 mm high and is inverted ($m < 0$), as in Fig. 34–9, and shown in our sketch, Fig. 34–11.

FIGURE 34–11
Example 34–1.
(Not to scale.)

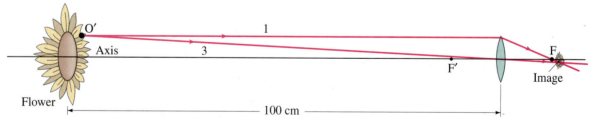

Flower 100 cm

EXAMPLE 34–2 **Object close to converging lens.** An object is placed 10 cm from a 15-cm-focal-length converging lens. Determine the image position and size (*a*) analytically, and (*b*) using a ray diagram.

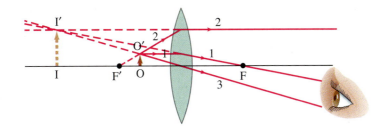

FIGURE 34–12 An object placed within the focal point of a converging lens produces a virtual image. Example 34–2.

SOLUTION (a) Given $f = 15\,\text{cm}$ and $d_o = 10\,\text{cm}$, then

$$\frac{1}{d_i} = \frac{1}{15\,\text{cm}} - \frac{1}{10\,\text{cm}} = -\frac{1}{30\,\text{cm}},$$

and $d_i = -30\,\text{cm}$. (Remember to take the reciprocal!) Because d_i is negative, the image must be virtual and on the same side of the lens as the object. The magnification

$$m = -\frac{d_i}{d_o} = -\frac{-30\,\text{cm}}{10\,\text{cm}} = 3.0.$$

➡ **PROBLEM SOLVING**

Don't forget to take the reciprocal

The image is three times as large as the object and is upright. This lens is being used as a simple magnifying glass, which we discuss in more detail in Section 34–7. (b) The ray diagram is shown in Fig. 34–12, and confirms the result in part (a). For point O′ on the top of the object, ray 1 is easy to draw. But ray 2 may take some thought: if we draw it heading toward F′, it is going the wrong way—so we have to draw it as if coming from F′ (and so dashed), striking the lens, and then going out parallel to the principal axis. We project it backward with a dashed line, as we must do also for ray 1, in order to find where they meet. Ray 3 is easy to draw, through the lens center, and it meets the other two at the image point, I′.

From this last Example and Fig. 34–12, we readily see that, whenever an object is placed between a converging lens and its focal point, the image is virtual.

EXAMPLE 34–3 **Diverging lens.** Where must a small insect be placed if a 25-cm-focal-length diverging lens is to form a virtual image 20 cm in front of the lens?

SOLUTION The ray diagram is basically that of Fig. 34–10 because our lens here is diverging and our image is in front of the lens within the focal distance. (It would be a valuable exercise to draw the ray diagram to scale, precisely, now.) Since $f = -25\,\text{cm}$ and $d_i = -20\,\text{cm}$, then Eq. 34–2 gives

$$\frac{1}{d_o} = \frac{1}{f} - \frac{1}{d_i} = -\frac{1}{25\,\text{cm}} + \frac{1}{20\,\text{cm}} = \frac{-4 + 5}{100\,\text{cm}} = \frac{1}{100\,\text{cm}}.$$

So the object must be 100 cm in front of the lens.

34–3 Combinations of Lenses

We now consider Examples illustrating how to deal with lenses used in combination. In general, when light passes through more than one lens, we find the image formed by the first lens as if it were alone. This image becomes the *object* for the second lens, and we find the image then formed by this second lens, which is the final image if there are only two lenses. The total magnification will be the product of the separate magnifications of each lens as we shall see.

Multiple lenses: image formed by first lens is object for second lens

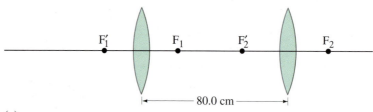

(a)

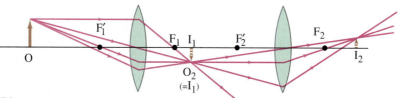

(b)

FIGURE 34–13 Example 34–4.

EXAMPLE 34–4 A two-lens system. Two converging lenses, with focal lengths $f_1 = 20.0\text{ cm}$ and $f_2 = 25.0\text{ cm}$, are placed 80.0 cm apart, as shown in Fig. 34–13a. An object is placed 60.0 cm in front of the first lens as shown in Fig. 34–13b. Determine (a) the position, and (b) the magnification, of the final image formed by the combination of the two lenses.

SOLUTION (a) The object is a distance $d_{o1} = +60.0\text{ cm}$ from the first lens, and this lens forms an image whose position can be calculated using the lens equation:

$$\frac{1}{d_{i1}} = \frac{1}{f_1} - \frac{1}{d_{o1}} = \frac{1}{20.0\text{ cm}} - \frac{1}{60.0\text{ cm}} = \frac{3 - 1}{60.0\text{ cm}} = \frac{1}{30.0\text{ cm}}.$$

Careful!
*Note that object distance for second lens is **not** equal to the image distance for first lens*

So the first image I_1 is at $d_{i1} = 30.0\text{ cm}$ behind the first lens. This image becomes the object for the second lens. It is a distance $d_{o2} = 80.0\text{ cm} - 30.0\text{ cm} = 50.0\text{ cm}$ in front of lens 2, as shown in Fig. 34–13b. The image formed by the second lens, again using the lens equation, is at a distance d_{i2} from the second lens:

$$\frac{1}{d_{i2}} = \frac{1}{f_2} - \frac{1}{d_{o2}} = \frac{1}{25.0\text{ cm}} - \frac{1}{50.0\text{ cm}} = \frac{4 - 2}{100.0\text{ cm}} = \frac{2}{100.0\text{ cm}}.$$

Hence $d_{i2} = 50.0\text{ cm}$ behind lens 2. This is the final image—see Fig. 34–13b.
(b) The first lens has a magnification (Eq. 34–3)

$$m_1 = -\frac{d_{i1}}{d_{o1}} = -\frac{30.0\text{ cm}}{60.0\text{ cm}} = -0.500.$$

Thus, the first image is inverted and is half as high as the object: again Eq. 34–3,

$$h_{i1} = m_1 h_{o1} = -0.500\, h_{o1}.$$

The second lens takes this image as object and changes its height by a factor

$$m_2 = -\frac{d_{i2}}{d_{o2}} = -\frac{50.0\text{ cm}}{50.0\text{ cm}} = -1.000.$$

The final image height is $\left(\text{remember } h_{o2} \text{ is the same as } h_{i1}\right)$:

$$h_{i2} = m_2 h_{o2} = m_2 h_{i1} = m_2 m_1 h_{o1}.$$

Total magnification is
$m = m_1 m_2$

We see from this equation that the total magnification is the product of m_1 and m_2, which here equals $(-1.000)(-0.500) = +0.500$, or $\frac{1}{2}$ the original height and upright.

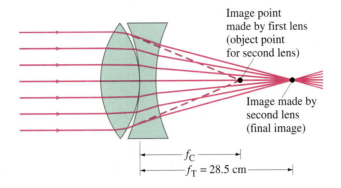

FIGURE 34–14 Determining the focal length of a diverging lens. Example 34–5.

EXAMPLE 34–5 **Measuring *f* for a diverging lens.** To measure the focal length of a diverging lens, a converging lens is placed in contact with it, as shown in Fig. 34–14. The Sun's rays are focused by this combination at a point 28.5 cm, behind the lenses as shown. If the converging lens has a focal length f_C of 16.0 cm, what is the focal length f of the diverging lens? Assume both lenses are thin and the space between them is negligible.

SOLUTION Rays from the Sun are focused 28.5 cm behind the combination, so the focal length of the total combination is $f_T = 28.5$ cm. If the diverging lens were absent, the converging lens would form the image at its focal point—that is, at a distance $f_C = 16.0$ cm behind it (dashed lines in Fig. 34–14). When the diverging lens is placed next to the converging lens, we treat the image formed by the first lens as the *object* for the second lens. Since this object lies to the right of the diverging lens, this is a situation where d_o is negative (see the sign conventions, page 841). Thus, for the diverging lens, the object is virtual and $d_o = -16.0$ cm. The diverging lens forms the image of this virtual object at a distance $d_i = 28.5$ cm away (this was given). Thus,

$$\frac{1}{f_D} = \frac{1}{d_o} + \frac{1}{d_i} = \frac{1}{-16.0 \text{ cm}} + \frac{1}{28.5 \text{ cm}} = -0.0274 \text{ cm}^{-1}.$$

We take the reciprocal to find $f_D = -1/(0.0274 \text{ cm}^{-1}) = -36.5$ cm. Note that the converging lens must be "stronger" than the diverging lens—that is, it must have a focal length whose magnitude is less than that of the diverging lens—if this technique is to work.

34–4 Lensmaker's Equation

In this Section, we will prove that parallel rays are brought to a focus at a *single* point for a thin lens. At the same time, we will also derive an equation that relates the focal length of a lens to the radii of curvature of its two surfaces, which is known as the lensmaker's equation.

In Fig. 34–15, a ray parallel to the axis of a lens is refracted at the front surface of the lens at point A_1 and is refracted at the back surface at point A_2. This ray then passes through point F, which we call the focal point for this ray. Point A_1 is a height h_1 above the axis, and point A_2 is height h_2 above the axis. C_1 and C_2 are the centers of curvature of the two lens surfaces; so the length $C_1 A_1 = R_1$, the radius of curvature of the front surface, and $C_2 A_2 = R_2$ is the radius of the second surface. The thickness of the lens has been grossly exaggerated so the various angles would be clear. But we will assume that the lens is actually very thin and that angles between the rays and the axis are small. In this approximation, $h_1 \approx h_2$, and the sines and tangents of all the angles will be equal to the angles themselves in radians. For example, $\sin \theta_1 \approx \tan \theta_1 \approx \theta_1$ (radians).

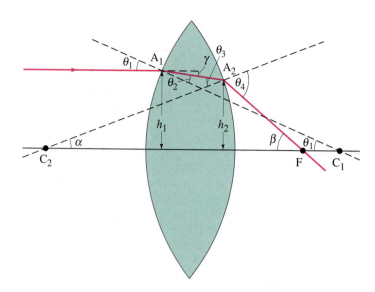

FIGURE 34–15 Diagram of a ray passing through a lens for derivation of the lensmaker's equation.

To this approximation, then, Snell's law tells us that

$$\theta_1 = n\theta_2$$

$$\theta_4 = n\theta_3$$

where n is the index of refraction of the glass, and we assume that the lens is surrounded by air ($n = 1$). Notice also in Fig. 34–15 that

$$\theta_1 \approx \sin\theta_1 = \frac{h_1}{R_1}$$

$$\alpha \approx \frac{h_2}{R_2}$$

$$\beta \approx \frac{h_2}{f}.$$

The last follows because the distance from F to the lens (assumed very thin) is f. From the diagram, the angle γ is

$$\gamma = \theta_1 - \theta_2.$$

A careful examination of Fig. 34–15 shows also that

$$\alpha = \theta_3 - \gamma.$$

This can be seen by drawing a horizontal line to the left from point A_2, which divides the angle θ_3 into two parts. The upper part equals γ and the lower part equals α. (The opposite angles between an oblique line and two parallel lines are equal.) Thus, $\theta_3 = \gamma + \alpha$. Finally, by drawing a horizontal line to the right from point A_2, we divide θ_4 into two parts. The upper part is α and the lower is β. Thus

$$\theta_4 = \alpha + \beta.$$

We now combine all these equations:

$$\alpha = \theta_3 - \gamma = \frac{\theta_4}{n} - (\theta_1 - \theta_2) = \frac{\alpha}{n} + \frac{\beta}{n} - \theta_1 + \theta_2,$$

or

$$\frac{h_2}{R_2} = \frac{h_2}{nR_2} + \frac{h_2}{nf} - \frac{h_1}{R_1} + \frac{h_1}{nR_1}.$$

Because the lens is thin, $h_1 \approx h_2$ and all h's can be canceled from all the numerators.

We then multiply through by n and rearrange to find that

$$\frac{1}{f} = (n - 1)\left(\frac{1}{R_1} + \frac{1}{R_2}\right).$$ (34–4)   *Lensmaker's equation*

This is called the **lensmaker's equation**. It relates the focal length of a lens to the radii of curvature of its two surfaces and its index of refraction. Notice that f does not depend on h_1 or h_2. Thus the position of the point F does not depend on where the ray strikes the lens. Hence, all rays parallel to the axis of a thin lens will pass through the same point F, which we wished to prove.

In our derivation, both surfaces are convex and R_1 and R_2 are considered positive.[†] Equation 34–4 also works for lenses with one or both surfaces concave; but for a concave surface, the radius must be considered *negative*.

Notice in Eq. 34–4 that the equation is symmetrical in R_1 and R_2. Thus, if a lens is turned around so that light impinges on the other surface, the focal length is the same even if the two lens surfaces are different.

EXAMPLE 34–6 **Calculating *f* for a converging lens.** A convex meniscus lens (Figs. 34–1a and 34–16) is made from glass with $n = 1.50$. The radius of curvature of the convex surface is 22.4 cm and that of the concave surface is 46.2 cm. (*a*) What is the focal length? (*b*) Where will it focus an object 2.00 m away?

SOLUTION (*a*) $R_1 = 22.4$ cm and $R_2 = -46.2$ cm; the latter is negative because it refers to the concave surface. Then

$$\frac{1}{f} = (1.50 - 1.00)\left(\frac{1}{22.4\text{ cm}} - \frac{1}{46.2\text{ cm}}\right)$$

$$= 0.0115\text{ cm}^{-1}.$$

So

$$f = \frac{1}{0.0115\text{ cm}^{-1}} = 87\text{ cm}$$

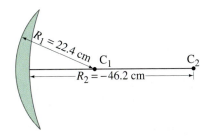

FIGURE 34–16 Example 34–6.

and the lens is converging. Notice that if we turn the lens around so that $R_1 = -46.2$ cm and $R_2 = +22.4$ cm, we get the same result. (*b*) From the lens equation, with $f = 0.87$ m and $d_o = 2.00$ m, we have

$$\frac{1}{d_i} = \frac{1}{f} - \frac{1}{d_o} = \frac{1}{0.87\text{ m}} - \frac{1}{2.00\text{ m}}$$

$$= 0.65\text{ m}^{-1},$$

so $d_i = 1/0.65\text{ m}^{-1} = 1.53$ m.

EXAMPLE 34–7 **Calculating *f* for a diverging lens.** A Lucite planoconcave lens (see Fig. 34–1b) has one flat surface and the other has $R = -18.4$ cm. What is the focal length?

SOLUTION From Table 33–1, n for Lucite is 1.51. A plane surface has infinite radius of curvature; if we call this R_1, then $1/R_1 = 0$. Therefore,

$$\frac{1}{f} = (1.51 - 1.00)\left(-\frac{1}{18.4\text{ cm}}\right).$$

So $f = (-18.4\text{ cm})/0.51 = -36$ cm, and the lens is diverging.

[†] Some books use a different convention—for example, R_1 and R_2 are considered positive if their centers of curvature are to the right of the lens, in which case a minus sign appears in their equivalent of Eq. 34–4.

In the remainder of this chapter, we discuss briefly some of the more common optical instruments, including the camera, our eyes and corrective lenses, the magnifying glass, telescopes, and microscopes. We start with the camera.

The basic elements of a **camera** are a lens, a light-tight box, a shutter to let light pass through the lens only briefly, and a sensitized plate or piece of film (Fig. 34–17). When the shutter is opened, light from external objects in the field of view is focused by the lens as an image on the film. The film contains light-sensitive chemicals that undergo change when light strikes them. In the development process, chemical reactions cause the changed areas to turn opaque so that the image is recorded on the film.[†] You can see the image yourself if you can remove the back of the camera and then view through a piece of tissue or wax paper (on which the image can form) placed at the position of the film with the shutter open.

There are three main adjustments on good-quality cameras: shutter speed, *f*-stop, and focusing (see Fig. 34–18), and we now discuss them.

Shutter speed. This refers to how long the shutter is open and the film exposed. It may vary from a second or more ("time exposures") to $\frac{1}{1000}$ s or less. To avoid blurring from camera movement, speeds faster than $\frac{1}{100}$ s are normally used. If the object is moving, faster shutter speeds are needed to "stop" the action. A shutter can be "behind the lens," as in Fig. 34–17, or a "focal plane" shutter which is a moveable curtain just in front of the film.

***f*-stop.** The amount of light reaching the film must be carefully controlled to avoid **underexposure** (too little light for any but the brightest objects to show up) or **overexposure** (too much light, so that all bright objects look the same, with a consequent lack of contrast and a "washed-out" appearance). To control the exposure, a "stop" or iris diaphragm, whose opening is of variable diameter, is placed behind the lens (Fig. 34–17). The size of the opening is varied to compensate for bright or dark days, the sensitivity of the film[‡] used, and for different shutter speeds. The size of the opening is specified by the ***f*-stop**, defined as

$$f\text{-stop} = \frac{f}{D},$$

where f is the focal length of the lens and D is the diameter of the opening. For example, when a 50-mm-focal-length lens has an opening $D = 25\,\text{mm}$, we say it is set at $f/2$. When the lens is set at $f/8$, the opening is only $6\frac{1}{4}\,\text{mm}$ $\left(50/6\frac{1}{4} = 8\right)$. The faster the shutter speed, or the darker the day, the greater the opening must be to get a proper exposure. This corresponds to a smaller *f*-stop number. The smaller the *f*-stop number, the more light passes through the lens to the film. The smallest *f*-number of a lens (largest opening) is referred to as the *speed* of the lens. It is common to find $f/2.0$ lenses today, and some even faster. The advantage of a fast lens is that it allows pictures to be taken under poor lighting conditions. Good quality lenses consist of several elements to reduce the defects present in simple thin lenses (Section 34–10). Standard *f*-stop markings on good lenses are 1.0, 1.4, 2.0, 2.8, 4.0, 5.6, 8, 11, 16, 22, and 32. Each of these stops corresponds to a diameter reduction by a factor of about $\sqrt{2} = 1.4$. Because the amount of light reaching the film is proportional to the *area* of the opening, and therefore proportional to the diameter squared, each standard *f*-stop corresponds to a factor of 2 in light intensity reaching the film.

The camera

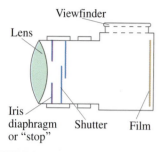

FIGURE 34–17 A simple camera.

FIGURE 34–18 On this camera, the *f*-stops and the focusing ring are on the camera lens. Shutter speeds are selected on the small wheel on top of the camera body.

[†]This is called a *negative*, because the black areas correspond to bright objects and vice versa. The same process occurs during printing to produce a black-and-white "positive" picture from the negative. Color film makes use of three dyes corresponding to the primary colors.

[‡]Different films have different sensitivities to light, referred to as the "film speed," and specified as an "ASA" number; a "faster" film is more sensitive and needs less light to produce a good image.

(a)   (b)

FIGURE 34–19 Photos with camera focused (a) on nearby object with distant object blurry, and (b) on more distant object with nearby object blurry.

Focusing. Focusing is the operation of placing the lens at the correct position relative to the film for the sharpest image. The image distance is a minimum for objects at infinity (the symbol ∞ is used for infinity) and is equal to the focal length. For closer objects, the image distance is greater than the focal length, as can be seen from the lens equation, $1/f = 1/d_o + 1/d_i$. To focus on nearby objects, the lens must therefore be moved away from the film, and this is usually done by turning a ring on the lens.

Focusing

If the lens is focused on a nearby object, a sharp image of it will be formed, but distant objects may be blurry (Fig. 34–19). The rays from a point on the distant object will be out of focus—they will form a circle on the film as shown (exaggerated) in Fig. 34–20. The distant object will thus produce an image consisting of overlapping circles and will be blurred. These circles are called **circles of confusion**. To include near and distant objects in the same photo, you can try setting the lens focus at an intermediate position. For a given distance setting, there is a range of distances over which the circles of confusion will be small enough that the images will be reasonably sharp. This is called the **depth of field**. For a particular choice of circle of confusion diameter as upper limit (typically taken to be 0.03 mm for 35-mm cameras), the depth of field varies with the lens opening. If the lens opening is smaller, only rays through the central part of the lens are accepted, and these form smaller circles of confusion for a given object distance. Hence, at smaller lens openings, a greater range of object distances will fit within the circle of confusion criterion, so that the depth of field is greater.

Depth of field

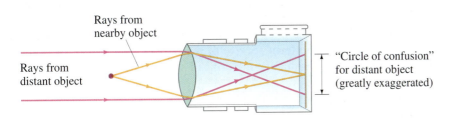

Rays from nearby object

Rays from distant object

"Circle of confusion" for distant object (greatly exaggerated)

FIGURE 34–20 When the lens is positioned to focus on a nearby object, points on a distant object produce circles and are therefore blurred. (The effect is shown greatly exaggerated.)

Camera lenses are categorized into normal, telephoto, and wide angle, according to focal length and film size. A **normal lens** is one that covers the film with a field of view that corresponds approximately to that of normal vision. A normal lens for 35-mm film has a focal length in the vicinity of 50 mm.[†] A **telephoto lens**, as its name implies, acts like a telescope to magnify images. They have longer focal lengths than a normal lens. As we saw in Section 34–2 (Eq. 34–3), the height of the image for a given object distance is proportional to the image distance, and the image distance will be greater for a lens with longer focal length. For distant objects, the image height is very nearly proportional to the focal length (can you prove this?). Thus a 200-mm telephoto lens for use with a 35-mm camera gives a 4× magnification over the normal 50-mm lens. A **wide-angle lens** has a shorter focal length than normal: a wider field of view is included and objects appear smaller. A **zoom lens** is one whose focal length can be changed so that you seem to zoom up to, or away from, the subject as you change the focal length.

Telephoto and wide-angle lenses

CONCEPTUAL EXAMPLE 34–8 | **Shutter speed.** To improve the depth of field, you stop down your camera lens by two *f*-stops (say, from *f*/4 to *f*/8). What should you do to the shutter speed to maintain the same exposure?

RESPONSE The amount of light admitted by the lens is proportional to the area of the lens opening. Reducing the lens opening by one *f*-stop reduces the diameter by a factor of $\sqrt{2}$, and the area by a factor of 2. Stopping down by two *f*-stops reduces the area of the lens opening by a factor of 4. To maintain the same exposure, the shutter must be open 4 times as long. So if the shutter speed had been $\frac{1}{250}$ s, you have to increase it to $\frac{1}{60}$ s.

34–6 | The Human Eye; Corrective Lenses

The eye

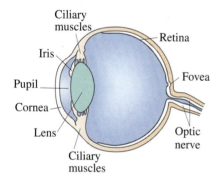

FIGURE 34–21 Diagram of a human eye.

The human eye resembles a camera in its basic structure (Fig. 34–21), but is far more sophisticated. The eye is an enclosed volume into which light passes through a lens. A diaphragm, called the **iris** (the colored part of your eye), adjusts automatically to control the amount of light entering the eye. The hole in the iris through which light passes (the **pupil**) is black because no light is reflected from it (it's a hole), and very little light is reflected back out from the interior of the eye. The **retina**, which plays the role of the film in a camera, is on the curved rear surface. It consists of a complex array of nerves and receptors known as *rods* and *cones* which act to change light energy into electrical signals that travel along the nerves. The reconstruction of the image from all these tiny receptors is done mainly in the brain, although some analysis is apparently done in the complex interconnected nerve network at the retina itself. At the center of the retina is a small area called the **fovea**, about 0.25 mm in diameter, where the cones are very closely packed and the sharpest image and best color discrimination are found.

Unlike a camera, the eye contains no shutter. The equivalent operation is carried out by the nervous system, which analyzes the signals to form images at the rate of about 30 per second. This can be compared to motion picture or television cameras, which operate by taking a series of still pictures at a rate of 24 (movies) or 30 (U.S. television) per second. Their rapid projection on the screen gives the appearance of motion.

The lens of the eye does little of the bending of the light rays. Most of the refraction is done at the front surface of the **cornea** (index of refraction = 1.376), which also acts as a protective covering. The lens acts as a fine adjustment for

Focusing the eye

[†]Note that a "35-mm camera" uses film that is 35 mm wide; that 35 mm is not to be confused with a focal length.

focusing at different distances. This is accomplished by the ciliary muscles (Fig. 34–21), which change the curvature of the lens so that its focal length is changed. To focus on a distant object, the muscles are relaxed and the lens is thin, Fig. 34–22a, and parallel rays focus at the focal point (on the retina). To focus on a nearby object, the muscles contract, causing the center of the lens to be thicker, Fig. 34–22b, thus shortening the focal length so that images of nearby objects can be focused on the retina, behind the focal point. This focusing adjustment is called **accommodation**.

The closest distance at which the eye can focus clearly is called the **near point** of the eye. For young adults it is typically 25 cm, although younger children can often focus on objects as close as 10 cm. As people grow older, the ability to accommodate is reduced and the near point increases. A given person's **far point** is the farthest distance at which an object can be seen clearly. For some purposes it is useful to speak of a **normal eye** (a sort of average over the population), which is defined as one having a near point of 25 cm and a far point of infinity. To check your own near point, place this book close to your eye and slowly move it away until the type is sharp.

The "normal" eye is more of an ideal than a commonplace. A large part of the population have eyes that do not accommodate within the normal range of 25 cm to infinity, or have some other defect. Two common defects are nearsightedness and farsightedness. Both can be corrected to a large extent with lenses—either eyeglasses or contact lenses.

Nearsightedness, or *myopia*, refers to an eye that can focus only on nearby objects. The far point is not infinity but some shorter distance, so distant objects are not seen clearly. It is usually caused by an eyeball that is too long, although sometimes it is the curvature of the cornea that is too great. In either case, images of distant objects are focused in front of the retina. A diverging lens, because it causes parallel rays to diverge, allows the rays to be focused at the retina (Fig. 34–23a) and thus corrects this defect.

Farsightedness, or *hyperopia*, refers to an eye that cannot focus on nearby objects. Although distant objects are usually seen clearly, the near point is somewhat greater than the "normal" 25 cm, which makes reading difficult. This defect is caused by an eyeball that is too short or (less often) by a cornea that is not sufficiently curved. It is corrected by a converging lens, Fig. 34–23b. Similar to hyperopia is *presbyopia*, which refers to the lessening ability of the eye to accommodate as one ages, and the near point moves out. Converging lenses also compensate for this.

Astigmatism is usually caused by an out-of-round cornea or lens so that point objects are focused as short lines, which blurs the image. Astigmatism is corrected with the use of a cylindrical lens which, for eyes that are nearsighted or farsighted, is superimposed on the spherical surface, so that the radius of curvature of the correcting lens is different in different planes.

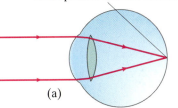

Focal point of lens and cornea

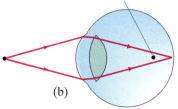

Focal point of lens and cornea

FIGURE 34–22 Accommodation by a normal eye: (a) lens relaxed, focused at infinity; (b) lens thickened, focused on a nearby object.

➡ **PHYSICS APPLIED**

Corrective lenses

Nearsightedness

Farsightedness

Astigmatism

FIGURE 34–23 Correcting eye defects with lenses: (a) a nearsighted eye, which cannot focus clearly on distant objects, can be corrected by use of a diverging lens; (b) a farsighted eye, which cannot focus clearly on nearby objects, can be corrected by use of a converging lens.

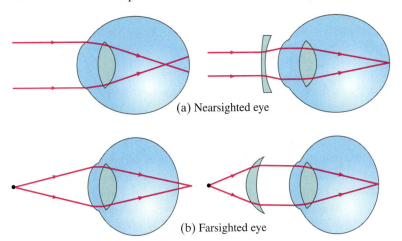

(a) Nearsighted eye

(b) Farsighted eye

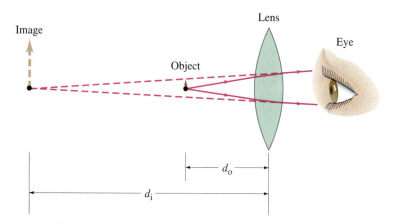

FIGURE 34–24 Lens of reading glasses (Example 34–9).

EXAMPLE 34–9 **Farsighted eye.** A particular farsighted person has a near point of 100 cm. Reading glasses must have what lens power so that this person can read a newspaper at a distance of 25 cm? Assume the lens is very close to the eye.

SOLUTION When the object is placed 25 cm from the lens, we want the image to be 100 cm away on the *same* side of the lens, and so it will be virtual, Fig. 34–24. Thus, $d_o = 25$ cm, $d_i = -100$ cm, and the lens equation gives

$$\frac{1}{f} = \frac{1}{25 \text{ cm}} + \frac{1}{-100 \text{ cm}} = \frac{4 - 1}{100 \text{ cm}} = \frac{1}{33 \text{ cm}}.$$

So $f = 33$ cm $= 0.33$ m. The power P of the lens is $P = 1/f = +3.0$ D. The plus sign indicates that it is a converging lens.

EXAMPLE 34–10 **Nearsighted eye.** A nearsighted eye has near and far points of 12 cm and 17 cm, respectively. (a) What lens power is needed for this person to see distant objects clearly, and (b) what then will be the near point? Assume that the lens is 2.0 cm from the eye (typical for eyeglasses).

SOLUTION (a) First we determine the power of the lens needed to focus objects at infinity, when the eye is relaxed. For a distant object ($d_o = \infty$), as shown in Fig. 34–25, the lens must put the image 17 cm from the eye (its far point), which is 15 cm in front of the lens; hence $d_i = -15$ cm. We use the lens equation to solve for the focal length of the needed lens:

$$\frac{1}{f} = -\frac{1}{15 \text{ cm}} + \frac{1}{\infty} = -\frac{1}{15 \text{ cm}}.$$

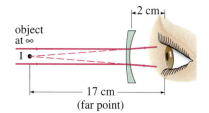

FIGURE 34–25 Example 34–10a.

FIGURE 34–26 Example 34–10b.

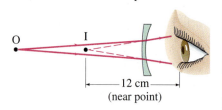

So $f = -15$ cm $= -0.15$ m or $P = 1/f = -6.7$ D. The minus sign indicates that it must be a diverging lens.
(b) To determine the near point when wearing the glasses, we note that a sharp image will be 12 cm from the eye (its near point, see Fig. 34–26), which is 10 cm from the lens; so $d_i = -0.10$ m and the lens equation gives

$$\frac{1}{d_o} = \frac{1}{f} - \frac{1}{d_i} = -\frac{1}{0.15 \text{ m}} + \frac{1}{0.10 \text{ m}} = \frac{1}{0.30 \text{ m}}.$$

So $d_o = 30$ cm, which means the near point when the person is wearing glasses is 30 cm in front of the lens.

Contact lenses

Contact lenses could be used to correct the eye in Example 34–10. Since contacts are placed directly on the cornea, we would not subtract out the 2.0 cm for the image distances. That is, for distant objects $d_i = -17$ cm, so $P = 1/f = -5.9$ D (diopters). The new near point would be 41 cm. Thus we see that a contact lens and an eyeglass lens will require slightly different powers, or focal lengths, for the same eye because of their different placements relative to the eye.

34–7 Magnifying Glass

Much of the remainder of this chapter will deal with optical devices that are used to produce magnified images of objects. We first discuss the **simple magnifier**, or **magnifying glass**, which is simply a converging lens (see chapter opening photo).

How large an object appears, and how much detail we can see on it, depends on the size of the image it makes on the retina. This, in turn, depends on the angle subtended by the object at the eye. For example, a penny held 30 cm from the eye looks twice as tall as one held 60 cm away because the angle it subtends is twice as great (Fig. 34–27). When we want to examine detail on an object, we bring it up close to our eyes so that it subtends a greater angle. However, our eyes can accommodate only up to a point (the near point), and we will assume a standard distance of 25 cm as the near point in what follows.

A magnifying glass allows us to place the object closer to our eye so that it subtends a greater angle. As shown in Fig. 34–28a, the object is placed at the focal point or just within it. Then the converging lens produces a virtual image, which must be at least 25 cm from the eye if the eye is to focus on it. If the eye is relaxed, the image will be at infinity, and in this case the object is exactly at the focal point. (You make this slight adjustment yourself when you "focus" on the object by moving the magnifying glass.)

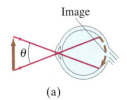

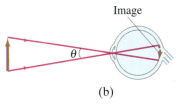

FIGURE 34–27 When the same object is viewed at a shorter distance, the image on the retina is greater, so the object appears larger and more detail can be seen. The angle θ that the object subtends in (a) is greater than in (b).

FIGURE 34–28 Leaf viewed (a) through a magnifying glass, and (b) with the unaided eye, with the eye focused at its near point.

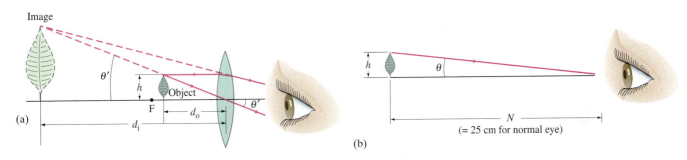

A comparison of part (a) of Fig. 34–28 with part (b), in which the same object is viewed at the near point with the unaided eye, reveals that the angle the object subtends at the eye is much larger when the magnifier is used. The **angular magnification** or **magnifying power**, M, of the lens is defined as the ratio of the angle subtended by an object when using the lens, to the angle subtended using the unaided eye, with the object at the near point N of the eye ($N = 25$ cm for the normal eye):

$$M = \frac{\theta'}{\theta}, \tag{34–5}$$

where θ and θ' are shown in Fig. 34–28. We can write M in terms of the focal length by noting that $\theta = h/N$ (Fig. 34–28b) and $\theta' = h/d_o$ (Fig. 34–28a), where h is the height of the object and we assume the angles are small so θ and θ' equal their sines and tangents. If the eye is relaxed (for least eye strain), the image will be at infinity and the object will be precisely the focal point; see Fig. 34–29. Then $d_o = f$ and $\theta' = h/f$. Thus

$$M = \frac{\theta'}{\theta} = \frac{h/f}{h/N} = \frac{N}{f}. \quad \begin{bmatrix} \text{eye focused at } \infty; \\ N = 25 \text{ cm for normal eye} \end{bmatrix} \tag{34–6a}$$

We see that the shorter the focal length of the lens, the greater the magnification.

FIGURE 34–29 With the eye relaxed, the object is placed at the focal point, and the image is at infinity. Compare to Fig. 34–28a where the image is at the eye's near point.

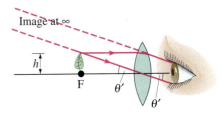

Magnification of a simple magnifier

[However, simple single-lens magnifiers are limited to about 2 or 3× because of distortion due to spherical aberration (Section 34–10).] The magnification of a given lens can be increased a bit by moving the lens and adjusting your eye so it focuses on the image at the eye's near point. In this case, $d_i = -N$ (see Fig. 34–28a) if your eye is very near the magnifier. Then the object distance d_o is given by

$$\frac{1}{d_o} = \frac{1}{f} - \frac{1}{d_i} = \frac{1}{f} + \frac{1}{N}.$$

We see from this equation that $d_o < f$, as shown in Fig. 34–28a, since $N/(f + N)$ must be less than 1. With $\theta' = h/d_o$ the magnification is

$$M = \frac{\theta'}{\theta} = \frac{h/d_o}{h/N} = \frac{N}{d_o} = N\left(\frac{1}{f} + \frac{1}{N}\right)$$

or

$$M = \frac{N}{f} + 1. \qquad \left[\begin{array}{c}\text{eye focused at near point, } N; \\ N = 25 \text{ cm for normal eye}\end{array}\right] \quad \textbf{(34–6b)}$$

We see that the magnification is slightly greater when the eye is focused at its near point, rather than relaxed.

EXAMPLE 34–11 **ESTIMATE** **A jeweler's loupe.** An 8-cm-focal-length converging lens is used as a "jeweler's loupe," which is a magnifying glass. Estimate (a) the magnification when the eye is relaxed, and (b) the magnification if the eye is focused at its near point $N = 25$ cm.

SOLUTION (a) With the relaxed eye focused at infinity, $M = N/f = 25 \text{ cm}/8 \text{ cm} \approx 3\times$. (b) The magnification when the eye is focused at its near point ($N = 25$ cm), and the lens is near the eye, is:

$$M = 1 + \frac{N}{f} = 1 + \frac{25}{8} \approx 4\times.$$

FIGURE 34–30 (a) Objective lens (mounted now in an ivory frame) from the telescope with which Galileo made his world-shaking discoveries, including the moons of Jupiter. (b) Later telescopes made by Galileo.

(a)

(b)

* ## 34–8 Telescopes

A telescope is used to magnify objects that are very far away. In most cases, the object can be considered to be at infinity.

Galileo, although he did not invent it,[†] developed the telescope into a usable and important instrument. He was the first to examine the heavens with the telescope (Fig. 34–30), and he made world-shaking discoveries (the moons of Jupiter, the phases of Venus, sunspots, the structure of the Moon's surface, that the Milky Way is made up of a huge number of individual stars, among others).

Several types of **astronomical telescope** exist. The common **refracting** type, sometimes called **Keplerian**, contains two converging lenses located at opposite ends of a long tube, as diagrammed in Fig. 34–31. The lens closest to the object is called the **objective lens** and forms a real image I_1 of the distant object in the plane of its focal point F_o (or near it if the object is not at infinity). Although this image, I_1, is smaller than the original object, it subtends a greater angle and is very close to the second lens, called the **eyepiece**, which acts as a magnifier. That is, the eye-

[†]Galileo built his first telescope in 1609 after having heard of such an instrument existing in Holland. The first telescopes magnified only 3 to 4 times, but Galileo soon made a 30-power instrument. The first Dutch telescope seems to date from about 1604, but there is a reference suggesting it may have been copied from an Italian telescope built as early as 1590. Kepler gave a ray description (1611) of the Keplerian telescope, which is named for him because he first described it, although he did not build it.

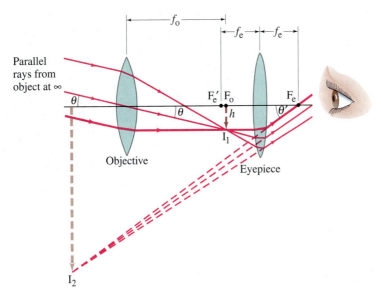

Objective

Eyepiece

FIGURE 34–31 Astronomical telescope (refracting). Parallel light from one point on a distant object $(d_o = \infty)$ is brought to a focus by the objective lens in its focal plane. This image (I_1) is magnified by the eyepiece to form the final image I_2. Only two of the rays shown are standard rays described in Fig. 34–6.

piece magnifies the image produced by the objective to produce a second, greatly magnified image, I_2, which is virtual, and inverted. If the viewing eye is relaxed, the eyepiece is adjusted so the image I_2 is at infinity. Then the real image I_1 is at the focal point F'_e of the eyepiece, and the distance between the lenses is $f_o + f_e$ for an object at infinity.

To find the total magnification of this telescope, we note that the angle an object subtends as viewed by the unaided eye is just the angle θ subtended at the telescope objective. From Fig. 34–31 we can see that $\theta \approx h/f_o$, where h is the height of the image I_1 and we assume θ is small so that $\tan\theta \approx \theta$. Note, too, that the thickest of the three rays drawn in Fig. 34–31 is parallel to the axis before it strikes the eyepiece and therefore passes through the eyepiece focal point F_e. Thus, $\theta' \approx h/f_e$ and the total magnifying power (angular magnification) of this telescope is

$$M = \frac{\theta'}{\theta} = -\frac{f_o}{f_e}, \qquad (34\text{--}7)$$

Telescope magnification

where we have inserted a minus sign to indicate that the image is inverted. To achieve a large magnification, the objective lens should have a long focal length and the eyepiece a short focal length.

FIGURE 34–32 This large refracting telescope was built in 1897 and is housed at Yerkes Observatory in Wisconsin. The objective lens is 102 cm (40 inches) in diameter, and the telescope tube is about 19 m long.

EXAMPLE 34–12 **Telescope magnification.** The largest refracting telescope in the world is located at the Yerkes Observatory in Wisconsin, Fig. 34–32. It is referred to as a "40-inch" telescope, meaning that the diameter of the objective is 40 inches, or 102 cm. The objective has a focal length of 19 m, and the eyepiece has a focal length of 10 cm. (*a*) Calculate the total magnifying power of this telescope. (*b*) Estimate the length of the telescope.

SOLUTION (*a*) From Eq. 34–7 we find,

$$M = -\frac{f_o}{f_e} = -\frac{19\text{ m}}{0.10\text{ m}} = -190\times.$$

(*b*) For a relaxed eye, the image I_1 is at the focal point of both the eyepiece and the objective lenses. The distance between the two lenses is thus $f_o + f_e \approx 19$ m, which is essentially the length of the telescope.

For an astronomical telescope to produce bright images of distant stars, the objective lens must be large to allow in as much light as possible. Indeed, the diameter of the objective (and hence its "light-gathering power") is an important parameter for an astronomical telescope, which is why the largest ones are specified by giving the objective diameter (such as the 200-inch Hale telescope on Palomar Mountain). The construction and grinding of large lenses is very difficult. Therefore,

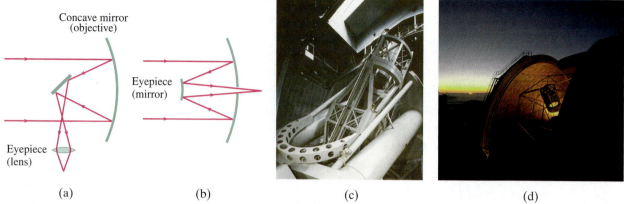

FIGURE 34–33 A concave mirror can be used as the objective of an astronomical telescope. Either a lens (a) or a mirror (b) can be used as the eyepiece. Arrangement (a) is called the Newtonian focus and (b) the Cassegrainian focus. Other arrangements are also possible. (c) The 200-inch (mirror diameter) Hale telescope on Palomar Mountain in California. (d) The 10-meter Keck telescope on Mauna Kea, Hawaii. The Keck combines thirty-six 1.8 meter six-sided mirrors into the equivalent of a very large (10-m diameter) single reflector.

Reflecting telescopes the largest telescopes are **reflecting telescopes** that use a curved mirror as the objective, Fig. 34–33, since a mirror has only one surface to be ground and can be supported along its entire surface[†] (a large lens, supported at its edges, would sag under its own weight). Normally, the eyepiece lens or mirror (see Fig. 34–33) is removed so that the real image formed by the objective mirror can be recorded directly on film.

A **terrestrial telescope** (for use in viewing objects on Earth), unlike its astronomical counterpart, must provide an upright image. Two designs are shown in Fig. 34–34. The **Galilean** type shown in part (a), which Galileo used for his great astronomical discoveries, has a diverging lens as eyepiece which intercepts the converging rays from the objective lens before they reach a focus, and acts to form a virtual upright image. This design is often used in opera glasses. The tube is reasonably short, but the field of view is small. The second type, shown in Fig. 34–34b, is often called a **spyglass** and makes use of a third lens ("field lens") that acts to make the image upright as shown. A spyglass must be quite long. The most practical design today is the **prism binocular** which was shown in Fig. 33–31. The objective and eyepiece are converging lenses. The prisms reflect the rays by total internal reflection and shorten the physical size of the device, and they also act to produce an upright image. One prism reinverts the image in the vertical plane, the other in the horizontal plane.

FIGURE 34–34 Terrestrial telescopes that produce an upright image: (a) Galilean; (b) spyglass, or field-lens, type.

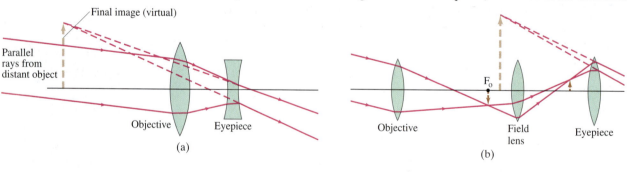

* **34–9** **Compound Microscope**

* 34–9 Compound Microscope

PHYSICS APPLIED

Microscopes

The compound **microscope**, like the telescope, has both objective and eyepiece (or ocular) lenses, Fig. 34–35. The design is different from that for a telescope because a microscope is used to view objects that are very close, so the object distance is

[†]Another advantage of mirrors is that they exhibit no chromatic aberration because the light doesn't pass through them; and they can be ground in a parabolic shape to correct for spherical aberration (Section 34–10). The reflecting telescope was first proposed by Newton.

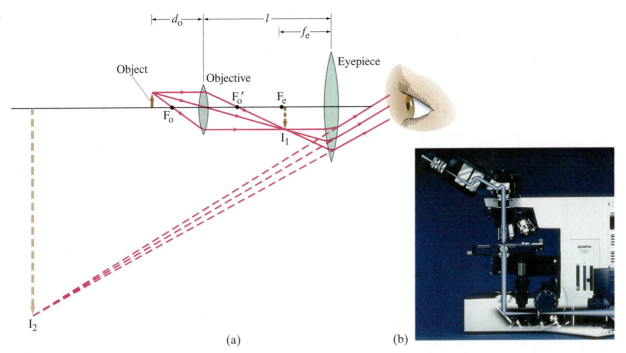

FIGURE 34–35 Compound microscope: (a) ray diagram, (b) photograph (illumination comes from the lower right, then up through the slide holding the object).

very small. The object is placed just beyond the objective's focal point as shown in Fig. 34–35a. The image I_1 formed by the objective lens is real, quite far from the lens, and much enlarged. This image is magnified by the eyepiece into a very large virtual image, I_2, which is seen by the eye and is inverted.

The overall magnification of a microscope is the product of the magnifications produced by the two lenses. The image I_1 formed by the objective is a factor m_o greater than the object itself. From Fig. 34–35a and Eq. 34–3 for the lateral magnification of a simple lens, we have

$$m_o = \frac{h_i}{h_o} = \frac{d_i}{d_o} = \frac{l - f_e}{d_o}, \qquad (34–8)$$

where d_o and d_i are the object and image distances for the objective lens, l is the distance between the lenses (equal to the length of the barrel), and we ignored the minus sign in Eq. 34–3 which only tells us that the image is inverted. We set $d_i = l - f_e$, which is true only if the eye is relaxed, so that the image I_1 is at the eyepiece focal point F_e. The eyepiece acts like a simple magnifier. If we assume that the eye is relaxed, the eyepiece angular magnification M_e is (from Eq. 34–6a)

$$M_e = \frac{N}{f_e}, \qquad (34–9)$$

where the near point $N = 25\,\text{cm}$ for the normal eye. Since the eyepiece enlarges the image formed by the objective, the overall angular magnification M is the product of the lateral magnification of the objective lens, m_o, times the angular magnification, M_e, of the eyepiece lens (Eqs. 34–8 and 34–9):

$$M = M_e m_o = \left(\frac{N}{f_e}\right)\left(\frac{l - f_e}{d_o}\right) \qquad (34–10a)$$

$$\approx \frac{Nl}{f_e f_o}. \qquad (34–10b)$$

Magnification

of

microscope

The approximation, Eq. 34–10b, is accurate when f_e and f_o are small compared to l, so $l - f_e \approx l$ and $d_o \approx f_o$ (Fig. 34–35a). This is a good approximation for large magnifications, since these are obtained when f_o and f_e are very small (they are in the denominator of Eq. 34–10b). In order to make lenses of very short focal length, which can be done best for the objective, compound lenses involving several elements must be used to avoid serious aberrations, as discussed in the next Section.

EXAMPLE 34–13 Microscope. A compound microscope consists of a 10× eyepiece and a 50× objective 17.0 cm apart. Determine (a) the overall magnification, (b) the focal length of each lens, and (c) the position of the object when the final image is in focus with the eye relaxed. Assume a normal eye, so $N = 25$ cm.

SOLUTION (a) The overall magnification is $(10\times)(50\times) = 500\times$.
(b) The eyepiece focal length is (Eq. 34–9) $f_e = N/M_e = 25$ cm$/10 = 2.5$ cm. For the objective lens, it is easier to next find d_o (part (c)) before we find f_o because we can use Eq. 34–8. Solving for d_o, we find

$$d_o = \frac{l - f_e}{m_o} = \frac{(17.0 \text{ cm} - 2.5 \text{ cm})}{50} = 0.29 \text{ cm}.$$

Then, from the lens equation with $d_i = l - f_e = 14.5$ cm (see Fig. 34–35a),

$$\frac{1}{f_o} = \frac{1}{d_o} + \frac{1}{d_i} = \frac{1}{0.29 \text{ cm}} + \frac{1}{14.5 \text{ cm}} = 3.52;$$

so $f_o = 0.28$ cm.
(c) We just calculated $d_o = 0.29$ cm, which is very close to f_o.

---

*$\boxed{34\text{–}10}$ Aberrations of Lenses and Mirrors

Earlier in this chapter, we developed a theory of image formation by a thin lens. We found, for example, that all rays from each point on an object are brought to a single point as the image point. This, and other results, were based on approximations such as that all rays make small angles with one another and we can use $\sin \theta \approx \theta$. Because of these approximations, we expect deviations from the simple theory and these are referred to as **lens aberrations**. There are several types of aberration; we will briefly discuss each of them separately but all may be present at one time.

Spherical aberration   Consider an object at any point (even at infinity) on the axis of a lens. Rays from this point that pass through the outer regions of the lens are brought to a focus at a different point from those that pass through the center of the lens. This is called **spherical aberration**, and is shown exaggerated in Fig. 34–36. Consequently, the image seen on a piece of film (for example) will not be a point but a tiny circular patch of light. If the film is placed at the point C, as indicated, the circle will have its smallest diameter, which is referred to as the **circle of least confusion**. Spherical aberration is present whenever spherical surfaces are used. It can be corrected by using nonspherical lens surfaces, but to grind such lenses is difficult and expensive. It can be minimized with spherical surfaces by choosing the curvatures so that equal amounts of bending occur at each lens surface; a lens can be designed like this for only one particular object distance. Spherical aberration is usually corrected (by which we mean reduced greatly) by the use of several lenses in combination, and by using only the central part of lenses.

FIGURE 34–36 Spherical aberration (exaggerated). Circle of least confusion is at C.

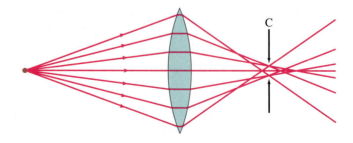

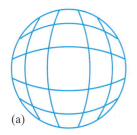
(a)

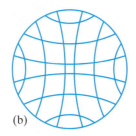
(b)

FIGURE 34–37 Distortion. Lenses may image a square grid of perpendicular lines to produce (a) barrel distortion or (b) pincushion distortion. These distortions can be seen in the photographs of Fig. 34–1d.

For object points off the lens axis, additional aberrations occur. Rays passing through the different parts of the lens cause spreading of the image that is noncircular. We won't go into the details but merely point out that there are two effects: **coma** (because the image of a point is comet-shaped rather than a circle) and **off-axis astigmatism.**[†] Furthermore, the image points for objects off the axis but at the same distance from the lens do not fall on a flat plane but on a curved surface—that is, the focal plane is not flat. (We expect this because the points on a flat plane, such as the film in a camera, are not equidistant from the lens.) This aberration is known as **curvature of field** and is obviously a problem in cameras and other devices where the film is placed in a flat plane. In the eye, however, the retina is curved, which compensates for this effect. Another aberration, known as **distortion**, is a result of variation of magnification at different distances from the lens axis. Thus a straight line object some distance from the axis may form a curved image. A square grid of lines may be distorted to produce "barrel distortion," or "pincushion distortion," Fig. 34–37. The latter is common in extreme wide-angle lenses.

Off-axis aberrations

All the above aberrations occur for monochromatic light and hence are referred to as *monochromatic aberrations*. Normal light is not monochromatic, and there will also be **chromatic aberration**. This aberration arises because of dispersion—the variation of index of refraction of transparent materials with wavelength (Section 33–6). For example, blue light is bent more than red light by glass. So if white light is incident on a lens, the different colors are focused at different points, Fig. 34–38, and there will be colored fringes in the image. Chromatic aberration can be eliminated for any two colors (and reduced greatly for all others) by the use of two lenses made of different materials with different indices of refraction and dispersion. Normally one lens is converging and the other diverging, and they are often cemented together (Fig. 34–39). Such a lens combination is called an **achromatic doublet** (or "color-corrected" lens).

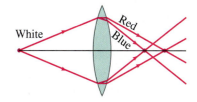

FIGURE 34–38 Chromatic aberration. Different colors are focused at different points.

FIGURE 34–39 Achromatic doublet.

It is not possible to fully correct all aberrations. Combining two or more lenses together can reduce them. High-quality lenses used in cameras, microscopes, and other devices are **compound lenses** consisting of many simple lenses (referred to as **elements**). A typical high-quality camera lens may contain six to eight (or more) elements.

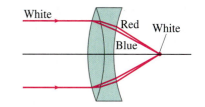

For simplicity we will normally indicate lenses in diagrams as if they were simple lenses. But it must be remembered that good-quality lenses are compound.

The human eye is also subject to aberrations, but they are minimal. Spherical aberration, for example, is minimized because (1) the cornea is less curved at the edges than at the center, and (2) the lens is less dense at the edges than at the center. Both effects cause rays at the outer edges to be bent less strongly, and thus help to reduce spherical aberration. Chromatic aberration is partially compensated for because the lens absorbs the shorter wavelengths appreciably and the retina is less sensitive to the blue and violet wavelengths. This is just the region of the spectrum where dispersion—and thus chromatic aberration—is greatest (Fig. 33–26).

Spherical mirrors (Section 33–4) also suffer aberrations including spherical aberration (see Fig. 33–13). Mirrors can be ground in a parabolic shape to correct for spherical aberration, but they are much harder to make and therefore very expensive. Spherical mirrors do not, however, exhibit chromatic aberration because the light does not pass through them.

[†] Although the effect is the same as for astigmatism in the eye (Section 34–6), the cause is different. Off-axis astigmatism is no problem in the eye because objects are clearly seen only at the fovea, on the lens axis.

Summary

A lens uses refraction to produce a real or virtual image. Parallel rays of light are focused to a point, called the **focal point**, by a converging lens. The distance of the focal point from the lens is called the **focal length** f of the lens. After parallel rays pass through a diverging lens, they appear to diverge from a point, its focal point; and the corresponding focal length is considered negative.

The **power** P of a lens, which is $P = 1/f$, is given in diopters, which are units of inverse meters (m^{-1}).

For a given object, the position and size of the image formed by a lens can be found by ray tracing. Algebraically, the relation between image and object distances, d_i and d_o, and the focal length f, is given by the **lens equation**:

$$\frac{1}{d_o} + \frac{1}{d_i} = \frac{1}{f}.$$

The ratio of image height to object height, which equals the magnification m, is

$$m = \frac{h_i}{h_o} = -\frac{d_i}{d_o}.$$

When using the various equations of geometric optics, it is important to remember the **sign conventions** for all quantities involved: carefully review them when doing problems.

A **camera** lens forms an image on film by allowing light in through a shutter. The lens is focused by moving it relative to the film, and its f-stop (or lens opening) must be adjusted for the brightness of the scene and the chosen shutter speed. The f-stop is defined as the ratio of the focal length to the diameter of the lens opening.

The human **eye** also adjusts for the available light—by opening and closing the iris. It focuses not by moving the lens, but by adjusting the shape of the lens to vary its focal length. The image is formed on the retina, which contains an array of receptors known as rods and cones.

Diverging eyeglass or contact lenses are used to correct the defect of a nearsighted eye, which cannot focus well on distant objects. Converging lenses are used to correct for defects in which the eye cannot focus on close objects.

A **simple magnifier** is a converging lens that forms a virtual image of an object placed at (or within) the focal point. The **angular magnification**, when viewed by a relaxed normal eye, is

$$M = \frac{N}{f},$$

where f is the focal length of the lens and N is the near point of the eye (25 cm for a "normal" eye).

An **astronomical telescope** consists of an **objective** lens or mirror and an **eyepiece** that magnifies the real image formed by the objective. The **magnification** is equal to the ratio of the objective and eyepiece focal lengths, and the image is inverted:

$$M = -\frac{f_o}{f_e}.$$

A compound **microscope** also uses objective and eyepiece lenses, and the final image is inverted. The total magnification is the product of the magnifications of the two lenses and is approximately

$$M \approx \left(\frac{N}{f_e}\right)\left(\frac{l}{f_o}\right),$$

where l is the distance between the lenses, N is the near point of the eye, and f_o and f_e are the focal lengths of objective and eyepiece, respectively.

Microscopes, telescopes, and other optical instruments are limited in the formation of sharp images by **lens aberrations**. These include **spherical aberration**, in which rays passing through the edge of a lens are not focused at the same point as those that pass near the center; and **chromatic aberration**, in which different colors are focused at different points. Compound lenses, consisting of several elements, can largely correct for aberrations.

Questions

1. Where must the film be placed if a camera lens is to make a sharp image of an object far away?

2. A photographer moves closer to his subject and then refocuses. Does the camera lens move farther away from or nearer to the film?

3. Can a diverging lens form a real image under any circumstances? Explain.

4. Show that a real image formed by a thin lens is always inverted, whereas a virtual image is always upright if the object is real.

5. Light rays are said to be "reversible." Is this consistent with the lens equation?

6. Can real images be projected on a screen? Can virtual images? Can either be photographed? Discuss carefully.

7. A thin converging lens is moved closer to a nearby object. Does the real image formed change (*a*) in position, (*b*) in size? If yes, describe how.

8. Compare the mirror equation with the lens equation. Discuss similarities and differences, especially the sign conventions for the quantities involved.

9. A lens is made of a material with an index of refraction $n = 1.30$. In air, it is a converging lens. Will it still be a converging lens if placed in water? Explain, using a ray diagram.

10. A dog stands facing a converging lens with its tail in the air. If the nose and the tail are each focused on a screen in turn, which will have the greater magnification?

11. A cat stands facing a converging lens with its tail in the air. Under what circumstances (if any) would the image of the nose be virtual and the image of the tail be real? Where would the image of the rest of the cat be?

12. Why, in Example 34–5, must the converging lens have a shorter focal length than the diverging lens if the latter's focal length is to be determined by combining them?

13. An unsymmetrical lens (say, planoconvex) forms an image of a nearby object. Does the image point change if the lens is turned around?

14. The thicker a double convex lens is in the center as compared to its edges, the shorter its focal length for a given lens diameter. Explain.

15. Does the focal length of a lens depend on the fluid in which it is immersed? What about the focal length of a spherical mirror? Explain.

16. An underwater lens consists of a carefully shaped thin-walled plastic container filled with air. What shape should it have in order to be (a) converging (b) diverging? Use ray diagrams to support your answer.

* 17. Why is the depth of field greater, and the image sharper, when a camera lens is "stopped down" to a larger f-number? Ignore diffraction.

* 18. Explain why swimmers with good eyes see distant objects as blurry when they are underwater. Use a diagram and also show why goggles correct this problem.

* 19. Will a nearsighted person who wears corrective lenses be able to see clearly underwater when wearing glasses? Use a diagram to show why or why not.

* 20. You can tell whether a person is nearsighted or farsighted by looking at the width of the face through their glasses. If the person's face appears narrower through the glasses, is the person farsighted or nearsighted?

* 21. The human eye is much like a camera—yet, when a camera shutter is left open and the camera moved, the image will be blurred; but when you move your head with your eyes open, you still see clearly. Explain.

* 22. In attempting to discern distant details, people will sometimes squint. Why does this help?

* 23. Is the image formed on the retina of the human eye upright or inverted? Discuss the implications of this for our perception of objects.

* 24. Reading glasses use converging lenses. A simple magnifier is also a converging lens. Are reading glasses therefore magnifiers? Discuss the similarities and differences between converging lenses as used for these two different purposes.

* 25. Spherical aberration in a thin lens is minimized if rays are bent equally by the two surfaces. If a planoconvex lens is used to form a real image of an object at infinity, which surface should face the object? Use ray diagrams to show why.

Problems

Sections 34–1 and 34–2

1. (I) A sharp image is located 88.0 mm behind a 65.0-mm-focal-length converging lens. Find the object distance (a) using a ray diagram, (b) by calculation.

2. (I) A certain lens focuses an object 2.85 m away as an image 48.3 cm on the other side of the lens. What type of lens is it and what is its focal length? Is the image real or virtual?

3. (I) (a) What is the power of a 27.5-cm-focal-length lens? (b) What is the focal length of a −6.25-diopter lens? Are these lenses converging or diverging?

4. (II) A stamp collector uses a converging lens with focal length 24 cm to view a stamp 18 cm in front of the lens. (a) Where is the image located? (b) What is the magnification?

5. (II) How large is the image of the Sun on the film used in a camera with (a) a 28-mm-focal-length lens, (b) a 50-mm-focal-length lens, and (c) a 200-mm-focal-length lens? The Sun's diameter is 1.4×10^6 km and it is 1.5×10^8 km away.

6. (II) A 35-mm slide (picture size is actually 24 by 36 mm) is to be projected on a screen 1.80 by 2.70 m placed 8.00 m from the projector. What focal-length lens should be used if the image is to cover the screen?

7. (II) An 80-mm-focal-length lens is used to focus an image on the film of a camera. The maximum distance allowed between the lens and the film plane is 120 mm. (a) How far ahead of the film should the lens be if the object to be photographed is 10.0 m away? (b) 3.0 m away? (c) 1.0 m away? (d) What is the closest object this lens could photograph sharply?

8. (II) A −6.0-diopter lens is held 12.5 cm from an ant 1.0 mm high. What is the position, type, and height of the image?

9. (II) It is desired to magnify reading material by a factor of 2.5× when a book is placed 8.0 cm behind a lens. (a) Draw a ray diagram and describe the type of image this would be. (b) What type of lens is needed for this? (c) What is the power of the lens in diopters?

10. (II) (a) How far from a 50.0-mm-focal-length lens must an object be placed if its image is to be magnified 2.00× and be real? (b) What if the image is to be virtual and magnified 2.00×?

11. (II) (a) An object 37.5 cm in front of a certain lens is imaged 8.20 cm in front of that lens (on the same side as the object). What type of lens is this and what is its focal length? Is the image real or virtual? (b) If the image were located, instead, 46.0 cm in front of the lens, what type of lens would it be and what focal length would it have?

12. (II) (a) A 2.20-cm-high insect is 1.10 m from a 135-mm-focal-length lens. Where is the image, how high is it, and what type is it? (b) What if $f = -135$ mm?

13. (II) A bright object and a viewing screen are separated by a distance of 76.0 cm. At what location(s) between the object and the screen should a lens of focal length 16.0 cm be placed in order to produce a crisp image on the screen? [*Hint*: First draw a diagram.]

14. (II) How far apart are an object and an image formed by a 75-cm-focal-length converging lens if the image is 2.75× larger than the object and is real?

15. (II) Show analytically that the image formed by a converging lens is real and inverted if the object is beyond the focal point $(d_o > f)$, and is virtual and upright if the object is within the focal point $(d_o < f)$. Describe the image if the object is itself an image formed by another lens, so its position is beyond the lens (on the side of the lens opposite from the incoming light), for which $-d_o > f$, and for which $0 < -d_o < f$.

16. (III) (a) Show that the lens equation can be written in the *Newtonian form*:

$$xx' = f^2,$$

where x is the distance of the object from the focal point on the front side of the lens, and x' is the distance of the image to the focal point on the other side of the lens. Calculate the location of an image if the object is placed 45.0 cm in front of a convex lens with a focal length of 32.0 cm using (b) the standard form of the thin lens formula, and (c) the Newtonian form, derived above.

Section 34–3

17. (II) Two 27.0-cm-focal-length converging lenses are placed 16.5 cm apart. An object is placed 35.0 cm in front of one. Where will the final image formed by the second lens be located? What is the total magnification?

18. (II) A diverging lens with $f = -31.5$ cm is placed 14.0 cm behind a converging lens with $f = 20.0$ cm. Where will an object at infinity be focused?

19. (II) The two converging lenses of Example 34–4 are placed so they are now only 20.0 cm apart. The object is still 60.0 cm in front of the first lens as in Fig. 34–13. In this case, determine (a) the position of the final image, and (b) the overall magnification. (c) Sketch the ray diagram for this system.

20. (II) A 31.0-cm-focal-length converging lens is 24.0 cm behind a diverging lens. Parallel light strikes the diverging lens. After passing through the converging lens, the light is again parallel. What is the focal length of the diverging lens? [*Hint*: First draw a ray diagram.]

21. (II) A diverging lens is placed next to a converging lens of focal length f_C, as in Fig. 34–14. If f_T represents the focal length of the combination, show that the focal length of the diverging lens, f_D, is given by

$$\frac{1}{f_D} = \frac{1}{f_T} - \frac{1}{f_C}.$$

Section 34–4

22. (I) A double concave lens has surface radii of 33.4 cm and 23.8 cm. What is the focal length if $n = 1.58$?

23. (I) Both surfaces of a double convex lens have radii of 31.0 cm. If the focal length is 28.9 cm, what is the index of refraction of the lens material?

24. (I) Show that if the lens of Example 34–7 is reversed so the light enters the curved face, the focal length is unchanged.

25. (I) A planoconvex lens (Fig. 34–1a) is to have a focal length of 17.5 cm. If made from fused quartz, what must be the radius of curvature of the convex surface?

26. (I) A glass ($n = 1.50$) planoconcave lens has a focal length of -21.5 cm. What is the radius of the concave surface?

27. (II) A prescription for a corrective lens calls for +2.50 diopters. The lensmaker grinds the lens from a "blank" with $n = 1.56$ and a preformed convex front surface of radius of curvature of 20.0 cm. What should be the radius of curvature of the other surface?

28. (II) An object is placed 90.0 cm from a glass lens ($n = 1.56$) with one concave surface of radius 22.0 cm and one convex surface of radius 18.5 cm. Where is the final image? What is the magnification?

29. (III) A glass lens ($n = 1.50$) in air has a power of +4.5 diopters. What would its power be if it were submerged in water?

* Section 34–5

* 30. (I) A 45-mm-focal-length lens has f-stops ranging from $f/1.4$ to $f/22$. What is the corresponding range of lens diaphragm diameters?

* 31. (I) A television camera lens has a 14-cm focal length and a lens diameter of 6.0 cm. What is its f-number?

* 32. (I) A properly exposed photograph is taken at $f/16$ and $\frac{1}{60}$ s. What lens opening would be required if the shutter speed were $\frac{1}{1000}$ s?

* 33. (II) A "pinhole" camera uses a tiny pinhole instead of a lens. Show, using ray diagrams, how reasonably sharp images can be formed using such a pinhole camera. In particular, consider two point objects 2.0 cm apart that are 1.0 m from a 1.0-mm-diameter pinhole. Show that on a piece of film 7.0 cm behind the pinhole the two objects produce two separate circles that do not overlap.

* 34. (II) Suppose that a correct exposure is $\frac{1}{250}$ s at $f/11$. Under the same conditions, what exposure time would be needed for a pinhole camera if the pinhole diameter is 1.0 mm and the film is 7.0 cm from the hole?

* 35. (II) A nature photographer wishes to photograph a 32-m tall tree from a distance of 55 m. What focal-length lens should be used if the image is to fill the 24-mm height of the film?

Section 34-6

36. (I) A human eyeball is about 2.0 cm long and the pupil has a maximum diameter of about 5.0 mm. What is the "speed" of this lens?

37. (II) A person struggles to read by holding a book at arm's length, a distance of 50 cm away. What power of reading glasses should be prescribed for him, assuming they will be placed 2.0 cm from the eye and he wants to read at the "normal" near point of 25 cm?

38. (II) Reading glasses of what power are needed for a person whose near point is 120 cm, so that he can read a computer screen at 50 cm? Assume a lens–eye distance of 1.8 cm.

39. (II) Show that if the nearsighted person in Example 34–10 wore contact lenses corrected for the far point ($= \infty$), that the near point would be 41 cm. (Would glasses be better in this case?)

40. (II) A person's left eye is corrected by a -4.0-diopter lens, 2.0 cm from the eye. (a) Is this person near- or farsighted? (b) What is this person's far point without glasses?

41. (II) A person's right eye can see objects clearly only if they are between 25 cm and 75 cm away. (a) What power of contact lens is required so that objects far away are sharp? (b) What will be the near point with the lens in place?

42. (II) About how much longer is the nearsighted eye in Example 34–10 than the 2.0 cm of a normal eye?

43. (II) One lens of a nearsighted person's eyeglasses has a focal length of -25.0 cm and the lens is 1.8 cm from the eye. If the person switches to contact lenses that are placed directly on the eye, what should be the focal length of the corresponding contact lens?

44. (II) What is the focal length of the eye lens system when viewing an object (a) at infinity, and (b) 30 cm from the eye? Assume that the lens–retina distance is 2.0 cm.

45. (II) A nearsighted person has near and far points of 10.0 and 20.0 cm respectively. If she puts on a pair of glasses with power $P = -4.0$ D, what are her new near and far points? Neglect the distance between her eye and the lens.

Section 34-7

46. (I) What is the magnification of a lens used with a relaxed eye if its focal length is 11 cm?

47. (I) What is the focal length of a magnifying glass of 3.0× magnification for a relaxed normal eye?

48. (I) A magnifier is rated at 2.5× for a normal eye focusing on an image at the near point. (a) What is its focal length? (b) What is its focal length if the 2.5× refers to a relaxed eye?

49. (II) Sherlock Holmes is using an 8.50-cm-focal-length lens as his magnifying glass. To obtain maximum magnification, where must the object be placed (assume a normal eye), and what will be the magnification?

50. (II) A 3.70-mm-wide bolt is viewed with a 9.00-cm-focal-length lens. A normal eye views the image at its near point. Calculate (a) the angular magnification, (b) the width of the image, and (c) the object distance from the lens.

51. (II) A small insect is placed 5.35 cm from a $+6.00$-cm-focal-length lens. Calculate (a) the position of the image, and (b) the angular magnification.

52. (II) A magnifying glass with a focal length of 9.5 cm is used to read print placed at a distance of 8.5 cm. Calculate: (a) the position of the image; (b) the linear magnification; and (c) the angular magnification.

53. (II) A magnifying glass is rated at 3.0× for a normal eye that is relaxed. What would be the magnification for a relaxed eye whose near point is (a) 60 cm, and (b) 18 cm? Explain the differences.

* Section 34-8

* 54. (I) What is the magnification of an astronomical telescope whose objective lens has a focal length of 72 cm, and whose eyepiece has a focal length of 2.8 cm? What is the overall length of the telescope when adjusted for a relaxed eye?

* 55. (I) The overall magnification of an astronomical telescope is desired to be 25×. If an objective of 80 cm focal length is used, what must be the focal length of the eyepiece? What is the overall length of the telescope when adjusted for use by the relaxed eye?

* 56. (I) An 8.0× binocular has 3.0-cm-focal-length eyepieces. What is the focal length of the objective lenses?

* 57. (II) An astronomical telescope has an objective with focal length 95 cm and a $+35$ D eyepiece. What is the total magnification?

* 58. (II) An astronomical telescope has its two lenses spaced 76.0 cm apart. If the objective lens has a focal length of 74.5 cm, what is the magnification of this telescope? Assume a relaxed eye.

* 59. (II) A Galilean telescope adjusted for a relaxed eye is 33.0 cm long. If the objective lens has a focal length of 36.0 cm, what is the magnification?

* 60. (II) What is the magnifying power of an astronomical telescope using a reflecting mirror whose radius of curvature is 6.2 m and an eyepiece whose focal length is 2.8 cm?

* 61. (II) The Moon's image appears to be magnified 120× by a reflecting astronomical telescope with an eyepiece having a focal length of 3.3 cm. What are the focal length and radius of curvature of the main mirror?

* 62. (II) A 130× astronomical telescope is adjusted for a relaxed eye when the two lenses are 1.25 m apart. What is the focal length of each lens?

* 63. (III) A 7.0× pair of binoculars has an objective focal length of 26 cm. If the binoculars are focused on an object 4.0 m away (from the objective), what is the magnification? (The 7.0× refers to objects at infinity; Eq. 34–7 holds only for objects at infinity and not for nearby ones.)

* Section 34-9

* 64. (I) A microscope uses an eyepiece with a focal length of 1.6 cm. Using a normal eye with a final image at infinity, the tube length is 17.5 cm and the focal length of the objective lens is 0.65 cm. What is the magnification of the microscope?

* 65. (I) A 650× microscope uses a 0.40-cm-focal-length objective lens. If the tube length is 17.5 cm, what is the focal length of the eyepiece? Assume a normal eye and that the final image is at infinity.

*66. (II) A microscope has a 12.0× eyepiece and a 56.0× objective 20.0 cm apart. Calculate (a) the total magnification, (b) the focal length of each lens, and (c) where the object must be for a normal relaxed eye to see it in focus.

*67. (II) A microscope has a 1.8-cm-focal-length eyepiece and 0.80-cm objective. Assuming a relaxed normal eye calculate (a) the position of the object if the distance between the lenses is 16.0 cm, and (b) the total magnification.

*68. (II) Repeat Problem 69 assuming that the final image is located 25 cm from the eyepiece (near point of a normal eye).

*69. (II) The eyepiece of a compound microscope has a focal length of 2.70 cm and the objective has $f = 0.740$ cm. If an object is placed 0.790 cm from the objective lens, calculate (a) the distance between the lenses when the microscope is adjusted for a relaxed eye, and (b) the total magnification.

* Section 34–10

*70. (II) An achromatic lens is made of two very thin lenses placed in contact that have focal lengths of $f_1 = -28$ cm and $f_2 = +23$ cm. (a) Is the combination converging or diverging? (b) What is the net focal length?

*71. (III) Let's examine spherical aberration in a particular situation. A planoconvex lens of index of refraction 1.50 and a radius of curvature $R = 12.0$ cm is shown in Fig. 34–40. Consider an incoming ray parallel to the principle axis and a height h above it as shown. Determine the distance d, from the flat face of the lens, to where this ray crosses the principle axis if (a) $h = 1.0$ cm, and (b) $h = 6.0$ cm. (c) How far apart are these "focal points"? (d) How large is the "circle of confusion" produced by the $h = 6.0$ cm ray at the "focal point" for $h = 1.0$ cm?

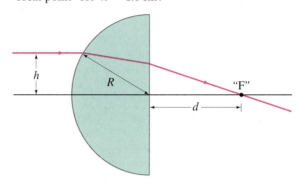

FIGURE 34–40 Problem 71.

General Problems

72. If a 135-mm telephoto lens is designed to cover object distances from 1.40 m to ∞, over what distance must the lens move relative to the plane of the film?

73. A 200-mm-focal-length lens can be adjusted so that it is 200.0 mm to 206.0 mm from the film. For what range of object distances can it be adjusted?

74. Show that for objects very far away (assume infinity), the magnification of a camera lens is proportional to its focal length.

75. For a camera equiped with a 50-mm-focal-length lens, what is the object distance if the image height equals the object height? How far is the object from the image on the film?

76. What is the magnifying power of a +8.0-D lens used as a magnifier? Assume a relaxed normal eye.

77. A lucite planoconvex lens (Fig. 34–1a) has one flat surface and one with $R = 18.4$ cm. It is used to view an object, located 66.0 cm away from the lens, which is a mixture of red and yellow. The index of refraction of the lucite is 1.5106 for red light and 1.5226 for yellow light. What are the locations of the red and yellow images formed by the lens?

78. A movie star catches a reporter shooting pictures of her at home. She claims the reporter was trespassing and to prove her point, she gives as evidence the film she seized. Her 1.75-m height is 8.25 mm high on the film, and the focal length of the camera lens was 200 mm. How far away from the subject was the reporter standing?

79. In a slide or movie projector, the film acts as the object whose image is projected on a screen (Fig. 34–41). If a 100-mm-focal-length lens is to project an image on a screen 7.50 m away, how far from the lens should the slide be? If the slide is 36 mm wide, how wide will the picture be on the screen?

Slide

Lens

Screen

FIGURE 34–41 Slide projector, Problem 79.

80. A converging lens with focal length of 12.0 cm is placed in contact with a diverging lens with a focal length of 20.0 cm. What is the focal length of the combination, and is the combination converging or diverging?

81. Show analytically that a diverging lens can never form a real image of a real object. Can you describe a situation in which a diverging lens can form a real image?

82. A lighted candle is placed 30 cm in front of a converging lens of focal length $f_1 = 15$ cm, which in turn is 50 cm in front of another converging lens of focal length $f_2 = 10$ cm (see Fig. 34–42). (a) Draw a ray diagram and estimate the location and the size of the final image. (b) Calculate the position and size of the final image.

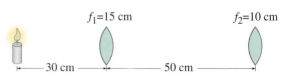

$f_1 = 15$ cm $f_2 = 10$ cm

30 cm   50 cm

FIGURE 34–42 Problem 82.

83. A bright object is placed on one side of a converging lens of focal length f, and a white screen for viewing the image is on the opposite side. The distance $d_T = d_i + d_o$ between the object and the screen is kept fixed, but the lens can be moved. Show that (a) if $d_T > 4f$, there will be two positions where the lens can be placed and a sharp image will be produced on the screen, and (b) if $d_T < 4f$, there will be no lens position where a sharp image is formed. (c) Determine the distance between the two lens positions in part (a), and the ratio of the image sizes.

84. (a) Show that if two thin lenses of focal lengths f_1 and f_2 are placed in contact with each other, the focal length of the combination is given by $f_T = f_1 f_2/(f_1 + f_2)$. (b) Show that the power P of the combination of two lenses is the sum of their separate powers, $P = P_1 + P_2$.

85. A lens whose index of refraction is n is submerged in a material whose index of refraction is n' ($n' \neq 1$). Derive the equivalent of Eqs. 34–2, 34–3 and 34–4 for this lens.

86. Sam purchases +2.5-diopter eyeglasses which correct his faulty vision to put his near point at 25 cm. (Assume he wears the lenses 2.0 cm from his eyes.) (a) Is Sam nearsighted or farsighted? (b) Calculate the focal length of Sam's glasses. (c) Calculate Sam's near point without glasses. (d) Pam, who has normal eyes with near point at 25 cm, puts on Sam's glasses. Calculate Pam's near point with Sam's glasses on.

87. A 50-year-old man uses +2.5-diopter lenses to be able to read a newspaper 25 cm away. Ten years later, he finds that he must hold the paper 35 cm away to see clearly with the same lenses. What power lenses does he need now? (Distances are measured from the lens.)

88. A woman can see clearly with her right eye only when objects are between 40 cm and 150 cm away. Prescription bifocals should have what powers so that she can see distant objects clearly (upper part) and be able to read a book 25 cm away (lower part)? Assume that the glasses will be 2.0 cm from the eye.

89. A child has a near point of 15 cm. What is the maximum magnification the child can obtain using an 8.0-cm-focal-length magnifier? Compare to that for a normal eye.

*90. As early morning passed toward midday, and the sunlight got more intense, a photographer who was taking repeated shots of the same subject noted that, if she kept her shutter speed constant, she had to change the f-number from $f/5.6$ to $f/22$. By how much had the sunlight intensity increased during that time?

*91. A person tries to make a telescope using the lenses from reading glasses. They have powers of +2.0 D and +5.0 D, respectively. (a) What maximum magnification telescope is possible? (b) Which lens should be used as the eyepiece?

*92. Exposure times must be increased for pictures taken at very short distances, because of the increased distance of the lens from the film for a focused image. (a) Show that when the object is so close to the camera that the image height equals the object height, the exposure time must be four times longer than when the object is a long distance away (say, ∞), given the same illumination and f-stop. (b) Show that if d_o is at least four or five times the focal length f of the lens, the exposure time is increased negligibly relative to the same object being a great distance away.

*93. The objective lens and the eyepiece of a telescope are spaced 85 cm apart. If the eyepiece is +25 diopters, what is the total magnification of the telescope?

The wave nature of light nicely explains how light reflected from the front and back surfaces of this very thin film of soapy water interferes constructively to produce the bright colors. Which color we see depends on the thickness of the film at that point—the film is very thin at the top and gets thicker going down toward the bottom (because of gravity). Can you use the clues (colors of light) to determine the thickness of the film at any point? You should be able to after reading this chapter.

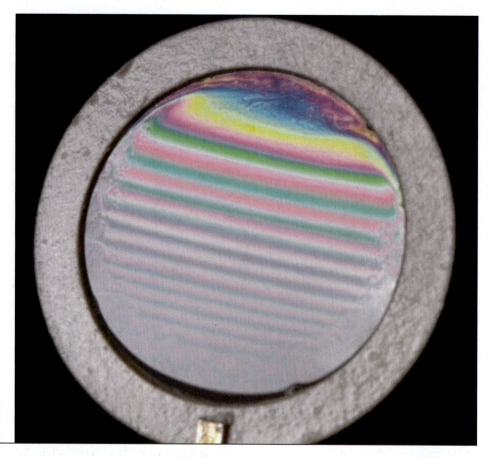

The Wave Nature of Light; Interference

That light carries energy is obvious to anyone who has focused the Sun's rays with a magnifying glass on a piece of paper and burned a hole in it. But how does light travel, and in what form is this energy carried? In our discussion of waves in Chapter 15, we noted that energy can be carried from place to place in basically two ways: by particles or by waves. In the first case, material bodies or particles can carry energy, such as an avalanche or rushing water. In the second case, water waves and sound waves, for example, can carry energy over long distances even though mass itself does not travel these distances. Then, what can we say about the nature of light: does light travel as a stream of particles away from its source, or does it travel in the form of waves that spread outward from the source?

Historically, this question has turned out to be a difficult one. For one thing, light does not reveal itself in any obvious way as being made up of tiny particles nor do we see tiny light waves passing by as we do water waves. The evidence seemed to favor first one side and then the other until about 1830, when most physicists had accepted the wave theory. By the end of the nineteenth century, light was considered to be an *electromagnetic wave* (Chapter 32). In the early twentieth century, light was shown to have a particle nature as well, as we shall discuss in Chapter 38. Nonetheless, the wave theory of light remains valid and has proved very successful. We now investigate the evidence for the wave theory and how it has explained a wide range of phenomena.

35–1 | Huygens' Principle and Diffraction

The Dutch scientist Christiaan Huygens (1629–1695), a contemporary of Newton, proposed a wave theory of light that had much merit. Still useful today is a technique he developed for predicting the future position of a wave front when an earlier position is known. This is known as **Huygens' principle** and can be stated as follows: *Every point on a wave front can be considered as a source of tiny wavelets that spread out in the forward direction at the speed of the wave itself. The new wave front is the envelope of all the wavelets—that is, the tangent to all of them.*

As a simple example of the use of Huygens' principle, consider the wave front AB in Fig. 35–1, which is traveling away from a source S. We assume the medium is *isotropic*—that is, the speed v of the waves is the same in all directions. To find the wave front a short time t after it is at AB, tiny circles are drawn with radius $r = vt$. The centers of these tiny circles are on the original wave front AB and the circles represent Huygens' (imaginary) wavelets. The tangent to all these wavelets, the line CD, is the new position of the wave front.

Huygens' principle is particularly useful when waves impinge on an obstacle and the wave fronts are partially interrupted. Huygens' principle predicts that waves bend in behind an obstacle, as shown in Fig. 35–2. This is just what water waves do, as we saw in Chapter 15 (Figs. 15–32 and 15–33). The bending of waves behind obstacles into the "shadow region" is known as **diffraction**. Since diffraction occurs for waves, but not for particles, it can serve as one means for distinguishing the nature of light.

Does light actually exhibit diffraction? In the mid-seventeenth century, a Jesuit priest, Francesco Grimaldi (1618–1663), had observed that when sunlight entered a darkened room through a tiny hole in a screen, the spot on the opposite wall was larger than would be expected from geometric rays. He also observed that the border of the image was not clear but was surrounded by colored fringes. Grimaldi attributed this to the diffraction of light.

Note that the ray model (Chapter 33) cannot account for diffraction, and it is important to be aware of the limitations of the ray model. Geometric optics using rays is so successful in its limited sphere because normal openings and obstacles are much larger than the wavelength of the light, and so relatively little diffraction or bending occurs.

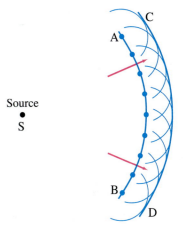

Huygens' principle

FIGURE 35–1 Huygens' principle used to determine the wave front CD when the wave front AB is given.

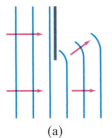 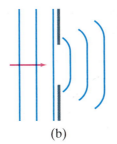 

(a) (b) (c)

FIGURE 35–2 Huygens' principle is consistent with diffraction (a) around the edge of an obstacle, (b) through a large hole, (c) through a small hole whose size is on the order of the wavelength of the wave.

35–2 | Huygens' Principle and the Law of Refraction

The laws of reflection and refraction were well known in Newton's time. The law of reflection could not distinguish between the two theories: waves versus particles, that we discussed on the previous page. For when waves reflect from an obstacle, the angle of incidence equals the angle of reflection (Fig. 15–22). The same is true of particles—think of a tennis ball without spin striking a flat surface.

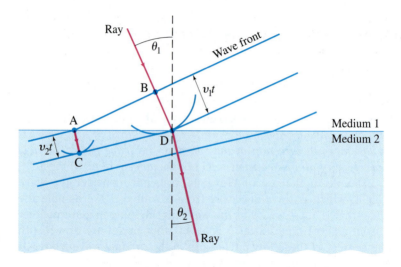

FIGURE 35–3 Refraction explained, using Huygens' principle.

The law of refraction is another matter. Consider light entering a medium where it is bent toward the normal, as when it travels from air into water. As shown in Fig. 35–3, this effect can be constructed using Huygens' principle if we assume the speed of light is less in the second medium $(v_2 < v_1)$. In time t, the point B on wave front AB goes a distance $v_1 t$ to reach point D. Point A, on the other hand, travels a distance $v_2 t$ to reach point C, and $v_2 t < v_1 t$. Huygens' principle is applied to points A and B to obtain the curved wavelets shown at C and D. The wave front is tangent to these two wavelets, so the new wave front is the line CD. Hence the rays, which are perpendicular to the wave fronts, bend toward the normal if $v_2 < v_1$, as drawn.[†] Newton favored a particle theory of light which predicted the opposite result, that the speed of light would be greater in the second medium $(v_2 > v_1)$. Thus the wave theory predicts that the speed of light in water, for example, is less than in air; and Newton's particle theory predicts the reverse. An experiment to actually measure the speed of light in water was performed in 1850 by the French physicist Jean Foucault, and it confirmed the wave-theory prediction. By then, however, the wave theory was already fully accepted, as we shall see in the next Section.

It is easy to show that Snell's law of refraction follows directly from Huygens' principle, given that the speed of light v in any medium is related to the speed in a vacuum, c, and the index of refraction, n, by Eq. 33–1, $v = c/n$. From the Huygens' construction of Fig. 35–3, angle ADC is equal to θ_2 and angle BAD is equal to θ_1. Then for the two triangles that have the common side AD, we have

$$\sin\theta_1 = \frac{v_1 t}{AD}, \qquad \sin\theta_2 = \frac{v_2 t}{AD}.$$

We divide these two equations and obtain:

$$\frac{\sin\theta_1}{\sin\theta_2} = \frac{v_1}{v_2}.$$

Then, since $v_1 = c/n_1$ and $v_2 = c/n_2$,

$$n_1 \sin\theta_1 = n_2 \sin\theta_2,$$

which is Snell's law of refraction, Eq. 33–5. (The law of reflection can be derived from Huygens' principle in a similar way, and this is given as Problem 1 at the end of the chapter.)

[†]This is basically the same as the discussion around Fig. 15–31.

When a light wave travels from one medium to another, its frequency does not change, but its wavelength does. This can be seen from Fig. 35–3, where we assume each of the blue lines representing a wave front corresponds to a crest (peak) of the wave. Then

Wavelength depends on n

$$\frac{\lambda_2}{\lambda_1} = \frac{v_2 t}{v_1 t} = \frac{v_2}{v_1} = \frac{n_1}{n_2},$$

where, in the last step, we used Eq. 33–1, $v = c/n$. If medium 1 is a vacuum (or air), so $n_1 = 1$, $v_1 = c$, and we call λ_1 simply λ, then the wavelength in another medium of index of refraction $n(= n_2)$ will be

$$\lambda_n = \frac{\lambda}{n}. \tag{35–1}$$

This result is consistent with the frequency f being unchanged since $c = f\lambda$. Combining this with $v = f\lambda_n$ in a medium where $v = c/n$ gives $\lambda_n = v/f = c/nf = f\lambda/nf = \lambda/n$, which checks.

Wave fronts can be used to explain how mirages are produced by refraction of light. Let us explain why, for example, on a hot day motorists sometimes see a mirage of water on the highway ahead of them, with distant vehicles seemingly reflected in it (Fig. 35–4a). On a hot day, there can be a layer of very hot air next to the roadway (made hot by the sun beating on the road). Hot air is less dense than cooler air, so the index of refraction is slightly lower in the hot air. In Fig. 35–4b, we see a diagram of light coming from one point on a distant car (on the right) heading left toward the observer. Wave fronts and two rays are shown. Ray A heads directly at the observer and follows a straight-line path, and represents the normal view of the car. Ray B is a ray initially directed slightly downward. But it does not hit the ground. Instead, ray B is bent slightly as it moves through layers of air of different index of refraction. The wave fronts, shown in blue in Fig. 35–4b, move slightly faster in the layers of air nearer the ground (see Fig. 35–3, and also the soldier analogy in Fig. 15–31). Thus ray B is bent as shown, and seems to the observer to be coming from below (dashed line) as if reflected off the road. Hence the mirage.

⮞ **PHYSICS APPLIED**

Highway mirages

FIGURE 35–4 (a) A highway mirage. (b) Drawing (greatly exaggerated) showing wave fronts and rays to explain highway mirages. Note how the sections of the wavefronts near the ground move faster and so are further apart.

(a)

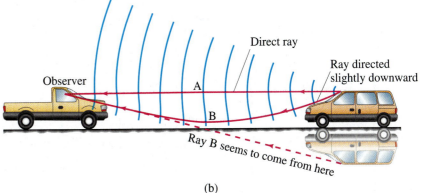

(b)

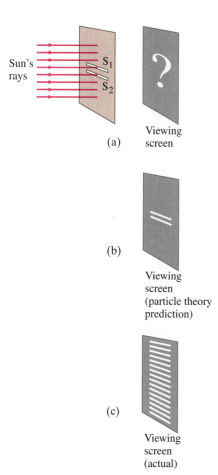

Sun's
rays

S₁
S₂

(a)
Viewing
screen

(b)
Viewing
screen
(particle theory
prediction)

(c)
Viewing
screen
(actual)

FIGURE 35–5 (a) Young's double-slit experiment. (b) If light consists of particles, we would expect to see two bright lines on the screen behind the slits. (c) Young observed many lines.

35–3 Interference—Young's Double-Slit Experiment

In 1801, the Englishman Thomas Young (1773–1829) obtained convincing evidence for the wave nature of light and was even able to measure the wavelengths for visible light. Figure 35–5a shows a schematic diagram of Young's famous double-slit experiment. Light from a single source (Young used the Sun) falls on a screen containing two closely spaced slits S_1 and S_2. If light consists of tiny particles, we might expect to see two bright lines on a screen placed behind the slits as in (b). But Young observed instead a series of bright lines as in (c). Young was able to explain this result as a **wave-interference** phenomenon. To see this, imagine plane waves of light of a single wavelength—called **monochromatic**, meaning "one color"—falling on the two slits as shown in Fig. 35–6. Because of diffraction, the waves leaving the two small slits spread out as shown. This is equivalent to the interference pattern produced when two rocks are thrown into a lake (Fig. 15–24), or when sound from two loudspeakers interferes (Fig. 16–16).

FIGURE 35–6 If light is a wave, light passing through one of two slits should interfere with light passing through the other slit.

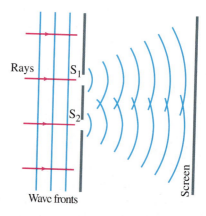

Rays
S₁
S₂
Wave fronts
Screen

To see how an interference pattern is produced on the screen, we make use of Fig. 35–7. Waves of wavelength λ are shown entering the slits S_1 and S_2, which are a distance d apart. The waves spread out in all directions after passing through the slits, but they are shown only for three different angles θ. In Fig. 35–7a, the waves reaching the center of the screen are shown ($\theta = 0$). The waves from the two slits travel the same distance, so they are in phase: a crest of one wave arrives at the same time as a crest of the other wave. Hence the amplitudes of the two waves add to

FIGURE 35–7 How the wave theory explains the pattern of lines seen in the double-slit experiment. (a) At the center of the screen the waves from each slit have traveled the same distance and are in phase. (b) At this angle θ, the lower wave travels an extra distance of one whole wavelength, and the waves are in phase; note from the shaded triangle that the extra distance equals $d \sin \theta$. (c) For this angle θ, the lower wave travels an extra distance equal to one-half wavelength, so the two waves arrive at the screen fully out of phase. (d) A more detailed diagram showing the geometry for parts (b) and (c).

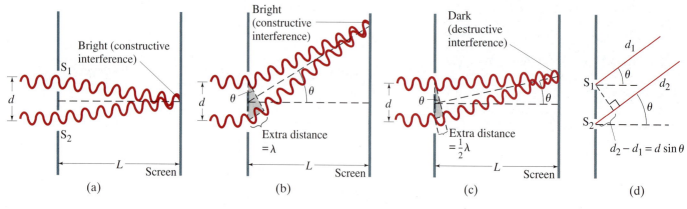

(a) Bright (constructive interference) — S₁, S₂, d, L, Screen

(b) Bright (constructive interference) — d, θ, θ, Extra distance = λ, L, Screen

(c) Dark (destructive interference) — d, θ, θ, Extra distance = ½λ, L, Screen

(d) S₁, S₂, d₁, d₂, θ, θ, $d_2 - d_1 = d \sin \theta$

form a larger amplitude as shown in Fig. 35–8a. This is **constructive interference**, and there is a bright spot at the center of the screen. Constructive interference also occurs when the paths of the two rays differ by one wavelength (or any whole number of wavelengths), as shown in Fig. 35–7b. But if one ray travels an extra distance of one-half wavelength (or $\frac{3}{2}\lambda$, $\frac{5}{2}\lambda$, and so on), the two waves are exactly out of phase when they reach the screen: the crests of one wave arrive at the same time as the troughs of the other wave, and so they add to produce zero amplitude (Fig. 35–8b). This is **destructive interference**, and the screen is dark, Fig. 35–7c. Thus, there will be a series of bright and dark lines (or **fringes**) on the viewing screen.

To determine exactly where the bright lines fall, first note that Fig. 35–7 is somewhat exaggerated; in real situations, the distance d between the slits is very small compared to the distance L to the screen. The rays from each slit for each case will therefore be essentially parallel and θ is the angle they make with the horizontal, as shown in Fig. 35–7d. From the right triangle shown in Fig. 35–7, we can see that the extra distance traveled by the lower ray is $d \sin\theta$. Constructive interference will occur, and a bright fringe will appear on the screen, when the *path difference, $d \sin\theta$*, equals a whole number of wavelengths:

$$d \sin\theta = m\lambda, \qquad m = 0, 1, 2, \cdots. \qquad \left[\begin{array}{c}\text{constructive} \\ \text{interference}\end{array}\right] \quad \textbf{(35–2a)}$$

The value of m is called the **order** of the interference fringe. The first order ($m = 1$), for example, is the first fringe on each side of the central fringe (at $\theta = 0$). Destructive interference occurs when the path difference, $d \sin\theta$, is $\frac{1}{2}, \frac{3}{2}$, and so on, wavelengths:

$$d \sin\theta = \left(m + \tfrac{1}{2}\right)\lambda, \qquad m = 0, 1, 2, \cdots. \qquad \left[\begin{array}{c}\text{destructive} \\ \text{interference}\end{array}\right] \quad \textbf{(35–2b)}$$

The intensity of the bright fringes is greatest for the central fringe ($m = 0$) and decreases for higher orders, as shown in Fig. 35–9.

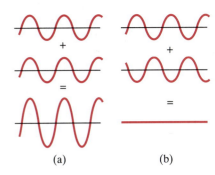

FIGURE 35–8
(a) Constructive interference.
(b) Destructive interference. (See also Section 15–8.)

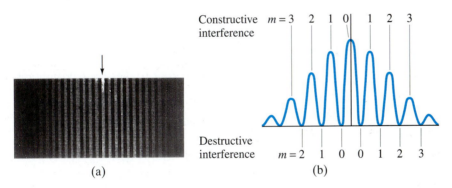

(a)

Constructive interference $m = 3$ 2   1   0| 1   2   3

Destructive interference $m = 2$ 1   0   0   1   2   3

(b)

FIGURE 35–9 (a) Interference fringes produced by a double-slit experiment and detected by photographic film placed on the viewing screen. The arrow marks the central fringe. (b) Intensity of light in the interference pattern. Also shown are values of m for Eq. 35–2a (constructive interference) and Eq. 35–2b (destructive interference).

| **CONCEPTUAL EXAMPLE 35–1** | **Interference pattern lines.** (*a*) Will there be an infinite number of points on the viewing screen where constructive and destructive interference occur, or only a finite number of points? (*b*) Are neighboring points of constructive interference uniformly spaced, or is the spacing between neighboring points of constructive interference not uniform?

RESPONSE (*a*) When you look at Eqs. 35–2a and b you might be tempted to say, given the statement $m = 0, 1, 2, \cdots$ beside the equations, that there are an infinite number of points of constructive and destructive interference. However, recall that $\sin\theta$ cannot exceed 1. Thus, there is an upper limit to the values of m that can be used in these equations. For Eq. 35–2a, the maximum value of m is the integer closest in value but smaller than d/λ. So there are a finite number of points of constructive and destructive interference spread out over an even infinitely tall screen. (*b*) The spacing between neighboring points of constructive or destructive interference is not uniform: The spacing gets larger as θ gets larger, and you can verify this statement mathematically. For small values of θ the spacing is nearly uniform as you will see in Example 35–2.

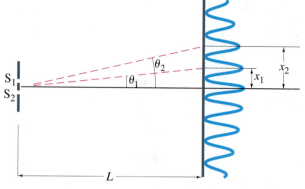

FIGURE 35–10 For small angles, the interference fringes occur at distance $x = \theta L$ above the center fringe ($m = 0$); θ_1 and x_1 are for the first order fringe ($m = 1$), θ_2 and x_2 are for $m = 2$.

EXAMPLE 35–2 **Line spacing for double-slit interference.** A screen containing two slits 0.100 mm apart is 1.20 m from the viewing screen. Light of wavelength $\lambda = 500$ nm falls on the slits from a distant source. Approximately how far apart will the bright interference fringes be on the screen?

SOLUTION Given $d = 0.100$ mm $= 1.00 \times 10^{-4}$ m, $\lambda = 500 \times 10^{-9}$ m, and $L = 1.20$ m, the first-order fringe ($m = 1$) occurs at an angle θ given by

$$\sin \theta_1 = \frac{m\lambda}{d} = \frac{(1)(500 \times 10^{-9}\,\text{m})}{1.00 \times 10^{-4}\,\text{m}} = 5.00 \times 10^{-3}.$$

This is a very small angle, so we can take $\sin \theta = \theta$, with θ in radians. The first-order fringe will occur a distance x_1 above the center of the screen (see Fig. 35–10) given by $x_1/L = \tan \theta_1 = \theta_1$, so

$$x_1 = L\theta_1 = (1.20\,\text{m})(5.00 \times 10^{-3}) = 6.00\,\text{mm}.$$

The second-order fringe ($m = 2$) will occur at

$$x_2 = L\theta_2 = L\frac{2\lambda}{d} = 12.0\,\text{mm}$$

above the center, and so on. Thus the lower order fringes are 6.00 mm apart.

CONCEPTUAL EXAMPLE 35–3 **Changing the wavelength.** (*a*) What happens to the interference pattern shown in Fig. 35–10, Example 35–2, if the incident light (500 nm) is replaced by light of wavelength 700 nm? (*b*) What happens instead if the slits are moved farther apart?

RESPONSE (*a*) When λ increases in Eq. 35–2a but d stays the same, then the angle θ for maxima increases and the interference pattern spreads out. (*b*) Increasing the slit spacing d reduces θ for each order, so the lines are closer together.

From Eqs. 35–2 we can see that, except for the zeroth-order fringe at the center, the position of the fringes depends on wavelength. Consequently, when white light falls on the two slits, as Young found in his experiments, the central fringe is white, but the first- (and higher-) order fringes contain a spectrum of colors like a *Wavelength (or frequency) determines color* rainbow; θ was found to be smallest for violet light and largest for red. By measuring the position of these fringes, Young was the first to determine the wavelengths of visible light (using Eqs. 35–2). In doing so, he showed that what distinguishes different colors physically is their wavelength, an idea put forward earlier by Grimaldi in 1665.

EXAMPLE 35–4 **Wavelengths from double-slit interference.** White light passes through two slits 0.50 mm apart and an interference pattern is observed on a screen 2.5 m away. The first-order fringe resembles a rainbow with violet and red light at either end. The violet light falls about 2.0 mm and the red 3.5 mm from the center of the central white fringe (Fig. 35–11). Estimate the wavelengths of the violet light and the red light.

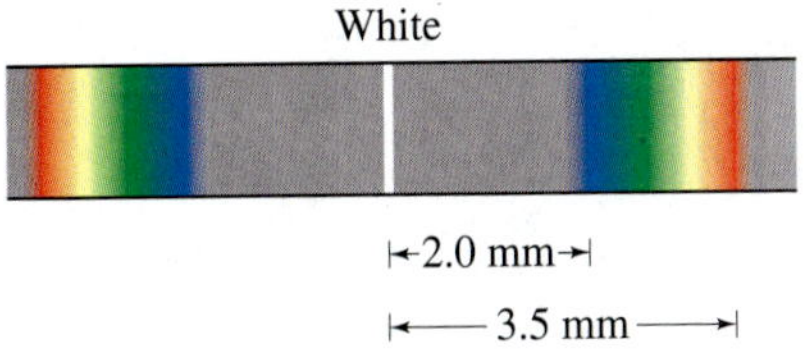

SOLUTION We use Eq. 35–2a with $m = 1$ and $\sin\theta = \theta$. Then for violet light, $x = 2.0$ mm, so (see also Fig. 35–10)

$$\lambda = \frac{d\theta}{m} = \frac{d}{m}\frac{x}{L} = \left(\frac{5.0 \times 10^{-4}\,\text{m}}{1}\right)\left(\frac{2.0 \times 10^{-3}\,\text{m}}{2.5\,\text{m}}\right) = 4.0 \times 10^{-7}\,\text{m},$$

or 400 nm. For red light, $x = 3.5$ mm, so

$$\lambda = \frac{d}{m}\frac{x}{L} = \left(\frac{5.0 \times 10^{-4}\,\text{m}}{1}\right)\left(\frac{3.5 \times 10^{-3}\,\text{m}}{2.5\,\text{m}}\right) = 7.0 \times 10^{-7}\,\text{m} = 700\,\text{nm}.$$

├—2.0 mm—┤

├——— 3.5 mm ———┤

FIGURE 35–11 Example 35–4.

35–4 Coherence

The two slits in Fig. 35–7 act as if they were two sources of radiation. They are called **coherent sources** because the waves leaving them bear the same phase relationship to each other at all times. This happens because the waves come from a single source to the left of the two slits in Fig. 35–7. An interference pattern is observed only when the sources are coherent. If two tiny lightbulbs replaced the two slits, an interference pattern would not be seen. The light emitted by one lightbulb would have a random phase with respect to the second bulb, and the screen would be more or less uniformly illuminated. Two such sources, whose output waves have phases that bear no fixed relationship to each other over time, are called **incoherent sources**.

Coherent and incoherent sources

The subject of coherence is rather complicated, and we discuss it only briefly. Two light beams do not have to be in phase to be coherent and produce an interference pattern. For example, suppose a piece of glass were placed in front of the lower slit in Fig. 35–7 and suppose the glass is just thick enough to slow down the light so that it enters the lower slit a half wavelength behind the light entering the upper slit. The two beams would be a constant 180° out of phase, but there would still be an interference pattern on the screen. (Can you guess what it would look like? Hint: The central point would be dark instead of bright.) Two beams can be coherent whether they are in phase or out of phase; the important thing is that they have a *constant* phase relation to each other over time.

Coherent sources of water or sound waves are easier to obtain than are coherent sources of light—two loudspeakers receiving the same pure frequency signal from an amplifier will be coherent sources. And two antennas connected to the same LC oscillator can be coherent sources of low-frequency electromagnetic waves. But LC oscillators at the high frequencies of visible light $(10^{15}\,\text{Hz})$ don't exist since L and C can't be made small enough. For sources of visible light we have to rely on the oscillations (or acceleration) of electric charge within atoms. In an incandescent light bulb, for example, the atoms in the filament are excited by heating, and give off "wave trains" of light, each of which lasts only about $10^{-8}\,\text{s}$. The light we see is the sum of a great many such wave trains that bear a random phase relation to each other. Two light bulbs are thus not coherent, and an interference pattern would not be seen. It wasn't until the 1950s that a really coherent source of light was developed, the *laser*. Because of its coherence, laser light on a double slit produces a very "clean" interference pattern.

Coherence is a relative concept. Perfect coherence of a beam would correspond to light that is perfectly sinusoidal (of one frequency) for all times, whereas complete incoherence occurs when there are waves whose phase relations are completely random over time. A measure of the "relative coherence" can be defined in terms of the sharpness of a two-slit interference pattern.

35–5 Intensity in the Double-Slit Interference Pattern

We saw in Section 35–3 that the interference pattern produced by the coherent light from two slits, S_1 and S_2 (Figs. 35–7 and 35–9), produces a series of bright and dark fringes. If the two monochromatic waves of wavelength λ are in phase at the slits, the maxima (brightest points) occur at angles θ given by

$$d \sin \theta = m\lambda,$$

and the minima (darkest points) when

$$d \sin \theta = \left(m + \tfrac{1}{2}\right)\lambda,$$

where m is an integer ($m = 0, 1, 2, ...$).

We now determine the intensity of the light at all points in the pattern (that is, for all θ). For simplicity, let us assume that if either slit were covered, the light passing through the other would diffract sufficiently to illuminate a large portion of the screen uniformly. The intensity I of the light at any point is proportional to the square of its wave amplitude (Section 15–3). Treating light as an electromagnetic wave, I is proportional to the square of the electric field E or magnetic field B (Section 32–7). Since E and B are proportional to each other, it doesn't matter which we use, but it is conventional to use E and write the intensity as $I \propto E^2$. The electric field $\mathbf{E}$ at any point P (see Fig. 35–12) will be the sum of the electric field vectors of the waves coming from each of the two slits, $\mathbf{E}_1$ and $\mathbf{E}_2$. Since $\mathbf{E}_1$ and $\mathbf{E}_2$ are essentially parallel (on a screen far away compared to the slit separation), the magnitude of the electric field at angle θ (that is, at point P) will be

$$E_\theta = E_1 + E_2.$$

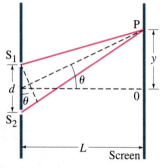

FIGURE 35–12 Determining the intensity in a double-slit interference pattern. Not to scale: in fact $L \gg d$, and the two rays become essentially parallel.

Both E_1 and E_2 vary sinusoidally with frequency $f = c/\lambda$, but they differ in phase, depending on their different travel distances from the slits. The electric field at P can then be written for the light from each of the two slits, using $\omega = 2\pi f$, as

$$E_1 = E_{10} \sin \omega t$$
$$E_2 = E_{20} \sin(\omega t + \delta) \qquad \textbf{(35–3)}$$

where E_{10} and E_{20} are their respective amplitudes and δ is the phase difference. The value of δ depends on the angle θ, so let us now determine δ as a function of θ.

At the center of the screen (point 0), $\delta = 0$. If the difference in path length from P to S_1 and S_2 is $d \sin \theta = \lambda/2$, the two waves are exactly out of phase so $\delta = \pi$ (or 180°). If $d \sin \theta = \lambda$, the two waves differ in phase by $\delta = 2\pi$. In general, then, δ is related to θ by

$$\frac{\delta}{2\pi} = \frac{d \sin \theta}{\lambda}$$

or

$$\delta = \frac{2\pi}{\lambda} d \sin \theta. \qquad \textbf{(35–4)}$$

To determine $E_\theta = E_1 + E_2$, we add the two scalars E_1 and E_2 which are sine functions differing by the phase δ. One way to determine the sum of E_1 and E_2 is to use a **phasor diagram**. (We used this technique before, in Chapter 31.) As shown in

Fig. 35–13, we draw an arrow of length E_{10} to represent the amplitude of E_1 (Eq. 35–3); and the arrow of length E_{20}, which we draw to make a fixed angle δ with E_{10}, represents the amplitude of E_2. When the diagram rotates at angular frequency ω about the origin, the projections of E_{10} and E_{20} on the vertical axis represent E_1 and E_2 as a function of time (see Eq. 35–3). We let $E_{\theta 0}$ be the "vector" sum[†] of E_{10} and E_{20}; it is the amplitude of the sum $E_\theta = E_1 + E_2$, and the projection of $E_{\theta 0}$ on the vertical axis is just E_θ. If the two slits provide equal illumination, so that $E_{10} = E_{20} = E_0$, then from Fig. 35–13 the angle $\phi = \delta/2$, and we can write

$$E_\theta = E_{\theta 0} \sin\left(\omega t + \frac{\delta}{2}\right). \qquad (35\text{–}5a)$$

From Fig. 35–13 we can also see that

$$E_{\theta 0} = 2E_0 \cos\phi = 2E_0 \cos\frac{\delta}{2}. \qquad (35\text{–}5b)$$

Combining Eqs. 35–5a and b, we obtain

$$E_\theta = 2E_0 \cos\frac{\delta}{2} \sin\left(\omega t + \frac{\delta}{2}\right), \qquad (35\text{–}5c)$$

where δ is given by Eq. 35–4. We can also derive Eq. 35–5c, when $E_{10} = E_{20}$, using the trigonometric identity

$$\sin A + \sin B = 2\sin\left(\frac{A+B}{2}\right)\cos\left(\frac{A-B}{2}\right).$$

That is, we add the two Eqs. 35–3 and obtain Eq. 35–5c:

$$E_\theta = E_1 + E_2 = 2E_0 \sin\left(\omega t + \frac{\delta}{2}\right)\cos\frac{\delta}{2}.$$

We are not really interested in E_θ as a function of time, since for visible light the frequency $(10^{14}$ to $10^{15}\,\text{Hz})$ is much too high to be noticeable. We are interested in the average intensity, which is proportional to the amplitude squared, $E_{\theta 0}^2$. We now drop the word "average," and we let $I_\theta (I_\theta \propto E_\theta^2)$ be the intensity at any point P at an angle θ to the horizontal. We let I_0 be the intensity at point 0, the center of the screen, where $\theta = \delta = 0$, so $I_0 \propto (E_{10} + E_{20})^2 = (2E_0)^2$. Then the ratio I_θ/I_0 is equal to the ratio of the squares of the electric-field amplitudes at these two points, so

$$\frac{I_\theta}{I_0} = \frac{E_{\theta 0}^2}{(2E_0)^2} = \cos^2\frac{\delta}{2}$$

where we used Eq. 35–5b. Thus the intensity I_θ at any point is related to the maximum intensity at the center of the screen by

$$I_\theta = I_0 \cos^2\frac{\delta}{2}$$
$$= I_0 \cos^2\left(\frac{\pi d \sin\theta}{\lambda}\right) \qquad (35\text{–}6)$$

where δ was given by Eq. 35–4. This is the relation we sought. From Eq. 35–6 we see that maxima occur where $\cos\delta/2 = \pm 1$, which corresponds to $\delta = 0, 2\pi, 4\pi, \cdots$; from Eq. 35–4, δ has these values when

$$d\sin\theta = m\lambda, \qquad m = 0, 1, 2, \cdots.$$

Minima occur where $\delta = \pi, 3\pi, 5\pi, \ldots$, which corresponds to

$$d\sin\theta = (m + \tfrac{1}{2})\lambda, \qquad m = 0, 1, 2, \cdots.$$

These are the same results we obtained in Section 35–3. But now we know not only the position of maxima and minima, but from Eq. 35–6 we can determine the intensity at all points.

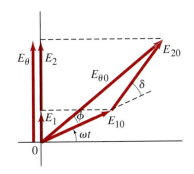

FIGURE 35–13 Phasor diagram for double-slit interference pattern.

[†]We are not adding the actual electric field vectors; instead we are adding the "phasors" in order to get the amplitude, taking into account the phase difference of the two waves.

In the usual situation where the distance L to the screen from the slits is large compared to the slit separation d ($L \gg d$), if we consider only points P whose distance y from the center (point 0) is small compared to L ($y \ll L$)—see Fig. 35–12—then

$$\sin \theta = \frac{y}{L}.$$

From this it follows (see Eq. 35–4) that

$$\delta = \frac{2\pi}{\lambda} \frac{d}{L} y.$$

Equation 35–6 then becomes

$$I_\theta = I_0 \left[\cos \left(\frac{\pi d}{\lambda L} y \right) \right]^2. \qquad\qquad [y \ll L, d \ll L] \quad \textbf{(35–7)}$$

The intensity I_θ as a function of the phase difference δ is plotted in Fig. 35–14. In the approximation of Eq. 35–7, the horizontal axis could as well be y, the position on the screen.

The intensity pattern expressed in Eqs. 35–6 and 35–7, and plotted in Fig. 35–14, shows a series of maxima of equal height, and is based on the assumption that each slit (alone) would illuminate the screen uniformly. This is never quite true, as we shall see when we discuss diffraction in the next chapter. We will see that the center maximum is strongest and each succeeding maximum to each side is less strong.

FIGURE 35–14 Intensity I as a function of phase difference δ and position on screen y (assuming $y \ll L$).

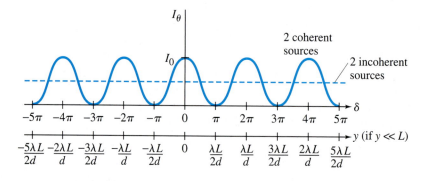

FIGURE 35–15 Example 35–5. The two dots represent the antennas.

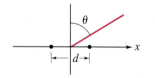

EXAMPLE 35–5 Radar antenna intensity. Two radio antennas are located close to each other as shown in Fig. 35–15, separated by a distance d. The antennas radiate in phase with each other, emitting waves of intensity I_0 at wavelength λ. (a) Calculate the net intensity as a function of θ for points very far from the antennas. (b) For $d = \lambda$, determine I and find in which directions I is a maximum and a minimum. (c) Repeat part b when $d = \lambda/2$.

SOLUTION (a) This setup is similar to Young's double-slit experiment. Points of constructive and destructive interference are still given by Eqs. 35–2a and b, and the net intensity as a function of position is given by Eq. 35–6 or 35–7. (b) We let $d = \lambda$ in Eq. 35–6, and find for the intensity,

$$I = I_0 \cos^2 (\pi \sin \theta).$$

I is a maximum, equal to I_0, when $\sin \theta = 0$, 1 or -1, meaning $\theta = 0$, 90°, 180° and 270°. I is zero when $\sin \theta = \frac{1}{2}$ and $-\frac{1}{2}$, for which $\theta = 30°$, 150°, 210° and 330°. (c) For $d = \lambda/2$ I is maximized for $\theta = 0$ and 180° and minimized for 90° and 270°.

Antenna arrays are important when a directional signal needs to be trasmitted such as in radar. We saw in Example 35–5 b and c that the intensity is maximized along particular directions and falls off slowly to zero at other angles. A narrow beam would be useful to concentrate radar energy on a target, and can be achieved using an array of more than two antennas. If the phases of the antennas are varied relative to each other, then the radiation pattern will shift direction, without having to physically rearrange the antennas. Such an antenna array is called an *electronically steered phase array*.

* Intensity for Incoherent Sources

Let us consider, briefly, the intensity pattern if the two slits of Fig. 35–12 were replaced by two *incoherent* sources of equal strength $\left(E_{10} = E_{20} = E_0\right)$. There would be no interference pattern, but rather a uniform illumination of the screen. The phase difference δ between the two waves varies randomly at any point P. Thus we must use the time average over $\cos^2 \delta/2$ in Eq. 35–6, which is $\frac{1}{2}$. The intensity, I_{inc}, of the two incoherent sources, will be

$$I_{inc} = \tfrac{1}{2} I_{coh}$$

where I_{coh} is the intensity due to two coherent sources at a maximum $\left(I_{coh}\right.$ is the I_0 of Eq. 35–6). This result, which is shown as the dashed line in Fig. 35–14, is confirmed by experiment and can be obtained in another way: when the sources are coherent, we add their wave amplitudes and then square to get the intensity

$$I_{coh} \propto \left(E_{10} + E_{20}\right)^2.$$

But if the sources are incoherent, each wave produces an intensity unrelated to the other, and the two intensities add up:

$$I_{inc} \propto E_{10}^2 + E_{20}^2.$$

That is, we square the amplitudes and *then* add them. If $E_{10} = E_{20} = E_0$, then the two relations above become

$$I_{coh} \propto \left(2E_0\right)^2 = 4E_0^2$$

$$I_{inc} \propto E_0^2 + E_0^2 = 2E_0^2$$

so

$$I_{inc} = \tfrac{1}{2} I_{coh}$$

which is our result above.

35–6 | Interference in Thin Films

Interference of light gives rise to many everyday phenomena such as the bright colors reflected from soap bubbles and from thin oil or gasoline films on water, Fig. 35–16. In these and other cases, the colors are a result of constructive interference between light reflected from the two surfaces of the thin film.

FIGURE 35–16 Thin film interference patterns seen in (a) soap bubbles, (b) thin films of soapy water, and (c) a thin layer of gasoline on water.

(a)

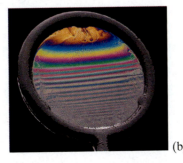
(b)

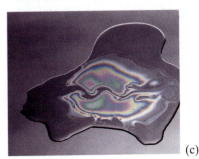
(c)

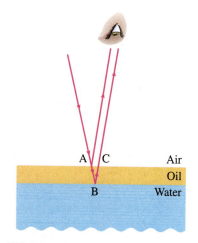

FIGURE 35-17 Light reflected from the upper and lower surfaces of a thin film of oil lying on water. Analysis assumes the light strikes the surface perpendicularly, but is shown here at a slight angle for clarity.

Newton's rings

To see how this happens, consider a smooth surface of water on top of which is a thin uniform layer of another substance, say an oil whose index of refraction is less than that of water (we'll see why we assume this in a moment); see Fig. 35-17. Assume for the moment that the incident light is of a single wavelength. Part of the incident light is reflected at A on the top surface, and part of the light transmitted is reflected at B on the lower surface. The part reflected at the lower surface must travel the extra distance ABC. If this *path difference* ABC is equal to one or a whole number of wavelengths in the film (λ_n), the two waves will reach the eye in phase and interfere constructively. Hence the region AC on the surface film will appear bright. But if ABC equals $\frac{1}{2}\lambda_n, \frac{3}{2}\lambda_n$, and so on, the two waves will be exactly out of phase, destructive interference occurs, and the area AC on the film will be dark. The wavelength λ_n is *the wavelength in the film*: $\lambda_n = \lambda/n$ where n is the index of refraction in the film and λ is the wavelength in vacuum.

When white light falls on such a film, the path difference ABC will equal λ_n (or $m\lambda_n$, with $m =$ an integer) for only one wavelength at a given viewing angle. The color corresponding to λ (λ in air) will be seen as very bright. For light viewed at a slightly different angle, the path difference ABC will be longer or shorter and a different color will undergo constructive interference. Thus, for an extended (non-point) source emitting white light, a series of bright colors will be seen next to one another. Variations in thickness of the film will also alter the path difference ABC and therefore affect the color of light that is most strongly reflected.

When a curved glass surface is placed in contact with a flat glass surface, Fig. 35-18, a series of concentric rings is seen when illuminated from above by monochromatic light. These are called **Newton's rings**[†] and they are due to interference between rays reflected by the top and bottom surfaces of the very thin *air gap* between the two pieces of glass. Because this gap (which is equivalent to a thin film) increases in width from the central contact point out to the edges, the extra path length for the lower ray (equal to BCD) varies; where it equals $0, \frac{1}{2}\lambda, \lambda, \frac{3}{2}\lambda, 2\lambda$, and so on, it corresponds to constructive and destructive interference; and this gives rise to the series of bright and dark lines seen in Fig. 35-18b.

The point of contact of the two glass surfaces (A in Fig. 35-18a) is dark in Fig. 35-18b. Since the path difference is zero here, we expect the rays reflected from each surface to be in phase and so this central point ought to be bright. But it is dark, which tells us the two rays must be completely out of phase. This can happen only because one of the waves undergoes a change in phase of 180° upon reflection, corresponding to $\frac{1}{2}$ cycle. Indeed, this and other experiments reveal that *a beam of light reflected by a material whose index of refraction is greater than that of the material in which it is traveling, changes phase by 180° or $\frac{1}{2}$ cycle.*

[†]Although Newton gave an elaborate description of them, they had been first observed and described by his contemporary, Robert Hooke.

FIGURE 35-18 Newton's rings.

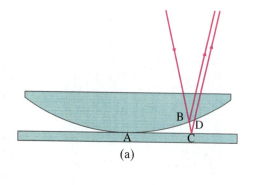

(a)

(b)

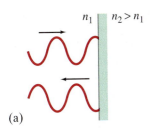 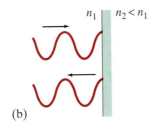

(a)   (b)

FIGURE 35–19 (a) Reflected ray changes phase by 180° or $\frac{1}{2}$ cycle if $n_2 > n_1$, but (b) does not if $n_2 < n_1$.

See Fig. 35–19. If the refractive index is less than that of the material in which the light is traveling, no phase change occurs.[†] [This can be derived from Maxwell's equations. It corresponds to the reflection of a wave traveling along a rope when it reaches the end; as we saw in Fig. 15–19, if the end is tied down, the wave changes phase and the pulse flips over, but if the end is free, no phase change occurs.] Thus the ray reflected by the curved surface above the air gap in Fig. 35–18a undergoes no change in phase. The ray reflected at the lower surface, where the beam in air strikes the glass, undergoes a 180° ($\frac{1}{2}\lambda$) phase change. Thus the two rays reflected at the point of contact A of the two glass surfaces (where the air gap approaches zero thickness) will be 180° ($\frac{1}{2}\lambda$) out of phase, and a dark spot occurs. Other dark bands will occur when the path difference BCD in Fig. 35–18a is equal to an integral number of wavelengths. Bright bands will occur when the path difference is $\frac{1}{2}\lambda, \frac{3}{2}\lambda$, and so on, because the phase change at one surface effectively adds another $\frac{1}{2}\lambda$.

FIGURE 35–20 (a) Light rays reflected from the upper and lower surfaces of a thin wedge of air interfere to produce bright and dark bands. (b) Pattern observed when glass plates are optically flat; (c) pattern when plates are not so flat. See Example 35–6.

EXAMPLE 35–6 **Thin film of air, wedge-shaped.** A very fine wire 7.35×10^{-3} mm in diameter is placed between two flat glass plates as in Fig. 35–20a. Light whose wavelength in air is 600 nm falls (and is viewed) perpendicular to the plates, and a series of bright and dark bands is seen, Fig. 35–20b. How many light and dark bands will there be in this case? Will the area next to the wire be bright or dark?

SOLUTION The thin film is the wedge of air between the two glass plates. Because of the phase change at the lower surface, there will be a dark band when the path difference is 0, λ, 2λ, 3λ, and so on. Since the light rays are perpendicular to the plates, the extra path length equals $2t$, where t is the thickness of the air gap at any point; so dark bands occur where

$$2t = m\lambda, \qquad m = 0, 1, 2, \cdots.$$

Bright bands occur when $2t = \left(m + \frac{1}{2}\right)\lambda$, where m is an integer. At the position of the wire, $t = 7.35 \times 10^{-6}$ m. At this point there will be $2t/\lambda = (2)(7.35 \times 10^{-6}\,\text{m})/(6.00 \times 10^{-7}\,\text{m}) = 24.5$ wavelengths. This is a "half integer," so the area next to the wire will be bright. There will be a total of 25 dark lines along the plates, corresponding to path lengths of $0\lambda, 1\lambda, 2\lambda, 3\lambda, \cdots, 24\lambda$, including the one at the point of contact A ($m = 0$). Between them, there will be 24 bright lines plus the one at the end, or 25. The bright and dark bands will be straight only if the glass plates are extremely flat. If they are not, the pattern is uneven, as in Fig. 35–20c. Thus we see a very precise way of testing a glass surface for flatness. Curved lens surfaces can be tested for precision similarly by placing the lens on a flat glass surface and observing Newton's rings (Fig. 35–18b) for perfect circularity.

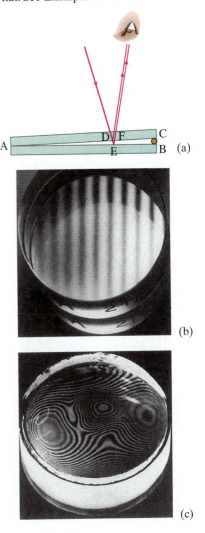

(a)

(b)

(c)

[†] Note that in Fig. 35–17, the light reflecting at both interfaces, air–oil and oil–water, underwent a phase change of $\frac{1}{2}\lambda$, since we assumed $n_{\text{water}} > n_{\text{oil}} > n_{\text{air}}$. Since the phase changes were equal, they didn't affect our analysis.

SECTION 35–6   Interference in Thin Films   **879**

If the wedge between the two glass plates of Example 35–6 is filled with some transparent substance other than air—say, water—the pattern shifts because the wavelength of the light changes. In a material where the index of refraction is n, the wavelength is $\lambda_n = \lambda/n$ where λ is the wavelength in vacuum (see Eq. 35–1). For instance, if the thin wedge of Example 35–6 were filled with water, $\lambda_n = 600\,\text{nm}/1.33 = 450\,\text{nm}$; instead of 25 dark lines, there would be 33.

When white light (rather than monochromatic light) is incident on the thin wedge of Figs. 35–18a or 35–20a, a colorful series of fringes is seen. This is because constructive interference occurs in the reflected light at different locations along the wedge for different wavelengths. Such a difference in thickness is part of the reason bright colors appear when light is reflected from a soap bubble or a thin layer of oil or gasoline on a puddle or lake (Fig. 35–16). Which wavelengths appear brightest also depends on the viewing angle, as we saw earlier.

➡ **PROBLEM SOLVING**

A formula is not enough: you must also check for phase changes at surfaces

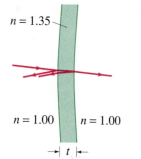

$n = 1.35$

$n = 1.00$ $n = 1.00$

t

FIGURE 35–21 Example 35–7.

➡ **PHYSICS APPLIED**

Lens coatings

EXAMPLE 35–7 **Thickness of soap bubble skin.** A soap bubble appears green ($\lambda = 540\,\text{nm}$) at the point on its front surface nearest the viewer. What is its minimum thickness? Assume $n = 1.35$.

SOLUTION The light is reflected perpendicularly from the point on a spherical surface nearest the viewer, Fig. 35–21. Therefore the path difference is $2t$, where t is the thickness of the soap film. Light reflected from the first (outer) surface undergoes a $\frac{1}{2}\lambda$ phase change (index of refraction of soap is greater than that of air), whereas that at the second (inner) surface does not. Therefore, green light is bright when the minimum path difference equals $\frac{1}{2}\lambda_n$. Thus, $2t = \lambda/2n$, so

$$t = \frac{\lambda}{4n} = \frac{(540\,\text{nm})}{(4)(1.35)} = 100\,\text{nm}.$$

This is the minimum thickness. The front surface would also appear green if $2t = 3\lambda/2n$, and, in general, if $2t = (2m + 1)\lambda/2n$ where m is an integer. Note that green is seen in air, so $\lambda = 540\,\text{nm}$ (not λ/n).

An important application of thin-film interference is in the coating of glass to make it "nonreflecting," particularly for lenses. A glass surface reflects about 4 percent of the light incident upon it. Good-quality cameras, microscopes, and other optical devices may contain six to ten thin lenses. Reflection from all these surfaces can reduce the light level considerably, and multiple reflections produce a background haze that reduces the quality of the image. By reducing reflection, transmission is increased. A very thin coating on the lens surfaces can reduce reflections considerably: the thickness of the film is chosen so that light (at least for one wavelength) reflecting from the front and rear surfaces of the film destructively interferes. The amount of reflection at a boundary depends on the difference in index of refraction between the two materials. Ideally, the coating material should have an index of refraction which is the geometric mean of those for air and glass, so that the amount of reflection at each surface is about equal. Then destructive interference can occur nearly completely for one particular wavelength depending on the thickness of the coating. Nearby wavelengths will at least partially destructively interfere, but it is clear that a single coating cannot eliminate reflections for all wavelengths. Nonetheless, a single coating can reduce total reflection from 4 percent to 1 percent of the incident light. Often the coating is designed to eliminate the center of the reflected spectrum (around 550 nm). The extremes of the spectrum—red and violet—will not be reduced as much. Since a mixture of red and violet produces purple, the light seen reflected from such coated lenses is purple (Fig. 35–22). Lenses containing two or three separate coatings can more effectively reduce a wider range of reflecting wavelengths.

FIGURE 35–22 A coated lens. Note color of light reflected from the front lens surface.

EXAMPLE 35–8 **Nonreflective coating.** What is the thickness of an optical coating of MgF_2, whose index of refraction is $n = 1.38$, which is designed to eliminate reflected light at wavelengths centered at 550 nm when incident normally on glass for which $n = 1.50$?

SOLUTION Figure 35–23 shows an incoming ray and two rays reflected from the front and rear surfaces of the coating on the lens. The rays are drawn not quite perpendicular to the lens so we can see each of them. To eliminate reflection, we want the reflected rays 1 and 2 to be $\frac{1}{2}$ wavelength out of phase with each other so they destructively interfere. Rays 1 and 2 *both* undergo a change of phase by $\frac{1}{2}\lambda$ when they reflect, respectively, from the front and rear surfaces of the coating. Therefore, we want the extra distance traveled by ray 2 $(= 2t)$ to be a half integral number of wavelengths. That is, $2t = (m + \frac{1}{2})\lambda_n$, where m is an integer and λ_n is the wavelength inside the MgF_2 coating. The minimum thickness $(m = 0)$ is usually chosen because destructive interference will then occur over the widest angle. Then

$$t = \frac{\lambda_n}{4} = \frac{\lambda}{4n} = \frac{(550\text{ nm})}{(4)(1.38)} = 99.6\text{ nm.}$$

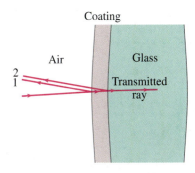

FIGURE 35–23 Incident ray of light is partially reflected at the front surface of a lens coating (ray 1) and again partially reflected at the rear surface of the coating (ray 2), with most of the energy passing as the transmitted ray into the glass.

* 35–7 Michelson Interferometer

A useful instrument involving wave interference is the **Michelson interferometer** (Fig. 35–24),[†] invented by the American Albert A. Michelson (Section 33–2). Monochromatic light from a single point on an extended source is shown striking a half-silvered mirror M_S. This **beam splitter** mirror M_S has a thin layer of silver that reflects only half the light that hits it, so that half of the beam passes through to a fixed mirror M_2, where it is reflected back. The other half is reflected by M_S up to a mirror M_1 that is movable (by a fine-thread screw), where it is also reflected back. Upon its return, part of beam 1 passes through M_S and reaches the eye; and part of beam 2, on its return, is reflected by M_S into the eye. If the two path lengths are identical, the two coherent beams entering the eye constructively interfere and brightness will be seen. If the movable mirror is moved a distance $\lambda/4$, one beam will travel an extra distance equal to $\lambda/2$ (because it travels back and forth over the distance $\lambda/4$). In this case, the two beams will destructively interfere and darkness will be seen. As M_1 is moved farther, brightness will recur (when the path difference is λ), then darkness, and so on.

Very precise length measurements can be made with an interferometer. The motion of mirror M_1 by only $\frac{1}{4}\lambda$ produces a clear difference between brightness and darkness. For $\lambda = 400$ nm, this means a precision of 100 nm or 10^{-4} mm! If mirror M_1 is tilted very slightly, the bright or dark spots are seen instead as a series of bright and dark lines or "fringes." By counting the number of fringes, or fractions thereof, extremely precise length measurements can be made.

Michelson saw that the interferometer could be used to determine the length of the standard meter in terms of the wavelength of a particular light. In 1960, that standard was chosen to be a particular orange line in the spectrum of krypton-86 (krypton atoms with atomic mass 86). Careful repeated measurements of the old standard meter (the distance between two marks on a platinum–iridium bar kept in Paris) were made to establish 1 meter as being 1,650,763.73 wavelengths of this light, which was *defined* to be the meter. In 1983, the meter was redefined in terms of the speed of light (Section 1–4).

FIGURE 35–24 Michelson interferometer.

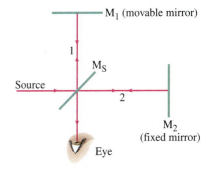

[†]There are other types of interferometer, but Michelson's is the best known.

Luminous Intensity

Although the *intensity* of light, as for any electromagnetic wave, is measured by the Poynting vector in W/m^2, and the total power output of a source can be measured in watts (the *radiant flux*), these are not adequate for measuring the visual sensation we call brightness. The reason is that we are really interested here only in the visible spectrum, whereas the two quantities just mentioned would take into account all wavelengths present. It is also important to take into account the eye's sensitivity to different wavelengths—the eye is most sensitive in the central, 550-nm (yellow), portion of the spectrum; so a yellow source would appear brighter than a red or blue source of the same power output.

These factors are taken into account in the quantity **luminous flux**, F_l, whose unit is the **lumen** (lm). One lumen is defined experimentally as the brightness of $\frac{1}{60}$ cm^2 of platinum surface at its melting temperature (1770°C). It is equivalent to $\frac{1}{683}$ watts of 555-nm light.

Since the luminous flux from a source may not be uniform over all directions, we define the **luminous intensity** I_l as the luminous flux per unit solid angle (steradian). Its unit is the **candela** (cd) where $1\,\text{cd} = 1\,\text{lm/sr}$.

Candela (unit)

The **illuminance**, E_l, is the luminous flux incident on a surface per unit area of the surface: $E_l = F_l/A$. Its unit is the lumen per square meter (lm/m^2) and is a measure of the illumination falling on a surface.[†]

We will not go into this subject in any more detail. We have introduced it here for completeness since the *luminous intensity*, as measured by the candela in SI units, is one of the seven basic quantities in the SI. (See Section 1–4 and Table 1–5.) The other six (and their units) we have already met: length (m), time (s), mass (kg), electric current (A), temperature (K), and amount of substance (mol).

> **EXAMPLE 35–9** **Light bulb illuminance.** The brightness of a particular type of 100-W light bulb is rated at 1700 lm. Determine (*a*) the luminous intensity and (*b*) the illuminance at a distance of 2.0 m. Assume the light output is uniform in all directions.
>
> **SOLUTION** (*a*) A full sphere corresponds to 4π sr. Hence, $I_l = 1700\,\text{lm}/4\pi\,\text{sr} = 135\,\text{cd}$. It does not depend on distance. (*b*) At $d = 2.0\,\text{m}$ from the source, the luminous flux per unit area is
>
> $$E_l = \frac{F_l}{4\pi d^2} = \frac{1700\,\text{lm}}{(4\pi)(2.0\,\text{m})^2} = 34\,\text{lm/m}^2.$$
>
> The illuminance decreases as the square of the distance.

PROBLEM SOLVING — Interference

1. Interference effects depend on the simultaneous arrival of two or more waves at the same point in space.

2. Constructive interference occurs when waves arrive in phase with each other: a crest of one wave arrives at the same time as a crest of the other wave. The amplitudes of the waves then add to form a larger amplitude. Constructive interference also occurs when the phase difference is exactly one full wavelength or any integer multiple of a full wavelength: $1\lambda, 2\lambda, 3\lambda, \cdots$.

3. Destructive interference occurs when a crest of one wave arrives at the same time as a trough of the other wave. The amplitudes add, but they are of opposite sign, so the total amplitude is reduced to zero if the two amplitudes are equal. Destructive interference occurs whenever the phase difference is a half-integral number of wavelengths. Thus, the total amplitude will be zero if two identical waves arrive one-half wavelength out of phase, or $(m + \frac{1}{2})\lambda$ out of phase where m is an integer.

4. For thin-film interference, don't forget an extra one-half wavelength phase shift that occurs when light reflects from an optically more dense medium (going from a medium of lesser toward greater index of refraction).

[†]The British unit is the foot-candle, or lumen per square foot.

Summary

The wave theory of light is strongly supported by the observations that light exhibits **interference** and **diffraction**. Wave theory also explains the refraction of light and the fact that light travels more slowly in transparent solids and liquids than it does in air. The wavelength of light in a medium with index of refraction n is

$$\lambda_n = \frac{\lambda}{n},$$

where λ is the wavelength in vacuum. The frequency is not changed.

Young's double-slit experiment clearly demonstrated the interference of light. The observed bright spots of the interference pattern were explained as constructive interference between the beams coming through the two slits, where the beams differ in path length by an integral number of wavelengths. The dark areas in between are due to destructive interference when the path lengths differ by $\frac{1}{2}\lambda, \frac{3}{2}\lambda$, and so on. The angles θ at which **constructive interference** occurs are given by

$$\sin\theta = m\frac{\lambda}{d},$$

where λ is the wavelength of the light, d the separation of the slits, and m an integer $(0, 1, 2, \cdots)$. **Destructive interference** occurs at angles θ given by

$$\sin\theta = \left(m + \tfrac{1}{2}\right)\frac{\lambda}{d}$$

where m is an integer $(0, 1, 2, \cdots)$.

The light intensity I_θ at any point in a double-slit interference pattern can be calculated using a phasor diagram, which predicts that

$$I_\theta = I_0\cos^2\frac{\delta}{2}$$

where I_0 is the intensity at $\theta = 0$ and the phase angle δ is

$$\delta = \frac{2\pi d}{\lambda}\sin\theta.$$

Two sources of light are perfectly **coherent** if the waves leaving them are sinusoidal, of the same single frequency, and maintain the same phase relationship at all times. If the light waves from the two sources have a random phase with respect to each other over time (as for two incandescent light bulbs) the two sources are **incoherent**.

Light reflected from the front and rear surfaces of a thin film of transparent material can interfere constructively or destructively, depending on the optical path difference. A phase change of 180° or $\frac{1}{2}\lambda$ occurs when the light reflects at a surface where the index of refraction increases. Such **thin-film interference** has many practical applications, such as lens coatings and using Newton's rings to check the uniformity of glass surfaces.

Questions

1. Does Huygens' principle apply to sound waves? To water waves?

2. What is the evidence that light carries energy?

3. Why is light sometimes described as rays and sometimes as waves?

4. We can hear sounds around corners but we cannot see around corners, yet both sound and light are waves. Explain the difference.

5. Can the wavelength of light be determined from reflection or refraction measurements?

6. Monochromatic red light is incident on a double slit and the interference pattern is viewed on a screen some distance away. Explain how the fringe pattern would change if the red light source is replaced by a blue light source.

7. Suppose white light falls on the two slits of Fig. 35–7, but one slit is covered by a red filter (700 nm) and the other by a blue (450 nm) filter. Describe the pattern on the screen.

8. Compare a double-slit experiment for sound waves to that for light waves. Discuss the similarities and differences.

9. Two rays of light from the same source destructively interfere if their path lengths differ by how much?

10. If Young's double-slit experiment were submerged in water, how would the fringe pattern be changed?

11. Why doesn't the light from the two headlights of a distant car produce an interference pattern?

12. Why are interference fringes noticeable only for a *thin* film like a soap bubble and not for a thick piece of glass, say?

13. Why are Newton's rings (Fig. 35–18) closer together farther from the center?

14. Some coated lenses appear greenish yellow when seen by reflected light. What wavelengths do you suppose the coating is designed to eliminate completely?

15. A drop of oil on a pond appears bright at its edges where its thickness is much less than the wavelengths of visible light. What can you say about the index of refraction of the oil?

* 16. Describe how a Michelson interferometer could be used to measure the index of refraction of air.

Problems

Section 35–2

1. (II) Derive the law of reflection—namely, that the angle of incidence equals the angle of reflection from a flat surface—using Huygens' principle for waves.

Section 35–3

2. (I) Monochromatic light falling on two slits 0.016 mm apart produces the fifth-order fringe at a 9.8° angle. What is the wavelength of the light used?

3. (I) The third-order fringe of 610 nm light is observed at an angle of 18° when the light falls on two narrow slits. How far apart are the slits?

4. (II) Monochromatic light falls on two very narrow slits 0.048 mm apart. Successive fringes on a screen 5.00 m away are 6.5 cm apart near the center of the pattern. What is the wavelength and frequency of the light?

5. (II) A parallel beam of light from a He-Ne laser, with a wavelength 656 nm, falls on two very narrow slits 0.060 mm apart. How far apart are the fringes in the center of the pattern if the screen is 3.6 m away?

6. (II) Light of wavelength 680 nm falls on two slits and produces an interference pattern in which the fourth-order fringe is 38 mm from the central fringe on a screen 2.0 m away. What is the separation of the two slits?

7. (II) If 720-nm and 660-nm light passes through two slits 0.58 mm apart, how far apart are the second-order fringes for these two wavelengths on a screen 1.0 m away?

8. (II) Suppose a thin piece of glass were placed in front of the lower slit in Fig. 35–7 so that the two waves enter the slits 180° out of phase (Fig. 35–25). Describe in detail the interference pattern on the screen.

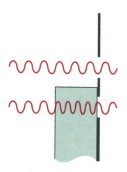

FIGURE 35–25
Problem 8.

9. (II) In a double-slit experiment it is found that blue light of wavelength 460 nm gives a second-order maximum at a certain location on the screen. What wavelength of visible light would have a minimum at the same location?

10. (II) Light of wavelength 480 nm in air falls on two slits 6.00×10^{-2} mm apart. The slits are immersed in water, as is a viewing screen 40.0 cm away. How far apart are the fringes on the screen?

11. (III) A very thin sheet of plastic ($n = 1.60$) covers one slit of a double-slit apparatus illuminated by 640-nm light. The center point on the screen, instead of being a maximum, is dark. What is the (minimum) thickness of the plastic?

Section 35–5

12. (I) If one slit in Fig. 35–12 is covered, by what factor does the intensity at the center of the screen change?

13. (II) Show that the angular full width at half maximum of the central peak in a double-slit interference pattern is given by $\Delta\theta = \lambda/2d$ if $\lambda \ll d$.

14. (II) Suppose that one slit of a double-slit apparatus is wider than the other so that the intensity of light passing through it is twice as great. Determine the intensity I as a function of position (θ) on the screen for coherent light.

15. (III) (a) Consider three equally spaced and equal-intensity coherent sources of light (such as adding a third slit to the two slits of Fig. 35–12). Use the phasor method to obtain the intensity as a function of the phase difference δ (Eq. 35–4). (b) Determine the positions of maxima and minima.

16. (III) Apply the phasor method to four parallel slits separated by equal distances d. Assume coherent light and determine the intensity as a function of position on the screen and find the positions of maxima and minima.

Section 35–6

17. (I) If a soap bubble is 120 nm thick, what color will appear at the center when illuminated normally by white light? Assume that $n = 1.34$.

18. (I) How far apart are the dark fringes in Example 35–6 if the glass plates are each 26.5 cm long?

19. (II) What is the minimum thickness (>0) of a soap film ($n = 1.34$) that would appear black if illuminated with 480-nm light? Assume there is air on both sides of the soap film.

20. (II) A lens appears greenish yellow ($\lambda = 570$ nm is strongest) when white light reflects from it. What minimum thickness of coating ($n = 1.28$) do you think is used on such a (glass) lens, and why?

21. (II) A total of 28 bright and 28 dark Newton's rings (not counting the dark spot at the center) are observed when 650-nm light falls normally on a planoconvex lens resting on a flat glass surface (Fig. 35–18). How much thicker is the center than the edges?

22. (II) A fine metal foil separates one end of two pieces of optically flat glass, as in Fig. 35–20. When light of wavelength 670 nm is incident normally, 25 dark lines are observed (with one at each end). How thick is the foil?

23. (II) How thick (minimum) should the air layer be between two flat glass surfaces if the glass is to appear bright when 480-nm light is incident normally? What if the glass is to appear dark?

24. (II) A thin film of alcohol ($n = 1.36$) lies on a flat glass plate ($n = 1.51$). When monochromatic light, whose wavelength can be changed, is incident normally, the reflected light is a minimum for $\lambda = 512$ nm and a maximum for $\lambda = 640$ nm. What is the thickness of the film?

25. (II) When a Newton's ring apparatus (Fig. 35–18) is immersed in a liquid, the diameter of the eighth dark ring decreases from 2.92 cm to 2.60 cm. What is the refractive index of the liquid?

26. (II) A planoconvex lucite lens 3.4 cm in diameter is placed on a flat piece of glass as in Fig. 35–18. When 580-nm light is incident normally, 48 bright rings are observed, the last one right at the edge. What is the radius of curvature of the lens surface, and the focal length of the lens?

27. (II) Show that the radius r of the mth dark Newton's ring, as viewed from directly above (Fig. 35–18), is given by $r = \sqrt{m\lambda R}$ where R is the radius of curvature of the curved glass surface and λ is the wavelength of light used. Assume that the thickness of the air gap is much less than R at all points and that $r \ll R$.

28. (III) Use the result of Problem 27 to show that the distance between adjacent dark Newton's rings is

$$\Delta r \approx \sqrt{\frac{\lambda R}{4\,m}}$$

for the mth ring, assuming $m \gg 1$.

29. (III) A single optical coating reduces reflection to zero for $\lambda = 550$ nm. By what factor is the intensity reduced by the coating for $\lambda = 450$ nm and $\lambda = 650$ nm as compared to no coating? Assume normal incidence.

* Section 35–7

* 30. (II) What is the wavelength of the light entering an interferometer if 344 bright fringes are counted when the movable mirror moves 0.125 mm?

* 31. (II) How far must the mirror M_1 in a Michelson interferometer be moved if 750 fringes of 589-nm light are to pass by a reference line?

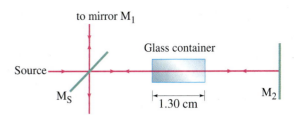

FIGURE 35–26 Problem 32.

* 32. (III) One of the beams of an interferometer (Fig. 35–26) passes through a small glass container containing a cavity 1.30 cm deep. When a gas is allowed to slowly fill the container, a total of 186 dark fringes are counted to move past a reference line. The light used has a wavelength of 610 nm. Calculate the index of refraction of the gas at its final density, assuming that the interferometer is in vacuum.

* 33. (III) The yellow sodium D lines have wavelengths of 589.0 and 589.6 nm. When they are used to illuminate a Michelson interferometer, it is noted that the interference fringes disappear and reappear periodically as the mirror M_1 is moved. Why does this happen? How far must the mirror move between one disappearance and the next?

* Section 35–8

* 34. (I) The illuminance of direct sunlight on Earth is about 10^5 lm/m^2. Estimate the luminous flux and luminous intensity of the Sun.

* 35. (II) The *luminous efficiency* of a light bulb is the ratio of luminous flux to electric power input. (*a*) What is the luminous efficiency of a 100-W, 1700-lm bulb? (*b*) How many 40-W, 60-lm/W fluorescent lamps would be needed to provide an illuminance of 250 lm/m^2 on a factory floor of area 25 m × 30 m? Assume the lights are 10 m above the floor and that half their flux reaches the floor.

General Problems

36. Light of wavelength λ strikes a screen containing two slits a distance d apart at an angle θ_i to the normal. Determine the angle θ_m at which the mth-order maximum occurs.

37. Television and radio waves can reflect from nearby mountains or from airplanes, and the reflections can interfere with the direct signal from the station. (*a*) Determine what kind of interference will occur when 75-MHz television signals arrive at a receiver directly from a distant station, and are reflected from an airplane 118 m directly above the receiver. (Assume $\frac{1}{2}\lambda$ change in phase of the signal upon reflection.) (*b*) What kind of interference will occur if the plane is 22 m closer to the receiver?

38. A radio station operating at 102.1 MHz broadcasts from two identical antennae at the same elevation but separated by a 7.0-m horizontal distance d, Fig. 35–27. A maximum signal is found along the midline, perpendicular to d at its midpoint and extending horizontally in both directions. If the midline is taken as 0°, at what other angle(s) θ is a maximum signal detected? A minimum signal? Assume all measurements are made much farther than 7 m from the antenna towers.

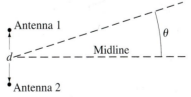

FIGURE 35–27
Problem 38.

39. Light of wavelength 690 nm passes through two narrow slits 0.60 mm apart. The screen is 1.50 m away. A second source of unknown wavelength produces its second-order fringe 1.13 mm closer to the central maximum than the 690-nm light. What is the wavelength of the unknown light?

40. What is the index of refraction of a clear material if a minimum of 150 nm thickness of it, when laid on glass, is needed to reduce reflection to nearly zero when light of 600 nm is incident normally upon it? Do you have a choice for an answer?

41. Monochromatic light of variable wavelength is incident normally on a thin sheet of plastic film in air. The reflected light is a minimum only for $\lambda = 510$ nm and $\lambda = 680$ nm in the visible spectrum. What is the thickness of the film ($n = 1.58$)?

42. In a water tank experiment, water waves are generated with their crests 3.5 cm apart and parallel. They pass through two openings 6.0 cm apart in a long wooden board. If the end of the tank is 2.0 m beyond the boards, where would you stand, relative to the "straight-through" direction, so that you received little or no wave action?

43. Compare the minimum thickness needed for an antireflective coating ($n = 1.38$) applied to a glass lens in order to eliminate (*a*) blue (450 nm), or (*b*) red (700 nm) reflections for light at normal incidence.

44. Suppose you viewed the light *transmitted* through a thin film on a flat piece of glass. Draw a diagram, similar to Fig. 35–17 or 35–23, and describe the conditions required for maxima and minima. Consider all possible values of index of refraction. Discuss the relative size of the minima compared to the maxima and to zero.

45. Stealth aircraft are designed to not reflect radar, whose wavelength is typically 2 cm, by using an antireflecting coating. Ignoring any change in wavelength in the coating, estimate its thickness.

46. Very highly reflective mirrors for a particular wavelength can be made by alternating many layers of *transparent* materials of indices of refraction n_1 and $n_2 (1 < n_1 < n_2)$. What should the minimum thicknesses d_1 and d_2 of Fig. 35–28 be, in terms of the incident wavelength λ, to maximize reflection?

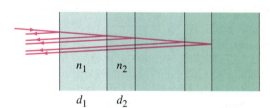

FIGURE 35–28 Problem 46.

47. When a charged particle, such as a proton, travels through a transparent medium at a speed v_p faster than the speed of light in that medium ($v = c/n$), it emits electromagnetic radiation (light) called **Čerenkov radiation**. It is the electromagnetic equivalent of a shock wave (see Section 16–8), and is confined to a particular angle that depends on v_p and v. Determine this angle for a proton traveling 2.21×10^8 m/s in a plastic whose index of refraction is 1.52.

48. What is the minimum (non-zero) thickness for the air layer between two flat glass surfaces if the glass is to appear dark when 640-nm light is incident normally? What if the glass is to appear bright?

49. *Lloyd's mirror* provides one way of obtaining a double-slit interference pattern from a single source so the light is coherent; as shown in Fig. 35–29, the light that reflects from the plane mirror appears to come from the virtual image of the slit. Describe in detail the interference pattern on the screen.

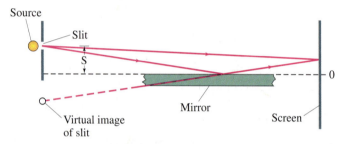

FIGURE 35–29 Problem 49.

50. Consider the antenna array of Example 35–5, Fig. 35–15. Let $d = \lambda/2$, and suppose that the two antennas are now 180° out of phase with each other. Find the directions for constructive and destructive interference, and compare with the case when the sources are in phase. (These results illustrate the basis for directional antennas.)

* **51.** Suppose the mirrors in a Michelson interferometer are perfectly aligned and the path lengths to mirrors M_1 and M_2 are identical. With these initial conditions, an observer sees a bright maximum at the center of the viewing area. Now one of the mirrors is moved a distance x. Determine a formula for the intensity at the center of the viewing area as a function of x, the distance the movable mirror is moved from the initial conditions.

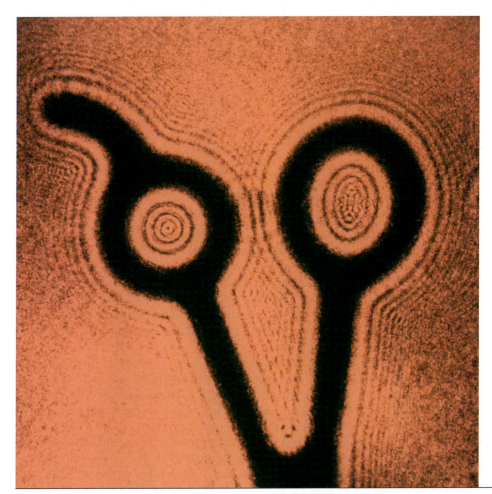

Parallel coherent light from a laser, which acts as nearly a point source, illuminates these shears. Instead of a clean shadow, there is a dramatic diffraction pattern, which is a strong confirmation of the wave theory of light. Diffraction patterns are washed out when typical extended sources of light are used, and hence are not seen, although a careful examination of shadows will reveal fuzziness. We will examine diffraction by a single slit, and how it affects the double-slit pattern, as well as diffraction gratings and x-rays. We will see how diffraction affects the resolution of optical instruments, and that the ultimate resolution can never be greater than the wavelength of the radiation used. Finally we study the polarization of light.

CHAPTER **36**

Diffraction and Polarization

Young's double-slit experiment put the wave theory of light on a firm footing. But full acceptance came only with studies on diffraction more than a decade later.

We have already discussed diffraction briefly with regard to water waves (Section 15–11) as well as for light (Section 35–1) and we have seen that it refers to the spreading or bending of waves around edges. Now we look at diffraction more closely, including its important practical effects of limiting the amount of detail, or *resolution*, that can be obtained with any optical instrument including telescopes, cameras, and the eye.

A part of the history of the wave theory of light belongs to Augustin Fresnel (1788–1827) who in 1819 presented to the French Academy a wave theory of light that predicted and explained interference and diffraction effects. Almost immediately Siméon Poisson (1781–1840) pointed out a counter-intuitive inference: according to Fresnel's wave theory, if light from a point source were to fall on a solid disk, part of the incident light would be diffracted around the edges and would constructively interfere at the center of the shadow (Fig. 36–1). That prediction seemed very unlikely. But when the experiment was actually carried out by François Arago, the bright spot was seen at the very center of the shadow (Fig. 36–2a). This was strong evidence for the wave theory.

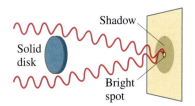

FIGURE 36–1 If light is a wave, a bright spot will appear at the center of the shadow of a solid disk illuminated by a point source of monochromatic light.

Figure 36–2a is a photograph of the shadow cast by a coin using a (nearly) point source of light, a laser in this case. The bright spot is clearly present at the center. Note that there also are bright and dark fringes beyond the shadow. These resemble the interference fringes of a double slit. Indeed, they are due to interference of waves diffracted around the disk, and the whole is referred to as a **diffraction pattern**. A diffraction pattern exists around any sharp-edged object illuminated by a point source, as shown in Fig. 36–2b and c. We are not always aware of diffraction because most sources of light in everyday life are not points, so light from different parts of the source washes out the pattern.

36–1 **Diffraction by a Single Slit**

To see how a diffraction pattern arises, we will analyze the important case of monochromatic light passing through a narrow slit. We will assume that parallel rays (plane waves) of light fall on the slit of width a, and pass through to a viewing screen very far away. If the viewing screen is not far away, lenses can be used to make the rays parallel.[†] As we know from studying water waves and from Huygens' principle, the waves passing through the slit spread out in all directions. We will now examine how the waves passing through different parts of the slit interfere with each other.

Parallel rays of monochromatic light pass through the narrow slit as shown in Fig. 36–3a. The light falls on a screen which is assumed to be very far away, so the rays heading for any point are essentially parallel. First we consider rays that pass straight through as in Fig. 36–3a. They are all in phase, so there will be a central bright spot on the screen. In Fig. 36–3b, we consider rays moving at an angle θ such that the ray from the top of the slit travels exactly one wavelength farther than the ray from the bottom edge. The ray passing through the very center of the slit will travel one-half wavelength farther than the ray at the bottom of the slit. These two rays will be exactly out of phase with one another and so will destructively interfere. Similarly, a ray slightly above the bottom one will cancel a ray that is the same distance above the central one. Indeed, each ray passing through the lower half of the slit will cancel with a corresponding ray passing through the upper half. Thus, all the

---

[†] Such a diffraction pattern, involving parallel rays, is called *Fraunhofer diffraction*. If the screen is close and no lenses are used, it is called *Fresnel diffraction*. The analysis in the latter case is rather involved, so we consider only the limiting case of Fraunhofer diffraction.

FIGURE 36–2 Diffraction pattern of (a) a circular disk (a coin), (b) a razor blade, (c) a single slit, each illuminated by a (nearly) point source of monochromatic light.

(a) (b) (c)

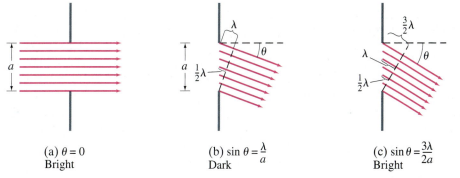

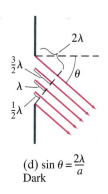

(a) $\theta = 0$
Bright

(b) $\sin\theta = \dfrac{\lambda}{a}$
Dark

(c) $\sin\theta = \dfrac{3\lambda}{2a}$
Bright

(d) $\sin\theta = \dfrac{2\lambda}{a}$
Dark

FIGURE 36–3 Analysis of diffraction pattern formed by light passing through a narrow slit.

rays destructively interfere in pairs, and so no light will reach the viewing screen at this angle. The angle θ at which this takes place can be seen from the diagram to occur when $\lambda = a\sin\theta$, so

$$\sin\theta = \frac{\lambda}{a}. \qquad\qquad \text{[first minimum]} \quad (36\text{–}1)$$

The light intensity is a maximum at $\theta = 0°$ and decreases to a minimum (intensity = zero) at the angle θ given by Eq. 36–1.

Now consider a larger angle θ such that the top ray travels $\frac{3}{2}\lambda$ farther than the bottom ray, as in Fig. 36–3c. In this case, the rays from the bottom third of the slit will cancel in pairs with those in the middle third because they will be $\lambda/2$ out of phase. However, light from the top third of the slit will still reach the screen, so there will be a bright spot centered near $\sin\theta \approx 3\lambda/2a$, but it will not be nearly as bright as the central spot at $\theta = 0°$. For an even larger angle θ such that the top ray travels 2λ farther than the bottom ray, Fig. 36–3d, rays from the bottom quarter of the slit will cancel with those in the quarter just above it because the path lengths differ by $\lambda/2$. And the rays through the quarter of the slit just above center will cancel with those through the top quarter. At this angle there will again be a minimum of zero intensity in the diffraction pattern. A plot of the intensity as a function of angle is shown in Fig. 36–4. This corresponds well with the photo of Fig. 36–2c. Notice that minima (zero intensity) occur at

$$a\sin\theta = m\lambda, \qquad m = 1, 2, 3, \cdots, \qquad \text{[minima]} \quad (36\text{–}2)$$

but *not* at $m = 0$ where there is the strongest maximum. Between the minima, smaller intensity maxima occur at approximately (but not exactly) $m \approx \frac{3}{2}, \frac{5}{2}, \cdots$. [Note that the *minima* for a diffraction pattern, Eq. 36–2, satisfy a criterion very similar to that for the *maxima* (bright spots) for double-slit interference, Eq. 35–2a.]

> **EXAMPLE 36–1** **Single-slit diffraction maximum.** Light of wavelength 750 nm passes through a slit 1.0×10^{-3} mm wide. How wide is the central maximum (a) in degrees, and (b) in centimeters, on a screen 20 cm away?
>
> **SOLUTION** (a) The first minimum occurs at
>
> $$\sin\theta = \frac{\lambda}{a} = \frac{7.5 \times 10^{-7}\,\text{m}}{1.0 \times 10^{-6}\,\text{m}} = 0.75.$$
>
> So $\theta = 49°$. This is the angle between the center and the first minimum, Fig. 36–5. The angle subtended by the whole central maximum, between the minima above and below the center, is twice this, or $98°$.
> (b) The width of the central maximum is $2x$, where $\tan\theta = x/20\,\text{cm}$. So $2x = 2(20\,\text{cm})(\tan 49°) = 46\,\text{cm}$. A large width of the screen will be illuminated, but it will not normally be very bright since the amount of light that passes through such a small slit will be small and it is spread over a large area.

Diffraction equation (angular width of central spot)

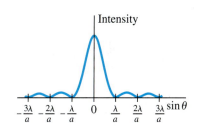

FIGURE 36–4 Intensity in the diffraction pattern of a single slit as a function of $\sin\theta$. Note that the central maximum is not only much higher than the maxima to each side, but it is also twice as wide ($2\lambda/a$ wide) as any of the others (only λ/a wide each).

Single slit diffraction minima

FIGURE 36–5 Example 36–1.

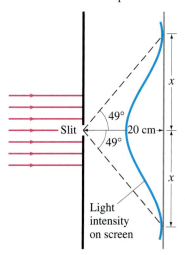

FIGURE 36–6 Example 36–2.

CONCEPTUAL EXAMPLE 36–2 **Diffraction spreads.** Light shines through a rectangular hole that is narrower in the vertical direction than the horizontal, Fig. 36–6. (a) Would you expect the diffraction pattern to be more spread out in the vertical direction or in the horizontal direction? (b) Should a rectangular loudspeaker horn at a stadium be high and narrow, or wide and flat?

RESPONSE (a) From Eq. 36–2 we can see that if we make the slit (width a) narrower, the pattern spreads out more. This is consistent with our earlier study of waves in Chapter 15. Hence the diffraction through the rectangular hole will be wider vertically, since the aperture in that direction is smaller.

(b) For the loudspeaker, the sound pattern desired is one spread out horizontally, so the horn should be tall and narrow (rotate Fig. 36–6 by 90°).

36–2 Intensity in Single-Slit Diffraction Pattern

FIGURE 36–7 Slit of width a divided into N strips of width Δy.

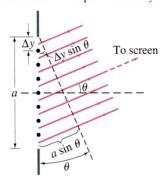

We have determined the positions of the minima in the diffraction pattern produced by light passing through a single slit, Eq. 36–2. We now discuss a method for predicting the amplitude and intensity at any point in the pattern using the phasor technique already discussed in Section 35–5.

Let us consider the slit divided into N very thin strips of width Δy as indicated in Fig. 36–7. Each strip sends light in all directions toward a screen on the right. Again we take the rays heading for any particular point on the distant screen to be parallel, all making an angle θ with the horizontal as shown. We choose the strip width Δy to be much smaller than the wavelength λ of the monochromatic light falling on the slit, so all the light from a given strip is in phase. The strips are of equal size, and if the whole slit is uniformly illuminated, we can take the electric field wave amplitudes ΔE_0 from each thin strip to be equal as long as θ is not too large. However, the separate amplitudes from the different strips will differ in phase. The phase difference in the light coming from adjacent strips will be (see Section 35–5, Eq. 35–4)

$$\Delta\beta = \frac{2\pi}{\lambda}\,\Delta y \sin\theta \tag{36–3}$$

since the difference in path length is $\Delta y \sin\theta$.

The total amplitude on the screen at any angle θ will be the sum of the separate wave amplitudes due to each strip. These wavelets have the same amplitude ΔE_0 but differ in phase. To obtain the total amplitude, we can use a phasor diagram as we did in Section 35–5 (Fig. 35–13). The phasor diagrams for four different angles θ are shown in Fig. 36–8. At the center of the screen, $\theta = 0$, the waves from each strip are all in phase ($\Delta\beta = 0$, Eq. 36–3), so the arrows representing each ΔE_0 line up as shown in Fig. 36–8a. The total amplitude of the light arriving at the center of the screen is then $E_0 = N\Delta E_0$. At a small angle θ, for a point on the distant screen not far from the center, Fig. 36–8b shows how the wavelets of amplitude ΔE_0 add up to give E_θ, the total amplitude on the screen at this angle θ. Note that each wavelet differs in phase from the adjacent one by $\Delta\beta$. The phase difference between the wavelets from the top and bottom edges of the slit is

$$\beta = N\,\Delta\beta = \frac{2\pi}{\lambda}N\,\Delta y \sin\theta = \frac{2\pi}{\lambda}a\sin\theta \tag{36–4}$$

where $a = N\,\Delta y$ is the total width of the slit. Although the "arc" in Fig. 36–8b has length $N\,\Delta E_0$, and so equals E_0 (total amplitude at $\theta = 0$), the amplitude of the total wave E_θ at angle θ is the *vector* sum of each wavelet amplitude and so is equal to the length of the chord as shown. The chord is shorter than the arc, so $E_\theta < E_0$.

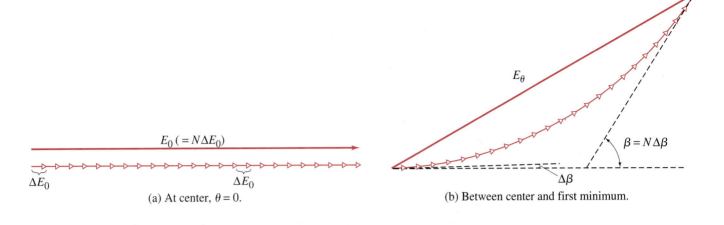

(a) At center, $\theta = 0$.

(b) Between center and first minimum.

(c) First minimum, $E_\theta = 0$ $(\beta = 360°)$.

(d) Near secondary maximum.

FIGURE 36–8 Phasor diagram for single-slit diffraction, giving the total amplitude E_θ at four different angles θ.

For greater θ, we eventually come to the case, illustrated in Fig. 36–8c, where the chain of arrows closes on itself. In this case the vector sum is zero, so $E_\theta = 0$ for this angle θ. This corresponds to the first minimum. Since $\beta = N\,\Delta\beta$ is 360° or 2π in this case, we have from Eq. 36–3,

$$2\pi = N\,\Delta\beta = N\left(\frac{2\pi}{\lambda}\,\Delta y\sin\theta\right)$$

or, since the slit width $a = N\,\Delta y$,

$$\sin\theta = \frac{\lambda}{a}.$$

Thus the first minimum $(E_\theta = 0)$ occurs where $\sin\theta = \lambda/a$, which is the same result we obtained in the previous Section, Eq. 36–1.

For even greater values of θ, the chain of arrows spirals beyond 360°. Figure 36–8d shows the case near the secondary maximum next to the first minimum. Here $\beta = N\,\Delta\beta \approx 360° + 180° = 540°$ or 3π. (Note that although β may be 540°, θ can still be a very small angle, depending on the values of a and λ.) When $\beta = 4\pi$, we have a double circle and again a minimum, where $\sin\theta = 2\lambda/a$, corresponding to $m = 2$ in Eq. 36–2. When greater angles θ are considered, new maxima and minima occur. But since the total length of the coil remains constant, equal to $N\,\Delta E_0 (= E_0)$, each succeeding maximum is smaller and smaller as the coil winds in on itself.

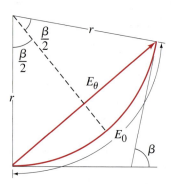

FIGURE 36–9 Determining amplitude E_θ as a function of θ for single-slit diffraction.

To obtain a quantitative expression for the amplitude (and intensity) for any point on the screen (that is, for any angle θ), we now consider the limit $N \to \infty$ so Δy become the infinitesimal width dy. In this case, the diagrams of Fig. 36–8 become smooth curves, one of which is shown in Fig. 36–9. For any angle θ, the wave amplitude on the screen is E_θ, equal to the chord in Fig. 36–9. The length of the arc is E_0, as before. If r is the radius of curvature of the arc, then

$$\frac{E_\theta}{2} = r \sin \frac{\beta}{2}.$$

Using radian measure for $\beta/2$, we also have

$$\frac{E_0}{2} = r \frac{\beta}{2}.$$

We combine these to obtain

$$E_\theta = E_0 \frac{\sin \beta/2}{\beta/2}. \tag{36–5}$$

The angle β is the phase difference between the waves from the top and bottom edges of the slit. The path difference for these two rays is $a \sin \theta$ (see Fig. 36–7 as well as Eq. 36–4), so

$$\beta = \frac{2\pi}{\lambda} a \sin \theta. \tag{36–6}$$

Intensity is proportional to the square of the wave amplitude, so the intensity I_θ at any angle θ is, from Eq. 36–5,

$$I_\theta = I_0 \left(\frac{\sin \beta/2}{\beta/2} \right)^2 \tag{36–7}$$

where $I_0 (\propto E_0^2)$ is the intensity at $\theta = 0$ (the central maximum). We can combine Eqs. 36–7 and 36–6 (although it is often simpler to leave them as separate equations) to obtain

Intensity in single-slit diffraction pattern

$$I_\theta = I_0 \left[\frac{\sin \left(\dfrac{\pi a \sin \theta}{\lambda} \right)}{\left(\dfrac{\pi a \sin \theta}{\lambda} \right)} \right]^2. \tag{36–8}$$

According to Eq. 36–8, minima $(I_\theta = 0)$ occur where $\sin (\pi a \sin \theta/\lambda) = 0$, which means $\pi a \sin\theta/\lambda$ must be $\pi, 2\pi, 3\pi$, and so on, or

$$a \sin\theta = m\lambda, \qquad m = 1, 2, 3, \cdots \qquad\qquad \text{[minima]}$$

which is what we have obtained previously, Eq. 36–2. Notice that m cannot be zero: when $\beta/2 = \pi a \sin\theta/\lambda = 0$, the denominator as well as the numerator in Eqs. 36–7 or 36–8 vanishes. We can evaluate the intensity in this case by taking the limit as $\theta \to 0$ (or $\beta \to 0$); for very small angles, $\sin\beta/2 \approx \beta/2$, so $(\sin\beta/2)/(\beta/2) \to 1$ and $I_\theta = I_0$, the maximum at the center of the pattern.

The intensity I_θ as a function of θ, as given by Eq. 36–8, corresponds to the diagram of Fig. 36–4.

EXAMPLE 36–3 **ESTIMATE** **Intensity at secondary maxima.** Estimate the intensities of the first two secondary maxima to either side of the central maximum.

SOLUTION The secondary maxima occur close to halfway between the minima, at about

$$\frac{\beta}{2} = \frac{\pi a \sin\theta}{\lambda} \approx \left(m + \tfrac{1}{2}\right)\pi. \qquad m = 1, 2, 3, \cdots.$$

The actual maxima are not quite at these points—their positions can be determined by differentiating Eq. 36–7 (see Problem 10)—but we are only seeking an estimate. Using these values for β in Eq. 36–7 or 36–8, with $\sin\left(m + \tfrac{1}{2}\right)\pi = 1$, gives

$$I_\theta = \frac{I_0}{\left(m + \tfrac{1}{2}\right)^2 \pi^2}. \qquad m = 1, 2, 3, \cdots.$$

For $m = 1$ and 2, we get

$$I_\theta = \frac{I_0}{22.2} = 0.045\, I_0 \qquad\qquad [m = 1]$$

$$I_\theta = \frac{I_0}{61.7} = 0.016\, I_0. \qquad\qquad [m = 2]$$

The first maximum to the side of the central peak has only 1/22, or 4.5 percent, the intensity of the central intensity, and succeeding ones are smaller still, just as we can see in Fig. 36–4 and the photo of Fig. 36–2c.

Diffraction by a circular opening produces a similar pattern (though circular rather than rectangular) and is of great practical importance, since lenses are essentially circular apertures through which light passes. We will discuss this in Section 36–4 and see how diffraction limits the resolution (or sharpness) of images.

*36–3 Diffraction in the Double-Slit Experiment

When we analyzed Young's double-slit experiment in Section 35–5, we assumed that the central portion of the screen was uniformly illuminated. This is equivalent to assuming the slits are infinitesimally narrow, so that the central diffraction peak is spread out over the whole screen. This can never be the case for real slits; diffraction reduces the intensity of the bright interference fringes to the side of center so they are not all of the same height as they were shown in Fig. 35–14. (They were shown more correctly in Fig. 35–9b.)

To calculate the intensity in a double-slit interference pattern, including diffraction, let us assume the slits have equal widths a and their centers are separated by a distance d. Since the distance to the screen is large compared to the slit separation d, the wave amplitude due to each slit is essentially the same at each point on the screen. Then the total wave amplitude at any angle θ will no longer be

$$E_{\theta 0} = 2E_0 \cos \frac{\delta}{2},$$

as was given by Eq. 35–5b. Rather, it must be modified, because of diffraction, by Eq. 36–5, so that

$$E_{\theta 0} = 2E_0 \left(\frac{\sin \beta/2}{\beta/2} \right) \cos \frac{\delta}{2}.$$

(Note that up to now in this chapter we have dealt only with the amplitude of E, and so we merely wrote E_θ rather than $E_{\theta 0}$ as was done in Chapter 35.) Thus the intensity will be given by

$$I_\theta = I_0 \left(\frac{\sin \beta/2}{\beta/2} \right)^2 \left(\cos \frac{\delta}{2} \right)^2 \tag{36–9}$$

where, from Eqs. 36–6 and 35–4,

$$\frac{\beta}{2} = \frac{\pi}{\lambda} a \sin \theta \qquad \text{and} \qquad \frac{\delta}{2} = \frac{\pi}{\lambda} d \sin \theta.$$

The first term in parentheses in Eq. 36–9 is sometimes called the "diffraction factor" and the second one the "interference factor." These two factors are plotted in Fig. 36–10a and b for the case when $d = 6a$, and $a = 10\lambda$. (Figure 36–10b is essentially the same as Fig. 35–14.) Figure 36–10c shows the product of these two curves, times I_0, which is the actual intensity as a function of θ (or as a function of position on the screen for θ not too large) as given by Eq. 36–9. As indicated by the dashed lines in Fig. 36–10c, the diffraction factor acts as a sort of envelope that limits the interference peaks.

EXAMPLE 36–4 **Diffraction plus interference.** Show why the central diffraction peak in Fig. 36–10c contains 11 interference fringes.

SOLUTION The first minimum in the diffraction pattern occurs where

$$\sin \theta = \frac{\lambda}{a}.$$

Since $d = 6a$,

$$d \sin \theta = 6a \left(\frac{\lambda}{a} \right) = 6\lambda.$$

From Eq. 35–2a, interference peaks (maxima) occur for $d \sin \theta = m\lambda$ where m can be 0, 1, $\cdots$ or any integer. Thus the diffraction minimum ($d \sin \theta = 6\lambda$) coincides with $m = 6$ in the interference pattern, so the $m = 6$ peak won't appear. Hence the central diffraction peak encloses the central interference peak ($m = 0$) and five peaks ($m = 1$ to 5) on each side for a total of 11. Since the sixth order doesn't appear, it is said to be a "missing order."

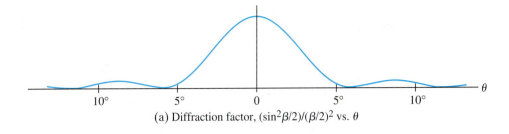

(a) Diffraction factor, $(\sin^2 \beta/2)/(\beta/2)^2$ vs. θ

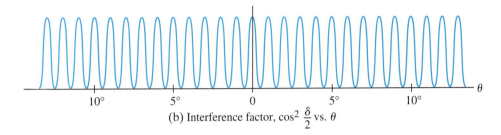

(b) Interference factor, $\cos^2 \dfrac{\delta}{2}$ vs. θ

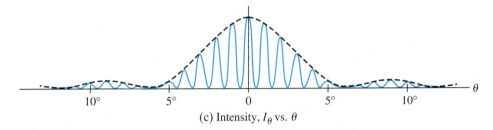

(c) Intensity, I_θ vs. θ

FIGURE 36–10 (a) Diffraction factor, (b) interference factor, and (c) the resultant intensity I_θ, plotted as a function of θ for $d = 6a = 60\lambda$.

Notice from this last Example that the number of interference fringes in the central diffraction peak depends only on the ratio d/a. It does not depend on wavelength λ. The actual spacing (in angle, or in position on the screen) does depend on λ. For the case illustrated, $a = 10\lambda$, and so the first diffraction minimum occurs at $\sin \theta = \lambda/a = 0.10$ or about $6°$.

The decrease in intensity of the interference fringes away from the center, as graphed in Fig. 36–10, is shown in Fig. 36–11.

The patterns due to interference and diffraction arise from the same phenomenon—the superposition of coherent waves of different phase. The distinction between them is thus not so much physical as for convenience of description, as in this Section where we analyzed the two-slit pattern in terms of interference and diffraction separately. In general, we use the word "diffraction" when referring to an analysis by superposition of many infinitesimal and usually contiguous sources, such as when we subdivide a source into infinitesimal parts. We use the term "interference" when we superpose the wave from a finite (and usually small) number of coherent sources.

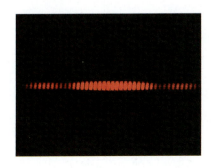
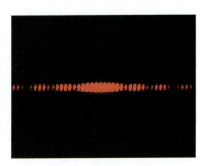

FIGURE 36–11 Photographs of a double-slit interference pattern showing effects of diffraction. In both cases $d = 0.50\,\text{mm}$, whereas $a = 0.040\,\text{mm}$ in (a) and $0.080\,\text{mm}$ in (b).

Limits of Resolution; Circular Apertures

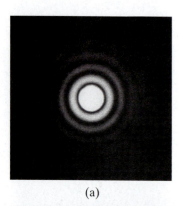

(a)

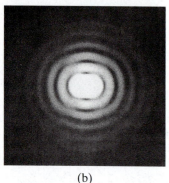

(b)

FIGURE 36–12 Photographs of images (greatly magnified) formed by a lens, showing the diffraction pattern of an image for: (a) a single point object; (b) two point objects whose images are barely resolved.

The ability of a lens to produce distinct images of two point objects very close together is called the **resolution** of the lens. The closer the two images can be and still be seen as distinct (rather than overlapping blobs), the higher the resolution. The resolution of a camera lens, for example, is often specified as so many lines per millimeter,[†] and can be determined by photographing a standard set of parallel lines on fine-grain film. The minimum spacing of lines distinguishable on film using the lens gives the resolution.

Two principal factors limit the resolution of a lens. The first is lens aberrations. As we saw, because of spherical and other aberrations, a point object is not a point on the image but a tiny blob. Careful design of compound lenses can reduce aberrations significantly, but they cannot be eliminated entirely. The second factor that limits resolution is *diffraction*, which cannot be corrected for because it is a natural result of the wave nature of light. We discuss it now.

In Section 36–1, we saw that because light travels as a wave, light from a point source passing through a slit is spread out into a diffraction pattern (Figs. 36–2 and 36–4). A lens, because it has edges, acts like a slit. When a lens forms the image of a point object, the image is actually a tiny diffraction pattern. Thus *an image would be blurred even if aberrations were absent.*

In the analysis that follows, we assume that the lens is free of aberrations, so that we can focus our attention on diffraction effects and how much they limit the resolution of a lens. In Fig. 36–4 we saw that the diffraction pattern produced by light passing through a rectangular slit has a central maximum in which most of the light falls. This central peak falls to a minimum on either side of its center at an angle $\theta \approx \sin\theta = \lambda/a$ (this is Eq. 36–1) where a is the width of the slit, λ is the wavelength of light used, and we assume θ is small. There are also low-intensity fringes beyond. For a lens, or any circular hole, the image of a point object will consist of a *circular* central peak (called the *diffraction spot* or *Airy disk*) surrounded by faint circular fringes, as shown in Fig. 36–12a. The central maximum has an angular half width given by

$$\theta = \frac{1.22\lambda}{D}$$

where D is the diameter of the circular opening.

This formula differs from that for a slit (Eq. 36–1) by the factor 1.22. This factor appears because the width of a circular hole is not uniform (like a rectangular slit) but varies from its diameter D to zero. A careful analysis shows that the "average" width is $D/1.22$. Hence we get the equation above rather than Eq. 36–1.

[†]This may be specified at the center of the field of view as well as at the edges, where it is usually less because of off-axis aberrations.

FIGURE 36–13 Intensity of light across the diffraction pattern of a circular hole.

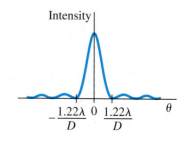

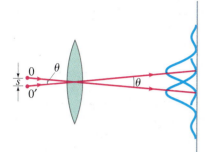

FIGURE 36–14 The *Rayleigh criterion.* Two images are just resolvable when the center of the diffraction peak of one is directly over the first minimum in the diffraction pattern of the other. The two point objects 0 and 0′ subtend an angle θ at the lens; only one ray is drawn for each object, to indicate the center of the diffraction pattern of its image.

The intensity of light in the diffraction pattern of light from a point source passing through a circular opening is shown in Fig. 36–13. The image for a non-point source is a superposition of such patterns. For most purposes we need consider only the central spot, since the concentric rings are so much dimmer.

If two point objects are very close, the diffraction patterns of their images will overlap as shown in Fig. 36–12b. As the objects are moved closer, a separation is reached where you can't tell if there are two overlapping images or a single image. The separation at which this happens may be judged differently by different observers. However, a generally accepted criterion is one proposed by Lord Rayleigh (1842–1919). This **Rayleigh criterion** states that *two images are just resolvable when the center of the diffraction disk of one image is directly over the first minimum in the diffraction pattern of the other*. This is shown in Fig. 36–14. Since the first minimum is at an angle $\theta = 1.22\lambda/D$ from the central maximum, Fig. 36–14 shows that two objects can be considered just resolvable if they are separated by this angle θ:

$$\theta = \frac{1.22\lambda}{D}.$$ (36–10)

Rayleigh criterion (resolution limit)

This is the limit on resolution set by the wave nature of light due to diffraction.

EXAMPLE 36–5 Hubble space telescope. The Hubble Space Telescope (HST) is a reflecting telescope that was placed in orbit above the Earth's atmosphere, so its resolution would not be limited by turbulence in the atmosphere (Fig. 36–15). Its objective diameter is 2.4 m. For visible light, say $\lambda = 550\,\text{nm}$, estimate the improvement in resolution the Hubble offers over Earth-bound telescopes, which are limited in resolution by movement of the Earth's atmosphere to about half an arc second (each degree is divided into 60 minutes each containing 60 seconds, so $1° = 3600\,\text{arc seconds}$).

SOLUTION Earth-bound telescopes are limited to an angular resolution of

$$\theta = \tfrac{1}{2}\left(\frac{1}{3600}\right)^{\circ}\left(\frac{2\pi\,\text{rad}}{360°}\right) = 2.4 \times 10^{-6}\,\text{rad}.$$

The Hubble, on the other hand, is limited by diffraction (Eq. 36–10) which for $\lambda = 550\,\text{nm}$ is

$$\theta = \frac{1.22(550 \times 10^{-9}\,\text{m})}{2.4\,\text{m}} = 2.8 \times 10^{-7}\,\text{rad},$$

which is almost ten times better resolution.

FIGURE 36–15 Hubble Space Telescope, with Earth in the background. The flat orange panels are solar cells that collect energy from the Sun.

Resolution of Telescopes and Microscopes; the λ Limit

You might think that a microscope or telescope could be designed to produce any desired magnification, depending on the choice of focal lengths and quality of the lenses. But this is not possible, because of diffraction. An increase in magnification above a certain point merely results in magnification of the diffraction patterns. This could be highly misleading since we might think we are seeing details of an object when we are really seeing details of the diffraction pattern. To examine this problem, we apply the Rayleigh criterion: two objects (or two nearby points on one object) are just resolvable if they are separated by an angle θ (Fig. 36–14) given by Eq. 36–10:

Resolution

$$\theta = \frac{1.22\lambda}{D}.$$

This is valid for either a microscope or a telescope, where D is the diameter of the objective lens. For a telescope, the resolution is specified by stating θ as given by this equation.[†]

For a microscope, it is more convenient to specify the actual distance, s, between two points that are just barely resolvable, Fig. 36–14. Since objects are normally placed near the focal point of the microscope objective, $\theta = s/f$, or $s = f\theta$. If we combine this with Eq. 36–10, we obtain for the **resolving power (RP)**:

Resolving power (microscope)

$$\mathrm{RP} = s = f\theta = \frac{1.22\lambda f}{D}. \tag{36–11}$$

This distance s is called the resolving power of the lens because it is the minimum separation of two object points that can just be resolved, assuming the highest quality lens since this limit is imposed by the wave nature of light.

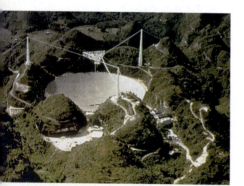

FIGURE 36–16 The 300-meter radiotelescope in Arecibo, Puerto Rico, uses radio waves (Fig. 32–12) instead of visible light.

EXAMPLE 36–6 **Telescope resolution (radio wave vs. visible light).** What is the theoretical minimum angular separation of two stars that can just be resolved by: (*a*) the 200-inch telescope on Palomar Mountain (Fig. 34–33c); and (*b*) the Arecibo radiotelescope (Fig. 36–16), whose diameter is 300 m and whose radius of curvature is also 300 m. Assume $\lambda = 550$ nm for the visible-light telescope in part (*a*), and $\lambda = 4$ cm (the shortest wavelength at which the radiotelescope has been operated) in part (*b*).

SOLUTION (*a*) Since $D = 200$ inch $= 5.1$ m, we have from Eq. 36–10 that

$$\theta = \frac{1.22\lambda}{D} = \frac{(1.22)(5.50 \times 10^{-7}\,\mathrm{m})}{(5.1\,\mathrm{m})} = 1.3 \times 10^{-7}\,\mathrm{rad},$$

or 0.75×10^{-5} deg. (Note that this is equivalent to resolving two points less than 1 cm apart from a distance of 100 km!) This is the limit set by diffraction. The resolution is not this good because of aberrations and, more importantly, turbulence in the atmosphere. In fact, large-diameter objectives are not justified by increased resolution, but by their greater light-gathering ability—they allow more light in, so fainter objects can be seen.

(*b*) Radiotelescopes are not hindered by atmospheric turbulence, and for radio waves with $\lambda = 0.04$ m the resolution is

$$\theta = \frac{(1.22)(0.04\,\mathrm{m})}{(300\,\mathrm{m})} = 1.6 \times 10^{-4}\,\mathrm{rad}.$$

[†]Telescopes with large-diameter objectives are usually limited not by diffraction but by other effects such as turbulence in the atmosphere. The resolution of a high-quality microscope, on the other hand, normally *is* limited by diffraction because microscope objectives are complex compound lenses containing many elements of small diameter (since f is small).

Diffraction sets an ultimate limit on the detail that can be seen on any object. In Eq. 36–11 we note that the focal length of a lens cannot be made less than (approximately) the radius of the lens, and even that is very difficult—see the lens-maker's equation (Eq. 34–4). In this best case, Eq. 36–11 gives, with $f \approx D/2$,

$$\text{RP} \approx \frac{\lambda}{2}. \tag{36–12}$$

Thus we can say, to within a factor of 2 or so, that

it is not possible to resolve detail of objects smaller than the wavelength of the radiation being used.

Resolution limited to λ

This is an important and useful rule of thumb.

Compound lenses are now designed so well that the actual limit on resolution is often set by diffraction—that is, by the wavelength of the light used. To obtain greater detail, one must use radiation of shorter wavelength. The use of UV radiation can increase the resolution by a factor of perhaps 2. Far more important, however, was the discovery in the early twentieth century that electrons have wave properties (Chapter 38) and that their wavelengths can be very small. The wave nature of electrons is utilized in the electron microscope (Section 38–7), which can magnify 100 to 1000 times more than a visible-light microscope because of the much shorter wavelengths. X-rays, too, have very short wavelengths and are often used to study objects in great detail (Section 36–10).

*36–6 Resolution of the Human Eye and Useful Magnification

The resolution of the human eye is limited by several factors, all of roughly the same order of magnitude. The resolution is best at the fovea, where the cone spacing is smallest, about $3\,\mu\text{m}$ (= 3000 nm). The diameter of the pupil varies from about 0.1 cm to about 0.8 cm. So for $\lambda = 550\,\text{nm}$ (where the eye's sensitivity is greatest), the diffraction limit is about $\theta \approx 1.22\lambda/D \approx 8 \times 10^{-5}\,\text{rad}$ to $6 \times 10^{-4}\,\text{rad}$. Since the eye is about 2 cm long, this corresponds to a resolving power of $s \approx (8 \times 10^{-5}\,\text{rad})(2 \times 10^{-2}\,\text{m}) \approx 2\,\mu\text{m}$ at best, to about $15\,\mu\text{m}$ at worst (pupil small). Spherical and chromatic aberration also limit the resolution to about $10\,\mu\text{m}$. The net result is that the eye can resolve objects whose angular separation is about $5 \times 10^{-4}\,\text{rad}$ at best. This corresponds to objects separated by 1 cm at a distance of about 20 m.

The typical near point of a human eye is about 25 cm. At this distance, the eye can just resolve objects that are $(25\,\text{cm})(5 \times 10^{-4}\,\text{rad}) \approx 10^{-4}\,\text{m} = \frac{1}{10}\,\text{mm}$ apart. Since the best light microscopes can resolve objects no smaller than about 200 nm at best (Eq. 36–12 for violet light, $\lambda = 400\,\text{nm}$), the useful magnification [= (resolution by naked eye)/(resolution by microscope)] is limited to about

$$\frac{10^{-4}\,\text{m}}{200 \times 10^{-9}\,\text{m}} = 500\times.$$

In practice, magnifications of about 1000× are often used to minimize eye-strain. Any greater magnification would simply make visible the diffraction pattern produced by the microscope objective.

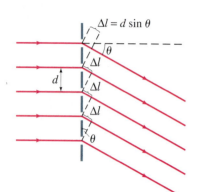

FIGURE 36–17 Diffraction grating.

36–7 Diffraction Grating

A large number of equally spaced parallel slits is called a **diffraction grating**, although the term "interference grating" might be as appropriate. Gratings can be made by precision machining of very fine parallel lines on a glass plate. The untouched spaces between the lines serve as the slits. Photographic transparencies of an original grating serve as inexpensive gratings. Gratings containing 10,000 lines per centimeter are common today, and are very useful for precise measurements of wavelengths. A diffraction grating containing slits is called a **transmission grating**. **Reflection gratings** are also used, which are made by ruling fine lines on a metallic or glass surface from which light is reflected and analyzed. The analysis is basically the same as for a transmission grating, which we now discuss.

The analysis of a diffraction grating is much like that of Young's double-slit experiment. We assume parallel rays of light are incident on the grating as shown in Fig. 36–17. We also assume that the slits are narrow enough so that diffraction by each of them spreads light over a very wide angle on a distant screen behind the grating, and interference can occur with light from all the other slits. Light rays that pass through each slit without deviation ($\theta = 0°$) interfere constructively to produce a bright line at the center of the screen. Constructive interference also occurs at an angle θ such that rays from adjacent slits travel an extra distance of $\Delta l = m\lambda$, where m is an integer. Thus, if d is the distance between slits, then we see from Fig. 36–17 that $\Delta l = d \sin\theta$, and

Diffraction grating maxima (m = order)

$$\sin\theta = \frac{m\lambda}{d}, \qquad m = 0, 1, 2, \cdots \qquad \text{[principal maxima]} \quad \textbf{(36–13)}$$

is the criterion to have a brightness maximum. This is the same equation as for the double-slit situation, and again m is called the order of the pattern.

There is an important difference between a double-slit and a multiple-slit pattern, however. The bright maxima are much *sharper* and *narrower* for a grating. Why this happens can be seen as follows. Suppose that the angle θ is increased just slightly beyond that required for a maximum. In the case of only two slits, the two waves will be only slightly out of phase, so nearly full constructive interference occurs. This means the maxima are wide (see Fig. 35–9). For a grating, the waves from two adjacent slits will also not be significantly out of phase. But waves from one slit and those from a second one a few hundred slits away may be exactly out of phase; all or nearly all the light will cancel in pairs in this way. For example, suppose the angle θ is different from its first-order maximum so that the extra path length for a pair of adjacent slits is not exactly λ but rather 1.0010λ. The wave through one slit and another one 500 slits below will be out of phase by 1.5000λ, or exactly $1\frac{1}{2}$ wavelengths, so the two will cancel. A pair of slits, one below each of these, will also cancel. That is, the light from slit 1 cancels with that from slit 501; light from slit 2 cancels with that from slit 502, and so on. Thus even for a tiny angle[†] corresponding to an extra path length of $\frac{1}{1000}\lambda$, there is much destructive interference, and so the maxima are very narrow. The more lines there are in a grating, the sharper will be the peaks (see Fig. 36–18). Because a grating produces much sharper (and brighter) lines than two slits alone, it is a far more precise device for measuring wavelengths.

Suppose the light striking a diffraction grating is not monochromatic, but rather consists of two or more distinct wavelengths. Then for all orders other than $m = 0$, each wavelength will produce a maximum at a different angle (Fig. 36–19a), just as for a double slit. If white light strikes a grating, the central ($m = 0$) maximum will be a sharp white peak. But for all other orders, there will be a distinct

Why more slits yield sharper peaks

FIGURE 36–18 Intensity as a function of viewing angle θ (or position on the screen) for (a) two slits, (b) six slits. For a diffraction grating, the number of slits is very large ($\sim 10^4$) and the peaks are narrower still.

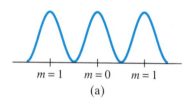

$m = 1 \qquad m = 0 \qquad m = 1$
(a)

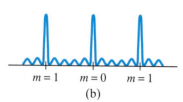

$m = 1 \qquad m = 0 \qquad m = 1$
(b)

[†]Depending on the total number of slits, there may or may not be complete cancellation for such an angle, so there will be very tiny peaks between the main maxima (see Fig. 36–18b), but they are usually much too small to be seen.

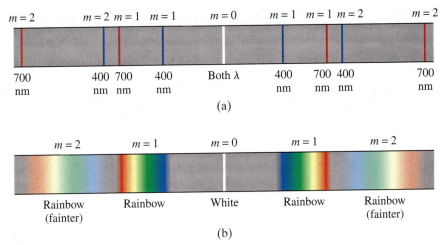

(a)

FIGURE 36–19 Spectrum produced by a grating: (a) two wavelengths, 400 nm and 700 nm; (b) white light. The second order will normally be dimmer than the first order. (Higher orders are not shown.) If the grating spacing is small enough, the second and higher orders will be missing.

(b)

spectrum of colors spread out over a certain angular width, Fig. 36–19b. Because a diffraction grating spreads out light into its component wavelengths, the resulting pattern is called a **spectrum**.

EXAMPLE 36–7 **Diffraction grating: lines.** Calculate the first- and second-order angles for light of wavelength 400 nm and 700 nm if the grating contains 10,000 lines/cm.

SOLUTION The grating contains 10^4 lines/cm $= 10^6$ lines/m, which means the separation between slits is $d = (1/10^6)\,\text{m} = 1.0 \times 10^{-6}\,\text{m}$. In first order $(m = 1)$, the angles are

$$\sin \theta_{400} = \frac{m\lambda}{d} = \frac{(1)(4.0 \times 10^{-7}\,\text{m})}{1.0 \times 10^{-6}\,\text{m}} = 0.400$$

$$\sin \theta_{700} = 0.700$$

so $\theta_{400} = 23.6°$ but $\theta_{700} = 44.4°$. In second order,

$$\sin \theta_{400} = \frac{(2)(4.0 \times 10^{-7}\,\text{m})}{1.0 \times 10^{-6}\,\text{m}} = 0.800$$

$$\sin \theta_{700} = 1.40$$

so $\theta_{400} = 53.1°$, but the second order does not exist for $\lambda = 700\,\text{nm}$ because $\sin \theta$ cannot exceed 1. No higher orders will appear.

The diffraction grating is the essential component of a spectroscope, a device for precise measurement of wavelengths, and we discuss it next.

*36–8 The Spectrometer and Spectroscopy

A **spectrometer** or **spectroscope**, Fig. 36–20, is a device[†] to measure wavelengths accurately using a diffraction grating, or a prism, to separate different wavelengths of light. Light from a source passes through a narrow slit S in the collimator. The slit is at the focal point of the lens L, so parallel light falls on the grating. The movable telescope can bring the rays to a focus. Nothing will be seen in the viewing telescope unless it is positioned at an angle θ that corresponds to a diffraction peak (first order is usually used) of a wavelength emitted by the source. The angle θ can be measured

FIGURE 36–20 Spectrometer or spectroscope.

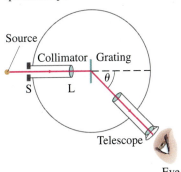

[†] If the spectrum of a source is recorded (say, on film) rather than viewed by the eye, the device is called a **spectrometer** or spectrograph, as compared to a **spectroscope**, which is for viewing only; but these terms are often used interchangeably. Devices that can also measure the intensity of light of a given wavelength are called **spectrophotometers**.

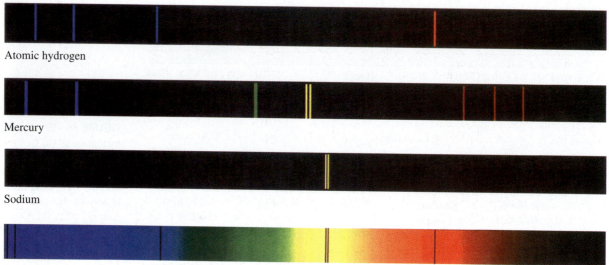

Atomic hydrogen

Mercury

Sodium

Solar absorption spectrum

FIGURE 36–21 Line spectra for the gases indicated, and spectrum from the Sun showing absorption lines.

to very high accuracy, so the wavelength of a line can be determined to high accuracy using Eq. 36–13:

$$\lambda = \frac{d}{m}\sin\theta,$$

where m is an integer representing the order, and d is the distance between grating lines. The line you see in a spectrometer corresponding to each wavelength is actually an image of the slit S. The narrower the slit, the narrower—but dimmer—the line is, and the more precisely we can measure its angular position. If the light contains a continuous range of wavelengths, then a continuous spectrum is seen in the spectroscope.

Line spectra
An important use of a spectrometer is for the identification of atoms or molecules. When a gas is heated or a large electric current is passed through it, the gas emits a characteristic **line spectrum**. That is, only certain discrete wavelengths of light are emitted, and these are different for different elements and compounds. Figure 36–21 shows the line spectra for a number of elements in the gas state. Line spectra occur only for gases at high temperatures and low pressure. The light from heated solids, such as a lightbulb filament, and even from a dense gaseous object such as the Sun, produces a **continuous spectrum** including a wide range of wavelengths.

Figure 36–21 also shows the Sun's "continuous spectrum," which contains a number of *dark* lines (only the most prominent are shown), called **absorption lines**. Atoms and molecules can absorb light at the same wavelengths at which they emit light. The Sun's absorption lines are due to absorption by atoms and molecules in the cooler outer atmosphere of the Sun, as well as by atoms and molecules in the Earth's atmosphere. A careful analysis of all these thousands of lines reveals that at least two-thirds of all elements are present in the Sun's atmosphere. The presence of elements in the atmosphere of other planets, in interstellar space, and in stars is also determined by spectroscopy.

*36–9 Peak Widths and Resolving Power for a Diffraction Grating

We now look at the pattern of maxima produced by a multiple-slit grating using phasor diagrams. We can determine a formula for the width of each peak, and we will see why there are tiny maxima between the principal maxima, as indicated in Fig. 36–18b. First of all, it should be noted that the two-slit and six-slit patterns shown in Fig. 36–18 were drawn assuming very narrow slits so that diffraction does not limit the height of the peaks. For real diffraction gratings, this is not normally the case: the slit width a is often not much smaller than the slit separation d, and diffraction thus limits the intensity of the peaks so the central peak ($m = 0$) is brighter than the side peaks. We won't worry about this effect on intensity except to note that if a diffraction minimum coincides with a particular order of the interference pattern, that order will not appear. (For example, if $d = 2a$, all the even orders, $m = 2, 4, \cdots$, will be missing. Can you see why? Hint: See Example 36–4.)

Figures 36–22 and 36–23 show phasor diagrams for a two-slit and a six-slit grating, respectively. Each short arrow represents the amplitude of a wave from a single slit, and their vector sum (as phasors) represents the total amplitude for a given viewing angle θ. Part (a) of each Figure shows the phasor diagram at $\theta = 0°$, at the center of the pattern, which is the central maximum ($m = 0$). Part (b) of each Figure shows the condition for the adjacent minimum: where the arrows first close on themselves (add to zero) so the amplitude E_θ is zero. For two slits, this occurs when the two separate amplitudes are 180° out of phase. For six slits, it occurs when each amplitude makes a 60° angle with its neighbor. For two slits, the minimum occurs when the phase between slits is $2\pi/2$ (in radians); for six slits it occurs when the phase δ is $2\pi/6$; and in the general case of N slits, the minimum occurs for a phase difference between adjacent slits of

$$\delta = \frac{2\pi}{N}. \qquad (36\text{–}14)$$

What does this correspond to in θ? First note that δ is related to θ by

$$\frac{\delta}{2\pi} = \frac{d \sin \theta}{\lambda} \quad \text{or} \quad \delta = \frac{2\pi}{\lambda} d \sin \theta \qquad (36\text{–}15)$$

just as in Eq. 35–4. Let us call $\Delta\theta_0$ the angular position of the minimum next to the peak at $\theta = 0$. This corresponds to an extra path length between adjacent slits (see Fig. 36–17) of $\Delta l = d \sin \Delta\theta_0$, so

$$\frac{\delta}{2\pi} = \frac{\Delta l}{\lambda} = \frac{d \sin \Delta\theta_0}{\lambda}.$$

We insert Eq. 36–14 for δ and find

$$\sin \Delta\theta_0 = \frac{\lambda}{Nd}. \qquad (36\text{–}16a)$$

Since $\Delta\theta_0$ is usually small (N is usually very large for a grating), $\sin \Delta\theta_0 \approx \Delta\theta_0$, so in the small angle limit we can write

$$\Delta\theta_0 = \frac{\lambda}{Nd}. \qquad (36\text{–}16b) \qquad \textit{Peak half width (m = 0)}$$

It is clear from either of the last two relations that the larger N is, the narrower will be the central peak. (For $N = 2$, $\sin \Delta\theta_0 = \lambda/2d$, which is what we obtained earlier for the double slit, Eq. 35–2b with $m = 0$.)

(a) Central maximum
$\theta = 0, \delta = 0$

$E_\theta = 0$
$\delta = 180°$

(b) Minimum: $\delta = 180°$

FIGURE 36–22 Phasor diagram for two slits (a) at the central maximum, (b) at the nearest minimum.

FIGURE 36–23 Phasor diagram for six slits (a) at the central maximum, (b) at the nearest minimum.

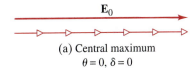

(a) Central maximum
$\theta = 0, \delta = 0$

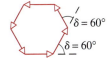

$\delta = 60°$

$\delta = 60°$

(b) Minimum: $\delta = 60°$

FIGURE 36–24 Phasor diagram for the secondary peak.

Either of Eqs. 36–16 shows why the peaks become narrower for larger N. The origin of the small secondary maxima between the principal peaks (see Fig. 36–18b) can be deduced from the diagram of Fig. 36–24. This is just a continuation of Fig. 36–23b (where $\delta = 60°$); but now the phase δ has been increased to almost 90°, where E_θ is a relative maximum. Note that E_θ is much less than E_0 (Fig. 36–23a), so the intensity in this secondary maximum is much smaller than in a principal peak. As δ (and θ) is increased further, E_θ again decreases to zero (a "double circle"), then reaches another tiny maximum, and so on. Eventually the diagram unfolds again and when $\delta = 360°$, all the amplitudes again lie in a straight line (as in Fig. 35–23a) corresponding to the next principal maximum ($m = 1$ in Eq. 36–13).

Equation 36–16b gives the half width of the central ($m = 0$) peak. To determine the half width of higher order peaks, $\Delta\theta_m$ for order m, we differentiate Eq. 36–15 so as to relate the change $\Delta\delta$ in δ, to the change $\Delta\theta$ in the angle θ:

$$\Delta\delta \approx \frac{d\delta}{d\theta} \Delta\theta = \frac{2\pi d}{\lambda} \cos\theta \, \Delta\theta.$$

If $\Delta\theta_m$ represents the half width of a peak of order m ($m = 1, 2, \cdots$)—that is, the angle between the peak maximum and the minimum to either side—then $\Delta\delta = 2\pi/N$ as given by Eq. 36–14. We insert this into the above relation and find

Peak half width (order m)

$$\Delta\theta_m = \frac{\lambda}{Nd \cos\theta_m}, \tag{36–17}$$

where θ_m is the angular position of the mth peak as given by Eq. 36–13. This derivation is valid, of course, only for small $\Delta\delta \,(= 2\pi/N)$ which is indeed the case for real gratings since N is on the order of 10^4 or more.

An important property of any diffraction grating used in a spectroscope is its ability to resolve two very closely spaced wavelengths. The **resolving power** R of a grating is defined as

Resolving power (grating)

$$R = \frac{\lambda}{\Delta\lambda}. \tag{36–18}$$

With a little work, using Eq. 36–17, we can show that $\Delta\lambda = \lambda N/m$ where N is the total number of grating lines and m is the order. Then we have

$$R = Nm. \tag{36–19}$$

The larger the value of R, the closer two wavelengths can be and still be resolvable. If R is given, the minimum separation $\Delta\lambda$ between two wavelengths near λ, is (Eq. 36–18)

$$\Delta\lambda = \frac{\lambda}{R}.$$

EXAMPLE 36–8 **Resolving two close lines.** Yellow sodium light, which consists of two wavelengths, $\lambda_1 = 589.00\,\text{nm}$ and $\lambda_2 = 589.59\,\text{nm}$, falls on a 7500-line/cm diffraction grating. Determine (a) the maximum order m that will be present for sodium light, (b) the width of grating necessary to resolve the two sodium lines, (c) the angular width of each sodium line.

SOLUTION (a) Since $d = 1\,\text{cm}/7500 = 1.33 \times 10^{-6}\,\text{m}$, then the maximum value of m for $\lambda = 589\,\text{nm}$ can be found from Eq. 35–13 with $\sin\theta \leq 1$:

$$m = \frac{d}{\lambda}\sin\theta \leq \frac{1.33 \times 10^{-6}\,\text{m}}{5.89 \times 10^{-7}\,\text{m}} = 2.25,$$

so $m = 2$ is the maximum order present.

(b) The resolving power needed is

$$R = \frac{\lambda}{\Delta\lambda} = \frac{589 \text{ nm}}{0.59 \text{ nm}} = 1000.$$

From Eq. 36–19, the total number N of lines needed is $N = R/m = 1000/2 = 500$, so the grating need only be $500/7500 \text{ cm}^{-1} = 0.0667 \text{ cm}$ wide. A typical grating is a few centimeters wide, and so will easily resolve the two lines.

(c) We need to use Eq. 36–17 for $m = 2$, but what is $\cos\theta_2$? From Eq. 36–13,

$$\sin\theta_2 = m\frac{\lambda}{d} = \frac{2(589 \times 10^{-9} \text{ m})}{1.33 \times 10^{-6} \text{ m}} = 0.886.$$

Then $\cos\theta_2 = (1 - \sin^2\theta_2)^{\frac{1}{2}} = 0.464$. For a grating with the minimum 500 lines, the angular width (Eq. 36–17) for $m = 2$ is

$$\Delta\theta_2 = \frac{\lambda}{Nd\cos\theta_2} = \frac{589 \times 10^{-9} \text{ m}}{(500)(1.33 \times 10^{-6} \text{ m})(0.464)} = 0.0019 \text{ rad},$$

or $0.11°$.

36–10 X-Rays and X-Ray Diffraction

In 1895, W. C. Roentgen (1845–1923) discovered that when electrons were accelerated by a high voltage in a vacuum tube and allowed to strike a glass (or metal) surface inside the tube, fluorescent minerals some distance away would glow, and photographic film would become exposed. Roentgen attributed these effects to a new type of radiation (different from cathode rays). They were given the name **X-rays** after the algebraic symbol x, meaning an unknown quantity. He soon found that X-rays penetrated through some materials better than through others, and within a few weeks he presented the first X-ray photograph (of his wife's hand). The production of X-rays today is usually done in a tube (Fig. 36–25) similar to Roentgen's, using voltages of typically 30 kV to 150 kV.

X-rays

FIGURE 36–25 X-ray tube. Electrons emitted by a heated filament in a vacuum tube are accelerated by a high voltage. When they strike the surface of the anode, the "target," X-rays are emitted.

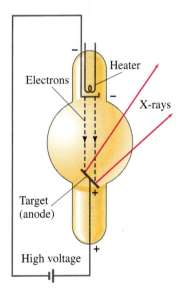

Investigations into the nature of X-rays indicated they were not charged particles (such as electrons) since they could not be deflected by electric or magnetic fields. It was suggested that they might be a form of invisible light. However, they showed no diffraction or interference effects using ordinary gratings. Of course, if their wavelengths were much smaller than the typical grating spacing of $10^{-6} \text{ m} (= 10^3 \text{ nm})$, no effects would be expected. Around 1912, it was suggested by Max von Laue (1879–1960) that if the atoms in a crystal were arranged in a regular array (see Fig. 17–2a), such a crystal might serve as a diffraction grating for very short wavelengths on the order of the spacing between atoms, estimated to be about $10^{-10} \text{ m} (= 10^{-1} \text{ nm})$. Experiments soon showed that X-rays scattered from a crystal did indeed show the peaks and valleys of a diffraction pattern (Fig. 36–26). Thus it was shown, in a single blow, that X-rays have a wave nature and that atoms are arranged in a regular way in crystals. Today, X-rays are recognized as electromagnetic radiation with wavelengths in the range of about 10^{-2} nm to 10 nm, the range readily produced in an X-ray tube.

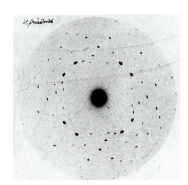

FIGURE 36–26 This X-ray diffraction pattern is one of the first observed by Max von Laue in 1912 when he aimed a beam of X-rays at a zinc sulfide crystal. The diffraction pattern was detected directly on a photographic plate.

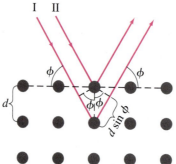

FIGURE 36–27 X-ray diffraction by a crystal.

We saw in Section 36–5 that light of shorter wavelength provides greater resolution when we are examining an object microscopically. Since X-rays have much shorter wavelengths than visible light, in principle they should offer much greater resolution. However, there seems to be no effective material to use as lenses for the very short wavelengths of X-rays. Instead, the clever but complicated technique of **X-ray diffraction** (or **crystallography**) has proved very effective for examining the microscopic world of atoms and molecules. In a simple crystal such as NaCl, the atoms are arranged in an orderly cubical fashion, Fig. 36–27, with atoms spaced a distance d apart. Suppose that a beam of X-rays is incident on the crystal at an angle ϕ to the surface, and that the two rays shown are reflected from two subsequent planes of atoms as shown. The two rays will constructively interfere if the extra distance ray I travels is a whole number of wavelengths farther than what ray II travels. This extra distance is $2d \sin \phi$. Therefore, constructive interference will occur when

$$m\lambda = 2d \sin \phi, \qquad m = 1, 2, 3, \cdots, \qquad (36\text{–}20)$$

Bragg equation

where m can be any integer. (Notice that ϕ is *not* the angle with respect to the normal to the surface.) This is called the **Bragg equation** after W. L. Bragg (1890–1971), who derived it and who, together with his father W. H. Bragg (1862–1942), developed the theory and technique of X-ray diffraction by crystals in 1912–1913. Thus, if the X-ray wavelength is known and the angle ϕ at which constructive interference occurs is measured, d can be obtained. This is the basis for X-ray crystallography.

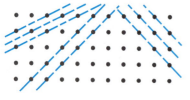

FIGURE 36–28 There are many possible planes existing within a crystal from which X-rays can be diffracted.

Actual X-ray diffraction patterns are quite complicated. First of all, a crystal is a three-dimensional object, and X-rays can be diffracted from different planes at different angles within the crystal, as shown in Fig. 36–28. Although the analysis is complex, a great deal can be learned about any substance that can be put in crystalline form. If the substance is not a single crystal but a mixture of many tiny crystals—as in a metal or a powder—then instead of a series of spots, as in Fig. 36–26, a series of circles is obtained, Fig. 36–29, each corresponding to diffraction of a certain order m from a particular set of parallel planes.

X-ray diffraction has been very useful in determining the structure of biologically important molecules. For example, it was with the help of X-ray diffraction that, in 1953, J. D. Watson and F. H. C. Crick worked out the double-helix structure of DNA.

FIGURE 36–29 (a) Diffraction of X-rays from a polycrystalline substance produces a set of circular rings as in (b), which is for polycrystalline sodium acetoacetate.

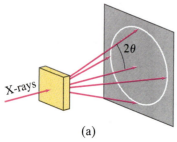

(a)

(b)

36–11 Polarization

An important and useful property of light is that it can be *polarized*. To see what this means, let us examine waves traveling on a rope. A rope can be set into oscillation in a vertical plane as in Fig. 36–30a, or in a horizontal plane as in Fig. 36–30b. In either case, the wave is said to be **linearly polarized** or **plane-polarized**—that is, the oscillations are in a plane.

If we now place an obstacle containing a vertical slit in the path of the wave, Fig. 36–31, a vertically polarized wave passes through, but a horizontally polarized wave will not. If a horizontal slit were used, the vertically polarized wave would be stopped. If both types of slit were used, both types of wave would be stopped. Note that polarization can exist *only* for *transverse waves*, and not for longitudinal waves such as sound. The latter oscillate only along the direction of motion, and neither orientation of slit would stop them.

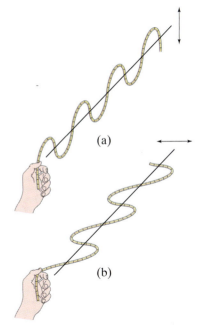

FIGURE 36–30 (above) Transverse waves on a rope polarized (a) in a vertical plane and (b) in a horizontal plane.

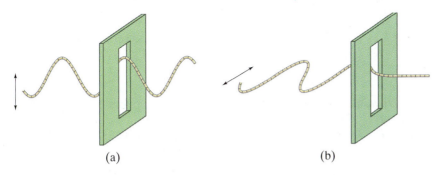

(a)   (b)

FIGURE 36–31 Vertically polarized wave passes through a vertical slit, but a horizontally polarized wave will not. [Note: This diagram would apply to the magnetic field **B** in an EM wave—see footnote below.]

Maxwell's theory of light as electromagnetic (EM) waves predicted that light can be polarized since an EM wave is a transverse wave. The direction of polarization in a plane-polarized EM wave is taken as the direction of the electric field vector **E**.

Light is not necessarily polarized. It can be **unpolarized**, which means that the source has oscillations in many planes at once, as shown in Fig. 36–32. An ordinary incandescent lightbulb emits unpolarized light, as does the Sun.

Polaroids

Plane-polarized light can be obtained from unpolarized light using certain crystals such as tourmaline. Or, more commonly today, we can use a **Polaroid sheet**. (Polaroid materials were invented in 1929 by Edwin Land.) A Polaroid sheet consists of complicated long molecules arranged parallel to one another. Such a Polaroid acts like a series of parallel slits to allow one orientation of polarization to pass through nearly undiminished (this direction is called the *axis* of the Polaroid), whereas a perpendicular polarization is absorbed almost completely.[†]

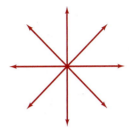

FIGURE 36–32 Oscillation of the electric field vectors in unpolarized light. The light is traveling into or out of the page.

[†]How this occurs can be explained at the molecular level. An electric field **E** that oscillates parallel to the long molecules can set electrons into motion along the molecules, thus doing work on them and transferring energy. Hence, if **E** is parallel to the molecules, it gets absorbed. An electric field **E** perpendicular to the long molecules does not have this possibility of doing work and transferring its energy, and so passes through freely. When we speak of the *axis* of a Polaroid, we mean the direction for which **E** is passed, so a Polaroid axis is perpendicular to the long molecules. (If we want to think of there being slits between the parallel molecules in the sense of Fig. 36–31, then Fig. 36–31 would apply for the **B** field in the EM wave, not the **E** field.)

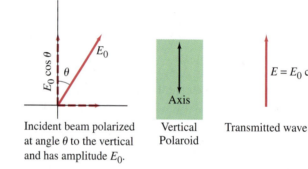

FIGURE 36–33 Vertical Polaroid transmits only the vertical component of a wave (electric field) incident upon it.

Incident beam polarized at angle θ to the vertical and has amplitude E_0.

Vertical Polaroid

Transmitted wave

If a beam of plane-polarized light strikes a Polaroid whose axis is at an angle θ to the incident polarization direction, the beam will emerge plane-polarized parallel to the Polaroid axis and its amplitude will be reduced by $\cos\theta$, Fig. 36–33. Thus, a Polaroid passes only that component of polarization (the electric field vector, **E**) that is parallel to its axis. Because the intensity of a light beam is proportional to the square of the amplitude (Sections 15–3 and 32–7), we see that the intensity of a plane-polarized beam transmitted by a polarizer is

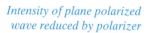
Intensity of plane polarized wave reduced by polarizer

$$I = I_0 \cos^2\theta, \tag{36–21}$$

where θ is the angle between the polarizer axis and the plane of polarization of the incoming wave, and I_0 is the incoming intensity.[†]

A Polaroid can be used as a **polarizer** to produce plane-polarized light from unpolarized light, since only the component of light parallel to the axis is transmitted. A Polaroid can also be used as an **analyzer** to determine (1) if light is polarized and (2) what is the plane of polarization. A Polaroid acting as an analyzer will pass the same amount of light independent of the orientation of its axis if the light is unpolarized; try rotating one lens of a pair of Polaroid sunglasses while looking through it at a lightbulb. If the light is polarized, however, when you rotate the Polaroid the transmitted light will be a maximum when the plane of polarization is parallel to the Polaroid's axis, and a minimum when perpendicular to it. If you do this while looking at the sky, preferably at right angles to the Sun's direction, you will see that skylight is polarized. (Direct sunlight is unpolarized, but don't look directly at the Sun, even through a polarizer, for damage to the eye may occur.) If the light trasmitted by an analyzer Polaroid falls to zero at one orientation, then the light is 100 percent plane-polarized. If it merely reaches a minimum, the light is *partially polarized*.

Unpolarized light consists of light with random directions of polarization. Each of these polarization directions can be resolved into components along two mutually perpendicular directions. On average, an unpolarized beam can be thought of as two plane-polarized beams of equal magnitude perpendicular to one another. When unpolarized light passes through a polarizer, one of the components is eliminated. So the intensity of the light passing through is reduced by half since half the light is eliminated, $I = \frac{1}{2}I_0$ (Fig. 36–34).

When two Polaroids are *crossed*—that is, their axes are perpendicular to one another—unpolarized light can be entirely stopped. As shown in Fig. 36–35, unpolarized light is made plane-polarized by the first Polaroid (the polarizer). The second Polaroid, the analyzer, then eliminates this component since its axis is perpendicular to the first. You can try this with Polaroid sunglasses (Fig. 36–36). Note that Polaroid sunglasses eliminate 50 percent of unpolarized light because of their polarizing property; they absorb even more because they are colored.

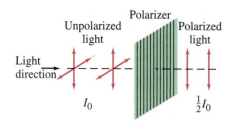

FIGURE 36–34 Unpolarized light has both vertical and horizontal components. After passing through a polarizer, one of these components is eliminated. The intensity of the light is reduced to half.

[†]Equation 36–21 is often referred to as **Malus' law**, after Etienne Malus, a contemporary of Fresnel.

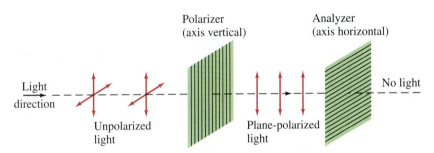

Polarizer (axis vertical)   Analyzer (axis horizontal)

Light direction

Unpolarized light   Plane-polarized light   No light

FIGURE 36–35 Crossed Polaroids completely eliminate light.

FIGURE 36–36 Crossed Polaroids. When the two polarized sunglass lenses overlap, with axes perpendicular, almost no light passes through.

EXAMPLE 36–9 **Two Polaroids at 60°.** Unpolarized light passes through two Polaroids; the axis of one is vertical and that of the other is at 60° to the vertical. What is the orientation and intensity of the transmitted light?

SOLUTION The first Polaroid eliminates half the light so the intensity is reduced by half: $I_1 = \frac{1}{2}I_0$. The light reaching the second polarizer is vertically polarized and so is reduced in intensity (Eq. 36–21) to

$$I_2 = I_1(\cos 60°)^2 = \frac{1}{4}I_1 .$$

Thus, $I_2 = \frac{1}{8}I_0$. The transmitted light has an intensity one-eighth that of the original and is plane-polarized at a 60° angle to the vertical.

CONCEPTUAL EXAMPLE 36–10 **Three Polaroids.** We saw in Fig. 36–35 that when unpolarized light falls on two crossed Polaroids (axes at 90°), no light passes through. What happens if a third Polaroid, with axis at 45° to each of the other two, is placed between them?

RESPONSE We start just as in Example 36–9. The first Polaroid changes the unpolarized light to plane-polarized and reduces the intensity from I_0 to $I = \frac{1}{2}I_0$. The second polarizer further reduces the intensity by $(\cos 45°)^2$, Eq. 36–21:

$$I_2 = I_1(\cos 45°)^2 = \frac{1}{2}I_1 = \frac{1}{4}I_0 .$$

The light leaving the second polarizer is plane polarized at 45° (Fig. 36–37) relative to the third polarizer, so the latter reduces the intensity to

$$I_3 = I_2(\cos 45°)^2 = \frac{1}{2}I_2$$

or $I_3 = \frac{1}{8}I_0$. Thus $\frac{1}{8}$ of the original intensity gets transmitted. But if we don't put the 45° Polaroid in at all, zero intensity results (Fig. 36–35). The 45° Polaroid must be inserted *between* the other two if transmission is to occur. Placing it before or after the other two results in zero intensity.

FIGURE 36–37 Example 36–10.

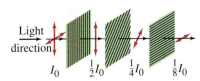

Light direction

I_0 $\frac{1}{2}I_0$ $\frac{1}{4}I_0$ $\frac{1}{8}I_0$

An important use of polarizers is in a **liquid crystal display** (LCD). LCDs are used in many displays, from digital watches to small TV sets. Each pixel (picture element) consists of a small liquid crystal contained between two glass plates whose outer surfaces have a thin film of polarizer. Liquid crystals are materials that, at room temperature, exist in a phase neither fully solid nor fully liquid. Useful ones are composed of rod-like molecules that interact only weakly with each other and tend to align with each other. The overall direction of alignment can be controlled using electric fields. Each pixel can be bright or dark depending on that alignment relative to the polarizers on the ends. Color displays can be obtained by using color filters.

PHYSICS APPLIED

Liquid crystals (LCD)

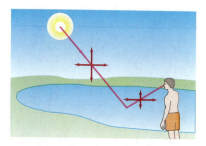

FIGURE 36–38 Light reflected from a nonmetallic surface, such as the smooth surface of water in a lake, is partially polarized parallel to the surface.

FIGURE 36–39 Photographs of a river, (a) allowing all light into the camera lens, and (b) using a polarizer which is adjusted to absorb most of the (polarized) light reflected from the water's surface, allowing the dimmer light from the bottom of the river, and any fish swimming there, to be seen more readily.

FIGURE 36–40 At θ_p the reflected light is plane-polarized parallel to the surface, and $\theta_p + \theta_r = 90°$, where θ_r is the refraction angle. (The large dots represent vibrations perpendicular to the page.)

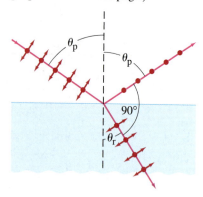

Polarization by Reflection

Another means of producing polarized light from unpolarized light is by reflection. When light strikes a nonmetallic surface at any angle other than perpendicular, the reflected beam is polarized preferentially in the plane parallel to the surface, Fig. 36–38. In other words, the component with polarization in the plane perpendicular to the surface is preferentially transmitted or absorbed. You can check this by rotating Polaroid sunglasses while looking through them at a flat surface of a lake or road. Since most outdoor surfaces are horizontal, Polaroid sunglasses are made with their axes vertical to eliminate the more strongly reflected horizontal component, and thus reduce glare. Fishermen wear Polaroids to eliminate reflected glare from the surface of a lake or stream and thus see beneath the water more clearly (Fig. 36–39).

(a)   (b)

The amount of polarization in the reflected beam depends on the angle, varying from no polarization at normal incidence to 100 percent polarization at an angle known as the **polarizing angle**, θ_p.[†] This angle is related to the index of refraction of the two materials on either side of the boundary by the equation

$$\tan\theta_p = \frac{n_2}{n_1}, \tag{36–22a}$$

where n_1 is the index of refraction of the material in which the beam is traveling, and n_2 is that of the medium beyond the reflecting boundary. If the beam is traveling in air, $n_1 = 1$, and Eq. 36–22a becomes

$$\tan\theta_p = n. \tag{36–22b}$$

The polarizing angle θ_p is also called **Brewster's angle**, and Eqs. 36–22 called *Brewster's law*, after the Scottish physicist David Brewster (1781–1868), who worked it out experimentally in 1812. Equations 36–22 can be derived from the electromagnetic wave theory of light. It is interesting that at Brewster's angle, the reflected and transmitted rays make a 90° angle to each other; that is, $\theta_p + \theta_r = 90°$, Fig. 36–40. This can be seen by substituting Eq. 36–22a, $n_2 = n_1 \tan\theta_p = n_1 \sin\theta_p/\cos\theta_p$, into Snell's law, $n_1 \sin\theta_p = n_2 \sin\theta_r$, and get $\cos\theta_p = \sin\theta_r$ which can only hold if $\theta_p = 90° - \theta_r$.

EXAMPLE 36–11 Polarizing angle. (a) At what incident angle is sunlight reflected from a lake plane-polarized? (b) What is the refraction angle?

SOLUTION (a) We use Eq. 36–22b with $n = 1.33$, so $\tan\theta_p = 1.33$ and $\theta_p = 53.1°$.
(b) $\theta_r = 90.0° - \theta_p = 36.9°$.

[†]Only a fraction of the incident light is reflected at the surface of the transparent medium. Although this reflected light is 100 percent polarized (if $\theta = \theta_p$), the remainder of the light, which is transmitted into the new medium, is only partially polarized.

36–12 | Scattering of Light by the Atmosphere

Sunsets are red, the sky is blue, and skylight is polarized (at least partially). These phenomena can be explained on the basis of the *scattering* of light by the molecules of the atmosphere. In Fig. 36–41 we see unpolarized light from the Sun impinging on a molecule of the Earth's atmosphere. The electric field of the EM wave sets the electric charges within the molecule into motion, and the molecule absorbs some of the incident radiation. But it quickly reemits this light since the charges are oscillating. As discussed in Section 32–4, oscillating electric charges produce EM waves. The electric field of these waves is in a plane that includes the line of oscillation. The intensity is strongest along a line perpendicular to the oscillation, and drops to zero along the line of oscillation (Section 32–4). In Fig. 36–41 the motion of the charges is resolved into two components. An observer at right angles to the direction of the sunlight, as shown, will see plane-polarized light because no light is emitted along the line of the other component of the oscillation. (Another way to understand this is to note that when viewing along the line of oscillation, one doesn't see the oscillation, and hence sees no waves made by it. Furthermore, the EM wave produces no oscillation along its own direction of motion.) At other viewing angles, both components will be present; one will be stronger, however, so the light appears partially polarized. Thus, the process of scattering explains the polarization of skylight.

Scattering of light by the Earth's atmosphere depends on λ. For particles much smaller than the wavelength of light (such as molecules of air), the particles will be less of an obstruction to long wavelengths than to short ones. The scattering decreases, in fact, as $1/\lambda^4$. Blue and violet light are thus scattered much more than red and orange, which is why the sky looks blue. At sunset, the Sun's rays pass through a maximum length of atmosphere. Much of the blue has been taken out by scattering. The light that reaches the surface of the Earth, and reflects off clouds and haze, is thus lacking in blue which is why sunsets appear reddish.

The dependence of scattering on $1/\lambda^4$ is valid only if the scattering objects are much smaller than the wavelength of the light. This is valid for oxygen and nitrogen molecules whose diameters are about 0.2 nm. Clouds, however, contain water droplets or crystals that are much larger than λ. They scatter all frequencies of light nearly uniformly. Hence clouds appear white (or gray, if shadowed).

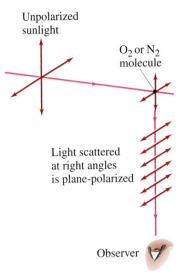

FIGURE 36–41 Unpolarized sunlight scattered by molecules of the air. An observer at right angles sees plane-polarized light, since the component of oscillation along the line of sight emits no light along that line.

➡ **PHYSICS APPLIED**

Why the sky is blue
Why sunsets are red

Summary

Diffraction refers to the fact that light, like other waves, bends around objects it passes, and spreads out after passing through narrow slits. This bending gives rise to a **diffraction pattern** due to interference between rays of light that travel different distances.

Light passing through a very narrow slit of width a will produce a pattern with a bright central maximum of half-width θ given by

$$\sin\theta = \frac{\lambda}{a},$$

flanked by fainter lines to either side.

The minima in the diffraction pattern occur at

$$a\sin\theta = m\lambda$$

where $m = 1, 2, 3, \cdots$, but not $m = 0$ (for which the pattern has its strongest maximum).

The **intensity** at any point in the single-slit diffraction pattern can be calculated using **phasor** diagrams. The same technique can be used to determine the intensity of the pattern produced by two slits.

The pattern for two-slit interference can be described as a series of maxima due to interference of light from the two slits, modified by an "envelope" due to diffraction at each slit.

The wave nature of light limits the sharpness or **resolution** of images. Because of diffraction, it is not possible to *discern details smaller than the wavelength* of the radiation being used. The useful magnification of a light microscope is limited by diffraction to about 1000×.

A **diffraction grating** consists of many parallel slits or lines, each separated from its neighbors by a distance d. The peaks of constructive interference occur at angles θ given by

$$\sin \theta = \frac{m\lambda}{d},$$

where $m = 0, 1, 2, \cdots$. The peaks are much brighter and sharper for a diffraction grating than for the simple two-slit apparatus. Peak width is inversely proportional to the total number of lines in the grating.

A diffraction grating (or a prism) is used in a **spectroscope** to separate different colors or to observe **line spectra**, since for a given order m, θ depends on λ. Precise determination of wavelength can be done with a spectroscope by careful measurement of θ.

In **unpolarized light**, the electric field vectors oscillate transversely at all angles. If the electric vector oscillates only in one plane the light is said to be **plane-polarized**. Light can also be partially polarized.

When an unpolarized light beam passes through a Polaroid sheet, the emerging beam is plane-polarized. When a light beam is polarized and passes through a Polaroid, the intensity varies as the Polaroid is rotated. Thus a Polaroid can act as a polarizer or as an analyzer.

The intensity of a plane-polarized beam incident on a Polaroid is reduced by the factor

$$\cos^2 \theta$$

where θ is the angle between the axis of the Polaroid and the initial plane of polarization.

Light can also be partially or fully **polarized by reflection**. If light traveling in air is reflected from a medium of index of refraction n, the reflected beam will be *completely* plane-polarized if the incident angle θ_p is given by $\tan \theta_p = n$. The fact that light can be polarized shows that it must be a transverse wave.

Questions

1. Radio waves and light are both electromagnetic waves. Why can we hear a radio behind a hill when we cannot see the transmitting antenna?

2. Hold one hand close to your eye and focus on a distant light source through a narrow slit between two fingers. (Adjust your fingers to obtain the best pattern.) Describe the pattern that you see.

3. A rectangular slit is twice as high as it is wide. Will the light spread more in the horizontal or in the vertical plane? Describe the pattern.

4. Explain why diffraction patterns are more difficult to observe with an extended light source than for a point source. Compare also a monochromatic source to white light.

5. For diffraction by a single slit, what is the effect of increasing (a) the slit width, (b) the wavelength?

6. Describe the single-slit diffraction pattern produced when white light falls on a slit having a width of (a) 50 nm, (b) 50,000 nm.

7. What happens to the diffraction pattern of a single slit if the whole apparatus is immersed in (a) water, (b) a vacuum, instead of in air.

8. Describe clearly the difference between the angles θ and β, as in Eq. 36–6.

9. In the single-slit diffraction pattern, why does the first off-center maximum not occur at exactly $\sin \theta = \frac{3}{2}\lambda/a$?

* 10. Figure 36–10 shows a two-slit interference pattern for the case when d is larger than a. Can the reverse case occur, when d is less than a?

* 11. When both diffraction and interference are taken into account in the double-slit experiment, discuss the effect of increasing (a) the wavelength, (b) the slit separation, (c) the slit width.

12. Discuss the similarities, and differences, of double-slit interference and single-slit diffraction.

13. Does diffraction limit the resolution of images formed by (a) spherical mirrors, (b) plane mirrors?

14. Do diffraction effects occur for virtual as well as real images?

15. What are the advantages (give at least two) for the use of large reflecting mirrors in astronomical telescopes?

16. Atoms have diameters of about 10^{-8} cm. Can visible light be used to "see" an atom? Why or why not?

17. Which color of visible light would give the best resolution in a microscope?

18. If monochromatic light were used in a microscope, would the color affect the resolution? Explain.

19. Could a diffraction grating just as well be called an interference grating? Discuss.

20. For light consisting of wavelengths between 400 nm and 700 nm, incident normally on a diffraction grating, for what orders (if any) would there be overlap in the observed spectrum? Does your answer depend on the slit width?

21. What is the difference in the interference patterns formed (i) by two slits 10^{-4} cm apart, (ii) by a diffraction grating containing 10^4 lines/cm?

22. White light strikes (a) a diffraction grating and (b) a prism. A rainbow appears on a wall just below the direction of the horizontal incident beam in each case. What is the color of the top of the rainbow in each case?

23. Explain why there are tiny peaks between the main peaks produced by a diffraction grating illuminated with monochromatic light. Why are the peaks so tiny?

24. What does polarization tell us about the nature of light?

25. How can you tell if a pair of sunglasses is polarizing or not?

* 26. What would be the color of the sky if the Earth had no atmosphere?

* 27. If the Earth's atmosphere were 50 times denser than it is, would sunlight still be white, or would it be some other color?

Problems

Section 36–1

1. (I) If 680-nm light falls on a slit 0.0345 mm wide, what is the angular width of the central diffraction peak?

2. (I) Monochromatic light falls on a slit that is 3.00×10^{-3} mm wide. If the angle between the first dark fringes on either side of the central maximum is 37.0° (dark fringe to dark fringe), what is the wavelength of the light used?

3. (II) Light of wavelength 550 nm falls on a slit that is 3.50×10^{-3} mm wide. Estimate how far from the central maximum the first diffraction maximum fringe is if the screen is 10.0 m away?

4. (II) Monochromatic light of wavelength 689 nm falls on a slit. If the angle between the first bright fringes on either side of the central maximum is 38°, estimate the slit width.

5. (II) If a slit diffracts 550-nm light so that the diffraction maximum is 8.0 cm wide on a screen 2.50 m away, what will be the width of the diffraction maximum for light with a wavelength of 400 nm?

6. (II) (a) For a given wavelength λ, what is the maximum slit width for which there will be no diffraction minima? (b) What is the maximum slit width so that no visible light exhibits a diffraction minimum?

7. (II) How wide is the central diffraction peak on a screen 3.50 m behind a 0.0655-mm-wide slit illuminated by 400-nm light?

8. (II) If parallel light falls on a single slit of width a at a 30° angle to the normal, describe the diffraction pattern.

Section 36–2

9. (II) If you double the width of a single slit, the intensity of the light passing through the slit is doubled. (a) Show, however, that the intensity at the center of the screen increases by a factor of 4. (b) Explain why this does not violate conservation of energy.

10. (III) (a) Explain why the secondary maxima in the single-slit diffraction pattern do not occur precisely at $\beta/2 = (m + \frac{1}{2})\pi$ where $m = 1, 2, 3, \cdots$. (b) By differentiating Eq. 36–7, show that the secondary maxima occur when $\beta/2$ satisfies the relation $\tan(\beta/2) = \beta/2$. (c) Carefully and precisely plot the curves $y = \beta/2$ and $y = \tan\beta/2$. From their intersections, determine the values of β for the first and second secondary maxima. What is the percent difference from $\beta/2 = (m + \frac{1}{2})\pi$?

11. (III) Determine, approximately, the angular width at half maximum (where $I = \frac{1}{2}I_0$) of the central diffraction peak for a single slit. [*Hint*: Use graphical methods, or trial and error; the Problem cannot be solved analytically.] To be concrete, assume $\lambda = 550$ nm and $a = 2.60 \times 10^{-3}$ mm.

*Section 36–3

*12. (II) Design a double-slit apparatus so that the central diffraction peak contains precisely fifteen fringes.

*13. (II) If a double-slit pattern contains exactly seven fringes in the central diffraction peak, what can you say about the slit width and separation?

*14. (II) Suppose $d = a$ in a double-slit apparatus, so that the two slits merge into one slit of width $2a$. Show that Eq. 36–9 reduces to the correct equation for single-slit diffraction.

*15. (II) Two 0.010-mm-wide slits are 0.030 mm apart (center to center). Determine (a) the spacing between interference fringes for 550 nm light on a screen 1.0 m away and (b) the distance between the two diffraction minima on either side of the central maximum of the envelope.

*16. (II) How many fringes are contained in the central diffraction peak for a double-slit pattern if (a) $d = 2.00a$, (b) $d = 12.0a$, (c) $d = 4.50a$, (d) $d = 7.20a$.

*17. (III) Draw phasor diagrams (as in Fig. 36–8) for the double-slit experiment, including both interference and diffraction. Make a diagram for each of several crucial points in the pattern (see Fig. 36–10c) such as at the center, first minimum, next maximum, and at $\sin\theta = \lambda/a$.

*18. (III) (a) Derive an expression for the intensity in the interference pattern for three equally spaced slits. Express in terms of $\delta = 2\pi d \sin\theta/\lambda$ where d is the distance between adjacent slits and assume the slit width $a \approx \lambda$. (b) Show that there is only one secondary maximum between principal peaks.

Sections 36–4 and 36–5

19. (I) What is the angular resolution limit set by diffraction for the 100-inch (mirror diameter) Mt. Wilson telescope ($\lambda = 500$ nm)?

20. (II) Two stars 10 light-years away are barely resolved by a 90 cm (mirror diameter) telescope. How far apart are the stars? Assume $\lambda = 550$ nm and that the resolution is limited by diffraction.

21. (II) The normal lens on a 35-mm camera has a focal length of 50 mm. Its aperture diameter varies from a maximum of 25 mm ($f/2$) to a minimum of 3.0 mm ($f/16$). Determine the resolution limit set by diffraction for $f/2$ and $f/16$. Specify as the number of lines per millimeter resolved on the film. Take $\lambda = 500$ nm.

22. (II) Suppose that you wish to construct a telescope that can resolve features 7.0 km across on the moon, 384,000 km away. You have a 2.0-m-focal-length objective lens whose diameter is 11.0 cm. What focal-length eyepiece is needed if your eye can resolve objects 0.10 mm apart at a distance of 25 cm? What is the resolution limit set by the size of the objective lens (that is, by diffraction)? Use $\lambda = 500$ nm.

Section 36–7

23. (I) At what angle will 440-nm light produce a third-order maximum when falling on a grating whose slits are 1.35×10^{-3} cm apart?

24. (I) How many lines per centimeter does a grating have if the third order occurs at a 13.0° angle for 650-nm light?

25. (I) A grating has 6600 lines/cm. How many spectral orders can be seen when it is illuminated by white light?

26. (I) A 3500-line/cm grating produces a third-order fringe at a 22.0° angle. What wavelength of light is being used?

27. (I) A source produces first-order lines when incident normally on a 10,000-line/cm diffraction grating at angles 29.8°, 37.7°, 39.6°, and 48.9°. What are the wavelengths?

28. (II) White light containing wavelengths from 400 nm to 750 nm falls on a grating with 7800 lines/cm. How wide is the first-order spectrum on a screen 2.80 m away?

29. (II) Show that the second- and third-order spectra of white light produced by a diffraction grating always overlap. What wavelengths overlap?

30. (II) Two first-order spectrum lines are measured by a 9550 line/cm spectroscope at angles, on each side of center, of $+26°38'$, $+41°02'$ and $-26°18'$, $-40°27'$. What are the wavelengths?

31. (II) Suppose the angles measured in Problem 30 were produced when the spectrometer (but not the source) was submerged in water. What then would be the wavelengths (in air)?

32. (II) The first-order line of 589-nm light falling on a diffraction grating is observed at a 15.5° angle. How far apart are the slits? At what angle will the third order be observed?

33. (II) Monochromatic light falls on a transmission diffraction grating at an angle ϕ to the normal. Show that Eq. 36–13 for diffraction maxima must be replaced by

$$d(\sin \phi \pm \sin \theta) = m\lambda. \qquad m = 0, 1, 2, \cdots.$$

Explain the $\pm$ sign.

* Section 36–9

* 34. (II) Missing orders occur for a diffraction grating when a diffraction minimum coincides with an interference maximum. Let a be the width of each slit and d the separation of slits and show (a) that if $d = 2a$, all even orders $(m = 2, 4, 6, \cdots)$ are missing. (b) Show that there will be missing orders whenever

$$\frac{d}{a} = \frac{m_1}{m_2}$$

where m_1 and m_2 are integers. (c) Discuss the case $d = a$, the limit in which the space between slits becomes negligible.

* 35. (II) Let 580-nm light be incident normally on a diffraction grating for which $d = 3.00a = 1200$ nm. (a) How many orders (principal maxima) are present? (b) If the grating is 1.80 cm wide, what is the full angular width of each principal maximum?

* 36. (II) A 6500-line/cm diffraction grating is 3.61 cm wide. If light with wavelengths near 624 nm falls on the grating, how close can two wavelengths be if they are to be resolved in any order? What order gives the best resolution?

* 37. (II) A diffraction grating has 16,000 rulings in its 2.4 cm width. Determine (a) its resolving power in first and second orders, and (b) the minimum wavelength resolution $(\Delta\lambda)$ it can yield for $\lambda \approx 410$ nm.

* 38. (II) For a fixed wavelength and diffraction angle, show that the resolving power of a diffraction grating depends on the total width, Nd, where N is the total number of slits each of width d.

* 39. (II) Determine a formula for the minimum difference in frequency, Δf, that a diffraction grating can resolve when two frequencies, $f_1 \approx f_2 = f$, are incident on it.

* Section 36–10

* 40. (II) X-rays of wavelength 0.138 nm fall on a crystal whose atoms, lying in planes, are spaced 0.265 nm apart. At what angle must the X-rays be directed if the first diffraction maximum is to be observed?

* 41. (II) First-order Bragg diffraction is observed at 26.2° from a crystal with spacing between atoms of 0.24 nm. (a) At what angle will second order be observed? (b) What is the wavelength of the X-rays?

* 42. (II) If X-ray diffraction peaks corresponding to the first three orders $(m = 1, 2, \text{and } 3)$ are measured, can both the X-ray wavelength λ and lattice spacing d be determined? Prove your answer.

Section 36–11

43. (I) Two polarizers are oriented at 75° to one another. Unpolarized light falls on them. What fraction of the light intensity is transmitted?

44. (I) Two Polaroids are aligned so that the light passing through them is a maximum. At what angle should one of them be placed so the intensity is subsequently reduced by half?

45. (I) What is Brewster's angle for an air–glass $(n = 1.56)$ surface?

46. (I) What is Brewster's angle for a diamond submerged in water if the light is hitting the diamond while traveling in the water?

47. (II) At what angle should the axes of two Polaroids be placed so as to reduce the intensity of the incident unpolarized light to $(a) \frac{1}{3}$, $(b) \frac{1}{10}$?

48. (II) The critical angle for total internal reflection at a boundary between two materials is 52°. What is Brewster's angle at this boundary?

49. (II) What would Brewster's angle be for reflections off the surface of water for light coming from beneath the surface? Compare to the angle for total internal reflection, and to Brewster's angle from above the surface.

50. (II) Two polarizers are oriented at 34.0° to one another. Light polarized at a 17.0° angle to each polarizer passes through both. What reduction in intensity takes place?

51. (II) Unpolarized light passes through five successive Polaroid sheets each of whose axis makes a 45° angle with the previous one. What is the intensity of the transmitted beam?

52. (II) Describe how to rotate the plane of polarization of a plane-polarized beam of light by 90° and produce only a 10 percent loss in intensity, using polarizers.

53. (III) The percent polarization P of a partially polarized beam of light is defined as

$$P = \frac{I_{max} - I_{min}}{I_{max} + I_{min}} \times 100$$

where I_{max} and I_{min} are the maximum and minimum intensities that are obtained when the light passes through a polarizer that is slowly rotated. Such light can be considered as the sum of two unequal plane-polarized beams of intensities I_{max} and I_{min} perpendicular to each other. Show that the light transmitted by a polarizer, whose axis makes an angle ϕ to the direction in which I_{max} is obtained, has intensity

$$\frac{1 + p \cos 2\phi}{1 + p} I_{max}$$

where $p = P/100$ is the "fractional polarization."

General Problems

54. When violet light of wavelength 415 nm falls on a single slit, it creates a central diffraction peak that is 9.20 cm wide on a screen that is 2.55 m away. How wide is the slit?

55. A teacher stands well back from an outside doorway 0.88 m wide, and blows a whistle of frequency 750 Hz. Ignoring reflections, estimate at what angle(s) it is *not* possible to hear the whistle clearly on the playground outside the doorway.

56. The wings of a certain beetle have a series of parallel lines across them. When normally incident 460-nm light is reflected from the wing, the wing appears bright when viewed at an angle of 50°. How far apart are the lines?

57. How many lines per centimeter must a grating have if there is to be no second-order spectrum for any visible wavelength?

58. Light is incident on a diffraction grating with 7500 lines per centimeter and the pattern is viewed on a screen located 2.5 m from the grating. The incident light beam consists of two wavelengths, $\lambda_1 = 4.4 \times 10^{-7}$ m and $\lambda_2 = 6.3 \times 10^{-7}$ m. Calculate the linear distance between the first-order bright fringes of these two wavelengths on the screen.

59. If parallel light falls on a single slit of width a at a 20° angle to the normal, describe the diffraction pattern.

60. When yellow sodium light, $\lambda = 589$ nm, falls on a diffraction grating, its first-order peak on a screen 60.0 cm away falls 3.32 cm from the central peak. Another source produces a line 3.71 cm from the central peak. What is its wavelength? How many lines/cm are on the grating?

61. What is the highest spectral order that can be seen if a grating with 6000 lines per cm is illuminated with 633-nm laser light? Assume normal incidence.

62. Two and only two full spectral orders can be seen on either side of the central maximum when white light is sent through a diffraction grating. What is the maximum number of lines per cm for the grating?

63. Two of the lines of the atomic hydrogen spectrum have wavelengths of 656 nm and 410 nm. If these fall at normal incidence on a grating with 7600 lines per cm, what will be the angular separation of the two wavelengths in the first-order spectrum?

64. Light falling normally on a 9850 line/cm grating is revealed to contain three lines in the first-order spectrum at angles of 31.2°, 36.4°, and 47.5°. What wavelengths are these?

65. (*a*) How far away can a human eye distinguish two car headlights 2.0 m apart? Consider only diffraction effects and assume an eye diameter of 5.0 mm and a wavelength of 500 nm. (*b*) What is the minimum angular separation an eye could resolve when viewing two stars, considering only diffraction effects? In reality, it is about 1′ of arc. Why is it not equal to your answer in (*b*)?

66. At what angle above the horizon is the Sun when light reflecting off a smooth lake is polarized most strongly?

67. Unpolarized light falls on two polarizer sheets whose axes are at right angles. (*a*) What fraction of the incident light intensity is transmitted? (*b*) What fraction is transmitted if a third polarizer is placed between the first two so that its axis makes a 60° angle with the axis of the first polarizer? (*c*) What if the third polarizer is in front of the other two?

68. Four polarizers are placed in succession with their axes vertical, at 30° to the vertical, at 60° to the vertical, and at 90° to the vertical. (*a*) Calculate what fraction of the incident unpolarized light is transmitted by the four polarizers. (*b*) Can the transmitted light be *decreased* by removing one of the polarizers? If so, which one? (*c*) Can the transmitted light intensity be extinguished by removing polarizers? If so, which one(s)?

69. At what angle should the axes of two Polaroids be placed so as to reduce the intensity of the incident unpolarized light by an additional factor (after the first Polaroid cuts it in half) of (*a*) 25 percent, (*b*) 10 percent, (*c*) 1 percent?

70. Two polarizers are oriented at 40° to each other and plane-polarized light is incident on them. If only 15 percent of the light gets through both of them, what was the initial polarization direction of the incident light?

71. Show that if two equally intense sources of light produce light that is plane-polarized, but with their planes of polarization perpendicular to each other, then they cannot produce an interference pattern even if they are in phase at all moments.

72. Spy planes fly at extremely high altitudes (25 km) to avoid interception. Their cameras are reportedly able to discern features as small as 5 cm. What must be the minimum aperture of the camera lens to afford this resolution? (Use $\lambda = 550$ nm.)

* 73. X-rays of wavelength 0.0973 nm are directed at an unknown crystal. The second diffraction maximum is recorded at 23.4°. What is the spacing between crystal planes?

* 74. X-rays of wavelength 0.10 nm fall on a microcrystalline powder sample as in Fig. 36–29. The sample is located 10 cm from the photographic film. The crystal structure of the sample has an atomic spacing of 0.25 nm. Calculate the radii of the diffraction rings corresponding to first- and second-order scattering.

Albert Einstein (1879–1955), one of the great minds of the twentieth century, was the creator of the special and general theories of relativity.

In this chapter we examine the special theory of relativity, which includes a number of non-classical results, including length contraction and time dilation for moving reference frames, and new formulas for momentum and energy.

CHAPTER 37

Special Theory of Relativity

FIGURE 37–1 Albert Einstein and his second wife.

Classical vs. modern physics

Physics at the end of the nineteenth century looked back on a period of great progress. The theories developed over the preceding three centuries had been very successful in explaining a wide range of natural phenomena. Newtonian mechanics beautifully explained the motion of objects on Earth and in the heavens. Furthermore, it formed the basis for successful treatments of fluids, wave motion, and sound. Kinetic theory explained the behavior of gases and other materials. Maxwell's theory of electromagnetism not only brought together and explained electric and magnetic phenomena, but it predicted the existence of electromagnetic (EM) waves that would behave in every way just like light—so light came to be thought of as an electromagnetic wave. Indeed, it seemed that the natural world, as seen through the eyes of physicists, was very well explained. A few puzzles remained, but it was felt that these would soon be explained using already known principles.

But it did not turn out so simply. Instead, these few puzzles were to be solved only by the introduction, in the early part of the twentieth century, of two revolutionary new theories that changed our whole conception of nature: the *theory of relativity* and *quantum theory*.

Physics as it was known at the end of the nineteenth century (what we've covered up to now in this book) is referred to as **classical physics**. The new physics that grew out of the great revolution at the turn of the twentieth century is now called **modern physics**. In this chapter, we present the special theory of relativity, which was first proposed by Albert Einstein (1879–1955; Fig. 37–1) in 1905. In the following chapter, we introduce the equally momentous quantum theory.

37–1 Galilean–Newtonian Relativity

Einstein's special theory of relativity deals with how we observe events, particularly how objects and events are observed from different frames of reference. This subject had, of course, already been explored by Galileo and Newton.

The special theory of relativity deals with events that are observed and measured from so-called **inertial reference frames** which (Sections 4–2 and 11–9) are reference frames in which Newton's first law is valid: if an object experiences no net force the object either remains at rest or continues in motion with constant speed in a straight line. It is easiest to analyze events when they are observed and measured from inertial frames, and the Earth, though not quite an inertial frame (it rotates), is close enough that for most purposes we can consider it an inertial frame. Rotating or otherwise accelerating frames of reference are noninertial frames, and won't concern us in this chapter (they are dealt with in Einstein's general theory of relativity).

A reference frame that moves with constant velocity with respect to an inertial frame is itself also an inertial frame, since Newton's laws hold in it as well. When we say that we observe or make measurements from a certain reference frame, it means that we are at rest in that reference frame.

Both Galileo and Newton were aware of what we now call the **relativity principle** applied to mechanics: that *the basic laws of physics are the same in all inertial reference frames.* You may have recognized its validity in everyday life. For example, objects move in the same way in a smoothly moving (constant-velocity) train or airplane as they do on Earth. (This assumes no vibrations or rocking—for they would make the reference frame noninertial.) When you walk, drink a cup of soup, play Ping-Pong, or drop a pencil on the floor while traveling in a train, airplane, or ship moving at constant velocity, the bodies move just as they do when you are at rest on Earth. Suppose you are in a car traveling rapidly along at constant velocity. If you release a coin from above your head inside the car, how will it fall? It falls straight downward with respect to the car, and hits the floor directly below the point of release, Fig. 37–2a. (If you drop the coin out the car's window, this won't happen because the moving air drags the coin backward relative to the car.) This is just how objects fall on the Earth—straight down—and thus our experiment in the moving car is in accord with the relativity principle.

Note in this example, however, that to an observer on the Earth, the coin follows a curved path, Fig. 37–2b. The actual path followed by the coin is different as viewed from different frames of reference. This does not violate the relativity principle because this principle states that the *laws* of physics are the same in all inertial frames. The same law of gravity, and the same laws of motion, apply in both reference frames. And the acceleration of the coin is the same in both reference frames. The difference in Figs. 37–2a and b is that in the Earth's frame of reference, the coin has an initial velocity (equal to that of the car). The laws of physics therefore predict it will follow a parabolic path like any projectile. In the car's ref-

Relativity principle: the laws of physics are the same in all inertial reference frames

FIGURE 37–2 A coin is dropped by a person in a moving car. (a) In the reference frame of the car, the coin falls straight down. (b) In a reference frame fixed on the Earth, the coin follows a curved (parabolic) path. The upper views show the moment of the coin's release, and the lower views are a short time later.

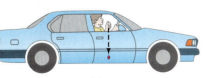
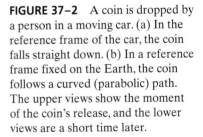

(a)
Reference frame = car

(b)
Reference frame = Earth

erence frame, there is no initial velocity, and the laws of physics predict that the coin will fall straight down. The laws are the same in both reference frames, although the specific paths are different.

Galilean–Newtonian relativity involves certain unprovable assumptions that make sense from everyday experience. It is assumed that the lengths of objects are the same in one reference frame as in another, and that time passes at the same rate in different reference frames. In classical mechanics, then, space and time are considered to be **absolute**: their measurement doesn't change from one reference frame to another. The mass of an object, as well as all forces, are assumed to be unchanged by a change in inertial reference frame.

The position of an object is, of course, different when specified in different reference frames, and so is velocity. For example, a person may walk inside a bus toward the front with a speed of 5 km/h. But if the bus moves 40 km/h with respect to the Earth, the person is then moving with a speed of 45 km/h with respect to the Earth. The acceleration of a body, however, is the same in any inertial reference frame according to classical mechanics. This is because the change in velocity, and the time interval, will be the same. For example, the person in the bus may accelerate from 0 to 5 km/h in 1.0 seconds, so $a = 5$ km/h/s in the reference frame of the bus. With respect to the Earth, the acceleration is $(45 \text{ km/h} - 40 \text{ km/h})/(1.0 \text{ s}) = 5$ km/h/s, which is the same.

Since neither F, m, nor a changes from one inertial frame to another, then Newton's second law, $F = ma$, does not change. Thus Newton's second law satisfies the relativity principle. It is easily shown that the other laws of mechanics also satisfy the relativity principle.

That the laws of mechanics are the same in all inertial reference frames implies that no one inertial frame is special in any sense. We express this important conclusion by saying that **all inertial reference frames are equivalent** for the description of mechanical phenomena. No one inertial reference frame is any better than another. A reference frame fixed to a car or an aircraft traveling at constant velocity is as good as one fixed on the Earth. When you travel smoothly at constant velocity in a car or airplane, it is just as valid to say you are at rest and the Earth is moving as it is to say the reverse. There is no experiment you can do to tell which frame is "really" at rest and which is moving. Thus, there is no way to single out one particular reference frame as being at absolute rest.

All inertial reference frames are equally valid

A complication arose, however, in the last half of the nineteenth century. When Maxwell presented his comprehensive and very successful theory of electromagnetism (Chapter 32), he showed that light can be considered an electromagnetic wave. Maxwell's equations predicted that the velocity of light c would be 3.00×10^8 m/s; and this is just what is measured, within experimental error. The question then arose: in what reference frame does light have precisely the value predicted by Maxwell's theory? For it was assumed that light would have a different speed in different frames of reference. For example, if observers were traveling on a rocket ship at a speed of 1.0×10^8 m/s away from a source of light, we might expect them to measure the speed of the light reaching them to be 3.0×10^8 m/s $- 1.0 \times 10^8$ m/s $= 2.0 \times 10^8$ m/s. But Maxwell's equations have no provision for relative velocity. They predicted the speed of light to be $c = 3.0 \times 10^8$ m/s. This seemed to imply there must be some special reference frame where c would have this value.

We discussed in Chapters 15 and 16 that waves travel on water and along ropes or strings, and sound waves travel in air and other materials. Nineteenth-century physicists viewed the material world in terms of the laws of mechanics, so it was natural for them to assume that light too must travel in some *medium*. They called this transparent medium the **ether** and assumed it permeated all space.[†] It was therefore assumed that the velocity of light given by Maxwell's equations must be with respect to the ether.

The "ether"

However, it appeared that Maxwell's equations did *not* satisfy the relativity principle. They were not the same in all inertial reference frames. They were simplest in the frame where $c = 3.00 \times 10^8$ m/s; that is, in a reference frame at rest in the ether. In any other reference frame, extra terms would have to be added to take into account the relative velocity. Thus, although most of the laws of physics obeyed the relativity principle, the laws of electricity and magnetism apparently did not. Instead, they seemed to single out one reference frame that was better than any other—a reference frame that could be considered absolutely at rest.

Scientists soon set out to determine the speed of the Earth relative to this absolute frame, whatever it might be. A number of clever experiments were designed. The most direct were performed by A. A. Michelson and E. W. Morley in the 1880s. The details of their experiment are discussed in the next Section. Briefly, what they did was measure the difference in the speed of light in different directions. They expected to find a difference depending on the orientation of their apparatus with respect to the ether. For just as a boat has different speeds relative to the land when it moves upstream, downstream, or across the stream, so too light would be expected to have different speeds depending on the velocity of the ether past the Earth.

Strange as it may seem, they detected no difference at all. This was a great puzzle. A number of explanations were put forth over a period of years, but they led to contradictions or were otherwise not generally accepted.

Then in 1905, Albert Einstein proposed a radical new theory that reconciled these many problems in a simple way. But at the same time, as we shall soon see, it completely changed our ideas of space and time.

* | 37–2 | The Michelson–Morley Experiment

The Michelson–Morley experiment was designed to measure the speed of the *ether*—the medium in which light was assumed to travel—with respect to the Earth. The experimenters thus hoped to find an absolute reference frame, one that could be considered to be at rest.

One of the possibilities nineteenth-century scientists considered was that the ether is fixed relative to the Sun, for even Newton had taken the Sun as the center of the universe. If this were the case (there was no guarantee, of course), the Earth's speed of about 3×10^4 m/s in its orbit around the Sun would produce a change of 1 part in 10^4 in the speed of light $(3.0 \times 10^8$ m/s$)$. Direct measurement of the speed of light to this accuracy was not possible. But A. A. Michelson, later with the help of E. W. Morley, was able to use his interferometer (Section 35–7) to measure the difference in the speed of light in different directions to this accuracy.

[†]The medium for light waves could not be air, since light travels from the Sun to Earth through nearly empty space. Therefore, another medium was postulated, the ether. The ether was not only transparent, but, because of difficulty in detecting it, was assumed to have zero density.

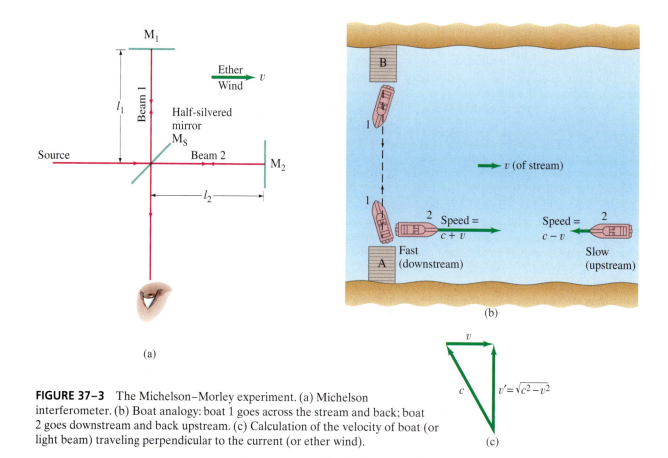

FIGURE 37–3 The Michelson–Morley experiment. (a) Michelson interferometer. (b) Boat analogy: boat 1 goes across the stream and back; boat 2 goes downstream and back upstream. (c) Calculation of the velocity of boat (or light beam) traveling perpendicular to the current (or ether wind).

This famous experiment is based on the principle shown in Fig. 37–3. Part (a) is a diagram of the Michelson interferometer, and it is assumed that the "ether wind" is moving with speed v to the right. (Alternatively, the Earth is assumed to move to the left with respect to the ether at speed v.) The light from the source is split into two beams by the half-silvered mirror M_S. One beam travels to mirror M_1 and the other to mirror M_2. The beams are reflected by M_1 and M_2 and are joined again after passing through M_S. The now superposed beams interfere with each other and the resultant is viewed by the observer's eye as an interference pattern (discussed in Section 35–7).

Whether constructive or destructive interference occurs at the center of the interference pattern depends on the relative phases of the two beams after they have traveled their separate paths. To examine this let us consider an analogy of a boat traveling up and down, and across, a river whose current moves with speed v, as shown in Fig. 37–3b. In still water, the boat can travel with speed c (not the speed of light in this case).

First we consider beam 2 in Fig. 37–3a, which travels parallel to the "ether wind." In its journey from M_S to M_2, we expect the light to travel with speed $c + v$, just as a boat traveling downstream (see Fig. 37–3b) adds the speed of the river to its own speed. Since the beam travels a distance l_2, the time it takes to go from M_S to M_2 is $t = l_2/(c + v)$. To make the return trip from M_2 to M_S, the light must move against the ether wind (like the boat going upstream), so its relative speed is expected to be $c - v$. The time for the return trip is $l_2/(c - v)$. The total time required for beam 2 to travel from M_S to M_2 and back to M_S is

$$ t_2 = \frac{l_2}{c + v} + \frac{l_2}{c - v} = \frac{2l_2}{c(1 - v^2/c^2)}. $$

Now let us consider beam 1, which travels crosswise to the ether wind. Here the boat analogy (part b) is especially helpful. The boat is to go from wharf A to wharf B directly across the stream. If it heads directly across, the stream's current will drag it downstream. To reach wharf B, the boat must head at an angle upstream. The precise angle depends on the magnitudes of c and v, but is of no interest to us in itself. Part (c) of Fig. 37–3 shows how to calculate the velocity v' of the boat relative to Earth as it crosses the stream. Since c, v, and v' form a right triangle, we have that $v' = \sqrt{c^2 - v^2}$. The boat has the same speed when it returns. If we now apply these principles to light beam 1 in Fig. 37–3a, we see that the beam travels with a speed $\sqrt{c^2 - v^2}$ in going from M_S to M_1 and back again. The total distance traveled is $2l_1$, so the time required for beam 1 to make the round trip is $2l_1/\sqrt{c^2 - v^2}$, or

$$t_1 = \frac{2l_1}{c\sqrt{1 - v^2/c^2}}.$$

Notice that the denominator in this equation for t_1 involves a square root, whereas that for t_2 does not.

If $l_1 = l_2 = l$, we see that beam 2 will lag behind beam 1 by an amount

$$\Delta t = t_2 - t_1 = \frac{2l}{c}\left(\frac{1}{1 - v^2/c^2} - \frac{1}{\sqrt{1 - v^2/c^2}}\right).$$

If $v = 0$, then $\Delta t = 0$, and the two beams will return in phase since they were initially in phase. But if $v \neq 0$, then $\Delta t \neq 0$, and the two beams will return out of phase. If this change of phase from the condition $v = 0$ to that for $v = v$ could be measured, then v could be determined. But the Earth cannot be stopped. Furthermore, we should not be too quick to assume that lengths are not affected by motion and therefore to assume $l_1 = l_2$.

Michelson and Morley realized that they could detect the difference in phase (assuming that $v \neq 0$) if they rotated their apparatus by 90°, for then the interference pattern between the two beams should change. In the rotated position, beam 1 would now move parallel to the ether and beam 2 perpendicular to it. Thus the roles could be reversed, and in the rotated position the times (designated by primes) would be

$$t_1' = \frac{2l_1}{c(1 - v^2/c^2)} \quad \text{and} \quad t_2' = \frac{2l_2}{c\sqrt{1 - v^2/c^2}}.$$

The time lag between the two beams in the nonrotated position (unprimed) would be

$$\Delta t = t_2 - t_1 = \frac{2l_2}{c(1 - v^2/c^2)} - \frac{2l_1}{c\sqrt{1 - v^2/c^2}}.$$

In the rotated position, the time difference would be

$$\Delta t' = t_2' - t_1' = \frac{2l_2}{c\sqrt{1 - v^2/c^2}} - \frac{2l_1}{c(1 - v^2/c^2)}.$$

When the rotation is made, the fringes of the interference pattern (Section 35–7) will shift an amount determined by the difference:

$$\Delta t - \Delta t' = \frac{2}{c}(l_1 + l_2)\left(\frac{1}{1 - v^2/c^2} - \frac{1}{\sqrt{1 - v^2/c^2}}\right).$$

This expression can be considerably simplified if we assume that $v/c \ll 1$. For in this case we can use the binomial expansion (Appendix A), so

$$\frac{1}{1 - v^2/c^2} \approx 1 + \frac{v^2}{c^2} \quad \text{and} \quad \frac{1}{\sqrt{1 - v^2/c^2}} \approx 1 + \frac{1}{2}\frac{v^2}{c^2}.$$

Then

$$\Delta t - \Delta t' \approx \frac{2}{c}(l_1 + l_2)\left(1 + \frac{v^2}{c^2} - 1 - \frac{1}{2}\frac{v^2}{c^2}\right)$$

$$\approx (l_1 + l_2)\frac{v^2}{c^3}.$$

Now we take $v = 3.0 \times 10^4$ m/s, the speed of the Earth in its orbit around the Sun. In Michelson and Morley's experiments, the arms l_1 and l_2 were about 11 m long. The time difference would then be about

$$\frac{(22 \text{ m})(3.0 \times 10^4 \text{ m/s})^2}{(3.0 \times 10^8 \text{ m/s})^3} \approx 7.0 \times 10^{-16} \text{ s}.$$

For visible light of wavelength $\lambda = 5.5 \times 10^{-7}$ m, say, the frequency would be $f = c/\lambda = (3.0 \times 10^8 \text{ m/s})/(5.5 \times 10^{-7} \text{ m}) = 5.5 \times 10^{14}$ Hz, which means that wave crests pass by a point every $1/(5.5 \times 10^{14} \text{ Hz}) = 1.8 \times 10^{-15}$ s. Thus, with a time difference of 7.0×10^{-16} s, Michelson and Morley should have noted a movement in the interference pattern of $(7.0 \times 10^{-16} \text{ s})/(1.8 \times 10^{-15} \text{ s}) = 0.4$ fringe. They could easily have detected this, since their apparatus was capable of observing a fringe shift as small as 0.01 fringe.

The null result

But they found *no significant fringe shift whatever*! They set their apparatus at various orientations. They made observations day and night so that they would be at various orientations with respect to the Sun (due to the Earth's rotation). They tried at different seasons of the year (the Earth at different locations due to its orbit around the Sun). Never did they observe a significant fringe shift.

This **null result** was one of the great puzzles of physics at the end of the nineteenth century. To explain it was a difficult challenge. One possibility to explain the null result was put forth independently by G. F. Fitzgerald and H. A. Lorentz (in the 1890s) in which they proposed that any length (including the arm of an interferometer) contracts by a factor $\sqrt{1 - v^2/c^2}$ in the direction of motion through the ether. According to Lorentz, this could be due to the ether affecting the forces between the molecules of a substance, which were assumed to be electrical in nature. This theory was eventually replaced by the far more comprehensive theory proposed by Albert Einstein in 1905—the special theory of relativity.

37–3 Postulates of the Special Theory of Relativity

The problems that existed at the turn of the century with regard to electromagnetic theory and Newtonian mechanics were beautifully resolved by Einstein's introduction of the theory of relativity in 1905. Einstein, however, was apparently not influenced directly by the null result of the Michelson–Morley experiment. What motivated Einstein were certain questions regarding electromagnetic theory and light waves. For example, he asked himself: "What would I see if I rode a light beam?" The answer was that instead of a traveling electromagnetic wave, he would see alternating electric and magnetic fields at rest whose magnitude changed in space, but did not change in time. Such fields, he realized, had never been detected and indeed were not consistent with Maxwell's electromagnetic theory. He argued, therefore, that it was unreasonable to think that the speed of light relative to any observer could be reduced to zero, or in fact reduced at all. This idea became the second postulate of his theory of relativity.

Einstein concluded that the inconsistencies he found in electromagnetic theory were due to the assumption that an absolute space exists. In his famous 1905 paper, he proposed doing away completely with the idea of the ether and the accompanying assumption of an absolute reference frame at rest. This proposal was embodied in two postulates. The first postulate was an extension of the Newtonian relativity principle to include not only the laws of mechanics but also those of the rest of physics, including electricity and magnetism:

> **First postulate (*the relativity principle*): The laws of physics have the same form in all inertial reference frames.**

The two postulates of special relativity

The second postulate is consistent with the first:

> **Second postulate (*constancy of the speed of light*): Light propagates through empty space with a definite speed c independent of the speed of the source or observer.**

These two postulates form the foundation of Einstein's **special theory of relativity**. It is called "special" to distinguish it from his later "general theory of relativity," which deals with noninertial (accelerating) reference frames. The special theory, which is what we discuss here, deals only with inertial frames.

The second postulate may seem hard to accept, for it violates commonsense notions. First of all, we have to think of light traveling through empty space. Giving up the ether is not too hard, however, for after all, it had never been detected. But the second postulate also tells us that the speed of light in vacuum is always the same, 3.00×10^8 m/s, no matter what the speed of the observer or the source. Thus, a person traveling toward or away from a source of light will measure the same speed for that light as someone at rest with respect to the source. This conflicts with our everyday notions, for we would expect to have to add in the velocity of the observer. Part of the problem is that in our everyday experience, we do not measure velocities anywhere near as large as the speed of light. Thus we can't expect our everyday experience to be helpful when dealing with such a high velocity. On the other hand, the Michelson–Morley experiment is fully consistent with the second postulate.[†]

Einstein's proposal has a certain beauty. For by doing away with the idea of an absolute reference frame, it was possible to reconcile classical mechanics with Maxwell's electromagnetic theory. The speed of light predicted by Maxwell's equations *is* the speed of light in vacuum in *any* reference frame.

Einstein's theory required giving up commonsense notions of space and time, and in the following Sections we will examine some strange but interesting consequences of special relativity. Our arguments for the most part will be simple ones. We will use a technique that Einstein himself did: we will imagine very simple experimental situations in which little mathematics is needed. In this way, we can see many of the consequences of relativity theory without getting involved in detailed calculations. Einstein called these "thought" experiments. Starting in Section 37–8 we will look more fully at the mathematics of relativity.

[†]The Michelson–Morley experiment can also be considered as evidence for the first postulate, for it was intended to measure the motion of the Earth relative to an absolute reference frame. Its failure to do so implies the absence of any such preferred frame.

37–4 | Simultaneity

One of the important consequences of the theory of relativity is that we can no longer regard time as an absolute quantity. No one doubts that time flows onward and never turns back. But, as we shall see in this Section and the next, the time interval between two events, and even whether two events are simultaneous, depends on the observer's reference frame.

Two events are said to occur simultaneously if they occur at exactly the same time. But how do we know if two events occur precisely at the same time? If they occur at the same point in space—such as two apples falling on your head at the same time—it is easy. But if the two events occur at widely separated places, it is more difficult to know whether the events are simultaneous since we have to take into account the time it takes for the light from them to reach us. Because light travels at finite speed, a person who sees two events must calculate back to find out when they actually occurred. For example, if two events are *observed* to occur at the same time, but one actually took place farther from the observer than the other, then the more distant one must have occurred earlier, and the two events were not simultaneous.

FIGURE 37–4 A moment after lightning strikes points A and B, the pulses of light are traveling toward the observer O, but O "sees" the lightning only when the light reaches O.

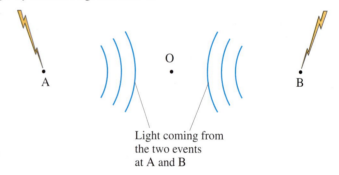

Light coming from the two events at A and B

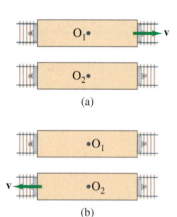

(a)

(b)

FIGURE 37–5 Observers O_1 and O_2, on two different trains (two different reference frames), are moving with relative velocity v. O_2 says that O_1 is moving to the right (a); O_1 says that O_2 is moving to the left (b). Both viewpoints are legitimate—it all depends on your reference frame.

We will now make use of a simple thought experiment. We assume an observer, called O, is located exactly halfway between points A and B where two events occur, Fig. 37–4. The two events may be lightning that strikes the points A and B, as shown, or any other type of events. For brief events like lightning, only short pulses of light will travel outward from A and B and reach O. O "sees" the events when the pulses of light reach point O. If the two pulses reach O at the same time, then the two events had to be simultaneous. This is because the two light pulses travel at the same speed (postulate 2), and since the distance OA equals OB, the time for the light to travel from A to O and B to O must be the same. Observer O can then definitely state that the two events occurred simultaneously. On the other hand, if O sees the light from one event before that from the other, then it is certain the former event occurred first.

The question we really want to examine is this: if two events are simultaneous to an observer in one reference frame, are they also simultaneous to another observer moving with respect to the first? Let us call the observers O_1 and O_2 and assume they are fixed in reference frames 1 and 2 that move with speed v relative to one another. These two reference frames can be thought of as trains (Fig. 37–5). O_2 says that O_1 is moving to the right with speed v, as in Fig. 37–5a; and O_1 says O_2 is moving to the left with speed v, as in Fig. 37–5b. Both viewpoints are legitimate according to the relativity principle. (There is, of course, no third point of view which will tell us which one is "really" moving.)

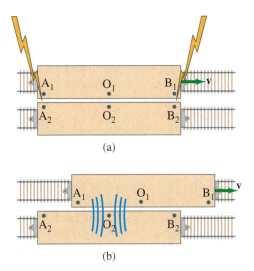

(a)

(b)

FIGURE 37–6 Thought experiment on simultaneity. To observer O_2, the reference frame of O_1 is moving to the right. In (a), one lightning bolt strikes the two reference frames at A_1 and A_2, and a second lightning bolt strikes at B_1 and B_2. (b) A moment later, the light from the two events reaches O_2 at the same time, so according to observer O_2, the two bolts of lightning strike simultaneously. But in O_1's reference frame, the light from B_1 has already reached O_1, whereas the light from A_1 has not yet reached O_1. So in O_1's reference frame, the event at B_1 must have preceded the event at A_1. Time is not absolute.

Now suppose two events occur that are observed and measured by both observers. Let us assume again that the two events are the striking of lightning and that the lightning marks both trains where it struck: at A_1 and B_1 on O_1's train, and at A_2 and B_2 on O_2's train. For simplicity, we assume that O_1 happens to be exactly halfway between A_1 and B_1, and that O_2 is halfway between A_2 and B_2. We now put ourselves in one reference frame or the other, from which we make our observations and measurements. Let us put ourselves in O_2's reference frame, so we observe O_1 moving to the right with speed v. Let us also assume that the two events occur *simultaneously* in O_2's frame, and just at the instant when O_1 and O_2 are opposite each other, Fig. 37–6a. A short time later, Fig. 37–6b, the light from A_2 and B_2 reaches O_2 at the same time (we assumed this). Since O_2 knows (or measures) the distances O_2A_2 and O_2B_2 as equal, O_2 knows the two events are simultaneous in the O_2 reference frame.

But what does observer O_1 observe and measure? From our (O_2) reference frame, we can predict what O_1 will observe. We see that O_1 moves to the right during the time the light is traveling to O_1 from A_1 and B_1. As shown in Fig. 37–6b, we can see from our O_2 reference frame that the light from B_1 has already passed O_1, whereas the light from A_1 has not yet reached O_1. Therefore, it is clear that O_1 will observe the light coming from B_1 before he observes the light coming from A_1. Now O_1's frame is as good as O_2's. Light travels at the same speed c for O_1 as for O_2 (the second postulate)[†]; and in the O_1 reference frame, this speed c is of course the same for light traveling from A_1 to O_1 as it is for light traveling from B_1 to O_1. Furthermore the distance O_1A_1 equals O_1B_1. Hence, since O_1 observes the light from B_1 before he observes the light from A_1 (we established this above, looking from the O_2 reference frame, Fig. 37–6b), then observer O_1 can only conclude that the event at B_1 occurred before the event at A_1. The two events are not simultaneous for O_1, even though they are for O_2.

We thus find that two events which are simultaneous to one observer are not necessarily simultaneous to a second observer.

It may be tempting to ask: "Which observer is right, O_1 or O_2?" The answer, according to relativity, is that they are *both* right. There is no "best" reference frame we can choose to determine which observer is right. Both frames are equally good. We can only conclude that *simultaneity is not an absolute concept*, but is relative. We are not aware of it in everyday life, however, because the effect is noticeable only when the relative speed of the two reference frames is very large (near c), or the distances involved are very large.

Simultaneity is relative

[†]Note that O_1 does not see himself catching up with one light beam and running away from the other (that is O_2's viewpoint of what happens for O_1). O_1 sees both light beams traveling at the same speed, c.

Because of the principle of relativity, the argument we gave for the thought experiment of Fig. 37–6 can be done from O_1's reference frame as well. In this case, O_1 will be at rest and will see event B_1 occur before A_1. But O_1 will recognize (by drawing a diagram equivalent to Fig. 37–6—try it and see!) that O_2, who is moving with speed v to the left, will see the two events as simultaneous.

37–5 Time Dilation and the Twin Paradox

The fact that two events simultaneous to one observer may not be simultaneous to a second observer suggests that time itself is not absolute. Could it be that time passes differently in one reference frame than in another? This is, indeed, just what Einstein's theory of relativity predicts, as the following thought experiment shows.

Figure 37–7 shows a spaceship traveling past Earth at high speed. The point of view of an observer on the spaceship is shown in part (a), and that of an observer on Earth in part (b). Both observers have accurate clocks. The person on the spaceship (a) flashes a light and measures the time it takes the light to travel across the spaceship and return after reflecting from a mirror. The light travels a distance $2D$ at speed c, so the time required, which we call Δt_0, is

$$\Delta t_0 = \frac{2D}{c}.$$

This is the time interval as measured by the observer on the spaceship.

The observer on Earth, Fig. 37–7b, observes the same process. But to this observer, the spaceship is moving. So the light travels the diagonal path shown going across the spaceship, reflecting off the mirror, and returning to the sender. Although the light travels at the same speed to this observer (the second postulate), it travels a greater distance. Hence the time required, as measured by

FIGURE 37–7 Time dilation can be shown by a thought experiment: the time it takes for light to travel over and back on a spaceship is longer for the observer on Earth (b) than for the observer on the spaceship (a).

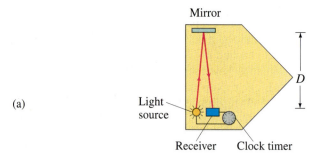

(a)

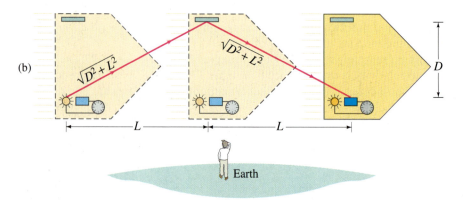

(b)

the observer on Earth, will be *greater* than that measured by the observer on the spaceship. The time interval, Δt, observed by the observer on Earth can be calculated as follows. In the time Δt, the spaceship travels a distance $2L = v\,\Delta t$ where v is the speed of the spaceship (Fig. 37–7b). Thus, the light travels a total distance on its diagonal path of $2\sqrt{D^2 + L^2}$, and therefore

$$c = \frac{2\sqrt{D^2 + L^2}}{\Delta t} = \frac{2\sqrt{D^2 + \dfrac{v^2(\Delta t)^2}{4}}}{\Delta t}.$$

We square both sides, and then solve for Δt, to find

$$c^2 = \frac{4D^2}{(\Delta t)^2} + v^2,$$

$$\Delta t = \frac{2D}{c\sqrt{1 - v^2/c^2}}.$$

We combine this with the formula above for $\Delta t_0 \left(\Delta t_0 = 2D/c\right)$ and find:

$$\Delta t = \frac{\Delta t_0}{\sqrt{1 - v^2/c^2}}.$$

(37–1) *Time-dilation formula*

Since $\sqrt{1 - v^2/c^2}$ is always less than 1, we see that $\Delta t > \Delta t_0$. That is, the time interval between the two events (the sending of the light, and its reception on the spaceship) is *greater* for the observer on Earth than for the observer on the spaceship. This is a general result of the theory of relativity, and is known as **time dilation**. Stated simply, the time-dilation effect says that

> **clocks moving relative to an observer are measured by that observer to run more slowly (as compared to clocks at rest).**

Time dilation: moving clocks run slow

However, we should not think that the clocks are somehow at fault. Time is actually measured to pass more slowly in any moving reference frame as compared to your own. This remarkable result is an inevitable outcome of the two postulates of the theory of relativity.

The concept of time dilation may be hard to accept, for it violates our commonsense understanding. We can see from Eq. 37–1 that the time dilation effect is negligible unless v is reasonably close to c. If v is much less than c, then the term v^2/c^2 is much smaller than the 1 in the denominator of Eq. 37–1, and then $\Delta t \approx \Delta t_0$ (see Example 37–2). The speeds we experience in everyday life are much smaller than c, so it is little wonder we don't ordinarily notice time dilation. Experiments have tested the time-dilation effect, and have confirmed Einstein's predictions. In 1971, for example, extremely precise atomic clocks were flown around the world in jet planes. The speed of the planes $\left(10^3\,\text{km/h}\right)$ was much less than c, so the clocks had to be accurate to nanoseconds $\left(10^{-9}\,\text{s}\right)$ in order to detect any time dilation. They were this accurate, and they confirmed Eq. 37–1 to within experimental error. Time dilation had been confirmed decades earlier, however, by observation on "elementary particles" which have very small masses (typically 10^{-30} to $10^{-27}\,\text{kg}$) and so require little energy to be accelerated to speeds close to the speed of light, c. Many of these elementary particles are not stable and decay after a time into lighter particles. One example is the muon, whose mean lifetime is $2.2\,\mu\text{s}$ when at rest. Careful experiments showed that when a muon is traveling at high speeds, its lifetime is measured to be longer than when it is at rest, just as predicted by the time-dilation formula.

Why we don't usually notice time dilation

EXAMPLE 37–1 **Lifetime of a moving muon.** (*a*) What will be the mean lifetime of a muon as measured in the laboratory if it is traveling at $v = 0.60\,c = 1.8 \times 10^8\,\text{m/s}$ with respect to the laboratory? Its mean life at rest is $2.2 \times 10^{-6}\,\text{s}$. (*b*) How far does a muon travel in the laboratory, on average, before decaying?

SOLUTION (*a*) If an observer were to move along with the muon (the muon would be at rest to this observer), the muon would have a mean life of $2.2 \times 10^{-6}\,\text{s}$. To an observer in the lab, the muon lives longer because of time dilation. From Eq. 37–1 with $v = 0.60\,c$, we have

$$\Delta t = \frac{\Delta t_0}{\sqrt{1 - \dfrac{v^2}{c^2}}} = \frac{2.2 \times 10^{-6}\,\text{s}}{\sqrt{1 - \dfrac{0.36\,c^2}{c^2}}} = \frac{2.2 \times 10^{-6}\,\text{s}}{\sqrt{0.64}} = 2.8 \times 10^{-6}\,\text{s}.$$

(*b*) At a speed of $1.8 \times 10^8\,\text{m/s}$, classical physics would tell us that with a mean life of $2.2\,\mu\text{s}$, an average muon would travel $d = vt = (1.8 \times 10^8\,\text{m/s})(2.2 \times 10^{-6}\,\text{s}) = 400\,\text{m}$. But relativity predicts an average distance of $(1.8 \times 10^8\,\text{m/s})(2.8 \times 10^{-6}\,\text{s}) = 500\,\text{m}$, and it is this longer distance that is measured experimentally.

Proper time

We need to make a comment about the use of Eq. 37–1 and the meaning of Δt and Δt_0. The equation is true only when Δt_0 represents the time interval between the two events in a reference frame where the two events occur at *the same point in space* (as in Fig. 37–7a where the two events are the light flash being sent and being received). This time interval, Δt_0, is called the **proper time**. Then Δt in Eq. 37–1 represents the time interval between the two events as measured in a reference frame moving with speed v with respect to the first. In Example 37–1 above, Δt_0 (and not Δt) was set equal to $2.2 \times 10^{-6}\,\text{s}$ because it is only in the rest frame of the muon that the two events ("birth" and "decay") occur at the same point in space.

EXAMPLE 37–2 **Time dilation at 100 km/h.** Let's check time dilation for everyday speeds. A car traveling 100 km/h covers a certain distance in 10.00 s according to the driver's watch. What does an observer on Earth measure for the time interval?

SOLUTION The car's speed relative to Earth is $100\,\text{km/h} = (1.00 \times 10^5\,\text{m})/(3600\,\text{s}) = 27.8\,\text{m/s}$. We set $\Delta t_0 = 10.00\,\text{s}$ in the time-dilation formula (the driver is at rest in the reference frame of the car), and then Δt is

$$\Delta t = \frac{\Delta t_0}{\sqrt{1 - \dfrac{v^2}{c^2}}} = \frac{10.00\,\text{s}}{\sqrt{1 - \left(\dfrac{27.8\,\text{m/s}}{3.00 \times 10^8\,\text{m/s}}\right)^2}} = \frac{10.00\,\text{s}}{\sqrt{1 - 8.59 \times 10^{-15}}}.$$

If you put these numbers into a calculator, you will obtain $\Delta t = 10.00\,\text{s}$, since the denominator differs from 1 by such a tiny amount. Indeed, the time measured by an observer on Earth would be no different from that measured by the driver, even with the best of today's instruments. A computer that could calculate to a large number of decimal places could reveal a difference between Δt and Δt_0. But we can estimate the difference quite easily using the binomial expansion (Appendix A),

➡ PROBLEM SOLVING
_____

Use of the binomial expansion

$$(1 \pm x)^n \approx 1 \pm nx. \qquad\qquad [\text{for } x \ll 1]$$

In our time-dilation formula, we have the factor $\left(1 - v^2/c^2\right)^{-\frac{1}{2}}$. Thus

$$\Delta t = \Delta t_0 \left(1 - \frac{v^2}{c^2}\right)^{-\frac{1}{2}} \approx \Delta t_0 \left(1 + \frac{1}{2}\frac{v^2}{c^2}\right)$$

$$\approx 10.00\,\text{s} \left[1 + \frac{1}{2}\left(\frac{27.8\,\text{m/s}}{3.00 \times 10^8\,\text{m/s}}\right)^2\right] \approx 10.00\,\text{s} + 4 \times 10^{-15}\,\text{s}.$$

So the difference between Δt and Δt_0 is predicted to be 4×10^{-15} s, an extremely small amount.

Time dilation has aroused interesting speculation about space travel. According to classical (Newtonian) physics, to reach a star 100 light-years away would not be possible for ordinary mortals (1 light-year is the distance light can travel in 1 year $= 3.0 \times 10^8$ m/s $\times 3.15 \times 10^7$ s $= 9.5 \times 10^{15}$ m). Even if a spaceship could travel at close to the speed of light, it would take over 100 years to reach such a star. But time dilation tells us that the time involved would be less for an astronaut. In a spaceship traveling at $v = 0.999c$, the time for such a trip would be only about $\Delta t_0 = \Delta t \sqrt{1 - v^2/c^2} = (100\,\text{yr})\sqrt{1 - (0.999)^2} = 4.5$ yr. Thus time dilation allows such a trip, but the enormous practical problems of achieving such speeds will not be overcome in the near future.

Notice, in this example, that whereas 100 years would pass on Earth, only 4.5 years would pass for the astronaut on the trip. Is it just the clocks that would slow down for the astronaut? The answer is no. All processes, including aging and other life processes, run more slowly for the astronaut according to the Earth observer. But to the astronaut, time would pass in a normal way. The astronaut would experience 4.5 years of normal sleeping, eating, reading, and so on. And people on Earth would experience 100 years of ordinary activity.

Not long after Einstein proposed the special theory of relativity, an apparent paradox was pointed out. According to this **twin paradox**, suppose one of a pair of 20-year-old twins takes off in a spaceship traveling at very high speed to a distant star and back again, while the other twin remains on Earth. According to the Earth twin, the traveling twin will age less. Whereas 20 years might pass for the Earth twin, perhaps only 1 year (depending on the spacecraft's speed) would pass for the traveler. Thus, when the traveler returns, the earthbound twin could expect to be 40 years old whereas the traveling twin would be only 21.

Twin paradox

This is the viewpoint of the twin on the Earth. But what about the traveling twin? If all inertial reference frames are equally good, won't the traveling twin make all the claims the Earth twin does, only in reverse? Can't the astronaut twin claim that since the Earth is moving away at high speed, time passes more slowly on Earth and the twin on Earth will age less? This is the opposite of what the Earth twin predicts. They cannot both be right, for after all the spacecraft returns to Earth and a direct comparison of ages and clocks can be made.

There is, however, not a paradox at all. The consequences of the special theory of relativity—in this case, time dilation—can be applied only by observers in inertial reference frames. The Earth is such a frame (or nearly so), whereas the spacecraft is not. The spacecraft accelerates at the start and end of its trip and, more importantly, when it turns around at the far point of its journey. During these acceleration periods, the twin on the spacecraft is not maintaining a steady inertial reference frame, so the spacecraft twin's predictions based on special relativity are not valid. The twin on Earth is in an inertial frame and can make valid predictions. Thus, there is no paradox. The traveling twin's point of view expressed above is not correct. The predictions of the Earth twin *are* valid according to the theory of relativity, and the prediction that the traveling twin returns having aged less is the proper one. Einstein's general theory of relativity, which deals with accelerating reference frames, confirms this result.

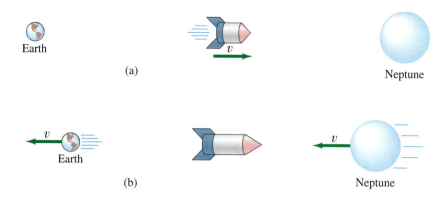

FIGURE 37–8 (a) A spaceship traveling at very high speed from Earth to Neptune, as seen from Earth's frame of reference. (b) As viewed by an observer on the spaceship, Earth and Neptune are moving at the very high velocity v: Earth leaves the spaceship, and a time Δt_0 later planet Neptune arrives at the spaceship. [Note that from the spaceship, part (b), each planet does not look shortened because at high speeds observers see the trailing edge (as in Fig. 37–10), and the net effect is to leave its appearance as a circle.]

37–6 Length Contraction

Not only time intervals are different in different reference frames. Space intervals—lengths and distances—are different as well, according to the special theory of relativity, and we illustrate this with a thought experiment.

Observers on Earth watch a spacecraft traveling at speed v from Earth to, say, Neptune, Fig. 37–8a. The distance between the planets, as measured by the Earth observers, is L_0. The time required for the trip, measured from Earth, is $\Delta t = L_0/v$. In Fig. 37–8b we see the point of view of observers on the spacecraft. In this frame of reference, the spaceship is at rest; Earth and Neptune move with speed v. (We assume v is much greater than the relative speed of Neptune and Earth, so the latter can be ignored.) The time between the departure of Earth and arrival of Neptune (as observed from the spacecraft) is the "proper time" (since the two events occur at the same point in space—i.e., on the spacecraft). Therefore the time interval is less for the spacecraft observers than for the Earth observers, because of time dilation. From Eq. 37–1, the time for the trip as viewed by the spacecraft is $\Delta t_0 = \Delta t\sqrt{1 - v^2/c^2}$. Since the spacecraft observers measure the same speed but less time between these two events, they must also measure the distance as less. If we let L be the distance between the planets as viewed by the spacecraft observers, then $L = v\,\Delta t_0$. We have already seen that $\Delta t_0 = \Delta t\sqrt{1 - v^2/c^2}$ and $\Delta t = L_0/v$, so we have $L = v\,\Delta t_0 = v\,\Delta t\sqrt{1 - v^2/c^2} = L_0\sqrt{1 - v^2/c^2}$. That is,

Length-contraction formula

$$L = L_0\sqrt{1 - \frac{v^2}{c^2}}. \tag{37–2}$$

This is a general result of the special theory of relativity and applies to lengths of objects as well as to distance. The result can be stated most simply in words as:

the length of an object is measured to be shorter when it is moving relative to the observer than when it is at rest.

Length contraction: moving objects are shorter (in the direction of motion)

This is called **length contraction**. The length L_0 in Eq. 37–2 is called the **proper length**. It is the length of the object—or distance between two points whose *positions are measured at the same time*—as determined by observers at rest with respect to it. Equation 37–2 gives the length L that will be measured by observers when the object travels past them at speed v. It is important to note, however, that length contraction occurs *only along the direction of motion*. For example, the moving spaceship in Fig. 37–8a is shortened in length, but its height is the same as when it is at rest.

Length contraction, like time dilation, is not noticeable in everyday life because the factor $\sqrt{1 - v^2/c^2}$ in Eq. 37–2 differs from 1.00 significantly only when v is very large.

EXAMPLE 37–3 **Painting's contraction.** A rectangular painting measures 1.00 m tall and 1.50 m wide. It is hung on the side wall of a spaceship which is moving past the Earth at a speed of 0.90 c. See Fig. 37–9a. (a) What are the dimensions of the picture according to the captain of the spaceship? (b) What are the dimensions as seen by an observer on the Earth?

SOLUTION (a) The painting (as well as everything else in the spaceship) looks perfectly normal to everyone on the spaceship, so the captain sees a 1.00 m by 1.50 m painting.
(b) Only the dimension in the direction of motion is shortened, so the height is unchanged at 1.00 m, Fig. 37–9b. The length, however, is contracted to

$$L = L_0 \sqrt{1 - \frac{v^2}{c^2}}$$
$$= (1.50 \text{ m}) \sqrt{1 - (0.90)^2} = 0.65 \text{ m}.$$

So the picture has dimensions 1.00 m × 0.65 m.

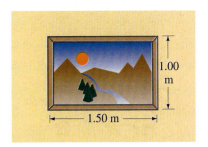

(a)

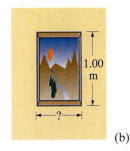

(b)

FIGURE 37–9 Example 37–3.

The appearance of moving objects

The Appearance of Objects Moving at Relativistic Speeds

Equation 37–2 tells us what the length of an object will be *measured* to be when traveling at speed v. The *appearance* of the object is another matter. Suppose, for example, you are traveling to the left past a small building at speed $v = 0.85 c$. This is equivalent to the building moving past you to the right at speed v. The building will look narrower (and the same height), but you will also be able to see the side of the building even if you are directly in front of it. This is shown in Fig. 37–10b—part (a) shows the building at rest. The fact that you see the side is not really a relativistic effect, but is due to the finite speed of light. To see how this occurs, we look at Fig. 37–10c which is a top view of the building, looking down. At the instant shown, the observer O is directly in front of the building. Light from points A and B reach O at the same time. If the building were at rest, light from point C could never reach O. But the building is moving at very high speed and does "get out of the way" so that light from C can reach O. Indeed, at the instant shown, light from point C when it was at an earlier location (C′ on the diagram) can reach O because the building has moved. In order to reach the observer at the same time as light from A and B, light from C had to leave at an earlier time since it must travel a greater distance. Thus it is light from C′ that reaches the observer at the same time as light from A and B. This, then, is how an observer might see both the front and side of an object at the same time even when directly in front of it.[†] It can be shown, by the same reasoning, that spherical objects will actually still have a circular outline even at high speeds. That is why the planets in Fig. 37–8b are drawn round rather than contracted.

[†]It would be an error to think that the building in Fig. 37–10b would look rotated. This is not correct since in that case side A would look shorter than side B. In fact, if the observer is directly in front, these sides appear equal in height. Thus the building looks contracted in its front face, but we also see the side, as described above. Also, though not shown in Fig. 37–10b, the walls of the building would appear curved, because of differing distances from the observer's eye of the various points from top to bottom along a vertical wall.

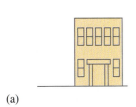

(a)

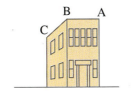

(b)

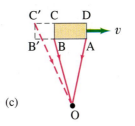

(c)

FIGURE 37–10 Building seen (a) at rest, and (b) moving at high speed. (c) Diagram explains why the side of the building is seen (see the text).

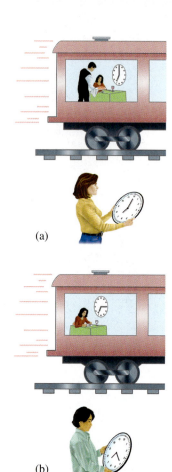

(a)

(b)

FIGURE 37–11 According to an accurate clock on a fast-moving train, a person (a) begins dinner at 7:00 and (b) finishes at 7:15. At the beginning of the meal, observers on Earth set their watches to correspond with the clock on the train. These observers measure the eating time as 20 minutes.

37–7 Four-Dimensional Space–Time

Let us imagine a person is on a train moving at a very high speed, say $0.65c$, Fig. 37–11. This person begins a meal at 7:00 and finishes at 7:15, according to a clock on the train. The two events, beginning and ending the meal, take place at the same point on the train. So the proper time between these two events is 15 min. To observers on Earth, the meal will take longer—20 min according to Eq. 37–1. Let us assume that the meal was served on a 20-cm-diameter plate. To observers on the Earth, the plate is only 15 cm wide (length contraction). Thus, to observers on the Earth, the meal looks smaller but lasts longer.

In a sense these two effects, time dilation and length contraction, balance each other. When viewed from the Earth, what the meal seems to lose in size it gains in length of time it lasts. Space, or length, is exchanged for time.

Considerations like this led to the idea of **four-dimensional space–time**: space takes up three dimensions and time is a fourth dimension. Space and time are intimately connected. Just as when we squeeze a balloon we make one dimension larger and another smaller, so when we examine objects and events from different reference frames, a certain amount of space is exchanged for time, or vice versa.

Although the idea of four dimensions may seem strange, it refers to the idea that any object or event is specified by four quantities—three to describe where in space, and one to describe when in time. The really unusual aspect of four-dimensional space–time is that space and time can intermix: a little of one can be exchanged for a little of the other when the reference frame is changed.

It is difficult for most of us to understand the idea of four-dimensional space–time. Somehow we feel, just as physicists did before the advent of relativity, that space and time are completely separate entities. Yet we have found in our thought experiments that they are not completely separate. Our difficulty in accepting this is reminiscent of the situation in the seventeenth century at the time of Galileo and Newton. Before Galileo, the vertical direction, that in which objects fall, was considered to be distinctly different from the two horizontal dimensions. Galileo showed that the vertical dimension differs only in that it happens to be the direction in which gravity acts. Otherwise, all three dimensions are equivalent, a viewpoint we all accept today. Now we are asked to accept one more dimension, time, which we had previously thought of as being somehow different. This is not to say that there is no distinction between space and time. What relativity has shown is that space and time determinations are not independent of one another.

37–8 Galilean and Lorentz Transformations

We now examine in detail the mathematics of relating quantities in one inertial reference frame to the equivalent quantities in another. In particular, we will see how positions and velocities *transform* (that is, change) from one frame to the other.

We begin with the classical or Galilean viewpoint. Consider two reference frames S and S′ which are each characterized by a set of coordinate axes, Fig. 37–12.

FIGURE 37–12 Inertial reference frame S′ moves to the right at speed v with respect to frame S.

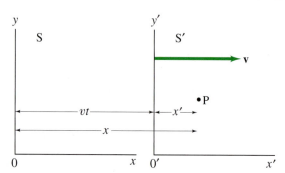

The axes x and y (z is not shown) refer to S and x' and y' to S'. The x' and x axes overlap one another, and we assume that frame S' moves to the right (in the x direction) at speed v with respect to S. And for simplicity let us assume the origins 0 and 0' of the two reference frames are superimposed at time $t = 0$.

Now consider an event that occurs at some point P (Fig. 37–12) represented by the coordinates x', y', z' in reference frame S' at the time t'. What will be the coordinates of P in S? Since S and S' overlap precisely initially, after a time t', S' will have moved a distance vt'. Therefore, at time t', $x = x' + vt'$. The y and z coordinates, on the other hand, are not altered by motion along the x axis; thus $y = y'$ and $z = z'$. Finally, since time is assumed to be absolute in Galilean–Newtonian physics, clocks in the two frames will agree with each other; so $t = t'$. We summarize these in the following **Galilean transformation equations**:

$$x = x' + vt'$$
$$y = y'$$
$$z = z'$$
$$t = t'.$$

[Galilean]   (37–3)   *Galilean transformation*

These equations give the coordinate of an event in the S frame when those in the S' frame are known. If those in the S frame are known, then the S' coordinates are obtained from

$$x' = x - vt, \qquad y' = y, \qquad z' = z, \qquad t' = t. \qquad \text{[Galilean]}$$

These four equations are the "inverse" transformation and are very easily obtained from Eqs. 37–3. Notice that the effect is merely to exchange primed and unprimed quantities and replace v by $-v$. This makes sense because from the S' frame, S moves to the left (negative x direction) with speed v.

Now suppose the point P in Fig. 37–12 represents a particle that is moving. Let the components of its velocity vector in S' be u'_x, u'_y, u'_z. (We use u to distinguish it from the relative velocity of the two frames, v.) Now $u'_x = dx'/dt'$, $u'_y = dy'/dt'$ and $u'_z = dz'/dt'$. The velocity of P as seen from S will have components u_x, u_y, and u_z. We can show how these are related to the velocity components in S' by differentiating Eqs. 37–3. For u_x we get

$$u_x = \frac{dx}{dt} = \frac{d(x' + vt')}{dt'} = u'_x + v$$

since v is assumed constant. For the other components, $u'_y = u_y$ and $u'_z = u_z$, so we have

$$u_x = u'_x + v$$
$$u_y = u'_y$$
$$u_z = u'_z.$$

[Galilean]   (37–4)   *Galilean velocity transformation*

These are known as the **Galilean velocity transformation equations**. We see that the y and z components of velocity are unchanged, but the x components differ by v: $u_x = u'_x + v$. This is just what we have used before when dealing with relative velocity.

The Galilean transformations, Eqs. 37–3 and 37–4, are valid only when the velocities involved are much less than c. We can see, for example, that the first of Eqs. 37–4 will not work for the speed of light; for light traveling in S' with speed $u'_x = c$ will have speed $c + v$ in S, whereas the theory of relativity insists it must be c in S. Clearly, then, a new set of transformation equations is needed to deal with relativistic velocities.

We will derive the required equations in a simple way, again looking at Fig. 37–12.

We assume that the transformation is linear and of the form

$$x = \gamma(x' + vt'), \qquad y = y', \qquad z = z'. \qquad \textbf{(i)}$$

That is, we modify the first of Eqs. 37–3 by multiplying by a constant γ which is yet to be determined ($\gamma = 1$ non-relativistically). But we assume the y and z equations are unchanged since there is no length contraction in these directions. We will not assume a form for t, but will derive it. The inverse equations must have the same form with v replaced by $-v$. (The principle of relativity demands it, since S′ moving to the right with respect to S is equivalent to S moving to the left with respect to S′.) Therefore

$$x' = \gamma(x - vt). \qquad \textbf{(ii)}$$

Now if a light pulse leaves the common origin of S and S′ at time $t = t' = 0$, after a time t it will have traveled a distance $x = ct$ or $x' = ct'$ along the x axis. Therefore, from Eqs. (i) and (ii) above,

$$ct = \gamma(ct' + vt') = \gamma(c + v)t', \qquad \textbf{(iii)}$$
$$ct' = \gamma(ct - vt) = \gamma(c - v)t. \qquad \textbf{(iv)}$$

We substitute t' from Eq. (iv) into Eq. (iii) and find $ct = \gamma(c + v)\gamma(c - v)(t/c) = \gamma^2(c^2 - v^2)t/c$. We cancel out the t on each side and solve for γ to find

$$\gamma = \frac{1}{\sqrt{1 - v^2/c^2}}.$$

Now that we have found γ, we need only find the relation between t and t'. To do so, we combine $x' = \gamma(x - vt)$ with $x = \gamma(x' + vt')$:

$$x' = \gamma(x - vt) = \gamma\big(\gamma[x' + vt'] - vt\big).$$

We solve for t and find $t = \gamma(t' + vx'/c^2)$. In summary,

Lorentz

transformation

equations

$$\begin{aligned} x &= \gamma(x' + vt') &&= \frac{1}{\sqrt{1 - v^2/c^2}}\,(x' + vt') \\[4pt] y &= y' \\[2pt] z &= z' && \qquad\qquad \textbf{(37–5)} \\[4pt] t &= \gamma\left(t' + \frac{vx'}{c^2}\right) &&= \frac{1}{\sqrt{1 - v^2/c^2}}\left(t' + \frac{vx'}{c^2}\right). \end{aligned}$$

These are called the **Lorentz transformation equations**. They were first proposed, in a slightly different form, by Lorentz in 1904 to explain the null result of the Michelson–Morley experiment and to make Maxwell's equations take the same form in all inertial reference frames. A year later Einstein derived them independently based on his theory of relativity. Notice that not only is the x equation modified as compared to the Galilean transformation, but so is the t equation; indeed, we see directly in this last equation how the space and time coordinates mix.

EXAMPLE 37–4 Deriving length contraction. Derive the length contraction formula, Eq. 37–2, from the Lorentz transformation equations.

SOLUTION Let an object of length L_0 be at rest on the x axis in S. The coordinates of its two end points are x_1 and x_2, so that $x_2 - x_1 = L_0$. At any instant in S′, the end points will be at x_1' and x_2' as given by the Lorentz transformation equations. The length measured in S′ is $L = x_2' - x_1'$. An observer in S′ measures this length by measuring x_2' and x_1' at the same time (in the S′ reference frame), so $t_2' = t_1'$. Then, from the first of Eqs. 37–5,

$$L_0 = x_2 - x_1 = \frac{1}{\sqrt{1 - v^2/c^2}}\,(x_2' + vt_2' - x_1' - vt_1').$$

Since $t_2' = t_1'$, we have

$$L_0 = \frac{1}{\sqrt{1 - v^2/c^2}}\,(x_2' - x_1') = \frac{L}{\sqrt{1 - v^2/c^2}},$$

or

$$L = L_0\sqrt{1 - v^2/c^2},$$

which is Eq. 37–2.

EXAMPLE 37–5 **Deriving time dilation.** Derive the time-dilation formula, Eq. 37–1, using the Lorentz transformation equations.

SOLUTION The time Δt_0 between two events that occur at the same place $(x'_2 = x'_1)$ in S′ is measured to be $\Delta t_0 = t'_2 - t'_1$. Since $x'_2 = x'_1$, then from the last of Eqs. 37–5, the time Δt between the events as measured in S is

$$\Delta t = t_2 - t_1 = \frac{1}{\sqrt{1 - v^2/c^2}} \left(t'_2 + \frac{vx'_2}{c^2} - t'_1 - \frac{vx'_1}{c^2} \right)$$

$$= \frac{1}{\sqrt{1 - v^2/c^2}} (t'_2 - t'_1) = \frac{\Delta t_0}{\sqrt{1 - v^2/c^2}},$$

which is Eq. 37–1. Notice that we chose S′ to be the frame in which the two events occur at the same place, so that $x'_1 = x'_2$ and the terms containing x'_1 and x'_2 cancel out.

The relativistically correct velocity equations are readily obtained by differentiating Eqs. 37–5 with respect to time. For example $\left(\text{using } \gamma = 1/\sqrt{1 - v^2/c^2}\right.$ and the chain rule for derivatives$\left.\right)$:

$$u_x = \frac{dx}{dt} = \frac{d}{dt} \left[\gamma(x' + vt') \right]$$

$$= \frac{d}{dt'} \left[\gamma(x' + vt') \right] \frac{dt'}{dt} = \gamma \left[\frac{dx'}{dt'} + v \right] \frac{dt'}{dt}.$$

But $dx'/dt' = u'_x$ and $dt'/dt = 1/(dt/dt') = 1/\left[\gamma(1 + vu'_x/c^2) \right]$ where we have differentiated the last of Eqs. 37–5 with respect to time. Therefore

$$u_x = \frac{\left[\gamma(u'_x + v) \right]}{\left[\gamma(1 + vu'_x/c^2) \right]} = \frac{u'_x + v}{1 + vu'_x/c^2}.$$

The others are obtained in the same way and we collect them here:

$$u_x = \frac{u'_x + v}{1 + vu'_x/c^2} \tag{37–6a}$$

Relativistic

$$u_y = \frac{u'_y \sqrt{1 - v^2/c^2}}{1 + vu'_x/c^2} \tag{37–6b}$$

velocity

$$u_z = \frac{u'_z \sqrt{1 - v^2/c^2}}{1 + vu'_x/c^2}. \tag{37–6c}$$

transformation

Note that even though the relative velocity **v** is in the x direction, the transformation of all the components of a particle's velocity are affected by v and the x component of the particle's velocity; this was not true for the Galilean transformation, Eqs. 37–4.

FIGURE 37–13 Rocket 2 is fired from rocket 1 with speed $u' = 0.60c$. What is the speed of rocket 2 with respect to the Earth?

EXAMPLE 37–6 **Adding velocities.** Calculate the speed of rocket 2 in Fig. 37–13 with respect to Earth.

SOLUTION Rocket 2 moves with speed $u' = 0.60c$ with respect to rocket 1. Rocket 1 has speed $v = 0.60c$ with respect to Earth. The velocities are along the same straight line which we take to be the x (and x') axis. We need use only the first of Eqs. 37–6. Then the speed of rocket 2 with respect to Earth is

$$u = \frac{u' + v}{1 + \dfrac{vu'}{c^2}} = \frac{0.60c + 0.60c}{1 + \dfrac{(0.60c)(0.60c)}{c^2}} = \frac{1.20c}{1.36} = 0.88c.$$

(The Galilean transformation would have given $u = 1.20c$.)

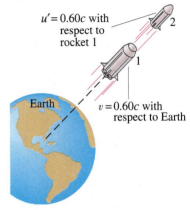

$u' = 0.60c$ with respect to rocket 1

$v = 0.60c$ with respect to Earth

Earth

Notice that Eqs. 37–6 reduce to the classical (Galilean) forms for velocities small compared to the speed of light, since $1 + vu'/c^2 \approx 1$ for v and $u' \ll c$. At the other extreme, let rocket 1 in Fig. 37–13 send out a beam of light, so that $u' = c$. Then Eq. 37–6a tells us the speed of light relative to Earth is

$$u = \frac{0.60\,c + c}{1 + \dfrac{(0.60\,c)(c)}{c^2}} = c,$$

which is consistent with the second postulate of relativity.

37–9 Relativistic Momentum and Mass

So far in this chapter, we have seen that two basic mechanical quantities, length and time intervals, need modification because they are relative—their value depends on the reference frame from which they are measured. We might expect that other physical quantities might need some modification according to the theory of relativity, such as momentum, energy, and mass.

Let us first examine momentum, which classically is defined as mass times velocity, $\mathbf{p} = m\mathbf{v}$. Classically, momentum is a conserved quantity. It would be a great benefit if the law of conservation of momentum were still valid in the relativistic domain, and we will try to insist that it does as we investigate any possible modification to the definition of momentum. A good guess as to how momentum might be altered would be the use of the factor $\gamma = 1/\sqrt{1 - v^2/c^2}$, as in Eqs. 37–1, 2, and 37–5. But let us be a little more general and let momentum be defined by $p = fmv$ where f is some function of v, $f(v)$. Now we consider a hypothetical collision between two objects—a thought experiment—and see what form $f(v)$ must take if momentum is to be conserved.

Our thought experiment involves the elastic collision of two *identical* balls. If the two balls travel at the same speed, v, we can safely say that they will have the same magnitude of momentum.

The collision in this thought experiment takes place as follows. We consider two inertial reference frames, S and S', moving along the x axis with a speed v with respect to each other. In reference frame S, a ball (call it ball A) is thrown with speed u along the y axis. In reference frame S', a second ball (B) is thrown with speed u along the negative y' axis. The two balls are thrown at just the right time so that they collide. We assume that they rebound elastically and, from symmetry, that each moves with the same speed u back along the y axis in its thrower's reference frame. Figure 37–14a shows the collision as seen by the person in reference frame S; and Fig. 37–14b shows the collision as seen from reference frame S'. In reference frame S ball A has velocity $+u$ along the y axis before the collision and $-u$ along the y axis after the collision. In frame S', ball A has, both before and after the collision, an x component of velocity equal to v, and a y component (see Eq. 37–6b with $u'_x = 0$) of magnitude

$$u\sqrt{1 - v^2/c^2}.$$

The same holds true for ball B, except in reverse. The velocity components are indicated in Fig. 37–14. Let us further assume that $u \ll v$, so that the speed of ball A is essentially v as seen in reference frame S', and thus A's momentum can be written $f(v)mv$. Similarly, the momentum of B in reference frame S is $f(v)mv$. We now apply the law of conservation of momentum, which we hope remains valid in relativity, even if momentum has to be redefined. That is, we assume that the total

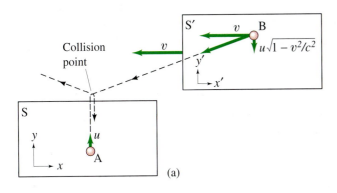

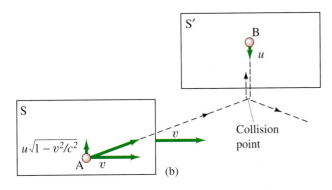

FIGURE 37–14 Deriving the momentum formula. Collision as seen (a) from reference frame S, (b) from reference frame S'.

momentum before the collision is equal to the total after the collision. We apply this to the y component of momentum in reference frame S (Fig. 37–14a):

$$f(u)mu - f(v)mu \sqrt{1 - v^2/c^2} = -f(u)mu + f(v)mu \sqrt{1 - v^2/c^2},$$

where we use $f(u)$ for ball A because its speed in S is only u.

We solve this for $f(v)$ and obtain

$$f(v) = \frac{f(u)}{\sqrt{1 - v^2/c^2}}.$$

This relation is valid for any u and v, (as long as $u \ll v$), and to simplify it so we can solve for f let us consider what happens if we let u become very small so that it approaches zero (this corresponds to a glancing collision with one of the balls essentially at rest and the other moving with speed v). Then the momentum terms $f(u)mu$ are in the nonrelativistic realm and take on the classical form, simply mu, meaning that $f(u) = 1$. So the previous equation becomes

$$f(v) = \frac{1}{\sqrt{1 - v^2/c^2}}.$$

We see that $f(v)$ comes out to be the factor we used before and called γ, and here has been shown to be valid for ball A. Using Fig. 37–14b we can derive the same relation for ball B. Thus we can conclude that we need to define the relativistic momentum of a particle as

Relativistic momentum

$$\mathbf{p} = \frac{m\mathbf{v}}{\sqrt{1 - v^2/c^2}} = \gamma m\mathbf{v}. \tag{37–7}$$

With this definition, as we saw in our thought experiment, the law of conservation of momentum will remain valid even in the relativistic realm. This relativistic momentum formula (Eq. 37–7) has been tested countless times on tiny elementary particles and been found valid.

This relativistic definition of momentum, Eq. 37–7, is sometimes interpreted as an increase in the mass of an object. That is, we can keep the form of our classical definition of momentum as

$$\mathbf{p} = m_{\text{rel}}\mathbf{v},$$

but only if we interpret m_{rel} to be the **relativistic mass**, which increases with speed according to

$$m_{\text{rel}} = \frac{m}{\sqrt{1 - v^2/c^2}}. \qquad \textbf{(37–8)}$$

Rest mass   In this "mass-increase" formula, m is referred to as the **rest mass** of the object—the mass it has as measured in a reference frame in which it is at rest; and m_{rel} is the object's relativistic mass in a reference frame in which it moves at speed v. With this interpretation, *the mass of an object appears to increase as its speed increases.* But we must be careful in the use of relativistic mass. We cannot just plug it into formulas like $F = ma$ or $K = \frac{1}{2}mv^2$ (as we can here for momentum, obtaining Eq. 37–7). For example, if we substitute Eq. 37–8 into $F = ma$, we obtain a formula that does not agree with experiment. If however, we write Newton's second law in its more general form, $\mathbf{F} = d\mathbf{p}/dt$, we do get a correct result. That is, Newton's second law, stated in its most general form is

$$\mathbf{F} = \frac{d\mathbf{p}}{dt} = \frac{d}{dt}(\gamma m\mathbf{v}) = \frac{d}{dt}\left(\frac{m\mathbf{v}}{\sqrt{1 - v^2/c^2}}\right) \qquad \textbf{(37–9)}$$

and is valid relativistically.[†]

Whenever we talk about the mass of an object, we will always mean its rest mass (a fixed value). On the rare occasion when we want to refer to an object's relativistic mass, we will say so explicitly.

37–10   The Ultimate Speed

A basic result of the special theory of relativity is that the speed of an object cannot equal or exceed the speed of light. That the speed of light is a natural speed limit in the universe can be seen from any of Eqs. 37–1, 37–2, 37–7, or the addition of velocities formula. It is perhaps easiest to see from Eq. 37–7. As an object is accelerated to greater and greater speeds, its momentum becomes larger and larger. Indeed, if v were to equal c, the denominator in this equation would be zero, and the momentum would be infinite. To accelerate an object up to $v = c$ would thus require infinite energy, and so is not possible.

37–11   Energy and Mass; $E = mc^2$

When a steady net force is applied to an object, the object increases in speed. Since the force is acting through a distance, work is done on the object and its energy increases. As the speed of the object approaches c, the speed cannot increase indefinitely since it cannot exceed c. On the other hand, the relativistic mass of the object increases with increasing speed. That is, the work done on an object not only increases its speed but also contributes to increasing its inertia. Normally, the work done on an object increases its energy. This new twist from the theory of relativity leads to the idea that mass is a form of energy, a crucial part of Einstein's theory of relativity.

[†] Clearly, Newton's second law written as $\mathbf{F} = ma$ is not valid, even if we use the relativistic mass. There is an extra term:

$$\mathbf{F} = \frac{d\mathbf{p}}{dt} = \frac{d(\gamma m\mathbf{v})}{dt} = \frac{d(m_{\text{rel}}\mathbf{v})}{dt} = m_{\text{rel}}\mathbf{a} + \frac{dm_{\text{rel}}}{dt}\mathbf{v}.$$

To find the mathematical relationship between mass and energy, we assume that the work-energy theorem is still valid in relativity, and we take the motion to be along the x axis. The work done to increase a particle's speed from zero to v is

$$W = \int_i^f F\,dx = \int_i^f \frac{dp}{dt}\,dx = \int_i^f \frac{dp}{dt}\,v\,dt = \int_i^f v\,dp$$

where i and f refer to the initial $(v = 0)$ and final $(v = v)$ states. Since $d(pv) = p\,dv + v\,dp$ we can write

$$v\,dp = d(pv) - p\,dv$$

so

$$W = \int_i^f d(pv) - \int_i^f p\,dv.$$

Since integration is the exact inverse of differentiation, the first term on the right becomes

$$\int_i^f d(pv) = pv\Big|_i^f = (\gamma mv)v = \frac{mv^2}{\sqrt{1 - v^2/c^2}}.$$

The second term in our equation for W above is easily integrated since

$$\frac{d}{dv}\left(\sqrt{1 - v^2/c^2}\right) = -(v/c^2)/\sqrt{1 - v^2/c^2},$$

and so becomes

$$-\int_i^f p\,dv = -\int_0^v \frac{mv}{\sqrt{1 - v^2/c^2}}\,dv = mc^2\sqrt{1 - v^2/c^2}\,\Big|_0^v$$

$$= mc^2\sqrt{1 - v^2/c^2} - mc^2.$$

Finally, we have for W:

$$W = \frac{mv^2}{\sqrt{1 - v^2/c^2}} + mc^2\sqrt{1 - v^2/c^2} - mc^2.$$

We multiply the second term by $\sqrt{1 - v^2/c^2}/\sqrt{1 - v^2/c^2} = 1$, and obtain

$$W = \frac{mc^2}{\sqrt{1 - v^2/c^2}} - mc^2.$$

By the work-energy theorem, the work done must equal the final kinetic energy K since the particle started from rest. Therefore

$$K = \frac{mc^2}{\sqrt{1 - v^2/c^2}} - mc^2 \qquad \textbf{(37–10a)}$$

$$= \gamma mc^2 - mc^2 = (\gamma - 1)mc^2. \qquad \textbf{(37–10b)}$$

Relativistic kinetic energy

Clearly, the kinetic energy K is not $\frac{1}{2}mv^2$ at high speeds. $\left[\frac{1}{2}mv^2\right.$ does not work for m as rest mass, nor for relativistic mass.$\left.\right]$

Equations 37–10 require some interpretation. First of all, what does the second term in Eq. 37–10a mean, the mc^2? Consistent with the idea that mass is a form of energy, Einstein called mc^2 the **rest energy** of the object. We can rearrange Eq. 37–10a to get

Rest energy, mc^2

$$E = K + mc^2, \qquad \textbf{(37–11a)}$$

where

$$E = \gamma mc^2 = \frac{mc^2}{\sqrt{1 - v^2/c^2}} \qquad \textbf{(37–11b)}$$

Total energy (defined)

is called the *total energy* E of the particle (assuming no potential energy), and equals the rest energy plus the kinetic energy.

For a particle at rest in a given reference frame, K is zero in Eq. 37–11a, so its total energy is its rest energy:

Mass related to energy

$$E = mc^2.$$ **(37–12)**

Here we have Einstein's famous formula, $E = mc^2$. This formula mathematically relates the concepts of energy and mass. But if this idea is to have any meaning from a practical point of view, then mass ought to be convertible to energy and vice versa. That is, if mass is just one form of energy, then it should be convertible to

Mass and energy interchangeable

other forms of energy just as other types of energy are interconvertible. Einstein suggested that this might be possible, and indeed changes of mass to other forms of energy, and vice versa, have been experimentally confirmed countless times. The interconversion of mass and energy is most easily detected in nuclear and elementary particle physics. For example, the neutral pion (π^0) of rest mass 2.4×10^{-28} kg is observed to decay into pure electromagnetic radiation (photons). The π^0 completely disappears in the process. The amount of electromagnetic energy produced is found to be exactly equal to that predicted by Einstein's formula, $E = mc^2$. The reverse process is also commonly observed in the laboratory: electromagnetic radiation under certain conditions can be converted into material particles such as electrons (see Section 38–4 on pair production). On a larger scale, the energy produced in nuclear power plants is a result of the loss in rest mass of the uranium fuel as it undergoes the process called fission. Even the radiant energy we receive from the Sun is an example of $E = mc^2$; the Sun's mass is continually decreasing as it radiates electromagnetic energy outward.

The relation $E = mc^2$ is now believed to apply to all processes, although the changes are often too small to measure. That is, when the energy of a system changes by an amount ΔE, the mass of the system changes by an amount Δm given by

$$\Delta E = (\Delta m)(c^2).$$

In a chemical reaction where heat is gained or lost, the masses of the reactants and the products will be different. Even when water is heated on a stove, the mass of the water increases very slightly.[†]

EXAMPLE 37–7 **Pion's kinetic energy.** A π^0 meson $(m = 2.4 \times 10^{-28}$ kg$)$ travels at a speed $v = 0.80\,c = 2.4 \times 10^8$ m/s. What is its kinetic energy? Compare to a classical calculation.

SOLUTION We substitute values into Eq. 37–10b

$$K = (\gamma - 1)mc^2$$

where

$$\gamma = \frac{1}{\sqrt{1 - v^2/c^2}} = \frac{1}{\sqrt{1 - (0.80)^2}} = 1.67.$$

Then

$$K = (1.67 - 1)(2.4 \times 10^{-28}\,\text{kg})(3.0 \times 10^8\,\text{m/s})^2$$
$$= 1.4 \times 10^{-11}\,\text{J}.$$

➥ **PROBLEM SOLVING**

Relativistic kinetic energy

Notice that the units of mc^2 are kg·m²/s², which is the joule. A classical calculation would give $K = \frac{1}{2}mv^2 = \frac{1}{2}(2.4 \times 10^{-28}\,\text{kg})(2.4 \times 10^8\,\text{m/s})^2 = 6.9 \times 10^{-12}$ J, about half as much, but this is not a correct result.

EXAMPLE 37–8 **Energy from nuclear decay.** The energy required or released in nuclear reactions and decays comes from a change in mass between the initial and final particles. In one type of radioactive decay, an atom of uranium $(m = 232.03714$ u$)$ decays to an atom of thorium $(m = 228.02873$ u$)$ plus an atom of helium $(m = 4.00260$ u$)$ where the masses (always rest masses) given are in atomic mass units $(1\,\text{u} = 1.6605 \times 10^{-27}$ kg$)$. Calculate the energy released in this decay.

[†]This last example is also easy to understand from the point of view of kinetic theory (Chapter 18): as heat is added, the temperature and therefore the average speed of the molecules increases, and Eq. 37–8 tells us that the relativistic mass of the molecules also increases.

SOLUTION The initial mass is 232.03714 u, and after the decay it is 228.02873 u + 4.00260 u = 232.03133 u, so there is a decrease in mass of 0.00581 u. This mass decrease, which equals $(0.00581 \text{ u})(1.66 \times 10^{-27} \text{ kg}) = 9.64 \times 10^{-30} \text{ kg}$, is changed into kinetic energy. Thus

Energy released in nuclear process

$$E = \Delta m c^2 = (9.64 \times 10^{-30} \text{ kg})(3.0 \times 10^8 \text{ m/s})^2 = 8.68 \times 10^{-13} \text{ J}.$$

Since $1 \text{ MeV} = 1.60 \times 10^{-13} \text{ J}$, the energy released is 5.4 MeV.

EXAMPLE 37–9 **Mass change in a chemical reaction.** When two moles of hydrogen and one mole of oxygen react to form two moles of water, the energy released is 484 kJ. How much does the mass decrease in this reaction?

SOLUTION Using Eq. 37–12 we have for the change in mass Δm:

$$\Delta m = \frac{\Delta E}{c^2} = \frac{(-484 \times 10^3 \text{ J})}{(3.00 \times 10^8 \text{ m/s})^2} = -5.38 \times 10^{-12} \text{ kg}.$$

The initial mass of the system is $0.002 \text{ kg} + 0.016 \text{ kg} = 0.018 \text{ kg}$. Thus the change in mass is relatively very tiny and can normally be neglected. [Conservation of mass is usually a reasonable principle to apply to chemical reactions.]

Equation 37–10a for the kinetic energy is

$$K = mc^2 \left(\frac{1}{\sqrt{1 - v^2/c^2}} - 1 \right).$$

At low speeds, $v \ll c$, we can expand the square root in the denominator using the binomial expansion, $(1 \pm x)^n = 1 \pm nx + n(n-1)x^2/2! + \cdots$. With $n = -\frac{1}{2}$, we get

$$K \approx mc^2 \left(1 + \frac{1}{2} \frac{v^2}{c^2} + \cdots - 1 \right)$$

$$\approx \tfrac{1}{2} mv^2,$$

where the dots in the first expression represent very small terms in the expansion which we have neglected since we assumed that $v \ll c$. Thus at low speeds, the relativistic form for kinetic energy reduces to the classical form, $K = \frac{1}{2}mv^2$. This is, of course, what we would like. It makes relativity a more viable theory in that it can predict accurate results at low speed as well as at high. Indeed, the other equations of special relativity also reduce to their classical equivalents at ordinary speeds: length contraction, time dilation, and modifications to momentum as well as kinetic energy, all disappear for $v \ll c$ since $\sqrt{1 - v^2/c^2} \approx 1$.

   A useful relation between the total energy E of a particle and its momentum p can also be derived. The momentum of a particle of mass m and speed v is given by Eq. 37–7

$$p = \gamma mv = \frac{mv}{\sqrt{1 - v^2/c^2}}.$$

Relativistic momentum

The total energy is

$$E = K + mc^2$$

or

$$E = \gamma mc^2 = \frac{mc^2}{\sqrt{1 - v^2/c^2}}.$$

We square this equation (and we insert "$v^2 - v^2$" which is zero, but will help us):

$$E^2 = \frac{m^2 c^2 (v^2 - v^2 + c^2)}{1 - v^2/c^2}$$

$$= p^2 c^2 + \frac{m^2 c^4 (1 - v^2/c^2)}{1 - v^2/c^2}$$

or

$$E^2 = p^2 c^2 + m^2 c^4. \tag{37–13}$$

Energy related to momentum

Thus, the total energy can be written in terms of the momentum p, or in terms of the kinetic energy (Eq. 37–11), where we have assumed there is no potential energy.

We can rewrite Eq. 37–13 as $E^2 - p^2c^2 = m^2c^4$. Since the rest mass m of a given particle is the same in any reference frame, we see that the quantity $E^2 - p^2c^2$ must also be the same in any reference frame. Thus, at any given moment the total energy E and momentum p of a particle will be different in different reference frames, but the quantity $E^2 - p^2c^2$ will have the same value in all inertial reference frames. We say that the quantity $E^2 - p^2c^2$ is **invariant** under a Lorentz transformation.

$E^2 - p^2c^2$
is invariant

* 37–12 Doppler Shift for Light

In Section 16–7 we discussed how the frequency and wavelength of sound are altered if the source of the sound and the observer are moving toward or away from each other. When a source is moving toward us, the frequency is higher than when the source is at rest. If the source moves away from us, the frequency is lower. We obtained four different equations for the Doppler shift (Eqs. 16–9a and b, Eqs. 16–10a and b), depending on the direction of the relative motion and whether the source or the observer is moving. The Doppler effect occurs also for light; but the shifted frequency or wavelength is given by slightly different equations, and there are only two of them, because for light—according to special relativity—we can make no distinction between motion of the source and motion of the observer. (Recall that sound travels in a medium such as air, whereas light does not—there is no evidence for an ether.)

To derive the Doppler shift for light, let us consider a light source and an observer that move toward each other, and let their relative velocity be v as measured in the reference frame of either the source or the observer. Figure 37–15a shows a source at rest emitting light waves of frequency f_0 and wavelength $\lambda_0 = c/f_0$. Two wavecrests are shown, a distance λ_0 apart, the second crest just having been emitted. In Fig. 37–15b, the source is shown moving at speed v toward a stationary observer who will see the wavelength λ being somewhat less than λ_0. (This is much like Fig. 16–20 for sound.) Let Δt represent the time between crests as detected by the observer, whose reference frame is shown in Fig. 37–15b. From Fig. 37–15b we see that

$$\lambda = c\,\Delta t - v\,\Delta t$$

where $c\,\Delta t$ is the distance crest 1 has moved in the time Δt after it was emitted, and $v\,\Delta t$ is the distance the source has moved in time Δt. So far our derivation has not differed from that for sound (Section 16–7). Now we invoke the theory of relativity. The time between emission of wavecrests has undergone time dilation:

$$\Delta t = \Delta t_0 / \sqrt{1 - v^2/c^2}$$

where Δt_0 is the time between emissions of wavecrests in the reference frame where the source is at rest (the "proper" time). In the source's reference frame (Fig. 37–15a), we have

$$\Delta t_0 = \frac{1}{f_0} = \frac{\lambda_0}{c}$$

(Eqs. 3–15 and 32–14). Thus

$$\lambda = (c - v)\,\Delta t = (c - v)\frac{\Delta t_0}{\sqrt{1 - v^2/c^2}} = \frac{(c - v)}{\sqrt{c^2 - v^2}}\lambda_0$$

or

$$\lambda = \lambda_0 \sqrt{\frac{c - v}{c + v}}. \qquad \begin{bmatrix} \text{source and observer} \\ \text{moving toward} \\ \text{each other} \end{bmatrix} \quad \textbf{(37–14a)}$$

Doppler

shift

for

light

The frequency f is (recall $\lambda_0 = c/f_0$)

$$f = \frac{c}{\lambda} = f_0 \sqrt{\frac{c + v}{c - v}}. \qquad \begin{bmatrix} \text{source and observer} \\ \text{moving toward} \\ \text{each other} \end{bmatrix} \quad \textbf{(37–14b)}$$

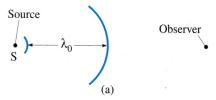

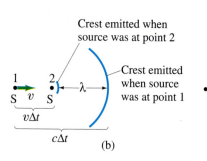

FIGURE 37–15 Doppler shift for light. (a) Reference frame of light source. (b) Reference frame of observer, toward whom source is moving.

Here f_0 is the frequency of the light as seen in the source's reference frame, and f is the frequency as measured by an observer moving toward the source or toward whom the source is moving. Equations 37–14 depend only on the relative velocity v. For relative motion *away* from each other we set $v < 0$ in Eqs. 37–14, and obtain

$$\lambda = \lambda_0 \sqrt{\frac{c + v}{c - v}}$$

$$f = f_0 \sqrt{\frac{c - v}{c + v}}.$$

$$\begin{bmatrix} \text{source and observer} \\ \text{moving away from} \\ \text{each other} \end{bmatrix}$$

(37–15a)

(37–15b)

From Eqs. 37–14 and 37–15 we see that light from a source moving toward us will have a higher frequency and shorter wavelength, whereas if a light source moves away from us, we will see a lower frequency and a longer wavelength. In the latter case, visible light will have its wavelength lengthened toward the red end of the visible spectrum (Fig. 33–24), an effect called a **redshift**. As we will see in the next chapter, all atoms have their own distinctive signature in terms of the frequencies of the light they emit. In 1929 the American astronomer Edwin Hubble (1889–1953) found that radiation from atoms in many galaxies is redshifted. That is, the frequencies of light emitted are lower than those emitted by stationary atoms on Earth, suggesting that the galaxies are receding from us. This is the origin of the idea that the universe is expanding.

37–13 The Impact of Special Relativity

A great many experiments have been performed to test the predictions of the special theory of relativity. Within experimental error, no contradictions have been found. Scientists have therefore accepted relativity as an accurate description of nature.

At speeds much less than the speed of light, the relativistic formulas reduce to the old classical ones, as we have discussed. We would, of course, hope—or rather, insist—that this be true since Newtonian mechanics works so well for objects moving with speeds $v \ll c$. This insistence that a more general theory (such as relativity) give the same results as a more restricted theory (such as classical mechanics which works for $v \ll c$) is called the **correspondence principle**. The two theories must correspond where their realms of validity overlap. Relativity thus does not contradict classical mechanics. Rather, it is a more general theory, of which classical mechanics is now considered to be a limiting case.

Correspondence principle

The importance of relativity is not simply that it gives more accurate results, especially at very high speeds. Much more than that, it has changed the way we view the world. The concepts of space and time are now seen to be relative, and intertwined with one another, whereas before they were considered absolute and separate. Even our concepts of matter and energy have changed: either can be converted to the other. The impact of relativity extends far beyond physics. It has influenced the other sciences, and even the world of art and literature; it has, indeed, entered the general culture.

From a practical point of view, we do not have much opportunity in our daily lives to use the mathematics of relativity. For example, the factor $\sqrt{1 - v^2/c^2}$, which appears in many relativistic formulas, has a value of 0.995 when $v = 0.10c$. Thus, for speeds even as high as $0.10c = 3.0 \times 10^7$ m/s, the factor $\sqrt{1 - v^2/c^2}$ in relativistic formulas gives a numerical correction of less than 1 percent. For speeds less than $0.10c$, or unless mass and energy are interchanged, we don't usually need to use the more complicated relativistic formulas, and can use the simpler classical formulas.

Summary

An **inertial reference frame** is one in which Newton's law of inertia holds. Inertial reference frames can move at constant velocity relative to one another; accelerating reference frames are noninertial.

The **special theory of relativity** is based on two principles: the **relativity principle**, which states that the laws of physics are the same in all inertial reference frames, and the principle of the **constancy of the speed of light**, which states that the speed of light in empty space has the same value in all inertial reference frames.

One consequence of relativity theory is that two events that are simultaneous in one reference frame may not be simultaneous in another. Other effects are **time dilation**: moving clocks are measured to run slow; and **length contraction**: the length of a moving object is measured to be shorter (in its direction of motion) than when it is at rest. Quantitatively,

$$L = L_0 \sqrt{1 - v^2/c^2}$$

$$\Delta t = \frac{\Delta t_0}{\sqrt{1 - v^2/c^2}}$$

where L and Δt are the length and time interval of objects (or events) observed as they move by at the speed v; L_0 and Δt_0 are the **proper length** and **proper time**—that is, the same quantities as measured in the rest frame of the objects or events.

The **Lorentz transformations** relate the positions and times of events in one inertial reference frame to their positions and times in a second inertial reference frame.

$$x = \gamma(x' + vt')$$
$$y = y'$$
$$z = z'$$
$$t = \gamma\left(t' + \frac{vx'}{c^2}\right)$$

where $\gamma = 1/\sqrt{1 - v^2/c^2}$.

Velocity addition also must be done in a special way. All these relativistic effects are significant only at high speeds, close to the speed of light, which itself is the ultimate speed in the universe.

The theory of relativity has changed our notions of space and time, and of momentum, energy, and mass. Space and time are seen to be intimately connected, with time being the fourth dimension in addition to the three dimensions of space.

The **momentum** of an object is given by

$$\mathbf{p} = \gamma m\mathbf{v} = \frac{m\mathbf{v}}{\sqrt{1 - v^2/c^2}}.$$

This formula can be interpreted as a *mass increase*, where the relativistic mass is $m_{\text{rel}} = \gamma m$ and m is the **rest mass** of the object ($v = 0$).

Mass and energy are interconvertible. The equation

$$E = mc^2$$

tells how much energy E is needed to create a mass m, or vice versa. Said another way, $E = mc^2$ is the amount of energy an object has because of its mass m. The law of conservation of energy must include mass as a form of energy.

The kinetic energy K of an object moving at speed v is given by

$$K = (\gamma - 1)mc^2 = \frac{mc^2}{\sqrt{1 - v^2/c^2}} - mc^2$$

where m is the rest mass of the object. The total energy E is

$$E = K + mc^2 = \gamma mc^2.$$

The momentum p of an object is related to its total energy E (assuming no potential energy) by

$$E^2 = p^2c^2 + m^2c^4.$$

Questions

1. You are in a windowless car in an exceptionally smooth train. Is there any physical experiment you can do in the train car to determine whether you are moving?

2. You might have had the experience of being at a red light when, out of the corner of your eye, you see the car beside you creep forward. Instinctively you stomp on the brake pedal, thinking that you are rolling backward. What does this say about absolute and relative motion?

3. A worker stands on top of a moving railroad car, and throws a heavy ball straight up (from his point of view). Ignoring air resistance, will the ball land on the car or behind it?

4. Does the Earth really go around the Sun? Or is it also valid to say that the Sun goes around the Earth? Discuss in view of the first principle of relativity (that there is no best reference frame).

5. If you were on a spaceship traveling at $0.5c$ away from a star, at what speed would the starlight pass you?

6. Will two events that occur at the same place and same time for one observer be simultaneous to a second observer moving with respect to the first?

7. Analyze the thought experiment of Section 37–4 from O_1's point of view. (Make a diagram analogous to Fig. 37–6.)

8. The time-dilation effect is sometimes expressed as "moving clocks run slowly." Actually, this effect has nothing to do with motion affecting the functioning of clocks. What then does it deal with?

9. Does time dilation mean that time actually passes more slowly in moving reference frames or that it only *seems* to pass more slowly?

10. A young-looking woman astronaut has just arrived home from a long trip. She rushes up to an old gray-haired man and in the ensuing conversation refers to him as her son. How might this be possible?

11. If you were traveling away from Earth at speed $0.5c$, would you notice a change in your heartbeat? Would your mass, height, or waistline change? What would observers on Earth using telescopes say about you?

12. Do time dilation and length contraction occur at ordinary speeds, say 90 km/h?

13. Suppose the speed of light were infinite. What would happen to the relativistic predictions of length contraction and time dilation?

14. Explain how the length-contraction and time-dilation formulas might be used to indicate that c is the limiting speed in the universe.

15. Does the equation $E = mc^2$ conflict with the conservation of energy principle? Explain.

16. If mass is a form of energy, does this mean that a spring has more mass when compressed than when relaxed?

17. It is not correct to say that "matter can neither be created nor destroyed." What must we say instead?

18. Is our intuitive notion that velocities simply add, as we did in Section 3–10, completely wrong?

Problems

Sections 37–4 to 37–6

1. (I) Lengths and time intervals depend on the factor
$$\sqrt{1 - v^2/c^2}$$
according to the theory of relativity (Eqs. 37–1 and 37–2). Evaluate this factor for speeds of: (*a*) $v = 20{,}000$ m/s (typical speed of a satellite); (*b*) $v = 0.0100c$; (*c*) $v = 0.100c$; (*d*) $v = 0.900c$; (*e*) $v = 0.990c$; (*f*) $v = 0.999c$.

2. (I) A spaceship passes you at a speed of $0.750c$. You measure its length to be 28.2 m. How long would it be when at rest?

3. (I) A certain type of elementary particle travels at a speed of 2.70×10^8 m/s. At this speed, the average lifetime is measured to be 4.76×10^{-6} s. What is the particle's lifetime at rest?

4. (I) If you were to travel to a star 100 light-years from Earth at a speed of 2.50×10^8 m/s, what would you measure this distance to be?

5. (II) What is the speed of a pion if its average lifetime is measured to be 4.10×10^{-8} s? At rest, its mean lifetime is 2.60×10^{-8} s.

6. (II) Suppose you decide to travel to a star 75 light-years away. How fast would you have to travel so the distance would be only 25 light-years?

7. (II) At what speed do the relativistic formulas for length and time intervals differ from classical values by 1.00 percent? (This is a reasonable way to estimate when to do relativistic calculations rather than classical.)

8. (II) Suppose a news report stated that starship *Enterprise* had just returned from a 5-year voyage while traveling at $0.84c$. (*a*) If the report meant 5.0 years of *Earth time*, how much time elapsed on the ship? (*b*) If the report meant 5.0 years of *ship time*, how much time passed on Earth?

9. (II) A certain star is 95.0 light-years away. How long would it take a spacecraft traveling $0.960c$ to reach that star from Earth, as measured by observers: (*a*) on Earth, (*b*) on the spacecraft? (*c*) What is the distance traveled according to observers on the spacecraft? (*d*) What will the spacecraft occupants compute their speed to be from the results of (*b*) and (*c*)?

10. (II) A friend of yours travels by you in her fast sports car at a speed of $0.660c$. It is measured in your frame to be 4.80 m long and 1.25 m high. (*a*) What will be its length and height at rest? (*b*) How many seconds would you say elapsed on your friend's watch when 20.0 s passed on yours? (*c*) How fast did you appear to be traveling according to your friend? (*d*) How many seconds would she say elapsed on your watch when she saw 20.0 s pass on hers?

11. (II) How fast must an average pion be moving to travel 15 m before it decays? The average lifetime, at rest, is 2.6×10^{-8} s.

Section 37–8

12. (I) Suppose in Fig. 37–12 that the origins of S and S' overlap at $t = t' = 0$ and that S' moves at speed $v = 30$ m/s with respect to S. In S', a person is resting at a point whose coordinates are $x' = 25$ m, $y' = 20$ m, and $z' = 0$. Calculate this person's coordinates in S (x, y, z) at (*a*) $t = 2.5$ s, (*b*) $t = 10.0$ s. Use the Galilean transformation.

13. (I) Repeat Problem 12 using the Lorentz transformation and a relative speed $v = 1.80 \times 10^8$ m/s, but choose the time to be (*a*) $2.5\,\mu$s and (*b*) $10.0\,\mu$s.

14. (I) A person on a rocket traveling at $0.50c$ (with respect to the Earth) observes a meteor come from behind and pass her at a speed she measures as $0.50c$. How fast is the meteor moving with respect to the Earth?

15. (II) Two spaceships leave Earth in opposite directions, each with a speed of $0.50c$ with respect to Earth. (a) What is the velocity of spaceship 1 relative to spaceship 2? (b) What is the velocity of spaceship 2 relative to spaceship 1?

16. (II) A spaceship leaves Earth traveling at $0.71c$. A second spaceship leaves the first at a speed of $0.87c$ with respect to the first. Calculate the speed of the second ship with respect to Earth if it is fired (a) in the same direction the first spaceship is already moving, (b) directly backward toward Earth.

17. (II) In Problem 12, suppose that the person moves with a velocity whose components are $u'_x = u'_y = 25.0$ m/s, What will be her velocity with respect to S? (Give magnitude and direction.)

18. (II) In Problem 13, suppose that the person moves with a velocity (with a rocket) whose components are $u'_x = u'_y = 2.0 \times 10^8$ m/s. What will be her velocity (magnitude and direction) with respect to S?

19. (II) A spaceship traveling at $0.66c$ away from Earth fires a module with a speed of $0.82c$ at right angles to its own direction of travel (as seen by the spaceship). What is the speed of the module, and its direction of travel (relative to the spaceship's direction), as seen by an observer on Earth?

20. (II) If a particle moves in the xy plane of system S (Fig. 37–12) with speed u in a direction that makes an angle θ with the x axis, show that it makes an angle θ' in S' given by $\tan \theta' = (\sin \theta)\sqrt{1 - v^2/c^2}/(\cos \theta - v/u)$.

21. (II) A stick of length L_0, at rest in reference frame S, makes an angle θ with the x axis. In reference frame S', which moves to the right with velocity $\mathbf{v} = v\mathbf{i}$ with respect to S, determine (a) the length L of the stick, and (b) the angle θ' it makes with the x' axis.

22. (III) In the old West, a marshal riding on a train traveling 50 m/s sees a duel between two men standing on the Earth 50 m apart parallel to the train. The marshal's instruments indicate that in his reference frame the two men fired simultaneously. (a) Which of the two men, the first one the train passes (A) or the second one (B) should be arrested for firing the first shot? That is, in the gunfighter's frame of reference, who fired first? (b) How much earlier did he fire? (c) Who was struck first?

23. (III) A farm boy studying physics believes that he can fit a 13.0-m long pole into a 10.0-m long barn if he runs fast enough, carrying the pole. Can he do it? Explain in detail. How does this fit with the idea that when he is running the barn looks even shorter than 10.0 m?

Section 37–9

24. (I) What is the momentum of a proton traveling at $v = 0.85c$?

25. (I) At what speed will an object's relativistic mass be twice its rest mass?

26. (II) A particle of rest mass m travels at a speed $v = 0.20c$. At what speed will its momentum be doubled?

27. (II) (a) A particle travels at $v = 0.10c$. By what percentage will a calculation of its momentum be wrong if you use the classical formula? (b) Repeat for $v = 0.50c$.

28. (II) What is the percent change in momentum of a proton that accelerates (a) from $0.45c$ to $0.90c$, (b) from $0.90c$ to $0.98c$?

Section 37–11

29. (I) A certain chemical reaction requires 4.82×10^4 J of energy input for it to go. What is the increase in mass of the products over the reactants?

30. (I) When a uranium nucleus at rest breaks apart in the process known as fission in a nuclear reactor, the resulting fragments have a total kinetic energy of about 200 MeV. How much mass was lost in the process?

31. (I) Calculate the rest energy of an electron in joules and in MeV $(1\ \text{MeV} = 1.60 \times 10^{-13}\ \text{J})$.

32. (I) Calculate the rest mass of a proton in MeV/c^2.

33. (I) The total annual energy consumption in the United States is about 8×10^{19} J. How much mass would have to be converted to energy to fuel this need?

34. (II) How much energy can be obtained from conversion of 1.0 gram of mass? How much mass could this energy raise to a height of 100 m?

35. (II) Show that when the kinetic energy of a particle equals its rest energy, the speed of the particle is about $0.866c$.

36. (II) At what speed will an object's kinetic energy be 25 percent of its rest energy?

37. (II) (a) How much work is required to accelerate a proton from rest up to a speed of $0.997c$? (b) What would be the momentum of this proton?

38. (II) Calculate the kinetic energy and momentum of a proton traveling 2.60×10^8 m/s.

39. (II) What is the momentum of a 750-MeV proton (that is, its kinetic energy is 750 MeV)?

40. (II) What is the speed of a proton accelerated by a potential difference of 95 MV?

41. (II) What is the speed of an electron whose kinetic energy is 1.00 MeV?

42. (II) What is the speed of an electron when it hits a television screen after being accelerated by the 25,000 V of the picture tube?

43. (II) Two identical particles of rest mass m approach each other at equal and opposite speeds, v. The collision is completely inelastic and results in a single particle at rest. What is the rest mass of the new particle? How much energy was lost in the collision? How much kinetic energy is lost in this collision?

44. (II) Calculate the speed of a proton $(m = 1.67 \times 10^{-27}\ \text{kg})$ whose kinetic energy is exactly half its total energy.

45. (II) What is the speed and momentum of an electron $(m = 9.11 \times 10^{-31}\ \text{kg})$ whose kinetic energy equals its rest energy?

46. (II) Suppose a spacecraft of mass 27,000 kg is accelerated to $0.21c$. (a) How much kinetic energy would it have? (b) If you used the classical formula for kinetic energy, by what percentage would you be in error?

47. (II) Calculate the kinetic energy and momentum of a proton $(m = 1.67 \times 10^{-27}\ \text{kg})$ traveling 8.4×10^7 m/s. By what percentages would your calculations have been in error if you had used classical formulas?

48. (II) The americium nucleus, $^{241}_{95}$Am, decays to a neptunium nucleus, $^{237}_{93}$Np, by emitting an alpha particle of mass 4.00260 u and kinetic energy 5.5 MeV. Estimate the mass of the neptunium nucleus, ignoring its recoil, given that the americium mass is 241.05682 u.

49. (II) An electron $(m = 9.11 \times 10^{-31} \text{ kg})$ is accelerated from rest to speed v by a conservative force. In this process, its potential energy decreases by 5.60×10^{-14} J. Determine the electron's speed, v.

50. (II) Make a graph of the kinetic energy versus momentum for (a) a particle of nonzero rest mass, and (b) a particle with zero rest mass.

51. (II) What magnetic field intensity is needed to keep 900-GeV protons revolving in a circle of radius 1.0 km (at, say, the Fermilab synchrotron)? Use the relativistic mass. The proton's rest mass is 0.938 GeV/c^2. (1 GeV = 10^9 eV.)

52. (II) A negative muon traveling at 33 percent the speed of light collides head on with a positive muon traveling at 50 percent the speed of light. The two muons (each of mass 105.7 MeV/c^2) annihilate, and produce how much electromagnetic energy?

53. (II) Show that the energy of a particle of charge e revolving in a circle of radius r in a magnetic field B is given by $E(\text{in eV}) = Brc$ in the relativistic limit $(v \approx c)$.

54. (II) Show that the kinetic energy K of a particle of rest mass m is related to its momentum p by the equation

$$p = \sqrt{K^2 + 2Kmc^2}/c.$$

55. (III) (a) In reference frame S, a particle has momentum $\mathbf{p} = p_x \mathbf{i}$ along the positive x axis. Show that in frame S', which moves with speed v as in Fig. 37–12, the momentum has components

$$p'_x = \frac{p_x - vE/c^2}{\sqrt{1 - v^2/c^2}}$$

$$p'_y = p_y$$

$$p'_z = p_z$$

$$E' = \frac{E - p_x v}{\sqrt{1 - v^2/c^2}}.$$

(These transformation equations hold, actually, for any direction of $\mathbf{p}$.) (b) Show that p_x, p_y, p_z, E/c transform according to the Lorentz transformation in the same way as x, y, z, ct.

* **Section 37–12**

* **56.** (II) A certain galaxy has a Doppler shift given by $f_0 - f = 0.797 f_0$. How fast is it moving away from us?

* **57.** (II) A quasar emits familiar hydrogen lines whose wavelengths are 3.0 times longer than what we measure in the laboratory. (a) What is the speed of this quasar? (b) What result would you obtain if you used the "classical" Doppler shift discussed in Chapter 16?

* **58.** (II) A spaceship moving toward Earth at $0.80c$ transmits radio signals at 95.0 MHz. At what frequency should Earth receivers be tuned?

* **59.** (II) Starting from Eq. 37–15a, show that the Doppler shift in wavelength is

$$\frac{\Delta \lambda}{\lambda} = \frac{v}{c}$$

if $v \ll c$.

General Problems

60. As a rule of thumb, anything traveling faster than about $0.1c$ is called *relativistic*—i.e., for which the correction using special relativity is a significant effect. Is the electron in a hydrogen atom (radius 0.5×10^{-10} m) relativistic? (Treat the electron as though it were in a circular orbit around the proton.)

61. An atomic clock is taken to the North Pole, while another stays at the Equator. How far will they be out of synchronization after a year has elapsed?

62. The nearest star to Earth is Proxima Centauri, 4.3 light-years away. (a) At what constant velocity must a spacecraft travel from Earth if it is to reach the star in 4.0 years, as measured by travelers on the spacecraft? (b) How long does the trip take according to Earth observers?

63. Derive a formula showing how the apparent density of an object changes with speed v relative to an observer. [*Hint*: Use relativistic mass.]

64. An airplane travels 1500 km/h around the world, returning to the same place, in a circle of radius essentially equal to that of the Earth. Estimate the difference in time to make the trip as seen by Earth and airplane observers. [*Hint*: Use the binomial expansion, Appendix A.]

65. (a) What is the speed of an electron whose kinetic energy is 10,000 times its rest energy? Such speeds are reached in the Stanford Linear Accelerator, SLAC. (b) If the electrons travel in the lab through a tube 3.0 km long (as at SLAC), how long is this tube in the electrons' reference frame?

66. How many grams of matter would have to be totally destroyed to run a 100-W lightbulb for 1 year?

67. What minimum amount of electromagnetic energy is needed to produce an electron and a positron together? A positron is a particle with the same mass as an electron, but has the opposite charge. (Note that electric charge is conserved in this process. See Section 38–4.)

68. A 1.68-kg mass oscillates on the end of a spring whose spring constant is $k = 48.7$ N/m. If this system is in a spaceship moving past Earth at $0.900\,c$, what is its period of oscillation according to (a) observers on the ship, and (b) observers on Earth?

69. An electron $(m = 9.11 \times 10^{-31}\,\text{kg})$ enters a uniform magnetic field $B = 1.8$ T, and moves perpendicular to the field lines with a speed $v = 0.92\,c$. What is the radius of curvature of its path?

70. An observer on Earth sees an alien vessel approach at a speed of $0.60\,c$. The *Enterprise* comes to the rescue (Fig. 37–16), overtaking the aliens while moving directly toward Earth at a speed of $0.90\,c$ relative to Earth. What is the relative speed of one vessel as seen by the other?

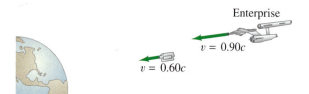

FIGURE 37–16   Problem 70.

71. A free neutron can decay into a proton, an electron, and a neutrino. Assume the neutrino's rest mass is zero, and the other masses can be found in the table inside the front cover. Determine the total kinetic energy shared among the three particles when a neutron decays at rest.

72. The Sun radiates energy at a rate of about 4×10^{26} W. (a) At what rate is the Sun's mass decreasing? (b) How long does it take for the Sun to lose a mass equal to that of Earth? (c) Estimate how long the Sun could last if it radiated constantly at this rate.

73. An unknown particle is measured to have a negative charge and a speed of 2.24×10^{8} m/s. Its momentum is determined to be 3.07×10^{-22} kg·m/s. Identify the particle by finding its mass.

74. How much energy would be required to break a helium nucleus into its constituents, two protons and two neutrons? The rest masses of a proton (including an electron), a neutron, and helium are, respectively, 1.00783 u, 1.00867 u, and 4.00260 u. (This is called the *total binding energy* of the $^{4}_{2}$He nucleus.)

75. What is the percentage increase in the (relativistic) mass of a car traveling 110 km/h as compared to at rest?

76. Two protons, each having a speed of $0.935\,c$ in the laboratory, are moving toward each other. Determine (a) the momentum of each proton in the laboratory, (b) the total momentum of the two protons in the laboratory, and (c) the momentum of one proton as seen by the other proton.

77. Show analytically that a particle with momentum p and energy E has a speed given by

$$v = \frac{pc^2}{E} = \frac{pc}{\sqrt{m^2c^2 + p^2}}.$$

78. A slab of glass moves to the right with speed v. A flash of light is emitted at point A (Fig. 37–17) and passes through the glass arriving at point B a distance L away. The glass has thickness d in the reference frame where it is at rest, and the speed of light in the glass is c/n. How long does it take the light to go from point A to point B according to an observer at rest with respect to points A and B? Check your answer for the cases $v = 0$, $n = 1$, and for $v = c$.

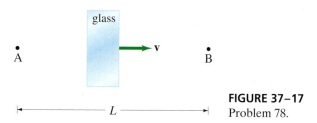

FIGURE 37–17
Problem 78.

79. Show that the space-time "distance" $(c\Delta t)^2 - (\Delta x)^2$ is invariant, meaning that all observers in all inertial reference frames calculate the same number for this quantity for any pair of events.

80. The fictional starship *Enterprise* obtains its power by combining matter and antimatter, achieving complete conversion of mass into energy. If the mass of the *Enterprise* is approximately 5×10^{9} kg, how much mass must be converted into kinetic energy to accelerate it from rest to one-tenth the speed of light?

81. A spacecraft (reference frame S′) moves past Earth (reference frame S) at velocity **v**, which points along the x and x' axes. The spacecraft emits light along its y' axis as shown in Fig. 37–18. (a) What angle does this light make with the x axis in the Earth's reference frame? (b) Show that the light moves with speed c also in the Earth's reference frame (i.e., given c in frame S′). (c) Compare these relativistic results to what you would have obtained classically (Galilean transformations).

FIGURE 37–18   Problem 81.

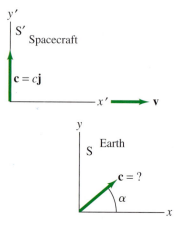

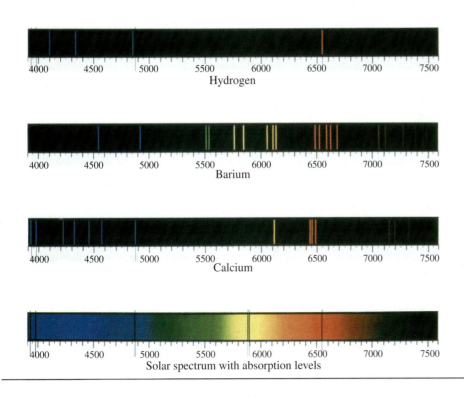

Atoms are tiny, and not observable using visible light. Trying to figure out the structure of atoms is a fascinating task. A major source of information about atoms comes from a study of the light emitted by the atoms of a pure material. Shown here are the line spectra (visible part) emitted by thin gases of hydrogen (top), barium, and calcium. At the bottom is the spectrum of the sun (visible light) in which are seen dark lines, indicating absorption at those wavelengths. What information do these spectra give us about atoms? We will see in this chapter that an entirely new theory is needed, the quantum theory.

CHAPTER **38**

Early Quantum Theory and Models of the Atom

The second aspect of the revolution that shook the world of physics in the early part of the twentieth century (the first part was Einstein's theory of relativity) was the quantum theory. Unlike the special theory of relativity, the revolution of quantum theory required almost three decades to unfold, and many scientists contributed to its development. It began in 1900 with Planck's quantum hypothesis, and culminated in the mid-1920s with the theory of quantum mechanics of Schrödinger and Heisenberg which has been so effective in explaining the structure of matter.

38–1 Planck's Quantum Hypothesis

One of the observations that was unexplained at the end of the nineteenth century was the spectrum of light emitted by hot objects. We saw in Chapter 19 that all objects emit radiation whose total intensity is proportional to the fourth power of the Kelvin temperature (T^4). At normal temperatures, we are not aware of this electromagnetic radiation because of its low intensity. At higher temperatures, there is sufficient infrared radiation that we can feel heat if we are close to the object. At still higher temperatures (on the order of 1000 K), objects actually glow, such as a red-hot electric stove burner or the element in a toaster. At temperatures above 2000 K, objects glow with a yellow or whitish color, such as white-hot iron and the filament of a lightbulb. As the temperature increases, the electromagnetic radiation emitted by bodies is most intense at higher and higher frequencies.

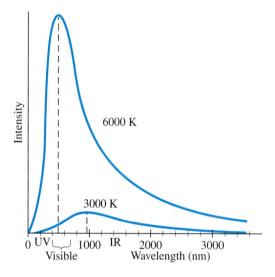

FIGURE 38–1 Spectrum of frequencies emitted by a blackbody at two different temperatures.

Blackbody radiation

The spectrum of light emitted by a hot dense object is shown in Fig. 38–1 for an idealized **blackbody**. A blackbody is a body that would absorb all the radiation falling on it (and so would appear black under reflection when illuminated from outside). The radiation such a blackbody would emit when hot and luminous, called **blackbody radiation** (though not necessarily black in color), is the easiest to deal with, and the radiation approximates that of many real objects. As can be seen, the spectrum contains a continuous range of frequencies. Such a continuous spectrum is emitted by any heated solid or liquid, and even by dense gases. The 6000-K curve in Fig. 38–1, corresponding to the temperature of the surface of the Sun, peaks in the visible part of the spectrum. For lower temperatures, the total radiation drops considerably and the peak occurs at longer wavelengths. Hence the blue end of the visible spectrum (and the UV) is relatively weaker. (This is why objects glow with a red color at around 1000 K.) It is found experimentally that the wavelength at the peak of the spectrum, λ_P, is related to the Kelvin temperature T by

$$\lambda_P T = 2.90 \times 10^{-3}\,\text{m}\cdot\text{K}. \qquad (38\text{–}1)$$

This is known as **Wien's law**.

EXAMPLE 38–1 **The Sun's surface temperature.** Estimate the temperature of the surface of our Sun, given that the Sun emits light whose peak intensity occurs in the visible spectrum at around 500 nm.

SOLUTION Wien's law gives

$$T = \frac{2.90 \times 10^{-3}\,\text{m}\cdot\text{K}}{\lambda_P} = \frac{2.90 \times 10^{-3}\,\text{m}\cdot\text{K}}{500 \times 10^{-9}\,\text{m}} \approx 6000\,\text{K}.$$

EXAMPLE 38–2 **Star color.** Suppose a star has a surface temperature of 32,500 K. What color would this star appear?

SOLUTION From Wien's law we have,

$$\lambda_P = \frac{2.90 \times 10^{-3}\,\text{m}\cdot\text{K}}{T} = \frac{2.90 \times 10^{-3}\,\text{m}\cdot\text{K}}{3.25 \times 10^{4}\,\text{K}} = 89.2\,\text{nm}.$$

The peak is in the UV range of the spectrum. In the visible region, the curve will be descending (see Fig. 38–1), so the shortest visible wavelengths will be strongest. Hence the star will appear bluish (or blue-white).

A major problem facing scientists in the 1890s was to explain blackbody radiation. Maxwell's electromagnetic theory had predicted that oscillating electric charges produce electromagnetic waves, and the radiation emitted by a hot object could be due to the oscillations of electric charges in the molecules of the material. Although this would explain where the radiation came from, it did not correctly predict the observed spectrum of emitted light. Two important theoretical curves based on classical ideas were those proposed by W. Wien (in 1896) and by Lord Rayleigh (in 1900). The latter was modified later by J. Jeans and since then has been known as the Rayleigh–Jeans theory. As experimental data came in, it became clear that neither Wien's nor the Rayleigh–Jeans formulations were in accord with experiment. Wien's was accurate at short wavelengths but deviated from experiment at longer wavelengths, whereas the reverse was true for the Rayleigh–Jeans theory (see Fig. 38–2).

The break came in late 1900 when Max Planck (1858–1947) proposed an empirical formula that nicely fit the data:

$$I(\lambda, T) = \frac{2\pi hc^2 \lambda^{-5}}{e^{hc/\lambda kT} - 1}.$$

$I(\lambda, T)$ is the radiation intensity as a function of wavelength λ at the temperature T; k is Boltzman's constant, c is the speed of light, and h is a new constant, now called **Planck's constant**. The value of h was estimated by Planck by fitting his formula for the blackbody radiation curve to experiment. The value accepted today is

$$h = 6.626 \times 10^{-34} \, \text{J} \cdot \text{s}.$$

Planck then sought a theoretical basis for the formula and within two months found that he could obtain the formula by making a new and radical (though not so recognized at the time) assumption: that the energy distributed among the oscillations of atoms within molecules is not continuous but instead consists of a finite number of very small discrete amounts, each related to the frequency of oscillation by

$$E_{min} = hf.$$

Planck's assumption suggests that the energy of any molecular vibration could be only some whole number multiple of hf:

$$E = nhf, \qquad n = 1, 2, 3, \cdots. \tag{38–2}$$

This idea is often called **Planck's quantum hypothesis** ("quantum" means "discrete amount" as opposed to "continuous"), although little attention was brought to this point at the time. In fact, it appears that Planck considered it more as a mathematical device to get the "right answer" rather than as a discovery comparable to those of Newton. Planck himself continued to seek a classical explanation for the introduction of h. The recognition that this was an important and radical innovation did not come until later, after about 1905 when others, particularly Einstein, entered the field.

The quantum hypothesis, Eq. 38–2, states that the energy of an oscillator can be $E = hf$, or $2hf$, or $3hf$, and so on, but there cannot be vibrations whose energy lies between these values. That is, energy would not be a continuous quantity as had been believed for centuries; rather it is **quantized**—it exists only in discrete amounts. The smallest amount of energy possible (hf) is called the **quantum of energy**. Recall from Chapter 14 that the energy of an oscillation is proportional to the amplitude squared. Thus another way of expressing the quantum hypothesis is that not just any amplitude of vibration is possible. The possible values for the amplitude are related to the frequency f.

A simple analogy may help. A stringed instrument such as a violin or guitar can be played over a continuous range of frequencies by moving your finger along the string. A flute or piano, on the other hand, is "quantized" in the sense that only certain frequencies (notes) can be played. Or compare a ramp, on which a box can be placed at any height, to a flight of stairs on which the box can have only certain discrete amounts of potential energy, as shown in Fig. 38–3.

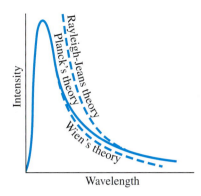

FIGURE 38–2 Comparison of the Wien and the Rayleigh–Jeans theories to that of Planck, which closely follows experiment.

Planck's quantum hypothesis

FIGURE 38–3 Ramp versus stair analogy. (a) On a ramp, a box can have continuous values of energy. (b) But on stairs, the box can have only discrete (quantized) values of energy.

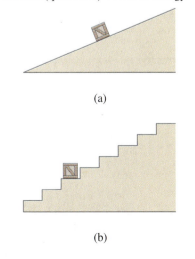

(a)

(b)

Photon Theory of Light and the Photoelectric Effect

In 1905, the same year that he introduced the special theory of relativity, Einstein made a bold extension of the quantum idea by proposing a new theory of light. Planck's work had suggested that the vibrational energy of molecules in a radiating object is quantized with energy $E = nhf$, where n is an integer. Einstein argued that therefore, when light is emitted by a molecular oscillator, the molecule's vibrational energy of nhf must decrease by an amount hf (or by $2hf$, etc.) to another integer times hf, namely $(n - 1)hf$. Then to conserve energy, the light ought to be emitted in packets or quanta, each with an energy

$$E = hf. \tag{38–3}$$

Again h is Planck's constant. Since all light ultimately comes from a radiating source, this suggests that perhaps *light is transmitted as tiny particles*, or **photons**, as they are now called, rather than as waves. This, too, was a radical departure from classical ideas. Einstein proposed a test of the quantum theory of light: quantitative measurements on the photoelectric effect.

Photoelectric effect

The **photoelectric effect** is the phenomenon that when light shines on a metal surface, electrons are emitted from the surface. (The photoelectric effect occurs in other materials, but is most easily observed with metals.) It can be observed using the apparatus shown in Fig. 38–4. A metal plate P and a smaller electrode C are placed inside an evacuated glass tube, called a **photocell**. The two electrodes are connected to an ammeter and a source of emf, as shown. When the photocell is in the dark, the ammeter reads zero. But when light of sufficiently high frequency illuminates the plate, the ammeter indicates a current flowing in the circuit. To explain completion of the circuit, we can imagine electrons, ejected by the impinging radiation, flowing across the tube from the plate to the "collector" C as shown in the diagram.

That electrons should be emitted when light shines on a metal is consistent with the electromagnetic (EM) wave theory of light: the electric field of an EM wave could exert a force on electrons in the metal and eject some of them. Einstein pointed out, however, that the wave theory and the photon theory of light give very different predictions on the details of the photoelectric effect. For example, one thing that can be measured with the apparatus of Fig. 38–4 is the maximum kinetic energy (K_{max}) of the emitted electrons. This can be done by using a variable voltage source and reversing the terminals so that electrode C is negative and P is positive. The electrons emitted from P will be repelled by the negative electrode, but if this reverse voltage is small enough, the fastest electrons will still reach C and there will be a current in the circuit. If the reversed voltage is increased, a point is reached where the current reaches zero—no electrons have sufficient kinetic energy to reach C. This is called the *stopping potential*, or *stopping voltage*, V_0, and from its measurement, K_{max} can be determined using conservation of energy:

$$K_{max} = eV_0.$$

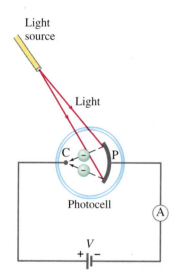

FIGURE 38–4 The photoelectric effect.

Now let us examine the details of the photoelectric effect from the point of view of the wave theory versus Einstein's particle theory. First the wave theory, assuming monochromatic light. The two important properties of a light wave are its intensity and its frequency (or wavelength). When these two quantities are varied, the wave theory makes the following predictions:

Wave

theory

predictions

1. If the light intensity is increased, the number of electrons ejected and their maximum kinetic energy should be increased because the higher intensity means a greater electric field amplitude, and the greater electric field should eject electrons with higher speed.

2. The frequency of the light should not affect the kinetic energy of the ejected electrons. Only the intensity should affect K_{max}.

The photon theory makes completely different predictions. First we note that in a monochromatic beam, all photons have the same energy ($= hf$). Increasing the intensity of the light beam means increasing the number of photons in the beam, but does not affect the energy of each photon as long as the frequency is not changed. According to Einstein's theory, an electron is ejected from the metal by a collision with a single photon. In the process, all the photon energy is transferred to the electron and the photon ceases to exist. Since electrons are held in the metal by attractive forces, some minimum energy W_0 (called the **work function**, which is on the order of a few electron volts for most metals) is required just to get an electron out through the surface. If the frequency f of the incoming light is so low that hf is less than W_0, then the photons will not have enough energy to eject any electrons at all. If $hf > W_0$, then electrons will be ejected and energy will be conserved in the process. That is, the input energy (of the photon), hf, will equal the outgoing kinetic energy K of the electron plus the energy required to get it out of the metal, W:

$$hf = K + W. \tag{38-4a}$$

The least tightly held electrons will be emitted with the most kinetic energy (K_{max}), in which case W in this equation becomes the work function W_0, and K becomes K_{max}:

$$hf = K_{max} + W_0. \tag{38-4b}$$

Many electrons will require more energy than the bare minimum (W_0) to get out of the metal, and thus the kinetic energy of such electrons will be less than the maximum.

From these considerations, the photon theory makes the following predictions:

1. An increase in intensity of the light beam means more photons are incident, so more electrons will be ejected; but since the energy of each photon is not changed, the maximum kinetic energy of electrons is not changed by an increase in intensity.

2. If the frequency of the light is increased, the maximum kinetic energy of the electrons increases linearly, according to Eq. 38–4b. That is,

$$K_{max} = hf - W_0.$$

This relationship is plotted in Fig. 38–5.

3. If the frequency f is less than the "cutoff" frequency f_0, where $hf_0 = W_0$, no electrons will be ejected at all, no matter how great the intensity.

These predictions of the photon theory are clearly very different from the predictions of the wave theory. In 1913–1914, careful experiments were carried out by R. A. Millikan. The results were fully in agreement with Einstein's photon theory.

One other aspect of the photoelectric effect also confirmed the photon theory. If extremely low light intensity is used, the wave theory predicts a time delay before electron emission so that an electron can absorb enough energy to exceed the work function. The photon theory predicts no such delay—it only takes one photon (if its frequency is high enough) to eject an electron—and experiments showed no delay. This too confirmed Einstein's photon theory.

Photon

theory

predictions

FIGURE 38–5 Photoelectric effect: the maximum kinetic energy of ejected electrons increases linearly with the frequency of incident light. No electrons are emitted if $f < f_0$.

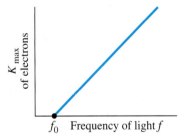

EXAMPLE 38–3 **Photon energy.** Calculate the energy of a photon of blue light, $\lambda = 450$ nm.

SOLUTION Since $f = c/\lambda$, we have

$$E = hf = \frac{hc}{\lambda} = \frac{(6.63 \times 10^{-34} \, \text{J·s})(3.0 \times 10^8 \, \text{m/s})}{(4.5 \times 10^{-7} \, \text{m})} = 4.4 \times 10^{-19} \, \text{J},$$

or $(4.4 \times 10^{-19} \, \text{J})/(1.6 \times 10^{-19} \, \text{J/eV}) = 2.7$ eV.

EXAMPLE 38–4 ESTIMATE Photons from a lightbulb. Estimate how many visible light photons a 100-W lightbulb emits per second.

SOLUTION Let's assume an average wavelength in the middle of the visible spectrum, $\lambda \approx 500$ nm. The energy emitted in one second ($= 100$ J) is $E = nhf$ when n is the number of photons emitted per second and $f = c/\lambda$. Hence

$$n = \frac{E}{hf} = \frac{E\lambda}{hc} = \frac{(100 \text{ J})(500 \times 10^{-9} \text{ m})}{(6.63 \times 10^{-34} \text{ J} \cdot \text{s})(3.0 \times 10^{8} \text{ m/s})} = 2.5 \times 10^{20}.$$

This is an overestimate since much of the 100 J of electric energy input is transformed into heat rather than light. If the efficiency is between 1 percent and 10 percent, then the number of photons emitted is on the order of 10^{19}, still an enormous number.

EXAMPLE 38–5 Photoelectron speed and energy. What is the maximum kinetic energy and speed of an electron ejected from a sodium surface whose work function is $W_0 = 2.28$ eV when illuminated by light of wavelength: (a) 410 nm; (b) 550 nm?

SOLUTION (a) For $\lambda = 410$ nm,

$$hf = \frac{hc}{\lambda} = 4.85 \times 10^{-19} \text{ J} \quad \text{or} \quad 3.03 \text{ eV}.$$

From Eq. 38–4b, $K_{max} = 3.03 \text{ eV} - 2.28 \text{ eV} = 0.75 \text{ eV}$, or 1.2×10^{-19} J. Since $K = \frac{1}{2}mv^2$ where $m = 9.1 \times 10^{-31}$ kg,

$$v = \sqrt{\frac{2K}{m}} = 5.1 \times 10^{5} \text{ m/s}.$$

Notice that we used the nonrelativistic equation for kinetic energy. If v had turned out to be more than about $0.1c$, our calculation would have been inaccurate by more than a percent or so, and we would probably prefer to redo it using the relativistic form (Eq. 37–10).

(b) For $\lambda = 550$ nm, $hf = 3.61 \times 10^{-19}$ J $= 2.26$ eV. Since this photon energy is less than the work function, no electrons are ejected.

Applications of the Photoelectric Effect

The photoelectric effect, besides playing an important historical role in confirming the photon theory of light, also has many practical applications. Burglar alarms and automatic door openers often make use of the photocell circuit of Fig. 38–4. When a person interrupts the beam of light, the sudden drop in current in the circuit activates a switch—often a solenoid—which operates a bell or opens the door. UV or IR light is sometimes used in burglar alarms because of its invisibility. Many smoke detectors use the photoelectric effect to detect tiny amounts of smoke that interrupt the flow of light and so alter the electric current. Photographic light meters use this circuit as well. Photocells are used in many other devices, such as absorption spectrophotometers, to measure light intensity. One type of film sound track is a variably shaded narrow section at the side of the film. Light passing through the film is thus "modulated," and the output electrical signal of the photocell detector follows the frequencies on the sound track. See Fig. 38–6. For many applications today, the vacuum-tube photocell of Fig. 38–4 has been replaced by a semiconductor device known as a **photodiode**. In these semiconductors, the absorption of a photon liberates a bound electron, which changes the conductivity of the material, so the current through a photodiode is altered.

FIGURE 38–6 Optical sound track on movie film. In the projector, light from a small source (different from that for the picture) passes through the sound track on the moving film. The light and dark areas on the sound track vary the intensity of the transmitted light which reaches the photocell, whose current output is then a replica of the original sound. This output is amplified and sent to the loudspeakers. High-quality projectors can show movies containing several parallel sound tracks to go to different speakers around the theater.

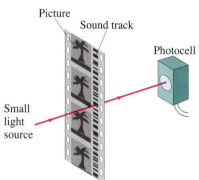

Picture
Sound track
Photocell
Small light source

EXAMPLE 38–6 Photosynthesis. In *photosynthesis*, the process by which pigments such as chlorophyll in plants capture the energy of sunlight to change CO_2 to useful carbohydrate, about nine photons are needed to transform one molecule of CO_2 to carbohydrate and O_2. Assuming light of wavelength $\lambda = 670$ nm (chlorophyll absorbs most strongly in the range 650 nm to 700 nm), how efficient is the photosynthetic process? The reverse chemical reaction releases an energy of 4.9 eV/molecule of CO_2.

SOLUTION The energy of nine photons, each of energy $hf = hc/\lambda$ is $(9)(6.6 \times 10^{-34}\,\text{J·s})(3.0 \times 10^8\,\text{m/s})/(6.7 \times 10^{-7}\,\text{m}) = 2.7 \times 10^{-18}$ J or 17 eV. Thus the process is $(4.9\,\text{eV}/17\,\text{eV}) = 29$ percent efficient.

38–3 Photons and the Compton Effect

A number of other experiments were carried out in the early twentieth century which also supported the photon theory. One of these was the **Compton effect** (1923) named after its discoverer, A. H. Compton (1892–1962). Compton scattered short wavelength light (actually X-rays) from various materials. He found that the scattered light had a slightly longer wavelength than did the incident light, and therefore a slightly lower frequency indicating a loss of energy. This finding, he showed, could be explained on the basis of the photon theory as incident photons colliding with electrons of the material, Fig. 38–7. He applied the laws of conservation of energy and momentum to such collisions and found that the predicted energies of scattered photons were in accord with experimental results.

Let us look at the Compton effect in detail and seek to predict the shifted wavelength using the photon theory of light and Fig. 38–7. First we note that the photon is truly a relativistic particle—it travels at the speed of light. Thus we must use relativistic formulas for dealing with its mass, energy, and momentum. The momentum of any particle of rest mass m is given by $p = \gamma mv = mv/\sqrt{1 - v^2/c^2}$. Since $v = c$ for a photon, the denominator is zero. So the rest mass of a photon must also be zero, or its momentum would be infinite. Of course a photon is never at rest. The momentum of a photon, from Eq. 37–13 with $m = 0$, is given by $E^2 = p^2 c^2$, or

$$p = \frac{E}{c}.$$

Since $E = hf$, the momentum of a photon is related to its wavelength by

$$p = \frac{hf}{c} = \frac{h}{\lambda}. \tag{38–5}$$

If the incoming photon in Fig. 38–7 has wavelength λ, then its total energy and momentum are

$$E = hf = \frac{hc}{\lambda} \quad \text{and} \quad p = \frac{h}{\lambda}.$$

After the collision of Fig. 38–7, the photon scattered at the angle ϕ has a wavelength which we call λ'. Its energy and momentum are

$$E' = \frac{hc}{\lambda'} \quad \text{and} \quad p' = \frac{h}{\lambda'}.$$

The electron, assumed at rest before the collision but free to move when struck is scattered at an angle θ as shown in Fig. 38–7.

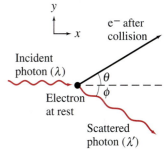

FIGURE 38–7 The Compton effect. A single photon of wavelength λ strikes an electron in some material, knocking it out of its atom. The scattered photon has less energy (since some is given to the electron) and hence has a longer wavelength λ'. Experiments found scattered X-rays of just the wavelengths predicted by conservation of energy and momentum using the photon model.

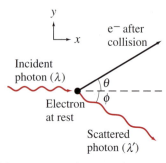

FIGURE 38–7 (repeated) The Compton effect.

FIGURE 38–8 Plots of intensity of radiation scattered from a target such as graphite (carbon), for three different angles. The values for λ' match Eq. 38–6. For (a) $\phi = 0°$, $\lambda' = \lambda_0$. In (b) and (c) a peak is found not only at λ' due to photons scattered from free electrons (or very nearly free) but also a peak at almost precisely λ_0. The latter is due to scattering from electrons very tightly bound to their atoms so the mass in Eq. 38–6 becomes very large (mass of the atom) and $\Delta\lambda$ becomes very small.

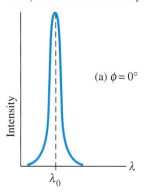

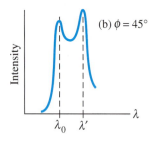

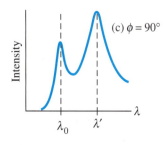

The electron's kinetic energy is (see Eq. 37–10):

$$K_e = \left(\frac{1}{\sqrt{1 - v^2/c^2}} - 1 \right) m_e c^2$$

where m_e is the rest mass of the electron and v is its speed. The electron's momentum is

$$p_e = \frac{1}{\sqrt{1 - v^2/c^2}} m_e v.$$

We apply conservation of energy to the collision (see Fig. 38–7):

$$\text{incoming photon} \rightarrow \text{scattered photon} + \text{electron}$$

$$\frac{hc}{\lambda} = \frac{hc}{\lambda'} + \left(\frac{1}{\sqrt{1 - v^2/c^2}} - 1 \right) m_e c^2.$$

And we apply conservation of momentum to the x and y components of momentum:

$$\frac{h}{\lambda} = \frac{h}{\lambda'} \cos\phi + \frac{m_e v \cos\theta}{\sqrt{1 - v^2/c^2}}$$

$$0 = \frac{h}{\lambda'} \sin\phi - \frac{m_e v \sin\theta}{\sqrt{1 - v^2/c^2}}.$$

We can combine these three equations to eliminate v and θ (the algebra is left as an exercise—see Problem 25) and we obtain, as Compton did, an equation for the wavelength of the scattered photon in terms of its scattering angle ϕ:

$$\lambda' = \lambda + \frac{h}{m_e c} (1 - \cos\phi).$$

For $\phi = 0$, the wavelength is unchanged (there is no collision for this case of the photon passing straight through). At any other angle, λ' is longer than λ. The difference in wavelength,

$$\Delta\lambda = \lambda' - \lambda = \frac{h}{m_e c} (1 - \cos\phi), \qquad \textbf{(38–6)}$$

is called the **Compton shift**. The quantity $h/m_e c$, which has the dimensions of length, is called the **Compton wavelength** λ_C of a free electron,

$$\lambda_C = \frac{h}{m_e c} = 2.43 \times 10^{-3}\,\text{nm} = 2.43\,\text{pm}. \qquad \text{[electron]}$$

Equation 38–6 predicts that λ' depends on the angle ϕ at which they are detected. Compton's measurements of 1923 were consistent with this formula, confirming the value of λ_C and the dependence of λ' on ϕ. See Fig. 38–8. The wave theory of light predicts no wavelength shift: an incoming electromagnetic wave of frequency f should set electrons into oscillation at the same frequency f, and such oscillating electrons should reemit EM waves of this same frequency f (Chapter 32), and would not change with the angle ϕ. Hence the Compton effect adds to the firm experimental foundation for the photon theory of light.

EXAMPLE 38–7 **X-ray scattering.** X-rays of wavelength 0.140 nm are scattered from a block of carbon. What will be the wavelengths of X-rays scattered at (a) 0°, (b) 90°, (c) 180°?

SOLUTION (a) For $\phi = 0°$, $\cos\phi = 1$, and Eq. 38–6 gives $\lambda' = \lambda = 0.140$ nm. This makes sense since for $\phi = 0°$, there really isn't any collision as the photon goes straight through without interacting.
(b) For $\phi = 90°$, $\cos\phi = 0$, so

$$\lambda' = \lambda + \frac{h}{m_e c} = 0.140\ \text{nm} + \frac{6.63 \times 10^{-34}\ \text{J·s}}{(9.11 \times 10^{-31}\ \text{kg})(3.00 \times 10^8\ \text{m/s})}$$
$$= 0.140\ \text{nm} + 2.4 \times 10^{-12}\ \text{m} = 0.142\ \text{nm};$$

that is, the wavelength is longer by one Compton wavelength (= 0.0024 nm for an electron).
(c) For $\phi = 180°$, which means the photon is scattered backward, returning in the direction from which it came (a direct "head-on" collision), $\cos\phi = -1$, so

$$\lambda' = \lambda + 2\frac{h}{m_e c} = 0.140\ \text{nm} + 2(0.0024\ \text{nm}) = 0.145\ \text{nm}.$$

The Compton effect has been used to diagnose bone disease such as osteoporosis. Gamma rays, which are photons of even shorter wavelength than X-rays, coming from a radioactive source are scattered off bone material. The total intensity of the scattered radiation is proportional to the density of electrons, which is in turn proportional to the bone density. Changes in the density of bone material can indicate the onset of osteoporosis.

➡ **PHYSICS APPLIED**

Measuring bone density

38–4 Photon Interactions; Pair Production

When a photon passes through matter, it interacts with the atoms and electrons. There are four important types of interactions that a photon can undergo:

1. The photon can be scattered from an electron (or a nucleus) and in the process lose some energy; this is the *Compton effect* (Fig. 38–7). But notice that the photon is not slowed down. It still travels with speed c, but its frequency will be lower.

2. The *photoelectric effect*: a photon may knock an electron out of an atom and in the process itself disappear.

3. The photon may knock an atomic electron to a higher energy state in the atom if its energy is not sufficient to knock the electron out altogether. In this process the photon also disappears, and all its energy is given to the atom. Such an atom is then said to be in an *excited state*, and we shall discuss this more later.

4. *Pair production*: A photon can actually create matter, such as the production of an electron and a positron, Fig. 38–9. (A positron has the same mass as an electron, but the opposite charge, $+e$.)

This last process is called **pair production**, and the photon disappears in the process of creating the electron–positron pair. This is an example of rest mass being created from pure energy, and it occurs in accord with Einstein's equation $E = mc^2$. Notice that a photon cannot create an electron alone since electric charge would not then be conserved. The inverse of pair production also occurs: if an electron collides with a positron, the two **annihilate** each other and their energy, including their mass, appears as electromagnetic energy of photons. Because of this process positrons usually do not last long in nature.

Photon

interactions

FIGURE 38–9 Pair production: a photon disappears and produces an electron and a positron.

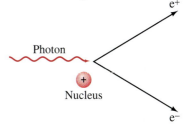

Pair production

EXAMPLE 38–8 **Pair production.** What is (*a*) the minimum energy of a photon that can produce an electron–positron pair? (*b*) What is this photon's wavelength?

SOLUTION (*a*) Because $E = mc^2$, and the mass created is equal to two electron masses, the photon must have energy

$$E = 2(9.11 \times 10^{-31}\,\text{kg})(3.0 \times 10^8\,\text{m/s})^2 = 1.64 \times 10^{-13}\,\text{J} = 1.02\,\text{MeV}.$$

A photon with less energy cannot undergo pair production.

(*b*) Since $E = hf = hc/\lambda$, the wavelength of a 1.02-MeV photon is

$$\lambda = \frac{hc}{E} = \frac{(6.6 \times 10^{-34}\,\text{J} \cdot \text{s})(3.0 \times 10^8\,\text{m/s})}{(1.64 \times 10^{-13}\,\text{J})} = 1.2 \times 10^{-12}\,\text{m},$$

which is 0.0012 nm. Such photons are in the gamma-ray (or very short X-ray) region of the electromagnetic spectrum.

Pair production cannot occur in empty space, for energy and momentum could not simultaneously be conserved. In Example 38–8, for instance, energy is conserved, but only enough energy was provided to create the electron–positron pair at rest and thus with no momentum to carry away the initial momentum of the photon. Indeed, it can be shown that at any energy, an additional massive object, such as an atomic nucleus, must take part in the interaction to carry off some of the momentum.

38–5 Wave–Particle Duality; the Principle of Complementarity

The photoelectric effect, the Compton effect, and other experiments have placed the particle theory of light on a firm experimental basis. But what about the classic experiments of Young and others (Chapters 35 and 36) on interference and diffraction which showed that the wave theory of light also rests on a firm experimental basis?

We seem to be in a dilemma. Some experiments indicate that light behaves like a wave; others indicate that it behaves like a stream of particles. These two theories seem to be incompatible, but both have been shown to have validity. Physicists have finally come to the conclusion that this duality of light must be accepted as a fact of life. It is referred to as the **wave–particle duality**. Apparently, light is a more complex phenomenon than just a simple wave or a simple beam of particles.

To clarify the situation, the great Danish physicist Niels Bohr (1885–1962, Fig. 38–10) proposed his famous **principle of complementarity**. It states that to understand any given experiment, we must use either the wave or the photon theory, but not both. Yet we must be aware of both the wave and particle aspects of light if we are to have a full understanding of light. Therefore these two aspects of light complement one another.

It is not possible to "visualize" this duality. We cannot picture a combination of wave and particle. Instead, we must recognize that the two aspects of light are different "faces" that light shows to experimenters.

Part of the difficulty stems from how we think. Visual pictures (or models) in our minds are based on what we see in the everyday world. We apply the concepts of waves and particles to light because in the macroscopic world we see that energy is transferred from place to place by these two methods. We cannot see directly whether light is a wave or particle—so we do indirect experiments. To explain the experiments, we apply the models of waves or of particles to the nature of light. But these are abstractions of the human mind. When we try to conceive of what light really "is," we insist on a visual picture. Yet there is no reason why light should conform to these models (or visual images) taken from the macroscopic

FIGURE 38–10 Niels Bohr, walking with Enrico Fermi along the Appian Way outside Rome.

Wave–particle duality

Principle of complementarity

world. The "true" nature of light—if that means anything—is not possible to visualize. The best we can do is recognize that our knowledge is limited to the indirect experiments, and that in terms of everyday language and images, light reveals both wave and particle properties.

It is worth noting that Einstein's equation $E = hf$ itself links the particle and wave properties of a light beam. In this equation, E refers to the energy of a particle; and on the other side of the equation, we have the frequency f of the corresponding wave.

38–6 Wave Nature of Matter

In 1923, Louis de Broglie (1892–1987) extended the idea of the wave–particle duality. He much appreciated the symmetry in nature, and argued that if light sometimes behaves like a wave and sometimes like a particle, then perhaps those things in nature thought to be particles—such as electrons and other material objects—might also have wave properties. De Broglie proposed that the wavelength of a material particle would be related to its momentum in the same way as for a photon, Eq. 38–5, $p = h/\lambda$. That is, for a particle having linear momentum p, the wavelength λ is given by

$$\lambda = \frac{h}{p}.$$

(38–7)   *de Broglie wavelength*

This is sometimes called the **de Broglie wavelength** of a particle.

EXAMPLE 38–9 **Wavelength of a ball.** Calculate the de Broglie wavelength of a 0.20-kg ball moving with a speed of 15 m/s. Discuss.

SOLUTION $\lambda = \dfrac{h}{p} = \dfrac{h}{mv} = \dfrac{(6.6 \times 10^{-34}\,\text{J}\cdot\text{s})}{(0.20\,\text{kg})(15\,\text{m/s})} = 2.2 \times 10^{-34}\,\text{m}.$

This is an incredibly small wavelength. Even if the speed were extremely small, say 10^{-4} m/s, the wavelength would be about 10^{-29} m. Indeed, the wavelength of any ordinary object is much too small to be measured and detected. The problem is that the properties of waves, such as interference and diffraction, are significant only when the size of objects or slits is not much larger than the wavelength. And there are no known objects or slits to diffract waves only 10^{-30} m long, so the wave properties of ordinary objects go undetected.

But tiny elementary particles, such as electrons, are another matter. Since the mass m appears in the denominator in Eq. 38–7, a very small mass should have a much larger wavelength.

EXAMPLE 38–10 **Wavelength of an electron.** Determine the wavelength of an electron that has been accelerated through a potential difference of 100 V.

SOLUTION We assume that the speed of the electron will be much less than c, so we use nonrelativistic mechanics. (If this assumption were to come out wrong, we would have to recalculate using relativistic formulas—see Sections 37–9 and 37–11.) The gain in kinetic energy will equal the loss in potential energy, so $\frac{1}{2}mv^2 = eV$ and

$$v = \sqrt{\frac{2\,eV}{m}} = \sqrt{\frac{(2)(1.6 \times 10^{-19}\,\text{C})(100\,\text{V})}{(9.1 \times 10^{-31}\,\text{kg})}} = 5.9 \times 10^6\,\text{m/s}.$$

Then

$$\lambda = \frac{h}{mv} = \frac{(6.6 \times 10^{-34}\,\text{J}\cdot\text{s})}{(9.1 \times 10^{-31}\,\text{kg})(5.9 \times 10^6\,\text{m/s})} = 1.2 \times 10^{-10}\,\text{m},$$

or 0.12 nm.

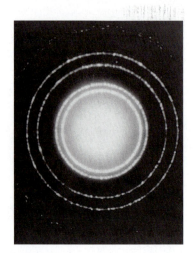

FIGURE 38–11 Diffraction pattern of electrons scattered from aluminum foil, as recorded on film.

From this Example, we see that electrons can have wavelengths on the order of 10^{-10} m. Although small, this wavelength can be detected: the spacing of atoms in a crystal is on the order of 10^{-10} m and the orderly array of atoms in a crystal could be used as a type of diffraction grating, as was done earlier for X-rays (see Section 36–10). C. J. Davisson and L. H. Germer performed the crucial experiment; they scattered electrons from the surface of a metal crystal and, in early 1927, observed that the electrons were scattered into a pattern of regular peaks. When they interpreted these peaks as a diffraction pattern, the wavelength of the diffracted electron wave was found to be just that predicted by de Broglie, Eq. 38–7. In the same year, G. P. Thomson (son of J. J. Thomson, who is credited with the discovery of the particle nature of electrons as we saw in Section 27–7), using a different experimental arrangement, also detected diffraction of electrons. (See Fig. 38–11. You can compare it to X-ray diffraction, Section 36–10.) Later experiments showed that protons, neutrons, and other particles also have wave properties.

Thus the wave–particle duality applies to material objects as well as to light. The principle of complementarity applies to matter as well. That is, we must be aware of both the particle and wave aspects in order to have an understanding of matter, including electrons. But again we must recognize that a visual picture of a "wave–particle" is not possible.

What is an electron?

We might ask ourselves: "What is an electron?" The early experiments of J. J. Thomson (Section 27–7) indicated a glow in a tube, and that glow moved when a magnetic field was applied. The results of these and other experiments were best interpreted as being caused by tiny negatively charged particles which we now call electrons. No one, however, has actually seen an electron directly. The drawings we sometimes make of electrons as tiny spheres with a negative charge on them are merely convenient pictures (now recognized to be inaccurate). Again we must rely on experimental results, some of which are best interpreted using the particle model and others using the wave model. These models are mere pictures that we use to extrapolate from the macroscopic world to the tiny microscopic world of the atom. And there is no reason to expect that these models somehow reflect the reality of an electron. We thus use a wave or a particle model (whichever works best in a situation) so that we can talk about what is happening. But we should not be led to believe that an electron *is* a wave or a particle. Instead we could say that an electron is the set of its properties that we can measure. Bertrand Russell said it well when he wrote that an electron is "a logical construction."

➥ PHYSICS APPLIED

Electron diffraction

EXAMPLE 38–11 Electron diffraction. The wave nature of electrons is manifested in experiments where an electron beam interacts with the atoms on the surface of a solid. By studying the angular distribution of the diffracted electrons, one can indirectly measure the geometrical arrangement of atoms. Assume that the electrons strike perpendicular to the surface of a solid (see Fig. 38–12), and that their energy is low, $K = 100$ eV, so that they interact only with the surface layer of atoms. If the smallest angle at which a diffraction maximum occurs is at 24°, what is the separation d between the atoms on the surface?

FIGURE 38–12 Exmple 38–11.

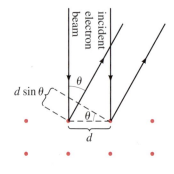

SOLUTION Treating the electrons as waves, we need to determine the condition where the difference in path traveled by the wave diffracted from adjacent atoms is an integer multiple of the de Broglie wavelength, so that constructive interference occurs. The path length difference is $d \sin \theta$; so for the smallest value of θ we must have

$$d \sin \theta = \lambda.$$

However, λ is related to the (non-relativistic) kinetic energy K by

$$K = \frac{p^2}{2m_e} = \frac{h^2}{2m_e \lambda^2}.$$

Thus

$$\lambda = \frac{h}{\sqrt{2m_e K}} = \frac{(6.63 \times 10^{-34}\,\text{J·s})}{\sqrt{2(9.11 \times 10^{-31}\,\text{kg})(100\,\text{eV})(1.6 \times 10^{-19}\,\text{J/eV})}} = 0.123\,\text{nm}.$$

The surface inter-atomic spacing is

$$d = \frac{\lambda}{\sin \theta} = \frac{0.123\,\text{nm}}{\sin 24°} = 0.30\,\text{nm}.$$

*38–7 Electron Microscopes

The idea that electrons have wave properties led to the development of the **electron microscope**, which can produce images of much greater magnification than a light microscope. Figures 38–13 and 38–14 are diagrams of two types, the **transmission electron microscope**, which produces a two-dimensional image, and the **scanning electron microscope**, which produces images with a three-dimensional quality. In each design, the objective and eyepiece lenses are actually magnetic fields that exert forces on the electrons to bring them to a focus. The fields are produced by carefully designed current-carrying coils of wire.

As discussed in Section 36–5, the best resolution of details on an object is on the order of the wavelength of the radiation used to view it. Electrons accelerated by voltages on the order of 10^5 V have wavelengths on the order of 0.004 nm. The maximum resolution obtainable would be on this order, but in practice, aberrations in the magnetic lenses limit the resolution in transmission electron microscopes to at best about 0.1 to 0.5 nm. This is still 10^3 times finer than that attainable with a light microscope, and corresponds to a useful magnification of about a million. Such magnifications are difficult to attain, and more common magnifications are 10^4 to 10^5. The maximum resolution attainable with a scanning electron microscope is somewhat less, about 5 to 10 nm at best.

PHYSICS APPLIED

Electron microscope

FIGURE 38–13 Transmission electron microscope. The magnetic-field coils are designed to be "magnetic lenses," which bend the electron paths and bring them to a focus, as shown.

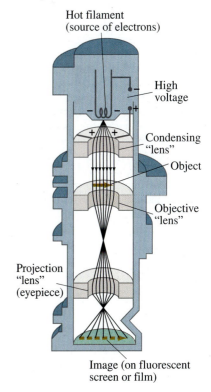

FIGURE 38–14 Scanning electron microscope. Scanning coils move an electron beam back and forth across the specimen. Secondary electrons produced when the beam strikes the specimen are collected and modulate the intensity of the beam in the CRT to produce a picture.

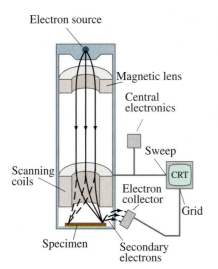

Early Models of the Atom

The idea that matter is made up of atoms was accepted by most scientists by 1900. With the discovery of the electron in the 1890s, scientists began to think of the atom itself as having a structure and electrons as part of that structure. We now introduce our modern approach to the atom, and of the quantum theory with which it is intertwined.[†]

A typical model of the atom in the 1890s visualized the atom as a homogeneous sphere of positive charge inside of which there were tiny negatively charged electrons, a little like plums in a pudding, Fig. 38–15. J. J. Thomson, soon after his discovery of the electron in 1897, argued that the electrons in this model should be moving.

Around 1911, Ernest Rutherford (1871–1937) and his colleagues performed experiments whose results contradicted Thomson's model of the atom. In these experiments a beam of positively charged "alpha (α) particles" was directed at a thin sheet of metal foil such as gold, Fig. 38–16a. (These newly discovered α particles were emitted by certain radioactive materials and were soon shown to be doubly ionized helium atoms—that is, having a charge of +2e.) It was expected from Thomson's model that the alpha particles would not be deflected significantly since electrons are so much lighter than alpha particles, and the alpha particles should not have encountered any massive concentration of positive charge in that model to strongly repel them. The experimental results completely contradicted these predictions. It was found that most of the alpha particles passed through the foil unaffected, as if the foil were mostly empty space. And of those deflected, a few were deflected at very large angles—some even nearly back in the direction from which they had come. This could happen, Rutherford reasoned, only if the positively charged alpha particles were being repelled by a massive positive charge concentrated in a very small region of space (see Fig. 38–16b). He theorized that the atom must consist of a tiny but massive positively charged nucleus, containing over 99.9 percent of the mass of the atom, surrounded by electrons some distance away. The electrons would be moving in orbits about the nucleus—much as the planets move around the Sun—because if they were at rest, they would fall into the nucleus

[†] Some readers may say: "Tell us the facts as we know them today, and don't bother us with the historical background and its outmoded theories." Such an approach would ignore the creative aspect of science and thus give a false impression of how science develops. Moreover, it is not really possible to understand today's view of the atom without insight into the concepts that led to it

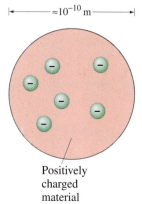

FIGURE 38–15 Plum-pudding model of the atom.

Rutherford's planetary model

FIGURE 38–16 (a) Experimental setup for Rutherford's experiment: α particles emitted by radon strike a metallic foil and some rebound backward; (b) backward rebound of α particles explained as the repulsion from a heavy positively charged nucleus.

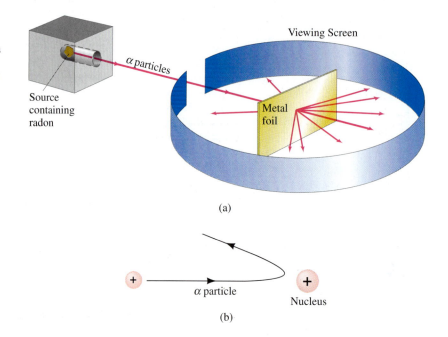

due to electrical attraction, Fig. 38–17. Rutherford's experiments suggested that the nucleus must have a radius of about 10^{-15} to 10^{-14} m. From kinetic theory, and especially Einstein's analysis of Brownian movement (see Section 17–1), the radius of atoms was estimated to be about 10^{-10} m. Thus the electrons would seem to be at a distance from the nucleus of about 10,000 to 100,000 times the radius of the nucleus itself (if the nucleus were the size of a baseball, the atom would have the diameter of a big city several kilometers across). So an atom would be mostly empty space.

Rutherford's "planetary" model of the atom (also called the "nuclear model of the atom") was a major step toward how we view the atom today. It was not, however, a complete model and presented some major problems, as we shall see.

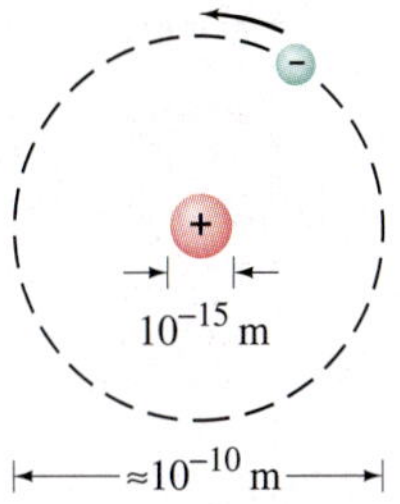

FIGURE 38–17 Rutherford's model of the atom, in which electrons orbit a tiny positive nucleus (not to scale). The atom is visualized as mostly empty space.

38–9 Atomic Spectra: Key to the Structure of the Atom

Earlier in this chapter we saw that heated solids (as well as liquids and dense gases) emit light with a continuous spectrum of wavelengths. This radiation is assumed to be due to oscillations of atoms and molecules, which are largely governed by the interaction of each atom or molecule with its neighbors.

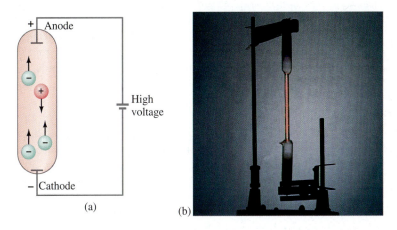

(a)   (b)

FIGURE 38–18 Gas-discharge tube: (a) is a diagram; (b) photo of actual discharge tube for hydrogen.

Rarefied gases can also be excited to emit light. This is done by intense heating, or more commonly by applying a high voltage to a "discharge tube" containing the gas at low pressure, Fig. 38–18. The radiation from excited gases had been observed early in the nineteenth century, and it was found that the spectrum was not continuous, but *discrete*. Since excited gases emit light of only certain wavelengths, when this light is analyzed through the slit of a spectroscope or spectrometer, a **line spectrum** is seen rather than a continuous spectrum. The line spectra emitted by a number of elements in the visible region are shown here in Fig. 38–19,

FIGURE 38–19 Emission spectra of the gases (a) atomic hydrogen, (b) helium, and (c) the *solar absorption* spectrum.

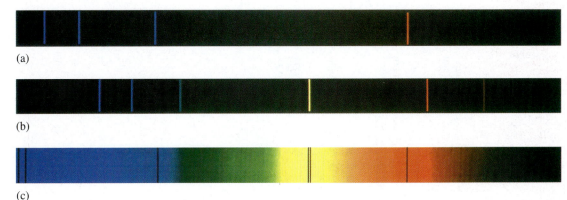

(a)

(b)

(c)

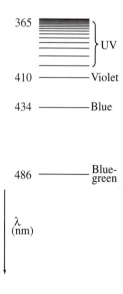

365 ⎤
     ⎬ UV
410 —————— Violet

434 —————— Blue

486 —————— Blue-green

λ
(nm)

656 —————— Red

FIGURE 38–20 Balmer series of lines for hydrogen.

and also at the start of this chapter and in Chapter 35. The **emission spectrum** is characteristic of the material and can serve as a type of "fingerprint" for identification of the gas.

We also saw (Chapter 35) that if a continuous spectrum passes through a gas, dark lines are observed in the emerging spectrum, at wavelengths corresponding to lines normally emitted by the gas. This is called an **absorption spectrum** (Fig. 38–19c), and it became clear that gases can absorb light at the same frequencies at which they emit. Using film sensitive to ultraviolet and to infrared light, it was found that gases emit and absorb discrete frequencies in these regions as well as in the visible.

For our purposes here, the importance of the line spectra is that they are emitted (or absorbed) by gases at low pressure and low density. In such low density gases, the atoms are far apart on the average and hence the light emitted or absorbed is assumed to be by *individual atoms* rather than through interactions between atoms, as in a solid, liquid, or dense gas. Thus the line spectra serve as a key to the structure of the atom: any theory of atomic structure must be able to explain why atoms emit light only of discrete wavelengths, and it should be able to predict what these wavelengths are.

Hydrogen is the simplest atom—it has only one electron orbiting its nucleus. It also has the simplest spectrum. The spectrum of most atoms shows little apparent regularity. But the spacing between lines in the hydrogen spectrum decreases in a regular way, Fig. 38–20. Indeed, in 1885, J. J. Balmer (1825–1898) showed that the four visible lines in the hydrogen spectrum (with measured wavelengths 656 nm, 486 nm, 434 nm, and 410 nm) fit the following formula

$$\frac{1}{\lambda} = R\left(\frac{1}{2^2} - \frac{1}{n^2}\right), \qquad n = 3, 4, \cdots, \tag{38–8}$$

where n takes on the values 3, 4, 5, 6 for the four lines, and R, called the **Rydberg constant**, has the value $R = 1.0974 \times 10^7 \, \text{m}^{-1}$. Later it was found that this **Balmer series** of lines extended into the UV region, ending at $\lambda = 365 \, \text{nm}$, as shown in Fig. 38–20. Balmer's formula, Eq. 38–8, also worked for these lines with higher integer values of n. The lines near 365 nm became too close together to distinguish, but the limit of the series at 365 nm corresponds to $n = \infty$ (so $1/n^2 = 0$ in Eq. 38–8).

Later experiments on hydrogen showed that there were similar series of lines in the UV and IR regions, and each series had a pattern just like the Balmer series, but at different wavelengths, Fig. 38–21. Each of these series was found to fit a formula with the same form as Eq. 38–8 but with the $1/2^2$ replaced by $1/1^2, 1/3^2, 1/4^2$, and so on. For example, the so-called **Lyman series** contains lines with wavelengths from 91 nm to 122 nm (in the UV region) and fits the formula

$$\frac{1}{\lambda} = R\left(\frac{1}{1^2} - \frac{1}{n^2}\right), \qquad n = 2, 3, \cdots.$$

And the wavelengths of the **Paschen series** (in the IR region) fit

$$\frac{1}{\lambda} = R\left(\frac{1}{3^2} - \frac{1}{n^2}\right), \qquad n = 4, 5, \cdots.$$

FIGURE 38–21 Line spectrum of atomic hydrogen. Each series fits the formula $\frac{1}{\lambda} = R\left(\frac{1}{n'^2} - \frac{1}{n^2}\right)$ where $n' = 1$ for the Lyman series, $n' = 2$ for the Balmer series, $n' = 3$ for the Paschen series, and so on; n can take on all integer values from $n = n' + 1$ up to infinity. The only lines in the visible region of the electromagnetic spectrum are part of the Balmer series.

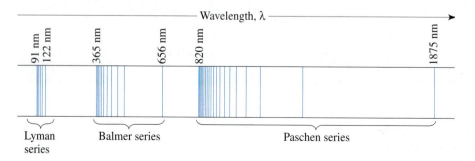

Wavelength, λ

91 nm 122 nm   365 nm   656 nm   820 nm   1875 nm

Lyman series   Balmer series   Paschen series

The Rutherford model, as it stood, was unable to explain why atoms emit line spectra. It had other difficulties as well. According to the Rutherford model, electrons orbit the nucleus, and since their paths are curved the electrons are accelerating. Hence they should give off light like any other accelerating electric charge (Chapter 32). Then, since energy is conserved, the electron's own energy must decrease to compensate. Hence electrons would be expected to spiral into the nucleus. As they spiraled inward, their frequency would increase in a short time and so too would the frequency of the light emitted. Thus the two main difficulties of the Rutherford model are these: (1) it predicts that light of a continuous range of frequencies will be emitted, whereas experiment shows line spectra; (2) it predicts that atoms are unstable—electrons should quickly spiral into the nucleus—but we know that atoms in general are stable, since the matter around us is stable.

Clearly Rutherford's model was not sufficient. Some sort of modification was needed, and it was Niels Bohr who provided it by adding an essential idea—the quantum hypothesis.

38–10 The Bohr Model

Bohr had studied in Rutherford's laboratory for several months in 1912 and was convinced that Rutherford's planetary model of the atom had validity. But in order to make it work, he felt that the newly developing quantum theory would somehow have to be incorporated in it. The work of Planck and Einstein had shown that in heated solids, the energy of oscillating electric charges must change discontinuously—from one discrete energy state to another, with the emission of a quantum of light. Perhaps, Bohr argued, the electrons in an atom also cannot lose energy continuously, but must do so in quantum "jumps." In working out his theory during the next year, Bohr postulated that electrons move about the nucleus in circular orbits, but that only certain orbits are allowed. He further postulated that an electron in each orbit would have a definite energy and would move in the orbit *without radiating energy* (even though this violated classical ideas since accelerating electric charges are supposed to emit EM waves; see Chapter 32). He thus called the possible orbits **stationary states**. Light is emitted, he hypothesized, only when an electron jumps from one stationary state to another of lower energy. When such a jump occurs, a single photon of light is emitted whose energy, since energy is conserved, is given by

$$hf = E_u - E_l, \tag{38–9}$$

where E_u refers to the energy of the upper state and E_l the energy of the lower state. See Fig. 38–22.

Bohr set out to determine what energies these orbits would have, since the spectrum of light emitted could then be predicted from Eq. 38–9. In the Balmer formula he had the key he was looking for. Bohr quickly found that his theory would be in accord with the Balmer formula if he assumed that the electron's angular momentum L is quantized and equal to an integer n times $h/2\pi$. As we saw in Chapter 10, angular momentum is given by $L = I\omega$, where I is the moment of inertia and ω is the angular velocity. For a single particle of mass m moving in a circle of radius r with speed v, $I = mr^2$ and $\omega = v/r$; hence, $L = I\omega = (mr^2)(v/r) = mvr$. Bohr's **quantum condition** is

$$L = mvr_n = n\frac{h}{2\pi}, \qquad n = 1, 2, 3, \cdots, \tag{38–10}$$

where n is an integer and r_n is the radius of the nth possible orbit. The allowed orbits are numbered 1, 2, 3, ..., according to the value of n, which is called the **quantum number** of the orbit.

Stationary state

Angular momentum quantized

Quantum number, n

FIGURE 38–22 An atom emits a photon (energy = hf) when its energy changes from E_u to a lower energy E_l.

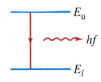

E_u

hf

E_l

Equation 38–10 did not have a firm theoretical foundation. Bohr had searched for some "quantum condition," and such tries as $E = hf$ (where E represents the energy of the electron in an orbit) did not give results in accord with experiment. Bohr's reason for using Eq. 38–10 was simply that it worked; and we now look at how. In particular, let us determine what the Bohr theory predicts for the measurable wavelengths of emitted light.

An electron in a circular orbit of radius r_n (Fig. 38–23) would have a centripetal acceleration v^2/r_n produced by the electrical force of attraction between the negative electron and the positive nucleus. This force is given by Coulomb's law,

$$F = \frac{1}{4\pi\epsilon_0} \frac{(Ze)(e)}{r^2}.$$

The charge on the nucleus is $+Ze$, where Z is the number of positive charges[†] (i.e., protons). For the hydrogen atom, $Z = +1$.

In Newton's second law, $F = ma$, we substitute $a = v^2/r_n$ and Coulomb's law for F, and obtain

$$F = ma$$

$$\frac{1}{4\pi\epsilon_0} \frac{Ze^2}{r_n^2} = \frac{mv^2}{r_n}.$$

We solve this for r_n, and then substitute for v from Eq. 38–10 (which says $v = nh/2\pi m r_n$):

$$r_n = \frac{Ze^2}{4\pi\epsilon_0 mv^2} = \frac{Ze^2 4\pi^2 m r_n^2}{4\pi\epsilon_0 n^2 h^2}.$$

We solve for r_n (it appears on both sides, so we cancel one of them) and find

$$r_n = \frac{n^2 h^2 \epsilon_0}{\pi m Z e^2} = \frac{n^2}{Z} r_1 \tag{38–11}$$

where

$$r_1 = \frac{h^2 \epsilon_0}{\pi m e^2}. \tag{38–12a}$$

Equation 38–11 gives the radii of all possible orbits. The smallest orbit is for $n = 1$, and for hydrogen ($Z = 1$) has the value

$$r_1 = \frac{(6.626 \times 10^{-34}\,\text{J·s})^2 (8.85 \times 10^{-12}\,\text{C}^2/\text{N·m}^2)}{(3.14)(9.11 \times 10^{-31}\,\text{kg})(1.602 \times 10^{-19}\,\text{C})^2}$$

or

$$r_1 = 0.529 \times 10^{-10}\,\text{m}. \tag{38–12b}$$

The radius of the smallest orbit in hydrogen, r_1, is sometimes called the **Bohr radius**. From Eq. 38–11, we see that the radii of the larger orbits increase as n^2, so

$$r_2 = 4r_1 = 2.12 \times 10^{-10}\,\text{m},$$

$$r_3 = 9r_1 = 4.76 \times 10^{-10}\,\text{m},$$

$$\vdots$$

$$r_n = n^2 r_1.$$

The first four orbits are shown in Fig. 38–24. Notice that, according to Bohr's model, an electron can exist only in the orbits given by Eq. 38–11. There are no allowable orbits in between.

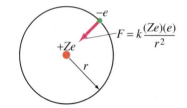

FIGURE 38–23 Electric force (Coulomb's law) keeps the negative electron in orbit around the positively charged nucleus.

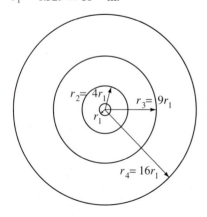

Bohr radius

FIGURE 38–24 Possible orbits in the Bohr model of hydrogen; $r_1 = 0.529 \times 10^{-10}$ m.

[†] We include Z in our derivation so that we can treat other single-electron ("hydrogenlike") atoms such as the ions He[+] ($Z = 2$) and Li[2+] ($Z = 3$). Helium in the neutral state has two electrons; if one electron is missing, the remaining He[+] ion consists of one electron revolving around a nucleus of charge $+2e$. Similarly, doubly ionized lithium, Li[2+], also has a single electron, and in this case $Z = 3$.

In each of its possible orbits, the electron would have a definite energy, as the following calculation shows. The total energy equals the sum of the kinetic and potential energies. The potential energy of the electron is given by $U = qV = -eV$, where V is the potential due to a point charge $+Ze$ as given by Eq. 23–5: $V = (1/4\pi\epsilon_0)(Q/r) = (1/4\pi\epsilon_0)(Ze/r)$. So

$$U = -eV = -\frac{1}{4\pi\epsilon_0}\frac{Ze^2}{r}.$$

The total energy E_n for an electron in the nth orbit of radius r_n is the sum of the kinetic and potential energies:

$$E_n = \tfrac{1}{2}mv^2 - \frac{1}{4\pi\epsilon_0}\frac{Ze^2}{r_n}.$$

When we substitute v from Eq. 38–10 and r_n from Eq. 38–11 into this equation, we obtain

$$E_n = -\left(\frac{Z^2e^4m}{8\epsilon_0^2 h^2}\right)\left(\frac{1}{n^2}\right) \qquad n = 1, 2, 3, \cdots \qquad \textbf{(38–13)} \qquad \textit{Energy levels}$$

or

$$E_n = \frac{Z^2}{n^2}E_1 \qquad n = 1, 2, 3, \cdots$$

where E_1 is the lowest energy level $(n = 1)$ for hydrogen $(Z = 1)$. The value of E_1 is

$$E_1 = -\frac{me^4}{8\epsilon_0^2 h^2} = -2.17 \times 10^{-18}\,\text{J} = -13.6\,\text{eV}, \qquad \textit{Ground state of hydrogen}$$

where we have converted joules to electron volts, as is customary in atomic physics. Since n^2 appears in the denominator of Eq. 38–13, the energies of the larger orbits in hydrogen $(Z = 1)$ are given by

$$E_n = \frac{-13.6\,\text{eV}}{n^2}.$$

For example,

$$E_2 = \frac{-13.6\,\text{eV}}{4} = -3.40\,\text{eV}, \qquad \textit{Excited states (first two)}$$

$$E_3 = \frac{-13.6\,\text{eV}}{9} = -1.51\,\text{eV}.$$

We see that not only are the orbit radii quantized, but from Eq. 38–13 so is the energy. The quantum number n that labels the orbit radii also labels the energy levels. The lowest **energy level** or **energy state** has energy E_1, and is called the **ground state**. The higher states, E_2, E_3, and so on, are called **excited states**.

Notice that although the energy for the larger orbits has a smaller numerical value, all the energies are less than zero. Thus, -3.4 eV is a higher energy than -13.6 eV. Hence the orbit closest to the nucleus (r_1) has the lowest energy. The reason the energies have negative values has to do with the way we defined the zero for potential energy (U). For two point charges, $U = kq_1q_2/r$ corresponds to zero potential energy when the two charges are infinitely far apart. Thus, an electron that can just barely be free from the atom by reaching $r = \infty$ (or, at least, far from the nucleus) with zero kinetic energy will have $E = 0$, corresponding to $n = \infty$ in Eq. 38–13. If an electron is free and has kinetic energy, then $E > 0$. To remove an electron that is part of an atom requires an energy input (otherwise atoms would not be stable). Since $E \geq 0$ for a free electron, then an electron bound to an atom must have $E < 0$. That is, energy must be added to bring its energy up, from a negative value, to at least zero in order to free it.

The minimum energy required to remove an electron from the ground state of an atom is called the **binding energy** or **ionization energy**. The ionization energy for hydrogen has been measured to be 13.6 eV, and this corresponds precisely to removing an electron from the lowest state, $E_1 = -13.6$ eV, up to $E = 0$ where it can be free.

Binding energy
Ionization energy

It is useful to show the various possible energy values as horizontal lines on an energy-level diagram.[†] This is shown for hydrogen in Fig. 38–25. The electron in a hydrogen atom can be in any one of these levels according to Bohr theory. But it could never be in between, say at -9.0 eV. At room temperature, nearly all H atoms will be in the ground state ($n = 1$). At higher temperatures, or during an electric discharge when there are many collisions between free electrons and atoms, many atoms can be in excited states ($n > 1$). Once in an excited state, an atom's electron can jump down to a lower state, and give off a photon in the process. This is, according to the Bohr model, the origin of the emission spectra of excited gases. The vertical arrows in Fig. 38–25 represent the transitions or jumps that correspond to the various observed spectral lines. For example, an electron jumping from the level $n = 3$ to $n = 2$ would give rise to the 656-nm line in the Balmer series, and the jump from $n = 4$ to $n = 2$ would give rise to the 486-nm line (see Fig. 38–20). We can predict wavelengths of the spectral lines emitted by combining Eq. 38–9 with Eq. 38–13. Since $hf = hc/\lambda$, we have from Eq. 38–9

Line spectra emission explained

$$\frac{1}{\lambda} = \frac{hf}{hc} = \frac{1}{hc}(E_n - E_{n'}),$$

and then using Eq. 38–13,

$$\frac{1}{\lambda} = \frac{Z^2 e^4 m}{8\epsilon_0^2 h^3 c}\left(\frac{1}{(n')^2} - \frac{1}{(n)^2}\right), \tag{38–14}$$

where n refers to the upper state and n' to the lower state. This theoretical formula

[†]Note that above $E = 0$, an electron is free and can have any energy (E is not quantized). Thus there is a continuum of energy states above $E = 0$, as indicated in the energy-level diagram of Fig. 38–25.

FIGURE 38–25 Energy-level diagram for the hydrogen atom, showing the origin of spectral lines for the Lyman, Balmer, and Paschen series.

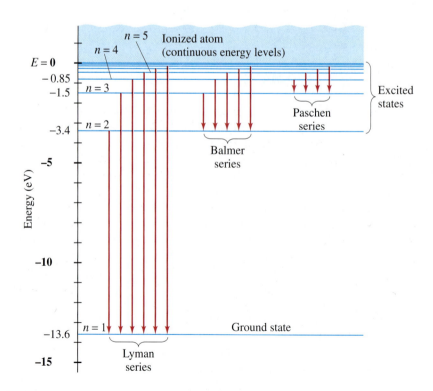

has the same form as the experimental Balmer formula, Eq. 38–8, with $n' = 2$. Thus we see that the Balmer series of lines corresponds to transitions or "jumps" that bring the electron down to the second energy level. Similarly, $n' = 1$ corresponds to the Lyman series and $n' = 3$ to the Paschen series (see Fig. 38–25). When the constant $(Z^2e^4m/8\epsilon_0^2h^3c)$ in Eq. 38–14 is evaluated with $Z = 1$, it is found to have the measured value of the Rydberg constant, $R = 1.0974 \times 10^7 \, \text{m}^{-1}$ in Eq. 38–8, in accord with experiment (see Problem 47).

The great success of Bohr's theory is that it explains why atoms emit line spectra and accurately predicts, for hydrogen, the wavelengths of emitted light. The Bohr theory also explains absorption spectra: photons of just the right wavelength can knock an electron from one energy level to a higher one. To conserve energy, only photons that have just the right energy (and frequency) will be absorbed. This explains why a continuous spectrum of light entering a gas will emerge with dark (absorption) lines at frequencies that correspond to emission lines.

Absorption lines explained

The Bohr theory also ensures the stability of atoms. It establishes stability by fiat: the ground state is the lowest state for an electron and there is no lower energy level to which it can go and emit more energy. Finally, as we saw above, the Bohr theory accurately predicts the ionization energy of 13.6 eV for hydrogen. However, the Bohr theory was not so successful for other atoms, as we shall see.

EXAMPLE 38–12 **Wavelength of a Lyman line.** Use Fig. 38–25 to determine the wavelength of the first Lyman line, the transition from $n = 2$ to $n = 1$. In what region of the electromagnetic spectrum does this lie?

SOLUTION In this case, $hf = E_2 - E_1 = 13.6 \, \text{eV} - 3.4 \, \text{eV} = 10.2 \, \text{eV} = 1.63 \times 10^{-18} \, \text{J}$. Since $\lambda = c/f$, we have

$$\lambda = \frac{c}{f} = \frac{hc}{E_2 - E_1} = \frac{(6.63 \times 10^{-34} \, \text{J·s})(3.00 \times 10^8 \, \text{m/s})}{1.63 \times 10^{-18} \, \text{J}} = 1.22 \times 10^{-7} \, \text{m},$$

or 122 nm, which is in the UV region. See Fig. 38–21.

EXAMPLE 38–13 **Wavelength of a Balmer line.** Determine the wavelength of light emitted when a hydrogen atom makes a transition from the $n = 6$ to the $n = 2$ energy level according to the Bohr model.

SOLUTION We can use Eq. 38–14 or its equivalent, Eq. 38–8, with $R = 1.097 \times 10^7 \, \text{m}^{-1}$. Thus

$$\frac{1}{\lambda} = (1.097 \times 10^7 \, \text{m}^{-1})\left(\frac{1}{4} - \frac{1}{36}\right) = 2.44 \times 10^6 \, \text{m}^{-1}.$$

So $\lambda = 1/(2.44 \times 10^6 \, \text{m}^{-1}) = 4.10 \times 10^{-7} \, \text{m}$ or 410 nm. This is the fourth line in the Balmer series, Fig. 38–20, and is violet in color.

EXAMPLE 38–14 **Absorption wavelength.** Use Fig. 38–25 to determine the maximum wavelength that hydrogen in its ground state can absorb. What would be the next smaller wavelength that would work?

SOLUTION Maximum λ corresponds to minimal energy, and thus the jump from the ground state to the first excited state (Fig. 38–25) for which the energy is $13.6 \, \text{eV} - 3.4 \, \text{eV} = 10.2 \, \text{eV}$; the required wavelength, as we saw in Example 38–12, is 122 nm. The next possibility is to jump from the ground state to the second excited state, which requires $13.6 \, \text{eV} - 1.5 \, \text{eV} = 12.1 \, \text{eV}$ and corresponds to a wavelength

$$\lambda = \frac{c}{f} = \frac{hc}{E_3 - E_1}$$

$$= \frac{(6.63 \times 10^{-34} \, \text{J·s})(3.00 \times 10^8 \, \text{m/s})}{(12.1 \, \text{eV})(1.60 \times 10^{-19} \, \text{J/eV})} = 103 \, \text{nm}.$$

EXAMPLE 38–15 Ionization energy. Use the Bohr model to determine the ionization energy of the He^+ ion, which has a single electron. Also calculate the minimum wavelength a photon must have to cause ionization.

SOLUTION We want to determine the minimum energy required to lift the electron from its ground state and to barely reach the free state at $E = 0$. The ground state energy of He^+ is given by Eq. 38–13 with $n = 1$ and $Z = 2$. Since all the symbols in Eq. 38–13 are the same as for the calculation for hydrogen, except that Z is 2 instead of 1, we see that E_1 will be $Z^2 = 2^2 = 4$ times the E_1 for hydrogen. That is,

$$E_1 = 4(-13.6\,\text{eV}) = -54.4\,\text{eV}.$$

Thus, to ionize the He^+ ion should require 54.4 eV, and this value agrees with experiment. The minimum wavelength photon that can cause ionization will have energy $hf = 54.4\,\text{eV}$ and wavelength $\lambda = c/f = hc/hf = (6.63 \times 10^{-34}\,\text{J·s})(3.00 \times 10^8\,\text{m/s})/(54.4\,\text{eV})(1.60 \times 10^{-19}\,\text{J/eV}) = 22.8\,\text{nm}$. If the atom absorbed a photon of greater energy (wavelength shorter than 22.8 nm), the atom could still be ionized and the freed electron would have kinetic energy of its own.

In this last Example, we saw that E_1 for the He^+ ion is four times more negative than that for hydrogen. Indeed the energy-level diagram for He^+ looks just like that for hydrogen, Fig. 38–25, except that the numerical values for each energy level are four times larger. It is important to note, however, that we are talking here about the He^+ *ion*. Normal (neutral) helium has two electrons and its energy level diagram is entirely different.

CONCEPTUAL EXAMPLE 38–16 Hydrogen at 20°C. Estimate the average kinetic energy of hydrogen atoms (or molecules) at room temperature, and use the result to explain why nearly all H atoms are in the ground state at room temperature, and hence emit no light.

RESPONSE According to kinetic theory (Chapter 18), the average kinetic energy of atoms or molecules in a gas is given by Eq. 18–4,

$$\bar{K} = \tfrac{3}{2}kT,$$

where $k = 1.38 \times 10^{-23}\,\text{J/K}$ is Boltzmann's constant, and T is the kelvin (absolute) temperature. Room temperature is about $T = 300\,\text{K}$, so

$$\bar{K} = \tfrac{3}{2}(1.38 \times 10^{-23}\,\text{J/K})(300\,\text{K}) = 6.2 \times 10^{-21}\,\text{J},$$

or, in electron volts:

$$\bar{K} = \frac{6.2 \times 10^{-21}\,\text{J}}{1.6 \times 10^{-19}\,\text{J/eV}} = 0.04\,\text{eV}.$$

The average kinetic energy is thus very small compared to the energy between the ground state and the next higher energy state ($13.6\,\text{eV} - 3.4\,\text{eV} = 10.2\,\text{eV}$). Any atoms in excited states emit light and eventually fall to the ground state. Once in the ground state, collisions with other atoms can transfer energy of only 0.04 eV on the average. A small fraction of atoms can have much more energy (see Section 18–2 on the distribution of molecular speeds), but even kinetic energy that is 10 times the average is not nearly enough to excite atoms above the ground state. Thus, at room temperature, nearly all atoms are in the ground state. Atoms can be excited to upper states by very high temperatures, or by passing a current of high energy electrons through the gas, as in a discharge tube (Fig. 38–18).

We should note that Bohr made some radical assumptions that were at variance with classical ideas. He assumed that electrons in fixed orbits do not radiate light even though they are accelerating (moving in a circle), and he assumed that angular momentum is quantized. Furthermore, he was not able to say how an electron moved when it made a transition from one energy level to another. On the other hand, there is no real reason to expect that in the tiny world of the atom electrons would behave as ordinary-sized objects do. Nonetheless, he felt that where quantum theory overlaps with the macroscopic world, it should predict classical results. This is the **correspondence principle**, already mentioned in regard to relativity (Section 37–13). This principle does work for Bohr's theory of the hydrogen atom. The orbit sizes and energies are quite different for $n = 1$ and $n = 2$, say. But orbits with $n = 100,000,000$ and $100,000,001$ would be very close in size and energy (see Fig. 38–25). Indeed, jumps between such large orbits, which would approach macroscopic sizes, would be imperceptible. Such orbits would thus appear to be continuously spaced, which is what we expect in the everyday world.

Correspondence principle

Finally, be careful not to believe that the well-defined orbits of the Bohr model actually exist. The Bohr model is only a model, not reality. The idea of electron orbits was rejected a few years later, and today electrons are better thought of as forming "clouds."

38–11 de Broglie's Hypothesis Applied to Atoms

Bohr's theory was largely of an *ad hoc* nature. Assumptions were made so that theory would agree with experiment. But Bohr could give no reason why the orbits were quantized, nor why there should be a stable ground state. Finally, ten years later, a reason was proposed by de Broglie.

We saw in Section 38–6 that in 1923, Louis de Broglie proposed that material particles, such as electrons, have a wave nature; and that this hypothesis was confirmed by experiment several years later.

One of de Broglie's original arguments in favor of the wave nature of electrons was that it provided an explanation for Bohr's theory of the hydrogen atom. According to de Broglie, a particle of mass m moving with a nonrelativistic speed v would have a wavelength (Eq. 38–7) of

$$\lambda = \frac{h}{mv}.$$

Each electron orbit in an atom, he proposed, is actually a standing wave. As we saw in Chapter 15, when a violin or guitar string is plucked, a vast number of wavelengths are excited. But only certain ones—those that have nodes at the ends—are sustained. These are the *resonant* modes of the string. Waves with other wavelengths interfere with themselves upon reflection and their amplitudes quickly drop to zero. With electrons moving in circles, according to Bohr's theory, de Broglie argued that the electron wave was a *circular* standing wave that closes on itself, Fig. 38–26. If the wavelength of a wave does not close on itself, as in Fig. 38–27, destructive interference takes place as the wave travels around the loop, and the wave quickly dies out. Thus, the only waves that persist are those for which the circumference of the circular orbit contains a whole number of wavelengths, Fig. 38–28. The circumference of a Bohr orbit of radius r_n is $2\pi r_n$, so we have

$$2\pi r_n = n\lambda, \qquad n = 1, 2, 3, \cdots.$$

When we substitute $\lambda = h/mv$, we get $2\pi r_n = nh/mv$, or

$$mvr_n = \frac{nh}{2\pi}.$$

This is just the *quantum condition* proposed by Bohr on an *ad hoc* basis, Eq. 38–10.

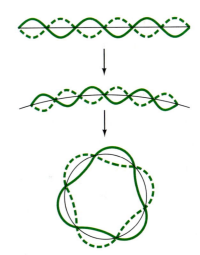

FIGURE 38–26 An ordinary standing wave compared to a circular standing wave.

FIGURE 38–27 When a wave does not close (and hence interferes destructively with itself), it rapidly dies out.

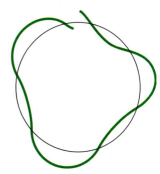

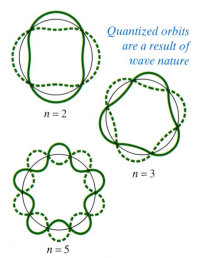

Quantized orbits are a result of wave nature

$n = 2$

$n = 3$

$n = 5$

FIGURE 38–28 Standing circular waves for two, three, and five wavelengths on the circumference; n, the number of wavelengths, is also the quantum number.

And it is from this equation that the discrete orbits and energy levels were derived. Thus we have an explanation for the quantized orbits and energy states in the Bohr model: they are due to the wave nature of the electron, and that only resonant "standing" waves can persist. This implies that the *wave–particle duality* is at the root of atomic structure.

It should be noted in viewing the circular electron waves of Fig. 38–28 that the electron is not to be thought of as following the oscillating wave pattern. In the Bohr model of hydrogen, the electron, considered as a particle, moves in a circle. The circular wave, on the other hand, represents the *amplitude* of the electron "matter wave," and in Fig. 38–28 the wave amplitude is shown superimposed on the circular path of the particle orbit for convenience.

Bohr's theory worked well for hydrogen and for one-electron ions. But it did not prove as successful for multielectron atoms. Bohr theory could not predict line spectra even for the next simplest atom, helium. It could not explain why some emission lines are brighter than others, nor why some lines are actually split into two or more closely spaced lines ("fine structure"). A new theory was needed and was indeed developed in the 1920s. This new and radical theory is called *quantum mechanics*, and finally solved the problem of atomic structure; but it gives us a very different view of the atom: the idea of electrons in well-defined orbits was replaced with the idea of electron "clouds." And this new theory of quantum mechanics has given us a wholly different view of the basic mechanisms underlying physical processes.

Summary

Quantum theory has its origins in **Planck's quantum hypothesis** that molecular oscillations are quantized: their energy E can only be integer (n) multiples of hf, where h is Planck's constant and f is the natural frequency of oscillation: $E = nhf$. This hypothesis explained the spectrum of radiation emitted by (black) bodies at high temperature.

Einstein proposed that for some experiments, light could be pictured as being emitted and absorbed as quanta (particles), which we now call **photons**, each with energy

$$E = hf.$$

He proposed the photoelectric effect as a test for the photon theory of light. In the **photoelectric effect**, the photon theory says that each incident photon can strike an electron in a material and eject it if it has sufficient energy. The maximum energy of ejected electrons is then linearly related to the frequency of the incident light. The photon theory is also supported by the **Compton effect** and the observation of electron–positron **pair production**.

The wave–particle duality refers to the idea that light and matter (such as electrons) have both wave and particle properties. The wavelength of a material object is given by

$$\lambda = \frac{h}{mv},$$

where mv is the momentum of the object. The **principle of complementarity** states that we must be aware of both the particle and wave properties of light and of matter for a complete understanding of them.

Early models of the atom include the plum-pudding model and Rutherford's planetary (or nuclear) model. Rutherford's model, which was created to explain the backscattering of alpha particles from thin metal foils, assumes that an atom consists of a tiny but massive positively charged nucleus surrounded (at a relatively great distance) by electrons.

To explain the line spectra emitted by atoms, as well as the stability of atoms, **Bohr theory** postulated (1) that electrons bound in an atom can only occupy orbits for which the angular momentum is quantized, which results in discrete values for the radius and energy; (2) that an electron in such a **stationary state** emits no radiation; (3) that, if an electron jumps to a lower state, it emits a photon whose energy equals the difference in energy between the two states; (4) that the angular momentum L of atomic electrons is quantized by the rule

$$L = \frac{nh}{2\pi},$$

where n is an integer called a **quantum number**. The $n = 1$ state in hydrogen is the **ground state**, which has an energy $E_1 = -13.6 \, \text{eV}$; higher values of n correspond to **excited states** and their energies are

$$E_n = -\frac{13.6 \, \text{eV}}{n^2}.$$

Atoms are excited to these higher states by collisions with other atoms or electrons or by absorption of a photon of just the right frequency.

De Broglie's hypothesis that electrons (and other matter) have a wavelength $\lambda = h/mv$ gave an explanation for Bohr's quantized orbits by bringing in the wave–particle duality: the orbits correspond to circular standing waves in which the circumference of the orbit equals a whole number of wavelengths.

Questions

1. What can be said about the relative temperature of whitish-yellow, reddish, and bluish stars?

2. If energy is radiated by all objects, why can we not see them in the dark? (See also Section 19–10.)

3. Does a lightbulb at a temperature of 2500 K produce as white a light as the Sun at 6000 K? Explain.

4. An ideal black body can be approximated by a small hole in an otherwise enclosed cavity. Explain. [*Hint*: The pupil of your eye is an approximate case.]

5. Darkrooms for developing black-and-white film are sometimes lit by a red bulb. Why red? Would such a bulb work in a darkroom for developing color photographs?

6. If the threshold wavelength in the photoelectric effect increases when the emitting metal is changed to a different metal, what can you say about the work functions of the two metals?

7. Explain why the existence of a cutoff frequency in the photoelectric effect more strongly favors a particle theory rather than a wave theory of light.

8. UV light causes sunburn, whereas visible light does not. Explain.

9. If an X-ray photon is scattered by an electron, does its wavelength change? If so, does it increase or decrease?

10. In both the photoelectric effect and in the Compton effect we have a photon colliding with an electron causing the electron to fly off. What then, is the difference between the two processes?

11. Show that the units of Planck's constant are those for angular momentum.

12. Consider a point source of light. How would the intensity of light vary with distance from the source according to (a) wave theory, (b) particle (photon) theory? Would this help to distinguish the two theories?

13. Explain how the photoelectric circuit of Fig. 38–4 could be used in (a) a burglar alarm, (b) a smoke detector, (c) a photographic light meter, and (d) a spectrophotometer (see Section 36–8).

14. Why do we say that light has wave properties? Why do we say that light has particle properties?

15. Why do we say that electrons have wave properties? Why do we say that electrons have particle properties?

16. What is the difference between a photon and an electron? Be specific: make a list.

17. If an electron and a proton travel at the same speed, which has the shorter wavelength?

18. In Rutherford's planetary model of the atom, what keeps the electrons from flying off into space?

19. Which of the following can emit a line spectrum: (a) gases, (b) liquids, (c) solids? Which can emit a continuous spectrum?

20. Why doesn't the O_2 gas in the air around us give off light?

21. How can you tell if there is oxygen near the surface of the Sun?

22. When a wide spectrum of light passes through hydrogen gas at room temperature, absorption lines are observed that correspond only to the Lyman series. Why don't we observe the other series?

23. Explain how the closely spaced energy levels for hydrogen near the top of Fig. 38–25 correspond to the closely spaced spectral lines at the top of Fig. 38–20.

24. Discuss the differences between Rutherford's and Bohr's theory of the atom.

25. Is it possible for the de Broglie wavelength of a "particle" to be greater than the dimensions of the particle? To be smaller? Is there any direct connection?

26. Approximately how large would h have to be so that quantum effects would be noticeable in the macroscopic world of everyday life?

27. Does it make sense that the potential energy of the electron in a hydrogen atom is negative and greater in magnitude than the kinetic energy, as given by the Bohr theory? What is the significance of this?

28. In a helium atom, which contains two electrons, do you think that on the average the electrons are closer to the nucleus or farther away than in a hydrogen atom? Why?

29. How can the spectrum of hydrogen contain so many lines when hydrogen contains only one electron?

30. The Lyman series is brighter than the Balmer series, because this series of transitions ends up in the most common state for hydrogen, the ground state. Why then was the Balmer series discovered first?

31. Use conservation of momentum to explain why photons emitted by hydrogen atoms have slightly less energy than that predicted by Eq. 38–9.

Problems

Section 38–1

1. (I) How hot is a metal being welded if it radiates most strongly at 440 nm?

2. (I) (a) What is the temperature if the peak of a blackbody spectrum is at 25.0 nm? (b) What is the wavelength at the peak of a blackbody spectrum if the body is at a temperature of 2800 K?

3. (I) An HCl molecule vibrates with a natural frequency of 8.1×10^{13} Hz. What is the difference in energy (in joules and electron volts) between possible values of the oscillation energy?

4. (I) Estimate the peak wavelength for radiation from (a) ice at 0° C, (b) a floodlamp at 3300 K, (c) helium at 4 K, assuming blackbody emission. In what region of the EM spectrum is each?

5. (II) The steps of a flight of stairs are 20.0 cm high (vertically). If a 58.0-kg person stands with both feet on the same step, what is the gravitational potential energy of this person, relative to the ground, on (a) the first step, (b) the second step, (c) the third step, (d) the nth step? (e) What is the change in energy as the person descends from step 6 to step 2?

6. (II) Estimate the peak wavelength of light issuing from the pupil of the human eye (which approximates a blackbody) assuming normal body temperature.

7. (III) Planck's radiation law is given by:

$$I(\lambda, T) = \frac{2\pi hc^2 \lambda^{-5}}{e^{hc/\lambda kT} - 1}$$

where $I(\lambda, T)$ is the rate energy is radiated per unit surface area per unit wavelength interval at wavelength λ and Kelvin temperature T. (a) Show that Wien's displacement law follows from this relationship. (b) Determine the value of h from the experimental value of $\lambda_P T$ given in the text. (c) Derive the Stefan-Boltzmann law (the T^4 dependence of the rate at which energy is radiated—Eq. 19–15), by integrating Planck's formula over all wavelengths; that is, show that

$$\int I(\lambda, T) \, d\lambda \propto T^4.$$

Section 38–2

8. (I) What is the energy range in eV, of photons in the visible spectrum, of wavelength 400 nm to 750 nm?

9. (I) What is the energy of photons (in eV) emitted by an 88.5-MHz FM radio station?

10. (I) A typical gamma ray emitted from a nucleus during radioactive decay may have an energy of 300 keV. What is its wavelength? Would we expect significant diffraction of this type of light when it passes through an everyday opening, like a door?

11. (I) About 0.1 eV is required to break a "hydrogen bond" in a protein molecule. What are the minimum frequency and maximum wavelength of a photon that can accomplish this?

12. (I) What minimum frequency of light is needed to eject electrons from a metal whose work function is 4.3×10^{-19} J?

13. (II) What is the longest wavelength of light that will emit electrons from a metal whose work function is 3.10 eV?

14. (II) The work functions for sodium, cesium, copper and iron are 2.3, 2.1, 4.7 and 4.5 eV respectively. Which of these metals will not emit electrons when visible light shines on it?

15. (II) In a photoelectric-effect experiment it is observed that no current flows unless the wavelength is less than 570 nm. (a) What is the work function of this material? (b) What is the stopping voltage required if light of wavelength 400 nm is used?

16. (II) What is the maximum kinetic energy of electrons ejected from barium $\left(W_0 = 2.48 \text{ eV}\right)$ when illuminated by white light, $\lambda = 400$ to 750 nm?

17. (II) When UV light of wavelength 255 nm falls on a metal surface, the maximum kinetic energy of emitted electrons is 1.40 eV. What is the work function of the metal?

18. (II) The threshold wavelength for emission of electrons from a given surface is 350 nm. What will be the maximum kinetic energy of ejected electrons when the wavelength is changed to (a) 280 nm, (b) 380 nm?

19. (II) A certain type of film is sensitive only to light whose wavelength is less than 660 nm. What is the energy (eV and kcal/mol) needed for the chemical reaction to occur which causes the film to change?

20. (II) When 230-nm light falls on a metal, the current through a photoelectric circuit (Fig. 38–4) is brought to zero at a reverse voltage of 1.64 V. What is the work function of the metal?

Section 38–3

21. (II) The quantity h/mc, which has the dimensions of length, is called the Compton wavelength. Determine the Compton wavelength for (a) an electron, (b) a proton. (c) Show that if a photon has wavelength equal to the Compton wavelength of a particle, the photon's energy is equal to the rest energy of the particle.

22. (II) X-rays of wavelength $\lambda = 0.120$ nm are scattered from a carbon block. What is the Compton wavelength shift for photons detected at angles (relative to the incident beam) of (a) 45°, (b) 90°, (c) 180°?

23. (II) For each of the scattering angles in the previous Problem, determine (a) the fractional energy loss of the photon, and (b) the energy given to the scattered electron.

24. (II) (a) Show that the fractional energy loss of scattered photons in the Compton effect can be written as $\Delta\lambda/\lambda$. (b) Is this an exact expression? If not, specify what quantity limits the accuracy and what its value can be so this formula is accurate to 0.1 percent.

25. (III) In the Compton effect (see Fig. 38–7), use the relativistic equations for conservation of energy and of linear momentum to show that the Compton shift in wavelength is given by Eq. 38–6.

26. (III) In the Compton effect, a 0.100-nm photon strikes a free electron in a head-on collision and knocks it into the forward direction. The rebounding photon recoils directly backward. Use conservation of (relativistic) energy and momentum to determine (a) the kinetic energy of the electron, and (b) the wavelength of the recoiling photon. (Note: Use Eq. 38–5, but not Eq. 38–6.)

Section 38–4

27. (I) How much total kinetic energy will an electron–positron pair have if produced by a 2.84-MeV photon?

28. (II) What is the longest wavelength photon that could produce a proton–antiproton pair? (Each has a mass of 1.67×10^{-27} kg.)

29. (II) What is the minimum photon energy needed to produce a $\mu^+ - \mu^-$ pair? The mass of each μ (muon) is 207 times the mass of the electron. What is the wavelength of such a photon?

30. (II) A gamma-ray photon produces an electron–positron pair, each with a kinetic energy of 345 keV. What was the energy and wavelength of the photon?

Section 38–6

31. (I) Calculate the wavelength of a 0.21-kg ball traveling at 0.10 m/s.

32. (I) What is the wavelength of a neutron $(m = 1.67 \times 10^{-27}$ kg) traveling at 5.5×10^4 m/s?

33. (I) Through how many volts of potential difference must an electron be accelerated to achieve a wavelength of 0.28 nm?

34. (II) Calculate the de Broglie wavelength of an electron in your TV picture tube if it is accelerated by 30,000 V. Is it relativistic? How does its wavelength compare to the size of the "neck" of the tube, typically 5 cm? Do we have to worry about diffraction problems blurring our picture on the screen?

35. (II) What is the wavelength of an electron of energy (a) 10 eV, (b) 100 eV, (c) 1.0 keV?

36. (II) Show that if an electron and a proton have the same nonrelativistic kinetic energy, the proton has the shorter wavelength.

37. (II) Calculate the ratio of the kinetic energy of an electron to that of a proton if their wavelengths are equal. Assume that the speeds are non-relativistic.

38. (II) What is the wavelength of an O_2 molecule in the air at room temperature? [*Hint*: See Chapter 18.]

39. (III) A Cadillac with a mass of 2000 kg approaches a freeway underpass that is 10 m across. At what speed must the car be moving, in order for it to have a wavelength such that it might somehow "diffract" after passing through this single "slit"? How do these conditions compare to normal freeway speeds of 30 m/s?

* **Section 38–7**

* **40.** (II) What voltage is needed to produce electron wavelengths of 0.10 nm? (Assume that the electrons are nonrelativistic.)
* **41.** (II) Electrons are accelerated by 2250 V in an electron microscope. What is the maximum possible resolution?

Sections 38–9 and 38–10

42. (I) For the three hydrogen transitions indicated below, with n being the initial state and n' being the final state, is the transition an absorption or an emission? Which is higher, the initial state energy or the final state energy of the atom? Finally, which of these transitions involves the largest energy photon? (a) $n = 1$, $n' = 3$ (b) $n = 6$, $n' = 2$ (c) $n = 4$, $n' = 5$.

43. (I) How much energy is needed to ionize a hydrogen atom in the $n = 2$ state?

44. (I) The third longest wavelength in the Paschen series in hydrogen (Fig. 38–25) corresponds to what transition?

45. (I) Calculate the ionization energy of doubly ionized lithium, Li^{2+}, which has $Z = 3$.

46. (I) (a) Determine the wavelength of the second Balmer line ($n = 4$ to $n = 2$ transition) using Fig. 38–25. Determine likewise (b) the wavelength of the second Lyman line and (c) the wavelength of the third Balmer line.

47. (I) Evaluate the Rydberg constant R using Bohr theory (compare Eqs. 38–8 and 38–14) and show that its value is $R = 1.0974 \times 10^7 \, m^{-1}$.

48. (II) What is the longest wavelength light capable of ionizing a hydrogen atom in the ground state?

49. (II) What wavelength photon would be required to ionize a hydrogen atom in the ground state and give the ejected electron a kinetic energy of 10.0 eV?

50. (II) In the Sun, an ionized helium (He^+) atom makes a transition from the $n = 6$ state to the $n = 2$ state, emitting a photon. Can that photon be absorbed by hydrogen atoms present in the Sun? If so, between what energy states will the hydrogen atom jump?

51. (II) Construct the energy-level diagram for the He^+ ion (see Fig. 38–25).

52. (II) Construct the energy-level diagram for doubly ionized lithium, Li^{2+}.

53. (II) What is the potential energy and the kinetic energy of an electron in the ground state of the hydrogen atom?

54. (II) An excited hydrogen atom could, in principle, have a radius of 1.00 mm. What would be the value of n for a Bohr orbit of this size? What would its energy be?

55. (II) Is the use of nonrelativistic formulas justified in the Bohr atom? To check, calculate the electron's velocity, v, in terms of c for the ground state of hydrogen, and then calculate $\sqrt{1 - v^2/c^2}$.

56. (II) Suppose an electron was bound to a proton, as in the hydrogen atom, by the gravitational force rather than by the electric force. What would be the radius, and energy, of the first Bohr orbit?

57. (II) Show that the magnitude of the potential energy of an electron in any Bohr orbit of a hydrogen atom is twice the magnitude of its kinetic energy in that orbit.

58. (III) *Correspondence principle*: Show that for large values of n, the difference in radius Δr between two adjacent orbits (with quantum numbers n and $n - 1$) is given by

$$\Delta r = r_n - r_{n-1} \approx \frac{2r_n}{n},$$

so $\Delta r/r_n \to 0$ as $n \to \infty$, in accordance with the correspondence principle. [Note that we can check the correspondence principle by either considering large values of $n(n \to \infty)$ or by letting $h \to 0$. Are these equivalent?]

59. (III) (a) For very large values of n, show that when an electron jumps from the level n to the level $n - 1$, the frequency of the light emitted is equal to

$$f = \frac{v}{2\pi r_n}.$$

(b) Show that this is also just the frequency predicted by classical theory for an electron revolving in a circular orbit of radius r_n with speed v. (c) Explain why this is consistent with the correspondence principle.

General Problems

60. The Big Bang theory states that the beginning of the Universe was accompanied by a huge burst of photons. Those photons are still present today and make up the so called Cosmic Microwave Background Radiation. The Universe radiates like a blackbody with a temperature of about 2.7 K. Calculate the peak wavelength of this radiation.

61. At low temperatures, nearly all the atoms in hydrogen gas will be in the ground state. What minimum frequency photon is needed if the photoelectric effect is to be observed?

62. A beam of 85-eV electrons is scattered from a crystal, as in X-ray diffraction, and a first-order peak is observed at $\theta = 38°$. What is the spacing between planes in the diffracting crystal? (See Section 36–10).

63. Show that the energy E (in electron volts) of a photon whose wavelength is λ (meters) is given by

$$E = 1.24 \times 10^{-6}/\lambda.$$

64. A microwave oven produces electromagnetic radiation at $\lambda = 12.2$ cm and produces a power of 760 W. Calculate the number of microwave photons produced by the microwave oven each second.

65. Sunlight reaching the Earth's surface has an intensity of about 1000 W/m^2. Estimate how many photons per square meter per second this represents. Take the average wavelength to be 550 nm.

66. A beam of red laser light ($\lambda = 633$ nm) hits a black wall and is fully absorbed. If this exerts a total force $F = 5.5$ nN on the wall, how many photons per second are hitting the wall?

67. If a 100-W light bulb emits 3.0 percent of the input energy as visible light (average wavelength 550 nm) uniformly in all directions, estimate how many photons per second of visible light will strike the pupil (4.0 mm diameter) of the eye of an observer 1.0 km away.

68. An electron and a positron collide head on, annihilate, and create two 0.90 MeV photons traveling in opposite directions. What were the initial kinetic energies of electron and positron?

69. In Compton's original experiment he saw X-rays scattered where the wavelength shifted from 0.0711 nm to 0.0735 nm. Through what angle did these X-rays scatter?

70. By what potential difference must (a) a proton ($m = 1.67 \times 10^{-27}$ kg), and (b) an electron ($m = 9.11 \times 10^{-31}$ kg), be accelerated to have a wavelength $\lambda = 5.0 \times 10^{-12}$ m?

71. In some of Rutherford's experiments (Fig. 38–16) the α particles (mass $= 6.64 \times 10^{-27}$ kg) had a kinetic energy of 4.8 MeV. How close could they get to a gold nucleus (charge $= +79e$)? Ignore the recoil motion of the nucleus.

72. By what fraction does the mass of an H atom decrease when it makes an $n = 3$ to $n = 1$ transition?

73. For what maximum kinetic energy is a collision between an electron and a hydrogen atom in its ground state definitely elastic?

74. Using Bohr theory, derive an equation for the angular velocity ω, and frequency f, of an electron in a hydrogen atom. Determine (a) for the ground state and (b) for the first excited state ($n = 2$).

75. Calculate the ratio of the gravitational to electric force for the electron in a hydrogen atom. Can the gravitational force be safely ignored?

76. A child's swing has a natural frequency of 0.75 Hz. (a) What is the separation between possible energy values (in joules)? (b) If the swing reaches a vertical height of 45 cm above its lowest point and has a mass of 20 kg (including the child), what is the value of the quantum number n? (c) What is the fractional change in energy between levels whose quantum numbers are n (as just calculated) and $n + 1$? Would quantization be measurable in this case?

77. Electrons accelerated by a potential difference of 12.3 V pass through a gas of hydrogen atoms at room temperature. What wavelengths of light will be emitted?

78. Atoms can be formed in which a muon (mass = 207 times the mass of an electron) replaces one of the electrons in an atom. Calculate, using Bohr theory, the energy of the photon emitted when a muon makes a transition from $n = 2$ to $n = 1$ in a muonic $^{208}_{82}$Pb atom (lead whose nucleus has a mass 208 times the proton mass and charge $+ 82e$).

79. In a particular photoelectric experiment a stopping potential of 2.10 volts is measured when ultraviolet light having a wavelength of 290 nm is incident on the metal. Using the same setup, what will the new stopping potential be if blue light with a wavelength of 440 nm is used, instead of the ultraviolet light.

80. In an X-ray tube (see Fig. 36–25 and discussion in Section 36–10), the high voltage between filament and target is V. After being accelerated through this voltage, an electron strikes the target where it is decelerated (by positively charged nuclei) and in the process one or more X-ray photons are emitted. (a) Show that the photon of shortest wavelength will have

$$\lambda_0 = \frac{hc}{eV}.$$

(b) What is the shortest wavelength of X-ray emitted when accelerated electrons strike the face of a 30-kV television picture tube?

81. Show that the wavelength of a particle of mass m with kinetic energy K is given by the relativistic formula $\lambda = hc/\sqrt{K^2 + 2mc^2K}$.

82. What is the kinetic energy and wavelength of a "thermal" neutron (one that is in equilibrium at room temperature–see Chapter 18)?

83. What is the theoretical limit of resolution for an electron microscope whose electrons are accelerated through 60 kV? (Relativistic formulas should be used.)

84. The intensity of the Sun's light in the vicinity of the Earth is about 1000 W/m^2. Imagine a spacecraft with a mirrored square sail of dimension 1.0 km. Estimate how much thrust (in newtons) this craft will experience due to collisions with the Sun's photons. [*Hint:* Assume the photons bounce off the sail with no change in magnitude of their momentum.]

85. The human eye can respond to as little as 10^{-18} J of light energy. If this comes with a wavelength at the peak of visual sensitivity, 550 nm, how many photons lead to an observable flash?

86. Light with a wavelength of 300 nm strikes a metal whose work function is 2.2 eV. What is the shortest de Broglie wavelength for the electrons that are produced as photoelectrons?

Participants in the 1927 Solvay Conference. Albert Einstein, seated at front center, had difficulty accepting that nature could behave according to the rules of quantum mechanics, which we study in this chapter starting with the wave function and the uncertainty principle. We examine the Schrödinger equation and its solutions for some simple cases: free particles, the square well, and tunnelling through a barrier.

A. Piccard E. Henriot P. Ehrenfest Ed. Herzen Th. De Donder E. Schrödinger E. Verschaffelt W. Pauli W. Heisenberg R. H. Fowler L. Brillouin
P. Debye M. Knudsen W. L. Bragg H. A. Kramers P. A. M. Dirac A. H. Compton L. de Broglie M. Born N. Bohr
I. Langmuir M. Planck Mme Curie H. A. Lorentz A. Einstein P. Langevin Ch. E. Guye C. T. R. Wilson O. W. Richardson

CHAPTER 39

Quantum Mechanics

Bohr's model of the atom gave us a first (though rough) picture of what an atom is like. It proposed an explanation for why there should be emission and absorption of light by atoms at discrete wavelengths, as well as for the stability of atoms. The wavelengths of the line spectra and the ionization energy for hydrogen (and one-electron ions) that are predicted are in excellent agreement with experiment. But the Bohr theory had important limitations. It was not able to predict the line spectra for more complex atoms—not even for the neutral helium atom, which has only two electrons. Nor could it explain why emission lines, when viewed with great precision, consist of two or more very closely spaced lines (referred to as *fine structure*). The Bohr theory also did not explain why some spectral lines were brighter than others. And it could not explain the bonding of atoms in molecules or in solids and liquids.

From a theoretical point of view, too, the Bohr theory was not satisfactory. For it was a strange mixture of classical and quantum ideas. Moreover, the wave–particle duality was still not really resolved.

We mention these limitations of the Bohr theory not to disparage it—for it was a landmark in the history of science. Rather, we mention them to show why, in the early 1920s, it became increasingly evident that a new, more comprehensive theory was needed. It was not long in coming. Less than two years after de Broglie

FIGURE 39–1 Erwin Schrödinger with Lise Meitner (see Chapter 43).

FIGURE 39–2 Werner Heisenberg (center) on Lake Como with Wolfgang Pauli (right) and Enrico Fermi (left).

gave us his matter–wave hypothesis, Erwin Schrödinger (1887–1961; Fig. 39–1) and Werner Heisenberg (1901–1976; Fig. 39–2) independently developed a new comprehensive theory. Their separate approaches were quite different but were soon shown to be fully compatible. In this chapter we look at this new theory, quantum mechanics. In the following two chapters we will discuss the results of quantum mechanics applied to the hydrogen atom, to other atoms, and to molecules and solids.

39–1 | Quantum Mechanics—A New Theory

The new theory, called **quantum mechanics**, unifies the wave–particle duality into a single consistent theory. As a theory, quantum mechanics has been extremely successful. It has successfully dealt with the spectra emitted by complex atoms, even the fine details. It explains the relative brightness of spectral lines and how atoms form molecules. It is also a much more general theory that covers all quantum phenomena from blackbody radiation to atoms and molecules. It has explained a wide range of natural phenomena and from its predictions many new practical devices have become possible. Indeed, it has been so successful that it is accepted today by nearly all physicists as the fundamental theory underlying physical processes.

Quantum mechanics deals mainly with the microscopic world of atoms and light. But in our everyday macroscopic world, we do perceive light and we accept that ordinary objects are made up of atoms. This new theory must therefore also account for the verified results of classical physics. That is, when it is applied to macroscopic phenomena, quantum mechanics must be able to produce the old classical laws. This, the **correspondence principle** (already mentioned in Section 37–13), is met fully by quantum mechanics. This doesn't mean we throw away classical theories such as Newton's laws. In the everyday world, the latter are far easier to apply and they give an accurate description. But when we deal with high speeds, close to the speed of light, we must use the theory of relativity; and when we deal with the tiny world of the atom, we use quantum mechanics.

Correspondence principle

Although we won't go into the detailed mathematics of quantum mechanics, we will discuss the main ideas and how they involve the wave and particle properties of matter to explain atomic structure and other applications.

39–2 The Wave Function and Its Interpretation; the Double-Slit Experiment

The important properties of any wave are its wavelength, frequency, and amplitude. For an electromagnetic wave, the wavelength determines whether the light is in the visible spectrum or not, and if so, what color it is. We also have seen that the wavelength (or frequency) is a measure of the energy of the corresponding photon $(E = hf)$. The amplitude or displacement of an electromagnetic wave at any point is the strength of the electric (or magnetic) field at that point, and is related to the intensity of the wave (the brightness of the light).

For material particles such as electrons, quantum mechanics relates the wavelength to momentum according to de Broglie's formula, $\lambda = h/p$. But what corresponds to the amplitude or displacement of a matter wave? In quantum mechanics, this role is played by the **wave function**, which is given the symbol Ψ (the Greek capital letter psi, pronounced "sigh"). Thus Ψ represents the displacement as a function of time and position, of a new kind of field which we might call a "matter" field or a matter wave.

Wave function

To calculate the wave function Ψ in a given situation (say, for an electron in an atom) is one of the basic tasks of quantum mechanics. Indeed, the development of an equation to do so was Schrödinger's great contribution. The *Schrödinger wave equation*, as it is called, is considered to be the basic equation for the description of nonrelativistic material particles. We will discuss the Schrödinger equation and its solutions for some simple cases later in this chapter. Now let us ask: What is the meaning of the wave function Ψ for a particle? One way to interpret Ψ is simply as the displacement at any point in space and time of a "matter wave," so that it plays the role that E (the electric field) does for an electromagnetic wave. Another interpretation is possible, however, based on the wave–particle duality. To understand this, we make an analogy with light.

We saw in Chapter 15 that the intensity I of any wave is proportional to the square of the displacement amplitude. This holds true for light waves as well, and, as we saw in Chapter 32, the energy density in a light wave is proportional to E^2, where E is the electric field strength. From the *particle* point of view, the energy density of a light beam is proportional to the number of photons per unit volume, n. The more photons (of a given wavelength) per unit volume there are, the greater the energy density. Hence

$$n \propto E^2.$$

That is, the density of photons is proportional to the square of the electric field strength.

If the light beam is very weak, only a few photons will be involved. Indeed, it is possible to "build up" a photograph on film using very weak light so that the effect of individual photons can be seen. If we are dealing with only one photon, the relationship above $(n \propto E^2)$ can be interpreted in a slightly different way. At any point the square of the electric field strength, E^2, is a measure of the *probability* that a photon will be found within a unit volume at that location. Said another way, if dV represents an infinitesimal volume element around that point, then $E^2 \, dV$ is proportional to the probability of finding a photon in the volume dV. At points where E^2 is large, there is a high probability of finding the photon; where E^2 is small, the probability of finding the photon is low.

We can interpret matter waves in the same way. The wave function Ψ may vary in magnitude from point to point in space and time. If Ψ describes a collection of many electrons, then $|\Psi|^2 \, dV$ will be proportional to the number of electrons expected to be found in the volume dV around any given point. When dealing with small numbers of electrons we can't make very exact predictions, so $|\Psi|^2$ takes on the character of a probability. If Ψ, which depends on time and position,[†] represents a single electron (say in an atom), then $|\Psi|^2$ is interpreted as follows: *$|\Psi|^2$ dV at a certain point in space and time represents the probability of finding the electron within the volume dV about the given position at that time.* Thus, $|\Psi|^2$ is often referred to as the **probability density** or **probability distribution**, since it is the probability of finding the particle per unit volume.

To understand this better, we take as a thought experiment the familiar double-slit experiment, and consider it both for light and for electrons.

Consider two slits whose size and separation are on the order of the wavelength of whatever we direct at them, either light or electrons, Fig. 39–3. We know very well what would happen in this case for light, since this is just Young's double-slit experiment (Section 35–3): an interference pattern would be seen on the screen behind. If light were replaced by electrons with wavelength comparable to the slit size, they too would produce an interference pattern (recall Fig. 38–11). In the case of light, the pattern would be visible to the eye or could be recorded on film. For electrons, a fluorescent screen could be used (it glows where an electron strikes).

Now, if we reduced the flow of electrons (or photons) so that they passed one at a time through the slits, we would see a flash each time one struck the screen. At first, the flashes would seem random. Indeed, there is no way to predict just where any one electron would hit the screen. If we let the experiment run for a long time, however, and kept track of where each electron hit the screen, we would soon see a pattern emerging—namely the interference pattern predicted by the wave theory; see Fig. 39–4. Thus, although we could not predict where a given electron would strike the screen, we could predict probabilities. (The same can be said for photons.) The probability, as mentioned before, is proportional to $|\Psi|^2$. Where $|\Psi|^2$ is zero, we would get a minimum in the interference pattern. And where $|\Psi|^2$ is a maximum, we would get a peak in the interference pattern.

Since the interference pattern would occur even when electrons (or photons) passed through the slits one at a time, it is clear that the interference pattern would not arise from the interaction of one electron with another. It is as if the electron passed through both slits at the same time, interfering with itself. This is possible because, remember, an electron is not precisely a particle. It is as much a wave as it is a particle, and a wave could certainly travel through both slits at once. But what would happen if we covered one of the slits so we knew that the electrons passed through the other one, and a little later we covered the second slit so that electrons had to have passed through the first? The result would be that no interference pattern would be seen. We would see, instead, two bright areas (or diffraction patterns) on the screen behind the slits. This confirms our idea that if both slits are open, the screen shows an interference pattern as if each electron passed through both slits, like a wave. Yet each electron would make a tiny spot on the screen as if it were a particle.

The main point of this discussion is this: If we treat electrons (and other matter) as if they were waves, then Ψ represents the wave amplitude. If we treat them as particles, then we must treat them on a *probabilistic* basis. The square of the wave function, $|\Psi|^2$, gives the probability per unit volume of finding a given electron at a given point. We cannot predict—or even follow—the path of a single electron precisely through space and time.

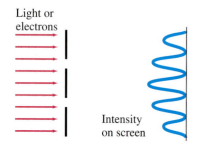

FIGURE 39–3 Parallel beam, of light or electrons, falls on two slits whose sizes are comparable to the wavelength. An interference pattern is observed.

FIGURE 39–4 Young's double-slit experiment done with electrons—note that the pattern is not evident with only a few electrons (top photo), but with more and more electrons (second and third photos), the familiar double-slit interference pattern (Chapter 35) is seen.

[†]The wave function Ψ is generally a complex quantity (that is, it involves $i = \sqrt{-1}$) and hence is not directly observable. On the other hand, $|\Psi|^2$, the absolute value of Ψ squared, is always a real quantity and it is to $|\Psi|^2$ that we can give a physical interpretation.

39–3 The Heisenberg Uncertainty Principle

Whenever a measurement is made, some uncertainty or error is always involved. For example, you cannot make an absolutely exact measurement of the length of a table. Even with a measuring stick that has markings 1 mm apart, there will be an inaccuracy of about $\frac{1}{2}$ mm or so. More precise instruments will produce more precise measurements. But there is always some uncertainty involved in a measurement no matter how good the measuring device. We expect that by using more precise instruments, the uncertainty in a measurement can be made indefinitely small.

But according to quantum mechanics, there is actually a limit to the accuracy of certain measurements. This limit is not a restriction on how well instruments can be made; rather, it is inherent in nature. It is the result of two factors: the wave–particle duality, and the unavoidable interaction between the thing observed and the observing instrument. Let us look at this in more detail.

Measurement uncertainty inherent in nature

To make a measurement on an object without somehow disturbing it, at least a little, is not possible. Consider trying to locate a Ping-pong ball in a completely dark room. You grope about trying to find its position; and just when you touch it with your finger, it bounces away. Whenever we measure the position of an object, whether it's a Ping-pong ball or an electron, we always touch it with something else that gives us the information about its position. To locate a lost Ping-pong ball in a dark room, you could probe about with your hand or a stick; or you could shine a light and detect the light reflecting off the ball. When you search with your hand or a stick, you find the ball's position when you touch it. But when you touch the ball you unavoidably bump it, and give it some momentum. Thus you won't know its *future* position. The same would be true, but to a much lesser extent, if you observe the Ping-pong ball using light. In order to "see" the ball, at least one photon must scatter from it, and the reflected photon must enter your eye or some other detector. When a photon strikes an ordinary-sized object, it does not appreciably alter the motion or position of the object. But when a photon strikes a very tiny object like an electron, it can transfer momentum to the object and thus greatly change the object's motion and position in an unpredictable way. The mere act of measuring the position of an object at one time makes our knowledge of its future position imprecise.

Now let us see where the wave–particle duality comes in. Imagine a thought experiment in which we are trying to measure the position of an object, say an electron, with photons, Fig. 39–5. (The arguments would be similar if we were using, instead, an electron microscope.) As we saw in Chapter 36, objects can be seen to an accuracy at best of about the wavelength of the radiation used. If we want an accurate position measurement, we must use a short wavelength. But a short wavelength corresponds to high frequency and large momentum $(p = h/\lambda)$; and the more momentum the photons have, the more momentum they can give the object when they strike it. If photons of longer wavelength, and correspondingly smaller momentum are used, the object's motion when struck by the photons will not be affected as much. But the longer wavelength means lower resolution, so the object's position will be less accurately known. Thus the act of observing produces a significant uncertainty in either the *position* or the *momentum* of the electron. This is the essence of the *uncertainty principle* first enunciated by Heisenberg in 1927.

Quantitatively, we can make an approximate calculation of the magnitude of this effect. If we use light of wavelength λ, the position can be measured at best to an accuracy of about λ. That is, the uncertainty in the position measurement, Δx, is approximately

$$\Delta x \approx \lambda.$$

Suppose that the object can be detected by a single photon. The photon has a

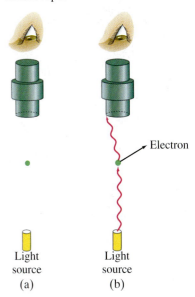

FIGURE 39–5 Thought experiment for observing an electron with a powerful light microscope. At least one photon must scatter from the electron (transferring some momentum to it) and enter the microscope.

Electron

Light source
(a)

Light source
(b)

momentum $p = h/\lambda$. When the photon strikes our object, it will give some or all of this momentum to the object, Fig. 39–5. Therefore, the final momentum of our object will be uncertain in the amount

$$\Delta p \approx \frac{h}{\lambda}$$

since we can't tell beforehand how much momentum will be transferred. The product of these uncertainties is

$$(\Delta x)(\Delta p) \approx h.$$

Of course, the uncertainties could be worse than this, depending on the apparatus and the number of photons needed for detection. In Heisenberg's more careful calculation, he found that at the very best

UNCERTAINTY PRINCIPLE
(Δx and Δp)

$$(\Delta x)(\Delta p_x) \gtrsim \frac{h}{2\pi},$$

(39–1)

where Δp_x is the uncertainty of the momentum in the x direction.[†] This is a mathematical statement of the **Heisenberg uncertainty principle**, or, as it is sometimes called, the **indeterminancy principle**. It tells us that we cannot measure both the position and momentum of an object precisely at the same time. The more accurately we try to measure the position, so that Δx is small, the greater will be the uncertainty in momentum, Δp_x. If we try to measure the momentum very precisely, then the uncertainty in the position becomes large. The uncertainty principle does not forbid individual exact measurements, however. For example, in principle we could measure the position of an object exactly. But then its momentum would be completely unknown. Thus, although we might know the position of the object exactly at one instant, we could have no idea at all where it would be a moment later.

Another useful form of the uncertainty principle relates energy and time, and we examine this as follows. The object to be detected has an uncertainty in position $\Delta x \approx \lambda$. Now the photon used to detect it travels with speed c, and it takes a time $\Delta t \approx \Delta x/c \approx \lambda/c$ to pass through the distance of uncertainty. Hence, the measured time when our object is at a given position is uncertain by about

$$\Delta t \approx \frac{\lambda}{c}.$$

Since the photon can transfer some or all of its energy ($= hf = hc/\lambda$) to our object, the uncertainty in energy of our object as a result is

$$\Delta E \approx \frac{hc}{\lambda}.$$

The product of these two uncertainties is

$$(\Delta E)(\Delta t) \approx h.$$

Heisenberg's more careful calculation gives

UNCERTAINTY PRINCIPLE
(ΔE and Δt)

$$(\Delta E)(\Delta t) \gtrsim \frac{h}{2\pi}.$$

(39–2)

This form of the uncertainty principle tells us that the energy of an object can be uncertain, or may even be nonconserved, by an amount ΔE for a time $\Delta t \approx h/(2\pi \, \Delta E)$.

[†]Note, however, that quantum mechanics does allow simultaneous precise measurements of p_x and y: that is $(\Delta y)(\Delta p_x) \gtrsim 0$.

The quantity $(h/2\pi)$ appears so often in quantum mechanics that for convenience it is given the symbol $\hbar$ ("h-bar"). That is

$$\hbar = \frac{h}{2\pi} = \frac{6.626 \times 10^{-34}\,\text{J}\cdot\text{s}}{2\pi} = 1.055 \times 10^{-34}\,\text{J}\cdot\text{s}.$$

By using this notation, Eqs. 39–1 and 39–2 for the uncertainty principle can be written

$$(\Delta x)(\Delta p_x) \gtrsim \hbar \quad \text{and} \quad (\Delta E)(\Delta t) \gtrsim \hbar.$$

We have been discussing the position and velocity of an electron as if it were a particle. But it isn't a particle. Indeed, we have the uncertainty principle because an electron—and matter in general—has wave as well as particle properties. What the uncertainty principle really tells us is that if we insist on thinking of the electron as a particle, then there are certain limitations on this simplified view—namely, that the position and velocity cannot both be known precisely at the same time; and that the energy can be uncertain (or nonconserved) in the amount ΔE for a time $\Delta t \approx \hbar/\Delta E$.

Because Planck's constant, h, is so small, the uncertainties expressed in the uncertainty principle are usually negligible on the macroscopic level. But at the level of the atom, the uncertainties are significant. Because we consider ordinary objects to be made up of atoms containing nuclei and electrons, the uncertainty principle is relevant to our understanding of all of nature. The uncertainty principle expresses, perhaps most clearly, the probabilistic nature of quantum mechanics. It thus is often used as a basis for philosophic discussion.

EXAMPLE 39–1 **Position uncertainty of electron.** An electron moves in a straight line with a constant speed $v = 1.10 \times 10^6\,\text{m/s}$ which has been measured to a precision of 0.10 percent. What is the maximum precision with which its position could be simultaneously measured?

SOLUTION The momentum of the electron is $p = mv = (9.11 \times 10^{-31}\,\text{kg}) \cdot (1.10 \times 10^6\,\text{m/s}) = 1.00 \times 10^{-24}\,\text{kg}\cdot\text{m/s}$. The uncertainty in the momentum is 0.10 percent of this, or $\Delta p = 1.0 \times 10^{-27}\,\text{kg}\cdot\text{m/s}$. From the uncertainty principle, the best simultaneous position measurement will have an uncertainty of

$$\Delta x = \frac{\hbar}{\Delta p} = \frac{1.06 \times 10^{-34}\,\text{J}\cdot\text{s}}{1.0 \times 10^{-27}\,\text{kg}\cdot\text{m/s}} = 1.1 \times 10^{-7}\,\text{m},$$

or 110 nm. This is about 1000 times the diameter of an atom.

EXAMPLE 39–2 **Position uncertainty of a baseball.** What is the uncertainty in position, imposed by the uncertainty principle, on a 150-g baseball thrown at $(93 \pm 2)\,\text{mph} = (42 \pm 1)\,\text{m/s}$?

SOLUTION The uncertainty in the momentum is

$$\Delta p = m\,\Delta v = (0.150\,\text{kg})(1\,\text{m/s}) = 0.15\,\text{kg}\cdot\text{m/s}.$$

Hence the uncertainty in a position measurement could be as small as

$$\Delta x = \frac{\hbar}{\Delta p} = \frac{1.06 \times 10^{-34}\,\text{J}\cdot\text{s}}{0.15\,\text{kg}\cdot\text{m/s}} = 7 \times 10^{-34}\,\text{m},$$

which is a distance incredibly smaller than any we could imagine observing or measuring. Indeed, the uncertainty principle sets no relevant limit on measurement for macroscopic objects.

EXAMPLE 39–3 **ESTIMATE** **J/ψ lifetime calculated.** The J/ψ meson, discovered in 1974, was measured to have an average mass of 3100 MeV/c^2 (note the use of energy units since $E = mc^2$) and an intrinsic width of 63 keV/c^2. By this we mean that the masses of different J/ψ mesons were actually measured to be slightly different from one another. This mass "width" is related to the very short lifetime of the J/ψ before it decays into other particles. Estimate its lifetime using the uncertainty principle.

SOLUTION The uncertainty of 63 keV/c^2 in the J/ψ's mass is an uncertainty in its rest energy, which in joules is

$$\Delta E = (63 \times 10^3 \, \text{eV})(1.60 \times 10^{-19} \, \text{J/eV}) = 1.01 \times 10^{-14} \, \text{J}.$$

Then we expect its lifetime τ ($= \Delta t$ here) to be

$$\tau \approx \frac{\hbar}{\Delta E} = \frac{1.06 \times 10^{-34} \, \text{J} \cdot \text{s}}{1.01 \times 10^{-14} \, \text{J}} \approx 1.0 \times 10^{-20} \, \text{s}.$$

Lifetimes this short are difficult to measure directly, and the assignment of very short lifetimes depends on this use of the uncertainty principle. (See Chapter 44.)

The uncertainty principle applies also for angular variables:

$$(\Delta L_z)(\Delta \phi) \gtrsim \hbar$$

where L is the component of angular momentum along a given axis (z) and ϕ is the angular position in a plane perpendicular to that axis.

39–4 Philosophic Implications; Probability versus Determinism

The classical Newtonian view of the world is a deterministic one (see Section 6–5). One of its basic ideas is that once the position and velocity of an object are known at a particular time, its future position can be predicted if the forces on it are known. For example, if a stone is thrown a number of times with the same initial velocity and angle, and the forces on it remain the same, the path of the projectile will always be the same. If the forces are known (gravity and air resistance, if any), the stone's path can be precisely predicted. This mechanistic view implies that the future unfolding of the universe, assumed to be made up of particulate bodies, is completely determined.

This classical deterministic view of the physical world has been radically altered by quantum mechanics. As we saw in the analysis of the double-slit experiment (Section 39–2), electrons all prepared in the same way will not all end up in the same place. According to quantum mechanics, certain probabilities exist that an electron will arrive at different points. This is very different from the classical view, in which the path of a particle is precisely predictable from the initial position and velocity and the forces exerted on it. According to quantum mechanics, the position and velocity of an object cannot even be known accurately at the same time. This is expressed in the uncertainty principle, and arises because basic entities, such as electrons, are not considered simply as particles: they have wave properties as well. Quantum mechanics allows us to calculate only the probability[†] that, say, an electron (when thought of as a particle) will be observed at various places. Quantum mechanics says there is some inherent unpredictability in nature.

[†]Note that these probabilities can be calculated precisely, just like exact predictions of probabilities at dice or playing cards, but unlike predictions of probabilities at sporting events or for natural or man-made disasters, which are only estimates.

Since matter is considered to be made up of atoms, even ordinary-sized objects are expected to be governed by probability, rather than by strict determinism. For example, quantum mechanics predicts a finite (but negligibly small) probability that when you throw a stone, its path will suddenly curve upward instead of following the downward-curved parabola of normal projectile motion. Quantum mechanics predicts with extremely high probability that ordinary objects will behave just as the classical laws of physics predict. But these predictions are considered probabilities, not certainties. The reason that macroscopic objects behave in accordance with classical laws with such high probability is due to the large number of molecules involved: when large numbers of objects are present in a statistical situation, deviations from the average (or most probable) approach zero. It is the average configuration of vast numbers of molecules that follows the so-called fixed laws of classical physics with such high probability, and gives rise to an apparent "determinism." Deviations from classical laws are observed when small numbers of molecules are dealt with. We can say, then, that although there are no precise deterministic laws in quantum mechanics, there are statistical laws based on probability.

It is important to note that there is a difference between the probability imposed by quantum mechanics and that used in the nineteenth century to understand thermodynamics and the behavior of gases in terms of molecules (Chapters 18 and 20). In thermodynamics, probability is used because there are far too many particles to keep track of. But the molecules are still assumed to move and interact in a deterministic way following Newton's laws. Probability in quantum mechanics is quite different; it is seen as *inherent* in nature, and not as a limitation on our abilities to calculate or to measure.

Although a few physicists have not given up the deterministic view of nature and have refused to accept quantum mechanics as a complete theory—one was Einstein—nonetheless, the vast majority of physicists do accept quantum mechanics and the probabilistic view of nature. This view, which as presented here is the generally accepted one, is called the **Copenhagen interpretation** of quantum mechanics in honor of Niels Bohr's home, since it was largely developed there through discussions between Bohr and other prominent physicists.

Copenhagen interpretation

Because electrons are not simply particles, they cannot be thought of as following particular paths in space and time. This suggests that a description of matter in space and time may not be completely correct. This deep and far-reaching conclusion has been a lively topic of discussion among philosophers. Perhaps the most important and influential philosopher of quantum mechanics was Bohr. He argued that a space–time description of actual atoms and electrons is not possible. Yet a description of experiments on atoms or electrons must be given in terms of space and time and other concepts familiar to ordinary experience, such as waves and particles. We must not let our *descriptions* of experiments lead us into believing that atoms or electrons themselves actually move in space and time as particles.

39–5 The Schrödinger Equation in One Dimension— Time-Independent Form

In order to describe physical systems quantitatively using quantum mechanics, we must have a means of determining the wave function Ψ mathematically. As mentioned in Section 39–2, the basic equation (in the nonrelativistic realm) for determining Ψ is the *Schrödinger equation*. We cannot, however, derive the Schrödinger equation from some higher principles, just as Newton's second law, for example, cannot be derived. The relation $\mathbf{F} = m\mathbf{a}$ was *invented* by Newton to describe how the motion of a body is related to the net applied force. As we saw early in this book, Newton's second law works exceptionally well. In the realm of classical physics it is the starting point for analytically solving a wide range of problems, and the solutions it yields are fully consistent with experiment. The validity of any fundamental equation resides in its agreement with experiment. The Schrödinger equation forms part of a new theory, and it too had to be *invented*—and then checked against experiment, a test that it passed splendidly.

The Schrödinger equation can be written in two forms: the time-dependent version and the time-independent version. We will mainly be interested in steady-state situations—that is, when there is no time dependence—and so we mainly deal with the time-independent version. (We briefly discuss the time-dependent version in the next, optional, section.) The time-independent version involves a wave function with only spatial dependence which we represent by lowercase psi, $\psi(x)$.

In classical mechanics, we solved problems using two approaches: via Newton's laws with the concept of force, and by using the energy concept with the conservation laws. The Schrödinger equation is based on the energy approach. Even though the Schrödinger equation cannot be derived, we can suggest what form it might take by using conservation of energy and considering a very simple case: that of a free particle on which no forces act, so that its potential energy U is constant. We assume that our particle moves along the x axis, and since no force acts on it, its momentum remains constant and its wavelength ($\lambda = h/p$) is fixed. To describe a wave for a free particle such as an electron, we expect that its wave function will satisfy a differential equation that is akin to (but not identical to) the classical wave equation. Let us see what we can infer about this equation. Consider a simple traveling wave of a single wavelength λ whose wave displacement, as we saw in Chapter 15 for mechanical waves and in Chapter 32 for electromagnetic waves, is given by $A \sin(kx - \omega t)$, or more generally as a superposition of sine and cosine: $A \sin(kx - \omega t) + B \cos(kx - \omega t)$. We are only interested in the spatial dependence, so we consider the wave at a specific moment, say $t = 0$. Thus we write as the wave function for our free particle

$$\psi(x) = A \sin kx + B \cos kx, \tag{39-3a}$$

where A and B are constants[†] and $k = 2\pi/\lambda$ (Eq. 15–11). For a particle of mass m and velocity v, the de Broglie wavelength is $\lambda = h/p$, where $p = mv$ is the particle's momentum. Hence

$$k = \frac{2\pi}{\lambda} = \frac{2\pi p}{h} = \frac{p}{\hbar}, \tag{39-3b}$$

One requirement for our wave equation, then, is that it have the wave function $\psi(x)$ as given by Eq. 39–3 as a solution for a free particle. A second requirement is that it be consistent with the conservation of energy, which we can express as

$$\frac{p^2}{2m} + U = E,$$

where E is the total energy, U is the potential energy, and (since we are considering the nonrelativistic realm) the kinetic energy K of our particle of mass m is $K = \frac{1}{2}mv^2 = p^2/2m$. Since $p = \hbar k$ (Eq. 39–3b), we can write the conservation of energy condition as

$$\frac{\hbar^2 k^2}{2m} + U = E. \tag{39-4}$$

Thus we are seeking a differential equation that satisfies conservation of energy

[†]In quantum mechanics, constants can be complex (i.e., with a real and/or imaginary part).

(Eq. 39–4) when $\psi(x)$ is its solution. Now, note that if we take two derivatives of our expression for $\psi(x)$, Eq. 39–3a, we get a factor $-k^2$ multiplied by $\psi(x)$:

$$\frac{d\psi(x)}{dx} = \frac{d}{dx}(A\sin kx + B\cos kx) = k(A\cos kx - B\sin kx)$$

$$\frac{d^2\psi(x)}{dx^2} = k\frac{d}{dx}(A\cos kx - B\sin kx) = -k^2(A\sin kx + B\cos kx) = -k^2\psi(x).$$

Can this last term be related to the k^2 term in Eq. 39–4? Indeed, if we multiply this last relation by $-\hbar^2/2m$, we obtain

$$-\frac{\hbar^2}{2m}\frac{d^2\psi(x)}{dx^2} = \frac{\hbar^2k^2}{2m}\psi(x).$$

The right side is just the first term on the left in Eq. 39–4 multiplied by $\psi(x)$. If we multiply Eq. 39–4 through by $\psi(x)$, and make this substitution, we obtain

$$-\frac{\hbar^2}{2m}\frac{d^2\psi(x)}{dx^2} + U(x)\psi(x) = E\psi(x). \tag{39–5}$$

SCHRÖDINGER EQUATION
(time-independent form)

This is, in fact, the one-dimensional **time-independent Schrödinger equation**, where for generality we have written $U = U(x)$. It is the basis for solving problems in non-relativistic quantum mechanics. For a particle moving in three dimensions there would be additional derivatives with respect to y and z (see Chapter 40).

But note that we have by no means derived the Schrödinger equation. Although we have made a good argument in its favor, other arguments could also be made which might or might not lead to the same equation. The Schrödinger equation as written (Eq. 39–5) is useful and valid only because it has given results in accord with experiment for a wide range of situations.

There are some requirements we impose on any wave function that is a solution of the Schrödinger equation in order that it be physically meaningful. First, we insist that it be a continuous function; after all, if $|\psi|^2$ represents the probability of finding a particle at a certain point, we expect that probability to be continuous from point to point and not to take discontinuous jumps. Second, we want the wave function to be *normalized*. By this we mean that for a single particle, the probability of finding the particle at one point or another (i.e., the probabilities summed over all space) must be exactly 1 (or 100 percent). For a single particle, $|\psi|^2$ represents the probability of finding the particle in unit volume. Then

$$|\psi|^2\,dV$$

is the probability of finding the particle within a volume dV, where ψ is the value of the wave function in this infinitesimal volume dV. For the one-dimensional case, $dV = dx$. Then the sum of the probabilities over all space—that is, the probability of finding the particle at one point or another—becomes

$$\int_{\text{all space}} |\psi|^2\,dV = \int |\psi|^2\,dx = 1. \tag{39–6}$$

Normalization

This is called the **normalization condition**, and the integral is taken over whatever region of space in which the particle has a chance of being found, which is often all of space, from $x = -\infty$ to $x = \infty$.

The more general form of the Schrödinger equation, including time dependence, for a particle of mass m moving in one dimension, is

Schrödinger equation (time-dependent form)

$$-\frac{\hbar^2}{2m}\frac{\partial^2\Psi(x,t)}{\partial x^2} + U(x)\Psi(x,t) = i\hbar\frac{\partial\Psi(x,t)}{\partial t}. \tag{39–7}$$

This is the **time-dependent Schrödinger equation**; here $U(x)$ is the potential energy of the particle as a function of position, and i is the imaginary number $i = \sqrt{-1}$. For a particle moving in three dimensions, there would be additional derivatives with respect to y and z, just as for the classical wave equation discussed in Section 15–5. Indeed, it is worth noting the similarity between the Schrödinger wave equation for zero potential energy ($U = 0$) and the classical wave equation: $\partial^2 D/\partial t^2 = v^2\partial^2 D/\partial x^2$, where D is the wave displacement (equivalent of the wave function). In both equations there is the second derivative with respect to x; but in the Schrödinger equation there is only the first derivative with respect to time, whereas the classical wave equation has the second derivative for time.

As we pointed out in the preceding Section, we cannot derive the time-dependent Schrödinger equation. But we can show how the time-independent Schrödinger equation (Eq. 39–5) is obtained from it. For many problems in quantum mechanics, it is possible to write the wave function as a product of separate functions of space and time:

$$\Psi(x,t) = \psi(x)f(t).$$

Substituting this into the time-dependent Schrödinger equation (Eq. 39–7), we get:

$$-\frac{\hbar^2}{2m}f(t)\frac{d^2\psi(x)}{dx^2} + U(x)\psi(x)f(t) = i\hbar\psi(x)\frac{df(t)}{dt}.$$

We divide both sides of this equation by $\psi(x)f(t)$ and obtain an equation that involves only x on one side and only t on the other:

Separation of variables

$$-\frac{\hbar^2}{2m}\frac{1}{\psi(x)}\frac{d^2\psi(x)}{dx^2} + U(x) = i\hbar\frac{1}{f(t)}\frac{df(t)}{dt}.$$

This *separation of variables* is very convenient. Since the left side is a function only of x, and the right side is a function only of t, the equality can be valid for all values of x and all values of t only if each side is equal to a constant (the same constant, of course), which we call C:

$$-\frac{\hbar^2}{2m}\frac{1}{\psi(x)}\frac{d^2\psi(x)}{dx^2} + U(x) = C \tag{39–8a}$$

$$i\hbar\frac{1}{f(t)}\frac{df(t)}{dt} = C. \tag{39–8b}$$

We multiply the first of these (Eq. 39–8a) through by $\psi(x)$ and obtain

$$-\frac{\hbar^2}{2m}\frac{d^2\psi(x)}{dx^2} + U(x)\psi(x) = C\psi(x). \tag{39–8c}$$

This we recognize immediately as the time-independent Schrödinger equation, Eq. 39–5, where the constant C equals the total energy E. Thus we have obtained the time-independent form of Schrödinger's equation from the time-dependent form.

Equation 39–8b is easy to solve. Putting $C = E$, we rewrite Eq. 39–8b as

$$\frac{df(t)}{dt} = -i\frac{E}{\hbar}f(t)$$

(note that since $i^2 = -1$, $i = -1/i$), and then as

$$\frac{df(t)}{f(t)} = -i\frac{E}{\hbar}\,dt.$$

We integrate both sides to obtain

$$\ln f(t) = -i\frac{E}{\hbar}t$$

or

$$f(t) = e^{-i\left(\frac{E}{\hbar}\right)t}.$$

Thus the total wave function is

$$\Psi(x, t) = \psi(x)e^{-i\left(\frac{E}{\hbar}\right)t}, \tag{39–9}$$

where $\psi(x)$ satisfies Eq. 39–5. It is, in fact, the solution of the time-independent Schrödinger equation (Eq. 39–5) that is the major task of nonrelativistic quantum mechanics. Nonetheless, we should note that in general the wave function $\Psi(x, t)$ is a complex function since it involves $i = \sqrt{-1}$. It has both a real and an imaginary part.[†] Since $\Psi(x, t)$ is not purely real, it cannot itself be physically measurable. Rather it is only $|\Psi|^2$, which *is* real, that can be measured physically.

Note also that

$$\left|f(t)\right| = \left|e^{-i\left(\frac{E}{\hbar}\right)t}\right| = 1,$$

so $\left|f(t)\right|^2 = 1$. Hence the probability density in space does not depend on time:

$$\left|\Psi(x, t)\right|^2 = \left|\psi(x)\right|^2.$$

We thus will be interested only in the time-independent Schrödinger equation, Eq. 39–5, which we will now examine for a number of simple situations.

39–7 Free Particles; Plane Waves and Wave Packets

A **free particle** is one that is not subject to any force, and we can therefore take its potential energy to be zero. (Although we dealt with the free particle in Section 39–5 in arguing for Schrödinger's equation, here we treat it directly using Schrödinger's equation as the basis.) Schrödinger's equation (Eq. 39–5) with $U(x) = 0$ becomes

$$-\frac{\hbar^2}{2m}\frac{d^2\psi(x)}{dx^2} = E\psi(x),$$

which can be written

$$\frac{d^2\psi}{dx^2} + \frac{2mE}{\hbar^2}\psi = 0.$$

This is a familiar equation that we encountered in Chapter 14 (Eq. 14–3) in connection with the simple harmonic oscillator. The solution to this equation, but with

---

[†] Recall that $e^{-i\theta} = \cos\theta - i\sin\theta$.

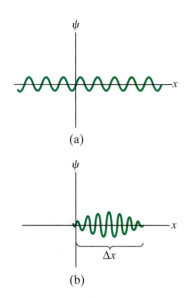

(a)

(b)

FIGURE 39–6 (a) A plane wave describing a free particle. (b) A wave packet of "width" Δx.

Wave packet

appropriate variable changes[†] for our case here, is

$$\psi = A \sin kx + B \cos kx, \qquad \text{[free particle]} \quad \textbf{(39–10)}$$

where

$$k = \sqrt{\frac{2mE}{\hbar^2}}. \qquad \textbf{(39–11a)}$$

Since $U = 0$, the total energy E of the particle is $E = \frac{1}{2}mv^2 = p^2/2m$ (where p is the momentum); thus

$$k = \frac{p}{\hbar} = \frac{h}{\lambda\hbar} = \frac{2\pi}{\lambda}. \qquad \textbf{(39–11b)}$$

So a free particle of momentum p and energy E can be represented by a plane wave that varies sinusoidally. If we are not interested in the phase, we can choose $B = 0$ in Eq. 39–10, and we show this sine wave in Fig. 39–6a.

Note in Fig. 39–6a that the sine wave will extend indefinitely[‡] in the $+x$ and $-x$ directions. Thus, since $|\psi|^2$ represents the probability of finding the particle, the particle could be anywhere between $x = -\infty$ and $x = \infty$. This is fully consistent with the uncertainty principle (Section 39–3): the momentum of the particle was given and hence is known precisely ($p = \hbar k$), so the particle's position must be totally unpredictable. Mathematically, if $\Delta p = 0$, $\Delta x \gtrsim \hbar/\Delta p = \infty$.

To describe a particle whose position is well localized—that is, it is known to be within a small region of space—we can use the concept of a **wave packet**. Figure 39–6b shows an example of a wave packet whose width is about Δx as shown, meaning that the particle is most likely to be found within this region of space. A well-localized particle moving through space can thus be represented by a moving wave packet.

A wave packet can be represented mathematically as the sum of many plane waves (sine waves) of slightly different wavelengths. That this will work can be seen by looking carefully at Fig. 16–18. There we combined only two nearby frequencies (to explain why there are "beats") and found that the sum of two sine waves looked like a series of wave packets. If we add additional waves with other nearby frequencies, we can eliminate all but one of the packets and arrive at Fig. 39–6b. Thus a wave packet consists of waves of a *range* of wavelengths; hence it does not have a definite momentum $p(= h/\lambda)$, but rather, a range of momenta. This is, of course, consistent with the uncertainty principle: we have made Δx small, so the momentum cannot be precise; that is, Δp cannot be zero. Instead, our particle can be said to have a range of momenta, Δp, or to have an uncertainty in its momentum, Δp. It is not hard to show, even for this simple situation (see Problem 15), that $\Delta p \approx h/\Delta x$, in accordance with the uncertainty principle.

39–8 Particle in an Infinitely Deep Square Well Potential (a Rigid Box)

The Schrödinger equation can be solved analytically only for a few possible forms of the potential energy U. We consider two simple cases here which at first may not seem realistic, but have simple solutions that can be used as approximations to understand a variety of phenomena.

[†] In Eq. 14–3, t becomes x and ω becomes $k = \sqrt{2mE/\hbar^2}$. (Don't confuse this k with the spring constant k of Chapter 14.)

[‡] Such an infinite wave makes problems for normalization since $\int_{-\infty}^{\infty} |\psi|^2\, dx = A^2 \int_{-\infty}^{\infty} \sin^2 kx\, dx$ is infinite for any nonzero value for A. For practical purposes we can usually normalize the waves ($A \neq 0$) by assuming that the particle is in a large but finite region of space. The region can be chosen large enough so that momentum is still rather precisely fixed.

In our first case, we assume that a particle of mass m is confined to a one-dimensional box of width L whose walls are perfectly rigid. (This can serve as a very crude approximation for an electron in a metal, for example.) The particle is trapped in this box and collisions with the walls are perfectly elastic. The potential energy for this situation, which is commonly known as an **infinitely deep square well potential** or **rigid box**, is shown in Fig. 39–7. We can write the potential energy $U(x)$ as

$$U(x) = 0 \qquad 0 < x < L$$
$$U(x) = \infty \qquad x \le 0 \quad \text{and} \quad x \ge L.$$

For the region $0 < x < L$, where $U(x) = 0$, we already know the solution of the Schrödinger equation from our discussion in Section 39–7: it is just Eq. 39–10,

$$\psi(x) = A \sin kx + B \cos kx,$$

where (from Eq. 39–11a)

$$k = \sqrt{\frac{2mE}{\hbar^2}}.$$

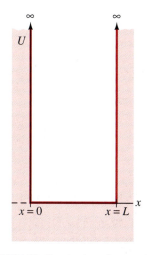

Outside the well $U(x) = \infty$, so $\psi(x)$ must be zero. (If it weren't, the product $U\psi$ in the Schrödinger equation wouldn't be finite; besides, if $U = \infty$, we can't expect a particle of finite total energy to be in such a region.) So we are concerned only with the wave function within the well, and we must determine the constants A and B as well as any restrictions on the value of k (and hence on the energy E).

We have insisted that the wave function must be continuous. Hence, if $\psi = 0$ outside the well, it must be zero at $x = 0$ and at $x = L$:

$$\psi(0) = 0 \quad \text{and} \quad \psi(L) = 0.$$

Boundary conditions

These are the **boundary conditions** for this problem. At $x = 0$, $\sin kx = 0$ but $\cos kx = 1$, so

$$0 = \psi(0) = A \sin 0 + B \cos 0 = 0 + B.$$

Thus B must be zero. Our solution is reduced to

$$\psi(x) = A \sin kx.$$

Now we apply the other boundary condition, $\psi = 0$ at $x = L$:

$$0 = \psi(L) = A \sin kL.$$

We don't want $A = 0$ or we won't have a particle at all ($|\psi|^2 = 0$ everywhere). Therefore, we set

$$\sin kL = 0,$$

which, since the sine is zero for angles of $0, \pi, 2\pi, 3\pi \cdots$ radians, can happen only if $kL = 0, \pi, 2\pi, 3\pi, \cdots$. In other words,

$$kL = n\pi \qquad n = 1, 2, 3, \cdots, \qquad \textbf{(39–12)}$$

where n is an integer. We eliminate the case $n = 0$ since this, too, produces $x = 0$ everywhere. Thus k, and hence E, cannot have just any value; rather, k is limited to values

$$k = \frac{n\pi}{L}.$$

Putting this expression in Eq. 39–11a (and substituting $h/2\pi$ for $\hbar$), we find that E can have only the values

$$E = n^2 \frac{h^2}{8mL^2}, \qquad n = 1, 2, 3, \cdots. \qquad \textbf{(39–13)}$$

Energy levels, quantized
(∞ potential well)

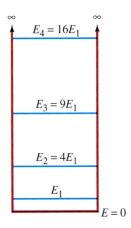

FIGURE 39–8 Possible energy levels for a particle in a box with perfectly rigid walls (infinite square well potential).

Zero-point energy

FIGURE 39–9 Wave functions corresponding to the quantum number n being 1, 2, 3 and 10 for a particle confined to a rigid box.

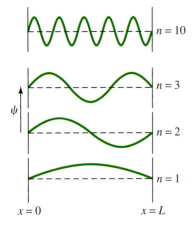

FIGURE 39–10 (below) The probability distribution for a particle in a rigid box for the states with $n = 1, 2, 3,$ and 10.

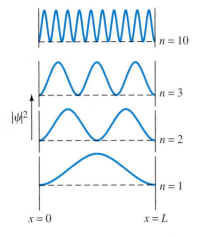

A particle trapped in a rigid box thus can only have certain *quantized energies*. The lowest energy (the ground state) has $n = 1$ and is given by

$$E_1 = \frac{h^2}{8mL^2}. \qquad \text{[ground state]}$$

The next highest energy ($n = 2$) is

$$E_2 = 4E_1,$$

and for higher energies (see Fig. 39–8),

$$E_3 = 9E_1$$
$$\vdots$$
$$E_n = n^2E_1.$$

The integer n is called the **quantum number** of the state. That the lowest energy, E_1, is not zero means that the particle in the box can never be at rest. This is contrary to classical ideas, according to which a particle can have $E = 0$. E_1 is called the **zero-point energy**. One outcome of this result is that even at a temperature of absolute zero (0 K), quantum mechanics predicts that particles in a box would not be at rest but would have a zero-point energy.

We can also note that the energy E_1 and momentum $p_1 = \hbar k = \hbar \pi / L$ in the ground state are related inversely to the width of the box. The smaller the width L, the larger the momentum (and energy). This can be considered a direct result of the uncertainty principle (see Problem 21).

The wave function $\psi = A \sin kx$ for each of the quantum states is (since $k = n\pi/L$)

$$\psi_n = A \sin\left(\frac{n\pi}{L} x\right). \qquad (39\text{–}14)$$

We can determine the constant A by imposing the normalization condition (Eq. 39–6):

$$1 = \int_{-\infty}^{\infty} \psi^2 \, dx = \int_0^L A^2 \sin^2\left(\frac{n\pi}{L} x\right) dx, \qquad (39\text{–}15)$$

where the integral needs to be done only over the range $0 < x < L$ because outside these limits $\psi = 0$. The integral (see Example 39–5) is equal to $A^2L/2$, so we have

$$A = \sqrt{\frac{2}{L}} \quad \text{and} \quad \psi_n = \sqrt{\frac{2}{L}} \sin\left(\frac{n\pi}{L} x\right).$$

The amplitude A is the same for all the quantum numbers. Figure 39–9 shows the wave functions (Eq. 39–14) for $n = 1, 2, 3,$ and 10. They look just like standing waves on a string—see Fig. 15–27. This is not surprising since the wave function solutions, Eq. 39–14, are the same as for the standing waves on a string, and the condition $kL = n\pi$ is the same in the two cases.

Figure 39–10 shows the probability distribution, $|\psi|^2$, for the same states ($n = 1, 2, 3, 10$) for which ψ is shown in Fig. 39–9. We see immediately that the particle is more likely to be found in some places than in others. For example, in the ground state ($n = 1$), the electron is much more likely to be found near the center of the box than near the walls. This is clearly at variance with classical ideas, which predict a uniform probability density—the particle would be as likely to be found at one point in the box as at any other. The quantum-mechanical probability densities for higher states are even more complicated, with areas of low probability not only near the walls but also at regular intervals in between.

EXAMPLE 39–4 **Electron in an infinite potential well.** (a) Calculate the three lowest energy levels for an electron trapped in an infinitely deep square well potential of width $L = 1.00 \times 10^{-10}$ m (about the diameter of a hydrogen atom in its ground state). (b) If a photon were emitted when the electron jumps from the $n = 2$ state to the $n = 1$ state, what would its wavelength be?

SOLUTION (a) The ground state ($n = 1$) has energy

$$E_1 = \frac{h^2}{8mL^2} = \frac{(6.63 \times 10^{-34}\,\text{J}\cdot\text{s})^2}{8(9.11 \times 10^{-31}\,\text{kg})(1.00 \times 10^{-10}\,\text{m})^2} = 6.03 \times 10^{-18}\,\text{J}.$$

In electron volts this is

$$E_1 = \frac{6.03 \times 10^{-18}\,\text{J}}{1.60 \times 10^{-19}\,\text{J/eV}} = 37.7\,\text{eV}.$$

Then

$$E_2 = (2)^2 E_1 = 151\,\text{eV}$$
$$E_3 = (3)^2 E_1 = 339\,\text{eV}.$$

(b) The energy difference is $E_2 - E_1 = 151\,\text{eV} - 38\,\text{eV} = 113\,\text{eV}$ or 1.81×10^{-17} J, and this would equal the energy of the emitted photon (energy conservation). Its wavelength would be

$$\lambda = \frac{c}{f} = \frac{hc}{E} = \frac{(6.63 \times 10^{-34}\,\text{J}\cdot\text{s})(3.00 \times 10^8\,\text{m/s})}{1.81 \times 10^{-17}\,\text{J}} = 1.10 \times 10^{-8}\,\text{m}$$

or 11.0 nm, which is in the ultraviolet region of the spectrum.

EXAMPLE 39–5 **Calculating a normalization constant.** Show that the normalization constant A for all wave functions describing a particle in an infinite potential well of width L has a value of $A = \sqrt{2/L}$.

SOLUTION The wave functions are

$$\psi = A \sin \frac{n\pi x}{L}.$$

To normalize ψ, we must have (Eq. 39–15)

$$1 = \int_0^L |\psi|^2 \, dx = \int_0^L A^2 \sin^2 \frac{n\pi x}{L} \, dx.$$

We need integrate only from 0 to L since $\psi = 0$ for all other values of x. To evaluate this integral we let $\theta = n\pi x/L$ and use the trigonometric identity $\sin^2 \theta = \frac{1}{2}(1 - \cos 2\theta)$. Then, with $dx = L\, d\theta/n\pi$, we have

$$1 = A^2 \int_0^{n\pi} \sin^2 \theta \left(\frac{L}{n\pi}\right) d\theta = \frac{A^2 L}{2n\pi} \int_0^{n\pi} (1 - \cos 2\theta)\, d\theta$$

$$= \frac{A^2 L}{2n\pi} \left(\theta - \tfrac{1}{2} \sin 2\theta\right) \Big|_0^{n\pi}$$

$$= \frac{A^2 L}{2}.$$

Thus $A^2 = 2/L$ and

$$A = \sqrt{\frac{2}{L}}.$$

EXAMPLE 39–6 | **ESTIMATE** | **Confined bacterium.** A tiny bacterium with a mass of about 10^{-14} kg is confined between two rigid walls 0.1 mm apart. (*a*) Estimate its minimum speed. (*b*) If, instead, its speed is about 1 mm in 100 s, estimate the quantum number of its state.

SOLUTION (*a*) The minimum speed occurs in the ground state, $n = 1$; since $E = \frac{1}{2}mv^2$, we have

$$v = \sqrt{\frac{2E}{m}} = \sqrt{\frac{h^2}{4m^2L^2}} = \frac{h}{2mL} = \frac{6.6 \times 10^{-34}\,\text{J}\cdot\text{s}}{2(10^{-14}\,\text{kg})(10^{-4}\,\text{m})} \approx 3 \times 10^{-16}\,\text{m/s}.$$

This is a speed so small that we could not measure it and the object would seem at rest, consistent with classical physics.
(*b*) Given $v = 10^{-3}\,\text{m}/100\,\text{s} = 10^{-5}\,\text{m/s}$, the kinetic energy of the bacterium is

$$E = \tfrac{1}{2}mv^2 = \tfrac{1}{2}(10^{-14}\,\text{kg})(10^{-5}\,\text{m/s})^2 = 0.5 \times 10^{-24}\,\text{J}.$$

From Eq. 39–13, the quantum number of this state is

$$n = \sqrt{E\left(\frac{8mL^2}{h^2}\right)} = \sqrt{\frac{(0.5 \times 10^{-24}\,\text{J})(8)(10^{-14}\,\text{kg})(10^{-4}\,\text{m})^2}{(6.6 \times 10^{-34}\,\text{J}\cdot\text{s})^2}}$$

$$\approx \sqrt{1 \times 10^{21}} \approx 3 \times 10^{10}.$$

This number is so large that we could never distinguish between adjacent energy states (between $n = 3 \times 10^{10}$ and $3 \times 10^{10} + 1$). The energy states would appear to form a continuum. Thus, even though the energies involved here are small ($\ll 1$ eV), we are still dealing with a macroscopic object (though visible only under a microscope) and the quantum result is not distinguishable from a classical one. This is in accordance with the correspondence principle.

* 39–9 Finite Potential Well

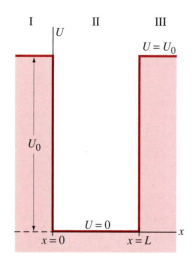

FIGURE 39–11 Potential energy U vs x for a finite one-dimensional square well.

Let us now look at a particle in a box whose walls are not perfectly rigid. That is, the potential energy outside the box or well is not infinite, but rises to some level U_0, as shown in Fig. 39–11. This is called a **finite potential well**. It can serve as an approximation for, say, a neutron in a nucleus. There are some significant new features that arise for the finite well as compared to the infinite well. We divide the well into three regions as shown in Fig. 39–11. In region II, inside the well, the Schrödinger equation is the same as before ($U = 0$), although the boundary conditions will be different. So we write the solution for region II as

$$\psi_{\text{II}} = A \sin kx + B \cos kx \qquad (0 < x < L)$$

but we don't immediately set $B = 0$ or assume that k is given by Eq. 39–12.
In regions I and III, the Schrödinger equation, now with $U(x) = U_0$, is

$$-\frac{\hbar^2}{2m}\frac{d^2\psi}{dx^2} + U_0\psi = E\psi.$$

We rewrite this as

$$\frac{d^2\psi}{dx^2} - \left[\frac{2m(U_0 - E)}{\hbar^2}\right]\psi = 0.$$

Let us assume that E is less than U_0, so the particle is "trapped" in the well (at least classically). There might be only one such **bound state**, or several, or even none, as we shall discuss later. We define the constant G by

$$G^2 = \frac{2m(U_0 - E)}{\hbar^2} \qquad \textbf{(39–16)}$$

and rewrite the Schrödinger equation as

$$\frac{d^2\psi}{dx^2} - G^2\psi = 0.$$

This equation has the general solution

$$\psi_{\text{I, III}} = Ce^{Gx} + De^{-Gx},$$

which can be confirmed by direct substitution, since

$$\frac{d^2}{dx^2}\left(e^{\pm Gx}\right) = G^2 e^{\pm Gx}.$$

In region I, x is always negative, so D must be zero (otherwise, $\psi \to \infty$ as $x \to -\infty$, giving an unacceptable result). Similarly in region III, where x is always positive, C must be zero. Hence

$$\psi_{\text{I}} = Ce^{Gx} \qquad (x < 0)$$
$$\psi_{\text{III}} = De^{-Gx} \qquad (x > L).$$

In regions I and III, the wave function decreases exponentially with distance from the well. The mathematical forms of the wave function inside and outside the well are different, but we insist that the wave function be continuous even at the two walls. We also insist that the slope of ψ, which is its first derivative, be continuous at the walls. Hence we have the boundary conditions:

$$\psi_{\text{I}} = \psi_{\text{II}} \quad \text{and} \quad \frac{d\psi_{\text{I}}}{dx} = \frac{d\psi_{\text{II}}}{dx} \quad \text{at } x = 0$$

$$\psi_{\text{II}} = \psi_{\text{III}} \quad \text{and} \quad \frac{d\psi_{\text{II}}}{dx} = \frac{d\psi_{\text{III}}}{dx} \quad \text{at } x = L.$$

At the left-hand wall ($x = 0$) these boundary conditions become

$$Ce^0 = A\sin 0 + B\cos 0 \qquad \text{or} \qquad C = B$$

and

$$GCe^0 = kA\cos 0 - kB\sin 0 \qquad \text{or} \qquad GC = kA.$$

These are two of the relations that link the constants A, B, C, D and the energy E. We get two more relations from the boundary conditions at $x = L$, and a fifth relation from normalizing the wave functions over all space, $\int_{-\infty}^{\infty}|\psi|^2\, dx = 1$. These five relations allow us to solve for the five unknowns, including the energy E. We will not go through the detailed mathematics, but we will discuss some of the results.

Figure 39–12a shows the wave function ψ for the three lowest possible states, and Fig. 39–12b shows the probability distributions $|\psi|^2$. We see that the wave functions are smooth at the walls of the well. Within the well ψ has the form of a sinusoidal wave; for the ground state, there is less than a half wavelength. Compare this to the infinite well (Fig. 39–9), where the ground-state wave function is exactly a half wavelength: $\lambda = 2L$. For our finite well, $\lambda > 2L$. Thus for a finite well the momentum of a particle ($p = h/\lambda$), and hence its ground-state energy, will be less than for an infinite well of the same width L.

Outside the finite well we see that the wave function drops off exponentially on either side of the walls. That ψ is not zero beyond the walls means that the particle can sometimes be found outside the well. This completely contradicts classical ideas. Outside the well, the potential energy of the particle is greater than its total energy: $U_0 > E$. This violates conservation of energy. But we clearly see in Fig. 39–12b that the particle can spend some time outside the well, where $U_0 > E$ (although the penetration into this classically forbidden region is generally not far since $|\psi|^2$ decreases exponentially with distance from either wall). The penetration

FIGURE 39–12 The wave functions (a), and probability distributions (b), for the three lowest possible states of a particle in a finite potential well. Each of the ψ and $|\psi|^2$ curves has been superposed on its energy level (dashed lines) for convenience.

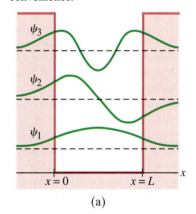

(a)

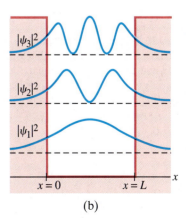

(b)

of a particle into a classically forbidden region is a very important result of quantum mechanics. But how can it be? How can we accept this nonconservation of energy? We can look to the uncertainty principle, in the form

$$\Delta E \, \Delta t \gtrsim \hbar.$$

It tells us that the energy is uncertain, and thus can even be nonconserved, by an amount ΔE for very short times $\Delta t \sim \hbar / \Delta E$.

Now let us consider the situation when the total energy E of the particle is greater than U_0. In this case the particle is a free particle and everywhere its wave function is sinusoidal, Fig. 39–13. Its wavelength is different outside the well than inside, as shown. Since $K = \frac{1}{2}mv^2 = p^2/2m$, the wavelength in region II is

$$\lambda = \frac{h}{p} = \frac{h}{\sqrt{2mE}} \qquad 0 < x < L,$$

whereas in regions I and III, where $p^2/2m = K = E - U_0$, the wavelength is

$$\lambda = \frac{h}{p} = \frac{h}{\sqrt{2m(E - U_0)}} \qquad x < 0 \quad \text{and} \quad x > L.$$

For $E > U_0$, any energy E is possible. But for $E < U_0$, as we saw above, the energy is quantized and only certain states are possible.

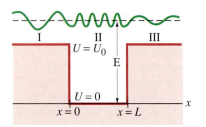

FIGURE 39–13 Particle of energy E traveling over a potential well whose depth U_0 is less than E (measured in the well).

39–10 | Tunneling through a Barrier

We saw in Section 39–9 that according to quantum mechanics, a particle such as an electron can penetrate a barrier into a region forbidden by classical mechanics. There are a number of important applications of this phenomenon, particularly as applied to penetration of a thin barrier.

We consider a particle of mass m and energy E traveling to the right along the x axis in free space where the potential energy $U = 0$ so the energy is all kinetic energy ($E = K$). The particle encounters a narrow potential barrier whose height U_0 (in energy units) is greater than E, and whose thickness is L (distance units); see Fig. 39–14a. Since $E < U_0$, we would expect from classical physics that the particle could not penetrate the barrier but would simply be reflected and would return in the opposite direction. Indeed, this is what happens for macroscopic objects. But quantum mechanics predicts a nonzero probability for finding the particle on the other side of the barrier. We can see how this can happen in part (b) of Fig. 39–14, which shows the wave function. The approaching particle has a sinusoidal wave function. Within the barrier the solution to the Schrödinger equation is a decaying exponential just as for the finite well of Section 39–9. However, before the exponential dies away to zero, the barrier ends (at $x = L$),

FIGURE 39–14 (a) A potential barrier of height U_0 and thickness L. (b) The wave function for a particle of energy $E\,(< U_0)$ that approaches from the left. The curve for ψ is superposed, for convenience, on the energy level line (dashed).

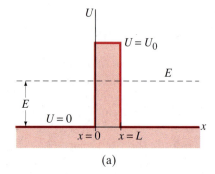

(a)

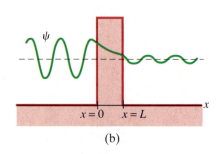

(b)

and for $x > L$ there is again a sinusoidal wave function, since $U = 0$ and $E = K > 0$. But it is a sine wave of greatly reduced amplitude. Nonetheless, because $|\psi|^2$ is nonzero beyond the barrier, we see that there is a nonzero probability that the particle penetrates the barrier. This process is called **tunneling** through the barrier, or **barrier penetration**. Although we cannot observe the particle within the barrier (it would violate conservation of energy), we can detect it after it has penetrated the barrier.

Quantitatively, we can describe the tunneling probability with a *transmission coefficient, T*, and a *reflection coefficient, R*. Suppose, for example, that $T = 0.03$ and $R = 0.97$; then if 100 particles struck the barrier, on the average 3 would tunnel through and 97 would be reflected. Note that $T + R = 1$, since an incident particle must either reflect or tunnel through. The transmission coefficient can be determined by writing the wave function for each of the three regions, just as we did for the finite well, and then applying the boundary conditions that ψ and $d\psi/dx$ must be continuous at the edges of the barrier ($x = 0$ and $x = L$). The calculation shows (see Problem 36) that if T is small ($\ll 1$), then

$$T \approx e^{-2GL}, \tag{39-17a}$$

where

$$G = \sqrt{\frac{2m(U_0 - E)}{\hbar^2}}. \tag{39-17b}$$

(This is the same G as in Section 39–9, Eq. 39–16.) We note that increasing the height of the barrier, U_0, or increasing its thickness, L, will drastically reduce T. Indeed, for macroscopic situations, T is extremely small, in accord with classical physics, which predicts no tunneling (again the correspondence principle).

EXAMPLE 39–7 **Barrier penetration.** A 50-eV electron approaches a square barrier 70 eV high and (*a*) 1.0 nm thick, (*b*) 0.10 nm thick. What is the probability that the electron will tunnel through?

SOLUTION (*a*) First we write the energy in SI units.

$$U_0 - E = (70\,\text{eV} - 50\,\text{eV})(1.6 \times 10^{-19}\,\text{J/eV}) = 3.2 \times 10^{-18}\,\text{J}.$$

Then, using Eq. 39–17, we have

$$2GL = 2\sqrt{\frac{2(9.11 \times 10^{-31}\,\text{kg})(3.2 \times 10^{-18}\,\text{J})}{(1.06 \times 10^{-34}\,\text{J}\cdot\text{s})^2}}(1.0 \times 10^{-9}\,\text{m}) = 46$$

and

$$T = e^{-2GL} = e^{-46} = 1.1 \times 10^{-20},$$

which is extremely small.
(*b*) For $L = 0.10$ nm, $2GL = 4.6$ and

$$T = e^{-4.6} = 0.010.$$

Thus the electron has a 1 percent chance of penetrating a 0.1-nm-thick barrier, but only 1 chance in 10^{20} to penetrate a 1-nm barrier. By reducing the barrier thickness by a factor of 10, the probability of tunneling through was increased 10^{18} times! Clearly the transmission coefficient is extremely sensitive to the values of L, $U_0 - E$, and m.

Tunneling is a result of the wave properties of material particles, and also occurs for classical waves. For example, we saw in Section 33–7 that, when light traveling in glass strikes a glass–air boundary at an angle greater than the critical angle, the light is 100 percent totally reflected. We studied this phenomenon of

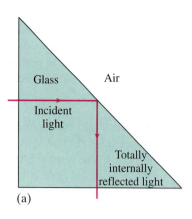

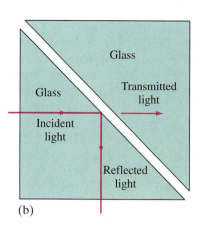

FIGURE 39–15 (a) Light traveling in glass strikes the interface with air at an angle greater than the critical angle, and is totally internally reflected. (b) A small amount of light tunnels through a narrow air gap between two pieces of glass.

(a)

(b)

FIGURE 39–16 Potential energy seen by an alpha particle (charge q) in presence of nucleus (charge Q), showing the wave function for tunneling out.

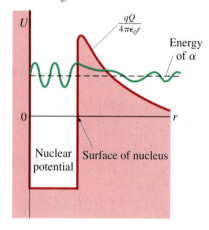

FIGURE 39–17 Probe tip of a scanning tunneling microscope moves up and down to maintain a constant tunneling current, producing an image of the surface.

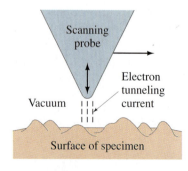

total internal reflection from the point of view of ray optics, and we show it here in Fig. 39–15a. The wave theory, however, predicts that waves actually penetrate the air for a few wavelengths—almost as if they "needed" to pass the interface to find out there is air beyond and hence need to be totally reflected. Indeed, if a second piece of glass is brought near the first as shown in Fig. 39–15b, a transmitted wave that has tunneled through the air gap can be experimentally observed. You can actually observe this for yourself by looking down into a glass of water at an angle such that light entering your eye has been totally internally reflected from the (outer) glass surface (it will look silvery). If you press a moistened fingertip against the glass, you can see the whorls of the ridges on your fingerprints, because at the ridges you have interfered with the total internal reflection at the outer surface of the glass. So you see light that has penetrated the gap and reflected off the ridges on your finger.

Applications of Tunneling

Tunneling thus occurs even for classical waves. What is new in quantum mechanics is that material particles have wave properties and hence can tunnel. Tunneling has provided the basis for a number of useful devices, as well as helped to explain a number of important phenomena, some of which we mention briefly now.

Some atomic nuclei undergo **radioactive decay** by the emission of an alpha (α) particle, which consists of two protons and two neutrons. Inside a radioactive nucleus, we can imagine that the protons and neutrons are moving about, and sometimes two of each come together and form this stable entity, the alpha particle. We will study alpha decay in more detail in Chapter 42, but for now we note that the potential energy diagram for the alpha particle inside this type of nucleus looks something like Fig. 39–16. The square well represents the attractive nuclear force that holds the nucleus together. To this is added the $1/r$ Coulomb potential energy of repulsion between the positive alpha particle and the remaining positively charged nucleus. The barrier that results is called the **Coulomb barrier**. The wave function for the tunneling particle shown must have energy greater than zero (or the barrier would be infinitely wide and tunneling could not occur), but less than the height of the barrier. If the alpha particle had energy higher than the barrier, it would always be free and the original nucleus wouldn't exist. Thus the barrier keeps the nucleus together, but occasionally a nucleus of this type can decay by the tunneling of an alpha particle. The probability of an α-particle escaping, and hence the "lifetime" of a nucleus, depends on the height and width of the barrier, and can take on a very wide range of values for only a limited change in barrier width as we saw in Example 39–7. Lifetimes of α-decaying radioactive nuclei range from less than 1 μs to 10^{10} yr.

A so-called **tunnel diode** is an electronic device made of two types of semiconductor carrying opposite-sign charge carriers, separated by a very thin neutral region. Current can tunnel through this thin barrier and can be controlled by the voltage applied to it, which affects the height of the barrier.

The **scanning tunneling electron microscope** (STM), developed in the 1980s, makes use of tunneling through a vacuum. A tiny probe, whose tip may be only one (or a few) atoms wide, is moved across the specimen to be examined in a series of linear passes, like those made by the electron beam in a TV tube or CRT. The tip, as it scans, remains very close to the surface of the specimen, about 1 nm above it, Fig. 39–17. A small voltage applied between the probe and the surface causes electrons to tunnel through the vacuum between them. This tunneling current is very sensitive to the gap width (see Example 39–7), so that a feedback mechanism can be used to raise and lower the probe to maintain a constant electron tunneling current. The probe's vertical motion, following the surface of the specimen, is then plotted as a function of position, producing a three-dimensional image of the surface. (See, for example, Fig. 39–18.) Surface features as fine as the size of an atom can be resolved: a resolution better than 0.1 nm laterally and 10^{-2} to 10^{-3} nm vertically. This kind of resolution was not available previously and has given a great impetus to the study of the surface structure of materials. The "topographic" image of a surface actually represents the distribution of electron charge (electron wave probability distributions).

The new **atomic force microscope** (AFM) is in many ways similar to an STM, but can be used on a wider range of sample materials. Instead of detecting an electric current, the AFM measures the force between a cantilevered tip and the sample, a force which depends strongly on the tip–sample separation at each point. The tip is moved as for the STM.

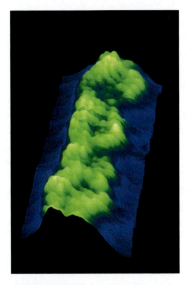

FIGURE 39–18 Image of cellular DNA, magnified about 2 million times, taken with a scanning tunneling microscope. Three turns of the DNA double helix can be seen in this false-color image.

Summary

In 1925, Schrödinger and Heisenberg separately worked out a new theory, **quantum mechanics**, which is now considered to be the basic theory at the atomic level. It is a statistical theory rather than a deterministic one.

An important aspect of quantum mechanics is the Heisenberg **uncertainty principle**. It results from the wave-particle duality and the unavoidable interaction between the observed object and the observer.

One form of the uncertainty principle states that the position x and momentum p_x of an object cannot both be measured precisely at the same time. The products of the uncertainties, $(\Delta x)(\Delta p_x)$, can be no less than $\hbar (= h/2\pi)$:

$$(\Delta p_x)(\Delta x) \gtrsim \hbar.$$

Another form states that the energy can be uncertain, or nonconserved, by an amount ΔE for a time Δt where

$$(\Delta E)(\Delta t) \gtrsim \hbar.$$

A particle such as an electron is represented by a **wave function** ψ. The square of the wave function, $|\psi|^2$, at any point in space represents the **probability** of finding the particle at that point. The wave function must be **normalized**, meaning that $\int |\psi|^2 \, dV$ over all space must equal 1, since the particle must be found at one place or another.

In nonrelativistic wave mechanics, ψ satisfies the **Schrödinger equation**:

$$-\frac{\hbar^2}{2m}\frac{d^2\psi}{dx^2} + U\psi = E\psi,$$

here in its one-dimensional time-independent form,

where U is the potential energy as a function of position and E is the total energy of the particle.

A **free particle** subject to no forces has a sinusoidal wave function $\psi = A \sin kx + B \cos kx$ with $k = p/\hbar$ and p is the particle's momentum. Such a wave of fixed momentum is spread out indefinitely in space as a plane wave.

A **wave packet**, localized in space, is a superposition of sinusoidal waves with a range of momenta.

For a particle confined to an **infinitely deep square well potential**, or **rigid box**, the Schrödinger equation gives the wave functions

$$\psi = A \sin kL,$$

where $A = \sqrt{2/L}$, as solutions inside the well. The energy is quantized,

$$E = \frac{\hbar^2 k^2}{2m} = \frac{n^2 h^2}{8mL^2},$$

where n is an integer.

In a **finite potential well**, the wave function extends into the classically forbidden region where the total energy is less than the potential energy. That this is possible is consistent with the uncertainty principle. The solutions to the Schrödinger equation in these areas are decaying exponentials.

Because quantum-mechanical particles can penetrate such classically forbidden areas, they can **tunnel** through thin barriers even though the potential energy in the barrier is greater than the total energy of the particle.

Questions

1. Compare a matter wave ψ to (a) a wave on a string, (b) an EM wave. Discuss similarities and differences.

2. Explain why Bohr's theory of the atom is not compatible with quantum mechanics, particularly the uncertainty principle.

3. Explain why it is that the more massive an object is, the easier it becomes to predict its future position.

4. In view of the uncertainty principle, why does a baseball seem to have a well-defined position and speed, whereas an electron does not?

5. Would it ever be possible to balance a very sharp needle precisely on its point? Explain.

6. When you check the pressure in a tire, doesn't some air inevitably escape? Is it possible to avoid this escape of air altogether? What is the relation to the uncertainty principle?

7. It has been said that the ground-state energy in the hydrogen atom can be precisely known but that the excited states have some uncertainty in their values (an "energy width"). Is this consistent with the uncertainty principle in its energy form? Explain.

8. If Planck's constant were much larger than it is, how would this affect our everyday life?

9. In what ways is Newtonian mechanics contradicted by quantum mechanics?

10. Discuss the connection between the zero-point energy for a particle in a box and the uncertainty principle.

11. The wave function for a particle in a box is zero at points within the box (except for $n = 1$). Does this mean that the probability of finding the particle at these points is zero? Does it mean that the particle cannot pass by these points? Explain.

12. What does the probability density look like for a particle in an infinite potential well for large values of n, say $n = 100$ or $n = 1000$? As n becomes very large, do your predictions approach classical predictions in accord with the correspondence principle?

13. For a particle in an infinite potential well the separation between energy states increases as n increases (see Eq. 39–13). But doesn't the correspondence principle require closer spacing between states as n increases so as to approach a classical (nonquantized) situation? Explain.

14. A particle is trapped in an infinite potential well. Describe what happens to the particle's ground-state energy and wave function as the potential walls become finite and get lower and lower until they finally reach zero ($U = 0$ everywhere).

15. A hydrogen atom and a helium atom, each with 4 eV of kinetic energy, approach a thin barrier 6 MeV high. Which has the greater probability of tunneling through?

Problems

Section 39–2

1. (II) The neutrons in a parallel beam, each having kinetic energy $\frac{1}{40}$ eV, are directed through two slits 1.0 mm apart. How far apart will the interference peaks be on a screen 1.0 m away?

2. (II) Bullets of mass 2.0 g are fired in parallel paths with speeds of 120 m/s through a hole 3.0 mm wide. How far from the hole must you be to detect a 1.0-cm spread in the beam?

Section 39–3

3. (I) A proton is traveling with a speed of $(4.825 \pm 0.012) \times 10^5$ m/s. With what maximum accuracy can its position be ascertained?

4. (I) An electron remains in an excited state of an atom for typically 10^{-8} s. What is the minimum uncertainty in the energy of the state (in eV)?

5. (I) If an electron's position can be measured to an accuracy of 1.6×10^{-8} m, how accurately can its velocity be known?

6. (II) A 12-g bullet leaves a rifle at a speed of 150 m/s. (a) What is the wavelength of this bullet? (b) If the position of the bullet is known to an accuracy of 0.55 cm (radius of the barrel), what is the minimum uncertainty in its vertical momentum? (c) If the accuracy of the bullet were determined only by the uncertainty principle (an unreasonable assumption), by how much might the bullet miss a pinpoint target 300 m away?

7. (II) An electron and a 150-g baseball are each traveling 75 m/s measured to an accuracy of 0.065 percent. Calculate and compare the uncertainty in position of each.

8. (II) Use the uncertainty principle to show that if an electron were present in the nucleus ($r \approx 10^{-15}$ m), its kinetic energy (use relativity) would be hundreds of MeV. (Since such electron energies are not observed, we conclude that electrons are not present in the nucleus.) [Hint: a particle can have energy as large as its uncertainty.]

9. (II) An electron in the $n = 2$ state of hydrogen remains there on the average about 10^{-8} s before jumping to the $n = 1$ state. (a) Estimate the uncertainty in the energy of the $n = 2$ state. (b) What fraction of the transition energy is this? (c) What is the wavelength, and width (in nm), of this line in the spectrum of hydrogen?

10. (II) How accurately can the position of a 2.50-keV electron be measured assuming its energy is known to 1.00 percent?

11. (III) In a double-slit experiment on electrons (or photons), suppose that we use indicators to determine which slit each electron went through (Section 39–2). These indicators must tell us the y coordinate to within $d/2$, where d is the distance between slits. Use the uncertainty principle to show that the interference pattern will be destroyed [Note: First show that the angle θ between maxima and minima of the interference pattern is given by $\frac{1}{2}\lambda/d$, Fig. 39–19.]

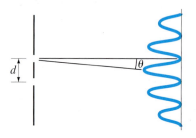

FIGURE 39–19
Problem 11.

*12. (III) (a) Show that $\Psi(x, t) = Ae^{i(kx-\omega t)}$ is a solution to the time-dependent Schrödinger equation for a free particle $[U(x) = U_0 = \text{constant}]$ but that $\Psi(x, t) = A \cos(kx - \omega t)$ and $\Psi(x, t) = A \sin(kx - \omega t)$ are not. (b) Show that the valid solution of part (a) satisfies conservation of energy if the de Broglie relations hold, $\lambda = h/p$, $\omega = E/\hbar$. That is, show that direct substitution in Eq. 39–7 gives

$$\hbar\omega = \frac{\hbar^2 k^2}{2m} + U_0.$$

Section 39–7

13. (I) Write the wave function for (a) a free electron and (b) a free proton, each having a constant velocity $v = 4.0 \times 10^5 \, \text{m/s}$.

14. (I) A free electron has a wave function $\psi(x) = A \sin(1.0 \times 10^{10} x)$, where x is given in meters. Determine the electron's (a) wavelength, (b) momentum, (c) speed, and (d) kinetic energy.

15. (II) Show that the uncertainty principle holds for a "wave packet" that is formed by two waves of similar wavelength λ_1 and λ_2. To do so, follow the argument leading up to Eq. 16–8, but use as the two waves $\psi_1 = A \sin k_1 x$ and $\psi_2 = A \sin k_2 x$. Then show that the width of each "wave packet" is $\Delta x = 2\pi/(k_1 - k_2) = 2\pi/\Delta k$ (from $t = 0.05 \, \text{s}$ to $t = 0.15 \, \text{s}$ in Fig. 16–18). Finally, show that $\Delta x \, \Delta p = h$ for this simple situation.

Section 39–8

16. (II) Show that for a particle in a perfectly rigid box, the wavelength of the wave function for any state is the de Broglie wavelength.

17. (II) What is the minimum speed of an electron trapped in a 0.10-nm-wide infinitely deep square well?

18. (II) An $n = 3$ to $n = 1$ transition for an electron trapped in a rigid box produces a 240-nm photon. What is the width of the box?

19. (II) An electron trapped in an infinitely deep square well has a ground-state energy $E = 8.0 \, \text{eV}$. (a) What is the longest wave-length photon this system can emit, and (b) what is the width of the well?

20. (II) The longest-wavelength line in the spectrum emitted by an electron trapped in an infinitely deep square well is 690 nm. What is the width of the well?

21. (II) For a particle in a box with rigid walls, determine whether our results for the ground state are consistent with the uncertainty principle by calculating the product $\Delta p \, \Delta x$. Take $\Delta x \approx L$, since the particle is known to be at least within the box. For Δp, note that although p is known ($= \hbar k$), the direction of $\mathbf{p}$ is not known, so the x component could vary from $-p$ to $+p$; hence take $\Delta p \approx 2p$.

22. (II) Write a formula for the positions of (a) the maxima and (b) the minima in $|\psi|^2$ for a particle in the nth state in an infinite square well.

23. (II) Determine the lowest four energy levels and wave functions for an electron trapped in an infinitely deep potential well of width 2.0 nm.

24. (II) An electron is trapped in a rigid box 0.50 nm wide. (a) Determine the energies and wave functions for the four lowest states. (b) Determine the wavelengths of photons emitted for all possible transitions between these states.

25. (II) Consider an atomic nucleus to be a rigid box of width 10^{-14} m. What would be the ground-state energy for (a) an electron, (b) a neutron, and (c) a proton?

26. (II) Suppose that the walls of an infinite square well are at $x = -L/2$ and $x = L/2$. What will be the wave functions for the four lowest energy levels? Sketch the wave functions and probability densities.

27. (II) An electron is trapped in an infinitely deep potential well of width L. Determine the probability of finding the electron within $\frac{1}{4}L$ of either wall if it is (a) in the ground state, (b) in the $n = 4$ state. [Hint: Evaluate $\int_0^{L/4}|\psi|^2 \, dx + \int_{3L/4}^{L}|\psi|^2 \, dx$.] (c) What is the classical prediction?

28. (II) An electron is trapped in a 1.00-nm-wide rigid box. Determine the probability of finding the electron within 0.10 nm of the center of the box (on either side of center) for (a) $n = 1$, (b) $n = 5$, and (c) $n = 20$. (d) Compare to the classical prediction.

*29. (II) Sketch the wave functions and the probability distributions for the $n = 4$ and $n = 5$ states for a particle trapped in a finite square well.

*30. (II) An electron with 100 eV of kinetic energy in free space passes over a finite potential well 50 eV deep that stretches from $x = 0$ to $x = 0.50$ nm. What is the electron's wave-length (a) in free space, (b) when over the well? (c) Draw a diagram showing the potential energy and total energy as a function of x, and on the diagram sketch a possible wave function.

*31. (II) An electron is trapped in a 0.10-nm-wide finite square well of height $U_0 = 1.0$ keV. Estimate at what distance outside the walls of the well the ground state wave function drops to 1 percent of its value at the walls.

*32. (II) Suppose that a particle of mass m is trapped in a finite potential well that has a rigid wall at $x = 0$ ($U = \infty$ for $x < 0$) and a finite wall of height $U = U_0$ at $x = L$, Fig. 39–20. (a) Sketch the wave functions for the lowest three states. (b) What is the form of the wave function in the ground state in the three regions $x < 0$, $0 < x < L$, $x > L$?

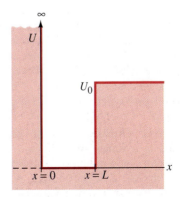

FIGURE 39–20 Problem 32.

33. (II) An electron approaches a potential barrier 10 eV high and 0.50 nm wide. If the electron has a 1.0 percent probability of tunneling through the barrier, what must its energy be?

34. (II) A 1.0-mA current of 1.0-MeV protons strikes a 2.0-MeV-high potential barrier 2.0×10^{-13} m thick. Estimate the transmitted current.

35. (II) For part (b) of Example 39–7, what effect will there be on the transmission coefficient if (a) the barrier height is raised 1 percent, (b) the barrier thickness is increased by 1 percent?

36. (III) Show that the transmission coefficient is given roughly by Eqs. 39–17 for a high or thick barrier, by calculating $|\psi(x = L)|^2 / |\psi(0)|^2$. [Hint: Assume that ψ is a decaying exponential inside the barrier.]

37. (III) A uranium-238 nucleus ($Z = 92$) lasts about 5×10^9 years before it decays by emission of an alpha particle ($Z = 2$, $M = 4M_{proton}$). (a) Assuming that the α particle is a point, and the nucleus is roughly 8 fm in radius, estimate the height of the Coulomb barrier (the peak in Fig. 39–16). (b) The alpha particle, when free, has kinetic energy ≈ 4 MeV. Estimate the width of the barrier. (c) Assuming that the square well has $U = 0$ inside (where $U = 0$ far from the nucleus), calculate the speed of the alpha particle and how often it hits the barrier, and from this (and Eq. 39–17) estimate its lifetime. [Hint: Replace the $1/r$ coulomb barrier with an "averaged" rectangular barrier (as in Fig. 39–14) of width equal to $\frac{1}{3}$ that calculated in (b).]

General Problems

38. If an electron's position can be measured to an accuracy of 2.0×10^{-8} m, how accurately can its velocity be known?

39. Estimate the lowest possible energy of a neutron contained in a typical nucleus of radius 1.0×10^{-15} m. [Hint: A particle can have an energy at least as large as its uncertainty.]

40. The Z^0 boson, discovered in 1985, is the mediator of the weak nuclear force, and it typically decays very quickly. Its average rest energy is 91.19 GeV, but its short lifetime shows up as an intrinsic width of 2.5 GeV. What is the lifetime of this particle?

41. What is the uncertainty in the mass of a muon ($m = 105.7$ MeV/c^2), specified in eV/c^2, given its average lifetime of 2.20 μs?

42. A free neutron ($m = 1.67 \times 10^{-27}$ kg) has a mean life of 900 s. What is the uncertainty in its mass (in kg)?

43. Use the uncertainty principle to estimate the position uncertainty for the electron in the ground state of the hydrogen atom. [Hint: Determine the momentum using the Bohr model of Section 38–10 and assume the momentum can be anywhere between this value and zero.] How does this compare to the Bohr radius?

44. A neutron is trapped in an infinitely deep potential well 2.0 fm in width. Determine (a) the four lowest possible energy states and (b) their wave functions. (c) What is the wavelength and energy of a photon emitted when the neutron makes a transition between the two lowest states? In what region of the EM spectrum does this photon lie? [Note: This is a rough model of an atomic nucleus.]

45. Estimate the kinetic energy and speed of an alpha particle ($q = 2e$, $m = 4$ proton masses) trapped in a nucleus 10^{-14} m wide. Assume an infinitely deep square well potential.

46. *Simple Harmonic Oscillator.* Suppose that an electron is trapped not in a square well, but one whose potential energy is that of a simple harmonic oscillator: $U(x) = \frac{1}{2}Cx^2$. That is, if the electron is displaced from $x = 0$, a restoring force $F = -Cx$ acts on it, where C is constant. (a) Sketch this potential energy. (b) Show that $x = Ae^{-Bx^2}$ is a solution to the Schrödinger equation and that the energy of this state is $E = \frac{1}{2}\hbar\omega$, where $\omega = 2\hbar B/m$, and A and B are constants. [Note: This is the ground state and this energy $\frac{1}{2}\hbar\omega$ is the zero-point energy for a harmonic oscillator. The energies of higher states are $E_n = (n + \frac{1}{2})\hbar\omega$, where n is an integer.]

47. A 10-gram pencil, 20 cm long, is balanced on its point. Classically, this is a configuration of (unstable) equilibrium, so the pencil could remain there forever if it were perfectly placed. A quantum mechanical analysis shows that the pencil must fall. (a) Why is this the case? (b) Estimate (within a factor of 2) how long it will take the pencil to hit the table if it is initially positioned as well as possible? [Hint: Use the uncertainty principle in its angular form to obtain an expression for the initial angle $\phi_0 \approx \Delta\phi$.]

48. The probability density for finding a particle at a particular location is $|\psi|^2$. The average value $\bar{x}$, the location of a particle averaged over many measurements, is given by

$$\bar{x} = \int_0^L x|\psi|^2 \, dx$$

for a particle in a rigid box. Calculate the average value of x for any value of n.

49. By how much does the tunneling current through the tip of an STM change if the tip rises 0.010 nm from some initial height above a sodium surface with a work function $W_0 = 2.28$ eV? [Hint: Let the work function (see Section 38–2) equal the energy needed to raise the electron to the top of the barrier.]

50. A small ball of mass 1.0×10^{-6} kg is dropped on a table from a height of 2.0 m. After each bounce the ball rises to 80 percent of its height before the bounce because of its inelastic collision with the table. Estimate how many bounces occur before the uncertainty principle plays a role in the problem. [Hint: Determine when the uncertainty in the ball's speed is comparable to its speed of impact on the table.]

A neon tube is a thin glass tube, moldable into various shapes, filled with neon (or other) gas that glows with a particular color when a current at high voltage passes through it. Gas atoms, excited to upper energy levels, jump down to lower energy levels and emit light (photons) whose wavelengths (color) are characteristic of the type of gas.

In this chapter we study what quantum mechanics tells us about atoms, their wave functions and energy levels, including the effect of the exclusion principle. We also discuss interesting applications such as lasers and holography.

CHAPTER 40

Quantum Mechanics of Atoms

At the beginning of Chapter 39 we discussed the limitations of the Bohr theory of atomic structure and why a new theory was needed. Although the Bohr theory had great success in predicting the wavelengths of light emitted and absorbed by the hydrogen atom, it could not do so for more complex atoms. Nor did it explain *fine structure*, the splitting of emission lines into two or more closely spaced lines. And, as a theory, it was an uneasy mixture of classical and quantum ideas.

Quantum mechanics came to the rescue in 1925 and 1926, and in this chapter we examine the quantum-mechanical theory of atomic structure, which is far more complete than the old Bohr theory.

40–1 Quantum-Mechanical View of Atoms

Although the Bohr model has been discarded as an accurate description of nature, nonetheless, quantum mechanics reaffirms certain aspects of the older theory, such as that electrons in an atom exist only in discrete states of definite energy, and that a photon of light is emitted (or absorbed) when an electron makes a transition from one state to another. But quantum mechanics is a much deeper theory, and has provided us with a very different view of the atom. According to quantum mechanics, electrons do not exist in well-defined circular orbits as in the Bohr theory. Rather, the electron (because of its wave nature) can be thought of as spread out in space as if it were a "cloud." The size and shape of the electron cloud can be calculated for a given state of an atom. For the ground state in the hydrogen atom,

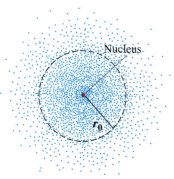

FIGURE 40–1 Electron cloud or "probability distribution" for the ground state of the hydrogen atom, as seen from afar. The circle represents the Bohr radius.

Probability distribution

the solution of the Schrödinger equation, as we will discuss in more detail in Section 40–3, gives

$$\psi(r) = \frac{1}{\sqrt{\pi r_0^3}} \, e^{-\frac{r}{r_0}}.$$

Here $\psi(r)$ is the wave function as a function of position, and it depends only on the radial distance r from the center, and not on angle θ or ϕ. (The constant r_0 has a value that happens to come out equal to the first Bohr radius.) Thus the electron cloud, whose density is $|\psi|^2$, for the ground state of hydrogen is spherically symmetric as shown in Fig. 40–1. The extent of the electron cloud at its higher densities roughly indicates the "size" of an atom, but just as a cloud may not have a distinct border, atoms do not have a precise boundary or a well-defined size. Not all electron clouds have a spherical shape, as we shall see later in this chapter. But note that $\psi(r)$, while becoming extremely small for large r (see the equation above), does not equal zero in any finite region. So quantum mechanics suggests that an atom is not mostly empty space. (Indeed, since $\psi \to 0$ only for $r \to \infty$, we might question the idea that there is any truly empty space in the universe.)

The electron cloud can be interpreted from either the particle or the wave viewpoint. Remember that by a particle we mean something that is localized in space—it has a definite position at any given instant. By contrast, a wave is spread out in space. The electron cloud, spread out in space as in Fig. 40–1, is a result of the wave nature of electrons. Electron clouds can also be interpreted as **probability distributions** for a particle. As we saw in Section 39–3, we cannot predict the path an electron will follow. After one measurement of its position we cannot predict exactly where it will be at a later time. We can only calculate the probability that it will be found at different points. If you were to make 500 different measurements of the position of an electron, considering it as a particle, the majority of the results would show the electron at points where the probability is high (dark area in Fig. 40–1). Only occasionally would the electron be found where the probability is low.

40–2 Hydrogen Atom: Schrödinger Equation and Quantum Numbers

The hydrogen atom is the simplest of all atoms, consisting of a single electron of charge $-e$ moving around a central nucleus (a single proton) of charge $+e$. It is with hydrogen that a study of atomic structure must begin.

The Schrödinger equation (see Eq. 39–5) includes a term containing the potential energy. For the hydrogen (H) atom, the potential energy is due to the Coulomb force between electron and proton:

$$U = -\frac{1}{4\pi\epsilon_0} \frac{e^2}{r}$$

where r is the radial distance from the proton (situated at $r = 0$) to the electron. See Fig. 40–2. The (time-independent) Schrödinger equation, which must now be written in three dimensions, is then

$$-\frac{\hbar^2}{2m}\left(\frac{\partial^2 \psi}{\partial x^2} + \frac{\partial^2 \psi}{\partial y^2} + \frac{\partial^2 \psi}{\partial z^2}\right) - \frac{1}{4\pi\epsilon_0}\frac{e^2}{r}\psi = E\psi, \qquad \textbf{(40–1)}$$

where $\partial^2\psi/\partial x^2$, $\partial^2\psi/\partial y^2$, and $\partial^2\psi/\partial z^2$ are partial derivatives with respect to x, y, and z. To solve the Schrödinger equation for the H atom, it is usual to write it in terms of spherical coordinates (r, θ, ϕ). We will not, however, actually go through the process of solving it. Instead, we look at the properties of the solutions, and (in the next Section) at the wave functions themselves.

FIGURE 40–2 Potential energy $U(r)$ for the hydrogen atom. The radial distance r of the electron from the nucleus is given in terms of the Bohr radius r_0.

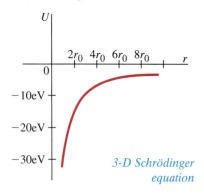

3-D Schrödinger equation

Recall from Chapter 39 that the solutions of the Schrödinger equation in one dimension for the infinite square well were characterized by a single quantum number, which we called n, which arises from applying the boundary conditions. In the three-dimensional problem of the H atom, the solutions of the Schrödinger equation are characterized by three quantum numbers corresponding to boundary conditions applied in the three dimensions. However, four different quantum numbers are actually needed to specify each state in the H atom, the fourth coming from a relativistic treatment. We now discuss each of these quantum numbers. Much of our analysis here will also apply to more complex atoms, which we discuss starting in Section 40–4.

Quantum mechanics predicts the same energy levels (Fig. 38–25) for the H atom as does the Bohr theory. That is,

$$E_n = -\frac{13.6 \text{ eV}}{n^2} \qquad n = 1, 2, 3, \ldots , \qquad \textbf{(40–2)}$$

where n is an integer called the **principal quantum number**. This is the same as the quantum number n that appeared in Bohr theory. It can have any integer value from 1 to ∞. The total energy of a state in the H atom depends on n, Eq. 40–2.

The **orbital quantum number**, l, is related to the magnitude of the orbital angular momentum of the electron; l can take on integer values from 0 to $(n - 1)$. For the ground state, $n = 1$, l can only be zero.[†] But for $n = 3$, say, l can be 0, 1, or 2. The actual magnitude of the orbital angular momentum L is related to the quantum number l by

$$L = \sqrt{l(l + 1)}\,\hbar. \qquad \textbf{(40–3)}$$

The value of l has almost no effect on the total energy in the hydrogen atom; only n does to any appreciable extent.[‡] But in atoms with two or more electrons, the energy does depend on l as well as n.

The **magnetic quantum number**, m_l, is related to the direction of the electron's orbital angular momentum, and it can take on integer values ranging from $-l$ to $+l$. For example, if $l = 2$, then m_l can be $-2, -1, 0, +1,$ or $+2$. Since angular momentum is a vector, it is not surprising that both its magnitude and its direction would be quantized. For $l = 2$, the five different directions allowed can be represented by the diagram of Fig. 40–3. This limitation on the direction of **L** is often called **space quantization**. In quantum mechanics, the direction of the angular momentum is usually specified by giving its component along the z axis (this choice is arbitrary). Then L_z is related to m_l by the equation

$$L_z = m_l \hbar. \qquad \textbf{(40–4)}$$

The values of L_x and L_y are not definite, however (see Problem 65). The name for m_l derives not from theory (which relates it to L_z) but from experiment. It was found that when a gas discharge tube (Fig. 38–18) was placed in a magnetic field, the spectral lines were split into several very closely spaced lines. This splitting, known as the **Zeeman effect**, implies that the energy levels must be split (Fig. 40–4), and thus that the energy of a state depends not only on n but also on m_l when a magnetic field is applied—hence the name "magnetic quantum number." (Why the energy should depend on the direction of **L** can be seen from a semiclassical view of a moving electron as an electric current which interacts with the magnetic field—see Chapter 27, Section 27–5, and Section 40–7.)

Finally, there is the **spin quantum number**, m_s, which for an electron can have only two values, $m_s = +\frac{1}{2}$ and $m_s = -\frac{1}{2}$. The existence of this quantum number did not come out of Schrödinger's original theory, as did n, l, and m_l. Instead, a subsequent modification by P. A. M. Dirac (1902–1984) explained its presence as a relativistic effect. The first hint that m_s was needed, however, came from experiment. A careful study of the spectral lines of hydrogen showed that each actually consisted of two (or more) very closely spaced lines even in the absence of an external

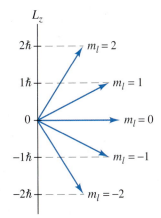

FIGURE 40–3 Quantization of angular momentum direction for $l = 2$.

FIGURE 40–4 When a magnetic field is applied, an $n = 3$, $l = 2$ energy level is split into five separate levels, corresponding to the five values of m_l (2, 1, 0, −1, −2). An $n = 2$, $l = 1$ level is split into three levels ($m_l = 1, 0, -1$). Transitions can occur between levels (not all transitions are shown), with photons of several slightly different frequencies being given off (the Zeeman effect).

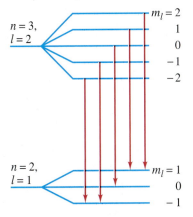

[†] Contrast this with the Bohr theory, which assigned $l = 1$ to the ground state (Eq. 38–10).

[‡] See discussion of *fine structure* starting at the bottom of this page.

magnetic field. It was at first hypothesized that this tiny splitting of energy levels, called **fine structure**, was due to angular momentum associated with a spinning of the electron. That is, the electron might spin on its axis as well as orbit the nucleus, just as the Earth spins on its axis as it orbits the Sun. The interaction between the tiny current of the spinning electron could then interact with the magnetic field due to the orbiting charge and cause the small observed splitting of energy levels. (So the energy depends slightly on m_l and m_s.) Today we consider this picture of a spinning electron as not legitimate. We cannot even view an electron as a localized object, much less a spinning one. What is important is that the electron can have two different states due to some intrinsic property that behaves as an angular momentum, and we still call this property "spin." The electron is said to have a spin quantum number $s = \frac{1}{2}$, which produces a spin angular momentum S given by

Spin angular momentum

$$ S = \sqrt{s(s + 1)}\,\hbar = \frac{\sqrt{3}}{2}\hbar. $$

(Compare Eq. 40–3.) This spin can have two different directions, $m_s = +\frac{1}{2}$ or $m_s = -\frac{1}{2}$, which are often said to be "spin up" and "spin down" (see Fig. 40–5). A state with spin down $\left(m_s = -\frac{1}{2}\right)$ has slightly lower energy than one with spin up. (Note that we include m_s, but not s, in our list of quantum numbers since s is the same for all electrons.)

The possible values of the four quantum numbers for an electron in the hydrogen atom are summarized in Table 40–1.

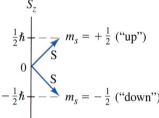

FIGURE 40–5 The spin angular momentum S can take on only two directions, $m_s = +\frac{1}{2}$ or $-\frac{1}{2}$, called "spin up" and "spin down."

TABLE 40–1   Quantum Numbers for an Electron

Name	Symbol	Possible Values
Principal	n	$1, 2, 3, \ldots, \infty$.
Orbital	l	For a given n: l can be $0, 1, 2, \ldots, n - 1$.
Magnetic	m_l	For given n and l: m_l can be $l, l - 1, \ldots, 0, \ldots, -l$.
Spin	m_s	For each set of n, l, and m_l: m_s can be $+\frac{1}{2}$ or $-\frac{1}{2}$.

CONCEPTUAL EXAMPLE 40–1   Possible states for $n = 3$. How many different states are possible for an electron whose principal quantum number is $n = 3$?

SOLUTION   For $n = 3$, l can have the values $l = 2, 1, 0$. For $l = 2$, m_l can be $2, 1, 0, -1, -2$, which is five different possibilities. For each of these, m_s can be either up or down ($+\frac{1}{2}$ or $-\frac{1}{2}$), so for $l = 2$ there are $2 \times 5 = 10$ states. For $l = 1$, m_l can be $1, 0, -1$, and since m_s can be $+\frac{1}{2}$ or $-\frac{1}{2}$ for each of these, we have 6 more possible states. Finally, for $l = 0$, m_l can only be 0, and there are only 2 states corresponding to $m_s = +\frac{1}{2}$ and $-\frac{1}{2}$. The total number of states is $10 + 6 + 2 = 18$, as detailed in the following table:

n	l	m_l	m_s	n	l	m_l	m_s
3	2	2	$\frac{1}{2}$	3	2	-2	$-\frac{1}{2}$
3	2	2	$-\frac{1}{2}$	3	1	1	$\frac{1}{2}$
3	2	1	$\frac{1}{2}$	3	1	1	$-\frac{1}{2}$
3	2	1	$-\frac{1}{2}$	3	1	0	$\frac{1}{2}$
3	2	0	$\frac{1}{2}$	3	1	0	$-\frac{1}{2}$
3	2	0	$-\frac{1}{2}$	3	1	-1	$\frac{1}{2}$
3	2	-1	$\frac{1}{2}$	3	1	-1	$-\frac{1}{2}$
3	2	-1	$-\frac{1}{2}$	3	0	0	$\frac{1}{2}$
3	2	-2	$\frac{1}{2}$	3	0	0	$-\frac{1}{2}$

EXAMPLE 40–2 **E and L for n = 3.** Determine (a) the energy and (b) the orbital angular momentum for each of the states in Example 40–1.

SOLUTION (a) The energy of a state depends only on n, except for the very small corrections mentioned above, which we will ignore. Since $n = 3$ for all these states, they all have the same energy,

$$E_3 = -\frac{13.6\,\text{eV}}{(3)^2} = -1.51\,\text{eV}.$$

(b) For $l = 0$,

$$L = \sqrt{l(l + 1)}\,\hbar = \sqrt{0(0 + 1)}\,\hbar = 0.$$

For $l = 1$,

$$L = \sqrt{1(1 + 1)}\,\hbar = \sqrt{2}\,\hbar$$
$$= 1.49 \times 10^{-34}\,\text{J} \cdot \text{s}.$$

Atomic angular momenta are generally given as a multiple of $\hbar$ ($\sqrt{2}\,\hbar$ in this case), rather than in SI units. But note that for $l = 1$, L is on the order of 10^{-34} J·s. This means that macroscopic angular momenta will have such extremely high quantum numbers that the quantization of angular momentum will not be detectable: L will appear continuous, in accordance with the correspondence principle. Finally, for $l = 2$,

$$L = \sqrt{2(2 + 1)}\,\hbar = \sqrt{6}\,\hbar.$$

Another prediction of quantum mechanics is that when a photon is emitted or absorbed, transitions can occur only between states with values of l that differ by one unit:

$$\Delta l = \pm 1.$$

Selection rule

According to this **selection rule**, an electron in an $l = 2$ state can jump only to a state with $l = 1$ or $l = 3$. It cannot jump to a state with $l = 2$ or $l = 0$. A transition such as $l = 2$ to $l = 0$ is called a **forbidden transition**. Actually, such a transition is not absolutely forbidden and can occur, but only with very low probability compared to **allowed transitions**—those that satisfy the selection rule $\Delta l = \pm 1$. Since the orbital angular momentum of an H atom must change by one unit when it emits a photon, conservation of angular momentum tells us that the photon must carry off angular momentum. Indeed, experimental evidence of many sorts shows that the photon can be assigned a spin quantum number of 1.

"forbidden" transition

40–3 | Hydrogen Atom Wave Functions

The solution of the Schrödinger equation for the ground state of hydrogen—the state with lowest energy E—has an energy $E_1 = -13.6$ eV, as we have seen. The wave function for the ground state depends only on r and so is spherically symmetric. As already mentioned in Section 40–1, its form is

$$\psi_{100} = \frac{1}{\sqrt{\pi r_0^3}}\, e^{-\frac{r}{r_0}} \tag{40–5a}$$

Ground state in hydrogen

where $r_0 = h^2\epsilon_0/\pi m e^2 = 0.0529$ nm is the Bohr radius (Section 38–10). The subscript 100 on ψ represents the quantum numbers n, l, m_l:

$$\psi_{nlm_l}.$$

For the ground state, $n = 1$, $l = 0$, $m_l = 0$, and there is only one wave function that serves for both $m_s = +\frac{1}{2}$ and $m_s = -\frac{1}{2}$ (the value of m_s does not affect the

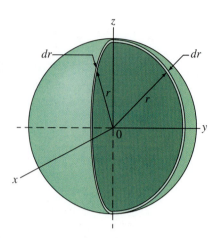

FIGURE 40–6 A spherical shell of thickness dr, inner radius r, and outer radius $r + dr$. Its volume is $dV = 4\pi r^2\, dr$.

spatial dependence of the wave function for any state, since spin is an *internal* or *intrinsic* property of the electron). The probability density for the ground state is

$$|\psi_{100}|^2 = \frac{1}{\pi r_0^3}\, e^{-\frac{2r}{r_0}} \tag{40–5b}$$

which falls off exponentially with r. Note that ψ_{100}, as well as all other wave functions we discuss, has been normalized:

$$\int_{\text{all space}} |\psi_{100}|^2\, dV = 1.$$

The quantity $|\psi|^2\, dV$ gives the probability of finding the electron in a volume dV about a given point. It is often more useful to specify the **radial probability distribution**, P_r, which is defined so that $P_r\, dr$ is the probability of finding the electron at a radial distance between r and $r + dr$ from the nucleus. That is, $P_r\, dr$ specifies the probability of finding the electron within a thin shell of thickness dr of inner radius r and outer radius $r + dr$, regardless of direction (see Fig. 40–6). The volume of this shell is the product of its surface area, $4\pi r^2$, and its thickness, dr:

$$dV = 4\pi r^2\, dr.$$

Hence

$$|\psi|^2\, dV = |\psi|^2 4\pi r^2\, dr$$

and the radial probability distribution is

$$P_r = 4\pi r^2 |\psi|^2. \tag{40–6}$$

Hydrogen ground state radial probability distribution

For the ground state of hydrogen, P_r becomes

$$P_r = 4\frac{r^2}{r_0^3}\, e^{-\frac{2r}{r_0}} \tag{40–7}$$

and is plotted in Fig. 40–7. The peak of the curve is the "most probable" value of r and occurs for $r = r_0$, the Bohr radius, which we now prove.

FIGURE 40–7 The radial probability distribution P_r for the ground state of hydrogen, $n = 1$, $l = 0$. The peak occurs at $r = r_0$, the Bohr radius.

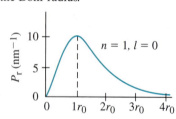

EXAMPLE 40–3 Most probable electron radius in hydrogen. Determine the most probable distance r from the nucleus at which to find the electron in the ground state of hydrogen.

SOLUTION The peak of the curve in Fig. 40–7 corresponds to the most probable value of r. At this point the curve has zero slope, so we take the derivative of Eq. 40–7 and set it equal to zero:

$$\frac{d}{dr}\left(4\frac{r^2}{r_0^3}\, e^{-\frac{2r}{r_0}}\right) = 0$$

$$\left(8\frac{r}{r_0^3} - \frac{8r^2}{r_0^4}\right)e^{-\frac{2r}{r_0}} = 0.$$

Since $e^{-\frac{2r}{r_0}}$ goes to zero only at $r = \infty$, it is the term in parentheses that must be zero:

$$8\frac{r}{r_0^3} - 8\frac{r^2}{r_0^4} = 0.$$

Therefore,

$$\frac{r}{r_0^3} = \frac{r^2}{r_0^4}$$

or

$$r = r_0.$$

The most probable radial distance of the electron from the nucleus according to quantum mechanics is at the Bohr radius, an interesting coincidence.

EXAMPLE 40–4 Calculating probability. Determine the probability of finding the electron in the ground state of hydrogen within two Bohr radii of the nucleus.

SOLUTION We need to integrate P_r from $r = 0$ out to $r = 2r_0$: that is, we want to find

$$P_r = \int_{r=0}^{2r_0} |\psi|^2 \, dV = \int_0^{2r_0} 4 \frac{r^2}{r_0^3} e^{-\frac{2r}{r_0}} \, dr.$$

We first make the substitution

$$x = 2\frac{r}{r_0}$$

and then integrate by parts $\left(\int u \, dv = uv - \int v \, du \right)$ letting $u = x^2$ and $dv = e^{-x} \, dx$ (and note that $dx = 2dr/r_0$):

$$P_r = \tfrac{1}{2} \int_{x=0}^4 x^2 e^{-x} \, dx = \tfrac{1}{2} \left[-x^2 e^{-x} + \int 2x e^{-x} \, dx \right] \Bigg|_0^4.$$

The second term we also integrate by parts with $u = 2x$ and $dv = e^{-x} \, dx$:

$$P_r = \tfrac{1}{2} \left[-x^2 e^{-x} - 2x e^{-x} + 2 \int e^{-x} \, dx \right] \Bigg|_0^4$$

$$= \left(-\tfrac{1}{2} x^2 - x - 1 \right) e^{-x} \Bigg|_0^4.$$

We evaluate this at $x = 0$ and at $x = 2(2r_0)/r_0 = 4$:

$$P_r = (-8 - 4 - 1) e^{-4} + e^0 = 0.76$$

or 76 percent. Thus the electron would be found 76 percent of the time within 2 Bohr radii of the nucleus and 24 percent of the time farther away. Note that this result depends on our wave function being properly normalized, which it is, as is readily shown by letting $r \to \infty$ and integrating over all space: $\int_0^\infty |\psi|^2 \, dV = 1$.

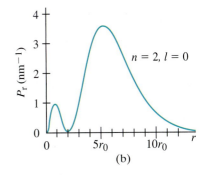

$|\psi_{200}|^2$

(a)

The first excited state in hydrogen has $n = 2$. For $l = 0$, the solution of the Schrödinger equation (Eq. 40–1) is a wave function that is again spherically symmetric:

$$\psi_{200} = \frac{1}{\sqrt{32\pi r_0^3}} \left(2 - \frac{r}{r_0} \right) e^{-\frac{r}{2r_0}}. \qquad (40\text{–}8)$$

Figure 40–8a shows the probability distribution $|\psi_{200}|^2$ and Fig. 40–8b shows[†] a plot of the radial probability distribution

$$P_r = \frac{1}{8} \frac{r^2}{r_0^3} \left(2 - \frac{r}{r_0} \right)^2 e^{-\frac{r}{r_0}}.$$

There are two peaks in this curve; the second, at $r \approx 5r_0$, is higher and corresponds to the most probable value for r in the $n = 2$, $l = 0$ state. We see that the

FIGURE 40–8 (a) The probability distribution $|\psi_{200}|^2$ and (b) the radial probability distribution P_r for the $n = 2$, $l = 0$ state in hydrogen.

[†] Just as for a particle in a deep square well potential (see Figs. 39–9 and 39–10), the higher the energy, the more nodes there are in ψ and $|\psi|^2$ also for the H atom.

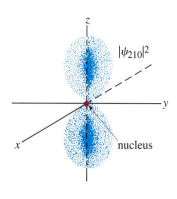

$|\psi_{210}|^2$

nucleus

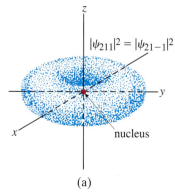

$|\psi_{211}|^2 = |\psi_{21-1}|^2$

nucleus

(a)

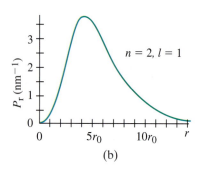

$n = 2, l = 1$

(b)

FIGURE 40–9 (a) The probability distribution for the three states with $n = 2$, $l = 1$. (b) Radial probability distribution for the sum of the three states with $n = 2$, $l = 1$, and $m_l = +1, 0,$ or -1.

electron tends to be somewhat farther from the nucleus in the $n = 2$, $l = 0$ state than in the $n = 1$, $l = 0$ state.

For the state with $n = 2$, $l = 1$, there are three possible wave functions, corresponding to $m_l = +1$, 0, or -1:

$$\psi_{210} = \frac{z}{\sqrt{32\pi r_0^5}} e^{-\frac{r}{2r_0}}$$

$$\psi_{211} = \frac{x + iy}{\sqrt{64\pi r_0^5}} e^{-\frac{r}{2r_0}} \tag{40–9}$$

$$\psi_{21-1} = \frac{x - iy}{\sqrt{64\pi r_0^5}} e^{-\frac{r}{2r_0}}.$$

where i is the imaginary number $i = \sqrt{-1}$. These wave functions are *not* spherically symmetric. The probability distributions, $|\psi|^2$, are shown in Fig. 40–9a, where we can see their directional orientation.

You may wonder how such non-spherically symmetric wave functions arise when the potential energy in the Schrödinger equation has spherical symmetry. Indeed, how could an electron select one of these states? In the absence of any external influence, such as a magnetic field in a particular direction, all three of these states are equally likely, and they all have the same energy. Thus an electron can be considered to spend one-third of its time in each of these states. The net effect, then, is the sum of these three wave functions squared:

$$\left|\psi_{210}\right|^2 + \left|\psi_{211}\right|^2 + \left|\psi_{21-1}\right|^2,$$

which is spherically symmetric, since $x^2 + y^2 + z^2 = r^2$. The radial probability distribution for this sum is shown in Fig. 40–9b.

Although the spatial distributions of the electron can be calculated for the various states, it is difficult to measure them experimentally. Indeed, most of the experimental information about atoms has come from a careful examination of the emission spectra under various conditions.

40–4 Complex Atoms; the Exclusion Principle

We have discussed the hydrogen atom in detail because it is the simplest to deal with. Now we briefly discuss more complex atoms, those that contain more than one electron, and whose energy levels can be determined experimentally from an analysis of the emission spectra. The energy levels are *not* the same as in the H atom, since the electrons interact with each other as well as with the nucleus. For atoms with more than one electron, the energy levels depend on both n and l.

Atomic number

The number of electrons in a neutral atom is called its **atomic number**, Z; Z is also the number of positive charges (protons) in the nucleus, and determines what kind of atom it is. That is, Z determines most of the properties that distinguish one atom from another.

Although modifications of the Bohr theory had been attempted in order to deal with complex atoms, the development of quantum mechanics in the years after 1925 proved far more successful. The mathematics becomes very difficult, however, since in multi-electron atoms, each electron is not only attracted to the nucleus but is repelled by the other electrons.

The simplest approach has been to treat each electron in an atom as occupying a particular state characterized by the quantum numbers n, l, m_l, and m_s. But to understand the possible arrangements of electrons in an atom, a new principle was needed. It was introduced by Wolfgang Pauli (1900–1958; Fig. 39–2) and is called the **Pauli exclusion principle**. It states:

No two electrons in an atom can occupy the same quantum state.

Pauli exclusion principle

Thus, no two electrons in an atom can have exactly the same set of the quantum numbers n, l, m_l, and m_s. The Pauli exclusion principle[†] forms the basis not only for understanding complex atoms, but also for understanding molecules and bonding, and other phenomena as well.

Let us now look at the structure of some of the simpler atoms when they are in the ground state. After hydrogen, the next simplest atom is *helium*, with two electrons. Both electrons can have $n = 1$, since one can have spin up $\left(m_s = +\tfrac{1}{2}\right)$ and the other spin down $\left(m_s = -\tfrac{1}{2}\right)$, thus satisfying the exclusion principle. Of course, since $n = 1$, l and m_l must be zero (Table 40–1). Thus the two electrons have the quantum numbers indicated in the table in the margin.

Lithium has three electrons, two of which can have $n = 1$. But the third cannot have $n = 1$ without violating the exclusion principle. Hence the third electron must have $n = 2$. Since it happens that the $n = 2$, $l = 0$ level has a lower energy than $n = 2$, $l = 1$, the electrons in the ground state have the quantum numbers indicated in the table in the margin. Of course, the quantum numbers of the third electron could also be, say, $\left(3, 1, -1, \tfrac{1}{2}\right)$. But the atom in this case would be in an excited state since it would have greater energy. It would not be long before it jumped to the ground state with the emission of a photon. At room temperature, unless extra energy is supplied (as in a discharge tube), the vast majority of atoms are in the ground state because the average thermal energy $\left(K = \tfrac{3}{2}kT \approx 0.04 \text{ eV at } T = 300 \text{ K}\right)$ is much less than the energy needed to excite atoms (1 to 10 eV).

We can continue in this way to describe the quantum numbers of each electron in the ground state of larger and larger atoms. That for sodium, with its eleven electrons, is shown in the table in the margin.

Figure 40–10 shows a simple energy level diagram where occupied states are shown as up or down arrows $\left(m_s = +\tfrac{1}{2} \text{ or } -\tfrac{1}{2}\right)$, and possible empty states are shown as small circles.

The ground-state configuration for all atoms is given in the **periodic table**, which is displayed inside the back cover of this book, and discussed in the next Section.

[†] The exclusion principle applies to identical particles whose spin quantum number is a half-integer $\left(\tfrac{1}{2}, \tfrac{3}{2}, \text{ and so on}\right)$, including electrons, protons, and neutrons; such particles are called **fermions** (after E. Fermi who derived a statistical theory describing them—see Section 41–6.). The exclusion principle does not apply to particles with integer spin quantum number (0, 1, 2, and so on), such as the photon and π meson, all of which are referred to as **bosons** (after S. N. Bose, who derived a statistical theory for them).

Helium, $Z = 2$

n	l	m_l	m_s
1	0	0	$-\tfrac{1}{2}$
1	0	0	$\tfrac{1}{2}$

Lithium, $Z = 3$

n	l	m_l	m_s
1	0	0	$-\tfrac{1}{2}$
1	0	0	$\tfrac{1}{2}$
2	0	0	$-\tfrac{1}{2}$

Sodium, $Z = 11$

n	l	m_l	m_s
1	0	0	$-\tfrac{1}{2}$
1	0	0	$\tfrac{1}{2}$
2	0	0	$-\tfrac{1}{2}$
2	0	0	$\tfrac{1}{2}$
2	1	1	$-\tfrac{1}{2}$
2	1	1	$\tfrac{1}{2}$
2	1	0	$-\tfrac{1}{2}$
2	1	0	$\tfrac{1}{2}$
2	1	-1	$-\tfrac{1}{2}$
2	1	-1	$\tfrac{1}{2}$
3	0	0	$-\tfrac{1}{2}$

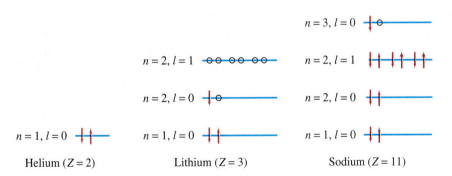

FIGURE 40–10 Energy level diagram showing occupied states (arrows) and unoccupied states (○) for He, Li, and Na. Note that we have shown the $n = 2$, $l = 1$ level of Li even though it is empty.

TABLE 40–2 Values of *l*

Value of *l*	Letter Symbol	Maximum Number of Electrons in Subshell
0	s	2
1	p	6
2	d	10
3	f	14
4	g	18
5	h	22
⋮	⋮	⋮

TABLE 40–3 Electron Configuration of Some Elements

Z (No. of Electrons)	Element[†]	Ground State Configuration (outer electrons)
1	H	$1s^1$
2	He	$1s^2$
3	Li	$2s^1$
4	Be	$2s^2$
5	B	$2s^2 2p^1$
6	C	$2s^2 2p^2$
7	N	$2s^2 2p^3$
8	O	$2s^2 2p^4$
9	F	$2s^2 2p^5$
10	Ne	$2s^2 2p^6$
11	Na	$3s^1$
12	Mg	$3s^2$
13	Al	$3s^2 3p^1$
14	Si	$3s^2 3p^2$
15	P	$3s^2 3p^3$
16	S	$3s^2 3p^4$
17	Cl	$3s^2 3p^5$
18	Ar	$3s^2 3p^6$
19	K	$4s^1$
20	Ca	$4s^2$
21	Sc	$3d^1 4s^2$
22	Ti	$3d^2 4s^2$
23	V	$3d^3 4s^2$
24	Cr	$3d^5 4s^1$
25	Mn	$3d^5 4s^2$
26	Fe	$3d^6 4s^2$

[†]Names of elements can be found in Appendix D.

40–5 The Periodic Table of Elements

More than a century ago, Dmitri Mendeleev (1834–1907) arranged the then known elements into what we now call the **periodic table** of the elements. The atoms were arranged according to increasing mass, but also so that elements with similar chemical properties would fall in the same column. Today's version is shown inside the back cover. Each square contains the atomic number Z, the symbol for the element, and the atomic mass (in atomic mass units). Finally, in the lower left corner the configuration of the ground state of the atom is given. This requires some explanation. Electrons with the same value of n are referred to as being in the same **shell**. Electrons with $n = 1$ are in one shell (the K shell), those with $n = 2$ are in a second shell (the L shell), those with $n = 3$ are in the third (M) shell, and so on. Electrons with the same values of n and l are referred to as being in the same **subshell**. Letters are often used to specify the value of l as shown in Table 40–2. That is, $l = 0$ is the s subshell; $l = 1$ is the p subshell; $l = 2$ is the d subshell; beginning with $l = 3$, the letters follow the alphabet, f, g, h, i, and so on. (The first letters s, p, d, and f were originally abbreviations of "sharp," "principal," "diffuse," and "fundamental," experimental terms referring to the spectra.)

The Pauli exclusion principle limits the number of electrons possible in each shell and subshell. For any value of l, there are $2l + 1$ different m_l values (m_l can be any integer from 1 to l, from -1 to $-l$, or zero), and two different m_s values. There can be, therefore, at most $2(2l + 1)$ electrons in any l subshell. For example, for $l = 2$, five m_l values are possible $(2, 1, 0, -1, -2)$, and for each of these, m_s can be $+\frac{1}{2}$ or $-\frac{1}{2}$ for a total of $2(5) = 10$ states. Table 40–2 lists the maximum number of electrons that can occupy each subshell.

Since the energy levels depend almost entirely on the values of n and l, it is customary to specify the electron configuration simply by giving the n value and the appropriate letter for l, with the number of electrons in each subshell given as a superscript. The ground-state configuration of sodium, for example, is written as $1s^2 2s^2 2p^6 3s^1$. This is simplified in the periodic table by specifying the configuration only of the outermost electrons and any other nonfilled subshells (see Table 40–3 here, and the periodic table inside the back cover).

CONCEPTUAL EXAMPLE 40–5 Electron configurations. Which of the following electron configurations are possible, and which forbidden? (*a*) $1s^2 2s^2 2p^6 3s^3$; (*b*) $1s^2 2s^2 2p^6 3s^2 3p^5 4s^2$; (*c*) $1s^2 2s^2 2p^6 2d^1$.

RESPONSE (*a*) This is not allowed, because there are too many electrons in the s subshell in the M ($n = 3$) shell. The s subshell has $m_l = 0$, with two slots only: for "spin up" and "spin down" electrons. (*b*) This is allowed, but it is an excited state. One of the electrons in the 3p subshell has jumped up to the 4s subshell. Since there are 19 electrons, the element is potassium. (*c*) This is not allowed, because there is no d ($l = 2$) subshell in the $n = 2$ shell (Table 40–1). The outermost electron will have to be (at least) in the $n = 3$ shell.

The grouping of atoms in the periodic table is according to increasing atomic number, Z. There is also a strong regularity according to chemical properties. And although this is treated in chemistry textbooks, we discuss it here briefly because it is a result of quantum mechanics. See the periodic table on the inside back cover.

All the noble gases (in the last column of the periodic table) have completely filled shells or subshells. That is, their outermost subshell is completely full, and the electron distribution is spherically symmetric. With such full spherical symmetry, other electrons are not attracted nor are electrons readily lost (ionization energy is high). This is why the noble gases are nonreactive (more on this when we discuss molecules and bonding in Chapter 41). Column seven contains the **halogens**, which lack one electron from a filled shell. Because of the shapes of the orbits (see Section 41–1), an additional electron can be accepted from another atom, and hence these elements are quite reactive. They have a valence of −1, meaning that when an extra electron is acquired, the resulting ion has a net charge of −1e. At the left of the periodic table, column I contains the **alkali metals**, all of which have a single outer s electron. This electron spends most of its time outside the inner closed shells and subshells which shield it from most of the nuclear charge. Indeed, it is relatively far from the nucleus and is attracted to it by a net charge of only about +1e, because of the shielding effect of the other electrons. Hence this outer electron is easily removed and can spend much of its time around another atom, forming a molecule. This is why the alkali metals are highly reactive and have a valence of +1. The other columns of the periodic table can be treated similarly.

The presence of the transition elements in the center of the table, as well as the lanthanides (rare earths) and actinides below, is a result of incomplete inner shells. For the lowest Z elements, the subshells are filled in a simple order: first 1s, then 2s, followed by 2p, 3s, and 3p. You might expect that 3d ($n = 3, l = 2$) would be filled next, but it isn't. Instead, the 4s level actually has a slightly lower energy than the 3d (due to electrons interacting with each other), so it fills first (K and Ca). Only then does the 3d shell start to fill up, beginning with Sc. (The 4s and 3d levels are close, so some elements have only one 4s electron, such as Cr.) Most of the chemical properties of these **transition elements** are governed by the relatively loosely held 4s electrons and hence they usually have valences of +1 or +2. See also Table 40–3. A similar effect is responsible for the rare earths, which are shown at the bottom of the periodic table for convenience. All have very similar chemical properties, which are determined by their two outer 6s or 7s electrons, whereas the different numbers of electrons in the unfilled inner shells have little effect.

40–6 X-Ray Spectra and Atomic Number

The line spectra of atoms in the visible, UV, and IR regions of the EM spectrum are mainly due to transitions between states of the outer electrons. Much of the charge of the nucleus is shielded from these electrons by the negative charge on the inner electrons. But the innermost electrons in the $n = 1$ shell "see" the full charge of the nucleus. Since the energy of a level is proportional to Z^2 (see Eq. 38–13), for an atom with $Z = 50$, we would expect wavelengths about $50^2 = 2500$ times shorter than those found in the Lyman series of hydrogen (around 100 nm), or 10^{-2} to 10^{-1} nm. Such short wavelengths lie in the X-ray region of the spectrum.

X-rays are produced when electrons accelerated by a high voltage strike the metal target inside the X-ray tube (Section 36–10). If we look at the spectrum of wavelengths emitted by an X-ray tube, we see that the spectrum consists of two parts: a continuous spectrum with a cutoff at some λ_0 which depends only on the voltage across the tube, and a series of peaks superimposed. A typical example is shown in Fig. 40–11. The smooth curve and the cutoff wavelength λ_0 move to the left as the voltage across the tube increases. The peaks (labeled K_α and K_β in Fig. 40–11), however, remain at the same wavelength when the voltage is changed, although they are located at different wavelengths when different target materials are used. This observation suggests that the peaks are characteristic of the material used. Indeed, we can explain them by imagining that the electrons accelerated by the high voltage of the tube can reach sufficient energies that when they collide with the atoms of the target, they can knock out one of the very tightly held inner electrons. Then we explain these **characteristic X-rays** (the peaks in Fig. 40–11) as photons emitted when an electron in an upper state drops down to fill the vacated lower state. The K lines result from transitions *into* the K shell ($n = 1$). The K_α line is a transition that originates from the $n = 2$ (L) shell, the K_β line from the $n = 3$ (M) shell. An L line is due to a transition into the L shell, and so on.

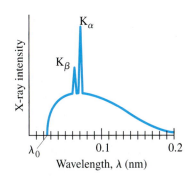

FIGURE 40–11 Spectrum of X-rays emitted from a molybdenum target in an X-ray tube operated at 50 kV.

Characteristic X-rays

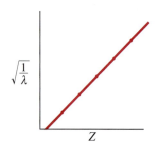

FIGURE 40–12 Plot of $\sqrt{1/\lambda}$ vs. Z for K_α X-ray lines.

Measurement of the characteristic X-ray spectra has allowed a determination of the inner energy levels of atoms. It has also allowed the determination of Z values for many atoms, since (as we have seen) the wavelength of the shortest X-rays emitted will be inversely proportional to Z^2. Actually, for an electron jumping from, say, the $n = 2$ to the $n = 1$ level, the wavelength is inversely proportional to $(Z - 1)^2$ because the nucleus is shielded by the one electron that still remains in the 1s level. In 1914, H. G. J. Moseley (1887–1915) found that a plot of $\sqrt{1/\lambda}$ versus Z produced a straight line, Fig. 40–12. The Z values of a number of elements were determined by fitting them to such a **Moseley plot**. The work of Moseley put the concept of atomic number on a firm experimental basis.

EXAMPLE 40–6 **X-ray wavelength.** Estimate the wavelength for an $n = 2$ to $n = 1$ transition in molybdenum ($Z = 42$). What is the energy of such a photon?

SOLUTION We use the Bohr formula, Eq. 38–14, with Z^2 replaced by $(Z - 1)^2 = (41)^2$. Or, more simply, we can use the result of Example 38–12 for the $n = 2$ to $n = 1$ transition in hydrogen ($Z = 1$). Since

$$\lambda \propto \frac{1}{(Z - 1)^2},$$

we will have

$$\lambda = \frac{(1.22 \times 10^{-7}\,\text{m})}{(41)^2} = 0.073\ \text{nm}.$$

This is close to the measured value (Fig. 40–11) of 0.071 nm. Each of these photons would have energy (in eV) of:

$$E = hf = \frac{hc}{\lambda} = \frac{(6.63 \times 10^{-34}\,\text{J}\cdot\text{s})(3.00 \times 10^8\,\text{m/s})}{(7.3 \times 10^{-11}\,\text{m})(1.60 \times 10^{-19}\,\text{J/eV})} = 17\ \text{keV}.$$

EXAMPLE 40–7 **Determining atomic number.** High-energy photons are used to bombard an unknown material. The strongest peak is found for X-rays emitted with an energy of 66 keV. Guess what the material is.

SOLUTION The strongest X-rays are generally for the K_α line (see Fig. 40–11) which occurs when photons knock out K shell electrons (the innermost orbit) and their place is taken by electrons from the L shell. We use the Bohr model, and assume the electrons "see" a nuclear charge of $Z - 1$ (screened by one electron). The hydrogen transition $n = 2$ to $n = 1$ would yield about 10.2 eV (see Fig. 38–25 or Example 38–12). Then since energy E is proportional to Z^2 (Eq. 38–13), or rather $(Z - 1)^2$ as we've just discussed, we can write

$$\frac{(Z - 1)^2}{1^2} = \frac{66 \times 10^3\ \text{eV}}{10.2\ \text{eV}} = 6.5 \times 10^3,$$

so $Z - 1 = \sqrt{6500} = 81$, and $Z = 82$, which makes it lead.

FIGURE 40–13 Bremsstrahlung photon produced by an electron decelerated by interaction with a target atom.

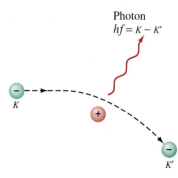

Now we briefly analyze the continuous part of an X-ray spectrum (Fig. 40–11) based on the photon theory of light. When electrons strike the target, they collide with atoms of the material and give up most of their energy as heat (about 99 percent, so X-ray tubes must be cooled). However, electrons can also give up energy by emitting a photon of light. An electron can be decelerated by interaction with atoms of the target (Fig. 40–13). But an accelerating charge can emit radiation (Chapter 32), and in this case it is called **bremsstrahlung** (German for "braking radiation"). Because energy is conserved, the energy of the emitted photon, hf, must equal the loss of kinetic energy K of the electron, $\Delta K = K - K'$, so

$$hf = \Delta K.$$

An electron may lose all or a part of its energy in such a collision. The continuous X-ray spectrum (Fig. 40–11) is explained as being due to such bremsstrahlung collisions in which varying amounts of energy are lost by the electrons. The shortest-wavelength X-ray (the highest frequency) must be due to an electron that gives up *all* its kinetic energy to produce one photon in a single collision. Since the initial kinetic energy of an electron is equal to the energy given it by the accelerating voltage, V, then $K = eV$. In a single collision in which the electron is brought to rest, we have

$$hf_0 = eV$$

or

$$\lambda_0 = \frac{hc}{eV}, \qquad\qquad (40\text{–}10)$$

where $\lambda_0 = c/f_0$ is the cutoff wavelength, Fig. 40–11. This prediction for λ_0 corresponds precisely with that observed experimentally. This result is further evidence that X-rays are a form of electromagnetic radiation (light)[†] and that the photon theory of light is valid.

EXAMPLE 40–8 Cutoff wavelength. What is the shortest-wavelength X-ray photon emitted in an X-ray tube subjected to 50 kV?

SOLUTION From Eq. 40–10,

$$\lambda_0 = \frac{(6.6 \times 10^{-34}\,\text{J·s})(3.0 \times 10^{8}\,\text{m/s})}{(1.6 \times 10^{-19}\,\text{C})(5.0 \times 10^{4}\,\text{V})} = 2.5 \times 10^{-11}\,\text{m},$$

or 0.025 nm. This agrees well with experiment, Fig. 40–11.

---

*40–7 Magnetic Dipole Moments; Total Angular Momentum

An electron orbiting the nucleus of an atom can be considered as a current loop, classically, and thus might be expected to have a **magnetic dipole moment** as discussed in Chapter 27. Indeed, in Example 27–9 we did a classical calculation of the magnetic dipole moment of the electron in the ground state of hydrogen based, essentially, on the Bohr model, and found it to give

$$\mu = IA = \tfrac{1}{2}evr.$$

Here v is the orbital velocity of the electron, and for a particle moving in a circle of radius r, its angular momentum is

$$L = mvr.$$

So we can write

$$\mu = \frac{1}{2}\frac{e}{m}L.$$

The direction of the angular momentum **L** is perpendicular to the plane of the current loop. So is the direction of the magnetic dipole moment vector $\boldsymbol{\mu}$, although in

---

[†] If X-rays were not photons but rather neutral particles with rest mass m_0, Eq. 40–10 would not hold.

the opposite direction since the electron's charge is negative. Hence we can write the vector equation

$$\boldsymbol{\mu} = -\frac{1}{2}\frac{e}{m}\mathbf{L}. \qquad (40\text{--}11)$$

This rough semiclassical derivation was based on the Bohr theory. The same result (Eq. 40–11), is obtained using quantum mechanics. Since $\mathbf{L}$ is quantized in quantum mechanics, the magnetic dipole moment, too, must be quantized.

As we saw in Section 27–5, a magnetic dipole moment in a magnetic field $\mathbf{B}$ experiences a torque, and the potential energy U of such a system depends on $\mathbf{B}$ and the orientation of $\boldsymbol{\mu}$ relative to $\mathbf{B}$ (Eq. 27–12):

$$U = -\boldsymbol{\mu} \cdot \mathbf{B}.$$

If the magnetic field $\mathbf{B}$ is in the z direction, then $U = -\mu_z B_z$ and from Eq. 40–4 $(L_z = m_l \hbar)$ and Eq. 40–11, we have

$$\mu_z = -\frac{e\hbar}{2m}m_l.$$

(Be careful here not to confuse the electron mass m with the magnetic quantum number, m_l.) It is useful to define the quantity

$$\mu_B = \frac{e\hbar}{2m} \qquad (40\text{--}12)$$

which is called the **Bohr magneton** and has the value $\mu_B = 9.27 \times 10^{-24}\,\text{J/T}$ (joule/tesla). Then we can write

$$\mu_z = -\mu_B m_l, \qquad (40\text{--}13)$$

where m_l has integer values from 0 to $\pm l$ (see Table 40–1). An atom placed in a magnetic field would have its energy split into levels that differ by $\Delta U = \mu_B B$; this is the *Zeeman effect*, and was shown in Fig. 40–4.

* Stern-Gerlach Experiment

The first evidence of this *space quantization* (Section 40–2) came in 1922 in a famous experiment known as the **Stern-Gerlach experiment**. Silver atoms (and later others) were heated in an oven from which they escaped as shown in Fig. 40–14. The atoms were made to pass through a collimator, which eliminated all but a narrow beam. The beam then passed into a *nonhomogeneous* magnetic field. The field was deliberately made nonhomogeneous so that it would exert a force on atomic magnetic moments: remember that the potential energy (in this case $-\boldsymbol{\mu} \cdot \mathbf{B}$) must change in space if there is to be a force ($F_x = -dU/dx$, etc., Section 8–2). If $\mathbf{B}$ has a gradient along the z axis, as in Fig. 40–14, then the force is along z:

$$F_z = -\frac{dU}{dz} = \mu_z \frac{dB_z}{dz}.$$

FIGURE 40–14 The Stern-Gerlach experiment, which is done inside a vacuum chamber.

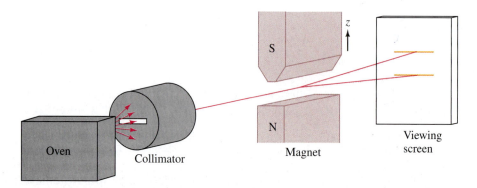

Thus the silver atoms would be deflected up or down depending on the value of μ for each atom. Classically, we would expect to see a continuous distribution on the viewing screen, since we would expect the atoms to have randomly oriented magnetic moments. But Stern and Gerlach saw instead two distinct lines for silver (and for other atoms sometimes more than two lines). These observations were the first evidence for space quantization, though not fully explained until a few years later. If the lines were due to orbital angular momentum, there should have been an odd number of them, corresponding to the possible values of m_l (since $\mu_z = -\mu_B m_l$). For $l = 0$, there is only one possibility, $m_l = 0$. For $l = 1$, m_l can be 1, 0, or −1, and we would expect three lines, and so on. Why there are only two lines was eventually explained by the concept of electron spin. With a spin of $\frac{1}{2}$, the electron spin can have only two orientations in space, as we saw in Fig. 40–5. Hence a magnetic dipole moment associated with spin would have only two positions. Thus, the two states for silver seen in the Stern-Gerlach experiment must be due to the spin of its one valence electron. Silver atoms must thus have zero orbital angular momentum but a total spin of $\frac{1}{2}$ due to this one valence electron (of its 47 electrons, the spins of the first 46 cancel). Two lines were found also for the H atom in its ground state, again due to the spin $\frac{1}{2}$ of its electron since the orbital angular momentum is zero.

The Stern-Gerlach deflection is proportional to the magnetic dipole moment, μ_z, and for a spin $\frac{1}{2}$ particle we expect $\mu_z = -\mu_B m_s = -(\frac{1}{2})(e\hbar/2m)$ as for the case of orbital angular momentum, Eq. 40–13. Instead, μ_z for spin was found to be about twice as large:

$$\mu_z = -g\mu_B m_s, \tag{40–14}$$

Magnetic dipole moment for spin

where g, called the **g-factor** or **gyromagnetic ratio**, has been measured to be slightly larger than 2: $g = 2.0023 \cdots$ for a free electron. This unexpected factor of (about) 2 clearly indicates that spin cannot be viewed as a classical angular momentum. It is a purely quantum-mechanical effect. Equation 40–14 is the same as Eq. 40–13 for orbital angular momentum with m_s replacing m_l. But for the orbital case, $g = 1$.

* Total Angular Momentum J

An atom can have both orbital and spin angular momenta. For example, in the 2p state of hydrogen $l = 1$ and $s = \frac{1}{2}$. In the 4d state, $l = 2$ and $s = \frac{1}{2}$. The **total angular momentum** is the vector sum of the orbital angular momentum $\mathbf{L}$ and the spin $\mathbf{S}$:

$$\mathbf{J} = \mathbf{L} + \mathbf{S}.$$

Total angular momentum

According to quantum mechanics, the magnitude of the total angular momentum $\mathbf{J}$ is quantized:

$$J = \sqrt{j(j + 1)}\,\hbar. \tag{40–15}$$

For the single electron in the H atom, quantum mechanics gives the result that j can be

$$j = l + s = l + \tfrac{1}{2}$$

or

$$j = l - s = l - \tfrac{1}{2}$$

but never less than zero, just as for l and s. For the 1s state, $l = 0$ and $j = \frac{1}{2}$ is the only possibility. For p states, say the 2p state, $l = 1$ and j can be either $\frac{3}{2}$ or $\frac{1}{2}$. The z component for j is quantized in the usual way:

$$m_j = j, j - 1, \cdots, -j.$$

For a 2p state with $j = \frac{1}{2}$, m_j can be $\frac{1}{2}$ or $-\frac{1}{2}$; for $j = \frac{3}{2}$, m_j can be $\frac{3}{2}, \frac{1}{2}, -\frac{1}{2}, -\frac{3}{2}$, for a total of four states. Note that the state of a single electron can be specified by giving n, l, m_l, m_s, or by giving n, j, l, m_j (only one of these descriptions at a time).

The interaction of magnetic fields with atoms, as in the Zeeman effect and the Stern-Gerlach experiment, involves the *total* angular momentum. Thus the Stern-Gerlach experiment on H atoms in the ground state shows two lines (for $m_j = +\frac{1}{2}$ and $-\frac{1}{2}$), but for the first excited state it shows four lines corresponding to the four possible m_j values $\left(\frac{3}{2}, \frac{1}{2}, -\frac{1}{2}, -\frac{3}{2}\right)$.

* Spectroscopic Notation

We can specify the state of an atom, including the total angular momentum quantum number j, using the following **spectroscopic notation**. For a single electron state we can write

$$nL_j,$$

where the value of L is specified using the same letters as in Table 40–2, but in upper case:

$$L = 0 \quad 1 \quad 2 \quad 3 \quad 4 \cdots$$

$$\text{letter} = S \quad P \quad D \quad F \quad G \cdots.$$

So the $2P_{3/2}$ state has $n = 2$, $l = 1$, $j = \frac{3}{2}$, whereas $1S_{1/2}$ specifies the ground state in hydrogen.

* Fine Structure; Spin-Orbit Interaction

It is also a magnetic effect that produces the *fine structure* splitting mentioned in Section 40–2, which occurs in the absence of any external field. Instead, it is due to a magnetic field produced by the atom itself. We can see how it occurs by putting ourselves in the reference frame of the electron, in which case we see the nucleus revolving about us as a moving charge or electric current that produces a magnetic field, B_n. The electron has an intrinsic magnetic dipole moment μ_s (Eq. 40–14) and hence its energy will be altered by an amount (Eq. 27–12)

$$\Delta U = -\boldsymbol{\mu}_s \cdot \mathbf{B}_n.$$

Since μ_s takes on quantized values according to the values of m_s, the energy of a single electron state will split into two closely spaced energy levels (for $m_s = \frac{1}{2}$ and $-\frac{1}{2}$). This tiny splitting of energy levels produces a tiny splitting in spectral lines. For example, in the H atom, the $2P \rightarrow 1S$ transition is split into two lines corresponding to $2P_{1/2} \rightarrow 1S_{1/2}$ and $2P_{3/2} \rightarrow 1S_{1/2}$. The difference in energy between these two is only about $5 \times 10^{-5}\,\text{eV}$, which is very small compared to the $2P \rightarrow 1S$ transition energy of $13.6\,\text{eV} - 3.4\,\text{eV} = 10.2\,\text{eV}$.

The magnetic field $\mathbf{B}_n$ produced by the orbital motion is proportional to the orbital angular momentum $\mathbf{L}$, and since $\boldsymbol{\mu}_s$ is proportional to the spin $\mathbf{S}$, then $\Delta U = -\boldsymbol{\mu}_s \cdot \mathbf{B}_n$ can be written

Spin-orbit interaction

$$\Delta U \propto \mathbf{L} \cdot \mathbf{S}.$$

This interaction, which produces the fine structure, is thus called the **spin-orbit interaction**. Its magnitude is related to a dimensionless constant known as the **fine structure constant**,

Fine-structure constant

$$\alpha = \frac{e^2}{2\epsilon_0 hc} \approx \frac{1}{137},$$

which also appears elsewhere in atomic physics.

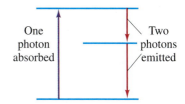

FIGURE 40–15 Fluorescence.

FIGURE 40–16 When UV light illuminates these various "fluorescent" rocks, they fluoresce in the visible region of the spectrum.

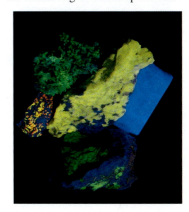

* | 40–8 | Fluorescence and Phosphorescence

When an atom is excited from one energy state to a higher one by the absorption of a photon, it may return to the lower level in a series of two (or more) jumps if there is an energy level in between (Fig. 40–15). The photons emitted will consequently have lower energy and frequency than the absorbed photon. When the absorbed photon is in the UV and the emitted photons are in the visible region of the spectrum, this phenomenon is called **fluorescence** (Fig. 40–16).

The wavelength for which fluorescence will occur depends on the energy levels of the particular atoms. Because the frequencies are different for different substances, and because many substances fluoresce readily, fluorescence is a powerful tool for identification of compounds. It is also used for assaying—determining how much of a substance is present—and for following substances along a natural pathway as in plants and animals. For detection of a given compound, the stimulating light must be monochromatic, and solvents or other materials present must not fluoresce in the same region of the spectrum. Sometimes the observation of fluorescent light being emitted is sufficient to detect a compound. In other cases, spectrometers are used to measure the wavelengths and intensities of the emitted light.

Fluorescent lightbulbs work in a two-step process. The applied voltage accelerates electrons that strike atoms of the gas in the tube and cause them to be excited. When the excited atoms jump down to their normal levels, they emit UV photons which strike a fluorescent coating on the inside of the tube. The light we see is a result of this material fluorescing in response to the UV light striking it.

Materials such as those used for luminous watch dials are said to be **phosphorescent**. When an atom is raised to a normal excited state, it drops back down within about 10^{-8} s. In phosphorescent substances, atoms can be excited by photon absorption to energy levels, said to be **metastable**, which are states that last much longer—even a few seconds or longer. In a collection of such atoms, many of the atoms will descend to the lower state fairly soon, but many will remain in the excited state for over an hour. Hence light will be emitted even after long periods. When you put your watch dial close to a bright lamp, it excites many atoms to metastable states, and you can see the glow a long time after.

* | 40–9 | Lasers

A laser is a device that can produce a very narrow intense beam of monochromatic coherent light. (By *coherent*, we mean that across any cross section of the beam, all parts have the same phase.) The emitted beam is a nearly perfect plane wave. An ordinary light source, on the other hand, emits light in all directions (so the intensity decreases rapidly with distance), and the emitted light is incoherent (the different parts of the beam are not in phase with each other). The excited atoms that emit the light in an ordinary lightbulb act independently, so each photon emitted can be considered as a short wave train that lasts about 10^{-8} s. These wave trains bear no phase relation to one another. Just the opposite is true of lasers.

The action of a laser is based on quantum theory. We have seen that a photon can be absorbed by an atom if (and only if) its energy hf corresponds to the energy difference between an occupied energy level of the atom and an available excited state, Fig. 40–17a. This is, in a sense, a resonant situation. If the atom is already in the excited state, it may of course jump spontaneously (i.e., no apparent stimulus) to the lower state with the emission of a photon. However, if a photon with this same energy strikes the excited atom, it can stimulate the atom to make the transition sooner to the lower state, Fig. 40–17b. This phenomenon is called **stimulated emission**, and it can be seen that not only do we still have the original photon, but also a second one of the same frequency as a result of the atom's transition. And these two photons are

FIGURE 40–17 (a) Absorption of a photon. (b) Stimulated emission.

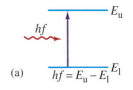

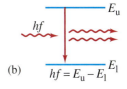

Stimulated emission

exactly *in phase*, and they are moving in the same direction. This is how coherent light is produced in a laser. Hence the name "laser," which is an acronym for **l**ight **a**mplification by **s**timulated **e**mission of **r**adiation.

The natural population of atoms in thermal equilibrium at any temperature T (in K) is given by the **Boltzmann distribution** (or **Boltzmann factor**):

$$N_n = Ce^{-\frac{E_n}{kT}}, \tag{40–16a}$$

where N_n is the number of atoms in the state with energy E_n. For two states n and n', the ratio of the number of atoms in the two states is

$$\frac{N_n}{N_{n'}} = e^{-\left(\frac{E_n - E_{n'}}{kT}\right)} \tag{40–16b}$$

Thus most atoms are in the ground state unless the temperature is very high. In the two-level system of Fig. 40–17, most atoms are normally in the lower state, so the majority of incident photons will be absorbed. In order to obtain the coherent light from stimulated emission, two conditions must be satisfied. First, atoms must be excited to the higher state, so that an **inverted population** is produced, one in which more atoms are in the upper state than in the lower one (Fig. 40–18). Then *emission* of photons will dominate over absorption. Hence the system will not be in thermal equilibrium. And second, the higher state must be a **metastable state**— a state in which the electrons remain longer than usual[†] so that the transition to the lower state occurs by stimulated emission rather than spontaneously. How these conditions are achieved for different lasers will be discussed shortly. For now, we assume that the atoms have been excited to an upper state. Figure 40–19 is a schematic diagram of a laser: the "lasing" material is placed in a long narrow tube at the ends of which are two mirrors, one of which is partially transparent (perhaps 1 or 2 percent). Some of the excited atoms drop down fairly soon after being excited. One of these is the atom shown on the far left in Fig. 40–19. If the emitted photon strikes another atom in the excited state, it stimulates this atom to emit a photon of the *same* frequency, moving in the *same* direction, and *in phase* with it. These two photons then move on to strike other atoms causing more stimulated emission. As the process continues, the number of photons multiplies. When the photons strike the end mirrors, most are reflected back, and as they move in the opposite direction, they continue to stimulate more atoms to emit photons. As the photons move back and forth between the mirrors, a small percentage passes through the partially transparent mirror at one end. These photons make up the narrow coherent external laser beam.

Inside the tube, some spontaneously emitted photons will be emitted at an angle to the axis, and these will merely go out the side of the tube and not affect the narrowness of the main beam. In a well-designed laser, the spreading of the beam is limited only by diffraction, so the angular spread is $\approx \lambda/a$ (see Eq. 36–1) where a is the diameter of the end mirror. The diffraction spreading can be incred-

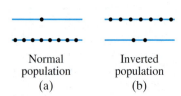

FIGURE 40–18 Two energy levels for a collection of atoms. Each dot represents the energy state of one atom. (a) A normal situation; (b) an inverted population.

Normal population (a)   Inverted population (b)

[†] An excited atom may land in such a state and can jump to a lower state only by a so-called forbidden transition (discussed in Section 40–2), which is why its lifetime is longer than normal.

FIGURE 40–19 Schematic laser diagram, showing excited atoms stimulated to emit light.

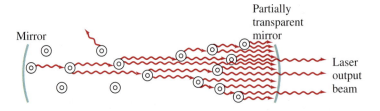

Mirror

Partially transparent mirror

Laser output beam

ibly small. The light energy, instead of spreading out in space as it does for an ordinary light source, is directed in a pencil-thin beam.

The excitation of the atoms in a laser can be done in several ways to produce the necessary inverted population. In a ruby laser, the lasing material is a ruby rod consisting of Al_2O_3 with a small percentage of aluminum (Al) atoms replaced by chromium (Cr) atoms. The Cr atoms are the ones involved in lasing. The atoms are excited by strong flashes of light of wavelength 550 nm, which corresponds to a photon energy of 2.2 eV. As shown in Fig. 40–20, the atoms are excited from state E_0 to state E_2. This process is called **optical pumping**. The atoms quickly decay either back to E_0 or to the intermediate state E_1, which is metastable with a lifetime of about 3×10^{-3} s (compared to 10^{-8} s for ordinary levels). With strong pumping action, more atoms can be forced into the E_1 state than are in the E_0 state. Thus we have the inverted population needed for lasing. As soon as a few atoms in the E_1 state jump down to E_0, they emit photons that produce stimulated emission of the other atoms and the lasing action begins. A ruby laser thus emits a beam whose photons have energy 1.8 eV and a wavelength of 694.3 nm (or "ruby-red" light).

In a helium–neon (He–Ne) laser, the lasing material is a gas, a mixture of about 15 percent He and 85 percent Ne. The atoms are excited by applying a high voltage to the tube so that an electric discharge takes place within the gas. In the process, some of the He atoms are raised to the metastable state E_1 shown in Fig. 40–21, which corresponds to a jump of 20.61 eV, almost exactly equal to an excited state in neon, 20.66 eV. The He atoms do not quickly return to the ground state by spontaneous emission, but instead often give their excess energy to a Ne atom when they collide—see Fig. 40–21. In such a collision, the He drops to the ground state and the Ne atom is excited to the state E_3' (the prime refers to neon states). The slight difference in energy (0.05 eV) is supplied by the kinetic energy of the moving molecules ($\frac{3}{2}kT \approx 0.04$ eV at $T = 300$ K). In this manner, the E_3' state in Ne—which is metastable—becomes more populated than the E_2' level. This inverted population between E_3' and E_2' is what is needed for lasing.

Other types of laser include: chemical lasers, in which the energy input comes from the chemical reaction of highly reactive gases; dye lasers, whose frequency is tunable; CO_2 gas lasers, capable of high power output in the infrared; rare-earth solid-state lasers such as the high-power Nd:Yag laser; and the *pn* junction laser in which the transitions occur between the bottom of the conduction band and the upper part of the valence band (Section 41–7).

The excitation of the atoms in a laser can be done continuously or in pulses. In a **pulsed laser**, the atoms are excited by periodic inputs of energy. The multiplication of photons continues until all the atoms have been stimulated to jump down

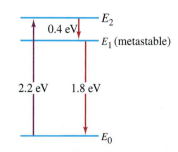

FIGURE 40–20 Energy levels of chromium in a ruby crystal. Photons of energy 2.2 eV "pump" atoms from E_0 to E_2, followed by decay to the metastable state E_1. Lasing action occurs by stimulated emission of photons in transition from E_1 to E_0.

He–Ne laser

Other lasers

FIGURE 40–21 Energy levels for He and Ne. He is excited in the electric discharge to the E_1 state. This energy is transferred to the E_3' level of the Ne by collision. E_3' is metastable and decays to E_2' by stimulated emission, producing the output beam of the laser.

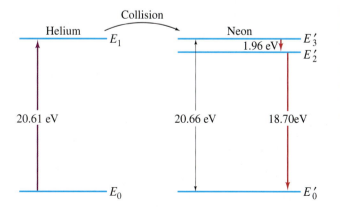

to the lower state, and the process is repeated with each input pulse. In a **continuous laser**, the energy input is continuous so that as atoms are stimulated to jump down to the lower level, they are soon excited back up to the upper level so that the output is a continuous laser beam. Any laser, of course, is not a source of energy. Energy must be put in, and the laser converts a part of this input energy into an intense narrow beam output.

➡ **PHYSICS APPLIED**

Medical and other uses of lasers

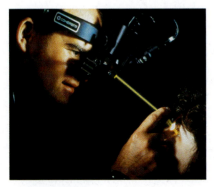

FIGURE 40–22 Laser being used in eye surgery.

The unique feature of light from a laser is, as mentioned before, that it is a coherent narrow beam of a single frequency (or several distinct frequencies). Because of this feature, the laser has found many applications. Lasers are a useful surgical tool. The narrow intense beam can be used to destroy tissue in a localized area, or to break up gallstones and kidney stones. Because of the heat produced, a laser beam can be used to "weld" broken tissue, such as a detached retina (Fig. 40–22). For some types of internal surgery, the laser beam can be carried by an optical fiber (Section 33–7) to the surgical point. An example is the removal of plaque clogging human arteries. Tiny organelles within a living cell have been destroyed using lasers by researchers studying how the absence of that organelle affects the behavior of the cell. Laser beams have been used to destroy cancerous and precancerous cells; at the same time, the heat seals off capillaries and lymph vessels, thus "cauterizing" the wound in the process to prevent spread of the disease. The intense heat produced in a small area by a laser beam is also used for welding and machining metals and for drilling tiny holes in hard materials. The beam of a laser is narrow in itself (typically, a few mm). But because the beam is coherent, monochromatic, and essentially parallel and narrow, lenses can be used to focus the light into incredibly small areas without the usual aberration problems. The limiting factor thus becomes diffraction, and the energy crossing unit area per unit time can be very large. The precise straightness of a laser beam is also useful to surveyors for lining up equipment precisely, especially in inaccessible places.

➡ **PHYSICS APPLIED**

CD players and bar codes

In everyday life, lasers are used as bar-code readers (at store checkout stands) and in compact disc (CD) players. The laser beam reflects off the stripes and spaces of a bar code, and off the tiny pits of a CD as shown in Fig. 40–23. The recorded information on a CD is a series of pits and spaces representing 0s and 1s (or "off" and "on") of a digitized code that is decoded electronically before being sent to the audio or video system. A bar-code reader is similar.

FIGURE 40–23 Reading a CD. The fine beam of a laser, focused even more finely with lenses, is directed at the undersurface of a rotating compact disc. The beam is reflected back from the areas between pits but reflects much less from pits. The reflected light is detected as shown, reflected by a half-reflecting mirror MS. The strong and weak reflections correspond to the 0s and 1s of the binary code representing the audio or video signal.

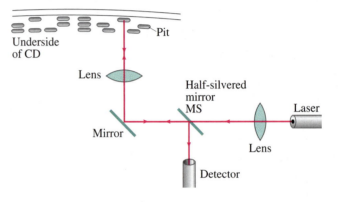

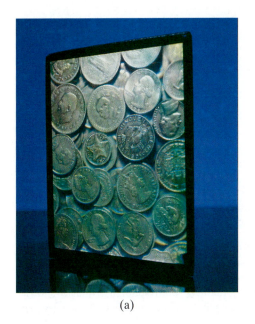

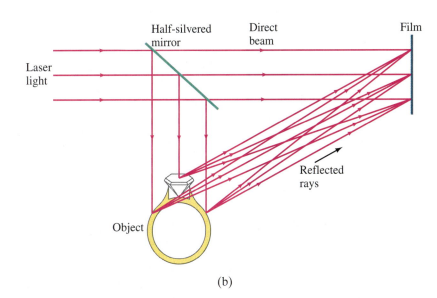

FIGURE 40–24   (a) Photo of a hologram of coins. (b) Making a hologram. Light reflected from various points on the object interferes (at the film) with light from the direct beam.

One of the most interesting applications of laser light is the production of three-dimensional images called **holograms** (see Fig. 40–24). In an ordinary photograph, the film simply records the intensity of light reaching it at each point. When the photograph or transparency is viewed, light reflecting from it or passing through it gives us a two-dimensional picture. In holography, three-dimensional images are formed by interference, without lenses. When a laser hologram is made on film, a broadened laser beam is split into two parts by a half-silvered mirror, Fig. 40–24b. One part goes directly to the film; the rest passes to the object to be photographed, from which it is reflected to the film. Light from every point on the object reaches each point on the film, and the interference of the two beams allows the film to record both the intensity and relative phase of the light at each point. It is crucial that the light be coherent—that is, in phase at all points—which is why a laser is used. After the film is developed, it is placed again in a laser beam and a three-dimensional image of the object is created. You can walk around such an image and see it from different sides as if it were the original object. Yet, if you try to touch it with your hand, there will be nothing material there.

The details of how the image is formed are quite complicated. But we can get the basic idea by considering one single point on the object. In Fig. 40–25a the rays OA and OB have reflected from one point on our object. The rays CA and DB come directly from the source and interfere with OA and OB at points A and B on the film. A set of interference fringes is produced as shown in Fig. 40–25b. The spacing between the fringes changes from top to bottom as shown. Why this

→ **PHYSICS APPLIED**

Holography

FIGURE 40–25   Light from point O on the object interferes with light of the direct beam (rays CA and DB).

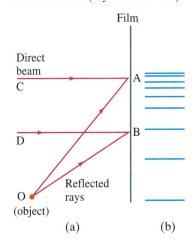

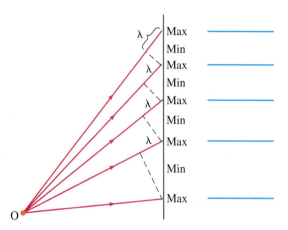

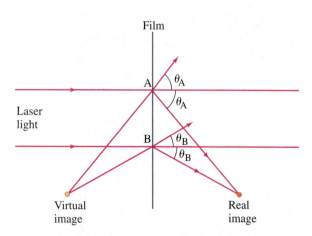

FIGURE 40–26 Each of the rays shown leaving point O is one wavelength shorter than the one above it. If the top ray is in phase with the direct beam (not shown), which has the same phase at all points on the screen, all the rays shown produce constructive interference. From this diagram it can be seen that the fringe spacing increases toward the bottom.

FIGURE 40–27 Reconstructing the image of one point on the object. Laser beam strikes film that is like a diffraction grating of variable spacing. Rays corresponding to the first diffraction maxima are shown emerging. The angle $\theta_A > \theta_B$ because the spacing at B is greater than at A ($\sin\theta = \lambda/d$). Hence real and virtual images of the point are reproduced as shown.

happens is explained in Fig. 40–26. Thus the hologram of a single point object would have the pattern shown in Fig. 40–25b. The film in this case looks like a diffraction grating with variable spacing. Hence, when coherent laser light is passed back through the developed film to reconstruct the image, the diffracted rays in the first order maxima occur at slightly different angles because the spacing changes. (Remember Eq. 36–13, $\sin\theta = \lambda/d$: where the spacing d is greater, the angle θ is less.) Hence, the rays diffracted upward (in first order) seem to diverge from a single point, Fig. 40–27. This is a virtual image of the original object, which can be seen with the eye. Rays diffracted in first order *downward* converge to make a real image, which can be seen and also photographed. (Note that the straight-through undiffracted rays are of no interest.) Of course real objects consist of many points, so a hologram will be a complex interference pattern which, when laser light is incident on it, will reproduce an image of the object. Each image point will be at the correct (three-dimensional) position with respect to other points, so the image accurately represents the original object. And it can be viewed from different angles as if viewing the original object. Holograms can be made in which a viewer can walk entirely around the image (360°) and see all sides of it.

White-light holograms   So-called **volume** or **white-light holograms** do not require a laser to see the image, but can be viewed with ordinary white light (preferably a nearby point source, such as the Sun or a clear bulb with a small bright filament). Such holograms must be made, however, with a laser. They are made not on thin film, but on a *thick* emulsion. The interference pattern, instead of being two-dimensional as for an ordinary hologram, is actually three-dimensional (hence the name "volume hologram"). The interference pattern in the film emulsion can be thought of as an array of bands or ribbons (consisting of the silver grains from the development process) where constructive interference occurred. This array, and the reconstruction of the image, can be compared to Bragg scattering of X-rays from the atoms in a crystal (see Section 36–10). White light can reconstruct the image because the Bragg condition (Eq. 36–20) selects out the appropriate single wavelength. If the hologram is originally produced by lasers emitting the three additive primary colors (red, green, and blue), the three-dimensional image can be seen in full color when viewed with white light.

Summary

In the quantum mechanical view of the atom, the electrons do not have well-defined orbits, but instead exist as a "cloud." Electron clouds can be interpreted as an electron wave spread out in space, or as a **probability distribution** for electrons considered as particles.

For the simplest atom, hydrogen, the Schrödinger equation contains the potential energy $U = -(1/4\pi\epsilon_0)(e^2/r)$. The solutions give the same values of energy as the old Bohr theory.

According to quantum mechanics, the state of an electron in an atom is specified by four **quantum numbers**: n, l, m_l, and m_s:

the principal quantum number, n, can take on any integer value $(1, 2, 3, \cdots)$ and corresponds to the quantum number of the old Bohr theory;

l can take on values from 0 up to $n - 1$;

m_l can take on integer values from $-l$ to $+l$;

m_s can be $+\frac{1}{2}$ or $-\frac{1}{2}$.

The energy levels in the hydrogen atom depend on n, whereas in other atoms they depend on n and l.

When an external magnetic field is applied, the spectral lines are split (the **Zeeman effect**), indicating that the energy depends also on m_l in this case.

Even in the absence of a magnetic field, precise measurements of spectral lines show a tiny splitting of the lines called **fine structure**, whose explanation is that the energy depends very slightly on m_l and m_s.

Transitions between states that obey the selection rule $\Delta l = \pm 1$ are far more probable than other so-called "forbidden" transitions.

The ground-state wave function in hydrogen has spherical symmetry, as do other $l = 0$ states. States with $l > 0$ have some directionality in space.

The probability density, $|\psi|^2$, and the *radial probability density*, $P_r = 4\pi r^2|\psi|^2$, are both useful to illustrate the spatial extent of the electron cloud.

The arrangement of electrons in multi-electron atoms is governed by the **Pauli exclusion principle**, which states that no two electrons can occupy the same quantum state—that is, they cannot have the same set of quantum numbers n, l, m_l, and m_s.

As a result, electrons in multi-electron atoms are grouped into **shells** (according to the value of n) and **subshells** (according to l).

Electron configurations are specified using the numerical values of n, and using letters for l: s, p, d, f, etc., for $l = 0, 1, 2, 3,$ and so on, plus a superscript for the number of electrons in that subshell. Thus, the ground state of hydrogen is $1s^1$, whereas that for oxygen is $1s^2 2s^2 2p^4$. The **periodic table** arranges the elements in horizontal rows according to increasing atomic number (number of electrons in the neutral atom). The shell structure gives rise to a periodicity in the properties of the elements, so that each vertical column can contain elements with similar chemical properties.

X-rays, which are a form of electromagnetic radiation of very short wavelength, are produced when high-speed electrons strike a target. The spectrum of X-rays so produced consists of two parts, a continuous spectrum produced when the electrons are decelerated by atoms of the target, and peaks representing photons emitted by atoms of the target after being excited by collision with the high-speed electrons. Measurement of these peaks allows determination of inner energy levels of atoms and determination of Z.

Questions

1. Discuss the differences between Bohr's view of the atom and the quantum-mechanical view.

2. The probability density $|\psi|^2$ is a maximum at the center of the H atom $(r = 0)$ for the ground state, whereas the radial probability density $P_r = 4\pi r^2|\psi|^2$ is zero at this point. Explain why.

3. Why do three quantum numbers come out of the Schrödinger theory (rather than, say, two or four)?

4. In Fig. 40–4, why do the upper and lower levels have different energy splittings in a magnetic field?

5. Which model of the hydrogen atom, the Bohr model or the quantum-mechanical model, predicts that the electron spends more time near the nucleus?

6. The size of atoms varies by only a factor of three or so from largest to smallest, yet the number of electrons varies from one to over 100. Why?

7. Excited hydrogen and excited helium atoms both radiate light as they jump down to the $n = 1$, $l = 0$, $m_l = 0$ state. Yet the two elements have very different emission spectra. Why?

8. The 589-nm yellow line in sodium is actually two very closely spaced lines. This splitting is due to an "internal" Zeeman effect. Can you explain this? [*Hint*: Put yourself in the reference frame of the electron.]

9. Which of the following electron configurations are forbidden? (*a*) $1s^2 2s^2 2p^4 3s^2 4p^2$; (*b*) $1s^2 2s^2 2p^8 3s^1$, (*c*) $1s^2 2s^2 2p^6 3s^2 3p^5 4s^2 4d^5 4f^1$.

10. Give the complete electron configuration for a uranium atom (careful scrutiny across the periodic table on the inside back cover will provide useful hints).

11. In what column of the periodic table would you expect to find the atom with each of the following configurations? (a) $1s^2 2s^2 2p^6 3s^2$; (b) $1s^2 2s^2 2p^6 3s^2 3p^6$; (c) $1s^2 2s^2 2p^6 3s^2 3p^6 4s^1$; (d) $1s^2 2s^2 2p^5$.

12. On what factors does the periodicity of the periodic table depend? Consider the exclusion principle, quantization of angular momentum, spin, and any others you can think of.

13. How would the periodic table look if there were no electron spin but otherwise quantum mechanics were valid? Consider the first 20 elements or so.

14. The ionization energy for neon ($Z = 10$) is 21.6 eV and that for sodium ($Z = 11$) is 5.1 eV. Explain the large difference.

15. Why do chlorine and iodine exhibit similar properties?

16. Explain why potassium and sodium exhibit similar properties.

17. Why are the chemical properties of the rare earths so similar?

18. Why do we not expect perfect agreement between measured values of X-ray line wavelengths and those calculated using Bohr theory, as in Example 40–6?

19. Why does the Bohr theory, which does not work at all well for normal transitions involving the outer electrons for He and more complex atoms, nevertheless predict reasonably well the atomic X-ray spectra for transitions deep inside the atom?

20. Why does the cutoff wavelength in Fig. 40–11 imply a photon nature for light?

21. How would you figure out which lines in an X-ray spectrum correspond to K_α, K_β, L, etc., transitions?

22. Why do the characteristic X-ray spectra vary in a systematic way with Z, whereas the visible spectra (Fig. 36–21) do not?

23. Why do we expect electron transitions deep within an atom to produce shorter wavelengths than transitions by outer electrons?

* 24. Why is the direction of the magnetic dipole moment of an electron opposite to that of its orbital angular momentum?

* 25. Why is a nonhomogeneous field used in the Stern-Gerlach experiment?

* 26. Compare spontaneous emission to stimulated emission.

* 27. Does the intensity of light from a laser fall off as the inverse square of the distance?

* 28. How does laser light differ from ordinary light? How is it the same?

* 29. Explain how a 0.0005-W laser beam, photographed at a distance, can seem much stronger than a 1000-W street lamp.

Problems

Section 40–2

1. (I) For $n = 6$, what values can l have?

2. (I) For $n = 5$, $l = 3$, what are the possible values of m_l and m_s?

3. (I) How many different states are possible for an electron whose principal quantum number is $n = 4$? Write down the quantum numbers for each state.

4. (I) If a hydrogen atom has $m_l = -3$, what are the possible values of n, l, m_s?

5. (I) A hydrogen atom is known to have $l = 4$. What are the possible values for n, m_l, and m_s?

6. (I) Calculate the magnitude of the angular momentum of an electron in the $n = 4$, $l = 2$ state of hydrogen.

7. (II) A hydrogen atom is in the 6g state. Determine (a) the principal quantum number, (b) the energy of the state, (c) the orbital angular momentum and its quantum number l, and (d) the possible values for the magnetic quantum number.

8. (II) (a) Show that the number of different states possible for a given value of l is equal to $2(2l + 1)$. (b) What is this number for $l = 0, 1, 2, 3, 4, 5$, and 6?

9. (II) Show that the number of different electron states possible for a given value of n is $2n^2$. (See Problem 8.)

10. (II) An excited H atom is in a 6d state. (a) Name all the states (n, l) to which the atom is "allowed" to jump with the emission of a photon. (b) How many different wavelengths are there (ignoring fine structure)?

Section 40–3

11. (I) Show that the ground-state wave function, Eq. 40–5, is normalized. [*Hint*: See Example 40–4.]

12. (II) For the ground state of hydrogen, what is the value of (a) ψ, (b) $|\psi|^2$, and (c) P_r at $r = r_0$?

13. (II) For the $n = 2$, $l = 0$ state of hydrogen, what is the value of (a) ψ, (b) $|\psi|^2$, and (c) P_r at $r = 5r_0$?

14. (II) Show that ψ_{200} as given by Eq. 40–8 is normalized.

15. (II) By what factor is it more likely to find the electron in the ground state of hydrogen at the Bohr radius (r_0) than at twice the Bohr radius $(2r_0)$?

16. (II) For the ground state of hydrogen, what is the probability of finding the electron within a spherical shell of inner radius $0.99r_0$ and outer radius $1.01r_0$?

17. (II) For the $n = 2$, $l = 0$ state of hydrogen, what is the probability of finding the electron within a spherical shell of inner radius $4.00r_0$ and outer radius $5.00r_0$?

18. (II) (a) Show that the probability of finding the electron in the ground state of hydrogen at less than one Bohr radius from the nucleus is 32 percent. (b) What is the probability of finding a 1s electron between $r = r_0$ and $r = 2r_0$?

19. (II) Determine the radius r of a sphere centered on the nucleus within which the probability of finding the electron for the ground state of hydrogen is (a) 50 percent, (b) 90 percent, (c) 99 percent.

20. (II) (a) Estimate the probability of finding an electron, in the ground state of hydrogen, within the nucleus assuming it to be a sphere of radius $r = 1.1$ fm. (b) What would be the probability if the electron were replaced with a muon, which is very similar to an electron (Chapter 44) except that its mass is 207 times greater?

21. (II) Determine the average radial probability distribution P_r for the $n = 2$, $l = 1$ state in hydrogen by calculating

$$P_r = 4\pi r^2 \left[\tfrac{1}{3}|\psi_{210}|^2 + \tfrac{1}{3}|\psi_{211}|^2 + \tfrac{1}{3}|\psi_{21-1}|^2 \right].$$

22. (II) Use the result of Problem 21 to show that the most probable distance r from the nucleus for an electron in the 2p state of hydrogen is $r = 4r_0$, which is just the second Bohr radius (Eq. 38–11, Fig. 38–24).

23. (II) Show that the mean value of r for an electron in the ground state of hydrogen is $\bar{r} = \tfrac{3}{2}r_0$, by calculating

$$\bar{r} = \int_{\text{all space}} r|\psi_{100}|^2 \, dV = \int_0^\infty r|\psi_{100}|^2 4\pi r^2 \, dr.$$

24. (III) Show that the probability of finding the electron within 1 Bohr radius of the nucleus in the hydrogen atom is (a) 3.4 percent for the $n = 2$, $l = 0$ state, and (b) 0.37 percent for the $n = 2$, $l = 1$ state.

25. (III) Show that ψ_{100} (Eq. 40–5a) satisfies the Schrödinger equation (Eq. 40–1) with the Coulomb potential, for energy $E = -me^4/8\epsilon_0^2 h^2$.

26. (III) For the $n = 2$, $l = 0$ state in hydrogen, what is the probability of finding the electron within the smaller peak in the radial probability distribution, Fig. 40–8b?

27. (III) The wave function for the $n = 3$, $l = 0$ state in hydrogen is

$$\psi_{300} = \frac{1}{\sqrt{27\pi r_0^3}} \left(1 - \frac{2r}{3r_0} + \frac{2r^2}{27r_0^2} \right) e^{-\frac{r}{3r_0}}.$$

(a) Determine the radial probability distribution P_r for this state, and (b) draw the curve for it on a graph. (c) Determine the most probable distance from the nucleus for an electron in this state.

Sections 40–4 and 40–5

28. (I) List the quantum numbers for each electron in the ground state of nitrogen ($Z = 7$).

29. (I) List the quantum numbers for each electron in the ground state of (a) carbon ($Z = 6$), (b) magnesium ($Z = 12$).

30. (I) How many electrons can be in the $n = 6$, $l = 3$ subshell?

31. (II) What is the full electron configuration for (a) selenium (Se), (b) gold (Au), (c) uranium (U)? [Hint: See the periodic table inside the back cover.]

32. (II) For each of the following atomic transitions, state whether the transition is *allowed* or *forbidden*, and why: (a) 4p → 3p; (b) 2p → 1s; (c) 3d → 2d; (d) 4d → 3s; (e) 4s → 2p.

33. (II) Using the Bohr formula for the radius of an electron orbit, estimate the average distance from the nucleus for an electron in the innermost ($n = 1$) orbit in uranium ($Z = 92$). Approximately how much energy would be required to remove this innermost electron?

34. (II) Estimate the binding energy of the third electron in lithium using Bohr theory. [Hint: This electron has $n = 2$ and "sees" a net charge of approximately $+1e$.] The measured value is 5.36 eV.

35. (II) Show that the total angular momentum is zero for a filled subshell.

36. (II) Let us apply the exclusion principle to an infinitely high square well (Section 39–8). Let there be five electrons confined to this rigid box whose width is L. Find the lowest energy state of this system, by placing the electrons in the lowest available levels, consistent with the Pauli exclusion principle.

Section 40–6

37. (I) What are the shortest-wavelength X-rays emitted by electrons striking the face of a 30-kV TV picture tube? What are the longest wavelengths?

38. (I) If the shortest-wavelength bremsstrahlung X-rays emitted from an X-ray tube have $\lambda = 0.029$ nm, what is the voltage across the tube?

39. (I) Show that the cutoff wavelength λ_0 is given by

$$\lambda_0 = \frac{1240 \text{ nm}}{V},$$

where V is the X-ray tube voltage in volts.

40. (II) Use the result of Example 40–6 to estimate the X-ray wavelength emitted when a cobalt atom ($Z = 27$) jumps from $n = 2$ to $n = 1$.

41. (II) Estimate the wavelength for an $n = 2$ to $n = 1$ transition in iron ($Z = 26$).

42. (II) Use Bohr theory to estimate the wavelength for an $n = 3$ to $n = 1$ transition in molybdenum. The measured value is 0.063 nm. Why do we not expect perfect agreement?

43. (II) A mixture of iron and an unknown material are bombarded with electrons. The wavelength of the K_α lines are 194 pm for iron and 229 pm for the unknown. What is the unknown material?

44. (II) Use conservation of energy and momentum to show that a moving electron cannot give off an X-ray photon unless there is a third body present, such as an atom or nucleus.

*Section 40–7

*45. (I) Verify that the Bohr magneton has the value $\mu_B = 9.27 \times 10^{-24}$ J/T (see Eq. 40–12).

*46. (II) Suppose that the splitting of energy levels shown in Fig. 40–4 was produced by a 2.0-T magnetic field. (a) What is the separation in energy between adjacent m_l levels for the same l? (b) How many different wavelengths will there be for 3d to 2p transitions, if m_l can change only by ± 1 or 0? (c) What is the wavelength for each of these transitions?

*47. (II) In a Stern-Gerlach experiment, Ag atoms issued from the oven with an average speed of 700 m/s and passed through a magnetic field gradient $dB/dz = 1.5 \times 10^3$ T/m for a distance of 4.0 cm. (a) What is the separation of the two beams as they emerge from the magnet? (b) What would the separation be if the g-factor were 1 for electron spin?

* **48.** (II) (a) Write down the quantum numbers for each electron in the aluminum atom. (b) Which subshells are filled? (c) The last electron is in the 3p state; what are the possible values of the total angular momentum quantum number, j, for this electron? (d) Explain why the angular momentum of this last electron also represents the total angular momentum for the entire atom (ignoring any angular momentum of the nucleus). (e) How could you use a Stern-Gerlach experiment to determine which value of j the atom has?

* **49.** (II) What are the possible values of j for an electron in (a) the 3p, (b) the 4f, and (c) the 4d state of hydrogen? What is J in each case?

* **50.** (II) For an electron in a 5g state, what are all the possible values of j, m_j, J, and J_z?

* **51.** (II) The difference between the $2P_{3/2}$ and $2P_{1/2}$ energy levels in hydrogen is about 5×10^{-5} eV, due to the spin-orbit interaction. (a) Taking the electron's (orbital) magnetic moment to be 1 Bohr magneton, estimate the internal magnetic field due to the electron's orbital motion. (b) Estimate the internal magnetic field using a simple model of the nucleus revolving in a circle about the electron.

* **Section 40–9**

* **52.** (II) A laser used to weld detached retinas puts out 25-ms-long pulses of 640-nm light which average 0.65 W output during a pulse. How much energy can be deposited per pulse and how many photons does each pulse contain?

* **53.** (II) Estimate the angular spread of a laser beam due to diffraction if the beam emerges through a 3.0-mm-diameter mirror. Assume that $\lambda = 694$ nm. What would be the diameter of this beam if it struck a satellite 300 km above the Earth?

* **54.** (II) Suppose that the energy level system in Fig. 40–20 is not being pumped and is in thermal equilibrium. Determine the fraction of atoms in levels E_2 and E_1 relative to E_0 at $T = 300$ K.

* **55.** (II) To what temperature would the system in Fig. 40–20 have to be raised (see Problem 54) so that in thermal equilibrium the level E_2 would have half as many atoms as E_0? (Note that pumping mechanisms do not maintain thermal equilibrium.)

* **56.** (II) Show that a population inversion for two levels (as in a pumped laser) corresponds to a negative Kelvin temperature in the Boltzmann distribution. Explain why such a situation does not contradict the idea that negative Kelvin temperatures cannot be reached in the normal sense of temperature.

General Problems

57. The ionization (binding) energy of the outermost electron in boron is 8.26 eV. (a) Use the Bohr model to estimate the "effective charge," Z_{eff}, seen by this electron. (b) Estimate the average orbital radius.

58. Show that there can be 18 electrons in a "g" subshell.

59. What is the full electron configuration in the ground state for elements with Z equal to (a) 27, (b) 36, (c) 38? [Hint: See the periodic table inside the back cover.]

60. What are the largest and smallest possible values for the angular momentum L of an electron in the $n = 5$ shell?

61. Estimate (a) the quantum number l for the orbital angular momentum of the Earth about the Sun, and (b) the number of possible orientations for the plane of Earth's orbit.

62. Use the Bohr theory (especially Eq. 38–14) to show that the Moseley plot (Fig. 40–12) can be written

$$\sqrt{\frac{1}{\lambda}} = a(Z - b),$$

where $b \approx 1$, and evaluate a.

63. Determine the most probable distance from the nucleus of an electron in the $n = 2$, $l = 0$ state of hydrogen.

64. In the so-called *vector model* of the atom, space quantization of angular momentum (Fig. 40–3) is illustrated as shown in Fig. 40–28. The angular momentum vector of magnitude $L = \sqrt{l(l + 1)}\hbar$ is thought of as precessing around the z axis (like a spinning top or gyroscope) in such a way that the z component of angular momentum, $L_z = m_l\hbar$, also stays constant. Calculate the possible values for the angle θ between $\mathbf{L}$ and the z axis (a) for $l = 1$, (b) $l = 2$, and (c) $l = 3$. (d) Determine the minimum value of θ for $l = 100$ and $l = 10^6$. Is this consistent with the correspondence principle?

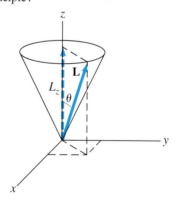

FIGURE 40–28 The vector model for orbital angular momentum. The orbital angular momentum vector $\mathbf{L}$ is imagined to precess about the z axis; L and L_z remain constant, but L_x and L_y continually change. Problems 64 and 65.

65. The vector model (Problem 64) gives some insight into the uncertainty principle for angular momentum, which is

$$\Delta L_z \, \Delta\phi \geq \hbar$$

for the z component. Here ϕ is the angular position measured in the plane perpendicular to the z axis. Once m_l for an atom is known, L_z is known precisely, so $\Delta L_z = 0$. (a) What does this tell us about ϕ? (b) What can you say about L_x and L_y, which are *not* quantized (only L and L_z are)? (c) Show that although L_x and L_y are not quantized, nonetheless $(L_x^2 + L_y^2)^{1/2} = [l(l+1) - m_l^2]^{1/2}\hbar$ is.

* 66. Show that the diffractive spread of a laser beam, $\approx \lambda/a$ as described in Section 40–9, is precisely what you might expect from the uncertainty principle. [*Hint*: Since the beam's width is constrained by the dimension of the aperture a, the component of the light's momentum perpendicular to the laser axis is uncertain.]

* 67. Silver atoms (spin $= \frac{1}{2}$) are placed in a 1.0-T magnetic field which splits the ground state into two close levels. (a) What is the difference in energy between these two levels, and (b) what wavelength photon could cause a transition from the lower level to the upper one? (c) How would your answer differ if the atoms were hydrogen?

* 68. Estimate the angular spread of a laser beam due to diffraction if the beam emerges through a 4.0-mm diameter mirror. Assume that $\lambda = 694$ nm. What would be the diameter of this beam if it struck (a) a satellite 1000 km above the Earth, or (b) the Moon?

* 69. *Populations in the H atom.* Use the Boltzmann factor (Eq. 40–16) to estimate the fraction of H atoms in the $n = 2$ and $n = 3$ levels (relative to the ground state) for thermal equilibrium at (a) $T = 300$ K and (b) $T = 6000$ K. [Note: Since there are eight states with $n = 2$ and only two with $n = 1$, multiply your result for $n = 2$ by $\frac{8}{2} = 4$; do similarly for $n = 3$.] (c) Given 1.0 g of hydrogen, estimate the number of atoms in each state at $T = 6000$ K. (d) Estimate the number of $n = 3$ to $n = 1$ and $n = 2$ to $n = 1$ photons that will be emitted per second at $T = 6000$ K. Assume that the lifetime of each excited state is 10^{-8} s.

70. A very simple model of a "one-dimensional" metal consists of N electrons confined to a rigid box of width L. We neglect the Coulomb interaction between the electrons (this is called the "independent electron model" and does remarkably well in predicting many properties of real metals). The Fermi energy of a metal is the energy of the most energetic electron when the metal is in its ground state ($T = 0$ K). (a) Calculate the Fermi energy for this one-dimensional metal, taking into account the Pauli exclusion principle. You can assume for simplicity that N is even. (b) What is the smallest amount of energy that this metal can absorb? Consider the limit of your answer for large N. (The relatively small value you should obtain is indicative of how metals can readily conduct).

71. If the principal quantum number n were limited to the range from 1 to 6, how many elements would we find in nature?

72. It is possible for atoms to be excited into states with very high values of the principal quantum number. Electrons in these so-called *Rydberg states* have very small ionization energies and huge orbital radii. This makes them particularly sensitive to external perturbation, as would be the case if the atom were in an electric field. Consider the $n = 50$ state of the hydrogen atom. Determine the binding energy, the radius of the orbit, and the effective cross sectional area of this Rydberg state.

* 73. In the analytical technique known as *electron spin resonance* (ESR), materials with atoms having one or more unpaired electrons are placed in a variable magnetic field plus a constant radio frequency (RF) electric field. The magnetic field splits the energy of the unpaired electrons according to the two possible spin orientations. If the splitting is the same as the photon energy which is available from the RF radiation, a resonant absorption takes place. This can be detected, and is a measure of the g-factor of the electrons in those atoms. The g-factor for molecules and solids can be different than for electrons in atoms, due to the chemical interactions in the material. Therefore, ESR provides a fingerprint of the system. In a particular ESR experiment, the RF radiation has a fixed wavelength of 2.0 cm. As the magnetic field is varied, a resonance is observed at 0.476 T. What is the g-factor for the unpaired electrons in the sample?

This Pentium chip is one of the leading processors used in computers today. It contains over 3 million transistors, not to mention diodes and other semiconductor electronic elements. Before discussing semiconductors and their applications later in this chapter, we begin with a study of how quantum theory describes how atoms bond together to form molecules, and how molecules behave. We then examine how atoms and molecules come together to form solids, with emphasis on metals and semi-conductors, and their use in electricity and electronics.

CHAPTER 41

Molecules and Solids

Since its development in the 1920s, quantum mechanics has had a profound influence on our lives, both intellectually and technologically. Even the way we view the world has changed, as we have seen in the last couple of chapters. In the present chapter, we will discuss how quantum mechanics has given us an understanding of the structure of molecules and matter in bulk, as well as a number of important applications including semiconductor devices. Our discussions will necessarily be qualitative for the most part.

41–1 | Bonding in Molecules

One of the great successes of quantum mechanics was to give scientists, at last, an understanding of the nature of chemical bonds. Since it is based in physics, and because this understanding is so important in many fields, we discuss it here.

By a molecule, we mean a group of two or more atoms that are strongly held together so as to function as a single unit. When atoms make such an attachment, we say that a chemical **bond** has been formed. There are two main types of strong chemical bond: covalent and ionic. Many bonds are actually intermediate between these two types.

Covalent Bonds

Covalent bond

To understand how **covalent bonds** are formed, we take the simplest case, the bond that holds two hydrogen atoms together to form the hydrogen molecule, H_2. The mechanism is basically the same for other covalent bonds. As two H atoms approach each other, the electron clouds begin to overlap, and the electrons from each atom can "orbit" both nuclei. (This is sometimes called "sharing" electrons.) If both electrons are in the ground state ($n = 1$) of their respective atoms, there are two possibilities: their spins can be parallel (both up or both down), in which case the total spin is $S = \frac{1}{2} + \frac{1}{2} = 1$; or their spins can be opposite ($m_s = +\frac{1}{2}$ for one, $m_s = -\frac{1}{2}$ for the other), so that the total spin $S = 0$. We shall now see that a bond is formed only for the $S = 0$ state, when the spins are opposite. First we consider the $S = 1$ state, for which the spins are the same. The two electrons cannot both be in the lowest energy state and be attached to the same atom, for then they would have identical quantum numbers in violation of the exclusion principle. The exclusion principle tells us that since no two electrons can occupy the same quantum state, if two electrons have the same quantum numbers, they must be different in some other way—namely, by being in different places in space (for example, attached to different atoms). When the two hydrogen atoms approach, the electrons will stay away from each other as shown by the probability distribution of Fig. 41–1. The positively charged nuclei then repel each other, and no bond is formed.

For the $S = 0$ state, on the other hand, the spins are opposite and the two electrons are consequently in different quantum states. Hence they can come close together spatially. In this case, the probability distribution looks like Fig. 41–2. As can be seen, the electrons spend much of their time between the two positively charged nuclei, and the latter are attracted to the negatively charged electron cloud between them. This attraction, which holds the two atoms together to form a molecule, constitutes a *covalent bond*.

The probability distributions of Figs. 41–1 and 41–2 can perhaps be better understood on the basis of waves. What the exclusion principle requires is that when the spins are the same, there is destructive interference of the electron wave functions in the region between the two atoms. But when the spins are opposite, constructive interference occurs in the region between the two atoms, resulting in a large amount of negative charge there. Thus a covalent bond can be said to be the result of constructive interference of the electron wave functions in the space between the two atoms, and of the electrostatic attraction of the two positive nuclei for the negative charge concentration between them.

Why a bond is formed can also be understood from the energy point of view. When the two H atoms approach close to one another, if the spins of their electrons are opposite, the electrons can occupy the same space, as discussed above. This means that each electron can now move about in the space of two atoms instead of in the volume of only one. Because each electron now occupies more space, it is less well localized. From the uncertainty principle, with Δx increased, the momentum, and hence the energy, can be less. Another way of understanding this is from the point of view of de Broglie waves (Section 38–11). Because each electron has a larger "orbit," its wavelength λ can be longer, so its momentum $p = h/\lambda$ (Eq. 38–7) can be less. With less momentum, each electron has less energy when the two atoms combine than when they are separate. That is, the molecule has less energy than the two separate atoms, and so is more stable. An energy input is required to break the H_2 molecule into two separate H atoms, so the H_2 molecule is a stable entity. This is what we mean by a *bond*. The energy required to break a bond is called the **bond energy**, the **binding energy**, or the **dissociation energy**. For the hydrogen molecule, H_2, the bond energy is 4.5 eV.

Ionic Bonds

An **ionic bond** is, in a sense, a special case of the covalent bond. Instead of the electrons being shared equally, they are shared unequally. For example, in sodium chloride (NaCl), the outer electron of the sodium spends nearly all its time around the chlorine (Fig. 41–3). The chlorine atom acquires a net negative charge as a result of the extra electron, whereas the sodium atom is left with a net positive charge. The electrostatic attraction between these two charged atoms holds them together. The resulting bond is called an *ionic bond* because it is created by the attraction between the two ions (Na^+ and Cl^-). But to understand the ionic bond,

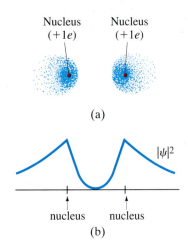

FIGURE 41–1 Electron probability distribution (electron cloud) for two H atoms when the spins are the same ($S = 1$): (a) electron cloud; (b) projection of $|\psi|^2$ along the line through the centers of the two atoms.

FIGURE 41–2 Electron probability distribution for two H atoms when the spins are opposite ($S = 0$): (a) electron cloud; (b) projection of $|\psi|^2$ along the line through the centers of the atoms. In this case a bond is formed because the positive nuclei are attracted to the concentration of negative charge between them. This is a hydrogen molecule, H_2.

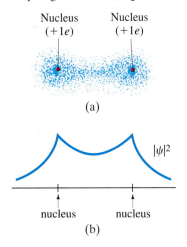

FIGURE 41–3 (below) Probability distribution $|\psi|^2$ for the last electron of Na in NaCl.

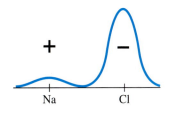

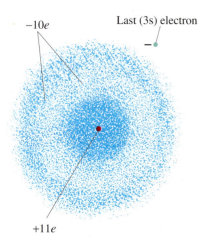

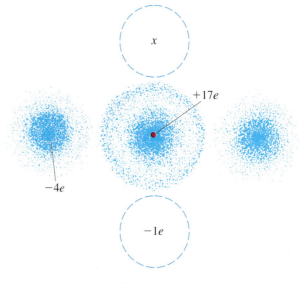

−10e

Last (3s) electron

+11e

FIGURE 41–4 In a neutral sodium atom, the 10 inner electrons shield the nucleus, so the single outer electron is attracted by a net charge of +1e.

FIGURE 41–5 Neutral chlorine atom. The +17e of the nucleus is shielded by the 12 electrons in the inner shells and subshells. Four of the five 3p electrons are shown in doughnut-shaped clouds, and the fifth is in the (dashed-line) cloud concentrated about the vertical axis. An extra electron at x will be attracted by a net charge that can be as much as +5e.

x

+17e

−4e

−1e

we must understand why the extra electron from the sodium spends so much of its time around the chlorine. After all, the chlorine is neutral; why should it attract another electron?

The answer lies in the probability distributions of the two neutral atoms. Sodium contains 11 electrons, 10 of which together form spherically symmetric closed shells (Fig. 41–4). The last electron spends most of its time beyond these closed shells. Because the closed shells have a total charge of −10e and the nucleus has charge +11e, the outermost electron in sodium "feels" a net attraction due to +1e. It is not held very strongly. On the other hand, 12 of chlorine's 17 electrons form closed shells, or subshells (corresponding to $1s^2 2s^2 2p^6 3s^2$). These 12 form a spherically symmetric shield around the nucleus. The other five electrons are in 3p states whose probability distributions are not spherically symmetric and have a form similar to those for the 2p states in hydrogen shown in Fig. 40–9a. Four of these 3p electrons can have "doughnut-shaped" distributions symmetric about the z axis, as shown in Fig. 41–5. The fifth can have a "barbell-shaped" distribution (as for $m_l = 0$ in Fig. 40–9a), which in Fig. 41–5 is shown only as a dashed outline because it is half empty. That is, the exclusion principle allows one more electron to be in this state (it will have spin opposite to that of the electron already there). If an extra electron—say from a Na atom—happens to be in the vicinity, it can be in this state, say at point x in Fig. 41–5. It could experience an attraction due to as much as +5e because the +17e of the chlorine nucleus is partly shielded at this point by the 12 inner electrons. Thus, the outer electron of a Na atom will be more strongly attracted by the +5e of the chlorine atom that it "sees" at certain points rather than by the +1e of its own atom. This, combined with the strong attraction between the two ions when the extra electron stays with the Cl⁻, produces the charge distribution of Fig. 41–3, and hence the ionic bond.

Partial Ionic Character of Covalent Bonds

A pure covalent bond in which the electrons are shared equally occurs mainly in symmetrical molecules such as H_2, O_2, and Cl_2. When the atoms involved are different from each other, it is usual to find that the shared electrons are more likely to be in the vicinity of one atom than the other. The extreme case is an ionic bond; in intermediate cases the covalent bond is said to have a *partial ionic character*. The molecules themselves are **polar**—that is, they have an electric dipole moment because one part (or parts) of the molecule has a net positive charge and other parts a net negative charge. An example is the water molecule, H_2O (Fig. 41–6). The shared electrons are more likely to be found around the oxygen atom than around the two hydrogens. The reason is similar to that discussed above in con-

FIGURE 41–6
The water molecule is polar.

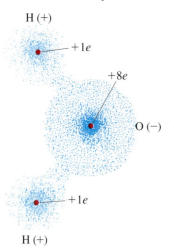

H (+)

+1e

+8e

O (−)

+1e

H (+)

nection with ionic bonds. Oxygen has eight electrons $(1s^2 2s^2 2p^4)$, of which four form a spherically symmetric core and the other four could have, for example, a doughnut-shaped distribution. The barbell-shaped distribution on the z axis (like that shown dashed in Fig. 41–5) could be empty, so electrons from hydrogen atoms can be attracted by a net charge of $+4e$. They are also attracted by the H nuclei, so they partly orbit the H atoms as well as the O atom. The net effect is that there is a net positive charge on each H atom (less than $+1e$), because the electrons spend only part of their time there. And, there is a net negative charge on the O atom.

41–2 Potential-Energy Diagrams for Molecules

It is useful to analyze the interaction between two objects—say, between two atoms or molecules—with the use of a potential-energy diagram, a plot of the potential energy versus the separation distance.

For the simple case of two point charges, q_1 and q_2, the potential energy U is given by (see Chapter 23):

$$U = \frac{1}{4\pi\epsilon_0}\frac{q_1 q_2}{r},$$

where r is the distance between the charges, and the constant $(1/4\pi\epsilon_0)$ is equal to $9.0 \times 10^9\,\text{N}\cdot\text{m}^2/\text{C}^2$. If the two charges have the same sign, the potential energy is positive for all values of r, and a graph of U versus r in this case is shown in Fig. 41–7a. The force is repulsive (the charges have the same sign) and the curve rises as r decreases; this makes sense since work is done to bring the charges together, thereby increasing their potential energy. If, on the other hand, the two charges are of opposite sign, the potential energy is negative because the product $q_1 q_2$ is negative. The force is attractive in this case and the graph of $U (\propto - 1/r)$ versus r looks like Fig. 41–7b. The potential energy becomes more *negative* as r decreases.

Now let us look at the potential energy diagram for the formation of a covalent bond, such as for the hydrogen molecule. The potential energy of one H atom in the presence of the other is plotted in Fig. 41–8. The potential energy U decreases as the atoms approach, because the electrons concentrate between the two nuclei (Fig. 41–2), so attraction occurs. However, at very short distances, the electrons would be "squeezed out"—there is no room for them between the two nuclei. Without the electrons between them, each nucleus would feel a repulsive force due to the other, so the curve rises as r decreases further. There is an optimum separation of the atoms, r_0 in Fig. 41–8, at which the energy is lowest. This is the point of greatest stability for the hydrogen molecule, and r_0 is the average separation of atoms in the H_2 molecule. The depth of this "well" is the *binding energy*,[†] as shown. This is how much energy must be put into the system to separate the two atoms to infinity, where the potential energy $U = 0$. For the H_2 molecule, the binding energy is about 4.5 eV, and $r_0 = 0.074\,\text{nm}$. In molecules made

[†]The binding energy corresponds not quite to the bottom of the potential-energy curve, but to the lowest energy state, slightly above it, as shown in Fig. 41–8. Compare to a square well, Fig. 39–8.

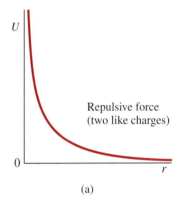

(a)

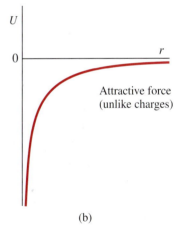

(b)

FIGURE 41–7 Potential energy U as a function of the separation for two point charges of (a) like sign and (b) opposite sign.

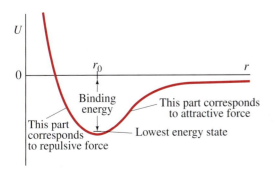

FIGURE 41–8 Potential-energy diagram for H_2 molecule; r is the separation of the two H atoms. The binding energy (the energy difference between $U = 0$ and the lowest energy state near the bottom of the well) is 4.5 eV, and $r_0 = 0.074\,\text{nm}$.

of larger atoms, say, oxygen or nitrogen, repulsion also occurs at short distances, because the closed inner electron shells begin to overlap and the exclusion principle forbids their coming too close. The repulsive part of the curve rises even more steeply than $1/r$. A reasonable approximation to the potential energy, at least in the vicinity of r_0, is

$$ U = -\frac{A}{r^m} + \frac{B}{r^n}, \tag{41-1} $$

where A and B are constants associated with the attractive and repulsive parts of the potential energy and m and n are small integers. For ionic and some covalent bonds, the attractive term can often be written with $m = 1$ (Coulomb potential).

For many bonds, the potential-energy curve has the shape shown in Fig. 41–9. There is still an optimum distance r_0 at which the molecule is stable. But when the atoms approach from a large distance, the force is initially repulsive rather than attractive. The atoms thus do not interact spontaneously. Instead, some additional energy must be injected into the system to get it over the "hump" (or barrier) in the potential-energy diagram. This required energy is called the **activation energy**.

The curve of Fig. 41–9 is much more common than that of Fig. 41–8. The activation energy often reflects a need to break other bonds, before the one under discussion can be made. For example, to make water from O_2 and H_2, the H_2 and O_2 molecules must first be broken into H and O atoms by an input of energy; this is what the activation energy represents. Then the H and O atoms can combine to form H_2O with the release of a great deal more energy than was put in initially. The initial activation energy can be provided by applying an electric spark to a mixture of H_2 and O_2, breaking a few of these molecules into H and O atoms. The resulting explosive release of energy when these atoms combine to form H_2O quickly provides the activation energy needed for further reactions, so additional H_2 and O_2 molecules are broken up and recombined to form H_2O.

The potential-energy diagrams for ionic bonds can have similar shapes. In NaCl, for example, the Na^+ and Cl^- ions attract each other at distances a bit larger than some r_0, but at shorter distances the overlapping of inner electron shells gives rise to repulsion. The two atoms thus are most stable at some intermediate separation, r_0, and there often is an activation energy.

FIGURE 41–9 Potential energy diagram for a bond requiring an activation energy.

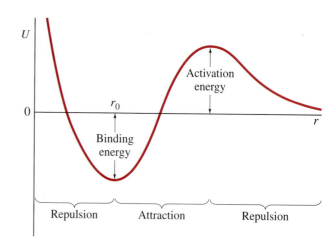

EXAMPLE 41-1 **ESTIMATE** **Sodium chloride bond.** A potential-energy diagram for the NaCl ionic bond is shown in Fig. 41–10, where we have set $U = 0$ for free Na and Cl neutral atoms (which are represented on the right in Fig. 41–10). Measurements show that 5.14 eV are required to remove an electron from a neutral Na atom to produce the Na^+ ion; and 3.61 eV of energy is released when an electron is "grabbed" by a Cl atom to form the Cl^- ion. Thus, forming Na^+ and Cl^- ions from neutral Na and Cl atoms requires 5.14 eV − 3.61 eV = 1.53 eV of energy, a form of activation energy. This is shown as the "bump" in Fig. 41–10. But note that the potential-energy diagram from here out to the right is not really a function of distance—it is drawn dashed to remind us that it only represents the energy difference between the ions and the neutral atoms (for which we have chosen $U = 0$). (a) Calculate the separation distance, r_1, at which the potential of the Na^+ and Cl^- ions drops to zero (measured value is $r_1 = 0.94$ nm). (b) Estimate the binding energy of the NaCl bond, which occurs at a separation $r_0 = 0.24$ nm. Ignore the repulsion of the overlapping electron shells that occurs at this distance (and causes the rise of the potential-energy curve for $r < r_0$, Fig. 41–10).

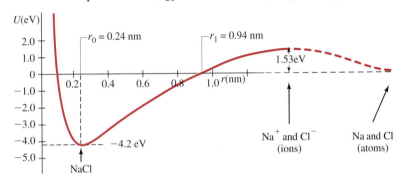

FIGURE 41–10 Potential-energy diagram for the NaCl bond. Beyond about $r = 1.2$ nm, the diagram is schematic only, and represents the energy difference between ions and neutral atoms. $U = 0$ is chosen for the two separated atoms Na and Cl (not for the ions). [For the two ions, Na^+ and Cl^-, the zero of potential energy at $r = \infty$, corresponds to $U \approx 1.53$ eV on this diagram.]

SOLUTION (a) The potential energy of two point charges is given by Coulombs' law:

$$U' = \frac{1}{4\pi\epsilon_0} \frac{q_1 q_2}{r},$$

where we distinguish U' from the U in Fig. 41–10. This formula works for our two ions if we set $U' = 0$ at $r = \infty$, which in the plot of Fig. 41–10 corresponds to $U = +1.53$ eV (Fig. 41–10 is drawn for $U = 0$ for the free *atoms*). The point r_1 in Fig. 41–10 corresponds to $U' = -1.53$ eV relative to the two free ions. Since $q_1 = +e$ and $q_2 = -e$, we have

$$r_1 = \frac{1}{4\pi\epsilon_0} \frac{q_1 q_2}{U'} = \frac{(9.0 \times 10^9\,\mathrm{N \cdot m^2/C^2})(-1.60 \times 10^{-19}\,\mathrm{C})(+1.60 \times 10^{-19}\,\mathrm{C})}{(-1.53\,\mathrm{eV})(1.60 \times 10^{-19}\,\mathrm{J/eV})}$$

$$= 0.94\,\mathrm{nm},$$

which is just the measured value.

(b) At $r_0 = 0.24$ nm, the potential energy of the two ions (relative to $r = \infty$ for the two ions) is

$$U' = \frac{1}{4\pi\epsilon_0} \frac{q_1 q_2}{r}$$

$$= \frac{(9.0 \times 10^9\,\mathrm{N \cdot m^2/C^2})(-1.60 \times 10^{-19}\,\mathrm{C})(+1.60 \times 10^{-19}\,\mathrm{C})}{(0.24 \times 10^{-9}\,\mathrm{m})(1.60 \times 10^{-19}\,\mathrm{J/eV})} = -6.0\,\mathrm{eV}.$$

Thus, we estimate that 6.0 eV of energy is given up when Na^+ and Cl^- ions form a NaCl bond. Put another way, it takes 6.0 eV to break the NaCl bond and form the Na^+ and Cl^- ions. To get the binding energy—the energy to separate the NaCl into Na and Cl atoms—we need to subtract out the 1.53 eV (the "bump" in Fig. 41–10) needed to ionize them:

$$\text{binding energy} = 6.0\,\mathrm{eV} - 1.53\,\mathrm{eV} = 4.5\,\mathrm{eV}.$$

The measured value (shown on Fig. 41–10) is 4.2 eV. The difference can be attributed to the energy associated with the repulsion of the electron shells at this distance.

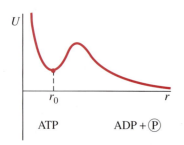

FIGURE 41–11 Potential energy diagram for the formation of ATP from ADP and phosphate (Ⓟ).

Sometimes the potential energy of a bond looks like that of Fig. 41–11. In this case, the energy of the bonded molecule, at a separation r_0, is greater than when there is no bond ($r = \infty$). That is, an energy *input* is required to make the bond (hence the binding energy is negative), and there is energy release when the bond is broken. Such a bond is stable only because there is the barrier of the activation energy. This type of bond is important in living cells, for it is in such bonds that energy can be stored efficiently in certain molecules, particularly ATP (adenosine triphosphate). The bond that connects the last phosphate group (designated Ⓟ in Fig. 41–11) to the rest of the molecule (ADP, meaning adenosine diphosphate, since it contains only two phosphates) is of the form shown in Fig. 41–11. Energy is actually stored in this bond. When the bond is broken (ATP → ADP + Ⓟ), energy is released and this energy can be used to make other chemical reactions "go."

41–3 Weak (van der Waals) Bonds

Once a bond between two atoms or ions is made, energy must normally be supplied to break the bond and separate the atoms. As mentioned in Section 41–1, this energy is called the *bond energy* or *binding energy*. The binding energy for covalent and ionic bonds is typically 2 to 5 eV. These bonds, which hold atoms together to form molecules, are often called **strong bonds** to distinguish them from so-called "weak bonds." The term **weak bond**, as we use it here, refers to an attachment between molecules due to simple electrostatic attraction—such as *between* polar molecules (not *within* a polar molecule, which is a strong bond). The strength of the attachment is much less than for the strong bonds. Binding energies are typically in the range 0.04 to 0.3 eV—hence their name "weak bonds."

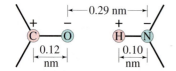

FIGURE 41–12 The C^+—O^- and H^+—N^- dipoles attract each other. (These dipoles may be part of larger molecules. See Fig. 41–13.)

Weak bonds are generally the result of attraction between dipoles. For example, Fig. 41–12 shows two molecules, which have permanent dipole moments, attracting one another. Besides such **dipole–dipole bonds**, there can also be **dipole–induced dipole bonds**, in which a polar molecule with a permanent dipole moment can induce a dipole moment in an otherwise electrically balanced (nonpolar) molecule, just as a single charge can induce a separation of charge in a nearby object (see Fig. 21–7). There can even be an attraction between two nonpolar molecules. Even though a molecule may not have a permanent dipole moment on the average, we can think of its electrons as moving about so that at any instant there may be a separation of charge. Such transient dipoles can induce a dipole moment in a nearby molecule, creating a weak attraction. All these weak bonds are referred to as **van der Waals bonds**, and the forces involved **van der Waals forces**. The potential energy has the general shape shown in Fig. 41–8, with the attractive van der Waals potential energy varying as $1/r^6$.

When one of the atoms in a dipole–dipole bond is hydrogen, as in Fig. 41–12, it is called a **hydrogen bond**. A hydrogen bond is generally quite strong for a so-called "weak bond." This is because the hydrogen atom is the smallest atom and thus can be approached more closely. A hydrogen bond also has a partial "covalent" character: that is, electrons between the two dipoles may be shared to a small extent, thus making a stronger, more lasting, bond.

Weak bonds are important in liquids and solids when strong bonds are absent (see Section 41–5). They are also very important for understanding the activities of cells, such as the double-helix shape of DNA (Fig. 41–13), and DNA replication. The average kinetic energy of molecules in a cell is around $\frac{3}{2}kT \approx 0.04$ eV, about the magnitude of weak bonds. This means that a weak bond can readily be broken just by a molecular collision. Hence weak bonds are not very permanent—they are, instead, brief attachments. But because of this, they play an important role in the cell. On the other hand, strong bonds—those that hold molecules together—are almost never broken simply by molecular collision.

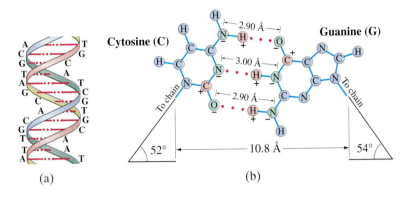

Thus they are relatively permanent. They can be broken by chemical action (the making of even stronger bonds), and this usually happens in the cell with the aid of an enzyme, which is a protein molecule.

EXAMPLE 41–2   Nucleotide energy. Calculate the interaction energy between the C=O dipole of thymine and the H—N dipole of adenine, assuming that the two dipoles are lined up as shown in Fig. 41–12. Dipole moment measurements (see Table 23–2) give $q_H = -q_N = 0.19e = 3.0 \times 10^{-20}$ C, and $q_C = -q_O = 0.42e = 6.7 \times 10^{-20}$ C.

SOLUTION The interaction energy will be equal to the potential energy of one dipole in the presence of the other, since this will be equal to the work needed to pull the two dipoles infinitely far apart. U will consist of four terms:

$$U = U_{CH} + U_{CN} + U_{OH} + U_{ON},$$

where U_{CH} means the potential energy of C in the presence of H, and similarly for the other terms. We do not have terms corresponding to C and O, or N and H, because the two dipoles are assumed to be stable entities (the bonds C=O and H—N remain intact). The potential energy for two point charges is $U_{12} = (1/4\pi\epsilon_0)(q_1 q_2/r)$, where r is the distance between them. So

$$U = \frac{1}{4\pi\epsilon_0}\left(\frac{q_C q_H}{r_{CH}} + \frac{q_C q_N}{r_{CN}} + \frac{q_O q_H}{r_{OH}} + \frac{q_O q_N}{r_{ON}}\right).$$

Using the distances shown in Fig. 41–12, we get:

$$U = (9.0 \times 10^9 \, \text{N·m}^2/\text{C})\left(\frac{(6.7)(3.0)}{0.31} + \frac{(6.7)(-3.0)}{0.41} + \frac{(-6.7)(3.0)}{0.19}\right.$$
$$\left. + \frac{(-6.7)(-3.0)}{0.29}\right)\frac{(10^{-20} \, \text{C})^2}{(10^{-9} \, \text{m})}$$
$$= -1.82 \times 10^{-20} \, \text{J} = -0.11 \, \text{eV}.$$

The potential energy is negative, meaning 0.11 eV of work (or energy input) is required to separate the molecules. That is, the binding energy of this "weak" or hydrogen bond is 0.11 eV. This is only an estimate, of course, since other charges in the vicinity would have an influence too.

41–4 Molecular Spectra

When atoms combine to form molecules, the wave functions (and probability distributions) of the outer electrons overlap, and this interaction alters the energy levels. Nonetheless, molecules can undergo transitions between electron energy levels just as atoms do. For example, the H_2 molecule can absorb a photon of just the

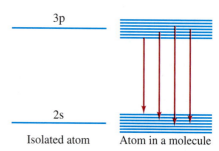

FIGURE 41–14 The individual energy levels of an isolated atom become bands of closely spaced levels in molecules, as well as in solids and liquids.

3p

2s

Isolated atom   Atom in a molecule

right frequency to excite one of its ground state electrons to an excited state. The excited electron can then return to the ground state, emitting a photon. The energy of photons emitted by molecules is of the same order of magnitude as for atoms, typically 1 to 10 eV.

Additional energy levels become possible for molecules (but not for atoms) because the molecule as a whole can rotate, and the atoms of the molecule can vibrate relative to each other. The energy levels for both rotational and vibrational levels are quantized, and are generally spaced much more closely $(10^{-3}$ to 10^{-1} eV) than the electronic levels. Each atomic energy level thus becomes a set of closely spaced levels corresponding to the vibrational and rotational motions, Fig. 41–14. Transitions from one level to another appear as many very closely spaced lines. In fact, the lines are not always distinguishable, and these spectra are called **band spectra**. Each type of molecule has its own characteristic spectrum, which can be used for identification and for determination of structure.

Rotational Energy Levels in Molecules

Let us now look in more detail at rotational and vibrational states in molecules. We begin with rotation, considering here only diatomic molecules, although the analysis can be extended to polyatomic molecules. When a diatomic molecule rotates about an axis through the center of mass perpendicular to the line joining the two atoms as shown in Fig. 41–15, its kinetic energy of rotation (see Section 10–10) is

$$E_{rot} = \frac{1}{2} I\omega^2 = \frac{(I\omega)^2}{2I},$$

where $(I\omega)$ is the angular momentum (Section 10–9). Quantum mechanics predicts quantization of angular momentum just as in atoms (see Eq. 40–3):

$$I\omega = \sqrt{L(L+1)}\,\hbar, \qquad L = 0, 1, 2, \cdots,$$

where L is an integer called the **rotational angular momentum quantum number**. Thus the rotational energy is quantized:

$$E_{rot} = \frac{(I\omega)^2}{2I} = L(L+1)\frac{\hbar^2}{2I}, \qquad L = 0, 1, 2, \cdots. \tag{41–2}$$

Transitions between rotational energy levels are subject to the *selection rule* (see the end of Section 40–2):

Selection rule

$$\Delta L = \pm 1.$$

The energy of a photon emitted or absorbed for a transition between rotational states with angular momentum quantum number L and $L - 1$ will be

$$\Delta E_{rot} = E_L - E_{L-1} = \frac{\hbar^2}{2I}L(L+1) - \frac{\hbar^2}{2I}(L-1)(L)$$

$$= \frac{\hbar^2}{I}L. \qquad \begin{bmatrix} L \text{ is for the} \\ \text{upper state} \end{bmatrix} \tag{41–3}$$

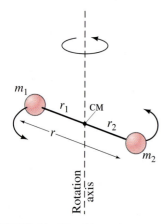

FIGURE 41–15 Diatomic molecule rotating about a vertical axis. The two atoms, of mass m_1 and m_2, are separated by a distance $r = r_1 + r_2$ where r_1 and r_2 are their respective distances from the center of mass.

We see that the transition energy increases directly with L. Figure 41–16 shows some of the allowed rotational energy levels and transitions. Measured absorption lines fall in the microwave or far-infrared regions of the spectrum, and their frequencies are generally $2, 3, 4, \cdots$ times higher than the lowest one, as predicted by Eq. 41–3.

The moment of inertia of the molecule in Fig. 41–15 rotating about its center of mass (Section 10–6) is

$$I = m_1 r_1^2 + m_2 r_2^2,$$

where r_1 and r_2 are the distances of each atom from their common center of mass. We show in Example 41–3 that I can be written

$$I = \frac{m_1 m_2}{m_1 + m_2} r^2 = \mu r^2, \qquad \textbf{(41–4)}$$

where $r = r_1 + r_2$ is the distance between the two atoms of the molecule and $\mu = m_1 m_2 / (m_1 + m_2)$ is called the **reduced mass**. If $m_1 = m_2$, then $\mu = \frac{1}{2} m_1 = \frac{1}{2} m_2$.

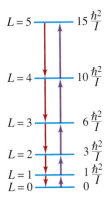

FIGURE 41–16 Rotational energy levels and allowed transitions (emission and absorption) for a diatomic molecule. Upward-pointing arrows represent absorption of a photon, and downward arrows represent emission.

EXAMPLE 41–3 Reduced mass. Show that the moment of inertia of a diatomic molecule rotating about its center of mass can be written

$$I = \mu r^2,$$

where

$$\mu = \frac{m_1 m_2}{m_1 + m_2}$$

is the reduced mass, Eq. 41–4, and r is the distance between the atoms.

SOLUTION The moment of inertia of a single particle of mass m a distance r from the rotation axis is $I = mr^2$. Hence for our diatomic molecule (Fig. 41–15)

$$I = m_1 r_1^2 + m_2 r_2^2.$$

Now $r = r_1 + r_2$ and $m_1 r_1 = m_2 r_2$ because the axis of rotation passes through the center of mass. Hence

$$r_1 = r - r_2 = r - \frac{m_1}{m_2} r_1,$$

so

$$r_1 = \frac{r}{1 + \dfrac{m_1}{m_2}} = \frac{m_2 r}{m_1 + m_2}.$$

Similarly,

$$r_2 = \frac{m_1 r}{m_1 + m_2}.$$

Then

$$I = m_1 \left(\frac{m_2 r}{m_1 + m_2} \right)^2 + m_2 \left(\frac{m_1 r}{m_1 + m_2} \right)^2 = \frac{m_1 m_2 (m_1 + m_2) r^2}{(m_1 + m_2)^2}$$

$$= \frac{m_1 m_2}{m_1 + m_2} r^2 = \mu r^2,$$

where

$$\mu = \frac{m_1 m_2}{m_1 + m_2},$$

which is what we wished to show.

EXAMPLE 41–4 Rotational transition. A rotational transition $L = 1$ to $L = 0$ for the molecule CO has a measured absorption wavelength $\lambda = 2.60\,\text{mm}$ (microwave region). Use this to calculate (a) the moment of inertia of the CO molecule, and (b) the CO bond length, r. (c) Calculate the wavelengths of the next three rotational transitions, and the energies of the photon emitted for each of these four transitions.

SOLUTION (a) From Eq. 41–3, we can write

$$\frac{\hbar^2}{I}L = \Delta E = hf = \frac{hc}{\lambda_1}.$$

With $L = 1$ (the upper state) in this case, we solve for I:

$$I = \frac{\hbar^2 L}{hc}\lambda_1 = \frac{h\lambda_1}{4\pi^2 c} = \frac{(6.63 \times 10^{-34}\,\text{J}\cdot\text{s})(2.60 \times 10^{-3}\,\text{m})}{4\pi^2(3.00 \times 10^8\,\text{m/s})}$$

$$= 1.46 \times 10^{-46}\,\text{kg}\cdot\text{m}^2.$$

(b) The masses of C and O are 12.0 and 16.0 u, respectively, where $1\,\text{u} = 1.66 \times 10^{-27}\,\text{kg}$. Thus the reduced mass is

$$\mu = \frac{m_1 m_2}{m_1 + m_2} = \frac{(12.0)(16.0)}{28.0}(1.66 \times 10^{-27}\,\text{kg}) = 1.14 \times 10^{-26}\,\text{kg}$$

or 6.86 u. Then, from Eq. 41–4, the bond length is

$$r = \sqrt{\frac{I}{\mu}} = \sqrt{\frac{1.46 \times 10^{-46}\,\text{kg}\cdot\text{m}^2}{1.14 \times 10^{-26}\,\text{kg}}} = 1.13 \times 10^{-10}\,\text{m} = 0.113\,\text{nm}.$$

(c) From Eq. 41–3, $\Delta E \propto L$. Hence $\lambda = c/f = hc/\Delta E$ is proportional to $1/L$. Thus, for $L = 2$ to $L = 1$ transitions, $\lambda_2 = \frac{1}{2}\lambda_1 = 1.30\,\text{mm}$. For $L = 3$ to $L = 2$, $\lambda_3 = \frac{1}{3}\lambda_1 = 0.87\,\text{mm}$. And for $L = 4$ to $L = 3$, $\lambda_4 = 0.65\,\text{mm}$. All are close to measured values. The energies of the photons, $hf = hc/\lambda$, are respectively $4.8 \times 10^{-4}\,\text{eV}$, $9.5 \times 10^{-4}\,\text{eV}$, $1.4 \times 10^{-3}\,\text{eV}$, and $1.9 \times 10^{-3}\,\text{eV}$.

Vibrational Energy Levels in Molecules

The potential energy of the two atoms in a typical diatomic molecule has the shape shown in Fig. 41–8 or 41–9, and Fig. 41–17 again shows the potential energy for the H_2 molecule. We note that the potential energy, at least in the vicinity of the equilibrium separation r_0, closely resembles the potential energy of a harmonic oscillator, $U = \frac{1}{2}kx^2$, which is shown superposed in dashed lines. Thus, for small displacements from r_0, each atom experiences a restoring force approximately proportional to the displacement, and the molecule vibrates as a simple harmonic oscillator (SHO). The classical frequency of vibration is

Classical vibration frequency

$$f = \frac{1}{2\pi}\sqrt{\frac{k}{\mu}}, \tag{41–5}$$

FIGURE 41–17 Potential energy for the H_2 molecule and for a simple harmonic oscillator ($U_{\text{SHO}} = \frac{1}{2}kx^2$, with $|x| = |r - r_0|$). The 0.50 eV energy height marked is for use in Example 41–5 to estimate k. [Note that $U_{\text{SHO}} = 0$ is not the same as $U = 0$ for the molecule.]

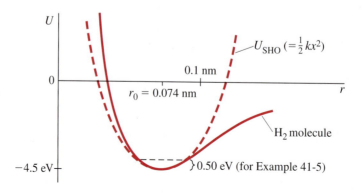

where k is the "stiffness constant" (as for a spring, Chapter 14) and instead of the mass m we must again use the reduced mass $\mu = m_1 m_2/(m_1 + m_2)$. (This is shown in Problem 59.) The Schrödinger equation for the SHO potential energy yields solutions for energy that are quantized according to

$$E_{vib} = (\nu + \tfrac{1}{2})hf \qquad \nu = 0, 1, 2, \cdots, \qquad \textbf{(41–6)}$$

where f is given by Eq. 41–5 and ν is an integer called the **vibrational quantum number**. The lowest energy state ($\nu = 0$) is not zero (as for rotation) but has $E = \tfrac{1}{2}hf$. This is called the **zero-point energy**.[†] Higher states have energy $\tfrac{3}{2}hf, \tfrac{5}{2}hf$, and so on, as shown in Fig. 41–18. Transitions are subject to the selection rule

$$\Delta\nu = \pm 1,$$

so allowed transitions occur only between adjacent states and all give off photons of energy

$$\Delta E_{vib} = hf. \qquad \textbf{(41–7)}$$

This is very close to experimental values for small ν, but for higher energies, the potential-energy curve (Fig. 41–17) begins to deviate from a perfect SHO curve, which affects the wavelengths and frequencies of the transitions. Typical transition energies are on the order of $10^{-1}\,eV$, about 10 to 100 times larger than for rotational transitions, with wavelengths in the infrared region of the spectrum $(\approx 10^{-5}\,m)$.[‡]

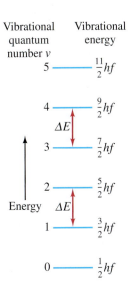

FIGURE 41–18 Allowed vibrational energies of a diatomic molecule, where f is the fundamental frequency of vibration, given by Eq. 41–5. Transitions are allowed only between adjacent levels ($\Delta\nu = \pm 1$), which are equally spaced.

EXAMPLE 41–5 ESTIMATE Wavelength for H_2. (a) Use the curve of Fig. 41–17 to estimate the value of the stiffness constant k for the H_2 molecule, and then (b) estimate the fundamental wavelength for vibrational transitions.

SOLUTION (a) We choose arbitrarily an energy height of 0.50 eV which is indicated in Fig. 41–17. By measuring directly on the graph, we find that this energy corresponds to a vibration on either side of $r_0 = 0.074\,nm$ of about $x = \pm 0.017\,nm$. For SHO, $U_{SHO} = \tfrac{1}{2}kx^2$ and $U_{SHO} = 0$ at $x = 0$ ($r = r_0$); then

$$k = \frac{2U_{SHO}}{x^2} \approx \frac{2(0.50\,eV)(1.6 \times 10^{-19}\,J/eV)}{(1.7 \times 10^{-11}\,m)^2} \approx 550\,N/m.$$

This value of k would also be reasonable for a macroscopic spring. (b) The reduced mass is $\mu = m_1 m_2/(m_1 + m_2) = m_1/2 = \tfrac{1}{2}(1.0\,u)(1.66 \times 10^{-27}\,kg) = 0.83 \times 10^{-27}\,kg$. Hence, using Eq. 41–5,

$$\lambda = \frac{c}{f} = 2\pi c \sqrt{\frac{\mu}{k}} = 2\pi (3.0 \times 10^8\,m/s) \sqrt{\frac{0.83 \times 10^{-27}\,kg}{550\,N/m}}$$

$$= 2300\,nm,$$

which is in the infrared region of the spectrum.

Experimentally, we do the inverse process: The wavelengths of vibrational transitions for a given molecule are measured, and from this the stiffness constant k can be calculated. The values of k calculated in this way are a measure of the strength of the molecular bond.

[†] Recall this phenomenon for a square well, Fig. 39–8.

[‡] Forbidden transitions with $\Delta\nu = 2$ are emitted somewhat more weakly, but their observation can be important in some cases, such as in astronomy.

Vibrational energy levels in hydrogen. Given that the hydrogen molecule emits infrared radiation of wavelength around 2300 nm, (*a*) what is the separation in energy between different vibrational levels, and (*b*) what is the lowest vibrational energy state?

SOLUTION (*a*)

$$\Delta E_{vib} = hf = \frac{hc}{\lambda} = \frac{(6.63 \times 10^{-34}\,J\cdot s)(3.00 \times 10^8\,m/s)}{(2300 \times 10^{-9}\,m)(1.60 \times 10^{-19}\,J/eV)} = 0.54\,eV.$$

(*b*) The lowest vibrational energy has $\nu = 0$ in Eq. 41–6: $E = \frac{1}{2}hf = 0.27\,eV$.

Rotational plus Vibrational Levels

When energy is imparted to a molecule, both the rotational and vibrational modes can be excited. Because rotational energies are an order of magnitude or so smaller than vibrational energies, which in turn are smaller than the electronic energy levels, we can represent the grouping of levels as shown in Fig. 41–19. Transitions between energy levels, with emission of a photon, are subject to the *selection rules*:

$$\Delta \nu = \pm 1$$

and

$$\Delta L = \pm 1.$$

Some allowed and forbidden (marked ✕) transitions are indicated in Fig. 41–19. Not all transitions and levels are shown, and the separation between vibrational levels, and (even more) between rotational levels, has been exaggerated. But we can clearly see the origin of the very closely spaced lines that give rise to the band spectra, as mentioned with reference to Fig. 41–14 earlier in this Section.

The spectra are quite complicated, so we consider briefly only transitions within the same electronic level, such as those at the top of Fig. 41–19. A transition from a state with quantum numbers ν and L, to one with quantum numbers $\nu + 1$ and $L \pm 1$ (see the selection rules above), will absorb[†] a photon of energy:

$$
\begin{aligned}
\Delta E &= \Delta E_{vib} + \Delta E_{rot} \\
&= hf + (L + 1)\frac{\hbar^2}{I} \quad \begin{bmatrix} L \to L + 1 \\ (\Delta L = +1) \end{bmatrix}, \quad L = 0, 1, 2, \cdots \\
&= hf - L\frac{\hbar^2}{I} \quad \begin{bmatrix} L \to L - 1 \\ (\Delta L = -1) \end{bmatrix}, \quad L = 1, 2, 3, \cdots,
\end{aligned}
$$

(41–8)

where we have used Eqs. 41–3 and 41–7. Note that for $L \to L - 1$ transitions, L cannot be zero since there is then no state with $L = -1$. Equations 41–8 predict an absorption spectrum like that shown schematically in Fig. 41–20, with transi-

[†] This is for absorption. For emission of a photon, the transition would be $\nu \to \nu - 1, L \to L \pm 1$.

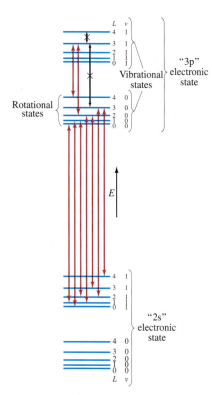

FIGURE 41–19 Combined electronic, vibrational, and rotational energy levels. Transitions marked with an ✕ are not allowed by the selection rules.

FIGURE 41–20 Expected spectrum for transitions between combined rotational and vibrational states.

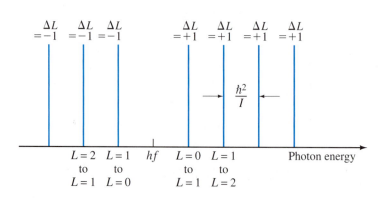

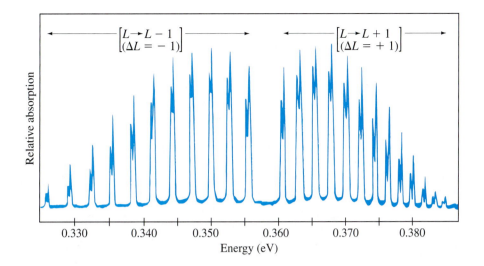

FIGURE 41–21 Absorption spectrum for HCl molecules. Lines on the left correspond to transitions where $L \to L - 1$; those on the right are for $L \to L + 1$. Each line has a double peak because chlorine has two isotopes of different mass and different moment of inertia.

tions $L \to L - 1$ on the left and $L \to L + 1$ on the right. Figure 41–21 shows the molecular absorption spectrum of HCl, which follows this pattern very well. (Each line in that spectrum is split into two because Cl consists of two isotopes of different mass; hence there are two kinds of HCl molecule with different moments of inertia I.)

EXAMPLE 41–7 **ESTIMATE** **The HCl molecule.** Estimate the moment of inertia of the HCl molecule using the absorption spectrum shown in Fig. 41–21. For the purposes of a rough estimate you can ignore the difference between the two isotopes.

SOLUTION The locations of the peaks in Fig. 41–21 should correspond to Eqs. 41–8. We don't know what value of L each peak corresponds to in Fig. 41–21, but we can estimate the energy difference between peaks to be about $\Delta E' = 0.0025$ eV. Then from Eqs. 41–8, the energy difference between two peaks is given by

$$\Delta E' = \Delta E_{L+1} - \Delta E_L$$
$$= \frac{\hbar^2}{I}.$$

Then

$$I = \frac{\hbar^2}{\Delta E'} = \frac{\left(6.626 \times 10^{-34} \, \text{J} \cdot \text{s}/2\pi\right)^2}{(0.0025 \, \text{eV})\left(1.6 \times 10^{-19} \, \text{J/eV}\right)} = 2.8 \times 10^{-47} \, \text{kg} \cdot \text{m}^2.$$

If we use the results of Ex. 41–3, we can get a better idea of what this number means. We write $I = \mu r^2$ and calculate μ:

$$\mu = \frac{m_1 m_2}{m_1 + m_2} = \frac{(1.0 \, \text{u})(35 \, \text{u})}{36 \, \text{u}} \left(1.66 \times 10^{-27} \, \text{kg/u}\right) = 1.6 \times 10^{-27} \, \text{kg}.$$

Then the bond length is given by (Eq. 41–4)

$$r = \left(\frac{I}{\mu}\right)^{\frac{1}{2}} = \left(\frac{2.8 \times 10^{-47} \, \text{kg} \cdot \text{m}^2}{1.6 \times 10^{-27} \, \text{kg}}\right)^{\frac{1}{2}} = 1.3 \times 10^{-10} \, \text{m},$$

which is the expected order of magnitude for a bond length.

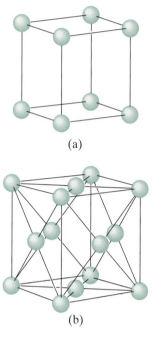

(a)

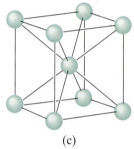

(b)

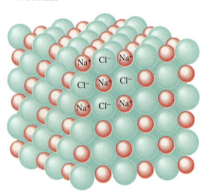

(c)

FIGURE 41–22 Arrangement of atoms in (a) a simple cubic crystal, (b) face-centered cubic crystal (note the atom at the center of each face) and (c) body-centered cubic crystal. Each shows the relationship of the bonds. Each of these "cells" is repeated in three dimensions to the edges of the macroscopic crystal.

FIGURE 41–23 Diagram of an NaCl crystal, showing the "packing" of atoms.

41–5 Bonding in Solids

Quantum mechanics has been a great tool for understanding the structure of solids. This active field of research today is called **solid-state physics**, or **condensed-matter physics** so as to include liquids as well. The rest of this chapter is devoted to this subject, and we begin with a brief look at the structure of solids and the bonds that hold them together.

Although some solid materials are *amorphous* in structure, in that the atoms and molecules show no long-range order, we will be interested here in the large class of *crystalline* substances whose atoms, ions, or molecules are generally believed to form an orderly array known as a *lattice*. Figure 41–22 shows three of the possible arrangements of atoms in a crystal: simple cubic, face-centered cubic, and body-centered cubic. The NaCl crystal is face-centered cubic (see Fig. 41–23), with one Na^+ ion or one Cl^- ion at each lattice point (i.e., considering Na and Cl separately).

The molecules of a solid are held together in a number of ways. The most common are by *covalent* bonding (such as between the carbon atoms of the diamond crystal) or *ionic* bonding (as in a NaCl crystal). Often the bonds are partially covalent and partially ionic. Our discussion of these bonds earlier in this chapter for molecules applies equally well here to solids.

Let us look for a moment at the NaCl crystal of Fig. 41–23. Each Na^+ ion feels an attractive Coulomb potential due to each of the six "nearest neighbor" Cl^- ions surrounding it. Note that one Na^+ does not "belong" exclusively to one Cl^-, so we must not think of ionic solids as consisting of individual molecules. Each Na^+ also feels a repulsive Coulomb potential due to other Na^+ ions, although this is weaker since the other Na^+ ions are farther away. Thus we expect a net attractive potential

$$U = -\alpha \frac{1}{4\pi\epsilon_0} \frac{e^2}{r}.$$

The factor α is called the *Madelung constant*. If each Na^+ were surrounded by only the six Cl^- ions, α would be 6, but the influence of all the other ions reduces it to a value $\alpha = 1.75$ for the NaCl crystal. The potential must also include a term representing the repulsive force when the wave functions of the inner shells and subshells overlap, and this has the form $U = B/r^m$, where m is a small integer. Thus

$$U = -\frac{\alpha}{4\pi\epsilon_0} \frac{e^2}{r} + \frac{B}{r^m}, \tag{41–9}$$

which has the same form as Eq. 41–1 for molecules (Section 41–2). It can be shown (Problem 19) that, at the equilibrium distance r_0,

$$U = U_0 = -\frac{\alpha}{4\pi\epsilon_0} \frac{e^2}{r_0} \left(1 - \frac{1}{m}\right).$$

This U_0 is known as the *ionic cohesive energy*; it is a sort of "binding energy"—the energy (per ion) needed to take the solid apart into separated ions.

A different type of bond occurs in metals. Metal atoms have relatively loosely held outer electrons. Present-day **metallic bond** theories propose that in a metallic solid, these outer electrons roam rather freely among all the metal atoms which, without their outer electrons, act like positive ions. The electrostatic attraction between the metal ions and this negative electron "gas" is what is believed, at least in part, to hold the solid together. The binding energy of metal bonds are typically 1 to 3 eV, somewhat weaker than ionic or covalent bonds (5 to 10 eV in solids). The "free electrons," according to this theory, are responsible for the high electrical and thermal conductivity of metals (see Sections 41–6 and 41–7). This theory also nicely accounts for the shininess of smooth metal surfaces: the electrons are free and can vibrate at any frequency, so when light of a range of frequencies falls on a metal, the electrons can vibrate in response and re-emit light of those same frequencies. Hence the reflected light will consist largely of the same frequencies as the incident light. Compare this to nonmetallic materials that have a distinct color—the atomic electrons exist only in certain energy states, and when white light falls on them, the atoms absorb at certain frequencies, and reflect other frequencies which make up the color we see.

The atoms or molecules of some materials, such as the noble gases, can form only **weak bonds** with each other. As we saw in Section 41–3, weak bonds have very low binding energies and would not be expected to hold atoms together as a liquid or solid at room temperature. The noble gases condense only at very low temperatures, where the atomic kinetic energy is small and the weak attraction can then hold the atoms together.

41–6 Free-Electron Theory of Metals

Let us look more closely at the free-electron theory of metals mentioned in the preceding Section. Let us imagine the electrons trapped within the metal as being in a potential well: inside the metal, the potential energy is zero, but at the edges of the metal there are high potential walls. Since very few electrons leave the metal at room temperature, we can imagine the walls as being infinitely high (as in Section 39–8). At higher temperatures, electrons do leave the metal (we know that thermionic emission occurs, Section 23–9), so we must recognize that the well is of finite depth. In this simple model, the electrons are trapped within the metal, but are free to move about inside the well whose size is macroscopic—the size of the piece of metal. The energy will be quantized, but the spacing between energy levels will be very tiny (see Eq. 39–13) since L is very large. Indeed, for a cube 1 cm on a side, the number of states with energy between, say, 5.0 and 5.5 eV, is on the order of 10^{22} (see Example 41–8).

To deal with such vast numbers of states, which are so closely spaced as to seem continuous, we need to use statistical methods. We define a quantity known as the **density of states**, $g(E)$, whose meaning is similar to the Maxwell distribution, Eq. 18–6 (see Section 18–2). That is, the quantity $g(E)\,dE$ represents the number of states per unit volume that have energy between E and $E + dE$. A careful calculation (see Problem 35), which must treat the potential well as three dimensional, shows that

$$g(E) = \frac{8\sqrt{2}\,\pi m^{\frac{3}{2}}}{h^3}\,E^{\frac{1}{2}}. \qquad (41-10)$$

This is plotted in Fig. 41–24.

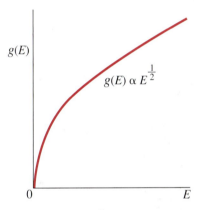

FIGURE 41–24 Density of states $g(E)$ as a function of energy E (Eq. 41–10).

Density of states

EXAMPLE 41–8 Electron states in copper. Estimate the number of states in the range 5.0 to 5.5 eV available to electrons in a 1.0-cm cube of copper metal.

SOLUTION Since $g(E)$ is the number of states per unit volume per unit energy interval, the number N of states is approximately (it is approximate because ΔE is not small)

$$N \approx g(E)V\,\Delta E,$$

where the volume $V = 1.0\,\text{cm}^3 = 1.0 \times 10^{-6}\,\text{m}^3$ and $\Delta E = 0.50\,\text{eV}$. We evaluate $g(E)$ at 5.25 eV, and find (Eq. 41–10):

$$N \approx g(E)V\,\Delta E = \frac{8\sqrt{2}\,\pi(9.1 \times 10^{-31}\,\text{kg})^{\frac{3}{2}}}{(6.63 \times 10^{-34}\,\text{J}\cdot\text{s})^3}\,\sqrt{(5.25\,\text{eV})(1.6 \times 10^{-19}\,\text{J/eV})}$$

$$\times\,(1.0 \times 10^{-6}\,\text{m}^3)(0.50\,\text{eV})(1.6 \times 10^{-19}\,\text{J/eV})$$

$$\approx 8 \times 10^{21}$$

in $1.0\,\text{cm}^3$. Note that the type of metal did not enter the calculation.

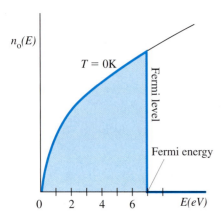

FIGURE 41–25 At $T = 0\,\text{K}$, all states up to energy E_F, called the Fermi energy, are filled.

Fermi level

Fermi energy

FIGURE 41–26 The Fermi–Dirac probability function for two temperatures, $T = 0\,\text{K}$ (black line) and $T = 1200\,\text{K}$ (blue curve). For $f(E) = 1$, a state with energy E is certainly occupied. For $f(E) = 0.5$, which occurs at $E = E_F$, the state with E_F has a 50% chance of being occupied.

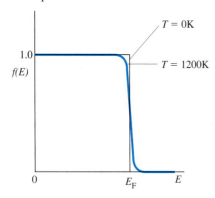

Fermi factor

Equation 41–10 gives us the density of states. Now we must ask: How are the states available to an electron gas actually populated? Let us first consider the situation at absolute zero, $T = 0\,\text{K}$. For a classical ideal gas, all the particles would be in the lowest state, with zero kinetic energy $\left(= \frac{3}{2}kT = 0\right)$. But the situation is vastly different for an electron gas because electrons obey the exclusion principle. Electrons do not obey classical statistics but rather a quantum statistics called **Fermi–Dirac statistics**[†] that takes into account the exclusion principle. All particles that have spin $\frac{1}{2}$ (or other half-integral spin: $\frac{3}{2}, \frac{5}{2}$, etc.), such as electrons, protons, and neutrons, obey Fermi–Dirac statistics and are referred to as **fermions**.[‡] The electron gas in a metal is often called a **Fermi gas**. According to the exclusion principle, no two electrons in the metal can have the same set of quantum numbers. Therefore, in each of the states of our potential well, there can be at most two electrons: one with spin up $\left(m_s = +\frac{1}{2}\right)$ and one with spin down $\left(m_s = -\frac{1}{2}\right)$. (This factor of 2 has already been included in Eq. 41–10.) Thus, at $T = 0\,\text{K}$, the possible energy levels will be filled, two electrons each, up to a maximum level called the **Fermi level**. This is shown in Fig. 41–25, where the vertical axis is labeled $n_o(E)$ for "density of occupied states." The energy of the state at the Fermi level is called the **Fermi energy**, E_F. To determine E_F, we integrate Eq. 41–10 from $E = 0$ to $E = E_F$ (all states up to E_F are filled at $T = 0\,\text{K}$):

$$\frac{N}{V} = \int_0^{E_F} g(E)\, dE, \tag{41–11}$$

where N/V is the number of conduction electrons per unit volume in the metal. Then, solving for E_F, the result (see Example 41–9) is

$$E_F = \frac{h^2}{8m}\left(\frac{3}{\pi}\frac{N}{V}\right)^{\frac{2}{3}}. \tag{41–12}$$

The average energy in this distribution (see Problem 31) is

$$\overline{E} = \frac{3}{5}E_F. \tag{41–13}$$

For copper, $E_F = 7.0\,\text{eV}$ (see Example 41–9) and $\overline{E} = 4.2\,\text{eV}$. This is very much greater than the energy of thermal motion at room temperature $\left(\frac{3}{2}kT \approx 0.04\,\text{eV}\right)$. Clearly, all motion does not stop at absolute zero.

Thus, at $T = 0$, all states with energy below E_F are occupied, and all states above E_F are empty. What happens for $T > 0$? We expect that some (at least) of the electrons will increase in energy due to thermal motion. Classically, the distribution of occupied states would be given by the Boltzmann factor, $e^{-E/kT}$ (see Eqs. 40–16). But for our electron gas, a quantum-mechanical system obeying the exclusion principle, the probability of a given state of energy E being occupied is given by the **Fermi–Dirac probability function** (or **Fermi factor**):

$$f(E) = \frac{1}{e^{(E - E_F)/kT} + 1}, \tag{41–14}$$

where E_F is the Fermi energy. This function is plotted in Fig. 41–26 for two temperatures, $T = 0\,\text{K}$ and $T = 1200\,\text{K}$ (just below the melting point of copper). At $T = 0$ (or as T approaches zero) the factor $e^{(E - E_F)/kT}$ in Eq. 41–14 is zero if $E < E_F$ and is ∞ if $E > E_F$. Thus

$$f(E) = \begin{cases} 1 & E < E_F \\ 0 & E > E_F \end{cases} \quad \text{at} \quad T = 0.$$

This is what is plotted in black in Fig. 41–26 for $T = 0$ and is consistent with Fig. 41–25: all states up to the Fermi level are occupied [probability $f(E) = 1$]

[†]Developed independently by Enrico Fermi (Figs. 39–2, 39–10, and 42–7) in early 1926 and by P.A.M. Dirac a few months later.

[‡]Particles with integer spin (0, 1, 2, etc.), such as the photon, obey *Bose–Einstein* statistics and are called *bosons*.

and all states above are unoccupied. For $T = 1200\,\text{K}$, the Fermi factor changes only a little, as shown in Fig. 41–26 as the blue curve. Note that at any temperature T, when $E = E_\text{F}$, then Eq. 41–14 gives $f(E) = 0.50$, meaning the state at $E = E_\text{F}$ has a 50% chance of being occupied. To see how $f(E)$ affects the actual distribution of electrons in energy states, we must weight the density of possible states, $g(E)$, by the probability that those states will be occupied, $f(E)$. The product of these two functions then gives the **density of occupied states**,

$$n_\text{o}(E) = g(E)f(E) = \frac{8\sqrt{2}\,\pi m^{\frac{3}{2}}}{h^3}\frac{E^{\frac{1}{2}}}{e^{(E-E_\text{F})/kT} + 1}. \qquad (41\text{–}15)$$

Density of occupied states

Then $n_\text{o}(E)\,dE$ represents the number of electrons per unit volume with energy between E and $E + dE$ in thermal equilibrium at temperature T. This is plotted in Fig. 41–27 for $T = 1200\,\text{K}$, a temperature at which a metal is so hot it would glow. We see immediately that the distribution differs very little from that at $T = 0$. We see also that the changes that do occur are concentrated about the Fermi level. A few electrons from slightly below the Fermi level move to energy states slightly above it. The average energy of the electrons increases only very slightly when the temperature is increased from $T = 0\,\text{K}$ to $T = 1200\,\text{K}$. This is very different from the behavior of an ideal gas, for which kinetic energy increases directly with T. Nonetheless, this behavior is readily understood as follows. Energy of thermal motion at $T = 1200\,\text{K}$ is about $\frac{3}{2}kT \approx 0.1\,\text{eV}$. The Fermi level, on the other hand, is on the order of several eV: for copper it is $E_\text{F} \approx 7.0\,\text{eV}$. An electron at $T = 1200\,\text{K}$ may have 7 eV of energy, but it can acquire at most only a few times 0.1 eV of energy by a (thermal) collision with the lattice. Only electrons very near the Fermi level would find vacant states close enough to make such a transition. Essentially none of the electrons could increase in energy by, say, 3 eV, so electrons farther down in the electron gas are unaffected. Only electrons near the top of the energy distribution can be thermally excited to higher states. And their new energy is on the average only slightly higher than their old energy.

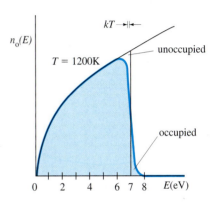

FIGURE 41–27 The density of occupied states for the electron gas in copper. The width kT represents thermal energy at $T = 1200\,\text{K}$.

EXAMPLE 41–9 **The Fermi level.** For the metal copper, determine (a) the Fermi energy, (b) the average energy of electrons, and (c) the speed of electrons at the Fermi level (this is called the *Fermi speed*).

SOLUTION (a) We combine Eqs. 41–10 and 41–11:

$$\frac{N}{V} = \frac{8\sqrt{2}\,\pi m^{\frac{3}{2}}}{h^3}\int_0^{E_\text{F}} E^{\frac{1}{2}}\,dE = \frac{8\sqrt{2}\,\pi m^{\frac{3}{2}}}{h^3}\frac{2}{3}E_\text{F}^{\frac{3}{2}}.$$

Solving for E_F, we obtain

$$E_\text{F} = \frac{h^2}{8m}\left(\frac{3}{\pi}\frac{N}{V}\right)^{\frac{2}{3}}$$

[note: $\left(\sqrt{2}\right)^{\frac{2}{3}} = 2\left(\sqrt{2}/2^{\frac{3}{2}}\right)^{\frac{2}{3}} = 2\left(\frac{1}{2}\right)^{\frac{2}{3}}$], and this is Eq. 41–12. We calculated N/V, the number of conduction electrons per unit volume in copper, in Example 25–12 to be $N/V = 8.4 \times 10^{28}\,\text{m}^{-3}$. Hence for copper

$$E_\text{F} = \frac{(6.63 \times 10^{-34}\,\text{J}\cdot\text{s})^2}{8(9.1 \times 10^{-31}\,\text{kg})}\left[\frac{3(8.4 \times 10^{28}\,\text{m}^{-3})}{\pi}\right]^{\frac{2}{3}}\frac{1}{1.6 \times 10^{-19}\,\text{J/eV}} = 7.0\,\text{eV}.$$

(b) From Eq. 41–13, $\overline{E} = \frac{3}{5}E_\text{F} = 4.2\,\text{eV}$.
(c) In our model, we took $U = 0$ inside the metal, so $E =$ kinetic energy $= \frac{1}{2}mv^2$. Therefore, at the Fermi level, the Fermi speed is

$$v_\text{F} = \sqrt{\frac{2E_\text{F}}{m}} = \sqrt{\frac{2(7.0\,\text{eV})(1.6 \times 10^{-19}\,\text{J/eV})}{9.1 \times 10^{-31}\,\text{kg}}} = 1.6 \times 10^6\,\text{m/s},$$

a very high speed. The temperature of a classical gas would have to be extremely high to produce an average particle speed this large.

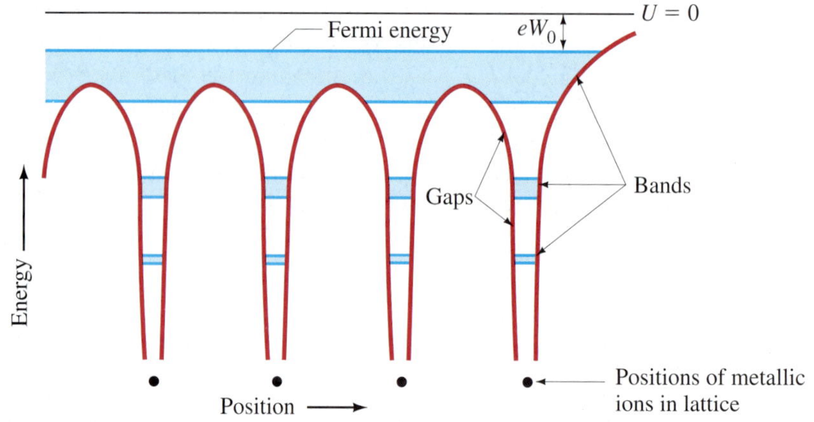

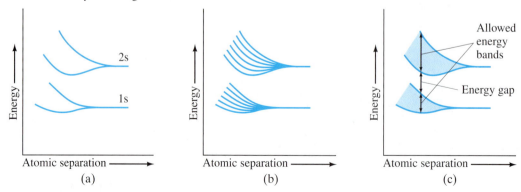

FIGURE 41–28 Potential energy for an electron in a metal crystal, with deep potential wells in the vicinity of each ion in the crystal lattice.

The simple model of the electron gas presented here provides good explanations for the electrical and thermal properties of conductors. But it does not explain why some materials are good conductors and others are good insulators. To provide an explanation, our model of a metal gas as a uniform potential well needs to be refined to include the effect of the lattice. Figure 41–28 shows a "periodic" potential that takes into account the attraction of electrons for each atomic ion in the lattice. Here we have taken $U = 0$ for an electron free of the metal; so within the metal, electron energies are less than zero (just as for molecules, or for the H atom, for example, in which the ground state has $E = -13.6\,\text{eV}$). The quantity eW_0 represents the minimum energy to remove an electron from the metal, where W_0 is the *work function* (see Section 38–2). The crucial outcome of putting a periodic potential (more easily approximated with narrow square wells) into the Schrödinger equation is that the allowed energy states are divided into *bands*, with energy gaps in between. Only electrons in the highest band, close to the Fermi level, are able to move about freely within the metal crystal. In the next Section we will see physically why there are bands and how they explain the properties of conductors, insulators, and semiconductors.

41–7 Band Theory of Solids

We saw in Section 41–1 that when two hydrogen atoms approach each other, the wave functions overlap, and the two 1s states (one for each atom) divide into two states of different energy. (As we saw, only one of these states, $S = 0$, has low enough energy to give a bound H_2 molecule.) Figure 41–29a shows this situation for 1s and 2s states for two atoms: as the atoms get closer, the 1s and 2s states split

FIGURE 41–29 The splitting of 1s and 2s atomic energy levels as (a) two atoms approach each other (the atomic separation decreases, moving toward the left); (b) the same for six atoms, and (c) for many atoms when they come together to form a solid.

into two levels. If six atoms come together, as in Fig. 41–29b, each of the states splits into six levels. If a large number of atoms come together to form a solid, then each of the original atomic levels becomes a **band** as shown in Fig. 41–29c. The energy levels are so close together in each band that they seem essentially continuous. This is why the spectrum of heated solids (Section 38–1) appears continuous.

The crucial aspect of a good **conductor** is that the highest energy band containing electrons is only partially filled. Consider sodium, for example, whose energy bands are shown in Fig. 41–30. The 1s, 2s, and 2p bands are full (just as in a Na atom) and don't concern us. The 3s band, however, is only half full. To see why, recall that the exclusion principle stipulates that in an atom, only two electrons can be in the 3s state, one with spin up and one with spin down. These two states have slightly different energy. For a solid consisting of N atoms, the 3s band will contain $2N$ possible energy states. Now a sodium atom has a single 3s electron, so in a sample of sodium metal containing N atoms, there are N electrons in the 3s band, and N unoccupied states. When a potential difference is applied across the metal, electrons can respond by accelerating and increasing their energy, since there are plenty of unoccupied states of slightly higher energy available. Hence, a current flows readily and sodium is a good conductor. The characteristic of all good conductors is that the highest energy band is only partially filled, or two bands overlap so that unoccupied states are available. An example of the latter is magnesium, which has two 3s electrons, so its 3s band is filled. But the unfilled 3p band overlaps the 3s band in energy, so there are lots of available states for the electrons to move into. Thus magnesium, too, is a good conductor.

In a material that is a good **insulator**, on the other hand, the highest band containing electrons, called the **valence band**, is completely filled. The next higher energy band, called the **conduction band**, is separated from the valence band by a "forbidden" **energy gap**, (or band gap), E_g, of typically 5 to 10 eV. So at room temperature (300 K), where thermal energies (that is, average kinetic energy—see Chapter 18) are on the order of $\frac{3}{2}kT \approx 0.04$ eV, almost no electrons can acquire the 5 eV needed to reach the conduction band. When a potential difference is applied across the material, no available states are accessible to the electrons, and no current flows. Hence, the material is a good insulator.

Figure 41–31 compares the relevant energy bands (a) for conductors, (b) for insulators, and also (c) for the important class of materials known as **semiconductors**. The bands for a pure (or **intrinsic**) semiconductor, such as silicon or germanium, are like those for an insulator, except that the unfilled conduction band is separated from the filled valence band by a much smaller energy gap, E_g, typically on the order of 1 eV. At room temperature, a few electrons can acquire enough thermal energy to reach the conduction band, and so a very small current may flow when a voltage is applied. At higher temperatures, more electrons have enough energy to jump the gap. This effect can often more than offset the effects of more frequent collisions due to increased disorder at higher temperature, so that the resistivity of semiconductors can *decrease* with increasing temperature (see Table 25–1). But this is not the whole story of semiconductor conduction. When a potential difference is applied to a semiconductor, the few electrons in the conduction band move toward the positive electrode. Electrons in the valence band try to do the same thing, and a few can because there are a small number of unoccupied

FIGURE 41–30 Energy bands for sodium.

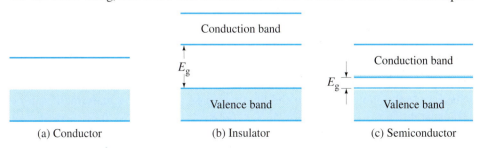

(a) Conductor (b) Insulator (c) Semiconductor

FIGURE 41–31 Energy bands for (a) a conductor, (b) an insulator, and (c) a semiconductor. Shading represents occupied states. Pale shading in part (c) represents electrons that can pass from the valence band to the conduction band due to thermal agitation at room temperature (exaggerated).

Holes (in semiconductor)

states which were left empty by the electrons reaching the conduction band. Such unfilled electron states are called **holes**. Each electron in the valence band that fills a hole in this way as it moves toward the positive electrode leaves behind a hole, so that the holes migrate toward the negative electrode. As the electrons tend to accumulate at one side of the material, the holes tend to accumulate on the opposite side. We will look at this phenomenon in more detail in the next Section.

EXAMPLE 41–10 **Calculating the energy gap.** It is found that the conductivity of a certain semiconductor increases when light of wavelength 345 nm or shorter strikes it, suggesting that electrons are being promoted from the valence band to the conduction band. What is the energy gap, E_g, for this semiconductor?

SOLUTION The longest wavelength, or lowest energy, photon to cause an increase in conductivity has $\lambda = 345$ nm, and it can transfer to an electron an energy

$$E_g = hf = \frac{hc}{\lambda} = \frac{(6.63 \times 10^{-34}\,\text{J·s})(3.00 \times 10^8\,\text{m/s})}{(345 \times 10^{-9}\,\text{m})(1.60 \times 10^{-19}\,\text{J/eV})} = 3.6\,\text{eV}.$$

EXAMPLE 41–11 **ESTIMATE** **Free electrons in semiconductors and insulators.** Use the Fermi–Dirac probability function, Eq. 41–14, to estimate the order of magnitude of the numbers of free electrons in the conduction band of a solid containing 10^{21} atoms, assuming the solid is at room temperature ($T = 300$ K) and is (a) a semiconductor with $E_g \approx 1.1$ eV, (b) an insulator with $E_g \approx 5$ eV. Compare to a conductor.

SOLUTION At $T = 0$, all states above the Fermi energy E_F are empty, and all those below are full. So for semiconductors and insulators we can take E_F to be about midway between the valence and conduction bands, Fig. 41–32, and it does not change significantly as we go to room temperature. (a) For the semiconductor, the gap $E_g \approx 1.1$ eV, so $E - E_F \approx 0.55$ eV for the lowest states in the conduction band. Since at room temperature we have $kT \approx 0.026$ eV, then $(E - E_F)/kT \approx 0.55\,\text{eV}/0.026\,\text{eV} \approx 21$ and

$$f(E) = \frac{1}{e^{(E-E_F)/kT} + 1} \approx \frac{1}{e^{21}} \approx 10^{-9}.$$

Thus about 1 atom in 10^9 can contribute an electron to the conductivity. (b) For the insulator with $E - E_F \approx 5.0\,\text{eV} - \frac{1}{2}(5.0\,\text{eV}) = 2.5$ eV, we get

$$f(E) \approx \frac{1}{e^{2.5/0.026} + 1} \approx \frac{1}{e^{96}} \approx 10^{-42}.$$

Thus in an ordinary sample containing 10^{21} atoms, there would be no free electrons in an insulator $(10^{21} \times 10^{-42} = 10^{-21})$, about $10^{12}\,(10^{21} \times 10^{-9})$ free electrons in a semiconductor, and about 10^{21} free electrons in a good conductor.

CONCEPTUAL EXAMPLE 41–12 **Which is transparent?** The energy gap for silicon is 1.14 eV at room temperature while that of zinc sulfide (ZnS) is 3.6 eV. Which one of these is opaque and which is transparent to visible light?

RESPONSE Visible light photons span energies from roughly 1.8 eV to 3.2 eV ($E = hf = hc/\lambda$ where $\lambda = 400$ nm to 700 nm and 1 eV $= 1.6 \times 10^{-19}$ J). Light is absorbed by the electrons in a material. Silicon's gap is small enough to absorb these photons, thus bumping electrons well up into the conduction band, and so silicon is opaque. On the other hand, zinc sulfide's energy gap is too wide to absorb visible photons, and so the light can pass through the material; it can be transparent.

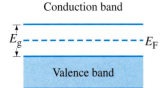

Conduction band

E_g ─────────────── E_F

Valence band

FIGURE 41–32 The Fermi energy is midway between the valence band and the conduction band.

41–8 Semiconductors and Doping

The most commonly used semiconductors in modern electronics are silicon (Si) and germanium (Ge). An atom of silicon or germanium has four outer electrons that act to hold the atoms in the regular lattice structure of the crystal, shown schematically in Fig. 41–33a. Germanium and silicon acquire useful properties for use in electronics only when a tiny amount of impurity is introduced into the crystal structure (perhaps 1 part in 10^6 or 10^7). This is called **doping** the semiconductor. Two kinds of doped semiconductor can be made, depending on the type of impurity used. If the impurity is an element whose atoms have five outer electrons, such as arsenic, we have the situation shown in Fig. 41–33b, with the arsenic atoms holding positions in the crystal lattice where normally silicon atoms would be. Only four of arsenic's electrons fit into the bonding structure. The fifth does not fit in and can move relatively freely, somewhat like the electrons in a conductor. Because of this small number of extra electrons, a doped semiconductor becomes slightly conducting. The density of conduction electrons in an intrinsic (pure) semiconductor is about 1 per 10^9 atoms, as we saw in Example 41–11. With an impurity concentration of 1 in 10^6 or 10^7 when doped, the conductivity will be much higher and it can be controlled with great precision. An arsenic-doped silicon crystal is called an **n-type semiconductor** because *n*egative charges (electrons) carry the electric current.

Doped semiconductors

n-type

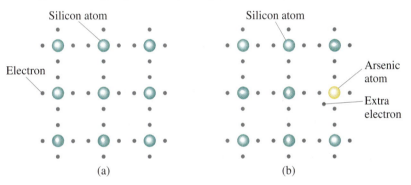

p-type

FIGURE 41–33 Two-dimensional representation of a silicon crystal. (a) Four (outer) electrons surround each silicon atom. (b) Silicon crystal doped with a few arsenic atoms: the extra electron doesn't fit into the crystal lattice and so is free to move about. This is an *n*-type semiconductor.

In a **p-type semiconductor**, a small percentage of semiconductor atoms are replaced by atoms with three outer electrons—such as gallium. As shown in Fig. 41–34a, there is a "hole" in the lattice structure near a gallium atom since it has only three outer electrons. Electrons from nearby silicon atoms can jump into this hole and fill it. But this leaves a hole where that electron had previously been, Fig. 41–34b. The vast majority of atoms are silicon, so holes are almost always next to a silicon atom. Since silicon atoms require four outer electrons to be neutral, this means that there is a net positive charge at the hole. Whenever an electron moves to fill a hole, the positive hole is then at the previous position of that electron. Another electron can then fill this hole, and the hole thus moves to a new location; and so on. This type of semiconductor is called *p-type* because it is the *p*ositive holes that seem to carry the electric current. Note, however, that both *p*-type and *n*-type semiconductors have *no net charge* on them.

Holes are positive

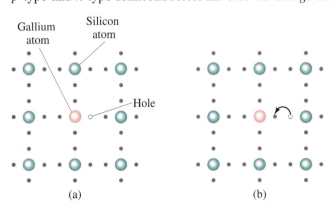

FIGURE 41–34 A *p*-type semiconductor, gallium-doped silicon. (a) Gallium has only three outer electrons, so there is an empty spot, or *hole*, in the structure. (b) Electrons from silicon atoms can jump into the hole and fill it. As a result, the hole moves to a new location (to the right in this figure), to where the electron used to be.

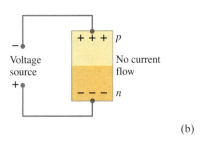

FIGURE 41–35 Impurity energy levels in doped semiconductors.

Donor level

Conduction band

Valence band

n-type

Conduction band

Acceptor level

Valence band

p-type

According to the band theory (Section 41–7), in a doped semiconductor, the impurity provides additional energy states between the bands, as shown in Fig. 41–35. In an *n*-type semiconductor, the impurity energy level lies just below the conduction band. Electrons in this energy level need only about 0.05 eV (in Si; even less in Ge) of energy to reach the conduction band; this is on the order of the thermal energy, $\frac{3}{2}kT$ ($= 0.04$ eV at 300 K), so transitions occur readily at room temperature. This energy level can thus supply electrons to the conduction band, so it is called a **donor** level. In *p*-type semiconductors, the impurity energy level is just above the valence band (Fig. 41–35). It is called an **acceptor** level because electrons from the valence band can easily jump into it. Positive holes are left behind in the valence band, and as other electrons move into these holes, the holes move about as discussed earlier.

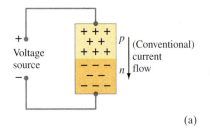

(a)

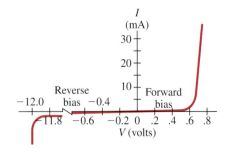

(b)

FIGURE 41–36 Schematic diagram showing how a semiconductor diode operates. Current flows when the voltage is connected in forward bias, as in (a), but not when connected in reverse bias, as in (b).

FIGURE 41–37 Current through a diode as a function of applied voltage.

* **41–9 Semiconductor Diodes**

Semiconductor diodes and transistors are essential components of modern electronic devices. The miniaturization achieved today allows many thousands of diodes, transistors, resistors, and so on, to be placed on a single *chip* only a millimeter on a side. We now discuss, briefly and qualitatively, the operation of diodes and transistors.

When an *n*-type semiconductor is joined to a *p*-type, a **p-n junction diode** is formed. Separately, the two semiconductors are electrically neutral. When joined, a few electrons near the junction diffuse from the *n*-type into the *p*-type semiconductor, where they fill a few of the holes. The *n*-type is left with a positive charge, and the *p*-type acquires a net negative charge. Thus a potential difference is established, with the *n* side positive relative to the *p* side, and this prevents further diffusion of electrons.

If a battery is connected to a diode with the positive terminal to the *p* side and the negative terminal to the *n* side as in Fig. 41–36a, the externally applied voltage opposes the internal potential difference and the diode is said to be **forward biased**. If the voltage is great enough (about 0.3 V for Ge, 0.6 V for Si at room temperature), a current will flow. The positive holes in the *p*-type semiconductor are repelled by the positive terminal of the battery and the electrons in the *n*-type are repelled by the negative terminal of the battery. The holes and electrons meet at the junction, and the electrons cross over and fill the holes. A current is flowing. Meanwhile, the positive terminal of the battery is continually pulling electrons off the *p* end, forming new holes, and electrons are being supplied by the negative terminal at the *n* end. Consequently, a large current flows through the diode.

When the diode is **reverse biased**, as in Fig. 41–36b, the holes in the *p* end are attracted to the battery's negative terminal and the electrons in the *n* end are attracted to the positive terminal. The current carriers do not meet near the junction and, ideally, no current flows.

A graph of current versus voltage for a typical diode is shown in Fig. 41–37. As can be seen, a real diode does allow a small amount of reverse current to flow due to thermal motion $\left(\frac{3}{2}kT\right)$ of electron pairs. For most practical purposes, this is negligible (at room temperature, a few μA in Ge, a few pA in Si; the reverse current increases rapidly with temperature, however, and may render a diode ineffective above 200°C).

EXAMPLE 41–13 **A diode.** The diode whose current–voltage characteristics are shown in Fig. 41–37 is connected in series with a 4.0-V battery and a resistor. If a current of 10 mA is to pass through the diode, what resistance must the resistor have?

SOLUTION In Fig. 41–37, we see that the voltage drop across the diode is about 0.7 V when the current is 10 mA. Therefore, the voltage drop across the resistor is 4.0 V − 0.7 V = 3.3 V, so $R = V/I = (3.3\ V)/(1.0 \times 10^{-2}\ A) = 330\ \Omega$.

If the voltage across a diode connected in reverse bias is increased greatly, a point is reached where breakdown occurs. The electric field across the junction becomes so large that ionization of atoms results. The electrons thus pulled off their atoms contribute to a larger and larger current as breakdown continues. The voltage remains constant over a wide range of currents. This is shown on the far left in Fig. 41–37. This property of diodes can be used to accurately regulate a voltage supply. A diode designed for this purpose is called a **zener diode**. When placed across the output of an unregulated power supply, a zener diode can maintain the voltage at its own breakdown voltage as long as the supply voltage is always above this point. Zener diodes can be obtained corresponding to voltages of a few volts to hundreds of volts.

Since a *p-n* junction diode allows current to flow only in one direction (as long as the voltage is not too high), it can serve as a **rectifier**—to change ac into dc. A simple rectifier circuit is shown in Fig. 41–38a where the arrow inside the symbol for a diode indicates the direction in which a diode conducts conventional (+) current. The ac source applies a voltage across the diode alternately positive and negative. Only during half of each cycle will a current pass through the diode; so only then is there a current through the resistor *R*. Hence, a graph of the voltage V_{ab} across *R* as a function of time looks like Fig. 41–38b. This **half-wave rectification** is not exactly dc, but it is unidirectional. More useful is a **full-wave rectifier** circuit, which uses two diodes (or sometimes four) as shown in Fig. 41–39a. At any given instant, either one diode or the other will conduct current to the right. Therefore, the output across the load resistor *R* will be as shown in Fig. 41–39b. Actually this is the voltage if the capacitor *C* were not in the circuit. The capacitor tends to store charge, and thus helps to smooth out the current as shown in Fig. 41–39c.

Rectifier circuits are important because most line voltage is ac, and most electronic devices require a dc voltage for their operation. Hence, diodes are found in nearly all electronic devices including radio and TV sets, calculators, and computers.

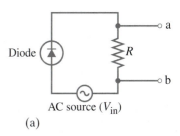

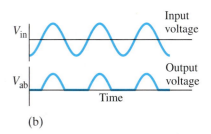

(a)

(b)

FIGURE 41–38 (a) A simple (half-wave) rectifier circuit using a semiconductor diode. (b) AC source input voltage, and output voltage across *R*, as functions of time.

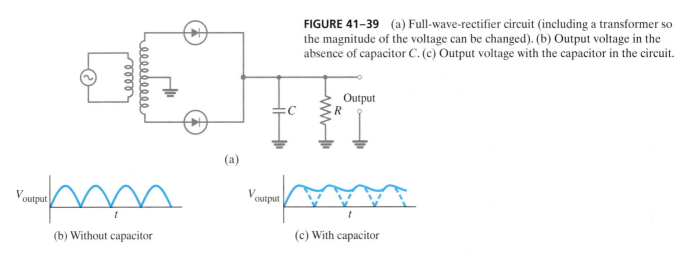

FIGURE 41–39 (a) Full-wave-rectifier circuit (including a transformer so the magnitude of the voltage can be changed). (b) Output voltage in the absence of capacitor *C*. (c) Output voltage with the capacitor in the circuit.

(a)

(b) Without capacitor

(c) With capacitor

Another useful device is a **light-emitting diode** (LED). When a *p-n* junction is forward biased, a current begins to flow. Electrons cross from the *n* region into the *p* region and combine with holes, and a photon can be emitted with an energy approximately equal to the band gap, E_g (see Figs. 41–31c and 41–35). Often the energy, and hence the wavelength, is in the red region of the visible spectrum, producing the familiar LED displays on VCRs, CD players, car instrument panels, clocks, and so on. Infrared (i.e., nonvisible) LEDs are used in remote controls for TV, VCRs, and stereos.

Solar cells

Photodiodes and **solar cells** are *p-n* junctions used in the reverse way. Photons are absorbed, creating electron–hole pairs if the photon energy is greater than the band gap energy, E_g. The created electrons and holes produce a current that, when connected to an external circuit, becomes a source of emf and power.

A diode is called a **nonlinear device** because the current is not proportional to the voltage; that is, a graph of current versus voltage (Fig. 41–37) is not a straight line as it is for a resistor (which ideally *is* linear). Transistors are also *nonlinear* devices.

* 41–10 Transistors and Integrated Circuits

Transistors

A simple **junction transistor** consists of a crystal of one type of doped semiconductor sandwiched between two crystals of the opposite type. Both *pnp* and *npn* transistors are made, and they are shown schematically in Fig. 41–40a. The three semiconductors are given the names *collector*, *base*, and *emitter*. The symbols for *npn* and *pnp* transistors are shown in Fig. 41–40b. The arrow is always placed on the emitter and indicates the direction of (conventional) current flow in normal operation.

The operation of a transistor can be analyzed qualitatively—very briefly—as follows. Consider an *npn* transistor connected as shown in Fig. 41–41. A voltage V_{CE} is maintained between the collector and emitter by the battery $\mathscr{E}_C$. The voltage applied to the base is called the *base bias voltage*, V_{BE}. If V_{BE} is positive, conduction electrons in the emitter are attracted into the base. Since the base region is very thin (perhaps 1 μm), most of these electrons flow right across into the collector, which is maintained at a positive voltage. A large current, I_C, flows between collector and emitter and a much smaller current, I_B, through the base. A small variation in the base voltage due to an input signal causes a large change in the collector current and therefore a large change in the voltage drop across the output resistor R_C. Hence a transistor can *amplify* a small signal into a larger one.

Amplifiers

FIGURE 41–40 (a) Schematic diagram of *npn* and *pnp* transistors. (b) Symbols for *npn* and *pnp* transistors.

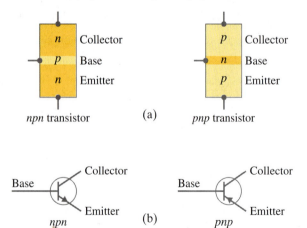

FIGURE 41–41 An *npn* transistor used as an amplifier.

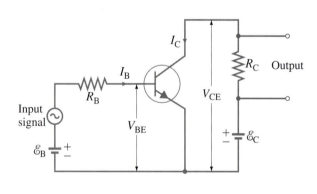

Normally a small ac signal is to be amplified, and when added to the base bias voltage causes the voltage and current at the collector to vary at the same rate but magnified. Thus, what is important for amplification is the *change* in collector current for a given input *change* in base current. The **current gain** is defined as the ratio

$$\beta_I = \frac{\text{output (collector) ac current}}{\text{input (base) ac current}} = \frac{i_C}{i_B}$$ *Current gain*

where i_B and i_C are the ac base current (input signal) and ac collector current (output). β_I is typically on the order of 10 to 100. Similarly, the **voltage gain** is

$$\beta_V = \frac{\text{output (collector) ac voltage}}{\text{input (base) ac voltage}}.$$ *Voltage gain*

Transistors are the basic elements in modern electronic **amplifiers** of all sorts.

A *pnp* transistor operates like an *npn*, except that holes move instead of electrons. The collector voltage is negative, and so is the base voltage in normal operation.

In Fig. 41–41 two batteries were shown: $\mathscr{E}_C$ supplied the collector voltage and $\mathscr{E}_B$ the base bias voltage. In practice, only one source is often used, and the base bias voltage can be obtained using a resistance voltage divider as in Fig. 41–42. Transistors can be connected in many other ways as well, and many new and innovative uses have been found for them.

Transistors were a great advance in miniaturization of electronic circuits. Although individual transistors are very small compared to the once used vacuum tubes, they are huge compared to **integrated circuits** or **chips** (see photo at start of this chapter). Tiny amounts of impurities can be placed at particular locations within a single silicon crystal. These can be arranged to form diodes, transistors, and resistors (undoped semiconductors). Capacitors and inductors can also be formed, although they are often connected separately. A tiny chip, only 1 mm on a side, may contain thousands of transistors and other circuit elements. Integrated circuits are now the heart of computers, television, calculators, cameras, and the electronic instruments that control aircraft, space vehicles, and automobiles. The "miniaturization" produced by integrated circuits not only allows extremely complicated circuits to be placed in a small space, but also has allowed a great increase in the speed of operation of, say, computers, because the distances the electronic signals travel are so tiny.

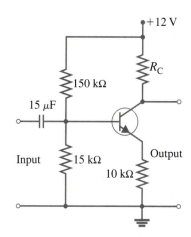

FIGURE 41–42 Typical transistor circuit involving an *npn* transistor. $\mathscr{E}_C = +12\text{ V}$ and $\mathscr{E}_B$ is determined by the resistors.

Summary

Quantum mechanics explains the bonding together of atoms to form **molecules**. In a **covalent bond**, the electron clouds of two or more atoms overlap because of constructive interference between the electron waves. The positive nuclei are attracted to this concentration of negative charge between them, forming the bond. An **ionic bond** is an extreme case of a covalent bond in which one or more electrons from one atom spend much more time around the other atom than around their own. The atoms then act as oppositely charged ions that attract each other, forming the bond.

These **strong bonds** hold molecules together, and also hold atoms and molecules together in solids. Also important are **weak bonds** (or **van der Waals bonds**), which are generally dipole attractions between molecules.

When atoms combine to form molecules, the energy levels of the outer electrons are altered because they now interact with each other. Additional energy levels also become possible because the atoms can vibrate with respect to each other, and the molecule as a whole can rotate. The energy levels for both vibrational and rotational motion are quantized, and are very close together (typically, 10^{-1} eV to 10^{-3} eV apart). Each atomic energy level thus becomes a set of closely spaced levels corresponding to the vibrational and rotational motions. Transitions from one level to another appear as many very closely spaced lines. The resulting spectra are called **band spectra**.

The quantized rotational energy levels are given by

$$E_{\text{rot}} = L(L+1)\frac{\hbar^2}{2I}, \qquad L = 0, 1, 2, \cdots,$$

where I is the moment of inertia of the molecule.

The energy levels for vibrational motion are given by

$$E_{vib} = \left(\nu + \tfrac{1}{2}\right)hf, \qquad \nu = 0, 1, 2, \cdots,$$

where f is the classical natural frequency of vibration for the molecule. Transitions between energy levels are subject to the selection rules $\Delta L = \pm 1$ and $\Delta \nu = \pm 1$.

Some **solids** are bound together by covalent and ionic bonds, just as molecules are. In metals, the electrostatic force between free electrons and positive ions helps form the **metallic bond**.

In the free-electron theory of metals, electrons occupy the possible energy states according to the exclusion principle. At $T = 0\,K$, all possible states are filled up to a maximum energy level called the *Fermi energy*, E_F, the magnitude of which is typically a few eV. All states above E_F are vacant at $T = 0\,K$. At normal temperatures $(300\,K)$ the distribution of occupied states is only slightly altered and is given by the **Fermi–Dirac probability function**

$$f(E) = \frac{1}{e^{(E-E_F)/kT} + 1}.$$

In a crystalline solid, the possible energy states for electrons are arranged in **bands**. Within each band the levels are very close together, but between the bands there may be forbidden **energy gaps**. Good conductors are characterized by the highest occupied band (the **conduction band**) being only partially full, so there are many accessible states available to electrons to move about and accelerate when a voltage is applied. In a good insulator, the highest occupied energy band (the **valence band**) is completely full, and there is a large energy gap (5 to 10 eV) to the next highest band, the *conduction band*. At room temperature, molecular kinetic energy (thermal energy) available due to collisions is only about 0.04 eV, so almost no electrons can jump from the valence to the conduction band. In a **semiconductor**, the gap between valence and conduction bands is much smaller, on the order of 1 eV, so a few electrons can make the transition from the essentially full valence band to the nearly empty conduction band.

In a **doped** semiconductor, a small percentage of impurity atoms with five or three valence electrons replace a few of the normal silicon atoms with their four valence electrons. A five-electron impurity produces an **n-type** semiconductor with negative electrons as carriers of current. A three-electron impurity produces a **p-type** semiconductor in which positive **holes** carry the current. The energy level of impurity atoms lies slightly below the conduction band in an *n*-type semiconductor, and acts as a **donor** from which electrons readily pass into the conduction band. The energy level of impurity atoms in a *p*-type semiconductor lies slightly above the valence band and acts as an **acceptor** level, since electrons from the valence band easily reach it, leaving holes behind to act as charge carriers.

A semiconductor **diode** consists of a **p-n junction** and allows current to flow in one direction only; it can be used as a **rectifier** to change ac to dc. Common **transistors** consist of three semiconductor sections, either as **pnp** or **npn**. Transistors can amplify electrical signals and find many other uses. An integrated circuit consists of a tiny semiconductor crystal or "chip" on which many transistors, diodes, resistors, and other circuit elements have been constructed using careful placement of impurities.

Questions

1. What type of bond would you expect for (*a*) the N_2 molecule, (*b*) the HCl molecule, (*c*) Fe atoms in a solid?
2. Describe how the molecule $CaCl_2$ could be formed.
3. Does the H_2 molecule have a permanent dipole moment? Does O_2? Does H_2O? Explain.
4. Although the molecule H_3 is not stable, the ion H_3^+ is. Explain, using the Pauli exclusion principle.
5. The energy of a molecule can be divided into four categories. What are they?
6. If conduction electrons are free to roam about in a metal, why don't they leave the metal entirely?
7. A silicon semiconductor is doped with phosphorus. Will these atoms be donors or acceptors? What type of semiconductor will this be?
8. Explain why the resistivity of metals increases with temperature whereas the resistivity of semiconductors may decrease with increasing temperature.
9. Discuss the differences between an ideal gas and a Fermi electron gas.
10. Which aspects of Fig. 41–27 are peculiar to copper, and which are valid in general for other metals?

* 11. Can a diode be used to amplify a signal?
* 12. Figure 41–43 shows a "bridge-type" full-wave rectifier. Explain how the current is rectified and how current flows during each half cycle.

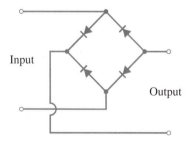

FIGURE 41–43 Question 12.

* 13. Compare the resistance of a *p-n* junction diode connected in forward bias to its resistance when connected in reverse bias.

* 14. Explain how a transistor could be used as a switch.

* **15.** If $\mathscr{E}_C$ were reversed in Fig. 41–41, how would the amplification be altered?

* **16.** Describe how a *pnp* transistor can operate as an amplifier.

* **17.** Do diodes and transistors obey Ohm's law?

* **18.** In a transistor, the base–emitter junction and the base–collector junction are essentially diodes. Are these junctions reverse-biased or forward-biased in the application shown in Fig. 41–41?

* **19.** What purpose does the capacitor in Fig. 41–42 serve?

Problems

Sections 41–1 to 41–3

1. (I) Estimate the binding energy of a KCl molecule by calculating the electrostatic potential energy when the K^+ and Cl^- ions are at their stable separation of 0.28 nm. Assume each has a charge of magnitude $1.0e$.

2. (I) Binding energies are often measured experimentally in kcal per mole, and then the binding energy in eV per molecule is calculated from that result. What is the conversion factor in going from kcal per mole to eV per molecule? What is the binding energy of KCl (= 4.43 eV) in kcal per mole?

3. (II) The measured binding energy of KCl is 4.43 eV. From the result of Problem 1, estimate the contribution to the binding energy of the repelling electron clouds at the equilibrium distance $r_0 = 0.28$ nm.

4. (II) Estimate the binding energy of the H_2 molecule, assuming the two H nuclei are 0.074 nm apart and the two electrons spend 33 percent of their time midway between them.

5. (III) Apply reasoning similar to that in the text for the $S = 0$ and $S = 1$ states in the formation of the H_2 molecule to show why the molecule He_2 is *not* formed. Show also why the He_2^+ molecular ion *is* formed (with a binding energy of 3.1 eV at $r_0 = 0.11$ nm).

Section 41–4

6. (I) Show that the quantity $\hbar^2/I$ has units of energy.

7. (I) What is the reduced mass of the molecules (*a*) NaCl; (*b*) N_2; (*c*) HCl?

8. (II) The so-called "characteristic rotational energy," $\hbar^2/2I$, for N_2 is 2.48×10^{-4} eV. Calculate the N_2 bond length.

9. (II) The fundamental vibration frequency for the CO molecule is 6.42×10^{13} Hz. Determine (*a*) the reduced mass, and (*b*) the effective value of the constant k. Compare to k for the H_2 molecule.

10. (II) Explain why there is no transition for $E = hf$ in Fig. 41–20 (and Fig. 41–21).

11. (II) (*a*) Calculate the characteristic rotational energy, $\hbar^2/2I$, for the O_2 molecule whose bond length is 0.121 nm. (*b*) What are the energy and wavelength of photons emitted in a $L = 2$ to $L = 1$ transition?

12. (II) The equilibrium separation of H atoms in the H_2 molecule is 0.074 nm (Fig. 41–8). Calculate the energies and wavelengths of photons for the rotational transitions (*a*) $L = 1$ to $L = 0$, (*b*) $L = 2$ to $L = 1$, and (*c*) $L = 3$ to $L = 2$.

13. (II) Calculate the bond length for the NaCl molecule given that three successive wavelengths for rotational transitions are 23.1 mm, 11.6 mm, and 7.71 mm.

14. (II) Derive Eqs. 41–8.

Section 41–5

15. (I) Estimate the ionic cohesive energy for NaCl taking $\alpha = 1.75$, $m = 8$, and $r_0 = 0.28$ nm.

16. (II) The spacing between "nearest neighbor" Na and Cl ions in a NaCl crystal is 0.24 nm. What is the spacing between two nearest neighbor Na ions?

17. (II) Common salt, NaCl, has a density of 2.165 g/cm^3. The molecular weight of NaCl is 58.44. Estimate the distance between nearest neighbor Na and Cl ions. [*Hint: Each* ion can be considered to have one "cube" or "cell" of side s (our unknown) extending out from it.]

18. (II) Repeat the previous Problem for KCl whose density is 1.99 g/cm^3.

19. (III) (*a*) Starting from Eq. 41–9, show that the ionic cohesive energy is given by $U_0 = -(\alpha e^2/4\pi\epsilon_0 r_0)(1 - 1/m)$. Determine U_0 for (*b*) NaI ($r_0 = 0.33$ nm) and (*c*) MgO ($r_0 = 0.21$ nm). Assume $m = 10$. (*d*) If you used $m = 8$ instead, how far off would your answers be?

20. (III) For a long one-dimensional chain of alternating positive and negative ions, show that the Madelung constant would be $\alpha = 2 \ln 2$ [*Hint:* Use a series expansion for $\ln(1 + x)$.]

Section 41–6

21. (I) Estimate the number of states available to electrons in a 1.0-cm^3 cube of copper between 6.90 and 7.00 eV.

22. (II) What, roughly, is the ratio of the density of molecules in an ideal gas at 300 K and 1 atm (say O_2) to the density of free electrons (assume one per atom) in a metal (copper) also at 300 K?

23. (II) Calculate the energy which has 90 percent occupancy probability for copper at (*a*) $T = 300$ K; (*b*) $T = 1200$ K.

24. (II) Calculate the energy which has 10 percent occupancy probability for copper at (*a*) $T = 300$ K; (*b*) $T = 1200$ K.

25. (II) What is the occupancy probability for a conduction electron in copper at $T = 300$ K for an energy $E = 1.010E_F$?

26. (II) Calculate the number of possible electron states in a 1.00-cm^3 cube of silver between $0.99E_F$ and $E_F (= 5.48$ eV).

27. (II) Calculate the Fermi energy and Fermi speed for sodium, which has a density of 0.97×10^3 kg/m^3 and has one conduction electron per atom.

28. (II) The atoms in zinc metal ($\rho = 7.1 \times 10^3$ kg/m^3) have two free electrons. Calculate (*a*) the density of conduction electrons, (*b*) their Fermi energy, and (*c*) their Fermi speed.

29. (II) Given that the Fermi energy of aluminum is 11.63 eV, (*a*) calculate the density of free electrons using Eq. 41–12, and (*b*) estimate the valence of aluminum using this model and the known density $(2.70 \times 10^3 \text{ kg/m}^3)$ and atomic weight (27.0) of aluminum.

30. (II) The neutrons in a neutron star (Chapter 45) can be treated as a Fermi gas with neutrons in place of the electrons in our model of an electron gas. Determine the Fermi energy for a neutron star of radius 10 km and mass twice that of our Sun. Assume that the star is made entirely of neutrons and is of uniform density.

31. (II) Show that the average energy of conduction electrons in a metal at $T = 0 \text{ K}$ is $\bar{E} = \frac{3}{5} E_\text{F}$ (Eq. 41–13) by calculating

$$\bar{E} = \frac{\displaystyle\int E\, n_\text{o}(E)\, dE}{\displaystyle\int n_\text{o}(E)\, dE}.$$

32. (II) Show that the probability for the state at the Fermi energy being occupied is exactly $\frac{1}{2}$, independent of temperature.

33. (II) (*a*) For copper at room temperature $(T = 300 \text{ K})$, calculate the Fermi factor, Eq. 41–14, for an electron with energy 0.10 eV above the Fermi energy. This represents the probability that this state is occupied. Is this reasonable? (*b*) What is the probability that a state 0.10 eV below the Fermi energy is occupied? (*c*) What is the probability that the state in part (*b*) is unoccupied?

34. (II) For a one-dimensional potential well, start with Eq. 39–13 and show that the number of states per unit energy interval for an electron gas is given by

$$g_L(E) = \sqrt{\frac{8mL^2}{h^2 E}}.$$

Remember that there can be two electrons (spin up and spin down) for each value of *n*. [*Hint*: Write the quantum number *n* in terms of *E*. Then $g_L(E) = 2\, dn/dE$ where *dn* is the number of energy levels between *E* and $E + dE$.]

35. (III) Proceed as follows to derive the density of states, $g(E)$, the number of states per unit volume per unit energy interval, Eq. 41–10. Let the metal be a cube of side *L*. Extend the discussion of Section 39–8 for an infinite well to three dimensions, giving energy levels

$$E = \frac{h^2}{8mL^2}\left(n_1^2 + n_2^2 + n_3^2\right).$$

(Explain the meaning of n_1, n_2, n_3.) Each set of values for the quantum numbers n_1, n_2, n_3 corresponds to one state. Imagine a space where n_1, n_2, n_3 are the axes, and each state is represented by a point on a cubic lattice in this space, each separated by 1 unit along an axis. Consider the octant $n_1 > 0$, $n_2 > 0$, $n_3 > 0$. Show that the number of states *N* within a radius $R = \left(n_1^2 + n_2^2 + n_3^2\right)^{\frac{1}{2}}$ is $2\left(\frac{1}{8}\right)\left(\frac{4}{3}\pi R^3\right)$. Then, to get Eq. 41–10, set $g(E) = (1/V)(dN/dE)$, where $V = L^3$ is the volume of the metal.

Section 41–7

36. (I) A semiconductor, bombarded with light of slowly increased frequency, begins to conduct when the wavelength of light is 640 nm; estimate the size of the energy gap E_g.

37. (I) Explain on the basis of energy bands why the sodium chloride crystal is a good insulator. [*Hint*: Consider the shells of Na^+ and Cl^- ions.]

38. (II) We saw that there are 2*N* possible electron states in the 3s band of Na, where *N* is the total number of atoms. How many possible electron states are there in the (*a*) 2s band, (*b*) 2p band, and (*c*) 3p band? (*d*) State a general formula for the total number of possible states in any given electron band.

39. (II) Calculate the longest-wavelength photon that can cause an electron in silicon $(E_\text{g} = 1.1 \text{ eV})$ to jump from the valence band to the conduction band.

40. (II) The energy gap E_g in germanium is 0.72 eV. When used as a photon detector, roughly how many electrons can be made to jump from the valence to the conduction band by the passage of a 710-keV photon that loses all its energy in this fashion?

Section 41–8

41. (II) Suppose that a silicon semiconductor is doped with phosphorus so that one silicon atom in 10^6 is replaced by a phosphorus atom. Assuming that the "extra" electron in every phosphorus atom is donated to the conduction band, by what factor is the density of conduction electrons increased? The density of silicon is 2330 kg/m^3, and the density of conduction electrons in pure silicon is about 10^{16} m^{-3} at room temperature.

* Section 41–9

* **42.** (I) At what wavelength will an LED radiate if made from a material with an energy gap $E_\text{g} = 1.4 \text{ eV}$?

* **43.** (I) If an LED emits light of wavelength $\lambda = 650 \text{ nm}$, what is the energy gap (in eV) between valence and conduction bands?

* **44.** (II) A silicon diode, whose current–voltage characteristics are given in Fig. 41–37, is connected in series with a battery and a 760-Ω resistor. What battery voltage is needed to produce a 12-mA current?

* **45.** (II) Suppose that the diode of Fig. 41–37 is connected in series to a 100-Ω resistor and a 2.0-V battery. What current flows in the circuit? [*Hint*: Draw a line on Fig. 41–37 representing the current in the resistor as a function of the voltage across the diode; the intersection of this line with the characteristic curve will give the answer.]

* **46.** (II) Sketch the resistance as a function of current, for $V > 0$, for the diode shown in Fig. 41–37.

* **47.** (II) An ac voltage of 120 V rms is to be rectified. Estimate very roughly the average current in the output resistor *R* (28 kΩ) for (*a*) a half-wave rectifier (Fig. 41–38), and (*b*) a full-wave rectifier (Fig. 41–39) without capacitor.

* **48.** (III) A silicon diode passes significant current only if the forward-bias voltage exceeds about 0.6 V. Make a rough estimate of the average current in the output resistor *R* of (*a*) a half-wave rectifier (Fig. 41–38), and (*b*) a full-wave rectifier (Fig. 41–39) without a capacitor. Assume that $R = 150\,\Omega$ in each case and that the ac voltage is 9.0 V rms in each case.

*49. (III) A 120-V rms 60-Hz voltage is to be rectified with a full-wave rectifier as in Fig. 41–39, where $R = 18\,k\Omega$, and $C = 25\,\mu F$. (a) Make a rough estimate of the average current. (b) What happens if $C = 0.10\,\mu F$? [Hint: See Section 26–4.]

*Section 41–10

*50. (II) Suppose that the current gain of the transistor in Fig. 41–41 is $\beta = i_C/i_B = 80$. If $R_C = 3.3\,k\Omega$, calculate the ac output voltage for a time-varying input current of $2.0\,\mu A$.

*51. (II) If the current gain of the transistor amplifier in Fig. 41–41 is $\beta = i_C/i_B = 100$, what value must R_C have if a 1.0-μA ac base current is to produce an ac output voltage of $0.40\,V$?

*52. (II) A transistor, whose current gain $\beta = i_C/i_B = 70$, is connected as in Fig. 41–41 with $R_B = 3.2\,k\Omega$ and $R_C = 6.8\,k\Omega$. Calculate (a) the voltage gain, and (b) the power amplification.

*53. (II) An amplifier has a voltage gain of 80 and a 15-$k\Omega$ load (output) resistance. What is the peak output current through the load resistor if the input voltage is an ac signal with a peak of $0.080\,V$?

General Problems

54. Estimate the binding energy of the H_2 molecule by calculating the difference in kinetic energy of the electrons between when they are in separate atoms and when they are in the molecule, using the uncertainty principle. Take Δx for the electrons in the separated atoms to be the radius of the first Bohr orbit, $0.053\,nm$, and for the molecule take Δx to be the separation of the nuclei, $0.074\,nm$.

55. The average translational kinetic energy of an atom or molecule is about $\bar{K} = \frac{3}{2}kT$ (see Chapter 18), where $k = 1.38 \times 10^{-23}\,J/K$ is Boltzmann's constant. At what temperature T will $\bar{K}$ be on the order of the bond energy (and hence the bond able to be broken by thermal motion) for (a) a covalent bond (say H_2) of binding energy $4.5\,eV$, and (b) a "weak" hydrogen bond of binding energy $0.15\,eV$?

56. In the ionic salt KF, the separation distance between ions is about $0.27\,nm$. (a) Estimate the electrostatic potential energy between the ions assuming them to be point charges (magnitude $1e$). (b) It is known that F releases $4.07\,eV$ of energy when it "grabs" an electron, and $4.34\,eV$ is required to ionize K. Find the binding energy of KF relative to free K and F atoms, neglecting the energy of repulsion.

57. The fundamental vibration frequency for the HCl molecule is $8.66 \times 10^{13}\,Hz$. Determine (a) the reduced mass, and (b) the effective value of the constant k. Compare to k for the H_2 molecule.

58. Imagine the two atoms of a diatomic molecule as if they were connected by a spring, Fig. 41–44. Show that the classical frequency of vibration is given by Eq. 41–5. [Hint: Let x_1 and x_2 be the displacements of each mass from initial equilibrium positions; then $m_1\,d^2x_1/dt^2 = -kx$, and $m_2\,d^2x_2/dt^2 = -kx$, where $x = x_1 + x_2$. Find another relationship between x_1 and x_2, assuming that the center of mass of the system stays at rest, and then show that $\mu\,d^2x/dt^2 = -kx$.]

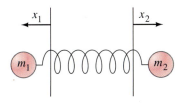

FIGURE 41–44 Problem 58.

59. Explain, using the Boltzmann factor (Eq. 40–16), why the heights of the peaks in Fig. 41–21 are different from one another. Explain also why the lines are not equally spaced. [Hint: Does the moment of inertia necessarily remain constant?]

60. Do we need to consider quantum effects for everyday rotating objects? Estimate the differences between rotational energy levels for a spinning baton compared to the energy of the baton. Assume the baton consists of a 30-cm-long bar with a mass of 200 g and two end masses (each points), each of mass 300 g, and that it rotates at 1.6 rev/s.

61. When solid argon melts at $-189°C$, its latent heat of fusion goes directly into breaking the bonds between the atoms. Solid argon is a weakly bound cubic lattice, with each atom connected to six neighbors, each bond having a binding energy of $3.9 \times 10^{-3}\,eV$. What is the latent heat of fusion for argon, in J/kg? [Hint: Show that in a simple cubic lattice (Fig. 41–45), there are three times as many bonds as there are atoms, when the number of atoms is large.]

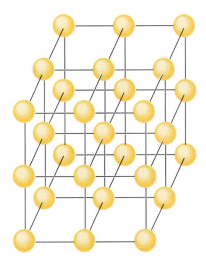

FIGURE 41–45 Problem 61.

62. A strip of silicon 1.5 cm wide and 1.0 mm thick is immersed in a magnetic field of strength 1.6 T perpendicular to the strip (Fig. 41–46). When a current of 0.20 mA is run through the strip, there is a resulting Hall effect voltage of 18 mV across the strip (Section 27–8). How many electrons per silicon atom are in the conduction band? The density of silicon is 2330 kg/m^3.

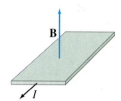

FIGURE 41–46   Problem 62.

63. Most of the Sun's radiation has wavelengths shorter than 1000 nm. For a solar cell to absorb all this, what energy gap ought the material have?

64. The energy gap between valence and conduction bands in germanium is 0.72 eV. What range of wavelengths can a photon have to excite an electron from the top of the valence band into the conduction band?

65. A TV remote control emits IR light. If the detector on the TV set is *not* to react to visible light, could it make use of silicon as a "window" with its energy gap $E_g = 1.14$ eV? What is the shortest-wavelength light that can strike silicon without causing electrons to jump from the valence band to the conduction band?

66. The *Fermi temperature* T_F is defined as that temperature at which the thermal energy kT (without the $\frac{3}{2}$) is equal to the Fermi energy: $kT_F = E_F$. (*a*) Determine the Fermi temperature for copper. (*b*) Show that for $T \gg T_F$, the Fermi factor (Eq. 41–14) approaches the Boltzmann factor. (Note: This last result is not very useful for understanding conductors. Why?)

67. For an arsenic donor atom in a doped silicon semiconductor, assume that the "extra" electron moves in a Bohr orbit about the arsenic ion. For this electron in the ground state, take into account the dielectric constant $K = 12$ of the Si lattice (which represents the weakening of the Coulomb force due to all the other atoms or ions in the lattice), and estimate (*a*) the binding energy, and (*b*) the orbit radius for this extra electron. [*Hint:* Substitute $\epsilon = K\epsilon_0$ in Coulomb's law.]

* 68. A full-wave rectifier (Fig. 41–39) uses two diodes to rectify a 75-V rms 60 Hz ac voltage. If $R = 8.8$ kΩ and $C = 30\,\mu F$, what will be the approximate percent variation in the output voltage? The variation in output voltage (Fig. 41–39c) is called *ripple voltage.* [*Hint:* See Section 26–4 and assume the discharge of the capacitor is approximately linear.]

* 69. A zener diode voltage regulator is shown in Fig. 41–47. Suppose that $R = 1.80$ kΩ, and that the diode breakdown voltage is 130 V: the diode is rated at a maximum current of 110 mA. (*a*) If $R_{load} = 16.0$ kΩ, over what range of supply voltages will the circuit maintain the voltage at 130 V? (*b*) If the supply voltage is 200 V, over what range of load resistance will the voltage be regulated?

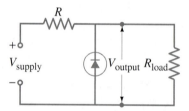

FIGURE 41–47   Problem 69.

This archeologist has unearthed the remains of a sea-turtle within an ancient man-made stone circle. Carbon dating of the remains can tell her when humans inhabited the site.

In this chapter we begin our discussion of nuclear physics including the properties of nuclei, the various forms of radioactivity, and how radioactive decay can be used in a variety of fields to determine the age of old objects, from bones and trees to rocks and other mineral substances, and obtain information on the history of the Earth.

CHAPTER 42

Nuclear Physics and Radioactivity

In the early part of the twentieth century, Rutherford's experiments led to the idea that at the center of an atom there is a tiny but massive nucleus. At the same time that the quantum theory was being developed and scientists were attempting to understand the structure of the atom and its electrons, investigations into the nucleus itself had also begun. In this chapter and the next, we take a brief look at *nuclear physics*.

42–1 Structure and Properties of the Nucleus

An important question to physicists in the early part of this century was whether the nucleus had a structure, and what that structure might be. It turns out that the nucleus is a complicated entity and is not fully understood even today. However, by the early 1930s, a model of the nucleus had been developed that is still useful.

According to this model, a nucleus is considered as an aggregate of two types of particles: protons and neutrons. (Of course, we must remember that these "particles" also have wave properties, but for ease of visualization and language, we often refer to them simply as "particles.") A **proton** is the nucleus of the simplest atom, hydrogen. It has a positive charge ($= +e = +1.60 \times 10^{-19}$ C) and a mass

Proton

$$m_{\text{p}} = 1.67262 \times 10^{-27} \text{ kg}.$$

The **neutron**, whose existence was ascertained only in 1932 by the Englishman James Chadwick (1891–1974), is electrically neutral ($q = 0$), as its name implies. Its mass, which is almost identical to that of the proton, is

Neutron

$$m_{\text{n}} = 1.67493 \times 10^{-27} \text{ kg}.$$

Nucleons

These two constituents of a nucleus, neutrons and protons, are referred to collectively as **nucleons**.

Although the hydrogen nucleus consists of a single proton alone, the nuclei of all other elements consist of both neutrons and protons. The different types of nuclei are often referred to as **nuclides**. The number of protons in a nucleus (or nuclide) is called the **atomic number** and is designated by the symbol Z. The total number of nucleons, neutrons plus protons, is designated by the symbol A and is called the **atomic mass number**. This name is used since the mass of a nucleus is very closely A times the mass of one nucleon. A nuclide with 7 protons and 8 neutrons thus has $Z = 7$ and $A = 15$. The **neutron number** N is $N = A - Z$.

Z and A

To specify a given nuclide, we need give only A and Z. A special symbol is commonly used which takes the form

$$^{A}_{Z}\text{X},$$

where X is the chemical symbol for the element (see Appendix D, and the periodic table inside the back cover), A is the atomic mass number, and Z is the atomic number. For example, $^{15}_{7}\text{N}$ means a nitrogen nucleus containing 7 protons and 8 neutrons for a total of 15 nucleons. In a neutral atom, the number of electrons orbiting the nucleus is equal to the atomic number Z (since the charge on an electron has the same magnitude but opposite sign to that of a proton). The main properties of an atom, and how it interacts with other atoms, are largely determined by the number of electrons. Hence Z determines what kind of atom it is: carbon, oxygen, gold, or whatever. It is redundant to specify both the symbol of a nucleus and its atomic number Z as described above. If the nucleus is nitrogen, for example, we know immediately that $Z = 7$. The subscript Z is thus sometimes dropped and $^{15}_{7}\text{N}$ is then written simply ^{15}N; in words we say "nitrogen fifteen."

For a particular type of atom (say, carbon), nuclei are found to contain different numbers of neutrons, although they all have the same number of protons. For example, carbon nuclei always have 6 protons, but they may have 5, 6, 7, 8, 9, or 10 neutrons. Nuclei that contain the same number of protons but different numbers of neutrons are called **isotopes**. Thus, $^{11}_{6}\text{C}$, $^{12}_{6}\text{C}$, $^{13}_{6}\text{C}$, $^{14}_{6}\text{C}$, $^{15}_{6}\text{C}$, and $^{16}_{6}\text{C}$ are all isotopes of carbon. Of course, the isotopes of a given element are not all equally common. For example, 98.9 percent of naturally occurring carbon (on Earth) is the isotope $^{12}_{6}\text{C}$, and about 1.1 percent is $^{13}_{6}\text{C}$. These percentages are referred to as the **natural abundances**.[†] Many isotopes that do not occur naturally can be produced in the laboratory by means of nuclear reactions (more on this later). Indeed, all elements beyond uranium ($Z > 92$) do not occur naturally and are only produced artificially, as are many nuclides with $Z \le 92$.

Isotopes

The approximate size of nuclei was determined originally by Rutherford from the scattering of charged particles. Of course, we cannot speak about a definite size for nuclei because of the wave–particle duality: their spatial extent must remain somewhat fuzzy. Nonetheless a rough "size" can be measured by scattering high-speed electrons off nuclei. It is found that nuclei have a roughly spherical

[†]The mass value for each element as given in the periodic table (inside back cover) is an average weighted according to the natural abundances of its isotopes.

shape with a radius that increases with A according to the approximate formula

$$r \approx (1.2 \times 10^{-15}\,\text{m})\left(A^{\frac{1}{3}}\right). \qquad \text{(42–1)}$$

Nuclear radii

Since the volume of a sphere is $V = \frac{4}{3}\pi r^3$, we see that the volume of a nucleus is proportional to the number of nucleons, $V \propto A$. This is what we would expect if nucleons were like impenetrable billiard balls: if you double the number of balls, you double the total volume. Hence, all nuclei have nearly the same density, and it is enormous (see Problem 5).

EXAMPLE 42–1 **ESTIMATE** **Nuclear sizes.** Estimate the diameter of the following nuclei: (a) ^1_1H, (b) $^{40}_{20}\text{Ca}$, (c) $^{208}_{82}\text{Pb}$, (d) $^{235}_{92}\text{U}$.

SOLUTION (a) For hydrogen, $A = 1$, Eq. 42–1 gives

$$d = \text{diameter} = 2r \approx 2.4 \times 10^{-15}\,\text{m}$$

since $A^{\frac{1}{3}} = 1^{\frac{1}{3}} = 1$.

(b) For calcium $d = 2r \approx (2.4 \times 10^{-15}\,\text{m})(40)^{\frac{1}{3}} = 8.2 \times 10^{-15}\,\text{m}$.

(c) For lead $d \approx (2.4 \times 10^{-15}\,\text{m})(208)^{\frac{1}{3}} = 14 \times 10^{-15}\,\text{m}$.

(d) For uranium $d \approx (2.4 \times 10^{-15}\,\text{m})(235)^{\frac{1}{3}} = 15 \times 10^{-15}\,\text{m}$.

The range of nuclear diameters is only from 2.4 fm to 15 fm.

The masses of nuclei can be determined, in one method, by measuring the radius of curvature of fast-moving nuclei in a magnetic field using a mass spectrometer, as discussed in Section 27–9. Indeed, as mentioned there, the existence of different isotopes of the same element was discovered using this device. Nuclear masses can be specified in **unified atomic mass units** (u). On this scale, a neutral $^{12}_{6}\text{C}$ atom is given the precise value 12.000000 u. A neutron then has a measured mass of 1.008665 u, a proton 1.007276 u, and a neutral hydrogen atom, ^1_1H (proton plus electron) 1.007825 u. The masses of many nuclides are given in Appendix D. It should be noted that the masses in this table, as is customary, are for the *neutral atom*, and not for a bare nucleus.

➥ **PROBLEM SOLVING**

Masses are for neutral atom

Masses are often specified using the electron-volt energy unit. This can be done because mass and energy are related, and the precise relationship is given by Einstein's equation $E = mc^2$ (Chapter 37). Since the mass of a proton is $1.67262 \times 10^{-27}\,\text{kg}$, or 1.007276 u, then

$$1.0000\,\text{u} = \left(\frac{1.0000\,\text{u}}{1.007276\,\text{u}}\right)(1.67262 \times 10^{-27}\,\text{kg})$$
$$= 1.66054 \times 10^{-27}\,\text{kg};$$

this is equivalent to an energy (see table inside front cover)

$$E = mc^2 = \frac{(1.66054 \times 10^{-27}\,\text{kg})(2.9979 \times 10^{8}\,\text{m/s})^2}{(1.6022 \times 10^{-19}\,\text{J/eV})}$$
$$= 931.5\,\text{MeV}.$$

Thus

$$1\,\text{u} = 1.6605 \times 10^{-27}\,\text{kg} = 931.5\,\text{MeV}/c^2.$$

Atomic mass unit

The rest masses of some of the basic particles are given in Table 42–1.

TABLE 42–1 Rest Masses in Kilograms, Unified Atomic Mass Units, and MeV/c^2

Object	Mass		
	kg	**u**	**MeV/c^2**
Electron	9.1094×10^{-31}	0.00054858	0.51100
Proton	1.67262×10^{-27}	1.007276	938.27
^1_1H atom	1.67353×10^{-27}	1.007825	938.78
Neutron	1.67493×10^{-27}	1.008665	939.57

Just as an electron has intrinsic spin and angular momentum quantum numbers, so too do nuclei and their constituents, the proton and neutron. Both the proton and the neutron are spin-$\frac{1}{2}$ particles. A nucleus, made up of protons and neutrons, has a **nuclear spin** quantum number, I, that can be either integer or half integer, depending on whether it is made up of an even or an odd number of nucleons. The *nuclear angular momentum* of a nucleus is given, as might be expected (see Section 40–2), by $\sqrt{I(I + 1)}\hbar$.

Nuclear magnetic moments are measured in terms of the **nuclear magneton**

$$\mu_N = \frac{e\hbar}{2m_p},$$

(42–2)

which is defined by analogy with the Bohr magneton for electrons ($\mu_B = e\hbar/2m_e$, Section 40–7). Since μ_N contains the proton mass, m_p, instead of the electron mass, it is about 2000 times smaller. The electron magnetic moment is about 1 Bohr magneton, so we might expect the proton to have a magnetic moment μ_p of about $1\,\mu_N$. Instead, it is

$$\mu_p = 2.7928\mu_N.$$

There is no satisfactory explanation for this large factor. Also surprising is the fact that the neutron has a magnetic moment:

$$\mu_n = -1.9135\mu_N.$$

This suggests that, although the neutron carries no net charge, there may be some sort of electric current within the neutron. The minus sign for μ_n indicates that its magnetic moment is opposite to its spin.

Important applications based on nuclear spin are nuclear magnetic resonance (NMR) and magnetic resonance imaging (MRI). They are discussed in the next chapter (Section 43–10).

42–2 Binding Energy and Nuclear Forces

The total mass of a stable nucleus is always less than the sum of the masses of its separate protons and neutrons, as the following Example shows.

EXAMPLE 42–2 $^4_2$**He mass compared to its constituents.** Compare the mass of a $^4_2$He nucleus to that of its constituent nucleons.

SOLUTION The mass of a neutral $^4_2$He atom, from Appendix D, is 4.002603 u. The mass of two neutrons and two protons (including the two electrons) is

$$2m_n = 2.017330\text{ u}$$
$$2m(^1_1\text{H}) = \underline{2.015650\text{ u}}$$
$$4.032980\text{ u}.$$

▶ PROBLEM SOLVING

Keep track of electron masses

We almost always deal with masses of neutral atoms—that is, nuclei with Z electrons—since this is how masses are measured. We must therefore be sure to balance out the electrons when we compare masses, which is why we used the mass of $^1_1$H in this Example rather than that of the proton alone.

Thus the mass of $^4_2$He is measured to be $4.032980\text{ u} - 4.002603\text{ u} = 0.030377\text{ u}$ less than the masses of its constituents. Where has this lost mass gone?

It has, in fact, gone into energy of another kind (such as radiation, or kinetic energy, for example). The mass (or energy) difference in the case of $^4_2$He, given in energy units, is $(0.030377\text{ u})(931.5\text{ MeV/u}) = 28.30\text{ MeV}$. This difference is

Binding energy

referred to as the **total binding energy** of the nucleus. The total binding energy represents the amount of energy that must be put into a nucleus in order to break it apart into its constituent protons and neutrons. If the mass of, say, a $^4_2$He nucleus were exactly equal to the mass of two neutrons plus two protons, the nucleus could fall apart without any input of energy. To be stable, the mass of a nucleus *must* be less than that of its constituent nucleons, so that energy input *is* needed to break it apart. Note that the binding energy is not something a nucleus has—it is energy it "lacks" relative to the total mass of its separate constituents.

Nuclear binding energy can be compared to the binding energy of electrons in an atom. We saw in Chapter 38 that the binding energy of the one electron in the hydrogen atom, for example, is 13.6 eV. The mass of a $_1^1$H atom is less than that of a single proton plus a single electron by 13.6 eV. Compared to the total mass of the atom (939 MeV), this is incredibly small (1 part in 10^8), and for practical purposes the mass difference can be ignored. The binding energies of nuclei are on the order of 10^6 times greater than the binding energies of electrons in atoms.

The **average binding energy per nucleon** is defined as the total binding energy of a nucleus divided by A, the total number of nucleons. For $_2^4$He, it is 28.3 MeV/4 = 7.1 MeV. Figure 42–1 shows the average binding energy per nucleon as a function of A for stable nuclei. The curve rises as A increases and reaches a plateau at about 8.7 MeV per nucleon above about $A \approx 40$. Beyond about $A \approx 80$, the curve decreases slowly, indicating that larger nuclei are held together a little less tightly than those in the middle of the periodic table. (We will see later that these characteristics allow the release of nuclear energy in the processes of fission and fusion.)

Binding energy per nucleon

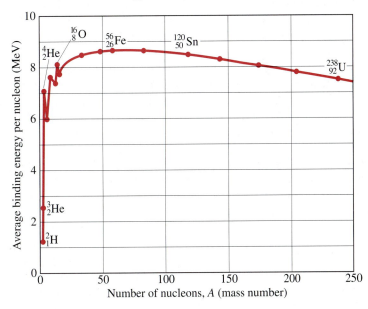

FIGURE 42–1 Average binding energy per nucleon as a function of mass number A for stable nuclei.

EXAMPLE 42–3 **Binding energy for iron.** Calculate the total binding energy and the average binding energy per nucleon for $_{26}^{56}$Fe, the most common stable isotope of iron.

SOLUTION $_{26}^{56}$Fe has 26 protons and 30 neutrons whose separate masses are

$$\begin{aligned}
(26)(1.007825\ u) &= 26.2035\ u\ \text{(includes electrons)} \\
(30)(1.008665\ u) &= \underline{30.2600\ u} \\
\text{Total} &= 56.4635\ u. \\
\text{Subtract mass of } _{26}^{56}\text{Fe:} &\quad -55.9349\ u\ \text{(Appendix D)} \\
\Delta m &= 0.5286\ u.
\end{aligned}$$

The total binding energy is thus

$$(0.5286\ u)(931.5\ \text{MeV}/u) = 492.4\ \text{MeV}$$

and the average binding energy per nucleon is

$$\frac{492.4\ \text{MeV}}{56\ \text{nucleons}} = 8.8\ \text{MeV}.$$

EXAMPLE 42–4 Binding energy of last neutron. What is the binding energy of the last neutron in $^{13}_{6}C$?

SOLUTION We compare the mass of $^{13}_{6}C$ to that of the atom with one less neutron, $^{12}_{6}C$, plus a free neutron (Appendix D):

$$\begin{aligned} \text{Mass } {}^{12}_{6}\text{C} &= \quad 12.000000 \text{ u} \\ \text{Mass } {}^{1}_{0}\text{n} &= \quad \underline{1.008665 \text{ u}} \\ \text{Total} &= \quad 13.008665 \text{ u}. \\ \text{Subtract mass of } {}^{13}_{6}\text{C:} &\quad \underline{-13.003355 \text{ u}} \\ \Delta m &= \quad 0.005310 \text{ u} \end{aligned}$$

which in energy is $(931.5 \text{ MeV/u})(0.005310 \text{ u}) = 4.95 \text{ MeV}$. That is, it would require 4.95 MeV input of energy to remove one neutron from $^{13}_{6}C$.

We can analyze nuclei not only from the point of view of energy, but also from the point of view of the forces that hold them together. We would not expect a collection of protons and neutrons to come together spontaneously, since protons are all positively charged and thus exert repulsive forces on each other. Indeed, the question arises as to how a nucleus stays together at all in view of the fact that the electric force between protons would tend to break it apart. Since stable nuclei *do* stay together, it is clear that another force must be acting. Because this new force is stronger than the electric force (which, in turn, is much stronger than gravity at the nuclear level) it is called the **strong nuclear force**. The strong nuclear force is an attractive force that acts between all nucleons—protons and neutrons alike. Thus protons attract each other via the nuclear force at the same time they repel each other via the electric force. Neutrons, since they are electrically neutral, only attract other neutrons or protons via the nuclear force.

The nuclear force turns out to be far more complicated than the gravitational and electromagnetic forces. A precise mathematical description is not yet possible. Nonetheless, a great deal of work has been done to try to understand the nuclear force. One important aspect of the strong nuclear force is that it is a **short-range** force: it acts only over a very short distance. It is very strong between two nucleons if they are less than about 10^{-15} m apart, but it is essentially zero if they are separated by a distance greater than this. Compare this to electric and gravitational forces, which can act over great distances and are therefore called **long-range** forces. The strong nuclear force has some strange quirks. For example, if a nuclide contains too many or too few neutrons relative to the number of protons, the binding of the nucleons is reduced; nuclides that are too unbalanced in this regard are unstable. As shown in Fig. 42–2, stable nuclei tend to have the same number of protons as neutrons ($N = Z$) up to about $A \approx 30$ or 40. Beyond this, stable nuclei contain more neutrons than protons. This makes sense since, as Z increases, the electrical repulsion increases, so a greater number of neutrons—which exert only the attractive nuclear force—are required to maintain stability. For very large Z, no number of neutrons can overcome the greatly increased electric repulsion. Indeed, there are no completely stable nuclides above $Z = 82$.

What we mean by a stable nucleus is one that stays together indefinitely. What then is an unstable nucleus? It is one that comes apart; and this results in radioactive decay. Before we discuss the important subject of radioactivity (the next Section), we note that there is a second type of nuclear force that is much weaker than the strong nuclear force. It is called the **weak nuclear force**, and we are aware of its existence only because it shows itself in certain types of radioactive decay. These two nuclear forces, the strong and the weak, together with the gravitational and electromagnetic forces, comprise the four known types of force in nature (more on this in Chapter 44).

Strong nuclear force

Long- and short-range forces

FIGURE 42–2 Number of neutrons versus number of protons for stable nuclides, which are represented by dots. The straight line represents $N = Z$.

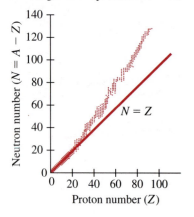

Weak force

42–3 | Radioactivity

Nuclear physics had its beginnings in 1896. In that year, Henri Becquerel (1852–1908) made an important discovery: in his studies of phosphorescence, he found that a certain mineral (which happened to contain uranium) would darken a photographic plate even when the plate was wrapped to exclude light. It was clear that the mineral emitted some new kind of radiation that, unlike X-rays, occurred without any external stimulus. This new phenomenon eventually came to be called **radioactivity**.

Discovery of radioactivity

Soon after Becquerel's discovery, Marie Curie (1867–1934) and her husband, Pierre Curie (1859–1906), isolated two previously unknown elements that were very highly radioactive (Fig. 42–3). These were named polonium and radium. Other radioactive elements were soon discovered as well. The radioactivity was found in every case to be unaffected by the strongest physical and chemical treatments, including strong heating or cooling and the action of strong chemical reagents. It was clear that the source of radioactivity must be deep within the atom, that it must emanate from the nucleus. And it became apparent that radioactivity is the result of the *disintegration* or *decay* of an unstable nucleus. Certain isotopes are not stable under the action of the nuclear force, and they decay with the emission of some type of radiation or "rays."

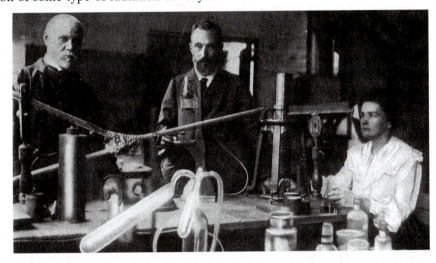

FIGURE 42–3 Marie and Pierre Curie in their laboratory (about 1906) where radium was discovered.

Many unstable isotopes occur in nature, and such radioactivity is called "natural radioactivity." Other unstable isotopes can be produced in the laboratory by nuclear reactions (Section 43–1); these are said to be produced "artificially" and to have "artificial radioactivity."

Rutherford and others began studying the nature of the rays emitted in radioactivity about 1898. They found that the rays could be classified into three distinct types according to their penetrating power. One type of radiation could barely penetrate a piece of paper. The second type could pass through as much as 3 mm of aluminum. The third was extremely penetrating: it could pass through several centimeters of lead and still be detected on the other side. They named these three types of radiation alpha (α), beta (β), and gamma (γ), respectively, after the first three letters of the Greek alphabet.

Each type of ray was found to have a different charge and hence is bent differently in a magnetic field, Fig. 42–4; α rays are positively charged, β rays are negatively charged, and γ rays are neutral. It was soon found that all three types of radiation consisted of familiar kinds of particles. Gamma rays are very high-energy photons whose energy is even higher than that of X-rays. Beta rays are electrons, identical to those that orbit the nucleus (but they are created within the nucleus itself). Alpha rays (or α particles) are simply the nuclei of helium atoms, $_2^4\text{He}$; that is, an α ray consists of two protons and two neutrons bound together.

FIGURE 42–4 Alpha and beta rays are bent in opposite directions by a magnetic field, whereas gamma rays are not bent at all.

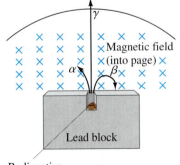

Radioactive sample (radium)

We now discuss each of these three types of radioactivity, or decay, in more detail.

42–4 Alpha Decay

When a nucleus emits an α particle ($^4_2$He), it is clear that the remaining nucleus will be different from the original: for it has lost two protons and two neutrons. Radium 226 ($^{226}_{88}$Ra), for example, is an α emitter. It decays to a nucleus with $Z = 88 - 2 = 86$ and $A = 226 - 4 = 222$. The nucleus with $Z = 86$ is radon (Rn) —see Appendix D or the periodic table. Thus the radium decays to radon with the emission of an α particle. This is written

α decay

$$^{226}_{88}\text{Ra} \rightarrow ^{222}_{86}\text{Rn} + ^4_2\text{He}.$$

See Fig. 42–5.

FIGURE 42–5 Radioactive decay of radium to radon with emission of an alpha particle.

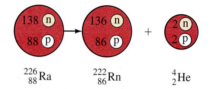

$^{226}_{88}$Ra $^{222}_{86}$Rn $^4_2$He

Daughter nucleus
Parent nucleus
Transmutation

It is clear that when α decay occurs, a new element is formed. The **daughter** nucleus ($^{222}_{86}$Rn in this case) is different from the **parent** nucleus ($^{226}_{88}$Ra in this case). This changing of one element into another is called **transmutation**.

Alpha decay can be written

$$^A_Z N \rightarrow ^{A-4}_{Z-2} N' + ^4_2\text{He} \qquad [\alpha \text{ decay}]$$

where N is the parent, N' the daughter, and Z and A are the atomic number and atomic mass number, respectively, of the parent.

Alpha decay occurs because the strong nuclear force is unable to hold very large nuclei together. Because the nuclear force is a short-range force, it acts only between neighboring nucleons. But the electric force can act all the way across a large nucleus. For very large nuclei, the large Z means the repulsive electric force becomes very large (Coulomb's law); and it acts between all protons. The strong nuclear force, since it acts only between neighboring nucleons, is overpowered and is unable to hold the nucleus together.

Why the strong nuclear force cannot hold a nucleus together

We can express the instability in terms of energy (or mass): the mass of the parent nucleus is greater than the mass of the daughter nucleus plus the mass of the α particle. The mass difference appears as kinetic energy, which is carried away by the α particle and the recoiling daughter nucleus. The total energy released is called the **disintegration energy**, Q, or the **Q-value** of the decay. From conservation of energy,

$$M_\text{P} c^2 = M_\text{D} c^2 + m_\alpha c^2 + Q,$$

so

Q-value

$$Q = M_\text{P} c^2 - (M_\text{D} + m_\alpha)c^2 \qquad \textbf{(42–3)}$$

where M_P, M_D, and m_α are the masses of the parent, daughter, and α particle, respectively. If the parent had *less* mass than the daughter plus the α particle (so $Q < 0$), the decay could not occur, for the conservation of energy law would be violated.

EXAMPLE 42–5 Uranium decay energy release. Calculate the disintegration energy when $^{232}_{92}U$ (mass = 232.037146 u) decays to $^{228}_{90}Th$ (228.028731 u) with the emission of an α particle. (As always, masses are for neutral atoms.)

SOLUTION Since the mass of the ^4_2He is 4.002603 u (Appendix D), the total mass in the final state is

$$228.028731 \text{ u} + 4.002603 \text{ u} = 232.031334 \text{ u}.$$

The mass lost when the $^{232}_{92}U$ decays is

$$232.037146 \text{ u} - 232.031334 \text{ u} = 0.005812 \text{ u}.$$

Since $1 \text{ u} = 931.5 \text{ MeV}$, the energy Q released is

$$Q = (0.005812 \text{ u})(931.5 \text{ MeV/u}) \approx 5.4 \text{ MeV},$$

and this energy appears as kinetic energy of the α particle and the daughter nucleus. (Using conservation of momentum, it can be shown that the α particle emitted by a $^{232}_{92}U$ nucleus at rest has a kinetic energy of about 5.3 MeV. Thus, the daughter nucleus—which recoils in the opposite direction from the emitted α particle—has about 0.1 MeV of kinetic energy. See Problem 68.)

If the mass of the daughter nucleus plus the mass of the α particle is less than the mass of the parent nucleus (so the parent is energetically allowed to decay), why are there any parent nuclei at all? That is, why haven't radioactive nuclei all decayed long ago, right after they were formed in the early stages of the universe? We can understand decay using a model of a nucleus inside of which there is an alpha particle (at least some of the time) bouncing around. The potential energy "seen" by the α particle would have a shape something like that shown in Fig. 42–6. The potential energy well (approximately square) between $r = 0$ and $r = R_0$ represents the short-range attractive nuclear force. Beyond the nuclear radius, R_0, the Coulomb repulsion dominates (since the nuclear force drops to zero) and we see the characteristic $1/r$ dependence of the Coulomb potential. The α particle, trapped within the nucleus, can be thought of as bouncing back and forth between the potential walls. Since the potential energy just beyond $r = R_0$ is greater than the energy of the α particle (dashed line), the α particle could not escape the nucleus if it were governed by classical physics. But according to quantum mechanics, there is a certain probability that the α particle can tunnel through the Coulomb barrier, from point A to point B in Fig. 42–6, as we discussed in Section 39–10. The height and width of the barrier affect the rate at which the nuclei decay (Section 42–8). Because of this barrier, the lifetimes of α-unstable nuclei can be quite long, from a fraction of a microsecond to over 10^{10} years. Note in Fig. 42–6 that the Q-value represents the total kinetic energy when the α particle is far from the nucleus.

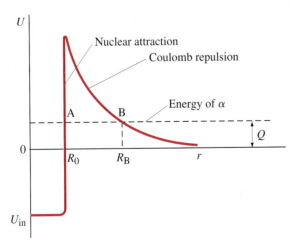

FIGURE 42–6 Potential energy for alpha particle and (daughter) nucleus, showing the Coulomb barrier through which the α particle must tunnel to escape. The Q-value of the reaction is also shown.

Why, you may wonder, do nuclei emit this combination of four nucleons called an α particle? Why not just four separate nucleons, or even one? The answer is that the α particle is very strongly bound, so that its mass is significantly less than that of four separate nucleons. As we saw in Example 42–2, two protons and two neutrons separately have a total mass of about 4.032980 u. The total mass of a $^{228}_{90}$Th nucleus plus four separate nucleons is 232.061711 u, which is greater than the mass of the parent nucleus. Such a decay could not occur because it would violate the conservation of energy. Similarly, it is almost always true that the emission of a single nucleon is energetically not possible. Or, to put it another way, for any nuclide for which it could be possible (mass of parent > mass of daughter + mass of nucleon), the decay happens so fast after formation of the parent that we don't see the parent in nature.

One widespread application of nuclear physics is present in nearly every home in the form of an ordinary **smoke detector**. The most common type of detector contains about 0.2 mg of the radioactive americium isotope, $^{241}_{95}$Am, in the form of AmO_2. The radiation ionizes the nitrogen and oxygen molecules in the air space between two oppositely charged plates. The resulting conductivity allows a small steady current. If smoke enters, the radiation is absorbed by the smoke particles rather than by the air molecules, thus reducing the current. The current drop is detected by the device's electronics and sets off the alarm. The radiation dose that escapes from an intact americium smoke detector is much less than the natural radioactive background, and so can be considered harmless. There is no question that smoke detectors save lives and reduce property damage.

42–5 | Beta Decay

Transmutation of elements also occurs when a nucleus decays by β decay—that is, with the emission of an electron or $β^-$ particle. The nucleus $^{14}_6$C, for example, emits an electron when it decays:

$$^{14}_6\text{C} \rightarrow \, ^{14}_7\text{N} + \text{e}^- + \text{a neutrino},$$

where e^- is the symbol for the electron. (The symbol $_{-1}^0$e is sometimes used for the electron whose charge corresponds to $Z = -1$ and, since it is not a nucleon and has very small mass, has $A = 0$.) The particle known as the neutrino, with rest mass $m = 0$ and charge $q = 0$, was not initially detected and was only later hypothesized to exist, as we shall discuss later in this Section. No nucleons are lost when an electron is emitted, and the total number of nucleons, A, is the same in the daughter nucleus as in the parent. But because an electron has been emitted from the nucleus itself, the charge on the daughter nucleus is $+1e$ greater than that on the parent. The parent nucleus had $Z = +6$, so from charge conservation the nucleus remaining behind must have a charge of $+7e$. So the daughter nucleus has $Z = 7$, which is nitrogen.

It must be carefully noted that the electron emitted in β decay is *not* an orbital electron. Instead, the electron is created *within the nucleus itself*. What happens is that one of the neutrons changes to a proton and in the process (to conserve charge) throws off an electron. Indeed, free neutrons actually do decay in this fashion:

$$\text{n} \rightarrow \text{p} + \text{e}^- + \text{a neutrino}.$$

Because of their origin in the nucleus, the electrons emitted in β decay are often referred to as "β particles," rather than as electrons, to remind us of their origin. They are, nonetheless, indistinguishable from orbital electrons.

EXAMPLE 42–6 **Energy release in $^{14}_{6}C$ decay.** How much energy is released when $^{14}_{6}C$ decays to $^{14}_{7}N$ by β emission? Use Appendix D.

SOLUTION The masses given in Appendix D are those of the neutral atom, and we have to keep track of the electrons involved. Assume the parent nucleus has six orbiting electrons so it is neutral, and its mass is 14.003242 u. The daughter, which in this decay is $^{14}_{7}N$, is not neutral since it has the same six electrons circling it but the nucleus has a charge of $+7e$. However, the mass of this daughter with its six electrons, plus the mass of the emitted electron (which makes a total of seven electrons), is just the mass of a neutral nitrogen atom. That is, the total mass in the final state is

(mass of $^{14}_{7}N$ nucleus + 6 electrons) + (mass of 1 electron),

and this is equal to

mass of neutral $^{14}_{7}N$ (includes 7 electrons),

which, from Appendix D is a mass of 14.003074 u. (Note that the neutrino doesn't contribute to either the mass or charge balance since it has $m = 0$ and $q = 0$.) Hence the mass after decay is 14.003074 u, whereas before decay, it was 14.003242 u. So the mass difference is 0.000168 u, which corresponds to 0.156 MeV or 156 keV.

According to this Example, we would expect the emitted electron to have a kinetic energy of 156 keV. (The daughter nucleus, because its mass is very much larger than that of the electron, recoils with very low velocity and hence gets very little of the kinetic energy.) Indeed, very careful measurements indicate that a few emitted β particles do have kinetic energy close to this calculated value. But the vast majority of emitted electrons have somewhat less energy. In fact, the energy of the emitted electron can be anywhere from zero up to the maximum value as calculated above. This range of electron kinetic energy was found for any β decay. It was as if the law of conservation of energy was being violated, and indeed Bohr actually considered this possibility. Careful experiments indicated that linear momentum and angular momentum also did not seem to be conserved. Physicists were troubled at the prospect of having to give up these laws, which had worked so well in all previous situations. In 1930, Wolfgang Pauli proposed an alternate solution: perhaps a new particle that was very difficult to detect was emitted during β decay in addition to the electron. This hypothesized particle could be carrying off the energy, momentum, and angular momentum required to maintain the conservation laws. This new particle was named the **neutrino**—meaning "little neutral one"—by the great Italian physicist Enrico Fermi (1901–1954; Fig. 42–7), who in 1934 worked out a detailed theory of β decay. (It was Fermi who, in this theory, postulated the existence of the fourth force in nature, which we call the weak nuclear force.) The electron neutrino has zero charge, spin of $\frac{1}{2}\hbar$, and seems to have zero rest mass, although we cannot yet rule out the possibility that it might have a very tiny rest mass. If its rest mass is zero, it is much like a photon in that it is neutral and travels at the speed of light. But the neutrino is far more difficult to detect. In 1956, complex experiments produced further evidence for the existence of the neutrino; but by then, most physicists had already accepted its existence.

The symbol for the neutrino is the Greek letter nu (ν). The correct way of writing the decay of $^{14}_{6}C$ is then

$$^{14}_{6}C \rightarrow\ ^{14}_{7}N + e^- + \bar{\nu}.$$

The bar ($^-$) over the neutrino symbol is to indicate that it is an "antineutrino." (Why this is called an antineutrino rather than simply a neutrino need not concern us now; it is discussed in Chapter 44.)

FIGURE 42–7 Enrico Fermi. Fermi contributed significantly to both theoretical and experimental physics, a feat almost unique in this century.

β^- *decay*

Many isotopes decay by electron emission. They are always isotopes that have too many neutrons compared to the number of protons. That is, they are isotopes that lie above the stable isotopes plotted in Fig. 42–2. But what about unstable isotopes that have too few neutrons compared to their number of protons—those that fall below the stable isotopes of Fig. 42–2? These, it turns out, decay by emitting a **positron** instead of an electron. A positron (sometimes called an e^+ or β^+ particle) has the same mass as the electron, but it has a positive charge of $+1e$. Because it is so like an electron, except for its charge, the positron is called the **antiparticle**[†] to the electron. An example of a β^+ decay is that of $^{19}_{10}\text{Ne}$:

Positron (β^+) decay

$$^{19}_{10}\text{Ne} \rightarrow {}^{19}_{9}\text{F} + e^+ + \nu,$$

where e^+ (or ^0_1e) stands for a positron. Note that the ν emitted here is a neutrino, whereas that emitted in β^- decay is called an antineutrino. Thus an antielectron (= positron) is emitted with a neutrino, whereas an antineutrino is emitted with an electron; this is discussed in Chapter 44.

We can write β^- and β^+ decay, in general, as follows:

$$^A_Z N \rightarrow {}^{A}_{Z+1}N' + e^- + \bar{\nu} \qquad [\beta^- \text{ decay}]$$

$$^A_Z N \rightarrow {}^{A}_{Z-1}N' + e^+ + \nu, \qquad [\beta^+ \text{ decay}]$$

where N is the parent nucleus and N' is the daughter.

Besides β^- and β^+ emission, there is a third related process. This is **electron capture** (abbreviated EC in Appendix D) and occurs when a nucleus absorbs one of its orbiting electrons. An example is ^7_4Be, which as a result becomes ^7_3Li. The process is written

Electron capture

$$^7_4\text{Be} + e^- \rightarrow {}^7_3\text{Li} + \nu,$$

or, in general,

$$^A_Z N + e^- \rightarrow {}^{A}_{Z-1}N' + \nu. \qquad [\text{electron capture}]$$

K-capture

Usually it is an electron in the innermost (K) shell that is captured, in which case it is called "K-capture." The electron disappears in the process and a proton in the nucleus becomes a neutron; a neutrino is emitted as a result. This process is inferred experimentally by detection of emitted X-rays (due to electrons jumping down to fill the empty state) of just the proper energy.

In β decay, it is the weak nuclear force that plays the crucial role. The neutrino is unique in that it interacts with matter only via the weak force, which is why it is so hard to detect.

42–6 | Gamma Decay

Gamma rays are photons having very high energy. They have their origin in the decay of a nucleus, much like emission of photons by excited atoms. Like an atom, a nucleus itself can be in an excited state. When it jumps down to a lower energy state, or to the ground state, it emits a photon which we call a γ ray. The possible energy levels of a nucleus are much farther apart than those of an atom: on the order of keV or MeV, as compared to a few eV for electrons in an atom. Hence, the emitted photons have energies that can range from a few keV to several MeV. For a given decay, the γ ray always has the same energy. Since a γ ray carries no charge, there is no change in the element as a result of a γ decay.

[†] Discussed in Chapter 44. Briefly, an antiparticle has the same mass as its corresponding particle, but opposite charge.

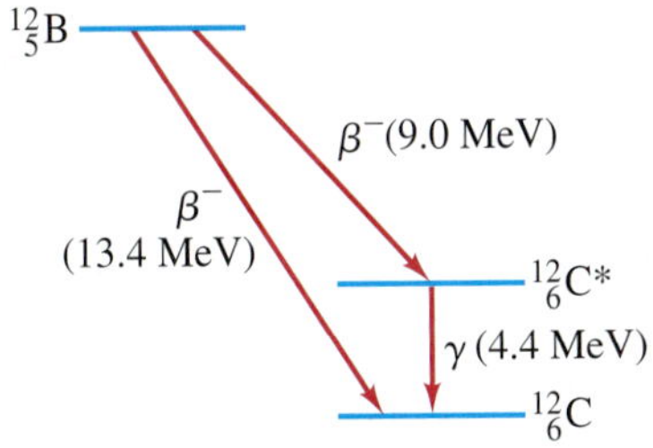

FIGURE 42–8 Energy-level diagram showing how $^{12}_5B$ can decay to the ground state of $^{12}_6C$ by β decay (total energy released = 13.4 MeV), or can instead β decay to an excited state of $^{12}_6C$ (indicated by *), which subsequently decays to its ground state by emitting a 4.4-MeV γ ray.

Internal conversion

How does a nucleus get into an excited state? It may occur because of a violent collision with another particle. More commonly, the nucleus remaining after a previous radioactive decay may be in an excited state. A typical example is shown in the energy-level diagram of Fig. 42–8. $^{12}_5B$ can decay by β decay directly to the ground state of $^{12}_6C$; or it can go by β decay to an excited state of $^{12}_6C$, which then decays by emission of a 4.4 MeV γ ray to the ground state.

We can write γ decay as

$$^A_ZN^* \rightarrow {}^A_ZN + \gamma, \qquad \text{[}\gamma \text{ decay]}$$

where the asterisk means "excited state" of that nucleus.

In some cases, a nucleus may remain in an excited state for some time before it emits a γ ray. The nucleus is then said to be in a **metastable state** and is called an **isomer**.

An excited nucleus can sometimes return to the ground state by another process known as **internal conversion** with no γ ray emitted. In this process, the excited nucleus interacts with one of the orbital electrons and ejects this electron from the atom with the same kinetic energy (minus the binding energy of the electron) that an emitted γ ray would have had.

What, you may wonder, is the difference between a γ ray and an X-ray? They both are electromagnetic radiation (photons) and, though γ rays usually have higher energy than X-rays, their range of energies overlap to some extent. The difference is not intrinsic. We use the term X-ray if the photon is produced by an electron–atom interaction, and γ ray if the photon is produced in a nuclear process.

42–7 Conservation of Nucleon Number and Other Conservation Laws

In all three types of radioactive decay, the classical conservation laws hold. Energy, linear momentum, angular momentum, and electric charge are all conserved. These quantities are the same before the decay as after. But a new conservation law is also revealed, the **law of conservation of nucleon number**. According to this law, the total number of nucleons (A) remains constant in any process, although one type can change into the other type (protons into neutrons or vice versa). This law holds in all three types of decay. Table 42–2 gives a summary of α, β, and γ decay.

TABLE 42–2 The Three Types of Radioactive Decay

α decay:
$$^A_ZN \rightarrow {}^{A-4}_{Z-2}N' + {}^4_2He$$

β decay:
$$^A_ZN \rightarrow {}^A_{Z+1}N' + e^- + \bar{\nu}$$
$$^A_ZN \rightarrow {}^A_{Z-1}N' + e^+ + \nu$$
$$^A_ZN + e^- \rightarrow {}^A_{Z-1}N' + \nu \text{ [EC]}^\dagger$$

γ decay:
$$^A_ZN^* \rightarrow {}^A_ZN + \gamma$$

*Indicates the excited state of a nucleus.
†Electron capture.

42–8 Half-Life and Rate of Decay

A macroscopic sample of any radioactive isotope consists of a vast number of radioactive nuclei. These nuclei do not all decay at one time. Rather, they decay one by one over a period of time. This is a random process: we can not predict exactly when a given nucleus will decay. But we can determine, on a probabilistic basis, approximately how many nuclei in a sample will decay over a given time period, by assuming that each nucleus has the same probability of decaying in each second that it exists.

The number of decays ΔN that occur in a very short time interval Δt is then proportional to Δt and to the total number N of radioactive nuclei present:

$$\Delta N = -\lambda N \, \Delta t. \qquad \textbf{(42–4)}$$

In this equation, λ is a constant of proportionality called the **decay constant**, which is different for different isotopes. The greater λ is, the greater the rate of decay and the more "radioactive" that isotope is said to be for a given number of nuclei. The number of decays that occur in the short time interval Δt is designated ΔN

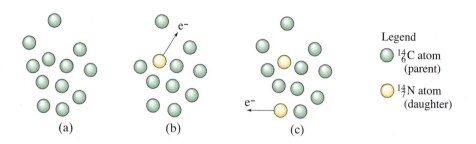

FIGURE 42–9 Radioactive nuclei decay one by one. Hence, the number of parent nuclei in a sample is continually decreasing. When a $^{14}_{6}C$ nucleus emits the electron, it becomes a $^{14}_{7}N$ nucleus.

(a)   (b)   (c)

Legend
$^{14}_{6}C$ atom (parent)
$^{14}_{7}N$ atom (daughter)

because each decay that occurs corresponds to a decrease by one in the number N of nuclei present. That is, radioactive decay is a "one-shot" process, Fig. 42–9. Once a particular parent nucleus decays into its daughter, it cannot do it again. The minus sign in Eq. 42–4 is needed to indicate that N is decreasing.

If we take the limit $\Delta t \rightarrow 0$ in Eq. 42–4, ΔN will be small compared to N, and we can write the equation in infinitesimal form as

$$dN = -\lambda N \, dt. \quad (42\text{–}5)$$

We can determine N as a function of t by rearranging this equation to

$$\frac{dN}{N} = -\lambda \, dt$$

and then integrating from $t = 0$ to $t = t$:

$$\int_{N_0}^{N} \frac{dN}{N} = -\int_{0}^{t} \lambda \, dt,$$

where N_0 is the number of parent nuclei present at $t = 0$ and N is the number remaining at time t. The integration gives

$$\ln \frac{N}{N_0} = -\lambda t$$

or

Radioactive decay law

$$N = N_0 e^{-\lambda t}. \quad (42\text{–}6)$$

Equation 42–6 is called the **radioactive decay law**. It tells us that the number of radioactive nuclei in a given sample decreases exponentially in time. This is shown in Fig. 42–10a, for the case of $^{14}_{6}C$ whose decay constant is $\lambda = 3.8 \times 10^{-12}\,\text{s}^{-1}$.

FIGURE 42–10 (a) The number N of parent nuclei in a given sample of $^{14}_{6}C$ decreases exponentially. (b) The number of decays per second also decreases exponentially. The half-life of $^{14}_{6}C$ is about 5730 yr, which means that the number of parent nuclei, N, and the rate of decay, dN/dt, decrease by half every 5730 yr.

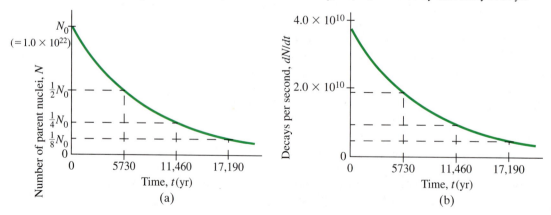

The rate of decay, or number of decays per second, in a pure sample is

$$\frac{dN}{dt}$$

which is called the **activity** of a given sample. From Eqs. 42–5 and 42–6,

$$\frac{dN}{dt} = -\lambda N = -\lambda N_0 e^{-\lambda t}. \qquad (42\text{–}7a)$$

At $t = 0$, the activity is

$$\left(\frac{dN}{dt} \right)_0 = -\lambda N_0. \qquad (42\text{–}7b)$$

Hence

$$\frac{dN}{dt} = \left(\frac{dN}{dt} \right)_0 e^{-\lambda t}, \qquad (42\text{–}7c) \qquad \textit{Activity}$$

so the activity also decreases exponentially in time at the same rate as for N (Fig. 42–10b).

The rate of decay of any isotope is often specified by giving its half-life rather than the decay constant λ. The **half-life** of an isotope is defined as the time it takes for half the original amount of isotope in a given sample to decay. For example, the half-life of $^{14}_{6}C$ is about 5730 yr. If at some time a piece of petrified wood contains, say, 1.00×10^{22} $^{14}_{6}C$ nuclei, then 5730 yr later it will contain only 0.50×10^{22} nuclei. After another 5730 yr it will contain 0.25×10^{22} nuclei, and so on. This is characteristic of the exponential function, and is shown in Fig. 42–10a. Since the rate of decay dN/dt is proportional to N, it too decreases by a factor of 2 every half-life, Fig. 42–10b.

The half-lives of known radioactive isotopes vary from as short as 10^{-22} s to about 10^{28} s (about 10^{21} yr). The half-lives of many isotopes[†] are given in Appendix D. It should be clear that the half-life (which we designate $T_{\frac{1}{2}}$) bears an inverse relationship to the decay constant. The longer the half-life of an isotope, the more slowly it decays, and hence λ is smaller. The precise relation is obtained from Eq. 42–6 by setting $N = N_0/2$ at $t = T_{\frac{1}{2}}$:

$$\frac{N_0}{2} = N_0 e^{-\lambda T_{\frac{1}{2}}} \qquad \text{or} \qquad e^{\lambda T_{\frac{1}{2}}} = 2.$$

We take natural logs of both sides ("ln" and "e" are inverse operations, meaning $\ln(e^x) = x$) and find

$$\ln\left(e^{\lambda T_{\frac{1}{2}}} \right) = \ln 2, \qquad \text{so} \qquad \lambda T_{\frac{1}{2}} = \ln 2$$

and

$$T_{\frac{1}{2}} = \frac{\ln 2}{\lambda} = \frac{0.693}{\lambda}. \qquad (42\text{–}8) \qquad \textit{Half-life}$$

We can then write Eq. 42–6 as

$$N = N_0 e^{-0.693t/T_{\frac{1}{2}}}.$$

Now let us use these concepts and equations in worked-out Examples.

[†] You may find the *mean life* of an isotope quoted. The mean life τ is defined as $\tau = 1/\lambda$ (see Problem 76), so that Eq. 42–6 can be written $N = N_0 e^{-t/\tau}$ just as for RC and LR circuits (Chapters 26 and 30) where τ is called the time constant. Since

$$\tau = \frac{1}{\lambda} = \frac{T_{\frac{1}{2}}}{0.693}$$

the mean life and half-life differ significantly in numerical value, so confusing them can cause serious error.

EXAMPLE 42–7 **Sample activity.** The isotope $^{14}_6C$ has a half-life of 5730 yr. If at some time a sample contains 1.00×10^{22} carbon-14 nuclei, what is the activity of the sample?

SOLUTION First we calculate the decay constant λ from Eq. 42–8, and obtain

$$\lambda = \frac{0.693}{T_{\frac{1}{2}}} = \frac{0.693}{(5730 \text{ yr})(3.156 \times 10^7 \text{ s/yr})} = 3.83 \times 10^{-12} \text{ s}^{-1},$$

since the number of seconds in a year is $(60)(60)(24)(365\frac{1}{4}) = 3.156 \times 10^7 \text{ s}$. From Eq. 42–5, the magnitude of the activity or rate of decay is

$$\frac{dN}{dt} = \lambda N = (3.83 \times 10^{-12} \text{ s}^{-1})(1.00 \times 10^{22})$$

$$= 3.83 \times 10^{10} \text{ decays/s}.$$

(The unit "decays/s" is often written simply as s^{-1} since "decays" is not a unit but refers only to the number.) Note that the graph of Fig. 42–10b starts at this value, corresponding to the original value of $N = 1.0 \times 10^{22}$ nuclei in Fig. 42–10a.

EXAMPLE 42–8 **A sample of radioactive $^{13}_7N$.** A laboratory has $1.49 \,\mu g$ of pure $^{13}_7N$, which has a half-life of 10.0 min (600 s). (a) How many nuclei are present initially? (b) What is the activity initially? (c) What is the activity after 1.00 h? (d) After approximately how long will the activity drop to less than one per second?

SOLUTION (a) Since the atomic mass is 13.0, then 13.0 g will contain 6.02×10^{23} nuclei (Avogadro's number). Since we have only $1.49 \times 10^{-6} \text{ g}$, the number of nuclei, N_0, that we have initially is given by the ratio

$$\frac{N_0}{1.49 \times 10^{-6} \text{ g}} = \frac{6.02 \times 10^{23}}{13.0 \text{ g}},$$

so $N_0 = 6.90 \times 10^{16}$ nuclei.
(b) From Eq. 42–8, $\lambda = (0.693)/(600 \text{ s}) = 1.16 \times 10^{-3} \text{ s}^{-1}$. Then, at $t = 0$ (Eq. 42–7b)

$$\left(\frac{dN}{dt}\right)_0 = \lambda N_0 = (1.16 \times 10^{-3} \text{ s}^{-1})(6.90 \times 10^{16}) = 8.00 \times 10^{13} \text{ s}^{-1}.$$

(c) After $1.00 \text{ h} = 3600 \text{ s}$, the activity will be (Eq. 42–7c)

$$\frac{dN}{dt} = \left(\frac{dN}{dt}\right)_0 e^{-\lambda t}$$

$$= (8.00 \times 10^{13} \text{ s}^{-1}) e^{-(1.16 \times 10^{-3} \text{ s}^{-1})(3600 \text{ s})} = 1.23 \times 10^{12} \text{ s}^{-1}.$$

This result can be obtained in another way: since 1.00 h represents six half-lives ($6 \times 10.0 \text{ min}$), the activity will decrease to $(\frac{1}{2})(\frac{1}{2})(\frac{1}{2})(\frac{1}{2})(\frac{1}{2})(\frac{1}{2}) = (\frac{1}{2})^6 = \frac{1}{64}$ of its original value, or $(8.00 \times 10^{13} \text{ s}^{-1})/64 = 1.25 \times 10^{12} \text{ s}^{-1}$. (The slight discrepancy between the two values arises because we kept only three significant figures.)
(d) We want to determine the time t when $dN/dt = 1.00 \text{ s}^{-1}$. From Eq. 42–7c, we have

$$e^{-\lambda t} = \frac{(dN/dt)}{(dN/dt)_0} = \frac{1.00 \text{ s}^{-1}}{8.00 \times 10^{13} \text{ s}^{-1}} = 1.25 \times 10^{-14}.$$

We take the natural log (ln) of both sides and divide by λ to find

$$t = -\frac{\ln(1.25 \times 10^{-14})}{\lambda} = \frac{32.0}{1.16 \times 10^{-3} \text{ s}^{-1}} = 2.76 \times 10^4 \text{ s} = 7.67 \text{ h}.$$

42–9 | Decay Series

It is often the case that one radioactive isotope decays to another isotope that is also radioactive. Sometimes this daughter decays to yet a third isotope which also is radioactive. Such successive decays are said to form a **decay series**. An important example is illustrated in Fig. 42–11. As can be seen, $^{238}_{92}$U decays by α emission to $^{234}_{90}$Th, which in turn decays by β decay to $^{234}_{91}$Pa. The series continues as shown, with several possible branches near the bottom. For example, $^{218}_{84}$Po can decay either by α decay to $^{214}_{82}$Pb or by β decay to $^{218}_{85}$At. The series ends at the stable lead isotope $^{206}_{82}$Pb. Other radioactive series also exist.

Because of such decay series, certain radioactive elements are found in nature that otherwise would not be. For when the solar system acquired its present form about 5 billion years ago, it is believed that nearly all nuclides were formed (by the fusion process, Sections 43–4 and 45–2). Many isotopes with short half-lives decayed quickly and no longer exist in nature today. But long-lived isotopes, such as $^{238}_{92}$U with a half-life of 4.5×10^9 yr, still do exist in nature today. Indeed, about half of the original $^{238}_{92}$U still remains (assuming that the origin of the solar system was about 5×10^9 yr ago). We might expect, however, that radium $\left(^{226}_{88}\text{Ra}\right)$, with a half-life of 1600 yr, would long since have disappeared from the Earth. Indeed, the original $^{226}_{88}$Ra nuclei must by now have all decayed. However, because $^{238}_{92}$U decays (in several steps) to $^{226}_{88}$Ra, the supply of $^{226}_{88}$Ra is continually replenished, which is why it is still found on Earth today. The same can be said for many other radioactive nuclides.

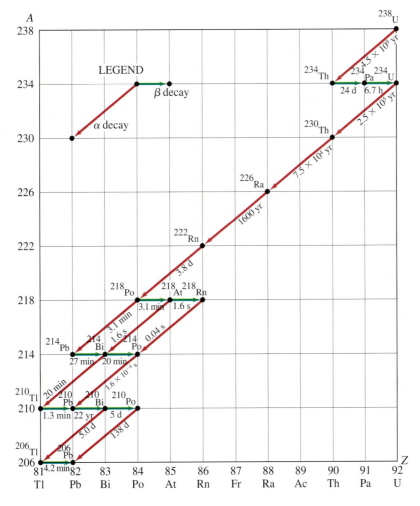

FIGURE 42–11 Decay series beginning with $^{238}_{92}$U. Nuclei in the series are specified by a dot representing A and Z values. Half-lives are given in seconds (s), minutes (min), hours (h), days (d), or years (yr). Note that a horizontal arrow represents β decay (A does not change), whereas a diagonal line represents α decay (A changes by 4, Z changes by 2).

CONCEPTUAL EXAMPLE 42–9 Decay chain. The decay chain starting with ^{234}U in Fig. 42–11 has nuclides with half-lives of 250,000 yr, 75,000 yr, 1600 yr, and a little under 4 days, respectively. Each decay in the chain has an alpha particle of a characteristic energy, and so we can monitor the radioactive decay rate of each nuclide. Given a sample that was pure ^{234}U a million years ago, which alpha decay would you expect to have the highest activity rate in the sample?

RESPONSE The first instinct is to say that the process with the shortest half-life would show the highest activity. Surprisingly, however, the activity rates in this sample are all the same! The reason is that in each case the decay of the parent acts as a bottleneck to the decay of the daughter. Compared to the 1600-yr half-life of ^{226}Ra, for example, its daughter ^{222}Rn decays almost immediately, but it cannot decay until it is made. (This is like an automobile assembly line: If worker A takes 20 minutes to do a task and then worker B takes only 1 minute to do the next task, worker B still does only one car every 20 minutes).

42–10 Radioactive Dating

Radioactive decay has many interesting applications. One is the technique of *radioactive dating* by which the age of ancient materials can be determined.

The age of any object made from once-living matter, such as wood, can be determined using the natural radioactivity of $^{14}_{6}$C. All living plants absorb carbon dioxide (CO_2) from the air and use it to synthesize organic molecules. The vast majority of these carbon atoms are $^{12}_{6}$C, but a small fraction, about 1.3×10^{-12}, is the radioactive isotope $^{14}_{6}$C. The ratio of $^{14}_{6}$C to $^{12}_{6}$C in the atmosphere has remained roughly constant over many thousands of years, in spite of the fact that $^{14}_{6}$C decays with a half-life of about 5730 yr. This is because energetic nuclei in the cosmic radiation, which impinges on the Earth from outer-space, strike nuclei of atoms in the atmosphere and break those nuclei into pieces, including free neutrons. The free neutrons can then collide with nitrogen nuclei in the atmosphere to produce the following nuclear transformation: $n + {}^{14}_{7}N \rightarrow {}^{14}_{6}C + p$. That is, a neutron strikes and is absorbed by a $^{14}_{7}$N nucleus, and a proton is knocked out in the process. The remaining nucleus is $^{14}_{6}$C. This continual production of $^{14}_{6}$C in the atmosphere roughly balances the loss of $^{14}_{6}$C by radioactive decay. As long as a plant or tree is alive, it continually uses the carbon from carbon dioxide in the air to build new tissue and to replace old. Animals eat plants, so they too are continually receiving a fresh supply of carbon for their tissues. Organisms cannot distinguish[†] $^{14}_{6}$C from $^{12}_{6}$C, and since the ratio of $^{14}_{6}$C to $^{12}_{6}$C in the atmosphere remains nearly constant, the ratio of the two isotopes within the living organism remains nearly constant as well. But when an organism dies, carbon dioxide is no longer absorbed and utilized. Because the $^{14}_{6}$C decays radioactively, the ratio of $^{14}_{6}$C to $^{12}_{6}$C in a dead organism decreases in time. Since the half-life of $^{14}_{6}$C is about 5730 yr, the $^{14}_{6}$C/$^{12}_{6}$C ratio decreases by half every 5730 yr. If, for example, the $^{14}_{6}$C/$^{12}_{6}$C ratio of an ancient wooden tool is half of what it is in living trees, then the object must have been made from a tree that was felled about 5700 years ago. Actually, corrections must be made for the fact that the $^{14}_{6}$C/$^{12}_{6}$C ratio in the atmosphere has not remained precisely constant over time. The determination of what this ratio has been over the centuries has required using techniques such as comparing the expected ratio to the actual ratio for objects whose age is known, such as very old trees whose annual rings can be counted.

[†] Organisms operate almost exclusively via chemical reactions—which involve only the outer orbital electrons of the atom; extra neutrons in the nucleus have essentially no effect.

EXAMPLE 42–10 **An ancient animal.** An animal bone fragment found in an archeological site has a carbon mass of 200 g. It registers an activity of 16 decays/s. What is the age of the bone?

➡ PHYSICS APPLIED

Archeological dating

SOLUTION When the animal was alive, the ratio of $^{14}_{6}C$ to $^{12}_{6}C$ in the 200-g piece of bone was 1.3×10^{-12}. The number of $^{14}_{6}C$ nuclei at that time was

$$N_0 = \left(\frac{6.02 \times 10^{23} \text{ atoms}}{12 \text{ g}} \right)(200 \text{ g})(1.3 \times 10^{-12}) = 1.3 \times 10^{13}.$$

From Eq. 42–7b, considering only the magnitude,

$$\left(\frac{dN}{dt} \right)_0 = \lambda N_0,$$

where $\lambda = 3.83 \times 10^{-12} \, \text{s}^{-1}$ (Example 42–7). So the original activity was

$$\left(\frac{dN}{dt} \right)_0 = (3.83 \times 10^{-12} \, \text{s}^{-1})(1.3 \times 10^{13}) = 50 \, \text{s}^{-1}.$$

From Eq. 42–7c,

$$\frac{dN}{dt} = \left(\frac{dN}{dt} \right)_0 e^{-\lambda t},$$

and we rewrite this as

$$e^{\lambda t} = \left[\frac{(dN/dt)_0}{(dN/dt)} \right].$$

Now we take the natural log (ln) of both sides to get

$$t = \frac{1}{\lambda} \ln \left[\frac{(dN/dt)_0}{(dN/dt)} \right] = \frac{1}{3.83 \times 10^{-12} \, \text{s}^{-1}} \ln \left[\frac{50 \, \text{s}^{-1}}{16 \, \text{s}^{-1}} \right]$$
$$= 2.98 \times 10^{11} \, \text{s} = 9400 \, \text{yr},$$

which is the time elapsed since the death of the animal.

Carbon dating is useful only for determining the age of objects less than about 60,000 years old. The amount of $^{14}_{6}C$ remaining in older objects is usually too small to measure accurately, although new techniques are allowing detection of even smaller amounts of $^{14}_{6}C$, pushing the time frame further back. On the other hand, radioactive isotopes with longer half-lives can be used in certain circumstances to obtain the age of older objects. For example, the decay of $^{238}_{92}U$, because of its long half-life of 4.5×10^9 years, is useful in determining the ages of rocks on a geologic time scale. When molten material solidified into rock as the temperature dropped, different compounds solidified according to the melting points, and thus different compounds separated to some extent. Uranium present in a material became fixed in position and the daughter nuclei that result from the decay of uranium were also fixed in that position. Thus, by measuring the amount of $^{238}_{92}U$ remaining in the material relative to the amount of daughter nuclei, the time when the rock solidified can be determined.

➡ PHYSICS APPLIED

Geological dating

Radioactive dating methods using $^{238}_{92}U$ and other isotopes have shown the age of the oldest Earth rocks to be about 4×10^9 yr. The age of rocks in which the oldest fossilized organisms are embedded indicates that life appeared at least 3 billion years ago. The earliest fossilized remains of mammals are found in rocks 200 million years old, and the first humanlike creatures seem to have appeared about 2 million years ago. Radioactive dating has been indispensable for the reconstruction of Earth's history.

➡ PHYSICS APPLIED

Oldest Earth rocks
and
earliest life

42–11 | Detection of Radiation

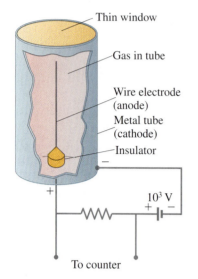

FIGURE 42–12 Diagram of a Geiger counter.

FIGURE 42–13 Scintillation counter with a photomultiplier tube.

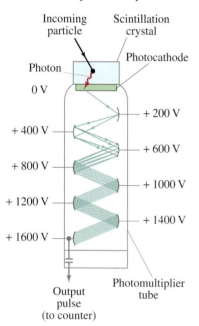

Individual particles such as electrons, protons, α particles, neutrons, and γ rays are not detected directly by our senses. Consequently, a variety of instruments have been developed to detect them.

One of the most common is the **Geiger counter**. As shown in Fig. 42–12, it consists of a cylindrical metal tube filled with a certain type of gas. A long wire runs down the center and is kept at a high positive voltage $(\approx 10^3\,\text{V})$ with respect to the outer cylinder. The voltage is just slightly less than that required to ionize the gas atoms. When a charged particle enters through the thin "window" at one end of the tube, it ionizes a few atoms of the gas. The freed electrons are attracted toward the positive wire and as they are accelerated they strike and ionize additional atoms. An "avalanche" of electrons is quickly produced, and when it reaches the wire anode, it produces a voltage pulse. The pulse, after being amplified, can be sent to an electronic counter, which counts how many particles have been detected. Or the pulses can be sent to a loudspeaker and each detection of a particle is heard as a "click."

A **scintillation counter** makes use of a solid, liquid, or gas known as a **scintillator** or **phosphor**. The atoms of a scintillator are easily excited when struck by an incoming particle and emit visible light when they return to their ground states. Typical scintillators are crystals of NaI and certain plastics. One face of a solid scintillator is cemented to a photomultiplier tube, and the whole is wrapped with opaque material to keep it light tight or is placed within a light-tight container. The **photomultiplier (PM) tube** converts the energy of the scintillator-emitted photon(s) into an electric signal. A PM tube is a vacuum tube containing several electrodes (typically 8 to 14), called *dynodes*, which are maintained at successively higher voltages as shown in Fig. 42–13. At its top surface is a photoelectric surface, called the *photocathode*, whose work function (Section 38–2) is low enough that an electron is easily released when struck by a photon from the scintillator. Such an electron is accelerated toward the first dynode. When it strikes the first dynode, the electron has acquired sufficient kinetic energy so that it can eject two to five more electrons. These, in turn, are accelerated to the second dynode, and a multiplication process begins. The number of electrons striking the last dynode may be 10^6 or more. Thus the passage of a particle through the scintillator results in an electric signal at the output of the PM tube that can be sent to an electronic counter just as for a Geiger tube. Because a scintillator crystal is a solid and therefore much more dense than the gas of a Geiger counter, it is a more efficient detector—especially for γ rays, which interact less with matter than do β rays.

In tracer work (Section 43–8), **liquid scintillators** are often used. Radioactive samples taken at different times or from different parts of an organism are placed directly in small bottles containing the liquid scintillator. This is particularly convenient for detection of β rays from ^3_1H and $^{14}_6\text{C}$, which have very low energies and have difficulty passing through the outer covering of a crystal scintillator or Geiger tube. A PM tube is still used to produce the electric signal.

A **semiconductor detector** consists of a reverse-biased *p-n* junction diode (Sections 41–8 and 9). A particle passing through the junction can excite electrons into the conduction band, leaving holes in the valence band. The freed charges produce a short electrical pulse that can be counted just as for Geiger and scintillation counters.

In Chapter 44 we will discuss additional detectors, those that allow visualization of particle tracks and are used in connection with nuclear reactions and elementary particles.

Summary

Nuclear physics is the study of atomic nuclei. Nuclei contain **protons** and **neutrons**, which are collectively known as **nucleons**. The total number of nucleons, A, is the **atomic mass number**. The number of protons, Z, is the **atomic number**. The number of neutrons equals $A - Z$. **Isotopes** are nuclei with the same Z, but with different numbers of neutrons. For an element X, an isotope of given Z and A is represented by

$$^A_Z X.$$

The nuclear radius is proportional to $A^{\frac{1}{3}}$, indicating that all nuclei have about the same density. Nuclear masses are specified in **unified atomic mass units** (u), where the mass of $^{12}_{6}C$ (including its 6 electrons) is defined as exactly 12.000000 u, or in terms of their energy equivalent (because $E = mc^2$), where

$$1\,u = 931.5\,\text{MeV}/c^2 = 1.66 \times 10^{-27}\,\text{kg}.$$

The mass of a stable nucleus is less than the sum of the masses of its constituent nucleons. The difference in mass (times c^2) is the **total binding energy**. It represents the energy needed to break the nucleus into its constituent nucleons. The **binding energy per nucleon** averages about 8 MeV per nucleon, and is lowest for low mass and high mass nuclei.

Unstable nuclei undergo **radioactive decay**; they change into other nuclei with the emission of an α, β, or γ particle. An α particle is a $^4_2$He nucleus; a β particle is an electron or positron; and a γ ray is a high-energy photon. In β decay, a **neutrino** is also emitted. The transformation of the parent into the daughter nucleus is called **transmutation** of the elements. Radioactive decay occurs spontaneously only when the rest mass of the products is less than the mass of the parent nucleus. The loss in mass appears as kinetic energy of the products.

Nuclei are held together by the **strong nuclear force**. The **weak nuclear force** makes itself apparent in β decay. These two forces, plus the gravitational and electromagnetic forces, are the four known types of force. Electric charge, linear and angular momentum, mass–energy, and **nucleon number** are **conserved** in all decays.

Radioactive decay is a statistical process. For a given type of radioactive nucleus, the number of nuclei that decay (ΔN) in a time Δt is proportional to the number N of parent nuclei present:

$$\Delta N = -\lambda N\,\Delta t.$$

The proportionality constant, λ, is called the **decay constant** and is characteristic of the given nucleus. The number N of nuclei remaining after a time t decreases exponentially

$$N \propto e^{-\lambda t}$$

as does the **activity**, dN/dt. The **half-life**, $T_{\frac{1}{2}}$, is the time required for half the nuclei of a radioactive sample to decay. It is related to the decay constant by $T_{\frac{1}{2}} = 0.693/\lambda$.

Questions

1. What do different isotopes of a given element have in common? How are they different?

2. What are the elements represented by the X in the following: (a) $^{232}_{92}$X; (b) $^{18}_{7}$X; (c) $^{1}_{1}$X; (d) $^{82}_{38}$X; (e) $^{247}_{97}$X?

3. How many protons and how many neutrons do each of the isotopes in Question 2 have?

4. Why are the atomic masses of many elements (see the periodic table) not close to whole numbers?

5. How do we know there is such a thing as the strong nuclear force?

6. What are the similarities and the differences between the strong nuclear force and the electric force?

7. What is the experimental evidence in favor of radioactivity being a nuclear process?

8. The isotope $^{64}_{29}$Cu is unusual in that it can decay by γ, β^-, and β^+ emission. What is the resulting nuclide for each case?

9. A $^{238}_{92}$U nucleus decays to a nucleus containing how many neutrons?

10. Describe, in as many ways as possible, the difference between α, β, and γ rays.

11. What element is formed by the radioactive decay of (a) $^{24}_{11}$Na (β^-); (b) $^{22}_{11}$Na (β^+); (c) $^{210}_{84}$Po (α)?

12. What element is formed by the decay of (a) $^{32}_{15}$P (β^-); (b) $^{35}_{16}$S (β^-); (c) $^{211}_{83}$Bi (α)?

13. Fill in the missing particle or nucleus:
    (a) $^{45}_{20}$Ca $\rightarrow$? $+ e^- + \bar{\nu}$
    (b) $^{58}_{29}$Cu $\rightarrow$? $+ \gamma$
    (c) $^{46}_{24}$Cr $\rightarrow$ $^{46}_{23}$V $+$?
    (d) $^{234}_{94}$Pu $\rightarrow$? $+ \alpha$
    (e) $^{239}_{93}$Np $\rightarrow$ $^{239}_{94}$Pu $+$?

14. Explain α decay by tunneling using the uncertainty principle.

15. Immediately after a $^{238}_{92}$U nucleus decays to $^{234}_{90}$Th $+ ^4_2$He, the daughter thorium nucleus still has 92 electrons circling it. Since thorium normally holds only 90 electrons, what do you suppose happens to the two extra ones?

16. When a nucleus undergoes either β^- or β^+ decay, what happens to the energy levels of the atomic electrons? What is likely to happen to these electrons following the decay?

17. The alpha particles from a given alpha-emitting nuclide are monoenergetic; that is, they all have the same kinetic energy. But the beta particles from a beta-emitting nuclide have a spectrum of energies. Explain the difference between these two cases.

18. Do isotopes that undergo electron capture generally lie above or below the line of stability in Fig. 42–2?

19. Can hydrogen or deuterium emit an α particle?

20. Why are many artificially produced radioactive isotopes rare in nature?

21. An isotope has a half-life of one month. After two months, will a given sample of this isotope have completely decayed? If not, how much remains?

22. Describe how the potential energy curve for an α particle in an α-emitting nucleus differs from that for a stable nucleus.

23. Explain the absence of β^+ emitters in the radioactive decay series of Fig. 42–11.

24. Can $^{14}_{6}C$ dating be used to measure the age of stone walls and tablets of ancient civilizations?

25. What assumptions are made in carbon-dating? What do you think could affect these assumptions?

Problems

Section 42–1

1. (I) What is the rest mass of an α particle in MeV/c^2?

2. (I) A pi meson has a mass of $139\ \text{MeV}/c^2$. What is this in atomic mass units?

3. (I) What is the approximate radius of an alpha particle (^4_2He)?

4. (I) By what percentage is the radius of the isotope $^{14}_{6}C$ greater than that of its sister $^{12}_{6}C$?

5. (II) (a) Determine the density of nuclear matter in kg/m^3, and show that it is essentially the same for all nuclei. (b) What would be the radius of the Earth if it had its actual mass but had the density of nuclei? (c) What would be the radius of a $^{238}_{92}U$ nucleus if it had the density of the Earth?

6. (II) (a) What is the approximate radius of a $^{64}_{29}\text{Cu}$ nucleus? (b) Approximately what is the value of A for a nucleus whose radius is $3.9 \times 10^{-15}\ \text{m}$?

7. (II) How much energy must an α particle have to just "touch" the surface of a $^{238}_{92}U$ nucleus?

8. (II) If an alpha particle were released from rest near the surface of a $^{243}_{95}\text{Am}$ nucleus, what would its kinetic energy be when far away?

9. (II) What stable nucleus has approximately half the radius of a uranium nucleus? [*Hint:* Find A and use Appendix D to get Z.]

Section 42–2

10. (I) Estimate the total binding energy for $^{40}_{20}\text{Ca}$, using Fig. 42–1.

11. (I) Use Fig. 42–1 to estimate the total binding energy of (a) $^{238}_{92}U$, and (b) $^{84}_{36}\text{Kr}$.

12. (II) Use Appendix D to calculate the binding energy of ^2_1H (deuterium).

13. (II) Calculate the binding energy per nucleon for a $^{14}_{7}N$ nucleus.

14. (II) Determine the binding energy of the last neutron in a $^{40}_{19}K$ nucleus.

15. (II) Calculate the total binding energy, and the binding energy per nucleon, for (a) ^6_3Li, (b) $^{208}_{82}\text{Pb}$. Use Appendix D.

16. (II) Calculate the binding energy of (a) the last proton, and (b) the last neutron, in a $^{12}_{6}C$ nucleus. Use Appendix D.

17. (II) Compare the binding energy of a neutron in $^{23}_{11}\text{Na}$ to that in $^{24}_{11}\text{Na}$.

18. (II) How much energy is required to remove (a) a proton, (b) a neutron, from $^{16}_{8}O$? Explain the difference in your answers.

19. (II) (a) Show that the nucleus ^8_4Be (mass = 8.005305 u) is unstable to decay into two α particles. (b) Is $^{12}_{6}C$ stable against decay into three α particles? Show why or why not.

Sections 42–3 to 42–7

20. (I) How much energy is released when tritium, ^3_1H, decays by β^- emission?

21. (I) What is the maximum kinetic energy of an electron emitted in the β decay of a free neutron?

22. (I) Show that the decay $^{11}_{6}C \rightarrow {}^{10}_{5}B + p$ is not possible because energy would not be conserved.

23. (II) $^{22}_{11}\text{Na}$ is radioactive. Is it a β^- or β^+ emitter? Write down the decay reaction, and estimate the maximum kinetic energy of the emitted β.

24. (II) Give the result of a calculation that shows whether or not the following decays are possible:
(a) $^{236}_{92}U \rightarrow {}^{235}_{92}U + n$;   (b) $^{16}_{8}O \rightarrow {}^{15}_{8}O + n$;
(c) $^{23}_{11}\text{Na} \rightarrow {}^{22}_{11}\text{Na} + n$.

25. (II) A $^{232}_{92}U$ nucleus emits an α particle with kinetic energy = 5.32 MeV. What is the daughter nucleus and what is the approximate atomic mass (in u) of the daughter atom? Ignore recoil of the daughter nucleus.

26. (II) When $^{23}_{10}\text{Ne}$ (mass = 22.9945 u) decays to $^{23}_{11}\text{Na}$ (mass = 22.9898 u), what is the maximum kinetic energy of the emitted electron? What is its minimum energy? What is the energy of the neutrino in each case?

27. (II) The nuclide $^{32}_{15}P$ decays by emitting an electron whose maximum kinetic energy can be 1.71 MeV. (a) What is the daughter nucleus? (b) What is its atomic mass (in u)?

28. (II) The isotope $^{218}_{84}\text{Po}$ can decay by either α or β^- emission. What is the energy release in each case? The mass of $^{218}_{84}\text{Po}$ is 218.008965 u.

29. (II) How much energy is released in electron capture by beryllium: $^7_4\text{Be} + {}^{\ 0}_{-1}e \rightarrow {}^7_3\text{Li} + \nu$?

30. (II) Decay series, such as that shown in Fig. 42–11, can be classified into four families, depending on whether the mass numbers have the form $4n$, $4n + 1$, $4n + 2$, or $4n + 3$, where n is an integer. Justify this statement and show that for a nuclide in any family, all its daughters will be in the same family.

31. (III) What is the energy of the α particle emitted in the decay $^{210}_{84}\text{Po} \rightarrow {}^{206}_{82}\text{Pb} + \alpha$? Take into account the recoil of the daughter nucleus.

32. (III) The α particle emitted when $^{238}_{92}U$ decays has 4.20 MeV of kinetic energy. Calculate the recoil kinetic energy of the daughter nucleus and the Q-value of the decay.

33. (III) (a) Show that when a nucleus decays by β^+ decay, the total energy released is equal to

$$(M_P - M_D - 2m_e)c^2,$$

where M_P and M_D are the masses of the parent and daughter atoms (neutral), and m_e is the mass of an electron or positron. (b) Determine the maximum kinetic energy of β^+ particles released when $^{11}_6C$ decays to $^{11}_5B$. What is the maximum energy the neutrino can have? What is its minimum energy?

34. (III) (a) Calculate the kinetic energy of the α particle emitted when $^{236}_{92}U$ decays. (b) Use Eq. 42–1 to estimate the radius of an α particle and $^{232}_{90}Th$ nucleus. Use this to estimate (c) the maximum height of the Coulomb barrier, and (d) its width AB in Fig. 42–6.

Sections 42–8 to 42–10

35. (I) A radioactive material produces 1280 decays per minute at one time, and 6 h later produces 320 decays per minute. What is its half-life?

36. (I) (a) What is the decay constant of $^{238}_{92}U$ whose half-life is 4.5×10^9 yr? (b) The decay constant of a given nucleus is $8.2 \times 10^{-5}\,s^{-1}$. What is its half-life?

37. (I) What is the activity of a sample of $^{14}_6C$ that contains 3.1×10^{20} nuclei?

38. (I) What fraction of a sample of $^{68}_{32}Ge$, whose half-life is about 9 months, will remain after 3.0 yr?

39. (I) How many nuclei of $^{238}_{92}U$ remain in a rock if the activity registers 875 decays per second?

40. (II) What fraction of a sample is left after (a) exactly 4 half-lives, (b) exactly $4\frac{1}{2}$ half-lives?

41. (II) In a series of decays, the nuclide $^{235}_{92}U$ becomes $^{207}_{82}Pb$. How many α and β^- particles are emitted in this series?

42. (II) The iodine isotope $^{131}_{53}I$ is used in hospitals for diagnosis of thyroid function. If $632\,\mu g$ are ingested by a patient, determine the activity (a) immediately, (b) 1.0 h later when the thyroid is being tested, and (c) 6 months later. Use Appendix D.

43. (II) $^{124}_{55}Cs$ has a half-life of 30.8 s. (a) If we have $9.8\,\mu g$ initially, how many Cs nuclei are present? (b) How many are present 2.0 min later? (c) What is the activity at this time? (d) After how much time will the activity drop to less than about 1 per second?

44. (II) Calculate the activity of a pure 7.7-μg sample of $^{32}_{15}P$ $(T_\frac{1}{2} = 1.23 \times 10^6\,s)$.

45. (II) The activity of a sample of $^{35}_{16}S$ $(T_\frac{1}{2} = 7.56 \times 10^6\,s)$ is 2.65×10^5 decays per second. What is the mass of the sample?

46. (II) A sample of $^{233}_{92}U$ $(T_\frac{1}{2} = 1.59 \times 10^5\,yr)$ contains 7.50×10^{19} nuclei. (a) What is the decay constant? (b) Approximately how many disintegrations will occur per minute?

47. (II) The activity of a sample drops by a factor of 10 in 8.6 minutes. What is its half-life?

48. (II) A 185-g sample of pure carbon contains 1.3 parts in 10^{12} (atoms) of $^{14}_6C$. How many disintegrations occur per second?

49. (II) A sample of $^{40}_{19}K$ is decaying at a rate of 8.70×10^2 decays/s. What is the mass of the sample?

50. (II) The rubidium isotope $^{87}_{37}Rb$, a β emitter with a half-life of 4.75×10^{10} yr, is used to determine the age of rocks and fossils. Rocks containing fossils of early animals contain a ratio of $^{87}_{38}Sr$ to $^{87}_{37}Rb$ of 0.0160. Assuming that there was no $^{87}_{38}Sr$ present when the rocks were formed, calculate the age of these fossils.

51. (II) Use Fig. 42–11 and calculate the relative decay rates for α decay of $^{218}_{84}Po$ and $^{214}_{84}Po$.

52. (II) ^7_4Be decays with a half-life of about 53 d. It is produced in the upper atmosphere, and filters down onto the Earth's surface. If a plant leaf is detected to have 250 decays/s of ^7_4Be, how long do we have to wait for the decay rate to drop to 10/s? Estimate the initial mass of ^7_4Be on the leaf.

53. (II) Which radioactive isotope of lead is being produced in a reaction where the measured activity of a sample drops to 1.050 percent of its original activity in 4.00 h?

54. (II) An ancient club is found that contains 190 g of carbon and has an activity of 5.0 decays per second. Determine its age assuming that in living trees the ratio of $^{14}C/^{12}C$ atoms is about 1.3×10^{-12}.

55. (II) At $t = 0$, a pure sample of radioactive nuclei contains N_0 nuclei whose decay constant is λ. Determine a formula for the number of daughter nuclei, N_D, as a function of time. Assume the daughter is stable and that $N_D = 0$ at $t = 0$.

56. (III) At $t = 0$, a pure sample of a radioactive nuclide (the parent) contains N_{P0} nuclei whose half-life is T_P. The daughter nuclide is also radioactive, with half-life T_D. (a) Determine the number of daughter nuclei, N_D, as a function of time, assuming that $N_D = 0$ at $t = 0$. (b) Plot N_D versus t for the cases $T_P = T_D$, $T_P = 3T_D$, $T_P = \frac{1}{3}T_D$. [Hint: Eq. 42–5 must be modified.]

General Problems

57. (a) What is the fraction of the hydrogen atom's mass that is in the nucleus? (b) What is the fraction of the hydrogen atom's volume that is occupied by the nucleus? (c) What is the density of nuclear matter? Compare this with water.

58. Using the uncertainty principle and the radius of a nucleus, estimate the minimum possible kinetic energy of a nucleon in, say, iron. Ignore relativistic corrections. [Hint: A particle can have an energy at least as large as its uncertainty.]

59. How much recoil energy does a $^{40}_{19}K$ nucleus get when it emits a 1.46 MeV gamma ray?

60. An old wooden tool is found to contain only 9.0 percent of $^{14}_6C$ that a sample of fresh wood would. How old is the tool?

61. A neutron star consists of neutrons at approximately nuclear density. Estimate, for a 10-km-diameter neutron star, (a) its mass number, (b) its mass (kg), and (c) the acceleration of gravity at its surface.

62. The $^3_1$H isotope of hydrogen, which is called *tritium* (because it contains three nucleons), has a half-life of 12.33 yr. It can be used to measure the age of objects up to about 100 yr. It is produced in the upper atmosphere by cosmic rays and brought to Earth by rain. As an application, determine approximately the age of a bottle of wine whose $^3_1$H radiation is about $\frac{1}{10}$ that present in new wine.

63. Some elementary particle theories (Section 44–11) suggest that the proton may be unstable, with a half-life $\geq 10^{32}$ yr. How long would you expect to wait for one proton in your body to decay (consider that your body is all water)?

64. When water is placed near an intense neutron source, the neutrons can be slowed down by collisions with the water molecules and eventually captured by a hydrogen nucleus to form the stable isotope called deuterium, $^2_1$H, giving off a gamma ray. What is the energy of the gamma ray?

65. How long must you wait (in half-lives) for a radioactive sample to drop to 1.00 percent of its original activity?

66. If the potassium isotope $^{40}_{19}$K gives 60 decays/s in a liter of milk, estimate how much $^{40}_{19}$K and regular $^{39}_{19}$K are in a liter of milk. Use Appendix D.

67. Show, using the decays given in Section 42–5, that the neutrino must have spin $\frac{1}{2}$.

68. In α decay of, say, a $^{226}_{88}$Ra nucleus, show that the nucleus carries away a fraction $1/(1 + A_D/4)$ of the total energy available, where A_D is the mass number of the daughter nucleus. [*Hint*: Use conservation of momentum as well as conservation of energy.] Approximately what percentage of the energy available is thus carried off by the α particle in the case cited?

69. Strontium-90 is produced as a nuclear fission product of uranium in both reactors and atomic bombs. Look at its location in the periodic table to see what other elements it might be similar to chemically, and tell why you think it might be dangerous to ingest. It has too many neutrons, and it decays with a half-life of about 29 yr. How long will we have to wait for the amount of $^{90}_{38}$Sr on the Earth's surface to reach 1 percent of its current level, assuming no new material is scattered about? Write down the decay reaction, including the daughter nucleus. Is the daughter radioactive? If so, write down its decay. Finish the decay scheme until you reach a stable nucleus.

70. The nuclide $^{191}_{76}$Os decays with β^- energy of 0.14 MeV accompanied by γ rays of energy 0.042 MeV and 0.129 MeV. (a) What is the daughter nucleus? (b) Draw an energy-level diagram showing the ground states of the parent and daughter and excited states of the daughter. To which of the daughter states does β^- decay of $^{191}_{76}$Os occur?

71. Use the uncertainty principle to argue why electrons are unlikely to be found in the nucleus. Use relativity. [*Hint*: A particle can have an energy at least as large as its uncertainty.]

72. Estimate the total binding energy for copper and then estimate the energy, in joules, needed to break a 3.0-g copper penny into its constituent nucleons.

73. Instead of giving atomic masses for nuclides as in Appendix D, some tables give the *mass excess*, Δ, defined as $\Delta = M - A$, where A is the atomic number and M is the mass in u. Determine the mass excess, in u and in MeV/c^2, for: (a) $^4_2$He; (b) $^{12}_6$C; (c) $^{107}_{47}$Ag; (d) $^{235}_{92}$U. (e) From a glance at Appendix D, can you make a generalization about the sign of Δ as a function of Z or A?

74. (a) A 100-gram sample of natural carbon contains the usual fraction of $^{14}_6$C. How long will it take on average before there is only one $^{14}_6$C nucleus left? (b) How does the answer in (a) change if the sample is 200 grams? What does this tell you about the limits of carbon dating?

75. If the mass of the proton were just a little closer to the mass of the neutron, the following reaction would be possible even at low collision energies:

$$e^- + p \rightarrow n + \nu.$$

Why would this situation be catastrophic? By what percentage would the proton's mass have to be increased to make this reaction possible?

76. (a) Show that the *mean life*, or *average lifetime* of a radioactive nuclide, defined as

$$\tau = \frac{\displaystyle\int_0^\infty N(t)t \, dt}{\displaystyle\int_0^\infty N(t) \, dt}$$

is $\tau = 1/\lambda$. (b) What fraction of the original number of nuclei remains after one mean life?

77. Two of the naturally occurring radioactive decay sequences start with $^{232}_{90}$Th, and $^{235}_{92}$U. The first five decays of these two sequences are:

$$\alpha, \beta, \beta, \alpha, \alpha$$

and

$$\alpha, \beta, \alpha, \beta, \alpha.$$

With this information determine the resulting intermediate daughter nuclei in each case.

78. What is the ratio of the kinetic energies for an alpha particle and a beta particle if both make tracks with the same radius of curvature in a magnetic field, oriented perpendicular to the paths of the particles.

79. In a rough theoretical model of α decay, the α particle moves back and forth within the parent nucleus at speed v_{in}, making a collision with the Coulomb barrier (Fig. 42–6) every $2R_0/v_{in}$ seconds, where R_0 is the radius of the nucleus. (a) Show that the decay constant λ in this model can be written roughly (using Eq. 39–17) as $\lambda = (v_{in}/2R_0)e^{-2GL}$, where L is the "average" barrier thickness (because the barrier is not square), and $G = \sqrt{2m(U_0 - E)}/\hbar^2$. (b) Use this result, plus estimates of the barrier height and width (as in Problem 34) to estimate the halflife of $^{236}_{92}$U. For the barrier thickness (Fig. 42–6) try first $L = \frac{1}{2}(R_B - R_0)$, and then (c) set $L = \frac{1}{3}(R_B - R_0)$ as an average thickness since the barrier is not square. Choose the potential U_{in} inside the well to be zero. (d) What barrier width would provide, in this simple model, the measured half-life of 2×10^7 yr?

Interior of the Tokamak Fusion Test Reactor at Princeton, shown with a scientist inside (in yellow). A plasma of electrons and light nuclei can be heated to high temperatures that rival the Sun. Confining the plasma by magnetic fields has proved difficult, and intense research is needed if the fusion of light nuclei is to fulfill its great promise as a source of abundant and relatively clean power. In this chapter we also discuss fission of large nuclei which is used in present-day nuclear power plants. We also examine nuclear reactions, dosimetry, and uses of radiation including therapy and imaging.

CHAPTER 43

Nuclear Energy; Effects and Uses of Radiation

We continue our study of nuclear physics in this chapter. We begin with a discussion of nuclear reactions, after which we examine the important energy-releasing processes of fission and fusion. The remainder of the chapter deals with the effects of nuclear radiation when it passes through matter, including us, and how radiation is used medically for therapy and diagnosis, including valuable imaging techniques.

43–1 Nuclear Reactions and the Transmutation of Elements

When a nucleus undergoes α or β decay, the daughter nucleus is that of a different element from the parent. The transformation of one element into another, called *transmutation*, also occurs by means of nuclear reactions. A **nuclear reaction** is said to occur when a given nucleus is struck by another nucleus, or by a simpler particle such as a γ ray or neutron, so that an interaction takes place. Ernest Rutherford was the first to report seeing a nuclear reaction. In 1919 he observed that some of the α particles passing through nitrogen gas were absorbed and protons emitted. He concluded that nitrogen nuclei had been transformed into oxygen nuclei via the reaction

$$\,_2^4\text{He} + \,_7^{14}\text{N} \rightarrow \,_8^{17}\text{O} + \,_1^1\text{H},$$

where $\,_2^4\text{He}$ is an α particle, and $\,_1^1\text{H}$ is a proton.

Since then, a great many nuclear reactions have been observed. Indeed, many of the radioactive isotopes used in the laboratory are made by means of nuclear reactions. Nuclear reactions can be made to occur in the laboratory, but they also occur regularly in nature. In Chapter 42 we saw an example of this: $^{14}_{6}\text{C}$ is continually being made in the atmosphere via the reaction $\text{n} + ^{14}_{7}\text{N} \rightarrow ^{14}_{6}\text{C} + \text{p}$.

Nuclear reactions are sometimes written in a shortened form: for example, the reaction

$$\text{n} + ^{14}_{7}\text{N} \rightarrow ^{14}_{6}\text{C} + \text{p}$$

is written

$$^{14}_{7}\text{N} \, (\text{n, p}) \, ^{14}_{6}\text{C}.$$

The symbols outside the parentheses on the left and right represent the initial and final nuclei, respectively. The symbols inside the parentheses represent the bombarding particle (first) and the emitted small particle (second).

In any nuclear reaction, both electric charge and nucleon number are conserved. These conservation laws are valuable in "balancing" nuclear reactions, as the following Example shows.

CONCEPTUAL EXAMPLE 43–1 **Deuterium reaction.** A neutron is observed to strike an $^{16}_{8}\text{O}$ nucleus and a deuteron is given off. (A **deuteron** is the nucleus of *Deuterium* **deuterium**, the isotope of hydrogen containing one proton and one neutron, $^{2}_{1}\text{H}$; it is sometimes given the symbol d or D.) What is the nucleus that results?

RESPONSE We have the reaction $\text{n} + ^{16}_{8}\text{O} \rightarrow ? + ^{2}_{1}\text{H}$. The total number of nucleons initially is $1 + 16 = 17$, and the total charge is $0 + 8 = 8$. The same totals apply to the right side of the reaction. Hence the product nucleus must have $Z = 7$ and $A = 15$. From the periodic table, we find that it is nitrogen that has $Z = 7$, so the nucleus produced is $^{15}_{7}\text{N}$. The reaction can be written $^{16}_{8}\text{O} \, (\text{n, d}) \, ^{15}_{7}\text{N}$, where d represents deuterium, $^{2}_{1}\text{H}$.

Energy (as well as momentum) is conserved in nuclear reactions, and we can use this to determine whether a given reaction can occur or not. For example, if the total mass of the products is less than the total mass of the initial particles, this decrease in mass (recall $E = mc^2$) appears as kinetic energy (K) of the outgoing particles. But if the total mass of the products is greater than the total mass of the initial reactants, the reaction requires an energy input. The reaction will then not occur unless the bombarding particle has sufficient kinetic energy. Consider a nuclear reaction of the general form

$$\text{a} + \text{X} \rightarrow \text{Y} + \text{b},$$

where a is a projectile particle (or small nucleus) that strikes nucleus X, producing nucleus Y and particle b (typically, p, n, α, γ). We define the **reaction energy**, or **Q-value**, in terms of the masses involved, as

Q-value $$Q = (M_\text{a} + M_\text{X} - M_\text{b} - M_\text{Y})c^2. \qquad (43\text{–}1)$$

Since energy is conserved, Q is equal to the change in kinetic energy (final minus initial):

$$Q = K_\text{b} + K_\text{Y} - K_\text{a} - K_\text{X}. \qquad (43\text{–}2)$$

Normally, $K_\text{X} = 0$ since X is the target nucleus at rest (or nearly so) struck by an incoming particle a. For $Q > 0$, the reaction is said to be *exothermic* or *exoergic*. Energy is released in the reaction, so the total kinetic energy is greater after the reaction than before. If Q is negative $(Q < 0)$, the reaction is said to be *endothermic* or *endoergic*. In this case the final total kinetic energy is less than the initial kinetic energy, and an energy input is required to make the reaction happen. The energy input comes from the kinetic energy of the initial colliding particles (a and X).

EXAMPLE 43–2 **A slow-neutron reaction.** The nuclear reaction

$$\text{n} + {}^{10}_{5}\text{B} \rightarrow {}^{7}_{3}\text{Li} + {}^{4}_{2}\text{He}$$

is observed to occur even when very slow-moving neutrons $(M_n = 1.0087 \, \text{u})$ strike a boron atom at rest. For a particular reaction in which $K_n \approx 0$, the helium $(M_{\text{He}} = 4.0026 \, \text{u})$ is observed to have a speed of $9.30 \times 10^6 \, \text{m/s}$. Determine (*a*) the kinetic energy of the lithium $(M_{\text{Li}} = 7.0160 \, \text{u})$, and (*b*) the Q-value of the reaction.

SOLUTION (*a*) Since the neutron and boron are both essentially at rest, the total momentum before the reaction is zero, and afterward is also zero. Therefore,

$$M_{\text{Li}} v_{\text{Li}} = M_{\text{He}} v_{\text{He}}.$$

We solve this for v_{Li} and substitute it into the equation for kinetic energy. We can use classical kinetic energy with little error, rather than relativistic formulas, because $v_{\text{He}} = 9.30 \times 10^6 \, \text{m/s}$ is not close to the speed of light c, and v_{Li} will be even less since $M_{\text{Li}} > M_{\text{He}}$. Thus we can write:

$$K_{\text{Li}} = \frac{1}{2} M_{\text{Li}} v_{\text{Li}}^2 = \frac{1}{2} M_{\text{Li}} \left(\frac{M_{\text{He}} v_{\text{He}}}{M_{\text{Li}}} \right)^2 = \frac{M_{\text{He}}^2 v_{\text{He}}^2}{2 M_{\text{Li}}}.$$

We put in numbers, changing the mass in u to kg and recalling that $1.60 \times 10^{-13} \, \text{J} = 1 \, \text{MeV}$:

$$K_{\text{Li}} = \frac{(4.0026 \, \text{u})^2 (1.66 \times 10^{-27} \, \text{kg/u})^2 (9.30 \times 10^6 \, \text{m/s})^2}{2(7.0160 \, \text{u})(1.66 \times 10^{-27} \, \text{kg/u})}$$

$$= 1.64 \times 10^{-13} \, \text{J} = 1.02 \, \text{MeV}.$$

(*b*) We are given the data $K_a = K_X = 0$ in Eq. 43–2, so $Q = K_{\text{Li}} + K_{\text{He}}$, where

$$K_{\text{He}} = \tfrac{1}{2} M_{\text{He}} v_{\text{He}}^2 = \tfrac{1}{2}(4.0026 \, \text{u})(1.66 \times 10^{-27} \, \text{kg/u})(9.30 \times 10^6 \, \text{m/s})^2$$

$$= 2.87 \times 10^{-13} \, \text{J} = 1.80 \, \text{MeV}.$$

Hence, $Q = 1.02 \, \text{MeV} + 1.80 \, \text{MeV} = 2.82 \, \text{MeV}.$

EXAMPLE 43–3 **Will the reaction "go"?** Can the reaction ${}^{13}_{6}\text{C}\,(\text{p, n})\,{}^{13}_{7}\text{N}$ occur when ${}^{13}_{6}\text{C}$ is bombarded by 2.0-MeV protons?

SOLUTION We use Eq. 43–1, looking up the masses of the nuclei in Appendix D. The total masses before and after the reaction are:

Before	After
$M({}^{13}_{6}\text{C}) = 13.003355$	$M({}^{13}_{7}\text{N}) = 13.005739$
$M({}^{1}_{1}\text{H}) = 1.007825$	$M(\text{n}) = 1.008665$
14.011180	14.014404

(We must use the mass of the ${}^{1}_{1}\text{H}$ atom rather than that of the bare proton because the masses of ${}^{13}\text{C}$ and ${}^{13}\text{N}$ include the electrons, and we must include an equal number of electrons on each side of the equation since none are created or destroyed.) The products have an excess mass of $(14.014404 - 14.011180)\text{u} = 0.003224 \, \text{u} \times 931.5 \, \text{MeV/u} = 3.00 \, \text{MeV}$. Thus $Q = -3.00 \, \text{MeV}$, and the reaction is endoergic. This reaction requires energy, and the 2.0 MeV protons do not have enough to make it go.

The proton in this last Example would have to have somewhat more than 3.00 MeV of kinetic energy to make this reaction go; 3.00 MeV would be enough to conserve

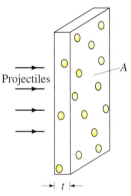

(a) Neutron captured by $^{238}_{92}$U.

(b) $^{239}_{92}$U decays by β decay to neptunium-239.

(c) $^{239}_{93}$Np itself decays by β decay to produce plutonium-239.

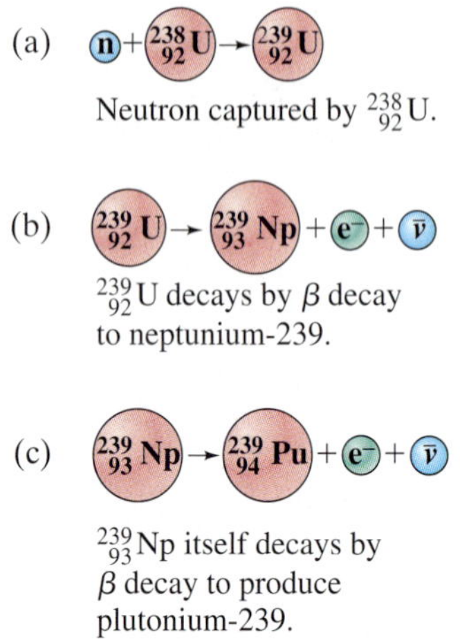

FIGURE 43–1 Neptunium and plutonium are produced in this series of reactions, after bombardment of $^{238}_{92}$U by neutrons.

energy, but a proton of this energy would produce the $^{13}_{7}$N and n with no kinetic energy and hence no momentum. Since an incident 3.0-MeV proton has momentum, conservation of momentum would be violated. A more complicated calculation shows that to conserve both energy and momentum, the minimum proton energy, called the **threshold energy**, is 3.23 MeV in this case (see Problem 14).

The artificial transmutation of elements took a great leap forward in the 1930s when Enrico Fermi realized that neutrons would be the most effective projectiles for causing nuclear reactions and in particular for producing new elements. Because neutrons have no net electric charge, they are not repelled by positively charged nuclei as are protons or alpha particles. Hence the probability of a neutron reaching the nucleus and causing a reaction is much greater than for charged projectiles,[†] particularly at low energies. Between 1934 and 1936, Fermi and his coworkers in Rome produced many previously unknown isotopes by bombarding different elements with neutrons. Fermi realized that if the heaviest known element, uranium, were bombarded with neutrons, it might be possible to produce new elements whose atomic numbers were greater than that of uranium. After several years of hard work, it was suspected that two new elements had been produced, neptunium ($Z = 93$) and plutonium ($Z = 94$). The full confirmation that such "transuranic" elements could be produced came several years later at the University of California, Berkeley. The reactions are shown in Fig. 43–1.

It was soon shown that what Fermi actually had observed when he bombarded uranium was an even stranger process—one that was destined to play an extraordinary role in the world at large. We discuss it in Section 43–3.

43–2 Cross Section

Some reactions have a higher probability of occurring than others. The reaction probability is specified by a quantity called the *cross section*. Although the size of a nucleus, like that of an atom, is not a clearly defined quantity since the edges are not distinct like those of a tennis ball or baseball, we can nonetheless define a **cross section** for nuclei undergoing collisions by using an analogy. Suppose that projectile particles strike a stationary target of total area A and thickness t, as shown in Fig. 43–2. Suppose the target is made up of identical objects (such as marbles or nuclei), each of which has a cross-sectional area σ. We assume that the objects are fairly far apart, so we don't have to worry about overlapping; this is often a reasonable assumption because nuclei have diameters on the order of 10^{-14} m but are at least 10^{-10} m (atomic size) apart even in solids. If there are n such objects per unit volume, the total area of all these tiny targets is

$$A' = nAt\sigma$$

since $nAt = (n)(\text{volume})$ is the total number of targets and σ is the area of each. If $A' \ll A$, most of the incident projectile particles will pass through the target without colliding. If R_0 is the rate at which the projectile particles strike the target (number/second), the rate at which collisions occur, R, is

$$R = R_0 \frac{A'}{A} = R_0 \frac{nAt\sigma}{A}$$

so

$$R = R_0 nt\sigma.$$

FIGURE 43–2 Projectile particles fall on a target of area A and thickness t made up of n nuclei per unit volume.

Projectiles

[†]That is, positively charged particles. Electrons rarely cause nuclear reactions because they do not interact via the strong nuclear force.

Thus, by measuring the collision rate, R, we can determine σ:

$$\sigma = \frac{R}{R_0\,nt}.$$

(43–3)   *Cross section*

If nuclei were simple billiard balls, and R the number of particles that are deflected per second, σ would represent the cross-sectional area of each ball. But nuclei are complicated objects that cannot be considered to have distinct boundaries. Furthermore, collisions can be either elastic or inelastic, and reactions can occur in which the nature of the particles can change. By measuring R for each possible process, we can determine an **effective cross section**, σ, for each process. None of these cross sections is necessarily related to a geometric cross-sectional area. Rather, σ is an "effective" target area. It is a *measure of the probability of a collision or of a particular reaction occurring* per target nucleus, independent of the dimensions of the entire target. The concept of cross section is useful because σ depends only on the properties of the interacting particles, whereas R depends on the thickness and area of the physical (macroscopic) target, on the number of particles in the incident beam, and so on.

We can define the elastic cross section σ_{el}, using Eq. 43–3, where R for a given experimental setup is the rate of elastic collisions (or **elastic scattering**), by which we mean collisions for which the final particles are the same as the initial particles ($a = b$, $X = Y$) and $Q = 0$. Similarly, the inelastic cross section, σ_{inel}, is related to the rate of inelastic collisions, or **inelastic scattering**, which involves the same final and initial particles but $Q \neq 0$, usually because excited states are involved. Then for each possible reaction there is a particular cross section. For protons (p) incident on $^{13}_{6}C$, for example, we could have various reactions, such as $p + {}^{13}_{6}C \rightarrow {}^{13}_{7}N + n$ or $p + {}^{13}_{6}C \rightarrow {}^{10}_{5}B + {}^{4}_{2}He$, and so on. The sum of all the separate reaction cross sections is called the **total reaction cross section**, σ_R. The **total cross section**, σ_T, is

Elastic scattering

Inelastic scattering

$$\sigma_T = \sigma_{el} + \sigma_{inel} + \sigma_R$$

Total cross section

and is a measure of all possible interactions or collisions. Said another way, σ_T is a measure of how many particles interact in some way and hence are eliminated from the incident beam. It is also possible to define "differential cross sections," which represent the probability of the deflected (or emitted) particles leaving at particular angles.

It is said that when one of the first nuclear cross sections was measured, a physicist, surprised that it was as large as it was ($\approx 10^{-28}\,m^2$), remarked, "it's as big as a barn." Ever since then nuclear cross sections have been measured in "barns," where 1 barn (bn) $= 10^{-28}\,m^2$.

The barn

The value of σ for a given reaction depends on, among other things, the incident kinetic energy. Typical nuclear cross sections are on the order of barns, but they can vary from millibarns to 1000 bn or more. Figure 43–3 shows the cross section for neutron capture in cadmium ($n + {}^{114}Cd \rightarrow {}^{115}Cd + \gamma$) as a function of neutron kinetic energy. Neutron cross sections for most materials are greater at low energies, as in Fig. 43–3. To produce nuclear reactions at a high rate it is therefore desirable that the bombarding neutrons have low energy. Neutrons that have been slowed down and have reached equilibrium with matter at room temperature ($\frac{3}{2}kT \approx 0.04\,eV$ at $T = 300\,K$) are called **thermal neutrons**.

Thermal neutrons

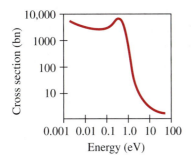

FIGURE 43–3 The neutron cross section for cadmium, which is extraordinarily large for $K \lesssim 1\,eV$. Note that both scales are logarithmic.

EXAMPLE 43–4 **Using cross section.** The reaction

$$p + {}^{56}_{26}Fe \rightarrow {}^{56}_{27}Co + n$$

has a cross section of 0.65 bn for a particular incident proton energy. Suppose the iron target has an area of $1.5\,cm^2$, and is $2.0\,\mu m$ thick. The density of iron is $7.8 \times 10^3\,kg/m^3$. If the protons are incident at a rate of 2.0×10^{13} particles/s, calculate the rate at which neutrons are produced.

SOLUTION We use Eq. 43–3 in the form:

$$R = R_0 n t \sigma.$$

Here $R_0 = 2.0 \times 10^{13}$ particles/s, $t = 2.0 \times 10^{-6}\,m$, and $\sigma = 0.65\,bn$. Recalling from Chapter 17 that one mole (mass = 56 g for iron) contains 6.02×10^{23} molecules, then the number of iron atoms per unit volume is

$$n = \frac{(7.8 \times 10^3\,kg/m^3)(6.02 \times 10^{23}\,atoms/mole)}{(56 \times 10^{-3}\,kg/mole)} = 8.4 \times 10^{28}\,atoms/m^3.$$

Then we find that the rate at which neutrons are produced is

$$R = (2.0 \times 10^{13}\,particles/s)(8.4 \times 10^{28}\,atoms/m^3)(2.0 \times 10^{-6}\,m)(0.65 \times 10^{-28}\,m^2)$$
$$= 2.2 \times 10^8\,particles/s.$$

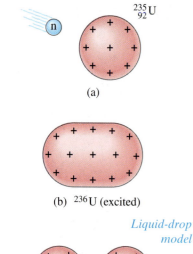

FIGURE 43–4 Fission of a ${}^{235}_{92}U$ nucleus after capture of a neutron, according to the liquid-drop model.

${}^{235}_{92}U$

(a)

(b) ${}^{236}U$ (excited)

Liquid-drop model

(c) *Fission fragments*

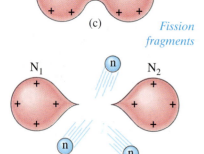

N_1 N_2

(d)

43–3 Nuclear Fission; Nuclear Reactors

In 1938, the German scientists Otto Hahn and Fritz Strassmann made an amazing discovery. Following up on Fermi's work, they found that uranium bombarded by neutrons sometimes produced smaller nuclei that were roughly half the size of the original uranium nucleus. Lise Meitner and Otto Frisch, two refugees from Nazi Germany working in Scandinavia, quickly realized what had happened: the uranium nucleus, after absorbing a neutron, actually had split into two roughly equal pieces. This was startling, for until then the known nuclear reactions involved knocking out only a tiny fragment (for example, n, p, or α) from a nucleus.

This new phenomenon was named **nuclear fission** because of its resemblance to biological fission (cell division). It occurs much more readily for ${}^{235}_{92}U$ than for the more common ${}^{238}_{92}U$. The process can be visualized by imaging the uranium nucleus to be like a liquid drop. According to this **liquid-drop model**, the neutron absorbed by the ${}^{235}_{92}U$ nucleus gives the nucleus extra internal energy (like heating a drop of water). This intermediate state, or **compound nucleus**, is ${}^{236}_{92}U$ (because of the absorbed neutron). The extra energy of this nucleus—it is in an excited state—appears as increased motion of the individual nucleons, which causes the nucleus to take on abnormal elongated shapes, Figure 43–4b. When the nucleus elongates (in this model) into the shape shown in Fig. 43–4c, the attraction of the two ends via the short-range nuclear force is greatly weakened by the increased separation distance, and the electric repulsive force becomes dominant. So the nucleus splits in two. The two resulting nuclei, N_1 and N_2, are called **fission fragments**, and in the process a number of neutrons (typically two or three) are also given off. The reaction can be written

$$n + {}^{235}_{92}U \rightarrow {}^{236}_{92}U \rightarrow N_1 + N_2 + neutrons. \qquad \textbf{(43–4)}$$

The compound nucleus, ${}^{236}_{92}U$, exists for less than $10^{-12}\,s$, so the process occurs very quickly. The two fission fragments more often split the original uranium mass as about 40 percent–60 percent rather than precisely half and half. A typical fission reaction is

$$n + {}^{235}_{92}U \rightarrow {}^{141}_{56}Ba + {}^{92}_{36}Kr + 3n,$$ (43–5)

although many others also occur. Figure 43–5 shows the distribution of fission fragments according to mass. Note that only rarely $(\sim 1 \text{ in } 10^4)$ does a fission result in equal mass fragments (small arrow in Fig. 43–5).

A tremendous amount of energy is released in a fission reaction because the mass of ${}^{235}_{92}U$ is considerably greater than the total mass of the fission fragments plus neutrons. This can be seen from the binding-energy-per-nucleon curve of Fig. 42–1; the binding energy per nucleon for uranium is about 7.6 MeV/nucleon, but for fission fragments that have intermediate mass (in the center portion of the graph, $A \approx 100$), the average binding energy per nucleon is about 8.5 MeV/nucleon. Since the fission fragments are more tightly bound, they have less mass. The difference in mass, or energy, between the original uranium nucleus and the fission fragments is about $8.5 - 7.6 = 0.9 \text{ MeV}$ per nucleon. Since there are 236 nucleons involved in each fission, the total energy released per fission is

$$(0.9 \text{ MeV/nucleon})(236 \text{ nucleons}) \approx 200 \text{ MeV}.$$

This is an enormous amount of energy for one single nuclear event. At a practical level, the energy from one fission is, of course, tiny. But if many such fissions could occur in a short time, an enormous amount of energy at the macroscopic level would be available. A number of physicists, including Fermi, recognized that the neutrons released in each fission (Eqs. 43–4 and 5) could be used to create a **chain reaction**. That is, one neutron initially causes one fission of a uranium nucleus; the two or three neutrons released can go on to cause additional fissions, so the process multiplies as shown schematically in Fig. 43–6. If a **self-sustaining chain reaction** was actually possible in practice, the enormous energy available in fission could be released on a larger scale. Fermi and his co-workers (at the University of

Chain reaction

FIGURE 43–5 Mass distribution of fission fragments from ${}^{235}_{92}U$ + n. The small arrow indicates equal mass fragments $(\frac{1}{2} \times (236 - 2) = 117)$. Note that the vertical scale is logarithmic.

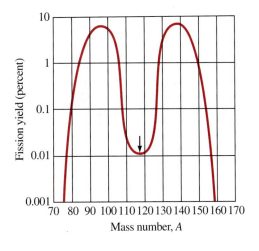

FIGURE 43–6 Chain reaction.

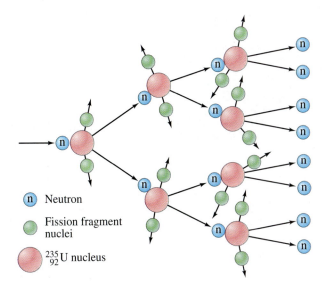

$\textcircled{n}$ Neutron

Fission fragment nuclei

${}^{235}_{92}U$ nucleus

FIGURE 43–7 Color painting of the first nuclear reactor, built by Fermi under the grandstand of Stagg Field at the University of Chicago. (There are no photographs of the original reactor because of military secrecy.) Natural uranium was used with graphite as moderator. On December 2, 1942, Fermi slowly withdrew the cadmium control rods and the reactor went critical. This first self-sustaining chain reaction was announced to Washington, by telephone, by Arthur Compton who witnessed the event and reported: "The Italian navigator has just landed in the new world."

Moderator

FIGURE 43–8 If the amount of uranium exceeds the critical mass, as in (b), a sustained chain reaction is possible. If the mass is less than critical, as in (a), most neutrons escape before additional fissions occur, and the chain reaction is not sustained.

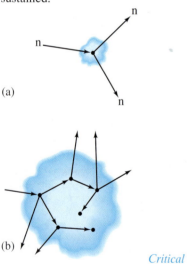

Critical mass

Chicago) showed it was possible by constructing the first **nuclear reactor** in 1942 (Fig. 43–7).

Several problems have to be overcome to make any nuclear reactor function. First, the probability that a $^{235}_{92}\text{U}$ nucleus will absorb a neutron is large only for slow neutrons, but the neutrons emitted during a fission, and which are needed to sustain a chain reaction, are moving very fast. A substance known as a **moderator** must be used to slow down the neutrons. The most effective moderator will consist of atoms whose mass is as close as possible to that of the neutrons. (To see why this is true, recall from Chapter 9, especially Examples 9–7 and 9–8, that a billiard ball striking an equal mass ball at rest can itself be stopped in one collision; but a billiard ball striking a heavy object bounces off with nearly the same speed it had.) The best moderator would thus contain ^1_1H atoms. Unfortunately, ^1_1H tends to absorb neutrons. But the isotope of hydrogen called *deuterium*, ^2_1H, does not absorb many neutrons and is thus an almost ideal moderator. Either ^1_1H or ^2_1H can be used in the form of water. In the latter case, it is **heavy water**, in which the hydrogen atoms have been replaced by deuterium. Another common moderator is *graphite*, which consists of $^{12}_6\text{C}$ atoms.

A second problem is that the neutrons produced in one fission may be absorbed and produce other nuclear reactions with other nuclei in the reactor, rather than produce further fissions. In a "light-water" reactor, the ^1_1H nuclei absorb neutrons, as does $^{238}_{92}\text{U}$ to form $^{239}_{92}\text{U}$ in the reaction $\text{n} + ^{238}_{92}\text{U} \rightarrow ^{239}_{92}\text{U} + \gamma$. Naturally occurring uranium[†] contains 99.3 percent $^{238}_{92}\text{U}$ and only 0.7 percent fissionable $^{235}_{92}\text{U}$. To increase the probability of fission of $^{235}_{92}\text{U}$ nuclei, natural uranium can be **enriched** to increase the percentage of $^{235}_{92}\text{U}$ using processes such as diffusion or centrifugation. Enrichment is not usually necessary for reactors using heavy water as moderator since heavy water doesn't absorb neutrons.

The third problem is that some neutrons will escape through the surface of the reactor core before they cause further fissions (Fig. 43–8). Thus the mass of fuel must be sufficiently large for a self-sustaining chain reaction to take place. The minimum mass of uranium needed is called the **critical mass**. The value of the critical mass depends on the moderator, the fuel ($^{239}_{94}\text{Pu}$ may be used instead of $^{235}_{92}\text{U}$), and how much the fuel is enriched, if at all. Typical values are on the order of a few kilograms (that is, not grams nor thousands of kilograms).

To have a self-sustaining chain reaction, it is clear that on the average at least one neutron produced in each fission must go on to produce another fission. The

[†] $^{238}_{92}\text{U}$ will fission, but only with fast neutrons ($^{238}_{92}\text{U}$ is more stable than $^{235}_{92}\text{U}$). The probability of absorbing a fast neutron and producing a fission is too low to produce a self-sustaining chain reaction.

average number of neutrons per each fission that do go on to produce further fissions is called the **multiplication factor**, f. For a self-sustaining chain reaction, we must have $f \geq 1$. If $f < 1$, the reactor is "subcritical." If $f > 1$, it is "supercritical." Reactors are equipped with movable **control rods** (usually of cadmium or boron), whose function is to absorb neutrons and maintain the reactor at just barely "critical," $f = 1$. The release of neutrons and subsequent fissions caused by them occurs so quickly that manipulation of the control rods to maintain $f = 1$ would not be possible if it weren't for the small percentage (~ 1 percent) of so-called **delayed neutrons**. They come from the decay of neutron-rich fission fragments (or their daughters) having lifetimes on the order of seconds—sufficient to allow enough reaction time to operate the control rods and maintain $f = 1$.

Critical reaction

Delayed neutrons

Types of nuclear reactor

Nuclear reactors have been built for use in research and to produce electric power. Fission produces many neutrons and a "research reactor" is basically an intense source of neutrons. These neutrons can be used as projectiles in nuclear reactions to produce nuclides not found in nature, including isotopes used as tracers and for therapy. A "power reactor" is used to produce electric power. The energy released in the fission process appears as heat, which is used to boil water and produce steam to drive a turbine connected to an electric generator (Fig. 43–9). The **core** of a nuclear reactor consists of the fuel and a moderator (water in most U.S. commercial reactors). The fuel is usually uranium enriched so that it contains 2 to 4 percent $^{235}_{92}$U. Water at high pressure or other liquid (such as liquid sodium) is allowed to flow through the core. The thermal energy it absorbs is used to produce steam in the heat exchanger, so the fissionable fuel acts as the heat input for a heat engine (Chapter 20).

Many problems are associated with nuclear power plants. Besides the usual thermal pollution associated with any heat engine (page 525), there is the problem of disposal of the radioactive fission fragments produced in the reactor, plus radioactive nuclides produced by neutrons interacting with the structural parts of the reactor. Fission fragments, like their uranium or plutonium parents, have about 50 percent more neutrons than protons. Nuclei with atomic number in the typical range for fission fragments ($Z \approx 30$ to 60) are stable only if they have more nearly equal numbers of protons and neutrons (see Fig. 42–2). Hence the highly neutron-rich fission fragments are very unstable and decay radioactively. The accidental release of highly radioactive fission fragments into the atmosphere poses a serious threat to human health (Section 43–5), as does possible leakage of the radioactive

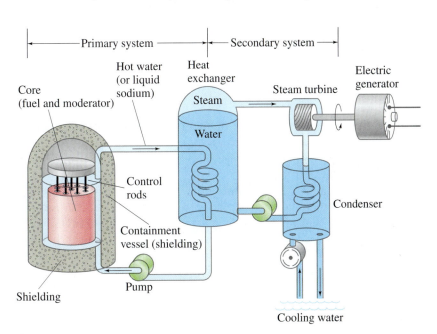

FIGURE 43–9 A nuclear reactor. The heat generated by the fission process in the fuel rods is carried off by hot water or liquid sodium and is used to boil water to steam in the heat exchanger. The steam drives a turbine to generate electricity and is then cooled in the condenser.

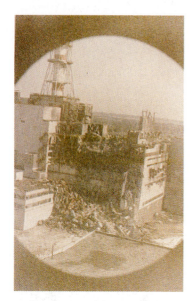

FIGURE 43-10 Devastation around Chernobyl in Russia, after the nuclear power plant disaster in 1986.

wastes when they are disposed of. The accidents at Three Mile Island (1979) and at Chernobyl (1986) have illustrated some of the dangers and have shown that nuclear plants must be constructed, maintained, and operated with great care and precision (Fig. 43–10). Finally, the lifetime of nuclear power plants is limited to 30-some years, due to buildup of radioactivity and the fact that the structural materials themselves are weakened by the intense conditions inside. Decommissioning of a power plant could take a number of forms, but the cost of any method of decommissioning a large plant will be very great.

So-called **breeder reactors** were proposed as a solution to the problem of limited supplies of fissionable uranium. A breeder reactor is one in which some of the neutrons produced in the fission of $^{235}_{92}U$ are absorbed by $^{238}_{92}U$, and $^{239}_{94}Pu$ is produced via the set of reactions shown in Fig. 43–1. $^{239}_{94}Pu$ is fissionable with slow neutrons, so after separation it can be used as a fuel in a nuclear reactor. Thus a breeder reactor "breeds" new fuel[†] $\left(^{239}_{94}Pu\right)$ from otherwise useless $^{238}_{92}U$. Since natural uranium is 99.3 percent $^{238}_{92}U$, this means that the supply of fissionable fuel could be increased by more than a factor of 100. But breeder reactors not only have the same problems as other reactors, but in addition present other serious problems. Not only is plutonium considered by many to be a serious health hazard in itself (radioactive with a half-life of 24,000 years), but plutonium produced in a reactor can readily be used in a bomb. Thus the use of a breeder reactor, even more than a conventional uranium reactor, presents the danger of nuclear proliferation, and the possibility of theft of fuel by terrorists who could produce a bomb.

It is clear that nuclear power presents many risks. Other large-scale energy-conversion methods, such as conventional coal-burning steam plants, also present health and environmental hazards, some of which were discussed in Chapter 20, including air pollution, oil spills, and the release of CO_2 gas which may be trapping heat as in a greenhouse and raising the Earth's temperature. The solution to the world's needs for energy is not only technological, but economic and political as well. A major factor surely is to "conserve"—to not waste energy and use as little as possible.

EXAMPLE 43–5 **ESTIMATE** **Uranium fuel amount.** Estimate the minimum amount of $^{235}_{92}U$ that needs to undergo fission in order to run a 1000-MW power reactor per year of continuous operation. Assume an efficiency (Chapter 20) of about 33 percent.

SOLUTION For 1000 MW output, the total power generation is 3000 MW, of which 2000 MW is dumped as "waste" heat. Thus the total energy release in 1 yr $\left(3 \times 10^7 \, s\right)$ from fission is about

$$\left(3 \times 10^9 \, J/s\right)\left(3 \times 10^7 \, s\right) \approx 10^{17} \, J.$$

If each fission releases 200 MeV of energy, the number of fissions required is

$$\frac{\left(10^{17} \, J\right)}{\left(2 \times 10^8 \, eV/fission\right)\left(1.6 \times 10^{-19} \, J/eV\right)} \approx 3 \times 10^{27} \, \text{fissions}.$$

The mass of a single uranium atom is about $(235 \, u)\left(1.66 \times 10^{-27} \, kg/u\right) \approx 4 \times 10^{-25} \, kg$, so the total mass needed is $\left(4 \times 10^{-25} \, kg\right)\left(3 \times 10^{27} \, \text{fissions}\right) \approx 1000 \, kg$, or about a ton. Since $^{235}_{92}U$ is only a fraction of normal uranium, and even when enriched it is never more than 10 percent of the total, the yearly requirement for uranium is on the order of tens of tons. This is orders of magnitude less than coal, both in mass and volume (see Problem 18 in Chapter 20).

[†] A breeder reactor does *not* produce more fuel than it uses.

The first use of fission, however, was not to produce electric power. Instead, it was first used as a fission bomb (the "atomic bomb"). In early 1940, with Europe already at war, Hitler banned the sale of uranium from the Czech mines he had recently taken over. Research into the fission process suddenly was enshrouded in secrecy. Physicists in the United States were alarmed. A group of them approached Einstein—a man whose name was a household word—to send a letter to President Roosevelt about the possibilities of using nuclear fission for a bomb far more powerful than any previously known, and inform him that Germany might already have begun development of such a bomb. Roosevelt responded by authorizing the program known as the Manhattan Project, to see if a bomb could be built. Work began in earnest after Fermi's demonstration in 1942 that a sustained chain reaction was possible. A new secret laboratory was developed on an isolated mesa in New Mexico known as Los Alamos. Under the direction of J. Robert Oppenheimer (1904–1967; Fig. 43–11), it became the home of famous scientists from all over Europe and the United States.

To build a bomb that was subcritical during transport but that could be made supercritical (to produce a chain reaction) at just the right moment, two pieces of uranium were used, each less than the critical mass but together greater than the critical mass. The two masses would be kept separate until the moment of detonation arrived. Then a kind of gun would force the two pieces together very quickly, a chain reaction of explosive proportions would occur, and a tremendous amount of energy would be released very suddenly. The first fission bomb was tested in the New Mexico desert in July 1945. It was successful. In early August, a fission bomb using uranium was dropped on Hiroshima and a second, using plutonium, was dropped on Nagasaki (Fig. 43–12). World War II ended shortly thereafter.

Besides its great destructive power, a fission bomb produces many highly radioactive fission fragments, as does a nuclear reactor. When a fission bomb explodes, these radioactive isotopes are released into the atmosphere and are known as *radioactive fallout.*

Testing of nuclear bombs in the atmosphere after World War II was a cause of concern, for the movement of air masses spread the fallout all over the globe. Radioactive fallout eventually settles to the Earth, particularly in rainfall, and is absorbed by plants and grasses and enters the food chain. This is a far more serious problem than the same radioactivity on the exterior of our bodies, since α and β particles are largely absorbed by clothing and the outer (dead) layer of skin. But once inside our bodies via food, the isotopes are in direct contact with living cells. One particularly dangerous radioactive isotope is $^{90}_{38}\mathrm{Sr}$, which is chemically much like calcium and becomes concentrated in bone, where it causes bone cancer and destruction of bone marrow. The 1963 treaty signed by over 100 nations that bans nuclear weapons testing in the atmosphere was motivated because of the hazards of fallout.

43–4 Fusion

The mass of every stable nucleus is less than the sum of the masses of its constituent protons and neutrons. For example, the mass of the helium isotope $^4_2\mathrm{He}$ is less than the mass of two protons plus the mass of two neutrons, as we saw in Example 42–2. Thus, if two protons and two neutrons were to come together to form a helium nucleus there would be a loss of mass. This mass loss is manifested in the release of a large amount of energy.

Atom bomb

FIGURE 43–11 J. Robert Oppenheimer, on the left, with General Leslie Groves, who was the administrative head of Los Alamos during the war. The photograph was taken at the Trinity site in the New Mexico desert, where the first atomic bomb was exploded.

FIGURE 43–12 Photo taken a month after the bomb was dropped on Nagasaki. The shacks were constructed afterwards from debris in the ruins.

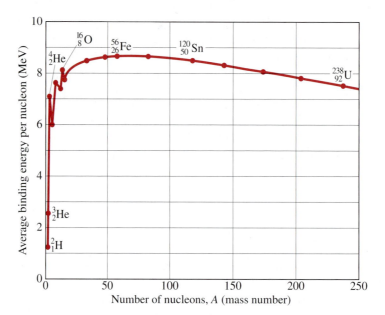

FIGURE 43–13 Average binding energy per nucleon as a function of mass number A for stable nuclei. Same as Fig. 42–1.

The process of building up nuclei by bringing together individual protons and neutrons, or building larger nuclei by combining small nuclei, is called **nuclear fusion**. A glance at Fig. 43–13 (same as Fig. 42–1) shows why small nuclei can combine to form larger ones with the release of energy: it is because the binding energy per nucleon is smaller for light nuclei than it is for those of increasing mass (up to about $A \approx 60$). It is believed that many of the elements in the universe were originally formed through the process of fusion (see Chapter 45), and that today, fusion is continually taking place within the stars, including our Sun, producing the prodigious amounts of radiant energy they emit.

EXAMPLE 43–6 **Fusion energy release.** One of the simplest fusion reactions involves the production of deuterium, ^2_1H, from a neutron and a proton: $^1_1\text{H} + \text{n} \rightarrow {}^2_1\text{H} + \gamma$. How much energy is released in this reaction?

SOLUTION From Appendix D, the initial rest mass is $1.007825\,\text{u} + 1.008665\,\text{u} = 2.016490\,\text{u}$, and after the reaction the mass is that of the ^2_1H, namely $2.014102\,\text{u}$. The mass difference is $0.002388\,\text{u}$, so the energy released is $(0.002388\,\text{u})(931.5\,\text{MeV/u}) = 2.22\,\text{MeV}$, and it is carried off by the ^2_1H nucleus and the γ ray.

The energy output of our Sun is believed to be due principally to the following sequence of fusion reactions:

Fusion reactions

$$^1_1\text{H} + {}^1_1\text{H} \rightarrow {}^2_1\text{H} + e^+ + \nu \qquad (0.42\,\text{MeV}) \quad \textbf{(43–6a)}$$

in the Sun

$$^1_1\text{H} + {}^2_1\text{H} \rightarrow {}^3_2\text{He} + \gamma \qquad (5.49\,\text{MeV}) \quad \textbf{(43–6b)}$$

(proton–proton cycle)

$$^3_2\text{He} + {}^3_2\text{He} \rightarrow {}^4_2\text{He} + {}^1_1\text{H} + {}^1_1\text{H} \qquad (12.86\,\text{MeV}) \quad \textbf{(43–6c)}$$

where the energy released (Q-value) for each reaction is given in parentheses. The net effect of this sequence, which is called the **proton–proton cycle**, is for four protons to combine to form one ^4_2He nucleus plus two positrons, two neutrinos, and two gamma rays:

$$4\,{}^1_1\text{H} \rightarrow {}^4_2\text{He} + 2e^+ + 2\nu + 2\gamma. \qquad \textbf{(43–7)}$$

Note that it takes two of each of the first two reactions (Eqs. 43–6a and b) to produce the two ^3_2He for the third reaction, so the total energy release for the net reaction, Eq. 43–7, is $(2 \times 0.42\,\text{MeV} + 2 \times 5.49\,\text{MeV} + 12.86\,\text{MeV}) = 24.7\,\text{MeV}$.

However, each of the two e^+ (formed in reaction Eq. 43–6a) quickly annihilates with an electron to produce $2m_e c^2 = 1.02\,\text{MeV}$, so the total energy released is $(24.7\,\text{MeV} + 2 \times 1.02\,\text{MeV}) = 26.7\,\text{MeV}$. The first reaction, the formation of deuterium from two protons (Eq. 43–6a), has a very low probability, and the infrequency of that reaction serves to limit the rate at which the Sun produces energy.

In stars hotter than the Sun, it is more likely that the energy output comes principally from the **carbon cycle**, which comprises the following sequence of reactions:

$$^{12}_{6}\text{C} + ^{1}_{1}\text{H} \rightarrow ^{13}_{7}\text{N} + \gamma$$

$$^{13}_{7}\text{N} \rightarrow ^{13}_{6}\text{C} + e^+ + \nu$$

$$^{13}_{6}\text{C} + ^{1}_{1}\text{H} \rightarrow ^{14}_{7}\text{N} + \gamma$$

$$^{14}_{7}\text{N} + ^{1}_{1}\text{H} \rightarrow ^{15}_{8}\text{O} + \gamma$$

$$^{15}_{8}\text{O} \rightarrow ^{15}_{7}\text{N} + e^+ + \nu$$

$$^{15}_{7}\text{N} + ^{1}_{1}\text{H} \rightarrow ^{12}_{6}\text{C} + ^{4}_{2}\text{He}.$$

Carbon

cycle

(some stars)

(See Problem 38.) The theory of the proton–proton cycle and of the carbon cycle as the source of energy for the Sun and stars was first worked out by Hans Bethe (1906–) in 1939.

CONCEPTUAL EXAMPLE 43–7 | **Stellar fusion.** What is the heaviest element likely to be produced in fusion processes in stars?

RESPONSE Fusion is possible as long as the final product has more binding energy (less mass) than the reactants, for then there is net release of energy. Since the binding energy curve in Fig. 43–13 (or 42–1) peaks near $A \approx 56$ to 58, which corresponds to iron or nickel, it would not be energetically favorable to produce elements heavier than that. Nevertheless, in the center of massive stars or in supernova explosions, there is enough energy available to drive endothermic reactions that produce heavier elements, as well.

The possibility of utilizing the energy released in fusion to make a power reactor is very attractive. The fusion reactions most likely to succeed in a reactor involve the isotopes of hydrogen, $^{2}_{1}\text{H}$ (deuterium) and $^{3}_{1}\text{H}$ (tritium), and are as follows, with the energy released given in parentheses:

Fusion reactor

$$^{2}_{1}\text{H} + ^{2}_{1}\text{H} \rightarrow ^{3}_{1}\text{H} + ^{1}_{1}\text{H} \qquad\qquad (4.03\,\text{MeV}) \quad \textbf{(43–8a)}$$

$$^{2}_{1}\text{H} + ^{2}_{1}\text{H} \rightarrow ^{3}_{2}\text{He} + n \qquad\qquad (3.27\,\text{MeV}) \quad \textbf{(43–8b)}$$

$$^{2}_{1}\text{H} + ^{3}_{1}\text{H} \rightarrow ^{4}_{2}\text{He} + n. \qquad\qquad (17.59\,\text{MeV}) \quad \textbf{(43–8c)}$$

Fusion reactions

for possible

reactor

Comparing these energy yields with that for the fission of $^{235}_{92}\text{U}$, we can see that the energy released in fusion reactions can be greater for a given mass of fuel than in fission. Furthermore, as fuel, a fusion reactor could use deuterium, which is very plentiful in the water of the oceans (the natural abundance of $^{2}_{1}\text{H}$ is 0.015 percent, or about 1 g of deuterium per 60 L of water). The simple proton–proton reaction of Eq. 43–6a, which could use a much more plentiful source of fuel, $^{1}_{1}\text{H}$, has such a small probability of occurring that it cannot be considered a possibility on Earth.

Although a successful fusion reactor has not yet been achieved, considerable progress has been made in overcoming the inherent difficulties. The problems are associated with the fact that all nuclei have a positive charge and thus repel each other. However, if they can be brought close enough together so that the short-range attractive nuclear force can come into play, the latter can pull the nuclei together and fusion will occur. Thus, in order for the nuclei to get close enough together, they must have rather large kinetic energy to overcome the electric repulsion. High kinetic energies are easily attainable with particle accelerators such as the cyclotron, but the number of particles involved is too small. To produce realistic amounts of energy, we must deal with matter in bulk, for which high kinetic energy means higher temperatures. Indeed, very high temperatures are required for fusion to occur, and fusion devices are often referred to as **thermonuclear devices**. The Sun and other stars are very hot, many millions of degrees, so the nuclei are moving fast enough for fusion to take place, and the energy released keeps the temperature high so that further fusion reactions can occur. The Sun and the stars represent huge self-sustaining thermonuclear reactors that stay together because of their great gravitational mass; but on Earth, containment of the plasma at the high temperatures and densities required has proven difficult.

It was realized after World War II that the temperature produced within a fission (or "atomic") bomb was close to 10^8 K. This suggested that a fission bomb could be used to ignite a fusion bomb (popularly known as a thermonuclear or hydrogen bomb) to release the vast energy of fusion. The uncontrollable release of fusion energy in an H-bomb was relatively easy to obtain. But to realize usable energy from fusion at a slow and controlled rate turned out to be difficult.

EXAMPLE 43–8 ESTIMATE Temperature needed for d–t fusion. Estimate the temperature required for deuterium–tritium fusion (d–t) to occur.

SOLUTION We assume the nuclei approach head-on, each with kinetic energy K, and that the nuclear force comes into play when the distance between their centers equals the sum of their radii as given by Eq. 42–1, namely $r_d \approx 1.5$ fm and $r_t \approx 1.7$ fm. The electrostatic potential energy (Chapter 23) of the two particles at this distance must equal the total kinetic energy of the two particles when far apart:

$$2K \approx \frac{1}{4\pi\epsilon_0} \frac{e^2}{(r_d + r_t)}$$

$$\approx \left(9.0 \times 10^9 \, \frac{N \cdot m^2}{C^2}\right) \frac{(1.6 \times 10^{-19} \, C)^2}{(3.2 \times 10^{-15} \, m)(1.6 \times 10^{-19} \, J/eV)}$$

$$\approx 0.45 \, MeV.$$

Thus, $K \approx 0.22$ MeV, and if we ask that the average kinetic energy be this high, then from Eq. 18–4, $\frac{3}{2}kT = \bar{K}$, we have

$$T = \frac{2K}{3k} = \frac{2(0.22 \, MeV)(1.6 \times 10^{-13} \, J/MeV)}{3(1.38 \times 10^{-23} \, J/K)} \approx 2 \times 10^9 \, K.$$

More careful calculations show that the temperature required for fusion is actually about an order of magnitude less than this rough estimate, partly because it is not necessary that the *average* kinetic energy be 0.22 MeV—a small percentage with this much energy (particles in the high-energy tail of the Maxwell distribution, Fig. 18–3) would be sufficient—and because of tunneling through the Coulomb barrier. Reasonable estimates for a usable fusion reactor are in the range $T \gtrsim 2$ to 4×10^8 K.

It is not only a high temperature that is required for a fusion reactor. There must also be a high density of nuclei to ensure a sufficiently high collision rate. A real difficulty with controlled fusion is to contain nuclei long enough and at a high enough density for sufficient reactions to occur that a usable amount of energy is obtained. At the temperatures needed for fusion, the atoms are ionized, and the resulting collection of nuclei and electrons is referred to as a **plasma**. Ordinary materials vaporize at a few thousand degrees at best, and hence cannot be used to contain a high-temperature plasma. Two major containment techniques are being investigated at present: *magnetic confinement* and *inertial confinement*.

Plasma

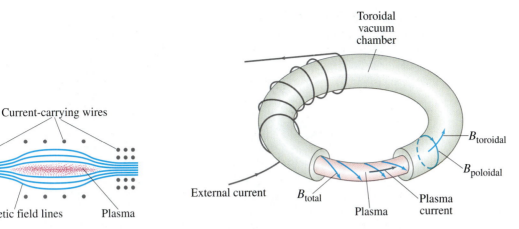

FIGURE 43–14 "Magnetic bottle" used to confine a plasma.

FIGURE 43–15 Tokamak configuration showing total **B** field due to external current plus current in the plasma.

Magnetic confinement

In **magnetic confinement**, magnetic fields are used to try to contain the hot plasma. One possibility is the "magnetic bottle" shown in Fig. 43–14. The paths of the charged particles in the plasma are bent by the magnetic field, and where the lines are close together, the force on the particles is such that they are reflected back toward the center by this "magnetic mirror." Unfortunately, magnetic bottles develop "leaks" and the charged particles slip out before sufficient fusion takes place. A more complex device called the **tokamak**, first developed in the USSR, shows considerable promise. A tokamak (Fig. 43–15) is toroid-shaped and involves complicated magnetic fields: first, current-carrying conductors produce a magnetic field directed along the axis of the toroid ("toroidal" field); an additional field is produced by currents within the plasma itself ("poloidal" field). The combination of these fields produces a helical field as shown in Fig. 43–15, confining the plasma, at least briefly, so it doesn't touch the vacuum chamber's metal walls.

In 1957, J. D. Lawson showed that the product of ion density n and confinement time τ must satisfy, at a minimum, approximately

$$n\tau \gtrsim 3 \times 10^{20} \, \text{s/m}^3.$$

Lawson criterion

This **Lawson criterion** must be reached to produce **ignition**, by which we mean a self-sustaining thermonuclear "burn" that continues after all external heating is turned off. To reach **break-even**, the point at which the energy output due to fusion is equal to the energy input to heat the plasma, requires an $n\tau$ about an order of magnitude less. The break-even point was very closely approached at the Tokamak Fusion Test Reactor (TFTR) at Princeton, and the very high temperature needed for ignition $(4 \times 10^8 \, \text{K})$ was exceeded—although not both of these at the same time. It is believed that a working fusion reactor might be a reality in the first decade of this century.

Ignition

Break-even

The second method for containing the fuel for fusion is **Inertial confinement**: a small pellet of deuterium and tritium is struck simultaneously from several directions by very intense laser beams. The intense influx of energy in such a laser fusion device heats and ionizes the pellet into a plasma. The outer layers evaporate, but the collisions they make with ions in the core of the pellet drive the latter inward. This implosion raises the density to about 10^3 times normal and further heats the core to temperatures at which fusion occurs. The confinement time is very short, on the order of 10^{-11} to 10^{-9} s, during which time the ions do not move appreciably because of their own inertia. Very soon thereafter fusion takes place and the pellet explodes.

Inertial confinement

(a)   (b)

FIGURE 43–16 (a) Target chamber (5 m in diameter) of NOVA laser at Lawrence Livermore Laboratory, into which 10 laser beams converge on a target. (b) A 1-mm-diameter DT (deuterium–tritium) target, on its support, at the center of the target chamber.

The NOVA laser used for fusion research at the Lawrence Livermore Laboratory (Fig. 43–16) can deposit 10^5 J into the pellet over a 10^{-9} s pulse. This is a power input of 10^{14} W, more than the total electric-generator capacity of the United States! Of course, it is sustained only over an extremely short interval of time. A future reactor might implode 100 such pellets per second, thus requiring an input of 10^5 J $\times$ $100\,s^{-1} = 10^7$ W on the average. Inertial confinement depends on high particle density, n, over very short time intervals, τ. Recently, the Lawson criterion has been achieved, but at a temperature less than that needed for ignition.

Besides the problems of confinement, the building of a practical fusion reactor will require the development of materials used in its construction that can withstand high temperatures and high levels of radiation.

43–5 Passage of Radiation Through Matter; Radiation Damage

When we speak of *radiation*, we include α, β, γ, and X-rays, as well as protons, neutrons, and other particles such as pions (see Chapter 44). Because charged particles can ionize the atoms or molecules of any material they pass through, they are referred to as **ionizing radiation**. And because radiation produces ionization, it can cause considerable damage to materials, particularly to biological tissue.

Charged particles, such as α and β rays and protons, cause ionization because of the electric force. That is, when they pass through a material, they can attract or repel electrons strongly enough to remove them from the atoms of the material. Since the α and β rays emitted by radioactive substances have energies on the order of a MeV (10^4 to 10^7 eV), whereas ionization of atoms and molecules requires on the order of 10 eV, it is clear that a single α or β particle can cause thousands of ionizations.

Neutral particles also give rise to ionization when they pass through materials. For example, X-ray and γ-ray photons can ionize atoms by knocking out electrons by means of the photoelectric and Compton effects (Chapter 38). Furthermore, if a γ ray has sufficient energy (greater than 1.02 MeV), it can undergo pair production: an electron and a positron are produced (Section 38–4). The charged particles produced in all three of these processes can themselves go on to produce further ionization. Neutrons, on the other hand, interact with matter mainly by collisions with nuclei, with which they interact strongly. Often the nucleus is broken apart by such a collision, altering the molecule of which it was a part. And the fragments produced can in turn cause ionization.

Radiation passing through matter can do considerable damage. Metals and other structural materials become brittle and their strength can be weakened if the radiation is very intense, as in nuclear reactor power plants and for space vehicles that must pass through areas of intense cosmic radiation.

The radiation damage produced in biological organisms is due primarily to ionization produced in cells. Several related processes can occur. Ions or radicals are produced that are highly reactive and take part in chemical reactions that interfere with the normal operation of the cell. All forms of radiation can ionize atoms by knocking out electrons. If these are bonding electrons, the molecule may break apart, or its structure may be altered so that it does not perform its normal function or performs a harmful function. If many cells die, the organism may not be able to recover. On the other hand, a cell may survive but be defective. It may go on dividing and produce many more defective cells, to the detriment of the whole organism. Thus radiation can cause cancer—the rapid production of defective cells.

Biological damage

Radiation damage to biological organisms is often separated into categories. *Somatic damage* refers to any part of the body except the reproductive organs, and affects that particular organism, causing cancer and, at high doses, radiation sickness (characterized by nausea, fatigue, loss of body hair, and other symptoms) or even death. *Genetic damage* refers to damage to reproductive cells, causing mutations, the majority of which are harmful, which are transmitted to future generations. The possible damage done by the medical use of X-rays and other radiation must be balanced against the medical benefits and prolongation of life as a result of their use.

43–6 Measurement of Radiation—Dosimetry

Although the passage of ionizing radiation through the human body can cause considerable damage, radiation can also be used to treat certain diseases, particularly cancer, often by using very narrow beams directed at the cancerous tumor to destroy it (Section 43–7). It is therefore important to be able to quantify the amount, or **dose**, of radiation. This is the subject of **dosimetry**.

The strength of a source can be specified at a given time by stating the **source activity**, or how many disintegrations occur per second. The traditional unit is the **curie** (Ci), defined as

Source activity

$$1 \text{ Ci} = 3.70 \times 10^{10} \text{ disintegrations per second.}$$

(This figure comes from the original definition as the activity of exactly 1 gram of radium.) Although the curie is still in common use, the proper SI unit for source activity is the **becquerel** (Bq), defined as

$$1 \text{ Bq} = 1 \text{ disintegration/s.}$$

Commercial suppliers of **radionuclides** (radioactive nuclides) specify the activity at a given time. Since the activity decreases over time, particularly for short-lived isotopes, it is important to take this into account.

The source activity (dN/dt) is related to the half-life, $T_{\frac{1}{2}}$, by (see Section 42–8):

$$\left| \frac{dN}{dt} \right| = \lambda N = \frac{0.693}{T_{\frac{1}{2}}} N.$$

EXAMPLE 43–9 **Radioactivity taken up by cells.** In a certain experiment, $0.016 \, \mu\text{Ci}$ of $^{32}_{15}\text{P}$ is injected into a medium containing a culture of bacteria. After 1 h, the cells are washed and a detector that is 70 percent efficient (counts 70 percent of emitted β particles) records 720 counts per minute from all the cells. What percentage of the original $^{32}_{15}\text{P}$ was taken up by the cells?

SOLUTION The halflife of $^{32}_{15}\text{P}$ being about 14 days, the loss of activity over the 1 hour can be ignored. The total number of disintegrations per second originally was $(0.016 \times 10^{-6})(3.7 \times 10^{10}) = 590$. The counter could be expected to count 70 percent of this, or 410 per second. Since it counted $720/60 = 12$ per second, then $12/410 = 0.029$ or 2.9 percent was incorporated into the cells.

Another type of measurement is the exposure or **absorbed dose**—that is, the effect the radiation has on the absorbing material. The earliest unit of dosage was the **roentgen** (R), which was defined in terms of the amount of ionization produced by the radiation (1.6×10^{12} ion pairs per gram of dry air at standard conditions). Today, 1 R is defined as the amount of X- or γ radiation that deposits 0.878×10^{-2} J of energy per kilogram of air. The roentgen was largely superseded by another unit of absorbed dose applicable to any type of radiation, the **rad**: *1 rad is that amount of radiation which deposits energy at a rate of 1.00×10^{-2} J/kg in any absorbing material.* (This is quite close to the roentgen for X- and γ rays.) The proper SI unit for absorbed dose is the **gray** (Gy):

$$1 \, \text{Gy} = 1 \, \text{J/kg} = 100 \, \text{rad}, \qquad \textbf{(43–9)}$$

and is now coming into use. The absorbed dose depends not only on the strength of a given radiation beam (number of particles per second) and the energy per particle, but also on the type of material that is absorbing the radiation. Since bone, for example, is denser than flesh and absorbs more of the radiation normally used, the same beam passing through a human body deposits a greater dose (in rads or grays) in bone than in flesh.

The gray and the rad are physical units of dose—the energy deposited per unit mass of material. They are, however, not the most meaningful units for measuring the biological damage produced by radiation. This is because equal doses of different types of radiation cause differing amounts of damage. For example, 1 rad of α radiation does 10 to 20 times the amount of damage as 1 rad of β or γ rays. This difference arises largely because α rays (and other heavy particles such as protons and neutrons) move much more slowly than β and γ rays of equal energy due to their greater mass. Hence, ionizing collisions occur closer together, so more irreparable damage can be done. The **relative biological effectiveness** (RBE) or **quality factor** (QF) of a given type of radiation is defined as the number of rads of X or γ radiation that produces the same biological damage as 1 rad of the given radiation. Table 43–1 gives the QF for several types of radiation. The numbers are approximate since they depend somewhat on the energy of the particles and on the type of damage that is used as the criterion.

The **effective dose** can be given as the product of the dose in rads and the QF, and this unit is known as the **rem** (which stands for *rad equivalent man*):

$$\text{effective dose (in rem)} \;=\; \text{dose (in rad)} \times \text{QF}. \qquad \textbf{(43–10a)}$$

This unit is being replaced by the SI unit for "effective dose," the **sievert** (Sv):

$$\text{effective dose (Sv)} \;=\; \text{dose (Gy)} \times \text{QF}. \qquad \textbf{(43–10b)}$$

By this definition, 1 rem of any type of radiation does approximately the same amount of biological damage. For example, 50 rem of fast neutrons does the same damage as 50 rem of γ rays. But note that 50 rem of fast neutrons is only 5 rads, whereas 50 rem of γ rays is 50 rads.

We are constantly exposed to low-level radiation from natural sources: cosmic rays, natural radioactivity in rocks and soil, and naturally occurring radioactive isotopes in our food, such as $^{40}_{19}\text{K}$. Radon, $^{222}_{86}\text{Rn}$, is of considerable concern today. It is the product of radium decay and is an intermediate in the decay series from uranium (see Fig. 42–11). Most intermediates remain in the rocks where formed, but radon is a gas that can escape from rock (and from building material like concrete) to enter the atmosphere we breathe. Although radon is inert chemically (it is a noble gas), it is not inert physically—it decays by alpha emission, and its products, also radioactive, are *not* chemically inert and can attach to the interior of the lung.

The natural radioactive background averages about 0.36 rem (3.6 mSv) per year per person, although there are large variations. From medical X-rays, the average person receives about 40 mrem (0.4 mSv) per year. The U.S. government

The rad (unit)

The gray (unit)

RBE or QF

Effective dose

The rem (unit)

The sievert (unit)

Natural radioactive
background

TABLE 43–1
Quality Factor (QF) of Different Kinds of Radiation

Type	QF
X and γ rays	≈ 1
β (electrons)	≈ 1
Fast protons	1
Slow neutrons	≈ 3
Fast neutrons	Up to 10
α particles and heavy ions	Up to 20

specifies the recommended upper limit of allowed radiation for an individual in the general populace at about 0.5 rem (5 mSv) per year, exclusive of natural sources. It is not known if low doses of radiation increase the chances of cancer or genetic defects, so the attitude today is to play safe and keep the radiation dose low.

The upper limit for people who work around radiation—in hospitals, power plants, research—has been set somewhat higher, on the order of 5 rem/yr whole-body dose (presumably because such people know what they are getting into). People who work around radiation generally carry some type of dosimeter, typically a **radiation film badge**, which is a piece of film wrapped in light-tight material. The passage of ionizing radiation through the film changes it so that the film is darkened upon development, and so indicates the received dose.

Large doses of radiation can cause a large number of unpleasant symptoms such as nausea, fatigue, and loss of body hair. Such effects are sometimes referred to as **radiation sickness**. Large doses can also be fatal, although the time span of the dose is important. A short dose of 1000 rem (10 Sv) is nearly always fatal. A 400-rem (4-Sv) dose in a short period of time is fatal in 50 percent of the cases. However, the body possesses remarkable repair processes, so that a 400-rem dose spread over several weeks is not usually fatal. It will, nonetheless, cause considerable damage to the body.

Radiation sickness

EXAMPLE 43–10 **Whole-body dose.** What whole-body dose is received by a 70-kg laboratory worker exposed to a 40 mCi $^{60}_{27}$Co source, assuming the person's body has cross-sectional area 1.5 m^2 and is, on average, 4.0 m from the source for 4.0 h per day? $^{60}_{27}$Co emits γ rays of energy 1.33 MeV and 1.17 MeV in quick succession. Approximately 50 percent of the γ rays interact in the body and deposit all their energy. (The rest pass through.)

SOLUTION The total γ-ray energy per decay is $(1.33 + 1.17)$MeV $= 2.50$ MeV, so the total energy emitted by the source is

$$(0.040 \text{ Ci})(3.7 \times 10^{10} \text{ decays/Ci·s})(2.50 \text{ MeV}) = 3.7 \times 10^{9} \text{ MeV/s}.$$

The proportion of this intercepted by the body is its 1.5 m^2 area divided by the area of a sphere of radius 4.0 m (Fig. 43–17)

$$\frac{1.5 \text{ m}^2}{4\pi r^2} = \frac{1.5 \text{ m}^2}{4\pi(4.0 \text{ m})^2} = 7.5 \times 10^{-3}.$$

So the rate energy is deposited in the body (remembering that only $\frac{1}{2}$ of the γ rays interact in the body) is

$$E = (\tfrac{1}{2})(7.5 \times 10^{-3})(3.7 \times 10^{9} \text{ MeV/s})(1.6 \times 10^{-13} \text{ J/MeV})$$
$$= 2.2 \times 10^{-6} \text{ J/s}.$$

Since 1 Gy = 1 J/kg, the whole-body dose rate for this 70-kg person is $(2.2 \times 10^{-6} \text{ J/s})/(70 \text{ kg}) = 3.1 \times 10^{-8}$ Gy/s. In the space of 4.0 h, this amounts to a dose of $(4.0 \text{ h})(3600 \text{ s/h})(3.1 \times 10^{-8} \text{ Gy/s}) = 4.5 \times 10^{-4}$ Gy. Since QF ≈ 1 for gammas, the effective dose (Eq. 43–10) is 450 μSv or (see Eq. 43–9):

$$(100 \text{ rad/Gy})(4.5 \times 10^{-4} \text{ Gy})(1) = 45 \text{ mrem}.$$

This 45-mrem effective dose is nearly 10 percent of the normal allowed dose for a whole year (500 mrem/yr), or 1 percent of the yearly allowance for radiation workers. This worker should not receive such a dose every day and should seek ways to reduce it (shield the source, vary the work, work farther away, etc.).

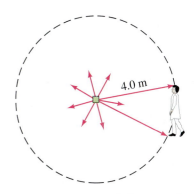

FIGURE 43–17 Radiation spreads out in all directions. A person 4.0 m away intercepts only a fraction: her cross-sectional area divided by the area of a sphere of radius 4.0 m.

The applications of radioactivity and radiation to human beings and other organisms is a vast field that has filled many books. In the medical field there are two basic aspects: (1) *radiation therapy*—the treatment of disease (mainly cancer)—which we discuss in this Section; and (2) the *diagnosis* of disease, which we discuss in the following Sections of this chapter.

Radiation can cause cancer. It can also be used to treat it. Rapidly growing cancer cells are especially susceptible to destruction by radiation. Nonetheless, large doses are needed to kill the cancer cells, and some of the surrounding normal cells are inevitably killed as well. It is for this reason that cancer patients receiving radiation therapy often suffer side effects characteristic of radiation sickness. To minimize the destruction of normal cells, a narrow beam of γ or X-rays is often used when the cancerous tumor is well localized. The beam is directed at the tumor, and the source (or body) is rotated so that the beam passes through various parts of the body to keep the dose at any one place as low as possible—except at the tumor and its immediate surroundings, where the beam passes at all times (Fig. 43–18). The radiation may be from a radioactive source such as $^{60}_{27}\text{Co}$, or it may be from an X-ray machine that produces photons in the range 200 keV to 5 MeV. Protons, neutrons, electrons, and pions, which are produced in particle accelerators (Section 44–2), are also being used in cancer therapy.

In some cases, a tiny radioactive source may be inserted directly inside a tumor, which will eventually kill the majority of the cells. A similar technique is used to treat cancer of the thyroid with the radioactive isotope $^{131}_{53}\text{I}$. The thyroid gland tends to concentrate any iodine present in the bloodstream; so when $^{131}_{53}\text{I}$ is injected into the blood, it becomes concentrated in the thyroid, particularly in any area where abnormal growth is taking place. The intense radioactivity emitted can then destroy the defective cells.

Radioactive isotopes are commonly used in biological, chemical, and medical research as **tracers**. A given compound is artificially synthesized using a radioactive isotope such as $^{14}_{6}\text{C}$ or $^{3}_{1}\text{H}$. Such "tagged" molecules can then be traced as they move through an organism or as they undergo chemical reaction. The presence of these tagged molecules (or parts of them, if they undergo chemical change) can be detected by a Geiger or scintillation counter (see Section 42–11).

For medical diagnosis, the radionuclide commonly used today is $^{99\text{m}}_{43}\text{Tc}$, a long-lived excited state of technetium-99 (the "m" in the symbol stands for "metastable" state). It is formed when $^{99}_{42}\text{Mo}$ decays. The great usefulness of $^{99\text{m}}_{43}\text{Tc}$ derives from its convenient half-life of 6 h (short, but not too short) and the fact that it can combine with a large variety of compounds. The compound to be labeled with the radionuclide is so chosen because it concentrates in the organ or region of the anatomy to be studied. Detectors outside the body then record, or image, the distribution of the radioactively labeled compound. The detection can be done by a single detector (Fig. 43–19) which is moved across the body, measuring the intensity of radioactivity at a large number of points. The image represents the relative intensity of radioactivity at each point. The relative radioactivity is a diagnostic tool. For example, high or low radioactivity may represent overactivity or underactivity of an organ or part of an organ, or in another case may represent a lesion or tumor. More complex *gamma cameras* make use of many detectors which simultaneously record the radioactivity at many points. The measured intensities can be displayed on a CRT (TV monitor or oscilloscope screen), and allow "dynamic" studies (that is, images that change in time) to be performed.

→ PHYSICS APPLIED

Radiation therapy

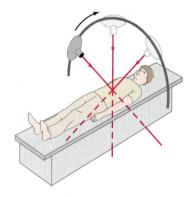

FIGURE 43–18 Radiation source rotates so that the beam always passes through the diseased tissue, but minimizing the dose in the rest of the body.

FIGURE 43–19 Collimated gamma-ray detector for scanning (moving) over patient. The collimator (to "collimate" means to "make straight") is necessary to select γ rays that come in a straight line from the patient. Without the collimator, γ rays from all parts of the body could strike the scintillator, producing a very poor image.

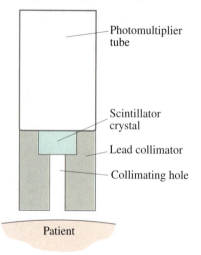

Photomultiplier tube

Scintillator crystal

Lead collimator

Collimating hole

Patient

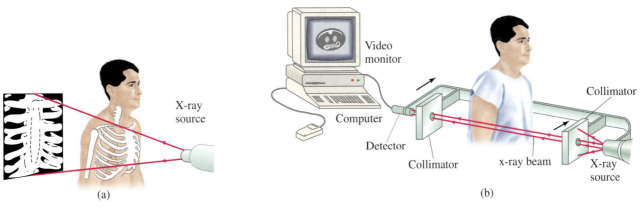

FIGURE 43–20   (a) Conventional X-ray imaging, which is essentially shadowing. (b) Tomographic imaging: the X-ray source and detector move together across the body, the transmitted intensity being measured at a large number of points. Then the source–detector assembly is rotated slightly (say, 1°) and another scan is made. This process is repeated for perhaps 180°. The computer reconstructs the image of the slice and it is presented on a TV monitor (cathode-ray tube).

*43–9   Imaging by Tomography: CAT Scans and Emission Tomography

Normal X-Ray Image

For a conventional medical (or dental) X-ray photograph, the X-rays emerging from the tube (Fig. 36–25, Section 36–10) pass through the body and are detected on photographic film or a fluorescent screen, Fig. 43–20a. The rays travel in very nearly straight lines through the body with minimal deviation since at X-ray wavelengths there is little diffraction or refraction. There is absorption (and scattering), however; and the difference in absorption by different structures in the body is what gives rise to the image produced by the transmitted rays. The less the absorption, the greater the transmission and the darker the film. The image is, in a sense, a "shadow" of what the rays have passed through.

Computerized Axial Tomography (CAT Scan)

In conventional X-ray images, the entire thickness of the body is projected onto the film; structures overlap and in many cases are difficult to distinguish. In the 1970s, a revolutionary new technique called **computerized tomography** (CT) using X-rays was developed, which produces an image of a *slice* through the body. (The word **tomography** comes from the Greek: *tomos* = slice, *graph* = picture.) Structures and lesions previously impossible to visualize can now be seen with remarkable clarity. The principle behind CT is shown in Fig. 43–20b: a thin collimated beam of X-rays passes through the body to a detector that measures the transmitted intensity. Measurements are made at a large number of points as the source and detector are moved past the body together. The apparatus is then rotated slightly about the body axis and again scanned; this is repeated at (perhaps) 1° intervals for 180°. The intensity of the transmitted beam for the many points of each scan, and for each angle, are sent to a computer that reconstructs the image of the slice. Note that the imaged slice is perpendicular to the long axis of the body. For this reason, CT is sometimes called **computerized axial tomography** (CAT), although the abbreviation CAT, as in CAT scan, can also be read as **computer-assisted tomography**.

   The use of a single detector as in Fig. 43–20b would require some time to make the many scans needed to form a complete image. Much faster scanners use

➡ **PHYSICS APPLIED**
CT images

CAT scans

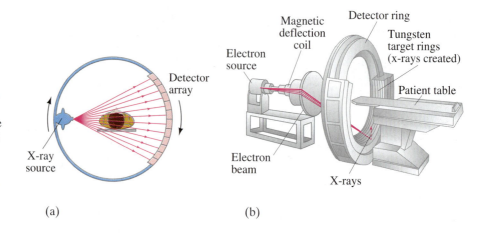

FIGURE 43–21 (a) Fan-beam scanner. Rays transmitted through the entire body are measured simultaneously at each angle. The source and detector rotate to take measurements at different angles. In another type of fan-beam scanner, there are detectors around the entire 360° of the circle which remain fixed as the source moves. (b) In another type, a beam of electrons from the source is directed by magnetic fields at tungsten targets surrounding the patient.

(a)   (b)

FIGURE 43–22 Two CT images, with different resolutions, each showing a cross section of a brain. Photo (a) is of low resolution; photo (b), of higher resolution, shows a brain tumor (dark area on the right).

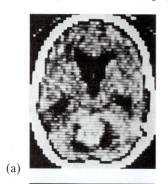

(a)

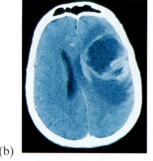

(b)

➡ PHYSICS APPLIED

Medical imaging

SPET
PET

a fan beam, Fig. 43–21a, in which beams passing through the entire cross section of the body are detected simultaneously by many detectors. The source and detectors are then rotated about the patient, and an image requires only a few seconds. Even faster, and therefore useful for heart scans, are fixed source machines wherein an electron beam is directed (by magnetic fields) to tungsten targets surrounding the patient, creating the X-rays. See Fig. 43–21b.

But how is the image formed? We can think of the slice to be imaged as being divided into many tiny picture elements (or **pixels**), which could be squares. For CT, the width of each pixel is chosen according to the width of the detectors and/or the width of the X-ray beams, and this determines the resolution of the image, which might be 1 mm. An X-ray detector measures the intensity of the transmitted beam. Subtracting this value from the intensity of the beam at the source, yields the total absorption (called a "projection") along that beam line. Complicated mathematical techniques, such as Fourier transforms, are used to analyze all the absorption projections for the huge number of beam scans measured, obtaining the absorption at each pixel and assigning each a "grayness value" according to how much radiation was absorbed. The image is made up of tiny spots (pixels) of varying shades of gray. Often the amount of absorption is color-coded. The colors in the resulting ("false-color") image have nothing to do, however, with the actual color of the object.

Figure 43–22 illustrates what actual CT images look like. It is generally agreed that CT scanning has revolutionized some areas of medicine by providing much less invasive, and/or more accurate, diagnosis.

Computerized tomography can also be applied to ultrasound imaging (Section 16–9) and to emissions from radioisotopes and nuclear magnetic resonance, which we discuss next.

Emission Tomography

It is possible to image the emissions of a radioactive tracer (see Section 43–8) in a single plane or slice through a body using computed tomography techniques. A basic gamma detector (Fig. 43–19) can be moved around the patient to measure the radioactive intensity from the tracer at many points and angles; the data are processed in much the same way as for X-ray CT scans. This technique is referred to as **single photon emission tomography** (SPET).[†]

Another important technique is **positron emission tomography** (PET), which makes use of positron emitters such as $^{11}_{6}C$, $^{13}_{7}N$, $^{15}_{8}O$, and $^{18}_{9}F$. These isotopes are incorporated into molecules that, when inhaled or injected, accumulate in the organ or region of the body to be studied. When such a nuclide undergoes β^+ decay, the emitted positron travels at most a few millimeters before it collides with a normal electron. In this collision, the positron and electron are annihilated, producing two γ rays $(e^+ + e^- \rightarrow 2\gamma)$, each having energy of 511 keV. The two γ rays fly off in oppo-

[†] Also known as SPECT, "single photon emission computed tomography."

site directions (180° ± 0.25°) since they must have almost exactly equal and opposite momenta to conserve momentum (the momenta of the e^+ and e^- are essentially zero compared to the momenta of the γ rays). Because the photons travel along the same line in opposite directions, their detection in coincidence by rings of detectors surrounding the patient (Fig. 43–23) readily establishes the line along which the emission took place. If the difference in time of arrival of the two photons could be determined accurately, the actual position of the emitting nuclide along that line could be calculated. Present-day electronics can measure times to at best ± 300 ps, so at the γ ray's speed $(c = 3 \times 10^8 \text{ m/s})$, the actual position could be determined to an accuracy on the order of about $vt \approx (3 \times 10^8 \text{ m/s})(300 \times 10^{-12} \text{ s}) \approx 10 \text{ cm}$, which is not very useful. Although there may be future potential for time-of-flight measurements to determine position, today computed tomography techniques are used instead, similar to those for X-ray CT, which can reconstruct PET images with a resolution on the order of 3–5 mm. One big advantage of PET is that no collimators are needed (as for detection of a single photon—see Fig. 43–19). Thus, fewer photons are "wasted" and lower doses can be administered to the patient with PET.

Both PET and SPET systems can give images related to biochemistry, metabolism, and function. This is to be compared to X-ray CT scans, whose images reflect shape and structure—that is, the anatomy of the imaged region.

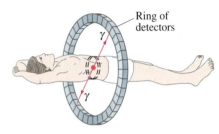

FIGURE 43–23 Positron emission tomography (PET) system showing a ring of detectors to detect the two annihilation γ rays $(e^+ + e^- \rightarrow 2\gamma)$ emitted at 180° to each other.

*43–10 Nuclear Magnetic Resonance (NMR) and Magnetic Resonance Imaging (MRI)

Nuclear magnetic resonance (NMR) is a phenomenon which soon after its discovery in 1946 became a powerful research tool in a variety of fields from physics to chemistry and biochemistry. It is also an important medical imaging technique. We first briefly discuss the phenomenon, and then look at its applications.

Nuclear Magnetic Resonance (NMR)

We saw in Chapter 40 that the energy levels in atoms are split when they are placed in a magnetic field B (the Zeeman effect) according to the angular momentum or spin of the state. The splitting is proportional to B and to the magnetic moment, μ. Nuclei too have magnetic moments (Section 42–1), and we examine only the simplest, the hydrogen $(_1^1\text{H})$ nucleus which consists of a single proton. Its spin angular momentum (and its magnetic moment), like that of the electron, can take on only two values when placed in a magnetic field: spin up (parallel to the field) and spin down (antiparallel to the field) as suggested in Fig. 43–24. When a magnetic field is present, an energy state splits into two levels as shown in Fig. 43–25 with the spin-up state (parallel to field) having the lower energy. The spin-down state acquires an additional energy μB_T and the spin-up state $-\mu B_T$ (Eq. 27–12 and Section 40–7), where B_T is the total magnetic field at the nucleus. The difference in energy between the two states (Fig. 43–25) is thus

$$\Delta E = 2\mu_p B_T,$$

where μ_p is the magnetic moment of the proton.

In a standard nuclear magnetic resonance (NMR) setup, the sample to be examined is placed in a static magnetic field. A radiofrequency (RF) pulse of electromagnetic radiation (that is, photons) is applied to the sample. If the frequency, f, of this pulse corresponds precisely to the energy difference between the two energy levels (Fig. 43–25), so that

$$hf = \Delta E = 2\mu_p B_T \qquad \textbf{(43–11)}$$

then the photons of the RF beam will be absorbed, exciting many of the nuclei from the lower state to the upper state. This is a resonance phenomenon since there is significant absorption only if f is very near $f = 2\mu_p B_T/h$. Hence the name "nuclear magnetic resonance." For free $_1^1\text{H}$ nuclei, the frequency is 42.58 MHz for a field $B_T = 1.0 \text{ T}$ (Example 43–11). If the H atoms are bound in a

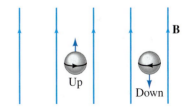

FIGURE 43–24 Schematic picture of a proton represented in a magnetic field **B** (pointing upward) with its two possible states of spin, up and down.

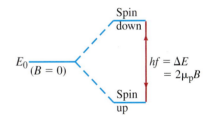

FIGURE 43–25 Energy E_0 in the absence of a magnetic field splits into two levels in the presence of a magnetic field.

molecule, the total magnetic field B_T at the H nuclei will be the sum of the external applied field (B_{ext}) plus the local magnetic field (B_{loc}) due to electrons and nuclei of neighboring atoms. Since f is proportional to B_T, the value of f for a given external field will be slightly different for bound H atoms than for free atoms:

$$hf = 2\mu_p(B_{ext} + B_{loc}).$$

This change in frequency, which can be measured, is called the "chemical shift." A great deal has been learned about the structure of molecules and bonds using this NMR technique.

EXAMPLE 43–11 **NMR for free protons.** Calculate the resonant frequency for free protons in a 1.000-T magnetic field.

SOLUTION We saw in Section 42–1 that the magnetic moment of the proton is

$$\mu_p = 2.7928\mu_N = 2.7928\left(\frac{e\hbar}{2m_p}\right).$$

Then

$$f = \frac{\Delta E}{h} = \frac{2\mu_p B}{h} = (2.7928)\left(\frac{eB}{2\pi m_p}\right) = 2.7928\left[\frac{(1.6022 \times 10^{-19}\,\text{C})(1.000\,\text{T})}{2\pi(1.6726 \times 10^{-27}\,\text{kg})}\right]$$

$$= 42.58\,\text{MHz}.$$

Magnetic Resonance Imaging (MRI)

→ **PHYSICS APPLIED**

NMR imaging (MRI)

For producing medically useful NMR images—now commonly called MRI, or **magnetic resonance imaging**—the element most used is hydrogen since it is the commonest element in the human body and gives the strongest NMR signals. The experimental apparatus is shown in Fig. 43–26. The large coils set up the static magnetic field, and the RF coils produce the RF pulse of electromagnetic waves (photons) that cause the nuclei to jump from the lower state to the upper one (Fig. 43–25). These same coils (or another coil) can detect the absorption of energy or the emitted radiation (also of frequency $f = \Delta E/h$) when the nuclei jump back down to the lower state.

The formation of a two-dimensional or three-dimensional image can be done using techniques similar to those for computed tomography (Section 43–9). The simplest thing to measure for creating an image is the intensity of absorbed and/or reemitted radiation from many different points of the body, and this would be a measure of the density of H atoms at each point. But how do we determine from what part of the body a given photon comes? One technique is to give the static

FIGURE 43–26 Typical MRI imaging setup: (a) diagram; (b) photograph.

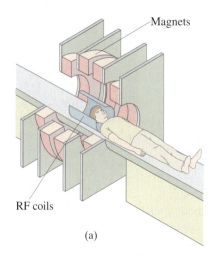
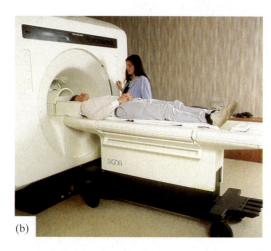

Magnets

RF coils

(a)

(b)

magnetic field a gradient; that is, instead of applying a uniform magnetic field, B_T, the field is made to vary with position across the width of the sample (or patient). Since the frequency absorbed by the H nuclei is proportional to B_T (Eq. 43–11), only one plane within the body will have the proper value of B_T to absorb photons of a particular frequency f. By varying f, absorption by different planes can be measured. Alternately, if the field gradient is applied *after* the RF pulse, the frequency of the emitted photons will be a measure of where they were emitted. See Fig. 43–27. If a magnetic field gradient in one direction is applied during excitation (absorption of photons) and photons of a single frequency are transmitted, only H nuclei in one thin slice will be excited. By applying a gradient in a different direction, perpendicular to the first, during reemission, the frequency f of the reemitted radiation will represent depth in that slice. Other ways of varying the magnetic field throughout the volume of the body can be used in order to correlate NMR frequency with position.

A reconstructed image based on the density of H atoms (that is, the intensity of absorbed or emitted radiation) is not very interesting. More useful are images based on the rate at which the nuclei decay back to the ground state, and such images can produce resolution of 1 mm or better. This NMR technique (sometimes called **spin-echo**) is producing images of great diagnostic value, both in the delineation of structure (anatomy) and in the study of metabolic processes. An NMR image is shown in Fig. 43–28.

NMR imaging is considered to be noninvasive. We can calculate the energy of the photons involved: as determined in Example 43–11, in a 1.0 T magnetic field, $f = 42.58\,\text{MHz}$ for ^1_1H. This corresponds to an energy of $hf = (6.6 \times 10^{-34}\,\text{J·s})(43 \times 10^6\,\text{Hz}) \approx 3 \times 10^{-26}\,\text{J}$ or about $10^{-7}\,\text{eV}$. Since molecular bonds are on the order of 1 eV, it is clear that the RF photons can cause little cellular disruption. This should be compared to X- or γ rays whose energies are 10^4 to $10^6\,\text{eV}$ and thus can cause significant damage. The static magnetic fields, though often large (~ 0.1 to 1 T), are believed to be harmless (except for people wearing heart pacemakers), although not a great deal of research has been done in this area.

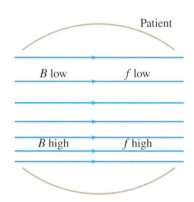

FIGURE 43–27 A static field that is stronger at the bottom than at the top. The frequency of absorbed or emitted radiation is proportional to B in NMR.

FIGURE 43–28 False-color NMR image (MRI) of a vertical section through the head showing structures in the normal brain.

Summary

A **nuclear reaction** occurs when two nuclei collide and two or more other nuclei (or particles) are produced. In this process, as in radioactivity, **transmutation** (change) of elements occurs.

The **reaction energy** or **Q-value** of a reaction $a + X \rightarrow Y + b$ is

$$Q = (M_a + M_X - M_Y - M_b)c^2$$
$$= K_b + K_Y - K_a - K_X.$$

The effective **cross section** for a reaction is a measure of the reaction probability per target nucleus.

In **fission** a heavy nucleus such as uranium splits into two intermediate-sized nuclei after being struck by a neutron. $^{235}_{92}\text{U}$ is fissionable by slow neutrons, whereas some fissionable nuclei require fast neutrons. Much energy is released in fission because the binding energy per nucleon is lower for heavy nuclei than it is for intermediate-sized nuclei, so the mass of a heavy nucleus is greater than the total mass of its fission products. The fission process releases neutrons, so that a **chain reaction** is possible. The **critical mass** is the minimum mass of fuel needed to sustain a chain reaction. In a **nuclear reactor** or nuclear bomb, a moderator is needed to slow down the released neutrons.

The **fusion** process, in which small nuclei combine to form larger ones, also releases energy. The energy from our Sun is believed to originate in the fusion reactions known as the **proton–proton cycle** in which four protons fuse to form a ^4_2He nucleus producing over 25 MeV of energy. A working fusion reactor for power generation has not yet proved possible because of the difficulty in containing the fuel (e.g., deuterium) long enough at the high temperature required. Nonetheless, great progress has been made in confining the collection of charged ions known as a **plasma**. The two main methods are **magnetic confinement**, using a magnetic field in a device such as the toroidal-shaped **tokamak**, and **inertial confinement**, in which intense laser beams compress a fuel pellet of deuterium and tritium.

Radiation can cause damage to materials, including biological tissue. Quantifying amounts of radiation is the subject of **dosimetry**. The **curie** (Ci) and the **becquerel** (Bq) are units that measure the **source activity** or rate of decay of a sample: $1\,\text{Ci} = 3.70 \times 10^{10}$ disintegrations per second, whereas $1\,\text{Bq} = 1$ disintegration/s. The **absorbed dose**, often specified in **rads**, measures the amount of energy deposited per unit mass of absorbing material: 1 rad is the amount of radiation that deposits energy at the rate of $10^{-2}\,\text{J/kg}$ of material. The SI unit of absorbed dose is the **gray**: $1\,\text{Gy} = 1\,\text{J/kg} = 100\,\text{rad}$. The **effective dose** is often specified by the **rem** = rad $\times$ QF, where QF is the "quality factor" of a given type of radiation; 1 rem of any type of radiation does approximately the same amount of biological damage. The average dose received per person per year in the United States is about 0.36 rem. The SI unit for effective dose is the **sievert**: $1\,\text{Sv} = 10^2\,\text{rem}$.

Questions

1. Fill in the missing particles or nuclei: (a) $^{137}_{56}\text{Ba}(n,\,\gamma)$?; (b) $^{137}_{56}\text{Ba}(n,\,?)^{137}_{55}\text{Cs}$; (c) $^2_1\text{H}(d,\,?)^4_2\text{He}$; (d) $^{197}_{79}\text{Au}(\alpha,\,d)$? where d stands for deuterium.

2. The isotope $^{32}_{15}\text{P}$ is produced by an (n, p) reaction. What must be the target nucleus?

3. When $^{22}_{11}\text{Na}$ is bombarded by deuterons (^2_1H), an α particle is emitted. What is the resulting nuclide?

4. Why are neutrons such good projectiles for producing nuclear reactions?

5. A proton strikes a $^{20}_{10}\text{Ne}$ nucleus, and an α particle is observed to emerge. What is the residual nucleus? Write down the reaction equation.

6. Are fission fragments β^+ or β^- emitters?

7. If $^{235}_{92}\text{U}$ released only 1.5 neutrons per fission on the average, would a chain reaction be possible? If so, what would be different?

8. $^{235}_{92}\text{U}$ releases an average of 2.5 neutrons per fission compared to 2.9 for $^{239}_{94}\text{Pu}$. Pure samples of which of these two nuclei do you think would have the smaller critical mass?

9. The energy from nuclear fission appears in the form of thermal energy—but the thermal energy of what?

10. Why can't uranium be enriched by chemical means?

11. How can a neutron, with practically no kinetic energy, excite a nucleus to the extent shown in Fig. 43–4?

12. Why would a porous block of uranium be more likely to explode if kept under water rather than in air?

13. A reactor that uses highly enriched uranium can use ordinary water (instead of heavy water) as a moderator and still have a self-sustaining chain reaction. Explain.

14. Why must the fission process release neutrons if it is to be useful?

15. Discuss the relative merits and disadvantages, including pollution and safety, of power generation by fossil fuels, nuclear fission, and nuclear fusion.

16. What is the reason for the "secondary system" in a nuclear reactor, Fig. 43–9? That is, why is the water heated by the fuel in a nuclear reactor not used directly to drive the turbines?

17. Discuss how the course of history might have been changed if, during World War II, scientists had refused to work on developing a nuclear bomb. Do you think it would have been possible to delay the building of a bomb indefinitely?

18. Research in molecular biology is moving toward the ability to perform genetic manipulations on human beings. The moral implications of future discoveries along these lines has led to a warning that this may be the molecular biologists' "Hiroshima." Discuss.

19. Does $E = mc^2$ apply in (a) fission, (b) fusion, (c) nuclear reactions?

20. Light energy emitted by the Sun and stars comes from the fusion process. What conditions in the interior of stars makes this possible?

21. How do stars, and our Sun, maintain confinement of the plasma for fusion?

22. What is the basic difference between fission and fusion?

23. People who work around metals that emit alpha particles are trained that there is little danger from proximity or even touching the material, but that they must take extreme precautions against ingesting it. Hence, there are strong rules against eating and drinking while working, and against machining the metal. Why?

24. Why is the recommended maximum radiation dose higher for women beyond the child-bearing age than for younger women?

25. A higher temperature is required for deuterium–deuterium ignition than for deuterium–tritium. Explain.

26. Radiation is sometimes used to sterilize medical supplies and even food. Explain how it works.

* 27. How might radioactive tracers be used to find a leak in a pipe?

Problems

Section 43–1

1. (I) Natural aluminum is all $^{27}_{13}\text{Al}$. If it absorbs a neutron, what does it become? Does it decay by β^+ or β^-? What will be the product nucleus?

2. (I) Determine whether the reaction $^2_1\text{H}(d,\,n)^3_2\text{He}$ requires a threshold energy. (d stands for deuterium, ^2_1H.)

3. (I) Is the reaction $^{238}_{92}\text{U}(n,\,\gamma)^{239}_{92}\text{U}$ possible with slow neutrons? Explain.

4. (II) Does the reaction $^7_3\text{Li}(p, \alpha)^4_2\text{He}$ require energy or does it release energy? How much energy?

5. (II) Calculate the energy released (or energy input required) for the reaction $^9_4\text{Be}(\alpha, n)^{12}_6\text{C}$.

6. (II) (a) Can the reaction $^{24}_{12}\text{Mg}(n, d)^{23}_{11}\text{Na}$ occur if the bombarding particles have 10.00 MeV of kinetic energy? (b) If so, how much energy is released?

7. (II) (a) Can the reaction $^7_3\text{Li}(p, \alpha)^4_2\text{He}$ occur if the incident proton has kinetic energy = 2500 keV? (b) If so, what is the total kinetic energy of the products?

8. (II) In the reaction $^{14}_7\text{N}(\alpha, p)^{17}_8\text{O}$, the incident α particles have 7.68 MeV of kinetic energy. (a) Can this reaction occur? (b) If so, what is the total kinetic energy of the products? The mass of $^{17}_8\text{O}$ is 16.999131 u.

9. (II) Calculate the Q-value for the "capture" reaction $^{16}_8\text{O}(\alpha, \gamma)^{20}_{10}\text{Ne}$.

10. (II) Calculate the total kinetic energy of the products of the reaction $^{13}_6\text{C}(d, n)^{14}_7\text{N}$ if the incoming deuteron (d) has $K = 36.3$ MeV.

11. (II) An example of a "stripping" nuclear reaction is $^6_3\text{Li}(d, p)\text{X}$. (a) What is X, the resulting nucleus? (b) Why is it called a "stripping" reaction? (c) What is the Q-value of this reaction? Is the reaction endothermic or exothermic?

12. (II) An example of a "pick-up" nuclear reaction is $^{12}_6\text{C}(^3_2\text{He}, \alpha)\text{X}$. (a) Why is it called a "pickup" reaction? (b) What is the resulting nucleus? (c) What is the Q-value of this reaction? Is the reaction endothermic or exothermic?

13. (II) (a) Complete the following nuclear reaction, $?(p, \gamma)^{32}_{16}\text{S}$. (b) What is the Q-value?

14. (III) Use conservation of energy and momentum to show that a bombarding proton must have an energy of 3.23 MeV in order to make the reaction $^{13}_6\text{C}(p, n)^{13}_7\text{N}$ occur. (See Example 43–3.)

Section 43–2

15. (I) What is the effective cross section for the collision of two hard spheres of radius R_1 and R_2.

16. (I) The cross section for the reaction $n + ^{10}_5\text{B} \rightarrow ^7_3\text{Li} + ^4_2\text{He}$ is about 40 bn for an incident neutron of low energy (kinetic energy ≈ 0). The boron is contained in a gas with $n = 1.7 \times 10^{21}$ nuclei/m^3 and the target has thickness $t = 12.0$ cm. What fraction of incident neutrons will be scattered?

17. (II) When the target is thick, the rate at which projectile particles collide with nuclei in the rear of the target is less than in the front of the target, since some scattering (i.e., collisions) takes place in the front layers. Let R_0 be the rate at which incident particles strike the front of the target, and R_x be the rate at a distance x into the target ($R_x = R_0$ at $x = 0$). Then show that the rate at which particles are scattered (and therefore lost from the incident beam) in a thickness dx is $-dR_x = R_x n\sigma dx$, where the minus sign means that R_x is decreasing and n is the number of nuclei per unit volume. Then show that $R_x = R_0 e^{-n\sigma x}$, where σ is the total cross section. If the thickness of the target is t, what does $R_x = R_0 e^{-n\sigma t}$ represent?

18. (II) A 1.0-cm-thick lead target reduces a beam of gamma rays to 30 percent of its original intensity. What thickness of lead will allow only one γ in 10^6 to penetrate (see Problem 17)?

19. (II) Use Fig. 43–3 to estimate what thickness of Cd $(\rho = 8650 \text{ kg/m}^3)$ will cause a 1 percent reaction rate $(R/R_0 = 0.01)$ for (a) 0.1-eV neutrons (b) 10-eV neutrons.

Section 43–3

20. (I) Calculate the energy released in the fission reaction $n + ^{235}_{92}\text{U} \rightarrow ^{88}_{38}\text{Sr} + ^{136}_{54}\text{Xe} + 12n$. Use Appendix D, and assume the initial kinetic energy of the neutron is very small.

21. (I) What is the energy released in the fission reaction of Eq. 43–5? (The masses of $^{141}_{56}\text{Ba}$ and $^{92}_{36}\text{Kr}$ are 140.91440 u and 91.92630 u, respectively.)

22. (I) How many fissions take place per second in a 200-MW reactor? Assume 200 MeV is released per fission.

23. (II) Suppose that the average electric power consumption, day and night, in a typical house is 300 W. What initial mass of $^{235}_{92}\text{U}$ would have to undergo fission to supply the electrical needs of such a house for a year? (Assume 200 MeV is released per fission, as well as 100% efficiency.)

24. (II) What initial mass of $^{235}_{92}\text{U}$ is required to operate a 500-MW reactor for 1 yr? Assume 40 percent efficiency.

25. (II) If a 1.0-MeV neutron emitted in a fission reaction loses one-half of its kinetic energy in each collision with moderator nuclei, how many collisions must it make to reach thermal energy $(\frac{3}{2}kT = 0.040 \text{ eV})$?

26. (II) Suppose that the neutron multiplication factor is 1.0004. If the average time between successive fissions in a chain of reactions is 1.0 ms, by what factor will the reaction rate increase in 1.0 s?

27. (II) Estimate the ratio of the height of the Coulomb barrier for α decay to that for fission of $^{236}_{92}\text{U}$. (Both are described by a potential energy diagram of the shape shown in Fig. 42–6.)

28. (II) Assuming a fission of $^{236}_{92}\text{U}$ into two roughly equal fragments, estimate the electric potential energy just as the fragments separate from each other. Assume that the fragments are spherical (see Eq. 42–1) and compare your calculation to the nuclear fission energy released, about 200 MeV.

Section 43–4

29. (I) What is the average kinetic energy of protons at the center of a star where the temperature is 10^7 K?

30. (II) Show that the energy released in the fusion reaction $^2_1\text{H} + ^3_1\text{H} \rightarrow ^4_2\text{He} + n$ is 17.59 MeV.

31. (II) Show that the energy released when two deuterium nuclei fuse to form ^3_2He with the release of a neutron is 3.27 MeV.

32. (II) Verify the Q-value stated for each of the reactions of Eqs. 43–6. [Hint: Be careful with electrons.]

33. (II) Calculate the energy release per gram of fuel for the reactions of Eqs. 43–8a, b, and c. Compare to the energy release per gram of uranium in fission.

34. (II) If a typical house requires 300 W of electric power on average, what minimum amount of deuterium fuel would have to be used in a year to supply these electrical needs? Assume the reaction of Eq. 43–8b.

35. (II) Show that the energies carried off by the ^4_2He nucleus and the neutron for the reaction of Eq. 43–8c are about 3.5 MeV and 14 MeV, respectively. Are these fixed values, independent of the plasma temperature?

36. (II) Suppose a fusion reactor ran on "d–d" reactions, Eqs. 43–8a and b. Estimate how much water, for fuel, would be needed per hour to run a 1000-MW reactor, assuming 30 percent efficiency.

37. (II) How much energy (J) is contained in 1.00 kg of water if its natural deuterium is used in the fusion reaction of Eq. 43–8a? Compare to the energy obtained from the burning of 1.0 kg of gasoline, about 5×10^7 J.

38. (III) The energy output of massive stars is believed to be due to the *carbon cycle* (see text). (a) Show that no carbon is consumed in this cycle and that the net effect is the same as for the proton–proton cycle. (b) What is the total energy release? (c) Determine the energy output for each reaction and decay. (d) Why does the carbon cycle require a higher temperature $(\approx 2 \times 10^7\,\text{K})$ than the proton–proton cycle $(\approx 1.5 \times 10^7\,\text{K})$?

39. (III) (a) Compare the energy needed for the first reaction of the carbon cycle to that for a deuteron–tritium reaction (Example 43–8). (b) If a deuteron–tritium reaction requires $T \approx 3 \times 10^8\,\text{K}$, estimate the temperature needed for the first carbon-cycle reaction.

40. (III) The deuterium–tritium pellet in a laser fusion device contains equal numbers of ^2_1H and ^3_1H atoms raised to a density of $200 \times 10^3\,\text{kg/m}^3$ by the laser pulses. Estimate (a) the density of particles in this compressed state, and (b) the length of time τ the pellet must be confined to meet the Lawson criterion for ignition.

Section 43–6

41. (I) A dose of 4.0 Sv of γ rays in a short period would be lethal to about half the people subjected to it. How many grays is this?

42. (I) Fifty rads of α-particle radiation is equivalent to how many rads of X-rays in terms of biological damage?

43. (I) How many rads of slow neutrons will do as much biological damage as 50 rads of fast neutrons?

44. (I) How much energy is deposited in the body of a 70-kg adult exposed to a 50-rad dose?

45. (II) A 0.025-μCi sample of $^{32}_{15}\text{P}$ is injected into an animal for tracer studies. If a Geiger counter intercepts 20 percent of the emitted β particles, what will be the counting rate, assumed 90 percent efficient.

46. (II) A 1.0-mCi source of $^{32}_{15}\text{P}$ (in NaHPO_4), a β^- emitter, is implanted in an organ where it is to administer 36 Gy. The half-life of $^{32}_{15}\text{P}$ is 14.3 days and 1 mCi delivers about 10 mGy/min. Approximately how long should the source remain implanted?

47. (II) About 35 eV is required to produce one ion pair in air. Show that this is consistent with the two definitions of the roentgen given in the text.

48. (II) $^{57}_{27}\text{Co}$ emits 122-keV γ rays. If a 70-kg person swallowed 1.85 μCi of $^{57}_{27}\text{Co}$, what would be the dose rate (Gy/day) averaged over the whole body? Assume that 50 percent of the γ-ray energy is deposited in the body.

49. (II) What is the mass of a 1.00-μCi $^{14}_6\text{C}$ source?

50. (II) Huge amounts of radioactive $^{131}_{53}\text{I}$ were released in the accident at Chernobyl in 1986. Chemically, iodine goes to the human thyroid. (Doctors can use it for diagnosis and treatment of thyroid problems.) In a normal thyroid, $^{131}_{53}\text{I}$ absorption can cause damage to the thyroid. (a) Write down the decay scheme for $^{131}_{53}\text{I}$. (b) Its half-life is 8.0 d; how long would it take for ingested $^{131}_{53}\text{I}$ to become 10 percent of the initial value? (c) Absorbing 1 mCi of $^{131}_{53}\text{I}$ can be harmful; what mass of iodine is this?

51. (II) Assume a liter of milk typically has an activity of 2000 pCi due to $^{40}_{19}\text{K}$. If a person drinks two glasses (0.5 L) per day, estimate the total dose (in Sv and in rem) received in a year. As a crude model, assume the milk stays in the stomach 12 hr and is then released. Assume also that very roughly 10 percent of the 1.5 MeV released per decay is absorbed by the body. Compare your result to the normal allowed dose per year. Make your estimate for (a) a 50-kg adult, and (b) a 5-kg baby.

52. (II) Radon gas, $^{222}_{86}\text{Rn}$, is considered a serious health hazard (see discussion in text). It decays by α-emission. (a) What is the daughter nucleus? (b) Is the daughter nucleus stable or radioactive? If the latter, how does it decay, and what is its lifetime? (c) Is the daughter nucleus also a noble gas, or is it chemically reacting? (d) Suppose 1.0 ng of $^{222}_{86}\text{Rn}$ seeps into a basement. What will be its activity? If the basement is then sealed, what will be the activity 1 month later?

*Section 43–9

* **53.** (II) (a) Suppose for a conventional X-ray image that the X-ray beam consisted of parallel rays. What would be the magnification of the image? (b) Suppose, instead, that the X-rays came from a point source (as in Fig. 43–20a) that is 15 cm in front of a human body 25 cm thick, and the film is pressed against the person's back. Determine and discuss the range of magnifications that result.

*Section 43–10

* **54.** (I) Calculate the wavelength of photons needed to produce NMR transitions in free protons in a 1.000-T field. In what region of the spectrum does it lie?

General Problems

55. J. Chadwick discovered the neutron by bombarding $^9_4$Be with the popular projectile of the day, alpha particles. (a) If one of the reaction products was the then unknown neutron, what was the other product? (b) What is the Q of this reaction?

56. Fusion temperatures are often given in keV. Determine the conversion factor from kelvins to keV using, as is common in this field, $\bar{K} = kT$ without the factor $\frac{3}{2}$.

57. One means of enriching uranium is by diffusion of the gas UF_6. Calculate the ratio of the speeds of molecules of this gas containing $^{235}_{92}$U and $^{238}_{92}$U, on which this process depends.

58. (a) What mass of $^{235}_{92}$U was actually fissioned in the first atomic bomb, whose energy was the equivalent of about 20 kilotons of TNT (1 kiloton of TNT releases 5×10^{12} J)? (b) What was the actual mass transformed to energy?

59. In a certain town the average yearly background radiation consists of 21 mrad of X-rays and γ rays plus 3.0 mrad of particles having a QF of 10. How many rem will a person receive per year on the average?

60. Deuterium makes up 0.015 percent of natural hydrogen. Make a rough estimate of the total deuterium in the Earth's oceans and estimate the total energy released if all of it were used in fusion reactors.

61. A shielded γ-ray source yields a dose rate of 0.052 rad/h at a distance of 1.0 m for an average-sized person. If workers are allowed a maximum dose rate of 5.0 rem/yr, how close to the source may they operate, assuming a 40-h work week? Assume that the intensity of radiation falls off as the square of the distance. (It actually falls off more rapidly than $1/r^2$ because of absorption in the air, so the answer above will give a better-than-permissible value.)

62. Radon gas, $^{222}_{86}$Rn, is formed by α decay. (a) Write the decay equation. (b) Ignoring the kinetic energy of the daughter nucleus (it's so massive), estimate the kinetic energy of the α particle produced. (c) Estimate the momentum of the alpha and of the daughter nucleus. (d) Estimate the kinetic energy of the daughter, and show that your approximation in (b) was valid.

63. The reaction $^{18}_{8}$O(p, n)$^{18}_{9}$F requires an input of energy equal to 2.453 MeV. What is the mass of $^{18}_{9}$F?

64. Consider a system of nuclear power plants that produce 3400 MW. (a) What total mass of $^{235}_{92}$U fuel would be required to operate these plants for 1 yr, assuming that 200 MeV is released per fission? (b) Typically 6 percent of the $^{235}_{92}$U nuclei that fission produce $^{90}_{38}$Sr, a β^- emitter with a half-life of 29 yr. What is the total radioactivity of the $^{90}_{38}$Sr, in curies, produced in 1 yr? (Neglect the fact that some of it decays during the 1-yr period.)

65. In the net reaction, Eq. 43–7, for the proton–proton cycle in the Sun, the neutrinos escape from the Sun with energy of about 0.5 MeV. The remaining energy, 26.2 MeV, is available within the Sun. Use this value to calculate the "heat of combustion" per kilogram of hydrogen fuel and compare it to the heat of combustion of coal, about 3×10^7 J/kg.

66. Energy reaches the Earth from the Sun at a rate of about 1400 W/m². Calculate (a) the total energy output of the Sun, and (b) the number of protons consumed per second in the reaction of Eq. 43–7, assuming that this is the source of all the Sun's energy. (c) Assuming that the Sun's mass of 2.0×10^{30} kg was originally all protons and that all could be involved in nuclear reactions in the Sun's core, how long would you expect the Sun to "glow" at its present rate?

67. Estimate how many solar neutrinos pass through a 100 m² ceiling of a room, at latitude 40°, in a year. [Hint: See Problems 65 and 66.]

68. Show, using the laws of conservation of energy and momentum, that for a nuclear reaction requiring energy, the minimum kinetic energy of the bombarding particle (the *threshold energy*) is equal to $[-Qm_{pr}/(m_{pr} - m_b)]$, where $-Q$ is the energy required (difference in total mass between products and reactants), m_b is the rest mass of the bombarding particle, and m_{pr} the total rest mass of the products. Assume the target nucleus is at rest before an interaction takes place, and that all particles are nonrelativistic.

69. The early scattering experiments performed around 1910 in Ernest Rutherford's laboratory in England produced the first evidence that an atom consists of a heavy nucleus surrounded by electrons. In one of these experiments, α particles struck a gold-foil target 4.0×10^{-5} cm thick in which there were 5.9×10^{28} gold atoms per cubic meter. Although most α particles either passed straight through the foil or were scattered at small angles, approximately 1.6×10^{-3} percent were scattered at angles greater than 90°—that is, in the backward direction. (a) Calculate the cross section, in barns, for backward scattering. (b) Rutherford concluded that such backward scattering could occur only if an atom consisted of a very tiny, massive, and positively charged nucleus with electrons orbiting some distance away. Assuming that backward scattering occurs for nearly direct collisions (i.e., $\sigma \approx$ area of nucleus), estimate the diameter of a gold nucleus.

70. Some stars, in a later stage of evolution, may begin to fuse two $^{12}_{6}$C nuclei into one $^{24}_{12}$Mg nucleus. (a) How much energy would be released in such a reaction? (b) What kinetic energy must two carbon nuclei each have when far apart, if they can then approach each other to within 6.0 fm, center-to-center? (c) Approximately what temperature would this require?

71. An average adult body contains about 0.10 μCi of $^{40}_{19}$K, which comes from food. (a) How many decays occur per second? (b) The potassium decays produce beta particles with energies of around 1.4 MeV. Calculate the dose per year in sieverts for a 50-kg adult. Is this a significant fraction of the 3.6 mSv/year background rate?

72. When the nuclear reactor accident occurred at Chernobyl in 1986, 2.0×10^7 Ci were released into the atmosphere. Assuming that this radiation was distributed uniformly over the surface of the Earth, what was the activity per square meter? (The actual activity was not uniform; even within Europe wet areas received more radioactivity from rainfall).

This computer-generated reconstruction of a proton–antiproton collision at Fermilab (Fig. 44–4) occurred at a combined energy of nearly 2 TeV. It is one of the events that provided evidence for the top quark, announced in 1995. The particle tracks are curved, due to a magnetic field, and the radius of curvature is a measure of each particle's momentum (Chapter 27). The top quark (t) has too brief a lifetime $(\approx 10^{-23}\,\text{s})$ to be detected itself, so we look for its possible decay products. Analysis indicates the following interaction and subsequent decays:

The tracks in the photo include jets (groups of particles moving in roughly the same direction), and a muon (μ^-) whose track is the pink one enclosed by a yellow rectangle to make it stand out. After reading this chapter, try to give the name for each symbol above and comment on whether all conservation laws hold.

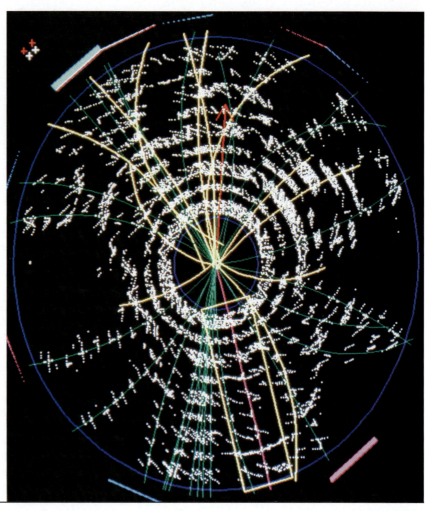

CHAPTER 44

Elementary Particles

I n the final two chapters of this book we discuss two of the most exciting areas of contemporary physics: (1) elementary particles, and (2) cosmology and astrophysics. These are subjects at the forefront of knowledge—one treats the smallest objects in the universe, the other the largest (and oldest) aspects of the universe. The reader who wants an understanding of the great beauties of present-day science—and/or wants to be a good citizen—will want to read these chapters, even if there is not time to cover them in a physics course.

In this penultimate chapter we discuss *elementary particle* physics, which represents the human endeavor to understand the basic building blocks of all matter. In the years after World War II, it was found that if the incoming particle in a nuclear reaction has sufficient energy, new types of particles can be produced. The earliest experiments used **cosmic rays**—particles that impinge on the Earth from space. But in order to produce high-energy particles in the laboratory, various types of particle accelerators have been constructed. Most commonly they accelerate protons or electrons, although heavy ions can also be accelerated, depending on the design. These high-energy accelerators have been used to probe the nucleus more deeply, to produce and study new particles, and to give us information about the basic forces and constituents of nature. Because the projectile particles are at high-energy, this field is sometimes called **high-energy physics**.

44–1 | High-Energy Particles

Particles accelerated to high energy are projectiles which can probe the interior of nuclei and nucleons that they strike. An important factor is that faster-moving projectiles can reveal more detail. The wavelength of projectile particles is given by de Broglie's wavelength formula (Eq. 38–7),

$$\lambda = \frac{h}{p}, \qquad\qquad (44\text{–}1)$$

de Broglie wavelength

showing that the greater the momentum p, of the bombarding particle, the shorter its wavelength. As discussed in Chapter 36 on diffraction, resolution of details in images is limited by the wavelength: the shorter the wavelength, the finer the detail that can be obtained. This is one reason why particle accelerators of higher and higher energy have been built in recent years.

EXAMPLE 44–1 **High resolution with electrons.** What is the wavelength, and hence the expected resolution, for a beam of 1.3-GeV electrons?

SOLUTION The $1.3\,\text{GeV} = 1300\,\text{MeV}$ refers to the kinetic energy, and is about 2500 times the rest energy (mass) of the electron $(mc^2 = 0.51\,\text{MeV})$. We are clearly dealing with relativistic speeds here, and it is easily shown (see Eqs. 37–10 and 37–13) that the speed of the electron is nearly $c = 3.0 \times 10^8\,\text{m/s}$. Therefore $K = (\gamma - 1)\,mc^2 \approx \gamma(mc^2) = 1300\,\text{MeV}$, and

$$\lambda = \frac{h}{p} = \frac{h}{\gamma m v} \approx \frac{h}{\gamma m c} = \frac{hc}{\gamma m c^2} \approx \frac{hc}{K}$$

$$= \frac{(6.6 \times 10^{-34}\,\text{J}\cdot\text{s})(3.0 \times 10^8\,\text{m/s})}{(1.3 \times 10^9\,\text{eV})(1.6 \times 10^{-19}\,\text{J/eV})} = 0.96 \times 10^{-15}\,\text{m},$$

or 0.96 fm. The maximum possible resolution of this beam of electrons is far greater than for a light beam in a light microscope $(\lambda \approx 500\,\text{nm})$. Indeed, this resolution of about 1 fm is on the order of the size of nuclei (see Eq. 42–1).

Another major reason for building high-energy accelerators is that new particles of greater mass can be produced at higher energies, as we will discuss shortly. Now we look briefly at several types of particle accelerator.

44–2 | Particle Accelerators and Detectors

Van de Graaff Accelerator

The key component of a Van de Graaff accelerator (Fig. 44–1) is a Van de Graaff generator, invented in 1931. A large, hollow, spherical conductor is charged to a high potential by a nonconducting moving belt in the following way: A high voltage of typically 50,000 V is applied to a pointed conductor A, which "sprays" positive charge onto the moving belt. (Actually, electrons are pulled off the belt onto the electrode A.) The belt carries the positive charge into the interior of the sphere, where it is "wiped off" the belt at B and races to the outer surface of the spherical conductor. (Remember that charge collects on the outer surface of any conductor, since the charges repel each other and try to get as far from each other as possible.) Very high potential differences can be obtained in this way, often giving the accelerated particles as much as 30 MeV of kinetic energy when generators are used in tandem. Connected to the Van de Graaff generator is an evacuated tube which serves as the particle accelerator. A source of H or He ions (p or α) is located inside the tube, and the large positive voltage repels them so they are accelerated toward the grounded target at the far end of the tube (Fig. 44–1).

FIGURE 44–1 Van de Graaff generator and accelerator.

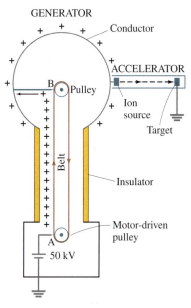

Cyclotron frequency

FIGURE 44–3 Diagram of a cyclotron. The magnetic field, applied by a large electromagnet, points into the page. A is the ion source. The field lines shown are for the electric field in the gap.

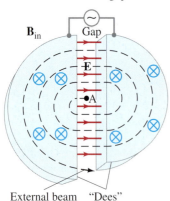

Cyclotron

The cyclotron, developed in 1930 by E. O. Lawrence (1901–1958; Fig. 44–2) at the University of California, Berkeley, uses a magnetic field to maintain charged ions—usually protons—in nearly circular paths (Chapter 27). The protons move within two D-shaped cavities, as shown in Fig. 44–3. Each time they pass into the gap between the "dees," a voltage accelerates them via the electric force, increasing their speed and increasing the radius of curvature of their path. After many revolutions, the protons acquire high kinetic energy and reach the outer edge of the cyclotron. They then either strike a target placed inside the cyclotron or leave the cyclotron with the help of a carefully placed "bending magnet" and are directed to an external target. Acceleration only occurs when the protons are in the gap *between* the dees, and the voltage must be alternating. When the protons are moving to the right across the gap in Fig. 44–3, the right dee must be electrically negative and the left one positive. A half-cycle later, the protons are moving to the left, so the left dee must be negative in order to accelerate them. The frequency, f, of the applied voltage must be equal to that of the circulating protons, and this is just the **cyclotron frequency** determined in Section 27–4, Eq. 27–6:

$$f = \frac{1}{T} = \frac{qB}{2\pi m}, \tag{44–2}$$

where q and m are the charge and mass of the particles moving in the magnetic field B.

EXAMPLE 44–2 **Cyclotron.** A small cyclotron of maximum radius $R = 0.25\,\text{m}$ accelerates protons in a 1.7-T magnetic field. Calculate (*a*) what frequency is needed for the applied alternating voltage, and (*b*) the kinetic energy of protons when they leave the cyclotron.

SOLUTION (*a*) From Eq. 44–2,

$$f = \frac{qB}{2\pi m}$$

$$= \frac{(1.6 \times 10^{-19}\,\text{C})(1.7\,\text{T})}{(6.28)(1.67 \times 10^{-27}\,\text{kg})} = 2.6 \times 10^7\,\text{Hz} = 26\,\text{MHz},$$

which is in the radio wave region of the EM spectrum (Fig. 32–12).
(*b*) The protons leave the cyclotron at $r = R = 0.25\,\text{m}$. From $F = ma$, we have (Eq. 27–5) $qvB = mv^2/r$, so $v = qBr/m$, and the kinetic energy is

$$K = \frac{1}{2}mv^2 = \frac{1}{2}m\frac{q^2B^2R^2}{m^2} = \frac{q^2B^2R^2}{2m}$$

$$= \frac{(1.6 \times 10^{-19}\,\text{C})^2(1.7\,\text{T})^2(0.25\,\text{m})^2}{(2)(1.67 \times 10^{-27}\,\text{kg})} = 1.4 \times 10^{-12}\,\text{J} = 8.7\,\text{MeV}.$$

Note that the magnitude of the voltage applied to the dees does not affect the final energy. But the higher this voltage, the fewer revolutions are required to bring the protons to full energy.

An important aspect of the cyclotron is that the frequency of the applied voltage, as given by Eq. 44–2, does not depend on the radius r. That is, the frequency does not have to be changed as the ions start from the source and are accelerated to paths of larger and larger radii. But this is only true at nonrelativistic energies. At higher speeds, the momentum is (Eq. 37–7) $p = mv/\sqrt{1 - v^2/c^2} = \gamma mv,$ so we must replace m in Eq. 44–2 with γm. One way to build a machine for higher speeds is to use a magnetic field that increases with radius so as to keep $B/\gamma m$ constant as the particles move to larger radius and momentum, thus allowing the frequency to be kept constant. Another approach is to keep B uniform while decreasing the frequency in time as a packet of ions increase in speed and mass, and reach larger orbits. Such a machine is called a **synchrocyclotron**.

Synchrocyclotron

Synchrotron

Another way to deal with the relativistic increase in effective mass (γm) with speed is to increase the magnetic field B in time as the particles speed up. Such devices are called **synchrotrons**, and today they can be enormous. The *Tevatron* accelerator at Fermilab (the Fermi National Accelerator Laboratory) at Batavia, Illinois, has a radius of 1.0 km, and that at CERN (European Center for Nuclear Research) in Geneva, Switzerland, is 4.3 km in radius. The Tevatron uses superconducting magnets to accelerate protons to about $1000 \text{ GeV} = 1 \text{ TeV}$ (hence its name; $1 \text{ TeV} = 10^{12} \text{ eV}$). These large synchrotrons use a narrow ring of magnets (see Fig. 44–4) with each magnet placed at the same radius from the center of the circle. The magnets are interrupted by gaps where high voltage accelerates the particles. Thus, once the particles are injected, they must move in a circle of constant radius. This is accomplished by giving them considerable energy initially in a much smaller accelerator, and then slowly increasing the magnetic field as they speed up in the large synchrotron.

Synchrotron

One problem of any accelerator is that accelerating electric charges radiate electromagnetic energy (see Chapter 32). Since ions or electrons are accelerated in an accelerator, we can expect considerable energy to be lost by radiation. The effect increases with speed and is especially important in circular machines where centripetal acceleration is present, particularly in synchrotrons, and hence is called **synchrotron radiation**. Synchrotron radiation can actually be useful, however. Intense beams of photons are sometimes needed, and they are usually obtained from an electron synchrotron.

FIGURE 44–4 (a) Aerial view of Fermilab at Batavia, Illinois; the accelerator is a circular ring 1.0 km in radius. (b) The interior of the tunnel of the main accelerator at Fermilab. The upper (rectangular-shaped) ring of magnets is for the older 500-GeV accelerator. Below it is the ring of superconducting magnets for the 1-TeV Tevatron.

(a)

(b)

Linear Accelerators

Linac

A Van de Graaff accelerator is essentially a linear accelerator since the ions move in a linear path. But the name *linear accelerator* is usually reserved for a more complex arrangement in which particles are accelerated many, many times along a straight-line path. Figure 44–5a is a diagram of a simple "linac." The ions pass through a series of tubular conductors. The voltage applied to the tubes must be alternating so that when positive ions (say) reach a gap, the tube in front of them is negative and the one they just left is positive. This assures that they are accelerated at each gap. As the ions increase in speed, they cover more distance in the same amount of time. Consequently, the tubes must be longer the farther they are from the source. Linear accelerators are of particular importance for accelerating electrons. Because of their small mass, electrons reach high speeds very quickly; an electron linac such as the one shown in Fig. 44–5a would have tubes nearly equal in length, since the electrons would be traveling close to $c = 3.0 \times 10^8 \, \text{m/s}$ for almost the entire distance. The amount of energy radiated away by electrons in a linear machine is much less than for a circular machine (see mention of synchrotron radiation above). The largest electron linear accelerator is that at Stanford (Stanford Linear Accelerator Center, or SLAC), Fig. 44–5b. It is about 3 km (2 miles) long and can accelerate electrons to 50 GeV. Many hospitals have 10-MeV electron linacs that produce photons to irradiate tumors.

FIGURE 44–5 (a) Diagram of a simple linear accelerator. (b) Photo of the Stanford Linear Accelerator (SLAC) in California.

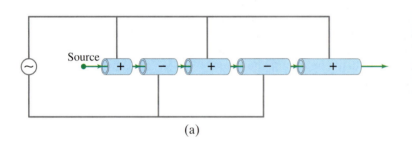

(a)

(b)

Colliding Beams

Colliders

Originally, high-energy physics experiments were carried out by allowing a beam of particles from an accelerator to strike a stationary target. A newer, very powerful technique allows us to obtain the maximum possible collision energy from a given accelerator: two beams of particles are accelerated to very high energy and are steered so that they collide head-on. One way to accomplish such **colliding beams** with a single accelerator is through the use of **storage rings**, in which the particles of one beam can continue to circulate while the second beam is being accelerated. The two beams are then steered so that they intersect and collide head-on. For example, in the experiments that provided strong evidence for the top quark (see chapter opening photo and Section 44–9), the Fermilab Tevatron accelerated protons and antiprotons each to 900 GeV, so that the combined energy of head-on collisions was 1.8 TeV. The largest collider accelerator today is the Large Electron-Positron (LEP) Collider at CERN, which has a circumference of 26.7 km (Fig. 44–6). It produces oppositely revolving beams of e^+ and e^- each of energy 95 GeV, for a total interaction energy of 190 GeV. The most powerful accelerator-collider will soon be the Large Hadron Collider (LHC) at CERN, scheduled to be completed about 2005. The two colliding beams will each carry 7 TeV protons for a total interaction energy of 14 TeV.

FIGURE 44–6 The large circle represents the position of the tunnel now containing the 8.5 km diameter LEP collider, about 100 m below the ground, at CERN (near Geneva) on the French-Swiss border. The LHC will use the same tunnel. The smaller circle shows the position of the Super Proton Synchrotron that will be used for accelerating protons in the LHC.

EXAMPLE 44–3 Protons at relativistic speeds. Determine the energy required to accelerate a proton in a high-energy accelerator (*a*) from rest to $v = 0.900c$, and (*b*) from $v = 0.900c$ to $v = 0.999c$. (*c*) What is the kinetic energy achieved by each proton?

SOLUTION We use the work-energy principle, which is still valid relativistically as we saw in Section 37–11: $W = \Delta K$. The kinetic energy of a proton is given by Eq. 37–10

$$K = (\gamma - 1)mc^2,$$

where the relativistic factor γ is

$$\gamma = \frac{1}{\sqrt{1 - v^2/c^2}}.$$

The work-energy theorem becomes

$$W = \Delta K = (\Delta\gamma)mc^2$$

since m (the rest mass) and c are constant.
(*a*) For $v = 0$, $\gamma = 1$; and for $v = 0.900c$

$$\gamma = \frac{1}{\sqrt{1 - (0.900)^2}} = 2.29.$$

Hence, with rest mass of the proton $mc^2 = 938 \text{ MeV}$, the work needed is

$$W = \Delta K = (\Delta\gamma)mc^2 = (2.29 - 1.00)(938 \text{ MeV}) = 1.21 \text{ GeV}.$$

(*b*) For $v = 0.999c$,

$$\gamma = \frac{1}{\sqrt{1 - (0.999)^2}} = 22.4.$$

So the work needed to accelerate a proton from $0.900c$ to $0.999c$ is

$$W = \Delta K = (\Delta\gamma)mc^2 = (22.4 - 2.29)(938 \text{ MeV}) = 18.9 \text{ GeV}.$$

(*c*) The kinetic energy reached by the proton in (*a*) is just equal to the work done on it, $K = 1.21 \text{ GeV}$. The final kinetic energy of the proton in (*b*), moving at $v = 0.999c$, is

$$K = (\gamma - 1)mc^2 = (21.4)(938 \text{ MeV}) = 20.1 \text{ GeV},$$

which makes sense since, starting from rest, we did work $W = 1.21 \text{ GeV} + 18.9 \text{ GeV} = 20.1 \text{ GeV}$ on it.

FIGURE 44–7 (a) In a cloud or bubble chamber, droplets or bubbles are formed around ions produced by the passage of a charged particle. (b) Bubble-chamber photo of particle tracks, here revealing evidence for the Ω^- particle. Note the curvature of the tracks in the magnetic field.

Path of particle

(a)

(b)

FIGURE 44–8 Photo of the large and complicated CDF detector at Fermilab which, utilizing wire drift chambers, detected the particles and produced the image of the tracks shown at the start of this chapter.

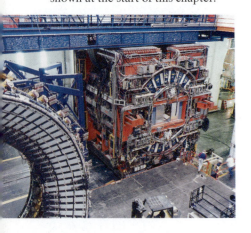

Particle Detectors

Detectors of particles such as scintillators, photomultipliers, and semiconductor detectors, which we discussed in Section 42–11, are also used in the study of nuclear reactions and in elementary particle physics. More sophisticated devices have been developed that allow the track of a charged particle to be *seen*. The simplest is the **photographic emulsion** (which, being small and simple and therefore portable, is now used mainly for cosmic-ray studies from balloons). A particle passing through a layer of photographic emulsion ionizes the atoms along its path. This results in a chemical change at these points, and when the emulsion is developed, the particle's path is revealed.

In a **cloud chamber**, a gas is cooled to a temperature slightly below its usual condensation point. (It is said to be "supercooled.") The gas molecules begin to condense on any ionized molecules present. Thus the ions produced when a charged particle passes through serve as centers on which tiny droplets form (Fig. 44–7a). Light scatters more from these droplets than from the gas background, so a photo of the cloud chamber at the right moment shows the track of the particle. An important instrument in the early days of nuclear physics, it is little used today.

The **bubble chamber**, invented in 1952 by D. A. Glaser (1926–), makes use of a superheated liquid. The liquid is kept close to its normal boiling point and the bubbles characteristic of boiling form around ions produced by the passage of a charged particle (Fig. 44–7b). A photograph of the interior of the chamber thus reveals the paths of particles that recently passed through. Because the bubble chamber uses a liquid—often liquid hydrogen—the density of atoms is much greater than in a cloud chamber. Hence it is a much more efficient device for observing the tracks of charged particles and their interactions with the nuclei of the liquid. Usually, a magnetic field is applied across the chamber and the momentum of the moving particles can be determined from the radius of curvature of their paths.

Much more used today is the **wire drift chamber**, which consists of a set of closely spaced fine wires immersed in a gas. Many wires are grounded, and others in between are kept at very high voltage. A charged particle passing through produces ions in the gas. The freed electrons drift toward the nearest high voltage wires, creating an "avalanche," and producing an electric pulse or signal at that wire. The positions of the particles can be determined both by the position of the wire and by the time it takes the pulses to reach detectors at the ends of the wires. The paths of the particles are reconstructed electronically with computers which can then "draw" a picture of the tracks, as shown in the photo at the start of the chapter: the white dots are the wires, and the colored lines are the particle tracks. The device that produced that photo is shown in Fig. 44–8. (Another example is shown in Fig. 44–12).

44–3 Beginnings of Elementary Particle Physics— Particle Exchange

By the mid-1930s, it was recognized that all atoms can be considered to be made up of neutrons, protons, and electrons. The basic constituents of the universe were no longer considered to be atoms but rather the proton, neutron, and electron. Besides these three *elementary particles*, several others were also known: the positron (a positive electron), the neutrino, and the γ particle (or photon), for a total of six elementary particles.

In the decades that followed, hundreds of other subnuclear particles were discovered. The properties and interactions of these particles, and which ones should be considered as fundamental or "elementary," became the substance of research in **elementary particle physics**.

Today, the standard model for elementary particles views even more fundamental entities, *quarks* and *leptons*, as the basic constituents of matter. In order to understand the standard model, we need to begin with the ideas leading up to its formulation.[†]

Elementary particle physics might be said to have begun in 1935 when the Japanese physicist Hideki Yukawa (1907–1981) predicted the existence of a new particle that would in some way mediate the strong nuclear force. To understand Yukawa's idea, we first consider the electromagnetic force. When we first discussed electricity, we saw that the electric force acts over a distance, without contact. To better perceive how a force can act over a distance, we saw that Faraday introduced the idea of a **field**. The force that one charged particle exerts on a second can be said to be due to the electric field set up by the first. Similarly, the magnetic field can be said to carry the magnetic force. Later (Chapter 32), we saw that electromagnetic (EM) fields can travel through space as waves. Finally, in Chapter 38, we saw that electromagnetic radiation (light) can be considered as either a wave or as a collection of particles called *photons*. Because of this wave–particle duality, it is possible to imagine that the electromagnetic force between charged particles is due to

(1) the EM field set up by one charged particle and felt by the other, or

(2) an exchange of photons or γ particles between them.

It is (2) that we want to concentrate on here, and a crude analogy for how an exchange of particles could give rise to a force is suggested in Fig. 44–9. In part (a), two children start throwing heavy pillows at each other; each catch results in the child being pushed backward by the impulse. This is the equivalent of a repulsive force. On the other hand, if the two children exchange pillows by grabbing them out of the other person's hand, they will be pulled toward each other, as when an attractive force acts.

For the electromagnetic force, it is photons that are exchanged between two charged particles that give rise to the force between them. A simple diagram describing this photon exchange is shown in Fig. 44–10. Such a diagram, called a **Feynman diagram** (after its inventor, the American physicist, Richard Feynman (1918–1988)), is based on the theory of **quantum electrodynamics** (QED).

Figure 44–10 represents the simplest case, in which a single photon is exchanged. One of the charged particles emits the photon and recoils somewhat as a result; and the second particle absorbs the photon. In any collision or *interaction*, energy and momentum are transferred from one charged particle to the other, carried by the photon. Because the photon is absorbed by the second particle very shortly after it is emitted by the first, it is not observable, and is referred to as a *virtual* photon, in contrast to one that is free and can be detected by instruments. The photon is said to *mediate*, or *carry*, the electromagnetic force.

(a) Repulsive force (children throwing pillows)

(b) Attractive force (children grabbing pillows from each other's hands)

FIGURE 44–9 Forces equivalent to particle exchange. (a) Repulsive force (children throwing pillows at each other). (b) Attractive force (children grabbing pillows from each other's hands).

FIGURE 44–10 Feynman diagram showing a photon acting as the carrier of the electromagnetic force between two electrons. This is sort of an *x* vs. *t* graph, with *t* increasing upward. Starting at the bottom, two electrons approach each other (the distance between them decreases in time). As they get close, momentum and energy get transferred from one to the other, carried by a photon (or, perhaps, by more than one), and the two electrons bounce apart.

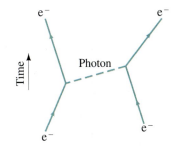

Particles that mediate or "carry" forces

[†] Just telling you how it is today would not be a scientific discussion but one of dogma—see footnote on p. 962.

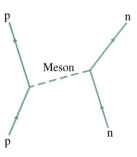

FIGURE 44–11 Meson exchange when a proton and neutron interact via the strong nuclear force.

By analogy with photon exchange that mediates the electromagnetic force, Yukawa argued in this early theory that there ought to be a particle that mediates the strong nuclear force—the force that holds nucleons together in the nucleus.

Mass estimate of exchange particle

Yukawa called this predicted particle a **meson** (meaning "of medium mass"), and Fig. 44–11 is a Feynman diagram of meson exchange. We can make a rough estimate of the mass of the meson as follows. Suppose the proton on the left in Fig. 44–11 is at rest. For it to emit a meson would require energy (to make the meson's mass) that would have to come from nowhere, which would violate conservation of energy. But the uncertainty principle allows nonconservation of energy by an amount ΔE if it occurs only for a time Δt given by $(\Delta E)(\Delta t) \approx h/2\pi$. We set ΔE equal to the energy needed to create the mass m of the meson: $\Delta E = mc^2$. Conservation of energy is violated only as long as the meson exists, which is the time Δt required for the meson to pass from one nucleon to the other. If we assume the meson travels at relativistic speed, close to the speed of light c, then Δt need be at most about $\Delta t = d/c$, where d is the maximum distance that can separate the interacting nucleons. Thus we can write

$$\Delta E \, \Delta t \approx \frac{h}{2\pi}$$

$$mc^2\left(\frac{d}{c}\right) \approx \frac{h}{2\pi}$$

or

Mass of exchange particle

$$mc^2 \approx \frac{hc}{2\pi d}. \qquad (44\text{–}3)$$

The range of the strong nuclear force (the maximum distance away it can be felt), is small—not much more than the size of a nucleon or small nucleus (see Eq. 42–1)—so let us take $d \approx 1.5 \times 10^{-15}$ m. Then from Eq. 44–3,

$$mc^2 \approx \frac{hc}{2\pi d} = \frac{(6.6 \times 10^{-34}\,\text{J}\cdot\text{s})(3.0 \times 10^8\,\text{m/s})}{(6.28)(1.5 \times 10^{-15}\,\text{m})}$$

$$\approx 2.1 \times 10^{-11}\,\text{J} = 130\,\text{MeV}.$$

The mass of the predicted meson is roughly $130\,\text{MeV}/c^2$ or about 250 times the electron mass of $0.51\,\text{MeV}/c^2$. [Note, incidentally, that since the electromagnetic force has infinite range ($d = \infty$), Eq. 44–3 tells us that the exchanged particle for the electromagnetic force, the photon, will have zero rest mass.]

Pion

The particle predicted by Yukawa was discovered in cosmic rays by C. F. Powell and G. Occhialini in 1947. It is called the "π" or pi meson, or simply the **pion**. It comes in three charge states: $+$, $-$, or 0. The π^+ and π^- have mass of $139.6\,\text{MeV}/c^2$ and the π^0 a mass of $135.0\,\text{MeV}/c^2$. All three interact strongly with matter. Reactions observed in the laboratory, using a particle accelerator, include

$$p + p \rightarrow p + p + \pi^0,$$
$$p + p \rightarrow p + n + \pi^+. \qquad (44\text{–}4)$$

The incident proton from the accelerator must have sufficient energy to produce the additional mass of the free pion.

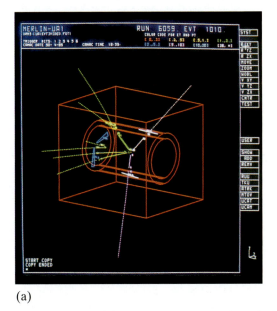

(a)

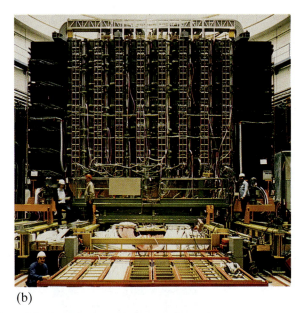

(b)

FIGURE 44–12 (a) Computer reconstruction of a Z-particle decay into an electron and a positron $(Z^0 \rightarrow e^+ + e^-)$ whose tracks are shown in white, which took place in the UA1 detector at CERN. (b) Photo of the UA1 detector at CERN as it was being built.

Yukawa's theory of pion exchange as carrier of the strong force is now out of date, and has been replaced by *quantum chromodynamics* in which the basic entities are *quarks*, and between them are exchanged *gluons* as the basic carriers of the strong force, as we shall discuss shortly. But the basic idea of the earlier theory, that forces can be understood as the exchange of particles, remains valid, as does Eq. 44–3 for estimating the mass of exchange particles.

There are four known types of force—or interaction—in nature. What about the other two: the weak nuclear force, and gravity? Theorists believe that these are also mediated by particles. The particles presumed to transmit the weak force are referred to as the W^+, W^-, and Z^0, and were only detected in 1983 (see Fig. 44–12). The quantum (or carrier) of the gravitational force is called the **graviton**, and if it exists it has not yet been observed. A comparison of the four forces is given in Table 44–1, where they are listed according to their (approximate) relative strengths. Notice that although gravity may be the most obvious force in daily life (because of the huge mass of the Earth), on a nuclear scale, it is much the weakest of the four forces and its effect at the nuclear or atomic level can nearly always be ignored.

Graviton

TABLE 44–1   The Four Forces in Nature

Type	Relative Strength (approx., for 2 protons in nucleus)	Field Particle
Strong nuclear	1	Gluons[†] (mesons)
Electromagnetic	10^{-2}	Photon
Weak nuclear	10^{-6}	$W^{\pm}$ and Z^0
Gravitational	10^{-38}	Graviton (?)

[†]Until the 1970s, thought to be mesons, but now gluons (see Section 44–10).

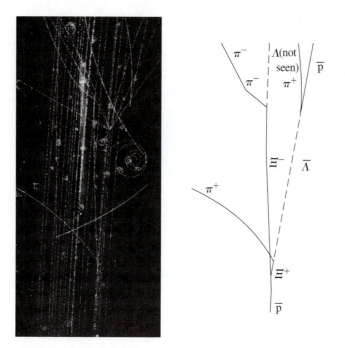

FIGURE 44–13 Liquid-hydrogen bubble-chamber photograph of an antiproton ($\bar{p}$) colliding with a proton, producing a Xi–anti-Xi pair ($\bar{p} + p \rightarrow \Xi^- + \bar{\Xi}^+$) that subsequently decay into other particles. The drawing indicates the assignment of particles to each track, which is based on how or if that particle decays, and on mass values estimated from measurement of momentum (curvature of track in magnetic field) and energy (heaviness of track, for example). Neutral particle paths are shown by dashed lines since neutral particles produce no bubbles and hence no tracks.

44–4 | Particles and Antiparticles

The positron, as we have seen, is basically a positive electron. That is, many of its properties are the same as for the electron, such as mass, but it has the opposite charge. The positron is said to be the **antiparticle** to the electron. After the positron was discovered in 1932, it was predicted that other particles also ought to have antiparticles. In 1955 the antiparticle to the proton was found, the **antiproton** ($\bar{p}$), which carries a negative charge; see Fig. 44–13. (The bar over the p is used to indicate antiparticle.) Soon after, the antineutron ($\bar{n}$) was found. All other particles also have antiparticles. But a few particles, like the photon and the π^0, do not have distinct antiparticles—we say that they are their own antiparticles.

Antiparticles are produced in nuclear reactions when there is sufficient energy available, and they do not live very long in the presence of matter. For example, a positron is stable when by itself, but if it encounters an electron, the two annihilate each other. The energy of their vanished mass, plus any kinetic energy they possessed, is converted to the energy of γ rays or of other particles. Annihilation also occurs for all other particle–antiparticle pairs.

44–5 | Particle Interactions and Conservation Laws

One of the important uses of high-energy accelerators is to study the interactions of elementary particles with each other. As a means of ordering this subnuclear world, the conservation laws are indispensable. The laws of conservation of energy, of momentum, of angular momentum, and of electric charge are found to hold precisely in all particle interactions.

A study of particle interactions has revealed a number of new conservation laws which (just like the old ones) are ordering principles: they help to explain why some reactions occur and others do not. For example, the following reaction has never been found to occur:

$$p + n \not\rightarrow p + p + \bar{p}$$

even though charge, energy, and so on, are conserved ($\bar{p}$ means an antiproton and $\nrightarrow$ means the reaction does not occur). To understand why such a reaction does not occur, physicists hypothesized a new conservation law, the conservation of **baryon number**. (Baryon number is a generalization of nucleon number, which we saw earlier is conserved in nuclear reactions and decays.) An important addition to this law is the proposal that whereas all nucleons have baryon number $B = +1$, all antinucleons (antiprotons, antineutrons) have $B = -1$. The reaction above does not conserve baryon number since on the left side we have $B = (+1) + (+1) = +2$, and on the right $B = (+1) + (+1) + (-1) = +1$. On the other hand, the following reaction does conserve B and *does* occur if the incoming proton has sufficient energy:

$$p + p \rightarrow p + p + \bar{p} + p,$$
$$B = +1 + 1 = +1 + 1 - 1 + 1.$$

As indicated, $B = +2$ on both sides of this equation. From these and other reactions, the conservation of baryon number has been established as a basic law of physics.

Also useful are the conservation laws for the three **lepton numbers**, associated with weak interactions including decays. In ordinary β decay, an electron or positron is emitted along with a neutrino or antineutrino. In a similar type of decay, a particle known as a **muon** can be emitted instead of an electron. A muon seems to be much like an electron, except its mass is 207 times as large ($106 \text{ MeV}/c^2$). The neutrino (ν_e) that accompanies an emitted electron is found to be different from the neutrino (ν_μ) that accompanies an emitted muon. Each of these neutrinos has an antiparticle: $\bar{\nu}_e$ and $\bar{\nu}_\mu$. In ordinary β decay we have, for example,

$$n \rightarrow p + e^- + \bar{\nu}_e$$

but never $n \rightarrow p + e^- + \bar{\nu}_\mu$ nor $n \rightarrow p + e^- + \bar{\nu}_e + \nu_e$. To explain why these do not occur, the concept of electron lepton number, L_e, was invented. If the electron (e^-) and the electron neutrino (ν_e) are assigned $L_e = +1$, and e^+ and $\bar{\nu}_e$ are assigned $L_e = -1$, whereas all other particles have $L_e = 0$, then all observed decays conserve L_e. For example, in $n \rightarrow p + e^- + \bar{\nu}_e$, $L_e = 0$ initially, and $L_e = 0 + (+1) + (-1) = 0$ after the decay. Decays that do not conserve L_e but would obey the other conservation laws are not observed to occur. Hence it is believed that L_e is conserved in all[†] interactions.

In a decay involving muons, such as

$$\pi^+ \rightarrow \mu^+ + \nu_\mu,$$

a second quantum number, muon lepton number (L_μ), is conserved. The μ^- and ν_μ are assigned $L_\mu = +1$, and μ^+ and $\bar{\nu}_\mu$ have $L_\mu = -1$, whereas other particles have $L_\mu = 0$. It is believed that L_μ is also conserved in all interactions or decays. Similar assignments can be made for a third lepton number, L_τ, associated with the more recently discovered τ lepton and its neutrino, ν_τ.

Keep in mind that antiparticles have not only opposite electric charge from their particles, but also opposite B, L_e, L_μ, and L_τ.

Baryon number

Lepton numbers

Antiparticles have opposite Q, B, L

| **CONCEPTUAL EXAMPLE 44–4** | **Lepton number in muon decay.** Which of the following decay schemes is possible for muon decay: (a) $\mu^- \rightarrow e^- + \bar{\nu}_e$; (b) $\mu^- \rightarrow e^- + \bar{\nu}_e + \nu_\mu$; (c) $\mu^- \rightarrow e^- + \nu_e$? All of these particles have $L_\tau = 0$.

RESPONSE A μ^- has $L_\mu = +1$ and $L_e = 0$. This is the initial state, and the final state (after decay) must also have $L_\mu = +1$, $L_e = 0$. In (a), the final state has $L_\mu = 0 + 0 = 0$, and $L_e = +1 - 1 = 0$; L_μ would not be conserved and indeed this decay is not observed to occur. The final state of (b) has $L_\mu = 0 + 0 + 1 = +1$ and $L_e = +1 - 1 + 0 = 0$, so both L_μ and L_e are conserved. This is in fact the most common decay mode of the μ^-. Finally, (c) does not occur because L_e ($= +2$ in the final state) is not conserved, nor is L_μ.

[†] Recent evidence hints that under certain circumstances, lepton number may not always be conserved.

EXAMPLE 44–5 **Energy and momentum are conserved.** In addition to the "number" conservation laws which help explain the decay schemes of particles, we can also apply the laws of conservation of energy and momentum. The decay of a Σ^+ particle with a rest mass of 1189 MeV/c^2 (see Table 44–2 in the next Section) commonly yields a proton (rest mass of 938 MeV/c^2) and a neutral pion, π^0 (rest mass of 135 MeV/c^2). What are the kinetic energies of the decay products, assuming the Σ^+ parent particle was at rest?

SOLUTION The energy released through the decay can be calculated from the change in the rest energies, as we did for nuclear processes (Eq. 42–3 or 43–1):

$$Q = \left[m_{\Sigma^+} - \left(m_p + m_{\pi^0}\right)\right]c^2 = \left[1189 - (938 + 135)\right] MeV = 116 \, MeV.$$

This energy Q becomes the kinetic energy of the resulting decay particles, p and π^0:

$$Q = K_p + K_{\pi^0}$$

with each particle's kinetic energy related to its momentum by (Eqs. 37–11 and 13):

$$K_p = E_p - m_p c^2 = \sqrt{\left(p_p c\right)^2 + \left(m_p c^2\right)^2} - m_p c^2,$$

and similarly for the pion. From momentum conservation, the proton and pion have the same magnitude of momentum since the original particle was at rest: $p_p = p_{\pi^0} = p$. Then

$$Q = 116 \, MeV = \left[\sqrt{(pc)^2 + (938 \, MeV)^2} - 938 \, MeV\right]$$
$$+ \left[\sqrt{(pc)^2 + (135 \, MeV)^2} - 135 \, MeV\right].$$

We solved this for pc, which gives $pc = 189 \, MeV$. Substituting into the expression for the kinetic energy, first for the proton, then for the pion we obtain $K_p = 19 \, MeV$ and $K_{\pi^0} = 97 \, MeV$.

44–6 Particle Classification

In the decades following the discovery of the π meson in the late 1940s, a great many other subnuclear particles were discovered, now numbering in the hundreds. It is useful to arrange the particles in categories according to their properties. One way of doing this is according to their interactions, since not all particles interact by means of all four of the forces known in nature (though all interact via gravity). Table 44–2 lists some of the more common particles classified in this way along with many of their properties. The particles listed are those that are stable, and many that are unstable. At the top of the table are the **gauge bosons** (so-named[†] after the theory that describes them, "gauge theory"), which include the *photon*, and the W and Z particles, that mediate the electromagnetic and weak interactions, respectively.

Gauge bosons

Leptons

Next in Table 44–2 are the **leptons**, which are particles that do not interact via the strong force but do interact via the weak nuclear force (as well as the much weaker gravitational force); those that carry electric charge also interact via the electromagnetic force. The leptons include the electron, the muon, and the tau (or τ, discovered in 1976 and more than 3000 times heavier than the electron), and three types of neutrino: the electron neutrino (ν_e), the muon neutrino (ν_μ), and the tau neutrino (ν_τ). They each have antiparticles, as indicated in Table 44–2.

Hadrons

The third category of particle in Table 44–2 is the **hadron**. Hadrons are those particles that interact via the strong nuclear force. Hence they are said to be **strongly interacting particles**. They also interact via the other forces, but the strong

[†] Bosons are particles that are not governed by the Pauli exclusion principle (see Sections 40–4, 41–6, and 44–10).

TABLE 44–2 Particles (stable under strong decay)†

Category	Forces involved	Particle name	Symbol	Anti-particle	Spin	Rest Mass (MeV/c^2)	B	L_e	L_μ	L_τ	S	Lifetime (s)	Principal Decay Modes
Gauge bosons	em	Photon	γ	Self	1	0	0	0	0	0	0	Stable	
	w, em	W	W^+	W^-	1	80.41×10^3	0	0	0	0	0	3×10^{-25}	$e\nu_e, \mu\nu_\mu, \tau\nu_\tau$, hadrons
	w	Z	Z^0	Self	1	91.19×10^3	0	0	0	0	0	3×10^{-25}	$e^+e^-, \mu^+\mu^-, \tau^+\tau^-$, hadrons
Leptons	w, em‡	Electron	e^-	e^+	$\frac{1}{2}$	0.511	0	+1	0	0	0	Stable	
		Neutrino (e)	ν_e	$\bar{\nu}_e$	$\frac{1}{2}$	$0(<4 \times 10^{-6})$‡	0	+1	0	0	0	Stable	
		Muon	μ^-	μ^+	$\frac{1}{2}$	105.7	0	0	+1	0	0	2.20×10^{-6}	$e^-\bar{\nu}_e\nu_\mu$
		Neutrino (μ)	ν_μ	$\bar{\nu}_\mu$	$\frac{1}{2}$	$0(<0.17)$‡	0	0	+1	0	0	Stable	
		Tau	τ^-	τ^+	$\frac{1}{2}$	1777	0	0	0	+1	0	2.91×10^{-13}	$\mu^-\bar{\nu}_\mu\nu_\tau, e^-\bar{\nu}_e\nu_\tau$, hadrons $+ \nu_\tau$
		Neutrino (τ)	ν_τ	$\bar{\nu}_\tau$	$\frac{1}{2}$	$0(<18)$‡	0	0	0	+1	0	Stable	
Hadrons (selected)													
Mesons	s, em, w	Pion	π^+	π^-	0	139.6	0	0	0	0	0	2.60×10^{-8}	$\mu^+\nu_\mu$
			π^0	Self	0	135.0	0	0	0	0	0	0.84×10^{-16}	2γ
		Kaon	K^+	K^-	0	493.7	0	0	0	0	+1	1.24×10^{-8}	$\mu^+\nu_\mu, \pi^+\pi^0$
			K^0_S	$\overline{K}^0_S$	0	497.7	0	0	0	0	+1	0.89×10^{-10}	$\pi^+\pi^-, 2\pi^0$
			K^0_L	$\overline{K}^0_L$	0	497.7	0	0	0	0	+1	5.17×10^{-8}	$\pi^\pm e^{\mp}\overset{(-)}{\nu}_e, \pi^\pm \mu^\mp \overset{(-)}{\nu}_\mu, 3\pi$
		Eta	η^0	Self	0	547.3	0	0	0	0	0	5×10^{-19}	$2\gamma, 3\pi^0, \pi^+\pi^-\pi^0$
		and others											
Baryons	s, em, w	Proton	p	$\bar{\text{p}}$	$\frac{1}{2}$	938.3	+1	0	0	0	0	Stable	
		Neutron	n	$\bar{\text{n}}$	$\frac{1}{2}$	939.6	+1	0	0	0	0	887	$\text{p}e^-\bar{\nu}_e$
		Lambda	Λ^0	$\overline{\Lambda}^0$	$\frac{1}{2}$	1115.7	+1	0	0	0	−1	2.63×10^{-10}	$\text{p}\pi^-, \text{n}\pi^0$
		Sigma	Σ^+	$\overline{\Sigma}^-$	$\frac{1}{2}$	1189.4	+1	0	0	0	−1	0.80×10^{-10}	$\text{p}\pi^0, \text{n}\pi^+$
			Σ^0	$\overline{\Sigma}^0$	$\frac{1}{2}$	1192.6	+1	0	0	0	−1	7.4×10^{-20}	$\Lambda^0\gamma$
			Σ^-	$\overline{\Sigma}^+$	$\frac{1}{2}$	1197.4	+1	0	0	0	−1	1.48×10^{-10}	$\text{n}\pi^-$
		Xi	Ξ^0	$\overline{\Xi}^0$	$\frac{1}{2}$	1314.9	+1	0	0	0	−2	2.90×10^{-10}	$\Lambda^0\pi^0$
			Ξ^-	$\overline{\Xi}^+$	$\frac{1}{2}$	1321.3	+1	0	0	0	−2	1.64×10^{-10}	$\Lambda^0\pi^-$
		Omega	Ω^-	Ω^+	$\frac{1}{2}$	1672.5	+1	0	0	0	−3	0.82×10^{-10}	$\Xi^0\pi^-, \Lambda^0 K^-, \Xi^-\pi^0$
		and others											

† See also Table 44–4 for particles with charm and bottomness.
‡ Neutrinos partake only in the weak interaction. Experimental upper limits on neutrino masses are given in parentheses. Recent evidence (2000) suggests that at least one of the neutrinos may have mass, although probably quite small ($<1\,\text{eV}$).

force predominates at short distances. The hadrons include nucleons, pions, and a large number of other particles. They are divided into two subgroups: **baryons**, which are those particles that have baryon number +1 (or −1 in the case of their antiparticles); and **mesons**, which have baryon number = 0.

Baryons

Mesons

Notice that the baryons $\Lambda, \Sigma, \Xi,$ and Ω all decay to lighter-mass baryons, and eventually to a proton or neutron. All these processes conserve baryon number. Since there is no lighter particle than the proton with $B = +1$, if baryon number is strictly conserved, the proton itself cannot decay and is stable. (But see Section 44–11.)

44–7 Particle Stability and Resonances

Many of the particles listed in Table 44–2 are unstable. The lifetime of an unstable particle depends on which force is most active in causing the decay. When we say the strong nuclear force is stronger than the electromagnetic, we mean that two particles will interact more strongly and more quickly if this force is acting. When a stronger force influences a decay, that decay occurs more quickly. Decays caused by the weak force typically have lifetimes of $10^{-13}\,\text{s}$ or longer. Particles that decay via the electromagnetic force have much shorter lifetimes, typically about 10^{-16} to $10^{-19}\,\text{s}$. (Exceptions to this scheme are the W and Z particles, which decay via the

Lifetime depends on which force is acting

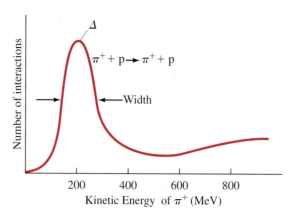

FIGURE 44–14 Number of π^+ particles scattered by a proton target as a function of the incident π^+ kinetic energy. The resonance shape represents the formation of a short-lived particle, the Δ, which has a charge in this case of $+2e(\Delta^{++})$.

weak interaction but have very short lifetimes due to their special nature as exchange particles.) The unstable particles listed in Table 44–2 decay either via the weak or the electromagnetic interaction. Decays that involve a γ (photon) are electromagnetic (such as $\pi^0 \to 2\gamma$). The other decays shown take place via the weak interaction, often accompanied by a neutrino which interacts only via the weak interaction: examples are $\pi^- \to \mu^- \nu_\mu$ and $\Sigma^- \to n\pi^-$.

A great many particles have been found that decay via the strong interaction, and these are not listed in Table 44–2. Such particles decay into other strongly interacting particles (say, n, p, π, but not involving γ, e, ν, and so on) and their lifetimes are very short, typically about 10^{-23} s. In fact their lifetimes are so short that they do not travel far enough to be detected before decaying. Their decay products can be detected, however, and it is from them that the existence of such short-lived particles is inferred. To see how this is done, we consider the first such particle discovered (by Fermi). Fermi used a beam of π^+ directed through a hydrogen target (protons) with varying amounts of energy. A graph of the number of interactions (π^+ scattered) versus the pion's kinetic energy is shown in Fig. 44–14. The large peak around 200 MeV was much higher than expected and certainly much higher than the number of interactions at neighboring energies. This led Fermi to conclude that the π^+ and proton combined momentarily to form a short-lived particle before coming apart again, or at least that they resonated together for a short time. Indeed, the large peak in Fig. 44–14 resembles a resonance curve (see Figs. 14–24, 14–27, and 31–9), and this new "particle"—now called the Δ—is referred to as a **resonance**. Hundreds of other resonances have been found in a similar way. Many resonances are regarded as excited states of other particles such as of the nucleon (proton or neutron).

The width of a resonance—in Fig. 44–14 the width of the Δ peak is on the order of 100 MeV—is an interesting application of the uncertainty principle. If a particle lives only 10^{-23} s, then its mass (i.e., its rest energy) will be uncertain by an amount $\Delta E \approx h/2\pi\Delta t \approx (6.6 \times 10^{-34}\,\text{J}\cdot\text{s})/(6)(10^{-23}\,\text{s}) \approx 10^{-11}\text{J} \approx 100\,\text{MeV}$, which is what is observed. Actually, the lifetimes of $\approx 10^{-23}$ s for such resonances are inferred by the reverse process: from the measured width being ≈ 100 MeV.

Very short-lived particles are inferred from their decay products

Resonance

44–8 Strange Particles

In the early 1950s, certain of the newly found particles, namely, the K, Λ, and Σ, were found to behave rather strangely in two ways. First, they were always produced in pairs. For example, the reaction

$$\pi^- + p \rightarrow K^0 + \Lambda^0$$

occurred with high probability, but the reaction $\pi^- + p \rightarrow K^0 + n$ was never observed to occur. This seemed strange because the unobserved reaction would not have violated any known conservation law, and plenty of energy was available. The second feature of these **strange particles** (as they came to be called) was that, although they were clearly produced via the strong interaction (that is, at a high rate), they did not decay at a rate characteristic of the strong interaction even though they decayed into strongly interacting particles (for example, $K \rightarrow 2\pi$, $\Sigma^+ \rightarrow p + \pi^0$). Instead of lifetimes of 10^{-23} s as expected for strongly interacting particles, strange particles have lifetimes of 10^{-10} to 10^{-8} s, which are characteristic of the weak interaction.

Strangeness and its conservation

To explain these observations, a new quantum number, **strangeness**, and a new conservation law, conservation of strangeness, were introduced. By assigning the strangeness numbers (S) indicated in Table 44–2, the production of strange particles in pairs was readily explained. Antiparticles were assigned opposite strangeness from their particles: one of each pair was assigned $S = +1$ and the other $S = -1$ (see Table 44–2). For example, in the reaction $\pi^- + p \rightarrow K^0 + \Lambda^0$, the initial state has strangeness $S = 0 + 0 = 0$, and the final state has $S = +1 - 1 = 0$, so strangeness is conserved. But for $\pi^- + p \rightarrow K^0 + n$, the initial state has $S = 0$ and the final state has $S = +1 + 0 = +1$, so strangeness would not be conserved; and this reaction is not observed.

Strangeness conserved in strong interactions but not in weak

To explain the decay of strange particles, it is assumed that strangeness is conserved in the strong interaction but is *not* conserved in the weak interaction. Thus, although strange particles were forbidden by strangeness conservation to decay to nonstrange particles of lower mass via the strong interaction, they could undergo such decay by means of the weak interaction. This would occur much more slowly, of course, which accounts for their longer lifetimes of 10^{-10} to 10^{-8} s.

The conservation of strangeness was the first example of a "partially conserved" quantity. In this case, the quantity strangeness is conserved by strong interactions but not by weak.

CONCEPTUAL EXAMPLE 44–6 **Guess the missing particle.** Using the conservation laws for particle interactions, determine the other particle as a result of the reaction

$$\pi^- + p \rightarrow K^0 + ?,$$

in addition to $K^0 + \Lambda^0$.

SOLUTION We write equations for the conserved numbers in this reaction, with B, L_e, S, and Q as unknowns whose determination will reveal what the possible particle might be:

Baryon number:	$0 + 1 = 0 + B$
Lepton number:	$0 + 0 = 0 + L_e$
Charge:	$-1 + 1 = 0 + Q$
Strangeness:	$0 + 0 = 1 + S.$

The unknown product particle would have to have these characteristics:

$$B = +1 \qquad L_e = 0 \qquad Q = 0 \qquad S = -1.$$

In addition to Λ^0, a neutral sigma particle, Σ^0, is also consistent with these numbers.

44–9 Quarks

We saw in our discussion of Table 44–2 that all particles, except the gauge bosons, fall into either of two categories: leptons and hadrons. The principal difference between these two groups is that the hadrons interact via the strong interaction, whereas the leptons do not. Another important difference that physicists had to deal with in the 1960s was that there were only four known leptons (e^-, μ^-, ν_e, ν_μ; the τ and ν_τ were not yet discovered), but there were well over a hundred hadrons.

The leptons are considered to be truly elementary particles since they do not seem to break down into smaller entities, do not show any internal structure, and have no measurable size. (Attempts to determine the size of leptons have put an upper limit of about 10^{-18} m.)

The hadrons, on the other hand, are more complex. Experiments indicate they do have an internal structure. And the fact that there are so many of them suggests that they can't all be elementary. To deal with this problem, M. Gell-Mann and G. Zweig in 1963 independently proposed that none of the hadrons so far observed, not even the proton and neutron, was elementary. Instead, they proposed that the hadrons are made up of combinations of three, more fundamental, pointlike entities called **quarks**.[†] Today, the quark theory is well-accepted, and quarks are considered the truly elementary particles, like leptons. The three quarks were labeled u, d, s, and given the names *up*, *down*, and *strange*. They were assumed to have fractional charge ($\frac{1}{3}$ or $\frac{2}{3}$ the charge on the electron—that is, less than the previously thought smallest charge). Other properties of quarks and antiquarks are indicated in Table 44–3. All hadrons known at the time could be constructed in theory from these three types of quark. Mesons would consist of a quark–antiquark pair. For example, a π^+ meson is considered a u$\bar{\text{d}}$ pair (note that

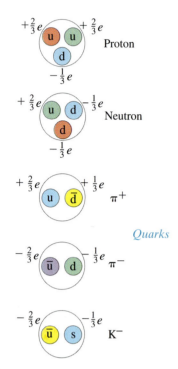

Quarks

FIGURE 44–15 Quark compositions for several particles.

[†] Gell-Mann chose the word from a phrase in James Joyce's *Finnegans Wake*.

TABLE 44–3 Properties of Quarks and Antiquarks

				Quarks				
Name	**Symbol**	**Spin**	**Charge**	**Baryon Number**	**Strangeness**	**Charm**	**Bottomness**	**Topness**
Up	u	$\frac{1}{2}$	$+\frac{2}{3}e$	$\frac{1}{3}$	0	0	0	0
Down	d	$\frac{1}{2}$	$-\frac{1}{3}e$	$\frac{1}{3}$	0	0	0	0
Strange	s	$\frac{1}{2}$	$-\frac{1}{3}e$	$\frac{1}{3}$	-1	0	0	0
Charmed	c	$\frac{1}{2}$	$+\frac{2}{3}e$	$\frac{1}{3}$	0	$+1$	0	0
Bottom	b	$\frac{1}{2}$	$-\frac{1}{3}e$	$\frac{1}{3}$	0	0	-1	0
Top	t	$\frac{1}{2}$	$+\frac{2}{3}e$	$\frac{1}{3}$	0	0	0	$+1$
				Antiquarks				
Name	**Symbol**	**Spin**	**Charge**	**Baryon Number**	**Strangeness**	**Charm**	**Bottomness**	**Topness**
Up	$\bar{\text{u}}$	$\frac{1}{2}$	$-\frac{2}{3}e$	$-\frac{1}{3}$	0	0	0	0
Down	$\bar{\text{d}}$	$\frac{1}{2}$	$+\frac{1}{3}e$	$-\frac{1}{3}$	0	0	0	0
Strange	$\bar{\text{s}}$	$\frac{1}{2}$	$+\frac{1}{3}e$	$-\frac{1}{3}$	$+1$	0	0	0
Charmed	$\bar{\text{c}}$	$\frac{1}{2}$	$-\frac{2}{3}e$	$-\frac{1}{3}$	0	-1	0	0
Bottom	$\bar{\text{b}}$	$\frac{1}{2}$	$+\frac{1}{3}e$	$-\frac{1}{3}$	0	0	$+1$	0
Top	$\bar{\text{t}}$	$\frac{1}{2}$	$-\frac{2}{3}e$	$-\frac{1}{3}$	0	0	0	-1

TABLE 44–4 Partial List of Hadrons Associated with Charm and Bottomness ($L_e = L_\mu = L_\tau = 0$)

Category	Particle	Anti-particle	Spin	Rest Mass (MeV/c^2)	Baryon Number	Strange-ness	Charm	Bottom-ness	Lifetime (s)	Principal Decay Modes
Mesons	D^+	D^-	0	1869.4	0	0	+1	0	10.6×10^{-13}	K + others, e + others
	D^0	$\overline{D}^0$	0	1864.6	0	0	+1	0	4.2×10^{-13}	K + others, μ or e + others
	D_S^+	D_S^-	0	1969	0	+1	+1	0	4.7×10^{-13}	K + others
	J/ψ (3097)	Self	1	3096.9	0	0	0	0	0.8×10^{-20}	Hadrons, e^+e^-, $\mu^+\mu^-$
	Υ (9460)	Self	1	9460.4	0	0	0	0	1.3×10^{-20}	Hadrons, $\mu^+\mu^-$, e^+e^-, $\tau^+\tau^-$
	B^-	B^+	0	5279	0	0	0	−1	1.5×10^{-12}	D^0 + others
	B^0	$\overline{B}^0$	0	5279	0	0	0	−1	1.5×10^{-12}	D^0 + others
Baryons	Λ_c^+	Λ_c^-	$\frac{1}{2}$	2285	+1	0	+1	0	2.0×10^{-13}	Hadrons (e.g., Λ + others)
	Σ_c^{++}	Σ_c^{--}	$\frac{1}{2}$	2453	+1	0	+1	0	?	$\Lambda_c^+ \pi^+$
	Σ_c^+	Σ_c^-	$\frac{1}{2}$	2454	+1	0	+1	0	?	$\Lambda_c^+ \pi^0$
	Σ_c^0	$\overline{\Sigma}_c^0$	$\frac{1}{2}$	2452	+1	0	+1	0	?	$\Lambda_c^+ \pi^-$
	Λ_b^0	$\overline{\Lambda}_b^0$	$\frac{1}{2}$	5640	+1	0	0	−1	1.1×10^{-12}	$J/\psi \Lambda^0$, $pD^0\pi^-$, $\Lambda_c^+\pi^+\pi^-\pi^-$

for the $u\bar{d}$ pair, $Q = \frac{2}{3}e + \frac{1}{3}e = +1e$, $B = \frac{1}{3} - \frac{1}{3} = 0$, $S = 0 + 0 = 0$, as they must for a π^+); and a $K^+ = u\bar{s}$, with $Q = +1$, $B = 0$, $S = +1$. Baryons, on the other hand, would consist of three quarks. For example, a neutron is n = ddu, whereas an antiproton is $\bar{p} = \bar{u}\bar{u}\bar{d}$. See Fig. 44–15.

Soon after the quark theory was proposed, physicists began looking for these fractionally charged particles, but direct detection has not been successful. Indeed, current models suggest that quarks may be so tightly bound together that they may not ever exist singly in the free state.

In 1964 several physicists proposed that there ought to be a fourth quark. Their argument was based on the expectation that there exists a deep symmetry in nature, including a connection between quarks and leptons. If there are four leptons (as was thought in the 1960s), then symmetry in nature would suggest there should also be four quarks. The fourth quark was said to be *charmed*. Its charge would be $+\frac{2}{3}e$ and it would have another property to distinguish it from the other three quarks. This new property, or quantum number, was called **charm** (see Table 44–3). Charm was assumed to be like strangeness: it would be conserved in strong and electromagnetic interactions, but would not be conserved by the weak. The new charmed quark would have charm $C = +1$ and its antiquark $C = -1$. The first charmed particle, the J/ψ meson, was discovered in 1974.

Charm

Charmed quark

Also in the 1970s strong evidence appeared for the tau (τ) lepton, with a mass of 1777 MeV/c^2. This lepton, like the electron and muon, presumably has a neutrino associated with it. Thus, the family of leptons is at present believed to have six members. This would upset the balance between leptons and quarks, the presumed basic building blocks of matter, unless two additional quarks also exist. Indeed, theoretical physicists postulated the existence of a fifth and sixth quark, named **top** and **bottom**. The names apply also to the new properties (quantum numbers) that distinguish the new quarks from the old quarks (see Table 44–3), and which (like strangeness) are conserved in strong, but not weak, interactions. New mesons involving b quarks were soon detected (Table 44–4). Convincing evidence for the top quark came in 1995 (see photo at start of this chapter) after years of searching. Its extremely high mass of about 175 GeV/c^2 contributed to its elusiveness because of the huge energy required to produce it.

Two more leptons

t and b quarks

TABLE 44–5 The Elementary Particles[†] as Seen Today

	First generation	Second generation	Third generation
Quarks	u, d	s, c	b, t
Leptons	e, ν_e	μ, ν_μ	τ, ν_τ
Gauge bosons	γ(photon)	$W^\pm, Z^0$	gluons

[†]Note that the quarks and leptons are arranged into three generations each, and the gauge particles are arranged in groups for the forces they mediate.

The elementary particles

Today, the truly elementary particles are considered to be the six quarks, the six leptons, and the gauge bosons that carry the fundamental forces. See Table 44–5, where the quarks and leptons are arranged in three groups (generations).

CONCEPTUAL EXAMPLE 44–7 Quark combinations. Find the baryon number, charge, and strangeness for the following quark combinations and identify the hadron particle that is made up of these quark combinations: (*a*) udd, (*b*) u$\bar{\text{u}}$, (*c*) uss, (*d*) sdd, and (*e*) b$\bar{\text{u}}$.

RESPONSE We use Table 44–3 to get the properties of the quarks, then Table 44–2 or 44–4 to find the particle that has these properties. (*a*) udd has

$$Q = +\tfrac{2}{3}e - \tfrac{1}{3}e - \tfrac{1}{3}e = 0,$$
$$B = \tfrac{1}{3} + \tfrac{1}{3} + \tfrac{1}{3} = 1,$$
$$S = 0 + 0 + 0 = 0,$$

as well as $C = 0$, bottomness $= 0$, topness $= 0$. The only baryon ($B = +1$) that has $Q = 0$, $S = 0$, etc., is the neutron (Table 44–2).
(*b*) u$\bar{\text{u}}$ has $Q = \tfrac{2}{3}e - \tfrac{2}{3}e = 0$, $B = 0$, and all other quantum numbers $= 0$. Sounds like a π^0.
(*c*) uss has $Q = 0$, $B = +1$, $S = -2$, others $= 0$. This is a Ξ^0.
(*d*) sdd has $Q = -1$, $B = +1$, $S = -1$, so must be a Σ^-.
(*e*) b$\bar{\text{u}}$ has $Q = -1$, $B = 0$, $S = 0$, $C = 0$, bottomness $= -1$, topness $= 0$. This must be a B$^-$ meson (Table 44–4).

44–10 The "Standard Model": Quantum Chromodynamics (QCD) and the Electroweak Theory

Not long after the quark theory was proposed, it was suggested that quarks have another property (or quality) called **color**. The distinction between the six quarks (u, d, s, c, b, t) was referred to as **flavor**. According to theory, each of the flavors of quark can have three colors, usually designated red, green, and blue. (These are the three primary colors which, when added together in equal amounts, as on a TV screen, produce white.) Note that the names "color" and "flavor" have nothing to do with our senses, but are purely whimsical—as are other names, such as charm, in this new field. (We did, however, "color" the quarks in Fig. 44–15.) The antiquarks are colored antired, antigreen, and antiblue. Baryons are made up of three quarks, one of each color. Mesons consist of a quark–antiquark pair of a particular color and its anticolor. Thus both baryons and mesons are colorless.

Originally, the idea of quark color was proposed to preserve the Pauli exclusion principle (Section 40–4). Not all particles obey the exclusion principle. Those that do, such as electrons, protons, and neutrons, are called **fermions**. Those that don't are called **bosons**. These two categories are distinguished also in their spin (Section 40–2): bosons have integer spin (0, 1, 2, etc.) whereas fermions have half-integer spin, usually $\tfrac{1}{2}$, as for electrons and nucleons, but can also be $\tfrac{3}{2}, \tfrac{5}{2}$, etc. Matter is made up mainly of fermions, but the carriers of the forces (γ, W, Z, and gluons, as we'll see) are all bosons. Quarks are fermions (they have spin $\tfrac{1}{2}$) and

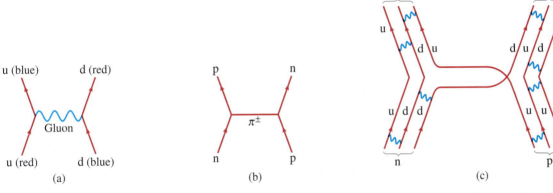

FIGURE 44–16 (a) The force between two quarks holding them together as part of a proton, for example, is carried by a gluon, which in this case involves a change in color. (b) Strong interaction n + p → n + p with the exchange of a charged π meson (+ or −, depending on whether it is considered moving to the left or to the right). (c) Quark representation of the same interaction n + p → n + p. The wavy lines between quarks represent gluon exchanges holding the hadrons together.

therefore should obey the exclusion principle. Yet for three particular baryons (uuu, ddd, and sss), all three quarks would have the same quantum numbers, and at least two of them have their spin in the same direction (since there are only two choices, spin up $[m_s = +\frac{1}{2}]$ or spin down $[m_s = -\frac{1}{2}]$). This would seem to violate the exclusion principle; but if quarks have an additional quantum number (color), which could be different for each quark, it would serve to distinguish them and the exclusion principle would hold. Although quark color, and the resulting threefold increase in the number of quarks, was thus originally an *ad hoc* idea, it also served to bring the theory into better agreement with experiment, such as predicting the correct lifetime of the π^0 meson. The idea of color soon became, in addition, a central feature of the theory as determining the force binding quarks together in a hadron. Each quark is assumed to carry a *color charge*, analogous to electric charge, and the strong force between quarks is often referred to as the **color force**. This new theory of the strong force is called **quantum chromodynamics** (*chroma* = color in Greek), or **QCD**, to indicate that the force acts between color charges (and not between, say, electric charges). The strong force between two hadrons[†] is considered to be a force between the quarks that make them up, as suggested in Fig. 44–16. The particles that transmit the color force (analogous to photons for the EM force) are called **gluons** (a play on "glue"). They are included in Table 44–5. There are eight gluons, according to the theory, all massless, and six of them have color charge.[‡] Thus gluons have replaced mesons (Table 44–1) as the particles carrying the strong (color) force.

QCD

Gluons

The color force has the interesting property that its strength increases with increasing distance (as does the force exerted by a coiled spring, Chapter 14); as two quarks approach each other very closely (equivalently, have high energy), the force between them becomes very small. This aspect is referred to as **asymptotic freedom**.

The weak force, as we have seen, is thought to be mediated by the W^+, W^-, and Z^0 particles. It acts between the "weak charges" that each particle has. Each elementary particle can thus have electric charge, weak charge, color charge, and gravitational mass, although one or more of these could be zero. For example, all leptons have color charge of zero, so they do not interact via the strong force.

[†] The strong force between hadrons appears feeble, however, in comparison to the force directly between quarks within hadrons.

[‡] Compare to the EM interaction, where the photon has no electric charge. Because gluons have color charge, they could attract each other and form composite particles (photons cannot). Such "glueballs" are being searched for.

To summarize, the standard model says that the truly elementary particles (Table 44–5) are the leptons, the quarks, and the gauge bosons (photon, W and Z, and the gluons). Some theories suggest there may be other bosons as well. The photon and the leptons are observed in experiments, and finally so too were the W^+, W^-, and Z^0. But so far only combinations of quarks (baryons and mesons) have been observed, and it seems likely that free quarks and gluons are unobservable.

One important aspect of new theoretical work is the attempt to find a unified basis for the different forces in nature. This was a long-held hope of Einstein, which he was never able to fulfill. A so-called **gauge theory** that unifies the weak and electromagnetic interactions was put forward in the 1960s by S. Weinberg, S. Glashow, *Electroweak theory* and A. Salam. In this **electroweak theory**, the weak and electromagnetic forces are seen as two different manifestations of a single, more fundamental, *electroweak* interaction. The electroweak theory has had many successes, including the prediction of the $W^\pm$ particles as carriers of the weak force, with masses of $81 \pm 2\ \mathrm{GeV}/c^2$ in excellent agreement with the measured values of $80.41 \pm 0.10\ \mathrm{GeV}/c^2$ (and similar accuracy for the Z^0). The electroweak theory plus QCD for the strong interac-*Standard Model* tion are often referred to today as the **Standard Model**.

Theoreticians have wondered why the W and Z have large masses rather than being massless like the photon. Electroweak theory suggests an explanation by means of a new **Higgs field** and its particle, the **Higgs boson**, which interact with the W and Z to "slow them down." In being forced to go slower than the speed of light, they must acquire mass.

44–11 | Grand Unified Theories

GUT
With the success of the unified electroweak theory, attempts have recently been made to incorporate it and QCD for the strong (color) force into a so-called **grand unified theory** (GUT). One type of such a grand unified theory of the electromagnetic, weak, and strong forces has been worked out in which there is only one class of particle—leptons and quarks belong to the same family and are able to change freely from one type to the other—and the three forces are different *Unification* aspects of a single underlying force. The unity is predicted to occur, however, only *of forces* on a scale of less than about $10^{-30}\ \mathrm{m}$. If two elementary particles (leptons or quarks) approach each other to within this **unification scale**, the apparently fundamental distinction between them would not exist at this level, and a quark could readily change to a lepton, or vice versa. Baryon and lepton numbers would not be conserved. The weak, electromagnetic, and strong (color) force would blend to a force of a single strength.

How could a lepton become a quark, or vice versa? The theory predicts the existence of particles, called X bosons, that can be exchanged between a quark and a lepton allowing one to change into the other, somewhat like the charged pion exchanged between the p and n in Fig. 44–16b allows the proton on the right to become a neutron. The mass of the proposed X boson, consistent with the uncertainty principle as applied earlier in this chapter (see Eq. 44–3 with $d \approx 10^{-30}\ \mathrm{m}$), would be about $10^{14}\ \mathrm{GeV}/c^2$, or 10^{14} times the proton mass. With such an incredibly large mass, there is little hope of seeing them in the laboratory. It is also this huge mass that would keep baryon and lepton numbers conserved in observed reactions, since the likelihood of producing such a massive particle, even as a virtual exchange particle, is extremely small at even the highest laboratory energies.

Symmetry
breaking
What happens between the unification distance of $10^{-30}\ \mathrm{m}$ and more normal (larger) distances is referred to as **symmetry breaking**. As an analogy, consider an atom in a crystal. Deep within the atom, there is much symmetry—in the innermost regions the electron cloud is spherically symmetric (Chapter 40). Farther out, this symmetry breaks down—the electron clouds are distributed preferentially along the lines (bonds) joining the atoms in the crystal. In a similar way, at $10^{-30}\ \mathrm{m}$ the force between elementary particles is theorized to be a single force—it is symmetrical

and does not single out one type of "charge" over another. But at larger distances, that symmetry is broken and we see three distinct forces. (In the "standard model" of electroweak interactions, Section 44–10, the symmetry breaking between the electromagnetic and the weak interactions occurs at about 10^{-18} m.)

CONCEPTUAL EXAMPLE 44–8 **Symmetry.** The table in Fig. 44–17 has four identical place settings. Four people sit down to eat. Describe the symmetry of this table and what happens to it when someone starts the meal.

RESPONSE The table has several kinds of symmetry. It is symmetric to rotations of 90°: that is, the table will look the same if everyone moved one chair to the left or to the right. It is also north–south symmetric and east–west symmetric, so that swaps across the table don't affect the way the table looks. It also doesn't matter whether any person picks up the fork to the left of the plate or the fork to the right. But once that first person picks up either fork, the choice is set for all the rest at the table as well. The symmetry has been *broken*. The underlying symmetry is still there—the water glasses could still be chosen either way—but some choice must get made and at that moment the symmetry of the diners is broken.

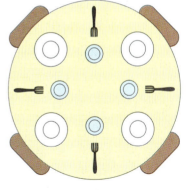

FIGURE 44–17 Symmetry around a table. Example 44–8.

Since unification occurs at such tiny distances and huge energies, the theory is difficult to test experimentally. But it is not completely impossible. One possibly testable prediction is the basis for the idea that the proton might decay (via, for example, $p \rightarrow \pi^0 + e^+$) and violate conservation of baryon number. This could happen if two quarks approached to within 10^{-31} m of each other. But it is very unlikely at normal temperature and energy, so the decay of a proton can only be an unlikely process. In the simplest form of GUT, the theoretical estimate of the proton lifetime for the decay mode $p \rightarrow \pi^0 + e^+$ is $\approx 10^{31}$ yr, and this has just come within the realm of testability. Proton decays have still not been seen and experiments put the lower limit on the proton lifetime for the above mode to be about 2.5×10^{32} yr, an order of magnitude greater than this prediction. This may seem a disappointment, but on the other hand, it presents a challenge. Indeed more complex GUTs are not affected by this result.

Proton decay?

EXAMPLE 44–9 **ESTIMATE** **Proton decay.** An experiment uses 3300 tons of water waiting to see a proton decay of the type $p \rightarrow \pi^0 + e^+$. If the experiment is run for four years without detecting a decay, estimate the lower limit on the proton half-life.

SOLUTION As with radioactive decay, the number of decays is proportional to the number of parent species (N), the time interval (Δt), and the decay constant (λ) which is related to the half-life $T_{\frac{1}{2}}$ by (see Eqs. 42–4 and 42–8):

$$\Delta N = -\lambda N \, \Delta t = -\frac{\ln 2}{T_{\frac{1}{2}}} N \, \Delta t.$$

Thus, dealing only with magnitudes,

$$T_{\frac{1}{2}} = \frac{N}{\Delta N} \Delta t \ln 2$$

which tells us that, for $\Delta N < 1$ over the four-year trial:

$$T_{\frac{1}{2}} > N(4 \, \text{yr})(0.69)$$

where N is the number of protons in 3300 tons of water. To determine N, we note that each molecule of H_2O contains $(2 + 8 =)$ 10 protons. So one mole of water (18 g) contains $10 \times 6 \times 10^{23}$ protons in 18 g of water, or about 3×10^{26} protons per kilogram. One ton is 10^3 kg, so the chamber contains $(3.3 \times 10^6 \, \text{kg}) \cdot (3 \times 10^{26} \, \text{protons/kg}) \approx 1 \times 10^{33}$ protons. Then our very rough estimate for a lower limit on the proton half-life is $T_{\frac{1}{2}} > (10^{33})(4 \, \text{yr})(0.7) \approx 3 \times 10^{33}$ yr.

An interesting prediction of unified theories relates to cosmology (see Chapter 45). It is thought that during the first 10^{-35} s after the theorized big bang that created the universe, the temperature was so extremely high that particles had energies corresponding to the unification scale. Baryon number would not have been conserved then, thus allowing an imbalance that might account for the observed predominance of matter ($B > 0$) over antimatter ($B < 0$) in the universe.

This last example is interesting, for it illustrates a deep connection between investigations at either end of the size scale: theories about the tiniest objects (elementary particles) have a strong bearing on the understanding of the universe as a whole. We will look at this more in the next chapter.

Even more ambitious than grand unified theories are attempts to also incorporate gravity, and thus unify all four forces in nature into a single theory. (Such theories are sometimes referred to misleadingly as **theories of everything**.) The only consistent theory so far that attempts to unify all four forces is called **string theory**, in which the elementary particles (Table 44–5) are imagined not as points but as one-dimensional strings about 10^{-35} m long.

A related idea is **supersymmetry**, which applied to strings is known as **superstring theory**. Supersymmetry predicts that interactions exist that would change fermions into bosons and vice versa, and that all known fermions have supersymmetric boson partners. Thus, for every quark there would be a *squark*, and for every lepton there would be a *slepton*. Likewise, for every known boson (photons and gluons, for example), there would be a supersymmetric fermion (*photinos* and *gluinos*). But why hasn't this "missing half" of the universe ever been detected? The best guess is that supersymmetric particles might be heavier than their conventional counterparts, perhaps too heavy to have been produced in today's accelerators. Until a supersymmetric particle is found, however, supersymmetry is just an elegant guess.

The world of elementary particles is opening new vistas. What happens in the near future is bound to be exciting.

Summary

Particle accelerators are used to accelerate charged particles, such as electrons and protons, to very high energy. High-energy particles have short wavelength and so can be used to probe the structure of matter at very small distances in great detail. High kinetic energy also allows the creation of new particles through collision (via $E = mc^2$).

Van De Graaff and linear accelerators use high voltage to accelerate particles along a line. Cyclotrons and synchrotrons use a magnetic field to keep the particles in a circular path and accelerate them at intervals by high voltage. **Colliding beams** allow higher interaction energy.

Particle **detectors** that can image particle tracks include photographic emulsions, cloud chambers, bubble chambers, and today wire drift chambers.

An **antiparticle** has the same mass as a particle but opposite charge. Certain other properties may also be opposite: for example, the antiproton has **baryon number** (nucleon number) opposite to that for the proton.

In all nuclear and particle reactions, the following conservation laws hold: momentum, angular momentum, mass-energy, electric charge, baryon number, and the three **lepton numbers**.

Certain particles have a property, called **strangeness**, which is conserved by the strong force but not by the weak force. The more recently noted properties, **charm**, **bottomness**, and **topness**, also are believed to be conserved by the strong force but not by the weak.

Just as the electromagnetic force can be said to be due to an exchange of photons, the strong nuclear force was first thought to be carried by *mesons* that have rest mass, or, according to more recent theory, by massless **gluons**. The W and Z particles carry the weak force. These fundamental force carriers (photon, W and Z, gluons) are called **gauge bosons**.

Other particles can be classified as either *leptons* or *hadrons*. **Leptons** participate in the weak and electromagnetic interactions. **Hadrons**, which today are considered to be made up of **quarks**, participate in the strong interaction as well. The hadrons can be classified as **mesons**, with baryon number zero, and **baryons**, with nonzero baryon number.

All particles, except for the photon, electron, neutrinos, and proton, decay with measurable half-lives varying from 10^{-25} s to 10^3 s. The half-life depends on which force is predominant in the decay. Weak decays usually have half-lives greater than about 10^{-13} s. Electromagnetic decays have half-lives on the order of 10^{-16} to 10^{-19} s. The shortest lived particles, called **resonances**, decay via the strong interaction and live typically for only about 10^{-23} s.

Today's standard model of elementary particles considers **quarks** as the basic building blocks of the hadrons. The six quarks are called **up**, **down**, **strange**, **charmed**, **bottom**, and **top** quarks. It is expected that there are the same number of quarks as leptons (six of each), and that quarks and leptons are the truly elementary particles along with the gauge bosons (γ, W, Z, gluons). Quarks are said to have **color**, and, according to **quantum chromodynamics** (QCD), the strong color force acts between their color charges and is transmitted by **gluons**. **Electroweak theory** views the weak and electromagnetic forces as two aspects of a single underlying interaction. QCD plus the electroweak theory are referred to as the **Standard Model**.

Grand unified theories of forces suggest that at very short distances $(10^{-30}$ m$)$ and very high energy, the weak, electromagnetic, and strong forces appear as a single force, and the fundamental difference between quarks and leptons disappears.

Questions

1. Give a reaction between two nucleons, similar to Eq. 44–4, that could produce a π^-.
2. If a proton is moving at very high speed, so that its kinetic energy is much greater than its rest energy (m_0c^2), can it then decay via p $\rightarrow$ n + π^+?
3. What would an "antiatom," made up of the antiparticles to the constituents of normal atoms, consist of? What might happen if *antimatter*, made of such antiatoms, came in contact with our normal world of matter?
4. What particle in a decay signals the electromagnetic interaction?
5. Does the presence of a neutrino among the decay products of a particle necessarily mean that the decay occurs via the weak interaction? Do all decays via the weak interaction produce a neutrino?
6. Why is it that a neutron decays via the weak interaction even though the neutron and one of its decay products (proton) are strongly interacting?
7. Which of the four interactions (strong, electromagnetic, weak, gravitational) does an electron take part in? A neutrino? A proton?

8. Check that charge and baryon number are conserved in each of the decays in Table 44–2.
9. Which of the particle decays in Table 44–2 occur via the electromagnetic interaction?
10. Which of the particle decays in Table 44–2 occur by the weak interaction?
11. By what interaction, and why, does $\Sigma^\pm$ decay to Λ^0? What about Σ^0 decaying to Λ^0?
12. The Δ baryon has spin $\frac{3}{2}$, baryon number 1, and charge $Q = +2$, $+1$, 0, or -1. Why is there no charge state $Q = -2$?
13. Which of the particle decays in Table 44–4 occur via the electromagnetic interaction?
14. Which of the particle decays in Table 44–4 occur by the weak interaction?
15. Quarks have spin $\frac{1}{2}$. How do you account for the fact that baryons have spin $\frac{1}{2}$ or $\frac{3}{2}$, and mesons have spin 0 or 1?
16. Suppose there were a kind of "neutrinolet" that was massless, had no color charge or electrical charge, and did not feel the weak force. Could you say that this particle even exists?

Problems

Sections 44–1 and 44–2

1. (I) What is the total energy of a proton whose kinetic energy is 6.35 GeV?
2. (I) Calculate the wavelength of 35-GeV electrons.
3. (I) What strength of magnetic field is used in a cyclotron in which protons make 2.8×10^7 revolutions per second?
4. (I) What is the time for one complete revolution for a very high-energy proton in the 1.0-km-radius Fermilab accelerator?
5. (I) If α particles are accelerated by the cyclotron of Example 44–2, what must be the frequency of the voltage applied to the dees?
6. (II) (a) If the cyclotron of Example 44–2 accelerated α particles, what maximum energy could they attain? What would their speed be? (b) Repeat for deuterons $(^2_1$H$)$. (c) In each case, what frequency of voltage is required?

7. (II) Which is better for picking out details of the nucleus: 30-MeV alpha particles or 30-MeV protons? Compare each of their wavelengths with the size of a nucleon in a nucleus.
8. (II) The voltage across the dees of a cyclotron is 55 kV. How many revolutions do protons make to reach a kinetic energy of 25 MeV?
9. (II) What is the wavelength, and maximum resolving power attainable, using 900-GeV protons at Fermilab?
10. (II) A cyclotron with a radius of 1.0 m is to accelerate deuterons $(^2_1$H$)$ to an energy of 10 MeV. (a) What is the required magnetic field? (b) What frequency is needed for the voltage between the dees? (c) If the potential difference between the dees averages 22 kV, how many revolutions will the particles make before exiting? (d) How much time does it take for one deuteron to go from start to exit, and (e) how far does it travel during this time?

11. (II) Protons are injected into the 1.0-km-radius Fermilab Tevatron with an energy of 8.0 GeV. If they are accelerated by 2.5 MV each revolution, how far do they travel and approximately how long does it take for them to reach 900 GeV?

12. (II) The Fermilab Tevatron takes about 20 seconds to bring the energies of the stored protons from 150 GeV to 900 GeV. The acceleration is done once per turn. Estimate the energy given to the protons on each turn. (You can assume that the speed of the protons is essentially c the whole time.)

13. (II) Show that the energy of a particle (charge e) in a synchrotron, in the relativistic limit ($v \approx c$), is given by $E(\text{in eV}) = Brc$, where B is magnetic field strength and r the radius of the orbit (SI units).

14. (II) What magnetic field intensity is needed at the 1.0-km-radius Fermilab synchrotron for 900-GeV protons?

Sections 44–3 to 44–6

15. (I) How much energy is released in the decay
$$\pi^+ \to \mu^+ + \nu_\mu?$$

16. (I) About how much energy is released when a Λ^0 decays to $n + \pi^0$? (See Table 44–2.)

17. (I) How much energy is required to produce a neutron–antineutron pair?

18. (I) Estimate the range of the strong force if the mediating particle were the kaon instead of the pion.

19. (II) Two protons are heading toward each other with equal speeds. What minimum kinetic energy must each have if a π^0 meson is to be created in the process? (See Table 44–2.)

20. (II) Which of the following decays are possible? For those that are forbidden, explain which laws are violated.
(a) $\Xi^0 \to \Sigma^+ + \pi^-$ (b) $\Omega^- \to \Sigma^0 + \pi^- + \nu$
(c) $\Sigma^0 \to \Lambda^0 + \gamma + \gamma$.

21. (II) Estimate the range of the weak force using Eq. 44–3, given the masses of the W and Z particles as about 80 to 90 GeV/c^2.

22. (II) What are the wavelengths of the two photons produced when a proton and antiproton at rest annihilate?

23. (II) (a) Show, by conserving momentum and energy, that it is impossible for an isolated electron to radiate only a single photon. (b) With this result in mind, how can you defend the photon exchange diagram in Fig. 44–10?

24. (II) What would be the wavelengths of the two photons produced when an electron and a positron, each with 420 keV of kinetic energy, annihilate?

25. (II) In the rare decay $\pi^+ \to e^+ + \nu_e$, what is the kinetic energy of the positron? Assume the π^+ decays from rest.

26. (II) What minimum kinetic energy must a neutron and proton each have if they are traveling at the same speed toward each other, collide, and produce a K^+K^- pair in addition to themselves? (See Table 44–2.)

27. (II) Calculate the kinetic energy of each of the two products in the decay $\Xi^- \to \Lambda^0 + \pi^-$. Assume the Ξ^- decays from rest.

28. (III) Could a π^+ meson be produced if a 100-MeV proton struck a proton at rest? What minimum kinetic energy must the incoming proton have?

29. (III) Calculate the maximum kinetic energy of the electron in the decay $\mu^- \to e^- + \bar{\nu}_e + \nu_\mu$. [Hint: In what direction do the two neutrinos move relative to the electron in order to give the latter the maximum kinetic energy? Both energy and momentum are conserved; use relativistic formulas.]

Sections 44–7 to 44–11

30. (I) Use Fig. 44–14 to estimate the energy width and then the lifetime of the Δ resonance using the uncertainty principle.

31. (I) The measured width of the J/ψ meson is 88 keV. Estimate its lifetime.

32. (I) The measured width of the ψ(3685) meson is 277 keV. Estimate its lifetime.

33. (I) What is the energy width (or uncertainty) of (a) η^0, and (b) Σ^0?

34. (I) The B$^-$ meson is presumed to be a b$\bar{u}$ quark combination. (a) Show that this is consistent for all quantum numbers. (b) What are the quark combinations for B$^+$, B^0, $\bar{B}^0$?

35. (II) What are the quark combinations that can form (a) a neutron, (b) an antineutron, (c) a Λ^0, (d) a $\bar{\Sigma}^0$?

36. (II) What particles do the following quark combinations produce? (a) uud, (b) $\bar{u}\bar{u}s$, (c) $\bar{u}s$, (d) d$\bar{u}$, (e) $\bar{c}s$.

37. (II) What is the quark combination needed to produce a D^0 meson ($Q = B = S = 0$, $C = +1$)?

38. (II) The D$_S^+$ meson has $S = C = +1$, $B = 0$. What quark combination would produce it?

39. (II) Draw possible Feynman diagrams using quarks (as in Fig. 44–16c) for the reactions (a) $\pi^- + p \to \pi^0 + n$, (b) $\bar{p} + p \to 2\pi^0$.

40. (II) Draw a possible quark Feynman diagram (see Fig. 44–16c) for reaction $K^- + p \to K^- + p$.

General Problems

41. What is the total energy of a proton whose kinetic energy is 25 GeV? What is its wavelength?

42. Assume there are 5×10^{13} protons at 900 GeV stored in the 1.0-km-radius ring of the Tevatron. (a) How much current (amperes) is carried by this beam? (b) How fast would a 1500 kg car have to move to carry the same kinetic energy as this beam?

43. The 4.25-km-radius LEP tunnel will be reused to house the magnets for the Large Hadron Collider (LHC). If the design calls for proton beams of energy 7.0 TeV, what magnetic field will be required?

44. (a) How much energy is released when an electron and a positron annihilate each other? (b) How much energy is released when a proton and an antiproton annihilate each other?

45. Which of the following reactions are possible, and by what interaction could they occur? For those forbidden, explain why.
    (a) $\pi^- + p \rightarrow K^+ + \Sigma^-$
    (b) $\pi^+ + p \rightarrow K^+ + \Sigma^+$
    (c) $\pi^- + p \rightarrow \Lambda^0 + K^0 + \pi^0$
    (d) $\pi^+ + p \rightarrow \Sigma^0 + \pi^0$
    (e) $\pi^- + p \rightarrow p + e^- + \bar{\nu}_e$
    (f) $\pi^- + p \rightarrow K^0 + p + \pi^0$
    (g) $K^- + p \rightarrow \Lambda^0 + \pi^0$
    (h) $K^+ + n \rightarrow \Sigma^+ + \pi^0 + \gamma$
    (i) $K^+ \rightarrow \pi^0 + \pi^0 + \pi^+$
    (j) $\pi^+ \rightarrow e^+ + \nu_e$

46. For the decay $\Lambda^0 \rightarrow p + \pi^-$, calculate (a) the Q-value (energy released), and (b) the kinetic energy of the p and π^-, assuming the Λ^0 decays from rest. (Use relativistic formulas.)

47. Symmetry breaking occurs in the electroweak theory at about 10^{-18} m. Show that this corresponds to an energy that is on the order of the mass of the $W^\pm$.

48. The mass of a π^0 can be measured by observing the reaction $\pi^- + p \rightarrow \pi^0 + n$ at very low incident π^- kinetic energy (assume it is zero). The neutron is observed to be emitted with a kinetic energy of 0.60 MeV. Use conservation of energy and momentum to determine the π^0 mass.

49. Calculate the Q-value for each of the reactions, Eq. 44–4, for producing a pion.

50. Calculate the Q-value for the reaction $\pi^- + p \rightarrow \Lambda^0 + K^0$, when negative pions strike stationary protons. Estimate the minimum pion kinetic energy needed to produce this reaction.

51. How many fundamental fermions are there in a water molecule?

52. Determine the maximum kinetic energy of (a) the positron, and (b) the π^-, in the decay $K^0 \rightarrow \pi^- + e^+ + \nu_e$.

53. (a) Show that the so-called unification distance of 10^{-30} m in grand unified theory is equivalent to an energy of about 10^{14} GeV. Use either the uncertainty principle or de Broglie's wavelength formula, and explain how they apply. (b) Calculate what temperature this corresponds to.

54. For the reaction $p + p \rightarrow 3p + \bar{p}$, where one of the initial protons is at rest, use relativistic formulas to show that the threshold energy is $6m_p c^2$, equal to three times the Q-value of the reaction, where m_p is the proton mass.

55. The lifetimes listed in Table 44–2 are in terms of *proper time*, measured in a reference frame where the particle is at rest. If a tau-lepton is created with a kinetic energy of 450 MeV, how long would its track be as measured in the lab, on average, ignoring any collisions?

56. A particle at rest, with a rest-energy of $m_p c^2$, decays into two fragments with rest energies of $m_{D1} c^2$ and $m_{D2} c^2$. Show that the kinetic energy of fragment D1 is

$$K_{D1} = \frac{1}{2m_p c^2}\left[\left(m_p c^2 - m_{D1} c^2\right)^2 - \left(m_{D2} c^2\right)^2\right]$$

57. Use the quark model to describe the reaction

$$\bar{p} + n \rightarrow \pi^- + \pi^0$$

58. What fraction of the speed of light c is the speed of a 7.0 TeV proton?

Hubble Space Telescope photo of columns of gas and dust inside M16, the Eagle Nebula. The Eagle Nebula is thought to be a site of recent star formation. To discuss the nature of the universe as we understand it today, we first discuss stars and galaxies and how they form and evolve, including the role of nucleosynthesis. We briefly discuss Einstein's general theory of relativity, which deals with gravity and curvature of space. Finally we look at the evidence for the expansion of the universe, and the standard moded of the univese evolving from an initial Big Bang.

CHAPTER 45

Astrophysics and Cosmology

In the previous chapter, we studied the tiniest objects in the universe—the elementary particles. Now we leap to the largest—stars and galaxies. These two extreme realms, elementary particles and the cosmos, are among the most intriguing and exciting subjects in science. And, surprising though it may seem, these distant realms are related in a fundamental way, as already hinted in Chapter 44.

Use of the techniques and ideas of physics to study the heavens is often referred to as **astrophysics**. At the base of our present theoretical understanding of the universe is Einstein's *general theory of relativity* and its theory of gravitation—for in the large-scale structure of the universe, gravity is the dominant force. General relativity serves also as the foundation for modern **cosmology**, which is the study of the universe as a whole. Cosmology deals especially with the search for a theoretical framework to understand the observed universe, its origin, and its future. The questions posed by cosmology are complex and difficult; the possible answers are often unimaginable. They are questions like "Has the universe always existed, or did it have a beginning in time?" Either alternative is difficult to imagine: time going back indefinitely into the past, or an actual moment when the universe began (but, then, what was there before?). And what about the size of the universe? Is it infinite in size? It is hard to imagine infinity. Or is it finite in size? This is also hard to imagine, for if the universe is finite, it does not make sense to ask what is beyond it, because the universe is all there is.

Our survey of astrophysics and cosmology will be necessarily brief and qualitative, but we will nonetheless touch on the major ideas. We begin with a look at what can be seen beyond the Earth.

45–1 Stars and Galaxies

According to the ancients, the stars, except for the few that seemed to move (the planets), were fixed on a sphere beyond the last planet. The universe was neatly self-contained, and we on Earth were at or near its center. But in the centuries following Galileo's first telescopic observations of the heavens in 1610, our view of the universe has changed dramatically. We no longer place ourselves at the center, and we view the universe as vastly larger. The distances involved are so great that we specify them in terms of the time it takes light to travel the given distance: for example, 1 light-second $= (3.0 \times 10^8 \text{ m/s})(1.0 \text{ s}) = 3.0 \times 10^8 \text{ m} = 3.0 \times 10^5 \text{ km}$; 1 light-minute $= 18 \times 10^6 \text{ km}$; and 1 **light-year** (ly) is

$$
\begin{aligned}
1 \text{ ly} &= (2.998 \times 10^8 \text{ m/s})(3.156 \times 10^7 \text{ s/yr}) \\
&= 9.46 \times 10^{15} \text{ m} \approx 10^{13} \text{ km}.
\end{aligned}
$$

Light-year

For specifying distances to the Sun and Moon, we usually use meters or kilometers, but we could specify them in terms of light. The Earth–Moon distance is 384,000 km, which is 1.28 light-seconds. The Earth–Sun distance is 1.50×10^{11} m, or 150,000,000 km; this is equal to 8.3 light-minutes. The most distant planet in the solar system, Pluto, is about 6×10^9 km from the Sun, or 6×10^{-4} ly. The nearest star to us, other than the Sun, is Proxima Centauri, about 4.3 ly away. (Note that the nearest star is 10,000 times farther from us than the farthest planet.)

On a clear moonless night, thousands of stars of varying degrees of brightness can be seen, as well as the elongated cloudy stripe known as the Milky Way (Fig. 45–1). It was Galileo who first observed, about 1610, that the Milky Way is comprised of countless individual stars. A century and a half later (about 1750), Thomas Wright suggested that the Milky Way was a flat disc of stars extending to great distances in a plane, which we call the **Galaxy** (Greek for "milky way").

FIGURE 45–1 A section of the Milky Way. The thin line is the trail of an artificial Earth satellite.

FIGURE 45–2 Our Galaxy, as it would appear from the outside: (a) "end view," in the plane of the disc; (b) "top view," looking down on the disc. (If only we could see it like this— from the outside!) (c) Infrared photograph of the inner reaches of the Milky Way, showing the central bulge of our Galaxy. This very wide angle photo extends over 180° of sky, and to be viewed properly it should be wrapped in a semicircle with your eyes at the center. The white dots are nearby stars.

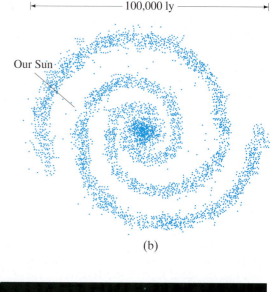

100,000 ly

Our Sun

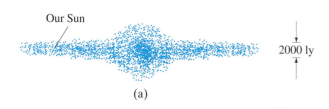

Our Sun

2000 ly

(a)

(b)

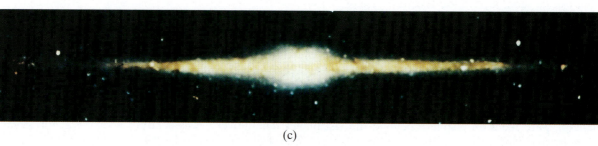

(c)

Our Galaxy has a diameter of almost 100,000 light-years and a thickness of very roughly 2000 light-years. It has a bulging central nucleus and spiral arms (Fig. 45–2). Our Sun, which seems to be just another star, is located more than halfway from the center to the edge, about 28,000 ly from the center. Our Galaxy contains about 10^{11} stars. The Sun orbits the galactic center approximately once every 200 million years, so its speed is about 250 km/s relative to the center of the Galaxy. The total mass of all the stars in our Galaxy is about 3×10^{41} kg.

EXAMPLE 45–1 ESTIMATE Our Galaxy's mass. Estimate the total mass of our Galaxy using the orbital data of the Sun (including our solar system) about the center of the Galaxy. Assume that most of the mass of the Galaxy can be approximated as a uniform sphere of mass (the central bulge, Fig. 45–2a).

SOLUTION Our Sun and solar system orbit the center of the Galaxy, according to the best measurements as mentioned above, with a speed of about $v = 250$ km/s at a distance from the Galaxy center of about $r = 28,000$ ly. We use Newton's second law, $F = ma$, with a being the centripetal acceleration, $a = v^2/r$, and F being the universal law of gravitation (Chapter 6)

$$F = ma$$
$$G\frac{Mm}{r^2} = m\frac{v^2}{r}$$

where M is the mass of the Galaxy and m is the mass of our Sun and solar system. Solving this, we find

$$M = \frac{rv^2}{G} \approx \frac{(28{,}000\ \text{ly})(10^{16}\ \text{m/ly})(2.5 \times 10^5\ \text{m/s})^2}{6.67 \times 10^{-11}\ \text{N}\cdot\text{m}^2/\text{kg}^2} \approx 3 \times 10^{41}\ \text{kg}.$$

In terms of *numbers* of stars, if they are like our Sun ($m = 2.0 \times 10^{30}$ kg), there would be about $(3 \times 10^{41}\ \text{kg})/(2 \times 10^{30}\ \text{kg}) \approx 10^{11}$ or about 100 billion stars.

FIGURE 45–3 This globular star cluster is located in the constellation Hercules.

We can see by telescope, in addition to stars both within and outside the Milky Way, many faint cloudy patches in the sky which were all referred to once as "nebulae" (Latin for "clouds"). A few of these, such as those in the constellations Andromeda and Orion, can actually be discerned with the naked eye on a clear night. Some are **star clusters** (Fig. 45–3), groups of stars that are so numerous they appear to be a cloud. Others are glowing clouds of gas or dust (Fig. 45–4), and it is for these that we now mainly reserve the word **nebula**. Most fascinating are those that belong to a third category: they often have fairly regular elliptical shapes and seem to be a great distance beyond our Galaxy. Immanuel Kant (about 1755) seems to have been the first to suggest that these latter might be circular discs, but appear elliptical because we see them at an angle, and are faint because they are so distant. At first it was not universally accepted that these objects were **extragalactic**—that is, outside our Galaxy. The very large telescopes constructed in the twentieth century revealed that individual stars could be resolved within these extragalactic objects and that many contained spiral arms. Edwin Hubble (1889–1953), who did much of this observational work in the 1920s using the 2.5-m (100-inch) telescope[†] on Mt. Wilson near Los Angeles, California, was also able to demonstrate that these objects were indeed extragalactic because of their great distances. The distance to the Andromeda nebula (a galaxy), for example, is over 2 million light-years, a distance 20 times greater than the diameter of our Galaxy. Thus it was determined that these nebulae are **galaxies** similar to ours. Today, the largest telescopes can see about 10^{11} galaxies. See Fig. 45–5. (Note that it is usual to capitalize the word galaxy only when it refers to our own.)

Galaxies tend to be grouped in **galaxy clusters**, with anywhere from a few to many thousands of galaxies in each cluster. Furthermore, clusters themselves seem to be organized into even larger aggregates: clusters of clusters of galaxies, or **superclusters**. The galaxies nearest us are about 2 million light-years away. The farthest detectable galaxies are thousands of times farther away, on the order of 10^{10} ly. (See Table 45–1.)

[†]2.5 m (= 100 inches) refers to the diameter of the curved objective mirror. The bigger the mirror, the more light it collects and the less diffraction there is, so more and fainter stars can be seen. See Chapters 34 and 36.

FIGURE 45–4 This gaseous nebula, found in the constellation Carina, is about 9000 light-years from us.

**TABLE 45–1
Heavenly Distances**

Object	Approx. Distance from Earth (ly)
Moon	4×10^{-8}
Sun	1.6×10^{-5}
Farthest planet in our solar system (Pluto)	6×10^{-4}
Nearest star (Proxima Centauri)	4.3
Center of our Galaxy	3×10^{4}
Nearest galaxy	2×10^{6}
Farthest galaxies	10^{10}

FIGURE 45–5 Photographs of galaxies. (a) Spiral galaxy in the constellation Hydra. (b) Two galaxies: the larger and more dramatic one is known as the Whirlpool galaxy. (c) The same galaxies as in (b), but this is a false-color infrared image which shows the arms of the spiral as being more regular than in the visible light photo (b); the different colors correspond to different light intensities.

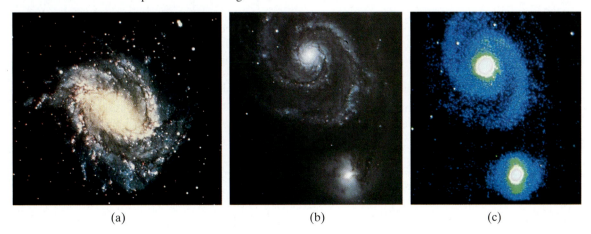

(a)　　　　　　(b)　　　　　　(c)

CONCEPTUAL EXAMPLE 45–2 **Looking back in time.** Astronomers often think of their telescopes as time machines, looking back toward the origin of the universe. How far back do they look?

RESPONSE The distance in light-years measures exactly how long in years the light has been traveling to reach us, so Table 45–1 tells us also how far back in time we are looking. For example, if we saw Proxima Centauri explode into a supernova today, then the event would have really occurred 4.3 years ago. The most distant objects, galaxies 10^{10} ly away, emitted the light we see now 10^{10} years ago, so what we see was how they were then, close to the beginning of the universe.

Besides the usual stars, clusters of stars, galaxies, and clusters and superclusters of galaxies, the universe contains a number of other interesting objects. Among these are stars known as *red giants*, *white dwarfs*, *neutron stars*, *black holes* (at least theoretically), and exploding stars called *novae* and *supernovae*. In addition there are *quasars* ("quasistellar radio sources"), which, if we judge their distance correctly, are galaxies thousands of times brighter than ordinary galaxies. Furthermore, there is radiation that reaches the Earth but does not emanate from the bright pointlike objects we call stars: it is a background radiation that seems to arrive uniformly from all directions in the universe. We discuss all these phenomena in this chapter.

How astronomical distances are measured

We have talked about the vast distances of objects in the universe. But how do we measure these distances? One basic technique employs simple geometry to measure the **parallax** of a star. By parallax we mean the apparent motion of a star, against the background of more distant stars, due to the Earth's motion about the Sun. As shown in Fig. 45–6, the sighting angle of a star relative to the plane of Earth's orbit (angle θ) is measured at different times of the year. Since we know the distance d from Earth to Sun, we can reconstruct the right triangles shown in the figure and can determine[†] the distance D to the star.

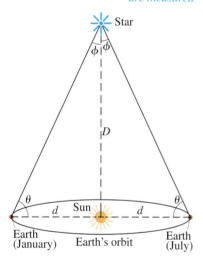

FIGURE 45–6 Distance to a star determined by parallax. (Not to scale.)

EXAMPLE 45–3 **ESTIMATE** **Distance to a star using parallax.** Estimate the distance D to a star if the angle θ in Fig. 45–6 is measured to be 89.99994°.

SOLUTION The angle ϕ in Figure 45–6 is 0.00006°, or about 1.0×10^{-6} radians. Since $\tan \phi = d/D$, then the distance D to the star is

$$D = \frac{d}{\tan \phi} = \frac{d}{\phi} = \frac{1.5 \times 10^8 \text{ km}}{1.0 \times 10^{-6} \text{ rad}} = 1.5 \times 10^{14} \text{ km},$$

or about 15 ly. (We can use $\tan \phi \approx \phi$ since ϕ is very small.)

The distances to stars are often specified in terms of parallax angle given in seconds of arc: 1 second (1″) is $\frac{1}{60}$ of a minute (′) of arc, which is $\frac{1}{60}$ of a degree, so $1″ = \frac{1}{3600}$ of a degree. The distance is then specified in parsecs (meaning *par*allax angle in *sec*onds of arc), where the **parsec** (pc) is defined as $1/\phi$ and ϕ is given in seconds. In the Example we just did, $\phi = (6 \times 10^{-5})°(3600) = 0.22″$ of arc, so we would say the star is at a distance of $1/0.22″ = 4.5$ pc. It is easy to show that the parsec is given by

Parsec

$$1 \text{ pc} = 3.26 \text{ ly}$$
$$= (3.26 \text{ ly})(9.46 \times 10^{15} \text{ m}) = 3.08 \times 10^{16} \text{ m}.$$

Parallax can be used to determine the distance to stars as far away as

---

[†]This is essentially the way the heights of mountains are determined, by "triangulation." See Example 1–7.

100 light-years ($\approx$30 parsecs). Beyond that distance, parallax angles are too small to measure. For greater distances, more subtle techniques must be employed. We can, for example, compare the apparent brightnesses of galaxies, and—using the inverse square law (intensity drops off as the square of the distance)—estimate their relative distances. This method cannot be very precise since we cannot expect all galaxies to have the same intrinsic brightness. A better estimate compares the brightest stars in galaxies or the brightest galaxies in galaxy clusters. It is reasonable to assume, for example, that the brightest stars in all galaxies are similar and have about the same intrinsic luminosity. Consequently, their *apparent* brightness would be a measure of how far away they are.

Another important technique for estimating distance is via the "redshift" in the line spectra of elements, which is related to the expansion of the universe, as will be discussed in Section 45–4.

As we look farther and farther away, the measurement techniques are less and less reliable, so there is more and more uncertainty in the measurements of large distances.

45–2 | Stellar Evolution: The Birth and Death of Stars

The stars appear unchanging. Night after night the heavens reveal no significant variations. Indeed, on a human time scale, the vast majority of stars (except novae and supernovae) change very little. Although stars *seem* fixed in relation to each other, many move sufficiently for the motion to be detected. Indeed, the speeds of stars relative to neighboring stars can be hundreds of km/s, but at their great distance from us, this motion is detectable only by careful measurement. Furthermore, there is a great range of brightness among stars. The differences in brightness are due both to differences in the amount of light stars emit and to their distances from us.

A useful parameter for a star or galaxy is its **absolute luminosity**, L, by which we mean the total power radiated in watts. Also important is the **apparent brightness**, l, defined as the power crossing unit area perpendicular to the path of the light at the Earth. Given that energy is conserved, and ignoring any absorption in space, the total emitted power L at a distance d from the star will be spread over a sphere of surface area $4\pi d^2$. If we let d be the distance from the star to the Earth, then L must be equal to $4\pi d^2$ times l (power per unit area at Earth). That is,

Luminosity and brightness of stars

$$l = \frac{L}{4\pi d^2}. \qquad\qquad (45\text{–}1)$$

EXAMPLE 45–4 Apparent brightness. Suppose a particular star has absolute luminosity equal to that of the Sun but is 10 pc away from Earth. By what factor will it appear dimmer than the Sun?

SOLUTION The luminosity L is the same for both stars, so the apparent brightness depends only on their relative distances. We use the inverse square law as stated above in Eq. 45–1: the star appears dimmer by a factor

$$\frac{l_1}{l_2} = \frac{d_2^2}{d_1^2} = \frac{\left(1.5 \times 10^8\,\text{km}\right)^2}{(10\,\text{pc})^2(3.26\,\text{ly/pc})^2\left(10^{13}\,\text{km/ly}\right)^2} \approx 2 \times 10^{-13},$$

where d_1 and d_2 are the distances from Earth to the star and to the Sun, respectively.

Careful study of nearby stars has shown that for most stars, the absolute luminosity depends on the mass:[†] *the more massive the star, the greater its luminosity.* Another important parameter of a star is its surface temperature, which can be determined from the spectrum of electromagnetic frequencies it emits, just as for a blackbody (Section 38–1). As we saw in Chapter 38, the spectrum of hotter and hotter bodies shifts from predominantly lower frequencies (and longer wavelengths, such as red) to higher frequencies (and shorter wavelengths such as blue). Quantitatively, the relation is given by Wien's law (Eq. 38–1): the peak wavelength λ_P in the spectrum of light emitted by a blackbody (and stars are fairly good approximations to blackbodies) is inversely proportional to its kelvin temperature T; that is, $\lambda_P T = 2.90 \times 10^{-3}\,\text{m}\cdot\text{K}$. The surface temperatures of stars typically range from about 3500 K (reddish) to perhaps 50,000 K (bluish).

An important astronomical discovery, made around 1900, was that for most stars, the color is related to the absolute luminosity and therefore to the mass. A useful way to present this relationship is by the so-called Hertzsprung–Russell (H–R) diagram. On the H–R diagram, the horizontal axis shows the temperature T whereas the vertical axis is the luminosity L, and each star is represented by a point on the diagram, Fig. 45–7. Most stars fall along the diagonal band termed the **main sequence**. Starting at the lower right we find the coolest stars, reddish in color; they are the least luminous and therefore of low mass. Farther up toward the left we find hotter and more luminous stars that are yellowish-white, like our Sun. Still farther up we find still more massive and more luminous stars, bluish in color. Stars that fall on this diagonal band are called *main-sequence stars*. There are also stars that fall outside the main sequence. Above and to the right we find extremely large stars, with high luminosities but with low (reddish) color temperature: these are called *red giants*. At the lower left, there are a few stars of low luminosity but with high temperature: these are the *white dwarfs*.

[†] The mass of a star can be determined by observing its gravitational effects. Most stars are part of a cluster, the simplest being a binary star in which two stars orbit around each other, allowing their masses to be determined using Newtonian mechanics.

FIGURE 45–7
Hertzsprung–Russell (H–R) diagram.

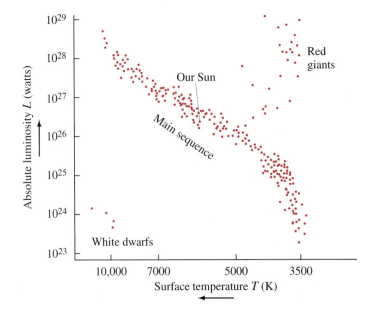

Before considering the different types of stars, let us first look at a couple of Examples that show how physical principles can be used to gain information about stars.

EXAMPLE 45–5 **Determining star temperatures and size.** Suppose that the distances to two nearby stars can be reasonably estimated and this data, together with their measured apparent brightnesses, suggests that the two stars have about the same absolute luminosity, L. The spectrum of one of the stars peaks at about 700 nm (so it is reddish). The spectrum of the other peaks at about 350 nm (bluish). Use Wien's law (Eq. 38–1) and the Stefan-Boltzmann equation (Section 19–10) to determine (*a*) the surface temperature of each star, and (*b*) how much larger one star is than the other.

SOLUTION (*a*) Wien's law states that $\lambda_P T = 2.90 \times 10^{-3}\,\text{m}\cdot\text{K}$. So the temperature of the reddish star is

$$T_r = \frac{2.90 \times 10^{-3}\,\text{m}\cdot\text{K}}{700 \times 10^{-9}\,\text{m}} = 4140\,\text{K}.$$

The temperature of the bluish star will be double this since its peak wavelength is half (350 nm vs. 700 nm); just to check:

$$T_b = \frac{2.90 \times 10^{-3}\,\text{m}\cdot\text{K}}{350 \times 10^{-9}\,\text{m}} = 8280\,\text{K}.$$

(*b*) The Stefan-Boltzmann equation, which we discussed in Chapter 19 (see Eq. 19–15), states that the power radiated *per unit area* of surface from a body is proportional to the fourth power of the kelvin temperature, T^4. Now the temperature of the bluish star is double that of the reddish star, so the bluish one must radiate $(2^4) = 16$ times as much energy per unit area. But we are given that they have the same luminosity (the same total power output), so the surface area of the blue star must be $\frac{1}{16}$ that of the red one. Since the surface area of a sphere is $4\pi r^2$, we conclude that the radius of the reddish star is $\sqrt{16} = 4$ times larger than the radius of the bluish star (or $4^3 = 64$ times the volume).

EXAMPLE 45–6 **ESTIMATE** **Distance to a star using H–R and color.** Suppose that detailed study of a certain star suggests that it most likely fits on the main sequence of an H–R diagram. Its measured apparent brightness is $l = 1.0 \times 10^{-12}\,\text{W/m}^2$, and the peak wavelength of its spectrum is $\lambda_P \approx 600\,\text{nm}$. Estimate its distance from us.

SOLUTION The star's temperature, from Wien's law (Eq. 38–1), is

$$T \approx \frac{2.90 \times 10^{-3}\,\text{m}\cdot\text{K}}{600 \times 10^{-9}\,\text{m}} \approx 4800\,\text{K}.$$

A star on the main sequence of an H–R diagram at this temperature has absolute luminosity of about $L \approx 1 \times 10^{26}\,\text{W}$, read off of Fig. 45–7. Then, from Eq. 45–1,

$$d = \sqrt{\frac{L}{4\pi l}} \approx \sqrt{\frac{1 \times 10^{26}\,\text{W}}{4(3.14)(1.0 \times 10^{-12}\,\text{W/m}^2)}} \approx 3 \times 10^{18}\,\text{m}.$$

Its distance from us in light-years and parsecs is

$$d = \frac{3 \times 10^{18}\,\text{m}}{10^{16}\,\text{m/ly}} \approx 300\,\text{ly} \approx \frac{300\,\text{ly}}{3.26\,\text{ly/pc}} \approx 90\,\text{pc}.$$

Why are there different types of stars, such as red giants and white dwarfs, as well as main-sequence stars? Were they all born this way, in the beginning? Or might each different type represent a different age in the life cycle of a star? Astronomers and astrophysicists today believe the latter is most likely the case. Note, however, that we cannot actually follow any but the tiniest part of the life cycle of any given star since they live for ages vastly greater than ours, on the order of millions or billions of years. Nonetheless, let us follow the process of **stellar evolution** from the birth to the death of a star, as astrophysicists have theoretically reconstructed it today.

Birth of a star

Contraction due to gravity

Stars are born, it is believed, when gaseous clouds (mostly hydrogen) contract due to the pull of gravity. A huge gas cloud might fragment into numerous contracting masses, each mass centered in an area where the density was only slightly greater than that at nearby points. Once such "globules" formed, gravity would cause each to contract in toward its center of mass. As the particles of such a *protostar* accelerate inward, their kinetic energy increases. When the kinetic energy is sufficiently high, the Coulomb repulsion that keeps the hydrogen nuclei apart can be overcome and nuclear fusion can take place. In a star like our Sun, the "burning" of hydrogen[†] occurs via the proton–proton cycle (Section 43–4, Eqs. 43–6), in which four protons fuse to form a $^4_2$He nucleus with the release of γ rays and neutrinos. These reactions require a temperature of about 10^7 K, corresponding to an average kinetic energy (kT) of about 1 keV. In more massive stars, the carbon cycle produces the same net effect: Four $^1_1$H produce a $^4_2$He—see Section 43–4. The fusion reactions take place primarily in the core of a star, where T is sufficiently high. (The surface temperature is, of course, much lower—on the order of a few thousand kelvins.) The tremendous release of energy in these fusion reactions produces a pressure sufficient to halt the gravitational contraction, and our protostar, now really a young *star*, stabilizes on the main sequence. Exactly where the star falls along the main sequence depends on its mass. The more massive the star, the farther up (and to the left) it falls on the diagram. To reach the main sequence requires perhaps 30 million years, if it is a star like our Sun, and (assuming our theory is right) it will remain there[‡] about 10 billion years (10^{10} yr). Although most stars are billions of years old, there is evidence that stars are actually being born at this moment.

Fusion begins when T (and $\bar{K}$) is large enough

Reaching the main sequence

As hydrogen "burns"—that is, fuses to form helium—the helium that is formed is denser and tends to accumulate in the central core where it was formed. As the core of helium grows, hydrogen continues to "burn" in a shell around it, Fig. 45–8. When much of the hydrogen within the core has been consumed, the production of energy decreases and is no longer sufficient to prevent the huge gravitational forces from once again causing the core to contract and heat up. The hydrogen in the shell around the core then "burns" even more fiercely because of this rise in temperature, causing the outer envelope of the star to expand and to cool. The surface temperature, thus reduced, produces a spectrum of light that peaks at longer wavelength. By this time the star has left the main sequence. It has become redder, and as it has grown in size, it has become more luminous. So it will have moved to the right and upward on the H–R diagram, as shown in Fig. 45–9. As it moves upward, it enters the **red giant** stage. Thus, theory explains the origin of red giants as a natural step in a star's evolution. Our Sun, for example, has been on the main sequence for about $4\frac{1}{2}$ billion years. It will probably remain there another 4 or 5 billion years. (We can take comfort in that!) When our Sun leaves the main sequence, it is expected to grow in size (as a red giant) until it occupies all the volume out to approximately the present orbit of Earth.

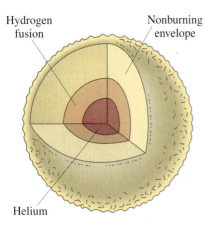

Hydrogen fusion

Nonburning envelope

Helium

FIGURE 45–8 A shell of "burning" hydrogen (fusing to become helium) surrounds the core where the newly formed helium gravitates.

Red giants

[†] The word "burn" is put in quotation marks because these high-temperature fusion reactions occur via a *nuclear* process, and must not be confused with ordinary burning (of, say, paper, wood, or coal) in air, which is a chemical reaction at the *atomic* level (and at a much lower temperature).

[‡] More massive stars, since they are hotter and the Coulomb repulsion is more easily overcome, "burn" much more quickly, and so use up their fuel faster, resulting in shorter lives. A star 10 times more massive than our Sun, for example, will remain on the main sequence only for about 10^7 years. Stars less massive than our Sun live much longer than our Sun's 10^{10} yr.

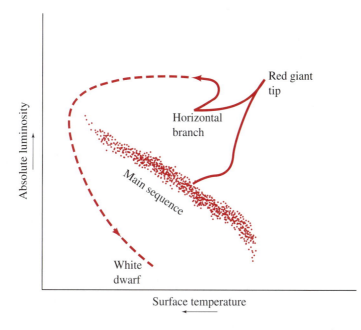

FIGURE 45–9 Evolution of a star like our Sun represented on an H–R diagram.

As a star's envelope expands, the core is shrinking and heating up. When the temperature reaches about 10^8 K, even helium nuclei, in spite of their greater charge and hence greater electrical repulsion, can then undergo fusion. The reactions are

$$\begin{aligned} {}_2^4\text{He} + {}_2^4\text{He} &\longrightarrow {}_4^8\text{Be} + \gamma \quad (Q = -95\,\text{keV}) \\ {}_2^4\text{He} + {}_4^8\text{Be} &\longrightarrow {}_6^{12}\text{C} + \gamma. \quad (Q = 7.4\,\text{MeV}) \end{aligned} \qquad (45\text{–}2)$$

Nucleosynthesis

The first reaction is (slightly) endothermic, and the ${}_4^8$Be formed is unstable and can decay quickly $(10^{-16}\,\text{s})$ back to two alpha particles. So the second reaction must occur very quickly after the first, and this is aided by the particularly large cross section for the second reaction. The net effect is

$$3\,{}_2^4\text{He} \longrightarrow {}_6^{12}\text{C}. \quad (Q = 7.3\,\text{MeV}).$$

The burning of helium causes a rapid and major change in the star, and the star moves rapidly to the "horizontal branch" on the H–R diagram (see Fig. 45–9). Further reactions are possible, creating elements of higher and higher Z, up to $Z = 10$ or 12. If the mass of the star is greater than about 0.7 solar masses, then still higher Z elements can be formed. The star can get even hotter and in the range $T = 2.5\text{–}5 \times 10^9$ K, nuclei as heavy as ${}_{26}^{56}$Fe and ${}_{28}^{56}$Ni can be made. But here the process of **nucleosynthesis**, the formation of heavy nuclei from lighter ones by fusion, ends. As we saw in Fig. 42–1, the average binding energy per nucleon begins to decrease for A greater than about 60. Further fusions would *require* energy, rather than release it. (Elements heavier than Ni are probably formed mainly by neutron capture, in supernova explosions, as we shall discuss shortly.)

Production of heavy elements

What happens next depends on the mass of the star. If the star has a residual mass less than about 1.4 solar masses[†] (known as the *Chandrasekhar limit*), no further fusion energy can be obtained and the star collapses under the action of gravity. As it shrinks, the star cools and typically follows the route shown in Fig. 45–9, according to theory, descending from the upper left downward, becoming a **white dwarf**. A white dwarf with a mass equal to that of the Sun would be about the size of the Earth. What determines its size is that it collapses to the point at which the electron clouds of the atoms start to overlap: at this point it collapses no further because, as the Pauli exclusion principle claims, no two electrons can be in the same quantum state. A white dwarf continues to lose internal energy, decreasing in temperature and becoming dimmer and dimmer until its light goes out. It has then become a *black dwarf*, a dark cold chunk of ash.

White dwarfs

[†]This cutoff refers to the mass of the star remaining *after* the red giant stage.

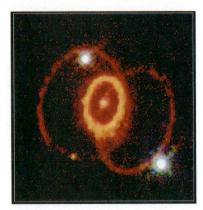

FIGURE 45–10 These glowing, expanding gas rings, a light-year in diameter, were observed by the Hubble Space Telescope a decade after the explosion of supernova SN1987a.

Neutron stars

Supernovae

FIGURE 45–11 SN1987a, the supernova that appeared in February 1987, is the very bright star on the right. The bright nebulous object is the Tarantula Nebula.

More massive stars (residual mass greater than 1.4 solar masses) are thought to follow a quite different scenario. A star with this great a mass can contract under gravity and heat up even further, reaching extremely high temperatures. The kinetic energy of the nuclei is then so high that fusion of elements heavier than iron is possible even though the reactions require energy input ($Q < 0$). Furthermore the high-energy collisions can also cause the breaking apart of iron and nickel nuclei into He nuclei, and eventually into protons and neutrons:

$$^{56}_{26}\text{Fe} \rightarrow 13\,^4_2\text{He} + 4\text{n}$$

$$^4_2\text{He} \rightarrow 2\text{p} + 2\text{n}.$$

These are energy-requiring (endoergic) reactions, but at such extremely high temperature and pressure there is plenty of energy available, enough even to force electrons and protons together to form neutrons in inverse β decay:

$$\text{e}^- + \text{p} \rightarrow \text{n} + \nu.$$

As the core contracts under the huge gravitational forces, the tremendous mass becomes essentially an enormous nucleus made up almost exclusively of neutrons. The size of the star is no longer limited by the exclusion principle applied to electrons, but rather applied to neutrons, and the star begins to contract rapidly toward forming an enormously dense **neutron star**. The contraction of the core would mean a great reduction in gravitational potential energy. Somehow this energy would have to be released. Indeed, it was suggested in the 1930s that the final core collapse to a neutron star may be accompanied by a catastrophic explosion whose tremendous energy could form virtually all elements of the periodic table and blow away the entire outer envelope of the star (Fig. 45–10), spreading its contents into interstellar space. Such explosions are believed to produce one type of observed **supernova** (there may be other mechanisms as well). Indeed, the presence of heavy elements on Earth and in our solar system suggests that our solar system may have formed from the debris from at least one supernova.

In a supernova explosion, a star's brightness is observed to suddenly increase billions of times in a period of just a few days and then fade away over the next few months. In February 1987 such a supernova occurred in an appendage to our Galaxy known as the Large Magellanic Cloud, only 170,000 ly away, and was visible to the naked eye in the southern hemisphere (Fig. 45–11). Supernovae are seen relatively often in distant galaxies, but this recent one, called SN1987a, is the first to have occurred close enough to be visible to the naked eye since the famous one of 1604 observed by Kepler and Galileo (and thus SN1987a is the first since the invention of the telescope). A supernova visible to the naked eye was observed by Chinese astronomers in A.D. 1054, and its remains are still visible in the sky (in the Crab Nebula), in the midst of which there is now a **pulsar**. Pulsars are astronomical objects that emit sharp pulses of radiation at regular intervals on the order of a second. They are now believed to be neutron stars which, because of conservation of angular momentum, increase greatly in rotational speed as their moment of inertia decreases during their contraction. The intense magnetic field of such a rapidly rotating star (perhaps as high as 10^6 to 10^{10} T) could trap and accelerate charged particles, which then give off radiation in a beam that rotates with the star. The discovery of pulsars in 1967 has lent support to the theory that neutron stars are the result of supernova explosions.

The core of a neutron star contracts to the point at which all neutrons are as close together as they are in a nucleus. That is, the density of a neutron star is on the order of 10^{14} times that of normal solids and liquids on Earth. A neutron star that has a mass 1.5 times that of our Sun would have a diameter of only about 10 km.

If the mass of a neutron star is less than about two or three solar masses, its subsequent evolution is thought to be similar to that of a white dwarf. If the mass is greater than this, the gravitational force may become so strong that the star contracts to an even smaller diameter and an even greater density, overcoming, in a sense, even the neutron exclusion principle. Gravity would then be so strong that even light emitted from it could not escape—it would be pulled back in by the force of gravity. In other words, the escape velocity from such a star is greater than c, the speed of light. Since no radiation could escape from such a star, we could not see it—it would be black. A body may pass by it and be deflected by its gravitational field, but if it came too close, it would be swallowed up, never to escape. This is a **black hole**. Black holes are predicted by theory to exist. Experimentally, evidence for their existence is strong but not everyone agrees it is sufficient to fully confirm the existence of black holes. One possibility is there may be a giant black hole at the center of our Galaxy—its mass is estimated at several million times that of the Sun—and it seems possible that many or all galaxies have black holes at their centers.

Black holes

45-3 General Relativity: Gravity and the Curvature of Space

We have seen that the force of gravity plays a dominant role in the processes that occur in stars and, in general, in the evolution of the universe as a whole. The reasons gravity, and not one of the other of the four forces in nature, plays the dominant role in the universe are (1) it is long-range and (2) it is always attractive. The strong and weak nuclear forces act over very short distances only, on the order of the size of a nucleus; hence they do not act over astronomical distances (although of course they act between nuclei and nucleons in stars to produce nuclear reactions). The electromagnetic force, like gravity, acts over great distances; but it can be either attractive or repulsive. And since the universe does not seem to contain large areas of net electric charge, a large net force does not occur. But gravity acts as an attractive force between *all* masses, and there are large accumulations in the universe of only the one "sign" of mass (not + and − as with electric charge).

However, the force of gravity as Newton described it in his law of universal gravitation shows discrepancies on a cosmological scale. Einstein, in his general theory of relativity, developed a theory of gravity that resolves these problems and forms the basis of cosmological dynamics.

In the *special theory of relativity* (Chapter 37), Einstein concluded that there is no way for an observer to determine whether a given frame of reference is at rest or is moving at constant velocity in a straight line. Thus the laws of physics must be the same in different inertial reference frames. But what about the more general case of motion where reference frames can be *accelerating*?

It is in the **general theory of relativity** that Einstein tackled the problem of accelerating reference frames and developed a theory of gravity. The mathematics of general relativity is very complex, so our discussion will be mainly qualitative.

We begin with Einstein's famous **principle of equivalence** (briefly discussed in Chapter 6), which states that

> **No experiment can be performed that could distinguish between a uniform gravitational field and an equivalent uniform acceleration.**

If observers sensed that they were accelerating (as in a vehicle speeding around a sharp curve), they could not prove by any experiment that in fact they weren't simply experiencing the pull of a gravitational field. Conversely, we might think we are being pulled by gravity when in fact we are undergoing an "inertial" acceleration having nothing to do with gravity.

As a thought experiment, consider a person in a freely falling elevator near the Earth's surface. If our observer held out a book and let go of it, what would happen? Gravity would pull it downward toward the Earth, but at the same rate $\left(g = 9.8\,\text{m/s}^2\right)$ at which the person and elevator were falling. So the book would hover right next to the person's hand (Fig. 45–12). The effect is exactly the same as if this reference frame were at rest and *no* forces were acting. On the other hand, suppose the elevator were far out in space where there is no gravitational field. If the person released the book, it would float, just as it does in Fig. 45–12.

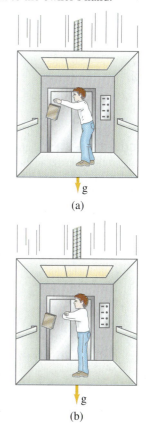

FIGURE 45–12 A freely falling elevator. The released book hovers next to the owner's hand.

(a)

(b)

If, instead, the elevator (out in space) were accelerating upward at an acceleration of $9.8\,\text{m/s}^2$, the book as seen by our observer would fall to the floor with an acceleration of $9.8\,\text{m/s}^2$, just as if it were falling because of gravity. According to the principle of equivalence, the observer could do no experiment to determine whether the book fell because the elevator was accelerating upward at $a = 9.8\,\text{m/s}^2$ in the absence of gravity, or because a gravitational field with $g = 9.8\,\text{m/s}^2$ was acting downward and he was at rest (say on the Earth). The two descriptions are equivalent.

Mass: inertial vs. gravitational

The principle of equivalence is related to the concept that there are two types of mass. Newton's second law, $F = ma$, uses **inertial mass**. We might say that inertial mass represents "resistance" to any type of force whatever. The second type of mass is **gravitational mass**. When one body attracts another by the gravitational force (Newton's law of universal gravitation, $F = Gm_1 m_2/r^2$, Chapter 6), the strength of the force is proportional to the product of the gravitational masses of the two bodies. This is much like the electric force between two bodies which is proportional to the product of their electric charges. The electric charge of a body is not related to its inertial mass; so why should we expect that a body's gravitational mass (call it gravitational charge if you like) be related to its inertial mass? We have assumed they were the same. Why? Because no experiment—not even high-precision experiments—has been able to discern any measurable difference between inertial and gravitational mass. This is another way to state the equivalence principle: *gravitational mass is equivalent to inertial mass.*

Light bending

The principle of equivalence can be used to show that light ought to be deflected due to the gravitational force of a massive body. Consider another thought experiment, in which an elevator is in free space where no gravity acts. If a light beam enters a hole in the side of the elevator, the beam travels straight across the elevator and makes a spot on the opposite side if the elevator is at rest (Fig. 45–13a). If the elevator is accelerating upward as in Fig. 45–13b, the light beam still travels straight across in a reference frame at rest. In the upwardly accelerating elevator, however, the beam is observed to curve downward. Why? Because during the time the light travels from one side of the elevator to the other, the elevator is moving upward at ever-increasing speed. Now according to the equivalence principle, an upwardly accelerating reference frame is equivalent to a downward gravitational field. Hence, we can picture the curved light path in Fig. 45–13b as being the effect of a gravitational field. Thus we expect gravity to exert a force on a beam of light and to bend it out of a straight-line path.

That light is affected by gravity is an important prediction of Einstein's general theory of relativity. And it can be tested. The amount a light beam would be

FIGURE 45–13 (a) Light beam goes straight across an elevator not accelerating. (b) The light beam bends (exaggerated) in an elevator accelerating in an upward direction.

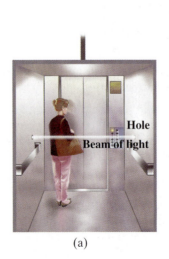

(a)

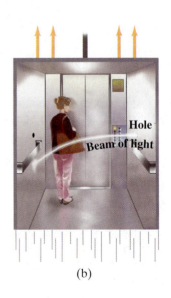

(b)

deflected from a straight-line path must be small even when passing a massive body. (For example, light near the Earth's surface after traveling 1 km is predicted to drop only about 10^{-10} m, which is equal to the diameter of a small atom and not detectable.) The most massive body near us is the Sun, and it was calculated that light from a distant star would be deflected by $1.75''$ of arc (tiny but detectable) as it passed near the Sun (Fig. 45–14). However, such a measurement could be made only during a total eclipse of the Sun, so that the Sun's tremendous brightness would not overwhelm the starlight passing near its edge. An opportune eclipse occurred in 1919 and scientists journeyed to the South Atlantic to observe it. Their photos of stars around the Sun revealed shifts in accordance with Einstein's prediction. (See Fig. 45–15.)

A light beam must travel by the shortest, most direct, path between two points. If it didn't, some other object could travel between the two points in a shorter time and thus have a greater speed than the speed of light—a clear contradiction of the special theory of relativity. If a light beam can follow a curved path (as discussed above), then this curved path must be the shortest distance between the two points—which suggests that *space itself is curved* and that it is the gravitational field that causes the curvature. Indeed, the curvature of space—or rather, of four-dimensional space-time—is a basic aspect of general relativity.

What is meant by *curved space*? To understand, let us recall that our normal method of viewing the world is via Euclidean plane geometry. In Euclidean geometry, there are many axioms and theorems we take for granted, such as that the sum of the angles of any triangle is $180°$. Other geometries, non-Euclidean which involve curved space, have also been imagined by mathematicians. Now it is hard enough to imagine three-dimensional curved space, much less curved four-dimensional space-time. So let us explain the idea of curved space by using two-dimensional surfaces.

Consider, for example, the two-dimensional surface of a sphere. It is clearly curved, Fig. 45–16, at least to us who view it from the outside—from our three-dimensional world. But how would hypothetical two-dimensional creatures determine whether their two-dimensional space were flat (a plane) or curved? One way would be to measure the sum of the angles of a triangle. If the surface is a plane, the sum of the angles is $180°$, as we learn in plane geometry. But if the space is curved, and a sufficiently large triangle is constructed, the sum of the angles will *not* be $180°$. To construct a triangle on a curved surface, say a sphere, we must use the equivalent of a straight line: that is, the shortest distance between two points, which is called a **geodesic**. On a sphere, a geodesic is an arc of a great circle (an arc contained in a plane passing through the center of the sphere) such as the Earth's equator and the Earth's longitude lines. Consider, for example, the large triangle of Fig. 45–16 whose sides are two "longitude" lines passing from the "north pole" to the equator, a part of which forms the third side. The two longitude lines make $90°$ angles with the equator (look at a world globe to see this more clearly). If they make, say, a $90°$ angle with each other at the north pole, the sum of these angles is $90° + 90° + 90° = 270°$. This is clearly *not* a Euclidean space. Note, however, that if the triangle is small in comparison to the radius of the sphere, the angles will add up to nearly $180°$, and the triangle (and space) will seem flat.

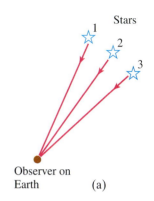

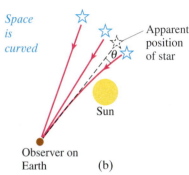

FIGURE 45–14 (a) Three stars in the sky. (b) If the light from one of these stars passes very near the Sun, whose gravity bends the rays, the star will appear higher than it actually is.

Geodesic

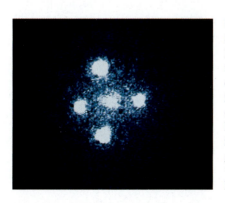

FIGURE 45–15 Photograph, taken by the Hubble Space Telescope, of the so-called "Einstein cross" (or Huchra's cross after its discoverer, John Huchra). It is thought to represent "gravitational lensing": The central spot is a relatively nearby galaxy, whereas the four other spots are thought to be images of a single quasar *behind* the galaxy, and the galaxy bends the light coming from the quasar to produce the four images. See also Fig. 45–14. (If the shape of the nearby galaxy were perfectly symmetric, we would expect the "image" of the distant quasar to be a circular ring or halo, instead of the four spots seen here.)

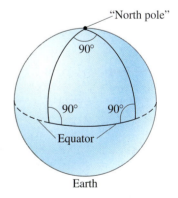

FIGURE 45–16 On a two-dimensional curved surface, the sum of the angles of a triangle may not be $180°$.

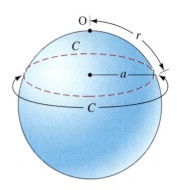

FIGURE 45–17 On a spherical surface, a circle of circumference C is drawn about point O as the center. The radius of the circle (not the sphere) is the distance r along the surface. (Note that in our three-dimensional view, we can tell that $2\pi a = C$; since $r > a$, then $2\pi r > C$.)

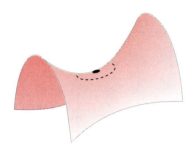

FIGURE 45–18 Example of a two-dimensional surface with negative curvature.

The universe: open or closed?

FIGURE 45–19 Rubber-sheet analogy for space-time curved by matter.

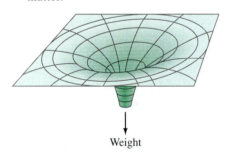

Weight

Another way to test the curvature of space is to measure the radius r and circumference C of a large circle. On a plane surface, $C = 2\pi r$. But on a two-dimensional spherical surface, C is *less* than $2\pi r$, as can be seen in Fig. 45–17. The proportionality constant between C and r is *less* than 2π. Such a surface is said to have *positive curvature*. On the saddlelike surface of Fig. 45–18, the circumference of a circle is greater than $2\pi r$, and the sum of the angles of a triangle is less than 180°. Such a surface is said to have a *negative curvature*.

Now, what about our universe? On a large scale (not just near a large mass), what is the overall curvature of the universe? Does it have positive curvature, negative curvature, or is it flat (zero curvature)? In the nineteenth century, Carl Friedrich Gauss (1777–1855) tried to determine whether our natural three-dimensional space deviated from Euclidean space by measuring the angles of a triangle formed by three mountain peaks using light rays as sides of the triangle. He was unable to detect any deviation from 180°, nor have experiments today detected any deviation.

Nonetheless, the question of the curvature of space is an important one in cosmology. If the universe has a positive curvature, then the universe would be *finite*, or *closed*. This does not mean that in such a universe the stars and galaxies would extend out to a certain boundary, beyond which there is empty space. Rather, galaxies would be spread throughout the space, and the space would fold back and "close on itself." There is no boundary or edge in such a universe. If a particle were to move in a straight line in a particular direction, it would eventually return to the starting point—albeit eons of time later. To ask "What is beyond such a closed universe?" is futile, since the space of the universe is all there is. If the curvature of space is zero or negative, on the other hand, the universe would be *open*. It would just go on and on and never fold back on itself. An open universe would be *infinite*. Whether the universe is open or closed depends, in part, on how much total mass there is in the universe, as we will discuss in Section 45–7. If the mass is great enough, it would bend space into a positively curved, closed, and finite space.

According to Einstein's theory, space-time is curved, especially near massive bodies. To visualize this, we might think of space as being like a thin rubber sheet; if a heavy weight is hung from it, it curves as shown in Fig. 45–19. The weight corresponds to a huge mass that causes space (space itself!) to curve. Thus, in Einstein's theory[†] we do not speak of the "force" of gravity acting on bodies. Instead we say that bodies and light rays move along geodesics (the equivalent of straight lines in plane geometry) in curved space-time. Thus, a body at rest or moving slowly near the great mass of Fig. 45–19 would follow a geodesic toward that body.

[†] Alexander Pope (1688–1744) wrote an epitaph for Newton:
"Nature, and Nature's laws lay hid in night:
God said, *Let Newton be!* and all was light."
Sir John Squire (1884–1958), perhaps uncomfortable with Einstein's profound thoughts, added:
"It did not last: the Devil howling '*Ho!*
Let Einstein be!' restored the status quo."

The extreme curvature of space-time shown in Fig. 45–19 could be produced by a **black hole**. A black hole, as we saw in the previous Section, is so dense that even light cannot escape from it. To become a black hole, a body of mass M must undergo **gravitational collapse**, contracting by gravitational self-attraction to within a radius called the **Schwarzschild radius**:

Black holes

$$R = \frac{2GM}{c^2}$$

where G is the gravitational constant and c the speed of light. If a body collapses to within this radius, it is predicted by general relativity to rapidly $(\approx 10^{-5}\,\text{s})$ collapse to a point at $r = 0$, forming an infinitely dense **singularity**. This prediction is uncertain, however, since in this realm we need to combine quantum mechanics with gravity, a unification of theories not yet fully achieved.

The Schwarzschild radius also represents the **event horizon** of a black hole. By event horizon we mean the surface beyond which no signals can ever reach us, and thus inform us of events that happen. As a star collapses toward a black hole, the light it emits is pulled harder and harder by gravity, but we can still see it. Once the matter passes within the event horizon the emitted light cannot escape, but is pulled back in by gravity.[†]

Event horizon

All we can know about a black hole is its mass, its angular momentum (there could be rotating black holes), and its electric charge. No other information, no details of its structure or the kind of matter it was formed of, can be known.

How might we observe black holes? We cannot see them because no light can escape from them. They would be black objects against a black sky. But they do exert a gravitational force on nearby bodies. The suspected black hole at the center of our Galaxy was discovered by examining the motion of matter in its vicinity. Another technique is to examine stars which appear to be rotating as if they were members of a *binary system* (two stars rotating about their common center of mass), although the companion is invisible. If the unseen star is a black hole, it might be expected to pull off gaseous material from its visible companion (Fig. 45–20). As this matter approached the black hole, it would be highly accelerated and should emit X-rays of a characteristic type. One of the candidates for a black hole is in the binary-star system Cygnus X-1.

[†] According to quantum mechanics, matter and radiation could escape from a black hole by tunneling, but the rate at which this could happen would be expected to be extremely low.

FIGURE 45–20 Artist's conception of how a black hole, which is one star of a binary pair, might pull matter from the other (more normal) star.

Star

Black hole

45–4 The Expanding Universe

We discussed in Section 45–2 how individual stars evolve from their birth to their death as white dwarfs, neutron stars, and black holes. But what about the universe as a whole: is it static, or does it evolve? The evolution of stars suggests that the universe as a whole evolves. One of the most important scientific ideas of this century proposes that distant galaxies are racing away from us, and that the farther they are from us, the faster they are moving away. How astronomers arrived at this astonishing idea and what it means for the past history of the universe as well as its future, will occupy us for the remainder of the book.

Hubble and the redshift

The idea that the universe is expanding was first put forth by Edwin Hubble in 1929. It was based on observations of the Doppler shift of light emitted by stars. In Chapter 16 we discussed how the frequency and wavelength of sound are altered if the source is moving toward or away from an observer. If the source moves toward us, the frequency is higher and the wavelength is shorter. If the source moves away from us, the frequency is lower and the wavelength is longer. The **Doppler effect** occurs also for light, but the shifted wavelength or frequency is given by a formula slightly different from that for sound because for light (according to special relativity), we can make no distinction between motion of the source and motion of the observer. (Recall that sound travels in a medium, such as air, but light does not; according to relativity, there is no ether.) According to special relativity, the formula is (see Section 37–12)

$$\lambda = \lambda_0 \sqrt{\frac{1 + v/c}{1 - v/c}}, \qquad \left[\begin{array}{l}\text{source and observer moving}\\\text{away from each other}\end{array}\right] \quad \textbf{(45–3)}$$

where λ_0 is the emitted wavelength as seen in a reference frame at rest with respect to the source, and λ is the wavelength measured in a frame moving with velocity v away from the source along the line of sight. Note that this relation depends only on the relative velocity v. (For relative motion toward each other, $v < 0$ in this formula.) Thus, when a source emits light of a particular wavelength and the source is moving away from us, the wavelength appears longer to us: the color of the light (if it is visible) is shifted toward the red end of the visible spec-

Redshift

trum, an effect known as a **redshift**. If the source moves toward us, the color shifts toward the blue (shorter wavelength) end of the spectrum. The amount of shift depends on the velocity of the source (Eq. 45–3). For speeds not too close to the speed of light, it is easy to show (Problem 30) that the fractional change in wavelength is proportional to the speed of the source to or away from us (as was the case for sound):

$$\frac{\Delta\lambda}{\lambda_0} = \frac{\lambda - \lambda_0}{\lambda_0} \approx \frac{v}{c}. \qquad\qquad [v \ll c] \quad \textbf{(45–4)}$$

In the spectra of stars and galaxies, lines are observed that correspond to lines in the known spectra of particular atoms (see Section 38–9). What Hubble found was that the lines seen in the spectra of galaxies were generally *redshifted*, and that the amount of shift seemed to be approximately proportional to the distance of the galaxy from us. That is, the velocity, v, of a galaxy moving away from us is proportional to its distance, d, from us:

Hubble's law

$$v = Hd. \qquad\qquad\qquad\qquad\qquad \textbf{(45–5)}$$

This is **Hubble's law**, one of the most important astronomical ideas of this century. The constant H is called the **Hubble parameter**. Hubble's law does not work well for nearby galaxies—in fact some are actually moving toward us ("blueshifted"); but this is believed to merely represent random motion of the galaxies. For more distant galaxies, the velocity of recession (Hubble's law) is much greater than that of random motion, and so is dominant.

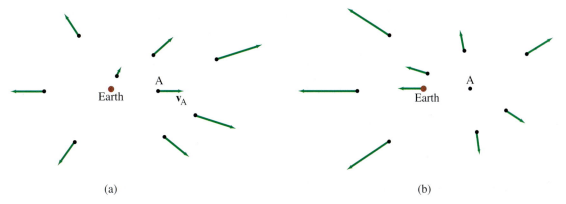

(a) (b)

FIGURE 45–21 Expansion of the universe looks the same from any point in the universe.

The value of H is not known very precisely. It is now estimated to be about

$$H \approx 70 \text{ km/s/Mpc}$$

Hubble's constant

(that is, 70 km/s per megaparsec of distance) with an uncertainty of about 20 percent. If we use light-years for distance, then $H \approx 20$ km/s per million light-years of distance.

What does it mean that distant galaxies are all moving away from us, and with ever greater speed the farther they are from us? It is as if there had been a great explosion at some distant time in the past. And at first sight we seem to be in the middle of it all. But we aren't, necessarily. The expansion appears the same from any other point in the universe. To understand why, see Fig. 45–21. In Fig. 45–21a we have the view from Earth (or from our Galaxy). The velocities of surrounding galaxies are indicated by arrows, pointing away from us, and greater for galaxies more distant from us. Now, what if we were on the galaxy labeled A in Fig. 45–21a? From Earth, galaxy A appears to be moving to the right at a velocity, call it $\mathbf{v}_A$, represented by the arrow pointing to the right. If we were *on* galaxy A, Earth would appear to be moving to the left at velocity $-\mathbf{v}_A$. To determine the velocities of other galaxies relative to A, we vectorially add the velocity vector, $-\mathbf{v}_A$, to all the velocity arrows shown in Fig. 45–21a. This yields Fig. 45–21b, where we see clearly that the universe is expanding away from galaxy A as well; and the velocities of galaxies receding from A are proportional to their distance from A.

Thus the expansion of the universe can be stated as follows: All galaxies are racing away from *each other* at an average rate of about 70 km/s per megaparsec of distance between them. The ramifications of this idea are profound, and we discuss them in a moment.

There is, however, a class of objects called **quasars**, or "quasistellar radio sources," that do not seem to conform to Hubble's law. Quasars are as bright as nearby stars but display very large redshifts. If quasars are normal participants in the general expansion of the universe according to Hubble's law, their large redshifts suggest they are very distant galaxies. If they are so far away, they must be incredibly bright—sometimes thousands of times brighter than normal galaxies. On the other hand, it has been suggested that some quasars, at least, are not of abnormal brightness because they are nearer than their redshifts suggest. In the first case we would have an unresolved brightness problem, in the second case an unresolved redshift problem. An interesting piece of evidence is that the population density of quasars seems to increase with distance from us. If they are at great distances from us—as their redshifts suggest—this would merely mean they were more common in the early universe than they are now. But if instead they are much closer to us—as their brightness suggests—it would seem that *we* are in a special place in the universe, the place where quasars are least populous. Most

Quasars

astronomers are unwilling to accept the latter possibility, for it would violate a basic assumption of uniformity, the so-called *cosmological principle*, which we discuss in the next paragraph. A recent suggestion that could resolve the quasar question is that these mysterious objects may be cores of galaxies that emit enormous amounts of energy when they draw in nearby material, for example when galaxies "collide." If the center of a galaxy is a black hole, a large amount of matter accelerated toward it would give off a prodigious amount of energy. It seems possible, then, that quasars are powered by black holes.

Cosmological principle

A basic assumption in cosmology has been that on a large scale, the universe looks the same to observers at different places at the same time. In other words, the universe is both isotropic (looks the same in all directions) and homogeneous (would look the same if we were located elsewhere, say in another galaxy). This assumption is called the **cosmological principle**. On a local scale, say in our solar system or within our Galaxy, it clearly does not apply (the sky looks different in different directions). But it has long been thought to be valid if we look on a large enough scale, so that the average population density of galaxies and clusters of galaxies ought to be the same in different areas of the sky. The expansion of the universe (Fig. 45–21) is consistent with the cosmological principle; and the uniformity of the cosmic microwave background radiation (discussed in the next Section) supports it. But matter (i.e., galaxies), even on the largest scale, seems not to be homogeneous but tends to clump, as already discussed. Although astronomers are reluctant to give up the cosmological principle, there is now some doubt about its validity. One possible resolution might be that over 90 percent of the universe may be nonluminous dark matter, which might be uniformly distributed. In any case, the cosmological principle has aided us in treating the universe as a single evolving entity and not as a random collection of material bodies. Another way of stating the cosmological principle is this: There is nothing special about the Earth on a cosmological scale; our large-scale observations are no different from those that might be made elsewhere in the universe.

The expansion of the universe, as described by Hubble's law, strongly suggests that galaxies must have been closer together in the past than they are now. Further, Hubble's law is consistent with all galaxies having been quite close together at some time in the past. This is, in fact, the basis of the *Big Bang* theory of the origin of the universe (discussed in the next Section) which pictures the beginning of the universe as a great explosion. Let us see what can then be said about the age of the universe.

Age of the universe

We can estimate the age of the universe using the Hubble parameter. With $H \approx 20$ km/s per million light-years, then the time required for the galaxies to arrive at their present separations would be approximately (starting with $v = d/t$ and using Eq. 45–5):

$$t = \frac{d}{v} = \frac{d}{Hd} = \frac{1}{H} \approx \frac{(10^6\,\text{ly})(10^{13}\,\text{km/ly})}{(20\,\text{km/s})(3 \times 10^7\,\text{s/y})} \approx 15 \times 10^9\,\text{yr},$$

or roughly 15 billion years. The age of the universe calculated in this way is called the *characteristic expansion time* or "Hubble age." It is not very precise, in part because we don't know the value of H precisely.

There are two other independent checks on the age of the universe. The first is determination of the age of the Earth (and solar system) from radioactivity, primarily using uranium, which places the age of the solar system at about $4\frac{1}{2}$ billion years. Second, by using the theory of stellar evolution, the ages of stars have been estimated to be about 10 to 15 billion years. These independent and unrelated determinations are consistent with a Big Bang occurring, according to our best estimates today, 10 to 15 billion years ago. The lower value determined from radioactivity is consistent since we would expect the origin of the Earth (and the solidification of rocks) to have occurred somewhat after the origin of the universe as a whole.

Before discussing the Big Bang theory in more detail, we briefly discuss one of the alternatives to the Big Bang—the **steady-state model**—which assumes that the universe is infinitely old and on the average looks the same now as it always has.[†] According to the steady-state model, no large-scale changes have taken place in the universe as a whole, particularly no Big Bang. To maintain this view in the face of the recession of galaxies away from each other, mass–energy conservation must be violated. That is, matter is assumed to be created continuously, keeping the density of the universe constant. The rate of mass creation required is very small—about one nucleon per cubic meter every 10^9 years.

The steady-state model provided the Big Bang model with healthy competition a half century ago. However, the discovery of the cosmic microwave background radiation (next Section), as well as the observed expansion of the universe, has made the Big Bang model the almost universally accepted model.

Steady-state model

45–5 The Big Bang and the Cosmic Microwave Background

The expansion of the universe seems to suggest, as we have seen, that the matter of the universe was once much closer together than it is now. This is the basis for the idea that the universe began about 15 billion years ago with a huge explosion known affectionately as the **Big Bang**.

The Big Bang

If there was a Big Bang, it must have occurred simultaneously at all points in the universe. If the universe is *finite*, the explosion would have taken place in a tiny volume approaching a point. However, this point of extremely dense matter is not to be thought of as a concentrated mass in the midst of a much larger space around it. Rather, the initial dense point *was* the universe—the entire universe. There wouldn't have been anything else. If, on the other hand, the universe is *infinite*, then the explosion would have occurred at *all* points in the universe since an infinite universe, even if smaller at an earlier time, would still have been infinite. In either case, when we say that the universe some time after the Big Bang was smaller than it is now, we mean that the average separation between galaxies was less. Thus, it is the *size of the universe itself* that has increased since the Big Bang.

What is the evidence supporting the Big Bang? First, the age of the universe as calculated from the Hubble expansion, from stellar evolution, and from radioactivity all point to a consistent time of origin for the universe, as we saw in the last Section. Another, and crucial, piece of evidence was the discovery in the 1960s of the **cosmic microwave background** radiation (or CMB), which came about as follows.

In 1964, Arno Penzias and Robert Wilson were experiencing difficulty with what they assumed to be background noise, or "static," in their radio telescope (a large antenna device for detecting radio waves from the heavens, Fig. 45–22).

FIGURE 45–22 Robert Wilson (left) and Arno Penzias, and behind them their "horn antenna."

[†] That the universe should be uniform in time as well as in space is essentially an extension of the cosmological principle and is sometimes referred to as the *perfect cosmological principle*.

Eventually, they became convinced that it was real and that it was coming from outside our Galaxy. They made precise measurements at a wavelength $\lambda = 7.35$ cm, which is in the microwave region of the electromagnetic spectrum. (See Fig. 32–12; this radiation is called "microwave" because the wavelength, though much greater than that of visible light, is somewhat smaller than wavelengths for ordinary radio waves, which are typically meters or hundred of meters.) The intensity of this radiation was found initially not to vary by day or night or time of year, nor to depend on direction. It came from all directions in the universe with equal intensity (within less than one part per thousand). It could only be concluded that this radiation came from beyond our Galaxy, from the universe as a whole. The remarkable uniformity of the cosmic microwave background radiation was in accordance with the cosmological principle. But theorists also felt that there needed to be some small inhomogeneities in the CMB that would have provided "seeds" around which galaxy formation could have started. Such small inhomogeneities, on the order of parts per million, were finally detected in 1992.

The intensity of this radiation as measured at $\lambda = 7.35$ cm corresponds to blackbody radiation (see Section 38–1) at a temperature of about 3 K, now measured more precisely to be 2.73 ± 0.01 K. When radiation at other wavelengths was measured, the intensities were found to fall on a blackbody curve as shown in Fig. 45–23, corresponding to a temperature of 2.73 K.

The discovery of the CMB at a temperature of 2.7 K ranks as one of the two most significant cosmological discoveries of the twentieth century. (The other was Hubble's expanding universe.) It is highly significant because it provides strong evidence in support of the Big Bang, and it gives us some idea of conditions in the very early universe. In fact, in the late 1940s, George Gamow and his collaborators calculated that a Big Bang origin of the universe should have generated just such a microwave background radiation.

To understand why, let us look at what a Big Bang might have been like. There must have been a tremendous release of energy. The temperature must have been extremely high, so high that there could not have been any atoms in the very early stages of the universe. Instead, the universe must have consisted solely of radiation (photons) and elementary particles. The universe would have been opaque—the photons in a sense "trapped," since as soon as they were emitted they would have been scattered or absorbed, primarily by electrons. Indeed, the microwave background radiation is strong evidence that matter and radiation were once in equilibrium at a very high temperature. As the universe expanded, the energy would have spread out over an increasingly larger volume and the temperature would have dropped. Only when the temperature had reached about 3000 K some 300,000 years later, could nuclei and electrons have stayed together as atoms.

FIGURE 45–23 Spectrum of cosmic microwave background radiation, showing blackbody curve and experimental measurements including that of Penzias and Wilson. (Thanks to G. F. Smoot and D. Scott. The vertical bars represent the experimental uncertainty in a measurement.)

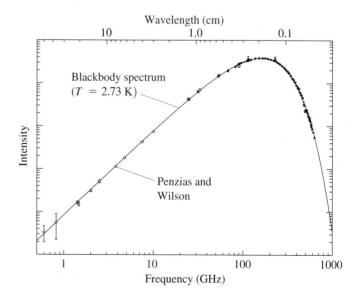

With the disappearance of free electrons, as they combined with nuclei to form atoms, the radiation would have been freed—"decoupled" from matter, we could say—to spread throughout the universe. As the universe expanded, so too the wavelengths of the radiation lengthened (you might think of standing waves, Section 15–9), thus redshifting to longer wavelengths that correspond to lower temperature (recall Wien's law, $\lambda_P T = $ constant, Section 38–1), until they would have reached the 2.7 K background radiation we observe today.

Photons "decoupled"

Although the total energy associated with the cosmic microwave background is much larger than that from other radiation sources (such as the light emitted by stars), it is small compared to the energy, and mass (remember $E = mc^2$), associated with matter. In fact today, radiation is believed to make up less than $\frac{1}{1000}$ of the energy of the universe: today the universe is *matter-dominated*. But it was not always so. The cosmic microwave background radiation strongly suggests that early in its history the universe was *radiation-dominated*.[†] But, as we shall see in the next Section, that period lasted less than $\frac{1}{10,000}$ of the history of the universe (thus far).

45–6 The Standard Cosmological Model: the Early History of the Universe

It is now generally agreed that the evolution of the universe must have been determined in the first few moments of the Big Bang. In the last decade or two, a convincing theory of the origin and evolution of the universe has developed, now known as the **standard model**. Much of this theory is based on recent theoretical and experimental advances in elementary particle physics. Indeed, cosmology and elementary particle physics have cross-fertilized to a surprising extent.

Let us go back now to the earliest of times—as close as possible to the Big Bang—and follow a standard model scenario of events as the universe expanded and cooled after the Big Bang. Initially, we will be talking of extremely small time intervals, as well as extremely high temperatures, far beyond anything in the universe today. Figure 45–24 is a graphical representation of the events, and it may be helpful to consult it as we go along.

We begin at a time only a minuscule fraction of a second after the Big Bang, 10^{-43} s. Although this is an unimaginably short time, predictions as early as this can be made based on present theory, albeit somewhat speculatively. Earlier than this instant, we can say nothing since we do not yet have a theory of quantum effects on gravity which would be needed for the incredibly high densities and temperatures then.

The standard-model "scenario" of the history of the universe after the Big Bang

[†] If there had not been such intense radiation in the first few minutes of the universe, nuclear reactions might have produced a much larger percentage of heavy nuclei than we see. Instead, nearly 75 percent of visible matter is hydrogen, presumably because the intense radiation immediately blasted apart any heavy nuclei that formed into their constituent protons and neutrons.

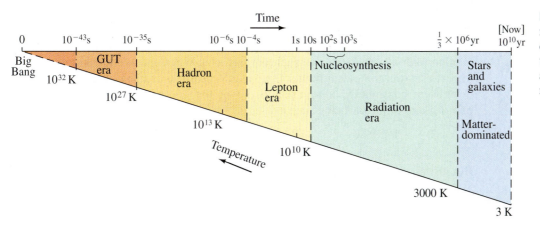

FIGURE 45–24 Graphical representation of the development of the universe after the Big Bang, according to the standard model.

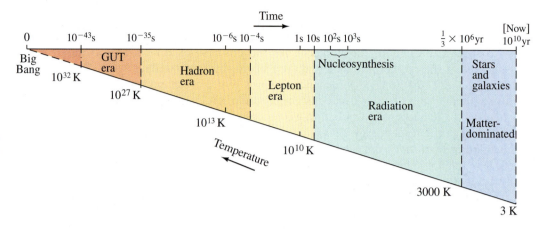

FIGURE 45–24 (repeated) Graphical representation of the development of the universe after the Big Bang, according to the standard model.

All four forces unified

It is imagined, however, that prior to 10^{-43} s, the four forces in nature were unified—there was only one force. The temperature would have been about 10^{32} K, corresponding to particles moving about every which way with an average kinetic energy K of 10^{19} GeV (see Eq. 18–4):

$$K \approx kT \approx \frac{(1.4 \times 10^{-23}\,\text{J/K})(10^{32}\,\text{K})}{1.6 \times 10^{-19}\,\text{J/eV}} \approx 10^{28}\,\text{eV} = 10^{19}\,\text{GeV}.$$

Symmetry broken (gravity condensed out)

(Note that the factor $\frac{3}{2}$ in Eq. 18–4 is usually ignored in such calculations.) At $t = 10^{-43}$ s, a kind of "phase transition" is believed to have occurred during which the gravitational force, in effect, "condensed out" as a separate force. This, and subsequent phase transitions, are somewhat analogous to the phase transitions water undergoes as it cools from a gas, condenses into a liquid, and with further cooling freezes into ice. The symmetry of the four forces was broken, but the strong, weak, and electromagnetic forces were still unified. At this point, the universe entered the so-called **grand unified era** (after grand unified theory—see Chapter 44). There was no distinction between quarks and leptons; baryon and lepton numbers were not conserved. Very shortly thereafter, as the universe expanded considerably and the temperature had dropped to about 10^{27} K, there was another phase transition during which the strong force condensed out. This probably occurred about 10^{-35} s after the Big Bang. Now the universe was filled with a soup of leptons and quarks. The leptons included electrons, muons, taus, neutrinos, and all their antiparticles. The quarks were initially free (something we have not seen in our present universe), but soon they began to "condense" into more normal particles: nucleons and the other hadrons and their antiparticles.

Quark confinement (hadron era)

With this **confinement of quarks**, the universe entered the **hadron era**.

We can think of this "soup" as a grand mixture of particles and antiparticles, as well as photons—all in roughly equal numbers—colliding with one another frequently and exchanging energy.

By the time the universe was only about a microsecond $(10^{-6}\,\text{s})$ old, it had cooled to about 10^{13} K, corresponding to an average kinetic energy of 1 GeV, and the vast majority of hadrons disappeared. To see why, let us focus on the most familiar hadrons: nucleons and their antiparticles. When the average kinetic energy of particles was somewhat higher than 1 GeV, protons, neutrons, and their antiparticles were continually being created out of the energies of collisions involving photons and other particles, such as

Most hadrons disappear

$$\text{photons} \rightarrow \text{p} + \bar{\text{p}},$$
$$\rightarrow \text{n} + \bar{\text{n}}.$$

But just as quickly, particles and antiparticles would annihilate: for example

$$\text{p} + \bar{\text{p}} \rightarrow \text{photons or leptons}.$$

So the processes of creation and annihilation of nucleons were in equilibrium. The numbers of nucleons and antinucleons were high—roughly as many as there were electrons, positrons, or photons. But as the universe expanded and cooled, and the average kinetic energy of particles dropped below about 1 GeV, which is the minimum energy needed to create nucleons and antinucleons (about 940 MeV each) in a typical collision, the process of nucleon creation could not continue. The process of annihilation could continue, however, with antinucleons annihilating nucleons, until there were almost no nucleons left. But not quite zero. To explain our present world, which consists mainly of matter (nucleons and electrons) with very little antimatter in sight, we must suppose that earlier in the universe, perhaps around 10^{-35} s after the Big Bang, a slight excess of quarks over antiquarks was formed.[†] This would have resulted in a slight excess of nucleons over antinucleons. And it is these "leftover" nucleons that we are made of today. The excess of nucleons over antinucleons was about one part in 10^9. Earlier, during the hadron era, there should have been about as many nucleons as photons. After it ended, the "leftover" nucleons thus numbered only about one nucleon per 10^9 photons, and this ratio has persisted to this day. Protons, neutrons, and all other heavier particles were thus tremendously reduced in number by about 10^{-6} s after the Big Bang. The lightest hadrons, the pions, disappeared as the nucleons had; because they are the lightest mass hadrons (140 MeV), they were the last hadrons to go, around 10^{-4} s after the Big Bang. Lighter particles, including electrons, positrons, neutrinos, photons—in roughly equal numbers—dominated, and the universe entered the **lepton era**.

Why is there matter now?

Lepton era

By the time the first full second had passed (clearly the most eventful second in history!), the universe had cooled to about 10 billion degrees, 10^{10} K. The average kinetic energy was about 1 MeV. This was still sufficient energy to create electrons and positrons and balance their annihilation reactions, since their masses correspond to about 0.5 MeV. So there were about as many e^+ and e^- as there were photons. But within a few more seconds, the temperature had dropped sufficiently so that e^+ and e^- could no longer be formed. Annihilation ($e^+ + e^- \rightarrow$ photons) continued. And, like nucleons before them, electrons and positrons all but disappeared from the universe—except for a slight excess of electrons over positrons (later to join with nuclei to form atoms). Thus, about $t = 10$ s after the Big Bang, the universe entered the **radiation era**. Its major constituents were now photons and neutrinos. But the neutrinos, partaking only in the weak force, rarely interacted. So the universe, until then experiencing an energy balance between matter and radiation, became **radiation-dominated**: much more energy was contained in radiation than in matter, a situation that would last hundreds of thousands of years (Fig. 45–24).

Radiation-dominated universe

Meanwhile, during the next few minutes, crucial events were taking place. Beginning about 2 or 3 minutes after the Big Bang, nuclear fusion began to occur. The temperature had dropped to about 10^9 K, corresponding to an average kinetic energy $\bar{K} \approx 100$ keV, where nucleons could strike each other and be able to fuse (Section 43–4), but now cool enough that newly formed nuclei would not be immediately broken apart by subsequent collisions. Deuterium, helium, and very tiny amounts of lithium nuclei were probably made. But the universe was cooling too quickly, and larger nuclei were not made. After only a few minutes, probably not even a quarter of an hour after the Big Bang, the temperature dropped far enough that nucleosynthesis stopped, not to start again for millions of years (in stars).

Making He nuclei

[†]An alternative possibility is that there was perfect symmetry between quarks and antiquarks, matter and antimatter, but that the universe somehow separated into domains, some containing only matter, others only antimatter. If this were true, we would expect antiparticles from such distant domains to reach us, at least occasionally, in cosmic rays; but none has ever been detected.

Thus, after the first hour or so of the universe, matter consisted mainly of bare nuclei of hydrogen (about 75 percent) and helium (about 25 percent)[†] and electrons. But radiation (photons) continued to dominate.

Our story is almost complete. The next important event is presumed to have occurred about 300,000 years later. The universe had expanded to about $\frac{1}{1000}$ of its present size, and the temperature had cooled to about 3000 K. The average kinetic energy of nuclei, electrons, and photons was less than an electron volt. Since ionization energies of atoms are on the order of eV, then as the temperature dropped below this point, electrons could orbit the bare nuclei and remain there (without *Birth of stable atoms* being ejected by collisions), thus forming atoms. With the birth of atoms, the photons—which had been continually scattering from the free electrons—now became much freer to spread nearly unhindered throughout the universe. We say that the photons become decoupled from matter. The total energy contained in radiation had been decreasing (lengthening in wavelength as the universe expanded), until at this point it was about equal to the total energy contained in matter. As the universe continued to expand, the radiation cooled further (to 2.7 K today, forming the cosmic microwave background radiation we detect from everywhere in the universe), and lost energy. But the mass of material particles did not decrease, so beginning at about this time, the energy of the universe became increasingly concentrated in matter rather than in radiation: the universe became *Matter-dominated* **matter-dominated**, as it remains today.[‡]
universe

After the birth of atoms, stars and galaxies could begin to form—probably by self-gravitation around mass concentrations (inhomogeneities). This began perhaps a million years after the Big Bang. The universe continued to evolve (see Section 45–2) until today, some 15 billion years later.

This scenario is by no means "proven" in any sense. Nor does it answer all questions. But it docs provide a tentative picture, for the first time, of how the universe may have begun and evolved. It does have problems, however, some of which have been resolved by a modification first proposed in the early 1980s *Inflation* known as the **inflationary scenario**. According to the inflationary scenario, at the earliest stages of cosmological evolution, around 10^{-35} s after the Big Bang, the universe underwent a very rapid exponential expansion associated with the phase transition (symmetry breaking) that separated the strong force from the electroweak (as discussed earlier in this Section). After the brief inflationary period, the scenario settles back to the standard expansion already discussed. The appeal of the inflationary scenario is that it provides natural explanations for a number of problems not resolved by the standard model, such as why the universe is as close to being as flat as it seems to be, and why the cosmic microwave background radiation is so uniform.

There are, however, questions we have not yet treated such as what is the future of the universe, which we will look at next, in the final Section of this book.

[†] This standard model prediction of a 25 percent primordial production of helium is fully in accord with what we observe today—the universe *does* contain about 25 percent He—and it is strong evidence in support of the standard Big Bang model. Furthermore, the theory says that 25 percent He abundance is fully consistent with there being three neutrino types, which is the number we observe so far. And it sets an upper limit of four to the maximum number of possible neutrino types. Actually, the fourth could be another type of low-mass particle, a *photino* or a *gravitino*, for example. Here we have a situation where cosmology actually makes a specific prediction about fundamental physics.

[‡] Although today matter contains more of the energy of the universe than does radiation, there are many more photons (perhaps 10^9 times more) than atoms, nuclei, and electrons. But each photon (at $T \approx 2.7$ K) has very little energy.

45–7 The Future of the Universe?

According to the standard Big Bang model, the universe is evolving and changing. Individual stars are evolving and dying as white dwarfs, neutron stars, black holes. At the same time, the universe as a whole is expanding. One important question is whether the universe will continue to expand forever. This question is connected to the curvature of space-time (Section 45–3) and to whether the universe is open (and infinite) or closed (and finite). There are three possibilities as shown in Fig. 45–25. If the curvature of the universe is *negative*, the expansion of the universe will never stop, although the rate of expansion might be expected to decrease due to the gravitational attraction of its parts. Such a universe would be *open* and infinite. If the universe is *flat* (no curvature), it would still be open and infinite but its expansion would slowly approach a zero rate. Finally, if the universe has *positive* curvature, it would be *closed* and finite. The effect of gravity in this case would be strong enough so that the expansion would eventually stop and the universe would then begin to contract. All matter eventually would collapse back onto itself in a **big crunch**. If, in this last case, the maximum expansion of the universe corresponded to, say, an intergalactic separation twice what it is now, the maximum expansion would occur about 30 or 40 billion years from now. Then, as the universe began to contract, the big crunch would occur about 100 billion years after the Big Bang.

An open or closed universe?

A big crunch?

Whether we live in an open and continually expanding universe, or a closed one that eventually will contract, is connected also to the average mass density in the universe. If the average mass density is above a critical value known as the **critical density**, estimated to be about

$$\rho_c \approx 10^{-26}\,\text{kg/m}^3$$

Critical density of the universe

(i.e., a few nucleons/m^3 on average throughout the universe), then gravity will prevent expansion from continuing forever, and will eventually pull the universe back into a big crunch. To say it another way, if $\rho > \rho_c$, there would be sufficient mass that gravity would give space-time a positive curvature. If instead the actual density is equal to the critical density, $\rho = \rho_c$, the universe will be flat and open. If the actual density is less than the critical density, $\rho < \rho_c$, the universe will have negative curvature and be open, expanding forever.

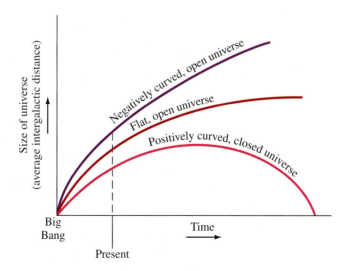

FIGURE 45–25 Three future possibilities for the universe.

EXAMPLE 45–7 ESTIMATE **Critical density of the universe.** Use energy conservation and the concept of escape velocity (Section 8–7) to estimate the critical density of the universe.

SOLUTION At the critical density, ρ_c, any given galaxy of mass m will just be able to "escape" away from our Galaxy. As we saw in Section 8–7, escape can just occur if the total energy E of the galaxy satisfies

$$E = K + U = \tfrac{1}{2}mv^2 - G\frac{mM}{R} = 0.$$

Here R is the distance of that galaxy m from our Galaxy, and we assume on the enormous scale of the universe that each galaxy is essentially a point. The total mass M that pulls inward on m is the total mass of all galaxies within a sphere of radius R (see Appendix C). If we assume the density of galaxies is roughly constant, then $M = \tfrac{4}{3}\pi\rho_c R^3$. Substituting this M into the above equation, and setting $v = HR$ (Hubble's law, Eq. 45–5), we obtain

$$\frac{GM}{R} = \tfrac{1}{2}v^2$$

or

$$\frac{G(\tfrac{4}{3}\pi\rho_c R^3)}{R} = \tfrac{1}{2}(HR)^2.$$

We solve for ρ_c:

$$\rho_c = \frac{3H^2}{8\pi G} \approx \frac{3\left[(70\times10^3\,\text{m/s/Mpc})(1\,\text{Mpc}/3.08\times10^{22}\,\text{m})\right]^2}{8(3.14)(6.67\times10^{-11}\,\text{N}\cdot\text{m}^2/\text{kg}^2)}$$

$$\approx 10^{-26}\,\text{kg/m}^3.$$

Great efforts have gone into measuring the actual density of the universe. Estimates of the amount of visible matter in the universe put the actual density at between one and two orders of magnitude less than the critical density, thereby suggesting an open universe. However, there is evidence for a significant amount of nonluminous matter in the universe, often referred to as the **missing mass** or **dark matter**, enough to bring the density to almost exactly ρ_c. For example, observations of the rotation of galaxies suggest that they rotate as if they had considerably more mass than we can see. And observations of the motion of galaxies within clusters also suggest that they have considerably more mass than can be seen. Furthermore, the highly regarded inflationary theory offers a strong argument in favor of ρ being closely equal to ρ_c, and thus that space on a grand scale is Euclidean. If there is nonluminous matter in the universe, what might it be? One suggestion is that the missing mass consists of previously unknown weakly interacting massive particles (WIMPS), possibly supersymmetric particles known as "neutralinos," or perhaps small primordial black holes made in the early stages of the universe. Another possibility for the missing mass (or part of it) are MACHOS (massive compact halo objects), which would be chunks of matter in the form of large planets (like Jupiter) or stars that are too small to sustain fusion; they might glow faintly due to energy released from gravitational contraction (thus they are sometimes referred to as *brown dwarfs*), but too faintly to be seen. Evidence for MACHOS was found in 1993, inferred from their (assumed) gravitational effect on light passing by them from a distant star.

Another proposal for the missing mass is that neutrinos, once believed to be massless, may actually have nonzero rest mass. Since the universe probably contains roughly as many neutrinos of each type as it does photons (that is, about 10^9 times the number of nucleons, although this neutrino background has yet to be detected), neutrino masses of only a few eV could help to bring the actual density

Missing mass or dark matter

of the universe up to the critical density. The supernova of 1987 offered an opportunity to estimate the electron neutrino mass. If neutrinos have nonzero rest mass, then their velocities are less than the speed of light ($v < c$). High-energy neutrinos emitted from the supernova should in this case have arrived at Earth earlier than lower-energy (and therefore slower) neutrinos emitted at the same instant. Since they traveled a distance of about 170,000 ly from SN1987a, the time difference ought to be measurable. To get an idea of the size of the effect, let us consider the following Example.

EXAMPLE 45–8 ESTIMATE Neutrino mass estimate from a supernova.
Suppose two neutrinos from SN1987a were detected on Earth (via the reaction $\bar{\nu}_e + p \rightarrow n + e^+$) 10 seconds apart, with measured kinetic energies of about 20 MeV and 10 MeV. (a) Estimate the rest mass of the neutrino, m_ν, using this data, assuming both neutrinos were emitted at the same time. (b) Theoretical models of supernova explosions suggest that the neutrinos are emitted in a burst that lasts from a second or two up to perhaps 10 s. If we assume the neutrinos are not emitted simultaneously but rather at any time over a 10-s interval, what then can we say about the neutrino mass based on the data given above?

SOLUTION (a) We expect the neutrino mass to be less than 100 eV (from laboratory measurements). Since our neutrinos have kinetic energy of 20 MeV and 10 MeV, we can make the approximation $m_\nu c^2 \ll E$, where E (the total energy) is essentially equal to the kinetic energy. From Eq. 37–11, we have

$$E = \frac{m_\nu c^2}{\sqrt{1 - v^2/c^2}}.$$

We solve this for v, the velocity of a neutrino with energy E:

$$v = c\left(1 - \frac{m_\nu^2 c^4}{E^2}\right)^{\frac{1}{2}} = c\left(1 - \frac{m_\nu^2 c^4}{2E^2} + \cdots\right),$$

where we have used the binomial expansion $(1 + x)^{\frac{1}{2}} = 1 + \frac{1}{2}x + \cdots$, and we ignore higher-order terms since $m_\nu^2 c^4 \ll E^2$. The time t for a neutrino to travel a distance $d (= 170,000 \text{ ly})$ is

$$t = \frac{d}{v} = \frac{d}{c\left(1 - \frac{m_\nu^2 c^4}{2E^2}\right)} \approx \frac{d}{c}\left(1 + \frac{m_\nu^2 c^4}{2E^2}\right),$$

where again we used the binomial expansion $\left[(1 + x)^{-1} = 1 - x + \cdots\right]$. The difference in arrival times for our two neutrinos of energies $E_1 = 20$ MeV and $E_2 = 10$ MeV is

$$t_2 - t_1 = \frac{d}{c}\frac{m_\nu^2 c^4}{2}\left(\frac{1}{E_2^2} - \frac{1}{E_1^2}\right).$$

We solve this for $m_\nu c^2$ and set $t_2 - t_1 = 10$ s:

$$m_\nu c^2 = \left[\frac{2c(t_2 - t_1)}{d}\frac{E_1^2 E_2^2}{E_1^2 - E_2^2}\right]^{\frac{1}{2}}$$

$$= \left[\frac{2(3 \times 10^8 \text{ m/s})(10 \text{ s})}{(1.7 \times 10^5 \text{ ly})(10^{16} \text{ m/ly})}\frac{(400 \text{ MeV}^2)(100 \text{ MeV}^2)}{(400 \text{ MeV}^2 - 100 \text{ MeV}^2)}\right]^{\frac{1}{2}}$$

$$= 22 \times 10^{-6} \text{ MeV} = 22 \text{ eV}.$$

We thus estimate the mass of the neutrino to be 22 eV/c^2, but there would of course be experimental uncertainties.
(b) If the two neutrinos of energy $E_1 = 20$ MeV and $E_2 = 10$ MeV were emitted at unknown times over a 10-s interval, then the 10-s difference in their arrival times could be due to a 10-s difference in their emission time. In this case our data are consistent with zero rest mass and it puts an approximate *upper limit* on the neutrino mass of 22 eV/c^2.

In the actual experiments, the most sensitive detector consisted of several thousand tons of water in an underground chamber. It detected 11 events in 12 seconds, probably via the reaction $\bar{\nu}_e + p \rightarrow n + e^+$. There was not a clear correlation between energy and time of arrival. Nonetheless, a careful analysis of this experiment and others, has set a rough upper limit on the electron anti-neutrino mass of about 4 eV. The upper limits for the masses of the other neutrinos are much higher (see Table 44–2). Very recent experiments that investigate mass differences among neutrinos suggest that at least one of the neutrinos may have nonzero mass.

The scenarios we have been describing are based on some well-founded assumptions. As of this writing (year 2000) one of those basic assumptions—that gravity should be slowing down the expansion of the universe—has been called into question, and there are suggestions that there might be a previously undetected repulsive component of gravity.

Latest experiments

Finally, very recent experiments on the early CMB are consistent with the universe being flat, and fully in accord with the inflation theory.

The questions raised by cosmology can seem absurd at times, they are so removed from everyday "reality." We can always say, "the Sun is shining, it's going to burn on for an unimaginably long time, all is well." Nonetheless, the questions of cosmology are deep ones that fascinate the human intellect. One aspect that is especially intriguing is this: calculations on the formation and evolution of the universe have been performed that deliberately varied the values—just slightly—of certain fundamental physical constants. The result? A universe in which life as we know it could not exist. [For example, if the difference in mass between proton and neutron were zero, or small (less than the mass of the electron, $0.511 \, \text{MeV}/c^2$), there would be no atoms: electrons would be captured by protons never to be freed again.] Such results have given rise to the so-called **Anthropic principle**, which says that if the universe were even a little different than it is, we could not be here. It seems as if the universe is exquisitely tuned, almost as if to accommodate us.

Summary

The night sky contains myriads of stars including those in the Milky Way, which is a "side view" of our **Galaxy** looking along the plane of the disc. Our Galaxy includes about 10^{11} stars. Beyond our Galaxy are billions of other galaxies.

Astronomical distances are measured in **light-years** $(1 \, \text{ly} \approx 10^{13} \, \text{km})$. The nearest star is about 4 ly away and the nearest other galaxy is 2 million ly away. Our galactic disc has a diameter of about 100,000 ly. Distances are often specified in **parsecs**, where 1 parsec = 3.26 ly.

Stars are believed to begin life as collapsing masses of hydrogen gas (protostars). As they contract, they heat up (potential energy is transformed to kinetic energy). When the temperature reaches 10 million degrees, nuclear fusion begins and forms heavier elements (**nucleosynthesis**), mainly helium at first. The energy released during these reactions balances the gravitational force, and the young star stabilizes as a **main-sequence** star. The tremendous luminosity of stars comes from the energy released during these thermonuclear reactions. After billions of years, as helium is collected in the core and hydrogen is used up, the core contracts and heats further. The envelope expands and cools, and the star becomes a **red giant** (larger diameter, redder color). The next stage of stellar evolution depends on the mass of the star. Stars of residual mass less than about 1.4 solar masses cool further and become **white dwarfs**, eventually fading and going out altogether. Heavier stars contract further due to their greater gravity: the density approaches nuclear density, the huge pressure forces electrons to combine with protons to form neutrons, and the star becomes essentially a huge nucleus of neutrons. This is a **neutron star**, and the energy released from its final core contraction is believed to produce **supernovae** explosions. If the star's residual mass is greater than two or three solar masses, it may contract even further and form a **black hole**, which is so dense that no matter or light can escape from it.

In the **general theory of relativity**, the **equivalence principle** states that an observer cannot distinguish acceleration from a gravitational field. Said another way, gravitational and inertial mass are the same. The theory predicts gravitational bending of light rays to a degree consistent with experiment. Gravity is treated as a curvature in space and time, the curvature being greater near massive bodies. The universe as a whole may be curved. If there is sufficient mass, the curvature of the universe is positive, and the universe is **closed** and **finite**; otherwise, it is **open** and **infinite**.

Distant galaxies display a **redshift** in their spectral lines, interpreted as a Doppler shift. The universe seems to be **expanding**, its galaxies racing away from each other at speeds (v) proportional to the distance (d) between them:

$$v = Hd,$$

which is known as **Hubble's law** (H is the **Hubble parameter**). This expansion of the universe suggests an explosive origin, the **Big Bang**, which probably occurred about 15 billion years ago.

Quasars are objects with a large redshift (suggesting great distance) and high luminosity (suggesting closeness or, more likely, extraordinary energy output).

The **cosmological principle** assumes that the universe, on a large scale, is homogeneous and isotropic.

Important evidence for the Big Bang model of the universe was the discovery of the **cosmic microwave background** radiation (CMB), which conforms to a blackbody radiation curve at a temperature of 2.7 K.

The **standard model** of the Big Bang provides a possible scenario as to how the universe developed as it expanded and cooled after the Big Bang. Starting at 10^{-43} seconds after the Big Bang, according to this model, there was a series of **phase transitions** during which previously unified forces of nature "condensed out" one by one. The **inflationary scenario** assumes that during one of these phase transitions, the universe underwent a brief but rapid exponential expansion. Until about 10^{-35} s, there was no distinction between quarks and leptons. Shortly thereafter, quarks were **confined** into hadrons (the **hadron era**). About 10^{-4} s after the Big Bang, the majority of hadrons disappeared, having combined with anti-hadrons, producing photons, leptons and energy, leaving mainly photons and leptons to freely move, thus introducing the **lepton era**. By the time the universe was about 10 s old, the electrons too had mostly disappeared, having combined with their antiparticles, and the universe became **radiation-dominated**. A couple of minutes later, nucleosynthesis began, but lasted only a few minutes. It then took several hundred thousand years before the universe was cool enough for electrons to combine with nuclei to form atoms. Also about this time, the background radiation had expanded and cooled so much that its total energy equaled the energy in matter. As the radiation cooled further, losing energy, the universe became **matter-dominated** (not in numbers, but in energy). Then stars and galaxies formed, producing a universe not much different than it is today—some 15 billion years later.

If the universe is open, it will continue to expand indefinitely. If it is closed, gravity is sufficiently strong to halt expansion and the universe will eventually begin to collapse back on itself, ending in a *big crunch*. Whether the universe is open or closed depends on whether its average mass density is above or below a critical density. The evidence today suggests that the universe is flat.

Questions

1. The Milky Way was once thought to be "cloudy," but no longer is considered so. Explain.

2. If you were measuring star parallaxes from the Moon instead of Earth, what corrections would you have to make? What changes would occur if you were measuring parallaxes from Mars?

3. Certain stars called *cepheid variables* change in luminosity with a period of typically several days. The period has been found to have a definite relationship with the absolute luminosity of the star. How could these stars be used to measure the distance to galaxies?

4. A star is in equilibrium when it radiates at its surface all the energy generated at its core. What happens when it begins to generate more energy than it radiates? Less energy? Explain.

5. Describe a red giant star. List some of its properties.

6. Select a point on the H–R diagram. Mark several directions away from this point. Now describe the changes that would take place in a star moving in each of these directions.

7. Does the H–R diagram reveal anything about the core of a star?

8. Why do some stars end up as white dwarfs, and others as neutron stars or black holes?

9. Can we tell, by looking at the population on the H–R diagram, that hotter suns have shorter lives? Explain.

10. What is a geodesic? What is its role in general relativity?

11. If it were discovered that the redshift of spectral lines of galaxies was due to something other than expansion, how might our view of the universe change? Would there be conflicting evidence? Discuss.

12. All galaxies appear to be moving away from us. Are we therefore at the center of the universe? Explain.

13. If you were located in a galaxy near the boundary of our observable universe, would galaxies in the direction of the Milky Way appear to be approaching you or receding from you? Explain.

14. What is the difference between the Hubble age of the universe and the actual age? Which is greater?

15. Compare an explosion on Earth to the Big Bang. Consider such questions as: Would the debris spread at a higher speed for more distant particles, as in the Big Bang? Would the debris come to rest? What type of universe would this correspond to, open or closed?

16. When the primordial nucleus exploded, thus creating the universe, into what did it expand? Discuss.

17. If nothing, not even light, escapes from a black hole, then how can we tell if one is there?

18. Explain what the 2.7 K cosmic microwave background radiation is. Where does it come from? Why is its temperature now so low?

19. The birth of atoms—that is, the combination of electrons with nuclei about which they orbit—occurred when the universe had cooled to about 3000 K and is generally called *recombination*. Why is this term misleading?

20. Why were atoms unable to exist until hundreds of thousands of years after the Big Bang?

21. Explain why today the universe is said to be matter-dominated, yet there are probably 10^9 times as many photons as there are massive particles.

22. If the universe is open, what will eventually happen to the cosmic microwave background radiation? If it is closed, what will happen to it?

23. Under what circumstances would the universe eventually collapse in on itself?

Problems

Sections 45–1 and 45–2

1. (I) Using the definitions of the parsec and the light-year, show that 1 pc $= 3.26$ ly.

2. (I) A star exhibits a parallax of 0.38 seconds of arc. How far away is it?

3. (I) The parallax angle of a star is $0.00019°$. How far away is the star?

4. (I) A star is 36 parsecs away. What is its parallax angle? State (a) in seconds of arc, and (b) in degrees.

5. (I) What is the parallax angle for a star that is 55 light-years away? How many parsecs is this?

6. (I) A star is 35 parsecs away. How long does it take for its light to reach us?

7. (I) If one star is twice as far away from us as a second star, will the parallax angle of the first star be greater or less than that of the second star? By what factor?

8. (II) We saw earlier (Chapter 19) that the rate energy reaches the Earth from the Sun (the "solar constant") is about 1.3×10^3 W/m². What is (a) the apparent brightness l of the Sun, and (b) the absolute luminosity L of the Sun?

9. (II) What is the apparent brightness of the Sun as seen on Jupiter? (Jupiter is 5.2 times farther from the Sun than the Earth.)

10. (II) Estimate the angular width that our Galaxy would subtend if observed from the nearest galaxy to us (Table 45–1). Compare to the angular width of the Moon from Earth.

11. (II) When our Sun becomes a red giant, what will be its average density if it expands out to the orbit of Earth (1.5×10^{11} m from the Sun)?

12. (II) When our Sun becomes a white dwarf, it is expected to be about the size of the Moon. What angular width will it subtend from Earth?

13. (II) Calculate the density of a white dwarf whose mass is equal to the Sun's and whose radius is equal to the Earth's. How many times larger than Earth's density is this?

14. (II) A neutron star whose mass is 1.5 solar masses has a radius of about 11 km. Calculate its average density and compare to that for a white dwarf (Problem 13) and to that of nuclear matter.

15. (II) Calculate the Q-values for the He burning reactions of Eq. 45–2. (The mass of ${}^8_4$Be is 8.005305 u.)

16. (II) In the later stages of stellar evolution, a star (if massive enough) will begin fusing carbon nuclei to form, for example, magnesium:

$$ {}^{12}_{6}\text{C} + {}^{12}_{6}\text{C} \rightarrow {}^{24}_{12}\text{Mg} + \gamma. $$

(a) How much energy is released in this reaction (see Appendix D). (b) How much kinetic energy must each carbon nucleus have (assume equal) in a head-on collision if they are just to touch (use Eq. 42–1) so that the strong force can come into play? (c) What temperature does this kinetic energy correspond to?

17. (II) Repeat Problem 16 for the reaction

$$ {}^{16}_{8}\text{O} + {}^{16}_{8}\text{O} \rightarrow {}^{28}_{14}\text{Si} + {}^{4}_{2}\text{He}. $$

18. (II) Suppose two stars of the same apparent brightness l are also believed to be the same size. The spectrum of one star peaks at 800 nm whereas that of the other peaks at 400 nm. Use Wien's law (Section 38–1) and the Stefan-Boltzmann equation (Eq. 19–15) to estimate their relative distances from us. [*Hint*: See Examples 45–5 and 45–6.]

19. (III) Stars located in a certain cluster are assumed to be about the same distance from us. Two such stars have spectra that peak at $\lambda_1 = 500$ nm and $\lambda_2 = 700$ nm, and the ratio of their apparent brightness is $l_1/l_2 = 0.091$. Estimate their relative sizes (give ratio of their diameters). [*Hint*: Use the Stefan-Boltzmann equation, Eq. 19–15.]

Section 45–3

20. (I) Describe a triangle, drawn on the surface of a sphere, for which the sum of the angles is (a) 360°, and (b) 180°.

21. (I) Show that the Schwarzschild radius for a star with mass equal to that (a) of our Sun is 2.95 km, and (b) of Earth is 8.9 mm.

22. (II) What is the Schwarzschild radius for a typical galaxy?

23. (II) What is the maximum sum-of-the-angles for a triangle on a sphere?

24. (II) Calculate the escape velocity, using Newtonian mechanics, from a body that has collapsed to its Schwarzschild radius.

Section 45–4

25. (I) If a galaxy is traveling away from us at 1.0 percent of the speed of light, roughly how far away is it?

26. (I) The redshift of a galaxy indicates a velocity of 3500 km/s. How far away is it?

27. (I) Estimate the speed of a galaxy (relative to us) that is near the "edge" of the universe, say 12 billion light-years away.

28. (II) Estimate the observed wavelength for the 656-nm line in the Balmer series of hydrogen in the spectrum of a galaxy whose distance from us is (a) 1.0×10^6 ly, (b) 1.0×10^8 ly, (c) 1.0×10^{10} ly.

29. (II) Estimate the speed of a galaxy, and its distance from us, if the wavelength for the hydrogen line at 434 nm is measured on Earth as being 610 nm.

30. (II) Starting from Eq. 45–3, show that the Doppler shift in wavelength is $\Delta\lambda/\lambda_0 \approx v/c$ (Eq. 45–4) for $v \ll c$. [*Hint:* Use the binomial expansion.]

Sections 45–5 to 45–7

31. (I) Calculate the wavelength at the peak of the blackbody radiation distribution at 2.7 K using the Wien law.

32. (II) The critical density for closure of the universe is $\rho_c \approx 10^{-26}$ kg/m³. State ρ_c in terms of the average number of nucleons per cubic meter.

33. (II) The size of the universe (the average distance between galaxies) at any one moment is believed to have been inversely proportional to the absolute temperature. Estimate the size of the universe, compared to today, at (a) $t = 10^6$ yr, (b) $t = 1$ s, (c) $t = 10^{-6}$ s, and (d) $t = 10^{-35}$ s.

34. (II) At approximately what time had the universe cooled below the threshold temperature for producing (a) kaons $(M \approx 500 \text{ MeV}/c^2)$, (b) Υ $(M \approx 9500 \text{ MeV}/c^2)$, and (c) muons $(M \approx 100 \text{ MeV}/c^2)$?

General Problems

35. Suppose that three main-sequence stars could undergo the three changes represented by the three arrows, A, B, and C, in the H–R diagram of Fig. 45–26. For each case, describe the changes in temperature, luminosity, and size.

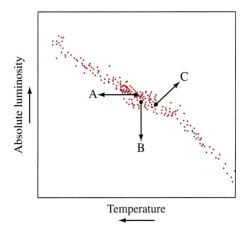

FIGURE 45–26 Problem 35.

36. Assume that the nearest stars to us have an intrinsic luminosity about the same as the Sun's. Their apparent brightness, however, is about 10^{11} times fainter than the Sun. From this, estimate the distance to the nearest stars. (Newton did this calculation, although he made a numerical error of a factor of 100.)

37. Use conservation of angular momentum to estimate the angular velocity of a neutron star which has collapsed to a diameter of 10 km, from a star whose radius was equal to that of our Sun $(7 \times 10^8 \text{ m})$, of mass 1.5 times that of the Sun, and which rotated (like our Sun) about once a month.

38. By what factor does the rotational kinetic energy change when the star in Problem 37 collapses to a neutron star?

39. A certain pulsar, believed to be a neutron star of mass 1.5 times that of the Sun, with diameter 10 km, is observed to have a rotation speed of 1.0 rev/s. If it loses rotational kinetic energy at the rate of 1 part in 10^9 per day, which is all transformed into radiation, what is the power output of the star?

40. Estimate the rate of which hydrogen atoms would have to be created, according to the steady-state model, to maintain the present density of the universe of about 10^{-27} kg/m³, assuming the universe is expanding with Hubble constant $H = 70$ km/s/Mpc.

41. Estimate what neutrino rest mass (in eV) would provide the critical density to close the universe. Assume the neutrino density is, like photons, about 10^9 times that of nucleons, and that nucleons make up only (a) 2 percent of the mass needed, or (b) 5 percent of the mass needed.

42. Two stars, whose spectra peak at 600 nm and 400 nm, respectively, both lie on the main sequence. Use Wien's law, the Stefan-Boltzmann equation, and the H–R diagram (Fig. 45–7) to estimate the ratio of their diameters. [*Hint:* See Examples 45–5 and 45–6.]

43. The farthest we can measure with parallax is about 30 parsecs. What is our minimum angular resolution (in degrees), based on this information?

44. Through some coincidence, the Balmer lines from singly ionized helium in a distant star happen to overlap with the Balmer lines from hydrogen (Fig. 38–20) in the Sun. How fast is the star receding from us?

45. What is the temperature that corresponds to the 1.8 TeV collisions at the Fermilab collider? To what era in cosmological history does this correspond?

46. Astronomers have recently measured the rotation of gas around what might be a supermassive black hole of about 2 billion solar masses at the center of a galaxy. If the radius from the galactic center to the gas clouds is 60 light-years, what Doppler shift $\Delta\lambda/\lambda_0$ do you estimate they saw?

47. Astronomers use an *apparent magnitude* (m) scale to describe apparent brightness. It uses a logarithmic scale, where a higher number corresponds to a less bright star. (For example, the Sun has magnitude -27, whereas most stars have positive magnitudes.) On this scale, a change in apparent magnitude by $+5$ corresponds to a decrease in apparent brightness by a factor of 100. If Venus has an apparent magnitude of -4.4, whereas Sirius has an apparent magnitude of -1.4, which is brighter? What is the ratio of the apparent brightness of these two objects?

48. How large would the Sun be if its density equaled the critical density of the universe, $\rho_c \approx 10^{-26}\,\text{kg/m}^3$? Express your answer in light-years and compare with the Earth–Sun distance and the size of our Galaxy.

49. Calculate the Schwarzschild radius using classical (Newtonian) gravitational theory, by calculating the minimum radius R for a sphere of mass M such that a photon (rest mass = 0) can escape from the surface. (General relativity gives $R = 2GM/c^2$.)

50. Estimate the radius of a white dwarf whose mass is equal to that of the Sun by the following method, assuming there are N nucleons and $\frac{1}{2}N$ electrons (why $\frac{1}{2}$?): (a) Use Fermi-Dirac statistics (Section 41–6) to show that the total energy of all the electrons is

$$E_e = \frac{3}{5}\left(\frac{1}{2}N\right)\frac{h^2}{8m_e}\left(\frac{3}{\pi}\frac{N}{2V}\right)^{\frac{2}{3}}.$$

[Hint: See Eqs. 41–12 and 41–13; we assume electrons fill energy levels from 0 up to the Fermi energy.] (b) The nucleons contribute to the total energy mainly via the gravitational force (note that the Fermi energy for nucleons is negligible compared to that for electrons—why?). Use a gravitational form of Gauss's law to show that the total gravitational potential energy of a uniform sphere of radius R is

$$-\frac{3}{5}\frac{GM^2}{R},$$

by considering the potential energy of a spherical shell of radius r due only to the mass inside it (why?) and integrate from $r = 0$ to $r = R$. (See also Appendix C.) (c) Write the total energy as a sum of these two terms, and set $dE/dR = 0$ to find the equilibrium radius, and evaluate it for a mass equal to the Sun's $(2.0 \times 10^{30}\,\text{kg})$.

51. Determine the radius of a neutron star using the same argument as in Problem 50 but for N neutrons only. Show that the radius of a neutron star, of 1.5 solar masses, is about 11 km.

52. (a) Use special relativity and Newton's law of gravitation to show that a photon of mass $m = E/c^2$ just grazing the Sun will be deflected by an angle $\Delta\theta$ given by

$$\Delta\theta = \frac{2GM}{c^2 R}$$

where G is the gravitational constant, R and M are the radius and mass of the Sun, and c is the speed of light. (b) Put in values and show $\Delta\theta = 0.87''$. (General relativity predicts an angle twice as large, $1.74''$.)

APPENDICES

Mathematical Formulas

A–1 Quadratic Formula

If

$$ax^2 + bx + c = 0$$

then

$$x = \frac{-b \pm \sqrt{b^2 - 4ac}}{2a}$$

A–2 Binomial Expansion

$$(1 \pm x)^n = 1 \pm nx + \frac{n(n-1)}{2!} x^2 \pm \frac{n(n-1)(n-2)}{3!} x^3 + \cdots$$

$$(x + y)^n = x^n \left(1 + \frac{y}{x}\right)^n = x^n \left(1 + n\frac{y}{x} + \frac{n(n-1)}{2!} \frac{y^2}{x^2} + \cdots\right)$$

A–3 Other Expansions

$$e^x = 1 + x + \frac{x^2}{2!} + \frac{x^3}{3!} + \cdots$$

$$\ln(1 + x) = x - \frac{x^2}{2} + \frac{x^3}{3} - \frac{x^4}{4} + \cdots$$

$$\sin\theta = \theta - \frac{\theta^3}{3!} + \frac{\theta^5}{5!} - \cdots$$

$$\cos\theta = 1 - \frac{\theta^2}{2!} + \frac{\theta^4}{4!} - \cdots$$

$$\tan\theta = \theta + \frac{\theta^3}{3} + \frac{2}{15}\theta^5 + \cdots \qquad |\theta| < \frac{\pi}{2}$$

In general:

$$f(x) = f(0) + \left(\frac{df}{dx}\right)_0 x + \left(\frac{d^2f}{dx^2}\right)_0 \frac{x^2}{2!} + \cdots$$

A–4 Areas and Volumes

Object	Surface area	Volume
Circle, radius r	πr^2	—
Sphere, radius r	$4\pi r^2$	$\frac{4}{3}\pi r^3$
Right circular cylinder, radius r, height h	$2\pi r^2 + 2\pi rh$	$\pi r^2 h$
Right circular cone, radius r, height h	$\pi r^2 + \pi r \sqrt{r^2 + h^2}$	$\frac{1}{3}\pi r^2 h$

A–5 Plane Geometry

1.

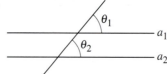

FIGURE A–1

If line a_1 is parallel to line a_2, then $\theta_1 = \theta_2$.

2.

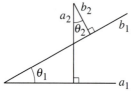

FIGURE A–2

If $a_1 \perp a_2$ and $b_1 \perp b_2$, then $\theta_1 = \theta_2$.

3. The sum of the angles in any plane triangle is 180°.

4. *Pythagorean theorem:*

FIGURE A–3

In any right triangle (one angle = 90°) of sides a, b, and c:

$$a^2 + b^2 = c^2$$

where c is the length of the hypotenuse (opposite the 90° angle).

A–6 Trigonometric Functions and Identities

(See Fig. A–4.)

FIGURE A–4

$$\sin \theta = \frac{o}{h} \qquad\qquad \csc \theta = \frac{1}{\sin \theta} = \frac{h}{o}$$

$$\cos \theta = \frac{a}{h} \qquad\qquad \sec \theta = \frac{1}{\cos \theta} = \frac{h}{a}$$

$$\tan \theta = \frac{o}{a} = \frac{\sin \theta}{\cos \theta} \qquad \cot \theta = \frac{1}{\tan \theta} = \frac{a}{o}$$

$$a^2 + o^2 = h^2 \qquad\qquad \text{[Pythagorean theorem]}.$$

Figure A–5 shows the signs (+ or −) that cosine, sine, and tangent take on for angles θ in the four quadrants (0° to 360°). Note that angles are measured counterclockwise from the x axis as shown; negative angles are measured from *below* the x axis, clockwise: for example, −30° = +330°, and so on.

FIGURE A–5

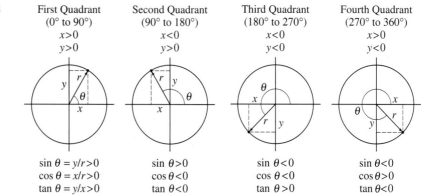

First Quadrant (0° to 90°)	Second Quadrant (90° to 180°)	Third Quadrant (180° to 270°)	Fourth Quadrant (270° to 360°)
$x > 0$	$x < 0$	$x < 0$	$x > 0$
$y > 0$	$y > 0$	$y < 0$	$y < 0$
$\sin \theta = y/r > 0$	$\sin \theta > 0$	$\sin \theta < 0$	$\sin \theta < 0$
$\cos \theta = x/r > 0$	$\cos \theta < 0$	$\cos \theta < 0$	$\cos \theta > 0$
$\tan \theta = y/x > 0$	$\tan \theta < 0$	$\tan \theta > 0$	$\tan \theta < 0$

The following are some useful identities among the trigonometric functions:

$$\sin^2\theta + \cos^2\theta = 1, \quad \sec^2\theta - \tan^2\theta = 1, \quad \csc^2\theta - \cot^2\theta = 1$$

$$\sin 2\theta = 2\sin\theta\cos\theta$$

$$\cos 2\theta = \cos^2\theta - \sin^2\theta = 2\cos^2\theta - 1 = 1 - 2\sin^2\theta$$

$$\tan 2\theta = \frac{2\tan\theta}{1 - \tan^2\theta}$$

$$\sin(A \pm B) = \sin A\cos B \pm \cos A\sin B$$

$$\cos(A \pm B) = \cos A\cos B \mp \sin A\sin B$$

$$\tan(A \pm B) = \frac{\tan A \pm \tan B}{1 \mp \tan A\tan B}$$

$$\sin(180° - \theta) = \sin\theta$$

$$\cos(180° - \theta) = -\cos\theta$$

$$\sin(90° - \theta) = \cos\theta$$

$$\cos(90° - \theta) = \sin\theta$$

$$\cos(-\theta) = \cos\theta$$

$$\sin(-\theta) = -\sin\theta$$

$$\tan(-\theta) = -\tan\theta$$

$$\sin\tfrac{1}{2}\theta = \sqrt{\frac{1 - \cos\theta}{2}}, \quad \cos\tfrac{1}{2}\theta = \sqrt{\frac{1 + \cos\theta}{2}}, \quad \tan\tfrac{1}{2}\theta = \sqrt{\frac{1 - \cos\theta}{1 + \cos\theta}}$$

$$\sin A \pm \sin B = 2\sin\left(\frac{A \pm B}{2}\right)\cos\left(\frac{A \mp B}{2}\right).$$

For any triangle (see Fig. A–6):

$$\frac{\sin\alpha}{a} = \frac{\sin\beta}{b} = \frac{\sin\gamma}{c} \qquad \text{[Law of sines]}$$

$$c^2 = a^2 + b^2 - 2ab\cos\gamma. \qquad \text{[Law of cosines]}$$

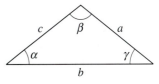

FIGURE A–6

A–7 Logarithms

The following identities apply to common logs (base 10), natural logs (base e) which are often abbreviated ln, or logs to any other base.

$$\log(ab) = \log a + \log b$$

$$\log\left(\frac{a}{b}\right) = \log a - \log b$$

$$\log a^n = n\log a.$$

A–8 Vectors

Vector addition is covered in Sections 3–2 to 3–5.
Vector multiplication is covered in Sections 3–3, 7–2 and 11–1.

Derivatives and Integrals

B–1 Derivatives: General Rules

(See also Section 2–3.)

$$\frac{dx}{dx} = 1$$

$$\frac{d}{dx}\big[af(x)\big] = a\frac{df}{dx} \qquad (a = \text{constant})$$

$$\frac{d}{dx}\big[f(x) + g(x)\big] = \frac{df}{dx} + \frac{dg}{dx}$$

$$\frac{d}{dx}\big[f(x)g(x)\big] = \frac{df}{dx}g + f\frac{dg}{dx}$$

$$\frac{d}{dx}\big[f(y)\big] = \frac{df}{dy}\frac{dy}{dx} \qquad \text{[chain rule]}$$

$$\frac{dx}{dy} = \frac{1}{\left(\dfrac{dy}{dx}\right)} \qquad \text{if } \frac{dy}{dx} \neq 0.$$

B–2 Derivatives: Particular Functions

$$\frac{da}{dx} = 0 \qquad (a = \text{constant})$$

$$\frac{d}{dx}x^n = nx^{n-1}$$

$$\frac{d}{dx}\sin ax = a\cos ax$$

$$\frac{d}{dx}\cos ax = -a\sin ax$$

$$\frac{d}{dx}\tan ax = a\sec^2 ax$$

$$\frac{d}{dx}\ln ax = \frac{1}{x}$$

$$\frac{d}{dx}e^{ax} = ae^{ax}$$

B–3 Indefinite Integrals: General Rules

(See also Section 7–3.)

$$\int dx = x$$

$$\int a f(x)\, dx = a \int f(x)\, dx \qquad (a = \text{constant})$$

$$\int [f(x) + g(x)]\, dx = \int f(x)\, dx + \int g(x)\, dx$$

$$\int u\, dv = uv - \int v\, du \qquad \text{[integration by parts]}$$

B–4 Indefinite Integrals: Particular Functions

(An arbitrary constant can be added to the right side of each equation.)

$$\int a\, dx = ax \qquad (a = \text{constant})$$

$$\int x^m\, dx = \frac{1}{m+1} x^{m+1} \qquad (m \neq -1)$$

$$\int \sin ax\, dx = -\frac{1}{a} \cos ax$$

$$\int \cos ax\, dx = \frac{1}{a} \sin ax$$

$$\int \tan ax\, dx = \frac{1}{a} \ln|\sec ax|$$

$$\int \frac{1}{x}\, dx = \ln x$$

$$\int e^{ax}\, dx = \frac{1}{a} e^{ax}$$

$$\int \frac{dx}{x^2 + a^2} = \frac{1}{a} \tan^{-1} \frac{x}{a}$$

$$\int \frac{dx}{x^2 - a^2} = \frac{1}{2a} \ln\left(\frac{x-a}{x+a}\right) \qquad (x^2 > a^2)$$

$$= -\frac{1}{2a} \ln\left(\frac{a+x}{a-x}\right) \qquad (x^2 < a^2)$$

$$\int \frac{dx}{\sqrt{x^2 \pm a^2}} = \ln(x + \sqrt{x^2 \pm a^2})$$

$$\int \frac{dx}{(x^2 \pm a^2)^{\frac{3}{2}}} = \frac{\pm x}{a^2 \sqrt{x^2 \pm a^2}}$$

$$\int \frac{x\, dx}{(x^2 \pm a^2)^{\frac{3}{2}}} = \frac{-1}{\sqrt{x^2 \pm a^2}}$$

$$\int \sin^2 ax\, dx = \frac{x}{2} - \frac{\sin 2ax}{4a}$$

$$\int xe^{-ax}\, dx = -\frac{e^{-ax}}{a^2} (ax + 1)$$

$$\int x^2 e^{-ax}\, dx = -\frac{e^{-ax}}{a^3} (a^2 x^2 + 2ax + 2)$$

B–5 A few Definite Integrals

$$\int_0^\infty x^n e^{-ax}\, dx = \frac{n!}{a^{n+1}}$$

$$\int_0^\infty e^{-ax^2}\, dx = \sqrt{\frac{\pi}{4a}}$$

$$\int_0^\infty xe^{-ax^2}\, dx = \frac{1}{2a}$$

$$\int_0^\infty x^2 e^{-ax^2}\, dx = \sqrt{\frac{\pi}{16a^3}}$$

$$\int_0^\infty x^3 e^{-ax^2}\, dx = \frac{1}{2a^2}$$

$$\int_0^\infty x^{2n} e^{-ax^2}\, dx = \frac{1 \cdot 3 \cdot 5 \cdots (2n-1)}{2^{n+1} a^n} \sqrt{\frac{\pi}{a}}$$

Gravitational Force due to a Spherical Mass Distribution

In Chapter 6 (Section 6–1), we stated that the gravitational force exerted by or on a uniform sphere acts as if all the mass of the sphere were concentrated at its center, if the other mass is outside the sphere. In other words, the gravitational force that a uniform sphere exerts on a particle outside it is

$$F = G\frac{mM}{r^2}, \qquad\qquad \text{[m outside sphere of mass M]}$$

where m is the mass of the particle, M the mass of the sphere, and r the distance of m from the center of the sphere. Now we will derive this result. We will use the concepts of infinitesimally small quantities and integration.

First we consider a very thin, uniform spherical shell (like a thin-walled basketball) of mass M whose thickness t is small compared to its radius R (Fig. C–1). The force on a particle of mass m at a distance r from the center of the shell can be calculated as the vector sum of the forces due to all the particles of the shell. We imagine the shell divided up into thin (infinitesimal) circular strips so that all points on a strip are equidistant from our particle m. One of these circular strips, labeled AB, is shown in Fig. C–1. It is $R\,d\theta$ wide, t thick, and has a radius $R\sin\theta$. The force on our particle m due to a tiny piece of the strip at point A is represented by the vector $\mathbf{F}_A$ shown. The force due to a tiny piece of the strip at point B, which is diametrically opposite A, is the force $\mathbf{F}_B$. We take the two pieces at A and B to be of equal mass, so $F_A = F_B$. The horizontal components of $\mathbf{F}_A$ and $\mathbf{F}_B$ are each equal to

$$F_A \cos\phi$$

and point toward the center of the shell. The vertical components of $\mathbf{F}_A$ and $\mathbf{F}_B$ are of equal magnitude and point in opposite directions, and so cancel. Since for every point on the strip there is a corresponding point diametrically opposite (as with A and B), we see that the net force due to the entire strip points toward the center of the shell. Its magnitude will be

$$dF = G\frac{m\,dM}{l^2}\cos\phi,$$

where dM is the mass of the entire circular strip and l is the distance from all

FIGURE C–1 Calculating the gravitational force on a particle of mass m due to a uniform spherical shell of radius R and mass M.

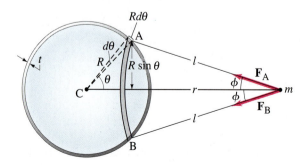

points on the strip to m, as shown. We write dM in terms of the density ρ; by density we mean the mass per unit volume (Section 13–1). Hence, $dM = \rho\, dV$, where dV is the volume of the strip and equals $(2\pi R \sin\theta)(t)(R\, d\theta)$. Then the force dF due to the circular strip shown is

$$dF = G \frac{m\rho 2\pi R^2 t \sin\theta\, d\theta}{l^2} \cos\phi. \qquad (C-1)$$

To get the total force F that the entire shell exerts on the particle m, we must integrate over all the circular strips: that is, from $\theta = 0°$ to $\theta = 180°$. But our expression for dF contains l and ϕ, which are functions of θ. From Fig. C–1 we can see that

$$l \cos\phi = r - R\cos\theta.$$

Furthermore, we can write the law of cosines for triangle CmA:

$$\cos\theta = \frac{r^2 + R^2 - l^2}{2rR}. \qquad (C-2)$$

With these two expressions we can reduce our three variables (l, θ, ϕ) to only one, which we take to be l. We do two things with Eq. C–2: (1) We put it into the equation for $l\cos\phi$ above:

$$\cos\phi = \frac{1}{l}(r - R\cos\theta) = \frac{r^2 + l^2 - R^2}{2rl};$$

and (2) we take the differential of both sides of Eq. C–2 (because $\sin\theta\, d\theta$ appears in the expression for dF, Eq. C–1):

$$-\sin\theta\, d\theta = -\frac{2l\, dl}{2rR} \qquad \text{or} \qquad \sin\theta\, d\theta = \frac{l\, dl}{rR},$$

since r and R are considered constants when summing over the strips. Now we insert these into Eq. C–1 for dF and find

$$dF = Gm\rho\pi t \frac{R}{r^2}\left(1 + \frac{r^2 - R^2}{l^2}\right) dl.$$

Now we integrate to get the net force on our thin shell of radius R. To integrate over all the strips ($\theta = 0°$ to $180°$), we must go from $l = r - R$ to $l = r + R$ (see Fig. C–1). Thus,

$$F = Gm\rho\pi t \frac{R}{r^2}\left[l - \frac{r^2 - R^2}{l}\right]_{l=r-R}^{l=r+R}$$

$$= Gm\rho\pi t \frac{R}{r^2}(4R).$$

The volume V of the spherical shell is its area $(4\pi R^2)$ times the thickness t. Hence the mass $M = \rho V = \rho 4\pi R^2 t$, and finally

$$F = G\frac{mM}{r^2}. \qquad \left[\begin{array}{c}\text{particle of mass } m \text{ outside a}\\ \text{thin uniform spherical shell of mass } M\end{array}\right]$$

This result gives us the force a thin shell exerts on a particle of mass m a distance r from the center of the shell, and *outside* the shell. We see that the force is the same as that between m and a particle of mass M at the center of the shell. In other words, for purposes of calculating the gravitational force exerted on or by a uniform spherical shell, we can consider all its mass concentrated at its center.

What we have derived for a shell holds also for a solid sphere, since a solid sphere can be considered as made up of many concentric shells, from $R = 0$ to $R = R_0$, where R_0 is the radius of the solid sphere. Why? Because if each shell has mass dM, we write for each shell, $dF = Gm\,dM/r^2$, where r is the distance from the center C to mass m and is the same for all shells. Then the total force equals the sum or integral over dM, which gives the total mass M. Thus the result

$$F = G\frac{mM}{r^2} \qquad \begin{bmatrix} \text{particle of mass } m \text{ outside} \\ \text{solid sphere of mass } M \end{bmatrix} \quad \text{(C–3)}$$

is valid for a solid sphere of mass M even if the density varies with distance from the center. (It is not valid if the density varies within each shell—that is, depends not only on R.) Thus the gravitational force exerted on or by spherical objects, including nearly spherical objects like the Earth, Sun, and Moon, can be considered to act as if the objects were point particles.

This result, Eq. C–3, is true only if the mass m is outside the sphere. Let us next consider a point mass m that is located inside the spherical shell of Fig. C–1. Here, r would be less than R, and the integration over l would be from $l = R - r$ to $l = R + r$, so

$$\left[l - \frac{r^2 - R^2}{l} \right]_{R-r}^{R+r} = 0.$$

Thus the force on any mass inside the shell would be zero. This result has particular importance for the electrostatic force, which is also an inverse square law. For the gravitational situation, we see that at points within a solid sphere, say 1000 km below the earth's surface, only the mass up to that radius contributes to the net force. The outer shells beyond the point in question contribute no net gravitational effect.

The results we have obtained here can also be reached using the gravitational analog of Gauss's law for electrostatics (Chapter 22).

APPENDIX D

Selected Isotopes

(1) Atomic Number Z	(2) Element	(3) Symbol	(4) Mass Number A	(5) Atomic Mass[†]	(6) % Abundance (or Radioactive Decay Mode)	(7) Half-life (if radioactive)
0	(Neutron)	n	1	1.008665	β^-	10.4 min
1	Hydrogen	H	1	1.007825	99.985%	
	Deuterium	d or D	2	2.014102	0.015%	
	Tritium	t or T	3	3.016049	β^-	12.33 yr
2	Helium	He	3	3.016029	0.000137%	
			4	4.002603	99.999863%	
3	Lithium	Li	6	6.015122	7.5%	
			7	7.016004	92.5%	
4	Beryllium	Be	7	7.016929	EC, γ	53.12 days
			9	9.012182	100%	
5	Boron	B	10	10.012937	19.9%	
			11	11.009305	80.1%	
6	Carbon	C	11	11.011434	β^+, EC	20.39 min
			12	12.000000	98.90%	
			13	13.003355	1.10%	
			14	14.003242	β^-	5730 yr
7	Nitrogen	N	13	13.005739	β^+	9.965 min
			14	14.003074	99.63%	
			15	15.000108	0.37%	
8	Oxygen	O	15	15.003065	β^+, EC	122.24 s
			16	15.994915	99.76%	
			18	17.999160	0.20%	
9	Fluorine	F	19	18.998403	100%	
10	Neon	Ne	20	19.992440	90.48%	
			22	21.991386	9.25%	
11	Sodium	Na	22	21.994437	β^+, EC, γ	2.6019 yr
			23	22.989770	100%	
			24	23.990963	β^-, γ	14.9590 h
12	Magnesium	Mg	24	23.985042	78.99%	
13	Aluminum	Al	27	26.981538	100%	

[†] The masses given in column (5) are those for the neutral atom, including the Z electrons. Updated 1999.

(1) Atomic Number Z	(2) Element	(3) Symbol	(4) Mass Number A	(5) Atomic Mass†	(6) % Abundance (or Radioactive Decay Mode)	(7) Half-life (if radioactive)
14	Silicon	Si	28	27.976927	92.23%	
			31	30.975363	β^-, γ	157.3 min
15	Phosphorus	P	31	30.973762	100%	
			32	31.973907	β^-	14.262 days
16	Sulfur	S	32	31.972071	95.02%	
			35	34.969032	β^-	87.32 days
17	Chlorine	Cl	35	34.968853	75.77%	
			37	36.965903	24.23%	
18	Argon	Ar	40	39.962383	99.600%	
19	Potassium	K	39	38.963707	93.2581%	
			40	39.963999	0.0117%	
					β^-, EC, γ, β^+	1.28×10^9 yr
20	Calcium	Ca	40	39.962591	96.941%	
21	Scandium	Sc	45	44.955910	100%	
22	Titanium	Ti	48	47.947947	73.8%	
23	Vanadium	V	51	50.943964	99.750%	
24	Chromium	Cr	52	51.940512	83.79%	
25	Manganese	Mn	55	54.938049	100%	
26	Iron	Fe	56	55.934942	91.72%	
27	Cobalt	Co	59	58.933200	100%	
			60	59.933822	β^-, γ	5.2714 yr
28	Nickel	Ni	58	57.935348	68.077%	
			60	59.930791	26.233%	
29	Copper	Cu	63	62.929601	69.17%	
			65	64.927794	30.83%	
30	Zinc	Zn	64	63.929147	48.6%	
			66	65.926037	27.9%	
31	Gallium	Ga	69	68.925581	60.108%	
32	Germanium	Ge	72	71.922076	27.66%	
			74	73.921178	35.94%	
33	Arsenic	As	75	74.921596	100%	
34	Selenium	Se	80	79.916522	49.61%	
35	Bromine	Br	79	78.918338	50.69%	
36	Krypton	Kr	84	83.911507	57.0%	
37	Rubidium	Rb	85	84.911789	72.17%	
38	Strontium	Sr	86	85.909262	9.86%	
			88	87.905614	82.58%	
			90	89.907737	β^-	28.79 yr
39	Yttrium	Y	89	88.905848	100%	
40	Zirconium	Zr	90	89.904704	51.45%	
41	Niobium	Nb	93	92.906377	100%	
42	Molybdenum	Mo	98	97.905408	24.13%	

† The masses given in column (5) are those for the neutral atom, including the Z electrons.

(1) Atomic Number Z	(2) Element	(3) Symbol	(4) Mass Number A	(5) Atomic Mass[†]	(6) % Abundance (or Radioactive Decay Mode)	(7) Half-life (if radioactive)
43	Technetium	Tc	98	97.907216	β^-, γ	4.2×10^6 yr
44	Ruthenium	Ru	102	101.904349	31.6%	
45	Rhodium	Rh	103	102.905504	100%	
46	Palladium	Pd	106	105.903483	27.33%	
47	Silver	Ag	107	106.905093	51.839%	
			109	108.904756	48.161%	
48	Cadmium	Cd	114	113.903358	28.73%	
49	Indium	In	115	114.903878	95.7%; β^-, γ	4.41×10^{14} yr
50	Tin	Sn	120	119.902197	32.59%	
51	Antimony	Sb	121	120.903818	57.36%	
52	Tellurium	Te	130	129.906223	33.80%	7.9×10^{20} yr
53	Iodine	I	127	126.904468	100%	
			131	130.906124	β^-, γ	8.0207 days
54	Xenon	Xe	132	131.904154	26.9%	
			136	135.907220	8.9%	
55	Cesium	Cs	133	132.905446	100%	
56	Barium	Ba	137	136.905821	11.23%	
			138	137.905241	71.70%	
57	Lanthanum	La	139	138.906348	99.9098%	
58	Cerium	Ce	140	139.905434	88.48%	
59	Praseodymium	Pr	141	140.907647	100%	
60	Neodymium	Nd	142	141.907718	27.13%	
61	Promethium	Pm	145	144.912744	EC, γ, α	17.7 yr
62	Samarium	Sm	152	151.919728	26.7%	
63	Europium	Eu	153	152.921226	52.2%	
64	Gadolinium	Gd	158	157.924101	24.84%	
65	Terbium	Tb	159	158.925343	100%	
66	Dysprosium	Dy	164	163.929171	28.2%	
67	Holmium	Ho	165	164.930319	100%	
68	Erbium	Er	166	165.930290	33.6%	
69	Thulium	Tm	169	168.934211	100%	
70	Ytterbium	Yb	174	173.938858	31.8%	
71	Lutecium	Lu	175	174.940767	97.4%	
72	Hafnium	Hf	180	179.946549	35.100%	
73	Tantalum	Ta	181	180.947996	99.988%	
74	Tungsten (wolfram)	W	184	183.950933	30.67%	
75	Rhenium	Re	187	186.955751	62.60%; β^-	4.35×10^{10} yr
76	Osmium	Os	191	190.960927	β^-, γ	15.4 days
			192	191.961479	41.0%	
77	Iridium	Ir	191	190.960591	37.3%	
			193	192.962923	62.7%	
78	Platinum	Pt	195	194.964774	33.8%	

[†]The masses given in column (5) are those for the neutral atom, including the Z electrons.

(1) Atomic Number Z	(2) Element	(3) Symbol	(4) Mass Number A	(5) Atomic Mass†	(6) % Abundance (or Radioactive Decay Mode)	(7) Half-life (if radioactive)
79	Gold	Au	197	196.966551	100%	
80	Mercury	Hg	199	198.968262	16.87%	
			202	201.970625	29.86%	
81	Thallium	Tl	205	204.974412	70.476%	
82	Lead	Pb	206	205.974449	24.1%	
			207	206.975880	22.1%	
			208	207.976635	52.4%	
			210	209.984173	β^-, γ, α	22.3 yr
			211	210.988731	β^-, γ	36.1 min
			212	211.991887	β^-, γ	10.64 h
			214	213.999798	β^-, γ	26.8 min
83	Bismuth	Bi	209	208.980383	100%	
			211	210.987258	α, γ, β^-	2.14 min
84	Polonium	Po	210	209.982857	α, γ	138.376 days
			214	213.995185	α, γ	164.3 μs
85	Astatine	At	218	218.008681	α, β^-	1.5 s
86	Radon	Rn	222	222.017570	α, γ	3.8235 days
87	Francium	Fr	223	223.019731	β^-, γ, α	21.8 min
88	Radium	Ra	226	226.025402	α, γ	1600 yr
89	Actinium	Ac	227	227.027746	β^-, γ, α	21.773 yr
90	Thorium	Th	228	228.028731	α, γ	1.9116 yr
			232	232.038050	100%; α, γ	1.405×10^{10} yr
91	Protactinium	Pa	231	231.035878	α, γ	3.276×10^4 yr
92	Uranium	U	232	232.037146	α, γ	68.9 yr
			233	233.039628	α, γ	1.592×10^5 yr
			235	235.043923	0.720%, α, γ	7.038×10^8 yr
			236	236.045561	α, γ	2.342×10^7 yr
			238	238.050782	99.2745%; α, γ	4.468×10^9 yr
			239	239.054287	β^-, γ	23.45 min
93	Neptunium	Np	237	237.048166	α, γ	2.144×10^6 yr
			239	239.052931	β^-, γ	2.3565 d
94	Plutonium	Pu	239	239.052157	α, γ	24,110 yr
			244	244.064197	α	8.08×10^7 yr
95	Americium	Am	243	243.061373	α, γ	7370 yr
96	Curium	Cm	247	247.070346	α, γ	1.56×10^7 yr
97	Berkelium	Bk	247	247.070298	α, γ	1380 yr
98	Californium	Cf	251	251.079580	α, γ	898 yr
99	Einsteinium	Es	252	252.082972	α, EC, γ	472 d
100	Fermium	Fm	257	257.095099	α, γ	101 d
101	Mendelevium	Md	258	258.098425	α, γ	51.5 d
102	Nobelium	No	259	259.10102	α, EC	58 min
103	Lawrencium	Lr	262	262.10969	α, EC, fission	216 min

† The masses given in column (5) are those for the neutral atom, including the Z electrons.

(1) Atomic Number Z	(2) Element	(3) Symbol	(4) Mass Number A	(5) Atomic Mass†	(6) % Abundance (or Radioactive Decay Mode)	(7) Half-life (if radioactive)
104	Rutherfordium	Rf	261	261.10875	α	65 s
105	Dubnium	Db	262	262.11415	α, fission, EC	34 s
106	Seaborgium	Sg	266	266.12193	α, fission	21 s
107	Bohrium	Bh	264	264.12473	α	0.44 s
108	Hassium	Hs	269	269.13411	α	9 s
109	Meitnerium	Mt	268	268.13882	α	0.07 s
110			271	271.14608	α	0.06 s
111			272	272.15348	α	1.5 ms
112			277	277	α	0.24 ms
114			289	289	α	20 s
116			289	289	α	0.6 ms
118			293	293	α	0.1 ms

† The masses given in column (5) are those for the neutral atom, including the Z electrons.

Answers to Odd-Numbered Problems

CHAPTER 1

1. (a) 1×10^{10} yr; (b) 3×10^{17} s.

3. (a) 1.156×10^3; (b) 2.18×10^1; (c) 6.8×10^{-3}; (d) 2.7635×10^1; (e) 2.19×10^{-1}; (f) 2.2×10^1.

5. 7.7%.

7. (a) 4%; (b) 0.4%; (c) 0.07%.

9. 1.0×10^5 s.

11. 9%.

13. (a) 0.286 6 m; (b) 0.000 085 V; (c) 0.000 760 kg; (d) 0.000 000 000 060 0 s; (e) 0.000 000 000 000 022 5 m; (f) 2,500,000,000 volts.

15. 1.8 m.

17. (a) 0.111 yd^2; (b) 10.76 ft^2.

19. (a) 3.9×10^{-9} in; (b) 1.0×10^8 atoms.

21. (a) 0.621 mi/h; (b) 1 m/s = 3.28 ft/s; (c) 0.278 m/s.

23. (a) 9.46×10^{15} m; (b) 6.31×10^4 AU; (c) 7.20 AU/h.

25. (a) 10^3; (b) 10^4; (c) 10^{-2}; (d) 10^9.

27. $\approx$20%.

29. 1×10^5 cm^3.

31. (a) $\approx$600 dentists.

33. $\approx 3 \times 10^8$ kg/yr.

35. 51 km.

37. $A = [L/T^4] = $ m/s^4, $B = [L/T^2] = $ m/s^2.

39. (a) 0.10 nm; (b) 1.0×10^5 fm; (c) 1.0×10^{10} Å; (d) 9.5×10^{25} Å.

41. (a) 3.16×10^7 s; (b) 3.16×10^{16} ns; (c) 3.17×10^{-8} yr.

43. (a) 1,000 drivers.

45. 1×10^{11} gal/yr.

47. 9 cm.

49. 4×10^5 t.

51. $\approx$4 yr.

53. 1.9×10^2 m.

55. (a) 3%, 3%; (b) 0.7%, 0.2%.

CHAPTER 2

1. 5.0 h.

3. 61 m.

5. 0.78 cm/s (toward $+x$).

7. $\approx$300 m/s.

9. (a) 10.1 m/s; (b) +3.4 m/s, away from trainer.

11. (a) 0.28 m/s; (b) 1.2 m/s; (c) 0.28 m/s; (d) 1.6 m/s; (e) -1.0 m/s.

13. (a) 13.4 m/s; (b) +4.5 m/s, away from master.

15. 24 s.

17. 55 km/h, 0.

19. 6.73 m/s.

21. 5.2 s

23. -7.0 m/s^2, 0.72.

25. (a) 4.7 m/s^2; (b) 2.2 m/s^2; (c) 0.3 m/s^2; (d) 1.6 m/s^2.

27. $v = (6.0 \text{ m/s}) + (17 \text{ m/s}^2)t$, $a = 17$ m/s^2.

29. 1.5 m/s^2, 99 m.

31. 1.7×10^2 m.

33. 4.41 m/s^2, $t = 2.61$ s.

35. 55.0 m.

37. (a) 2.3×10^2 m; (b) 31 s; (c) 15 m, 13 m.

39. (a) 103 m; (b) 64 m.

41. 31 m/s.

43. (b) 3.45 s.

45. 32 m/s (110 km/h).

47. 2.83 s.

49. (a) 8.81 s; (b) 86.3 m/s.

51. 1.44 s.

53. 15 m/s.

55. 5.44 s.

59. 0.035 s.

61. 1.8 m above the top of the window.

63. 52 m.

65. 19.8 m/s, 20.0 m.

67. (a) $v = (g/k)(1 - e^{-kt})$; (b) $v_{\text{term}} = g/k$.

69. $6h_{\text{Earth}}$.

71. 1.3 m.

73. (b) $H_{50} = 9.8$ m; (c) $H_{100} = 39$ m.

75. (a) 1.3 m; (b) 6.1 m/s; (c) 1.2 s.

77. (a) 3.88 s; (b) 73.9 m; (c) 38.0 m/s, 48.4 m/s.

79. (a) 52 min; (b) 31 min.

81. (a) $v_0 = 26$ m/s; (b) 35 m; (c) 1.2 s; (d) 4.1 s.

83. (a) 4.80 s; (b) 37.0 m/s; (c) 75.2 m.

85. She should decide to stop!

87. $\Delta v_{0\text{down}} = 0.8$ m/s, $\Delta v_{0\text{up}} = 0.9$ m/s.

89. 29.0 m.

CHAPTER 3

1. 263 km, 13° S of W.

3. $\mathbf{V}_{\text{wrong}} = \mathbf{V}_2 - \mathbf{V}_1$.

5. 13.6 m, 18° N of E,

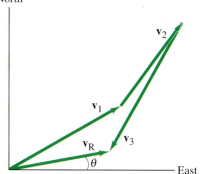

7. (a)

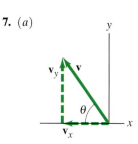

(b) $V_x = -11.7$, $V_y = 8.16$; (c) 14.3, 34.9° above $-x$-axis.

9. (a) $V_{\text{N}} = 476$ km/h, $V_{\text{W}} = 421$ km/h; (b) $d_{\text{N}} = 1.43 \times 10^3$ km, $d_{\text{W}} = 1.26 \times 10^3$ km.

11. (a) 4.2, 45° below $+x$-axis; (b) 5.1, 79° below $+x$-axis.

13. (a) 53.7, 1.40° above $-x$-axis; (b) 53.7, 1.40° below $+x$-axis.

15. (a) 94.5, 11.8° below $-x$-axis; (b) 150, 35.3° below $+x$-axis.

17. (a) $A_x = \pm 82.9$; (b) 166.6, 12.1° above $-x$-axis.

19. $(7.60 \text{ m/s})\mathbf{i} - (4.00 \text{ m/s})\mathbf{k}$; 8.59 m/s.

21. (a) Unknown; (b) 4.11 m/s^2, 33.2° north of east; (c) unknown.

23. (a) $\mathbf{v} = (4.0\,\text{m/s}^2)t\mathbf{i} + (3.0\,\text{m/s}^2)t\mathbf{j}$;
(b) $(5.0\,\text{m/s}^2)t$;
(c) $\mathbf{r} = (2.0\,\text{m/s}^2)t^2\mathbf{i} + (1.5\,\text{m/s}^2)t^2\mathbf{j}$;
(d) $\mathbf{v} = (8.0\,\text{m/s})\mathbf{i} + (6.0\,\text{m/s})\mathbf{j}$,
$|\mathbf{v}| = 10.0\,\text{m/s}$,
$\mathbf{r} = (8.0\,\text{m})\mathbf{i} + (6.0\,\text{m})\mathbf{j}$.

25. (a) $-(18.0\,\text{m/s})\sin(3.0\,\text{s}^{-1})t\mathbf{i}$
$+ (18.0\,\text{m/s})\cos(3.0\,\text{s}^{-1})t\mathbf{j}$;
(b) $-(54.0\,\text{m/s}^2)\cos(3.0\,\text{s}^{-1})t\mathbf{i}$
$- (54.0\,\text{m/s}^2)\sin(3.0\,\text{s}^{-1})t\mathbf{j}$;
(c) circle; (d) $a = (9.0\,\text{s}^{-2})r$, $180°$.

27. 44 m, 6.3 m.

29. $38°$ and $52°$.

31. 1.95 s.

33. 22 m.

35.

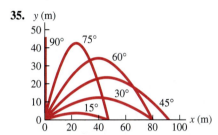

37. 5.71 s.

39. (a) 65.7 m; (b) 7.32 s; (c) 267 m;
(d) 42.2 m/s, $30.1°$ above the horizontal.

43. Unsuccessful, 34.7 m.

45. (a) $\mathbf{v}_0 = 3.42\,\text{m/s}$, $47.5°$ above the horizontal; (b) 5.32 m above the water; (c) $\mathbf{v}_f = 10.5\,\text{m/s}$, $77°$ below the horizontal.

47. $\theta = \tan^{-1}(gt/v_0)$ below the horizontal.

49. $\theta = \frac{1}{2}\tan^{-1}(-\cot\phi)$.

51. $7.29g$ up.

53. $5.9 \times 10^{-3}\,\text{m/s}^2$ toward the Sun.

55. $0.94g$.

59. 2.7 m/s, $22°$ from the river bank.

61. 23.1 s.

63. 1.41 m/s.

65. (a) 1.82 m/s; (b) 3.22 m/s.

67. (a) 60 m; (b) 75 s.

69. 58 km/h, $31°$, 58 km/h opposite to $\mathbf{v}_{12}$.

71. $0.0889\,\text{m/s}^2$.

73. $D_x = 60\,\text{m}$, $D_y = -35\,\text{m}$, $D_z = -12\,\text{m}$; 70 m; $\theta_h = 30°$ from the x-axis toward the −y-axis, $\theta_v = 9.8°$ below the horizontal.

75. 7.0 m/s.

77. $\pm 28.5°$, ± 25.2.

79. 170 km/h, $41.5°$ N of E.

81. $1.6\,\text{m/s}^2$.

83. 2.7 s, 1.9 m/s.

85. (a) $Dv/(v^2 - u^2)$;
(b) $D/(v^2 - u^2)^{1/2}$.

87. $54.6°$ below the horizontal.

89. (a) 464 m/s; (b) 355 m/s.

91. Row at an angle of $23°$ upstream and run 243 m in a total time of 20.7 min.

93. $1.8 \times 10^3\,\text{rev/day}$.

<hr/>

CHAPTER 4

3. $6.9 \times 10^2\,\text{N}$.

5. (a) $5.7 \times 10^2\,\text{N}$; (b) 99 N;
(c) $2.1 \times 10^2\,\text{N}$; (d) 0.

7. 107 N.

9. $-9.3 \times 10^5\,\text{N}$, 25% of the weight of the train.

11. $m > 1.9\,\text{kg}$.

13. $2.1 \times 10^2\,\text{N}$.

15. $-1.40\,\text{m/s}^2$ (down).

17. a (downward) $\geq 1.2\,\text{m/s}^2$.

19. $a_{\text{max}} = 0.557\,\text{m/s}^2$.

21. (a) $2.2\,\text{m/s}^2$; (b) 18 m/s; (c) 93 s.

23. $3.0 \times 10^3\,\text{N}$ downward.

25. (a) $1.4 \times 10^2\,\text{N}$; (b) 14.5 m/s.

27. Southwesterly direction.

29.

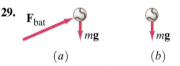

(a)   (b)

31. (a) $1.13\,\text{m/s}^2$, $52.2°$ below −x-axis; (b) $0.814\,\text{m/s}^2$, $42.3°$ above +x-axis.

33. (a) $m_1g - F_T = m_1a$, $F_T - m_2g = m_2a$.

35. (a) lower bucket = 34 N, upper bucket = 68N; (b) lower bucket = 40 N, upper bucket = 80 N.

37. $1.4 \times 10^3\,\text{N}$.

39. $F_B = 6890\,\text{N}$, $\mathbf{F}_A + \mathbf{F}_B = 8860\,\text{N}$.

41. (a) 2.2 m up the plane; (b) 2.2 s.

43. $\frac{5}{2}(F_0/m)t_0^2$.

47. (a)

(b) $a = m_2g/(m_1 + m_2)$,
$F_T = m_1m_2g/(m_1 + m_2)$.

49. $a = [m_2 + m_C(\ell_2/\ell)]g/$
$(m_1 + m_2 + m_C)$.

51. $1.74\,\text{m/s}^2$, $F_{T1} = 22.6\,\text{N}$, $F_{T2} = 20.9\,\text{N}$.

53. $(m + M)g\tan\theta$.

55. $F_{T1} = [4m_1m_2m_3/$
$(m_1m_3 + m_2m_3 + 4m_1m_2)]g$,
$F_{T3} = [8m_1m_2m_3/$
$(m_1m_3 + m_2m_3 + 4m_1m_2)]g$,
$a_1 = [(m_1m_3 - 3m_2m_3 + 4m_1m_2)/$
$(m_1m_3 + m_2m_3 + 4m_1m_2)]g$,
$a_2 = [(-3m_1m_3 + m_2m_3 + 4m_1m_2)/$
$(m_1m_3 + m_2m_3 + 4m_1m_2)]g$,
$a_3 = [(m_1m_3 + m_2m_3 - 4m_1m_2)/$
$(m_1m_3 + m_2m_3 + 4m_1m_2)]g$.

57. $v = \{[2m_2\ell_2 + m_C(\ell_2^2/\ell)]g/$
$(m_1 + m_2 + m_C)\}^{1/2}$.

59. $2.0 \times 10^{-2}\,\text{N}$.

61. 4.3 N.

63. $1.5 \times 10^4\,\text{N}$.

65. 1.2 s, no change.

67. (a) $2.45\,\text{m/s}^2$ (up the incline);
(b) 0.50 kg; (c) 7.35 N, 4.9 N.

69. $1.3 \times 10^2\,\text{N}$.

71. $8.8°$.

73. 82 m/s (300 km/h).

75. (a) $F = \frac{1}{2}Mg$;
(b) $F_{T1} = F_{T2} = \frac{1}{2}Mg$, $F_{T3} = \frac{3}{2}Mg$,
$F_{T4} = Mg$.

77. $-8.3 \times 10^2\,\text{N}$.

79. (a) $0.606\,\text{m/s}^2$; (b) 150 kN.

CHAPTER 5

1. 35 N, no force.

3. (a)

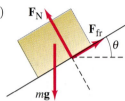

(b)

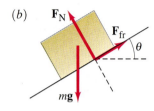

(c)

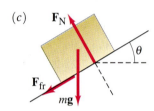

5. 0.20.

7. 69 N, $\mu_k = 0.54$.

9. 8.0 kg.

11. 1.3 m.

13. 1.3×10^3 N.

15. (a) 0.58; (b) 5.7 m/s; (c) 15 m/s.

17. (a) 1.4 m/s^2; (b) 5.4×10^2 N; (c) 1.41 m/s^2, 2.1×10^2 N.

19. (a) 86 cm up the plane; (b) 1.5 s.

21. (a) 2.8 m/s^2; (b) 2.1 N.

23. $a = \{\sin\theta - [(\mu_1 m_1 + \mu_2 m_2)/(m_1 + m_2)]\cos\theta\}g$, $F_T = [m_1 m_2(\mu_2 - \mu_1)/(m_1 + m_2)]g\cos\theta$.

25. (a) $\mu_k = (v_0^2/2gd\cos\theta) - \tan\theta$; (b) $\mu_s \geq \tan\theta$.

27. (a) 2.0 m/s^2 up the plane; (b) 5.4 m/s^2 up the plane.

29. $\mu_k = 0.40$.

31. (a) $c = 14$ kg/m; (b) 5.7×10^2 N.

33. $F_{\min} = (m + M)g(\sin\theta + \mu\cos\theta)/(\cos\theta - \mu_s\sin\theta)$.

35. $v_{\max} = 21$ m/s, independent of the mass.

37. (a) 0.25 m/s^2 toward the center; (b) 6.3 N toward the center.

39. Yes, $v_{\text{top, min}} = (gR)^{1/2}$.

41. 0.34.

43. 2.1×10^2 N.

45. 5.91°, 14.3 N.

47. (a) 5.8×10^3 N; (b) 4.1×10^2 N; (c) 31 m/s.

49. $F_T = 2\pi m R f^2$.

51. 66 km/h $< v <$ 123 km/h.

53. (a) $(1.6 \text{ m/s}^2)\mathbf{i}$; (b) $(0.98 \text{ m/s}^2)\mathbf{i} - (1.7 \text{ m/s}^2)\mathbf{j}$; (c) $-(4.9 \text{ m/s}^2)\mathbf{i} - (1.6 \text{ m/s}^2)\mathbf{j}$.

55. (a) 9.0 m/s^2; (b) 15 m/s^2.

57. $\tau = m/b$.

59. (a) $v = (mg/b) + [v_0 - (mg/b)]e^{-bt/m}$; (b) $v = -(mg/b) + [v_0 + (mg/b)]e^{-bt/m}$, $v \geq 0$.

61. (b) 1.8°.

63. 10 m.

65. $\mu_s = 0.41$.

67. 2.3.

69. 101 N, $\mu_k = 0.719$.

71. (b) Will slide.

73. Emerges with a speed of 13 m/s.

75. 27.6 m/s, 0.439 rev/s.

77. $\Sigma F_{\tan} = 3.3 \times 10^3$ N, $\Sigma F_R = 2.0 \times 10^3$ N.

79. (a) $F_{NC} > F_{NB} > F_{NA}$; (b) heaviest at C, lightest at A; (c) $v_{A\max} = (gR)^{1/2}$.

81. (a) 1.23 m/s; (b) 3.01 m/s.

83. $\phi = 31°$.

85. (a) $r = v^2/g\cos\theta$; (b) 92 m.

87. (a) 59 s; (b) greater normal force.

89. 29.2 m/s.

91. 302 m, 735 m.

93. $g(1 - \mu_s\tan\phi)/4\pi^2 f^2(\tan\phi + \mu_s) < r < g(1 + \mu_s\tan\phi)/4\pi^2 f^2(\tan\phi - \mu_s)$.

CHAPTER 6

1. 1.52×10^3 N.

3. 1.6 m/s^2.

5. $g_h = 0.91 g_{\text{surface}}$.

7. 1.9×10^{-8} N toward center of square.

9. $Gm^2\{(2/x_0^2) + [3x_0/(x_0^2 + y_0^2)^{3/2}]\}\mathbf{i} + Gm^2\{[3y_0/(x_0^2 + y_0^2)^{3/2}] + (4/y_0^2)\}\mathbf{j}$.

11. 1.26.

13. 3.46×10^8 m from Earth's center.

15. (b) g decreases with an increase in height; (c) 9.493 m/s^2.

19. 7.56×10^3 m/s.

21. 2.0 h.

23. (a) 56 kg; (b) 56 kg; (c) 75 kg; (d) 38 kg; (e) 0.

25. (a) 22 N (toward the Moon); (b) -1.7×10^2 N (away from the Moon).

27. (a) Gravitational force provides required centripetal acceleration; (b) 9.6×10^{26} kg.

29. 7.9×10^3 m/s.

31. $v = (Gm/L)^{1/2}$.

33. 0.0587 days (1.41 h).

35. 1.6×10^2 yr.

37. 2×10^8 yr.

39. $r_{\text{Europa}} = 6.71 \times 10^5$ km, $r_{\text{Ganymede}} = 1.07 \times 10^6$ km, $r_{\text{Callisto}} = 1.88 \times 10^6$ km.

41. 9.0 Earth-days.

43. (a) 2.1×10^2 A.U. $(3.1 \times 10^{13}$ m); (b) 4.2×10^2 A.U.; (c) 4.2×10^2.

45. (a) 5.9×10^{-3} N/kg; (b) not significant.

47. 2.7×10^3 km.

49. 6.7×10^{12} m/s^2.

51. 4.4×10^7 m/s^2.

53. $G' = 1 \times 10^{-4}$ N·m^2/kg$^2 \approx 10^6\, G$.

55. 5 h 35 min, 19 h 50 min.

57. (a) 10 h; (b) 6.5 km; (c) 4.2×10^{-3} m/s^2.

59. 5.4×10^{12} m, in the Solar System, Pluto.

61. $2.3 g_{\text{Earth}}$.

63. $m_P = g_P r^2/G$.

67. 7.9×10^3 m/s.

CHAPTER 7

1. 6.86×10^3 J.

3. 1.27×10^4 J.

5. 8.1×10^3 J.

7. 1 J $= 1 \times 10^7$ erg $= 0.738$ ft·lb.

9. 1.0×10^4 J.

13. (a) 3.6×10^2 N; (b) -1.3×10^3 J; (c) -4.6×10^3 J; (d) 5.9×10^3 J; (e) 0.

15. $W_{FN} = W_{mg} = 0$, $W_{FP} = -W_{fr} = 2.0 \times 10^2$ J.

21. (a) -16.1; (b) -238; (c) -3.9.

23. $\mathbf{C} = -1.3\mathbf{i} + 1.8\mathbf{j}$.

25. $\theta_x = 42.7°, \theta_y = 63.8°, \theta_z = 121°$.

27. 95°, $-35°$ from x-axis.

31. 0.089 J.

33. 2.3×10^3 J.

35. 2.7×10^3 J.

37. $(kX^2/2) + (aX^4/4) + (bX^5/5)$.

39. (a) 5.0×10^{10} J.

41. (a) $\sqrt{3}$; (b) $\frac{1}{4}$.

43. -5.02×10^5 J.

45. 3.0×10^2 N in the direction of the motion of the ball.

47. 24 m/s (87 km/h or 54 mi/h), the mass cancels.

49. (a) 72 J; (b) -35 J; (c) 37 J.

51. 10.2 m/s.

53. $\mu_k = F/2mg$.

55. (a) 6.5×10^2 J; (b) -4.9×10^2 J; (c) 0; (d) 4.0 m/s.

57. (a) 1.66×10^5 J; (b) 21.0 m/s; (c) 2.13 m.

59. $v_p = 2.0 \times 10^7$ m/s, $v_{pc} = 2.0 \times 10^7$ m/s; $v_e = 2.9 \times 10^8$ m/s, $v_{ec} = 8.4 \times 10^8$ m/s.

61. 1.74×10^3 J.

63. (a) 15 J; (b) 4.2×10^2 J; (c) -1.8×10^2 J; (d) -2.5×10^2 J; (e) 0; (f) 10 J.

65. (a) 12 J; (b) 10 J; (c) -2.1 J.

67. 86 kJ, $\theta = 42°$.

69. $(A/k)e^{-(0.10\,\mathrm{m})k}$.

71. 1.5 N.

73. 5.0×10^3 N/m.

75. (a) 6.6°; (b) 10.3°.

CHAPTER 8

1. 0.924 m.

3. 2.2×10^3 J.

5. (a) 51.7 J; (b) 15.1 J; (c) 51.7 J.

7. (a) Conservative; (b) $\frac{1}{2}kx^2 - \frac{1}{4}ax^4 - \frac{1}{5}bx^5 + $ constant.

9. (a) $\frac{1}{2}k(x^2 - x_0^2)$; (b) same.

11. 45.4 m/s.

13. 6.5 m/s.

15. (a) 1.0×10^2 N/m; (b) 22 m/s².

17. (a) 8.03 m/s; (b) 3.44 m.

19. (a) $v_{\max} = \left[v_0^2 + (kx_0^2/m)\right]^{1/2}$; (b) $x_{\max} = \left[x_0^2 + (mv_0^2/k)\right]^{1/2}$.

21. (a) 2.29 m/s; (b) 1.98 m/s; (c) 1.98 m/s; (d) $F_{Ta} = 0.87$ N, $F_{Tb} = 0.80$ N, $F_{Tc} = 0.80$ N; (e) $v_a = 2.59$ m/s, $v_b = 2.31$ m/s, $v_c = 2.31$ m/s.

23. $k = 12Mg/h$.

25. 4.5×10^6 J.

27. (a) 22 m/s; (b) 2.9×10^2 m.

29. 13 m/s.

31. 0.23.

33. 0.40.

35. (a) 0.13 m; (b) 0.77; (c) 0.46 m/s.

37. (a) $K = GM_E m_S/2r_S$; (b) $U = -GM_E m_S/r_S$; (c) $-\frac{1}{2}$.

39. (a) 6.2×10^5 m/s; (b) 4.2×10^4 m/s, $v_{esc}/v_{orbit} = \sqrt{2}$.

45. (a) 1.07×10^4 m/s; (b) 1.17×10^4 m/s; (c) 1.12×10^4 m/s.

47. (a) $dv_{esc}/dr = -\frac{1}{2}\left(2GM_E/r^3\right)^{1/2}$ $= -v_{esc}/2r$; (b) 1.09×10^4 m/s.

49. $GmM_E/12r_E$.

51. 1.1×10^4 m/s.

55. 5.4×10^2 N.

57. (a) 1.0×10^3 J; (b) 1.0×10^3 W.

59. 2.1×10^4 W, 28 hp.

61. 4.8×10^2 W.

63. 1.2×10^3 W.

65. 1.8×10^6 W.

67. (a) -25 W; (b) $+4.3 \times 10^3$ W; (c) $+1.5 \times 10^3$ W.

69. (a) 80 J; (b) 60 J; (c) 80 J; (d) 5.7 m/s at $x = 0$; (e) 32 m/s² at $x = \pm x_0$.

71. $a^2/4b$.

73. 8.0 m/s.

75. 32.5 hp.

77. (a) 28 m/s; (b) 1.2×10^2 m.

79. (a) $(2gL)^{1/2}$; (b) $(1.2gL)^{1/2}$.

81. (a) 1.1×10^6 J; (b) 60 W (0.081 hp); (c) 4.0×10^2 W (0.54 hp).

83. (a) 40 m/s; (b) 2.6×10^5 W.

87. (a) 29°; (b) 6.4×10^2 N; (c) 9.2×10^2 N.

89. (a) $-\dfrac{U_0}{r}\left(\dfrac{r_0}{r} + 1\right)e^{-r/r_0}$; (b) 0.030; (c) $F(r) = -C/r^2$, 0.11.

91. 6.7 hp.

93. (a) 2.8 m; (b) 1.5 m; (c) 1.5 m.

95. 76 hp.

97. (a) 5.00×10^3 m/s; (b) 2.89×10^3 m/s.

CHAPTER 9

1. 6.0×10^7 N, up.

3. (a) 0.36 kg·m/s; (b) 0.12 kg·m/s.

5. $(26\,\text{N·s})\mathbf{i} - (28\,\text{N·s})\mathbf{j}$.

7. (a) $(8h/g)^{1/2}$; (b) $(2gh)^{1/2}$; (c) $-(8m^2gh)^{1/2}$ (up); (d) mg (down), a surprising result.

9. 3.4×10^4 kg.

11. 4.4×10^3 m/s.

13. -0.667 m/s (opposite to the direction of the package).

15. 2, lesser kinetic energy has greater mass.

17. $\frac{3}{2}v_0\mathbf{i} - v_0\mathbf{j}$.

19. 1.1×10^{-22} kg·m/s, 36° from the direction opposite to the electron's.

21. (a) $(100\,\text{m/s})\mathbf{i} + (50\,\text{m/s})\mathbf{j}$; (b) 3.3×10^5 J.

23. 130 N, not large enough.

25. 1.1×10^3 N.

27. (a) $2mv/\Delta t$; (b) $2mv/t$.

29. (a)

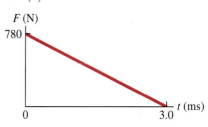

F (N)

780

0   3.0 t (ms)

(b) 1.2 N·s; (c) 1.2 N·s; (d) 3.9 g.

31. (a) $(0.84\,\text{N}) + (1.2\,\text{N/s})t$; (b) 18.5 N; (c) $(0.12\,\text{kg/s})\{\left[(49\,\text{m}^2/\text{s}^2) - (1.18\,\text{m}^2/\text{s}^3)t\right]^{1/2} + (9.80\,\text{m/s}^2)t\}$, 18.3 N.

33. $v_1' = -1.40$ m/s (rebound), $v_2' = 2.80$ m/s.

35. (a) 2.7 m/s; (b) 0.84 kg.

37. 3.2×10^3 m/s.

39. (a) 1.00; (b) 0.89; (c) 0.29; (d) 0.019.

41. (a) 0.32 m; (b) -3.1 m/s (rebound), 4.9 m/s; (c) Yes.

43. (a) $+M/(m + M)$; (b) 0.964.

45. 141°.

47. (b) $e = (h'/h)^{1/2}$.

49. (a) $v_1' = v_2' = 1.9$ m/s; (b) $v_1' = -1.6$ m/s, $v_2' = 7.9$ m/s; (c) $v_1' = 0$, $v_2' = 5.2$ m/s; (d) $v_1' = 3.1$ m/s, $v_2' = 0$; (e) $v_1' = -4.0$ m/s, $v_2' = 12$ m/s; result for (c) is reasonable, result for (d) is not reasonable, result for (e) is not reasonable.

51. 61° from first eagles's direction, 6.8 m/s.

53. (a) 30°; (b) $v_2' = v/\sqrt{3}$, $v_1' = v/\sqrt{3}$; (c) $\frac{2}{3}$.

55. $\theta_1' = 76°$, $v_n' = 5.1 \times 10^5$ m/s,
$v_{He}' = 1.8 \times 10^5$ m/s.

59. 6.5×10^{-11} m from the carbon atom.

61. 0.030 nm above center of H triangle.

63. $x_{CM} = 1.10$ m (East),
$y_{CM} = -1.10$ m (South).

65. $x_{CM} = 0$, $y_{CM} = 2r/\pi$.

67. $x_{CM} = 0$, $y_{CM} = 0$,
$z_{CM} = 3h/4$ above the point.

69. (a) 4.66×10^6 m.

71. (a) $x_{CM} = 4.6$ m; (b) 4.3 m;
(c) 4.6 m.

73. $mv/(M + m)$ up, balloon will also stop.

75. 55 m.

77. 0.899 hp.

79. (a) 2.3×10^3 N; (b) 2.8×10^4 N;
(c) 1.1×10^4 hp.

81. A "scratch shot".

83. 1.4×10^4 N, 43.3°.

85. 5.1×10^2 m/s.

87. $m_2 = 4.00m$.

89. 50%.

91. (a) No; (b) $v_1/v_2 = -m_2/m_1$;
(c) m_2/m_1; (d) does not move;
(e) center of mass will move.

93. 8.29 m/s.

95. (a) 2.5×10^{-13} m/s; (b) 1.7×10^{-17};
(c) 0.19 J.

97. $m \leq M/3$.

99. 29.6 km/s.

101. (a) 2.3 N·s; (b) 4.5×10^2 N.

103. (a) Inelastic collision; (b) 0.10 s;
(c) -1.4×10^5 N.

105. 0.28 m, 1.1 m.

CHAPTER 10

1. (a) $\pi/6$ rad = 0.524 rad;
(b) $19\pi/60 = 0.995$ rad;
(c) $\pi/2 = 1.571$ rad;
(d) $2\pi = 6.283$ rad;
(e) $7\pi/3 = 7.330$ rad.

3. 2.3×10^3 m.

5. (a) 0.105 rad/s;
(b) 1.75×10^{-3} rad/s;
(c) 1.45×10^{-4} rad/s; (d) zero.

7. (a) 464 m/s; (b) 185 m/s;
(c) 355 m/s.

9. (a) 262 rad/s;
(b) 46 m/s, 1.2×10^4 m/s² radial.

11. 7.4 cm.

---

13. (a) 1.75×10^{-4} rad/s²;
(b) $a_R = 1.17 \times 10^{-2}$ m/s²,
$a_{tan} = 7.44 \times 10^{-4}$ m/s².

15. (a) 0.58 rad/s2; (b) 12 s.

17. (a) $(1.67 \text{ rad/s}^4)t^3 - (1.75 \text{ rad/s}^3)t^2$;
(b) $(0.418 \text{ rad/s}^4)t^4 - (0.583 \text{ rad/s}^3)t^3$;
(c) 6.4 rad/s, 2.0 rad.

19. (a) ω_1 is in the $-x$-direction,
ω_2 is in the $+z$-direction;
(b) $\omega = 61.0$ rad/s, 35.0° above
$-x$-axis;
(c) $-(1.75 \times 10^3 \text{ rad/s}^2)\mathbf{j}$.

21. (a) 35 m·N; (b) 30 m·N.

23. 1.2 m·N (clockwise).

25. 3.5×10^2 N, 2.0×10^3 N.

27. 53 m·N.

29. (a) 3.5 kg·m²; (b) 0.024 m·N.

31. 2.25×10^3 kg·m², 8.8×10^3 m·N.

33. 9.5×10^4 m·N.

35. 10 m/s.

37. (a)

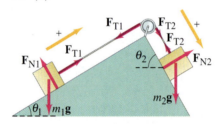

(b) $F_{T1} = 47$ N, $F_{T2} = 75$ N;
(c) 7.0 m·N, 1.7 kg·m².

39. Thin hoop (through center):
$k = R_0$;
Thin hoop (through diameter):
$k = [(R_0^2/2) + (w^2/12)]^{1/2}$;
Solid cylinder (through center):
$k = R/\sqrt{2}$;
Hollow cylinder (through center):
$k = [(R_1^2 + R_2^2)/2]^{1/2}$;
Uniform sphere (through center):
$k = (2r_0^2/5)^{1/2}$;
Rod (through center): $k = \ell/\sqrt{12}$;
Rod (through end): $k = \ell/\sqrt{3}$;
Plate (through center):
$k = [(\ell^2 + w^2)/12]^{1/2}$.

41. (a) 4.18 rad/s²; (b) 8.37 m/s²;
(c) 421 m/s²; (d) 3.07×10^3 N;
(e) 1.14°.

43. (a) $I_a = Ms^2/12$; (b) $I_b = Ms^2/12$.

45. (a) $5.30MR_0^2$; (b) -15%.

47. (a) $9MR_0^2/16$; (b) $MR_0^2/4$;
(c) $5MR_0^2/4$.

51. (b) $M\ell^2/12$, $Mw^2/12$.

---

53. 0.38 rev/s.

55. (a) As moment of inertia increases,
angular velocity must decrease;
(b) 1.6.

57. (a) 7.1×10^{33} kg·m²/s;
(b) 2.7×10^{40} kg·m²/s.

59. 0.45 rad/s, 0.80 rad/s.

61. 2.33×10^4 J.

63. 5×10^9, loss of gravitational
potential energy.

65. 1.4 m/s.

67. (a) 2.5 kg·m²; (b) 0.58 kg·m²;
(c) 0.35 s; (d) -72 J; (e) rotating.

69. 12.4 m/s.

71. 1.4×10^2 J.

73. (a) 4.48 m/s; (b) 1.21 J;
(c) $\mu_s \geq 0.197$.

75. $v = [10g(R_0 - r_0)/7]^{1/2}$.

77. (a) 4.5×10^5 J; (b) 0.18 (18%);
(c) 1.71 m/s²; (d) 6.4%.

79. (a) $12v_0^2/49 \mu_k g$;
(b) $v = 5v_0/7$, $\omega = 5v_0/7R$.

81. (a) 4.5 m/s², 19 rad/s²; (b) 5.8 m/s;
(c) 15.3 J; (d) 1.4 J;
(e) $K = 16.7$ J, $\Delta E = 0$;
(f) $a = 4.5$ m/s², $v = 5.8$ m/s, 14.1 J.

83. $\theta_{Sun} = 9.30 \times 10^{-3}$ rad (0.53°),
$\theta_{Moon} = 9.06 \times 10^{-3}$ rad (0.52°).

85. $\omega_1/\omega_2 = R_2/R_1$.

87. $\ell/2$, $\ell/2$.

89. (a) $-(I_W/I_P)\omega_W$ (down);
(b) $-(I_W/2I_P)\omega_W$ (down);
(c) $(I_W/I_P)\omega_W$ (up); (d) 0.

91. (a) $\omega_R/\omega_F = N_F/N_R$; (b) 4.0;
(c) 1.5.

93. (a) 1.5×10^2 rad/s²;
(b) 1.2×10^3 N.

95. (a) 0.070 rad/s²; (b) 40 rpm.

97. 7.9 N.

99. (b) 2.2×10^3 rad/s; (c) 24 min.

101. (a) 2.9 m; (b) 3.6 m.

103. (a) 1.2 rad/s; (b) 2.0×10^3 J,
1.2×10^3 J, loss of 8.0×10^2 J,
decrease of 40%.

105. (a) 1.7 m/s; (b) 0.84 m/s.

107. (a) $h_{min} = 2.7R_0$;
(b) $h_{min} = 2.7R_0 - 1.7r_0$.

109. (a) 0.84 m/s; (b) 0.96.

CHAPTER 11

7. (a) $-7.0\mathbf{i} - 14.0\mathbf{j} + 19.3\mathbf{k}$; (b) 164°.

11. $-(30.3 \text{ m·N})\mathbf{k}$ (in $-z$-direction).

13. $(18 \text{ m·kN})\mathbf{i} \pm (14 \text{ m·kN})\mathbf{j}$
$\mp (19 \text{ m·kN})\mathbf{k}$.

19. $(55\mathbf{i} - 90\mathbf{j} + 42\mathbf{k})\,\text{kg}\cdot\text{m}^2/\text{s}$.

21. $(a)\,[(7m/9) + (M/6)]\ell^2\omega^2$;
$(b)\,[(14m/9) + (M/3)]\ell^2\omega$.

23. $2.30\,\text{m/s}^2$.

25. $(a)\,L = [R_0 M_1 + R_0 M_2 + (I/R_0)]v$;
$(b)\,a = M_2 g/[M_1 + M_2 + (I/R_0^2)]$.

27. Rod rotates at 7.8 rad/s about the center of mass, which moves with constant velocity of 0.21 m/s.

31. $F_1 = [(d + r\cos\phi)/2d]m_1 r\omega^2 \sin\phi$,
$F_2 = [(d - r\cos\phi)/2d]m_1 r\omega^2 \sin\phi$.

33. $16\,\text{N}, -7.5\,\text{N}$.

35. $3m^2v^2/g(3m + 4M)(m + M)$.

37. $(1 - 4.7 \times 10^{-13})\omega_\text{E}$.

39. $(a)\,14\,\text{m/s};\ (b)\,6.8\,\text{rad/s}$.

41. $1.02 \times 10^{-3}\,\text{kg}\cdot\text{m}^2$.

43. $2.2\,\text{rad/s}\ (0.35\,\text{rev/s})$.

45. $\tan^{-1}(r\omega^2/g)$.

47. $(a)\,g$, along a radial line; $(b)\,0.998g$, $0.0988°$ south from a radial line; $(c)\,0.997g$, along a radial line.

49. North or south direction.

51. (a) South; $(b)\,\omega D^2 \sin\lambda/v_0$; $(c)\,0.46\,\text{m}$.

53. $(a)\,(-9.0\mathbf{i} + 12\mathbf{j} - 8.0\mathbf{k})\,\text{kg}\cdot\text{m}^2/\text{s}$; $(b)\,(9.0\mathbf{j} - 6.0\mathbf{k})\,\text{m}\cdot\text{N}$.

55. (a) Turn in the direction of the lean; $(b)\,\Delta L = 0.98\,\text{kg}\cdot\text{m}^2/\text{s}$, $\Delta L = 0.18 L_0$.

57. $(a)\,1.8 \times 10^3\,\text{kg}\cdot\text{m}^2/\text{s}^2$; $(b)\,1.8 \times 10^3\,\text{m}\cdot\text{N};\ (c)\,2.1 \times 10^3\,\text{W}$.

59. $v_\text{CM} = (3g\ell/4)^{1/2}$.

61. $(19\,\text{m/s})(1 - \cos\theta)^{1/2}$.

63. $(a)\,2.3 \times 10^4\,\text{rev/s}$; $(b)\,5.7 \times 10^3\,\text{rev/s}$.

CHAPTER 12

1. $379\,\text{N}, 141°$.

3. $1.6 \times 10^3\,\text{m}\cdot\text{N}$.

5. $6.52\,\text{kg}$.

7. $2.84\,\text{m}$ from the adult.

9. $0.32\,\text{m}$.

11. $F_{T1} = 3.4 \times 10^3\,\text{N}$,
$F_{T2} = 3.9 \times 10^3\,\text{N}$.

13. $F_1 = -2.94 \times 10^3\,\text{N}$ (down),
$F_2 = 1.47 \times 10^4\,\text{N}$.

15. Top hinge: $F_{Ax} = 55.2\,\text{N}$,
$F_{Ay} = 63.7\,\text{N}$; bottom hinge:
$F_{Bx} = -55.2\,\text{N}, F_{By} = 63.7\,\text{N}$.

17. (a)

$(b)\,1.5 \times 10^4\,\text{N};\ (c)\,6.7 \times 10^3\,\text{N}$.

19. $F_\text{T} = 1.4 \times 10^3\,\text{N}$ (up),
$F_\text{bone} = 2.1 \times 10^3\,\text{N}$ (down).

21. $2.7 \times 10^3\,\text{N}$.

23. $89.5\,\text{cm}$ from the feet.

25. $F_1 = 5.8 \times 10^3\,\text{N}, F_2 = 5.6 \times 10^3\,\text{N}$.

27. $(a)\,2.1 \times 10^2\,\text{N};\ (b)\,2.0 \times 10^3\,\text{N}$.

29. $7.1 \times 10^2\,\text{N}$.

31. $F_\text{T} = 2.5 \times 10^2\,\text{N}$,
$F_{AH} = 2.5 \times 10^2\,\text{N}$,
$F_{AV} = 2.0 \times 10^2\,\text{N}$.

33. $(a)\,1.00\,\text{N};\ (b)\,1.25\,\text{N}$.

35. $\theta_\text{max} = 40°$, same.

37. $(a)\,F_\text{T} = 182\,\text{N};\ (b)\,F_\text{N1} = 352\,\text{N}$,
$F_\text{N2} = 236\,\text{N};\ (c)\,F_\text{B} = 298\,\text{N}, 52.4°$.

39. $1.0 \times 10^2\,\text{N}$.

41. $(a)\,1.2 \times 10^5\,\text{N/m}^2;\ (b)\,2.4 \times 10^{-6}$.

43. $(a)\,1.3 \times 10^5\,\text{N/m}^2;\ (b)\,6.5 \times 10^{-7}$; $(c)\,0.0062\,\text{mm}$.

45. $9.6 \times 10^6\,\text{N/m}^2$.

47. $9.0 \times 10^7\,\text{N/m}^2, 9.0 \times 10^2\,\text{atm}$.

49. $2.2 \times 10^7\,\text{N}$.

51. $(a)\,1.1 \times 10^2\,\text{m}\cdot\text{N};\ (b)$ wall; (c) all three.

53. $3.9 \times 10^2\,\text{N}$, thicker strings, maximum strength is exceeded.

55. $(a)\,4.4 \times 10^{-5}\,\text{m}^2;\ (b)\,2.7\,\text{mm}$.

57. $1.2\,\text{cm}$.

61. $(a)\,F_\text{T} = 129\,\text{kN}$;
$F_\text{A} = 141\,\text{kN}, 23.5°$;
$(b)\,F_\text{DE} = 64.7\,\text{kN}$ (tension),
$F_\text{CE} = 32.3\,\text{kN}$ (compression),
$F_\text{CD} = 64.7\,\text{kN}$ (compression),
$F_\text{BD} = 64.7\,\text{kN}$ (tension),
$F_\text{BC} = 64.7\,\text{kN}$ (tension),
$F_\text{AC} = 97.0\,\text{kN}$ (compression),
$F_\text{AB} = 64.7\,\text{kN}$ (compression).

63. $(a)\,4.8 \times 10^{-2}\,\text{m}^2;\ (b)\,6.8 \times 10^{-2}\,\text{m}^2$.

65. $F_\text{AB} = 5.44 \times 10^4\,\text{N}$ (compression),
$F_{ACx} = 2.72 \times 10^4\,\text{N}$ (tension),
$F_\text{BC} = 5.44 \times 10^4\,\text{N}$ (tension),
$F_\text{BD} = 5.44 \times 10^4\,\text{N}$ (compression),
$F_\text{CD} = 5.44 \times 10^4\,\text{N}$ (tension),
$F_\text{CE} = 2.72 \times 10^4\,\text{N}$ (tension),
$F_\text{DE} = 5.44 \times 10^4\,\text{N}$ (compression).

67. $12\,\text{m}$.

69. $M_\text{C} = 0.191\,\text{kg}, M_\text{D} = 0.0544\,\text{kg}$,
$M_\text{A} = 0.245\,\text{kg}$.

71. $(a)\,Mg[h/(2R - h)]^{1/2}$;
$(b)\,Mg[h(2R - h)]^{1/2}/(R - h)$.

73. $\theta_\text{max} = 29°$.

75. $6, 2.0\,\text{m}$ apart.

77. 3.8.

79. $5.0 \times 10^5\,\text{N}, 3.2\,\text{m}$.

81. $(a)\,600\,\text{N};\ (b)\,F_\text{A} = 0, F_\text{B} = 1200\,\text{N}$;
$(c)\,F_\text{A} = 150\,\text{N}, F_\text{B} = 1050\,\text{N}$;
$(d)\,F_\text{A} = 750\,\text{N}, F_\text{B} = 450\,\text{N}$.

83. $6.5 \times 10^2\,\text{N}$.

85. $0.67\,\text{m}$.

87. Right end is safe, left end is not safe, $0.10\,\text{m}$.

89. $(a)\,F_\text{L} = 3.3 \times 10^2\,\text{N}$ up,
$F_\text{R} = 2.3 \times 10^2\,\text{N}$ down;
$(b)\,65\,\text{cm}$ from right hand;
$(c)\,123\,\text{cm}$ from right hand.

91. $\theta \geq 40°$.

93. (b) beyond the table;
$(c)\,D = L\displaystyle\sum_{i=1}^{n} \frac{1}{2i};\ (d)$ 32 bricks.

95. $F_\text{TB} = 134\,\text{N}, F_\text{TA} = 300\,\text{N}$.

97. $2.6w, 31°$ above horizontal.

CHAPTER 13

1. $3 \times 10^{11}\,\text{kg}$.

3. $4.3 \times 10^2\,\text{kg}$.

5. 0.8477.

7. $(a)\,3 \times 10^7\,\text{N/m}^2$;
$(b)\,2 \times 10^5\,\text{N/m}^2$.

9. $1.1\,\text{m}$.

11. $8.28 \times 10^3\,\text{kg}$.

13. $1.2 \times 10^5\,\text{N/m}^2$,
$2.3 \times 10^7\,\text{N}$ (down),
$1.2 \times 10^5\,\text{N/m}^2$.

15. $6.54 \times 10^2\,\text{kg/m}^3$.

17. $3.36 \times 10^4\,\text{N/m}^2\ (0.331\,\text{atm})$.

19. $(a)\,1.41 \times 10^5\,\text{Pa};\ (b)\,9.8 \times 10^4\,\text{Pa}$.

21. $(a)\,0.34\,\text{kg};\ (b)\,1.5 \times 10^4\,\text{N}$ (up).

23. $(c) \geq 0.38h$, no.

27. $4.70 \times 10^3\,\text{kg/m}^3$.

29. $8.5 \times 10^2\,\text{kg}$.

31. Copper.

33. $(a)\,1.14 \times 10^6\,\text{N};\ (b)\,4.0 \times 10^5\,\text{N}$.

35. (b) Above the center of gravity.

37. 0.88.

39. $7.9 \times 10^2\,\text{kg}$.

43. $4.1\,\text{m/s}$.

45. $9.5\,\text{m/s}$.

47. $1.5 \times 10^5\,\text{N/m}^2 = 1.5\,\text{atm}$.

49. $4.11 \times 10^{-3}\,\text{m}^3/\text{s}$.

51. $1.7 \times 10^6\,\text{N}$.

59. (a) $2[h_1(h_2 - h_1)]^{1/2}$;
  (b) $h_1' = h_2 - h_1$.
61. 0.072 Pa·s.
63. 4.0×10^3 Pa.
65. 11 cm.
67. (a) Laminar; (b) 3200, turbulent.
69. 1.9 m.
71. 9.1×10^{-3} N.
73. (a) $\gamma = F/4\pi r$; (b) 0.024 N/m.
75. (a) 0.88 m; (b) 0.55 m; (c) 0.24 m.
77. $1.5 \times 10^2 \,\text{N} \le F \le 2.2 \times 10^2 \,\text{N}$.
79. 0.051 atm.
81. 0.63 N.
83. 5 km.
85. 5.3×10^{18} kg.
87. 2.6 m.
89. 39 people.
91. 37 N, not float.
93. $d = D[v_0^2/(v_0^2 + 2gy)]^{1/4}$.
95. (a) 3.2 m/s; (b) 19 s.
97. 1.9×10^2 m/s.

CHAPTER 14

1. 0.60 m.
3. 1.15 Hz.
5. (a) 2.4 N/m; (b) 12 Hz.
7. (a) $0.866\, x_{max}$; (b) $0.500\, x_{max}$.
9. $0.866\, A$.
11. $[(k_1 + k_2)/m]^{1/2}/2\pi$.
13. (a) 8/7 s, 0.875 Hz; (b) 3.3 m,
  −10.4 m/s; (c) +18 m/s, −57 m/s^2.
15. 3.6 Hz.
19. (a) $y = -(0.220\,\text{m})\sin[(37.1\,\text{s}^{-1})t]$;
  (b) maximum extensions at 0.0423 s,
  0.211 s, 0.381 s,...; minimum
  extensions at 0.127 s, 0.296 s,
  0.465 s,....
21. $f = (3k/M)^{1/2}/2\pi$.
25. (a) $x = (12.0\,\text{cm})\cos[(25.6\,\text{s}^{-1})t$
  $+ 1.89\,\text{rad}]$;
  (b) $t_{max} = 0.294$ s, 0.539 s, 0.784 s,...;
  $t_{min} = 0.171$ s, 0.416 s, 0.661 s,...;
  (c) −3.77 cm; (d) +13.1 N (up);
  (e) 3.07 m/s, 0.110 s.
27. (a) 0.650 m; (b) 1.34 Hz; (c) 29.8 J;
  (d) $K = 25.0$ J, $U = 4.8$ J.
29. 9.37 m/s.
31. $A_1 = 2.24 A_2$.
33. (a) 4.2×10^2 N/m; (b) 3.3 kg.
35. 352.6 m/s.
39. 0.9929 m.
41. (a)0.248 m; (b)2.01 s.

43. (a) −12°; (b) +1.9°; (c) −13°.
45. $\frac{1}{3}$.
47. 1.08 s.
49. 0.31 g.
51. (a) 1.6 s.
53. 3.5 s.
55. (a) 0.727 s; (b) 0.0755;
  (c) $x = (0.189\,\text{m})e^{-(0.108/s)t}$
  $\sin[(8.64\,\text{s}^{-1})t]$.
57. (a) 8.3×10^{-4}%; (b) 39 periods.
59. (a) 5.03 Hz; (b) 0.0634 s^{-1};
  (c) 110 oscillations.
61.

$A_0 k / F_0$

[graph: resonance curve peaking near $\omega/\omega_0 = 1.0$, with $A_0 k/F_0$ axis from 0 to 4, ω/ω_0 axis from 0 to 2.0]

Answer to Problem 14-61.

65. (a) 198 s; (b) 8.7×10^{-6} W;
  (c) 8.8×10^{-4} Hz on either side of f_0.
69. (a) 0.63 Hz; (b) 0.65 m/s; (c) 0.077 J.
71. 151 N/m, 20.3 m.
73. 0.11 m.
75. 3.6 Hz.
77. (a) 1.1 Hz; (b) 13 J.
79. (a) 90 N/m; (b) 8.9 cm.
81. $k = \rho_{water}\, g A$.
83. Water will oscillate with SHM,
  $k = 2\rho g A$, the density and the cross
  section.
85. $T = 2\pi(ma/2k\,\Delta a)^{1/2}$.
87. (a) 1.64 s; (b) 0.67 m.

CHAPTER 15

1. 2.3 m/s.
3. 1.26 m.
5. 0.72 m.
7. 2.7 N.
9. (a) 75 m/s; (b) 7.8×10^3 N.
11. (a) 1.3×10^3 km;
  (b) cannot be determined.
13. (a) 0.25; (b) 0.50.
17. (a) 0.30 W; (b) 0.25 cm.
19. $D = D_M \sin[2\pi(x/\lambda + t/T) + \phi]$.
21. (a) 41 m/s; (b) 6.4×10^4 m/s^2;
  (c) 41 m/s, 8.2×10^3 m/s^2.

23. (a, c)

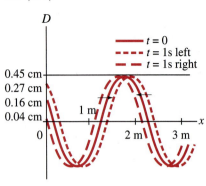

  (b) $D = (0.45\,\text{m})\cos[(3.0\,\text{m}^{-1})x$
  $- (6.0\,\text{s}^{-1})t + 1.2]$.
  (d) $D = (0.45\,\text{m})\cos[(3.0\,\text{m}^{-1})x$
  $+ (6.0\,\text{s}^{-1})t + 1.2]$.
25. $D = -(0.020\,\text{cm})\cos[(8.01\,\text{m}^{-1})x$
  $- (2.76 \times 10^3\,\text{s}^{-1})t]$.
27. The function is a solution.
31. (a) $v_2/v_1 = (\mu_1/\mu_2)^{1/2}$;
  (b) $\lambda_2/\lambda_1 = v_2/v_1 = (\mu_1/\mu_2)^{1/2}$;
  (c) lighter cord.
33. (c) $A_T = [2k_1/(k_2 + k_1)]A$
  $= [2v_2/(v_1 + v_2)]A$.
35. (b) $2D_M \cos(\frac{1}{2}\phi)$, purely sinusoidal;
  (d) $D = \sqrt{2}\, D_M \sin(kx - \omega t + \pi/4)$.
37. 440 Hz, 880 Hz, 1320 Hz, 1760 Hz.
39. $f_n = n(0.50\,\text{Hz}), n = 1, 2, 3,...$;
  $T_n = (2.0\,\text{s})/n, n = 1, 2, 3,...$.
41. 70 Hz.
45. 4.
47. (a) $D_2 = (4.2\,\text{cm})\sin[(0.71\,\text{cm}^{-1})x$
  $+ (47\,\text{s}^{-1})t + 2.1]$;
  (b) $D_{resultant} = (8.4\,\text{cm})$
  $\sin[(0.71\,\text{cm}^{-1})x + 2.1]$
  $\cos[(47\,\text{s}^{-1})t]$.
49. 308 Hz.
51. (a)

[graph: D, m vs t, s showing curves D_2, D_R, D_1]

(b)

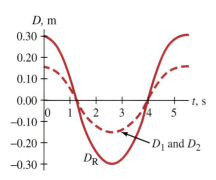

53. 5.4 km/s.

55. 29°.

57. 24°.

59. Speed will be greater in the less dense rod by a factor of $\sqrt{2}$.

61. (a) 0.050 m; (b) 2.3.

63. 0.99 m.

65. (a) solid curves,
    (c) dashed curves;

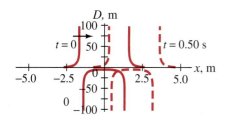

(b)
$D = (4.0 \text{ m}^3)/\{[x - (3.0 \text{ m/s})t]^2 - 2.0 \text{ m}^2\}$;

(d)

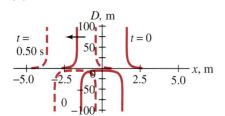

$D = (4.0 \text{ m}^3)/\{[x + (3.0 \text{ m/s})t]^2 - 2.0 \text{ m}^2\}$.

67. (a) 784 Hz, 1176 Hz, 880 Hz,
    1320 Hz; (b) 1.26; (c) 1.12; (d) 0.794.

69. $\lambda_n = 4L/(2n - 1), n = 1, 2, 3, \dots$.

71. $y = (3.5 \text{ cm}) \cos[(1.05 \text{ cm}^{-1})x - (1.39 \text{ s}^{-1})t]$.

73.

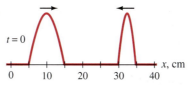

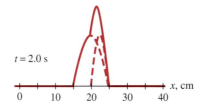

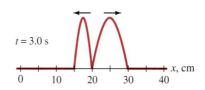

CHAPTER 16

1. 2.6×10^2 m.

3. 5.4×10^2 m.

5. 1200 m, 300 m.

7. (a) 1.1×10^{-8} m; (b) 1.1×10^{-10} m.

9. (a) $\Delta P = (4 \times 10^{-5} \text{ Pa})$
    $\sin[(0.949 \text{ m}^{-1})x - (315 \text{ s}^{-1})t]$;
    (b) $\Delta P = (4 \times 10^{-3} \text{ Pa})$
    $\sin[(94.9 \text{ m}^{-1})x - (3.15 \times 10^4 \text{ s}^{-1})t]$.

11. (a) 49 dB; (b) 3.2×10^{-10} W/m².

13. 150 Hz to 20,000 Hz.

15. (a) 9; (b) 9.5 dB.

17. (a) Higher frequency is greater by a factor of 2; (b) 4.

19. (a) 5.0×10^{-13} W; (b) 6.3×10^4 yr.

21. 87 dB.

23. (a) 5.10×10^{-5} m; (b) 29.8 Pa.

25. (a) 1.5×10^3 W; (b) 3.4×10^2 m.

27. (b) 190 dB.

29. (a) 570 Hz; (b) 860 Hz.

31. 8.6 mm $< L <$ 8.6 m.

33. (a) 110 Hz, 330 Hz, 550 Hz, 770 Hz;
    (b) 220 Hz, 440 Hz, 660 Hz, 880 Hz.

35. (a) 0.656 m; (b) 262 Hz, 1.31 m;
    (c) 1.31 m, 262 Hz.

37. −2.6%.

39. (a) 0.578 m; (b) 869 Hz.

41. 215 m/s.

43. 0.64, 0.20, −2 dB, −7 dB.

45. 28.5 kHz.

47. (a) 130.5 Hz, or 133.5 Hz;
    (b) ±2.3%.

49. (a) 343 Hz; (b) 1030 Hz, 1715 Hz.

53. 346 Hz.

57. (a) 1690 Hz; (b) 1410 Hz.

59. 30,890 Hz.

61. 120 Hz.

63. 91 Hz.

65. 90 beats/min.

67. (a) 570 Hz; (b) 570 Hz; (c) 570 Hz;
    (d) 570 Hz; (e) 594 Hz; (f) 595 Hz.

71. (a) 120; (b) 0.96°.

73. (a) 37°; (b) 1.7.

75. 0.278 s.

77. 55 m.

79. 410 km/h (255 mi/h).

81. 1, 0.444, 0.198, 0.0878, 0.0389.

83. 18.1 W.

85. 15 W.

87. 2.3 Hz.

89. $\Delta P_M/\Delta P_{M0} = D_M/D_{M0} = 10^6$.

91. 50 dB.

93. 17.5 m/s.

95. 2.3 kHz.

97. 550 Hz.

99. (a) 2.8×10^3 Hz; (b) 1.80 m;
    (c) 0.12 m.

101. (a) 2.2×10^{-7} m; (b) 5.4×10^{-5} m.

CHAPTER 17

1. 0.548.

3. (a) 20°C; (b) ≈3300°F.

5. 104.0°F.

7. −40°F = −40°C.

9. $\Delta L_{\text{Invar}} = 2.0 \times 10^{-6}$ m,
    $\Delta L_{\text{steel}} = 1.2 \times 10^{-4}$ m,
    $\Delta L_{\text{marble}} = 2.5 \times 10^{-5}$ m.

11. −69°C.

13. 5.1 mL.

15. 0.06 cm³.

19. −40 min.

21. -2.8×10^{-3} (0.28%).

23. 3.5×10^7 N/m².

25. (a) 27°C; (b) 4.3×10^3 N.

27. −459.7°F.

29. 1.07 m³.

31. 1.43 kg/m³.

33. (a) 14.8 m³; (b) 1.81 atm.

35. 1.80×10^3 atm.

37. 37°C.

39. 3.43 atm.

41. 0.588 kg/m³, water vapor is not an ideal gas.

43. 2.69×10^{25} molecules/m³.

45. 4.9×10^{22} molecules.

47. 7.7×10^3 N.

49. (a) 71.2 torr; (b) 157°C.

51. (a) 0.19 K; (b) 0.051%.

53. (a) Low; (b) 0.017%.

55. 1/6.

57. 5.1×10^{27} molecules, 8.4×10^3 mol.

59. 11 L, not advisable.

61. (a) 9.3×10^2 kg; (b) 1.0×10^2 kg.

63. 1.1×10^{44} molecules.

65. 3.3×10^{-7} cm.

67. 1.1×10^3 m.

69. 15 h.

71. 0.66×10^3 kg/m³.

73. ± 0.11 C°.

77. 3.6 m.

CHAPTER 18

1. (a) 5.65×10^{-21} J; (b) 7.3×10^3 J.

3. 1.17.

5. (a) 4.5; (b) 5.2.

7. $\sqrt{2}$.

11. (a) 461 m/s; (b) $19\ \mathrm{s}^{-1}$.

13. 1.00429.

17. Vapor.

19. (a) Gas, liquid, vapor;
(b) gas, liquid, solid, vapor.

21. 0.69 atm.

23. 11°C.

25. 1.96 atm.

27. 120°C.

29. (a) 5.3×10^6 Pa; (b) 5.7×10^6 Pa.

31. (b) $b = 4.28 \times 10^{-5}$ m³/mol,
$a = 0.365\ \mathrm{N \cdot m^4/mol^2}$.

33. (a) 10^{-7} atm; (b) 300 atm.

35. (a) 6.3 cm; (b) 0.58 cm.

37. 2×10^{-7} m.

39. (b) $4.7 \times 10^7\ \mathrm{s}^{-1}$.

43. 7.8 h.

45. (b) 4×10^{-11} mol/s; (c) 0.7 s.

47. 2.6×10^2 m/s,
$4 \times 10^{-17}\ \mathrm{N/m^2} \approx 4 \times 10^{-22}$ atm.

49. (a) 2.9×10^2 m/s; (b) 12 m/s.

51. Reasonable, 70 cm.

53. $mgh = 4.3 \times 10^{-5}(\frac{1}{2}mv_{\mathrm{rms}}^2)$,
reasonable.

55. $P_2/P_1 = 1.43,\ T_2/T_1 = 1.20$.

57. 1.4×10^5 K.

59. (a) 1.7×10^3 Pa; (b) 7.0×10^2 Pa.

61. 2×10^{13} m.

CHAPTER 19

1. 1.0×10^7 J.

3. 1.8×10^2 J.

5. 2.1×10^2 kg/h.

7. 83 kcal.

9. 4.7×10^6 J.

11. 40 C°.

13. 186°C.

15. 7.1 min.

17. (b) $mc_0[(T_2 - T_1) + a(T_2^2 - T_1^2)/2]$;
(c) $c_{\mathrm{mean}} = c_0[1 + \frac{1}{2}a(T_2 + T_1)]$.

19. 0.334 kg (0.334 L).

21. $\frac{2}{3}m$ steam and $\frac{4}{3}m$ water at 100°C.

23. 9.4 g.

25. 4.7×10^3 kcal.

27. 1.22×10^4 J/kg.

29. 360 m/s.

31. (a) 0; (b) 5.00×10^3 J.

33.

P(atm)

B · · · · · · · · · · A
1.0 ─ ─

0.5 ─ ─ ─ ─ ┤ C

0   1.0   2.0   *V*(L)

35. (a) 0; (b) −1300 kJ.

37. (a) 1.6×10^2 J; (b) $+1.6 \times 10^2$ J.

39. $W = 3.46 \times 10^3$ J, $\Delta U = 0$,
$Q = +3.46 \times 10^3$ J (into the gas).

41. +129 J.

45. (a) +25 J; (b) +63 J; (c) −95 J;
(d) −120 J; (e) −15 J.

47. $W = RT \ln\left(\dfrac{V_2 - b}{V_1 - b}\right) + a\left(\dfrac{1}{V_2} - \dfrac{1}{V_1}\right)$.

49. 22°C/h.

51. $4.98\ \mathrm{cal/mol \cdot K},\ 2.49\ \mathrm{kcal/kg \cdot K}$;
$6.97\ \mathrm{cal/mol \cdot K},\ 3.48\ \mathrm{kcal/kg \cdot K}$.

53. 83.7 g/mol, krypton.

55. 46 C°.

57. (a) 2.08×10^3 J; (b) 8.32×10^2 J;
(c) 2.91×10^3 J.

59. 0.379 atm, −51°C.

61. 1.33×10^3 J.

63. (a) $T_1 = 317$ K, $T_2 = 153$ K;
(b) -1.59×10^4 J; (c) -1.59×10^4 J;
(d) $Q = 0$.

65. (a)

P

P_1 ─ ─ ·1 T₁

T₃ 3 ← 2 T₂

0   *V*

(b) 231 K;
(c) $Q_{1 \to 2} = 0$,
$\Delta U_{1 \to 2} = -2.01 \times 10^3$ J,
$W_{1 \to 2} = +2.01 \times 10^3$ J;
$W_{2 \to 3} = -1.31 \times 10^3$ J,
$\Delta U_{2 \to 3} = -1.97 \times 10^3$ J,
$Q_{2 \to 3} = -3.28 \times 10^3$ J;
$W_{3 \to 1} = 0$,
$\Delta U_{3 \to 1} = +3.98 \times 10^3$ J,
$Q_{3 \to 1} = +3.98 \times 10^3$ J;
(d) $W_{\mathrm{cycle}} = +0.70 \times 10^3$ J,
$Q_{\mathrm{cycle}} = +0.70 \times 10^3$ J,
$\Delta U_{\mathrm{cycle}} = 0$.

67. (a) 64 W; (b) 22 W.

69. 4.8×10^2 W.

71. 31 h.

73. (a) 1.7×10^{17} W; (b) 278K (5°C).

75. (b) $\Delta Q/\Delta t = A(T_2 - T_1)/\Sigma(\ell_i/k_i)$.

77. 22%.

79. 4×10^{15} J.

81. 2.8 kcal/kg.

83. 30 C°.

85. 682 J.

87. 2.8 C°.

89. 2.58 cm, rod vaporizes.

91. 4.3 kg.

95. (a) 2.3 C°/s; (b) 84°C;
(c) convection, conduction, evaporation.

97. (a) $\rho = m/V = (mP/nR)/T$;
(b) $\rho = (m/nRT)P$.

99. (a) 1.9×10^5 J; (b) -1.4×10^5 J;

(c)

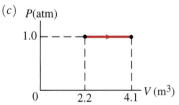

P(atm)

1.0 ─ ─ ·

0   2.2   4.1   *V* (m³)

101. 3.2×10^5 s = 3.7 d.

103. 10 C°.

CHAPTER 20

1. 24%.

3. 816 MW.

5. 18%.

7. 13 km³/day, 63 km².

9. 28.0%.

13. 1.2×10^{13} J/h.

15. 1.4×10^3 m/day.

17. 660°C.

19. 3.7×10^8 kg/h.

21. (a) $P_a = 5.15 \times 10^5$ Pa,
   $P_b = 2.06 \times 10^5$ Pa;
   (b) $V_c = 30.0$ L, $V_d = 12.0$ L;
   (c) 2.83×10^3 J; (d) -2.14×10^3 J;
   (e) 0.69×10^3 J; (f) 24%.

23. 5.7.

25. −21°C.

27. 2.9.

29. (a) 3.9×10^4 J; (b) 3.0 min.

31. 76 L.

33. 0.15 J/K.

35. +11 kcal/K.

37. +0.0104 cal/K · s.

39. 1.7×10^2 J/K.

43. (a) 0.312 kcal/K;
   (b) > −0.312 kcal/K.

45. (a) Adiabatic process;
   (b) $\Delta S_i = -nR \ln 2$, $\Delta S_a = 0$;
   (c) $\Delta S_{surr,i} = nR \ln 2$, $\Delta S_{surr,a} = 0$.

47. (a) Entropy is a state function.

51. (a) 5/16; (b) 1/64.

53. (b), (a), (c), (d).

55. 69%.

57. 2.6×10^3 J/K.

59. (a) 17; (b) 5.9×10^7 J/h.

61. (a) 5.3 C°; (b) +77 J/kg · K.

63. $\Delta S = K/T$.

65. (a)

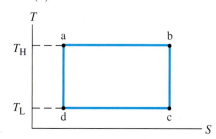

   (b) area = $Q_{net} = W_{net}$.

67. $e_{Stirling} = (T_H - T_L) \ln(V_b/V_a)/$
   $[T_H \ln(V_b/V_a)$
   $+ \tfrac{3}{2}(T_H - T_L)]$,
   $e_{Stirling} < e_{Carnot}$.

69. 0.091 hp.

71. (a) 1/379; (b) $1/1.59 \times 10^{11}$.

CHAPTER 21

1. 6.3×10^9 N.

3. 2.7×10^{-3} N.

5. 5.5×10^3 N.

7. 8.66 cm.

9. -5.4×10^7 C.

11. 83.8 N away from the center of the triangle.

13. 2.96×10^7 N toward the center of the square.

15. $\mathbf{F}_1 = (kQ^2/\ell^2)[(-2 + 3\sqrt{2}/4)\mathbf{i}$
   $+ (4 - 3\sqrt{2}/4)\mathbf{j}]$, $\mathbf{F}_2 = (kQ^2/\ell^2)$
   $[(2 + 2\sqrt{2})\mathbf{i} + (-6 + 2\sqrt{2})\mathbf{j}]$,
   $\mathbf{F}_3 = (kQ^2/\ell^2)[(-12 - 3\sqrt{2}/4)\mathbf{i}$
   $+ (6 + 3\sqrt{2}/4)\mathbf{j}]$, $\mathbf{F}_4 = (kQ^2/\ell^2)$
   $[(12 - 2\sqrt{2})\mathbf{i} + (-4 - 2\sqrt{2})\mathbf{j}]$.

17. (a) $Q_1 = Q_2 = \tfrac{1}{2}Q_T$;
   (b) Q_1 (or Q_2) = 0.

19. $0.402Q_0$, 0.366ℓ from Q_0.

21. 60.2×10^{-6} C, 29.8×10^{-6} C;
   -16.8×10^{-6} C, 106.8×10^{-6} C.

23. $\mathbf{F} = -(1.90kQ^2/\ell^2)(\mathbf{i} + \mathbf{j} + \mathbf{k})$.

25. 2.18×10^{-16} N (west).

27. 7.43×10^6 N/C (up).

29. $(1.39 \times 10^2$ N/C$)\mathbf{j}$.

33.

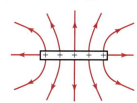

35. 8.26×10^{-10} N/C (south).

37. 4.5×10^6 N/C up, 1.2×10^7 N/C, 56° above the horizontal.

39. 5.61×10^4 N/C away from the opposite corner.

41. $Q_1/Q_2 = \tfrac{1}{4}$.

43. (a) $2Qy/4\pi\epsilon_0(y^2 + \ell^2)^{3/2}\,\mathbf{j}$.

45. $\dfrac{Q}{4\pi\epsilon_0}\left[\dfrac{x\mathbf{i} - (2a/\pi)\mathbf{j}}{(x^2 + a^2)^{3/2}}\right]$.

49. $\dfrac{-2\lambda \sin\theta_0}{4\pi\epsilon_0 R}\,\mathbf{i}$.

51. (a) $\dfrac{\lambda}{4\pi\epsilon_0 x(x^2 + L^2)^{1/2}}$
   $\{L\mathbf{i} + [x - (x^2 + L^2)^{1/2}]\mathbf{j}\}$.

53. $(\sigma/2\epsilon_0)\mathbf{k}$.

55. (a) $\mathbf{a} = -(3.5 \times 10^{15}$ m/s²$)\,\mathbf{i}$
   $-(1.41 \times 10^{16}$ m/s²$)\,\mathbf{j}$; (b) $\theta = -104°$.

57. $\theta = -28°$.

59. (b) $2\pi(4\pi\epsilon_0 mR^3/qQ)^{1/2}$.

61. (a) 3.4×10^{-20} C; (b) No;
   (c) 8.5×10^{-26} m · N;
   (d) 2.5×10^{-26} J.

63. (a) $\theta \ll 1$;
   (b) $(pE/I)^{1/2}/2\pi$.

65. (b) Direction of the dipole.

67. 6.8×10^3 C.

69. 5.7×10^{13} C.

71. $\mathbf{F}_1 = 0.30$ N, 265° from x-axis,
   $\mathbf{F}_2 = 0.26$ N, 139° from x-axis,
   $\mathbf{F}_3 = 0.26$ N, 30° from x-axis.

73. 4.2×10^5 N/C up.

75. $0.444Q_0$, 0.333ℓ from Q_0.

77. 5.60 m from the positive charge, and 3.60 m from the negative charge.

79. (a) In the direction of the velocity, to the right; (b) 2.1×10^2 N/C.

81. $\theta_0 = 18°$.

83. $(1.08 \times 10^7$ N/C$)/$
   $\{3.00 - \cos[(12.5$ s⁻¹$)t]\}^2$, up.

85. $E_A = 4.2 \times 10^4$ N/C (right),
   $E_B = -1.4 \times 10^4$ N/C (left),
   $E_C = -2.8 \times 10^3$ N/C (left),
   $E_D = -4.2 \times 10^4$ N/C (left).

87. $d(1 + \sqrt{2})$ from the negative charge, and $d(2 + \sqrt{2})$ from the positive charge.

CHAPTER 22

1. (a) 41 N · m²/C; (b) 29 N · m²/C;
   (c) 0.

3. $\Phi_{net} = 0$,
   $\Phi_{x=0} = -(6.50 \times 10^3$ N/C$)\ell^2$,
   $\Phi_{x=\ell} = +(6.50 \times 10^3$ N/C$)\ell^2$,
   $\Phi_{all\ others} = 0$.

5. 12.8 nC.

7. $-1.2\ \mu$C.

9. -3.75×10^{-11} C.

11. (a) -1.0×10^4 N/C (toward wire);
   (b) -2.5×10^4 N/C (toward wire).

13.

15. (a) 5.5×10^7 N/C (away from center);
   (b) 0.

17. (a) $-8.00\ \mu$C;
   (b) $+1.00\ \mu$C

19. (a) 0; (b) σ/ϵ_0;
   (c) unaffected.

21. (a) 0; (b) $\sigma_1 r_1^2/\epsilon_0 r^2$;
(c) $(\sigma_1 r_1^2 + \sigma_2 r_2^2)/\epsilon_0 r^2$;
(d) $\sigma_2/\sigma_1 = -(r_1/r_2)^2$; (e) $\sigma_1 = 0$.

23. (a) $q/4\pi\epsilon_0 r^2$;
(b) $(1/4\pi\epsilon_0)[Q(r^3 - r_1^3)$
$+ q(r_0^3 - r_1^3)]/(r_0^3 - r_1^3)r^2$;
(c) $(q + Q)/4\pi\epsilon_0 r^2$.

25. (a) $q/4\pi\epsilon_0 r^2$; (b) $(q + Q)/4\pi\epsilon_0 r^2$;
(c) $E(r < r_0) = Q/4\pi\epsilon_0 r^2$,
$E(r > r_0) = 2Q/4\pi\epsilon_0 r^2$;
(d) $E(r < r_0) = -Q/4\pi\epsilon_0 r^2$,
$E(r > r_0) = 0$.

27. (a) $\sigma R_0/\epsilon_0 r$; (b) 0; (c) same.

29. (a) 0; (b) $Q/2\pi\epsilon_0 Lr$;
(c) 0; (d) $eQ/4\pi\epsilon_0 L$.

31. (a) 0;
(b) -2.3×10^5 N/C (toward the axis);
(c) -1.8×10^4 N/C (toward the axis).

33. (a) $\rho_E r/2\epsilon_0$; (b) $\rho_E R_1^2/2\epsilon_0 r$;
(c) $\rho_E(r^2 + R_1^2 - R_2^2)/2\epsilon_0 r$;
(d) $\rho_E(R_3^2 + R_1^2 - R_2^2)/2\epsilon_0 r$;
(e)

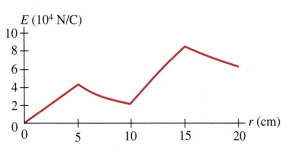

35. $Q/\epsilon_0\sqrt{2}$.

37. $\oint \mathbf{g} \cdot d\mathbf{A} = -4\pi GM$.

39. $Q_{\text{enclosed}} = \epsilon_0 b\ell^3$.

41. 3.95×10^2 N·m^2/C,
-1.69×10^2 N·m^2/C.

43. (a) 0; (b) $Q/25\pi\epsilon_0 r_0^2 \leq E \leq Q/\pi\epsilon_0 r_0^2$;
(c) not perpendicular;
(d) not useful.

45. (a) $0.677e = +1.08 \times 10^{-19}$ C;
(b) 3.5×10^{11} N/C.

47. (a) $\rho_E r_0/6\epsilon_0$ (right);
(b) $-17\rho_E r_0/54\epsilon_0$ (left).

49. (a) 0; (b) 5.65×10^5 N/C (right);
(c) 5.65×10^5 N/C (right);
(d) -5.00×10^{-6} C/m^2;
(e) $+ 5.00 \times 10^{-6}$ C/m^2.

CHAPTER 23

1. -4.2×10^{-5} J (done by the field).

3. 3.4×10^{-15} J.

5. $V_a - V_b = +72.8$ V.

7. 7.04 V.

9. 0.8 μC.

11. (a) $V_{\text{BA}} = 0$; (b) $V_{\text{CB}} = -2100$ V;
(c) $V_{\text{CA}} = -2100$ V.

13. (a) -9.6×10^8 V;
(b) $V(\infty) = +9.6 \times 10^8$ V.

15. (a) The same;
(b) $Q_2 = r_2 Q/(r_1 + r_2)$.

17. (a) $Q/4\pi\epsilon_0 r$;
(b) $(Q/8\pi\epsilon_0 r_0)[3 - (r^2/r_0^2)]$;
(c)

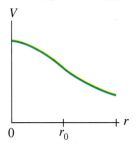

19. (a) $V_0 + (\sigma R_0/\epsilon_0) \ln(R_0/r)$;
(b) $V = V_0$; (c) $V \neq 0$.

21. (a) 29 V;
(b) -29 eV $(-4.6 \times 10^{-18}$ J).

23. $+0.19$ J.

25. 4.2 MV.

27. 2.33×10^7 m/s.

29. $V_{\text{BA}} = (1/2\pi\epsilon_0)q(2b - d)/b(d - b)$.

31. $\dfrac{\sigma}{2\epsilon_0}[(x^2 + R_2^2)^{1/2} - (x^2 + R_1^2)^{1/2}]$.

33. $\dfrac{Q}{8\pi\epsilon_0 L} \ln\left(\dfrac{x + L}{x - L}\right)$, $x > L$.

35. $\dfrac{a}{6\epsilon_0}[(x^2 + R^2)^{1/2}(R^2 - 2x^2) + 2x^3]$.

37. 3.2 mm.

39. (a) 8.5×10^{-30} C·m; (b) zero.

41. (a) -0.088 V; (b) 1%.

43. (a) 5.2×10^{-20} C.

47. $\mathbf{E} = 2y(2z - 1)\mathbf{i} - 2(y + x - 2xz)\mathbf{j}$
$+ (4xy)\mathbf{k}$.

49. (a) 9.6×10^4 eV; (b) 1.9×10^5 eV.

51. -2.4×10^4 V.

53. (a) $U = (1/4\pi\epsilon_0)(Q_1 Q_2/r_{12}$
$+ Q_1 Q_3/r_{13} + Q_1 Q_4/r_{14} + Q_2 Q_3/r_{23}$
$+ Q_2 Q_4/r_{24} + Q_3 Q_4/r_{34})$.
(b) $U = (1/4\pi\epsilon_0)(Q_1 Q_2/r_{12}$
$+ Q_1 Q_3/r_{13} + Q_1 Q_4/r_{14} + Q_1 Q_5/r_{15}$
$+ Q_2 Q_3/r_{23} + Q_2 Q_4/r_{24} + Q_2 Q_5/r_{25}$
$+ Q_3 Q_4/r_{34} + Q_3 Q_5/r_{35} + Q_4 Q_5/r_{45})$.

55. (a) 2.0 keV; (b) 42.8.

57. (a) $(-4 + \sqrt{2})Q^2/4\pi\epsilon_0 b$; (b) 0.

59. $3Q^2/20\pi\epsilon_0 r$.

61. 5.4×10^5 V/m.

63. 9×10^2 V.

65. (a) 1.1 MV; (b) 13 kg.

67. 7.2 MV.

69. 1.58×10^{12} electrons.

71. 1.7×10^{-12} V.

73. 1.03×10^6 m/s.

75. $V_a = -3.5\, Q/4\pi\epsilon_0 L$,
$V_b = -5.2\, Q/4\pi\epsilon_0 L$,
$V_c = -6.8\, Q/4\pi\epsilon_0 L$.

77. (a) 5.8×10^5 V; (b) 9.2×10^{-14} J.

79. $V_a - V_b = (\lambda/2\pi\epsilon_0) \ln(R_b/R_a)$.

83. (a) $\rho_E(r_2^3 - r_1^3)/3\epsilon_0 r$;
(b) $(\rho_E/6\epsilon_0)[3r_2^2 - r^2 - (2r_1^3/r)]$;
(c) $(\rho_E/2\epsilon_0)(r_2^2 - r_1^2)$, potential is
continuous at r_1 and r_2.

CHAPTER 24

1. 2.6 μF.

3. 6.3 pF.

5. 0.80 μF.

7. 2.0 C.

9. 1.8×10^2 m^2.

11. 7.1×10^{-4} F.

13. 23 nC.

17. 4.5×10^4 V/m.

19. (a) $\epsilon_0 A/(d - \ell)$; (b) 3.

21. 2880 pF, yes.

23. (a) $(C_1 C_2 + C_1 C_3 + C_2 C_3)/(C_2 + C_3)$;
(b) $Q_1 = 350$ μC, $Q_2 = 117$ μC.

25. (a) 3.71 μF; (b) $V_{\text{ab}} = V_1 = 26.0$ V,
$V_2 = 14.9$ V, $V_3 = 11.1$ V.

27. 18 nF (parallel), 1.6 nF (series).

29. (a) $3C/5$; (b) $Q_1 = Q_2 = CV/5$,
$Q_3 = 2CV/5$, $Q_4 = 3CV/5$,
$V_1 = V_2 = V/5$, $V_3 = 2V/5$,
$V_4 = 3V/5$.

31. $Q_1' = C_1 C_2 V_0/(C_1 + C_2)$,
$Q_2' = C_2^2 V_0/(C_1 + C_2)$.

33. (a) $Q_1 = Q_3 = 30\ \mu C$;
    $Q_2 = Q_4 = 60\ \mu C$;
    (b) $V_1 = V_2 = V_3 = V_4 = 3.75$ V;
    (c) 7.5 V.

35. 3.0 μF.

37. $C \approx (\epsilon_0 A/d)[1 - \frac{1}{2}(\theta\sqrt{A}/d)]$.

39. 2.0×10^{-3} J.

41. 2.3×10^{3} J.

43. 1.65×10^{-7} J.

45. (a) 2.5×10^{-5} J; (b) 6.2×10^{-6} J;
    (c) $Q_{par} = 4.2\ \mu C$, $Q_{ser} = 1.0\ \mu C$.

47. (a) 2.2×10^{-4} J; (b) 8.1×10^{-5} J;
    (c) -1.4×10^{-4} J; (d) stored
    potential energy is not conserved.

51. 1.5×10^{-10} F.

53. 0.46 μC.

55. 3.3×10^{2} J.

57. $C = 2\epsilon_0 A K_1 K_2/d(K_1 + K_2)$.

59. (a) $0.40Q_0$, $1.60Q_0$; (b) $0.40V_0$.

61. (a) 111 pF; (b) 1.66×10^{-8} C;
    (c) 1.84×10^{-8} C;
    (d) 1.17×10^{5} V/m; (e) 3.34×10^{4} V/m;
    (f) 150 V;
    (g) 172 pF; (h) 2.58×10^{-8} C.

63. 22%.

65. $Q = 4.41 \times 10^{-7}$ C,
    $Q_{ind} = 3.65 \times 10^{-7}$ C,
    $E_{air} = 2.69 \times 10^{4}$ V/m,
    $E_{glass} = 4.64 \times 10^{3}$ V/m;

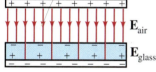

67. 11 μF.

69. (a) $4\times$; (b) $4\times$; (c) $\frac{1}{2}\times$.

71. 10.9 V.

73. (b) 1.5×10^{-10} F/m.

75. $U_2/U_1 = 1/2K$, $E_2/E_1 = 1/K$.

77. (a) 19 J; (b) 0.19 MW.

79. (a) 0.10 MV; (b) voltage will
    decrease exponentially.

81. 660 pF in parallel.

83. $Q_1 = 11\ \mu C$, $Q_2 = Q_3 = 13\ \mu C$,
    $V_1 = 11$ V, $V_2 = 6.5$ V, $V_3 = 4.4$ V.

85. $Q^2 x/2A\epsilon_0$.

87. (a) 7.4 nF, 0.33 μC, 1.5×10^{4} V/m,
    7.5×10^{-6} J; (b) 27 nF, 1.2 μC,
    1.5×10^{4} V/m, 2.7×10^{-5} J.

89. (a) 66 pF; (b) 30 μC; (c) 7.0 mm; (d)
    450 V.

CHAPTER 25

1. 9.38×10^{18} electron/s.

3. 2.1×10^{-11} A.

5. 7.5×10^{2} V.

7. 2.1×10^{21} electron/min.

9. (a) 16 Ω; (b) 6.8×10^{3} C.

11. 0.57 mm.

13. $R_{Al} = 1.2R_{Cu}$.

15. 1/6 the length, 8.3 Ω, 1.7 Ω.

17. 58.3°C.

19. 1.8×10^{3} °C.

21. $R_2 = \frac{1}{4}R_1$.

25. $R = (r_2 - r_1)/4\pi\sigma r_1 r_2$.

27. 3.2 W.

29. 37 V.

31. (a) 240 Ω, 0.50 A; (b) 96 Ω, 1.25 A.

33. 0.092 kWh, 22¢/month.

35. 1.1 kWh.

37. 3.

39. 5.3 kW.

41. 0.128 kg/s.

43. 0.094 A.

45. (a) Infinite; (b) 1.9×10^{2} Ω.

47. 636 V, 5.66 A.

49. 1.5 kW, 3.0 kW, 0.

51. (a) 7.8×10^{-10} m/s; (b) 10.5 A/m^2
    along the wire; (c) 1.8×10^{-7} V/m.

53. 2.7 A/m^2 north.

55. 12 h.

57. 6.67×10^{-2} S.

59. (a) 8.6 Ω, 1.1 W; (b) $4\times$.

61. (a) \$44; (b) 1.8×10^{3} kg/yr.

63. (a) -19.5%; (b) percentage decrease
    in the power output would be less.

65. (a) 1.44×10^{3} W; (b) 17 W; (c) 11 W;
    (d) 0.8¢/day.

67. (a) 1.5 kW; (b) 12.5 A.

69. 2.

71. 0.303 mm, 28.0 m.

73. (a) 1.2 kW; (b) 100 W.

75. 1.4×10^{12} protons.

77. $j_a = 2.8 \times 10^{5}$ A/m^2,
    $j_b = 1.6 \times 10^{5}$ A/m^2.

CHAPTER 26

1. (a) 8.39 V; (b) 8.49 V.

3. 0.060 Ω.

5. 360 Ω, 23 Ω.

7. 25 Ω, 70 Ω, 95 Ω, 18 Ω.

9. Series connection.

11. 4.6 kΩ.

13. 310 Ω, 3.7%.

15. 960 Ω in parallel.

17. 105 Ω.

19. (a) V_1 and V_2 increase; V_3 and V_4
    decrease; (b) I_1 (= I) and I_2
    increase; I_3 and I_4 decrease;
    (c) increases; (d) $I = I_1 = 0.300$ A,
    $I_2 = 0$, $I_3 = I_4 = 0.150$ A; $I = 0.338$ A,
    $I_2 = I_3 = I_4 = 0.113$ A.

21. 0.4 Ω.

23. 0.41 A.

25. $I_1 = 0.68$ A, $I_2 = -0.40$ A.

27. $I_1 = 0.18$ A right, $I_2 = 0.32$ A left,
    $I_3 = 0.14$ A up.

29. $I_1 = 0.274$ A, $I_2 = 0.222$ A,
    $I_3 = 0.266$ A, $I_4 = 0.229$ A,
    $I_5 = 0.007$ A, $I = 0.496$ A.

31. 52 V, –28 V.
    The negative value means the
    battery is facing the other direction.

33. 70 V.

35. $I_1 = 0.783$ A.

39. (a) $R(3R + 5R')/8(R + R')$; (b) $R/2$.

41. (a) 3.7 nF; (b) 22 μs.

43. $t = 1.23\tau$.

45. (a) $\tau = R_1 R_2 C/(R_1 + R_2)$;
    (b) $Q_{max} = \mathscr{E}R_2 C/(R_1 + R_2)$.

47. 2.1 μs.

49. 50 μA.

51. (a) 2.9×10^{-4} Ω in parallel;
    (b) 35 kΩ in series.

53. 22 V, 17 V, 14% low.

55. 0.85mA, 4.3V.

57. 9.6 V.

61. 3.6×10^{-2} C°.

63. Two resistors in series.

65. 2.2 V, 116 V.

67. 0.19 MΩ.

69. (a) 0.10 A; (b) 0.10 A; (c) 53 mA.

71. 2.5 V.

73. 46.1 V, 0.71 Ω.

75. (a) 72.0 W; (b) 14.2 W; (c) 3.76 W.

77. (a) 40 kΩ; (b) between b and c.

79. 375 cells, 3.8 m $\times$ 0.090 m.

81. (a) 0.50 A; (b) 0.17 A; (c) 3.3 μC;
    (d) 32 μs.

83. (a) + 6.8 V, 10.2 μC; (b) 28 μs.

CHAPTER 27

1. (a) 6.7 N/m; (b) 4.7 N/m.

3. 2.68 A.

5. 0.243 T.

7. (a) South pole; (b) 3.5×10^{2} A;
    (c) 5.22 N.

9. 5.5×10^3 A.

13. 1.05×10^{-13} N north.

15. (a) Down; (b) in; (c) right.

21. 1.6 m.

23. (a) 2.7 cm; (b) 3.8×10^{-7} s.

25. 1.034×10^8 m/s (west), gravity can be ignored.

27. $(6.4\mathbf{i} - 10.3\mathbf{j} - 0.24\mathbf{k})] \times 10^{-16}$ N.

29. 5.3×10^{-5} m, 3.3×10^{-4} m.

31. (a) 45°; (b) 3.5×10^{-3} m.

33. (a) $2\mu B$; (b) 0.

35. (a) 4.33×10^{-5} m · N; (b) north edge.

39. 29 μA.

41. 1.2×10^5 C/kg.

43. (a) 2.2×10^{-4} V/m; (b) 2.7×10^{-4} m/s; (c) 6.4×10^{28} electrons/m³.

45. (a)Determine polarity of the emf; (b) 0.43 m/s.

47. 1.53 mm, 0.76 mm.

51. 3.0 T up.

53. 1.1×10^{-6} m/s west.

55. 0.17 N, 68° above the horizontal toward the north.

57. 0.20 T, 26.6° from the vertical.

59. (c) 48 MeV.

61. Slower protons will deflect more, and faster protons will deflect less, $\theta = 12°$.

63. 2.0 A, down.

65. 7.3×10^{-3} T.

67. -2.1×10^{-20} J.

CHAPTER 28

1. 1.7×10^{-4} T, 3.1×.

3. 0.18 N attraction.

5.

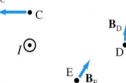

7. 8.9×10^{-5} T, 70° above horizontal.

9. 4.0×10^{-5} T, 15° below the horizontal.

11. (a) $(2.0 \times 10^{-5}$ T/A)$(15$ A $- I)$ up; (b) $(2.0 \times 10^{-5}$ T/A)$(15$ A $+ I)$ down.

13. 21 A down.

15. $[(\mu_0/4\pi)2I(d - 2x)/x(d - x)]\mathbf{j}$.

17. 4.12×10^{-5} T.

19. (b) $(\mu_0/4\pi)(2I/y)$.

21. 0.123 A.

23. (a) 6.4×10^{-3} T; (b) 3.8×10^{-3} T; (c) 2.1×10^{-3} T.

25. (a) 51 cm; (b) 1.3×10^{-2} T.

27. (a) $(\mu_0 I_0/2\pi R_1^2)r$ circular CCW; (b) $\mu_0 I_0/2\pi r$ circular CCW; (c) $(\mu_0 I_0/2\pi r)(R_3^2 - r^2)/(R_3^2 - R_2^2)$ circular CCW; (d) 0.

29. 3.6×10^{-6} T.

31. $\mu_0 I/8R$ out of the page.

33. (a) $\mu_0 I(R_1 + R_2)/4R_1 R_2$ into the page; (b) $\frac{1}{2}\pi I(R_1^2 + R_2^2)$ into the page.

35. (a) $\dfrac{Q\omega R^2}{4}\,\mathbf{i}$;

(b)

$$\frac{\mu_0 Q\omega}{2\pi R^2}\left[\frac{R^2 + 2x^2 - 2x\sqrt{R^2 + x^2}}{\sqrt{R^2 + x^2}}\right]\mathbf{i};$$
(c) yes.

37. (b) $B = \mu_0 IL/4\pi y(L^2 + y^2)^{1/2}$ circular.

39. (a) $(\mu_0 I_0/2\pi R)n \tan (\pi/n)$ into the page.

41. $B = \dfrac{\mu_0 I}{4\pi}\left\{\dfrac{(x^2 + y^2)^{1/2}}{xy}\right.$

$+ \dfrac{[(b - x)^2 + y^2]^{1/2}}{y(b - x)}$

$+ \dfrac{[(a - y)^2 + (b - x)^2]^{1/2}}{(a - y)(b - x)}$

$+ \left.\dfrac{[x^2 + (a - y)^2]^{1/2}}{x(a - y)}\right\},$

out of the page.

43. (a) 26 A · m²; (b) 31 m · N.

45. 30 T.

47. $F_M/L = 5.84 \times 10^{-5}$ up, $F_N/L = 3.37 \times 10^{-5}$ N/m 60° below the line toward P, $F_P/L = 3.37 \times 10^{-5}$ N/m 60° below the line toward N.

49. 0.27 mm, 1.4 cm.

51. $B = \mu_0 jt/2$ parallel to the sheet, perpendicular to the current (opposite directions on the two sides).

53. Between long, thin and short, fat.

55. 3×10^9 A.

57. B will decrease.

59. 2.1×10^{-6} g.

61. $2\mu_0 I/L\pi$ (left).

63. 4×10^{-6} T, about 10% of the Earth's field.

CHAPTER 29

1. -3.8×10^2 V.

3. Counterclockwise.

5. 0.026 V.

7. (a) Counterclockwise; (b) clockwise; (c) zero; (d) counterclockwise.

9. Counterclockwise.

11. (a) Clockwise; (b) 43 mV; (c) 17 mA.

13. 1.1×10^{-5} J.

15. 4.21 C.

17. (a) 5.2×10^{-2} A; (b) 0.32 mW.

19. 1.7×10^{-2} V.

21. $(\mu_0 Ia/2\pi) \ln 2$.

23. (a) 0.15 V; (b) 5.4×10^{-3} A; (c) 4.5×10^{-4} N.

25. (a) Will move at constant speed; (b) $v = v_0 e^{-B^2\ell^2 t/mR}$.

27. (a) $\dfrac{\mu_0 Iv}{2\pi} \ln\left(\dfrac{a + b}{b}\right)$ toward long wire; (b) $\dfrac{\mu_0 Iv}{2\pi} \ln\left(\dfrac{a + b}{b}\right)$ away from long wire.

31. 0.33 kV, 120 rev/s.

33. 100 V.

35. 13 A.

37. 3.54×10^4 turns.

39. 0.18.

41. (a) 5.2 V; (b) step-down transformer.

43. 549 V, 68.6 A.

45. 56.8 kW.

47. 0.188 V/m.

49. (b) Clockwise; (c) $dB/dt > 0$.

51. (a) IR/ℓ (constant); (b) $\dfrac{\mathscr{E}_0}{\ell}e^{-B^2\ell^2 t/mR}$.

53. 31 turns.

55. $v = 0.76$ m/s.

57. 184 kV.

59. 1.5×10^{17}.

61. (a) 23 A; (b) 90 V; (c) 6.9×10^2 W; (d) 75%.

63. (a) 0.85 A; (b) 8.2.

65. $\frac{1}{2}B\omega L^2$ toward the center.

71. $B\omega r$, radially out from the axis.

CHAPTER 30

1. $M = \mu_0 N_1 N_2 A/\ell$.

3. $M/\ell = \mu_0 n_1 n_2 \pi r_2^2$.

5. $M = (\mu_0 w/2\pi) \ln(\ell_2/\ell_1)$.

7. 1.2 H.

9. 2.5×10^{-6} H.

11. $r_1 \geq 2.5$ mm.

13. 3.

15. (a) $L_1 + L_2$; (b) $= L_1 L_2/(L_1 + L_2)$.

17. 15.9 J.

19. (a) $u_E = 4.4 \times 10^{-4}$ J/m^3,
$u_B = 1.6 \times 10^6$ J/m^3, $u_B \gg u_E$;
(b) $E = 6.0 \times 10^8$ V/m.

21. 4.4 J/m^3, 1.6×10^{-14} J/m^3.

23. $(\mu_0 I^2/4\pi) \ln(r_2/r_1)$.

25. $t/\tau = 4.6$.

27. $(dI/dt)_0 = V_0/L$.

29. (a) $(LV^2/2R^2)(1 - 2e^{-t/\tau} + e^{-2t/\tau})$;
(b) $t/\tau = 5.3$.

31. (a) 213 pF; (b) 46.5 μH.

35. (a) $Q = Q_0/\sqrt{2}$; (b) $T/8$.

37. $R = 2.30 \ \Omega$.

41. Decrease, 1.15 kΩ.

43. 20 mH, 95 turns.

45. (a) 21 mH; (b) 45 mA;
(c) 2.2×10^{-5} J.

47. 3.0×10^3 turns, 95 turns.

51. (b) Positioning one coil
perpendicular to the other;
(c) $L_1 L_2/(L_1 + L_2)$;
$(L_1 L_2 - M^2)/(L_1 + L_2 \mp 2M)$.

55. (a) $\frac{1}{2}(Q_0^2/C)e^{-Rt/L}$.

57.

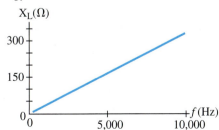

CHAPTER 31

1. (a) $3.7 \times 10^2 \ \Omega$; (b) $2.2 \times 10^{-2} \ \Omega$.

3. 9.90 Hz.

5.

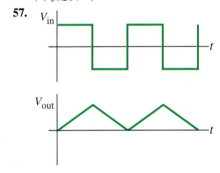

7. 0.13 H.

11. (a) 5.0%; (b) 98%.

13. (a) 9.0 kΩ; (b) 10.2 kΩ.

15. (a) 18 mA; (b) $-29°$; (c) 1.8 W;
(d) $V_R = 105$ V, $V_C = 58$ V.

17. (a) 0.38 A; (b) $-89°$; (c) 0.29 W.

19. 332 Ω.

21. (a) 0; (b) $\frac{2}{\pi} V_0$, $\overline{V}_{1/2} = \frac{2\sqrt{2}}{\pi} V_{\text{rms}}$.

23. 8.78 kΩ, $-7.66°$, 91.1 mA.

25. 265 Hz, 324 W.

27. 52.5 mA.

29. (b) $\omega'^2 = [(1/LC) - (R^2/2L^2)]$;
(c) $k \leftrightarrow 1/C, m \leftrightarrow L, b \leftrightarrow R$.

31. (a) $V_0^2 R/[2R^2 + 2(\omega L - 1/\omega C)^2]$;
(b) $\omega'^2 = 1/LC$; (c) $\Delta\omega = R/L$.

33. 4 Ω.

35. 9.76 nF.

37. 27.9 mH.

39. 1.6 kHz.

41. 14 Ω, 75 mH.

43. 2.2×10^3 Hz, 1.1×10^4 Hz.

45. (a) 23.6 kΩ, 10.8°; (b) 1.88×10^{-5} W;
(c) 2.8×10^{-5} A, 0.66 V, 4.7×10^{-4} V,
0.126 V.

49. $\{(R_1 + R_2)^2$
$+ [\omega(L_1 + L_2) -$
$(C_1 + C_2)/\omega C_1 C_2]^2\}^{1/2}$.

51. 19 Ω, 62 mH.

53. $I = \left(\frac{V_0}{Z}\right) \sin(\omega t + \phi)$,

$I_C = \left(\frac{V_0}{X_C}\right)\left[-\left(\frac{R}{Z}\right)\cos(\omega t + \phi) + \cos(\omega t)\right]$,

$I_L = \left(\frac{V_0}{X_L}\right)\left[\left(\frac{R}{Z}\right)\cos(\omega t + \phi) - \cos(\omega t)\right]$,

$Z = \sqrt{R^2 + \left(\frac{X_C X_L}{X_L - X_C}\right)^2}$,

$\tan\phi = \frac{X_C X_L}{(X_L - X_C)R}$.

1. 9.2×10^4 V/m · s.

3. 7.9×10^{14} V/m · s.

7. $\oint \mathbf{B} \cdot d\mathbf{A} = \mu_0 Q_m$,
$\oint \mathbf{E} \cdot d\boldsymbol{\ell} = \mu_0 \, dQ_m/dt - d\Phi_B/dt$.

9. 1.4×10^{-13} T.

11. (a) $B_0 = E_0/c$, $- y$-direction;
(b) $-z$-direction.

13. (a) 1.08 cm; (b) 3.0×10^{18} Hz.

15. 314 nm, ultraviolet.

17. (a) 4.3 min; (b) 71 min.

19. 1.77×10^{-6} W/m^2.

21. 7.82×10^{-7} J/h.

23. 4.50×10^{-6} J.

25. 3.8×10^{26} W.

29. $r < 3 \times 10^{-7}$ m.

31. 302 pF.

33. 2.59 nH $\leq L \leq$ 3.89 nH.

35. (a) 441 m; (b) 2.979 m.

37. 5.56 m, 0.372 m.

39. (a) 1.28 s; (b) 4.3 min.

43. 469 V/m.

45. Person at the radio hears the voice
0.14 s sooner.

47. (a) 0.40 W; (b) 12 V/m; (c) 12 V.

49. 1.5×10^{11} W.

51. (a) Parallel;
(b) 8.9 pF $\leq C \leq$ 80 pF;
(c) 1.05 mH $\leq L \leq$ 1.12 mH.

1. (a) 2.21×10^8 m/s;
(b) 1.99×10^8 m/s.

3. 8.33 min.

5. 3 m.

7. 3.4×10^3 rad/s.

9. I_3 is the desired image:

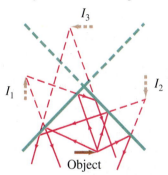

Depending on where you put your eye,
two other images may also be visible.

11. 5°.

15. 36.4 cm.

19. 4.5 m.

21. Convex, -20 m.

23. (a) Center of curvature; (b) real;
(c) inverted; (d) -1.

29. (a) Convex mirror; (b) 22 cm behind
surface; (c) -98 cm; (d) -196 cm.

33. 45.6°.

35. 24.9°.

37. 4.6 m.

43. 3.0%.

45. 0.22°.

47. 61.7°, lucite.

49. 93.5 cm.

51. $n_{\text{liquid}} \geq 1.5$.

55. 17.0 cm below the surface of the glass.

59. (a) 3.0 m, 4.0 m, 7.0 m; (b) toward you, away from you, toward you.

61. −3.80 m.

63. Chose different signs for the magnification; 13.3 cm, 26.7 cm.

65. ≥56.1°.

69. 81 cm (inside the glass).

CHAPTER 34

1. (a)

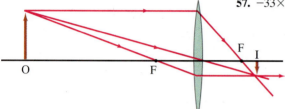

(b) 24.9 cm.

3. (a) 3.64 D, converging;
(b) −16.0 cm, diverging.

5. (a) −0.26 mm; (b) −0.47 mm;
(c) −1.9 mm.

7. (a) 81 mm; (b) 82 mm; (c) 87 mm;
(d) 24 cm.

9. (a) Virtual; (b) converging lens;
(c) 7.5 D.

11. (a) −10.5 cm (diverging), virtual;
(b) +203 cm (converging).

13. 22.9 cm, 53.1 cm.

15. Real and upright.

17. Real, 21.3 cm beyond second lens,
−0.708 (inverted).

19. (a) +7.14 cm; (b) −0.357 (inverted);
(c)

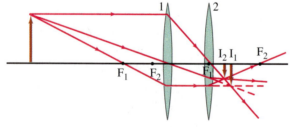

23. 1.54.

25. 8.1 cm.

27. −1.87 m (concave).

29. +1.15 D.

31. $f/2.3$.

35. 41 mm.

37. +2.3 D.

39. Glasses would be better.

41. (a) −1.33 D; (b) 38 cm.

43. −26.8 cm.

45. 17 cm, 100 cm.

47. 8.3 cm.

49. 6.3 cm from the lens, 3.9×.

51. (a) −49.4 cm; (b) 4.7×.

53. (a) 7.2×; (b) 2.2×.

55. 3.2 cm, 83 cm.

57. −33×.

59. 12×.

61. $f_o = 4.0$ m, $r = 8.0$ m.

63. 7.5×.

65. 1.7 cm.

67. (a) 0.85 cm; (b) 230×.

69. (a) 14.4 cm; (b) 137×.

71. (a) 15.9 cm; (b) 14.3 cm; (c) 1.6 cm;
(d) $r = 0.46$ cm.

73. $6.87 \text{ m} \leq d_o \leq \infty$.

75. 100 mm, 200 mm.

77. 79.4 cm, 75.5 cm.

79. 0.101 m, −2.7 m.

81. $0 < -d_o < -f$.

83. (c) $\Delta d = \sqrt{d_T^2 - 4d_T f}$,
$m = \left(\dfrac{d_T + \sqrt{d_T^2 - 4d_T f}}{d_T - \sqrt{d_T^2 - 4d_T f}} \right)^2$.

85. $1/f' = [(n/n') - 1][(1/R_1) + (1/R_2)]$;
$(1/d_o) + (1/d_i) = 1/f'$, where
$1/f' = [(n/n') - 1]/f(n - 1)$;
$m = -d_i/d_o$.

87. +3.6 D.

89. 2.9×, 4.1×.

91. (a) −2.5×; (b) 5.0 D.

93. −20×.

CHAPTER 35

3. 5.9 μm.

5. 3.9 cm.

7. 0.21 mm.

9. 613 nm.

11. 533 nm.

15. (a) $I_\theta/I_0 = (1 + 4\cos\delta + 4\cos^2\delta)/9$;
(b) $\sin\theta_{\max} = m\lambda/d$, $m = 0, \pm1, \pm2, \ldots$;
$\sin\theta_{\min} = (m + \frac{1}{3}k)\lambda/d$; $k = 1, 2$;
$m = 0, \pm1, \pm2, \ldots$.

17. Orange-red.

19. 179 nm.

21. 9.1 μm.

23. 120 nm, 240 nm.

25. 1.26.

29. 0.47, 0.23.

31. 0.221 mm.

33. 0.289 mm.

35. (a) 17 lm/W; (b) 156.

37. (a) Constructive; (b) destructive.

39. 464 nm.

41. 646 nm.

43. (a) 81.5 nm; (b) 127 nm.

45. 0.5 cm.

47. $\theta = 63.3°$.

49. $\sin\theta_{\max} = (m + \frac{1}{2})\lambda/2S$, $m = 0,1,2,\ldots$,
$\sin\theta_{\min} = m\lambda/2S$, $m = 0, 1, 2, \ldots$.

51. $I/I_0 = \cos^2(2\pi x/\lambda)$.

CHAPTER 36

1. 2.26°.

3. 2.4 m.

5. 5.8 cm.

7. 4.28 cm.

9. (b) Average intensity is 2×.

11. 10.7°.

13. $d = 4a$.

15. (a) 1.8 cm; (b) 11.0 cm.

17.

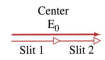

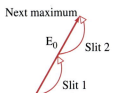

19. 2.4×10^{-7} rad $= (1.4 \times 10^{-5})°$
$= 0.050''$.

21. 820 lines/mm, 102 lines/mm.

23. 5.61°.

25. Two full orders.

27. 497 nm, 612 nm, 637 nm, 754 nm.

29. 600 nm to 750 nm of second order
overlaps with 400 nm to 500 nm of
third order.

31. 621 nm, 909 nm.

35. (a) Two orders; (b) 6.44×10^{-5} rad
$= 13.3''$, 7.36×10^{-5} rad $= 15.2''$,
2.52×10^{-4} rad $= 52.0''$.

37. (a) 1.60×10^4, 3.20×10^4;
(b) 0.026 nm, 0.013 nm.

39. $\Delta f = f/mN$.

41. (a) 62.0°; (b) 0.21 nm.

43. 0.033.

45. 57.3°.

47. (a) 35°; (b) 63°.

49. 36.9°, 53.1°.

51. $I_0/32$.

55. 31° on either side of the normal.

57. 12,500 lines/cm.

59. $\sin \theta = \sin 20° - (m\lambda/a)$,
$m = \pm1, \pm2, \ldots$.

61. Two orders.

63. 11.7°.

65. (a) 16 km; (b) 0.42'.

67. (a) 0; (b) $0.094 I_0$;
(c) no light gets transmitted.

69. (a) 30°; (b) 18°; (c) 5.7°.

73. 0.245 nm.

CHAPTER 37

1. (a) 1.00; (b) 0.99995; (c) 0.995;
(d) 0.436; (e) 0.141; (f) 0.0447.

3. 2.07×10^{-6} s.

5. $0.773c$.

7. $0.141c$.

9. (a) 99.0 yr; (b) 27.7 yr; (c) 26.6 ly;
(d) $0.960c$.

11. $0.89c = 2.7 \times 10^8$ m/s.

13. (a) (470 m, 20 m, 0);
(b) (1820 m, 20 m, 0).

15. (a) $0.80c$; (b) $-0.80c$.

17. 60 m/s, 24°.

19. 2.7×10^8 m/s, 43°.

21. (a) $L_0 \sqrt{1 - (v/c)^2 \cos^2\theta}$;
(b) $\tan \theta' = \tan \theta / \sqrt{1 - (v/c)^2}$.

23. Not possible in the boy's frame.

25. $0.866c$.

27. (a) 0.5%; (b) 13%.

29. 5.36×10^{-13} kg.

31. 8.20×10^{-14} J, 0.511 MeV.

33. 9×10^2 kg.

37. (a) 11.2 GeV (1.79×10^{-9} J);
(b) 6.45×10^{-18} kg·m/s.

39. 7.49×10^{-19} kg·m/s.

41. $0.941c$.

43. $M = 2m/\sqrt{1 - (v^2/c^2)}$,
$K_{loss} = 2mc^2\{[1/\sqrt{1 - (v^2/c^2)}] - 1\}$.

45. $0.866c$, 4.73×10^{-22} kg·m/s.

47. 39 MeV (6.3×10^{-12} J),
1.5×10^{-19} kg·m/s, -6%, -4%.

49. $0.804c$.

51. 3.0 T.

57. (a) $0.80c$; (b) $2.0c$.

61. 3.8×10^{-5} s.

63. $\rho = \rho_0/[1 - (v^2/c^2)]$.

65. (a) 1.5 m/s less than c; (b) 30 cm.

67. 1.02 MeV (1.64×10^{-13} J).

69. 2.2 mm.

71. 0.78 MeV.

73. Electron.

75. 5.19×10^{-13}%.

81. (a) $\alpha = \tan^{-1}[(c^2/v^2) - 1]^{1/2}$;
(c) $\tan \theta = c/v$, $u = \sqrt{v^2 + c^2}$.

CHAPTER 38

1. 6.59×10^3 K.

3. 5.4×10^{-20} J, 0.34 eV.

5. (a) 114 J; (b) 228 J; (c) 342 J;
(d) $114n$ J; (e) -456 J.

7. (b) $h = 6.63 \times 10^{-34}$ J·s.

9. 3.67×10^{-7} eV.

11. 2.4×10^{13} Hz, 1.2×10^{-5} m.

13. 400 nm.

15. (a) 2.18 eV; (b) 0.92 V.

17. 3.46 eV.

19. 1.88 eV, 43.3 kcal/mol.

21. (a) 2.43×10^{-12} m;
(b) 1.32×10^{-15} m.

23. (a) 5.90×10^{-3}, 1.98×10^{-2},
3.89×10^{-2};
(b) 60.8 eV, 204 eV, 401 eV.

27. 1.82 MeV.

29. 212 MeV, 5.85×10^{-15} m.

31. 3.2×10^{-32} m.

33. 19 V.

35. (a) 0.39 nm; (b) 0.12 nm;
(c) 0.039 nm.

37. 1.84×10^3.

39. 3.3×10^{-38} m/s, no diffraction.

41. 0.026 nm.

43. 3.4 eV.

45. 122 eV.

49. 52.5 nm.

51.

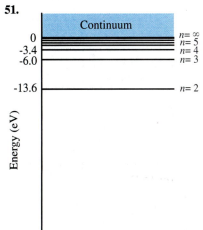

53. $U = -27.2$ eV, $K = +13.6$ eV.

55. Justified.

61. 3.28×10^{15} Hz.

65. 2.78×10^{21} photons/s·m².

67. 8.3×10^6 photons/s.

69. $\theta = 89.4°$.

71. 4.7×10^{-14} m.

73. 10.2 eV.

75. 4.4×10^{-40}, yes.

77. 653 nm, 102 nm, 122 nm.

79. 0.64 V.

83. 5×10^{-12} m.

85. 3.

CHAPTER 39

1. 1.8×10^{-7} m.

3. $\pm 1.3 \times 10^{-11}$ m.

5. 7.2×10^3 m/s.

7. 2.4×10^{-3} m, 1.4×10^{-32} m.

9. (a) 6.6×10^{-8} eV; (b) 6.5×10^{-9}; (c) 7.9×10^{-7} nm.

13. (a) $\psi = A \sin (3.5 \times 10^9 \text{ m}^{-1})x + B \cos (3.5 \times 10^9 \text{ m}^{-1})x$; (b) $\psi = A \sin (6.3 \times 10^{12} \text{ m}^{-1})x + B \cos (6.3 \times 10^{12} \text{ m}^{-1})x$.

17. 3.6×10^6 m/s.

19. (a) 52 nm; (b) 0.22 nm.

23. $E_1 = 0.094$ eV,
$\psi_1 = (1.00 \text{ nm}^{-1/2}) \sin (1.57 \text{ nm}^{-1} x)$;
$E_2 = 0.38$ eV,
$\psi_2 = (1.00 \text{ nm}^{-1/2}) \sin (3.14 \text{ nm}^{-1} x)$;
$E_3 = 0.85$ eV,
$\psi_3 = (1.00 \text{ nm}^{-1/2}) \sin (4.71 \text{ nm}^{-1} x)$;
$E_4 = 1.51$ eV,
$\psi_4 = (1.00 \text{ nm}^{-1/2}) \sin (6.28 \text{ nm}^{-1} x)$.

25. (a) 4 GeV; (b) 2 MeV; (c) 2 MeV.

27. (a) 0.18; (b) 0.50; (c) 0.50.

29.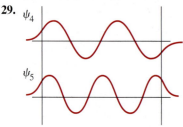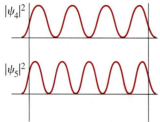

31. 0.03 nm.

33. 9.2 eV.

35. (a) Decreases by 8%; (b) decreases by 5%.

37. (a) 32 MeV; (b) 56 fm; (c) 8.8×10^{20} s^{-1}, 10^{10} yr.

39. 21 MeV.

41. 3.00×10^{-10} eV/c^2.

43. r_1 , the Bohr radius.

45. 0.5 MeV, 5×10^6 m/s.

47. (b) 4 s.

49. 14% decrease.

CHAPTER 40

1. $\ell = 0, 1, 2, 3, 4, 5.$

3. 32 states, $(4, 0, 0, -\frac{1}{2})$, $(4, 0, 0, +\frac{1}{2})$, $(4, 1, -1, -\frac{1}{2})$, $(4, 1, -1, +\frac{1}{2})$, $(4, 1, 0, -\frac{1}{2})$, $(4, 1, 0, +\frac{1}{2})$, $(4, 1, 1, -\frac{1}{2})$, $(4, 1, 1, +\frac{1}{2})$, $(4, 2, -2, -\frac{1}{2})$, $(4, 2, -2, +\frac{1}{2})$, $(4, 2, -1, -\frac{1}{2})$, $(4, 2, -1, +\frac{1}{2})$, $(4, 2, 0, -\frac{1}{2})$, $(4, 2, 0, +\frac{1}{2})$, $(4, 2, 1, -\frac{1}{2})$, $(4, 2, 1, +\frac{1}{2})$, $(4, 2, 2, -\frac{1}{2})$, $(4, 2, 2, +\frac{1}{2})$, $(4, 3, -3, -\frac{1}{2})$, $(4, 3, -3, +\frac{1}{2})$, $(4, 3, -2, -\frac{1}{2})$, $(4, 3, -2, +\frac{1}{2})$, $(4, 3, -1, -\frac{1}{2})$, $(4, 3, -1, +\frac{1}{2})$, $(4, 3, 0, -\frac{1}{2})$, $(4, 3, 0, +\frac{1}{2})$, $(4, 3, 1, -\frac{1}{2})$, $(4, 3, 1, +\frac{1}{2})$, $(4, 3, 2, -\frac{1}{2})$, $(4, 3, 2, +\frac{1}{2})$, $(4, 3, 3, -\frac{1}{2})$, $(4, 3, 3, +\frac{1}{2})$.

5. $n \geq 5, m_\ell = -4, -3, -2, -1, 0, 1, 2, 3, 4,$ $m_s = -\frac{1}{2}, +\frac{1}{2}.$

7. (a) 6; (b) -0.378 eV; (c) $\ell = 4$, $\sqrt{20}\hbar = 4.72 \times 10^{-34}$ kg$\cdot$m^2/s; (d) $m_\ell = -4, -3, -2, -1, 0, 1, 2, 3, 4.$

13. (a) $-[3/(32\pi r_0^3)^{1/2}] e^{-5/2}$; (b) $(9/32\pi r_0^3) e^{-5}$; (c) $(225/8 r_0) e^{-5}$.

15. 1.85.

17. 17.3%.

19. (a) $1.34 r_0$; (b) $2.7 r_0$; (c) $4.2 r_0$.

21. $\dfrac{r^4}{24 r_0^5} e^{-r/r_0}$.

27. (a) $\dfrac{4r^2}{27 r_0^3}\left(1 - \dfrac{2r}{3 r_0} + \dfrac{2r^2}{27 r_0^2}\right)^2 e^{-2r/3 r_0}$;

(b)

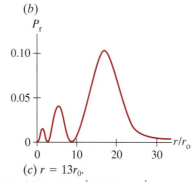

(c) $r = 13 r_0$.

29. (a) $(1, 0, 0, -\frac{1}{2})$, $(1, 0, 0, +\frac{1}{2})$, $(2, 0, 0, -\frac{1}{2})$, $(2, 0, 0, +\frac{1}{2})$, $(2, 1, -1, -\frac{1}{2})$, $(2, 1, -1, +\frac{1}{2})$; (b) $(1, 0, 0, -\frac{1}{2})$, $(1, 0, 0, +\frac{1}{2})$, $(2, 0, 0, -\frac{1}{2})$, $(2, 0, 0, +\frac{1}{2})$, $(2, 1, -1, -\frac{1}{2})$, $(2, 1, -1, +\frac{1}{2})$, $(2, 1, 0, -\frac{1}{2})$, $(2, 1, 0, +\frac{1}{2})$, $(2, 1, -1, -\frac{1}{2})$, $(2, 1, -1, +\frac{1}{2})$, $(3, 0, 0, -\frac{1}{2})$, $(3, 0, 0, +\frac{1}{2})$.

31. (a) $1s^2 2s^2 2p^6 3s^2 3p^6 3d^{10} 4s^2 4p^4$; (b) $1s^2 2s^2 2p^6 3s^2 3p^6 3d^{10} 4s^2 4p^6 4d^{10} 4f^{14} 5s^2 5p^6 5d^{10} 6s^1$; (c) $1s^2 2s^2 2p^6 3s^2 3p^6 3d^{10} 4s^2 4p^6 4d^{10} 4f^{14} 5s^2 5p^6 5d^{10} 6s^2 6p^6 5f^3 6d^1 7s^2$.

33. 5.8×10^{-13} m, 0.115 MeV.

37. 0.041 nm, 1 nm.

41. 0.19 nm.

43. Chromium.

47. (a) 0.25 mm; (b) 0.13 mm.

49. (a) $\frac{1}{2}, \frac{3}{2}, \frac{1}{2}\sqrt{3}\,\hbar, \frac{1}{2}\sqrt{15}\hbar$; (b) $\frac{5}{2}, \frac{7}{2}, \frac{1}{2}\sqrt{35}\hbar, \frac{1}{2}\sqrt{63}\hbar$; (c) $\frac{3}{2}, \frac{5}{2}, \frac{1}{2}\sqrt{15}\hbar, \frac{1}{2}\sqrt{35}\hbar$.

51. (a) 0.4 T; (b) 0.4 T.

53. 5.6×10^{-4} rad, 1.7×10^2 m.

55. 3.7×10^4 K.

57. (a) 1.56; (b) 1.4×10^{-10} m.

59. (a) $1s^2 2s^2 2p^6 3s^2 3p^6 3d^7 4s^2$; (b) $1s^2 2s^2 2p^6 3s^2 3p^6 3d^{10} 4s^2 4p^6$; (c) $1s^2 2s^2 2p^6 3s^2 3p^6 3d^{10} 4s^2 4p^6 5s^2$.

61. (a) 2.5×10^{74}; (b) 5.0×10^{74}.

63. $r = 5.24 r_0$.

65. (a) ϕ is unknown; (b) L_x and L_y are unknown.

67. (a) 1.2×10^{-4} eV; (b) 1.1 cm; (c) no difference.

69. (a) 3×10^{-171}, 7×10^{-203}; (b) 1.1×10^{-8}, 6.3×10^{-10}; (c) 6.6×10^{15}, 3.8×10^{14}; (d) 7×10^{23} photons/s, 4×10^{22} photons/s.

71. 182.

73. 2.25.

CHAPTER 41

1. 5.1 eV.

3. 0.7 eV.

7. (a) 13.941 u; (b) 7.0034 u; (c) 0.9801 u.

9. (a) 6.86 u; (b) 1.85×10^3 N/m.

11. (a) 1.79×10^{-4} eV; (b) 7.16×10^{-4} eV, 1.73 mm.

13. 2.36×10^{-10} m.

15. -7.9 eV.

17. 0.283 nm.

19. (b) -6.9 eV; (c) -10.8 eV; (d) 3%.

21. 1.8×10^{21}.

23. (a) 6.9 eV; (b) 6.8 eV.

25. 6.3%.

27. 3.2 eV, 1.05×10^6 m/s.

29. (a) 1.79×10^{29} m^{-3}; (b) 3.

33. (a) 0.021, reasonable; (b) 0.979; (c) 0.021.

37. A large energy is required to create a conduction electron by raising an electron from the valence band to the conduction band.

39. 1.1 μm.

41. 5×10^6.

43. 1.91 eV.

45. 13 mA.

47. (a) 2.1 mA; (b) 4.3 mA.

49. (a) 9.4 mA (smooth); (b) 6.7 mA (rippled).

51. 4.0 kΩ.

53. 0.43 mA.

55. (a) 3.5×10^4 K; (b) 1.2×10^3 K.

57. (a) 0.9801 u; (b) 482 N/m, 88% of the constant for H_2.

59. States with higher values of L are less likely to be occupied, so less likely to absorb a photon; I will depend on L.

61. 2.8×10^4 J/kg.

63. 1.24 eV.

65. 1.09 μm, could be used.

67. (a) 0.094 eV; (b) 0.63 nm.

69. (a) 145 V $\leq V \leq$ 343 V; (b) 3.34 k$\Omega \leq R_{load} < \infty$.

CHAPTER 42

1. 3729 MeV/c^2.

3. 1.9×10^{-15} m.

5. (a) 2.3×10^{17} kg/m^3; (b) 184 m; (c) 2.6×10^{-10} m.

7. 28 MeV.

9. $^{31}_{15}$P.

11. (a) 1.8×10^3 MeV; (b) 7.3×10^2 MeV.

13. 7.48 MeV.

15. (a) 32.0 MeV, 5.33 MeV; (b) 1636 MeV, 7.87 MeV.

17. 12.4 MeV, 7.0 MeV, neutron is more closely bound in ^{23}Na.

19. (b) Stable.

21. 0.782 MeV.

23. β^+ emitter, 1.82 MeV.

25. $^{228}_{90}$Th, 228.02883 u.

27. (a) $^{32}_{16}$S; (b) 31.97207 u.

29. 0.862 MeV.

31. 5.31 MeV.

33. (b) 0.961 MeV, 0.961 MeV to 0.

35. 3.0 h.

37. 1.2×10^9 decays/s.

39. 1.78×10^{20} nuclei.

41. 7 α particles, 4 β^- particles.

43. (a) 4.8×10^{16} nuclei; (b) 3.2×10^{15} nuclei; (c) 7.2×10^{13} decays/s; (d) 26 min.

45. 1.68×10^{-10} g.

47. 2.6 min.

49. 3.4 mg.

51. 8.6×10^{-7}.

53. $^{211}_{82}$Pb.

55. $N_D = N_0(1 - e^{-\lambda t})$.

57. (a) 0.99946; (b) 1.2×10^{-14}; (c) 2.3×10^{17} kg/m^3, $10^{14} \times$.

59. 28.6 eV.

61. (a) 7.2×10^{55}; (b) 1.2×10^{29} kg; (c) 3.2×10^{11} m/s^2.

63. 6×10^3 yr.

65. 6.64 $T_{1/2}$.

69. Calcium, stored by body in bones, 193 yr, $^{90}_{38}$SR $\rightarrow$ $^{90}_{39}$Y $+$ $^{0}_{-1}$e $+ \bar{\nu}$, $^{90}_{39}$Y is radioactive, $^{90}_{39}$Y $\rightarrow$ $^{90}_{40}$Zr $+$ $^{0}_{-1}$e $+ \bar{\nu}$, $^{90}_{40}$Zr is stable.

73. (a) 0.002603 u, 2.425 MeV/c^2; (b) 0; (c) -0.094909 u, -88.41 MeV/c^2; (d) 0.043924 u, 40.92 MeV/c^2; (e) $\Delta \geq 0$ for $0 \leq Z \leq 8$ and $Z \geq 85$, $\Delta < 0$ for $9 \leq Z \leq 84$.

75. 0.083%.

77. $^{228}_{88}$RA, $^{228}_{89}$Ac, $^{228}_{90}$Th, $^{224}_{88}$Ra, $^{220}_{87}$Rn, $^{231}_{90}$Th, $^{231}_{91}$Pa, $^{227}_{89}$Ac, $^{227}_{90}$Th, $^{223}_{88}$Ra.

79. (b) $\approx 10^{17}$ yr; (c) ≈ 60 yr; (d) 0.4.

CHAPTER 43

1. $^{28}_{13}$Al, β^- emitter, $^{28}_{14}$Si.

3. Possible.

5. 5.701 MeV is released.

7. (a) Can occur; (b) 19.85 MeV.

9. $+4.730$ MeV.

11. (a) $^{7}_{3}$Li; (b) neutron is stripped from the deuteron; (c) $+5.025$ MeV, exothermic.

13. (a) $^{31}_{15}$P(p, γ)$^{32}_{16}$S; (b) $+8.864$ MeV.

15. $\sigma = \pi(R_1 + R_2)^2$.

17. Rate at which incident particles pass through target without scattering.

19. (a) 0.7 μm; (b) 1 mm.

21. 173.2 MeV.

23. 0.116 g.

25. 25.

27. 0.11.

29. 1.3 keV.

33. 6.1×10^{23} MeV/g, 4.9×10^{23} MeV/g, 2.1×10^{24} MeV/g, 5.1×10^{23} MeV/g.

35. Not independent.

37. 3.23×10^9 J, $65 \times$.

39. (a) $4.9 \times$; (b) 1.5×10^9 K.

41. 400 rad.

43. 167 rad.

45. 1.7×10^2 counts/s.

49. 0.225 μg.

51. (a) 0.03 mrem $\approx$ 0.006% of allowed dose; (b) 0.3 mrem $\approx$ 0.06% of allowed dose.

53. (a) 1; (b) $1 \leq m \leq 2.7$.

55. (a) $^{12}_{6}$C; (b) $+5.70$ MeV.

57. 1.004.

59. 51 mrem/yr.

61. 4.6 m.

63. 18.000953 u.

65. 6.31×10^{14} J/kg, $\approx 10^7 \times$ the heat of combustion of coal.

67. 1×10^{24} neutrinos/yr.

69. (a) 6.8 bn; (b) 3×10^{-14} m.

71. (a) 3.7×10^3 decays/s; (b) 5.2×10^{-4} Sv/yr ≈ 0.15 background.

CHAPTER 44

1. 7.29 GeV.

3. 1.8 T.

5. 13 MHz.

7. $\lambda_\alpha = 2.6 \times 10^{-15}$ m $\approx$ size of nucleon, $\lambda_p = 5.2 \times 10^{-15}$ m $\approx$ 2(size of nucleon), α particle is better.

9. 1.4×10^{-18} m.

11. 2.2×10^6 km, 7.5 s.

15. 33.9 MeV.

17. 1.879 GeV.

19. 67.5 MeV.

21. 2.3×10^{-18} m.

23. (b) Uncertainty principle allows energy to not be conserved.

25. 69.3 MeV.

27. 8.6 MeV, 57.4 MeV.

29. 52.3 MeV.

31. 7.5×10^{-21} s.

33. (a) 1.3 keV; (b) 8.9 keV.

35. (a) n = d d u; (b) $\bar{n} = \bar{d} \bar{d} \bar{u}$; (c) Λ^0 = u d s; (d) $\Sigma^0 = \bar{u} \bar{d} \bar{s}$.

37. $D^0 = c\bar{u}$.

39. (*a*)

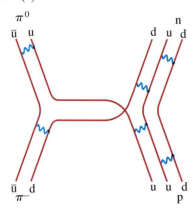

(*b*)

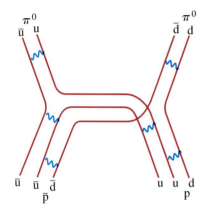

41. 26 GeV, 4.8×10^{-17} m.

43. 5.5 T.

45. (*a*) Possible, through the strong interaction; (*b*) possible, through the strong interaction; (*c*) possible, through the strong interaction; (*d*) forbidden, charge is not conserved; (*e*) possible, through the weak interaction; (*f*) forbidden, charge is not conserved; (*g*) possible, through the strong interaction; (*h*) forbidden, strangeness is not conserved; (*i*) possible, through the weak interaction; (*j*) possible, through the weak interaction.

49. -135.0 MeV, -140.9 MeV.

51. 64.

53. (*b*) 10^{27} K.

55. 6.58×10^{-5} m.

57. $\bar{u}\,\bar{u}\,\bar{d} + u\,d\,d \rightarrow \bar{u}\,d + d\,\bar{d}$.

CHAPTER 45

3. 4.8 ly.

5. 0.059″, 17 pc.

7. Less, $\phi_1/\phi_2 = \frac{1}{2}$.

9. 48 W/m².

11. 1.4×10^{-4} kg/m³.

13. 1.8×10^9 kg/m³, 3.3×10^5.

15. -92.2 keV, 7.366 MeV.

17. (*a*) 9.594 MeV is released; (*b*) 7.6 MeV; (*c*) 6×10^{10} K.

19. $d_2/d_1 = 6.5$.

23. 540°.

25. 1.4×10^8 ly.

27. 0.86*c*.

29. 0.328*c*, 4.6×10^9 ly.

31. 1.1 mm.

33. (*a*) 10^{-3}; (*b*) 10^{-10}; (*c*) 10^{-13}; (*d*) 10^{-27}.

35. (*a*) Temperature increases, luminosity is constant, size decreases; (*b*) temperature is constant, luminosity decreases, size decreases; (*c*) temperature decreases, luminosity increases, size increases.

37. 8×10^3 rev/s.

39. 7×10^{24} W.

41. (*a*) 46 eV; (*b*) 18 eV.

43. $\approx (2 \times 10^{-5})°$.

45. 1.4×10^{16} K, hadron era.

47. Venus is brighter, $\ell_V/\ell_S = 16$.

49. $R \geq GM/c^2$.

51. $R = \dfrac{h^2}{16m_n^{8/3}GM^{1/3}}\left(\dfrac{18}{\pi^2}\right)^{2/3}$.

Index

Amorphous solids, 1044
Ampère, Andrè, 637, 712, 722, 787
Ampère (unit), 637, 712
  operational definition of, 712
Ampère's law, 712–15, 788–90
Amplifier, 1054–55
Amplitude:
  pressure, 420
  of vibration, 363
  of wave, 390, 396, 419, 422
Amplitude modulation (AM), 804
Analog information, 749
Analyzer (of polarized light), 908
Andromeda, 1143
Aneroid gauge, 338–39
Angle:
  Brewster's, 910
  of dip, 688
  of incidence, 403, 408, 813, 822
  phase, 366, 398, 778
  polarizing, 910
  radian measure of, 240
  of reflection, 403, 813
  of refraction, 408–9, 822
Angstrom (unit), 824 *fn*
Angular acceleration, 241–42
Angular displacement, 246–47
Angular frequency, 366
Angular magnification, 853
Angular momentum, 256–60, 281–89
  in atoms, 965, 1005–7, 1015–18
  conservation, law of, 257–59,
      288–89
  of nuclei, 1064
  total, 1017–18
  vector, nature of, 259–60, 281–90
Angular position, 240
Angular quantities, 240–43, 246–47
Angular velocity, 241–44
Annihilation, 957, 1124, 1163
Anode, 605
Antenna, 787, 805, 876–77
Anthropic principle, 1169
Antilock brakes, 119
Antimatter, 1137 *pr*, 1163 *fn*
Antineutrino, 1071
Antineutron, 1124
Antinodes, 405, 427
Antiparticles, 1072, 1124–28, 1163 *fn*
Antiproton, 1124
Antiquark, 1130–31
Apparent brightness (stars and
    galaxies), 1145
Apparent magnitude, 1172 *pr*

Apparent weightlessness, 142
Approximate calculations, 9–12
Arago, F., 887
Arch, 319–21
Archimedes, 340–41
Archimedes' principle, 340–43
Area:
  formulas for, A-1
  under a curve or graph, 161–64
Aristotle, 2, 78–79
Armature, 698, 744
Asymptotic freedom, 1133
Astigmatism, 851
Astronomical telescope, 854–56
Astrophysics, 1140 *ff*
Atmosphere (unit), 337–40
Atmosphere, scattering of light by,
    911
Atmospheric pressure, 336–40
Atomic bomb, 1095
Atomic force microscope, 999
Atomic mass, 446
Atomic mass number, 1062
Atomic mass unit, 7, 446, 1063
Atomic number, 1010, 1012–15,
    1062
Atomic spectra, 863–65
Atomic structure:
  Bohr model of, 965–71, 977, 1003–5
  early models of, 862–63
  quantum mechanics of, 1003–24
Atomic weight, 446 *fn*
Atoms, 446–47, 965–73 (*see also*
    Atomic structure, Kinetic theory)
  angular momentum in, 967
  binding energy in, 968
  complex, 1010–15
  electric charge in, 547
  energy levels in, 967
  hydrogen, 964–71, 1004–10
  ionization energy in, 968, 970
  probability distribution in, 1004,
      1008–10
  shells and subshells in, 1012–13
  vector model of, 1028 *pr*
ATP, 1036
Atwood machine, 92, 277 *pr*, 286
Audible range, 418
Aurora borealis, 694
Autoclave, 482 *pr*
Autofocusing camera, 419
Average binding energy per nucleon,
    1065
Average lifetime, 1084 *pr*

Average speed, 18
Avogadro, Amedeo, 459
Avogadro's hypothesis, 459
Avogadro's number, 459
Axis, instantaneous, 246
Axis of lens, 837
Axis of rotation (*defn*), 240

B

Back, forces in, 331 *pr*
Back emf, 742–43
Background radiation, cosmic
    microwave, 1159–61, 1164
Balance, 308
Ballistic galvanometer, 755 *pr*
Ballistic pendulum, 218
Balmer, J. J., 964
Balmer formula, 964, 969
Balmer series, 964, 968–69
Band gap, 1049–50
Band spectra, 1038
Band theory of solids, 1048–50
Banking of curves, 118–20
Bar (unit), 337
Bar-code reader, 1022
Barn (unit), 1089
Barometer, 339
Barrier penetration, 996–99,
    1069–70, 1155 *fn*, 1168
Baryon (*defn*), 1127
Baryon number, 1124–26, 1132, 1162
Base (of transistor), 1054
Base and derived quantities and
    units, 8
Battery, 635–36, 659, 668
Beam splitter, 881
Beat frequency, 431
Beats, 429–32
Becquerel, H., 1067
Becquerel (unit), 1101
Bel, 421
Bell, Alexander Graham, 421
Bernoulli, Daniel, 345–46
Bernoulli's equation, 345–47
Bernoulli's principle, 345–49
Beta decay, 1067, 1070–73, 1125
Beta particle, 1067, 1070
Betatron, 754 *pr*
Biasing and bias voltage, 1052,
    1054–55
Big Bang, 1136, 1159–64
Big crunch, 1165

Bimetallic-strip thermometer, 448
Binary system, 1146, 1155
Binding energy:
  in atoms, 968
  in molecules, 1031, 1033, 1036
  in nuclei, 1064–66
  in solids, 1044
  total, 1064
Binding energy per nucleon, 1065
Binoculars, 827, 856
Binomial expansion, A-1
Biological damage by radiation,
  1101–2
Biot, Jean Baptist, 719
Biot-Savart law, 719–21
Blackbody, 950
Blackbody radiation, 950–51, 1160
Black dwarf, 1149
Black hole, 149, 1144, 1151, 1155,
  1158, 1166, 1168
Blood flow, 344–45
Blueshift, 1156
Blue sky, 911
Bohr, Niels, 958, 965, 969, 985
Bohr magneton, 1016, 1064
Bohr model of the atom, 965–71,
  977, 1003–5
Bohr radius, 590 pr, 966, 1004
Boiling, 475–76, 490–93 (see also
  Phase, changes of)
Boiling point, 475–76
Boltzmann, Ludwig, 535–36
Boltzmann's constant, 459
Boltzmann distribution, 1020
Boltzmann factor, 1020, 1046
Bomb calorimeter, 490
Bond:
  covalent, 1030–33, 1044
  dipole, 1036–37
  hydrogen, 1036
  ionic, 1031–33, 1044
  metallic, 1044
  molecular, 1030–33
  partially covalent/ionic, 1032–33,
    1044
  strong (defn), 1036
  van der Waals, 1036–37
  weak, 1036–37, 1045
Bond energy, 1031, 1036
Bonding:
  in molecules, 1030–33
  in solids, 1044–45
Bone density, 957
Bose, S. N., 1011 fn

Bose-Einstein statistics, 1046 fn
Bosons, 1011 fn, 1046 fn, 1126–27,
  1132–34
Bottomness and bottom quark,
  1130–31
Boundary conditions, 990, 995
Bound charge, 625
Bound state, 994
Bow wave, 436
Boyle, Robert, 455
Boyle's law, 455–56, 467
Bragg, W. H., 906
Bragg, W. L., 906
Bragg equation, 906
Brahe, Tycho, 143
Brake, hydraulic, 338
Braking distance of car, 29, 167
Brayton cycle, 544 pr
Breakdown voltage, 596
Break-even (fusion), 1099
Breaking point, 309
Breeder reactor, 1094
Bremsstrahlung, 1014–15
Brewster, D., 910
Brewster's angle and law, 910
Bridges, 315–18
Brightness, apparent, 1145
British system of units, 8
Broglie, Louis de (see de Broglie)
Bronchoscope, 827
Brown, Robert, 446
Brown dwarfs, 1166
Brownian movement, 446
Brushes, 698, 740
BSCCO, 650
Btu, 486
Bubble chamber, 1120, 1124
Bulk modulus, 310, 312
Buoyancy, 340–43
Buoyant force, 340–43
Burglar alarm, 954

C

Calculator errors, 5
Caloric theory, 486
Calorie (unit), 486–87
  related to the joule, 486
Calorimeter, 489–90
Calorimetry, 489–90
Camera, 848–50
  autofocusing, 419
  gamma, 1104

Camera flash unit, 621
Cancer, 1101, 1104
Candela (unit), 882
Cantilever, 304–5
Capacitance, 614–21
Capacitance bridge, 630 pr
Capacitive reactance, 775
Capacitor discharge, 672
Capacitor microphone, 677
Capacitors, 613–21
  charging of, 789–90
  in circuits, 669–73, 764–65, 774 ff
  as filter, 776
  reactance of, 775
  in series and parallel, 617–20
  uses of, 776
Capacity (see Capacitance)
Capillaries, 352
Capillarity, 351–52
Car, stopping distance, 29, 167
Carbon cycle, 1097, 1112 pr, 1148
Carbon dating, 1061, 1078–79
Carburetor of car, 348
Carnot, S., 520
Carnot cycle, 520–25
Carnot efficiency, 522–23
Carnot engine, 520–25
Carnot's theorem, 522–23
Carrier frequency, 804
Carrier of forces, 1121 ff
Cassegrainian focus, 856
Cathedrals, 319
Cathode, 605
Cathode rays, 605, 699 (see also
  Electron)
Cathode ray tube (CRT), 605–6, 700,
  805
CAT scan, 1105–6
Causality, 146
Cavendish, Henry, 135, 137, 148
CD player, 1022, 1054
Cell, electric, 635
Cellular telephone, 806
Celsius temperature scale, 448–49
Center of buoyancy, 357 pr
Center of gravity (CG), 223
Center of mass (CM), 221–24
  for human body, 221–22
  translational motion and, 225–27,
    262–65
Centigrade (see Celsius temperature
  scale)
Centipoise (unit), 350
Centrifugal (pseudo) force, 115, 291

Copier, electrostatic, 555
Core, of reactor, 1093
Coriolis force, 292–94
Cornea, 850
Corrective lenses, 850–52
Correspondence principle, 943, 971, 976 *pr*, 978
Cosmic microwave background radiation, 1159–61, 1164
Cosmic rays, 1102, 1114
Cosmological principle, 1157 *ff*
"perfect," 1159 *fn*
Cosmology, 1136, 1140–69
Coulomb, Charles, 549–50
Coulomb (unit), 550, 712
operational definition of, 712
Coulomb barrier, 998, 1098, 1148
Coulomb's law, 549–53, 575–86, 966
vector form of, 553
Coulomb potential, 1034
Counter emf, 742–43
Counter torque, 743
Covalent bond, 1030–33, 1044
Crane, 304
Creativity in science, 3
Crick, F.H.C., 906
Critical damping, 376
Critical density, of universe, 1165–68
Critical mass, 1092–95
Critical point, 473
Critical temperature, 473
Crossed Polaroids, 908
Cross product, 279–80
Cross section, 1088–90
CRT, 605–6, 700, 805
Crystal lattice, 446–47, 1044
Crystallography, 906
CT scan, 1105–6
Curie, M. and P., 726, 1067
Curie (unit), 1101
Curie temperature, 722
Curie's law, 726
Current (*see* Electric current)
Current density, 647–49
Current gain, 1055
Current sensitivity, 674
Curvature of field, 859
Curvature of universe (space-time), 149, 1151–55, 1165
Cycle (*defn*), 363
Cyclic universe, 1168
Cyclotron, 707 *pr*, 1116–17
Cyclotron frequency, 694, 1116

D

Damped harmonic motion, 374–77
Dark matter, 1166
Dating, radioactive, 1078–79
Daughter nucleus (*defn*), 1068
Davisson, C. J., 960
dB (unit), 421
DC circuits, 658–77
DC generator, 741
DC motor, 698
de Broglie, Louis, 959–61, 971, 977–78
de Broglie wavelength, 959, 971–72
Debye's law, 514 *pr*
Decay:
of elementary particles, 1124–34
radioactive (*see* Radioactivity)
Decay constant, 1073–76
Decay series, 1077–78
Deceleration (*defn*), 24
Deceleration parameter, 1168
Decibel (db), 421
Declination, magnetic, 688
Decommissioning nuclear plant, 1094
Dee, 1116
Defects of the eye, 851
Defibrillator, 632 *pr*, 651 *fn*
Degrees of freedom, 500–501
Del, 603 *fn*
Delayed neutron, 1093
Delta particle, 1128
Demagnetization, 725
Density, 332–33
and floating, 342
Density of occupied states, 1047
Density of states, 1045–48
Depth of field, 849
Derivatives, 21–22, A-4–A-5
partial, 399
Derived quantities, 8
Descartes, R., 146
Destructive interference, 404, 430, 871, 1031
Detection of radiation, 1080
Determinism, 147, 984–85
Deuterium, 1086, 1092, 1095–1100
Deuteron, 1086
Dew point, 476
Diamagnetism, 725–26
Diamond, 826–27
Dielectric constant, 621–22
Dielectrics, 621–26
molecular description of, 624–26

Dielectric strength, 622
Diesel engine, 540 *pr*
Differential cross section, 1089
Diffraction, 867, 887–911, 1020
by circular opening, 896–97
as distinguished from interference, 896
in double slit experiment, 896
of electrons, 960
Fraunhofer, 888 *fn*
Fresnel, 888 *fn*
of light, 867, 887–911
as limit to resolution, 896–97
by single slit, 888–93
of water waves, 410
X-ray, 905–6
Diffraction grating, 900–905
resolving power of, 903–5
Diffraction limit of lens resolution, 896–97
Diffraction patterns, 888 *ff*
of circular opening, 896–97
of single slit, 888–93
X-ray, 905–6
Diffraction spot or disc, 896
Diffuse reflection, 813
Diffusion, 479–80
Fick's law of, 480
Diffusion constant, 480
Diffusion equation, 480
Digital information, 749
Digital voltmeter, 675
Dimensional analysis, 12–13
Dimensions, 12–13
Diodes, 1052–54
Diopter, 838
Dip, angle of, 688
Dipole antenna, 793
Dipole bonds, 1036–37
Dipoles and dipole moments:
of atoms, 1015–18
electric, 565–67, 578, 601–3
magnetic, 695–97, 721
Dirac, P. A. M., 1005, 1046 *fn*
Direct current (dc) (*see* Electric current)
Discharge tube, 963
"Discovery" in science, 700
Disintegration energy (*defn*), 1068–69
Disorder and order, 533–34
Dispersion, 402, 824–25
Displacement, 17–18, 46–47
angular, 246–47

Induction, electromagnetic, 734 *ff*
Induction, Faraday's law, 736–38, 792
Inductive reactance, 774
Inductor, 758–59
    in circuits, 758–63, 772–83
    energy stored in, 760–61
    reactance of, 774
Inelastic collision, 214, 217–18
Inelastic scattering, 1089
Inertia, 79
    law of, 78–79
    moment of, 249–56, 374
    rotational (*defn*), 249–50
Inertial confinement, 1098–99
Inertial forces, 291–92
Inertial mass, 148, 1152
Inertial reference frame, 79, 291,
    917 *ff*
    equivalence of all, 918–19, 922
    transformations, 932–36
Inflationary universe, 1164
Infrared radiation (IR), 506, 799,
    825
Infrasonic waves, 419
Initial conditions, 365
Instantaneous axis, 246
Instruments, wind and stringed,
    424–29
Insulators:
    electrical, 547–48, 640, 1049–50
    thermal, 504, 1049–50
Integrals and integration, 36–37,
    161 *ff*, A-4–A-5, B-5
Integrated circuits, 1054–55
Intensity:
    of coherent and incoherent sources,
        877
    in interference and diffraction
        patterns, 874–77, 890–93
    of light, 877, 882
    of sound, 420–23
    of waves, 395–96, 420–23
Intensity level (*see* Sound level)
Interference, 404–5
    constructive, 404, 430, 871, 1031
    destructive, 404, 430, 871, 1031
    as distinguished from diffraction,
        895
    of light waves, 870–81, 894–96, 900
    of sound waves, 429–32
    by thin films, 877–81
    of water waves, 404–5
    of waves on a string, 404
Interference fringes, 871

Interference pattern:
    double slit, 870–73, 978–79
    double slit, including diffraction,
        893–95
    multiple slit, 900–905
Interferometers, 881
Internal combustion engine, 518–19
Internal conversion, 1073
Internal energy, 189–90, 487–88
Internal reflection, total, 415 *pr*,
    826–27
Internal resistance, 659
Intrinsic semiconductor (*defn*), 1049
Inverted population, 1020
Ion, 547
Ionic bonds, 1031–33, 1044
Ionic cohesive energy, 1044
Ionization energy, 968, 970
Ionizing radiation (*defn*), 1100
IR radiation, 506, 799, 825
Irreversible process, 520–25
Isobaric process (*defn*), 496–97
Isochoric process (*defn*), 496–97
Isolated system, 209, 493
Isomer, 1073
Isotherm, 495
Isothermal process (*defn*), 495
Isotopes, 702, 1062
    table of, A-9–A-13
Isotropic material, 867

J

J (total angular momentum),
    1017–18
J/ψ particle, 1131
Jeweler's loupe, 854
Joint, 315
Joule, James Prescott, 486
Joule (unit), 156, 167, 248 *fn*
    relation to calorie, 486
Jump start, 668–69
Junction diode, 1054–55
Junction transistor, 1054–55
Jupiter, moons of, 152, 854

K

Kant, Immanuel, 1143
Kaon, 1127
K-capture, 1072
K lines, 1013

Kelvin (unit), 455
Kelvin-Planck statement of second
    law of thermodynamics, 520, 523,
    533
Kelvin temperature scale, 455, 460,
    538
Kepler, Johannes, 143, 854 *fn*, 1150
Keplerian telescope, 854
Kepler's laws, 143–46, 288–89
Keyboard, computer, 616
Kilocalorie (unit), 486–87
Kilogram (unit), 7, 79
Kilowatt-hour (unit), 643
Kinematics, 16–67, 240–44 (*see also*
    Motion)
    for rotational motion, 240–44
    for translational motion, 16–67
    for uniform circular motion, 63–65
    vector kinematics, 53–55
Kinetic energy, 164–69, 182 *ff*,
    260–65
    in collisions, 214–17, 219
    molecular, relation to temperature,
        469
    relativistic, 938–42
    rotational, 260–65
    total, 262–65
    translational, 165–69
Kinetic friction, 106 *ff*
Kinetic theory, 445, 466 *ff*
Kirchhoff, G. R., 665
Kirchhoff's rules, 664–69

L

Ladder, 306–7
Laminar flow, 343–45
Land, Edwin, 907
Large Magellanic Cloud, 1150
Laser light, 873
Lasers, 873, 887, 1019–22, 1100
Latent heats, 490–93
Lateral magnification, 819, 841
Lattice structure, 446–47, 1044,
    1051–52
Laue, Max von, 905
Lawrence, E. O., 1116
Laws, 3–4 (*see also* specific name of
    law)
Lawson, J. D., 1099
Lawson criterion, 1099–1100
LC circuit, 764–65
LCD, 909

produces electric field and current, 747–49
of solenoid, 716–18
sources of, 709–26
of straight wire, 710, 720
of toroid, 708
Magnetic field lines, 687
Magnetic flux, 736, 791
Magnetic flux density (*see* Magnetic field)
Magnetic force, 686, 689–95
on electric current, 689–92
on moving electric charges, 692–95
Magnetic induction (*see* Magnetic field)
Magnetic lens, 961
Magnetic moment, 695–97, 1015–18, 1064
Magnetic permeability, 710, 724–25
Magnetic poles, 686–87
of Earth, 688
Magnetic quantum number, 1005–6, 1016
Magnetic resonance imaging, 1064, 1108–9
Magnetic tape and discs, 749
Magnetism, 686–726
Magnetization (vector), 726
Magnification:
angular, 853
of electron microscope, 961
lateral, 819, 841
of lens, 841–43
of lens combination, 843–45
longitudinal, 833 *pr*
of magnifying glass, 853–54
of microscope, 856–58, 898–99
of mirror, 819
sign conventions for, 819, 841
of telescope, 855, 898–99
useful, 899, 961
Magnifier, simple, 853–54
Magnifying glass, 836, 853–54
Magnifying power, 853, 855 (*see also* Magnification)
Magnitude (stars and galaxies), 1172 *pr*
Main-sequence, stars, 1146–49
Manhattan Project, 1095
Marconi, Guglielmo, 803
Manometer, 338
Mass, 7, 79, 82
atomic, 446
center of, 221–24

critical, 1092–95
gravitational vs. inertial, 148, 1152
inertial, 148, 1152
molecular, 446
in relativity theory, 936–38
rest, 938–40, 1063
standard of, 7
units of, 7, 79
variable, 227–29
Mass-energy transformation, 938–42
Mass excess, 1084 *pr*
Mass increase in relativity, 936–38
Mass spectrometer (spectrograph), 702
Matter, states of, 332, 446–47
Matter-dominated universe, 1161, 1164
Maxwell, James Clerk, 470, 787, 792, 797–98, 918
Maxwell distribution of molecular speeds, 470–72
Maxwell's equations, 787, 792–806, 918–19, 934
Mean free path, 478–79
Mean life, 1075, 1084 *pr*
Mean speed, 469
Measurement, 4–5, 981
Mechanical advantage, 93, 303
Mechanical energy, 182–89
Mechanical equivalent of heat, 486
Mechanical waves, 388–411
Mechanics (*defn*), 16
Medical imaging, 437, 1105–9
Meitner, L., 978, 1090
Melting point, 491 (*see also* Phase, changes of)
Mendeleev, D., 1012
Meson exchange, 1122
Mesons, 1122–36
Metallic bond, 1044
Metals, free electron theory of, 1045–48
Metastable state, 1019–20, 1073
Meter (unit), 6, 812 *fn*, 881
Meters, 674–76, 697
correction for resistance of, 676
Metric system, 6–8
MeV (million electron volts) (*see* Electron volt)
Mho (unit), 656 *pr*
Michelson, A., 811–12, 881, 919–22
Michelson interferometer, 881, 919–22

Michelson-Morley experiment, 919–23, 923 *fn*
Micrometer, 10
Microphones:
capacitor, 677
magnetic, 749
ribbon, 749
Microscope:
compound, 856–58
electron, 899, 961, 999
magnification of, 856–58, 898–99
resolving power of, 898–99
useful magnification, 899
Microscopic description of a system, 445
Microstate, 535–36
Microwaves, 798
Milky Way, 854, 1141
Millikan, R. A., 700, 953
Millikan oil-drop experiment, 700
Mirage, 869
Mirror equation, 819–22
Mirrors:
concave and convex, 819 *ff*
focal length of, 817, 821
plane, 812–15
spherical, 816–22
used in telescope, 856
Missing mass, 1166–68
MKS units (*see* SI units)
mm-Hg (unit), 339
Models, 3
Moderator, 1092–94
Modern physics (*defn*), 1, 916
Modulation, 804
Moduli of elasticity, 309–12
Molar specific heat, 498–99
Mole (*defn*), 456
volume of, for ideal gas, 457
Molecular mass, 446
Molecular rotation, 1038–40
Molecular spectra, 1037–43
Molecular speeds, 466–72
Molecular vibrations, 501, 1040–42
Molecules, 1030–43
bonding in, 1030–33
polar, 1032–33
P E diagrams for, 1033–36
spectra of, 1037–43
Moment arm, 247
Moment of a force, 247
Moment of inertia, 249–56, 374
Momentum, 206 *ff*, 936–38, 941
angular, 256–60, 281–89, 965

done by a gas, 496–98
in first law of thermodynamics, 494 *ff*
from heat engines, 517 *ff*
relation to energy, 164–69, 177–83, 190–91, 193, 195, 260–61
units of, 156
Work-energy principle, 165–68, 182, 190–91, 261
Work function, 953, 1048
Working substance (*defn*), 519
Wright, Thomas, 1141

X

X-ray crystallography, 906
X-ray diffraction, 905–6
X-rays, 799, 905–6, 1013–15, 1073, 1105–9

and atomic number, 1013–15
in electromagnetic spectrum, 799
spectra, 1013–15

Y

Young, Thomas, 870
Young's double-slit experiment, 870–73
Young's modulus, 309–10
Yukawa, H., 1121–22

Z

Z^0 particle, 1123, 1126–28, 1133
Zeeman effect, 708 *pr*, 1005, 1016, 1018, 1107
Zener diode, 1053

Zero, absolute, 455, 538
Zero-point energy, 992, 1041
Zeroth law of thermodynamics, 449–50
Zoom lens, 850
Zweig, G., 1130

Photo Credits

CO-1 NOAA/Phil Degginger/Color-Pic, Inc. **1-1a** Philip H. Coblentz/Tony Stone Images **1-1b** Richard Berenholtz/The Stock Market **1-1c** Antranig M. Ouzoonian, P.E./Weidlinger Associates, Inc. **1-2** Mary Teresa Giancoli **1-3a** Oliver Meckes/E.O.S./MPI-Tubingen/Photo Researchers, Inc. **1-3b** Douglas C. Giancoli **1-4** International Bureau of Weights and Measures, Sevres, France **1-5** Douglas C. Giancoli **1-6** Doug Martin/Photo Researchers, Inc. **1-9** David Parker/Science Photo Library/Photo Researchers, Inc. **1-10** The Image Works **CO-2** Joe Brake Photography/Crest Communications, Inc. **2-22** Justus Sustermans, painting of Galileo Galilei/The Granger Collection **2-23** Photograph by Dr. Harold E. Edgerton, © The Harold E. Edgerton 1992 Trust. Courtesy Palm Press, Inc. **CO-3** Michel Hans/Vandystadt Allsport Photography (USA), Inc **3-19** Berenice Abbott/Commerce Graphics Ltd., Inc. **3-21** Richard Megna/Fundamental Photographs **3-30a** Don Farrall/PhotoDisc, Inc. **3-30b** Robert Frerck/Tony Stone Images **3-30c** Richard Megna/Fundamental Photographs **CO-4** Mark Wagner/Tony Stone Images **4-1** AP/Wide World Photos **4-3** Central Scientific Company **4-5** Sir Godfrey Kneller, Sir Isaac Newton, 1702. Oil on canvass. The Granger Collection **4-6** Gerard Vandystadt/Agence Vandystadt/Photo Researchers, Inc. **4-8** David Jones/Photo Researchers, Inc. **4-11** Tsado/NASA/Tom Stack & Associates **4-31** Lars Ternblad/The Image Bank **4-34** Kathleen Schiaparelli **4-36** Jeff Greenberg/Photo Researchers, Inc. **CO-5a** Jess Stock/Tony Stone Images **CO-5b** Werner H. Muller/Peter Arnold, Inc. **5-12** Jay Brousseau/The Image Bank **5-17** Guido Alberto Rossi/The Image Bank **5-33** C. Grzimek/Okapia/Photo Researchers, Inc. **5-34** Photofest **5-38** Cedar Point Photo by Dan Feicht **CO-6** Earth Imaging/Tony Stone Images **6-9** NASA/Johnson Space Center **6-13** I. M. House/Tony Stone Images **6-14L** Jon Feingersh/The Stock Market **6-14M** © Johan Elbers 1995 **6-14R** Peter Grumann/The Image Bank **CO-7** Eric Miller/Reuters/Corbis **7-20a** Stanford Linear Accelerator Center/Science Photo Library/Photo Researchers, Inc. **7-20b** Account Phototake/Phototake NYC **7-24** Official U.S. Navy Photo **CO-8 and 8-10** Photograph by Dr. Harold E. Edgerton, © The Harold E. Edgerton 1992 Trust. Courtesy Palm Press **8-11** David Madison/David Madison Photography **8-16** AP/Wide World Photos **8-23** M. C. Escher's "Waterfall" © 1996 Cordon Art-Baarn-Holland. All rights reserved. **CO-9** Richard Megna/Fundamental Photographs **9-9** Photograph by Dr. Harold E. Edgerton, © The Harold E. Edgerton 1992 Trust. Courtesy Palm Press **9-16** D.J. Johnson **9-20** Courtesy Brookhaven National Laboratory **9-22** Berenice Abbott/Photo Researchers, Inc. **CO-10** Ch. Russeil/Kipa/Sygma Photo News **10-10b** Mary Teresa Giancoli **10-14a** Richard Megna/Fundamental Photographs **10-14b** Photoquest, Inc. **10-47** Jens Hartmann/AP/Wide World Photos **10-48** Focus on Sports, Inc. **10-49** Karl Weatherly/PhotoDisc, Inc **CO-11** AP/Wide World Photos **11-19c** NOAA/Phil Degginger/Color-Pic, Inc. **11-19d** NASA/TSADO/Tom Stack & Associates **11-34** Michael Kevin Daly/The Stock Market **CO-12** Steve Vidler/Leo de Wys, Inc. **12-2** AP/Wide World Photos **12-7** T. Kitchin/Tom Stack & Associates **12-10** The Stock Market **12-21** Douglas C. Giancoli **12-23** Mary Teresa Giancoli **12-27** Fabricius & Taylor/Liaison Agency, Inc. **12-36** Henryk T. Kaiser/Leo de Wys, Inc. **12-38** Douglas C. Giancoli **12-39** Galen Rowell/Mountain Light Photography, Inc. **12-41** Douglas C. Giancoli **12-43** Giovanni Paolo Panini (Roman, 1691-1765), Interior of the Pantheon, Rome c. 1734. Oil on canvas. 1.283 × .991 (50 1/2 × 39); framed: 1.441 × 1.143 (56 3/4 × 45). © 1995 Board of Trustees, National Gallery of Art, Washington. Samuel H. Kress Collection. Photo by Richard Carafelli. **12-44** Robert Holmes/Corbis **12-45** Italian Government Tourist Board **CO-13** Steven Frink/Tony Stone Images **13-12** Corbis **13-20** Department of Mechanical and Aerospace Engineering, Princeton University **13-31** Rod Planck/Thomas Stack & Associates **13-33** Alan Blank/Bruce Coleman Inc. **13-42** Douglas C. Giancoli **13-45** Galen Rowell/Mountain Light Photography, Inc. **13-51** NASA/Goddard Space Flight Center/Science Source/Photo Researchers, Inc. **CO-14** Richard Megna/Fundamental Photographs **14-4** Mark E. Gibson/Visuals Unlimited **14-14** Fundamental Photographs **14-15** Douglas C. Giancoli **14-22** Taylor Devices, Inc. **14-25** Martin Bough/Fundamental Photographs **14-26a** AP/Wide World Photos **14-26b** Paul X. Scott/Sygma Photo News **CO-15** Richard Megna/Fundamental Photographs **15-1** Douglas C. Giancoli **15-24** Douglas C. Giancoli **15-30** Martin G. Miller/Visuals Unlimited **15-32** Richard Megna/Fundamental Photographs **CO-16** Photographic Archives, Teatro alla Scala, Milan, Italy **16-6** Yoav Levy/Phototake NYC **16-9a** David Pollack/The Stock Market **16-9b** Andrea Brizzi/The Stock Market **16-10** Bob Daemmrich/The Image Works **16-13** Bildarchiv Foto Marburg/Art Resource, N.Y. **16-24a** Norman Owen Tomalin/Bruce Coleman Inc. **16-25b** Sandia National Laboratories, New Mexico **16-28a** P. Saada/Eurelios/Science Photo Library/Photo Researchers, Inc. **16-28b** Howard Sochurek/Medical Images Inc. **CO-17** Le Matin de Lausanne/Sygma Photo News **17-3** Bob Daemmrich/Stock Boston **17-5** Leonard Lessin/Peter Arnold, Inc. **17-14** Leonard Lessin/Peter Arnold, Inc. **17-17** Brian Yarvin/Photo Researchers, Inc. **CO-18** Tom Till/DRK Photo **18-9** Paul Silverman/Fundamental Photographs **18-14** Mary Teresa Giancoli **CO-19** Tom Bean/DRK Photo **19-25** Science Photo Library/Photo Researchers, Inc. **CO-20a** David Woodfall/Tony Stone Images **CO-20b** AP/Wide World Photos **20-7a** Sandia National Laboratories, New Mexico **20-7b** Martin Bond/Science Photo Library/Photo Researchers, Inc **20-7c** Lionel Delevingne/Stock Boston **20-13** Leonard Lessin/Peter Arnold, Inc. **CO-21** Fundamental Photographs **21-38** Michael J. Lutch/Boston Museum of Science **CO-23** Gene Moore/Phototake NYC **23-21** Jon Feingersh/The Stock Market **CO-24** Paul Silverman/Fundamental Photographs **CO-25** Mahaux Photography/The Image Bank **25-1** J.-L. Charmet/Science Photo Library/Photo Researchers, Inc. **25-10** T.J. Florian/Rainbow **25-13** Richard Megna/Fundamental Photographs **25-16** Barbara Filet/Tony Stone Images **25-26** Takeshi Takahara/Photo Researchers, Inc. **25-33** Liaison Agency, Inc. **CO-26a** Steve Weinrebe/Stock Boston **CO-26b** Sony Electronics, Inc. **26-22** Paul Silverman/Fundamental Photographs **26-26** Paul Silverman/Fundamental Photographs **CO-27** Richard Megna/Fundamental Photographs **27-4** Richard Megna/Fundamental Photographs **27-6** Mary Teresa Giancoli **27-8** Richard Megna/Fundamental Photographs **27-19** Pekka Parviainen/Science Photo Library/Photo Researchers, Inc. **CO-28** Manfred Kage/Peter Arnold, Inc. **28-21** Richard Megna/Fundamental Photographs **CO-29** Richard Megna/Fundamental Photographs **29-10** Werner H. Muller/Peter Arnold, Inc.

29-15 Tomas D.W. Friedmann/Photo Researchers, Inc. 29-20 Jon Feingersh/Comstock 29-28b National Earthquake Information Center, U.S.G.S. CO-30 Adam Hart-Davis/Science Photo Library/Photo Researchers, Inc. CO-31v1 Albert J. Copley/Visuals Unlimited CO-31v2 Mary Teresa Giancoli CO-32 Richard Megna/Fundamental Photographs 32-1 AIP Emilio Segre' Visual Archives 32-14 The Image Works CO-33 Douglas C. Giancoli 33-6 Douglas C. Giancoli 33-11a Mary Teresa Giancoli and Suzanne Saylor 33-11b Mary Teresa Giancoli 33-21 Mary Teresa Giancoli 33-25 David Parker/Science Photo Library/Photo Researchers, Inc. 33-28b Michael Giannechini/Photo Researchers, Inc. 33-34b S. Elleringmann/Bilderberg/Aurora & Quanta Productions 33-40 Douglas C. Giancoli 33-42 Mary Teresa Giancoli CO-34 Mary Teresa Giancoli 34-1c Douglas C. Giancoli 34-1d Douglas C. Giancoli 34-2 Douglas C. Giancoli and Howard Shugat 34-4 Douglas C. Giancoli 34-7a Douglas C. Giancoli 34-7b Douglas C. Giancoli 34-18 Mary Teresa Giancoli 34-19a Mary Teresa Giancoli 34-19b Mary Teresa Giancoli 34-26 Mary Teresa Giancoli 34-30a Franca Principe/Istituto e Museo di Storia della Scienza, Florence, Italy 34-32 Yerkes Observatory, University of Chicago 34-33c Palomar Observatory/California Institute of Technology 34-33d Joe McNally/Joe McNally Photography 34-35b Olympus America Inc. CO-35 Larry Mulvehill/Photo Researchers, Inc. 35-4a John M. Dunay IV/Fundamental Photographs 35-9a Bausch & Lomb Incorporated 35-16a Paul Silverman/ Fundamental Photographs 35-16b Richard Megna/Fundamental Photographs 35-16c Yoav Levy/Phototake NYC 35-18b Ken Kay/ Fundamental Photographs 35-20b Bausch & Lomb Incorporated 35-20c Bausch & Lomb Incorporated 35-22 Kristen Brochmann/ Fundamental Photographs CO-36 Richard Megna/Fundamental Photographs 36-2a Reprinted with permission from P.M. Rinard, American Journal of Physics, Vol. 44, #1, 1976, p. 70. Copyright 1976 American Association of Physics Teachers. 36-2b Ken Kay/ Fundamental Photographs 36-2c Ken Kay/Fundamental Photographs 36-11a Richard Megna/Fundamental Photographs 36-11b Richard Megna/Fundamental Photographs 36-12a and b Reproduced by permission from M. Cagnet, M. Francon, and J. Thrier, The Atlas of Optical Phenomena. Berlin: Springer-Verlag, 1962. 36-15 Space Telescope Science Institute 36-16 The Arecibo Observatory is part of the National Astronomy and Ionosphere Center which is operated by Cornell University under a cooperative agreement with the National Science Foundation. 36-21 Wabash Instrument Corp./Fundamental Photographs 36-26 Photo by W. Friedrich/Max von Laue. Burndy Library, Dibner Institute for the History of Science and Technology, Cambridge, Massachusetts. 36-29b Bausch & Lomb Incorporated 36-36 Diane Schiumo/Fundamental Photographs 36-39a Douglas C. Giancoli 36-39b Douglas C. Giancoli CO-37 Image of Albert Einstein licensed by Einstein Archives, Hebrew University, Jerusalem, represented by Roger Richman Agency, Beverly Hills, California 37-1 AIP Emilio Segre Visual Archives CO-38 Wabash Instrument Corp./Fundamental Photographs 38-10 Photo by S.A. Goudsmit, AIP Emilio Segre' Visual Archives 38-11 Education Development Center, Inc. 38-20b Richard Megna/Fundamental Photographs 38-21abc Wabash Instrument Corp./Fundamental Photographs CO-39 Institut International de Physique/American Institute of Physics/Emilio Segre Visual Archives 39-01 American Institute of Physics/Emilio Segre Visual Archives 39-02 F.D. Rosetti/American Institute of Physics/Emilio Segre Visual Archives 39-04 Advanced Research Laboratory/Hitachi Metals America, Ltd. 39-18 Driscoll, Youngquist, and Baldeschwieler, Caltech/Science Photo Library/Photo Researchers, Inc. CO-40 Patricia Peticolas/Fundamental Photographs 40-16 Paul Silverman/Fundamental Photographs 40-22 Yoav Levy/Phototake NYC 40-24b Paul Silverman/Fundamental Photographs CO-41 Charles O'Rear/Corbis CO-42 Reuters Newmedia Inc/Corbis 42-03 Chemical Heritage Foundation 42-07 University of Chicago, Courtesy of AIP Emilio Segre Visual Archives CO-43 AP/Wide World Photos 43-07 Gary Sheahan "Birth of the Atomic Age" Chicago (Illinois); 1957. Chicago Historical Society. 43-10 Liaison Agency, Inc. 43-11 LeRoy N. Sanchez/Los Alamos National Laboratory 43-12 Corbis 43-16a Lawrence Livermore National Laboratory/Science Source/Photo Researchers, Inc. 43-16b Gary Stone/Lawrence Livermore National Laboratory 43-22a Martin M. Rotker/Martin M. Rotker 43-22b Simon Fraser/Science Photo Library/Photo Researchers, Inc. 43-26b Southern Illinois University/Peter Arnold, Inc. 43-28 Mehau Kulyk/Science Photo Library/Photo Researchers, Inc. CO-44 Fermilab Visual Media Services 44-06 CERN/Science Photo Library/Photo Researchers, Inc. 44-08 Fermilab Visual Media Services CO-45 Jeff Hester and Paul Scowen, Arizona State University, and NASA 45-01 NASA Headquarters 45-02c NASA/Johnson Space Center 45-03 U.S. Naval Observatory Photo/NASA Headquarters 45-04 National Optical Astronomy Observatories 45-05a R.J. Dufour, Rice University 45-05b U.S. Naval Observatory 45-05c National Optical Astronomy Observatories 45-10 Space Telescope Science Institute/NASA Headquarters 45-11 National Optical Astronomy Observatories 45-15 European Space Agency/NASA Headquarters 45-22 Courtesy of Lucent Technologies/Bell Laboratories

Table of Contents Photos p. v (left) NOAA/Phil Degginger/Color-Pic, Inc. **p. v** (right) Mark Wagner/Tony Stone Images **p. vi** (left) Jess Stock/Tony Stone Images **p. vi** (right) Photograph by Dr. Harold E. Edgerton, © The Harold E. Edgerton 1992 Trust. Courtesy Palm Press **p. vii** (left) Ch. Russeil/Kipa/Sygma Photo News **p.vii** (right) Tibor Bognar/The Stock Market **p. viii** (left) Richard Megna/Fundamental Photographs **p. viii** (right) Douglas C. Giancoli **p. ix** (top) S. Feval/Le Matin de Lausanne/Sygma Photo News **p. ix** (bottom) Richard A. Cooke III/Tony Stone Images (right) David Woodfall/Tony Stone Images and AP/ Wide World Photos **p. x** (left) Fundamental Photographs **p. x** (right) Richard Kaylin/Tony Stone Images **p. xi** (left) Mahaux Photography/The Image Bank **p. xi** (right) Manfred Cage/Peter Arnold, Inc. **p. xii** Werner H. Muller/Peter Arnold, Inc. **p. xiii** (left) Mary Teresa Giancoli. **p. xiii** (right) Larry Mulvehill/Photo Researchers, Inc. **p. xiv** Image of Albert Einstein licensed by Einstein Archives, Hebrew University, Jerusalem, represented by Roger Richman Agency, Beverly Hills, California **p. xiv** (right) Donna McWilliam/AP/Wide World Photos **p. xv** (left) Charles O'Rear/Corbis (right) Fermilab **p. xvi** (left) Jeff Hester and Paul Scowen, Arizona State University, and NASA (right) R. J. Dufour, Rice University.

Periodic Table of the Elements§

Transition Elements

Legend:
Symbol — Cl 17 — Atomic Number
Atomic Mass§ — 35.4527
Electron Configuration (outer shells only) — $3p^5$

Group I	Group II		Group III	Group IV	Group V	Group VI	Group VII	Group VIII
H 1 — 1.00794 — $1s^1$								He 2 — 4.002602 — $1s^2$
Li 3 — 6.941 — $2s^1$	Be 4 — 9.012182 — $2s^2$		B 5 — 10.811 — $2p^1$	C 6 — 12.0107 — $2p^2$	N 7 — 14.00674 — $2p^3$	O 8 — 15.9994 — $2p^4$	F 9 — 18.9984032 — $2p^5$	Ne 10 — 20.1797 — $2p^6$
Na 11 — 22.989770 — $3s^1$	Mg 12 — 24.3050 — $3s^2$		Al 13 — 26.981538 — $3p^1$	Si 14 — 28.0855 — $3p^2$	P 15 — 30.973761 — $3p^3$	S 16 — 32.066 — $3p^4$	Cl 17 — 35.4527 — $3p^5$	Ar 18 — 39.948 — $3p^6$

Transition Elements block:

| Group I | Group II | Sc 21 / Ti 22 ... | ... |

| Sc 21 — 44.955910 — $3d^14s^2$ | Ti 22 — 47.867 — $3d^24s^2$ | V 23 — 50.9415 — $3d^34s^2$ | Cr 24 — 51.9961 — $3d^54s^1$ | Mn 25 — 54.938049 — $3d^54s^2$ | Fe 26 — 55.845 — $3d^64s^2$ | Co 27 — 58.933200 — $3d^74s^2$ | Ni 28 — 58.6934 — $3d^84s^2$ | Cu 29 — 63.546 — $3d^{10}4s^1$ | Zn 30 — 65.39 — $3d^{10}4s^2$ |

K 19 — 39.0983 — $4s^1$; Ca 20 — 40.078 — $4s^2$; Ga 31 — 69.723 — $4p^1$; Ge 32 — 72.61 — $4p^2$; As 33 — 74.92160 — $4p^3$; Se 34 — 78.96 — $4p^4$; Br 35 — 79.904 — $4p^5$; Kr 36 — 83.80 — $4p^6$

| Y 39 — 88.90585 — $4d^15s^2$ | Zr 40 — 91.224 — $4d^25s^2$ | Nb 41 — 92.90638 — $4d^45s^1$ | Mo 42 — 95.94 — $4d^55s^1$ | Tc 43 — (98) — $4d^55s^2$ | Ru 44 — 101.07 — $4d^75s^1$ | Rh 45 — 102.90550 — $4d^85s^1$ | Pd 46 — 106.42 — $4d^{10}5s^0$ | Ag 47 — 107.8682 — $4d^{10}5s^1$ | Cd 48 — 112.411 — $4d^{10}5s^2$ |

Rb 37 — 85.4678 — $5s^1$; Sr 38 — 87.62 — $5s^2$; In 49 — 114.818 — $5p^1$; Sn 50 — 118.710 — $5p^2$; Sb 51 — 121.760 — $5p^3$; Te 52 — 127.60 — $5p^4$; I 53 — 126.90447 — $5p^5$; Xe 54 — 131.29 — $5p^6$

| La 57–71† | Hf 72 — 178.49 — $5d^26s^2$ | Ta 73 — 180.9479 — $5d^36s^2$ | W 74 — 183.84 — $5d^46s^2$ | Re 75 — 186.207 — $5d^56s^2$ | Os 76 — 190.23 — $5d^66s^2$ | Ir 77 — 192.217 — $5d^76s^2$ | Pt 78 — 195.078 — $5d^96s^1$ | Au 79 — 196.96655 — $5d^{10}6s^1$ | Hg 80 — 200.59 — $5d^{10}6s^2$ |

Cs 55 — 132.90545 — $6s^1$; Ba 56 — 137.327 — $6s^2$; Tl 81 — 204.3833 — $6p^1$; Pb 82 — 207.2 — $6p^2$; Bi 83 — 208.98038 — $6p^3$; Po 84 — (209) — $6p^4$; At 85 — (210) — $6p^5$; Rn 86 — (222) — $6p^6$

| Ac 89–103‡ | Rf 104 — (261) — $6d^27s^2$ | Db 105 — (262) — $6d^37s^2$ | Sg 106 — (266) — $6d^47s^2$ | Bh 107 — (264) | Hs 108 — (269) | Mt 109 — (268) | 110 — (271) | 111 — (272) | 112 — (277) |

Fr 87 — (223) — $7s^1$; Ra 88 — (226) — $7s^2$; 114 — (289) ; 116 — (289) ; 118 — (293)

†Lanthanide Series

| La 57 — 138.9055 — $5d^16s^2$ | Ce 58 — 140.115 — $4f^15d^16s^2$ | Pr 59 — 140.90765 — $4f^35d^06s^2$ | Nd 60 — 144.24 — $4f^45d^06s^2$ | Pm 61 — (145) — $4f^55d^06s^2$ | Sm 62 — 150.36 — $4f^65d^06s^2$ | Eu 63 — 151.964 — $4f^75d^06s^2$ | Gd 64 — 157.25 — $4f^75d^16s^2$ | Tb 65 — 158.92534 — $4f^95d^06s^2$ | Dy 66 — 162.50 — $4f^{10}5d^06s^2$ | Ho 67 — 164.93032 — $4f^{11}5d^06s^2$ | Er 68 — 167.26 — $4f^{12}5d^06s^2$ | Tm 69 — 168.93421 — $4f^{13}5d^06s^2$ | Yb 70 — 173.04 — $4f^{14}5d^06s^2$ | Lu 71 — 174.967 — $4f^{14}5d^16s^2$ |

‡Actinide Series

| Ac 89 — (227.02775) — $6d^17s^2$ | Th 90 — 232.0381 — $6d^27s^2$ | Pa 91 — 231.03588 — $5f^26d^17s^2$ | U 92 — 238.0289 — $5f^36d^17s^2$ | Np 93 — (237) — $5f^46d^17s^2$ | Pu 94 — (244) — $5f^66d^07s^2$ | Am 95 — (243) — $5f^76d^07s^2$ | Cm 96 — (247) — $5f^76d^17s^2$ | Bk 97 — (247) — $5f^96d^07s^2$ | Cf 98 — (251) — $5f^{10}6d^07s^2$ | Es 99 — (252) — $5f^{11}6d^07s^2$ | Fm 100 — (257) — $5f^{12}6d^07s^2$ | Md 101 — (258) — $5f^{13}6d^07s^2$ | No 102 — (259) — $5f^{14}6d^07s^2$ | Lr 103 — (262) — $5f^{14}6d^17s^2$ |

§ Atomic mass values averaged over isotopes in percentages they occur on Earth's surface. For many unstable elements, mass of the longest-lived known isotope is given in parentheses. 1999 revisions. (See also Appendix D.)